Carpentry & Building Construction

A DO-IT-YOURSELF GUIDE

William P. Spence

Sterling Publishing Co., Inc.
New York, NY

Disclaimer

The author has made every attempt to present safe and sound building practices, but he makes no claim that the information in this book is complete or complies with every local building code.

The publisher does not warrant or guarantee any of the products described herein or perform any independent analysis in connection with any of the product information conatined herein. The publisher does not assume, and expressly disclaim any obligation to obtain and include information other than that provided by the manufacturer.

The reader is expressly warned to consider and adopt safety precautions that might be indicated by the activities herein and to avoid all potential hazards. By following the instructions conatined herein, the reader willingly assumes all risks in connection with such instructions.

The author and publisher make no representation or warranties of any kind, including but not limited to the warranties of fitness for a particular purpose or merchantability, nor are any such representations implied with respect to the material set forth herein, and the publisher takes no responsibility with respect to such material. The publisher shall not be liable for any special, consequential, or exemplary damage resulting, in whole or part, from the readers' use of, or reliance upon, this material.

Library of Congress Cataloging-in-Publication Data

Spence, William Perkins. 1925–
Carpentry & building construction : a do-it-yourself guide / by William P. Spence.
p. cm.
Includes index.
ISBN 0-8069-9845-8
1.Carpentry—Amateurs' manuals. 2.House construction—Amateurs' manuals.
I. Title. II. Title: Carpentry and building construction.
TH5606.S68 1999
694—dc21 98-40915
CIP

Edited by Rodman P. Neumann

1 3 5 7 9 10 8 6 4 2

Published by Sterling Publishing Company, Inc.
387 Park Avenue South, New York, N.Y. 10016

Distributed in Canada by Sterling Publishing
c/o Canadian Manda Group, One Atlantic Avenue, Suite 105
Toronto, Ontario, Canada M6K 3E7
Distributed in Great Britain and Europe by Cassell PLC
Wellington House, 125 Strand, London WC2R 0BB, England
Distributed in Australia by Capricorn Link (Australia) Pty Ltd.
P.O. Box 6651, Baulkham Hills, Business Centre, NSW 2153, Australia
Manufactured in the United States of America

Sterling ISBN 0-8069-9845-8

CONTENTS

Preface 5

SECTION I SAFETY, CODES & DRAWINGS 6

1. Safety on the Construction Site 6
2. Ladders, Scaffolding & Runways 18
3. Building Codes, Zoning Regulations, Building Permits & Inspections 30
4. Architectural Drawings & Specifications 34

SECTION II SITE & FOUNDATION PREPARATION 56

5. Using Leveling Instruments & Linear Measuring Tools 56
6. Locating the Building on the Site 72
7. Footings & Foundations 76
8. Concrete 100

SECTION III FLOOR, WALL & CEILING FRAMING 108

9. Floor Framing 108
10. Wall & Partition Framing 140
11. Framing the Ceiling 174

SECTION IV FRAMING THE ROOF & DORMERS 182

12. Roof Types & Design 182
13. Constructing a Gable Roof 192
14. Hip Roof Construction 214
15. Intersecting Roofs 226
16. Framing Flat, Shed, Gambrel & Mansard Roofs 238
17. Dormers 248
18. Trussed Roof Construction 252
19. Cornice Construction 266

SECTION V OTHER FRAMING METHODS 278

20. Post, Plank & Beam Construction 278
21. Heavy Timber 292
22. Pole Construction 298

SECTION VI DOORS & WINDOWS 304
23. Installing & Trimming Doors 304
24. Installing & Trimming Windows 338

SECTION VII FINISHING THE EXTERIOR 368
25. Finishing the Exterior Walls 368
26. Finishing the Roof 402
27. Decks & Porches 428

SECTION VIII PANELING & MOLDING 444
28. Moldings 444
29. Installing Base, Crown & Other Moldings 456
30. Paneling & Wainscoting 472

SECTION IX FLOORS & STAIRS 492
31. Installing Wood Floors 492
32. Other Types of Flooring 512
33. Stairs 522

SECTION X INSULATION & INTERIOR WALL FINISH 544
34. Thermal Insulation, Sound & Moisture Control 544
35. Installing Gypsum Drywall 560

SECTION XI CABINET CONSTRUCTION 576
36. Cabinets & Countertops 576

SECTION XII LIGHT-FRAME STEEL CONSTRUCTION 592
37. Cold-Formed Structural Steel Products 592
38. Cold-Formed Structural Steel Framing 598

SECTION XIII TOOLS & MATERIALS 610
39. Wood & Reconstituted Wood Products 610
40. Fasteners 632
41. The Carpenter's Tool Box 642
42. Using Power Tools 656

APPENDICES 676

INDEX 699

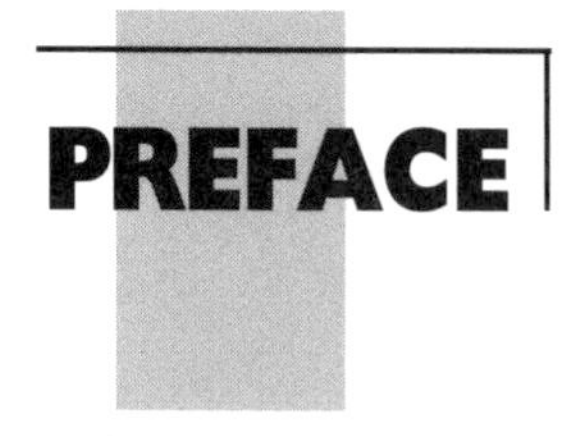

PREFACE

Carpentry & Building Construction provides information on the entire process of light frame construction. The homeowner will find it useful as repairs are made, additions are built, or an entire house is constructed. The apprentice carpenter and experienced carpenter will find extensive information about current construction materials and techniques. Students in vocational carpentry programs or apprenticeship programs will find it has the considerable detail they need.

Construction is one of the more dangerous occupational areas. Emphasis on safety on the construction site as well as the selection and use of safety equipment are included. The carpenter must be able to read the architectural working drawings and assemble the structure so it meets the building codes. Entire chapters are devoted to these areas.

A building must be accurately located on the site, and the footings, foundation, and framing must be level and plumb. The book shows several ways to do these, including using the latest surveying and leveling tools.

Considerable coverage of the materials of construction is included. This includes materials such as concrete, wood, reconstituted wood products, cold-formed steel, insulation, gypsum board, and fasteners.

The text thoroughly covers methods of constructing light frame buildings. The traditional wood stud and joist construction is completely detailed. Since there is an increasing use of post, plank, and beam construction for residential and small commercial buildings, an entire chapter covers this method. Pole construction, typically used for agricultural buildings, is shown when used for residential framing. The increasing use of cold-formed steel structural members makes it necessary for those in construction to be familiar with the framing materials available and techniques for assembling the framing. Two chapters introduce this important area.

Roof framing is difficult and can include a number of styles, each presenting specific layout, cutting, and assembly requirements. Seven chapters are devoted to roof framing including details on trussed roof construction.

After the framing is in place the exterior must be finished. Information on finishing the exterior walls includes many different materials as well as installing windows and exterior doors. These instructions will produce an airtight, watertight finished exterior.

Interior wall finishing materials include gypsum and wood products. The types and installation recommendations are detailed. Techniques for installing various finish flooring materials, such as wood, laminated wood, vinyl, and carpet, are presented. Special features such as stairs, insulation, cabinets, and countertops are useful to new construction as well as remodeling jobs. Finally, the hand and power tools needed and their safe and proper use are explained. Again, safety is of great importance.

The book contains more than two thousand drawings and photographs to help the reader understand the techniques presented. I have drafted a large number of these illustrations, but a great many were made available by the numerous companies noted on the credit lines and by their staff, who took time to respond to my requests. I appreciate their assistance.

William P. Spence

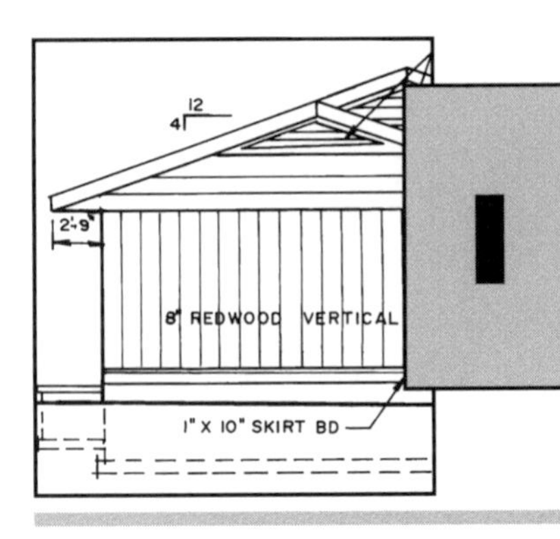

1 SAFETY, CODES & DRAWINGS

CHAPTER 1

SAFETY ON THE CONSTRUCTION SITE

Safety is everyone's business. Even though a contractor may have a safety supervisor on the construction site, this does not guarantee safe conditions. All those working on the job must understand the basic principles of safe behavior and act accordingly. Only then can accidents on a construction site be reduced. Everyone should be safety conscious. Safety consciousness is the awareness of good safety practices. This includes knowledge, attitude, obedience, and concern.

OCCUPATIONAL SAFETY & HEALTH ADMINISTRATION (OSHA)

OSHA was formed by the Occupational Safety and Health Administration Act of 1970. The act provides minimum safety and health standards for working conditions. OSHA is part of the U.S. Department of Labor. Its main duties are to:

1. Encourage employers and employees to reduce hazards in their workplaces.
2. Establish the responsibilities and rights of employers and employees.
3. Encourage new safety and health programs.
4. Establish record-keeping procedures to keep track of injuries and illnesses that happen on or because of the job.
5. Develop health and safety standards and enforce them.
6. Encourage states to establish safety and health programs.

There are two types of standards: those that apply to all industries and those that apply to one industry, such as construction.

Since the safety regulations are changed over the years it is important for the general contractor, the carpentry foreman, and the individual carpenters to keep current on these regulations.

The U.S. Bureau of Labor keeps a yearly record of on-the-job deaths in all industries. Transportation accidents account for 42 percent of the deaths. Since large construction sites have heavy on-site traffic clearly defined roads and traffic control signs are necessary. The construction workers must always be alert for moving vehicles such as trucks, forklifts, bulldozers, motorgraders, and mobile cranes. A summary of the major causes of accidents in all occupations is detailed in **1-1**. Notice that each area represents activities that do occur on the construction site. One classification,

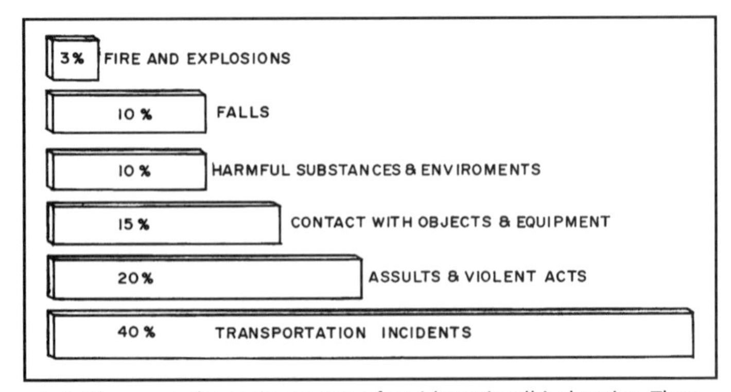

1-1 These are the major causes of accidents in all industries. They all apply to activities on the construction site. *(Courtesy U.S. Bureau of Labor)*

Falls, is a major concern for carpenters. While the total for all industries is 10 percent, about one-third of these occur on the construction site. Death and injury due to contact with various objects and equipment is another major accident area.

Carpenters should follow the recommendations of the manufacturer of power tools for safe use. They should wear adequate personal protection equipment, help keep the building and site clean, and maintain a positive attitude toward safe behavior.

CORRECT ATTITUDES

An important first step toward a safe working environment is to have workers develop a positive attitude toward safety. They can warn others of a dangerous situation, even pull a few protruding nails, or put some scrap in the dumpster. They also need to have a regard for their own personal safety. Wearing personal protection gear, using sharp quality tools, and not taking chances are important actions.

Safety is the responsibility of every person on the job.

GENERAL HOUSEKEEPING RULES

1. Keep walkways, runways, stairs, aisles, and work areas free from debris.
2. Store tools and materials in a safe manner.
3. Remove all nails from used lumber.
4. Keep oily rags and other combustible materials in nonflammable containers.
5. Clean up oil, grease, and other liquids that have been spilled on walkways. Clear snow and ice from walkways, scaffolding, and ladders.
6. Regularly remove trash and flammable waste materials from the site.
7. When dropping materials more than 20 feet, use an enclosed chute. When dropping materials short distances through holes on the inside, barricade off the area where they will land.
8. Keep the land around the building relatively level.
9. Establish trafficways on the site and maintain them in a level and smooth condition.

PERSONAL SAFETY

The employer is responsible for requiring the wearing of appropriate personal protective equipment in all operations where there is an exposure to hazardous conditions. Each individual should have the proper safety equipment. In some cases the contractor provides some equipment, but basic safety items are usually provided by each worker. Following are some things to observe:

1. Always wear a hard hat in areas designated as "hard hat areas." The hard hat lining must be adjusted so it sits firmly on the band inside **(see 1-2)**. Protective hats are available in two types. **Type 1** has a full brim not less than 1¼ inches wide and **Type 2** is brimless with a peak extending forward.

Three classes are recognized:

Class A. General service, limited voltage protection.

Class B. Utility service, high voltage protection.

Class C. Special service, no voltage protection.

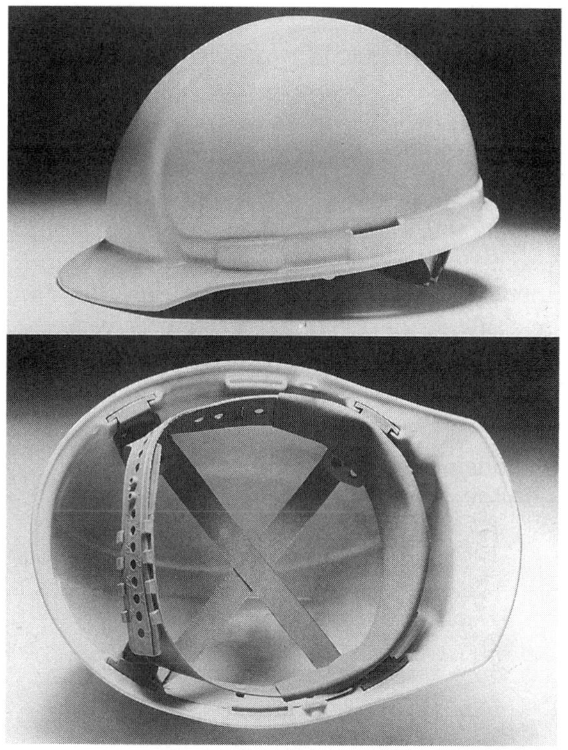

1-2 A safety hat is required on the construction site. It is typically called a "hard hat." The safety hat has a series of straps that rest on the head. This is a Type 2 hard hat. *(Courtesy Aearo Company)*

1-3 Safety glasses are required for most work on the construction site. *(Courtesy Willson Safety)*

Class A hard hats are intended for protection against impact hazards and are used in construction.

2. Purchase hard hats certified to meet the specifications in *American National Standards Institute,* Z89.1-1969.
3. Never wear a hard hat over another hat.
4. The hard hat must fit properly. The lining must be adjusted so that the hat sits squarely upon the head.
5. Eye and face protection equipment shall be required when machines or operations present potential eye or face damage. Eye and face protection equipment must meet the requirements specified in *American National Standards Institute,* Z87.1-1989, *Practice for Occupational and Educational Eye and Face Protection* **(see 1-3 and 1-4)**.

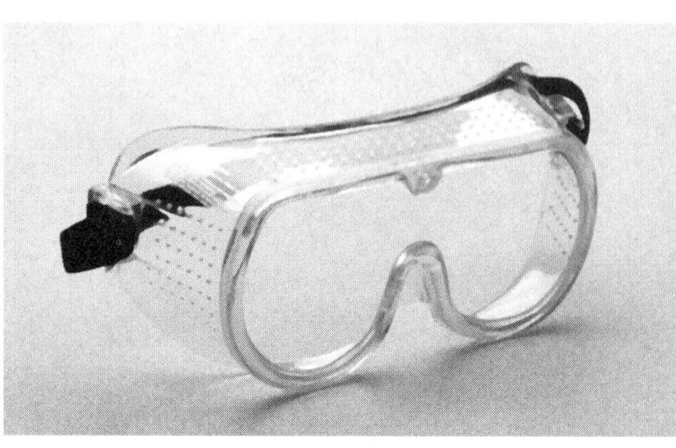

1-4 These high-impact safety goggles can be worn over regular eye glasses. *(Courtesy Willson Safety)*

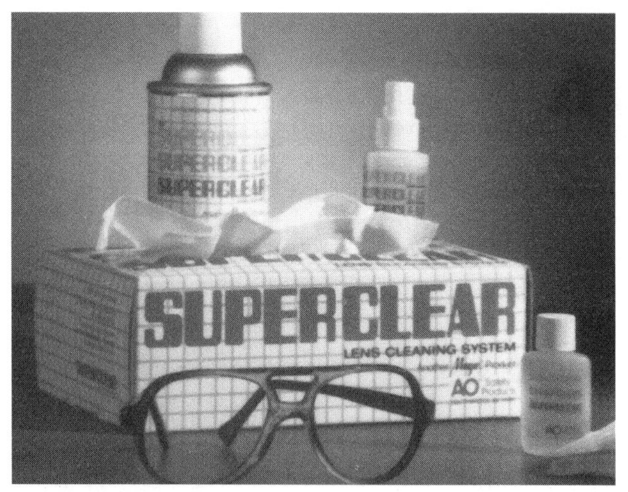

1-5 Eye safety equipment manufacturers have lens cleaning systems available. *(Courtesy Aearo Company)*

6. Face and eye protection equipment must be kept clean and in good repair **(see 1-5)**.
7. If there is danger to the face, a full-face shield should be worn **(see 1-6)**.
8. Wherever it is not feasible to reduce the noise levels or duration of exposure to that specified by OSHA regulations, ear protective devices shall be provided and used. Plain cotton is not an acceptable protection device.
9. Hearing-protection devices must be worn if working in areas subjected to noise levels of 90 decibels (dB) or more for several hours. Ear inserts or muff-type protectors are satisfactory **(see 1-7)**. Keep hearing-protection devices clean.

1-6 A full face shield is worn when there is the possibility of damage to the face. *(Courtesy Aearo Company)*

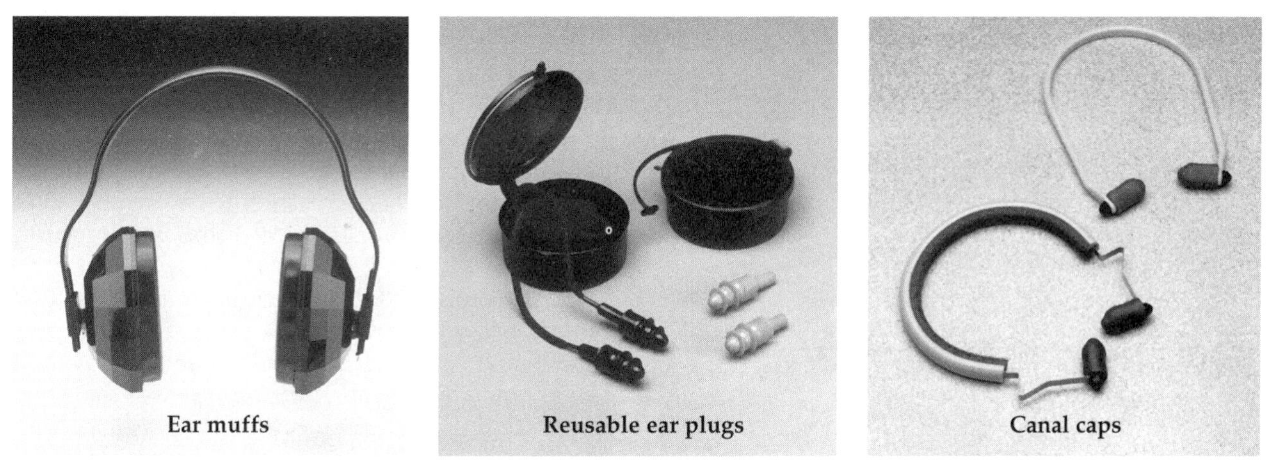

1-7 Typical ear protection devices are required for noise levels above those considered safe by OSHA regulations. *(Courtesy Aearo Company)*

10. When employees are exposed to harmful respiratory substances, respiratory protective devices must be used. Selection of a respirator should be made according to the guidelines in *American National Standard Practices for Respiratory Protection* Z88.2-1980 **(see 1-8 and 1-9)**.
11. Keep the respiration device clean. Change filters often.
12. Proper foot protection is needed at all times. Safety-toe shoes with a steel toe covering are recommended.
13. If working in wet conditions, rubber boots are needed.
14. Always wear a shirt and pants with long legs. Clothes should be loose enough to permit easy bending. They should not be too big because they can get caught.
15. Never wear jewelry of any kind. Rings, necklaces, bracelets, watches, ties, and other such articles must be removed.
16. When working aboveground where falls are likely, lifelines, safety belts, or lanyards should be used for employee safeguarding. These should be sized and secured as specified in OSHA requirements.
17. Stay alert when heavy equipment is being operated on the site. The beeping sound it emits means it is backing up. Since the operator has limited vision, be especially alert when you hear this warning.

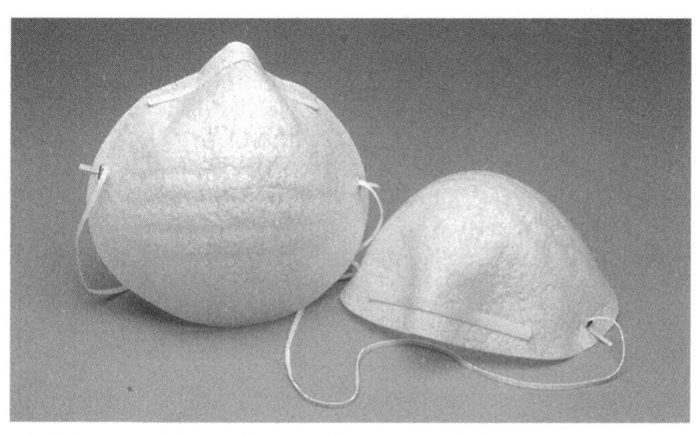

1-8 Disposable masks are effective for light dust conditions and should be discarded after a few hours' use. *(Courtesy Aearo Company)*

1-9 This is one type of respiratory protection device used when the air conditions are severe. It has replaceable filters which should be changed frequently. *(Courtesy Aearo Company)*

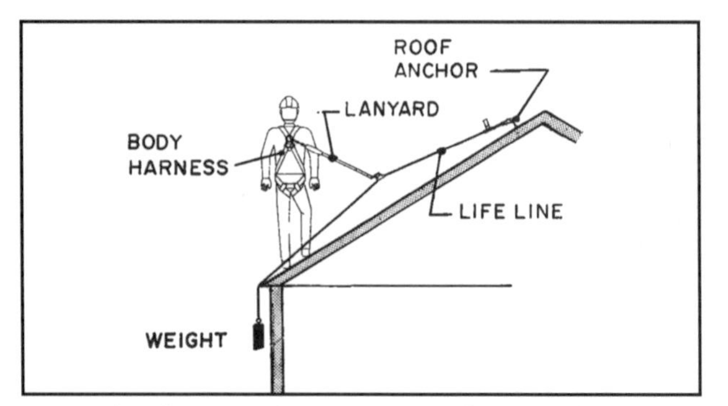

1-10 This illustrates the components of a typical fall arrest system. *(Courtesy DBI/SALA)*

FALL PROTECTION

Since falls are a major cause of accidents and deaths on the construction site, special attention needs to be given to safety procedures.

The Occupational Safety and Health Act (OSHA) requires the contractor to have various records on the site pertaining to safety, including a written fall protection plan, and to maintain accident records. Each employee should receive fall protection training, and approved fall prevention systems must be used. Various equipment manufacturers have developed fall protection systems that meet the standards of OSHA and the American National Standards Institute (ANSI). As the fall protection regulations change, both the manufacturer and construction personnel must be ready to make changes to meet the new requirements.

OSHA requires fall protection in areas such as ramps, runways, walkways, excavations, unprotected sides and edges, low-slope and steep slope roofs, holes, and wall openings as well as hoist areas, areas of formwork, reinforcing steel, leading edge work, bricklaying, and concrete erection. Any work area above six feet (1.8m) should have fall protection such as a personal fall arrest system, guardrails with toe boards, or a safety net system.

It is the duty of the employers to assess the workplace to determine if the walking and working surfaces have the strength and structural integrity to safely support the workers and any materials. Once this is assured they must select one of the protective options if a fall hazard is present.

1-11 This carpenter has on a body harness connected by a lanyard to a lifeline. *(Courtesy DBI/SALA)*

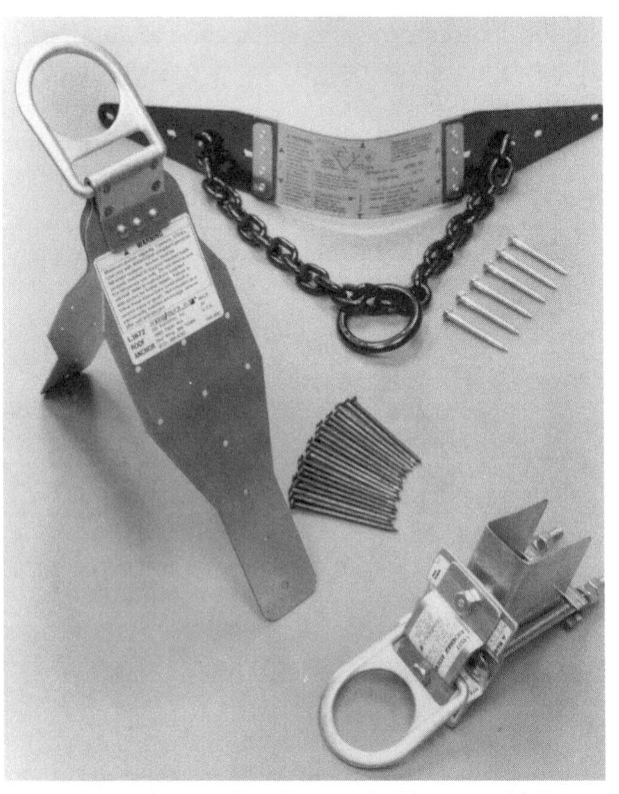

1-12 Some of the roof anchors used with personal fall arrest systems. *(Courtesy DBI/SALA)*

Personal Fall Arrest Systems

The personal fall arrest system consists of anchors, connectors, a lanyard, and a body harness and may include a lifeline, deceleration device, or some combination of these. Body belts are prohibited. In **1-10** and **1-11** such a system frequently used by carpenters and roofers is shown. It uses roof anchors to which the lifeline is connected. The carpenter's body harness is connected to the lifeline with a lanyard. The carpenter can slide the lanyard along the lifeline to reach the work area. However if a fall occurs the connecting device clamps tightly to the lifeline. Typical roof anchors are in **1-12**.

The personal fall arrest system in **1-13** holds the worker in a vertical position and connects the body harness to the metal forms. It limits the distance of a fall to a maximum of two feet (0.61m).

1-13 This fall protection device has a full body harness with chain connectors which in this case connect to the steel foundation forms. Notice the cable running from the back of the harness and above the worker's head. This holds the worker in the desired work location and limits a free fall to 2 feet (0.61m) or less. *(Courtesy DBI/SALA)*

The body harness is generally a high strength nylon belt and shoulder straps. It has a metal ring to which a lanyard is attached. The other end of the lanyard is secured to a lifeline.

For some types of work, such as working on high vertical surfaces, the lanyard is secured to a lifeline with a rope grab. The rope grab can be loosened and moved up and down the lifeline as the worker changes position. It locks to the lifeline if the worker falls. The worker is then suspended from the lifeline by the lanyard.

It should be pointed out that the safety harness will help reduce deaths from falls. However, occasionally the fall and the sudden stop may cause some physical injury. The worker should be extremely careful when working aboveground because the safety system will not totally protect from injury. Sometimes it gives a false sense of security.

Safety nets are used on jobs when safety belt use is limited by the nature of the situation. A good example is constructing a bridge over a river.

Warning line systems are barriers erected on a roof to warn employees that they are approaching an unprotected roof side or edge. Warning line systems also designate an area in which roofing work may take place without the use of a guardrail, body harness, or safety net system to protect employees in the area. They consist of ropes, wires, or chains, and the supporting stantions that hold them in place.

FIRST AID

A carpenter must understand basic first-aid techniques. Red Cross first-aid manuals and courses are available. A small first-aid kit could be carried by a carpenter. A large contractor will also have extensive first-aid equipment.

Some of the more common situations a carpenter should be able to handle are skin abrasions; sharp, deep cuts; puncture wounds; shock due to an accident or electrical shock; heat exhaustion; heat stroke; broken bones; burns; and frostbite.

Saving That Severed Finger

One accident carpenters face daily is the possibility of cutting into or severing a finger. First-aid manuals give the following advice:

1. Stop the bleeding. Wrap the stump with a firm, snug bandage, and elevate it above the rest of the limb. Move the person to a medical facility as soon as possible.
2. Retrieve the amputated part, rinse in cold water and wrap in a clean, damp cloth or paper towel, and place in a watertight plastic bag. Place the bag in ice water. Do not place directly on the ice. Carry it to the medical facility. Hopefully they will know where to send the patient to have the finger reattached.

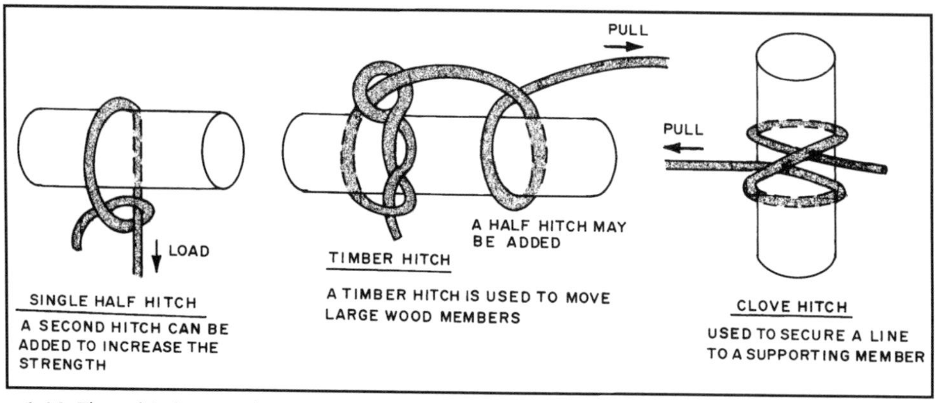

1-14 These hitches can be used to secure the end of a rope to a post or beam yet are easily untied.

TOOL SAFETY

Detailed safety instructions for using hand and power tools are given in Chapters 41 and 42. Following are some general safety hints for tool use.

1. All tools should be kept in good repair. Broken handles should be replaced. Dull saws and edge tools should be sharpened. A dull tool is a dangerous tool.
2. Keep tools clean.
3. Never remove guards from tools.
4. Use the right tool for the job. A chisel is a poor screwdriver. A screwdriver is a poor pry bar.
5. Read the manufacturer's instructions before using a tool. Obey them.
6. Store edge tools so that the cutting edges are covered.
7. When using wrenches, select the proper size. Do not overtighten threaded fasteners.
8. Use only a hammer for hammering. Pliers and other tools make poor hammers.
9. When sawing boards, be certain that they are held securely.
10. When using portable electric tools, be certain that they have a grounding wire and that it is connected to an approved ground.
11. Use only high-quality extension cords. A 100-foot cord of No. 12 wire will safely carry 20 amperes (A). A No. 8 wire will carry 35 A. A No. 4 wire will carry 60 A.
12. Examine electric extension cords for cuts and breaks.
13. Unplug any electric tool before you adjust or replace the cutting device.
14. Use only the attachments recommended for the tool by the manufacturer.
15. If working near welders, do not look at the arc.
16. Stay clear of all heavy equipment in operation. Stay alert for overhead and crane work as well as on-the-ground operations.
17. If demolition work is under way, observe the orders to seek safe quarters prior to the explosion.

SAFELY SECURING ROPES

The carpenter will occasionally have to tie a rope to a brace or post or join two ropes end to end. These connections must be made properly or an accident may occur. While various types of metal clips and rings are secured on the ends of lines for this purpose, it often becomes necessary to tie some type of hitch or bend to secure the end of a rope or to join the ends of two ropes.

Hitches

Hitches are used to fasten a rope to something, such as a timber, yet can be easily removed. They are shown in **1-14**.

The **clove hitch** is used when you want to temporarily connect a rope to a post or timber. The **half hitch** is another temporary connection. It is not a secure hitch and must be kept under tension. If a member is to be raised by a hoist, the **timber hitch** provides a more secure tie. It can have increased security by adding a half hitch.

1-15 Bends are used to tie together the ends of ropes of the same or different diameters.

Bends

Bends are used to join the ends of two ropes. These include the square knot and sheet bend **(see 1-15)**. The **square knot** is a strong connection and will not slip. It is used to connect ropes having the same diameter. The **sheet bend** is used to join ropes with different diameters. It can also be used to join a rope to a loop such as you find in the end of slings.

PREPARING EXCAVATIONS

While carpenters do not plan or dig excavations, such as those for a foundation, they do sometimes work in the excavated area. Excavations present a dangerous situation because of the possible collapse of the earth sides which could bury the worker. The carpenter must evaluate the effectiveness of the preparation that has been made to prevent the sides from collapsing. Remember, you do not have to have your head buried to die. If your chest is buried the pressure reduces your ability to breathe and you can suffocate because your lungs cannot expand and contract.

OSHA regulations require that in all excavations employees exposed to potential cave-ins must be protected by sloping or benching the sides of the excavation, or by placing a shield between the side of the excavation and the work area.

Excavations do not require a protective system if they are less than five feet (1.5m) deep or are made in stable rock.

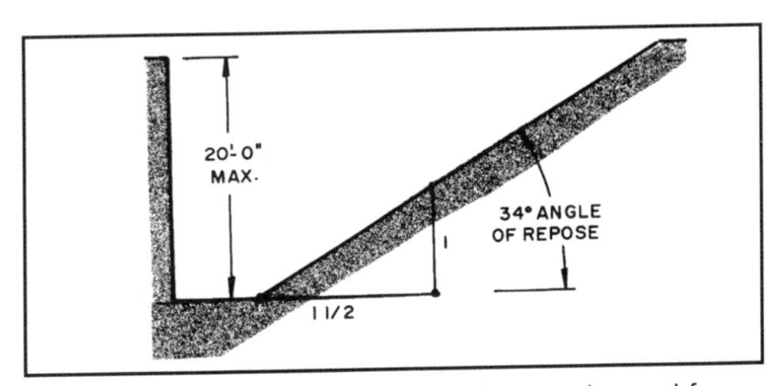

1-16 An excavation with a 1½ to 1 slope can be used for almost any soil for excavations up to 20 feet (6.1m) deep.

Sloping Requirements

One method for sloping the sides of the excavation is to have the sides on an angle not steeper than one and one-half horizontal units to one vertical unit, or 34 degrees **(see 1-16)**. This is referred to as the angle of repose. This can be used on excavations to a maximum of 20 feet (6.1m) deep. A slope of this grade or less is safe for any type of soil. Steeper slopes that may be acceptable in various soils must be tabulated and approved by a registered engineer **(see 1-17, page 14)**.

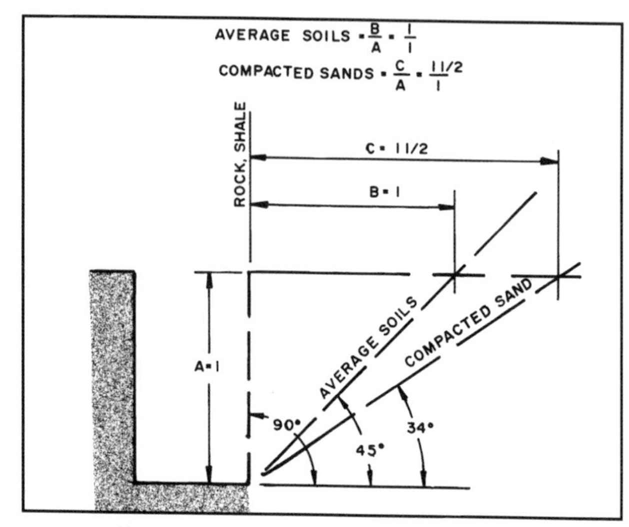

1-17 Different slopes can be used for various types of soil.

On very deep excavations the soil can be sloped by benching as shown in **1-18**.

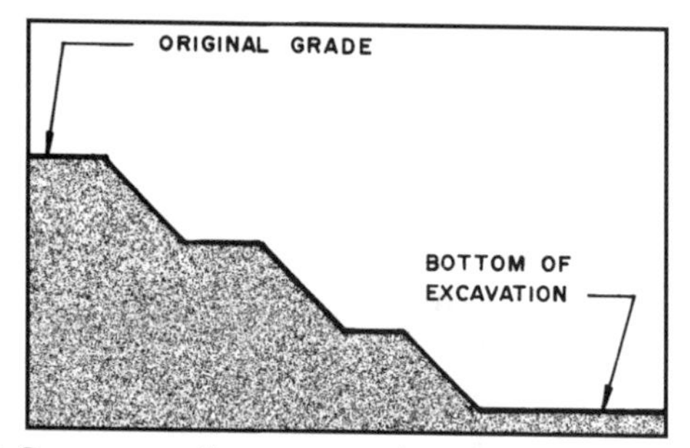

1-18 Deep excavations can use sloped sides by using the benching technique.

Trench Shields

Trench shields (sometimes called boxes) provide a protective work area for someone doing work in a limited space, such as welding a pipe. The shield **(see 1-19)** must be designed by a registered engineer and may be of timber, aluminum, or other suitable material. It must provide the same degree of protection offered by a complete shoring system.

Shoring Excavations

Shallow trenches can be shored using wood sheet piling braced by stringers and rakers as shown in **1-20**. Local and OSHA regulations control the depth, method of bracing, and size of the members.

Larger, deeper excavations can be shored with timber or steel sheeting piling or with steel soldier piles and wood lagging. The sheeting and soldier piles are driven into the soil with a pile-driving hammer.

The **timber sheet piling** is available in several types. The one shown in **1-21** uses square-edge planks. However, tongue-and-groove-edged planks are also used. These are driven into the soil side by side and supported with wales and rakers or soil or rock anchors.

Steel interlocking sheet piling sections are driven into the soil with a pile-driving hammer. They tend to be more watertight than timber sheet piling. They are braced with wales and rakers or soil or rock anchors **(see 1-22)**.

Sheet piling using **steel soldier piles and wood lagging** is shown in **1-23**. The soldier piles are driven into the soil, and wood lagging typically three inches thick is placed between them. Horizontal steel wales are placed across the soldier piles. The sheet piling may be secured using wales and bracing to the ground inside the excavation or with earth anchors or rock anchors **(see 1-24)**.

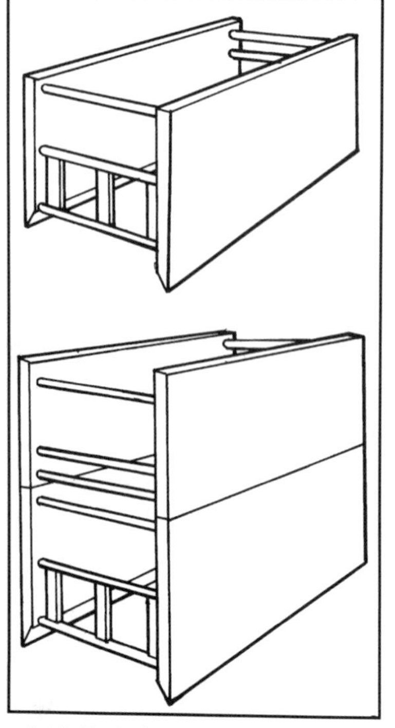
1-19 Typical trench shields.

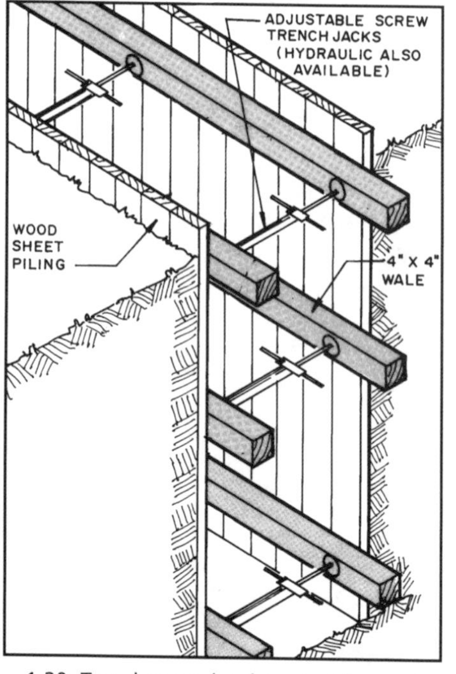

1-20 Trenches can be shored with wood sheet piling reinforced with wales and horizontal braces.

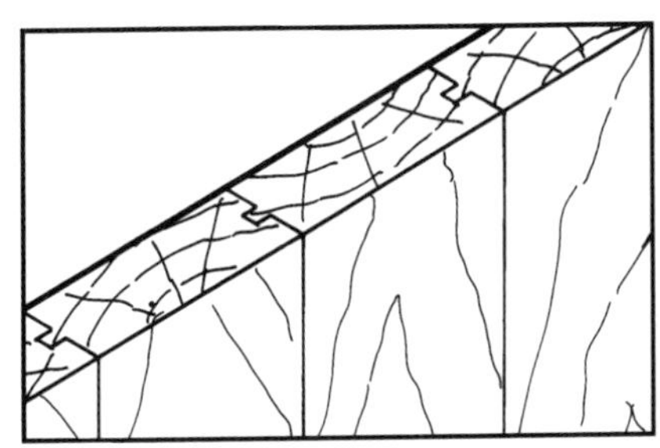

1-21 Timber sheet piling prevents the soil from collapsing into the excavation but is penetrated by ground water.

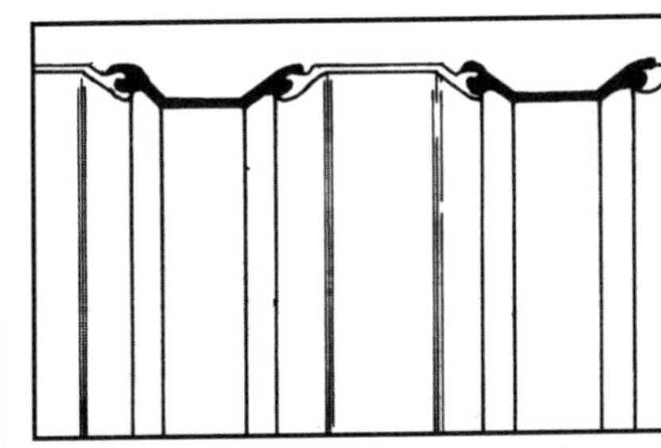

1-22 Steel interlocking sheet piling resists penetration by water and can be reused.

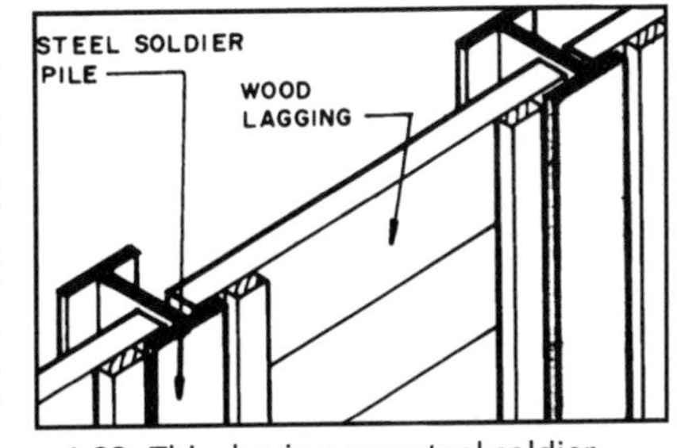

1-23 This shoring uses steel soldier piles and timber lagging.

Excavations over six feet (1.8m) deep must be protected by a guardrail system, fence, barricade, or cover to prevent workers from falling into it. Walkways over excavations over six feet (1.8m) deep also require guardrails. The guardrail should be at least 42 inches (107cm) high.

FLOOR & WALL OPENINGS

Codes establish requirements for protecting openings in walls and floors. Carpenters construct the required temporary railings and guardrails. In addition some jobs are located in populated areas and carpenters erect **barricades** to keep unauthorized people safely away from the site and control access by gates. Sometimes overhead platforms are built to protect passersby from materials that might be accidentally dropped.

Floor openings are protected by guardrails as shown in **1-25**. These include openings left where stairs will eventually be built. Small openings of 12 inches or less can cause workers to trip and allow debris to fall on those below. They should have a temporary cover. Following are some factors that must be considered.

1. Floor holes must be guarded with a standard railing or cover. If a cover is used, it must be able to carry twice the load passing over it. The cover should not be able to be accidentally moved out of place.
2. If the opening is a stairway or ladderway, it must have a swinging gate. The gate must be able to be locked.

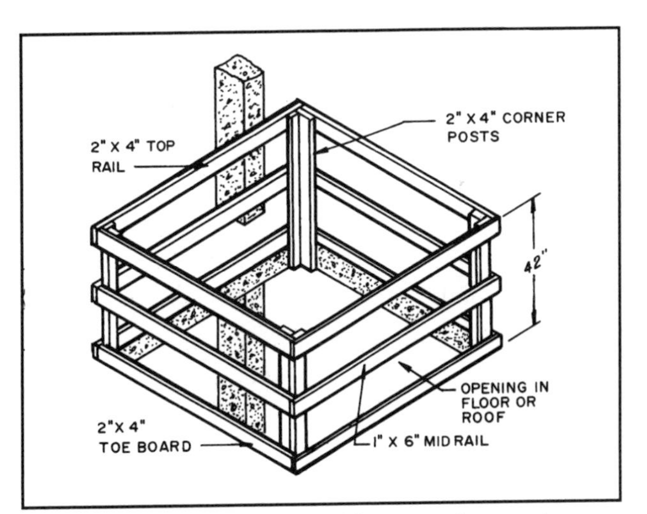

1-25 A recommended guardrail for protecting openings in the floor or roof.

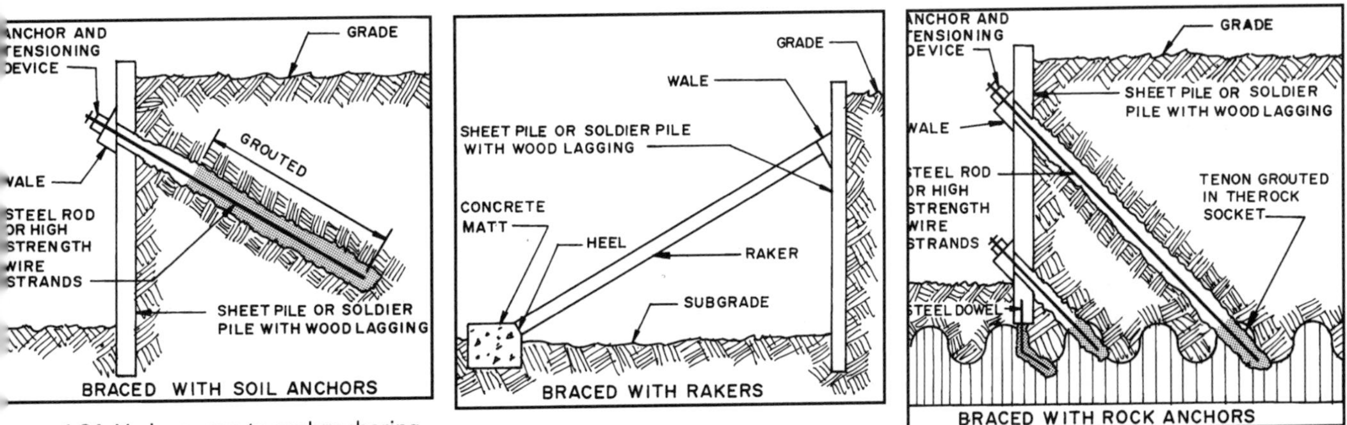

1-24 Various ways to anchor shoring.

3. Hatchways and chutes must have a standard railing or cover.
4. Wall openings, including window and door openings, with a drop of over four feet must have a standard railing **(see 1-26)**.
5. Temporary stairs should have a handrail.
6. Stairs should be kept clean.
7. Runways and ramps should have a standard rail if they are over four feet above the ground or floor.
8. Ramps should be securely fastened on both ends to prevent them from moving when used.

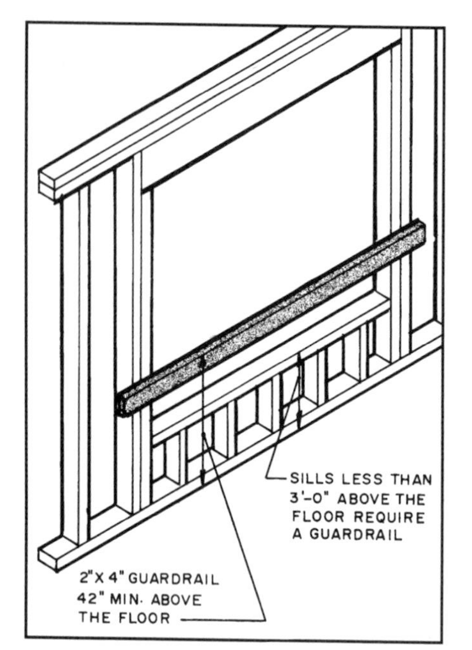

1-26 Wall openings should have temporary rails over the opening.

MATERIALS HANDLING SAFETY

1. Do not stack lumber and other materials too high. A maximum of 20 feet is normal for dressed lumber. When stacking, be certain to keep piles level and well supported.
2. Remove nails before handling used lumber.
3. When removing lumber from a stack, remove it so that the pile is kept level as it goes down.
4. When lifting materials, keep your back straight and lift with your legs **(see 1-27)**.
5. Stay out from under material being moved by a crane or forklift.
6. If a load is too heavy to lift alone, get some help.
7. Use mechanical aids to lift and move heavy objects. These include such things as wheeled carts, chain hoists, and forklifts.

1-27 Commonsense rules for lifting heavy materials.

FIRE PRECAUTIONS

The collection of materials, various trades, and many processes on a construction site make an accidental fire a likely possibility. Every construction worker should be constantly reminded to be aware of fire hazards and correct them or report them so action can be taken. Following are some preventative actions that should be taken.

1. Have adequate fire protection equipment placed strategically around the building and the construction site. Extinguishers suitable for the various types of fire anticipated should be available and they should be clearly marked.
2. Follow the rules for good housekeeping mentioned earlier in this chapter.
3. Control the location of smoking, especially when volatile materials are in use.
4. Take special precautions when hazardous activities, such as welding, are under way.
5. Cite the need for fire awareness in the weekly safety talks with all workers.
6. Always call the fire department if a fire exists. Do not rely solely on the possibility you may successfully extinguish it.

Extinguishers for Various Types of Fires

Fires are classified into four classes, and fire extinguishers are clearly marked to show the class of fire they will control.

Class A fires include those in wood, textiles, paper, and other fibrous flammable materials. Water is the proper extinguishing agent.

Class B fires include those caused by flammable liquids. They must be smothered with a chemical foam or carbon dioxide.

Class C fires are electrical fires. They are extinguished with nonconductive dry chemicals. Never use a Class A extinguisher on a Class C fire because water is a good electrical conductor and its use could lead to electrocution.

Class D fires involve combustible metal materials such as magnesium and sodium. They are extinguished with special powders that smother the fire. This type of fire should be extinguished only by trained personnel.

GROUND-FAULT CIRCUIT INTERRUPTERS

Injury and death from electrical shock is a real hazard on the construction site. The moving of materials and equipment can damage the temporary electrical lines. Working in times of snow or rain increases the hazard. Even very minor damage to an extension cord or piece of electrical equipment can cause shock. Most portable power tools now have double insulation which helps protect the user. However it does not protect from a damaged extension cord. Deficiencies in grounding and insulation can be overcome by using a **ground-fault circuit interrupter** (GFCI).

A GFCI is a fast-acting circuit breaker that senses small imbalances in a circuit caused by current leakage to the ground and in a fraction of a second shuts off the electricity. It also provides protection against fires, overheating, and damaged insulation.

OSHA requires the employer to provide approved ground-fault circuit interrupters for all 120-volt, single phase 15- and 20-ampere receptacle outlets that are in use by employees on construction sites and that are not a part of the permanent wiring of the building. GFCIs are available on extension cords and must be used on the job.

SAFETY AROUND MAJOR EQUIPMENT

A large construction site is the scene of considerable activity. The heavy equipment in use poses a constant source of danger.

Crane Safety

When working on site where cranes are in use observe the following:

1. Do not ride a load or a sling.
2. Do not work in the area below a crane when it is operating.

1-28 Operators of earthmoving equipment have limited vision. Those working on site must be constantly aware of the presence of this massive equipment. (Courtesy U.S. Department of the Army Corps of Engineers)

3. Be alert for the possibility of material falling as it is being raised or lowered.
4. Keep clear of cranes moving around the site.
5. When you are on the building, watch out for swinging booms, hooks, or loads.
6. Never try to serve as a crane signal person unless you are absolutely certain you know the hand signals.

Earthmoving Equipment Safety

When working on the site where earthmoving equipment is in use observe the following:

1. Stay alert when earthmoving equipment is being operated. The operator does not have a wide range of vision **(see 1-28)**.
2. Do not hitch rides on equipment.
3. Stay alert for an alarm indicating equipment is backing up.

ADDITIONAL INFORMATION

Fall Protection in Construction, Excavations

Materials Handling and Storage

Personnel Protective Equipment, Safety

Protection on Construction Sites

Standards for Scaffolds, Ground-Fault, and other publications related to safety on the job.

U.S. Department of Labor, Occupational Safety and Health Administration, 200 Constitution Ave. NW, Washington, DC 20210

CHAPTER 2

LADDERS, SCAFFOLDING & RUNWAYS

Much of the work performed by the carpenter occurs above the ground and requires the use of ladders and scaffolding. Working on ladders and scaffolding presents some hazards to the worker. However, if safety procedures are followed, the risk is greatly reduced. Procedures and equipment for the prevention of injury from falls is discussed in Chapter 1.

MANUFACTURED LADDERS

The manufactured ladders commonly used on light-frame construction include stepladders, straight ladders, and extension ladders **(see 2-1)**. The platform stepladder in **2-2** is used for interior work where a solid, smooth floor is available. Platforms that are raised and lowered either electrically or hydraulically are also frequently used **(see 2-3)**.

It is important to purchase high-quality ladders that meet OSHA requirements. Detailed information from OSHA on ladders and stairways is available in the publication *Stairways and Ladders, OSHA 3124*. National safety codes for portable ladders, available from the American National Standards Institute, include *ANSI-14.1-14.2-56, Safety Code for Portable Metal Ladders*.

GENERAL REQUIREMENTS FOR MANUFACTURED LADDERS

1. The rungs and steps of portable and fixed ladders should not be spaced less than 10 inches (254mm) apart and not more than 14 inches (355mm) apart. The steps on step stools must be not less than eight inches (20.3cm) apart or more than 12 inches (30.5cm) apart. The minimum clear distance between the side rails is 11.5 inches (292mm).

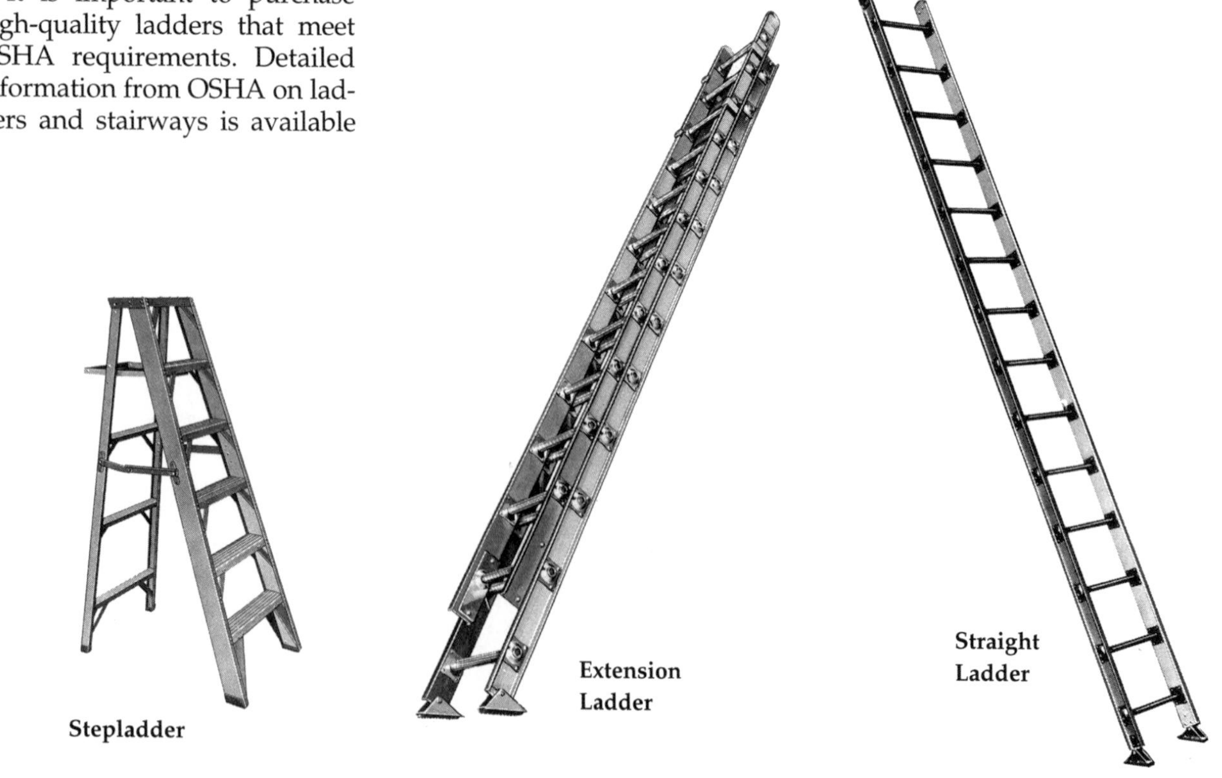

2-1 Commonly used types of ladders. *(Courtesy Aluminum Ladder Company)*

2. Straight ladders should not be tied together to create longer sections. Use longer extension ladders or build a platform partway up as a landing from one ladder to another.
3. Wood ladders must not be coated with an opaque coating because this hides defects.
4. Damaged ladders must be removed from the job and repaired or destroyed.
5. The rungs and steps must be corrugated, knurled, dimpled, or coated with a skid-resistant material to minimize slipping.
6. Ladders must have a load capacity equal to four times the maximum intended load.

2-2 A step platform ladder is moved on wheels and requires a firm, smooth base. *(Courtesy Aluminum Ladder Company)*

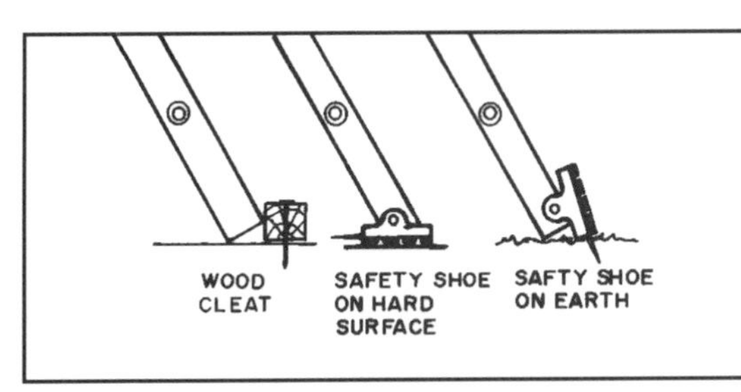

2-4 The legs of the ladder must have some type of foot or blocking to keep them from sliding.

USING STRAIGHT & EXTENSION LADDERS SAFELY

Following are recommendations to be observed when using ladders.

1. Never use a ladder that is defective. Examine regularly for cracked, broken, or missing parts.
2. Place ladders where they are out of the normal traffic pattern followed by workers on the job. This prevents them from accidentally being bumped or knocked down.
3. The feet of the ladder must be placed on a firm, nonslippery surface.
4. The ladder must have an approved type of nonskid feet **(see 2-4)**.
5. When necessary, nail blocking to keep the feet from sliding.
6. Clear the area of debris at the top and bottom of the ladder.

2-3 This aluminum aerial platform rises to 20 feet (6.1m) and uses a scissor design raising mechanism which reduces sway. *(Courtesy Mayville Engineering Company, Inc.)*

7. When raising a straight or extension ladder place the feet against something to keep them from sliding **(see 2-5)**.
8. The top of the ladder should extend three feet above the top of a wall to provides support for activities such as going up on a roof **(see 2-6)**.
9. Use a ladder that is the correct length for the job. Extension ladders are adjustable and make it easy to vary the length **(see 2-7)**.
10. The feet of the ladder should be away from a wall a distance equal to one-fourth of the working height of the ladder as shown in **2-6**.
11. The rungs and steps must be kept level.
12. When ascending or descending a ladder, keep your hands on both rails and face the ladder.
13. Move heavy materials to a roof or other high place with a forklift, crane, or hoist rather than trying to carry them up a ladder.
14. Keep the rungs and steps free of dirt and ice.
15. If a ladder enters an opening in a wall, nail a wood brace across the opening for support.
16. Avoid using ladders when exposed to high winds.
17. Move tools by carrying them in a carpenter's belt or raise them in a bucket on a hoist.
18. Observe the load limitations specified for the ladder. They must support at least four times the maximum intended load.
19. Ladders are not designed to carry loads when they are in a horizontal position, so do not use them as scaffolding.
20. Do not lean out away from the ladder. If you cannot reach something, move the ladder closer to it.

2-5 To raise a ladder, place the feet against something to keep them from sliding.

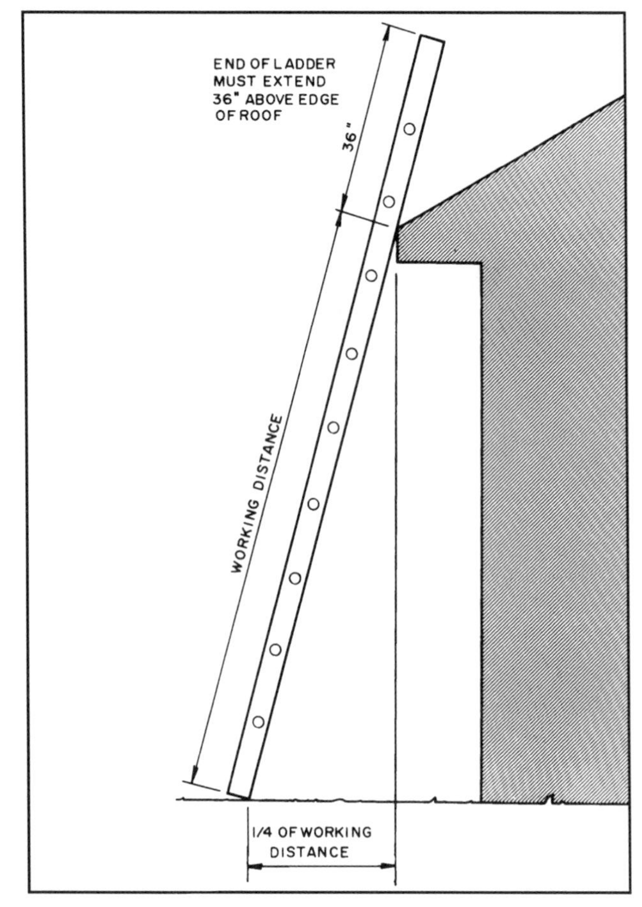

2-6 Set the feet of the ladder the proper distance from the wall. Make certain that it extends 36 inches (915mm) above the edge upon which it rests.

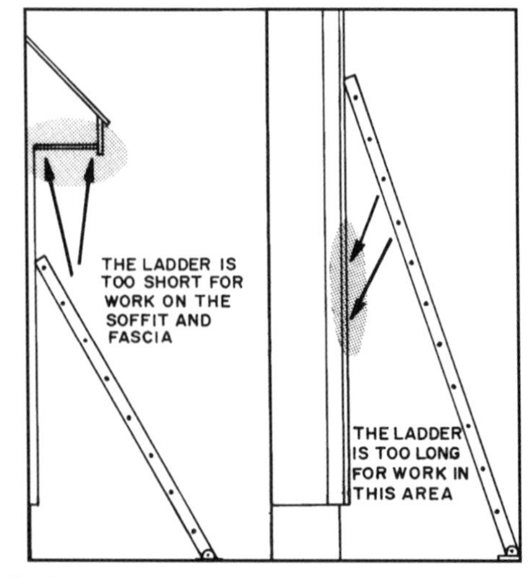

2-7 The ladder must be the proper length for reaching the work area.

21. Keep your shoes clean and dry.

22. Do not stand on the top three rungs of a straight or extension ladder.

23. Allow a 3-foot overlap between sections of a 36-foot extension ladder. A 4-foot overlap is required when using 48-foot extension ladders.

24. Metal ladders that could conduct electricity should not be used near energized lines or equipment. Conductive ladders must be prominently marked as conductive.

USING STEPLADDERS SAFELY

1. Observe the maximum load-carrying capacity specified.

2. Be certain the ladder is fully open and the braces are fully extended, locking the legs in place.

3. All four legs must be firmly set on a level surface.

4. Do not stand on the top two steps. These are usually marked "no step."

5. Keep the ladder in good repair or destroy it if damaged beyond repair.

SITE-BUILT LADDERS

Site-built ladders are made on the job using lumber sent in for construction of the building. The lumber used must be of good quality and pieces with large loose knots or splits must be discarded. The ladders must be built according to OSHA standards. A single-rung ladder provides one-way traffic while a double-rung ladder provides two-way traffic **(see 2-8)**.

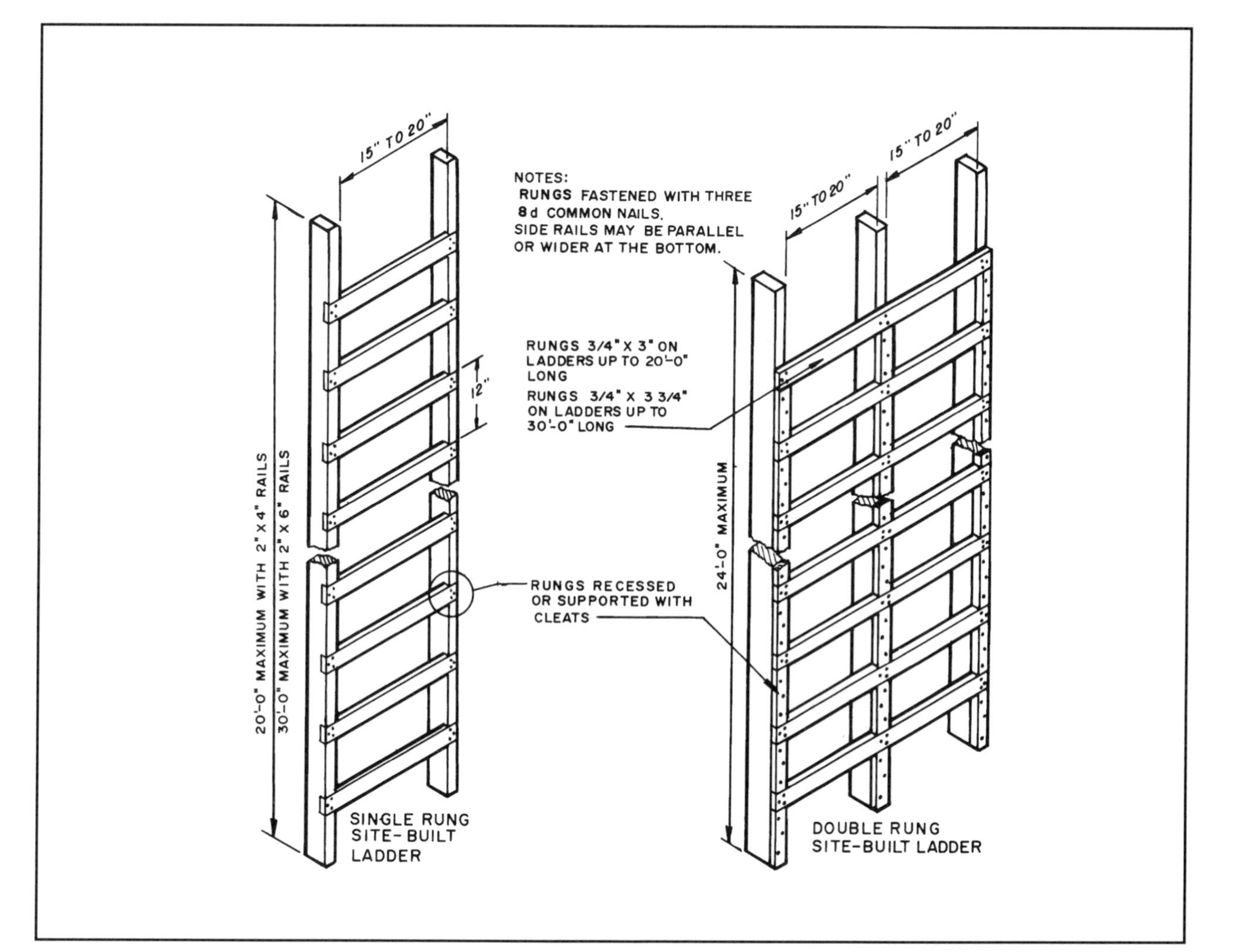

2-8 Typical site-built ladders.

Single-rung ladders cannot exceed 30 feet in length while double-rung ladders must not exceed 24 feet in length. The rungs are set into notches cut in the rails or are braced with cleats **(see 2-9)**. Recommended sizes of lumber for the various parts are shown in **Table 2-1**.

Double-rung ladders are used in each work area where 25 or more employers enter and leave or where two-way traffic is required. Two manufactured ladders can also serve the same area.

SCAFFOLDING

Scaffolds are temporary constructions built to provide a safe, elevated work platform. They enable the worker to walk on a horizontal platform and be protected from falls by a guardrail. This is much safer than working at high elevations using ladders.

Many contractors prefer to use factory manufactured metal scaffolding because it is safe, meets OSHA standards, and is easily erected and dismantled. It can be used for many years. Wood scaffolding is built on the job by carpenters and is often used on small jobs. It must be constructed to meet OSHA safety regulations. The wood should be of good quality, and members with large loose knots and splits should not be used. Proper nailing and bracing is important.

PLANKS

Planks may be laminated or solid wood, metal, or other approved materials. Both the Southern Pine Inspection and Lumber Inspection Bureaus provide plank material with their grade stamp which indicates it meets OSHA specifications. Solid wood planks meeting these requirements can span distances as shown in **Table 2-2**. Fabricated planks such as laminated veneer lumber are rated by the manufacturer based on the maximum intended load as shown in **Table 2-3**.

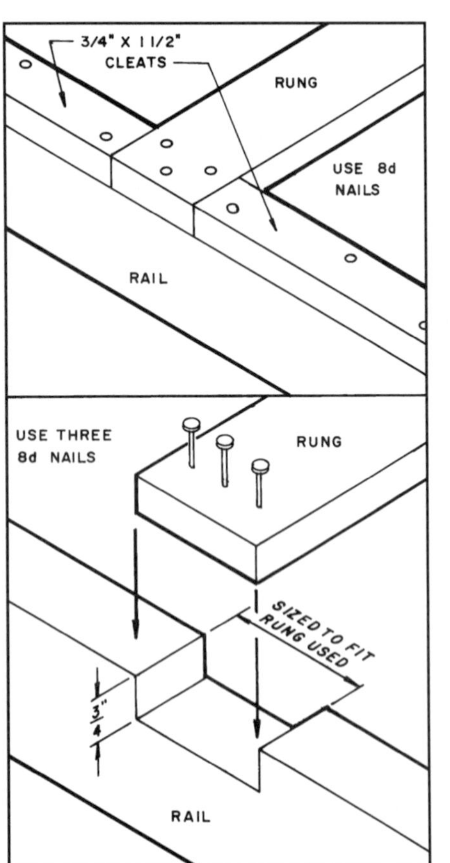

2-9 Site-built ladders may have the rungs recessed into the side rail or be supported with wood cleats.

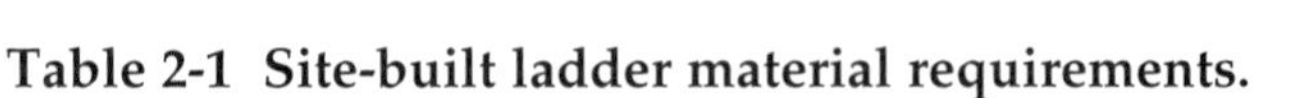

Table 2-1 Site-built ladder material requirements.

Ladder length	Minimum inside width at the base	Rung	Rails
Up to 12 ft	16 in	1 × 3 in	2 x 3 in
over 12 ft to 20 ft	18 in	1 × 4 in	2 x 4 in
over 20 ft to 30 ft	19 in	1 × 4 in	2 x 6 in

Table 2-2 Maximum spans for grade-stamped planks.*

Maximum intended nominal load (lb/sqft)	Maximum permissible span using full thickness undressed lumber (ft)	Maximum permissible span using nominal thickness undressed lumber (ft)
25	10	8
50	8	6
75	6	—

* OSHA *Report 29 CFR Part 126, Safety Standards for Scaffolds Used in the Construction Industry; Final Rule.*

SCAFFOLD CONSTRUCTION REQUIREMENTS

Scaffolding must meet the requirements established by the Occupational Safety and Health Administration. Manuals detailing these regulations may be secured from OSHA. Since these regulations are very detailed, only a few of the more commonly occurring situations facing carpenters are included in this chapter. Because these regulations are updated periodically, regular contact must be maintained to get information or revised regulations.

WOOD SINGLE-POLE SCAFFOLDS

Wood single-pole scaffolds use a single vertical pole placed about two to three feet away from the building and resting on a 2 × 6-inch footing board. The poles are connected with a 2 × 6-inch ribbon. Ledgers rest on the ribbon and are connected to the building with notched 2 × 6-inch blocks. The wood planking is supported by the ledgers **(see 2-10)**.

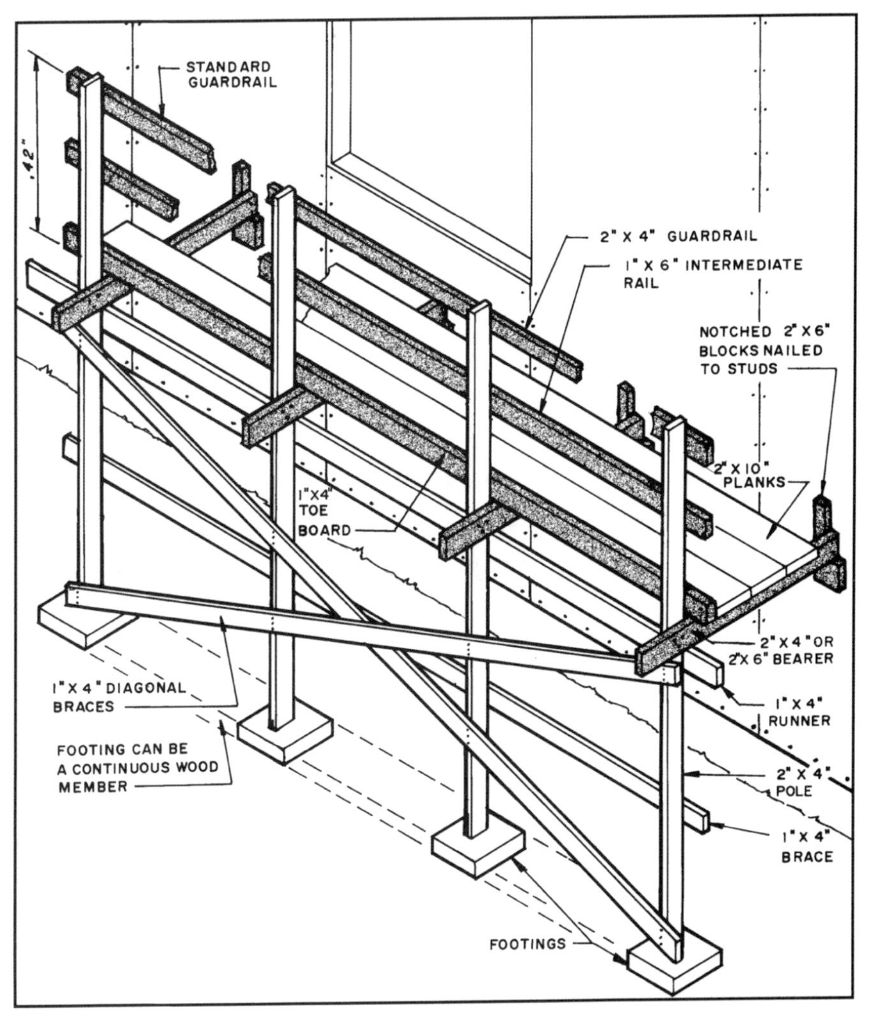

2-10 A job-built, light-duty, single-pole scaffold is fastened to the wall of the building.

Table 2-3 Loads used when fabricated planks are specified by the manufacturer.*

Rated load capacity	Intended load over the entire span area (lb/sqft)
Light duty	25
Medium duty	50
Heavy duty	75
	Intended load by the number of persons on the scaffold (lb/sqft)
One person	250 pounds at the center of the scaffold
Two persons	250 pounds 18 in. to the left and right of the center of the span
Three persons	One person at the center and one person 18 in. to the right and left of the center of the span. Each person equals 250 pounds.

* OSHA *Report 29 CFR Part 126, Safety Standards for Scaffolds Used in the Construction Industry; Final Rule.*

Design data for light-duty (25 lb/sqft) single-pole scaffolding is shown in **Table 2-4**. Larger members are required for medium (50 lb/sqft) and heavy-duty (75 lb/sqft) single-pole wood scaffolds. The scaffold platform should be at least 18 inches (457mm) wide. The scaffold should support its own weight and at least four times the intended load.

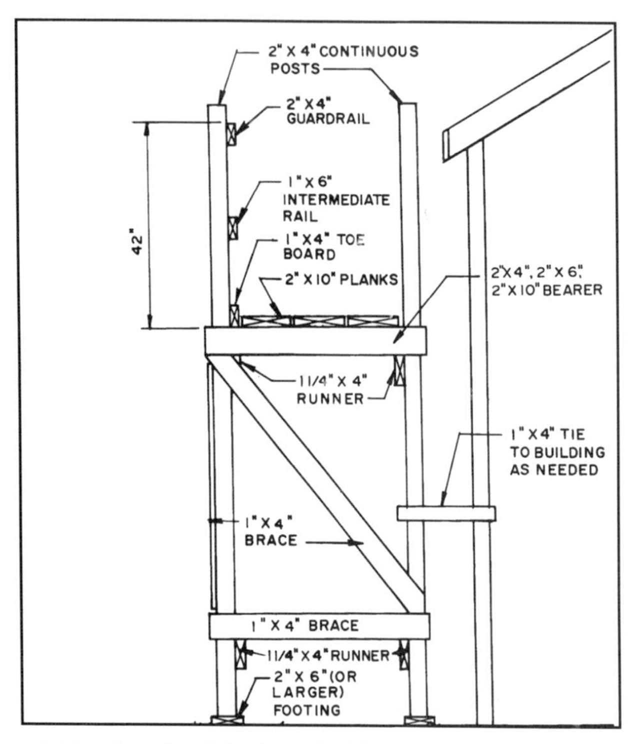

2-11 A low-rise, light-duty, double-pole wood scaffold.

Wood Double-Pole Scaffolding

Wood double-pole scaffolding uses two parallel rows of poles. One is next to the building and the other is several feet away. It stands free of the building and does not rely on the exterior wall to support any load. Typical construction of double-pole, low-rise scaffolding is shown in **2-11**. The same type of construction may be used for multi-story scaffolding. It is limited to a height of 60 feet **(see 2-12 and Table 2-5)**. Design data for light-duty (25 lb/sqft) and heavy-duty (75 lb/sqft) scaffolds are specified in OSHA publications.

Scaffold Platform Requirements

Each end of a platform should extend over the centerline of its support at least six inches (15cm) unless restrained by an approved method.

Each end of a platform 10 feet (3m) or less in length should not extend over its support more than 12 inches (30.5cm). A platform over 10 feet (3m) in length should not extend more than 18 inches (460mm) beyond its support.

Table 2-4 Single-pole wood scaffolds for light duty.*

		Up to 20 ft high	Up to 60 ft high
Maximum intended load (lb/sqft)		25	25
Poles		2 × 4 in	4 × 4 in
Max. pole span	Longitudinal	6 ft	10 ft
	Transverse	5 ft	5 ft
Runners		1 × 4 in	1¼ × 9 in
Bearers size and spacing		2 × 4 in – 3 ft	2 × 4 in – 3 ft
		2 × 6 in – 5 ft	2 × 6 in – 5 ft
Planking		1¼ × 9 in	2 × 10 in
Maximum vertical spacing of horizontal members		7 ft	9 ft
Bracing, horizontal, diagonal, and tie-ins		1 × 4 in	1 × 4 in

* U.S. Department of Labor, Occupational Safety and Health Administration publication, *Safety Standards for Scaffolds Used in the Construction Industry; Final Rule, 1996.*

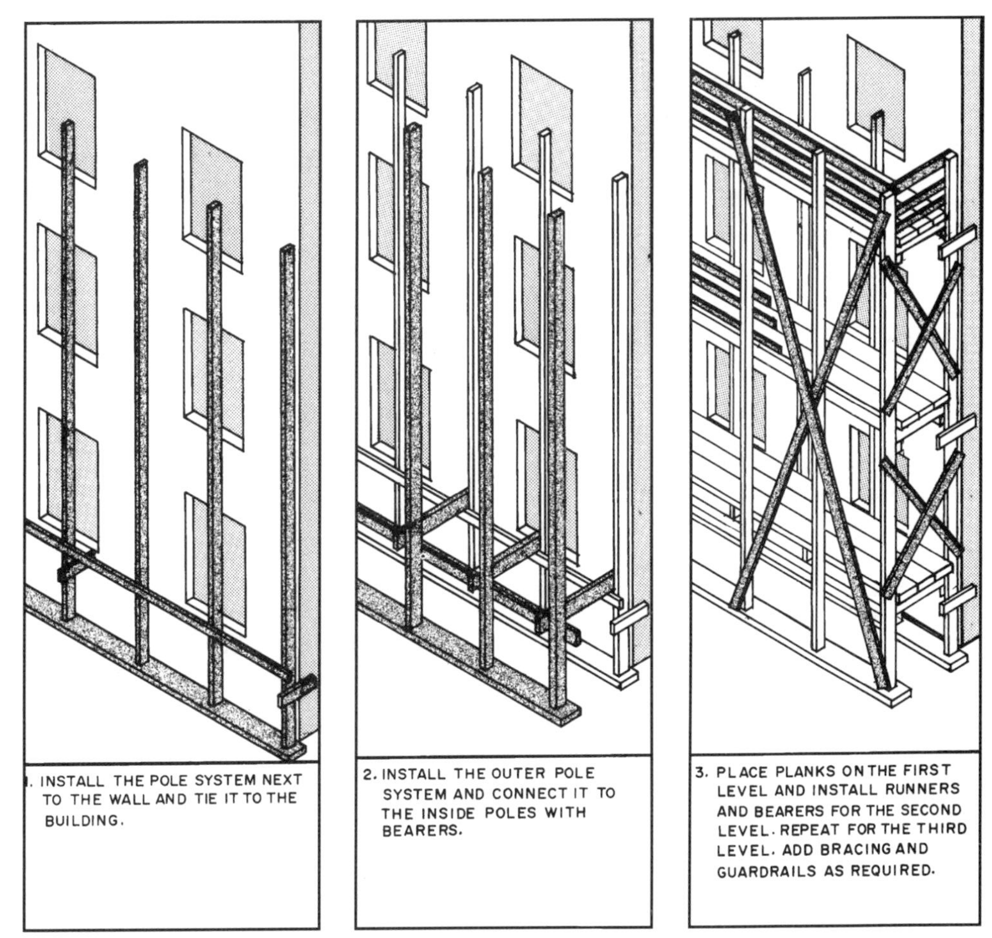

2-12 A typical light-duty, high-rise, double-pole scaffold.

Table 2-5 Double-pole wood scaffolds for light duty.*

		Up to 20 ft high	Up to 60 ft high
Maximum intended load (lb/ft2)		25	25
Poles		2 × 4 in	4 × 4 in
Max. pole span	Longitudinal	6 ft	10 ft
	Transverse	6 ft	10 ft
Runners		1¼ × 4 in	1¼ × 9 in
Bearer size and spacing		2 × 4 in – 3 ft	2 × 4 in – 3 ft
		2 × 6 in – 6 ft	2 × 10 in (rough) – 6 ft
Planking		1¼ × 9 in	2 × 10 in
Maximum vertical spacing of horizontal members		7 ft	7 ft
Bracing–horizontal, diagonal, and tie-ins		1 × 4 in	1 × 4 in

* U.S. Department of Labor, Occupational Safety and Health Administration publication, *Safety Standards for Scaffolds Used in the Construction Industry; Final Rule, 1996.*

When scaffold planks overlap to create a long platform the overlap must occur on supports and should be not less than 12 inches (30.5cm). This can be reduced if the planks are nailed together to prevent movement.

Wood platforms should not be covered with opaque finishes because this may cover cracks and other defects.

Metal Scaffolding

Metal tube and coupler scaffolding is a manufactured product made from steel or aluminum. The system is designed to easily fit together and studs are welded to the columns to which metal bracing is bolted. Likewise it is easily disassembled and moved to another job. The quality is good and the structural system is dependable. If any part is bent or broken it should be replaced. Straightening a bent part does not restore structural integrity. A typical installation is in **2-13**. The vertical posts are set on 2 × 10-inch (38 × 240-mm) lumber. The ground should be leveled and if necessary compacted. The scaffold should be tied to the building.

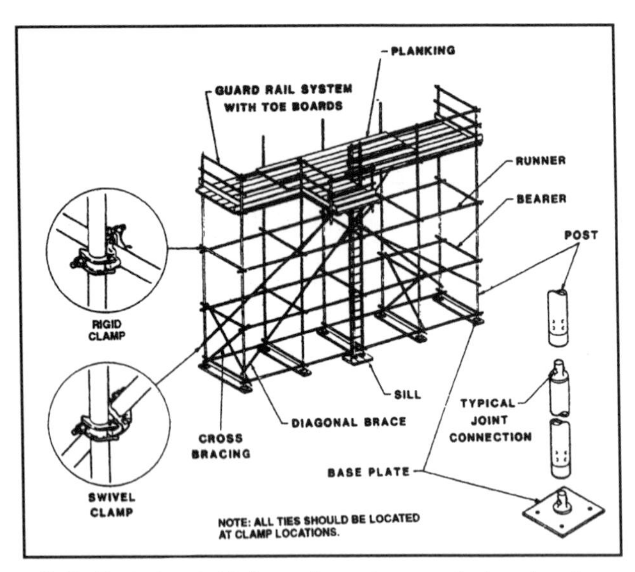

2-13 Typical metal tube and coupler manufactured scaffold. Notice the ladder built as part of the structure. *(From U.S. Department of Labor, Occupational Safety and Health Administration publication,* Safety Standards for Scaffolds Used in the Construction Industry; Final Rule*)*

2-14 A duplex nail has two heads.

Scaffolding Safety

1. A scaffold must be able to hold four times the load it is expected to carry.
2. The footing for a scaffold must be level and solid and must not have motion when weight is applied. The scaffold must be level and plumb.
3. Scaffolds must be erected by experienced carpenters.
4. The lumber used must be free of defects.
5. Wood scaffolding should be designed to meet OSHA standards.
6. Use duplex nails on wood scaffolding to help when taking it apart **(see 2-14)**.
7. If people must walk under scaffolding as they work, the space between the toe board and guardrail should be covered with screen. This prevents materials from falling on workers.
8. Typically scaffolding guardrails include a toe board which prevents tools from being knocked off onto those working below **(see 2-15)**.
9. Always use a ladder to climb up a scaffold. Never climb on the scaffold frame. Metal scaffolding will usually have a ladder built on it as part of the structure.
10. If work is going on above the scaffold, overhead protection is needed for those working on the scaffold.
11. Do not put more material on the scaffold than is needed for the immediate job. Keep it clear of debris.
12. The boards forming the scaffold platform should not have spaces between them. The minimum-size board is 2 × 10 inches. Discard cracked or damaged boards.

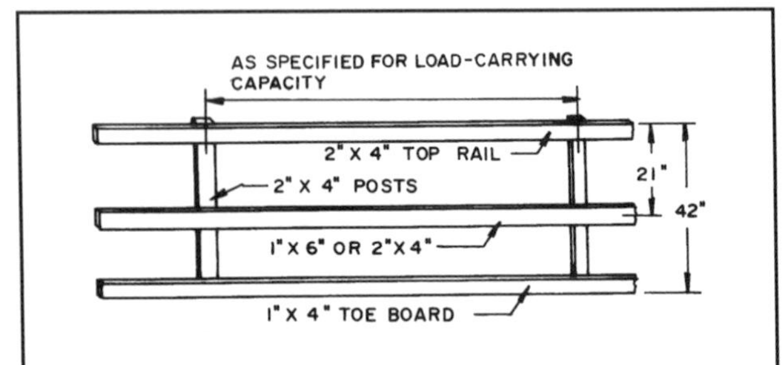

2-15 A standard wood guardrail.

13. The wood planks should overlap at least 12 inches and not extend beyond the ledger boards more than six inches.
14. When using manufactured scaffolding, observe the maximum design load.
15. Follow the manufacturer's instructions for assembling manufactured scaffolding.
16. Be certain that all hand screws and wing nuts are tight.
17. If the unit is on wheels, be certain the wheels are locked or blocked so that they do not roll before you climb to the platform.

Ladder Jacks

Ladder jacks are metal frames fastened to the rungs of ladders that support scaffold boards. Several types are available as shown in **2-16** and **2-17**. These are designed to support a worker doing light work. They should not be loaded with materials or several workers. This extra loading requires a scaffold.

Ladder jacks can be used safely provided these conditions are observed:

1. Use only for light duty with a maximum load of 25 lb/sqft.
2. Install only on quality manufactured ladders.
3. The platform should be 12 inches (30.5cm) wide.
4. Do not space the ladders over eight feet apart.
5. Ladders spaced closer than eight feet can support two people.

2-16 One type of ladder jack. *(Courtesy Aluminum Ladder Company)*

Roofing Brackets

A roofing bracket is a metal frame that goes over the roof ridge or is nailed in place on the roof. It has space for a wood plank **(see 2-18)**. Brackets can be adjusted to fit the pitch of the roof. When working with a roofing bracket, the worker should wear a safety belt.

2-17 Other variations of ladder jacks.

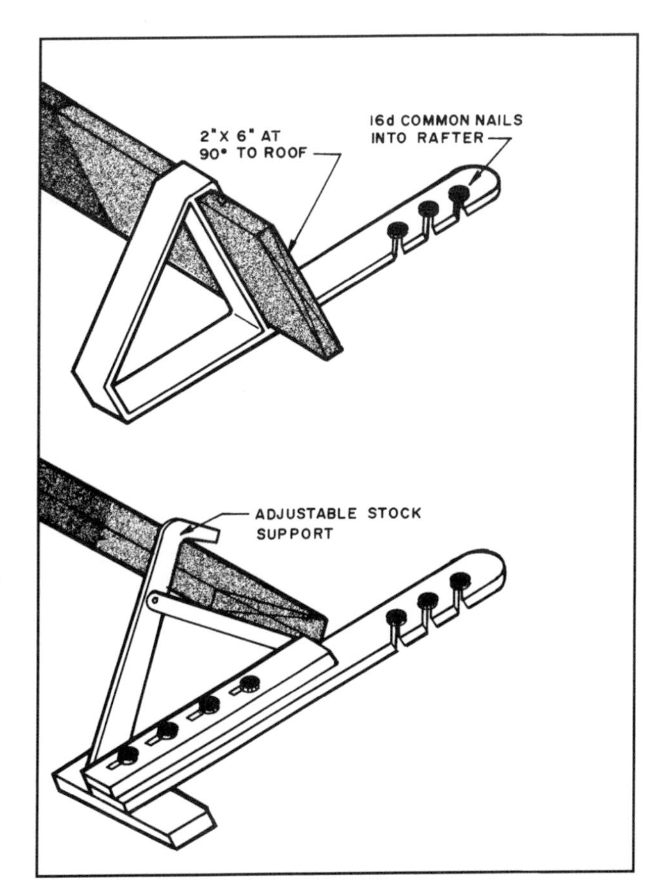

2-18 Two types of roofing bracket.

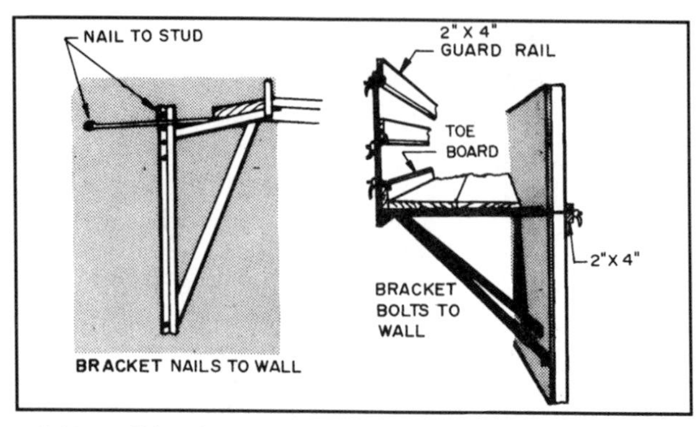

2-19 Wall brackets are nailed or bolted to the studs in the exterior wall.

Wall Brackets

A wall bracket is a metal frame that is either nailed to the building frame or is hooked around a stud **(see 2-19)**. These brackets must be installed carefully. Each nail should penetrate solid wood. Use 16d or 20d common nails. Nails that bend or have damaged heads should be replaced. Wall brackets are used for light duty work and should not be loaded with materials.

Trestle Jacks

Trestle jacks are used to support scaffolding boards for low work. They are most often used for interior jobs such as applying drywall sheets to a ceiling. Their height is adjustable **(see 2-20)**. The boards must be two-inch material and free from defects that would weaken them. The jacks are spaced to provide a firm support for the boards.

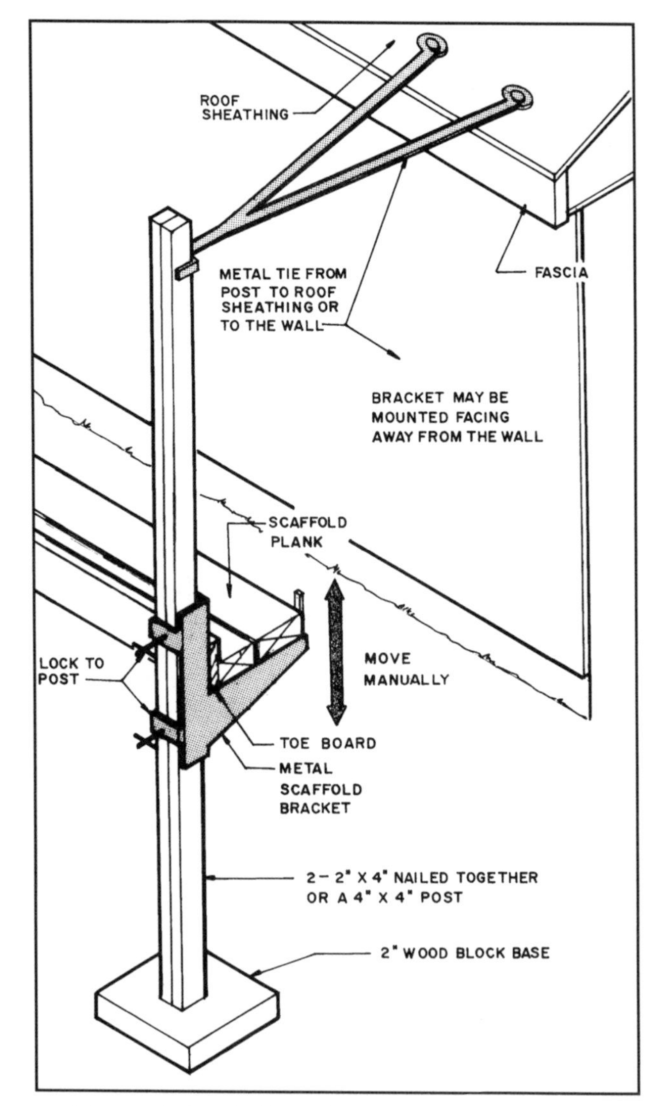

2-21 A typical scaffold bracket that is manually raised and lowered on a 4 × 4-inch pole.

Pump Jacks and Scaffold Brackets

Scaffold brackets are clamped to a 4 × 4-inch pole and support scaffold planks. They are raised and lowered manually on the post **(see 2-21)**. Pump jacks are much like scaffold brackets but are raised by pumping a pedal and lowered by turning a crank **(see 2-22)**. The 4 × 4-inch poles are connected to the roof sheathing or wall with metal ties.

The pole can be either two 2 × 4-inch wood members nailed together with 12d common nails staggered uniformly from opposite outside edges, or a solid 4 × 4-inch pole. The pole has a maximum height of 30 feet and a maximum load of 500 pounds between poles applied at the center of the span. Not more than two

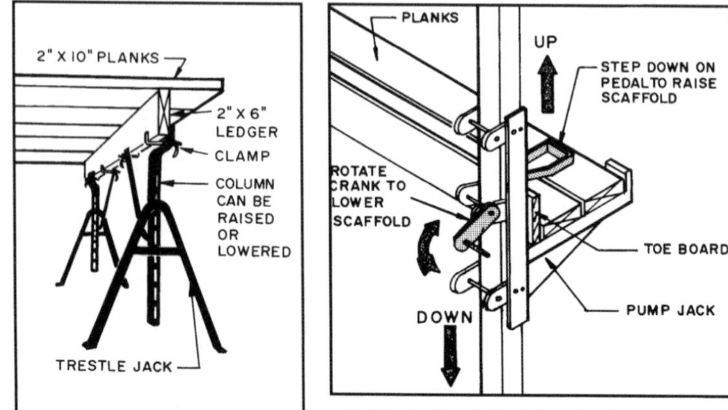

2-20 Trestle jacks rest on the ground or floor and support a low scaffold.

2-22 A plumb jack is raised on a 4 × 4-inch pole with a foot-operated pedal and lowered with a cranking mechanism.

people can be on a pump jack scaffold at the same time. The platform should be at least 12 inches (30.5cm) wide.

Elevated Platforms

Elevated platforms are available in a wide range of sizes. They are electrically or hydraulically operated and are controlled by the person on the platform. The platform has guardrails with toe boards as required by OSHA regulations. It can be used inside or outside a building. Units lifting workers above one story have outriggers to increase the stability of the unit. The manufacturer specifies the maximum lifting capacity and this must be posted on the unit and observed **(see 2-23)**.

RUNWAYS, RAMPS & TEMPORARY STAIRS

Temporary stairs must be constructed to carry the loads expected to be placed upon them. Wood ramps are widely used to move workers and materials up small elevations. Runways are temporary constructions used to move workers and materials such as power buggies that carry concrete from the mix truck up to where it will be placed **(see 2-23)**.

GUARDRAILS

A guardrail system designed for use wherever a safety system is needed is shown installed on a wood framed roof in **2-24**. It protects workers from falls. The basic unit is an aluminum post that has a base that is secured to the roof, floor, or other base material **(see 2-25)**. It is adjustable so it remains perpendicular to the ground for roof slopes of $4/12$, $8/12$, and $12/12$. The horizontal rails are standard 2 × 4-inch stock. The brackets are spaced 8 feet apart.

ADDITIONAL INFORMATION

U.S. Department of Labor, OSHA Publications, P.O. Box 37534, Washington, DC 20013-7535

American National Standards Institute, Inc., 1430 Broadway, New York, NY 10018

2-24 With this Guardrail 2000 safety system, metal brackets are nailed to the roof, providing the carpenter protection from falls. *(Courtesy Reichel Corporation)*

MAXIMUM SLOPE 1 IN 3 (20°)
4" WOOD CURB ON BOTH SIDES OF THE RAMP
GUARDRAIL
WOOD PLANK RUNWAY
FORMS FOR A CONCRETE WALL
RAMP OF WOOD PLANKS
SUPPORTS DESIGNED TO CARRY INTENDED LOADS
BRACE AS NECESSARY
POSTS 4X4 OR LARGER

2-23 Carpenters frequently have to build runways and ramps over which construction materials are moved.

2-25 This is the aluminum bracket that is secured to the floor or roof and supports horizontal 2 × 4-inch rails. *(Courtesy Reichel Corporation)*

CHAPTER 3

BUILDING CODES, ZONING REGULATIONS, BUILDING PERMITS & INSPECTIONS

Construction is regulated by state and local governments by the adoption of building codes. Building codes are regulations that establish the minimum standards of workmanship and quality of construction. Their purpose is to protect the safety, health, and welfare of those who will work or live in the building. Codes are recorded in detailed manuals and pertain to things such as foundations, structure, fire protection, finishes, means of egress, roofs, various materials of construction, electrical, mechanical and plumbing systems, and site work.

The architects and engineers designing the building must observe all codes, and the various construction trades must likewise perform their work so the actual construction meets the code regulations.

Some states have statewide codes that will apply to construction over the entire state. Local governmental units can move beyond the state codes and have additional requirements. The adopted code is the law and can be enforced by the local government.

MODEL BUILDING CODES

The preparation of a building code is a complex and technically demanding process. Therefore most governmental units adopt a model code and occasionally supplement it with special local requirements. They serve as models for local and state governments as they prepare their codes. Often the model code is adopted in its entirety.

Model codes are developed through the cooperative effort of building officials and construction industry representatives. The actual enforcement of the code is up to the building officials of the local government. Since model codes are revised on a regular basis to reflect new materials and techniques, architects, engineers, and those in the construction trades must keep up to date concerning the existing codes.

The major model codes include the Uniform Building Code, Standard Building Code, Basic Building Code, and National Building Code.

The Uniform Building Code is widely used in the western states while the Standard Building Code is typically adopted in the southern states. The Basic Building Code and the National Building Code are national model codes. In addition to the codes developed, each code organization publishes extensive code-related publications and numerous educational publications.

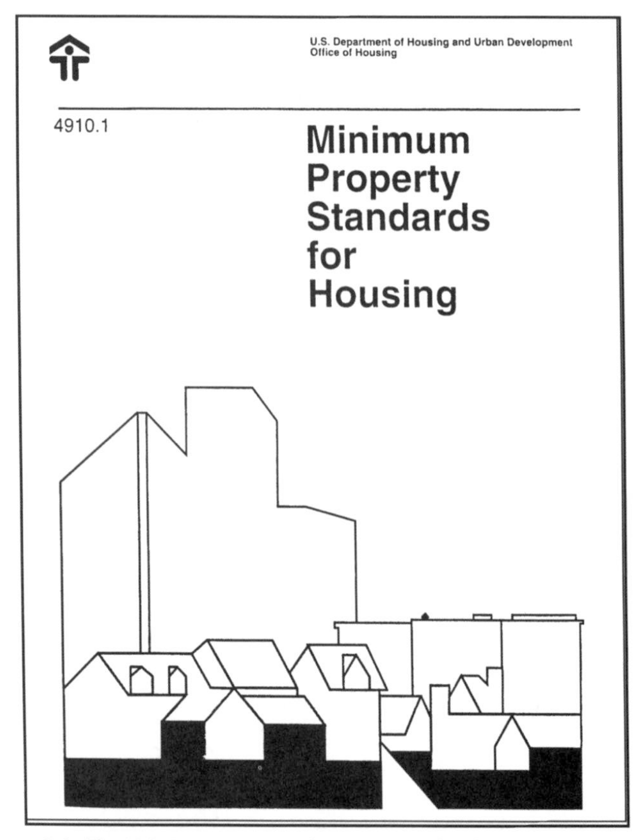

3-1 The U.S. Department of Housing and Urban Development publishes these standards for residential buildings.

The U.S. Department of Housing and Urban Development publishes the code in *Minimum Property Standards for Housing* **(see 3-1)**. Standards apply to buildings and sites designed and used for normal multifamily and care-type housing occupancy.

ZONING REGULATIONS

Zoning regulations specify the type of building that may be built in a specified area. These zones are recorded on a zoning map and are readily available to property owners and contractors at the local building official's office. The building official is an employee of the local government and is responsible for all activities related to construction. Zoning regulations are recorded in a document such as the one shown in **3-2**.

Typically zoning regulations will specify areas such as single-family residential, multifamily residential, recreational, agricultural, commercial, or industrial. The zoning ordinance carefully details exactly what may be built in each area. The property owner, architect, and contractor must be fully aware of the requirements for each piece of property. If the property owner feels an exception is needed and justified, appeals to the zoning regulations can be made to a governmental body, such as the Zoning Board of Adjustment.

The zoning regulations also specify a range of other things such as the minimum lot size, minimum building size, maximum height of a building, types of occupancy permitted, area that can be covered with impervious materials, and area in which a building can be built. The building area is established by setbacks. Setbacks are the distances the building must be from the sides of the lot. For example, a typical residential lot containing 15,000 square feet could have a front setback of 30 feet, side setbacks of 15 feet, and a rear setback of 30 feet. The building will be limited to a maximum height of 35 feet, not more than 25 percent of the lot covered by an impervious surface, and the building must have a minimum of 1800 square feet. A typical plan is shown in **3-3**.

3-2 Zoning ordinances are established by the local governing body.

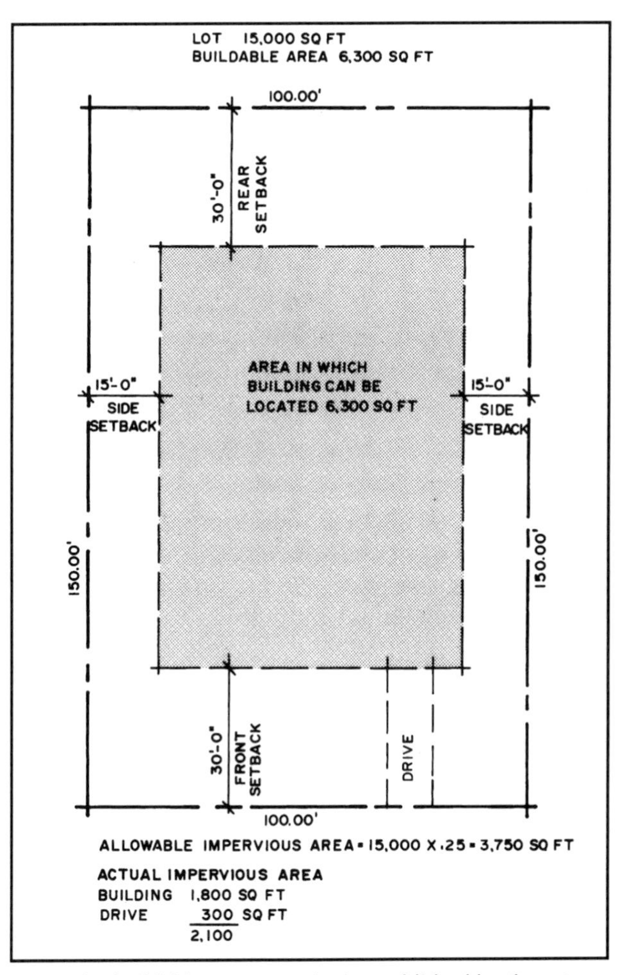

3-3 The buildable area on a site is established by the setbacks.

While the actual zoning districts will vary from one governmental jurisdiction to another, the following are typical.

Residential District. This regulates the land use where people will build homes and indicates the type of nonresidential buildings, such as churches and schools, that may be built in the district. Typically this district will allow only single-family dwellings.

Residential Multifamily District. This area is zoned to permit the construction of duplexes, apartments, townhouses, and condominiums.

Commercial District. This permits construction and occupancy of various businesses such as retail shops, theaters, and restaurants.

Industrial District. In this district companies producing products by some manufacturing process can be located. Often the type of activity is carefully limited.

Agricultural District. Use of the land in this area is permitted for growing crops and raising animals. Again, the type of crop and especially the types of animals allowed might be regulated.

Recreational District. This district reserves the land for recreational activities in an attempt to enhance the quality of life in the neighborhood. It could permit such things as parks, playgrounds, and community swimming pools.

The local government has some type of zoning and planning board. The board reviews and amends the zoning ordinances, conducts studies of local land use conditions, prepares a comprehensive land use plan, and holds public hearings on matters pertaining to zoning and land use. Architects, land developers, and general contractors must keep in close contact with the activities of this board because zoning changes do occur over time.

Building Permits

After the building has been designed and a general contractor hired, the contractor or owner applies for a building permit at the office of the local government. The application contains the information the local building official needs to consider **(see 3-4)**. In addition it is necessary to supply several sets of the architectural drawings. The building official examines this information to see if it meets the building codes and zoning ordinances. When the official is satisfied with the project, a building permit is issued **(see 3-5)**. The permit is posted in a conspicuous place on the construction site **(see 3-6)**. Some authorities require that a set of working drawings also be kept on the site.

Sometimes additional permits are required for selected aspects of the job such as electrical or plumbing work.

Inspection

As the construction proceeds, the contractor contacts the building official to get the required inspections and approval. For example, after the footings are dug but before they are poured the local building official will need to visit the site and approve them. This continues as the building is framed, wiring installed, plumbing installed, heat and air-conditioning installed, and other of the

VILLAGE OF PINEHURST, NORTH CAROLINA
APPLICATION/PERMIT TO BUILD

The undersigned hereby makes application to erect, alter, demolish or relocate a building at the location designated and agrees to comply with the plans as approved and with the provisions of all ordinances and laws of the State of North Carolina, the County of Moore, and the Village of Pinehurst regarding the work specified below. I hereby certify that I am seeking to acquire a vested right pursuant to G.S. 160A-385.1 and Chapter 9 of the Village Code.

A. PROPERTY LOCATION/OWNER
1. Address:
2. Within Town Limits: Yes No | 3. Lot Number:
4. Tax Map: Township: | 5. Unit or Subdivision:
6. Owner: | 7. Telephone (owner):

B. TYPE OF WORK
1. New Construction: | 2. Alteration or Addition:
3. Demolition: | 4. Relocation:

C. TYPE OF CONSTRUCTION (Wood Frame, Masonry, etc.):

D. BUILDING CHARACTERISTICS
1. Length: ft. | 2. Width: ft. | 3. Heated Area: sq. ft.; Garage sq. ft.
4. Number of Stories: | 5. Basement: Yes No | 6. Number of Rooms:

E. ESTIMATED CONSTRUCTION COST

F. ZONING REQUIREMENTS
1. Zoning District: | 2. Permitted Use:
3. Lot Area: sq. ft. | 4. Lot Width: ft. | 5. Building Height: ft.
6. Setbacks: Front- ft./Side- ft./Side- ft./Rear- ft.
7. Lot Coverage: % sq. ft. | 8. Overlay District:
9. Flood Zone: | 10. Watershed District:

G. UTILITIES
1. Well or Public Water: | 2. Septic Tank or Sewer:
3. Health Department Approval: Attached: No:

I. CONTRACTORS (Name, Address & Phone) (License Number)
1. Building:
2. Plumbing:
3. Heating:
4. Electrical:
5. Insul:
6. Gas Piping:

DATE ____________ APPLICANT ____________

(THIS SPACE FOR USE OF PLANNING & DEVELOPMENT DEPARTMENT ONLY)

This application for a building permit is (APPROVED,DISAPPROVED) subject to the requirements or for the reasons listed below. No changes shall be made from an approved plan without the consent of the Chairman of Architectural Advisory & Appearance Commission and/or the Building Inspector. No building may be occupied without first obtaining a CERTIFICATE OF COMPLIANCE issued by the Building Inspector.

REQUIREMENTS, CHANGES AND REMARKS

Building Inspector: ____________ Date: ____________
PERMIT N ____________ PERMIT FEE: $ ____________

3-4 An application for a permit to build is used by the local governing authority. *(Courtesy Village of Pinehurst, NC)*

BUILDING PERMIT

Contractor ______________________

Issued ______________ Permit No. ______________

Lot ________ Address ______________________

Nature of Work ______________________ Unit ________

BUILDING INSPECTIONS

Footing [ ] Slab [ ] Insulation [ ]

Foundation [ ] Framing [ ] Final [ ]

Heating Inspections **Plumbing Inspections** **Electrical Inspections**

Rough [ ] Sewer [ ] Rough [ ]

Gas Pipe [ ] Rough [ ]

Village of Pinehurst

NORTH CAROLINA

NOTE: This Permit, With A Set Of Plans Attached, MUST Be Displayed At The Address Shown Above, During The Entire Period Of Construction. Please Give One Working Day's Notice On Inspections Needed.

3-5 The building permit issued by the local building official grants approval for construction to begin. *(Courtesy Village of Pinehurst, NC)*

3-6 The general contractor must post the building permit in a conspicuous place on the construction site. The white plastic tube shown contains a set of the architectural drawings which must be kept on the site for inspection.

processes continue. The contractor must have the electrical work and plumbing inspected before they are covered up with gypsum wallboard, plaster, or other wall-finishing material. It is the responsibility of the contractor to request an inspection. This typically will require a day or two advanced notice. Therefore the contractor must carefully monitor the progress of the work.

The official record of inspections is recorded on the building permit posted on the site. The inspector notes the date of the inspection and signs the card as each inspection is satisfactorily completed. If the inspection notes unsatisfactory conditions the job will not progress until corrections are made and approved.

ADDITIONAL INFORMATION

Building Code Agencies

BOCA Building Officials and Code Administrators International
4051 West Fossmore Rd.
Country Club Hills, IL 60478

CABO Council of American Building Officials
5203 Leesburg Pike, Suite 708
Falls Church, VA 22041

IAPMO International Association of Plumbing and Mechanical Officials
20001 Walnut Drive South
Walnut, CA 91789

ICBO International Conference of Building Officials
5360 Workmen Mill Rd.
Whittier, CA 90601-2298

NCSBCS National Conference of States on Building Codes and Standards
505 Huntmar Park Dr., Suite 210
Herndon, VA 22070

SBCC Southern Building Code Congress International
900 Montclair Rd.
Birmingham, AL 35213-1206

CHAPTER 4

ARCHITECTURAL DRAWINGS & SPECIFICATIONS

Carpenters must be able to read architectural drawings and specifications. Together the drawings and specifications tell everything they need to know to construct the building.

A typical set of **architectural drawings** will include a site plan, foundation plan, floor plan, elevations of all exterior walls, required sections, and special details. Separate drawings are often used to show the electrical, plumbing, and heating and air-conditioning systems.

The **specifications** provide a word description of the materials, construction techniques, and equipment desired. They detail things that cannot be shown on the architectural drawings.

SPECIFICATIONS

Specifications are a written document providing detailed information about the required materials, finishes, and workmanship. This enables the architect to provide a clear description of things that cannot be easily shown on the architectural drawings. The specifications are grouped by trades. One system used by many architects is available from the Construction Specifications Institute. It is divided into 16 divisions. The major divisions in a typical specifications document are shown in the box at the right.

Construction Specifications

Division 1	General requirements (alternatives, quality control, etc.)
Division 2	Site work (clearing, drainage, etc.)
Division 3	Concrete (reinforcement, cast-in-place, etc.)
Division 4	Masonry (mortars, stone, etc.)
Division 5	Metals (joists, decking, etc.)
Division 6	Wood and plastics (rough carpentry, plastic fabrications, etc.)
Division 7	Thermal and moisture protection (waterproofing, etc.)
Division 8	Doors and windows (metal, wood, glazing, etc.)
Division 9	Finishes (tile, plaster, carpet, etc.)
Division 10	Specialties (fireplaces, lockers, etc.)
Division 11	Equipment (food service, laboratory, etc.)
Division 12	Furnishings (artwork, furniture, etc.)
Division 13	Special construction (clean room, vault, etc.)
Division 14	Conveying systems (elevator, hoists, etc.)
Division 15	Mechanical (plumbing, heat generation, etc.)
Division 16	Electrical (service, distribution, lighting, etc.)

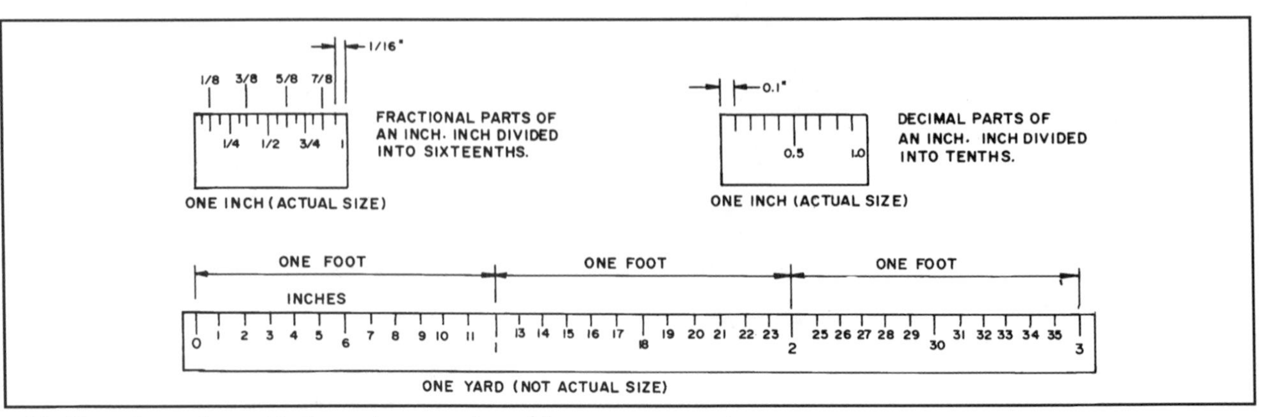

4-1 Measuring units used in the customary system of linear measurement.

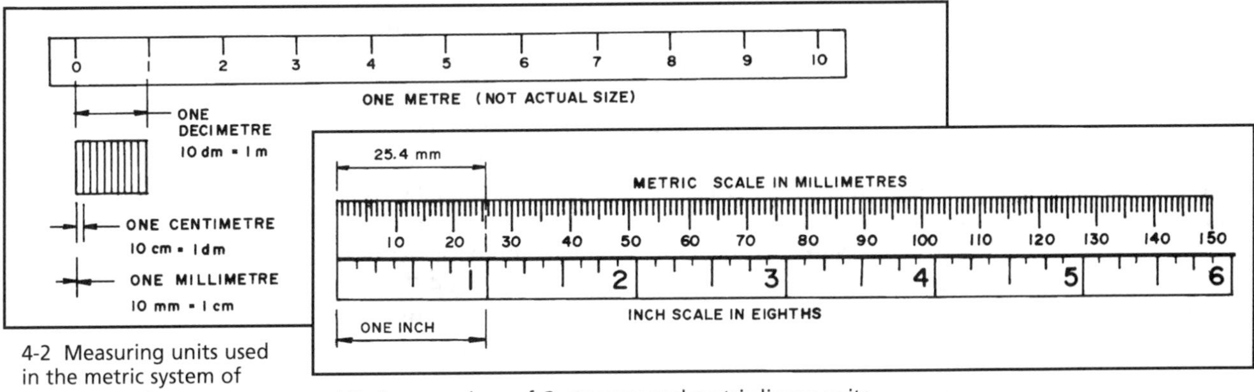

4-2 Measuring units used in the metric system of measurement.

4-3 A comparison of Customary and metric linear units.

METHODS OF LINEAR MEASUREMENT

To use architectural drawings the carpenter must be able to use both inch and metric systems of measurement. While inch systems are most commonly used, the metric system is slowly moving into use in construction as it already has in many other industries. The U.S. Federal Government has switched all of its construction projects to the metric system, so it is important for all construction trades to become familiar with the metric system. It is actually a lot easier to use than measuring in feet and inches.

Since the system is based on 10, it is possible to change from one metric linear unit to another by only moving the decimal point. For example, 2.575 meters equals 25.75 decameters, 257.5 centimeters, or 2575 millimeters **(see 4-2)**.

A comparison of an inch scale and a metric scale in millimeters is shown in **4-3**. Notice that one inch equals 25.4mm.

The metric units of linear measure, symbols used to show them on drawings, and conversion factors are shown in **4-4**.

The Customary System

The customary system (also referred to as the English or imperial system) of linear measurement uses the **yard** as the basic unit. The yard is divided in 3 units called **feet**. A foot is divided into 12 units called **inches**. An inch can be divided into various **fractional** parts such as ½, ⅜, ¼, 1⁄16, and 1⁄32 inch **(see 4-1)**.

The inch is also divided into decimal sizes based on 1⁄10 or 1⁄100 of an inch. For example, one-half of an inch is 0.5 inch.

The Metric System of Linear Measurement

The basic unit in the metric system of linear measurement is the **meter**, which is a little longer than the yard (39.37 inches).

The metric system of linear measurement is based on units of 10. A meter is divided into 10 decimeters (dm), 100 centimeters (cm), or 1000 millimeters (mm).

4-4 Metric measures of linear distances and conversion factors.

Metric Measures of Linear Distances

1 meter (m) = 10 decimeters (dm)

1 decimeter (dm) = 10 centimeters (cm)

1 centimeter (cm) = 10 millimeters (mm)

Linear Conversion Factors—Customary to Metric

inches to millimeters—multiply inches by 25.4

inches to centimeters—multiply inches by 2.54

inches to meters—multiply inches by 0.0254

feet to meters—multiply feet by 0.3048

Linear Conversion Factors—Metric to Customary

millimeters to inches—multiply millimeters by 0.0394

centimeters to inches—multiply centimeters by 0.394

meters to feet—multiply meters by 3.280

Other Metric Units

Other metric units frequently used on construction projects include mass, volume, and the area of a surface. The metric unit of **mass** is the **kilogram** (kg). The kilogram is the mass of the standard cylinder stored at the International Bureau of Weights and Measures in France. Although mass and weight are not exactly the same thing, for practical purposes mass is used as a measure of weight.

4-5 Metric measures of mass and conversion factors.

Metric Measures of Mass

1 kilogram (kg) = 1000 grams (g)

1 hectogram (hg) = 100 grams (g)

1 dekogram (dag) = 10 grams (g)

1 decigram (dg) = one-tenth of a gram (g)

1 centigram (cg) = one-hundreth of a gram (g)

1 milligram (mg) = one-thousandth of a gram (g)

Mass Conversion Factors

pounds to kilograms—multiply pounds by 0.454

kilograms to pounds—multiply kilograms by 2.205

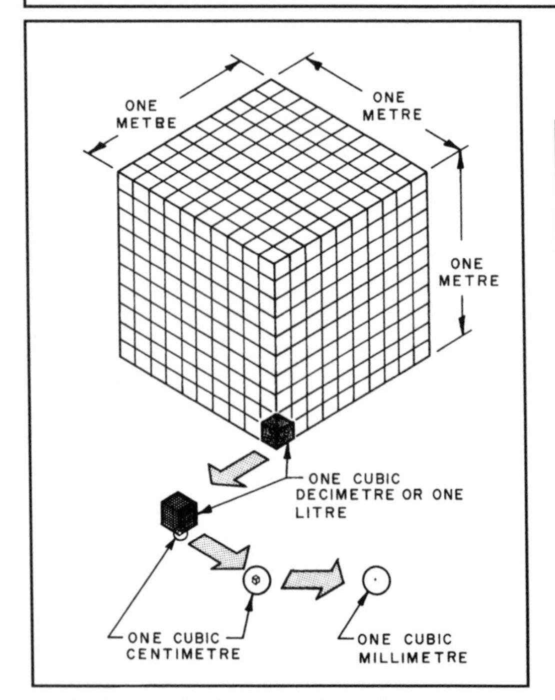

4-6 Metric units for measuring volume.

The metric units of mass, the symbols used to show them on drawings, and conversion factors are shown in **4-5**.

The basic metric unit of **volume** is the **cubic meter** (m^3). A cubic meter is a cube measuring one meter on each side **(see 4-6)**.

The metric units of volume, the symbols used to show them on drawings, and conversion factors are shown in **4-7**.

The metric unit of **area** is the **square meter** (m^2). A square meter is a square measuring one meter on each side **(see 4-8)**.

Large land areas are measured using the **square kilometer** (km^2). Smaller land areas are measured in square hectometers (hm^2). Metric measures of area, the symbols used to show them on drawings, and conversion factors are shown in **4-9**.

ARCHITECTURAL DRAWINGS

Architectural drawings show all the details needed to construct the building. A complete set for a typical residence contains the following drawings:

Site plan
Foundation plan
Floor plan
Exterior elevations
Framing plan
Various sections
Special details
Door and window schedules
Finish schedule

4-7 Metric units for measuring volume and conversion factors.

Metric Measures of Volume

1000 cubic millimeters (mm^3) = 1 cubic centimeter (cm^3)

1000 cubic centimeters (cm^3) = 1 cubic decimeter (dm^3)

1000 cubic decimeters (dm^3) = 1 cubic meter (m^3)

Conversion Factors for Measures of Volume

cubic yards to cubic meters (m^3)–multiply cubic yards by 0.764

cubic feet to cubic meters (m^3)–multiply cubic feet by 0.0028

gallons to cubic meters (m^3)–multiply gallons by 0.0038

quarts to cubic meters (m^3)–multiply quarts by 0.00095

ounces to cubic meters (m^3)–multiply ounces by 0.000029

Larger buildings will have separate electrical, plumbing and heating, air-conditioning, and ventilation plans. Every carpenter on the job must be able to read the architectural drawings.

Architectural Scales

Architectural drawings are drawn to **scale**. On drawings using **feet and inches** a fraction of an inch is used to represent a foot of the actual building. Thus the drawings are made much smaller than actual size. Scales used to draw residential and commercial buildings are shown in **4-10**. For example, residential floor plans generally are made to the scale of ¼ inch = 1 foot 0 inches of the actual house. The **metric scale** used for residential floor plans is usually 1:50. This means that one millimeter on the drawing represents 50 millimeters of the actual house.

4-9 Metric measures of area and conversion factors.

Metric Measures of Area

100 square millimeters (mm^2) = 1 square centimeter (cm^2)

100 square centimeters (cm^2) = 1 square decimeter (dm^2)

100 square decimeters (dm^2) = 1 square meter (m^2)

Designations for Large Areas

1 square kilometer (km^2) = 1,000,000 square meters (m^2)

1 square hectometer (hm^2) = 10,000 square meters (m^2)

Conversion Factors for Measurement of Area

square inches (in^2) to square millimeters (mm^2)—multiply in^2 by 645.4

square inches (in^2) to square centimeters (cm^2)—multiply in^2 by 6.454

square centimeters (cm^2) to square inches—multiply cm^2 by 0.155

square feet (ft^2) to square meters (m^2)—multiply ft^2 by 0.093

square meters (m^2) to square feet—multiply m^2 by 10.76

square yards (yd^2) to square meters (m^2)—multiply yd^2 by 0.836

square meters (m^2) to square yards—multiply m^2 by 1.197

acres to square meters (m^2)—multiply acres by 4046.87

square miles (mi^2) to square meters (m^2)—multiply sqmi by 2,589,998

square miles (mi^2) to square kilometers (km^2)—multiply mi^2 by 2.589

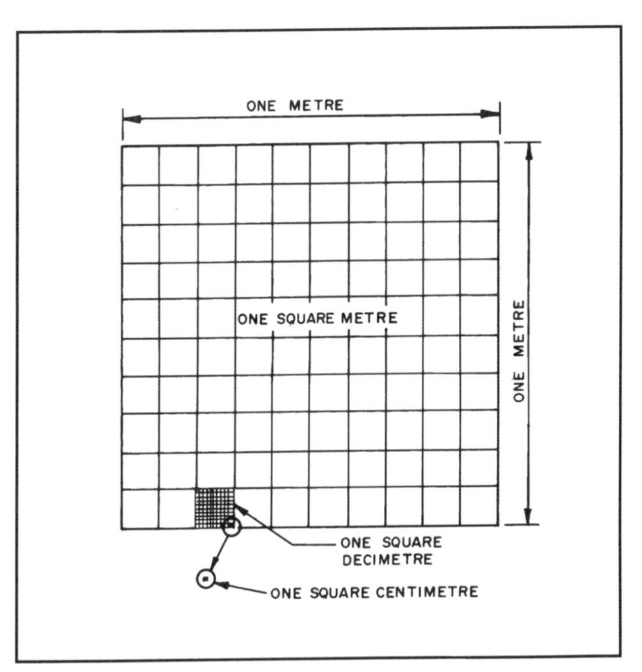

4-8 Metric units for measuring area.

4-10 Scales used for architectural drawings.

Customary (English) Architectural Scales	
Floor plan	¼" = 1'-0"
Foundation plan	¼" = 1'-0"
Elevations	¼" = 1'-0"
Construction details	¾" to 1½" = 1'-0"
Wall sections	¾" to 1½" = 1'-0"
Cabinet details	⅜" to ½" = 1'-0"
Plot plan	1" = 20' or 40'

Metric Architectural Scales	
Floor plan	1:50
Foundation plan	1:50
Elevations	1:50
Construction details	1:20 and 1:10
Wall sections	1:20 and 1:10
Cabinet details	1:50 and 1:25
Plot plan	1:100 and 1:500

Construction drawings are fully dimensioned. The dimensions shown are the actual size, even though the part is drawn to scale. A carpenter seldom has to measure a drawing to get a size. If the drawing is properly made, the dimension can be read off the drawing. Occasionally, a dimension is missing. The carpenter may have to measure the drawing to get it. This gives only an **approximate size** because the drawing may not be accurate and the carpenter's rule is not as finely subdivided as the architect's scale.

Land surveys are measured in **decimal feet** or **meters**. If the site plan is in feet, the lengths of the sides of the site are in decimal feet and are measured with a civil engineer's scale, and the location of the building and other features are located with feet and inches, which are measured with an architect's scale. If the site plan is metric all dimensions are measured with a metric scale.

Line Symbols

Architectural drawings have a variety of different types of lines which are used to give a special meaning. Some are used on the drawing of the building and others are related to dimensioning **(see 4-11)**. The use of these lines can be seen on the partial drawing in **4-12**. Notice in **4-11** the symbol marked G is used to indicate sections on foundations. The letter in the top of the symbol is the section identification letter and the letter-number in the bottom is the number of the sheet on which the section has been drawn.

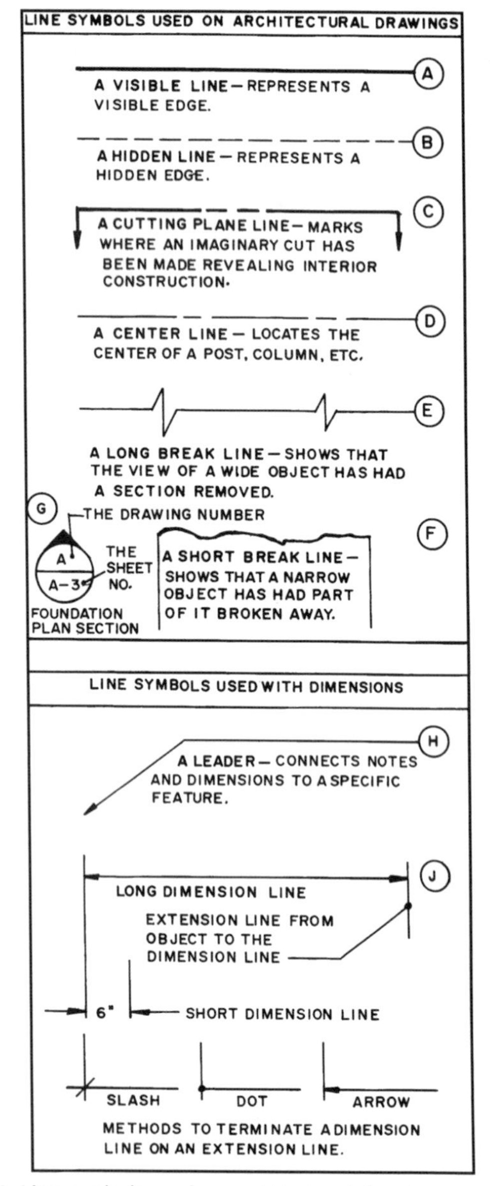

4-11 Line symbols used on architectural drawings.

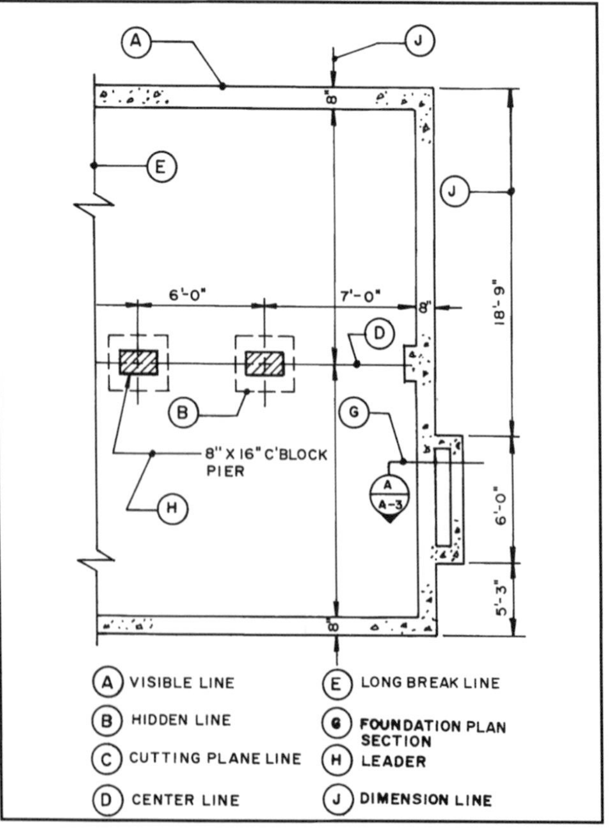

4-12 A partial architectural drawing showing the use of line symbols.

Architectural Symbols

Construction drawings are drawn to very small scales. This makes it impossible for the architect to show things as they really appear. Even if it were possible, the cost of the drafting time to do it would be excessive. **Architectural symbols** are used on drawings to represent these various elements. Selected symbols are shown in **4-13 through 4-18** on the following pages.

Another shortcut is to use abbreviations. This saves space and speeds up drafting time. Carpenters must know how to read these symbols and abbreviations **(see 4-19 on pages 42 and 43)**.

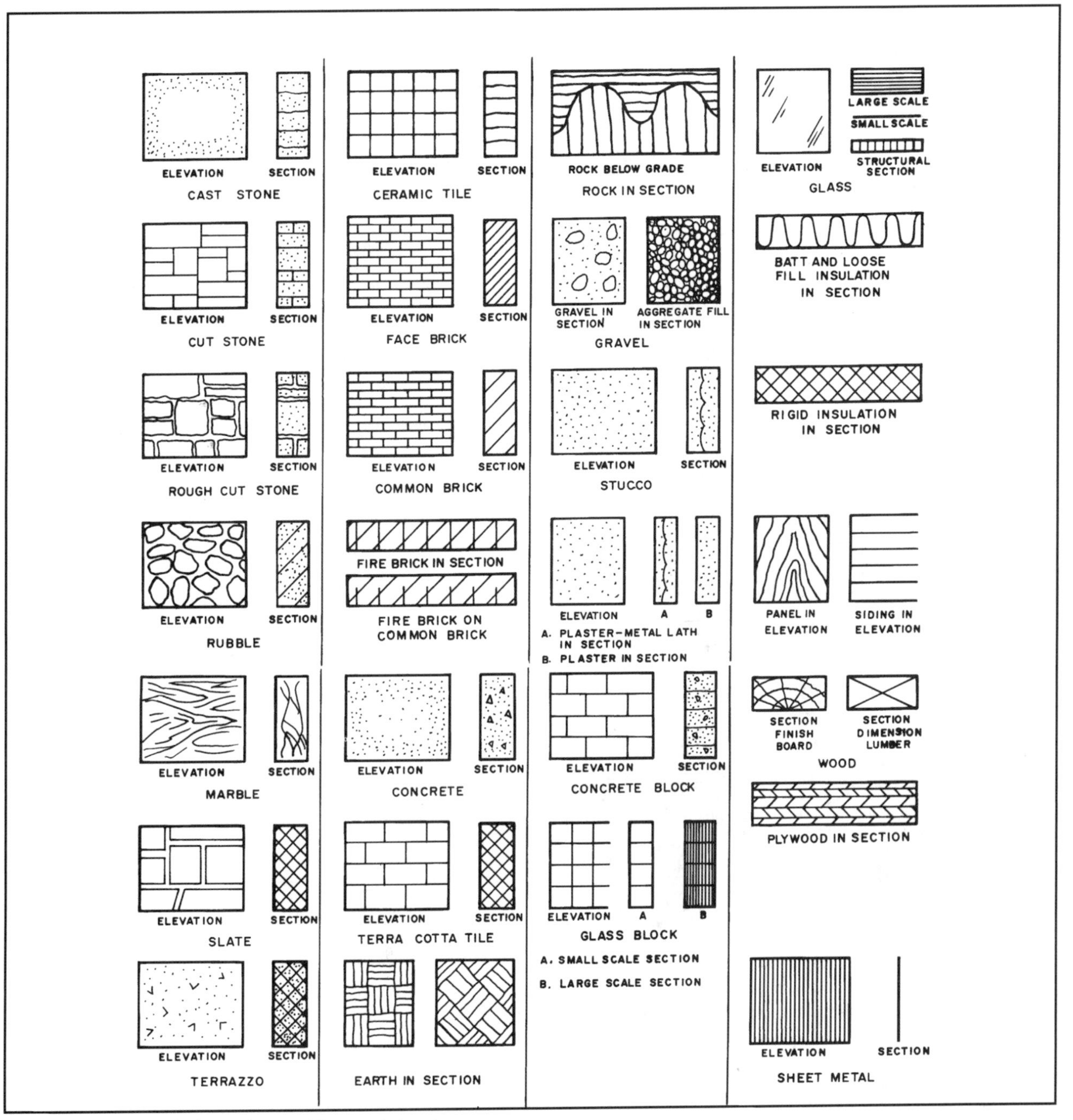

4-13 Materials symbols used on architectural drawings.

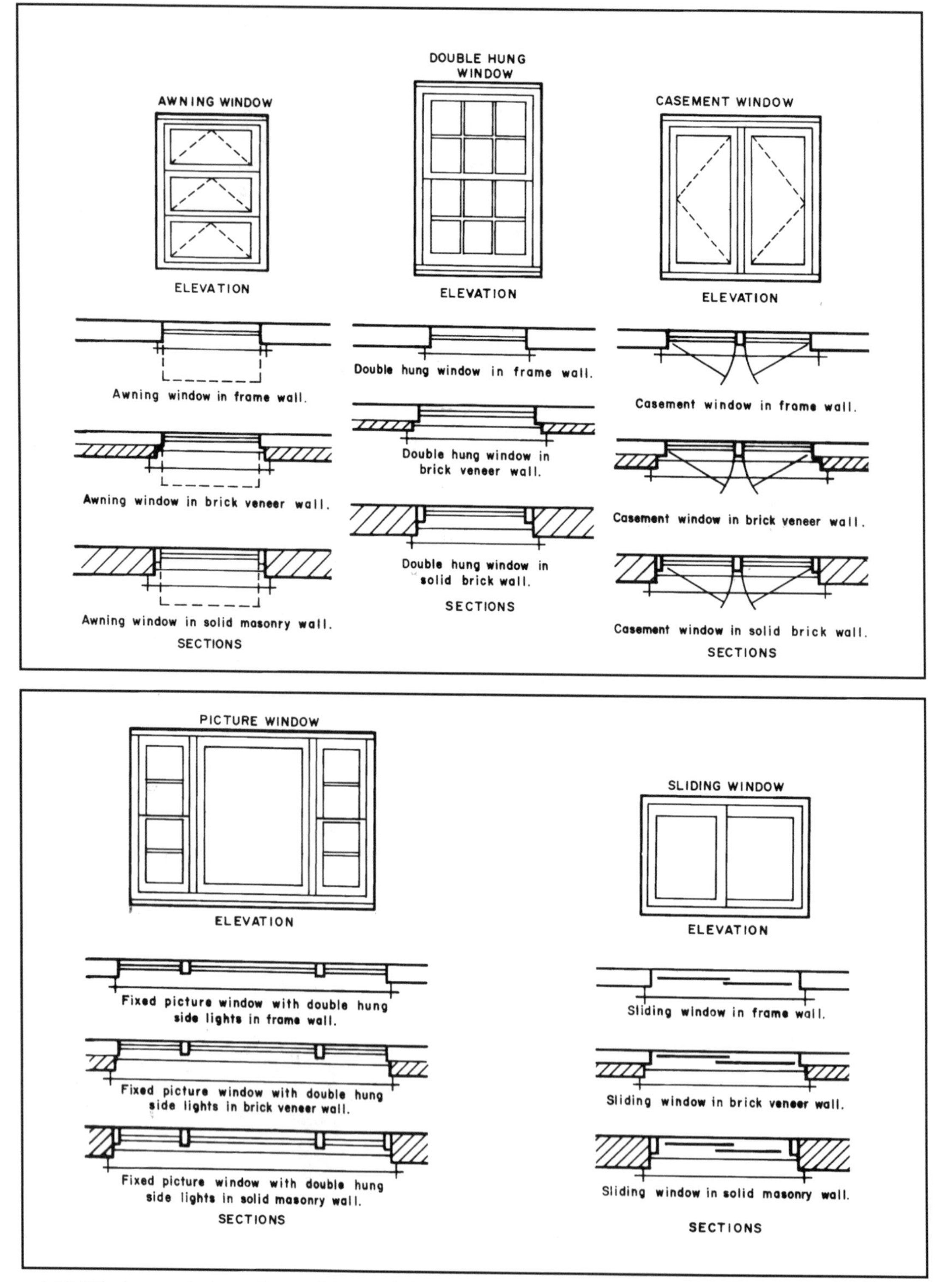

4-14 Window symbols used on architectural drawings.

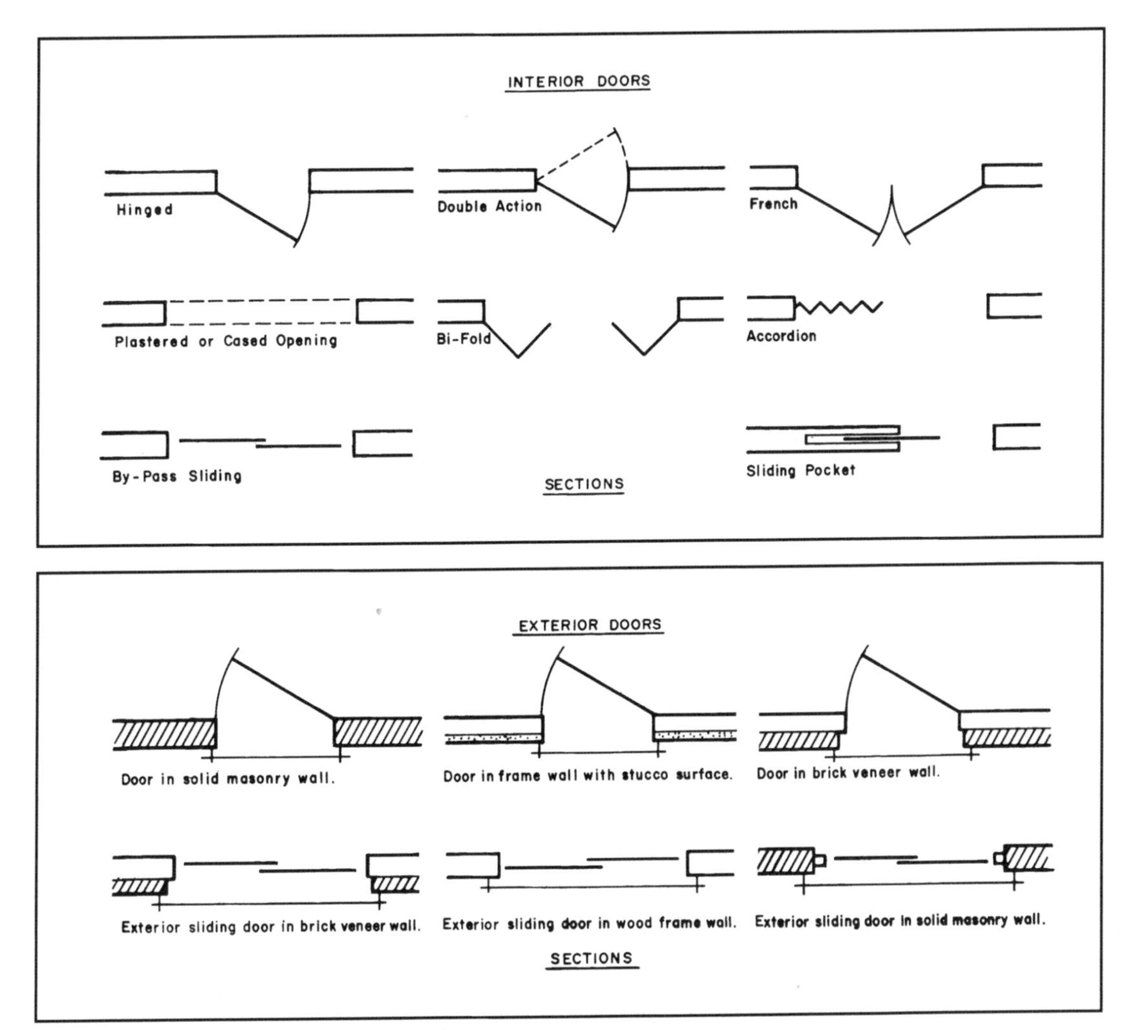

4-15 Door symbols used on architectural drawings.

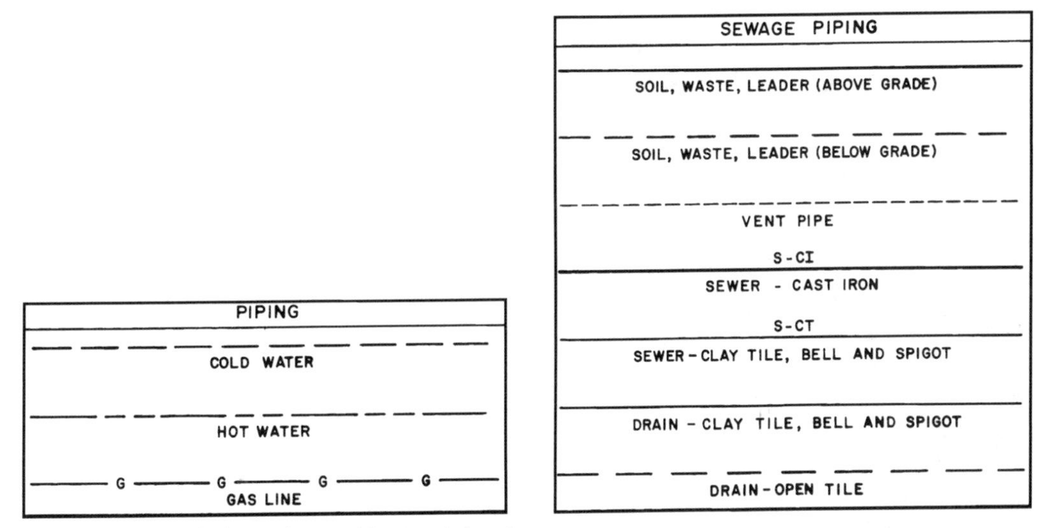

4-16 Piping symbols used on architectural drawings.

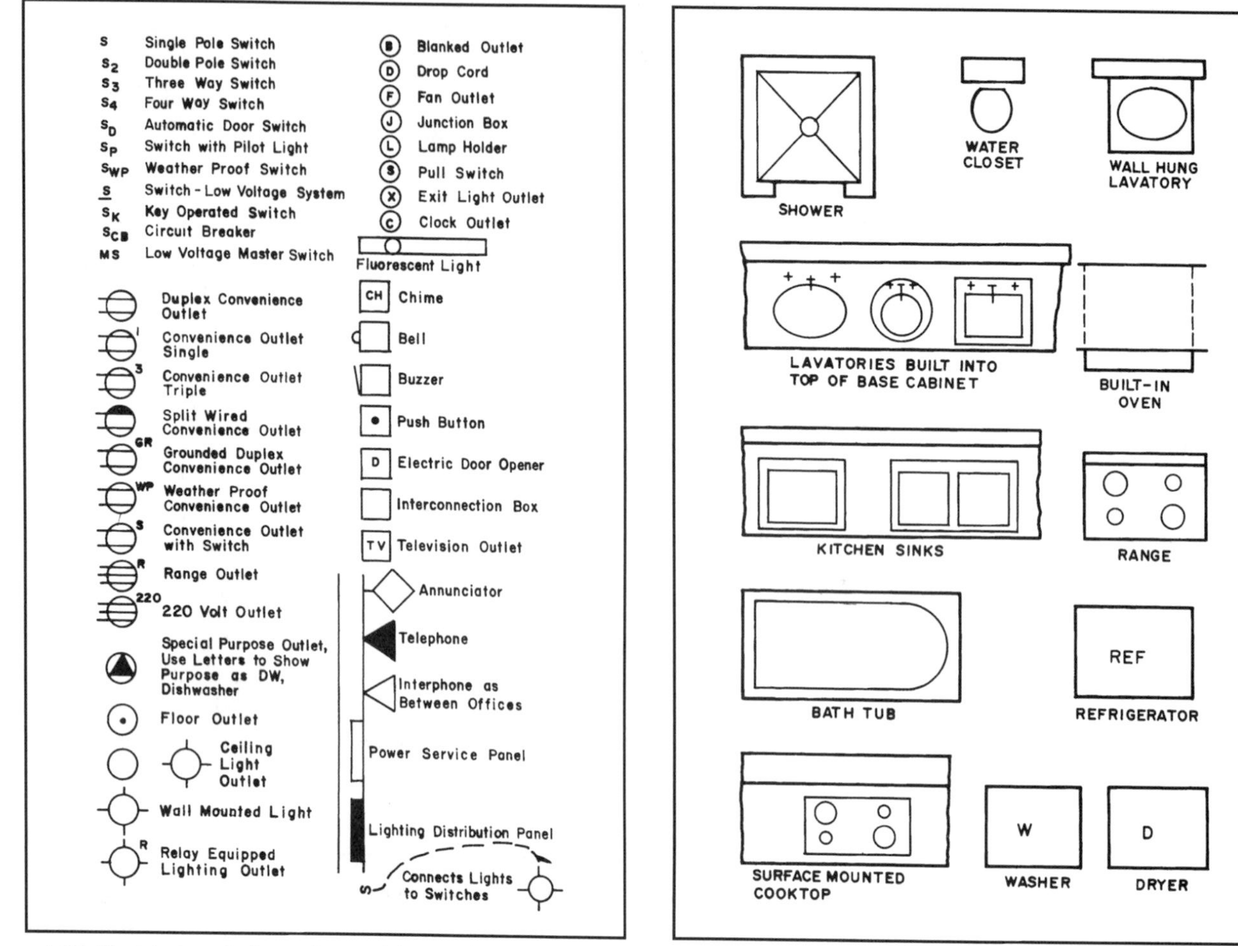

4-17 Electrical symbols used on architectural drawings.

4-18 Symbols for fixtures and appliances in kitchens and baths.

4-19 Some of the abbreviations used on architectural drawings.

Aluminum	AL
Anchor Bolt	AB
Area	A
Asphalt Tile	AT
Balcony	BALC
Bathroom	B
Bathtub	BT
Basement	BSMT
Beam	BM
Bearing	BRG
Bedroom	BR
Bench Mark	BM
Block	BLK
Blocking	BLKG
Brick	BRK
Board	BD
Building	BLDG
Building Line	BL
Cabinet	CAB
Casement	CSMT
Caulking	CLKG
Ceiling	CLG
Cement	CEM
Center	CTR
Center to Center	OC
Center Line	CL
Chimney	CHIM
Closet	CL
Column	COL
Concrete	CONC
Concrete Block	CONC BLK
Concrete Masonry Unit	CMU
Cornice	COR
Cubic Foot	CU FT
Cubic Inch	CU IN
Cubic Yard	CU YD
Detail	DET
Diameter	DIA
Dimension	DIM
Dining Room	DR
Dishwasher	DW
Ditto	DO
Door	DR
Dormer	DRM
Double Hung	DH
Drain	DR
Drywall	DW
East	E

Electric	ELEC
Elevation	EL
Excavate	EXC
Exterior	EXT
Existing	EXIST
Face Brick	FB
Feet	FT
Fill	F
Finish	FIN
Finish Floor	FIN FL
Fireplace	FPL
Fireproof	FPRF
Fixture	FIX
Flashing	FL
Floor	FLR
Flooring	FLG
Footing	FTG
Foundation	FDN
Furred Ceiling	FC
Galvanized Iron	GI
Girder	GDR
Glass	GL
Glue-Laminated	GLUE LAM
Grade	GR
Ground	GRD
Gypsum Board	GYP BD
Hardboard	HBD
Hardwood	HWD
Head	HD
Height	HT
Hose Bibb	HB
Hot Water Heater	HWH
Inch	IN
Insulation	INS
Interior	INT
Jamb	JMB
Joint	JT
Joist	JST
Kilogram	kg
Kitchen	KIT
Laminate	LAM
Landing	LC
Laundry	LAU
Lavatory	LAV
Length	LG
Level	LEV
Light	LT
Linen Closet	LC
Living Room	LR
Louver	LV
Lumber	LBR
Masonry Opening	MO
Material	MTL
Maximum	MAX
Metal	MET
Meter (metric)	M
Millimeter	mm
Minimum	MIN
Miscellaneous	MISC
Molding	MLDG
North	N
Not to Scale	NTS
On Center	OC
Opening	OPNG
Overhang	OH
Panel	PNL
Painted	PTD
Parallel	PAR
Partition	PTN
Plaster	PLAS
Plate	PL
Pounds per Square Inch	PSI
Precast	PRCST
Prefabricated	PREFAB
Pressure Treated	PT
Quantity	QTY
Rafter	RFTR
Refrigerator	REF
Reinforced	REINF
Reinforcement Bar	REBAR
Retaining Wall	RW
Ridge	RDG
Riser	R
Roof	RF
Roofing	RFG
Roof Drain	RD
Room	RM
Rough	RGH
Rough Opening	RO
Shake	SHK
Sheathing	SHTHG
Sheet Metal	SM
Shingle	SHGL
Siding	SDG
Sill	SL
Sink	SK
Skylight	SKL
Sliding Door	SL DR
Soffit	SOF
Soil Pipe	SP
Solar Panel	SLR PAN
South	S
Specification	SPEC
Square	SQ
Stairs	ST
Stairway	STWY
Steel	STL
Stone	ST
Suspended Ceiling	SUSP CLG
Tar and Gravel	T&G
Tee	T
Terra-Cotta	TC
Terrazzo	TER
Thick	THK
Tongue and Groove	T&G
Tread	TR
Typical	TYP
Unexcavated	UNEXC
Unfinished	UNFIN
Utility Room	UR
Vent	V
Ventilate	VENT
Vent Stack	VS
Vertical	VERT
Volume	VOL
Water	W
Water Closet	WC
Waterproof	WP
W-Beam	W
Weep Hole	WH
Welded Wire Fabric	WWF
West	W
Wide Flange	WF
Window	WDW
Wood	WD
Yard	YD
Yellow Pine	YP

BASIC TYPES OF HOUSES

The basic **shapes** of typical residences include square, rectangular, T-shaped, U-shaped, and L-shaped **(see 4-20)**. The square shape is the most economical because it has the lowest number of lineal feet of exterior wall and footing. As the shape becomes rectangular the number of feet of wall and footing increases.

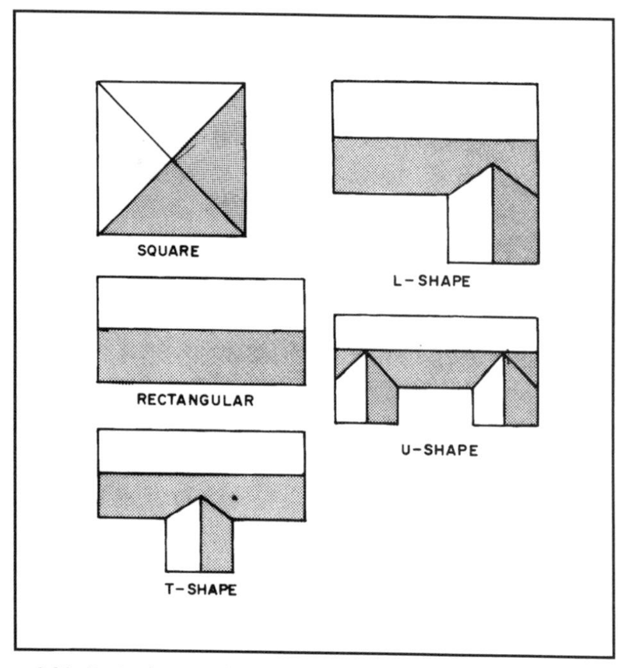

4-20 Basic shapes of residences.

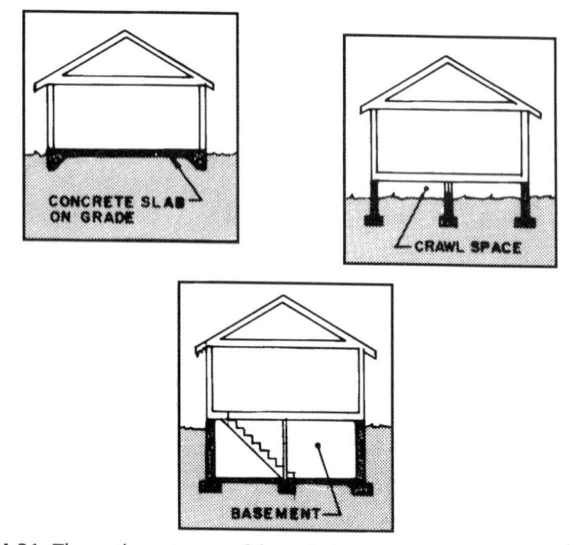

4-21 These three types of foundation construction are typical of those commonly used for residential construction.

The basic **types of houses** include one-story, one-and-one-half-story, two-story, and various split-level designs. These can be built using a concrete slab, crawl space, or basement foundation **(see 4-21)**.

The One-Story House

A one-story house has all the habitable rooms on the same level and above ground. The attic can be used for storage and a pull-down stair in the garage is usually used to provide access **(see 4-22)**. A one-and-one-half story house has some habitable area in the space between the roof and first floor ceiling **(see 4-23)**.

The Two-Story House

The two-story house has a full first floor and a full second floor above it. This is an economical house to build because the lineal feet of footing required are fewer than if the entire area were on one level. The attic can be used for storage **(see 4-24)**.

The Split-Level House

The split-level house is typically one story on one side and two stories on the other. The split can occur front-to-rear, rear-to-front or side-to-side. Some examples are in **4-25** through **4-27**. It can also have a basement under part of the building. Split-level houses are especially useful on lots having a fairly good slope.

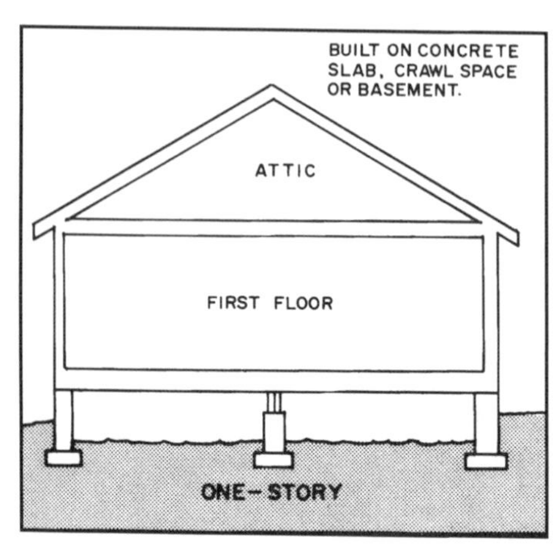

4-22 A one-story house has all habitable rooms on the first floor.

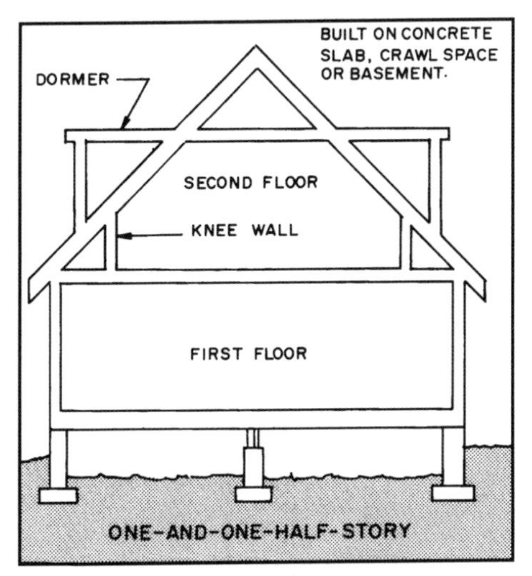

4-23 A one-and-one-half story house has some habitable area in the space between the roof and first floor ceiling.

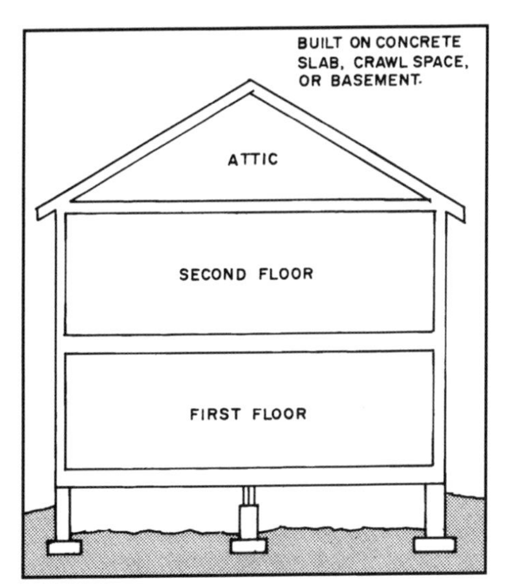

4-24 A two-story house has two full floors of habitable space below the roof.

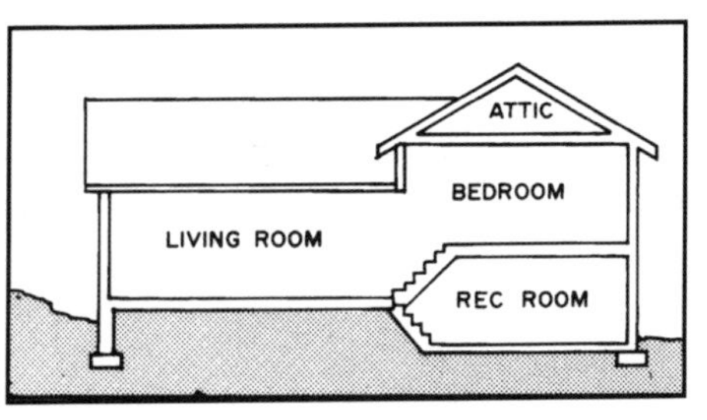

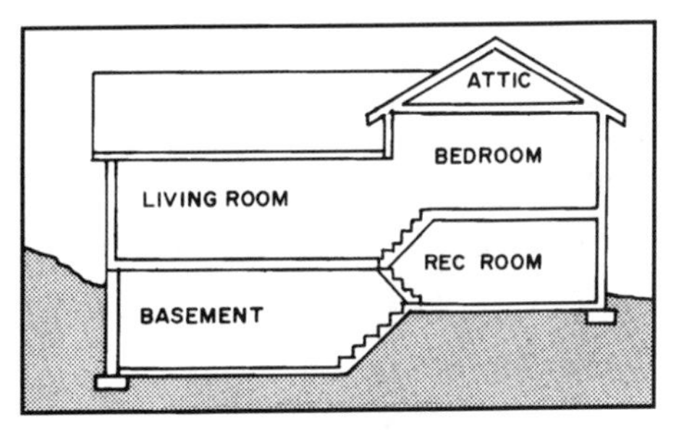

4-25 Typical side-to-side split-level houses.

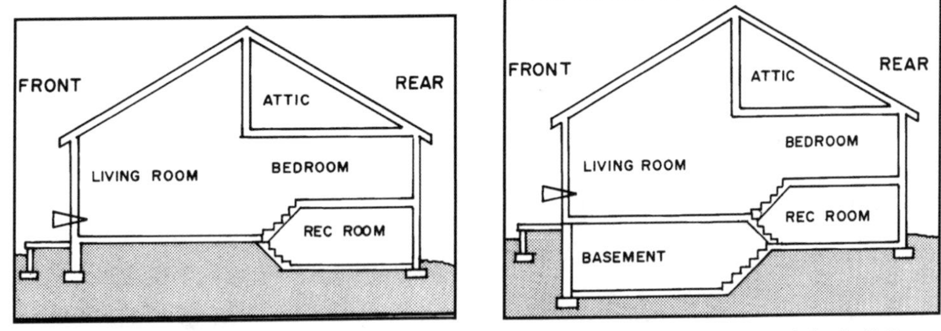

4-26 These split-level houses split the living levels from the front to the rear of the building.

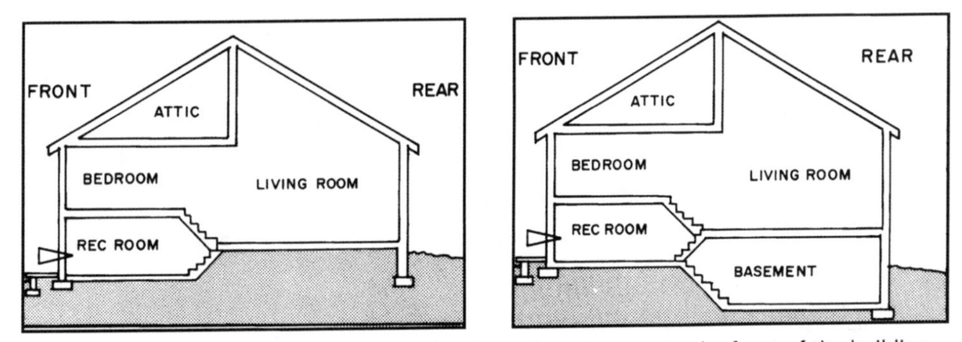

4-27 These split-level houses split the living levels from the rear to the front of the building.

READING A SITE PLAN

The site plan is a drawing of the building site showing the location of the building on the site, size of the site, contours, and other features, such as driveways and sidewalks. A typical site plan for a small residence is in **4-28**. A metric plan is shown in **4-29**. It should be noted that the plan may be surveyed and drawn in decimal feet or meters. Following are the things that are shown on a site plan. Relate these to the plans shown in **4-28** and **4-29**.

Property Lines

The length of each side of the lot is shown in decimal feet or meters. This will show the shape of the lot. Sometimes the surveyor's compass directions are also shown.

Compass Direction

Compass direction is indicated by an arrow pointing north. This helps the architect as the house is planned to orient it to take advantage of the sun. Typically the elevations (sides) of the house are referred to by compass directions as facing north, south, east, or west.

Setbacks

A **setback** is the minimum acceptable distance between the property line and the building as specified by the local building ordinances. No part of the building can be located in this area. Typically this includes not only the building but porches, decks, and exterior stairs. In some cases the roof overhang cannot encroach on the setback. The setback distances are specified for the front, rear, and side yards.

Building Location

An **outline** of the building with any protrusions such as porches, decks, or stairs is drawn on the site plan. It is sited with dimensions which are used when the foundation is built and verifies it is not encroaching on the setback. The elevation of the **finish floor** is recorded in the building outline.

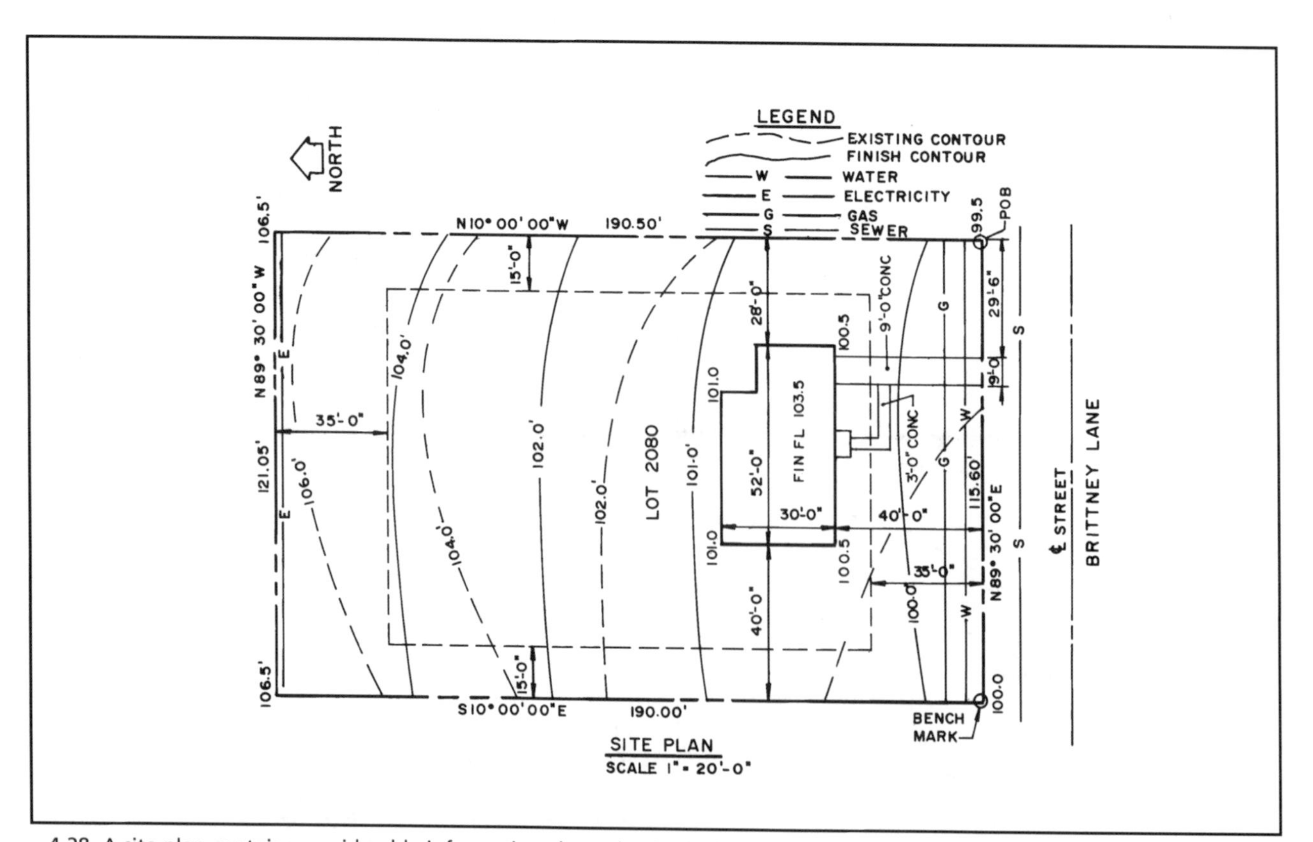

4-28 A site plan contains considerable information about the site before and after construction is completed.

Bench Mark

A **bench mark** is a marked reference point from which the elevation of other points on the site is taken. It is established by the surveyor, typically at one corner of the lot. It is specified as 100.0 feet and is shown on the site plan. A stake is driven to locate and establish this point. All elevations, such as the floor of the building and slope of the lot, are taken from this point. For example, a lot may slope away from the bench mark. If the rear of the lot were five feet lower than the bench mark it would be recorded on the site plan as 95.0 feet.

Site Elevations, Contours & Finished Grade

The site will often not be flat but have some areas higher than others. The elevation (height) of these areas is measured above and below the bench mark. As the site is surveyed, points of various elevations are recorded in even feet. The points that have the same elevation, 101.0 feet, are connected with a line called a contour line, those at 102.0 feet are connected, and this continues until the entire surface of the lot has contour lines **(see 4-29)**. If the lot is steep the contour lines may record elevations every two feet apart or even wider. The original contours are shown on the site plan with a dashed line.

As the building is designed it generally becomes obvious that the original site contours must be changed to accommodate the building, driveways, and sidewalks, and to direct surface water away from the building. These new grades are called the finished contour. It is shown with a solid line as seen in **4-29**. Again, these grades are established by the surveyor working from the bench mark. The finished grades are established after the exterior work on the building is finished. They may require moving around some of the soil on the lot, hauling away excess soil, or bringing in additional soil.

The finish grades are also indicated at the corners of the building, on the driveway, sidewalk, and other features influenced by them. For example, it may be desired to slope the driveway from the garage to the street. The finish grade at the garage door and at the street will show the slope required.

Other Features

There are countless other features that may appear on the site plan. Examples include a tree survey showing the location of trees to be saved, retaining walls, wells, septic tanks and drain fields, special slopes (called swales) to direct water away from the building or away from the lot next door, electrical power sources and connections to the house, city sewer lines and connections, city water lines, natural gas lines, and telephone and cable TV service.

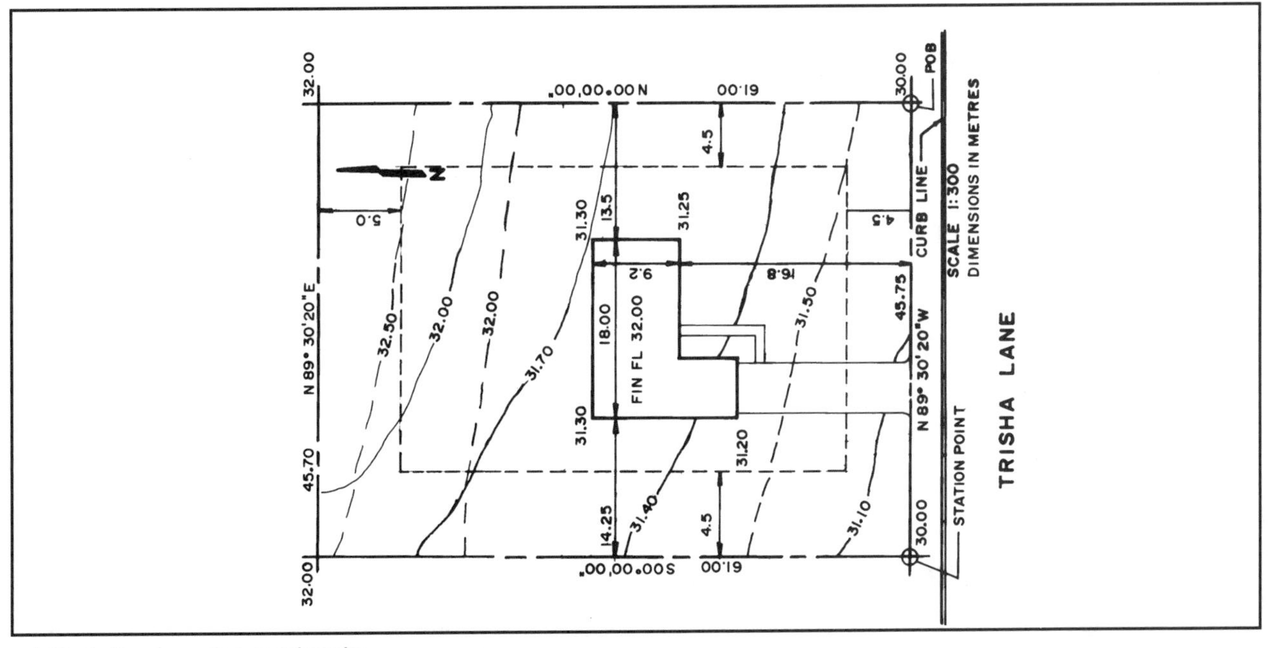

4-29 A site plan using metric units.

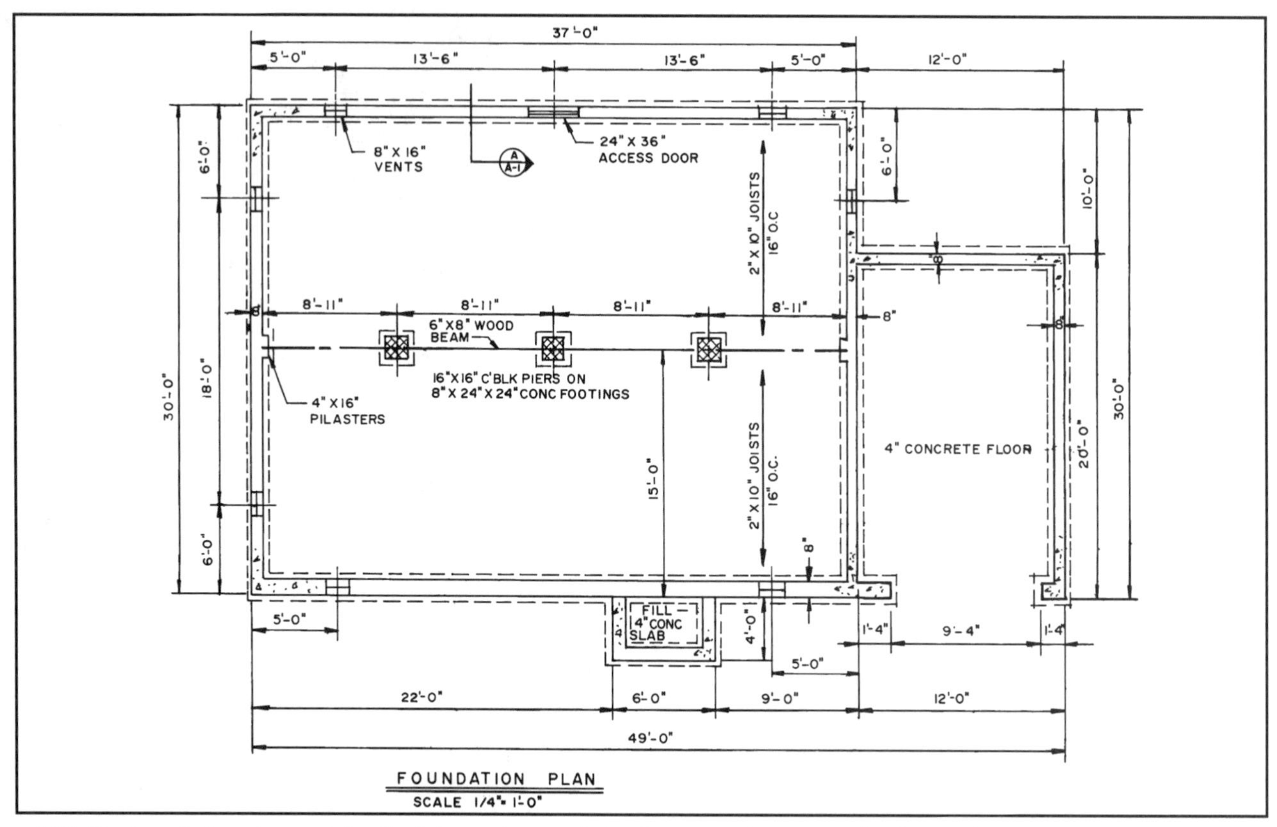

4-30 The foundation plan contains the information needed to lay out and construct the foundation.

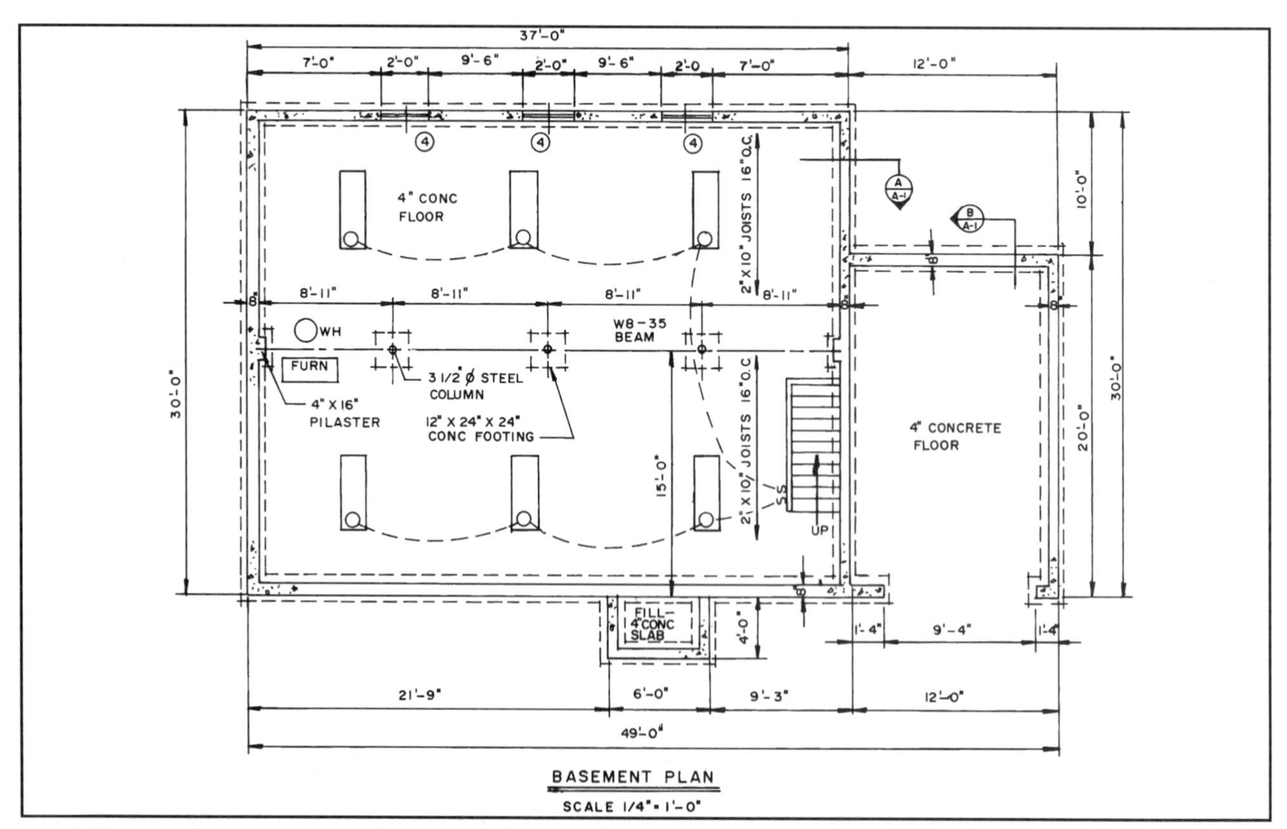

4-31 This basement plan contains foundation information as well as lighting layout and furnace and water heater locations.

READING FOUNDATION PLANS

The commonly used foundations include a full basement, a crawl space, or a slab-on-grade construction. Since the foundation carries the weight of the building plus other loads, such as snow, the drawing must clearly show the design sizes of all parts.

A typical **foundation plan** for a building with a **crawl space** will show the size of the foundation walls and footings, size and location of beams, joists, piers, columns, pilasters, and vents, and material symbols as required **(see 4-30)**.

If the building has a **basement** the drawing could have additional information such as stairs, interior partitions, doors, windows, a water heater, a furnace, electrical outlets and lights, an electrical panel, possible plumbing units such as lavatories and water closets, a floor drain, and notes about the floor and interior wall finishes **(see 4-31)**.

Notice the piers and columns are located by their centerlines. Joist information is given with a note. All openings in masonry walls are located by the sides of the openings. The wood framing for interior partitions is dimensioned to their centers and the exterior wall framing to the outside face of the stud wall.

A typical foundation plan for a building with a concrete slab floor is shown in **4-32**. While there are several ways this construction can be built, that in **4-32** is typical. The exterior dimensions of the slab are given and the footing is shown with hidden lines. Notice footings located on the interior of the slab. These are placed below interior loadbearing partitions which the carpenter frames and installs. The elevation of the finish floor and information about the slab construction are noted. Various sections are indicated and section detail drawings are made of each on the same drawing sheet containing the foundation plan.

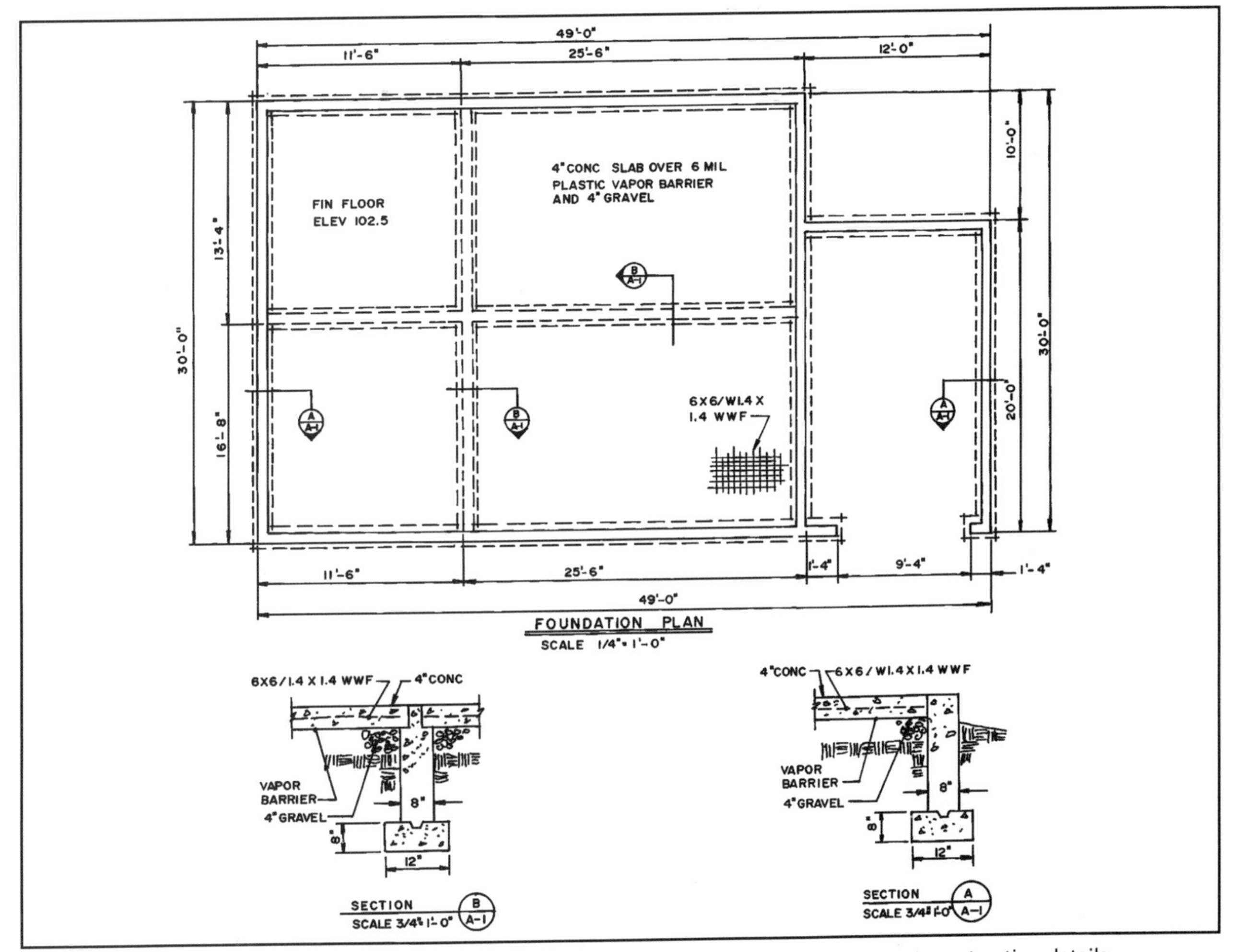

4-32 This foundation plan for a house with an on-grade concrete floor uses sections to present construction details.

READING FLOOR PLANS

The floor plan shows how the interior of the building would look if the roof were removed. Actually it is a section drawing cutting the building horizontally about halfway between the floor and ceiling. It shows the locations of interior partitions, doors, cabinets, plumbing fixtures, fireplaces, and other such items. It also shows the locations of windows and doors in the exterior wall. A separate floor plan is drawn for each floor of a house. Sometimes the locations of electric lights, outlets, switches, and other related electrical units are shown. Of major interest to the carpenter are the dimensions giving the sizes and locations of the walls and various openings.

The following paragraphs relate to the floor plan in **4-33**.

Exterior dimensions. Notice that the exterior dimensions indicate the overall length of each wall and the locations of doors and windows. The doors and windows are located by their centerlines.

Interior dimensions. The location of partitions is indicated within the plan. They are located by their center, and from the outside face of the exterior wall stud.

Doors and windows. Doors and windows are identified by a symbol indicating the type of window and a letter or number that appears on a schedule. The door and window schedule gives complete size and other information about each unit. Refer to the schedule in **4-34**; symbols are given earlier in **4-14** and **4-15** and abbreviations in **4-19**. Notice the hinged side of the door and the direction of swing are shown.

Identification. Various features, such as a water heater, are identified by symbols and usually a descriptive word or abbreviation. The rooms are identified, as are other features such as porches.

Materials. The materials used to construct the walls are shown with symbols identified earlier in **4-13**. On this plan the exterior walls are wood-framed with wood siding and all partitions are wood-framed.

Cabinets. In the bath is a single lavatory in a cabinet base. The kitchen cabinets are L-shaped with spaces for the refrigerator, sink, stove, and dishwasher indicated. Wall-hung cabinets are indicated with a dashed line. These symbols tell little about the cabinets so elevations must be drawn.

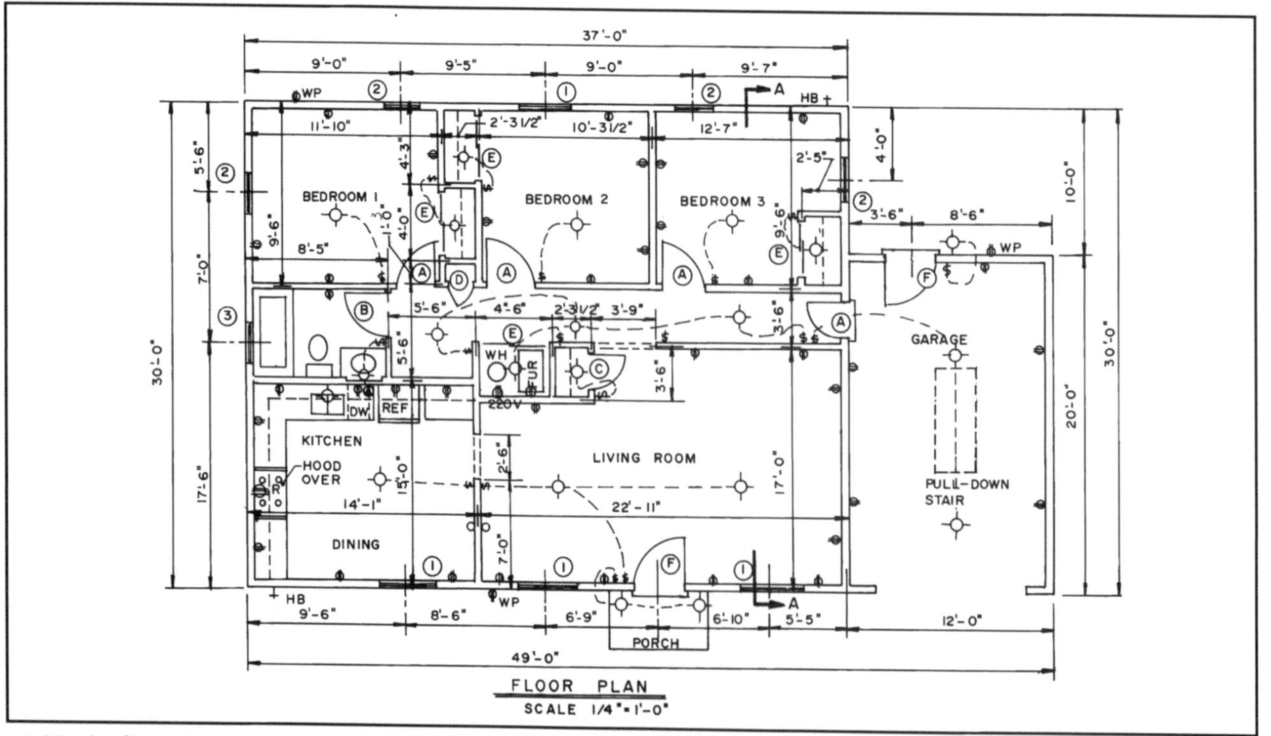

4-33 The floor plan is the most detailed of all the architectural drawings and is used constantly by carpenters as they frame the building.

Plumbing. The plumbing fixtures, water heater, and hose bibbs are shown with symbols **(refer to 4-16 and 4-18)**. Detailed information, such as color and brand, is given in the specifications. Typically a drawing showing the run of the piping is not made for small residential work. It is left to the plumber to make the best runs and meet codes.

Electrical. Lighting, duplex outlets, switches, connections to appliances, and other electrical features are indicated by symbols **(refer to 4-17)**. Notice the waterproof (WP) electrical outlets on the exterior building.

Usually the actual circuitry is not shown on small residences except for the connections between the switches and the lights. The run and development of circuits are usually left to the electricians, who must make certain the work meets the code.

Attic access. Some form of access door to the attic is necessary. On this plan it is a pull-down stair located in the garage ceiling. Other times it will be an opening in the ceiling with a removable cover providing access but requiring the use of a ladder.

Schedules. Schedules are used to show detailed and specific information about various features of a building. Door, window, and interior finish schedules are commonly used for residential construction. **Door schedules** contain the mark, size, material, number, and type of door.

4-34 Typical door and window schedules for residential construction.

Door Schedule

mark	size	material	number	remarks
A	2'-6" × 6'-8"	Birch	3	Flush, H.C.
B	2'-4" × 6'-8"	Birch	1	Flush, H.C.
C	2'-0" × 6'-8"	Birch	2	Flush, H.C.
D	1'-8" × 6'-8"	Birch	1	Flush, H.C.
E	2'-8" × 7'-0"	Pine	2	Exterior, 6 light
F	3'-0" × 7'-0"	Pine	1	Exterior, 6 light
G	2'-0" × 6'-8"	Pine	2	Bifold
H	3'-0" × 6'-8"	Birch	1	Flush, H.C., sliding
	2'-0" × 6'-8"	Birch	4	Flush, H.C., sliding

Window Schedule

mark	size	material	number	remarks
1	36 × 28	Wood	4	Double-hung, 6 light
2	24 × 24	Wood	4	Double-hung, 4 light
3	30 × 20	Wood	1	Sliding

Window schedules contain the mark, size, material, number, and type of window, plus any additional remarks deemed necessary **(see 4-34)**.

Interior finish schedules vary somewhat in what is presented but typically indicate the room name or number, wall and ceiling finish, and floor finish **(see 4-35)**. If colors are specified, a separate interior color schedule can be produced.

4-35 A typical interior finish schedule for residential construction.

Room	Finish Floor	Wall Finish		Ceiling Finish		Trim		Remarks
		Material	Finish	Material	Finish	Material	Finish	
Kitchen	vinyl	gypsum	latex paint	gypsum	latex paint	pine	semi-gloss enamel	
Living Rm	prefinished oak	do	do	do	do	do	do	(do = ditto)
Bedrm 1	carpet	—	do	do	do	do	do	
Bedrm 2	do	—	do	do	do	do	do	
Bedrm 3	do	—	do	do	do	do	do	
Hall	prefinished oak	do	do	do	do	do	do	
Bath	ceramic tile	do	do	do	do	do	do	tile tub surround & walls 4 ft above floor
Garage	concrete	do	do	do	do	do	do	

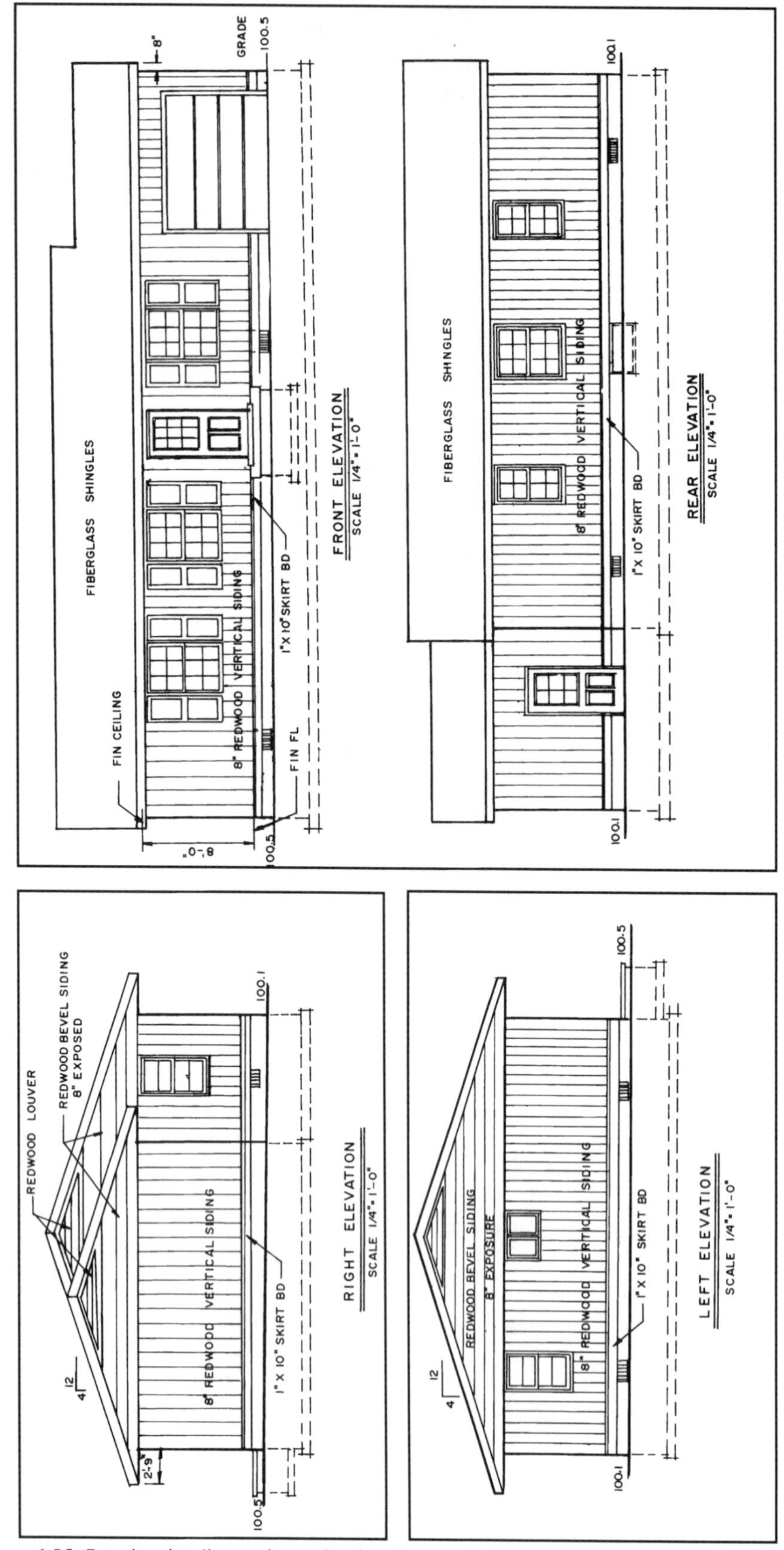

4-36 Exterior details are shown by drawing an elevation of each side of the house.

READING THE ELEVATIONS

An **elevation** is an exterior view of the building. A typical residence has four elevations. If a part of the building is on an angle, the elevation is drawn looking perpendicular to that wall. Typical elevation drawings are in **4-36**. Each view is identified by name. Typically they are labeled by their facing the points of the compass, North, South, East, or West, or by the type of view, Front, Rear, Right, or Left, as you face the front of the building.

The elevation shows the doors and windows indicated on the floor plan but has drawings of the actual windows and doors **(see 4-14 and 4-15)**. Shutters are shown if required. The materials used are lettered on the view and identified by line symbols **(see 4-13)**.

The height from the finished floor to the finished ceiling is shown along with other details such as porches, railings, decks, and stairs. The slope of the roof is indicated by a symbol such as

12
6⌈

This means there are 6 inches of rise in the roof for every 12 inches of horizontal run in the roof. This will be covered in more detail in the roof framing chapter. The foundation below grade is indicated with dashed lines.

READING SECTION VIEWS

Section views are cuts through a part of the building showing construction details. While there are many types that can be made, those for typical residential work include a typical wall section, foundation sections, and for more complex framing situations a section through the entire building. Any feature that is not clearly shown on the foundation and floor plan will benefit by having a section view through it. Section views are drawn to a larger scale than foundation and floor plans.

Typical Wall Sections

Wall sections are made running from the foundation through the details at the eave **(see 4-37)**. If more than one type of exterior siding is used, sections for each will be drawn. The section view identifies all materials, gives required sizes, and shows how the wall is to be built.

Building Sections

Building sections are drawn along planes that pass through the house from the front to the rear wall and the footing through the ridge of the roof **(see 4-38)**. While these are not usually required for a simple building as used in this chapter, they do clarify details when framing is unusual or more complex. All the members are identified and the sizes noted.

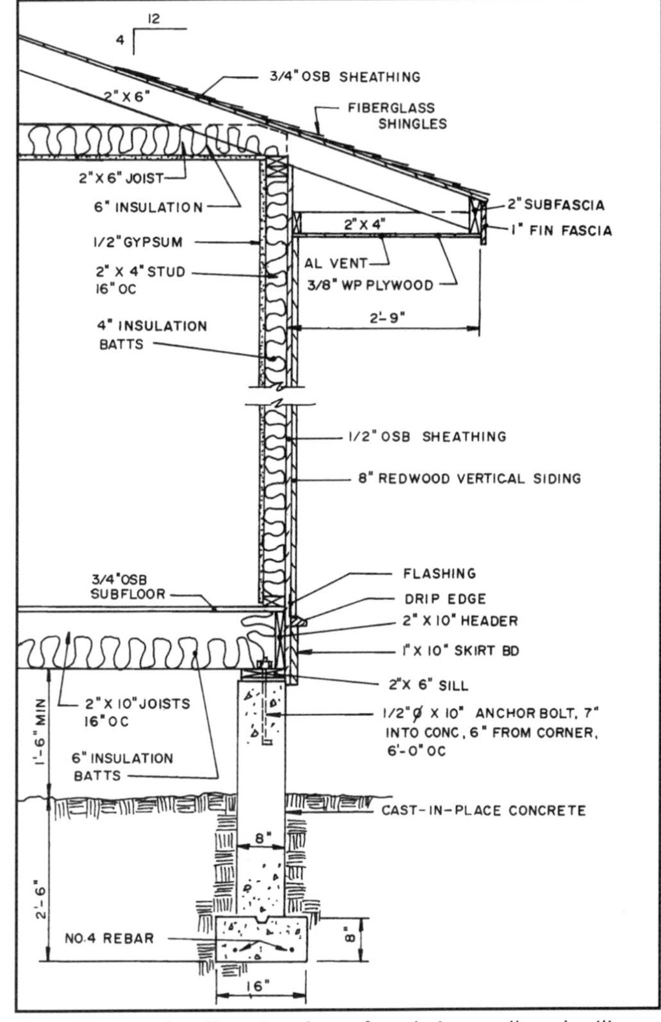

4-37 A typical wall section shows foundation, wall, and ceiling and roof framing details.

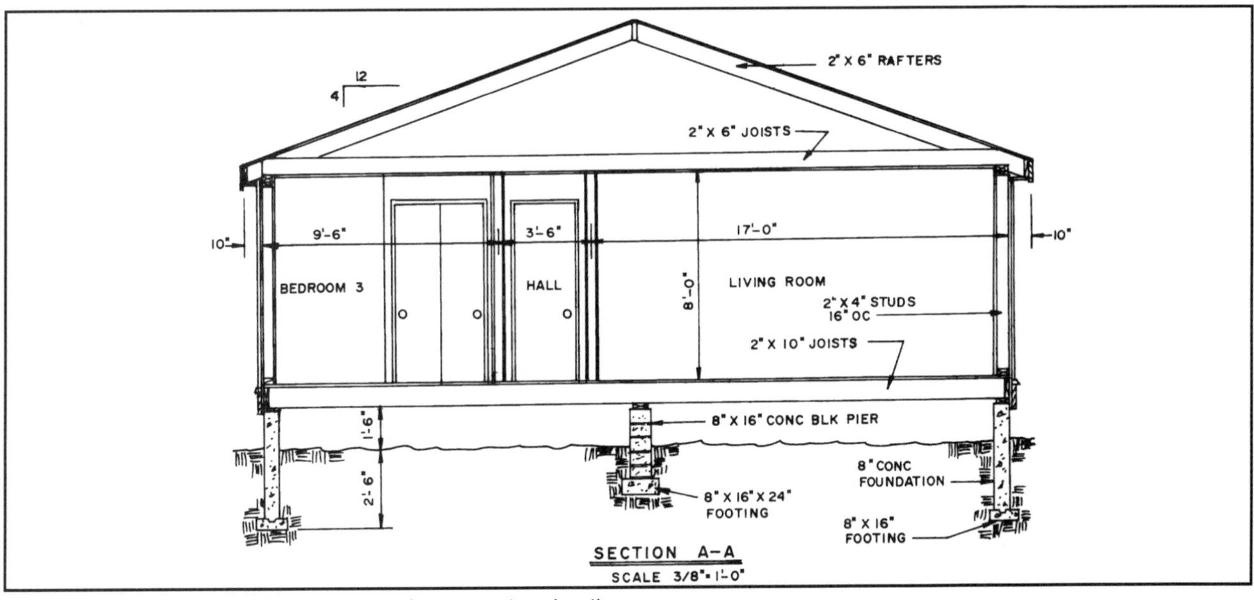

4-38 A building section shows structural construction details.

Foundation Sections

Foundation sections are located on the foundation plan as shown earlier in **4-30, 4-31,** and **4-32.** Most often they are drawn on the same sheet as the foundation plan. They show the sizes and materials required and how the foundation is to be built. Typical examples are shown below in **4-39** and **4-40**.

READING FRAMING PLANS

Some sets of construction drawings have **framing plans**. These show the location of each wood structural member. Typical examples are floor, ceiling, roof, and wall framing plans. The structural members are generally shown with a single line and the sizes and spacing of the members are noted. Examples are shown in **4-41** and **4-42**.

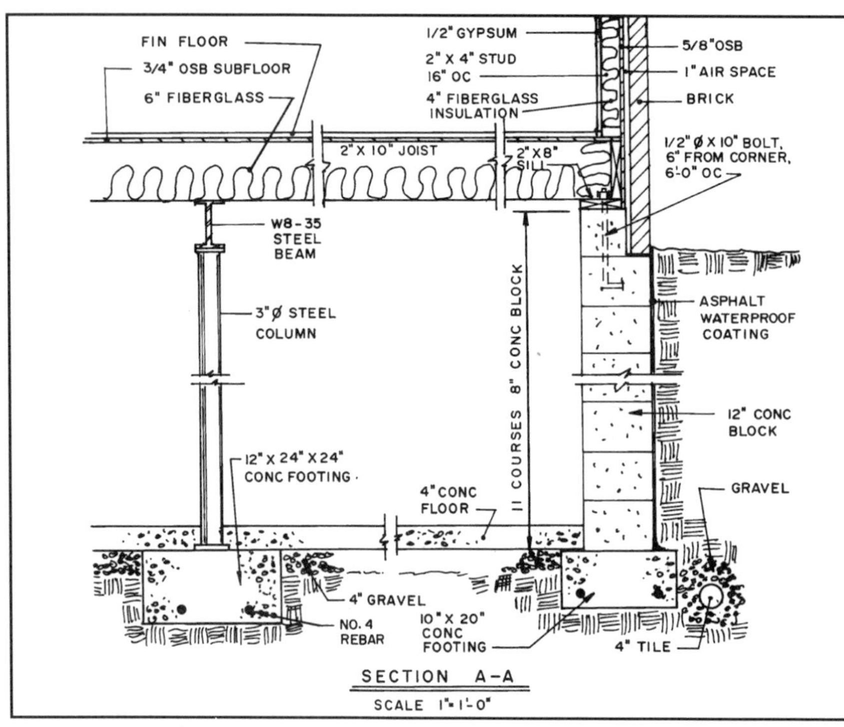

4-39 A typical foundation section for a house with a basement.

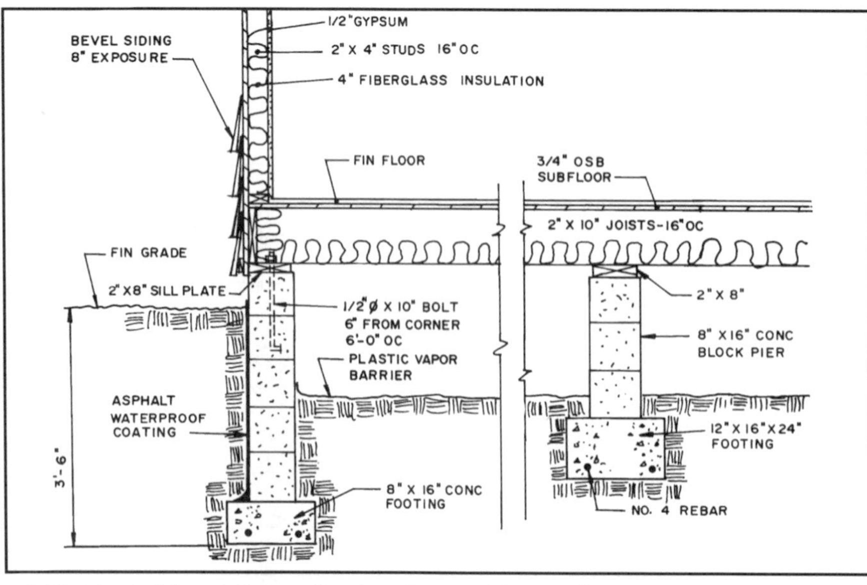

4-40 A typical foundation section for a house with a crawl space.

OTHER DRAWINGS

A set of architectural working drawings will contain a variety of detail drawings which the architect feels are necessary to accurately describe the project.

Stair Details

While the house plan used in this chapter does not have a stair, a typical drawing for one is shown in **4-43**. The plan view of the stair is drawn as part of the floor plan. The section view is drawn on a sheet devoted to various details. The architect must be certain the design meets local building codes. Critical measurements include the size of the tread and riser and the headroom.

Cabinet Details

Elevation views of the cabinets in the kitchen and bath, and of other units such as shelving in a den or living room, are drawn to show exactly how they are to be made. Often these are mass-produced stock units and this information may be given by lettering the manufacturer's item number on the cabinet. Refer to Section 11, Chapter 36, "Cabinet Construction."

Fireplace Details

While the plan of the house used in this chapter does not have a fireplace, fireplaces are typical in many homes. Custom built fireplaces require an elevation view showing details such as a mantel and surrounding material and a section giving the required dimensions. The section is very important because if left to the brick mason the firebox may not be designed to function properly **(see 4-44)**.

ADDITIONAL INFORMATION

Spence, William P., *Architectural Working Drawings: Residential and Commercial Buildings*, John Wiley and Sons, NY 1993

Spence, William P., *Architecture: Design, Engineering, Drawing*, Glencoe/McGraw-Hill, Mission Hills, CA. Sixth Edition, 1991

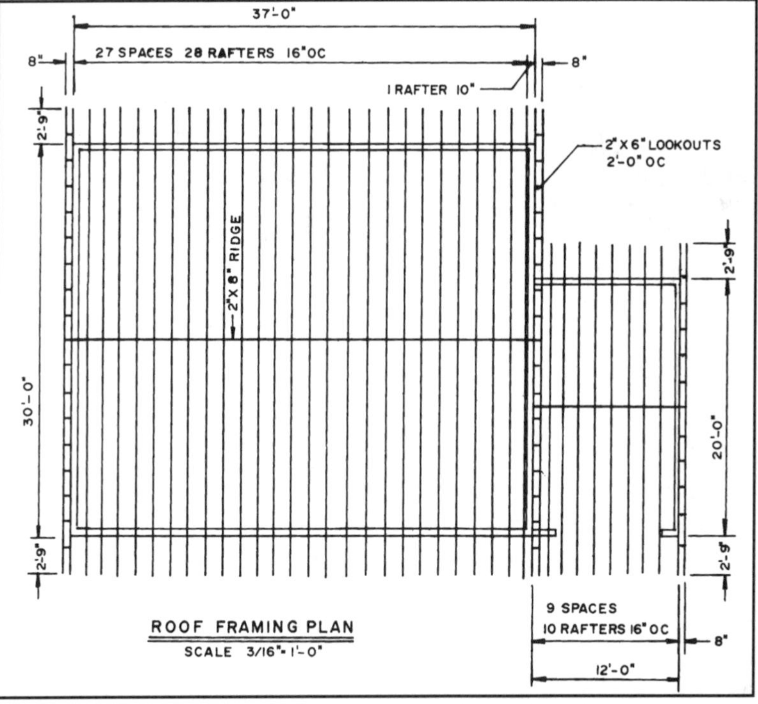

4-41 Carpenters use the roof framing plan to locate the rafters and establish the overhang at the eaves and rake.

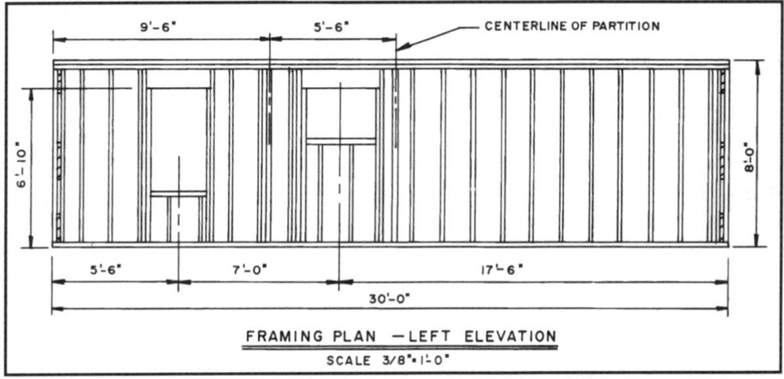

4-42 The wall framing plan indicates the framing of the wall and openings in the wall.

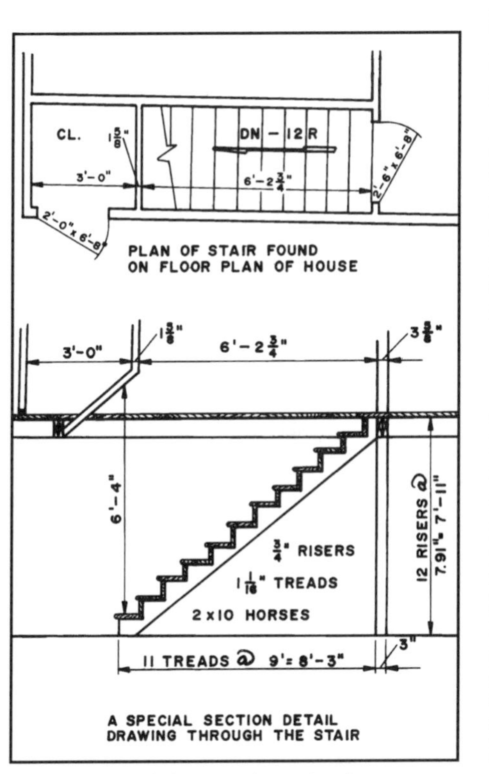

4-43 A detail drawing for a simple carpenter-built stair.

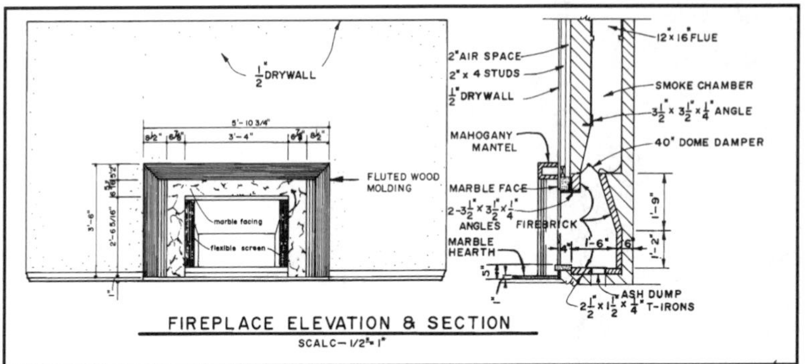

4-44 A fireplace detail specifies how the mason should size the firebox and gives information about the mantel.

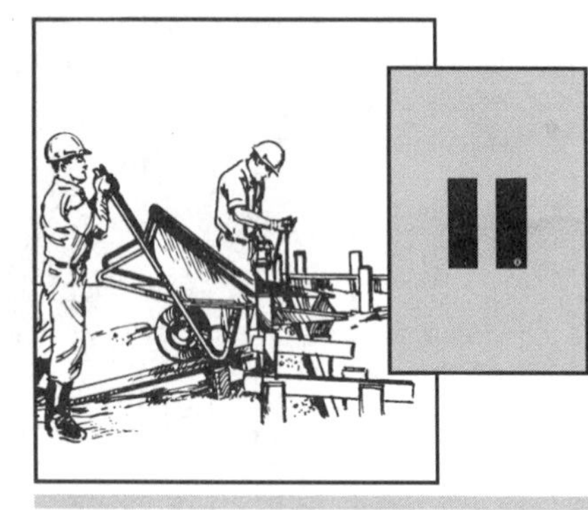

II SITE & FOUNDATION PREPARATION

CHAPTER 5

Using Leveling Instruments & Linear Measuring Tools

The use of leveling instruments and linear measuring tools is essential to various parts of the construction process, ranging from investigating the original site to locating the building on the site, and establishing grades and elevations of various parts of the structure.

LINEAR MEASURING TOOLS

Linear distances are measured with measuring tapes and various electronic distance-measuring (EDM) devices.

Measuring tapes are available on winding reels that store 50 to 300 feet of tape **(see 5-1)**. They are available with a variety of graduations. Some are graduated in feet, inches, and eighths of an inch or in feet, inches, and tenths of an inch, while metric tapes are graduated in meters, with subdivisions of centimeters and meters **(see 5-2)**. Stainless steel and fiberglass tapes are available.

EDM devices range from electronic tapes to laser devices that can measure long distances. One type of electronic tape is shown in **5-3**. This device is used in the same manner as the typical steel measuring tape. However, the distance is also shown by a digital readout on the top of the unit. It will read in feet and inches, inches only, or centimeters. It can be set to read outside-to-outside and inside-to-inside distances. The maximum length it can measure is 25 feet (762.5cm). It is battery operated. Another style of electronic tape is shown in **5-4**. The digital readout is on the side and can be converted to feet and inches, inches, and centimeters. These measuring devices are

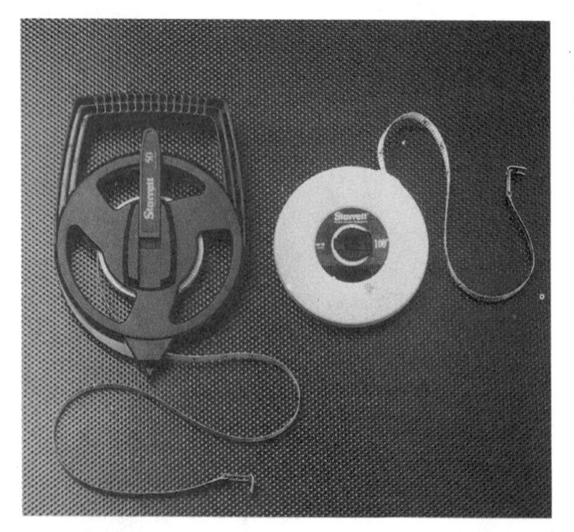

5-1 Tapes are used to measure linear distances. *(Courtesy L.S. Starrett Company)*

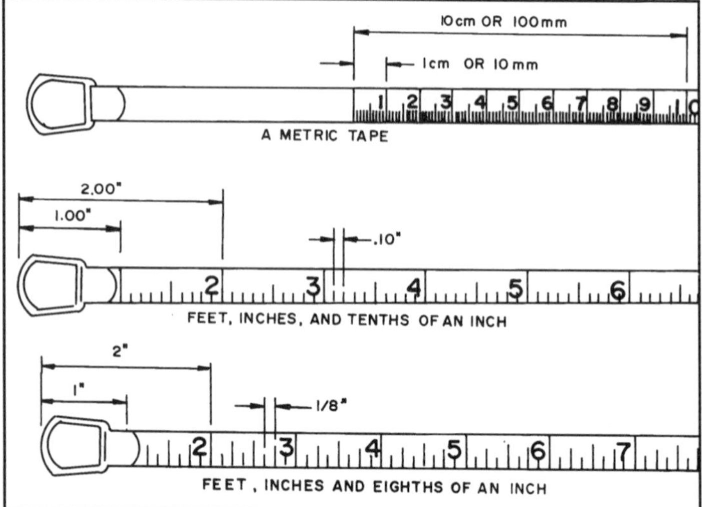

5-2 Tapes are available with customary and metric units.

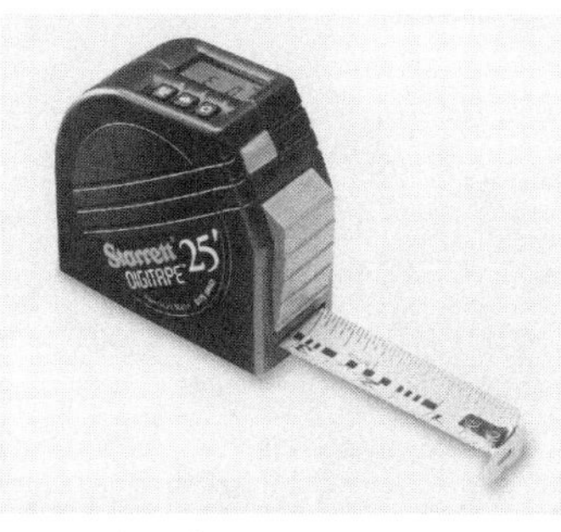

5-3 This electronic measuring device provides a steel tape and a digital readout. *(Courtesy L.S. Starrett Company)*

5-4 The digital readout on this electronic measuring tape can be converted to both customary and metric units. *(Courtesy Seiko Instruments, USA, Inc.)*

accurate to the nearest 1⁄16 inch (0.16cm). These units are useful for interior and exterior framing and trim since the distance they can measure is small. They are not useful for land surveying or locating a building on the lot.

Other EDM instruments derive distances by comparing transmitted and received signals and deriving the phase difference between them. The signal may be visible, infrared, or microwave wavelengths. The signal is reflected off an unmanned reflector (target) or by a second manned instrument located at the point to which the measurement is to be taken. The distance between the instrument and the target is shown on a digital readout.

LEVELING INSTRUMENTS

Leveling instruments are essential to the proper construction of a foundation, for wall, floor, and roof framing, and for many exterior and interior finishing processes. The tools available range from a simple bubble level to laser leveling instruments. The size, type, and capabilities vary considerably with the instrument. Manufacturers should be consulted for technical details.

OPTICAL LEVELING INSTRUMENTS

A carpenter is frequently required to use various optical leveling instruments. These include the level, automatic level, level transit, and automatic level transit.

The Level

A **level** (sometimes referred to as a builder's level) is a surveying instrument for measuring heights with respect to an established **horizontal line of sight**. The level is designed for use on light construction work, where it can be used for jobs such as leveling foundations, driveways, patios, floors, sills, curbs, and ditches. The telescope is mounted on a leveling plate which mounts on a tripod. Leveling screws and a level vial are used to set the telescope horizontally. The telescope rotates in a 360-degree horizontal circle but does not have a tilting mechanism allowing it to measure vertical angles. A **level transit** is used for these operations. The circle is graduated in single degrees and uses a vernier to measure angles to 15′ (minutes). The eyepiece has a sunshade to reduce interference from the sun **(see 5-5)**. A level mounted on a tripod accompanied by a leveling rod is shown in **5-6**.

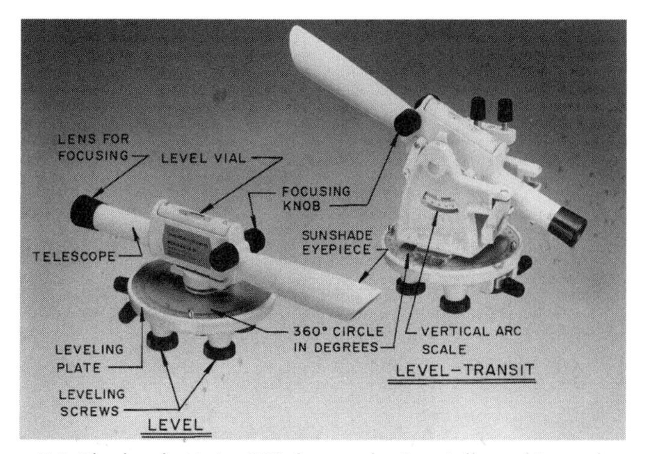

5-5 The level rotates 360 degrees horizontally and is used to lay out horizontal angles and check features for levelness. The level transit rotates 360 degrees horizontally and swings through a vertical arc 45 degrees above and 45 degrees below level. *(Courtesy David White, Inc.)*

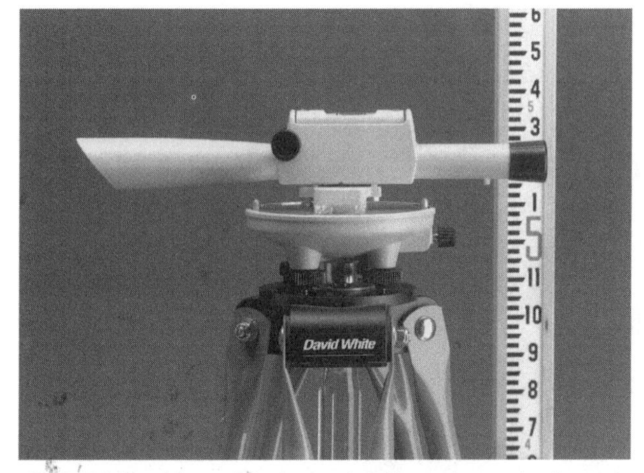

5-6 This level has a 20× (power) optics system, a horizontal circle divided into 360 degrees, a vernier divided into 15-minute graduations, and a manually adjusted leveling plate. Notice the leveling rod. *(Courtesy David White, Inc.)*

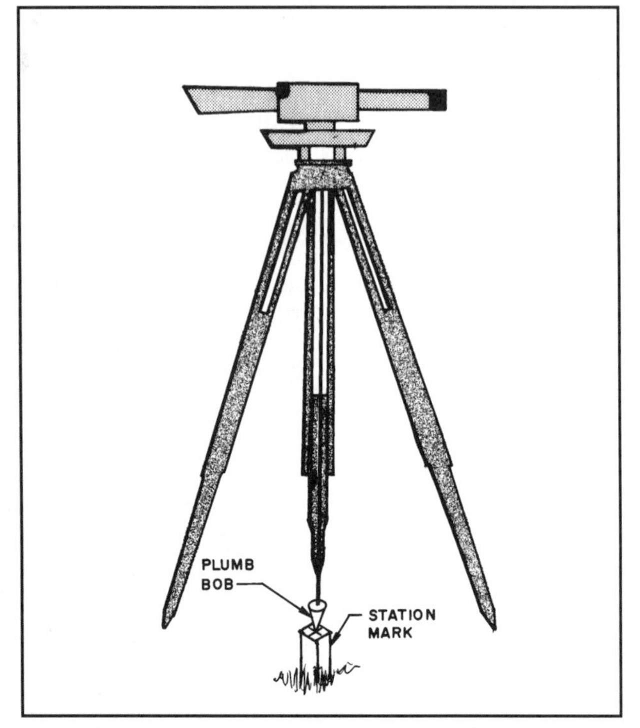

5-7 The level and level transit are positioned over an established survey point by dropping a plumb bob to a nail or mark on the station mark stake.

A plumb bob can be hung below the level, enabling it to be set directly above an established survey point **(see 5-7)**.

Manufacturers have available a range of models with telescopes typically from 12× to 26× (power). The **power** of the lenses controls how close the object viewed through the telescope appears. A target viewed through a 20× (power) telescope will appear 20 times closer than when viewed with the unaided eye.

The Automatic Level

The automatic level is a self-leveling instrument. It is set up in the same manner as the level except it has a compensator which will automatically correct it should it move above or below the established level. This instrument uses an optical pendulum that uses gravity to correct for variations from a true level line of sight. If the instrument is jarred slightly, the line of sight remains true **(see 5-8)**.

It is used for the same operations as the builder's level but has the advantage of not requiring manual releveling if it is jarred slightly.

The Level Transit

The level transit does the work of two instruments. It serves as a **level** as just described in previous sections and as a transit. A transit is a surveying instrument used for laying out and measuring **horizontal** and **vertical** angles, distances, directions, and differences in elevation. The laser-transit can shoot vertical angles up to 45 degrees and be set to plumb a vertical surface **(refer to 5-5)**. It can level foundations and other features as well as run straight lines for driveway curbs, building lines, fences, and setting stakes.

The level transit should be locked in the horizontal position, when used as a level, and rotated 360 degrees. When unlocked the telescope can be raised and lowered in a vertical position for laying out and measuring angles **(see 5-9)**.

Both manual and automatic leveling level transits are available. They are leveled in the same manner as described for levels. A level transit mounted on a tripod and a leveling rod are shown in **5-10**. An automatic leveling level transit is shown in **5-11**.

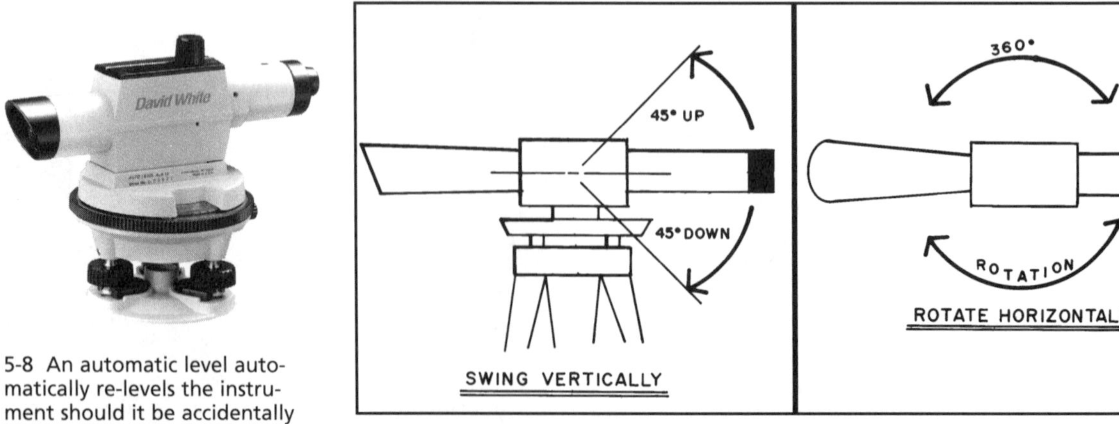

5-8 An automatic level automatically re-levels the instrument should it be accidentally jarred while being used. *(Courtesy David White, Inc.)*

5-9 The level transit can be rotated 360 degrees horizontally and move through a vertical arc of 45 degrees above and 45 degrees below level.

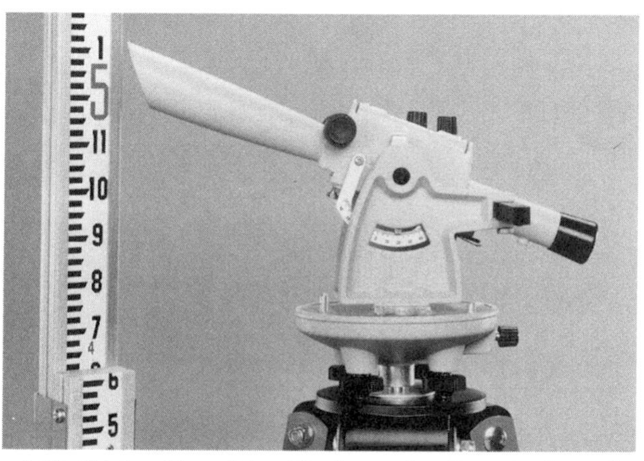

5-10 This level transit has 20× (power) optics, a horizontal circle divided into 360 degrees, a vernier divided into 15-minute graduations, a vertical arc ranging from 45 degrees above and 45 degrees below level, and a manually adjusted leveling plate. *(Courtesy David White, Inc.)*

5-11 This automatic leveling level transit has 18× (power) optics, a horizontal circle divided into 360 degrees, a vernier divided into 15-minute graduations, a vertical arc ranging from 45 degrees above to 45 degrees below level, and a compensator system that automatically keeps it level. *(Courtesy David White, Inc.)*

Tripods

A tripod has three legs that are either wood or metal **(see 5-12)**. The legs are adjustable in length to enable the level to be set horizontal on sloping or uneven ground **(see 5-13)**. The legs must be set firmly into the ground. Tripods set on hard surfaces, such as asphalt or concrete, can have their legs confined by a wood frame to prevent them from spreading and slipping on the surface **(see 5-14)**.

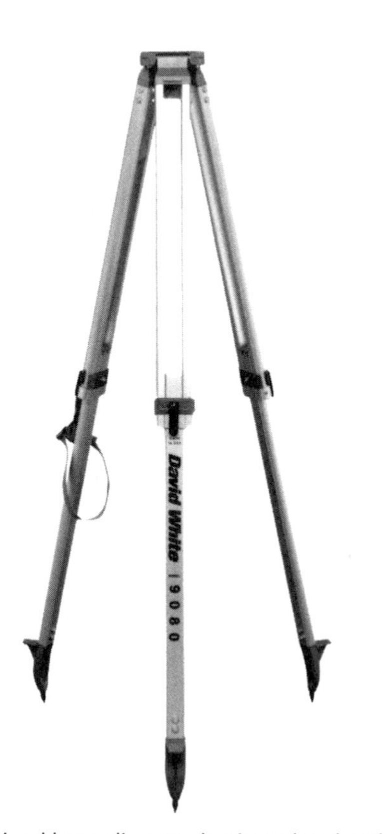

5-12 This tripod has a die-cast aluminum head and adjustable aluminum extension legs. Notice the carrying strap. *(Courtesy David White, Inc.)*

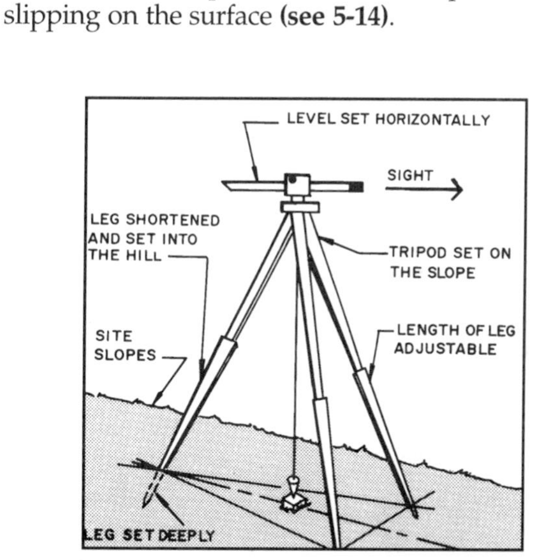

5-13 The length of tripod legs can be adjusted to assist in setting the instrument level. The legs are set firmly into the ground.

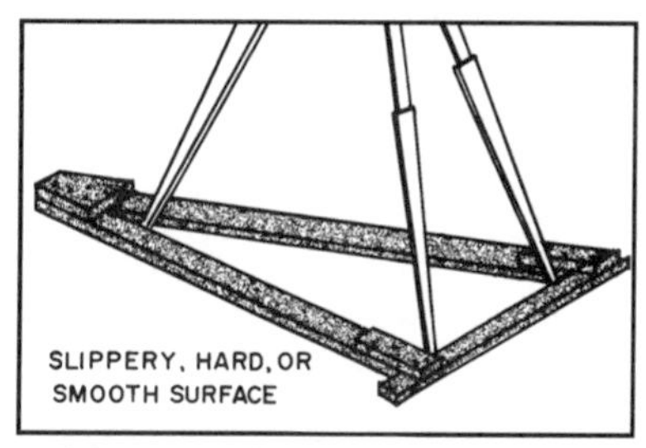

5-14 A wood frame can be used to keep the legs of the tripod from spreading or slipping on firm surfaces.

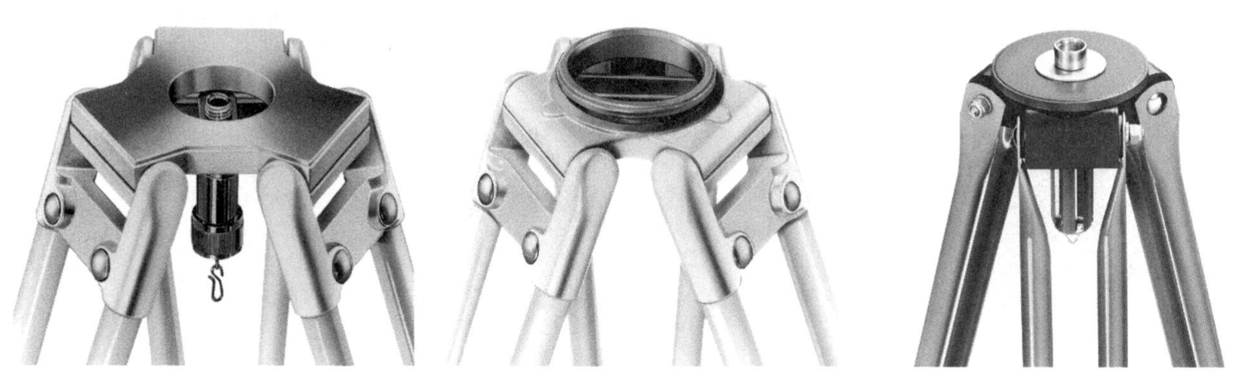

5-15 Typical threaded connections available on tripod heads. *(Courtesy David White, Inc.)*

The head has various types of threaded connections. The tripod selected must be one designed to accommodate the level **(see 5-15)**.

The Leveling Rod

Leveling rods are available made from hardwood, fiberglass, and aluminum. The leveling rod is placed at the point to be sighted by the level or level transit. It is in two sections and can be extended to 14 feet. A target can be attached to the rod to help locate the target center graduations **(see 5-16)**. Rods are available with graduations in feet and inches or in metric units. Those in feet and inches are the architect's rod and the engineer's rod. The **architect's rod** is subdivided into feet, inches, and eighths of an inch. This is typically used by carpenters and other construction workers. The **engineer's rod** is subdivided into feet, tenths of a foot, and hundredths of a foot. This is used by land surveyors and other engineering personnel. The **metric rod** is subdivided into meters, decimeters, and 5-millimeter units. (A decimeter is 1/1000th of a meter.) Metric construction projects use this for all purposes **(see 5-17)**.

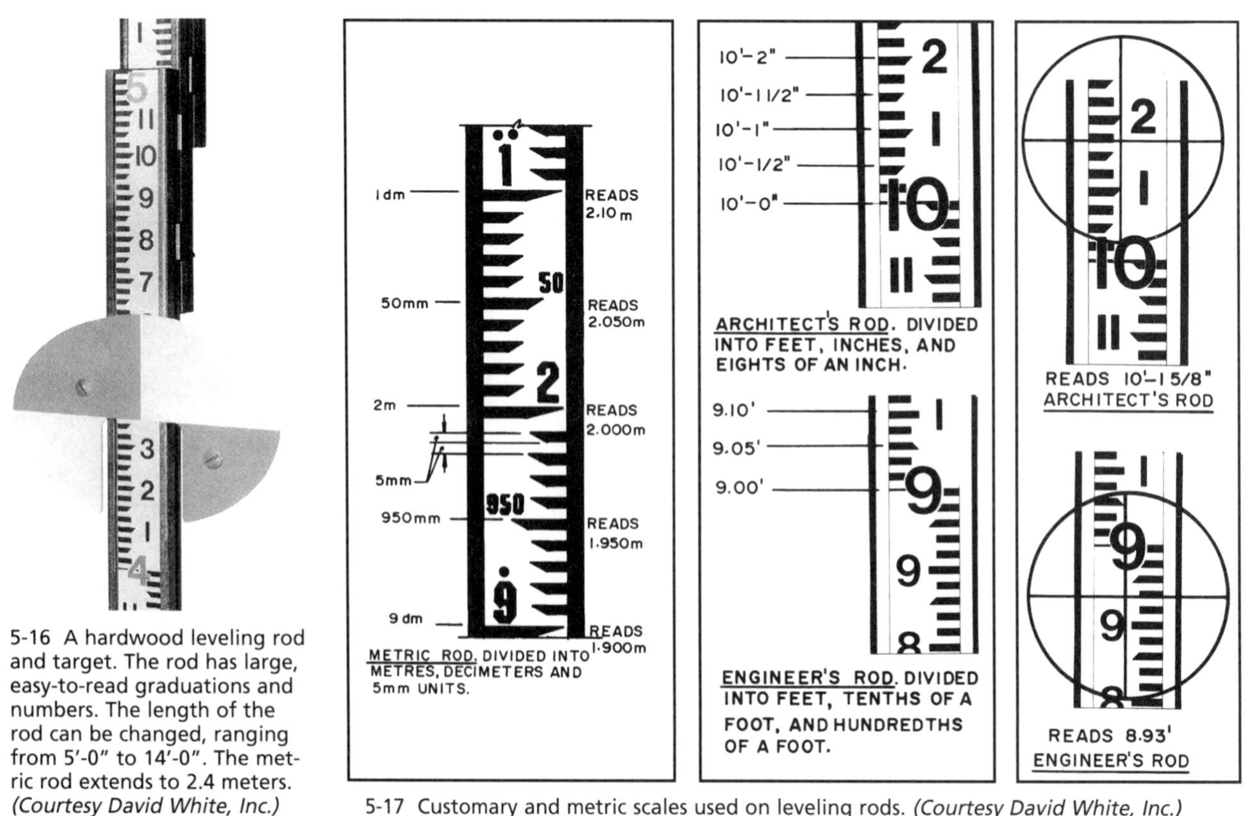

5-16 A hardwood leveling rod and target. The rod has large, easy-to-read graduations and numbers. The length of the rod can be changed, ranging from 5'-0" to 14'-0". The metric rod extends to 2.4 meters. *(Courtesy David White, Inc.)*

5-17 Customary and metric scales used on leveling rods. *(Courtesy David White, Inc.)*

Using the Horizontal Circle & Vernier Scale

Horizontal angles are measured in degrees and fractions of degrees. A circle is divided into 360 **degrees** (360º). A degree is divided into 60 **minutes** (60′). A minute is divided into 60 **seconds** (60″).

Levels and level transits use a **horizontal circle** divided into 360 degrees and a **vernier scale** marked every 5 minutes. The vernier scale is used to lay out **horizontal angles** that are not even degrees, such as 65 degrees 30 minutes. The horizontal circle remains in position as the instrument is rotated horizontally. The vernier scale is mounted onto the instrument frame that rotates as the telescope is moved. The horizontal scale is divided into quadrants, each of which includes a 90-degree arc. Each division is one degree. The vernier scale is divided into 15-minute intervals to the right and left of the zero mark **(see 5-18)**.

To swing a 90-degree angle, set the instrument on the first target and the horizontal circle on zero. Swing the telescope until the zero on the vernier lines up with the 90-degree mark on the circle, as shown in **5-19**. If you want an angle of 65 degrees 30 minutes, set the zero on the venier on 65 degrees and then move it an additional 30 minutes on the vernier **(see 5-20)**.

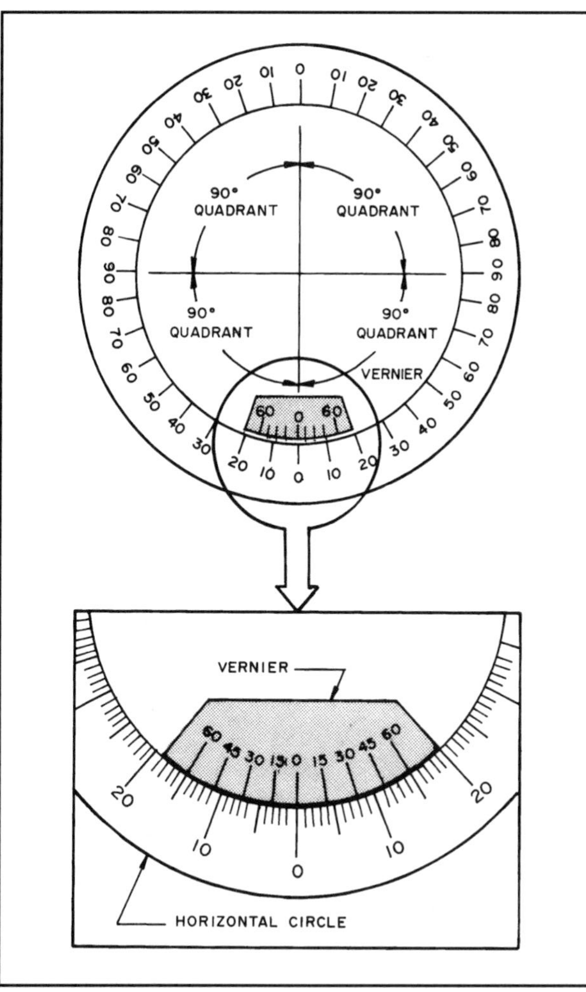

5-18 The horizontal circle is divided into 360 degrees and the vernier is divided into 15-minute intervals 60 degrees left and right of the zero mark.

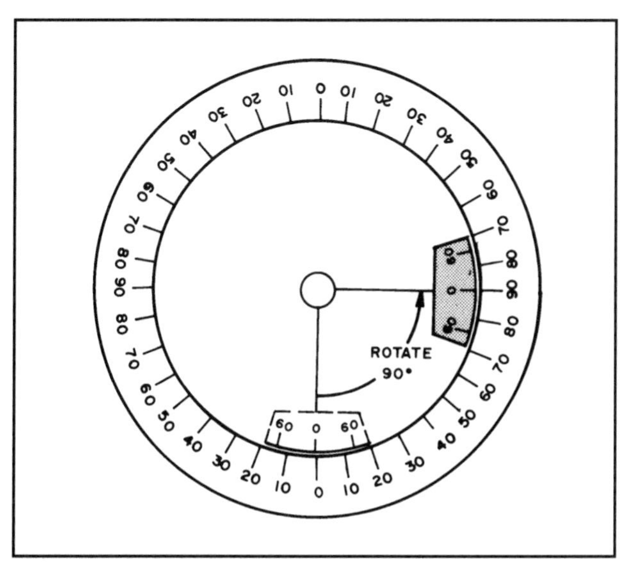

5-19 How to turn a 90-degree horizontal angle using the horizontal circle and the vernier scale.

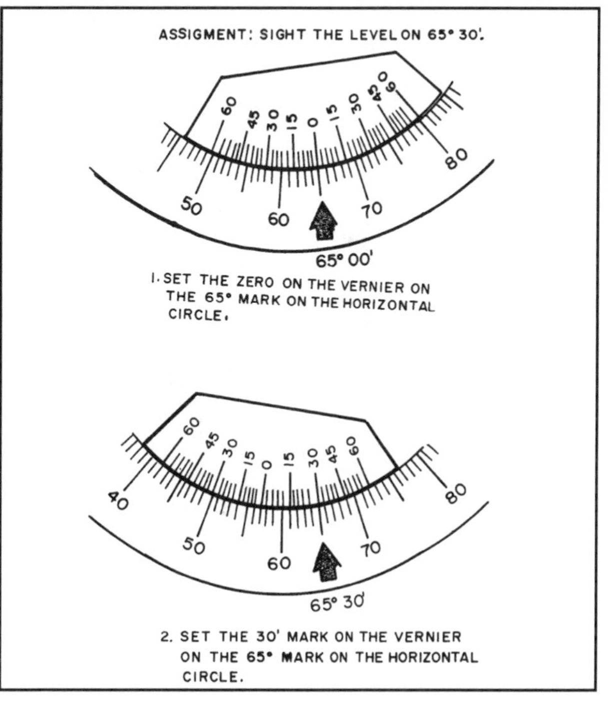

5-20 An angle that is not in even degrees locates the nearest even degree on the horizontal circle and the minutes with the vernier.

If you want to locate an arc greater than 90 degrees, it would be located as shown in **5-21**.

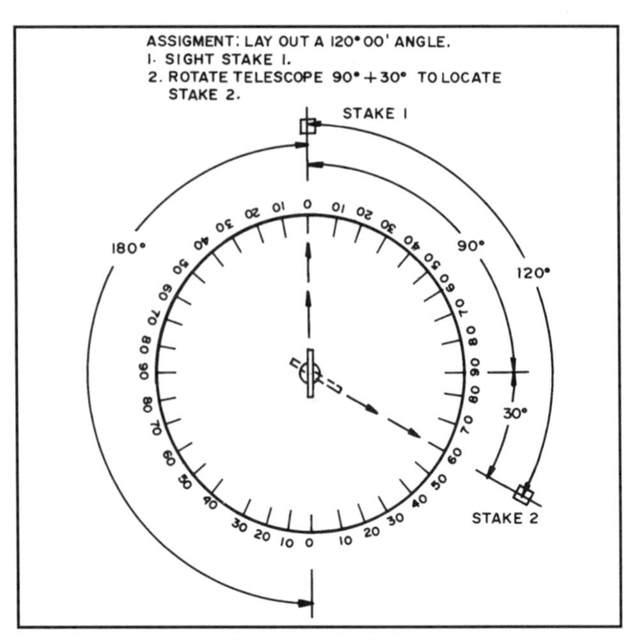

5-21 This is how to lay out an angle greater than 90 degrees.

Using the Vertical Arc & Vernier Scale

Level transits use a vertical arc marked in degrees to measure vertical angles. It serves the same purpose as the horizontal circle on a level. The vertical arc is marked in one-degree graduations. Some instruments do not have a vernier. Therefore they can only locate angles to the nearest degree. As the telescope is rotated vertically up or down the angle can be read on the vertical arc dial **(see 5-22)**.

SETTING UP THE LEVEL & LEVEL TRANSIT

The procedures for setting up the level and level transit are the same except the telescope on the level transit **must be locked** in its horizontal position before the instrument is leveled.

The tripod is set with the legs about 3 feet apart and firmly set into the ground. On rough or sloped ground one or more of the legs may have their length adjusted as shown earlier in **5-13**. The legs can be kept from spreading when placed on a hard surface by enclosing them in a wood triangle **(refer to 5-14)**.

Generally the instrument will be set over a known point which is located by a surveyor's stake or a permanent marker.

To set the instrument:

1. Place the tripod with its legs spread over the stake, visually positioning the center of the head directly over the stake.

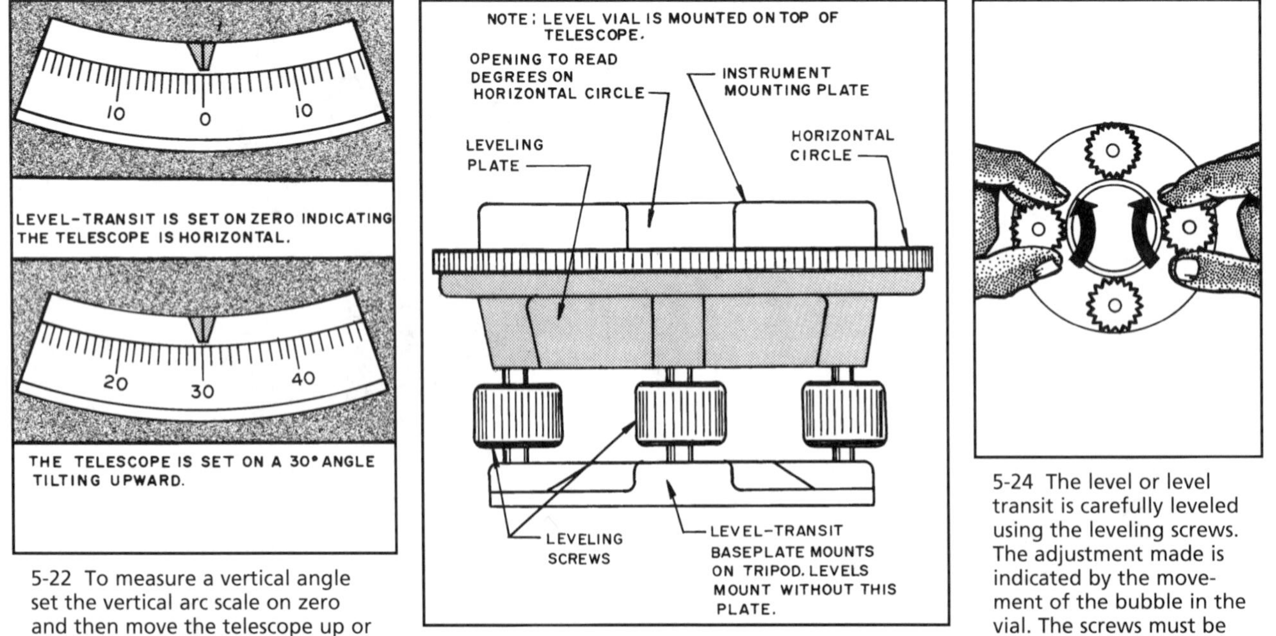

5-22 To measure a vertical angle set the vertical arc scale on zero and then move the telescope up or down until the pointer indicates the desired angle.

5-23 The telescope is leveled by turning the leveling screws located below the leveling plate.

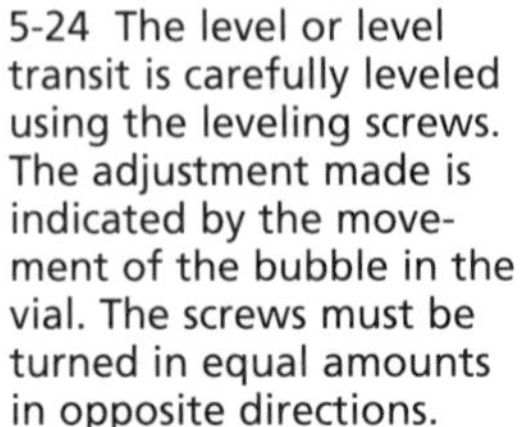

5-24 The level or level transit is carefully leveled using the leveling screws. The adjustment made is indicated by the movement of the bubble in the vial. The screws must be turned in equal amounts in opposite directions.

2. Adjust the legs so the tripod is level.

3. Fasten the instrument on the tripod. Lift it by its base and screw it onto the tripod. Tighten until firm but do not overtighten.

4. Attach the plumb bob to the instrument and adjust the length until it is just above the stake. It may be necessary to move the tripod a little to get the plumb bob centered on the stake.

5. If the level or level transit has a shifting center use it to bring the plumb bob directly on the mark on the stake. This involves moving the instrument on the leveling plate.

6. Now level the instrument with the leveling screws and the level vice. This is a preliminary leveling, so do not lock the screws.

7. Finally, fine-adjust the leveling screws and lock them in place.

Leveling the Level or Level Transit

Before these instruments can accurately lay out horizontal and vertical angles, the telescope must be absolutely level. After the instrument is set on the tripod, the leveling vial (a spirit level) and leveling screws are used to rough-level and fine-tune the instrument **(see 5-23)**.

The instrument has four leveling screws **(see 5-24)**. Screws opposite each other are moved simultaneously. To adjust for level, (A) the screws are moved in opposite directions, (B) they are moved at the same time, and (C) each is moved the same amount, as labeled in 5-24.

The steps to level the instrument are shown in **5-25**.

A. Point the telescope so it is over a pair of leveling screws. Turn the leveling screws below it. The direction the left thumb moves is the direction the bubble in the level vial will move. Turning the screws in will move the bubble to the right. Turning them out will move the bubble to the left. Adjust until the bubble is centered.

B. Rotate the telescope until it is directly above the other two leveling screws. Adjust as described until the bubble is centered in the vial.

C. Now go back to the first two leveling screws and repeat the leveling process. This is a fine-tuning operation and requires very small adjustments.

D. Finally rotate the telescope over each of the leveling screws and check for level. If necessary repeat the above leveling steps until the instrument is level in all directions.

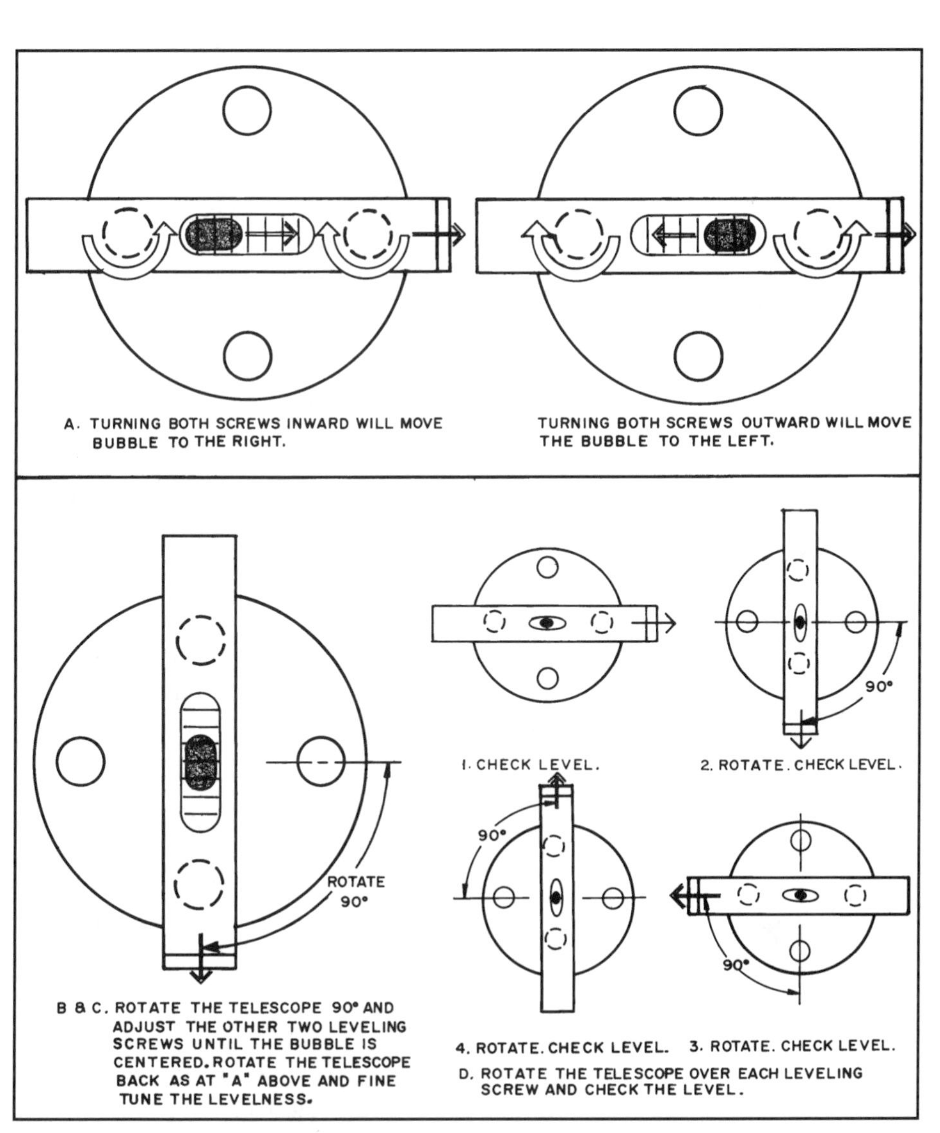

5-25 This is how to level the instrument and fine-tune the adjustments.

Move the rod to the left of the person holding it

Move the rod to the right of the person holding it.

Raise the target.

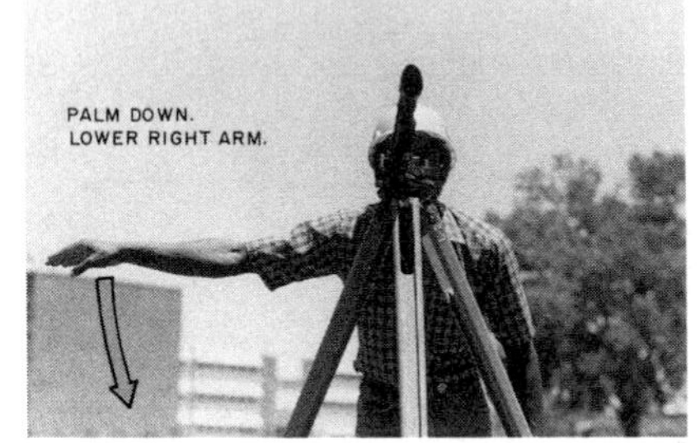

Lower the target.

The target is plumb.

The target is on grade.

5-26 Arm signals given by the surveyor to the rodman, directing the movement of the leveling rod into plumb and the target into the on-grade position.

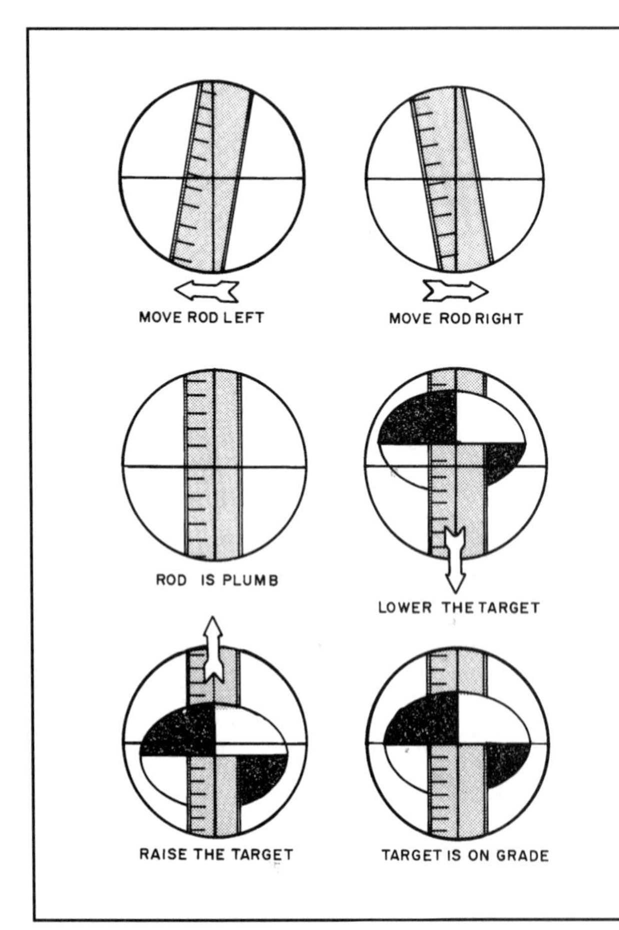

5-27 What the surveyor sees through the telescope and the directions he gives to the rodman are shown here.

Surveyor-to-Rodman Signals

The surveyor operates the level or level transit and the rodman holds the leveling rod at the point being established. Some teams use walkie-talkies to tell the rodman how the leveling rod and target should be moved. Others use arm signals, especially when distances between the two are relatively short. The standard signals are in **5-26**. Notice the surveyor signals the direction in which the rodman should move the rod (right or left) by pointing in the actual direction of movement. Since this is opposite the way the rodman faces, the surveyor points "move left" with the right arm and "move right" with the left arm. Likewise the "up" and "down" signals are used to locate the target on grade. The view of the rod and target as seen by the surveyor through the telescope is in **5-27**.

ON-SITE LEVELING OPERATIONS

Leveling operations can be performed with a level or level transit. When the level transit is used the telescope is locked in the horizontal position.

Often it is necessary to find the differences in elevation of the grade on the site. These elevations are needed before starting to establish the corner heights of the foundation and when moving soil to the finish grades indicated on the architectural drawings.

Finding the Difference in Grade Elevations

Typically a grade stake is driven at the high corner of the lot. This stake is usually indicated on the site plan. The level is placed over this point and the rod placed at intervals on the sloped lot. The height is read on the rod by the surveyor at each location as shown in **5-28**. These heights are recorded in the surveyor's log book. If the lot is relatively flat the sightings can be spaced farther apart than if the slope is steep. A steeply sloped lot may require sighting one point, then moving the level to that point and sighting the next point, until the total difference in grade has been measured. The amount of slope will determine the number of sightings required.

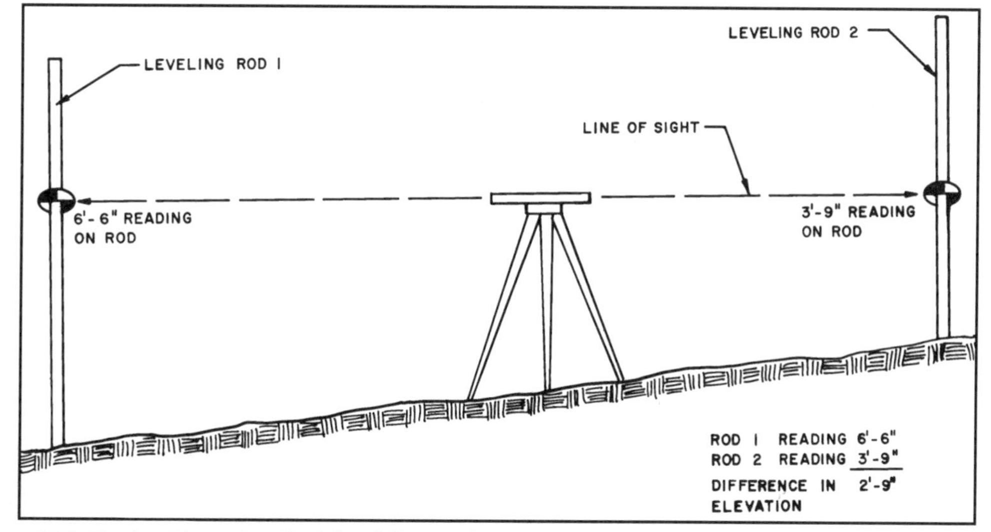

5-28 The level is used to find the difference between the elevations of two points.

Setting Footing Grade Stakes

The footing grade stakes locate the required elevation of the soil at each corner of the building. To set the grade stakes, place the level near the center of the building **(see 5-29)**. Sight a leveling rod at each corner and record the actual grade and note the difference between it and the desired grade. Sometimes some soil has to be removed or a low area filled to bring the grade to the desired height. The required grade is marked on the grade stake. This mark is the point from which cut-and-fill measurements are taken.

If an area requires a fill to get the grade to the proper height, the grade stake is marked F (fill) and the amount of fill required is marked on the stake. If the area requires soil be removed (a cut), the grade stake is marked C and the amount of cut is written on it **(see 5-30)**.

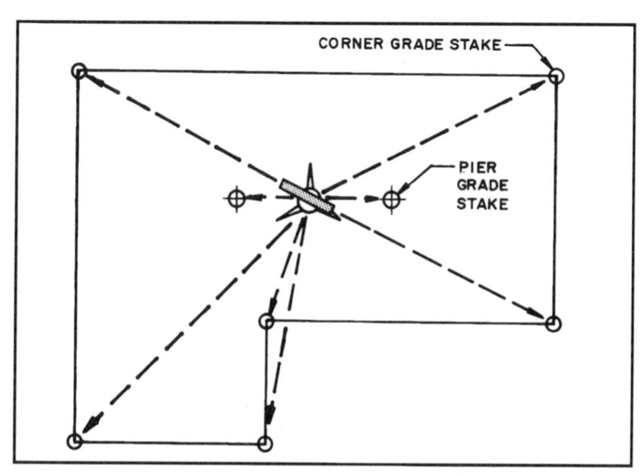

5-29 To establish the required elevation of the soil at each corner of a building, set the level in the center of the building and sight each corner.

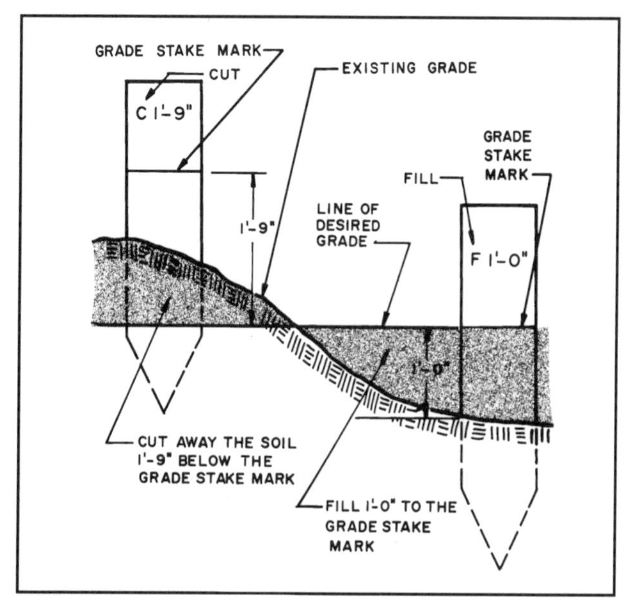

5-30 The required elevation of the soil is marked on the grade stake.

Finding the Elevation of Above-Grade Construction

Frequently it is necessary to verify the elevation of some part of a structure that is above grade, such as the second floor or a beam. This can be done with a level or level transit. Set the level over a known point which was established from the bench mark. This could be the top of the footing as shown in **5-31**. Sight the required height above the footing. Rotate the telescope and read the height on the leveling rod on the foundation form. If the part is above the height of the level, a rule or leveling rod can be suspended below the surface as shown in **5-32**.

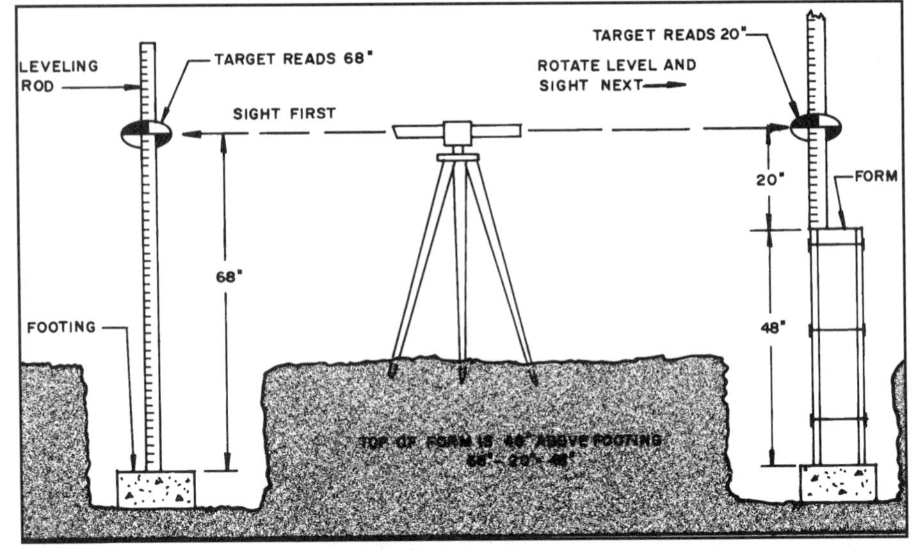

5-31 The elevation of part of a structure below a known point can be found with a level.

Contour Lines

Contour lines are lines that connect points having the same elevation. The surveyor establishes these and they are recorded on the site plan. They are essential as the architect locates the building and establishes grade lines and the height of the first floor of the building.

The distance between contour lines varies with the slope. Steeply sloped lots will have contour lines closer together than those on sites with moderate or little slope.

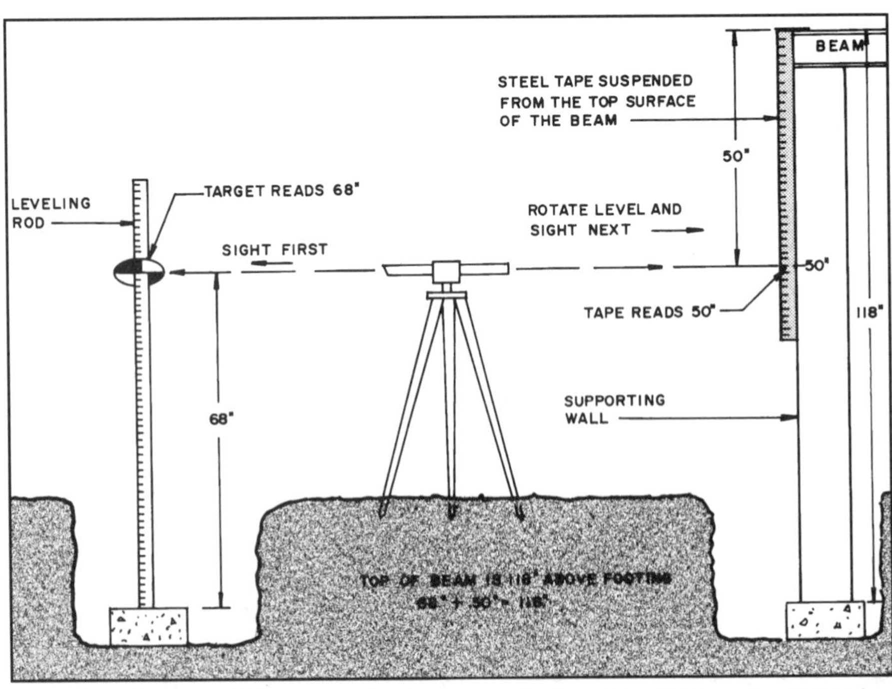

5-32 The elevation of high parts of a structure can be found by sighting a leveling rod or rule suspended from the feature.

Laying Out Horizontal Angles

Set the level over the corner stake and line up the plumb bob with the nail in the top of the stake. Level the instrument. Sight the leveling rod that is being held on one corner of the building.

Set the horizontal circle scale to zero to align with the zero on the vernier scale. Swing the level the number of degrees in the angle to be laid out and sight the leveling rod on that point. In **5-33** the angle to be laid out was 90 degrees. Drive a stake at the sighted point to locate this corner.

Measuring Vertical Angles

Vertical angles can be measured with the level transit. To lay out a vertical angle level the level-transit as previously described. Swing the level horizontally until the telescope is lined up on the member involved. Lock it in position. Rotate the level vertically to the angle desired. The angle will be found on the vertical arc. The angles are measured above and below the horizontal as shown in **5-34**.

A vertical member can be checked to see if it is plumb by sighting it as just described. When it is plumb from one side move the level transit 90 degrees and check from the other side.

Setting Stakes in a Row

It is often necessary to locate a series of stakes in a straight line, as is required when locating the edges of a driveway, sidewalk, or row of fence posts. The level transit is set and plumbed over the first stake in the row. The telescope is tilted and sighted on the edge of the end stake in the other end of the row. If necessary adjust the horizontal tangent screw to get the cross hair aligned with the edge of the stake and then lock the horizontal control. Locate each stake in the row by measuring the required distance they are to be apart and aiming the telescope at them. When the edge of each stake aligns with the cross hair, drive the stake into the ground.

CARE OF SURVEYING INSTRUMENTS

Leveling instruments must be handled carefully and stored in cases provided by the manufacturer. Dropping or jarring them can cause physical damage and possibly throw them out of alignment. They must be protected from dust and moisture. When on a tripod the instrument must be covered if it is not going to be used for a while. When fastening the instrument to the tripod grip it by the base. Do not grip or lift the instrument by the telescope. Be certain the tripod is firmly set in the ground or has the legs blocked so they will not slip if being used on hard surfaces. Be certain the legs are locked securely so they will not spread or unexpectedly change in length. Tighten the leveling screws and clamp screws firmly, but do not overtighten. Clean dust from the lens by blowing or dusting with a soft brush. Finally be certain everyone using the instrument has studied the manufacturer's manual.

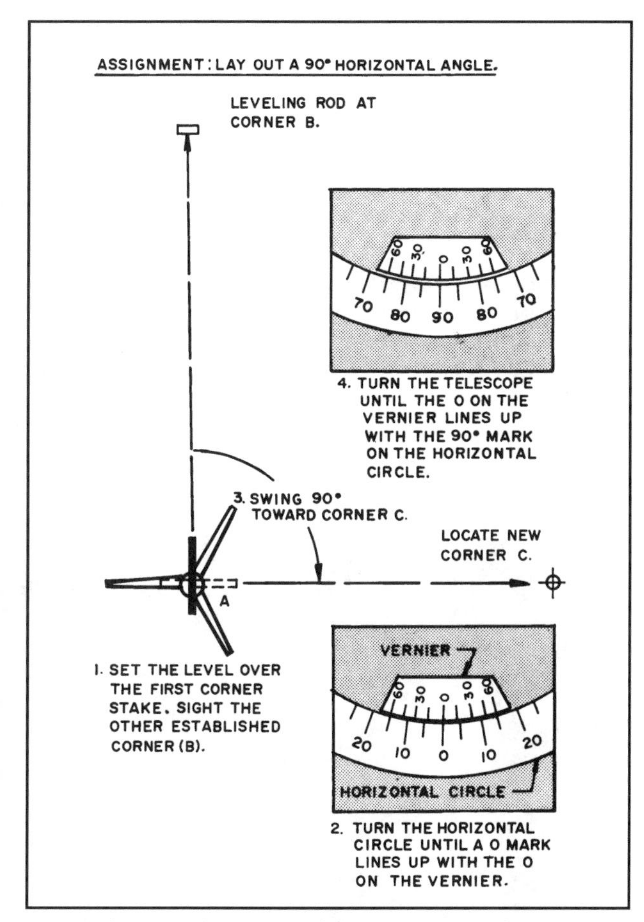

5-33 These are the steps to follow to lay out a horizontal angle.

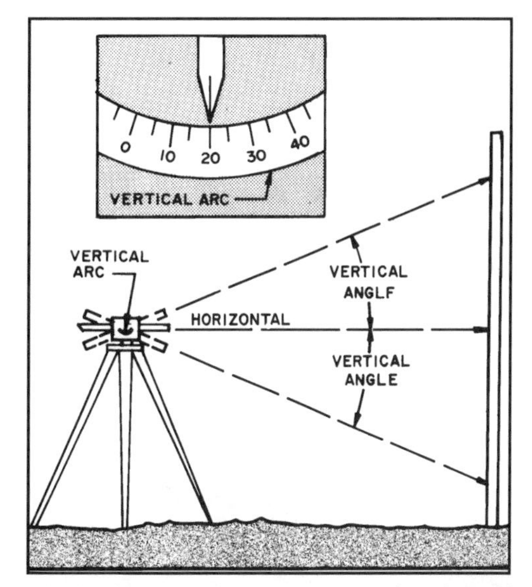

5-34 The level transit can measure vertical angles with its vertical arc scale.

Electronic Levels

One type of electronic level is shown in **5-35**. It is used to check members for levelness and plumb and can measure all angles within a 360-degree range. It has standard level bubble vials for quick checks such as are made with a standard level. For precision work a sensor module is used to provide digital readings. This battery-operated system reports the data on a digital display and is accurate to 1/10th of a degree. When the display reads 0.0 degrees, the member is level. When it reads 90.0 degrees, it is plumb. It also gives an audible beep when it is level or plumb. This is useful when it is difficult to see the digital display. Angles can be checked in degrees or percent slope (rise over run of a roof).

The level may be mounted on a tripod **(see 5-36)** or on a floor mounting **(see 5-37)**. It is used to lay out foundations, align window frames, locate and plumb partitions, locate horizontal lines, and perform many other leveling and plumbing functions. Some typical uses are shown in **5-38**.

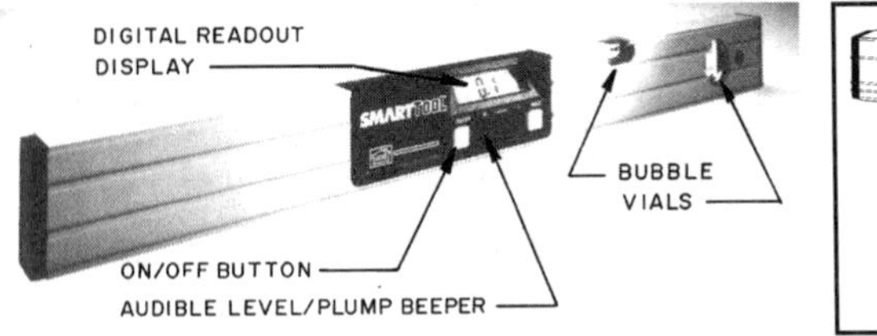

5-35 The electronic level checks members for levelness and plumb and can measure an angle within a 360-degree range. *(Courtesy Macklanburg-Duncan Company)*

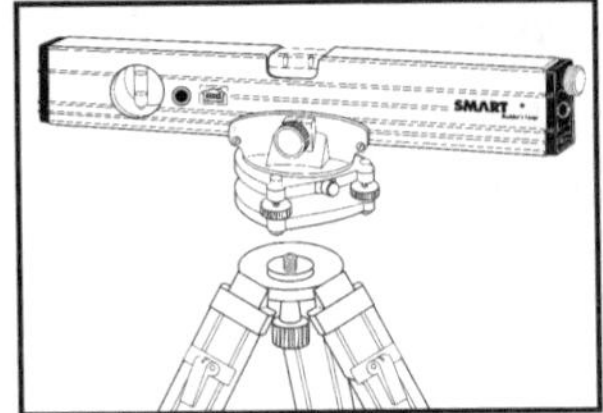

5-36 The electronic level can be mounted on a tripod. *(Courtesy Macklanburg-Duncan Company)*

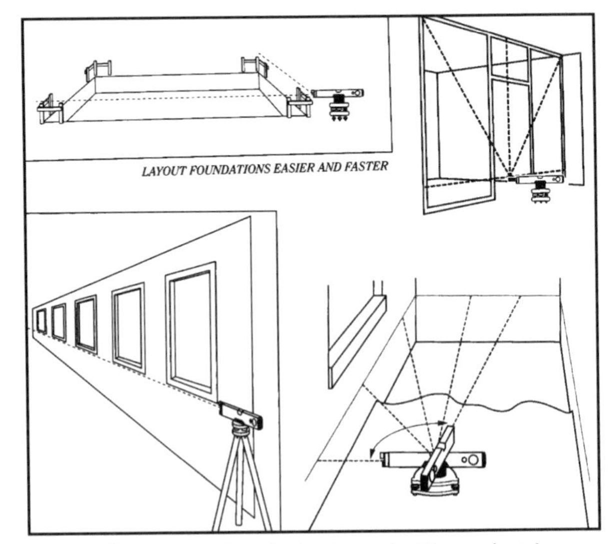

5-37 A floor-mounting device is used with an electric level for interior and exterior applications. *(Courtesy Macklanburg-Duncan Company)*

Lasers

A **laser** is a device that converts electromagnetic radiation of mixed frequencies to one or more discrete frequencies of highly amplified and coherent visible radiation. The emission is more intense, monochromatic, coherent, and directional than light emitted by a conventional source such as an incandescent lamp.

Some types of lasers operate in the **visible** part of the spectrum and can be seen against a visible reference. Others operate in the infrared part of the spectrum and have a beam that is **invisible** to the human eye, and a **detector** must be used to establish a reference line or plane.

5-38 Some typical uses for the electric level SmarTool®. *(Courtesy Macklanburg-Duncan Company)*

Laser detectors normally employ photoelectric cells or photodiodes. They are typically mounted on a leveling rod and are raised and lowered by the surveyor. The intersection of the beam with the detector produces a reading which gives the height of the earth below the beam.

Laser Levels

A laser level can be positioned to establish level (horizontal) and plumb (vertical) lines.

A small pocket laser level is shown in **5-39**. This unit is about 4 inches (101.6mm) square and 2 inches (50.8mm) thick. It is designed to be carried in a carrying case on the carpenter's belt. It is an automatic self-leveling laser that projects a beam vertically, providing instant plumb bob. In addition it establishes level (horizontal) lines and automatically establishes 90-degree corners **(see 5-40)**. Some applications are shown in **5-41**. It can be used indoors and outdoors.

A larger powerful laser level with the related sensor and leveling rod is shown in **5-42**. It has a rotating head which sends out a beam that may be visible when it strikes an object or a detector on a leveling rod **(see 5-43)**.

5-39 This pocket laser is small enough to be carried in a pouch on the carpenter's belt. *(Courtesy Levelite Technology, Inc.)*

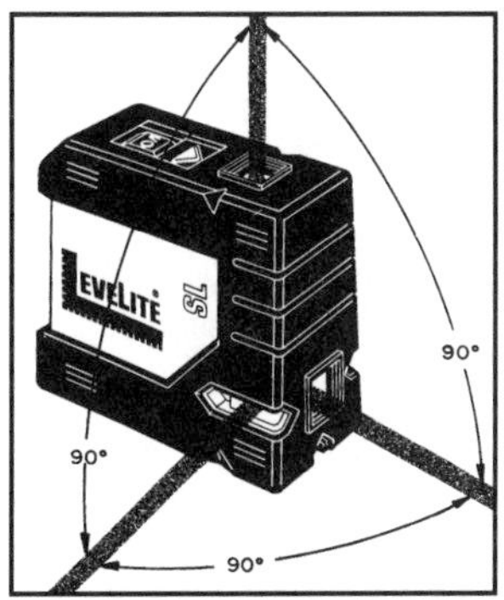

5-40 The pocket laser sends out vertical and level beams and establishes 90-degree corners. *(Courtesy Levelite Technology, Inc.)*

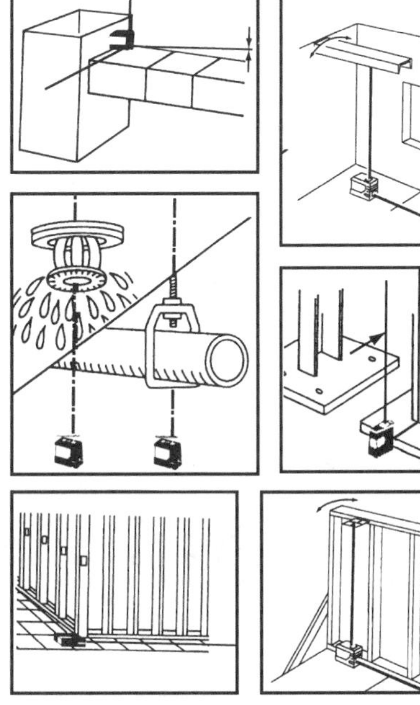

5-41 The pocket laser can quickly establish plumb, locate overhead points such as where to install a light, and run various level lines such as those needed for lining up studs on a partition. *(Courtesy Levelite Technology, Inc.)*

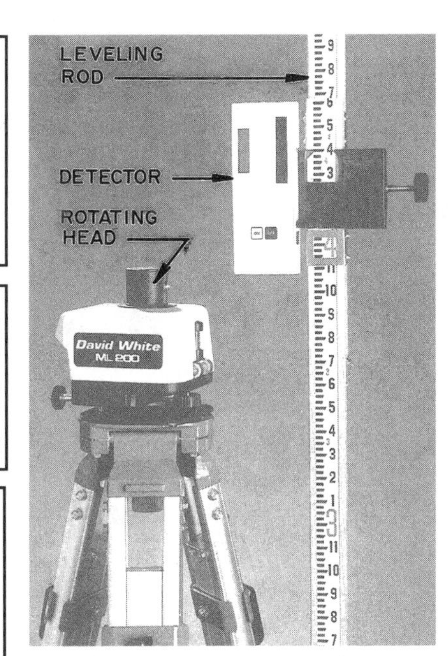

5-42 The laser level and the detector are powered by batteries. The detector uses a liquid crystal display (LCD) and has an audible signal when the laser hits the target. This instrument is leveled with leveling screws and sensitive spirit levels. *(Courtesy David White, Inc.)*

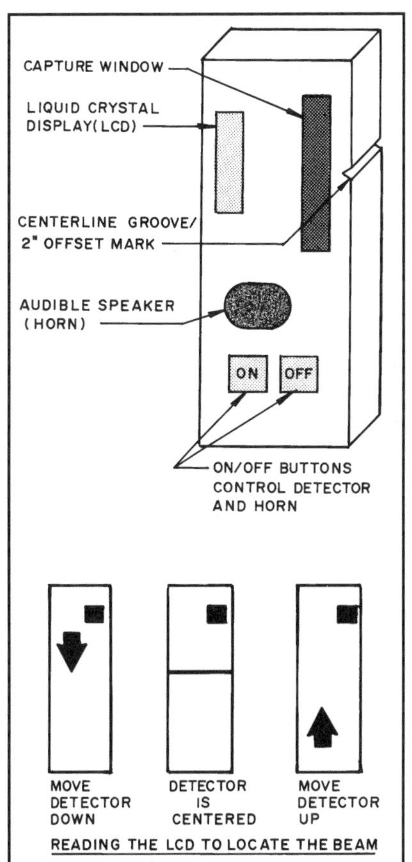

5-43 A typical electronic laser detector that includes an audible horn. The capture window receives the beam from the laser within a 90-degree arc. The centerline groove marks the center beam when the signal centerline is displayed on the LCD and the horn signals a continuous tone. *(Courtesy David White, Inc.)*

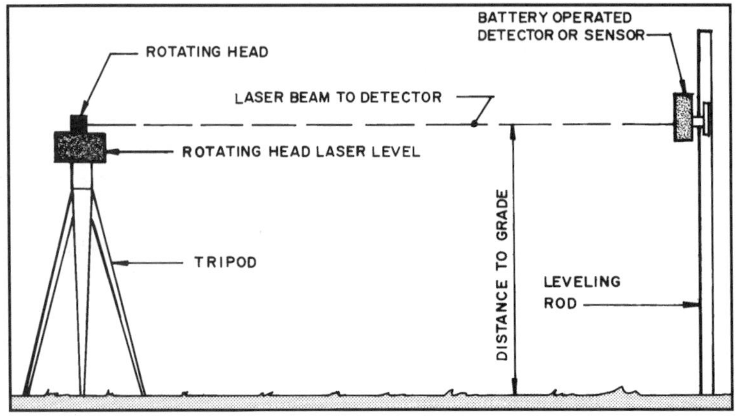

5-44 The laser level sends a laser beam to the detector.

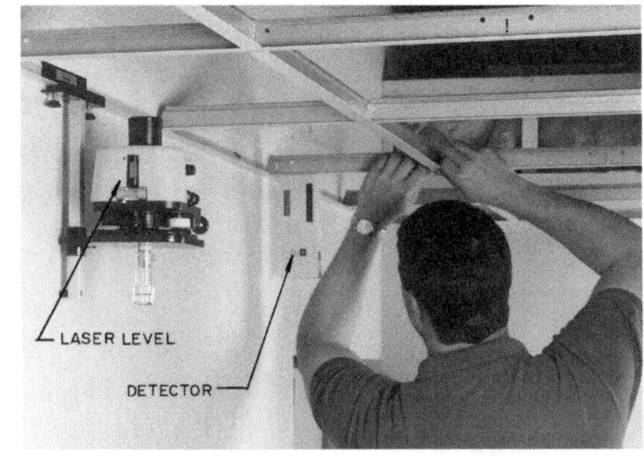

5-45 The laser level is hung on a wall mounting bracket when used to locate level lines near the ceiling. *(Courtesy David White, Inc.)*

A typical detector is shown in **5-44**. The rotating head sends a laser beam 360 degrees around the device. It does most of the same operations performed with the optical level transit. However only one person is needed to do the layout. These levels can have an accuracy of ⅛ inch (3mm) within 100 feet (30.5m). The range will vary with the laser model but is typically 200 to 800 feet (61 to 244m).

Laser levels may be manually leveled or be self-leveling. **Manually leveled** laser levels are mounted on a tripod and adjusted to the desired height. They are leveled with leveling screws and level bubbles similar to those on optical levels. **Electronically controlled self-leveling** laser levels have electro-levels and motors. Electrodes are positioned above the liquid in the electro-level bubble. When the bubble is not level, the liquid will touch the electrode, completing an electric circuit—this starts a small motor that rotates the assembly until the electrode is clear of the liquid, thus again stopping the motor. In this condition the laser level is horizontal.

5-46 The leveling rod with a detector is held by one worker who can check grades and levelness without the assistance of a second person. *(Courtesy David White, Inc.)*

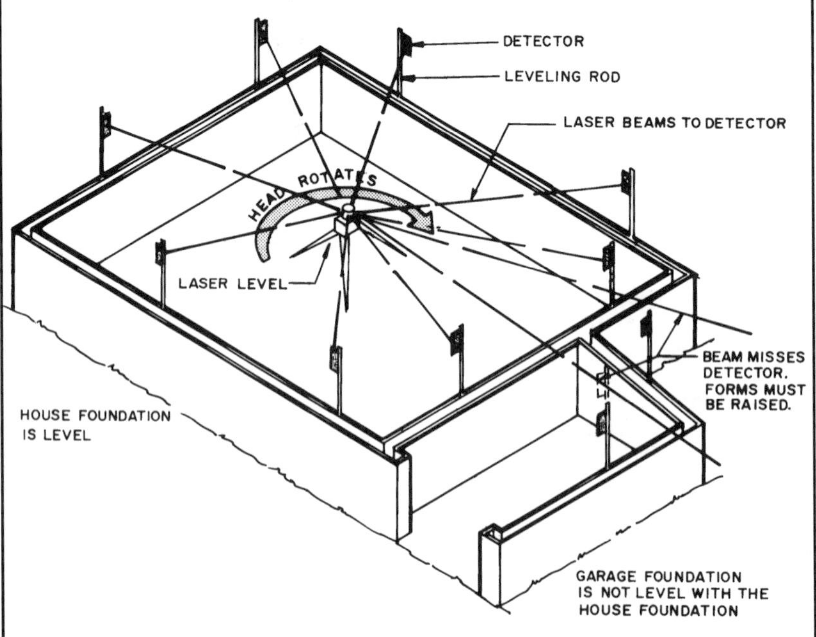

5-47 When the laser beam strikes all the detectors set on top of the foundation forms, the top of the forms is level. If the beam does not strike the detector capture window, the form in that location needs to be raised or lowered as required.

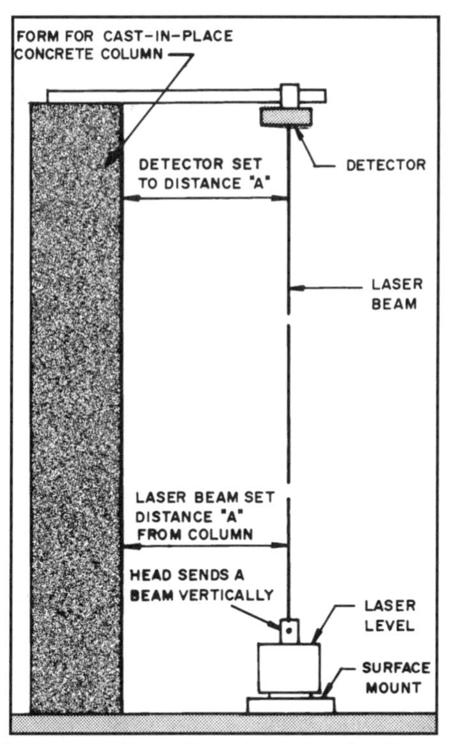

5-48 The laser level transit can be set to plumb vertical elements such as a wall or column. If the beam does not strike the detector, the column form must be moved until the target has been hit and firmly braced in that position.

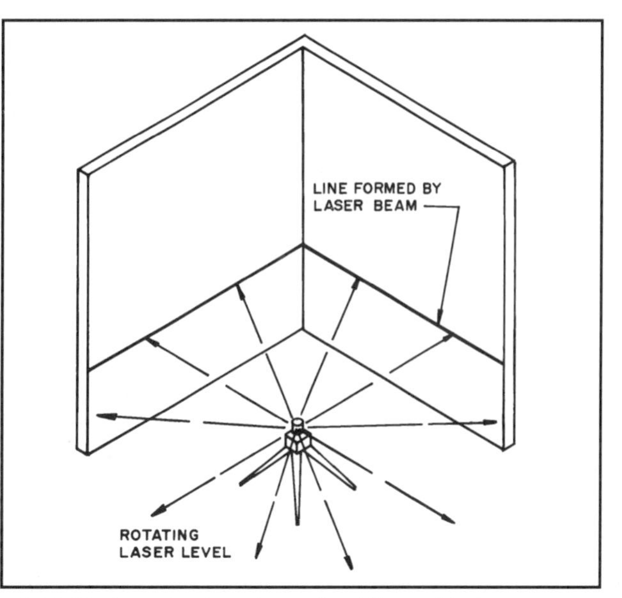

5-49 The laser level can locate level lines with the rotating head.

Setting Up a Laser Level

For work such as leveling a foundation or deck the laser level is mounted on a tripod. The legs of the tripod are firmly set into the ground. The unit is manually leveled to within 8 degrees and the automatic leveling device takes over and finalizes the leveling process. Other operations, such as locating dropped ceilings, require the laser level be hung on a **wall mounting bracket (see 5-45)**.

Typical Applications

The laser level can be used to establish grades (**see 5-46)** and level long features such as the forms for a foundation. In **5-47** the laser level has been set up in the center of the foundation. The rotating head sends out a beam to the targets set in various locations. The laser level is also used for leveling decks and patios.

Walls and columns can be accurately plumbed with the laser level. This instrument is set to project a plumb vertical line to a detector mounted on the wall or column **(see 5-48)**.

Interior work includes locating the wall line of a suspended ceiling and locating wood framed partitions or tracks for metal stud walls. The beam is visible and locations can be marked on the wall and floor **(see 5-49)**.

Laser Cleaning & Maintenance

Keep the unit clean and dry. It can be wiped off with a soft cloth dampened with solvent-free water. It should be stored in its carrying case when not in use. Keep the case clean and dry.

Safety

When using a laser level never expose your eyes to the beam, and if extended exposure occurs consult a physician immediately **(see 5-50)**.

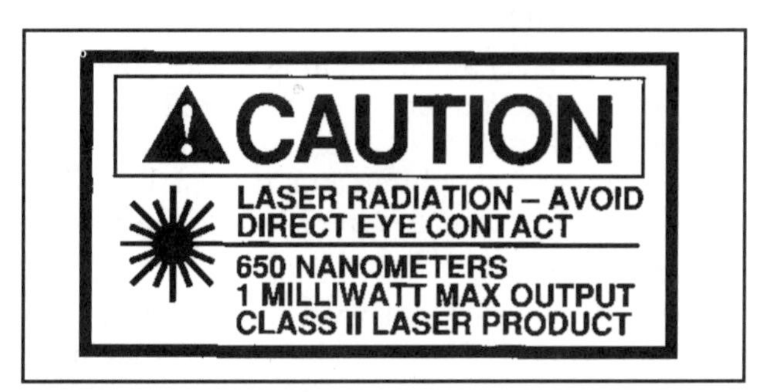

5-50 Laser radiation is harmful to your eyes.

CHAPTER 6

Locating the Building on the Site

The site plan shows the corners of the lot, required setbacks, and the location of the proposed building. In some areas the contractor is required to employ a licensed surveyor to locate the corners of the foundation. The general contractor's project supervisor may also be prepared to do this task. If the corners of the lot have not been staked, a surveyor will be employed to locate them. The deed or local records will give the needed information. The surveyor works from a known reference point to locate the corners of the lot. Each corner is then marked with a two-by-two-inch wood stake. These survey points are represented on the site plan by the corners of the lot. Once the corners of the lot are staked the corners of the foundation can be located **(see 6-1)**.

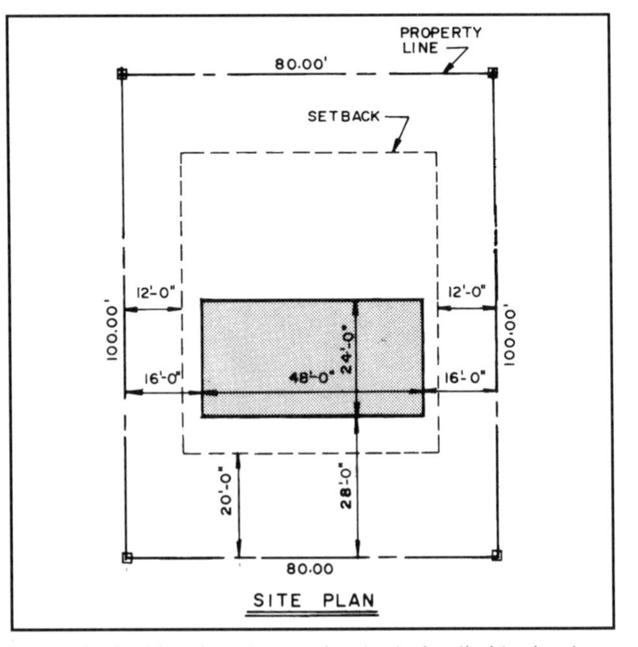

6-1 The building location on the site is detailed in the site plan.

LOCATING THE FOUNDATION

The following is a typical procedure for locating the corners of the foundation with a level transit **(see 6-2)**.

1. Set the level transit on corner 1 and sight corner 2. Measure the distance the right foundation corner is from corner 1 (16 feet) and stake it (stake 3). Then measure the length of the foundation (48 feet) and stake it at point 4.
2. Move the level transit to stake 3, sight stake 2, and swing 90 degrees toward point 5. Measure the distance from 3 to 5 (28 feet) to locate this corner of the foundation.
3. Raise the level transit sighting toward the rear corner of the foundation and sight beyond corner 5. Measure the distance from 5 to 6 (24 feet) along this sighting and drive stake 6. The distance from 5 to 6 is the width of the right side of the house (24 feet).
4. Move the level transit stake 4, sight on stake 3, then swing 90 degrees to locate the other side of the foundation. Measure the distance from 4 to 7 (28 feet). This locates the front left corner of the foundation.

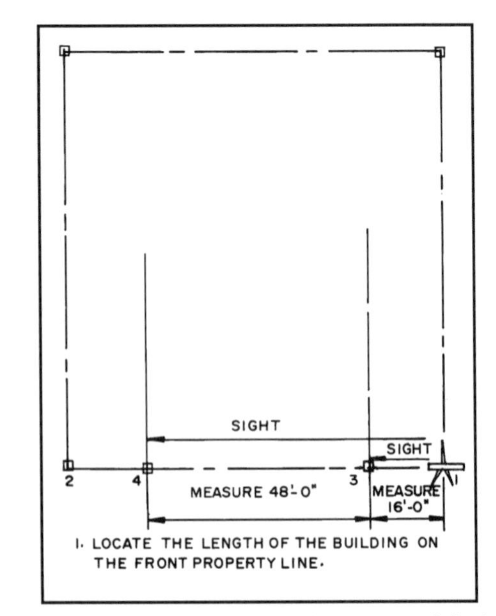

6-2 The steps to follow to locate the corners of the building. (Above and opposite page.)

5. Raise the level transit sighting along the left side of the foundation. Measure the width of the left side along this sighting and stake the rear corner 8 (24 feet). Connect stakes 5, 6, 7, and 8 to form the outline of the foundation.

The exact point locating each corner is located by a nail driven into the stake.

It should be noted that while this example shows a rectangular lot and house not all lots and houses are in this configuration. The angle of swing will vary if the lot is another shape, such as a triangle, or if the house is not composed of right-angled walls but rather of some angular shape.

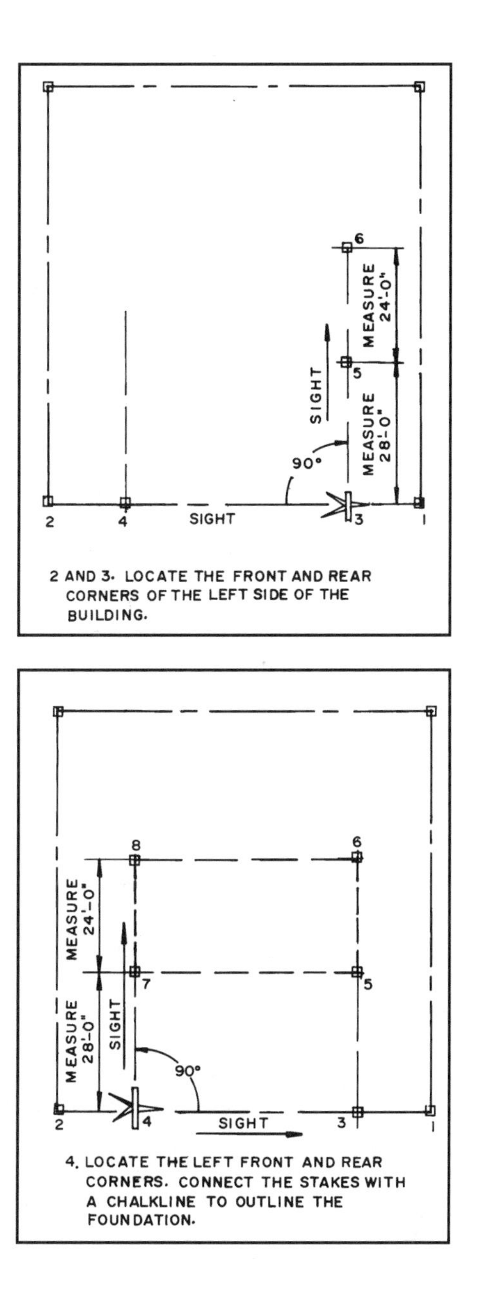

2 AND 3. LOCATE THE FRONT AND REAR CORNERS OF THE LEFT SIDE OF THE BUILDING.

4. LOCATE THE LEFT FRONT AND REAR CORNERS. CONNECT THE STAKES WITH A CHALKLINE TO OUTLINE THE FOUNDATION.

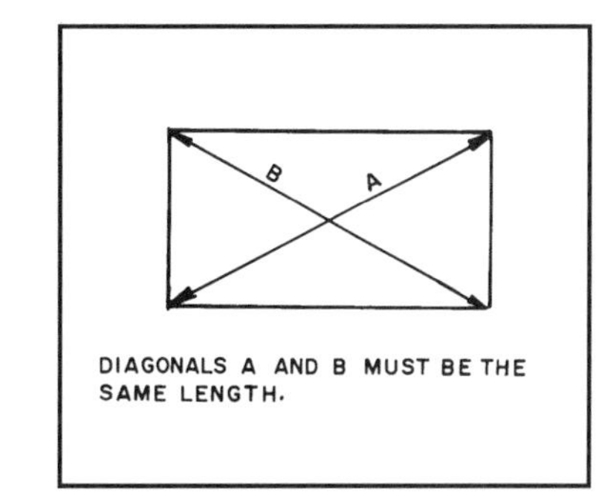

DIAGONALS A AND B MUST BE THE SAME LENGTH.

6-3 If the diagonals measure the same length the layout is square.

Checking the Layout

The stake locations are usually quite accurate when they have been laid out using the level transit or laser. The squareness can be checked by measuring the diagonals as shown in **6-3**. If the diagonals measure the same length the layout is square.

Individual corners can be checked to see that they are 90 degrees by using the 6-8-10 method **(see 6-4)**. Measure six feet on one chalkline and eight feet on the joining line. The distance between the ends (the hypotenuse of the triangle formed) must be exactly 10 feet The 90-degree angle can also be checked with laser instruments as shown in Chapter 5.

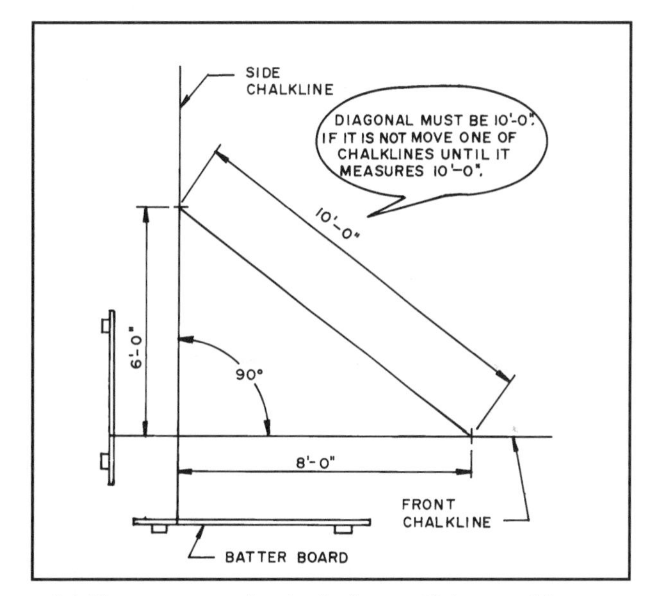

6-4 The corners can be checked to see if they are 90 degrees by using the 6-8-10 method.

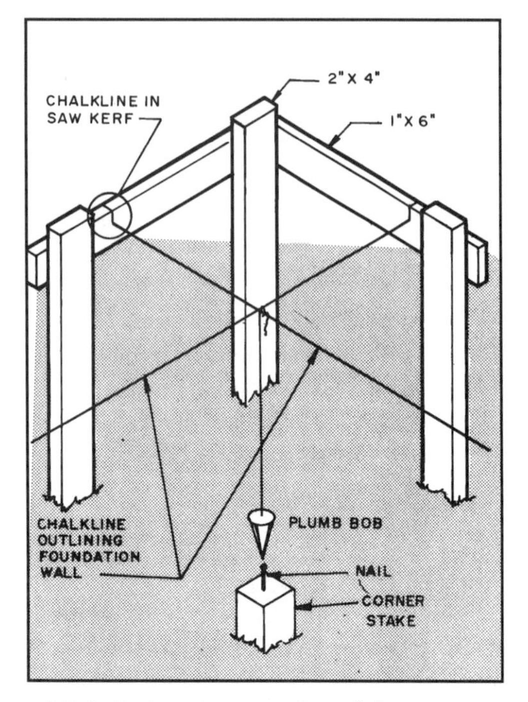

6-5 Batterboards are set clear of the area to be excavated and support a chalkline which locates the face of the foundation.

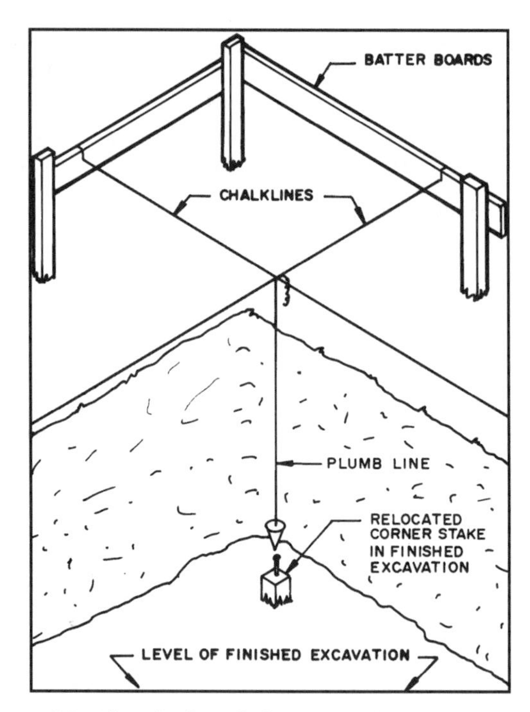

6-7 After the foundation excavation is complete the corner stake can be located by dropping a plumb line from the intersection of the chalklines.

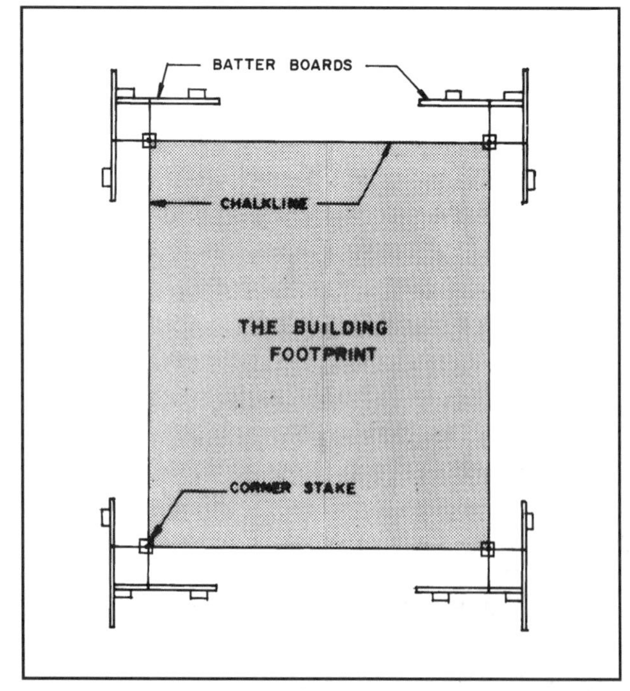

6-6 Batterboards are set after the corners of the building are located. The chalkline outlines the footprint of the building.

SETTING THE BATTERBOARDS

Once the corners of the foundation are staked the lines of the foundation are located using chalklines. These lines are used to locate the foundation walls and footings. When the excavations are dug for the footings the corner stakes are lost. Therefore batterboards are set up clear of any excavation and the chalklines are run from them. Notice in **6-5** how the batterboards are set before excavation begins. They are set at all corners of the building as shown in **6-6**. The intersection of the chalklines should occur directly above the nail in the corner stake. This intersection is used to relocate the corner of the foundation after the footings have been dug.

The tops of the batterboards are set to be the top of the finished foundation. They are used by the masons laying a concrete block foundation or those setting the forms for a cast-in-place concrete foundation.

EXCAVATING FOR THE FOOTINGS

The footing excavations are dug deep enough to place them below the frost line. This is established by local codes. If the site has been filled the excavation must continue until the footing is on undisturbed soil. Deep excavations are made wider than the footing to allow room to work for those who will install the footings and foundation walls. The footing excavations are typically dug with a backhoe. If an excavation is for a basement a bulldozer is often used. The depth of the excavation is carefully checked as digging proceeds. The final leveling is often done with hand shovels. After the excavation is completed the corner stake is relocated by dropping a plumb line as shown in **6-7**.

Footings near the surface of the soil, as used in warm climates, are usually dug with a hand shovel. Forms for the footings are not needed if the soil will stand firm after the footing has been dug. The concrete is poured directly into the excavation. Deep footings in wide excavations require the footing be formed **(see 6-8)**. The bottom of the footing excavation must be smooth, level, and in undisturbed soil. If dug too deep the soil cannot be replaced and extra concrete is required.

The depth of the footing is checked frequently during digging by measuring from the chalklines.

Detailed information on footings and foundations is in Chapter 7.

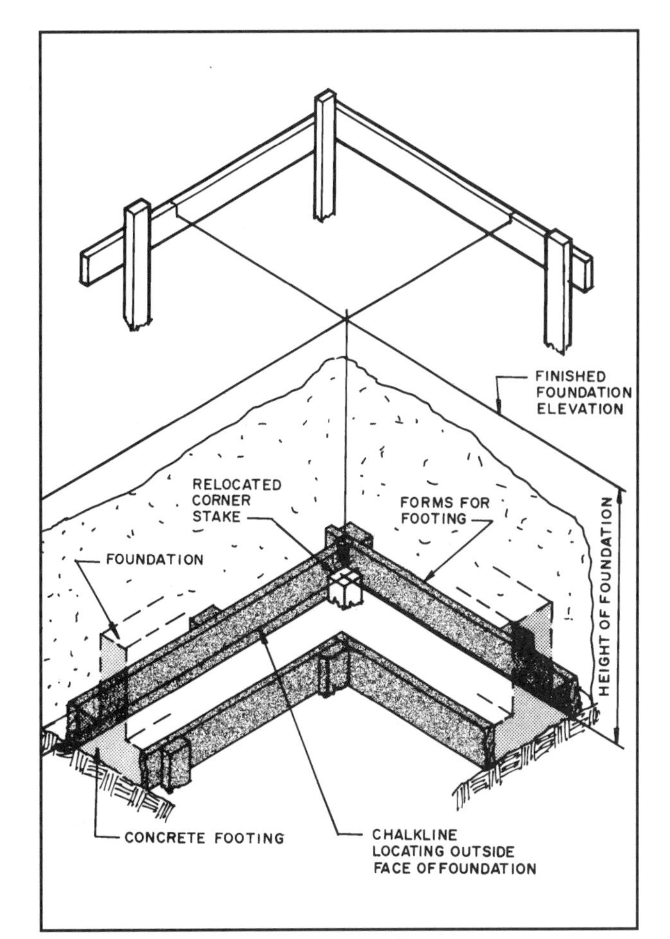

6-8 Some footings must be cast in forms.

CHAPTER 7

FOOTINGS & FOUNDATIONS

Building **concrete formwork** is one of the jobs a carpenter is expected to do. Forms are the units built to hold the concrete in place as it hardens. Some carpenters specialize in form construction. When small buildings are constructed, the carpenters generally do the formwork and all other carpentry on the building. When large multistory buildings are built, the carpenters are more likely to specialize in one area, such as formwork. Carpenters are needed anywhere concrete is to be formed and poured. This includes large projects such as bridges, dams, and sewage disposal projects **(see 7-1)**.

In addition to building forms that can be reused, carpenters build single-use forms.

A draft tube form for a hydroelectric powerhouse is shown in **7-2**. It is used only once but represents the ultimate in carpentry craftsmanship in heavy construction.

Properly constructed formwork is essential to the satisfactory construction of the building. The footings and foundation must be of the proper size and in the proper location. They must be level and of the proper height because they form the base upon which the building is built. The forms must be strong enough to withstand the pressures from the wet concrete and not bow or buckle.

A carpenter must carefully read the blueprints so that the forms are the proper size. The footings must be below the frost line of the ground. The **frost line** is the depth to which ground freezes in the winter. This varies from as little as one foot in warm climates to four to five feet in cold climates. The architect will indicate the depth of the footing on the drawings.

FOOTINGS

The **footing** is a concrete pad placed on the soil. The **foundation wall** is built on the footing. The footing holds the weight of the entire building and its contents and resists its sinking into the soil. It must be placed in the soil below the frost line and rest on undisturbed soil. The footing is cast-in-place concrete reinforced with steel reinforcing bars. Shallow footings in soil that remains standing as the trench is dug (cohesive soil) can be cast in the trench without forms **(see 7-3)**. Footings in soil that will not stand (noncohesive soil) or that are cast in an open area such as in an excavation for a basement require footing forms.

The size of the footings depends upon the load and the type of soil. The size and design are determined by the architect and shown on the

7-1 These massive concrete forms are framed using plywood panels and solid lumber bracing. *(Courtesy APA—The Engineered Wood Association)*

7-2 Single-use forms are often complex and require the services of very skilled carpenters. *(Courtesy Corp of Engineers, U.S. Department of the Army)*

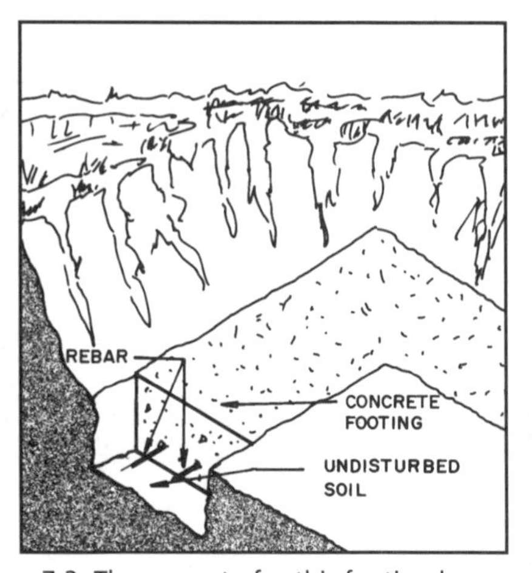

7-3 The concrete for this footing has been placed in an earth trench and did not require side forms.

architectural drawings. The load-bearing properties of the soil are determined by standard soil tests. Clay and sandy clay will support only 2000 pounds. per square foot while sandy gravel and gravel will support 5000 pounds per square foot. Local codes will often specify bearing capacities of soils. However, the only way to be certain is to have a soil engineer test actual soil samples.

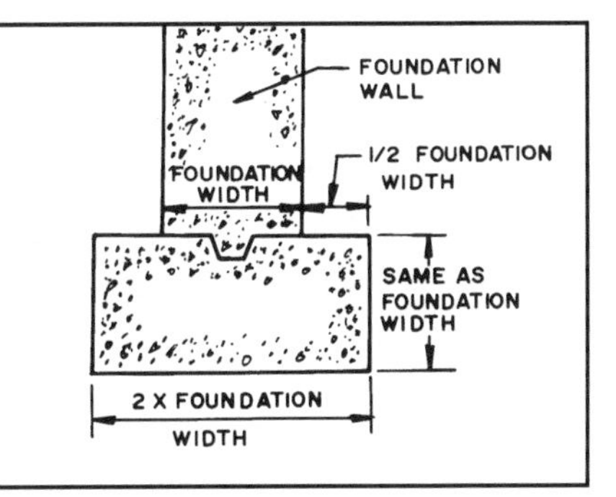

7-4 Typical footing proportions for light residential construction. Soil conditions and design loads can require the footing be larger.

A rectangular footing is possibly the most common type used **(see 7-4)**. While specific sizes will vary, the proportions shown are typical for normal residential construction and good soil. Following are steps typically followed to install footing forms from wood or prefabricated metal.

1. Drop a line with a plumb bob from the **batterboard** lines that locate the building. This locates the corner of the building **(see 7-5)**.
2. Drive a wood 2 × 2 x 16-inch stake below the plumb bob. Drive a nail in the stake at the point where the plumb bob touches it. This is the corner of the building.
3. Drive the stake into the ground until its top is the exact level of the top of the footing to be poured. Check this height with the point of beginning on the batterboards. The leveling instrument is used to make this check. This is an important step because the elevation of the entire building depends upon it.
4. Repeat these steps at the other corners of the building. Each corner will then have a stake locating it, and the top of the stake will be the exact height of the top of the footing.
5. Connect the nails in the corner stakes with a line. This forms the outer line of the building **(see 7-6)**.

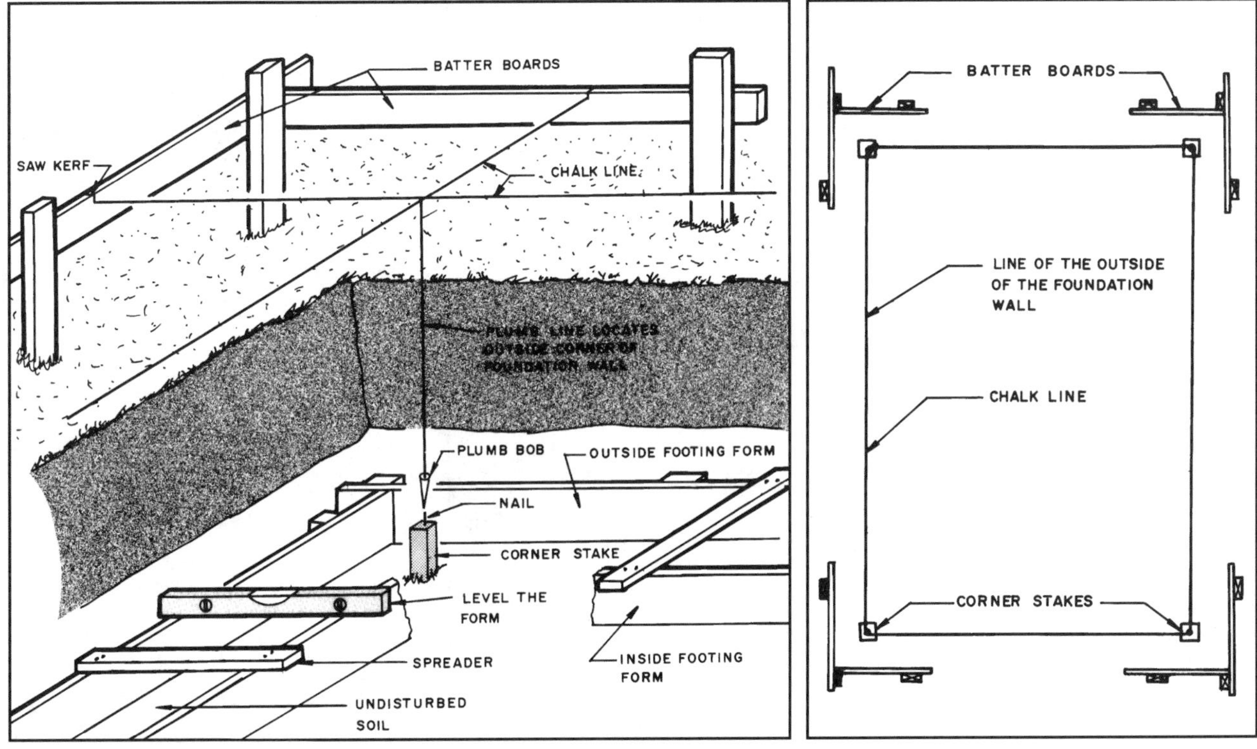

7-5 Batterboards, chalkline, and plumb bob are used to locate the corner stake in the excavation from which measurements are made to set the footing forms.

7-6 Run a chalkline between the corner stakes to locate the outside face of the foundation wall.

6. Now place the footing form boards. Their location is measured from the building line. The distance of the outside form board from the building line is the width of the footing projection. The inside form board is located by measuring the width of the footing from the outside form board. The inside form board is located using **spreaders**. A spreader is a wood member whose length equals the width of the footing **(see 7-7)**. The spreaders are removed as the concrete is poured.

   The tops of the form boards must be on the same level as the top of the corner stake. These can be lined up with a level. The form boards are supported by stakes driven into the ground every three or four feet. They are braced.

7. After the form boards have been set, the required reinforcing bars are placed in the form and raised above the soil the distance specified on the architectural drawings. These may be supported with wire, precast concrete, plastic, or fiber-reinforced cementitious bar supports **(see 7-8)**.

8. Before pouring the concrete, recheck the level of the form boards.

9. Now the footings can be poured. Care should be taken when placing the concrete so that the forms are not moved or damaged.

10. After the footings are poured and the concrete has begun to set, a **keyway** can be formed. A keyway is used to tie the foundation wall to the footing **(see 7-9)**. The keyway is usually made by pressing a 2 × 4-inch wood member into the concrete before it sets. The member is slanted on the sides to make removal easier.

    Another way to tie the foundation wall to the footing is to cast vertical rebars in the footing. These extend up into the foundation wall form and are tied to the reinforcing bars in the foundation **(see 7-10)**.

11. After the concrete is hard, the form boards are removed.

Metal footing forms are installed in much the same way.

If the width of the footing exceeds certain design limits it must be made thicker. A wide but thin footing will tend to crack. The thickness can be added by forming a step in the footing **(see 7-11)**. Typical wood forms for forming this spread footing are shown in **7-12**.

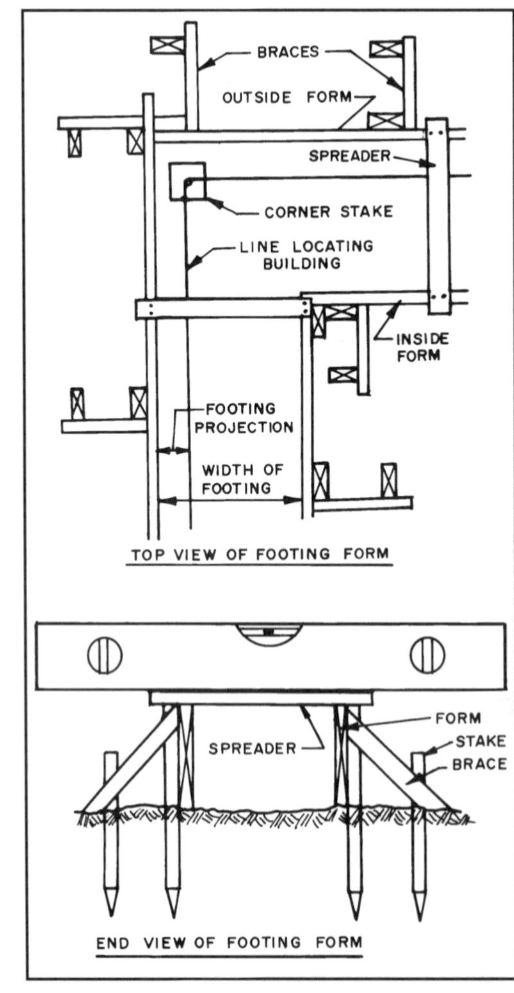

7-7 The footing form boards are located from the corner stake, leveled, and staked.

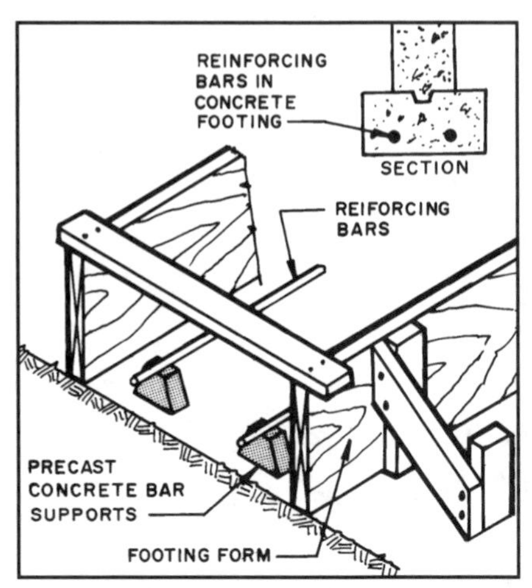

7-8 Reinforcing bars in footings are elevated with bar supports the distance specified on the architectural drawings.

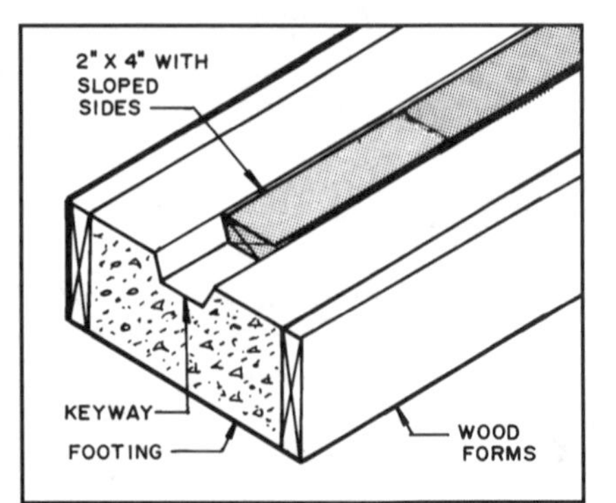

7-9 A keyway is formed in the footing to provide a tie to the foundation wall.

7-10 These rebars were cast in the footing and will be tied to the rebar in the foundation wall.

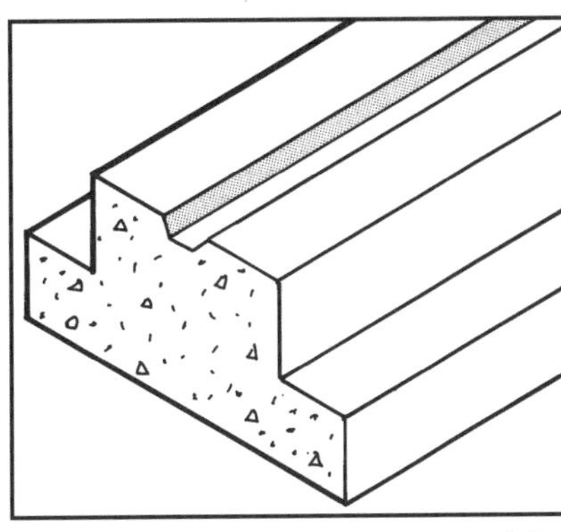

7-11 The footing may be stepped if the projection beyond the foundation wall exceeds the thickness of the footing.

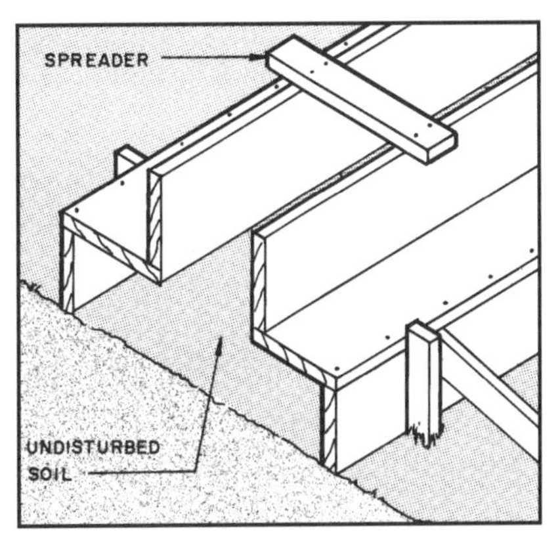

7-12 One way to build wood forms for a stepped spread footing.

If a building is built into the side of a hill, the footing might be stepped **(see 7-13)**. This reduces the amount of excavation needed and saves money because the foundation wall requires less forming and concrete. Stepped footings have horizontal and vertical steps. The vertical step is generally not higher than three-fourths of the horizontal distance between steps. The horizontal distance between steps should be not less than two feet zero inches **(see 7-14)**. The steps should overlap a distance equal to the thickness of the footing. A typical wood carpenter-built form is in **7-15**.

Another way to set footings on a sloped site is in **7-16**. The footing must be below the frost line and spaced so the maximum slope between adjacent footings is a 1 × 2 rise.

Piers and columns rest on footings. Generally each pier or column has a separate footing though there are times when a large footing might support several.

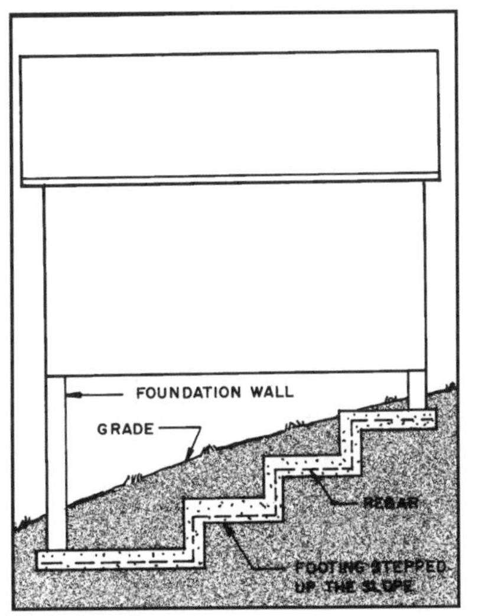

7-13 Buildings set into the side of a steep slope can step the footing up the slope.

7-14 Typical design details for a footing stepped up a slope.

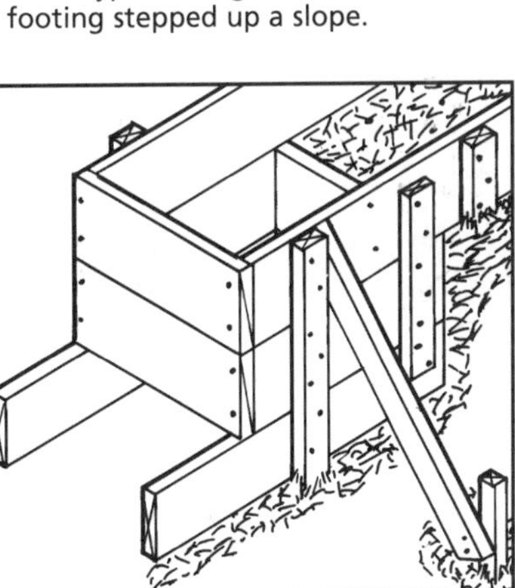

7-15 Typical ways to build wood forms for a stepped footing.

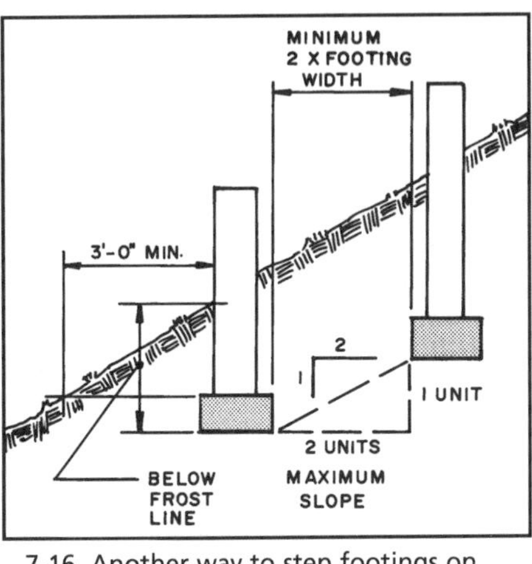

7-16 Another way to step footings on steeply sloped sites.

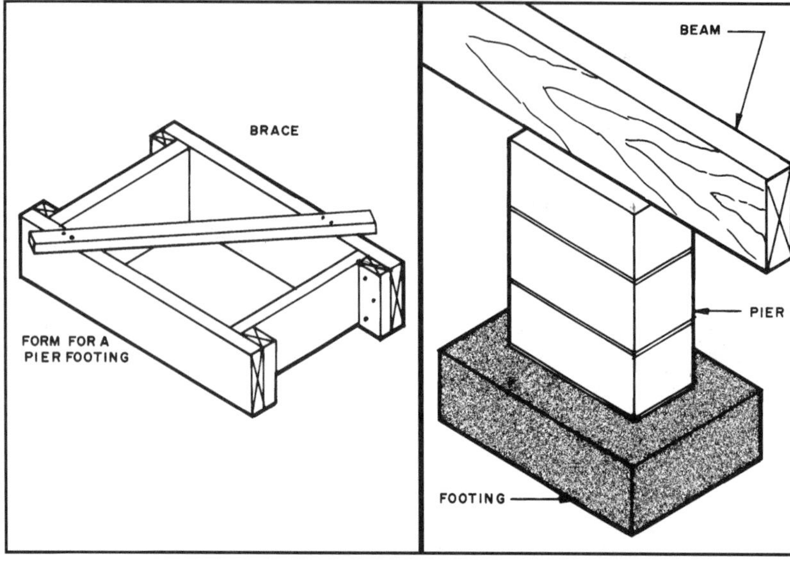

7-17 A typical pier and its footing.

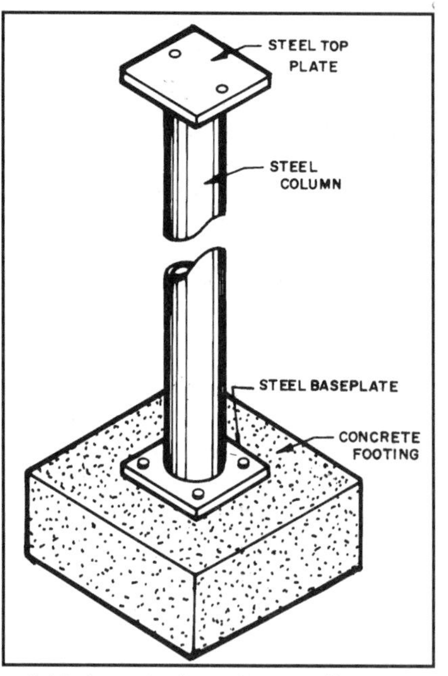

7-18 A steel column is generally secured to the footing with anchor bolts.

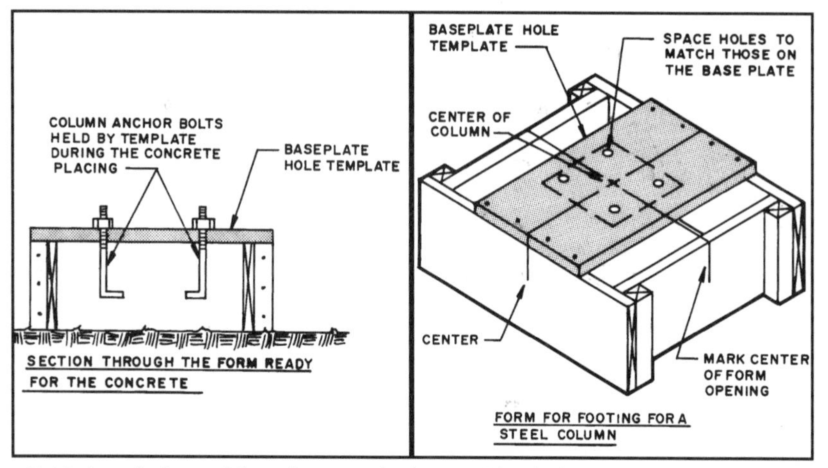

7-19 A typical wood form for a steel column with a hole template to hold the anchor bolts in the desired location as the concrete is placed in the form.

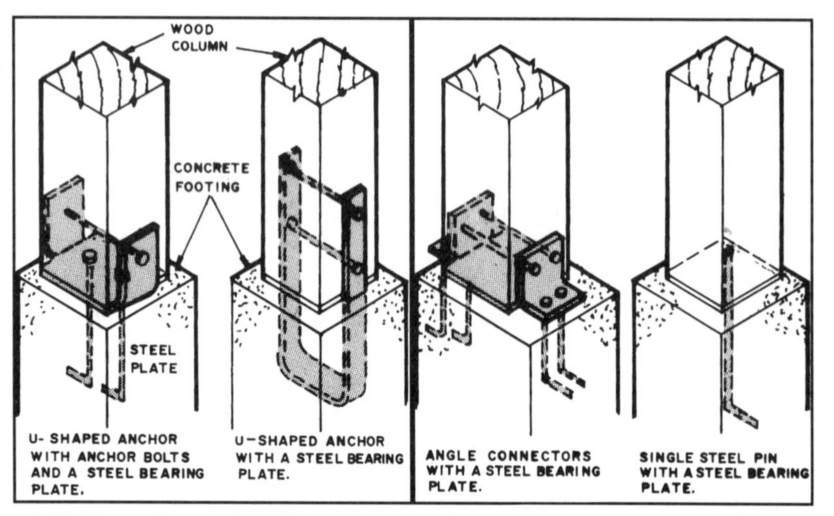

7-20 Several types of metal connectors used to secure wood columns to the concrete footing or foundation wall.

A **pier** is a masonry unit that is supported by the footing and typically carries a beam supporting a series of floor joists **(see 7-17)**. A **column** is either steel or wood and is supported by a footing as described for piers. Since the column must be secured to the footing in some way, provision is made for the form to hold the metal connections in place as the concrete is poured. For example, a steel column **(see 7-18)** has a metal baseplate that is secured to the footing with bolts. The bolts are held in the form with a hole template as shown in **7-19**. Wood columns are mounted with some type of metal anchor. Examples of some are in **7-20**. Notice the use of a steel baseplate below the wood column. This prevents moisture transfer from the concrete footing to the wood post.

Sometimes, a pier or column footing must be stepped. This is needed to get sufficient strength to carry the design load. A typical form for a stepped pier or column footing is shown in **7-21**.

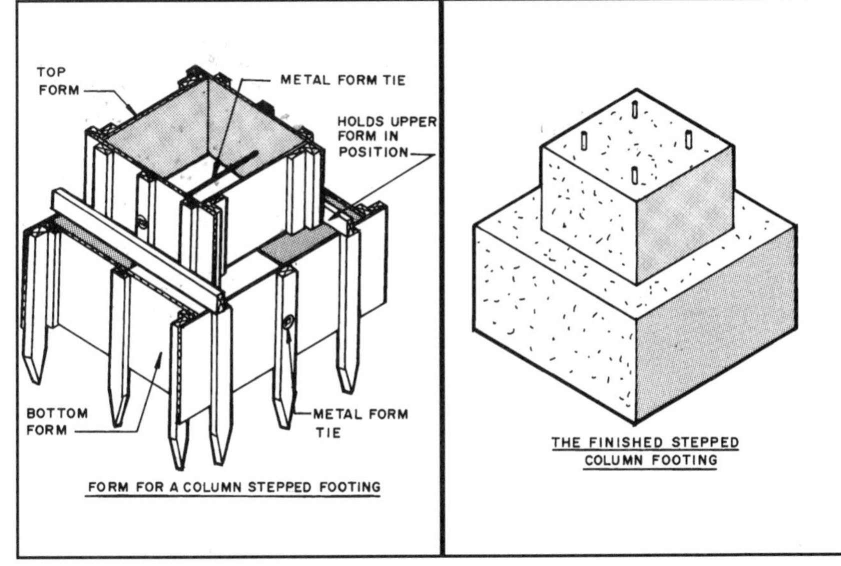

7-21 A typical form for a stepped column footing.

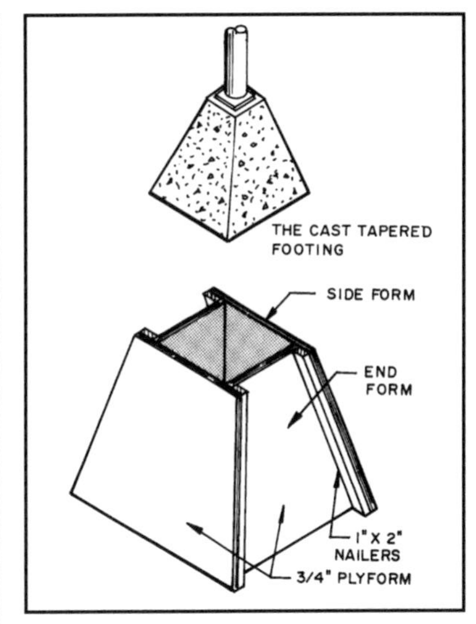

7-22 A form for a tapered footing. Anchor bolts or other steel column anchors are suspended in the form to secure the column to the footing.

Tapered column footings are formed as shown in **7-22**. The end forms are made the exact size of the footing. The side forms are larger so that the cleats can be attached.

Forms for piers and columns are located as shown on the architectural drawings. The area in which the footing will be located must be level and free of debris. The layout is shown in **7-23**. Each footing form is located by finding the exact center of the column or pier and running chalklines across the foundation. The form for the footing is placed with this point as its center and it is carefully staked in position. Care must be taken when pouring and screeding the concrete not to move the form. Anchor bolts are placed in the concrete as described earlier.

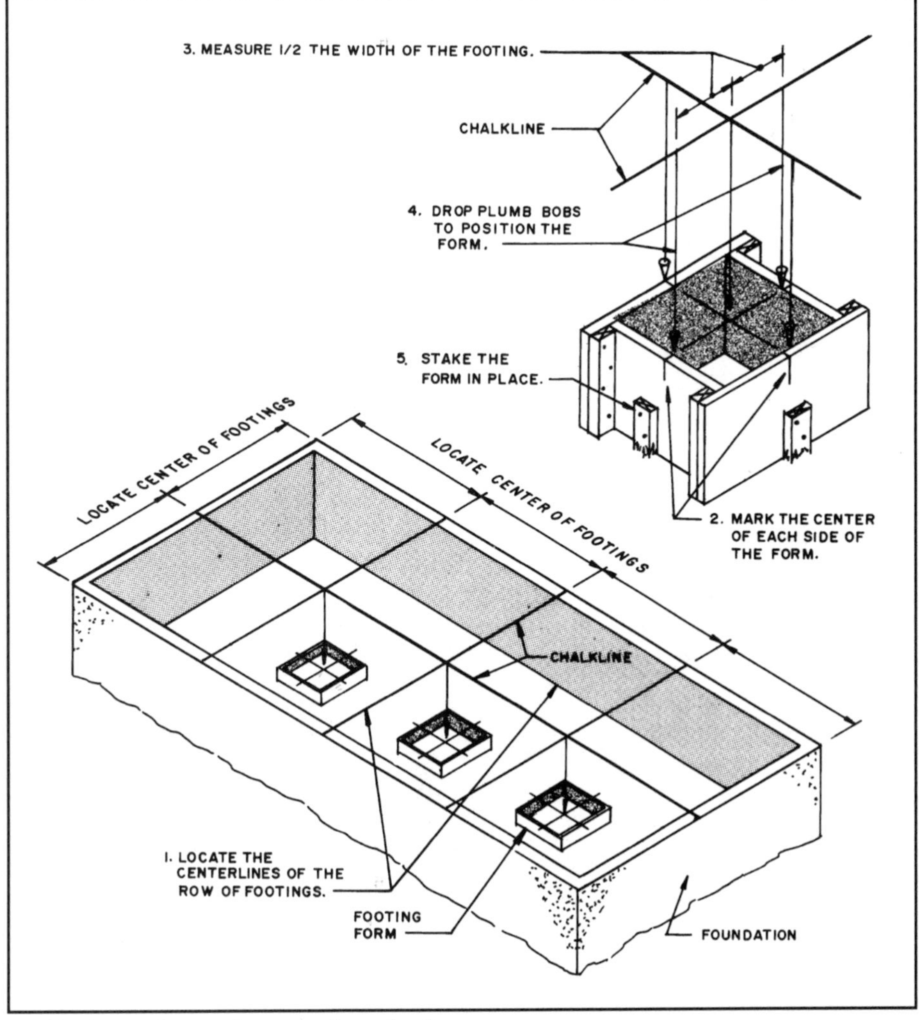

7-23 How to locate pier and column footings.

PIERS & GRADE BEAMS

Some structures are built using piers and a grade beam. The **pier** is a round concrete member cast in a hole bored in the earth. If the soil is noncohesive a round form is worked into the hole. A **grade beam** is a rectangular concrete structural member that is placed horizontally on top of a series of piers. This construction serves as the foundation for the building. The floor is framed on the grade beam in the same manner as a foundation wall **(see 7-24)**.

The soil below the grade beam is not allowed to touch it because freezing and thawing of the soil could cause the grade beam to heave. The area below the grade beam is filled with coarse gravel to allow subsurface water to drain away from the foundation. The bottom of the grade beam must be below the frost line.

Grade beams can be formed and poured on top of the piers or can be precast concrete units. This type of foundation is used when the soil near the surface of the earth is not stable and it is necessary to go through the unstable layer to reach soil that will support the building. On large multistory buildings piers will be drilled down until they reach a bearing rock layer.

MATERIALS IN FORMWORK

The materials the carpenter generally uses to build formwork are plywood, wood structural members, and metal ties.

A special plywood called **plyform** is made especially for the form facing material.

As forms are designed, the plywood panels must be of the proper thickness to support the pressure. In addition, the structural members, joists, studs, and walls must be of the proper size. For most residential work ⅝- or ¾-inch plyform is used. When a form is properly designed, the architect considers the rate at which concrete is poured, the job temperature, concrete slump, cement type, concrete density, method of vibration, and height of the form. The faster concrete is poured, the greater the pressure on the form. A pour of 1 to 2 feet in depth per hour is slow. A pour of 3 to 7 feet per hour is fast. Design tables for computing the structural design of concrete forms are available from APA—The Engineered Wood Association, P.O. Box 11700, Tacoma, WA 98411-1700.

Prefabricated metal framed panels are available. They use a structural metal frame covered with a plywood sheathing **(see 7-25)**. These forms can be reused after they have been stripped from

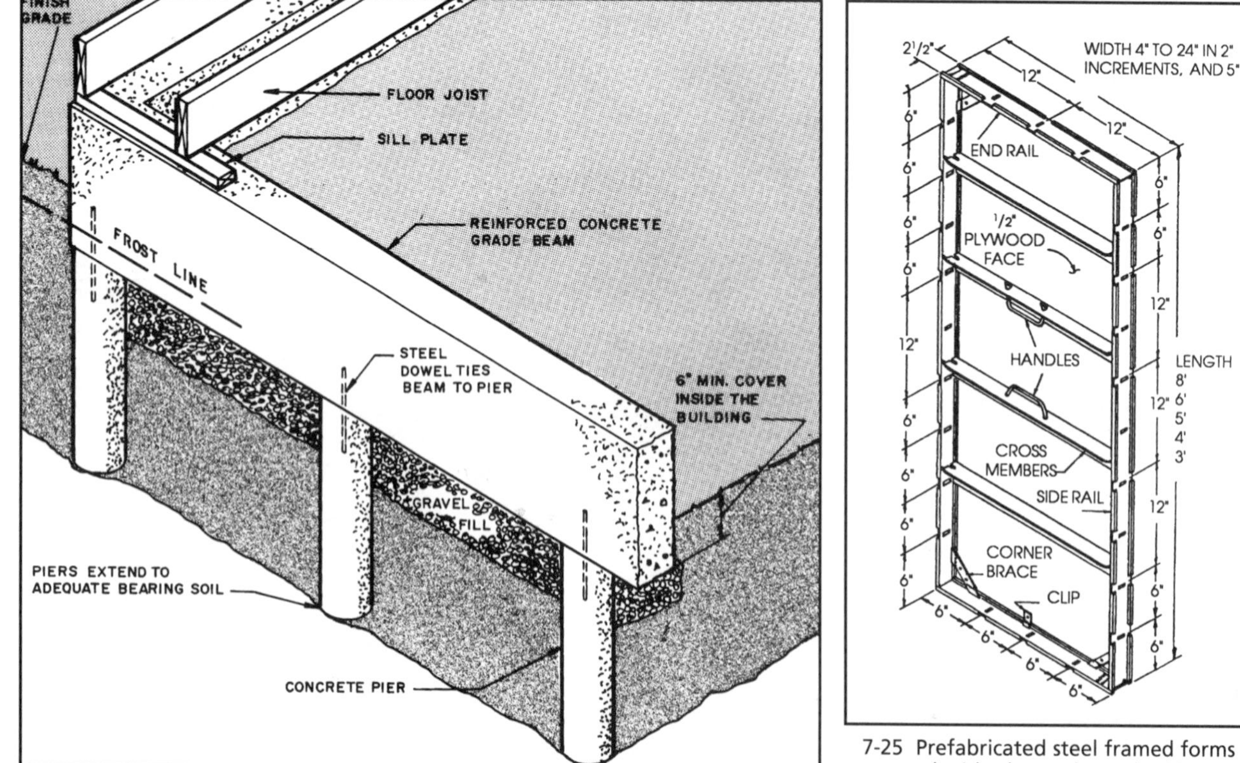

7-24 Grade beams serve as the foundation wall and are supported by piers.

7-25 Prefabricated steel framed forms covered with plywood panels. *(Courtesy Symons Corporation)*

the wall and cleaned. There are various metal ties used to hold the forms together. They use some form of wedge, screw, or cam to tighten them **(see 7-26)**.

Aluminum panels are available that produce a textured surface such as the brick pattern shown in **7-27**. The manufacturer provides a system for joining and bracing the forms, steel reinforcing bars are set in place, and then the forms are set on each side as shown in **7-28**.

Cellular plastic forms are also available for forming cast-in-place concrete walls. The forms remain in place after the concrete has cured and provide insulation for the wall **(see 7-29)**.

FORMING WALLS

The forms used for foundation walls are typically made up of preassembled wood or metal panels **(see 7-30)**. There are occasions when the design requires special forms be built by carpenters.

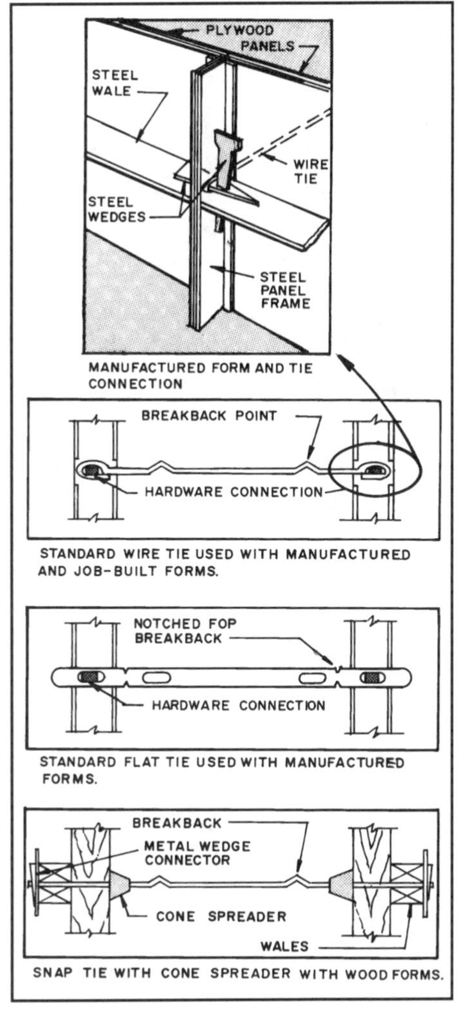

7-26 Typical metal ties used to space concrete forms.

7-27 A prefabricated aluminum form with a brick pattern stamped into the surface. *(Courtesy Precise Forms, Inc., Kansas City, MO)*

7-28 A footing with reinforcing bars set in it and extending up between the wall forms. Notice horizontal rebars are tied to the vertical rebars. *(Courtesy Precise Forms, Inc., Kansas City, MO)*

7-29 This insulated concrete forming system can be used above or below grade. The plastic form ties are used to space and hold the plastic form blocks in place. The plastic foam blocks are left on the wall after it cures and provide insulation. *(Courtesy AFM® Corporation)*

7-30 Metal forms are widely used to form concrete walls. *(Courtesy Precise Forms, Inc., Kansas City, MO)*

7-31 Metal forms with patterned surfaces are used to produce textured concrete surfaces. *(Courtesy Precise Forms, Inc., Kansas City, MO)*

7-32 Two of the possible patterns cast in concrete walls. One is a brick design and the other appears as board and batten siding. *(Courtesy Precise Forms, Inc., Kansas City, MO)*

Forms must be strong enough to support loads without bending or breaking. The forms also should not leak.

The finished surface of the concrete depends upon the surface of the inside of the form. Textures are produced using special plywood or metal forms with textured surfaces **(see 7-31)**.

Textured form surfaces tend to require an increase in the force that is needed to strip the forms. This limits the number of times the form can be used. Sometimes film coatings, such as lacquer, polyurethane, or epoxy, are used with a release agent to make stripping of plywood forms easier. Some finished concrete walls can be seen in **7-32**.

Another way to produce a textured surface is to attach a form liner to the inside of the wood form. One type is a plastic sheet containing a textured surface that is nailed to the surface of the wood form.

Most complex, specially built forms use plywood as the surface material. Since plywood has good insulating qualities, it helps level out temperatures during curing. This provides more consistent curing temperatures. It can be easily cut and formed to build special forms. No special equipment is needed.

After plywood forms are removed, they should immediately be cleaned and repaired. Holes should be patched. A fiber brush is used to clean the surface. The surface could be lightly coated with mill oil. A frequently used mill oil is a 100- or higher viscosity pale colored oil. A parting agent may be used instead of oil. Some parting agents, such as wax or silicone, affect the surface, so that painting the concrete is a problem. Metal forms also must be cleaned after they have been stripped. If they are oiled, then a special oil for metal forms must be used.

LOW WALL FORMS

Low wall forms are those under four feet in height. They are commonly made from ¾-inch plywood and 2 × 4-inch wood framing material. The studs are generally placed about two feet apart. It is important to brace the form at the bottom since the pressures from the liquid concrete are greatest there. Inclined braces are needed to support the top of the form.

A wood form tie is nailed to the studs at the top of the form. This holds the form at the proper width and prevents it from opening as the concrete is poured.

A typical wood form is shown in **7-33**. The footing is cast first and the wall is formed after it has hardened.

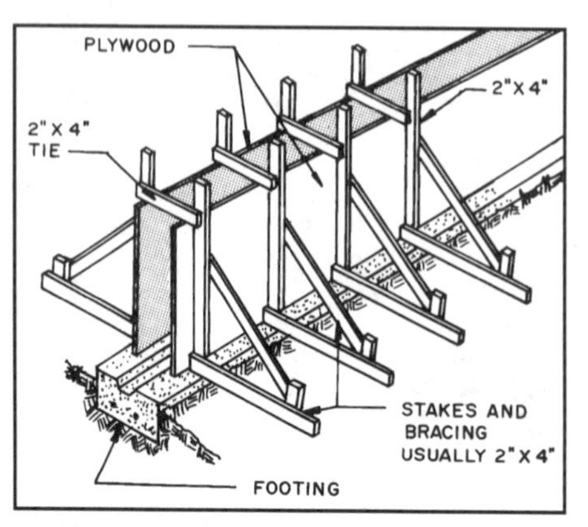

7-33 A typical wood form for a low wall.

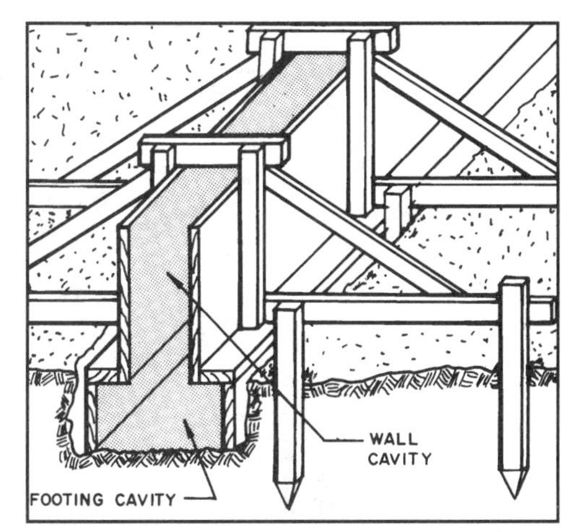

7-34 A wood form for monolithically casting a footing and a low wall.

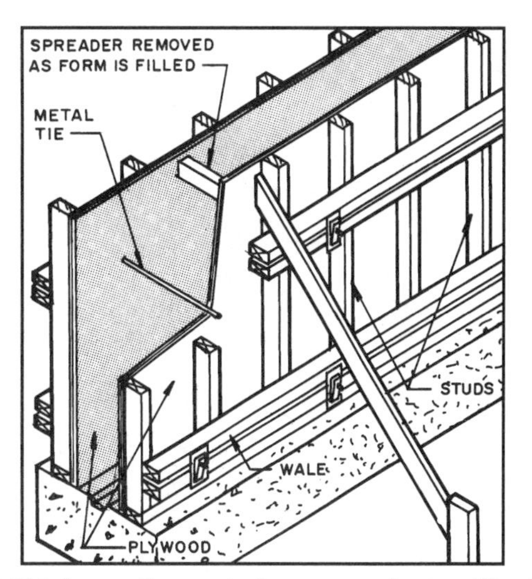

7-35 This low wall concrete form uses wales to stiffen it.

Forms for casting the footing and low wall **monolithically** are shown in **7-34**. "Monolithically" refers to casting them at the same time.

Wales can be used to keep low wall forms in line **(see 7-35)**. A finished low foundation wall with the forms removed is shown in **7-36**.

7-36 A low foundation with the forms removed. *(Courtesy Precise Forms, Inc., Kansas City, MO)*

FULL WALL FORMS

Forms for a conventional basement wall are usually made in 2 × 8- and 4 × 8-foot panels. Designs for wood panels are shown in **7-37**.

Forms are often held together with carriage bolts or duplex nails. This makes it easy to remove the forms **(see 7-38)**.

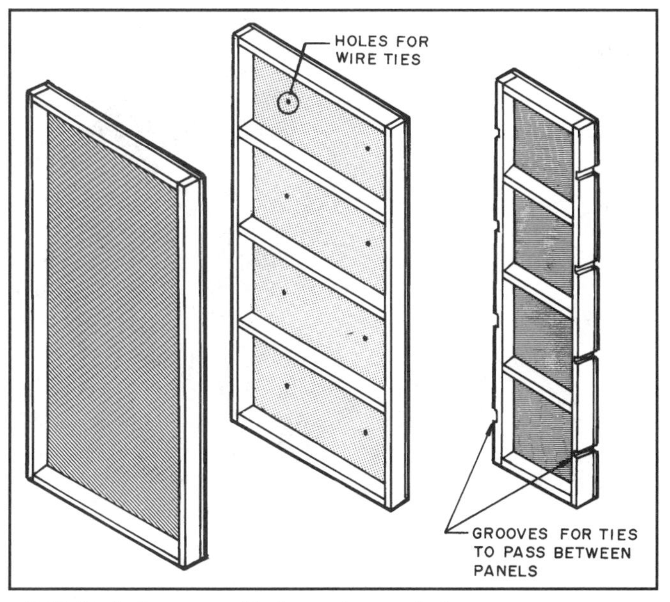

7-37 Typical preassembled wood concrete forms.

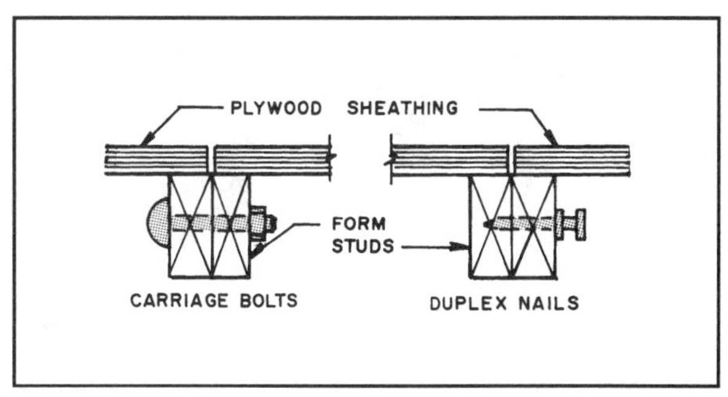

7-38 Typical ways of joining preassembled wood form panels.

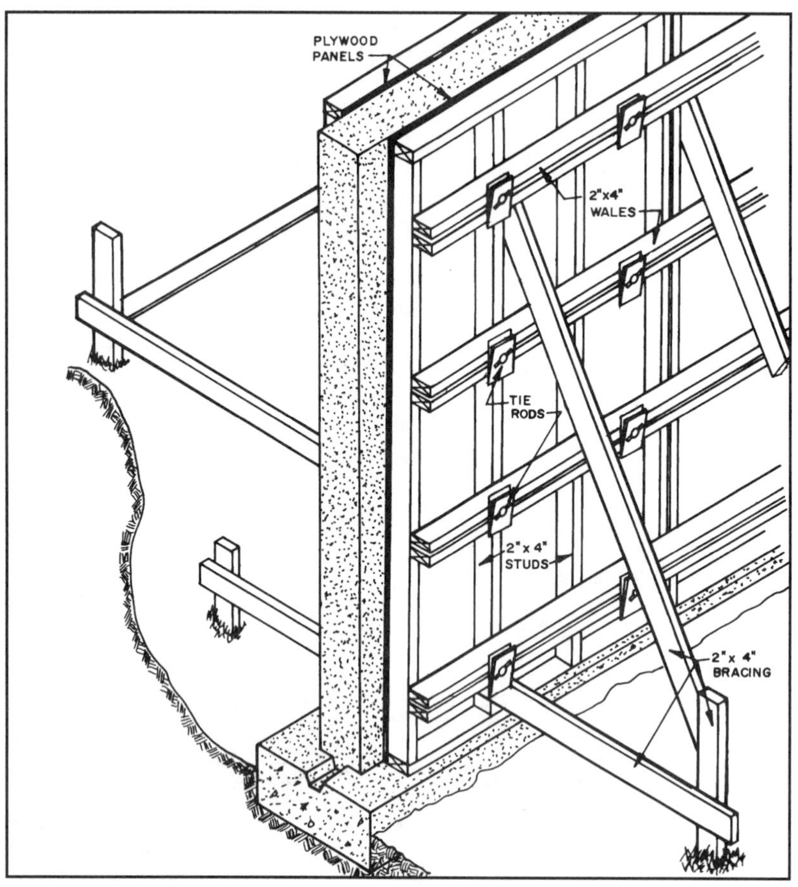

7-39 A full wall form using preassembled wood form panels. Notice the use of several rows of wales.

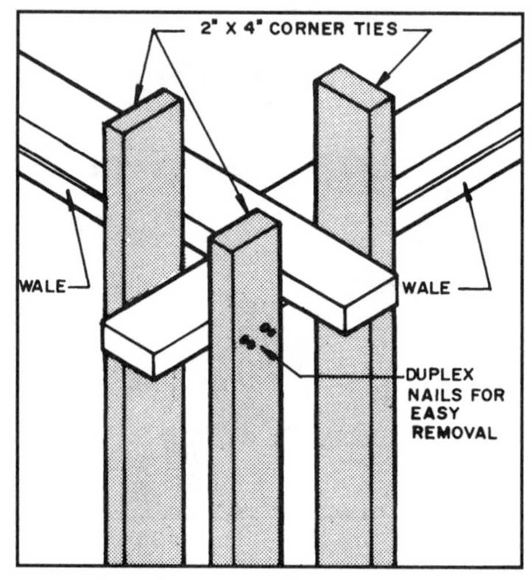

7-40 Wales are reinforced at the corners with vertical wood corner ties.

Wales are used to stiffen wood forms. A **wale** is a horizontal structural member placed outside the form. A wall form using these panels is shown in **7-39**. The wales are reinforced at the corner by corner ties **(see 7-40)**. The 2 × 4-inch members are nailed to the wales to hold the corner tight. Usually, **duplex nails** are used. A typical double wall form of this type is shown in **7-41**.

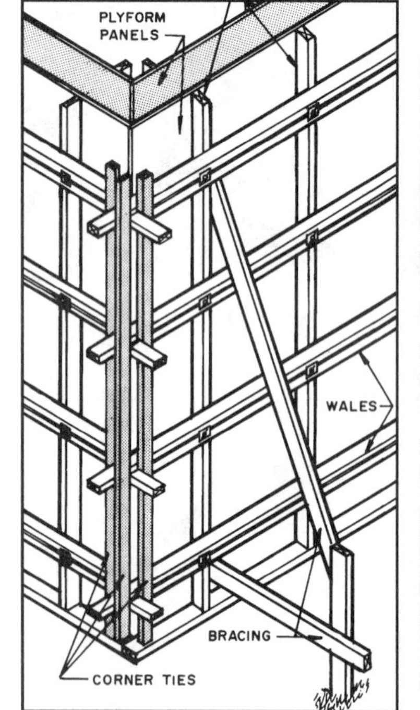

7-41 A typical double wall wood form with wales and corner ties.

7-42 Metal ties with cones are set in one side of this form, ready for the other side to be set on the footing.

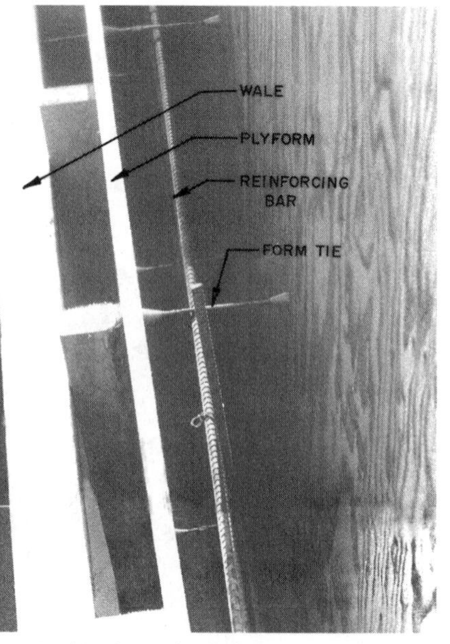

7-43 This shows both forms in place and the ties fastened to both. Notice the reinforcing rods have been placed in the wall cavity.

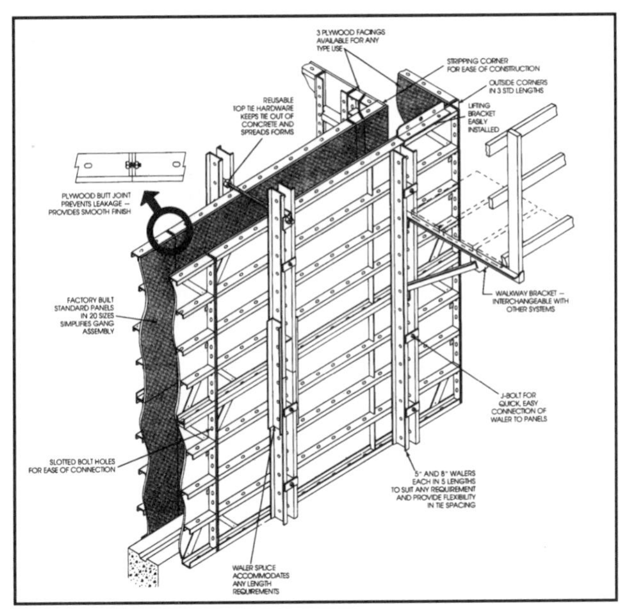

7-44 This form is built using standard metal framed panels and vertical wales. Notice the ties are connected to the wales. *(Courtesy Symons Corporation)*

7-45 These metal-framed concrete form panels are being stripped from the wall. *(Courtesy Precise Forms, Inc., Kansas City, MO)*

Metal form ties are used to help keep the forms the proper distance apart and strengthen them when concrete is poured. There are many different types available. One type is a rod with snap points **(see 7-42)**. The form tie goes through holes in the sheathing and studs. Some go through the wales or between double wales. This type is held tight by wedges pressing against the wales. The forms usually have the holes for the ties predrilled before they are set up on the footing **(see 7-43)**.

An example showing standard metal-framed full wall forms is in **7-44**. The ties are bolted to the vertical metal wales.

When the forms are stripped from the concrete wall, the ties then are broken at the snap point. This leaves small holes which need to be filled with mortar. A job that is being stripped is shown in **7-45**.

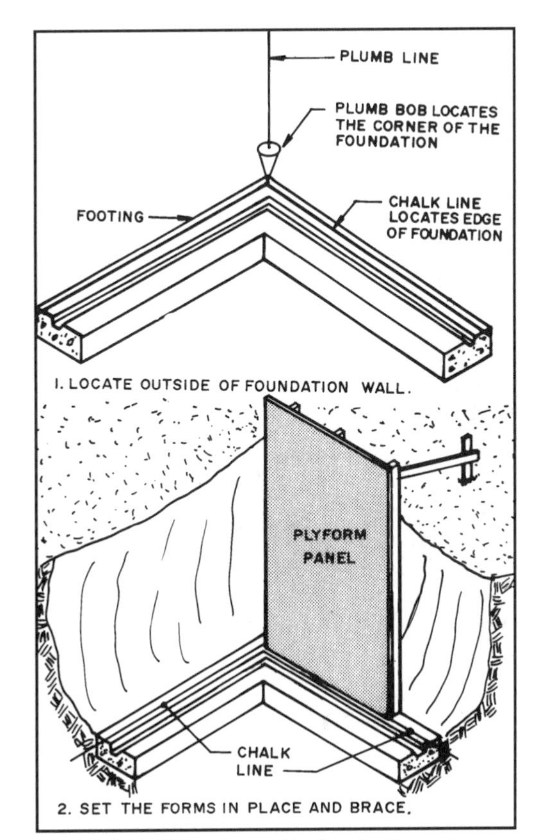

7-46 A chalkline is used to locate the outer line of the foundation on the footing.

Setting Up the Forms

Using a plumb bob, drop a line from the intersection of the building lines on the batterboards to the footing. Mark the location of each corner on the footing.

Run a chalkline from one corner mark to the next on the footing. Chalk the line and snap a chalkline on the footing. Repeat this around all sides of the building. The chalkline mark on the footing locates the outside face of the foundation wall **(see 7-46)**.

Next, stand a preassembled panel on the footing. Start by setting up a corner. Line up the face of the panel with the chalk mark. Nail the sole into the green concrete footing. Add a temporary brace. Then set the next panel. Join it to the first with bolts or duplex nails. Nail the sole to the footing and brace it.

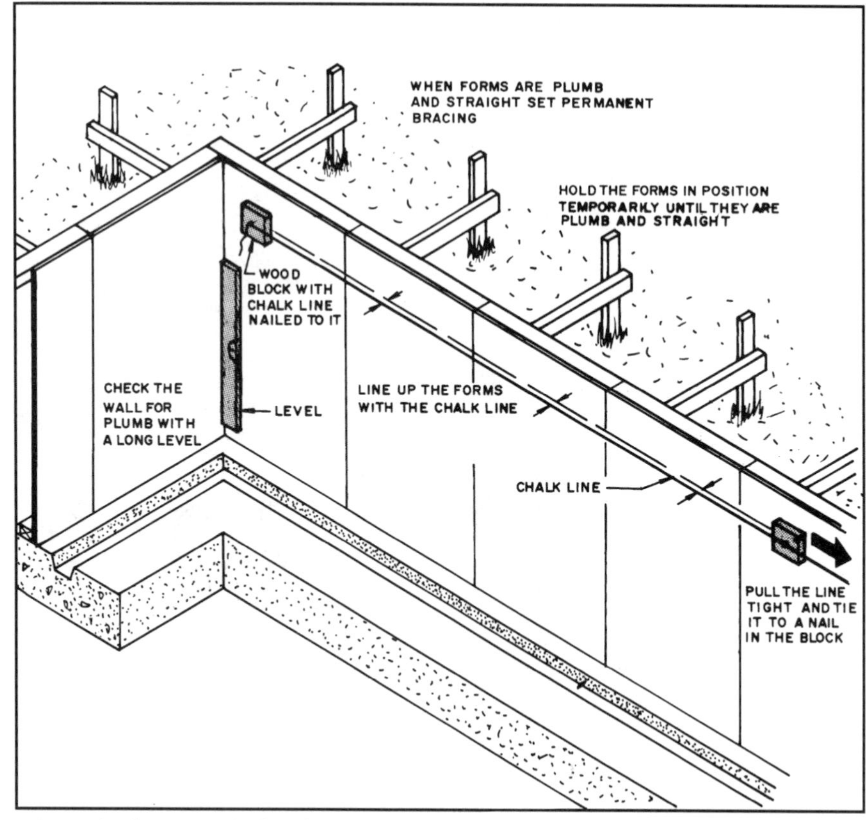

7-47 The forms can be lined up with a chalkline. Level transits (theodolites) are also widely used as discussed in Chapter 5.

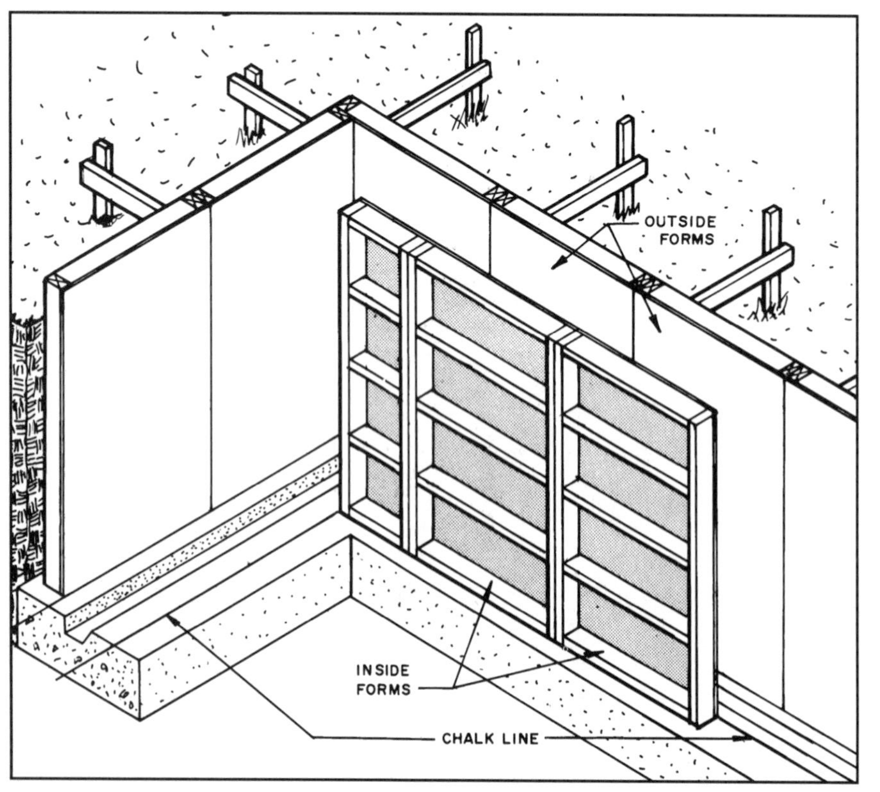

7-48 The inside wall forms are set and the ties are inserted.

The forms can be checked for **plumb** with a long builder's level. These levels are available up to 6 feet (183cm) long. Level transits (theodolites) are also used as explained in Chapter 5. The **straightness** of the form can be checked by nailing equally thick wood blocks at each end and running a chalkline between them as shown in **7-47**. Measure the distance between the line and the face of the forms. Adjust the braces until the chalkline is the same distance from the forms along the entire wall. Straightness is usually checked with a level transit as explained in Chapter 5. When the form is plumb and straight, permanently set the bracing.

When one wall is set, insert the form ties in the holes. Some forms have the ties pass through holes in the form. In this case the inner and outer panels must be placed so that the holes line up. Other forms have the ties pass in grooves between the panels. The panels must be placed so that their edges line up, to permit the ties to pass through them. Usually, the wall will require small panels to fill out the space evenly.

Set the forms for the inside face of the wall in the same manner as those for the outside. Insert the form ties through the holes in the form **(see 7-48)**. Now set the wales and secure them with some type of wale clamp as shown in **7-49**. Several types of connection hardware are available. Finally reinforce the corners by nailing vertical 2 × 4-inch corner ties to the wales **(refer back to 7-40 and 7-41)**.

Bracing the Forms

Typical bracing systems are shown in **7-39** through **7-41**. As the forms are put in place, temporary braces are often used. These are removed when the permanent bracing is installed. The bottom of the form has the greatest pressure and must be strongly braced to keep the form from slipping. The top of the form tends to spread and must also be braced. The top can be held with horizontal or diagonal braces. Often both are used **(see 7-50)**.

Adjustable braces are available **(see 7-51)**. They help to hold the form in place as it is assembled. This makes it easy to adjust the alignment of the form since its length can be easily changed with the screw.

The stakes in the ground should be 2 × 4-inch material and driven several feet deep. If the ground is soft, they must be driven even deeper. Watch for cracked or split material. If so weakened, they may break and permit a portion of the form to buckle.

7-49 The forms are stiffened and strengthened by the horizontal wales. Various types of metal wale connections are available.

7-50 Side and cross bracing have been used to support this form as the concrete is poured.

Forming Pilasters

Sometimes, a foundation wall will have pilasters. A **pilaster** is used to support the ends of beams and stiffen a long foundation wall **(see 7-52)**. Pilasters are usually formed when the wall is formed. In this way they become a part of the wall. A typical way to form pilasters is shown in **7-53**.

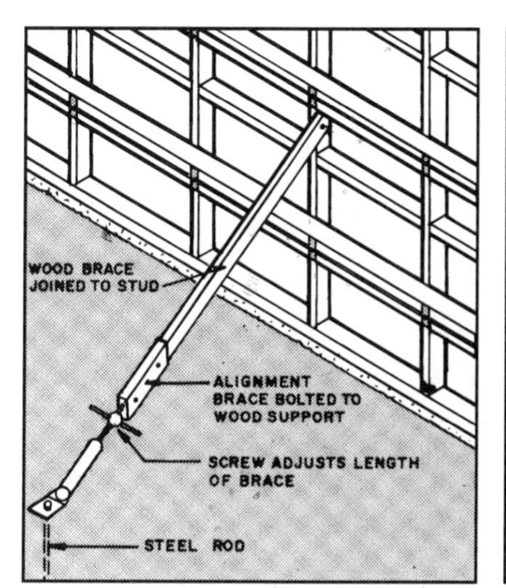

7-51 The length of this form brace can be adjusted as the form is plumbed and straightened.

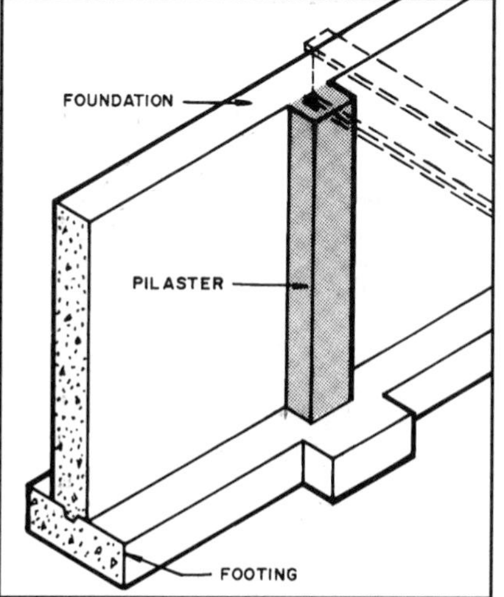

7-52 A pilaster strengthens a long foundation wall and can be used to support a beam.

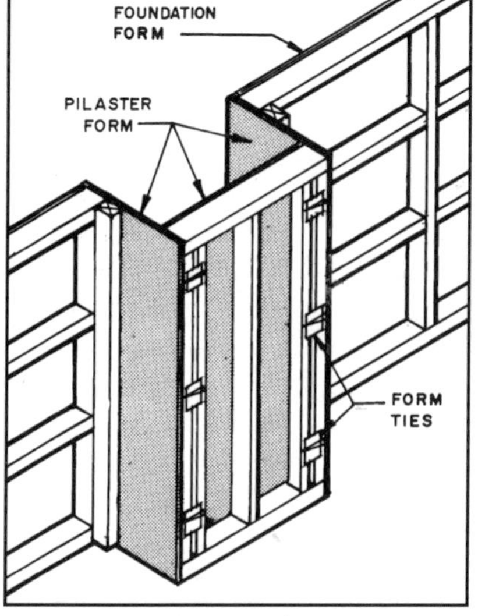

7-53 A typical pilaster form.

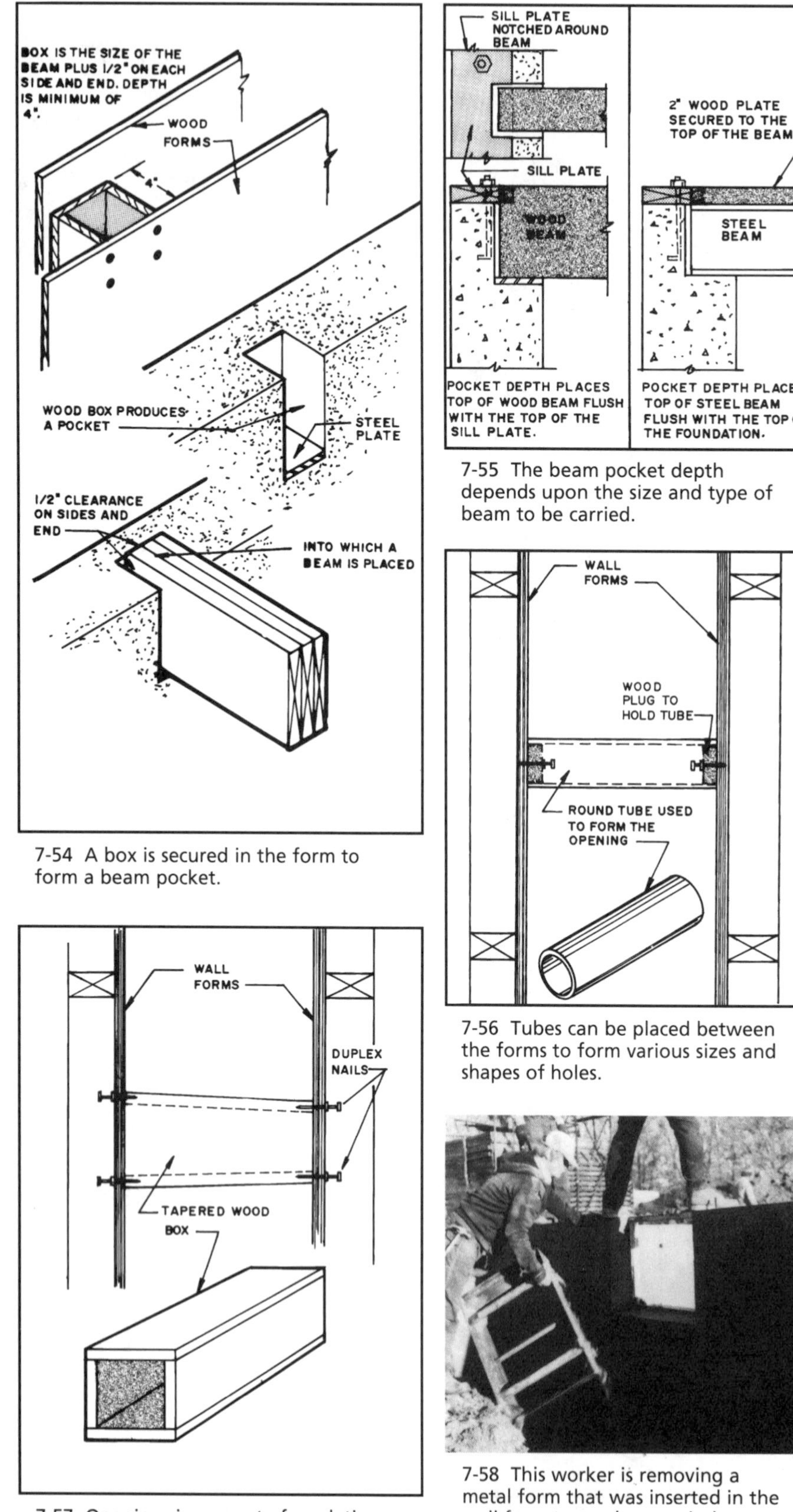

7-54 A box is secured in the form to form a beam pocket.

7-57 Openings in concrete foundations can be formed by inserting tapered wood forms.

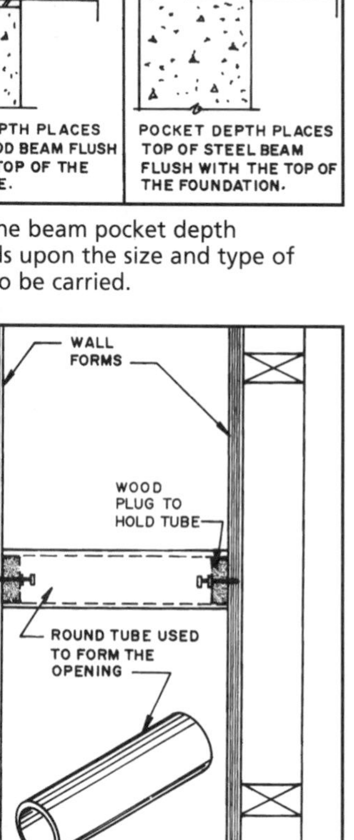

7-55 The beam pocket depth depends upon the size and type of beam to be carried.

7-56 Tubes can be placed between the forms to form various sizes and shapes of holes.

7-58 This worker is removing a metal form that was inserted in the wall form to produce a window opening. *(Courtesy Precise Forms, Inc., Kansas City, MO)*

Forming a Beam Pocket

Beam pockets are boxlike pockets set into the tops of foundations in which beams and girders are placed. They may also be placed above a pilaster. The form is a three-sided box as shown in **7-54**. The pocket left in the foundation should be about ½-inch (13mm) wider than the beam and deep enough to allow the top of a wooden beam or girder to be level with the top of the sill plate. The drawings will show the thickness of the steel plate to be placed below the beam.

If a steel beam is used it is usually set flush with the top of the foundation and a wood member is placed on top of it to bring it up to the level of the sill plate **(see 7-55)**.

Wall Openings

Concrete walls often have openings for doors, windows, and pipes. The work to form these must be done before the inner foundation wall form is erected.

Small openings, such as for pipes, can be made by placing fiber, plastic, or metal tubes between the forms. These tubes are held in place by wood blocks or plastic rosettes nailed to the forms **(see 7-56)**. Rectangular openings, such as for ventilation pipes, can be made using a wood boxlike form. The form is tapered to help remove it from the wall when the forms are stripped. It is nailed to the form with duplex nails **(see 7-57)**.

Large openings in concrete foundations, such as for windows, are made in several ways. An opening

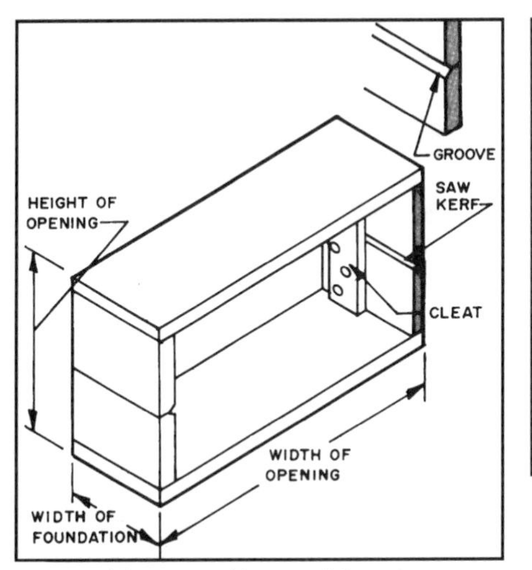

7-59 This removable wood form is placed inside the wall form to provide an opening in the wall.

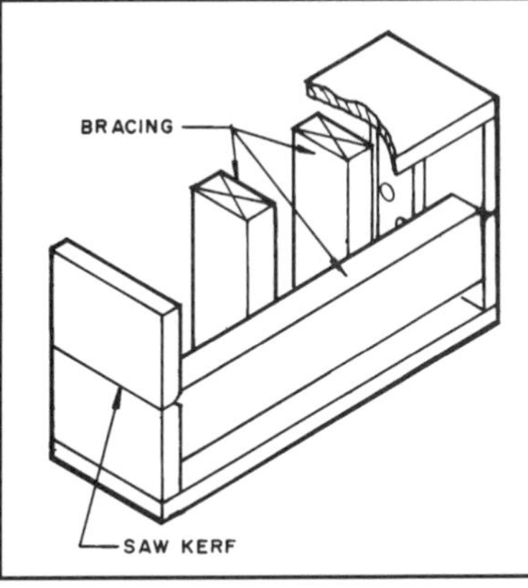

7-60 Internal bracing strengthens the wood form.

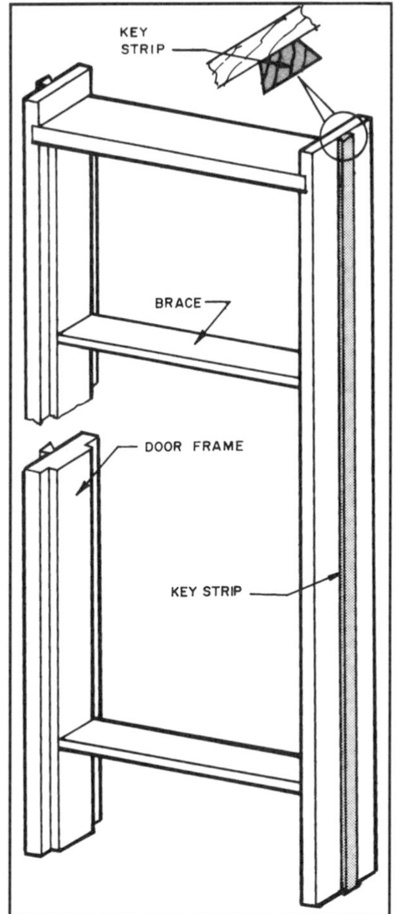

7-62 A wood door frame with a key strip designed to remain in the wall after the concrete hardens, thus securing the door frame to the foundation.

the size of the unit, such as a window, is marked out by building a box whose outside dimensions equal the wanted size of the opening. After the concrete has hardened, the form is removed **(see 7-58)**. To make it easy to remove the form, a saw kerf or notch can be cut in the sides **(see 7-59)**. A cleat is used to hold the sides in place. When the form is removed, the sides can be pried in until they collapse.

The form must have internal bracing. These are usually 2 × 4- or 2 × 6-inch members. The larger the opening, the more bracing is needed. If the form is poorly braced, the weight of the concrete will cause it to sag **(see 7-60)**.

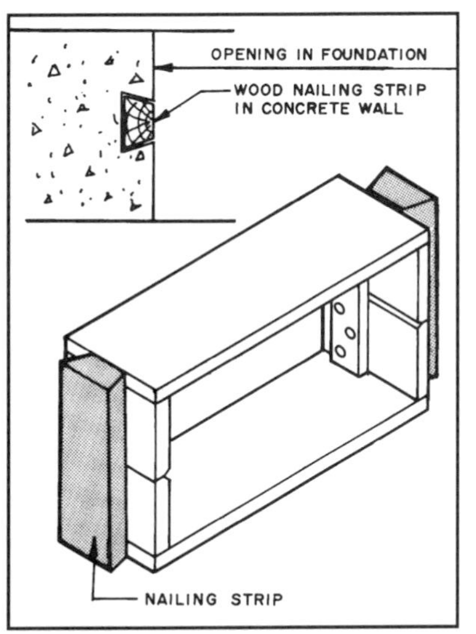

7-61 Wedge-shaped wood blocks are used for nailing strips in concrete walls.

Sometimes it is desired to leave a nailing strip in the concrete wall. This can be done by lightly nailing a wedge-shaped member to the box frame **(see 7-61)**. When the box is removed, the wedge remains in the concrete.

Sometimes, the door frame is installed in the foundation wall before the concrete is poured. A wood key strip is used to hold the frame in place **(see 7-62)**. If the width of the foundation is greater than that of the door frame, a filler strip must be used **(see 7-63)**. A door opening is being installed in **7-64**.

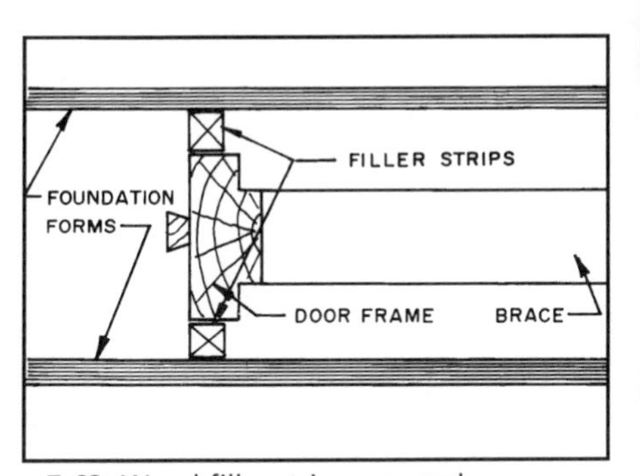

7-63 Wood filler strips are used between the wood door frame and the form.

7-64 The door frame is being placed in the foundation form. *(Courtesy Precise Forms, Inc., Kansas City, MO)*

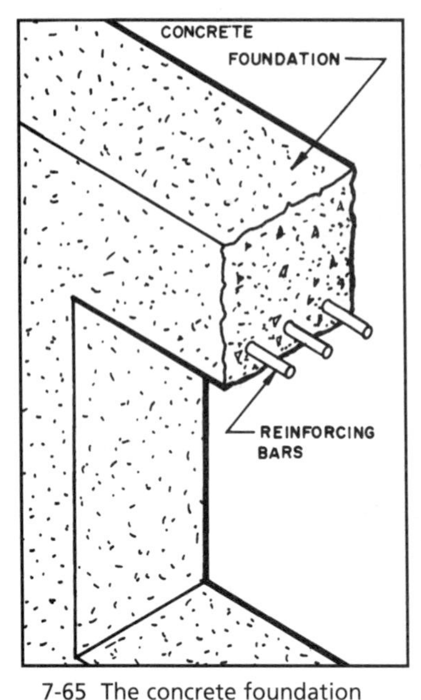

7-65 The concrete foundation above an opening must usually be reinforced with steel reinforcing bars.

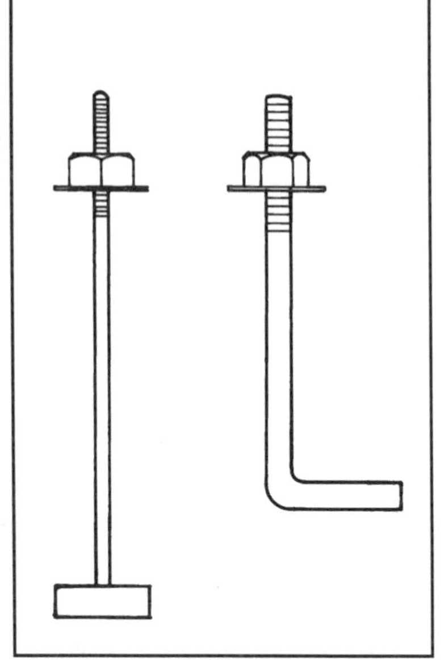

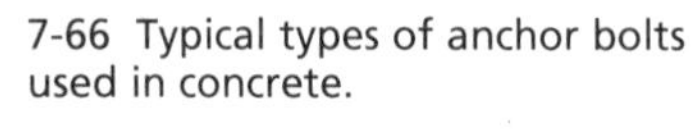

7-66 Typical types of anchor bolts used in concrete.

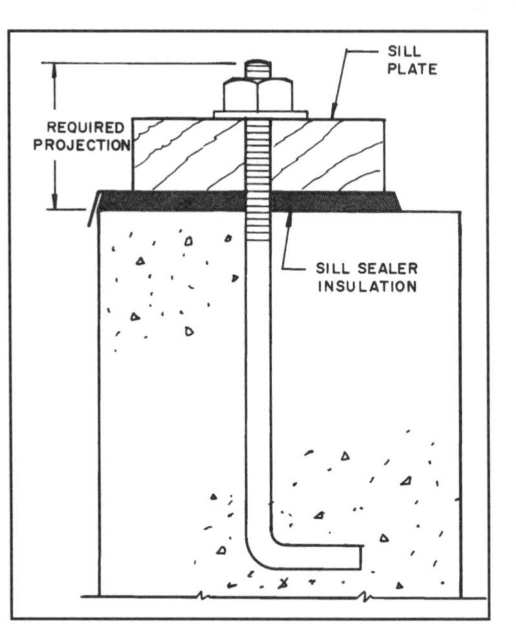

7-67 The anchor bolt must project above the top of the foundation far enough to pass through the items to be secured and allow the bolt to be secured.

Sometimes the concrete above an opening must be reinforced. This is often done by adding steel reinforcing rods above the opening **(see 7-65)**.

Sill Anchors

Various metal anchors are used to secure the wood sill to the concrete foundation after the wall has cured. **Anchor bolts** are commonly used for this purpose **(see 7-66)**. Anchor bolts are usually ½ inch (12mm) in diameter and are set into the concrete foundation at least 8 inches (20.3cm). Bolts are typically placed 12 inches (30.5cm) from each corner and a maximum of 6 feet (1.8m) apart. The spacing and placement are regulated by local building codes.

After the wall has been poured and leveled with the top of the form the anchor bolts are set into the wet concrete. Enough bolt is left protruding to go through the wood sill and accept a washer and nut **(see 7-67)**. The concrete is smoothed with a trowel around the bolt after it has been inserted.

Another anchor is the zinc-coated **steel anchor clip** shown in **7-68**. The clip is set into the concrete with the arms extending up. After the concrete has hardened the sill is set in place and the arms are bent around it and nailed to the sill.

Anchor bolts are set into the cavities of the

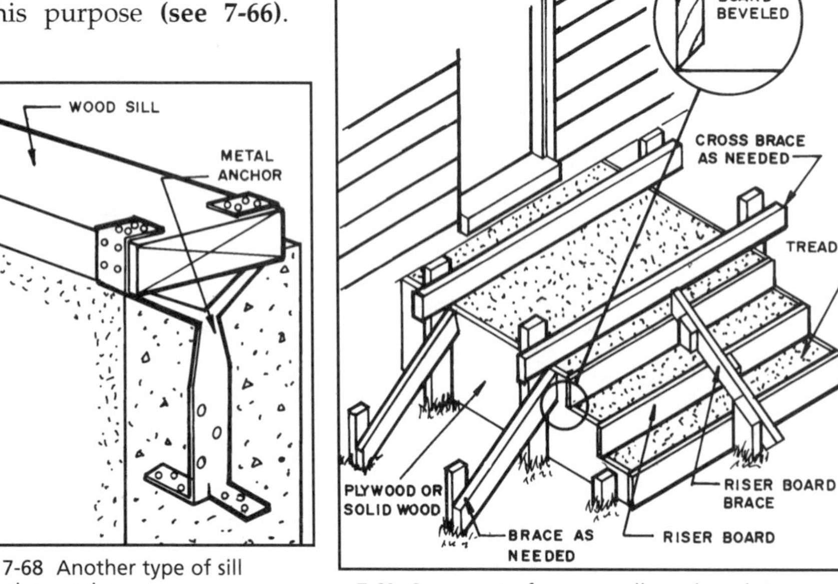

7-68 Another type of sill plate anchor.

7-69 One way to form a small porch and steps.

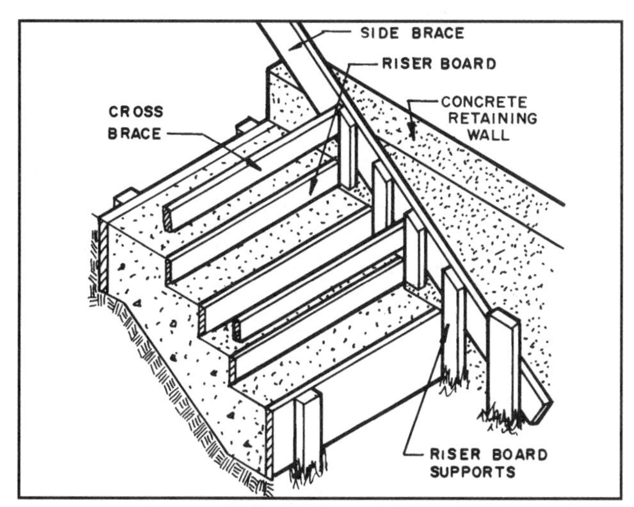

7-70 Form for a stair that butts a wall on one side.

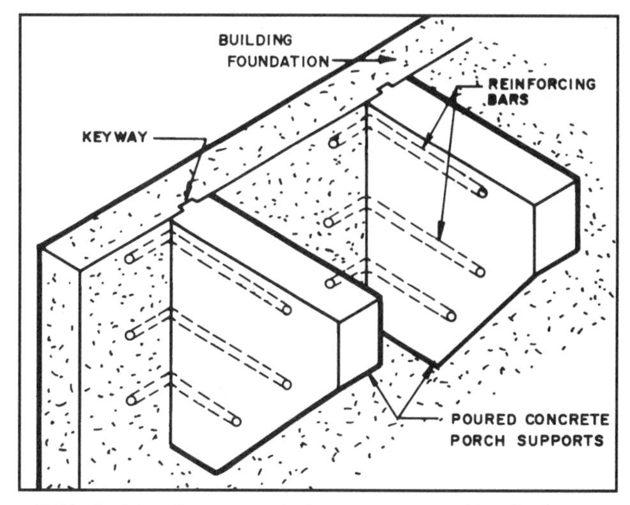

7-71 Cast-in-place concrete braces supported by the foundation wall can be used to support a porch floor.

concrete block foundations. The bolt should extend into the second row of blocks and the cavity filled with concrete. Anchor clips are installed in the same manner.

FORMING CONCRETE STAIRS

The forming for stairs that are open on the sides is shown in **7-69**. The side forms can be dimension lumber, such as 2 × 6-inch members or ¾-inch plyform. The side forms have the sizes of the landing, treads, and risers marked on them. They are cut to size. The riser boards are cut to length. This is the width of the steps plus material to nail them to the side forms.

Notice in **7-69** how the lower edge of the riser board is beveled. This provides room for the mason to trowel the step surface smooth.

Adequate bracing is necessary. One way to do this is shown in **7-69**.

Sometimes concrete steps are built where the sides are not open. This might be on an outside entrance to a basement. They are formed by using side braces to retain the riser boards **(see 7-70)**. Braces can be nailed to the top edges of the riser form boards if needed.

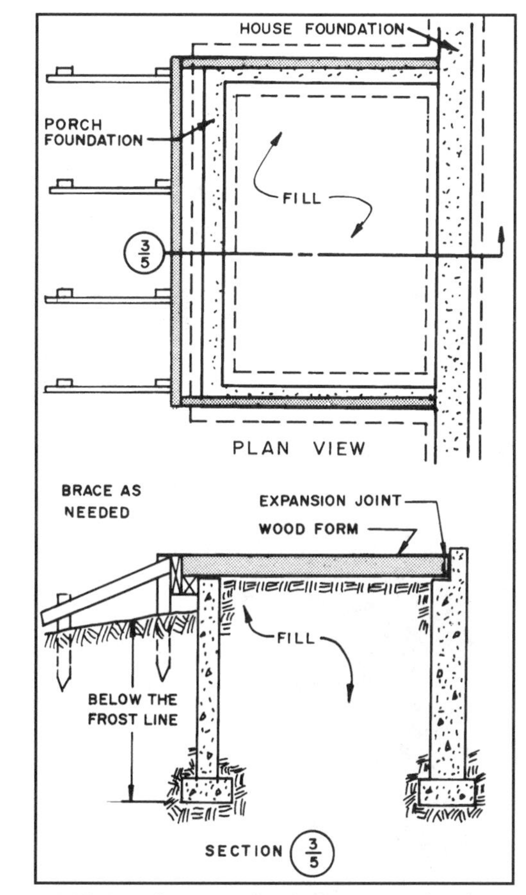

7-72 Forming a porch using a conventional footing and foundation.

FORMING ENTRANCE PLATFORMS

Many houses will have a small concrete entrance platform at each exterior door. This will consist of a landing and concrete steps. The platform needs support to keep it from settling and pulling away from the foundation. Often, concrete buttresses are poured and are cast as part of the foundation wall **(see 7-71)**. The floor slab is poured on top of these buttresses.

Porches that are larger are formed using a regular footing and foundation wall. The open area is filled with soil or gravel and serves as a base for the concrete. A wood form is built to form the floor **(see 7-72)**. **Reinforcing wire** is placed in the floor. If the span is large, steel reinforcing bars are used. Piers could be used to support the center of the slab if the porch is very large. A form for an on-grade entrance platform is shown in **7-73**. In cold climates the freezing and thawing will cause this platform to heave.

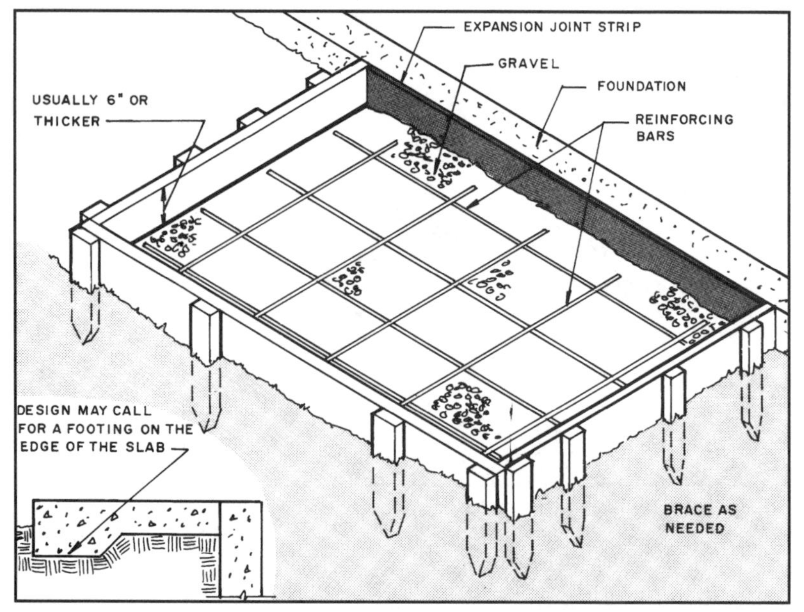

7-73 This is one way to form an on-grade entrance platform. The architectural drawings will give design details.

FORMING SIDEWALKS, PATIOS & DRIVES

Carpenters are often expected to build the forms for patios, sidewalks, and drives. Usually, these are built after the building is almost finished. This keeps them from being damaged.

The area to be covered with concrete is excavated to a depth needed to set the top of the slab at the level desired. Under normal conditions the concrete is placed on the soil. It should be firmly packed. If fill is needed to level the area it should be thoroughly packed. In areas where moisture may be a problem, four inches of rock or sand is used. This helps drainage, provides a firm underpinning, and helps in areas where cold may cause the walk to crack.

Stakes are driven at the corners of the area. They should be carefully located so that the size is correct. A chalkline is run between stakes on one side of the area. A leveling instrument is used to level the string. Mark the level on the stakes.

Stakes are driven into the ground along the chalkline. The side form boards are nailed to the stakes. These are usually 1 × 4-inch members. The top of the form board is lined up with the chalkline **(see 7-74)**.

Spreaders are cut the width of the sidewalk. They are used to locate the form boards on the other side. Normally the forms are set to provide a slope of ⅛ inch (3mm) per foot (30.5cm) on the finished surface of the slab. This helps drain the surface. This is usually done with a carpenter's level.

Wherever a slab meets a building, an **isolation joint** is needed. This is made of an asphalt-impregnated composition material. It is about ½ inch thick. This permits the slab to expand, contract, and settle without cracking. If it were firmly joined to the foundation—and the foundation then settled—the slab would break.

Slabs generally have reinforcing wire mesh. This strengthens them and helps reduce cracking. Reinforcing bars can be used when it appears extra strength is needed.

When the concrete is poured, a **screed** is used to level it. The forms are overfilled. The screed is placed on top of the side form boards. It is moved

7-74 Wood and metal side forms are used to form patios, sidewalks, and driveways.

7-75 After the concrete is placed in the form it is screeded.

from side to side as it is pulled along the form boards. This levels the concrete to the correct height **(see 7-75)**.

After the concrete is level and has partially set, the surface is troweled. Various degrees of smoothness can be had, depending upon the troweling done. Finally, control joints are made. These are cuts that go into the concrete about ½ inch. They provide weakened places for the concrete to crack if needed **(see 7-76)**. They are placed every four or five feet on sidewalks and every 10 feet on driveways.

7-76 The sidewalk is being finished with hand tools.

7-77 Large on-grade slabs are divided into sections by placing some form of expansion joint material at specified intervals.

The edges of the walk are rounded with an edging tool. This is done before the concrete hardens. It makes a rounded edge and helps prevent the edge from breaking off.

Slabs larger than 10 × 10 feet (3.05 x 3.05m) require the installation of some form of expansion joint material. Redwood strips 1 × 4 inches (25.4 × 101.6mm) are often used **(see 7-77)**. These are placed in the form before the concrete is poured.

FORMING CONCRETE SLAB FLOORS

Concrete slab floors may be poured directly on the compacted soil in some parts of the country. Other places require a 4-inch (101.6-mm) layer of compacted gravel below the slab. This helps drain subsurface water from below the slab.

As the slab is designed the method for heating the building must be considered. Heat can be provided by running copper tubing in the slab into which hot water will be pumped from a boiler. Electric heating cables can also be embedded in the slab. Hot air ducts can be laid and the slab poured over them. A down-flow furnace will move air through the ducts to vents cut through the slab.

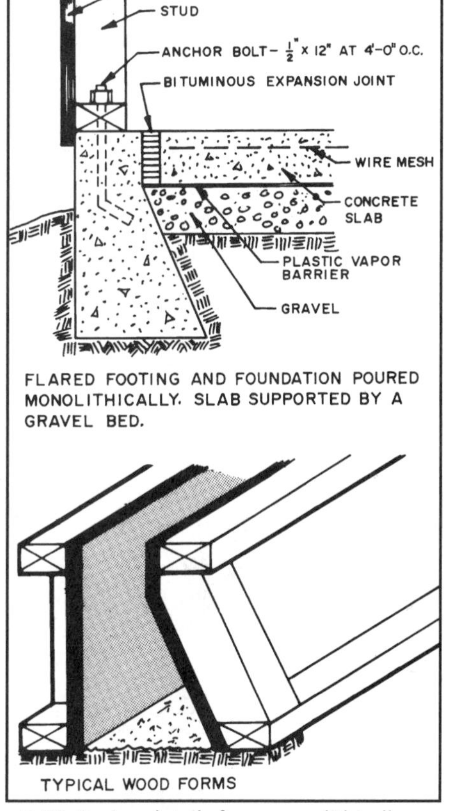

7-78 Design details for a monolithically cast flared footing and foundation for slab floor construction.

Following are the most commonly used types of concrete slab construction. All require a polyethylene-film moisture barrier be placed below the slab and insulation around the perimeter of the building. The slab is then reinforced with weld wire reinforcement which helps reduce the size of any cracks that may occur. Steel reinforcing bars are used in footings and any place in the slab that will be required to carry the design load. The slab is poured and troweled to a smooth finish.

In warm climates freezing is not a problem so the footing does not have to go deep into the soil. One type uses a flared footing with a floating slab **(see 7-78)**. In this design the slab and the foundation are separated. The foundation supports the weight of the building. The slab can rise and fall independently from the foundation.

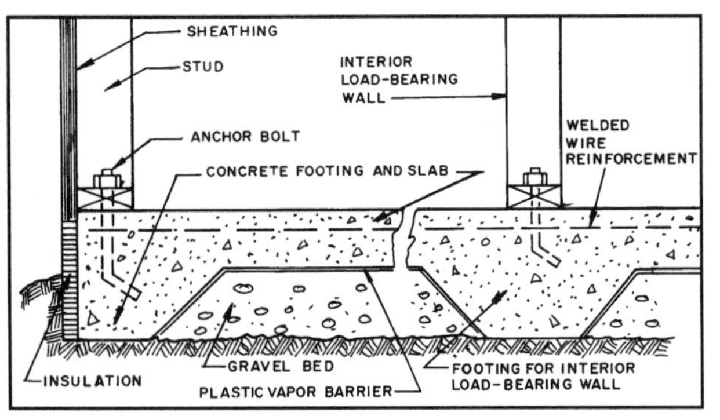

7-79 This is a monolithically cast concrete foundation and floor slab for use in warm climates.

Another type of slab used in warm climates has a shallow footing that is cast monolithically with the floor slab. Monolithically means they are cast together and form a single unit **(see 7-79)**. Notice that interior load-bearing walls require a footing be cast below them.

In climates where codes require the foundation extend below the frost line, a monolithically cast foundation and floor slab can be constructed using a grade beam as shown in **7-80**. When it is desired to have the floor independent of the foundation or when the foundation must extend deeply in the soil, a construction as shown in **7-81** can be used. When concrete block foundations are used the slab and foundation can be built much like those described for cast concrete foundations **(see 7-82)**. Notice the use of rigid insulation below the slab and on the face of the foundation wall.

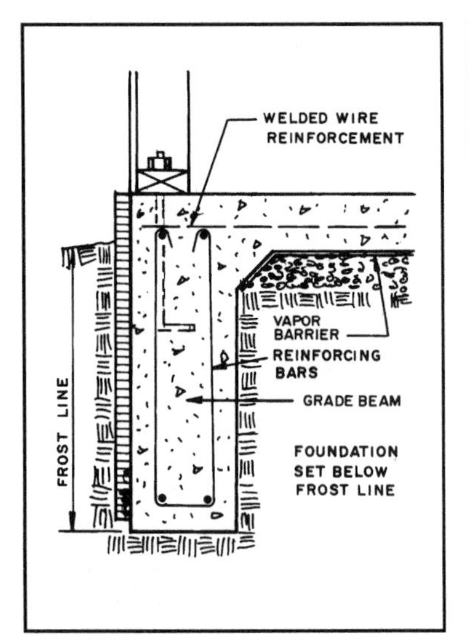

7-80 This monolithically cast grade beam extends to the frost line.

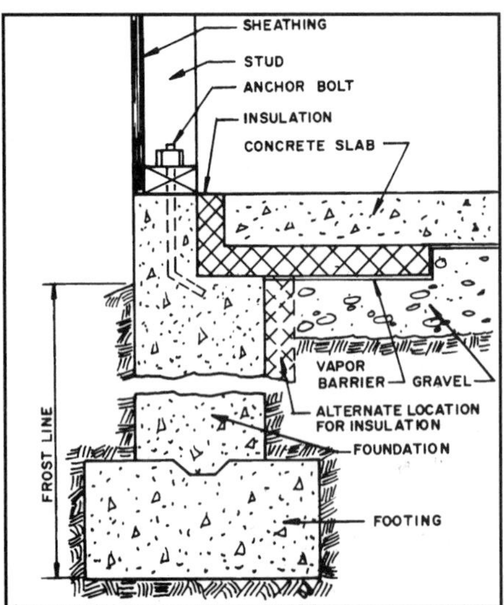

7-81 Design detail for a foundation for slab floor construction for use in a cold climate.

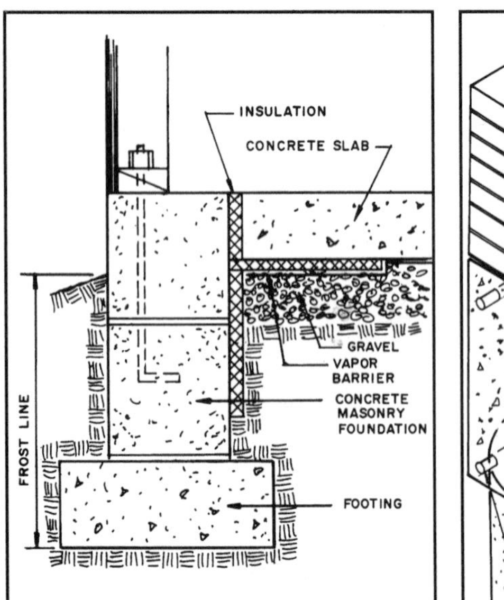

7-82 This detail shows the use of a concrete masonry foundation with a ground-supported concrete slab floor.

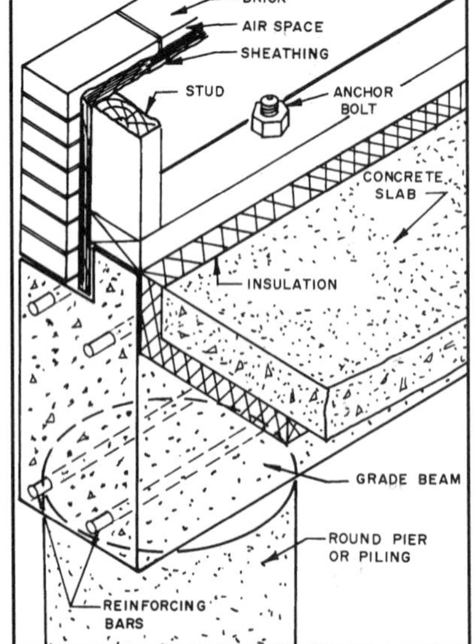

7-83 Details for a grade beam on a pier or piling with a ground-supported concrete slab floor.

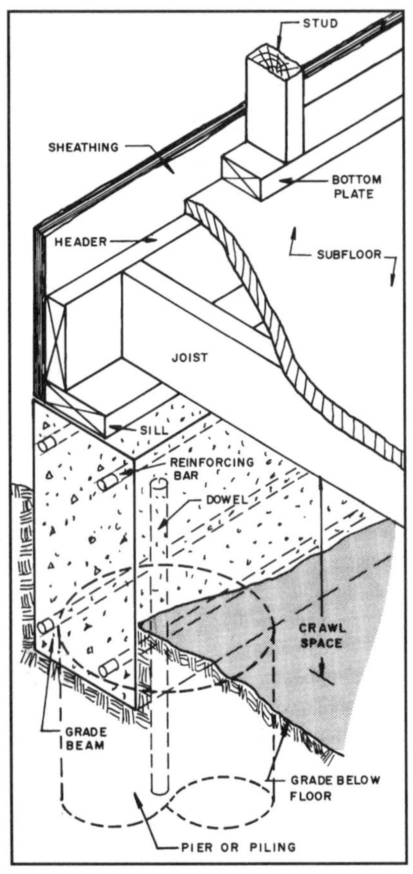

7-84 A typical detail using a grade beam to support a wood-framed floor.

FORMING GRADE BEAMS

Grade beams are structural members resting on piers or pilings that serve as the foundation wall **(refer back to 7-24)**. They may be formed and cast in place or precast and set with a crane. They are used with both concrete slab **(7-83)** and conventional wood-framed floors **(see 7-84)**.

The grade beam is formed the same as a low wall foundation. It rests on top of the piers. Steel reinforcing rods are placed at the bottom and top of the grade beam. The top of the form is the finished height of the foundation. It must be level. This is leveled in the same way as explained for foundation walls. A typical form for casting a beam is in **7-85**.

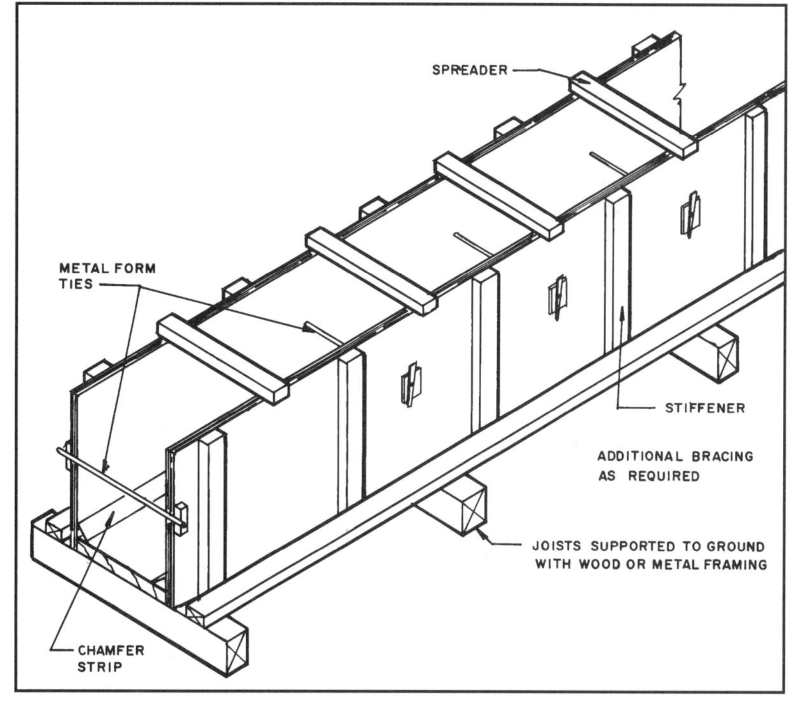

7-85 One way to prepare a form for a grade beam.

PERMANENT WOOD FOUNDATIONS

The **permanent wood foundation** is a load-bearing, wood-frame wall system designed for below grade use as a foundation for light frame construction. It is basically the same as aboveground frame construction except for several factors. The lumber and plywood used in framing are stress-graded to withstand the lateral soil pressures. The foundation also carries the usual live and dead loads. Vertical loads on the foundation are distributed to the supporting soil by a footing made of a wood footing plate and a structural gravel layer. A typical installation is shown in **7-86**.

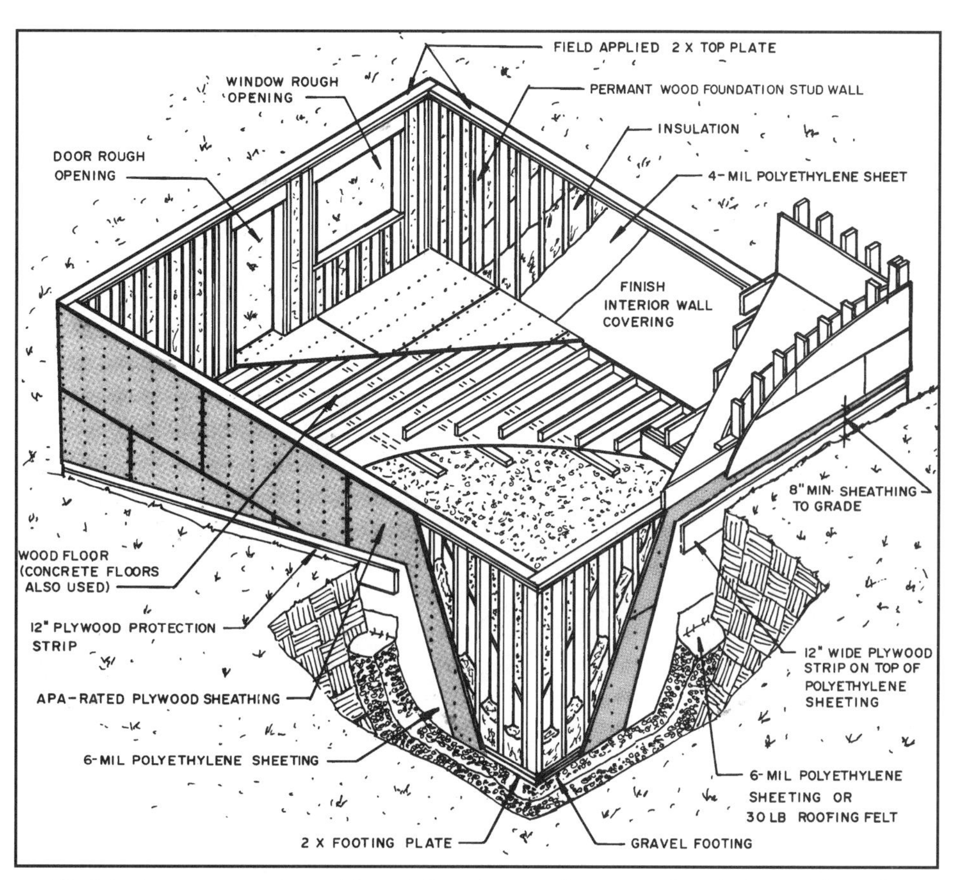

7-86 A typical permanent wood foundation. Approved wood studs and plywood form the wall, which is assembled with corrosion-resistant fasteners. The surfaces below grade are fully waterproofed. *(Courtesy Southern Pine Council)*

7-87 Foundation panels are assembled in a shop, trucked to the site, and set in place. This can occur in any type of weather. *(Courtesy Southern Pine Council)*

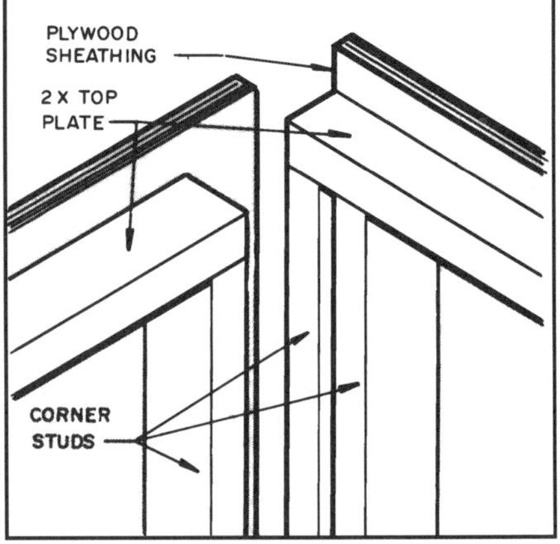

7-88 Framing an external corner on the permanent wood foundation. *(Courtesy Southern Pine Council)*

All lumber and plywood in contact with or close to the soil are pressure treated with wood preservatives. This protects against decay and insects. The preservative treatment must meet the requirements set by the American Wood Preservers Bureau. If lumber is cut after treatment, the cut surface should be brush-coated with preservative.

Extensive moisture control measures are used. Moisture reaching the upper part of the wall is deflected by polyethylene sheeting. This moisture flows into a porous gravel layer built around the lower part of the basement. The moisture flows through the gravel to a positively drained sump which removes the water.

There are specific requirements related to the species and grade of lumber used and the specifications for the plywood sheathing. Fasteners are corrosion resistant and usually silicon, bronze, copper, stainless steel, or zinc-coated steel.

The permanent wood foundation does not require the customary concrete footing. This reduces the overall weight and cost of construction. Since it can be assembled in a shop and erected on the site it permits builders to install

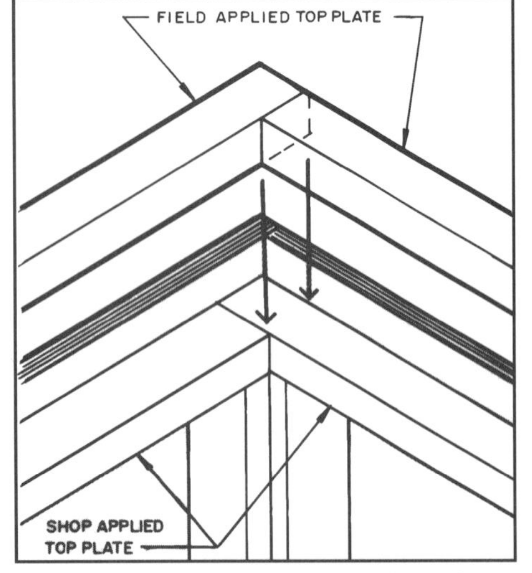

7-89 After the corner panels are butted and nailed, a field-applied top plate is added, helping tie the corner together. *(Courtesy Southern Pine Council)*

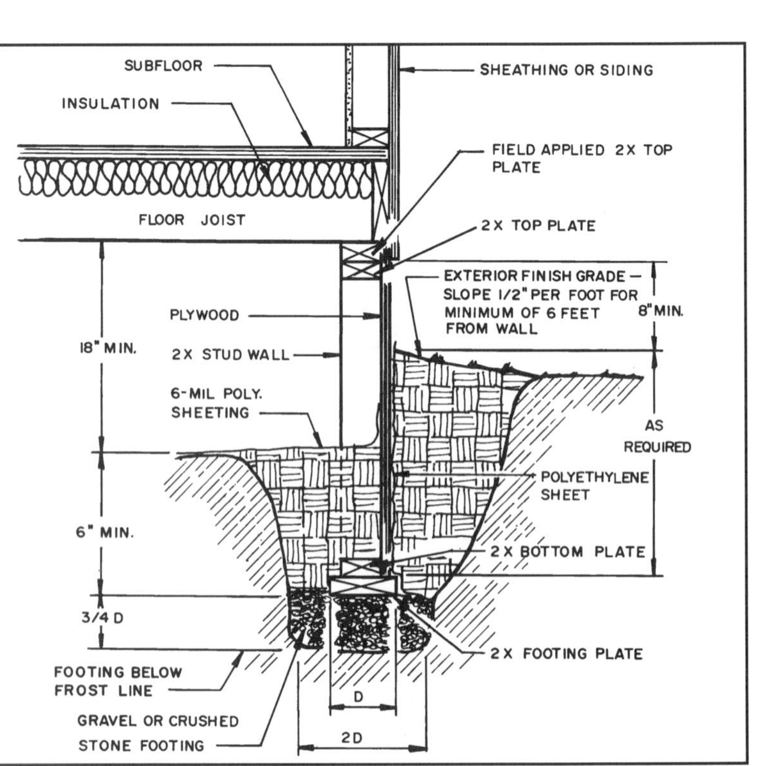

7-90 Construction details for a permanent wood foundation with frame construction over a crawl space. *(Courtesy Southern Pine Council)*

foundations in any kind of weather. Installation is fast and does not require the usual crew of concrete workers **(see 7-87)**.

The basement tends to be warmer. The wood wall provides good insulation. In addition, it can be insulated on the inside in the same manner as above-grade frame construction. This tends to reduce heating costs.

Since considerable attention is given to waterproofing and handling below-surface water, the basement produced is dry. This reduces the mildew often associated with basements.

The basement area is easier to finish into a living area. Normal procedures are used to apply wall finish over the insulation. It is easy to install electrical outlets around the basement walls.

Typical Construction Details

Exterior corners are framed as shown in **7-88**. After the panels are assembled a second top plate is nailed, tying the corner in place **(7-89)**.

Openings for doors and windows have the studs doubled and headers installed as shown earlier in **7-86**.

A section through a permanent wood foundation for a building with a crawl space is shown in **7-90** and for a building with a basement in **7-91**. Notice that concrete footings are not used. The frame wall has a pressure-treated wood footing plate that rests on a specially prepared gravel bed. The gravel bed should be below the frost line. The size and type of gravel is carefully specified.

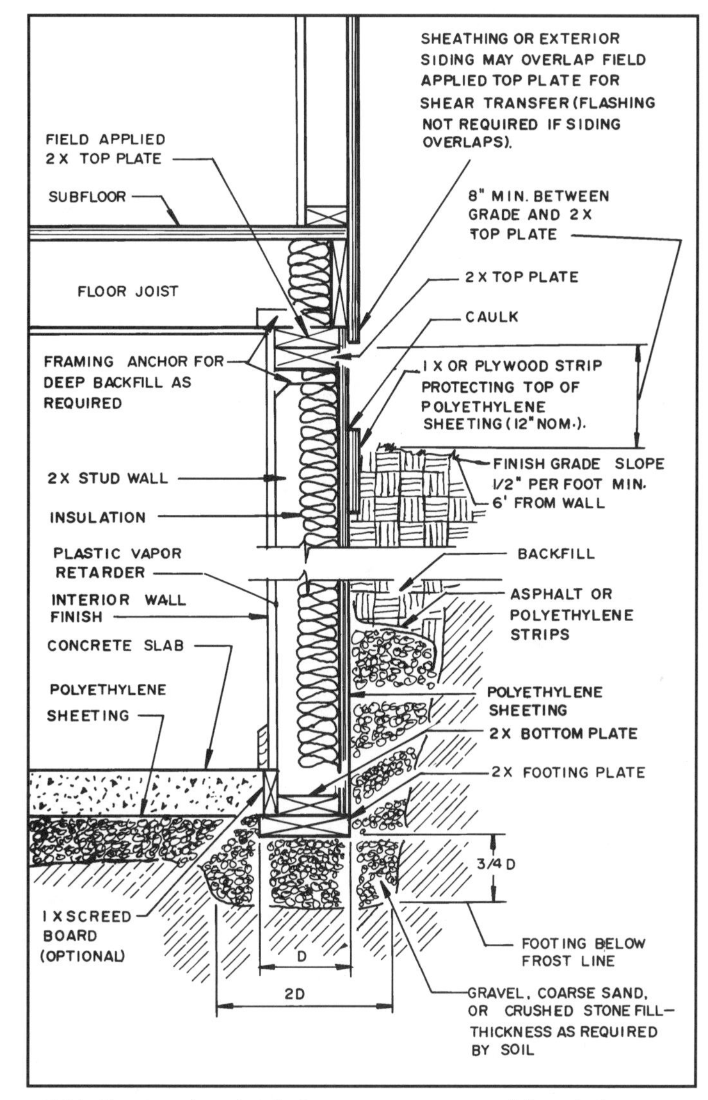

7-91 Construction details for a permanent wood foundation with frame construction over a full basement. *(Courtesy Southern Pine Council)*

CHAPTER 8

Concrete

Concrete is a mixture of cement, water, aggregates, and in some mixes one or more admixtures. Concrete is a very durable material when properly formulated, placed, and cured. It is poured into forms to produce footings, walls, piers, floors, beams, and other items. After the concrete has hardened the forms are removed **(see 8-1)**.

Portland Cement

Portland cements are made from raw materials containing alumina, lime, silica, and gypsum. Iron and magnesia are also present in most types of cement. These raw ingredients are mined, crushed, and ground to a fine powder. The ground materials, except for gypsum, are blended together and fired in a kiln at 2700 degrees F (1494 degrees C), forming a clinker. The clinker is ground and gypsum is added, forming Portland cement. The gypsum controls the setting time. The more gypsum in the blend the slower the cement will set.

Portland cements set and harden by a process called hydration. Hydration is a chemical reaction between the cement and water which produces heat, leading to hardening.

8-1 Concrete is used to form strong watertight foundations. This foundation has had the forms removed and is ready for the carpenters to begin construction of the floor. *(Courtesy Precise Forms, Inc., Kansas City, MO)*

Table 8-1 Types of Portland cement and air-entrained Portland cement.

United States Type Designation	Canadian Type Designation	
Types of Portland Cement		
I	10	Normal
II	20	Moderate
III	30	High early strength
IV	40	Low heat of hydration
V	50	Sulfate-resisting
Types of Air-Entrained Portland Cement		
IA	10A	Normal air-entrained
IIA	20A	Moderate air-entrained
IIIA	30A	High early strength air-entrained

The types of Portland cement are in **Table 8-1**. The United States and Canada use basically the same designations.

Type I. Normal is a general-purpose Portland cement used when the special properties offered by the other types are not needed. It is used for applications such as sidewalks, driveways, reinforced structural members, and concrete masonry units.

Type II. Moderate is used when protection against moderate sulfate attack is required. This typically occurs in underground situations. It is used for applications such as retaining walls and piers.

Type III. High early strength Portland cement is used when it is necessary to develop high early strength faster than normal cements. It reduces the curing time and is especially useful in freezing weather.

Type IV. Low heat of hydration Portland cement is used when pouring large masses of concrete because it reduces the heat produced, slowing the setting time—giving, of course, more time to place the concrete before it begins to set.

Type V. Sulfate-resisting Portland cement is used where concrete is exposed to severe surface conditions such as those that occur with below-ground applications.

Air-entrained Portland cement is available in three types, 1A, 11A, and 111A. Air-entraining additives in the mix produce millions of microscopic air bubbles in the concrete. This improves the ability of the concrete to resist the freezing and thawing cycles in the winter. It also improves the workability of the concrete and reduces the amount of water needed.

Storing Portland Cement

On small construction jobs the Portland cement is delivered to the site in water-resistant sacks. They must be stored on wood pallets which keep them above the earth or concrete floor in a storage building. They should be stacked closely together to reduce the flow of air between them and be covered with a plastic sheet or tarpaulin.

On large jobs the Portland cement is stored in large watertight bins from which it is withdrawn and moved to an on-site concrete batch plant.

Mixing Water

Almost any natural water that is safe to drink can be used to make concrete. Some waters, such as from lakes, may not be drinkable but if tested for impurities may prove to be usable. Seawater containing up to 35,000 PPM (parts per million) of dissolved salts is generally suitable as a mixing water for unreinforced concrete. If reinforcing is present the seawater may cause it to corrode.

Aggregates

Aggregates include sand, crushed stone, gravel, and other similar materials. They are mixed with Portland cement and water. Typically aggregates make up 60 to 75 percent of the volume of concrete, so their hardness and quality are very important to the strength of the finished concrete. The rest of the concrete is made up of the cement paste.

Aggregates can be tested to see if they have the properties needed to make quality concrete. Fine aggregates (sand) have particles smaller than ¼ inch (6mm). Larger aggregates used in concrete for most normal building construction range from ¾ inch to 1½ inches (19mm to 38mm). The aggregate used in a mix should contain particles having a wide range of sizes. Aggregates must not have more than one percent sea salt.

Admixtures

Admixtures are materials other than water, aggregates, and cement added to the mix during or before it is mixed. They are added to alter the properties. They are used sparingly, and the type and amount must be determined by concrete specialists. Admixtures in common use include:

1. Air-entraining additives—improve workability, resist freezing, thawing.
2. Accelerators—speed up set time.
3. Water reducers—reduce amount of water needed.
4. Retarders—slow setting time.
5. Pozzolans—increase cementing ability.
6. High-range water reducers—drastically reduce water needed.

Concrete Mixes

The proportions of the ingredients in a concrete mix vary with its intended use. The mix used should have the following properties:

1. Must be economical.
2. Can be placed and worked.
3. Can be finished.
4. Will have the required strength.
5. Will be durable and resist deterioration in the proposed application.

The mix to be used should be decided by a concrete specialist. On small jobs this person is usually in the employ of the concrete batch plant. On large jobs the mix is specified by an engineer—part of the overall specifications provided for the job.

A key to a successful mix is the water-cement ratio. The water-cement ratio is the proportion of water to the amount of cement. Most ratios range from 0.4 to 0.7 lb. of water to one pound of cement. If the amount of water added increases without increasing the amount of cement the strength is lowered. The various types of cement require different water-cement ratios. The water-cement ratio also influences the watertightness of the cured concrete.

Table 8-2 Suggested mixes for non-air-entrained concrete with a slump of 3 to 4 inches and a water–cement ratio of 4.5.*

Maximum Size Aggregate (inches)	Entrapped Air (%)	Water		Cement			Fine Aggregate		Coarse Aggregate	
		cu ft	gal	cu ft	lb	no. of sacks	lb	cu ft	lb	cu ft
⅜	3	6.6	46	8.3	968	10.3	1240	124	1260	126
½	2.5	6.3	44	7.9	921	9.8	1100	110	1520	152
¾	2.0	5.9	41	7.4	855	9.1	960	96	1800	180
1	1.5	5.6	39	7.0	817	8.7	910	91	1940	194
1½	1.0	5.1	36	6.5	752	8.0	880	88	2110	211

*Water-cement ratio (gallons per 94 lb cement) equals 4.5

Cement weighs 116 lb per cubic foot
One cubic foot of water equals 7 gallons
One sack of cement equals 94 pounds
Sand and gravel weigh 90 to 105 pounds per cubic foot
Water weighs 62.4 pounds per cubic foot
Basic data abstracted from the technical manual *No. 5–742, Concrete and Masonry,* U.S. Department of the Army.

The sizes and proportions of aggregates also influence the quality of the concrete. The most economical mix will use the most coarse aggregate that is permissible and have the specified water-cement ratio. A generalized example of possible mixes for non-air-entrained concrete is in **Table 8-2**. The various materials are measured in pounds or cubic feet and water is typically in gallons or cubic feet. If water is in cubic feet remember that one cubic foot of water contains seven gallons of water. The mix specifications also limit the maximum size of aggregate. It is important to have a range of aggregate sizes. The small aggregate becomes coated with the cement paste and fills the voids between the large aggregate. This produces a cured concrete with few or no voids between aggregates.

Sources of Concrete

On small projects, such as residences, the concrete is mixed off-site in a concrete batch plant **(8-2)**. This is a computer-controlled facility that produces the required mixes and deposits them in a transit-mix truck for delivery to the site **(see 8-3)**. On large projects the batch plant is erected on the site and the concrete is delivered where needed by transit-mix trucks. Generally, concrete must be placed within 1½ hours after the water has been added. If water is added after this time or at any time on the construction site, the water-cement

8-2 Part of an off-site concrete batch plant.

8-3 This ready-mix transit truck is at the batch plant, ready to receive a load for delivery to the site.

8-4 Wheelbarrows with pneumatic tires are used to move concrete short distances when the quantity of concrete to be poured is small.

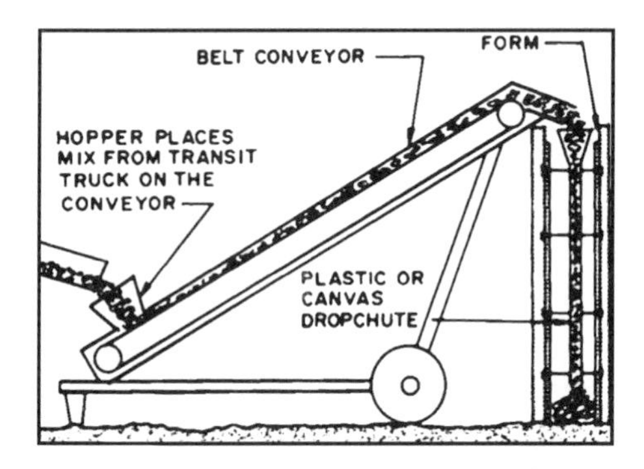

8-5 Concrete is placed in tall forms through a drop chute. The concrete is being moved to the drop chute with a belt conveyor.

ratio will be changed and the strength and durability of the cured material will be changed. No water should be added without first checking with the concrete specialist at the batch plant.

Placing Concrete

Concrete must have the recommended mix if it is to be placed properly. If it is too thin or too thick, placement will be difficult and the resulting concrete structure will be defective.

Before concrete is placed, the forms should be inspected to be certain they are complete and adequately braced. Also it is important that a sufficient amount of concrete has been ordered and delivery scheduled so the pour can be completed without interruption. If the amount of concrete runs short, that already poured will begin to harden before a new batch can be placed on it. This forms a seam called a cold joint. Concrete is mixed in cubic yard quantities.

The inside of the forms and soil under the footing are moistened so they do not absorb the mixing water from the concrete. The concrete should not be placed on frozen ground.

Place the concrete as near as possible to its final location **(see 8-4)**. When placed in the form or on the ground, such as for a slab, the concrete can be leveled with a vibrator—but not moved horizontally. On small forms, place the concrete at one end and move additional placements toward the other end.

Concrete should not be dropped into forms, such as for a wall. The fall plus the splash and agitation by rebars and form spacers will cause the aggregate to separate from the mix. Some form of drop chute must be used **(see 8-5)**.

Concrete can be placed using the chute on the transit-mix truck if the truck can be near the pour site **(see 8-6)**. Buggies are used to move concrete to less accessible areas and may require runways and ramps be built **(see 8-7)**. Power-operated conveyors, as shown in **8-8**, will lift concrete above the ground and fill chutes which direct the concrete into the form. Concrete pumps can move concrete long distances. The transit-mix truck fills the hopper on the pump. The concrete is pumped through a hose carried on a boom. A plastic or rubber tube-like chute on the end places the concrete down into the form **(see 8-9)**.

8-6 Whenever possible move the transit-mix truck close to the area that will be in the pour. (*Courtesy Corps of Engineers, U.S. Department of the Army*)

Curing Concrete

As explained earlier, concrete hardens due to a chemical reaction between the water and Portland cement (hydration) which produces heat and evaporates the water in the mix. The hydration of the cement depends upon the temperature, amount of time elapsed, and the presence of moisture. If the mix is properly formulated there is enough water to fully hydrate the cement. When the water is gone hydration stops and the strength of the concrete has been determined. The longer hydration can occur, the stronger the concrete produced. Therefore the concrete must be kept moist and not allowed to freeze until it has had time to reach the design strength. If the temperature is above 40 degrees F (4.5 degrees C) the curing must be continuous for at least seven days so it can reach 70 percent of the required compressive strength.

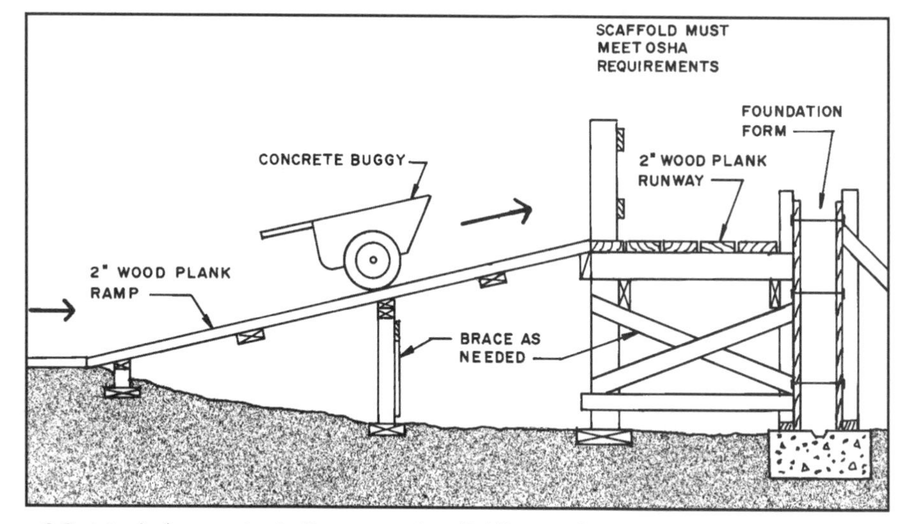

8-7 A typical carpenter-built ramp and scaffolding used to place concrete in a form that rises above grade.

8-8 This type of concrete belt conveyor is portable and is used for short reach and lift applications. *(Courtesy American Concrete Institute)*

8-9 This form is being filled using a concrete pump. Notice the flexible tube on the end which moves down into the form to place the concrete. *(Courtesy Morgen Manufacturing Company)*

After the concrete has been placed the forms should be kept covered and moist until they can be removed. After the forms are removed the surfaces of the concrete should be kept continuously wet with a spray or covered with a wet fabric that is kept wet. In cold weather the forms must be kept covered with blankets or some form of insulation. It may be necessary to heat the space in which the concrete has been poured.

Another way to help cure concrete is to spray a liquid membrane-forming curing compound on the surface. This wax and synthetic-resin film keeps the water from leaving the concrete, thus extending hydration. In some cases plastic sheathing can be placed over the surface to retard the loss of moisture.

8-11 Steel reinforcing bars are placed inside the forms. *(Courtesy Corps of Engineers, U.S. Department of the Army)*

Reinforcing Concrete

Concrete is very strong in compression and weak in tension **(see 8-10)**. This is why it is so widely used for foundation construction. When used to form beams or unsupported slabs, such as floors or roofs, it requires steel reinforcing to resist the forces of tension. Steel is excellent in its resistance to tension.

Concrete members are reinforced with steel reinforcing bars, welded wire fabric, and fibers mixed in the batch.

Reinforcing bars (also called rebars) are available smooth or with ridges on the surface. Those with ridges are called deformed bars. Deformed

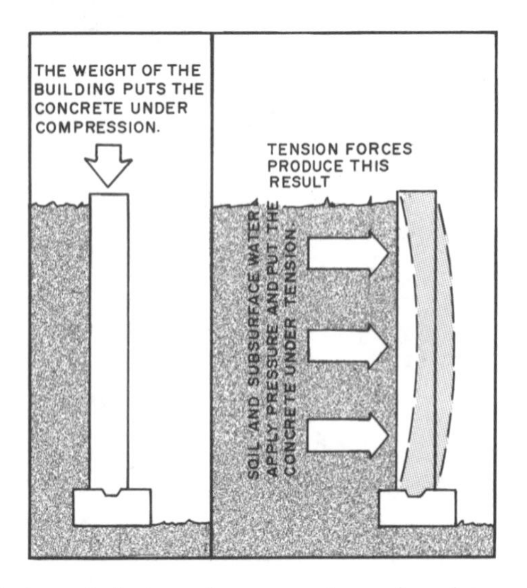

8-10 Concrete is strong in compression and weak in tension.

Table 8-3 Sizes of steel reinforcing bars

Inch-Size Steel Reinforcing Bars

Bar Size Designation	Cross-sectional Diameter (inches)
#3	0.375
#4	0.500
#5	0.625
#6	0.750
#7	0.875
#8	1.000
#9	1.128
#10	1.270
#11	1.410
#14	1.693
#18	2.257

Metric-Size Steel Reinforcing Bars

Bar Size Designation	Cross-sectional Diameter (millimeters)
#10M	11.3
#15M	16.0
#20M	19.5
#25M	25.2
#30M	29.9
#35M	35.7
#45M	43.7
#55M	56.4

bars are used for most reinforcing applications. Smooth bars are used in special situations.

The rebars are placed inside the forms and tied to each other and to the forms. The concrete is poured around them **(see 8-11)**.

Rebars are identified by number with Nos. 3 through 18 being most commonly used. The number indicates the number of ⅛ inches. For example the No. 3 rebar is ⅜ inches. Metric rebars are also available and are designated in millimeters. They are available with diameters from 10M through 55M. A 10M bar has a diameter of 10mm (actual size 11.3mm). **Table 8-3** shows the standard inch and metric sizes of reinforcing bars.

Welded wire reinforcement (WWR) (sometimes called welded wire fabric) is made of a wire mesh much like wire fencing **(see 8-12)**. The wires are welded together forming a rectangular pattern. It is used in flat slabs to reduce the amount of surface cracks and in some cases to control tension forces. It comes in sheets and rolls. It is specified by the wire diameter and the spacing of the wires. Various strength fabrics are available.

The style designation indicates the spacing of the wires and the wire diameter. For example, a 4 × 4-W1.4 × W1.4 style means the longitudinal and transverse wires are four inches apart and the wires are plain with a mean wire area of 0.014 square inches per foot. Fiber reinforcement involves mixing various fibers in the mix to add reinforcing properties. Fibers commonly used include glass fibers, polypropylene, polyethylene, polyester, acrylic, aramid, steel, asbestos, carbon, and various natural fibers such as wood, sisal, coconut, bamboo, jute, okwara, and elephant grass. Fibers are often added in addition to rebars or welded wire fabric.

The selection of the type and amount of reinforcement is the work of engineers and concrete specialists.

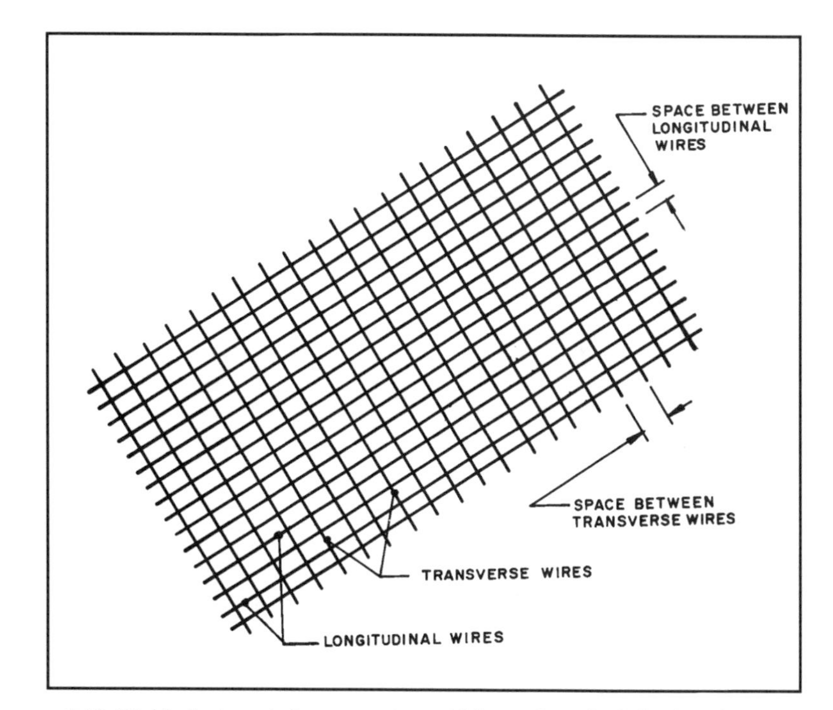

8-12 Welded wire reinforcement is a widely used steel reinforcing that is most often used in concrete slab construction.

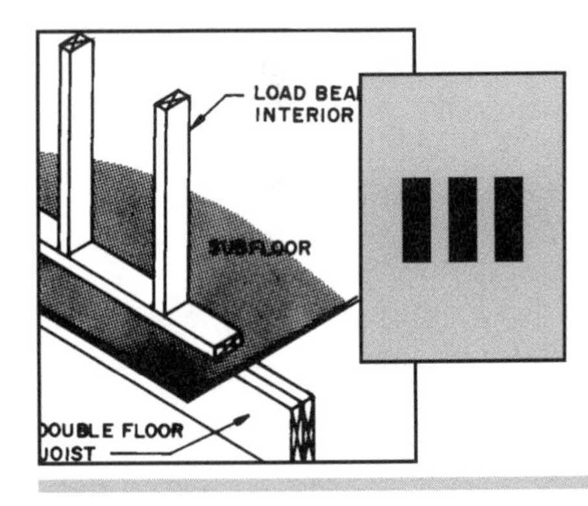

III FLOOR, WALL & CEILING FRAMING

CHAPTER 9

FLOOR FRAMING

Floor framing includes setting columns, beams, sills, headers, and joists and applying the subfloor. These framing elements form a strong, solid base on which the interior and exterior walls rest, and they carry other loads such as cabinets, stairs, appliances, bath fixtures, water heaters, furnaces, and eventually furniture. The floor also forms a diaphragm to resist the lateral pressure from the earth and subsurface water that is pressing on the foundation wall, tending to push it inward. The basic parts of a floor-framing system using solid wood joists and header are shown in **9-1**.

In the construction of a basement, steel columns are most commonly used, although wood posts can also provide adequate support. Beams may be made of steel or solid wood, they may be carpenter-built from two-inch lumber, or they may be some form of glued, laminated member. Second-floor loads are carried to the foundation through the exterior walls as well as through load-bearing walls on the interior of the building. These load-bearing walls rest on the floor structure and they require proper support below the floor.

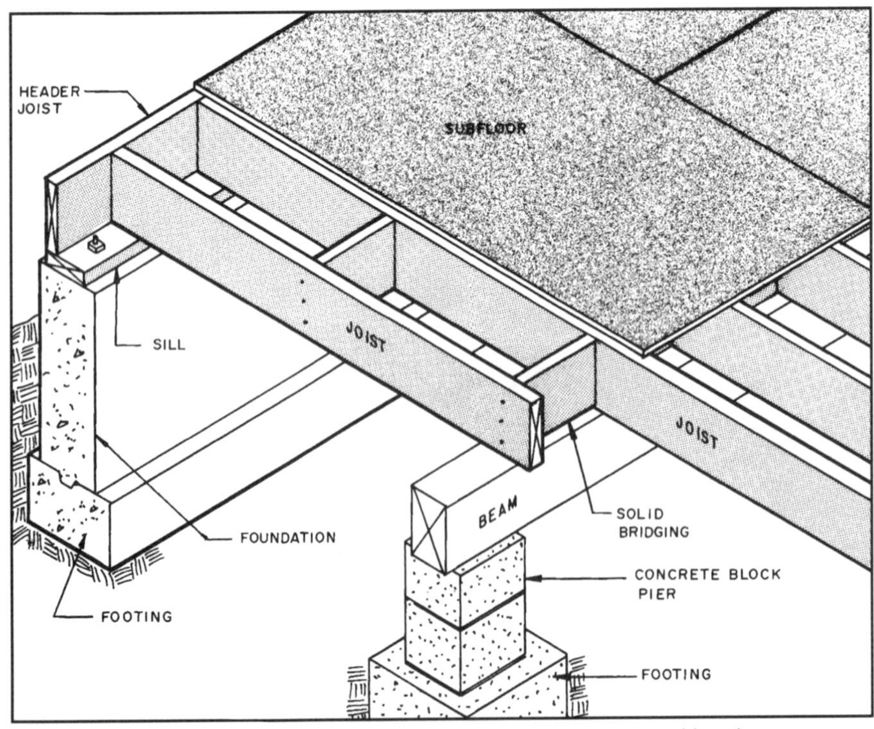

9-1 Parts of a typical wood framed floor using solid wood joists and headers.

FLOOR BEAMS

Usually the span from one foundation wall to the other is so long that floor joists cannot reach and carry the expected loads. Beams are installed in the foundation to shorten the distance joists must span. Confusion often exists when using the terms beam and girder. A **beam** is a straight structural member used to support transverse loads such as floor joists. A **girder** is a straight structural member used to support beams.

Beams in light construction could be a standard or wide-flange steel beam, solid wood, built-up wood, or some form of laminated wood.

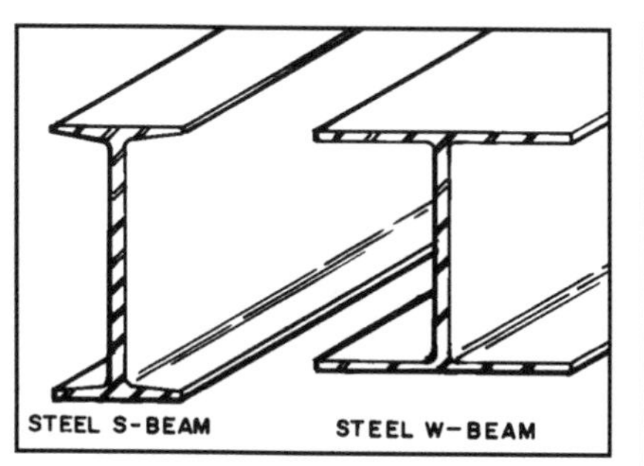

9-2 Steel S-beams (standard) and W-beams (wide-flange) beams are widely used in wood framed floor construction.

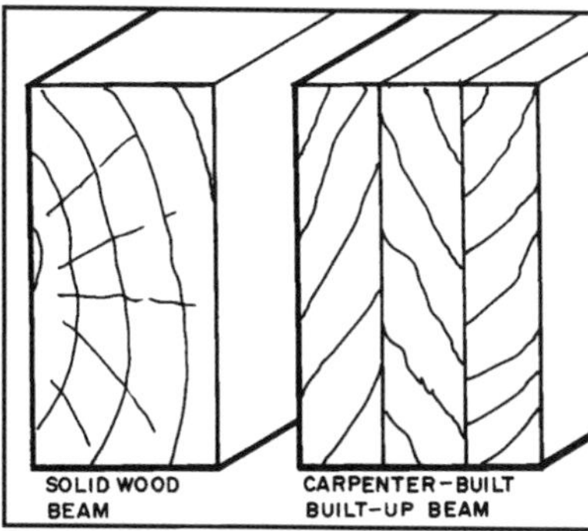

9-3 Solid wood and wood carpenter-built beams used in wood framed floor construction.

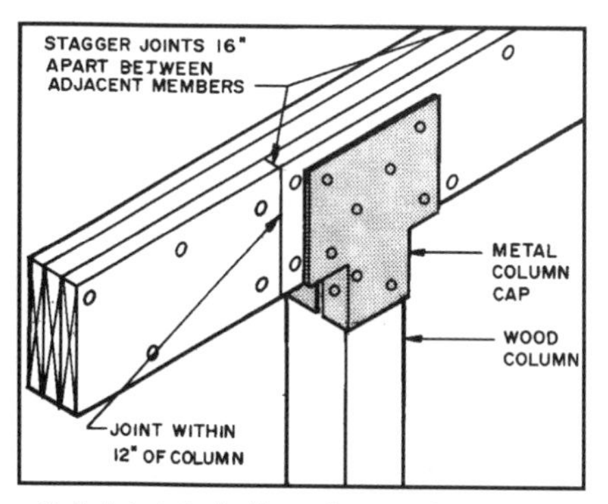

9-4 Joints in built-up beams should be over a column or post or within 12 inches of it, and the joints should be staggered.

STEEL BEAMS

Steel beams are frequently used because they can span long distances using members that are smaller in size than wood beams. This often helps give additional headroom in basements.

The two types used are standard (S) and wide-flange (W) **(see 9-2)**. The wide flange has a wider surface on which to rest the floor joists. Typical load-carrying capacities for two sizes of wide-flange beams are given in **Table 9-1**. The size of the beam will be determined by the architect and will be found on the drawing. If the framing carpenter installs the beam, it will be necessary to check and make certain it is as specified.

SOLID & BUILT-UP WOOD BEAMS

Wood beams can be solid wood members or made from assembled two-inch lumber, producing what is called a built-up beam. These are usually assembled by the carpenter **(see 9-3)**. The stock should be of high quality. The joints should be staggered and if possible occur over piers or columns or at least not more than 12 inches from a pier or column **(see 9-4)**. Solid wood beams are seldom used because they are difficult to secure in the large sizes required and are not seasoned in the interior cross section. They tend to check and crack as they dry after installation.

Allowable spans for built-up beams are found in Appendices H and R . The lumber must always be dry, and the lumber used must be the grade indicated in the table.

Built-up beams are often glued and nailed. The nails should follow the nailing patterns shown in **9-5**. Two-piece beams are joined with 10d nails, all driven from one side. Two nails are driven at each end. Three-piece beams are nailed from both sides using 20d nails. Two nails are driven at each end.

Span data for solid wood beams are found in Appendix R.

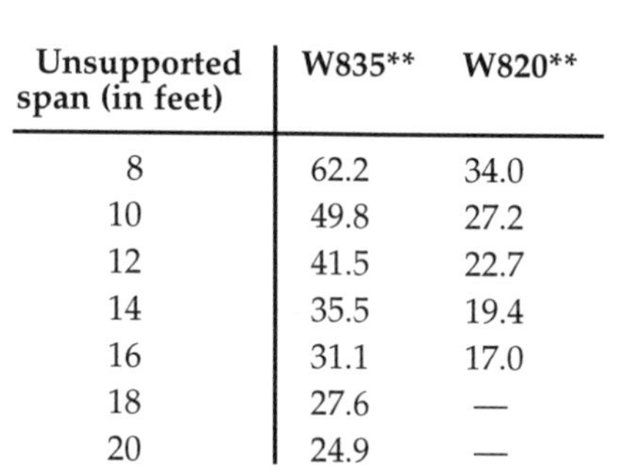

Table 9-1 Wide flange steel beam allowable loads in kips.*

Unsupported span (in feet)	W835**	W820**
8	62.2	34.0
10	49.8	27.2
12	41.5	22.7
14	35.5	19.4
16	31.1	17.0
18	27.6	—
20	24.9	—

* A kip is 1000 pounds.

** W means wide flange beam, 8 means it is 8 inches high, and 35 means it weighs 35 pounds per lineal foot.

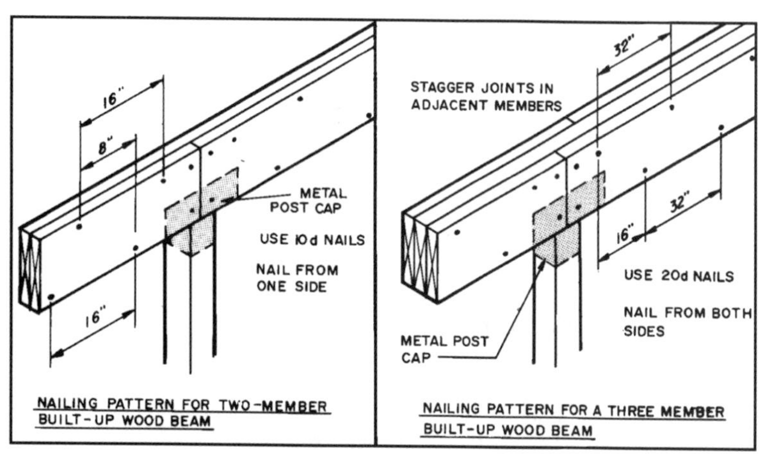

9-5 Recommended nailing patterns for built-up beams.

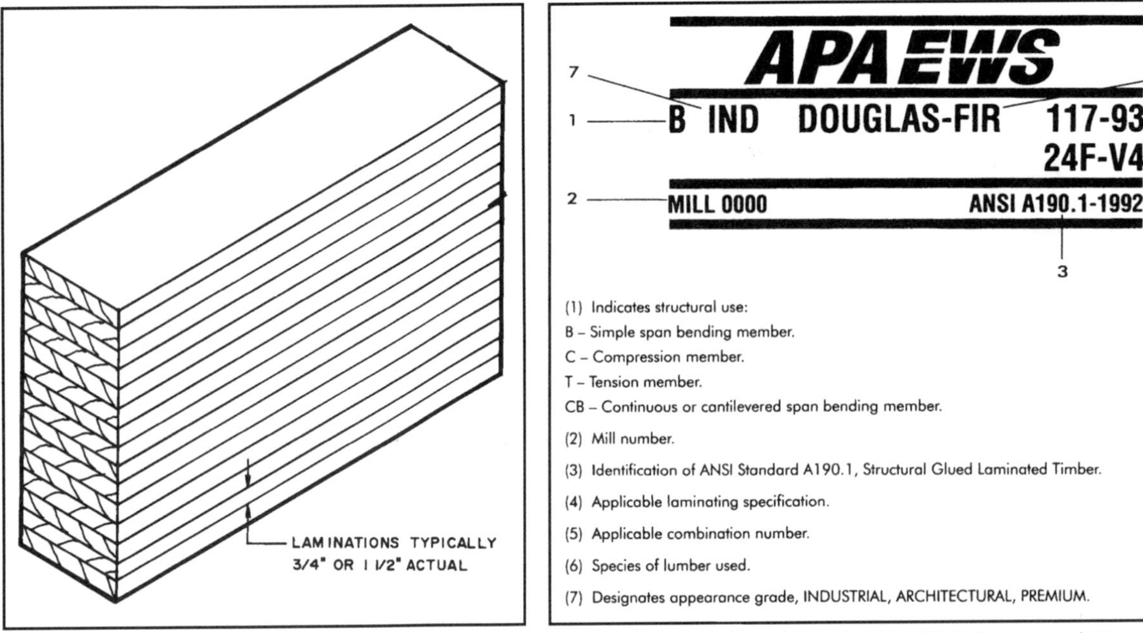

9-6 Glulam beams are made by bonding solid wood laminations.

9-7 A typical trademark for the APA-EWS (Engineered Wood Systems) products. *(Courtesy APA—The Engineered Wood Association)*

Building codes specify that beams should be of sufficient size to carry the dead and live loads without exceeding the allowable working stresses specified for various wood grades and species indicated in the code. When sizing beams consult the local codes.

GLUED LAMINATED BEAMS

Glued laminated beams are referred to as glulams. They are made by gluing, under carefully controlled conditions, high-quality wood strips that have defects cut away. The strips range from about ¾ to 1½ inches in thickness and are kiln-dried. This produces a high-quality beam that will carry heavy loads over long distances **(see 9-6)**.

9-8 These beams are made from parallel strand lumber.

These beams are more dimensionally stable than solid wood beams and carry more weight than the other types of wood beams over the same span. Typical span data for selected sizes are found in Appendices I and P. Consult manufacturers for specific data.

The American Plywood Association has expanded to incorporate this family of wood products including glulams. American Wood Systems was created by the APA to serve the needs of the engineered wood systems industry. A new trademark has been created, APA-EWS (Engineered Wood Systems). The APA-EWS trademark appears on glued laminated beams and may be used on prefabricated wood I-beams, structural composite lumber, and other engineered wood products **(see 9-7)**. Products bearing the APA-EWS trademark are supported with the same technical and promotional services APA provides to manufacturers of structural wood products.

PARALLEL STRAND LUMBER

Parallel strand lumber is used for beams, headers, columns, and posts. It is made from Douglas fir and southern pine **(see 9-8)**. The logs are peeled into veneer sheets 1⁄10 and ⅛ of an inch in thickness. These sheets are clipped into strands up to 8 feet in length. Defects are removed, and the strands are coated with a waterproof adhesive.

The oriented strands are fed into a rotary belt press and cured using microwave energy. This produces a billet that is cut into standard sizes up to 66 feet in length. One such product is marketed under the name Parallam®. It has been extensively tested for fire resistance, fastener-holding, moisture response, long-term loading, flexural response, stiffness, and internal bond. It has exceeded the performance of competing wood products. It can be sawed, drilled, and nailed like solid wood. It has been accepted by major building codes in the United States and Canada **(see Table 9-2)**.

Table 9-2 Parallam® beams are made in these sizes.

Thickness (inches)	Depth (inches)
1¾	7¼
3½	9¼
5¼	9½
7	11¼
	11½
	11⅞
	12
	12½
	14
	16
	18

9-9 Laminated veneer lumber (LVL) is made by bonding layers of wood veneer so that the grain of the veneers runs parallel to the axis of the member.

LAMINATED VENEER LUMBER

Laminated veneer lumber (LVL) is another form of manufactured structural member. It is a layered composite of wood veneers, usually southern pine, bonded with an exterior adhesive to form a structural member that has virtually no shrinkage, checking, twisting, or splitting **(see 9-9)**. It has more load-bearing capacity per pound than solid sawed lumber. It has a high allowable bending stress (3100 psi) and a modulus of elasticity (E) of 2.0. It is made in many depths and long lengths **(see 9-10)**. Typical depths range from 9¼ to 18 inches, and thicknesses are 1½ and 1¾ inches. Design data on spans are available from manufacturers.

9-10 Laminated veneer lumber is made in many depths and lengths. *(Courtesy Boise Cascade Corporation)*

STRUCTURAL-USE PLYWOOD-LUMBER BEAMS (BOX BEAMS)

Plywood-lumber beams (also called box beams) are made by combining lumber and structural panels in various combinations. They are made following specific design requirements as prescribed by APA—The Engineered Wood Association. They may be assembled by nailing, gluing, or gluing and nailing.

Glued beams require the lumber flanges to be glued together with rigid structural adhesive and the plywood webs to be glued to the lumber flanges under positive mechanical pressure, which is not practically achievable outside of a factory **(see 9-11)**.

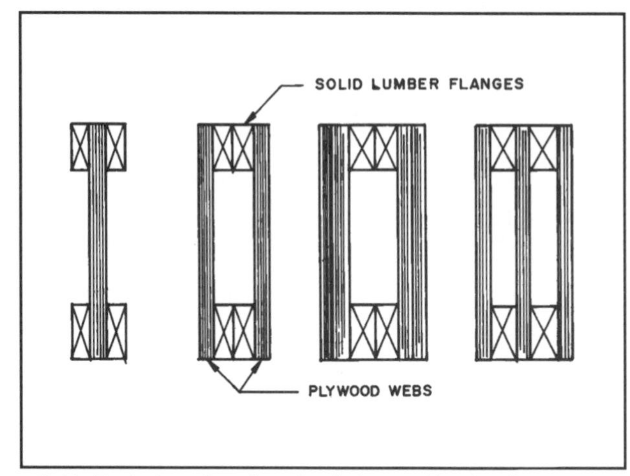

9-11 Typical sections for glued plywood-lumber beams. *(Courtesy APA—The Engineered Wood Association)*

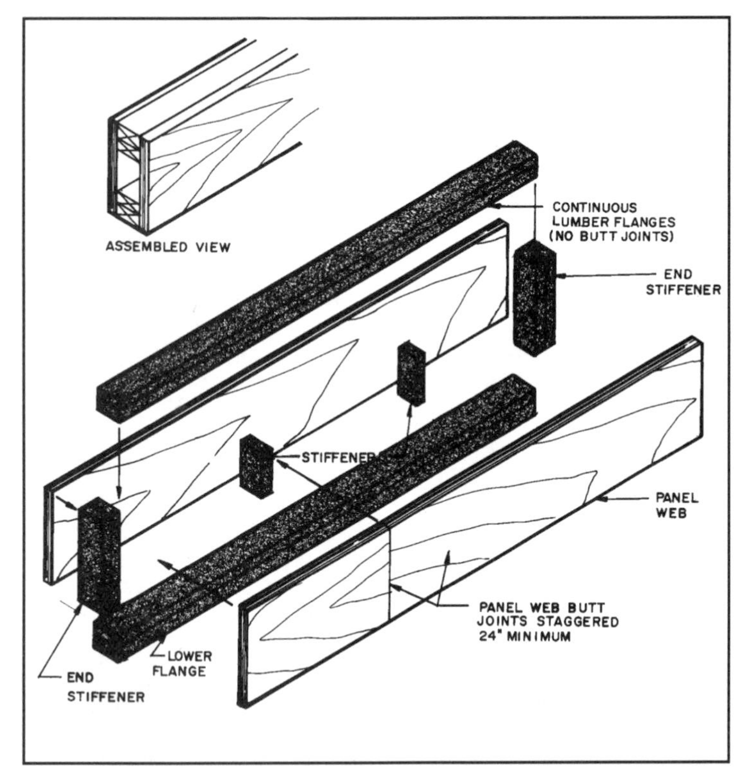

9-12 Plywood-lumber beams may be constructed by nailing an assembly of solid wood flanges and APA-rated panel products. *(Courtesy APA—The Engineered Wood Association)*

Nailed plywood-lumber beams can be constructed on the job site and glue may be added to provide additional stiffness. A typical use is a header over a garage door opening. Nailed plywood-lumber beams use 2 × 4-inch lumber flanges and APA-rated sheathing—Exposure 1, non-veneer Com-Ply®, or four- or five-ply plywood. These panels are all made with exterior adhesive **(see 9-12)**. Typical allowable loads for selected sizes of nailed plywood-lumber beams are given in **9-13**. These designs are based on the use of No. 1 Douglas fir or southern pine kiln-dried lumber. Reduce loads by 19 percent for No. 2 Douglas fir, or No. 2 KD15 southern pine.

To select a nailed plywood-lumber beam:

1. Figure the load on the beam. For this example assume a load of 260 pounds per lineal foot on an 18-foot beam.
2. Select the beam design. Examine the load design data in **9-13**. The 18-foot-length column shows that a 16-inch

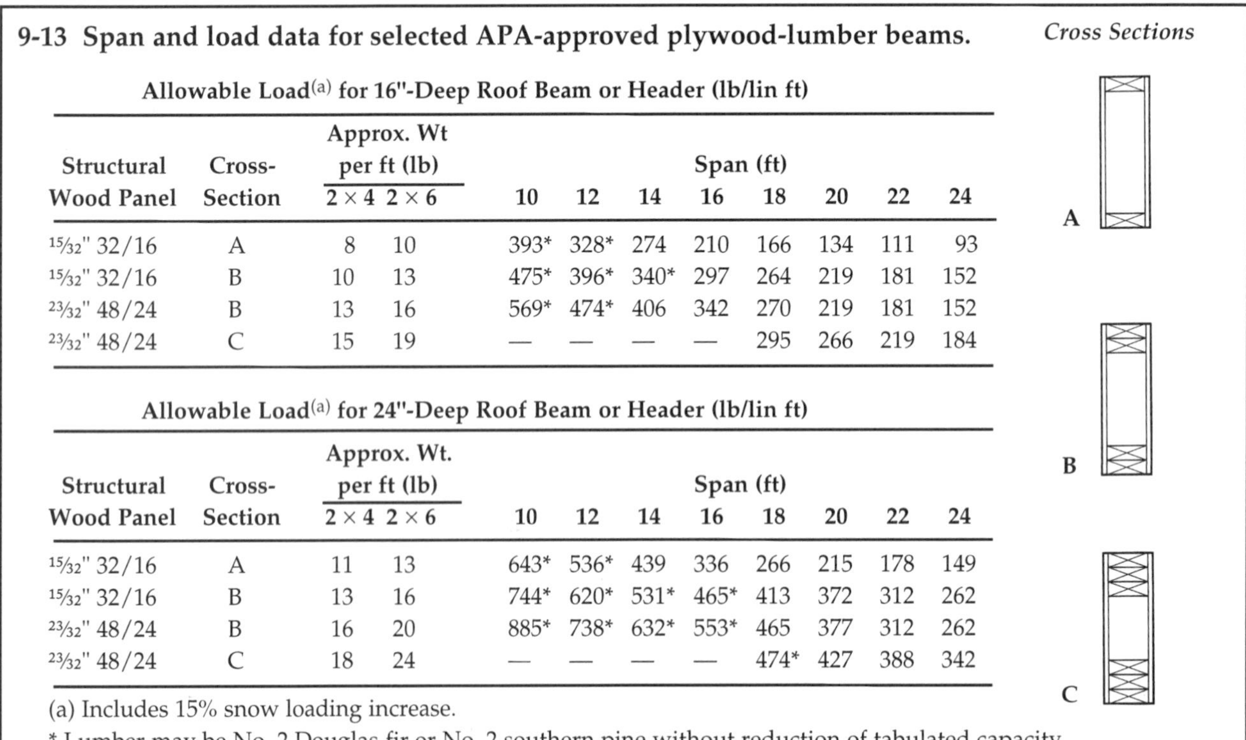

9-13 Span and load data for selected APA-approved plywood-lumber beams.

Cross Sections

Allowable Load(a) for 16"-Deep Roof Beam or Header (lb/lin ft)

Structural Wood Panel	Cross-Section	Approx. Wt per ft (lb) 2 × 4	2 × 6	Span (ft) 10	12	14	16	18	20	22	24
15/32" 32/16	A	8	10	393*	328*	274	210	166	134	111	93
15/32" 32/16	B	10	13	475*	396*	340*	297	264	219	181	152
23/32" 48/24	B	13	16	569*	474*	406	342	270	219	181	152
23/32" 48/24	C	15	19	—	—	—	—	295	266	219	184

Allowable Load(a) for 24"-Deep Roof Beam or Header (lb/lin ft)

Structural Wood Panel	Cross-Section	Approx. Wt. per ft (lb) 2 × 4	2 × 6	Span (ft) 10	12	14	16	18	20	22	24
15/32" 32/16	A	11	13	643*	536*	439	336	266	215	178	149
15/32" 32/16	B	13	16	744*	620*	531*	465*	413	372	312	262
23/32" 48/24	B	16	20	885*	738*	632*	553*	465	377	312	262
23/32" 48/24	C	18	24	—	—	—	—	474*	427	388	342

A

B

C

(a) Includes 15% snow loading increase.
* Lumber may be No. 2 Douglas fir or No. 2 southern pine without reduction of tabulated capacity.
(Courtesy APA—The Engineered Wood Association)

beam with double top and bottom flanges and $^{23}/_{32}$-inch panel webs with Cross Section B will carry 274 pounds per lineal foot. This is the design load. In some cases several designs would be appropriate.

To build this plywood-lumber beam:

1. Select the correct layout of stiffeners as shown in **9-14**. The panel joint locations require a minimum of a two-foot stagger between the panel butt joints on opposite sides of the beam. All vertical joints are located in the center half of the beam. Vertical stiffeners should be added in the layouts so that they are no more than four feet apart. Place vertical stiffeners behind the butt joints between panel webs. Six inches is added to the clear span to allow the double-bearing-end vertical stiffeners to rest on the double-load-carrying studs on each end of the beam.
2. Build the framework of lumber flanges and stiffeners. Use dry lumber having a moisture content of 19 percent or less for Douglas fir and 15 percent or less for southern pine. Lumber should be free of warp or characteristics that produce gaps greater than ⅛ inch between the lumber and the panel. Nail the flanges to the stiffeners with 8d common nails.
3. Fasten the panel webs to the framework with the face grain or strength axis (eight-foot dimension) in the long direction (same direction as the flanges) using 8d common nails. The required nailing pattern is shown in **9-15**.

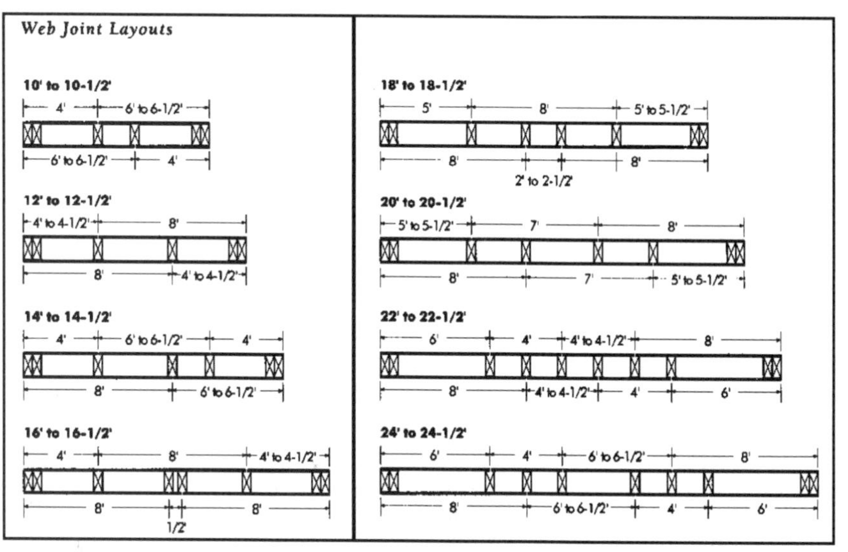

9-14 Recommended web joint layouts and stiffener locations for selected plywood-lumber beams. *(Courtesy APA—The Engineered Wood Association)*

Plywood-lumber beam headers are installed as shown in **9-16**. The beam rests on two 2 × 4-inch trimmer studs on each end and is face-nailed to a full-length stud with 16d common nails six inches O.C. (on center). For additional information contact APA—The Engineered Wood Association.

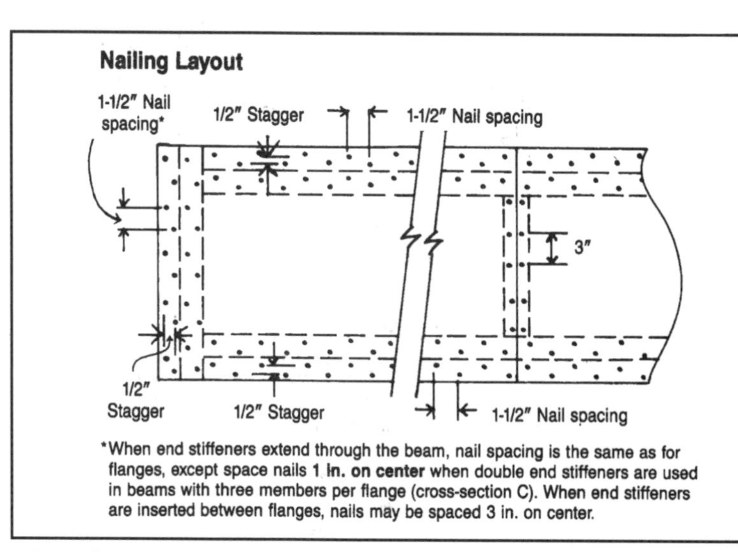

9-15 The required nailing pattern for carpenter-built plywood-lumber beams. *(Courtesy APA—The Engineered Wood Association)*

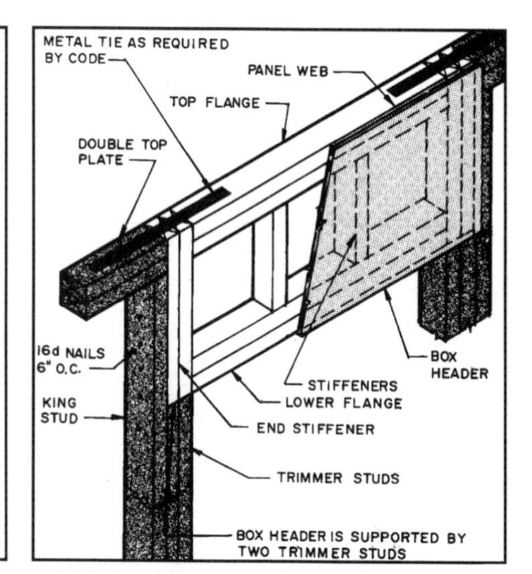

9-16 One way to install panel-lumber headers to the wall studs.

POSTS, COLUMNS & PIERS

Since the spans of beams below the floor are usually greater than the typical sizes of beams used, the beams are supported at intervals with posts, columns, or piers.

Piers

Buildings with crawl spaces usually use piers to support beams. These are usually concrete block with the cores filled with concrete. However, cast concrete piers also find some use **(see 9-17)**. Concrete block piers are usually 8 × 16 inches or 16 × 16 inches depending on the load to be carried. The size of the footing depends upon the load to be placed on the pier. A typical installation is in **9-18**.

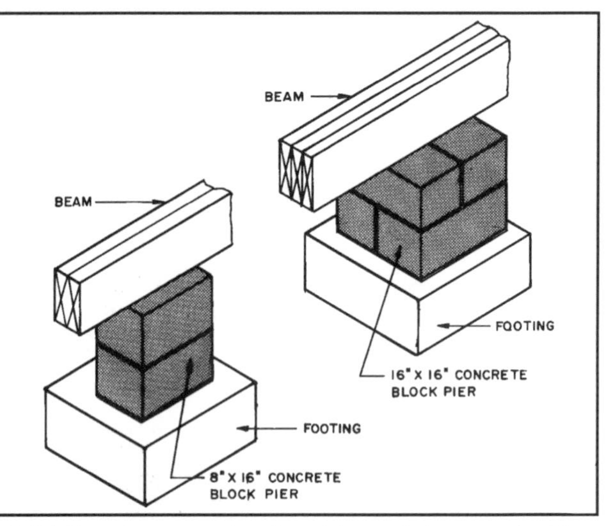

9-17 Floor beams can be given intermediate support with concrete block piers.

Wood Posts

Wood posts must be of high-quality material. They may be solid wood, glued laminated members (glulams), or parallel strand lumber (PSL). Beams up to four inches wide require a 4 × 4-inch or 4 × 6-inch post. Beams up to eight inches wide require a 6 × 8-inch or 8 × 8-inch post.

Wood posts are joined to wood beams and concrete footings with some type of metal connector. Typical connections are shown in **9-19** and **9-20**. The wood post cap connections are either nailed or bolted to the beam. The metal base connections are set in the concrete footing and bolted to the post.

9-18 A typical 16" × 16" concrete pier installation.

Steel Posts (Pipe Columns)

Steel columns are most commonly used to carry beams over basement space. They have a steel plate on top to carry the wood beam. If steel beams are used, they are often welded to the column **(see 9-21)**. Steel columns have a square steel plate on the bottom that may be bolted to the footing **(see 9-22)**.

INSTALLING BEAMS

Beams are set in pockets built in the foundation wall. The end of the beam should rest at least four inches directly on the foundation. Wood beam pockets should be large enough to allow a ½-inch air space around it on all sides. This helps provide circulation to keep the wood dry **(see 9-23)**. A damp

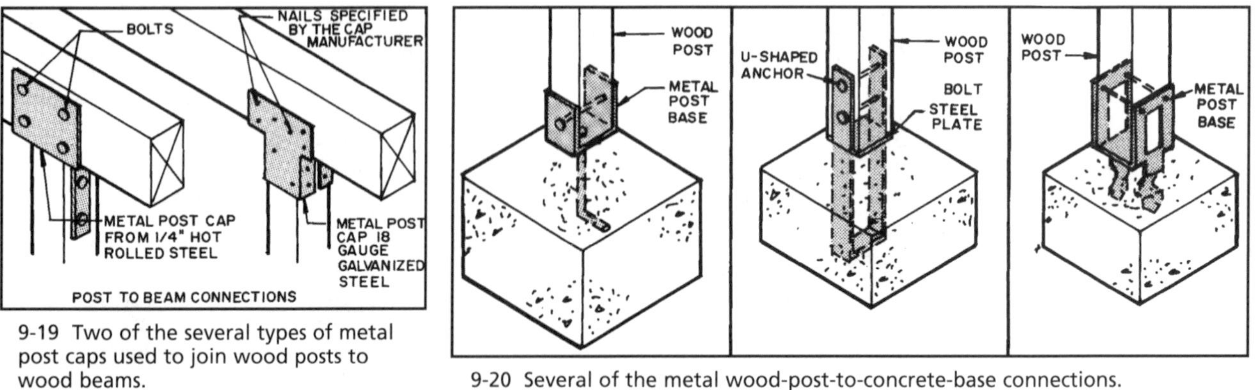

9-19 Two of the several types of metal post caps used to join wood posts to wood beams.

9-20 Several of the metal wood-post-to-concrete-base connections.

proofing material is placed in the pocket under the wood beam. Steel beams are set in the same way, except that the depth of the pocket must allow for the 2 × 4-inch or 2 × 6-inch wood member fastened to the top of the beam. The top of the beam plus the wood plate must be level with the top of the sill plate **(see 9-24)**.

If it is not possible to get adequate support for a beam on the foundation wall, a pilaster can be built as part of the foundation. While it can be any size required, a 4 × 12-inch pilaster is commonly used with concrete block construction.

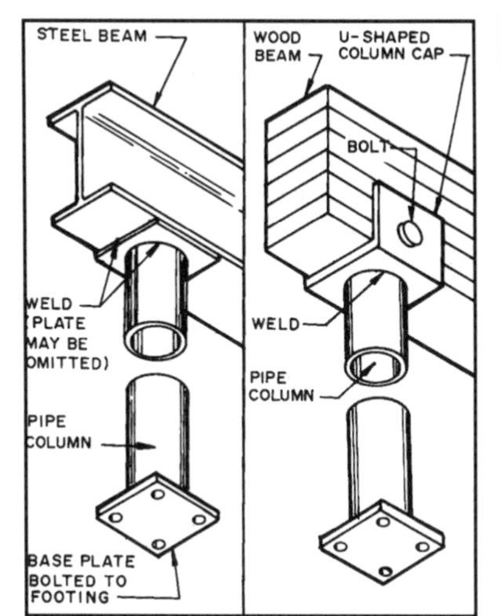

9-21 Metal posts are used to support steel and wood beams.

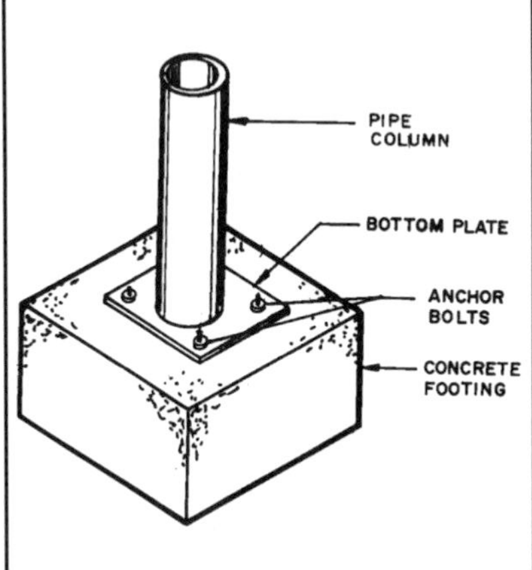

9-22 Steel columns are bolted to the concrete foundation.

THE SILL PLATE

The sill plate (often called a mudsill) is a pressure-treated wood member, usually 2 × 6-inches, that is bolted or strapped to the foundation. If bolts are used they should be ½ inch in diameter, embedded seven inches into the concrete or 15 inches into reinforced grouted masonry and spaced not more than 6 feet apart. Bolts should be located within 12 inches of the end of each piece of sill lumber, and each piece should have at least two bolts **(see 9-25)**.

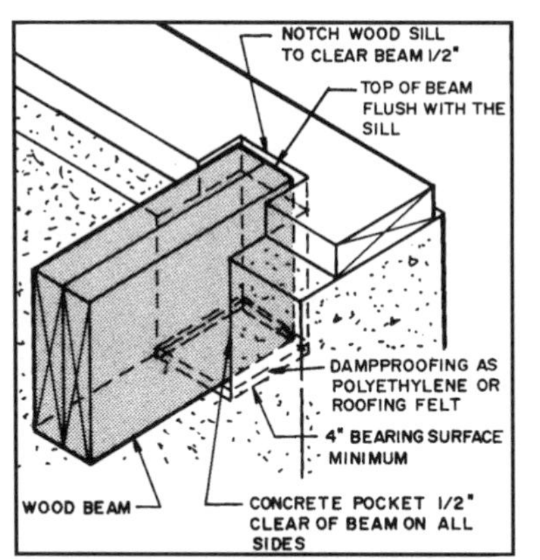

9-23 Wood beams are set in pockets in the masonry foundation wall with their tops above the foundation so that they are flush with the sill.

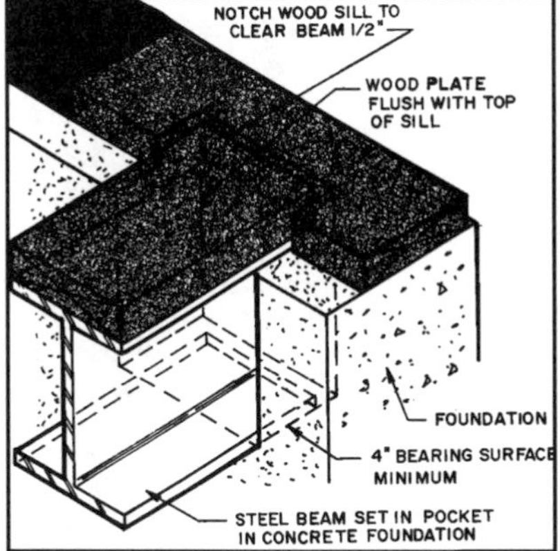

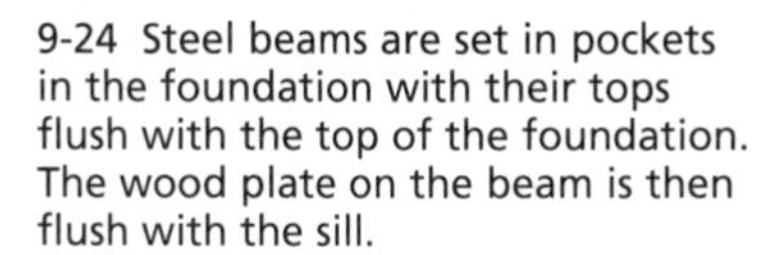

9-24 Steel beams are set in pockets in the foundation with their tops flush with the top of the foundation. The wood plate on the beam is then flush with the sill.

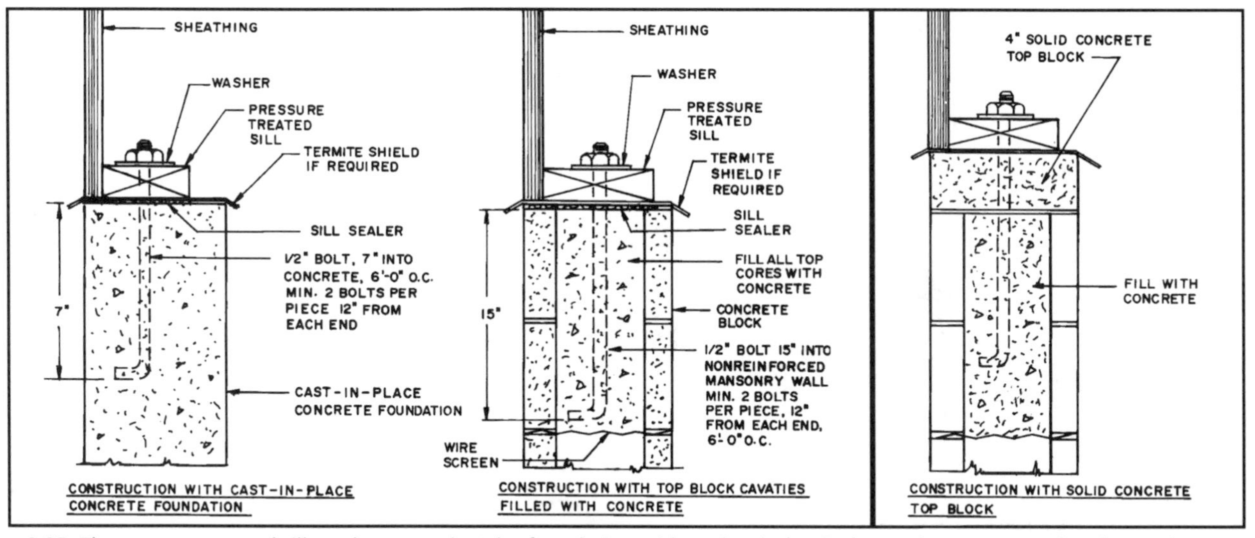

9-25 The pressure-treated sill can be secured to the foundation with anchor bolts. Code requirements must be observed.

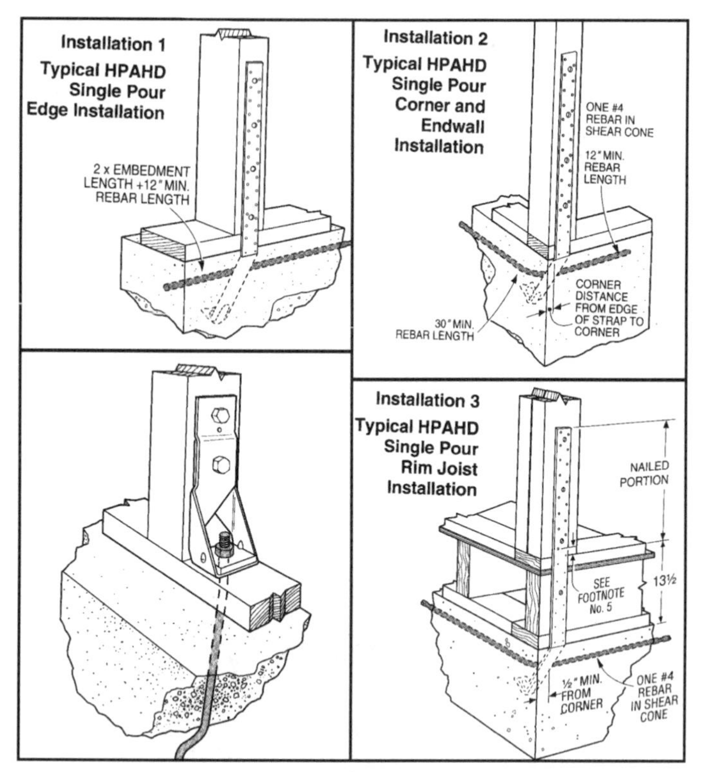

9-27 Strap anchors are used to hold the sill to the foundation or concrete floor. *(Courtesy Simpson Strong-Tie Company)*

The sill plate can be anchored with metal straps that are set into the concrete **(see 9-26)**. They are spaced six feet apart. They are bent around the wood plate and nailed to it. If used with concrete blocks, they must extend two blocks into the foundation. In some areas codes approve the use of power-driven fasteners which penetrate the concrete foundation.

Some codes permit the sill plate to be omitted. The floor joists then rest directly on top of the cast-in-place concrete foundation or concrete floor, or on the webs of the top course of concrete blocks that have the cores filled with concrete. Metal anchors tie the stud wall to the foundation **(see 9-27)**.

Installing the Sill Plate

1. Lay the sill plate on top of the foundation. It is best to use pieces 14 to 16 feet or longer. Position plates so that no joints occur over an opening, such as a vent, in the foundation. They should meet with a square butt joint.

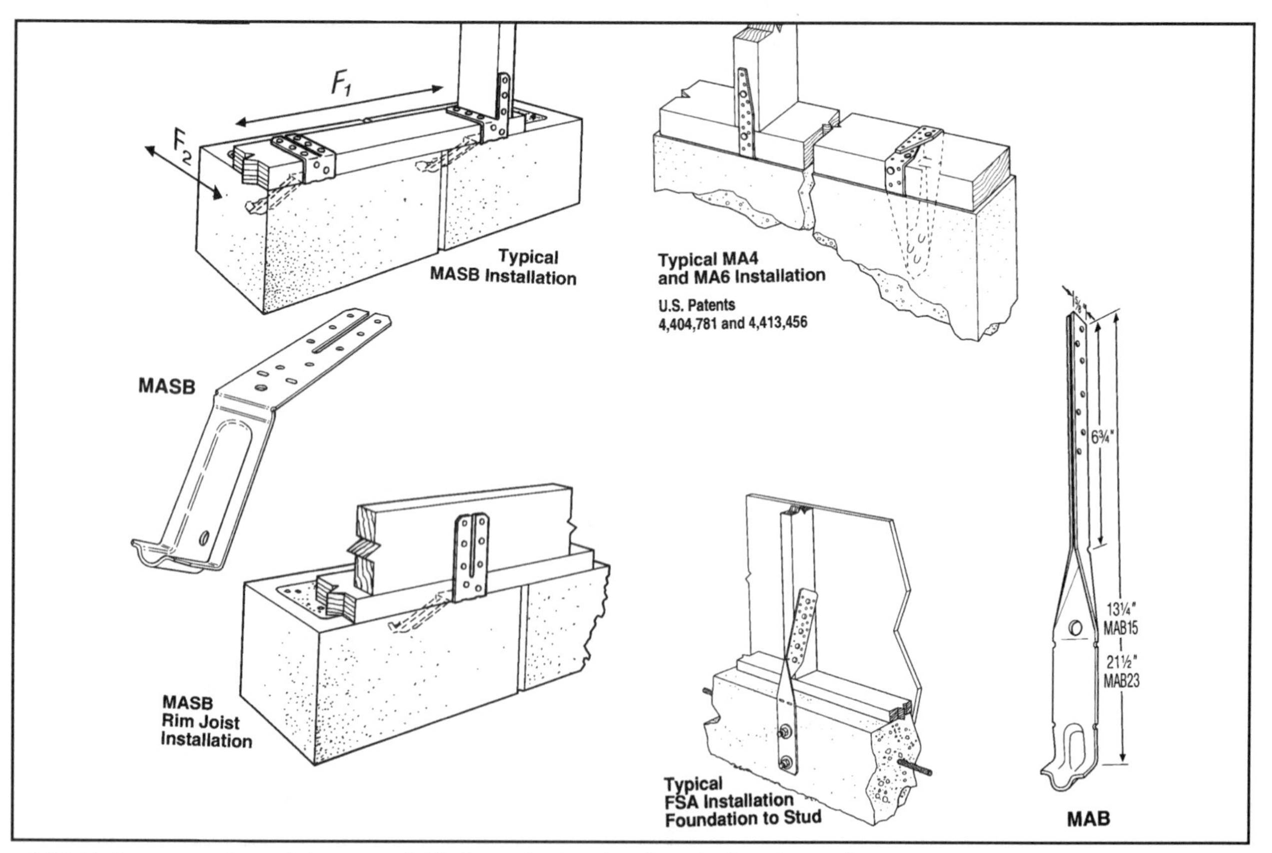

9-26 Various types of metal ties are available for anchoring the sill to the foundation. *(Courtesy Simpson Strong-Tie Company)*

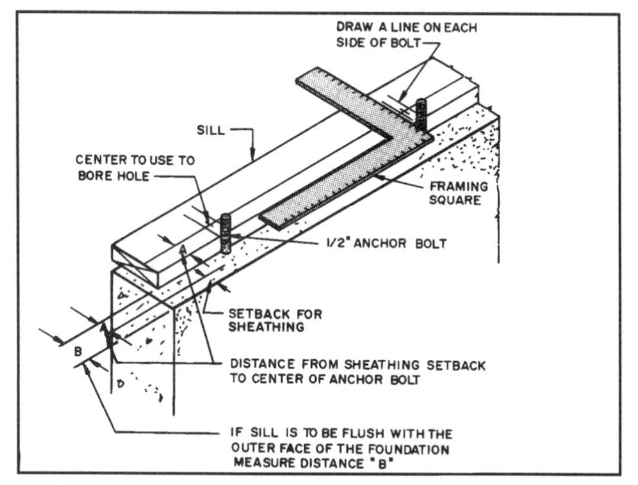

9-28 To locate the anchor bolt on the sill, lay the sill next to the bolts and mark their location with a square.

2. Draw square lines on the plate from each bolt **(see 9-28)**. Measure in from the outside edge of the foundation to the center of the bolt, and mark this. This locates the spot to bore the bolt hole. Examine the construction drawing to see if an additional setback allowance must be made for sheathing. Some have the sheathing set flush with the foundation **(see 9-29)**. Others prefer to let it overlap. Sill plate construction on concrete slab floors is shown in **9-30**.
3. Bore the bolt holes a little larger than ½ inch so that the sill can be moved a little, permitting it to be set square with abutting pieces.
4. If a termite shield and sill sealer are to be used, place these over the bolts. A termite shield is a metal piece that extends beyond the faces of the foundation. A sill sealer is a thin layer of insulation that seals the air spaces that occur between the sill plate and the foundation. Some prefer to caulk this joint.

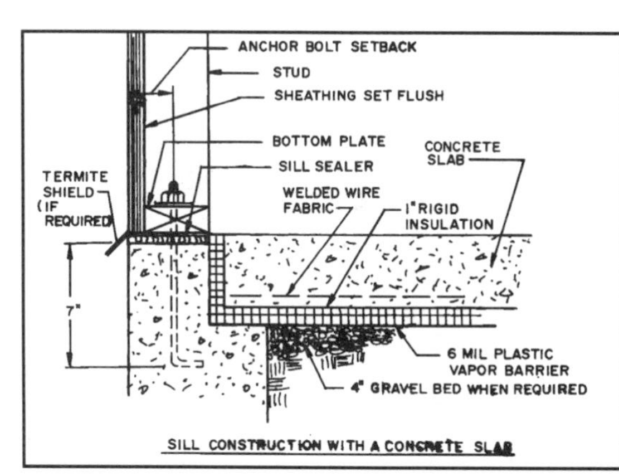

9-30 Typical sill plate construction for a building with a concrete slab floor.

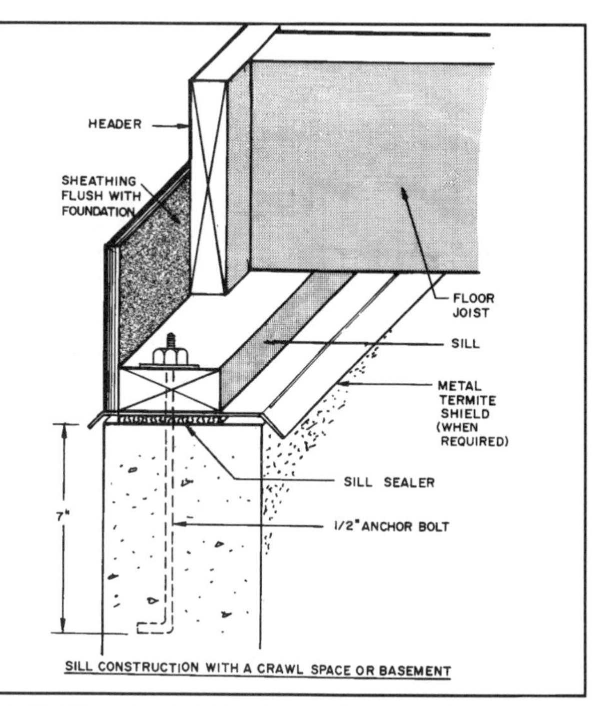

9-29 When the sheathing is to be flush with the face of the foundation, the sill is set in a distance equal to the thickness of the sheathing.

5. Place the sill plate over the bolts, put a washer on, and finger-tighten the bolt. Check to be certain the corners are square and adjust as necessary **(see 9-31)**.
6. Place a level on the sill plate. If it is not level, place metal shims between it and the foundation, and adjust until it is level. If the top of the foundation is very irregular, trowel a thin coat of mortar over it, and level the sill in the fresh mortar. Finger-tighten the bolts to hold the sill in place as the mortar sets.
7. Tighten the bolts with a wrench just enough to firm up the connection. Too much tightening might pull the bolt out of the concrete.

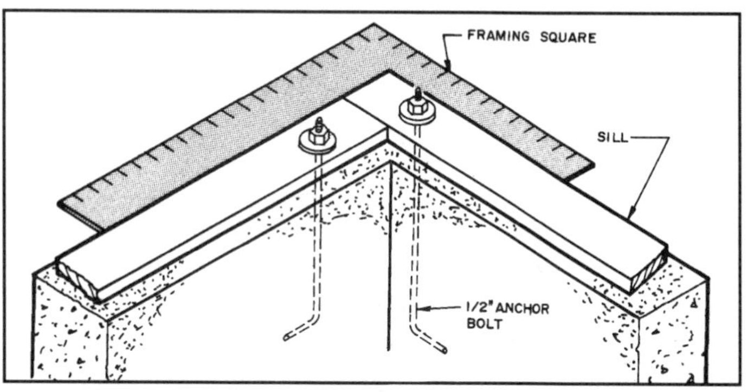

9-31 Use a framing square to check the sill for squareness.

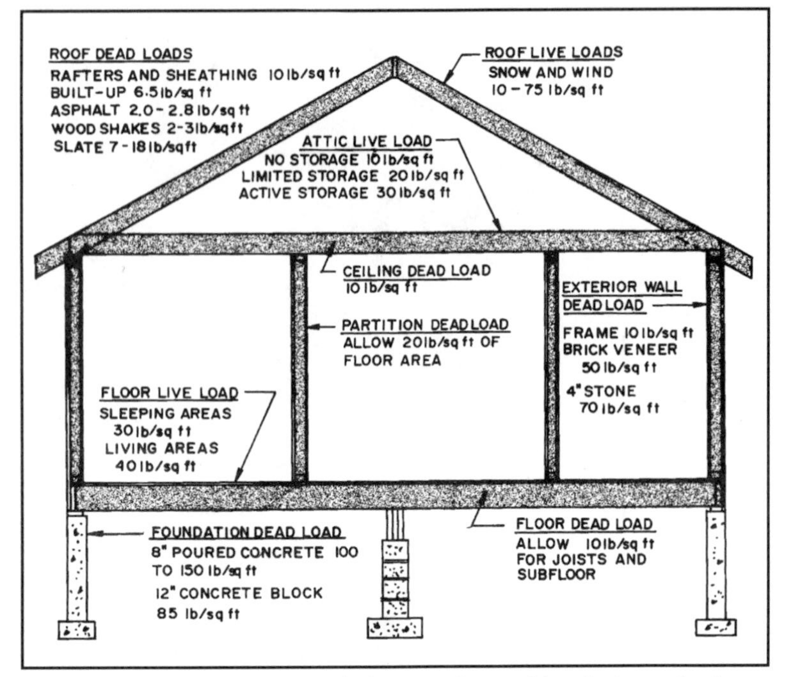

9-32 Typical live and dead loads that must be considered when selecting structural members for wood light frame construction.

FLOOR JOISTS

Floor joists must span the required distances and carry the design loads. For typical residential work joists must carry a total weight of around 100 pounds per square foot. This includes the weight of the floor materials, live load (furniture, people), partitions, ceiling materials, and attic storage **(see 9-32)**. These specifications will vary some with the actual design, but are typical. Generally, bedroom areas are figured to have a live load of 30 pounds per square foot, and living areas 40 pounds per square foot.

Joists must have sufficient strength to carry the design loads. In addition, stiffness must be considered. Sometimes designers choose a joist a size larger than required for strength, to get a firmer floor. Joists are generally two-inch-thick solid lumber or I-joists. Some types of wood floor truss joists are also used.

Solid Wood Floor Joists

The actual size of solid wood joist chosen depends therefore upon the loading, span, spacing between joists, and the grade and species of lumber used. The most common spacings used are 16 and 24 inches center to center. Span data were developed by the *In-Grade Testing Program* discussed in Chapter 39.

Tables giving allowable spans for lumber manufactured in the United States are found in Appendices J through O. To use these tables you must know the type shown, the size of the member, the joist spacing, the live and dead loads, and the grade of the lumber. For example, a 2 × 8-inch southern pine floor joist with a grade of No. 1 spaced 16 inches O.C. with a 40-pound live load and a 10-pound dead load will span 13′-1″.

Span data for other species of wood, such as those harvested in the western part of the United States, are available from various wood product associations, such as the Western Wood Products Association.

The Canadian Wood Council has developed extensive design tables for wood structural members in metric and English units. The selected span data found in Appendices F through I have been calculated in metric units for Spruce-Pine-Fir to meet the requirements of buildings covered in Part 9 of the *National Building Code of Canada (NBCC)* 1990. Floor joists are in many cases limited by both **deflection** and **vibration** performance, producing spans slightly less than those in span tables for products used in the United States, which only considers deflection.

Canadian metric lumber span ratings for floor joists vary with the species, whether the subfloor is nailed or nailed and glued, and no matter what

Table 9-3 Typical nail penetration requirements.

Nail size	Box nail penetration (inches)	Common nail penetration (inches)
6d	1⅛	1¼
8d	1¼	1½
10d	1½	1⅝
12d	1½	1⅝
16d	1½	1¾
20d	1⅝	2⅛
30d	1⅝	2¼
40d	1¾	2½

type of bridging and strapping are used. Data found in Appendices F through I give only a limited example.

Metric loads are given in kilopascals (kPa). Metric sizes are given in millimeters. Metric conversion factors are found in Appendix Q. Detailed data is available from the Canadian Wood Council, 1730 St. Laurent Blvd., Ottawa, Ontario, Canada, KIG 5L1.

Canadian lumber manufacturers also produce lumber to match the sizes and grades used in the United States. Span data for their Spruce-Pine-Fir classification are found in Appendices C through F. Refer to Chapter 39 for information on the *In-Grade Testing Program*.

9-33 These I-joists have laminated veneer lumber (LVL) flanges and oriented strand board webs. *(Courtesy Boise Cascade Corporation)*

I-Joists

Another manufactured wood structural member is the I-joist. It is made using an oriented strand board web with flanges of solid kiln-dried lumber or laminated veneer lumber **(see 9-33)**. I-joists are lightweight, need no midspan bridging, and allow greater on-center spacing than solid sawed joists.

They are useful for floor and ceiling joists, cantilever beams in balconies, cathedral ceiling joists, or other high-pitched roof construction. They are available in depths from 11⅞ to 24 inches and are 2¼ inches wide. Detailed design specifications are available from manufacturers.

NAIL SIZES TO USE

The size of nail to be used for structural framing depends on the lateral strength required. This is established by the building codes, which also specify the required penetration. Examples of typical penetration requirements are given in **Table 9-3**. The actual size and number of nails required for various assemblies will vary somewhat in different localities, but those shown in Appendices A and B are typical of those found in various codes. Consult your local codes for specific information.

WOOD JOIST-TO-BEAM CONNECTIONS

Wood framed floors can have the joists overlap at the beam, butt end-to-end, butt the beam and rest on a ledger, or be set flush with the top of the beam. When the joists overlap they should overlap four inches or as specified by local codes. Excessive overlaps should be avoided. The joists are toenailed to the wood beam from both sides **(see 9-34)**. When a steel beam is used it has a wood plate secured to it. It is often bolted to the beam, but other types of connections are also used **(see 9-35)**.

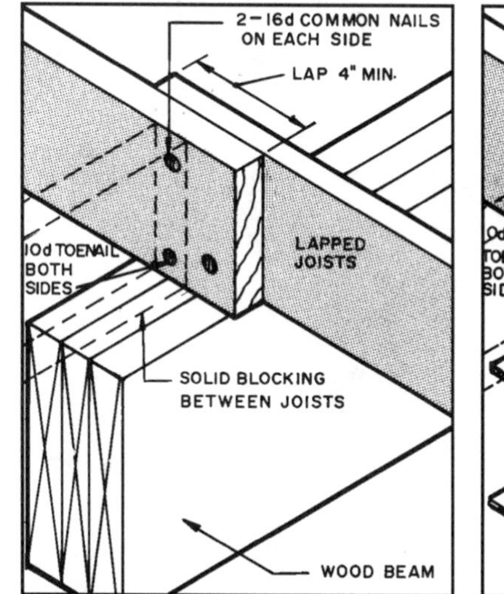

9-34 Floor joists are frequently installed by overlapping them on top of the wood beam. They are toenailed to the beam.

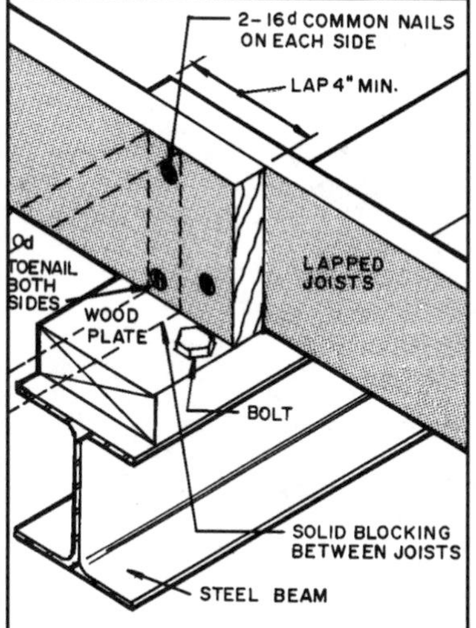

9-35 Overlapped floor joists on steel beams are secured to a wood plate bolted to the beam. They are toenailed to the plate.

If the joists are to rest on top of the beam and butt end-to-end, some form of wood or metal or wood splice plate is used to tie them together as shown in **9-36**. The gusset may be a two-inch-thick solid wood member, ¾-inch plywood, or a manufactured metal gussett plate. Each joist is toenailed to the beam from both sides.

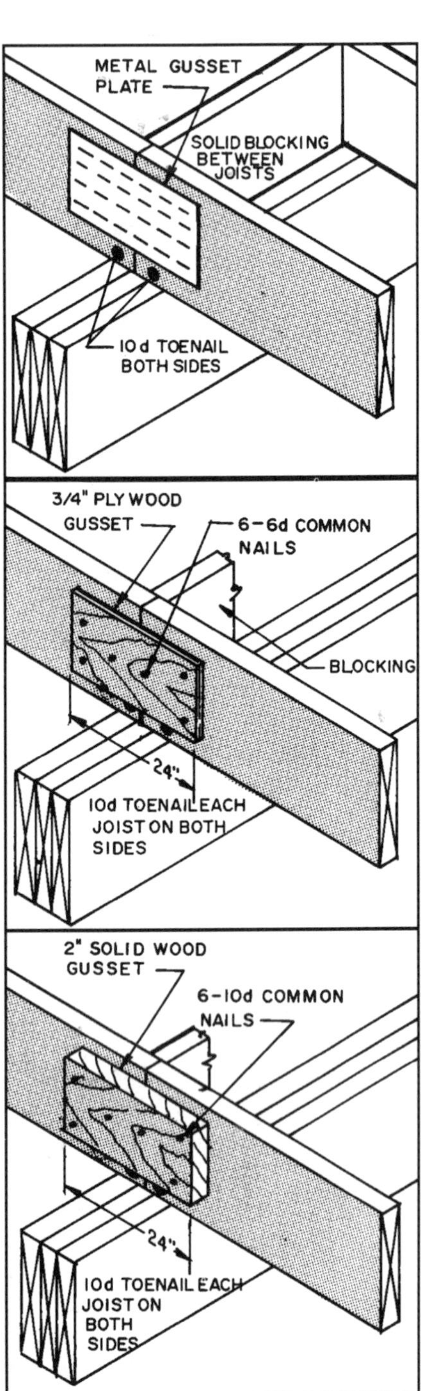

9-36 Floor joists can be installed by butting them on top of the beam. They are tied together with metal or wood splice plates.

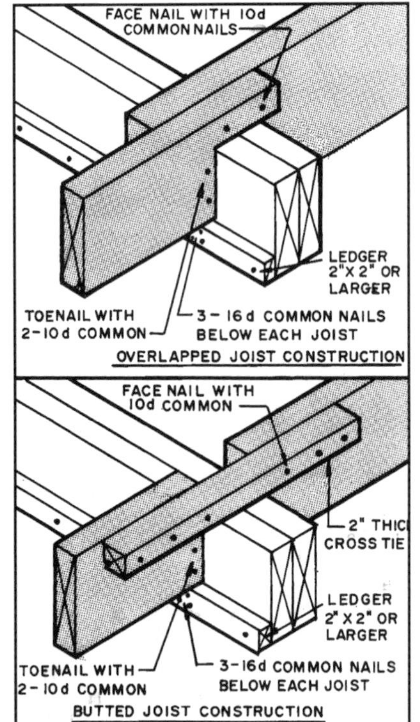

9-37 These joists overlap and are set on a wood ledger.

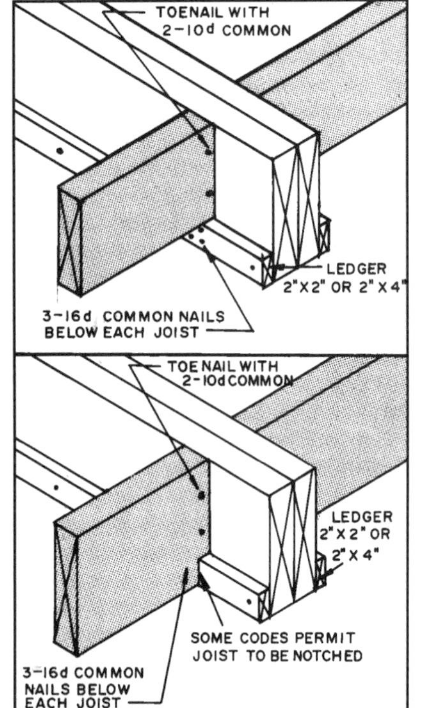

9-38 These joists are in line, butt the beam, and are flush with the top of the beam.

9-39 This view is looking down on a floor assembly having the joists set on a ledger and flush with the top of the wood beam.

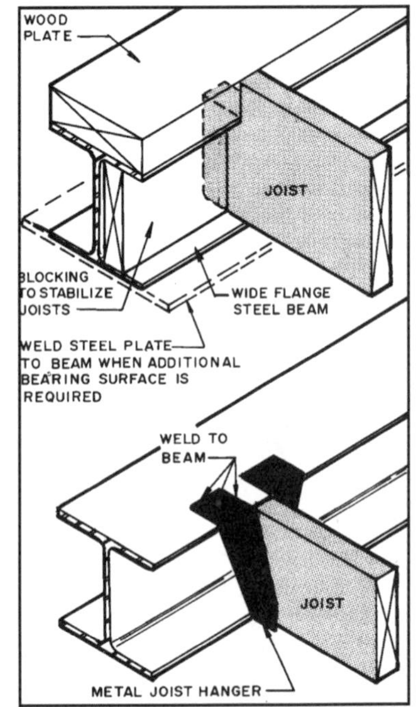

9-40 Two ways to butt wood joists to steel beams.

Joists can butt the beam and overlap and be set on a ledger as shown in **9-37**. Notice that a special nailing pattern is used in the ledger below each joist. If the joists are to be in line and butt a wood beam, they are framed as shown in **9-38** and **9-39**. Be certain to check your local code. Joists can be butted to steel beams and be flush with the wood plate as shown in **9-40**. Notice that some beams do not provide a wide enough flange for the joist to bear upon and a steel plate must be welded to the beam.

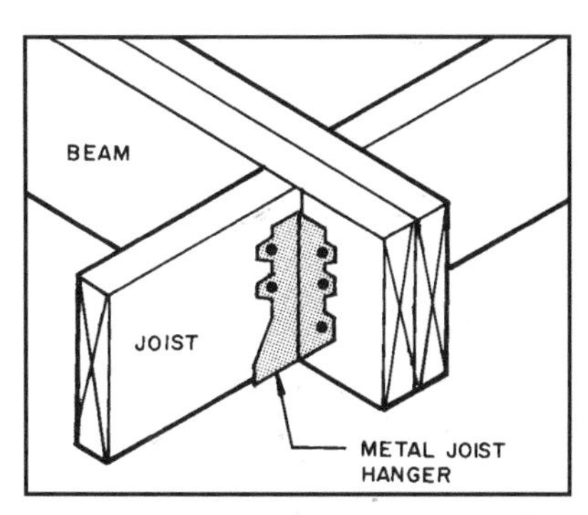

9-41 Metal joist hangers are widely used to butt wood joists to wood beams.

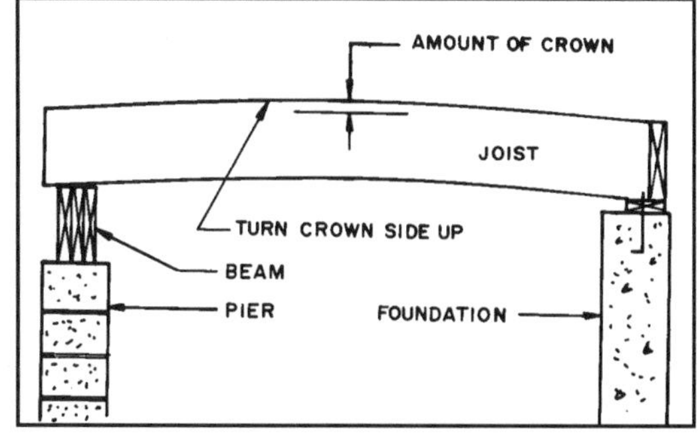

9-42 If a joist has a slight crown place the crown side up. If the crown is severe discard the member.

Another widely used method for butting joists to wood beams is to use metal joist hangers. The hangers are welded to steel beams **(see 9-40)** and nailed to wood beams as shown in **9-41**. It is important to use joist hangers recommended for use with the size floor joists being installed.

One way floors are framed is shown in **9-43**. The joists are lapped on the beam a minimum of four inches. The header joist is nailed to the ends of the joists with three 16d common nails. The header joist and the stringer joist are toenailed to the sill plate with 10d common nails, 16 inches O.C. Joists are toenailed to the beam and sill with two 10d common nails. The joist overlap is nailed with two 16d common nails on each side, at the end of the joist.

INSTALLING SOLID WOOD FLOOR JOISTS

Before any joists are actually installed on the foundation, each of the joists should be inspected for straightness. Those with any twist, cup, or large bow should not be used. If a joist has a slight crook, it can be used, but the crown should be placed up so that load will tend to straighten it out **(see 9-42)**. If a joist has a large knot near an edge, place this edge on top, because the top of the joist is in compression and the knot will have less effect on the strength.

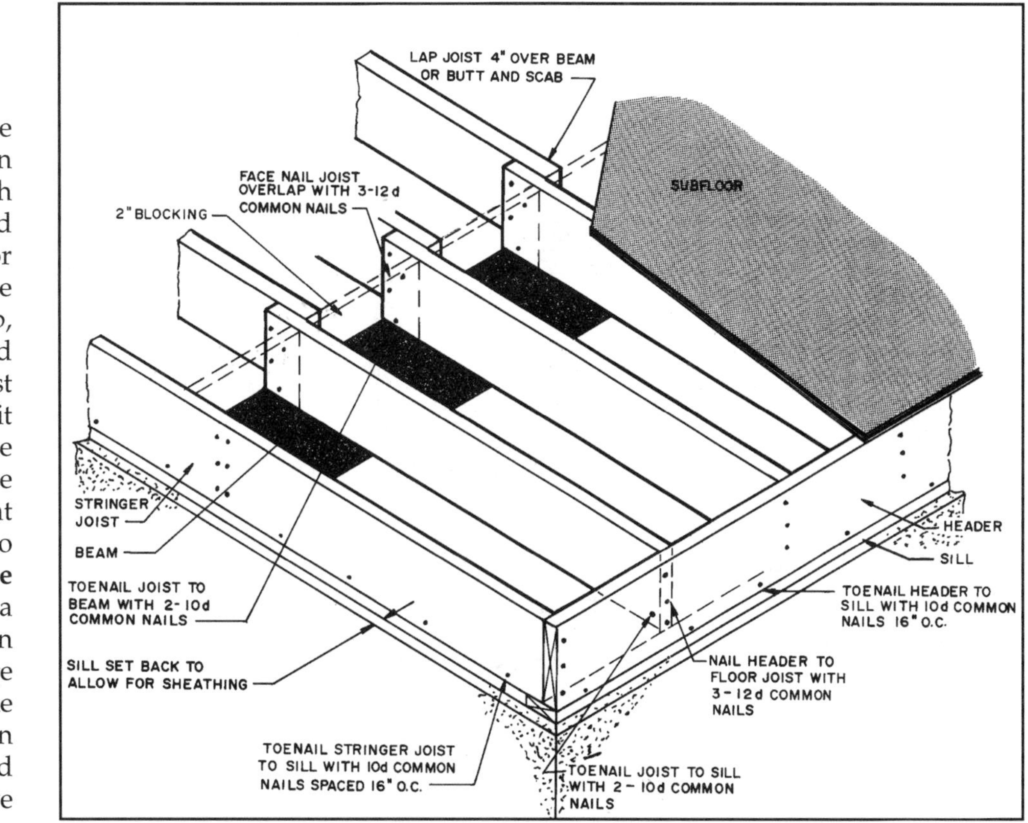

9-43 A typical platform framing system where the floor joists rest on top of the beam, overlap, and are side-nailed together.

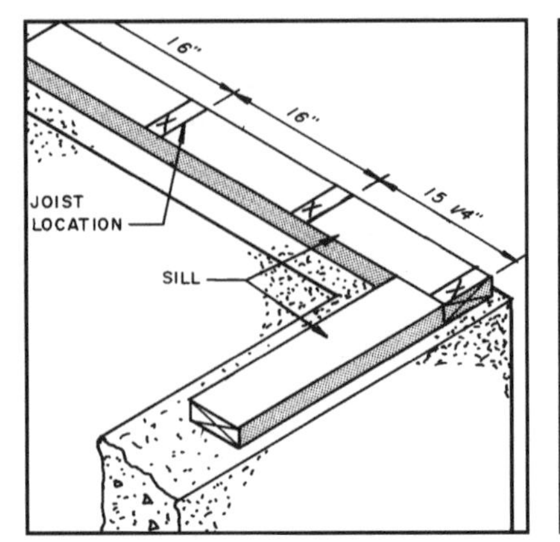

9-44 Some carpenters prefer to mark the joist locations on the sill.

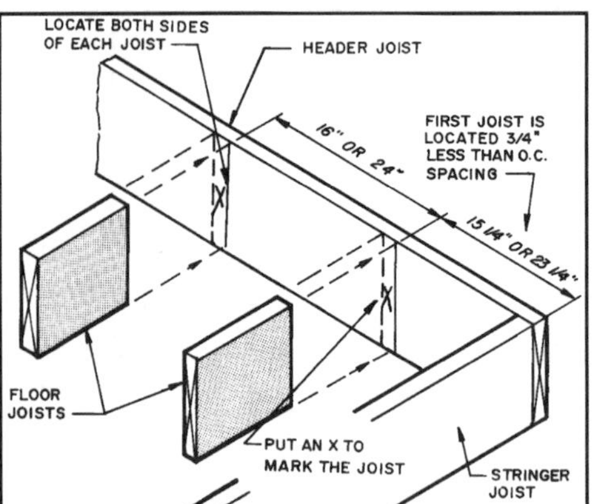

9-45 Locate each joist on the header joist. The first joist is ¾-inch less than the specified on-center (O.C.) spacing.

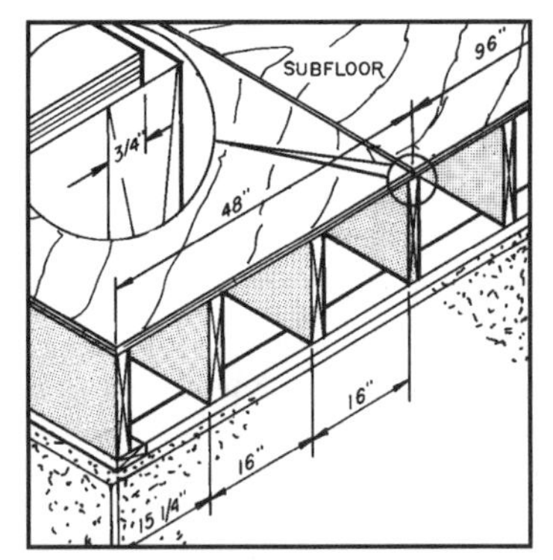

9-46 The joist spacing allows the subfloor panel to be flush with the outside of the stringer joist and leave a support on the third joist for the next panel.

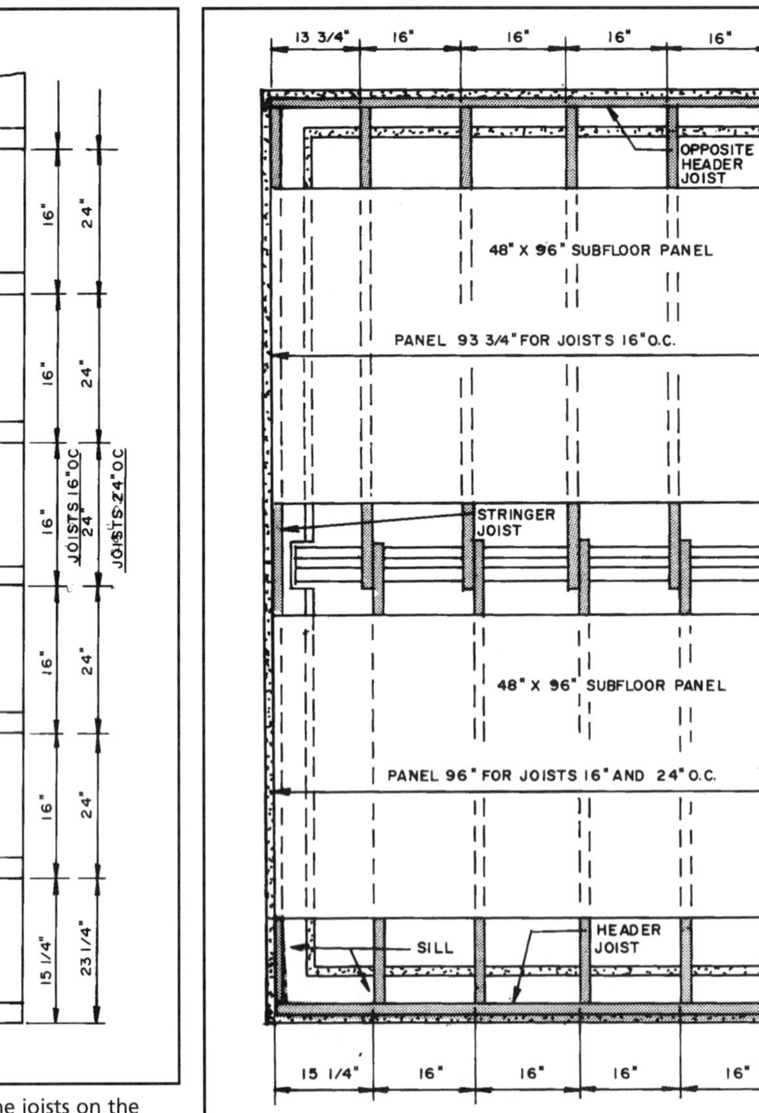

9-47 Locate the joists on the header joist. Mark both sides and identify the location with a large X.

9-48 A layout for floor joists when they are overlapped on the beam. Dimensions for 16-inch O.C. and 24-inch O.C. spacings are shown.

Locating the Floor Joists

The locations of the floor joists can be marked on the header joist or the sill plate. The following discussion relates to using the header joist. Marking the sill follows a similar procedure **(see 9-44)**.

The joist spacing is shown on the architectural drawings by a center-to-center distance. These are typically 16 inches or 24 inches O.C. Since this measurement is difficult to use in locating the joists, the edge of the joist is located. In **9-45** the stringer joist is set flush with the edge of the sill. The next joist is set in a distance ¾-inch less than the on-center spacing. For a 16-inch O.C. layout this is 15¼ inches. This allows the standard 4 × 8-foot sheets of subfloor to overlay the stringer joist and end on the center of the third regular joist **(9-46)**.

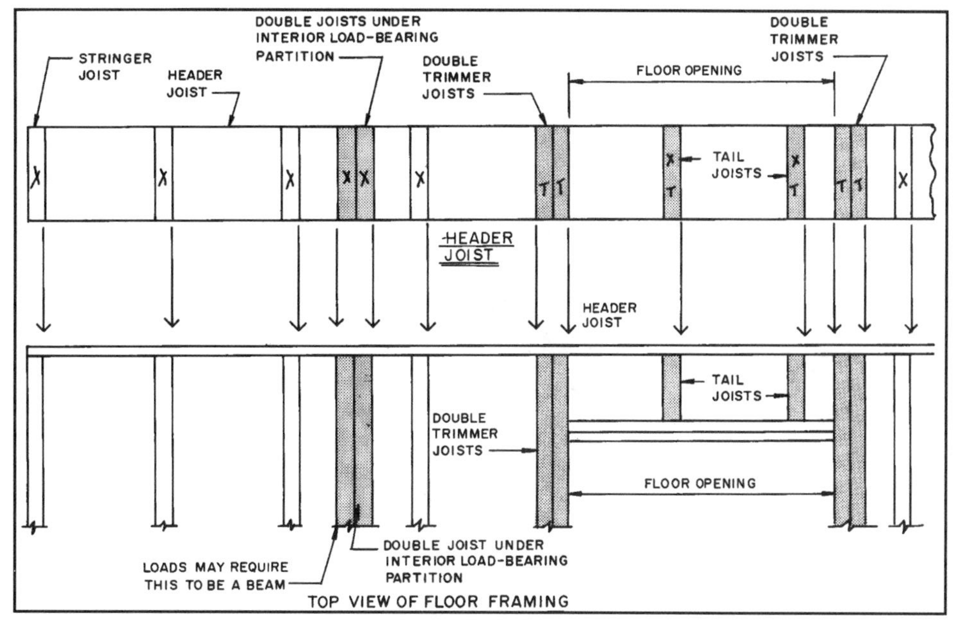

9-49 After the joists are located on the header joist, mark the other joists required, such as double and trimmer joists.

To layout the joist locations on the header joist, measure in 15¼ inches and mark the edge of the joist. Place an X to the left of this mark and locate the other side of the joist. Continue locating other joists in this manner, except space 16 inches as shown in **9-47**. If the joists are to be 24 inches O.C., set in the first joist 21¼ inches from the outside of the stringer joist and space the rest 24 inches O.C. Both sides of the joists are marked and an X is placed indicating the locations. This is necessary so the edge of the subfloor panels can all rest on a joist without cutting a panel. When floor joists are end-butted the layout on the header joists or sill is the same. When they are lapped over the beam the header on the opposite side is marked allowing for the joist thickness as shown in **9-48**.

9-50 The header joists are installed on top of the sill around the perimeter of the building.

There are many special situations where the joist layout requires additional joists. When an interior load-bearing wall carries only a ceiling load, the floor joists are doubled below the wall. If there are additional loads, such as a second floor, a beam sized to carry the load is used. If a load-bearing wall runs perpendicular to the floor joists and the load is minimum, every other joist below the wall can be doubled. Heavier loads require a beam below the joists. These extra joists must be marked on the header as shown in **9-49**. If there is an opening in the floor for a stair or fireplace the joists are doubled on each side and double headers are used, giving a layout like that in **9-50**.

Setting the Joists

Carpenters have various methods they prefer to set the joists. The following is a typical procedure.

After the header joist has been marked it is toenailed to the sill. Check it with a chalkline to be certain it is straight. Next set the stringer joists along each end of the foundation. Toenail them to the sill and nail to the header joist with three 16d common nails **(see 9-50)**.

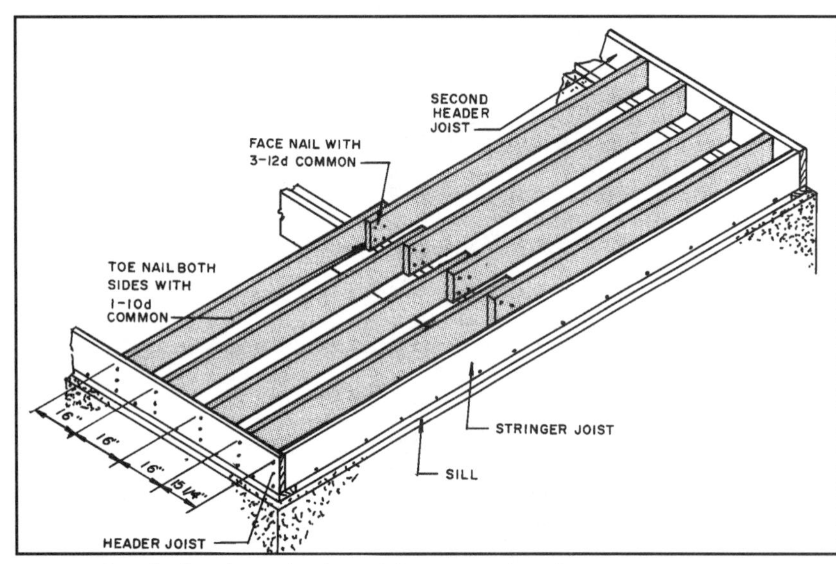

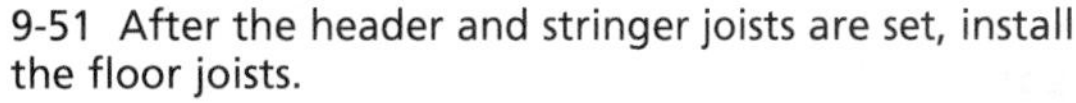
9-51 After the header and stringer joists are set, install the floor joists.

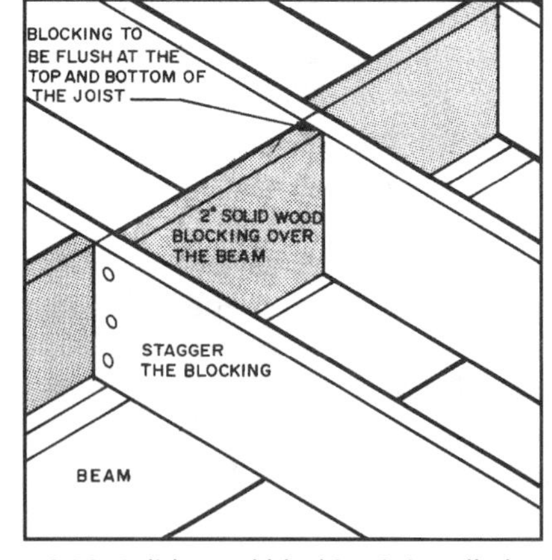

9-52 Solid wood blocking is installed between the joists at the beam. This spaces them the correct distance apart and serves as a fire block.

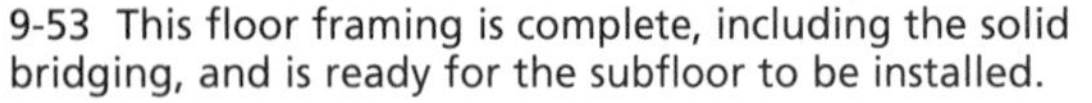
9-53 This floor framing is complete, including the solid bridging, and is ready for the subfloor to be installed.

After the joists have been cut to length, place them on the sill and the beam. Toenail to the sill and end-nail to the header joist. Then locate and toenail to the beam. When the lapping joists are installed toenail them to the sill and beam and nail the lapped ends with 16d common nails **(see 9-51)**.

At this point install the solid blocking between the joists at the beam **(see 9-52 and 9-53)**.

Some carpenters prefer not to nail the joists that are around an opening but leave the space open until all the regular joists are in place. Then they put in the required trimmer and header joists and cut and install the tail joists.

After the joists are installed they must be straightened before the subfloor panels can be installed. Begin by checking the stringer joist for straightness by using a chalkline as shown in **9-54**.

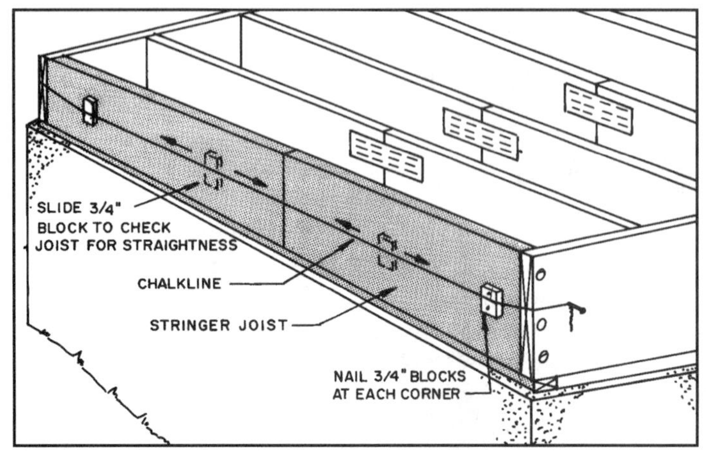

9-54 Stringer joists can be checked for straightness with a chalkline.

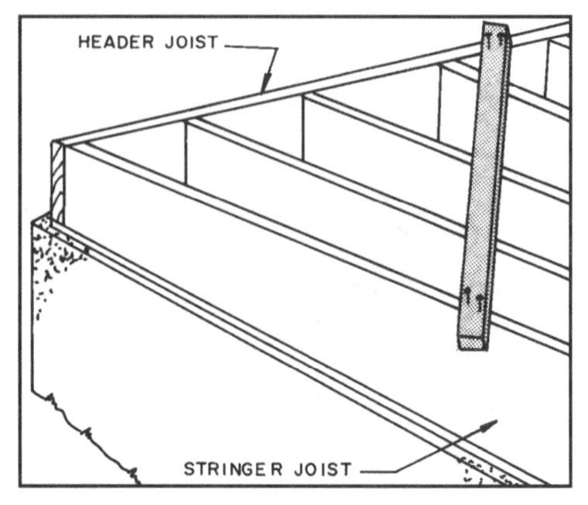

9-55 Straighten the stringer and header joists, set at a right angle, and nail on a temporary brace.

Shim it out with ¾-inch blocks. Move a ¾-inch block along the joist to see if it is straight. Adjust as necessary. This is important because the other joists are now to be lined up with it. Then straighten the header joist and nail a temporary diagonal brace between it and the stringer joist **(see 9-55)**.

The header joist, stringer joist, and floor joists can be held straight and properly spaced by holding them in the desired position with wood furring strips temporarily nailed across them. Mark the joist spaces on the strip and move the joists to align with the marks. Since the subfloor is now to be installed, nail the furring strip far enough in from the header joist (50 to 60 inches) to allow the subfloor panel to be placed and nailed. Leave the nail heads above the surface so the strips can be removed as the subfloor is installed **(see 9-56)**.

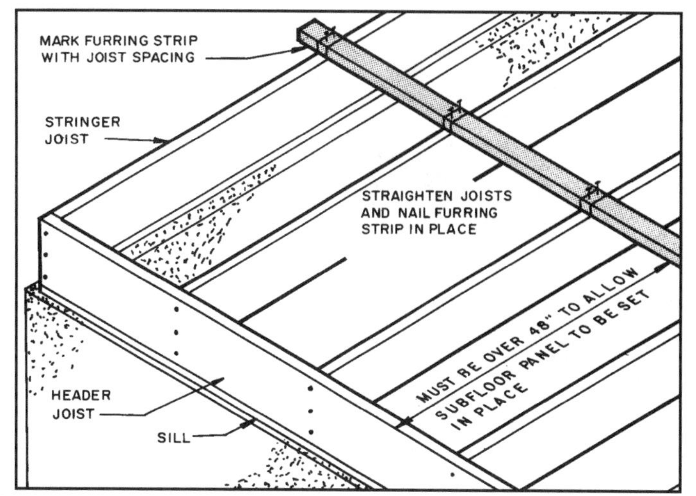

9-56 Use a temporary brace to keep floor joists spaced as required for the installation of the subfloor.

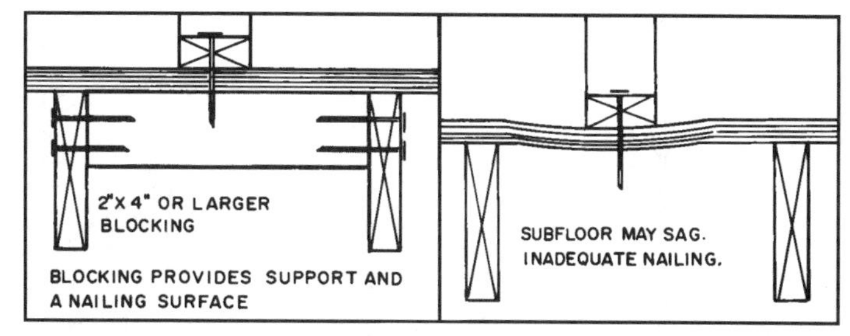

9-57 Blocking between the joists is required to support and provide a nailing surface for nonload-bearing partitions running parallel with the floor joists.

Supporting Interior Partitions

Nonload-bearing partitions running parallel with the joists are basically supported by the subfloor if it is ⅝ inch thick or thicker plywood or oriented strand board. However, blocking must be nailed between the joists every two feet to provide a solid nailing base for the wall plate as shown in **9-57**. This also helps support the partition because the subfloor will sometimes sag without this support. Any joists should be doubled that run under load-bearing interior partitions running parallel with the joists **(see 9-58)**. Solid blocking between the two joists is used when it is necessary to have access to the partition, e.g., to install heating ducts **(see 9-59)**.

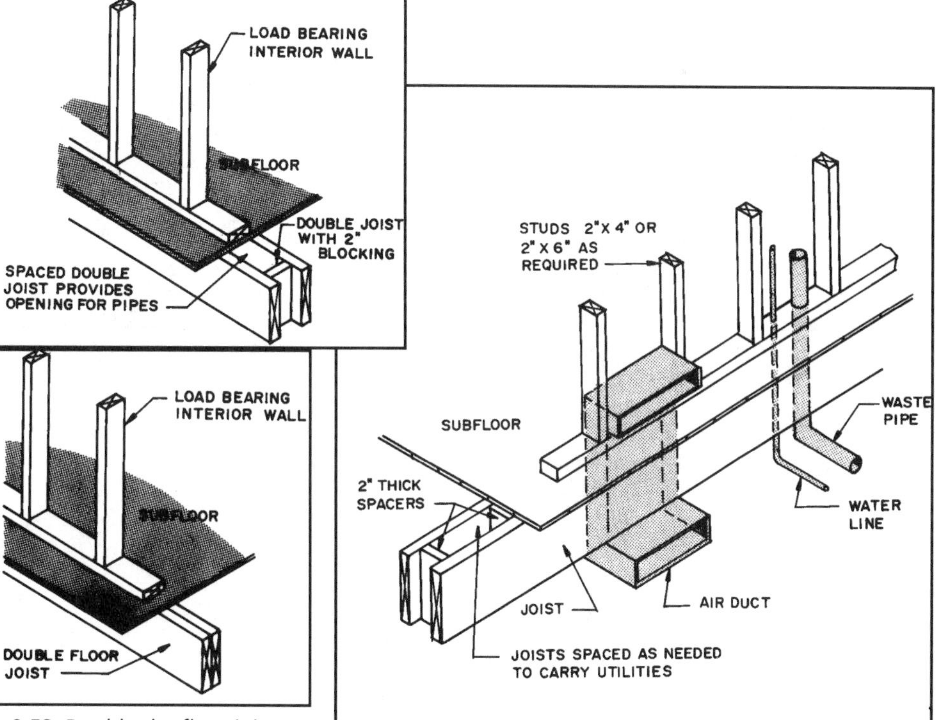

9-58 Double the floor joists under load-bearing interior partitions.

9-59 Joists can be spaced with two-inch solid wood spacers to allow plumbing and ducts to pass up into the wall.

Framing Floor Openings

It is often necessary to frame an opening in a floor for a stair or fireplace. The joists are doubled on each side of the opening and are called **trimmer joists**. The shortened joists abutting the opening are called **tail joists**. The tail joists are nailed to **headers**. If the opening is less than 4 feet wide, a single header is used. Double headers are required for openings 4′-0″ or larger **(see 9-60 and 9-61)**.

Tail joists shorter than six feet are nailed to the header with three 16d common end nails and two 10d nails toenailed. Tail joists longer than six feet are hung with metal joist hangers or ledger strips. Headers are joined to trimmers with three 16d common nails and two 10d toenails.

When nailing framed openings, first frame the opening with single headers and trimmers. Then facenail the second header and trimmers to those in place **(see 9-62)**. Tail joists are usually hung with metal joist hangers.

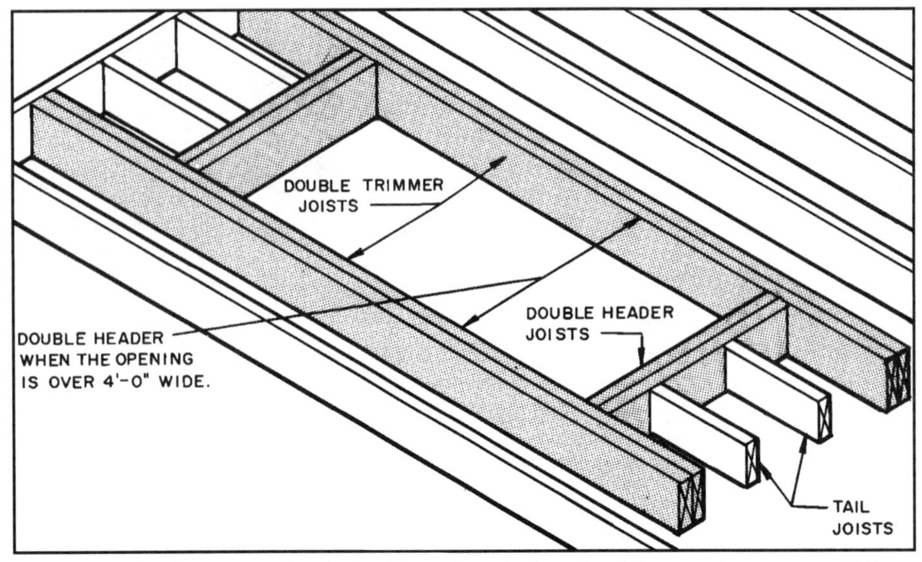

9-60 Framing for an opening in the floor when the length of the opening runs parallel with the floor joists.

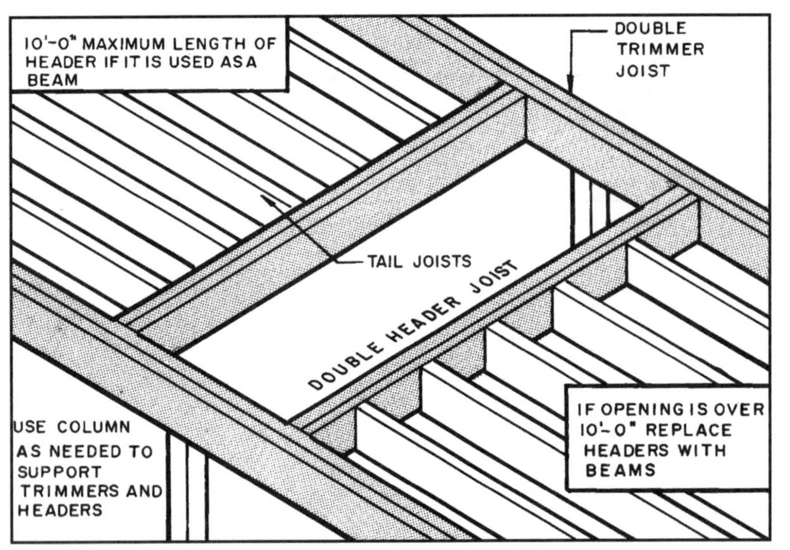

9-61 Framing for an opening in the floor when the length of the opening runs perpendicular to the floor joists.

Order of Assembly

A. End-nail trimmers 3 and 4 to header 1 with three 16d comon nails.
B. End-nail header 1 to tail joists with three 16d common nails and two 10d inside toenails.
C. Face-nail header 2 to header 1 with 12d common nails staggered 12 inches O.C.
D. End-nail trimmers 3 and 4 to header 1 with three 16d common nails.
E. Face-nail trimmers 5 and 6 to trimmers 3 and 4 with 12d common nails 12 inches O.C. staggered.

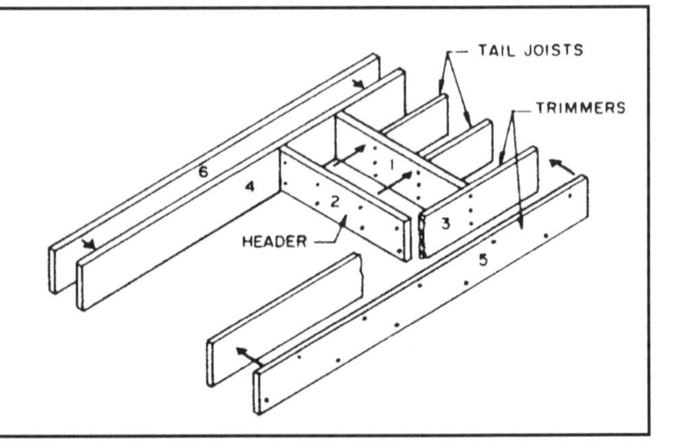

9-62 Typical nailing sequence for installing header and trimmer joists. Tail joists are frequently joined with metal joist hangers, eliminating the need to end-nail as shown in this illustration.

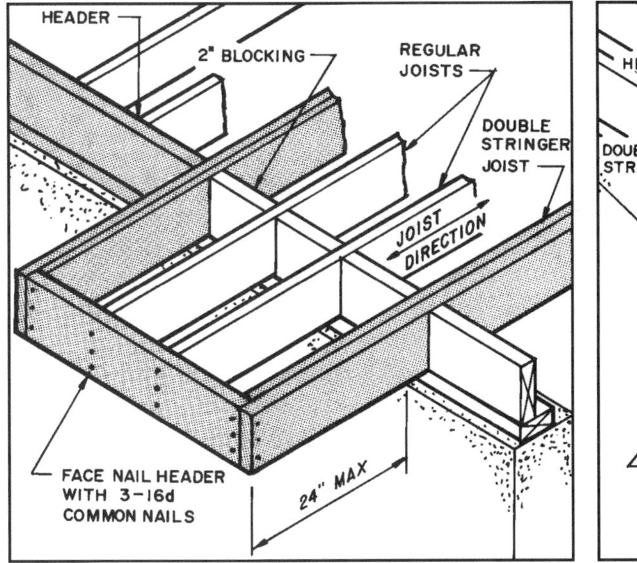

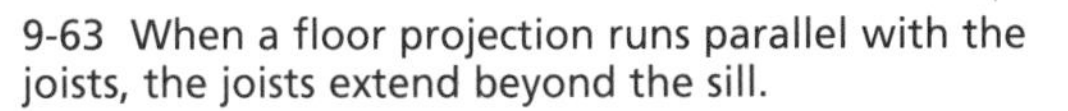
9-63 When a floor projection runs parallel with the joists, the joists extend beyond the sill.

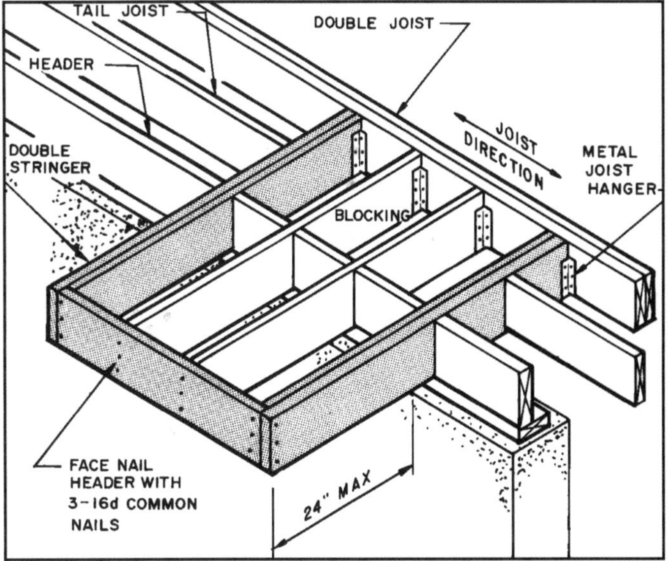

9-64 When a floor projection runs perpendicular to the joists, the joists must extend beyond the sill into the floor area.

Framing Floor Projections

Floors often cantilever over the foundation where bay windows or balconies are required. Some house styles have the second floor cantilever over the first floor. The framing details in **9-63** and **9-64** are generally adequate for cantilevers that are non-load-bearing, such as a bay window. If they are to support a wall, ceiling, and/or a roof, the structure requires an engineer to determine the design.

When joists run in the direction of the projection, they are extended beyond the foundation **(see 9-63)**. Under normal conditions a two-foot projection is maximum. Projections of a second floor seldom extend beyond 12 inches. Anything longer than this will need special design consideration. Notice that the joists on the outside edges are doubled and a single header is used. The subfloor extends out over the projection and is cut flush with the outside of the framing. The same nailing pattern that is used on the rest of the floor is used.

When the projection runs perpendicular to the joists, it must be framed by extending the joists in the projection back into the floor to the second joist **(see 9-64)**, which is doubled. The joists on the ends of the projection are also doubled. All joists are hung using metal joist hangers.

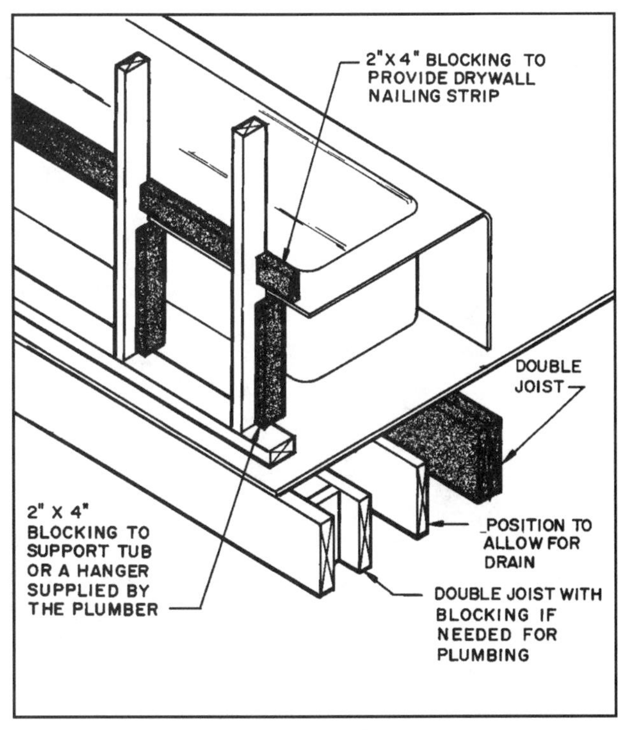

9-65 Framing to support a bathtub.

Framing for a Bath

The floor below a bathroom requires extra strength because of loads heavier than normal. When the floor is to have a wood subfloor and tile, carpet, or composition finish, the only extra framing is to double the joists under the edges of the bathtub **(see 9-65)**. If the wall is to hold plumbing, the joists are spaced apart with blocking.

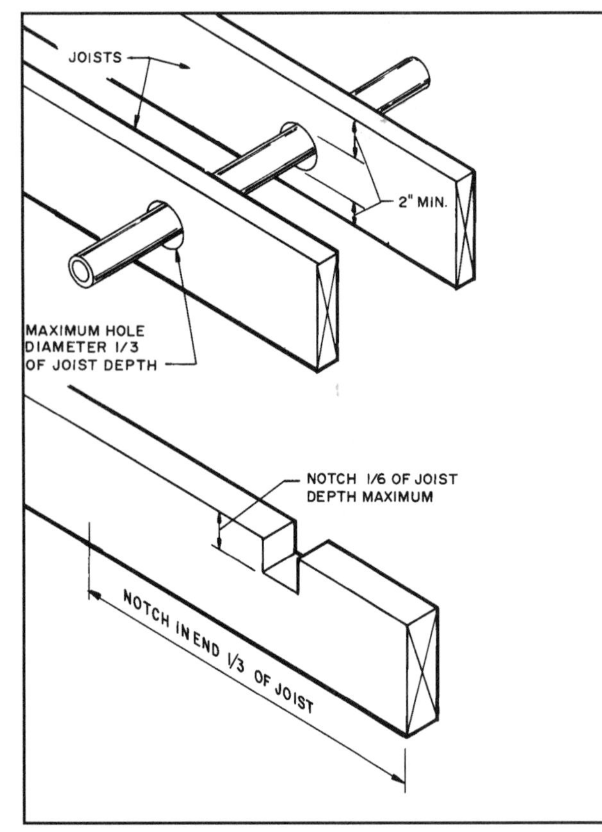

9-66 Recommended allowances for notching or boring holes in joists.

Notching Joists & Beams

Occasionally it is necessary to cut a notch or hole in a joist or beam to run through plumbing or other utilities. Notching of joists should be avoided whenever possible. Notches on the ends of joists should not exceed one fourth of the joist depth. Notches in the top or bottom of joists should not exceed one sixth of the depth and should not be located in the middle third of the joist **(see 9-66)**.

Holes bored in joists should not be within two inches of the top or bottom of the joist. The diameter of the hole should not exceed one third of the depth of the joist **(see 9-66)**.

If larger cuts are needed, the joist will need reinforcing. Sometimes sections of two-inch-thick lumber are nailed on both sides of the notched section, or the joist can be doubled. It is also possible to cut out the joist and frame the area in the same way as for an opening in the floor. Notching of I-joists, laminated veneer, parallel strand, and other manufactured beams and joists should be done according to the instructions of the manufacturer.

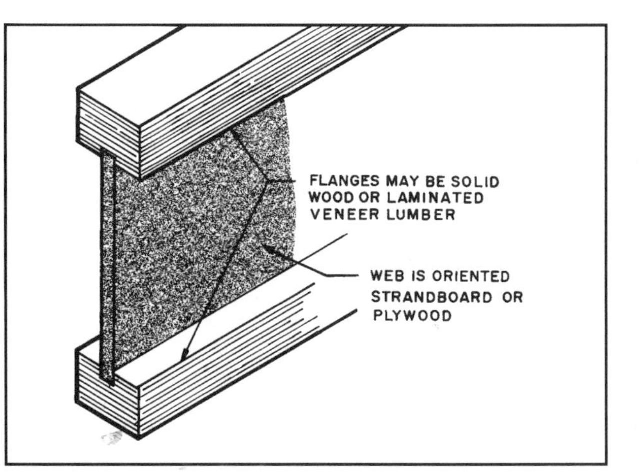

9-67 The structure of a typical I-joist.

I-JOIST FLOOR SYSTEM

An I-joist is a structural wood member used for floor, ceiling, and roof framing. It is manufactured using plywood or oriented strand board webs and solid wood or laminated veneer lumber flanges **(see 9-67)**. These joists have long span capabilities, and on some installations they eliminate the need for a center beam **(see 9-68)**. The framing is similar to that used with solid wood members, but certain important differences exist. The following construction details are typical of those recommended by companies manufacturing I-joists. The carpenter should be familiar with the details recommended by the manufacturer of the joists to be used. This is especially critical when loads, spacings, and spans are decided **(see 9-69)**.

9-68 The members on the left are insulated headers used over openings in the walls while those on the right are I-joists used for floor joists and roof rafters. *(Courtesy Superior Wood Systems, Inc.)*

9-69 I-joists provide straight joists of consistent quality that are easy to install. *(Courtesy Trus Joist MacMillan)*

Typically the joists rest on the wood sill on the foundation wall. The header joist (called a rim joist in the trade) may be an I-joist, an insulated rim joist, a reconstituted wood rim board, or CDX (exterior) plywood. These are detailed in **9-70**. The nailing of the I-joist should be done as recommended by the manufacturer or local code. A typical example is in **9-71**. I-joists are generally power-nailed, which greatly speeds the installation **(see 9-72)**.

The reconstituted wood rim board is a manufactured structural board made from wood fibers and appears similar to oriented strand board (OSB). It uses fibers longer than those in OSB, which are bonded with a high-performance adhesive.

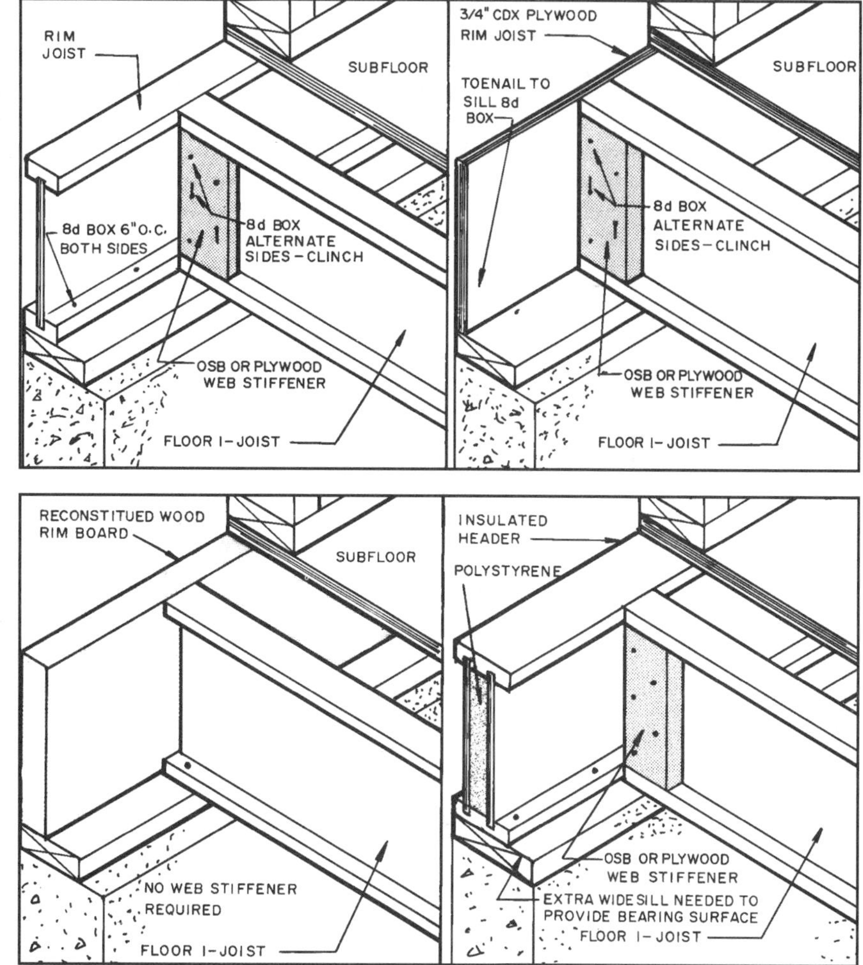

9-70 Typical I-beam construction at the sill.

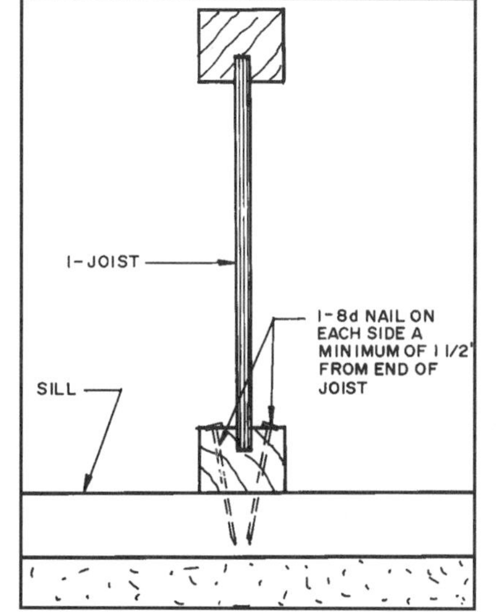

9-71 A recommended nailing pattern for securing the I-joist to the sill.

9-72 I-joists are rapidly secured to the sill with a power nailing tool. *(Courtesy Boise Cascade Corporation)*

9-73 Rim board ties together the ends of the I-joists and provides support for the subfloor and exterior wall. The product shown is TimberStrand (LSL)®. *(Courtesy Trus Joist MacMillan)*

When using rim board it is not necessary to use squash blocks or other filler material. The rim board supports the subfloor, exterior wall, and any exterior attachments in the same manner as solid wood header joists. One type is in **9-73**.

When concentrated loads are to be placed on the I-joists between the flanges, web stiffeners are required. Web stiffeners are needed for instances such as an intermediate wall or end-bearing wall where I-joists are used as headers or rim joists. Refer back to **9-70**.

I-joists can be doubled as needed under interior partitions. The flanges must be supported with solid wood and plywood blocking **(see 9-74)**.

I-joists butting a beam or another I-joist are hung with metal joist hangers **(see 9-75 and 9-76)**.

Some typical spans of I-joists are shown in **Table 9-4**.

Typical framing for I-joists such as ceiling or second floor joists can be seen in **9-77**. The blocking used on doubled joists around a stair opening to support the stair carriages is solid wood installed as shown in **9-78**.

Many of the previous illustrations show nailing recommendations. These are typical. However, manufacturers' recommendations and codes should be consulted. In some cases the recommendations of a structural engineer should be obtained.

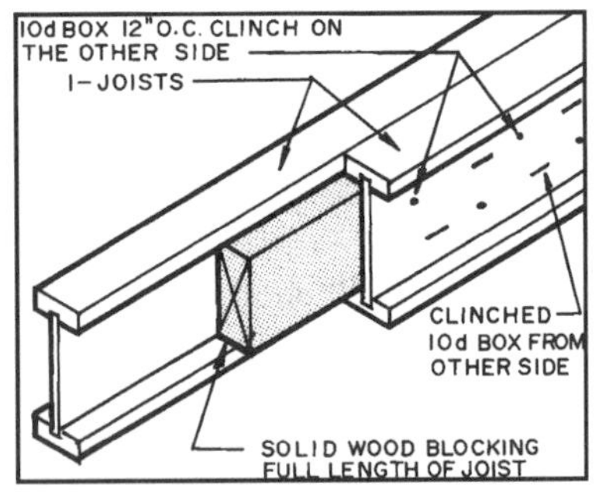

9-74 Double I-joists require solid wood blocking.

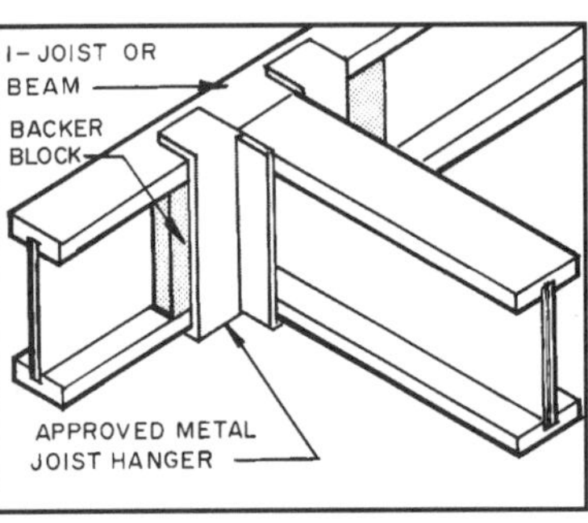

9-75 I-joists can be joined to beams or other I-joists with approved metal joist hangers.

Table 9-4 Typical spans for I-joists.*

Spacing, O.C. (inches)	Joist depth 9½	11⅞	14	16
12	18'-2"	21'-9"	24'-8"	27'-3"
16	16'-6"	19'-9"	22'-5"	24'-10"
19.2	15'-6"	18'-6"	21'-1"	23'-4"
24	14'-4"	17'-1"	19'-8"	21'-7"
32	11'-4"	14'-4"	—	—

* Based on a floor load of 40 psf (pounds per square foot) and 10 pounds dead load. Consult manufacturer for specific data.

9-76 This carpenter is joining I-joists to a wood beam with metal joist hangers. *(Courtesy Boise Cascade Corporation)*

In **9-79** are typical nailing recommendations for securing the I-joist and rim board or joist to a sill or top plate.

I-Joist Standards

I-joist standards for use by national and local building codes are available. Details can be obtained from APA—The Engineered Wood Association.

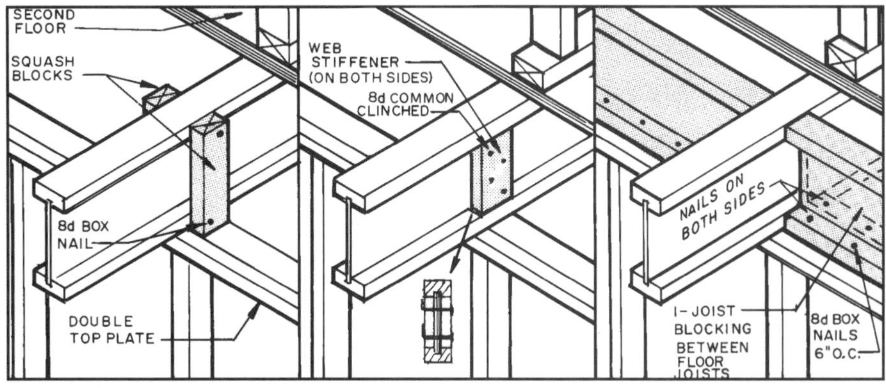

9-77 I-joists require some type of reinforcing when a load-bearing interior partition rests on them. This is typical construction for a second floor.

WOOD FLOOR TRUSSES

Wood floor trusses are widely used for floor joists **(see Table 9-5)**. They can carry reasonably high floor loads over longer spans than solid wood joists. These joists will span the entire width of a typical residence, eliminating the need for a center beam. Wood floor trusses are delivered to the construction site with several trusses strapped together and are removed from the truck with a hoist or forklift **(see 9-80)**.

Table 9-5 Typical span and load data for truss-type floor joists spaced 24" O.C.

Clear Span (in feet)	Joist depth 12	14	16	18	20	22
14	120*	118*	—	—	—	—
16	83	100	—	—	—	—
18	66	79	92*	105*	117*	120*
20	53	63	74	85	95	103
22	—	52	61	70	79	88
24	—	—	51	59	67	74
26	—	—	—	50	57	63
28	—	—	—	—	49	—
30	—	—	—	—	—	47

* Pounds per square foot (psf).
Consult manufacturer's catalog for specific data.

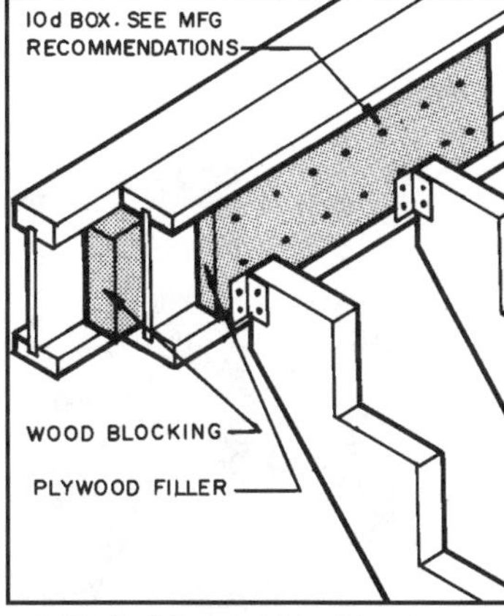

9-78 Solid wood blocking is added to provide a solid anchor for the metal angles holding the stair carriages to the double joist.

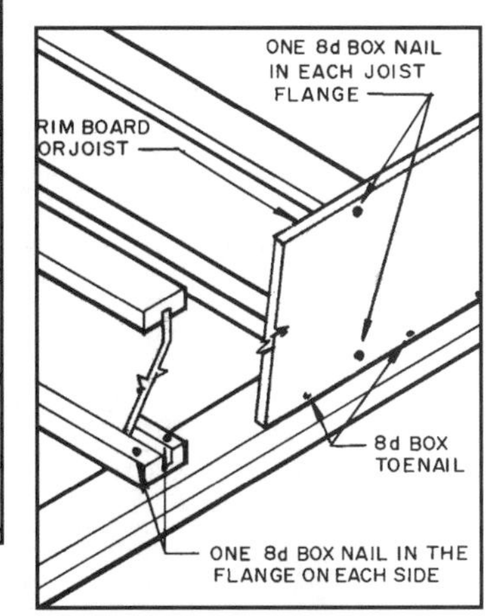

9-79 Typical nailing patterns for securing the I-joist to a sill or plate and a rim board or joist.

9-80 Wood floor trusses are delivered to the site with several of them bound together with metal strapping.

9-81 These wood-framed floor trusses are assembled in a factory with metal gusset plates.

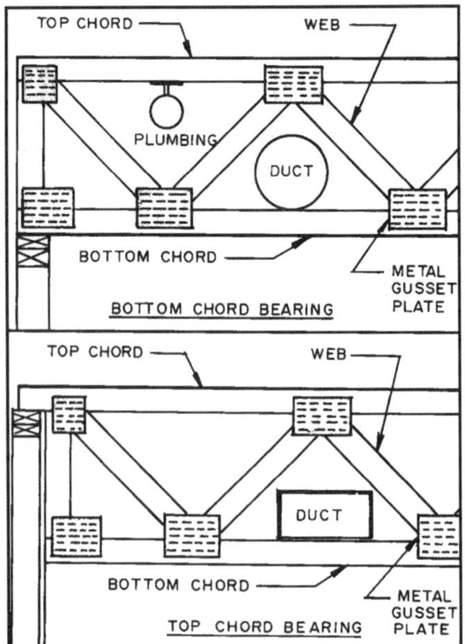

9-82 Wood floor trusses may be made to be bottom chord bearing or top chord bearing.

Wood floor trusses are usually made from small wood members, typically 2 × 4 inches. One type, shown in **9-81**, has wood flanges and webs. The members are joined with metal gusset plates which are pressed into place in the factory. Trusses may be bottom-chord-bearing or top-chord-bearing as shown in **9-82**. The bottom-chord-bearing type will give a higher basement ceiling since it rests on top of the foundation. The top-chord-bearing type lowers the height of the floor in relation to the top of the foundation.

Another type of wood floor truss uses metal webs. These are a manufactured unit that is nailed to the flanges **(see 9-83)**. Both types are typically spaced 24 inches O.C. so the subfloor used must be of a thickness that will span that distance and carry the required floor loads **(see 9-84)**.

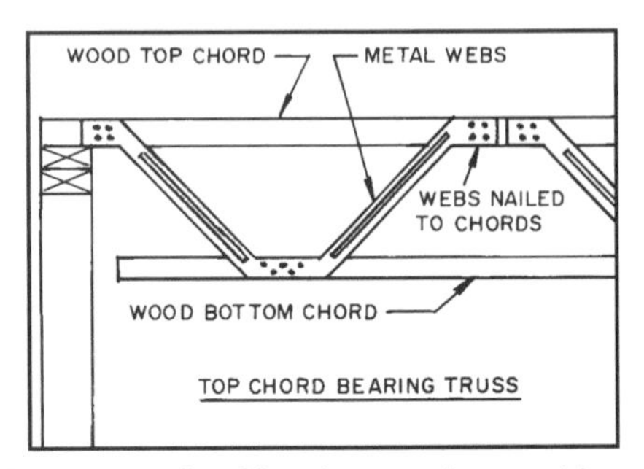

9-83 This is a top chord bearing wood truss with metal webs. Bottom chord bearing trusses are also available.

9-84 The wood floor trusses have metal webs and are spaced 24 inches O.C.

A major advantage to floor trusses is that they permit the passage of ducts, pipes, and electrical runs through the openings between the webs. In some cases round ducts as large as 12 to 16 inches will pass through the truss.

The floor trusses are engineered and manufactured in a factory to serve a specific use. The carpenter should never cut or alter a truss. If this must occur the manufacturer should be consulted.

As the floor trusses are installed they are kept spaced and stabilized with a temporary 1 × 4-inch furring nailed into each as described earlier for solid wood joists. Some type of permanent bridging is specified by the manufacturer. It may be diagonal bridging or a horizontal 1 × 4-inch member as shown in **9-85**. A completed installation after the subfloor has been installed is shown in **9-86**.

Bridging

Bridging is usually required when the ratio of joist depth to thickness is six to one or greater. It is used to stiffen the floor assembly, hold the joists in place, and help distribute a floor load to adjacent joists. Typical types of bridging are shown in **9-87**.

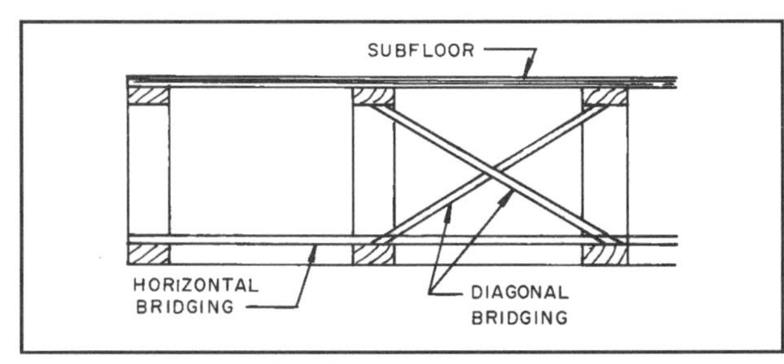

9-85 Some form of bridging is required when using wood floor trusses.

9-86 After the wood floor trusses are installed, the subfloor is glued and nailed to the top chord.

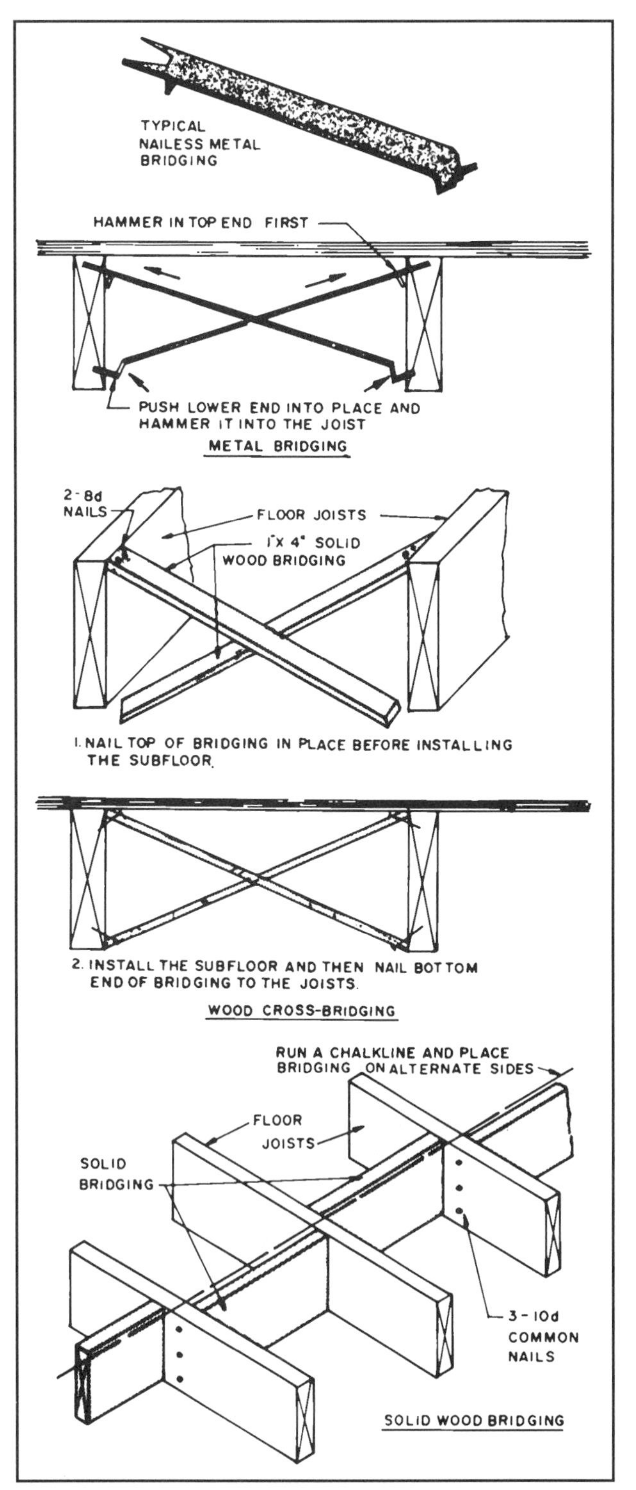

9-87 Commonly used types of bridging for wood-framed floors.

When installing diagonal wood bridging (usually 1 × 3-inch or 1 × 4-inch), drive two 8d common nails in each end of diagonal wood bridging before you start to install them. Notice they are on opposite sides of the piece. Run a chalkline across the joists to locate the line of bridging. Nail the top end of each piece with one member each side of the line before the subfloor is installed. Leave the bottom ends loose until the subfloor is in place **(see 9-88)**. Some prefer to wait until the wall and roof framing are in place and sheathed before nailing the bottom ends. This gives the joists time to settle and equalize.

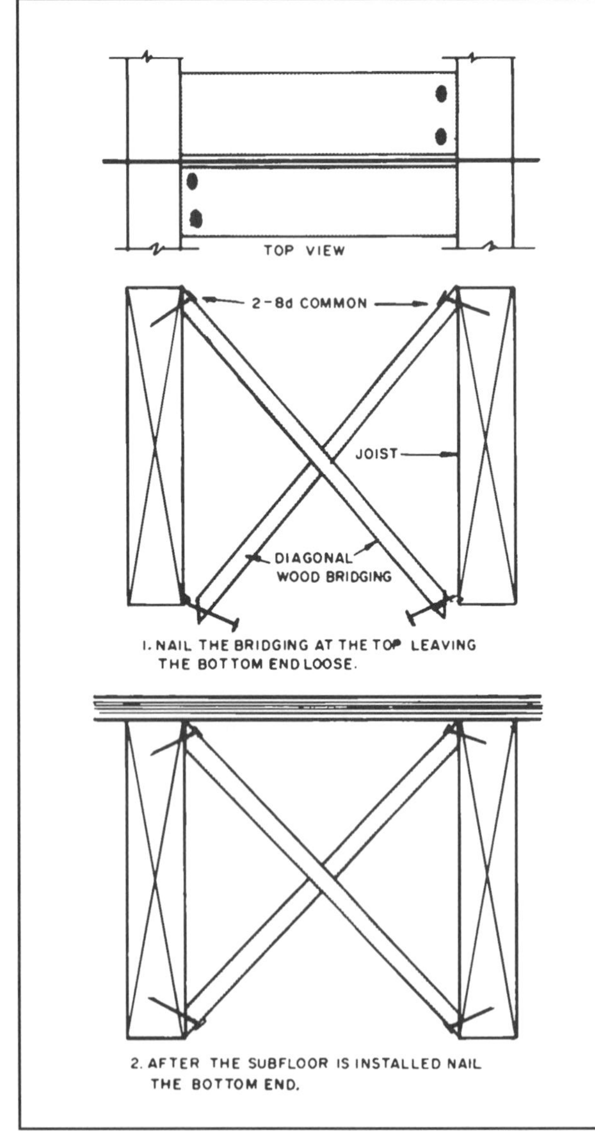

9-88 The top end of diagonal wood bridging is nailed before the subfloor is installed. The lower end is nailed after the subfloor is in place.

Solid wood bridging is cut to length (typically 14½ inches for 16-inch O.C. joists) a bit (1⁄16 inch) short of the actual distance between the joists. This helps get it to slide into place. This is especially helpful when the joists are a bit out of line. Usually bridging is not applied until the framing on the building is complete.

The bridging is staggered on each side of a chalkline. It is very difficult to nail the 10d common nails because of the limited space between the joists.

Metal bridging **(refer back to 9-87)** is installed by driving the pointed end into the joist near its top. After the subfloor is installed the bottom end is swung against the joist and the metal prongs on it are hammered into the joist.

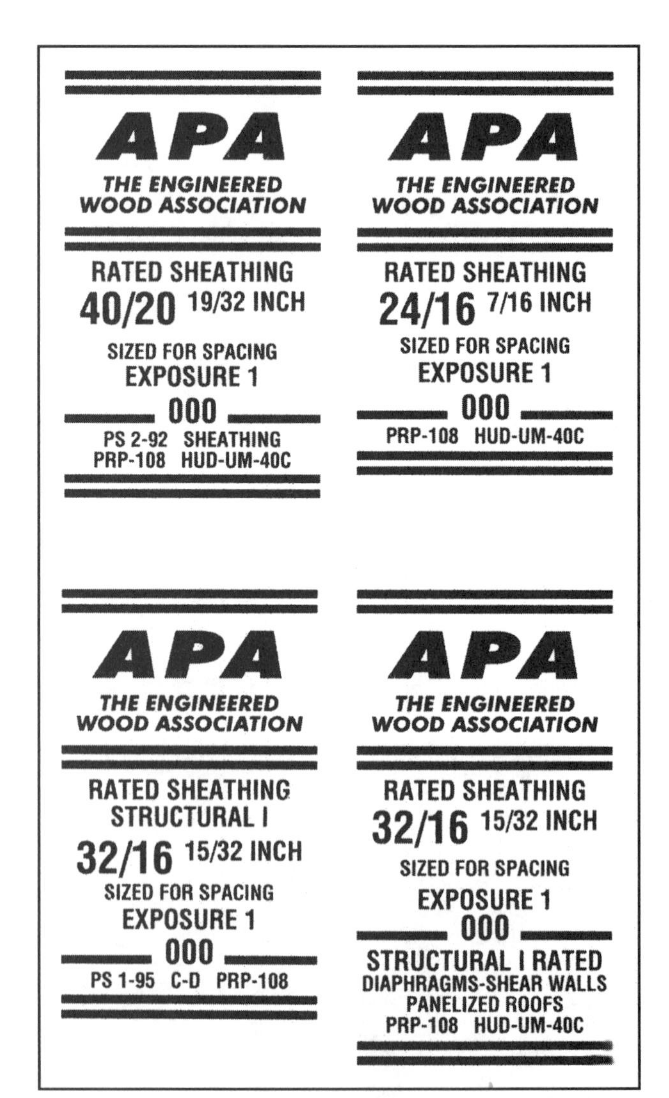

9-89 Grade marks for APA-Rated Sheathing and APA Structural Panels used for subflooring. *(Courtesy APA—The Engineered Wood Association)*

INSTALLING THE SUBFLOOR

The subfloor is secured to the joists, forming a strong, stiff platform. While solid wood boards nailed on a diagonal can be used, subfloors are now almost entirely plywood or some type of reconstituted wood panel.

Plywood Subfloor

A plywood subfloor may be APA-Rated Sheathing or APA-Rated Sturd-I-Floor. APA-Rated Sheathing and APA Structural I panels are used as a subfloor, if a separate underlayment panel or strip flooring is to be nailed over them. The stamp on the panel indicates the maximum span. For example, a span rating of 32⁄16 means the panel can be used as roof sheathing for rafters up to 32 inches O.C. and floor sheathing for joists up to 16 inches O.C. Generally ½- or ⅝-inch-thick plywood is used for joists spaced 16 inches O.C. and ¾ inch for joists 24 inches O.C. Underlayment plywood should be APA Underlayment Int or one of several other grades available **(see 9-89)**.

APA-Rated Sturd-I-Floor is a combination subfloor-underlayment providing the strength required and a smooth surface for the application of carpet or composition floor-covering materials. It is available with span ratings of 16, 20, 24, 32, and 48 inches O.C. The spans from 16 through 32 inches are framed with two-inch joists, I-joists, or floor trusses. These panels range in thickness from 19⁄32 to ⅞ inch **(see 9-90)**.

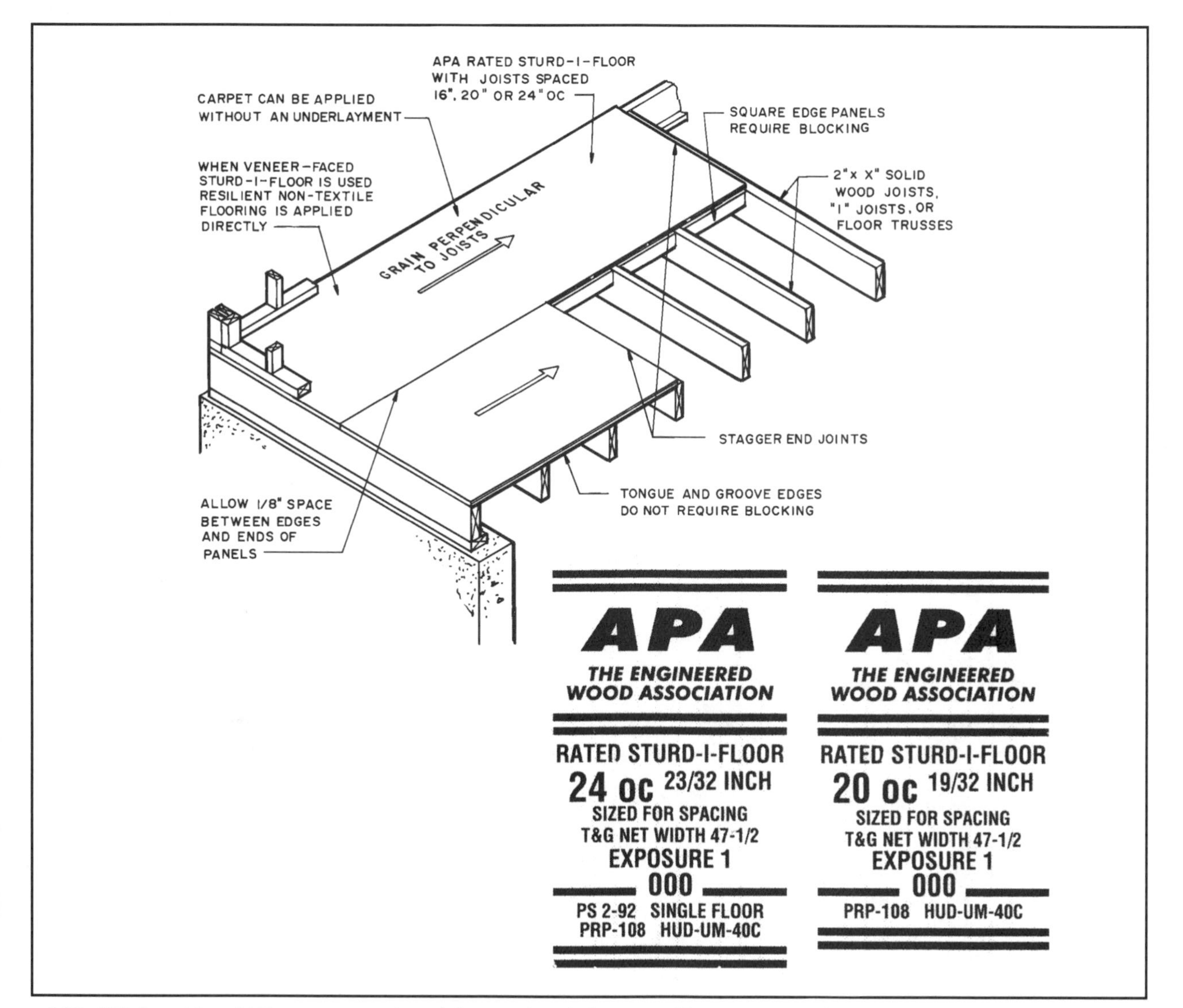

9-90 APA-Rated Sturd-I-Floor serves as the subfloor and underlayment. *(Courtesy APA—The Engineered Wood Association)*

Panels spanning 48 inches O.C. are placed on girders spaced 48 inches O.C. These girders may be 4-inch solid wood, 2-inch-thick stock nailed together, wood floor trusses, or lightweight steel beams **(see 9-91)**.

In **9-92** the carpenters are placing plywood floor panels on floor beams spaced 48 inches O.C. Notice the tongue on the long edge of the panel. It is set into a groove on the butting panel and a ⅛-inch space is left between the top veneers.

Plywood is installed with the face grain at right angles to the joists. Tongue-and-groove sheets are joined, leaving ⅛-inch spacing on all joints. If square-edged sheets are used, blocking must be installed below each joint **(see 9-93)**.

Plywood sheathing is nailed to floor joists using 6d ring- or screw-shank nails for panels 7⁄16 inch thick and 8d ring- or screw-shank nails for panels up through ¾ inch thick. Nails are spaced six inches apart on all edges and 12 inches apart on intermediate members for panels that are nailed only.

Underlayment is installed with the face grain of the panels perpendicular to the joists. End joints should be staggered and a 1⁄32-inch space left between the edges of panels. The underlayment is nailed six inches O.C. along the edges and eight inches O.C. on the interior, with 3d shank nails for panels ½ inch or thinner. Panels up to ¾ inch thick are nailed with 4d shank nails 6 inches O.C. on the edges and 12 inches O.C. on the interior. Fill and sand the joints, producing a smooth surface **(see 9-94)**.

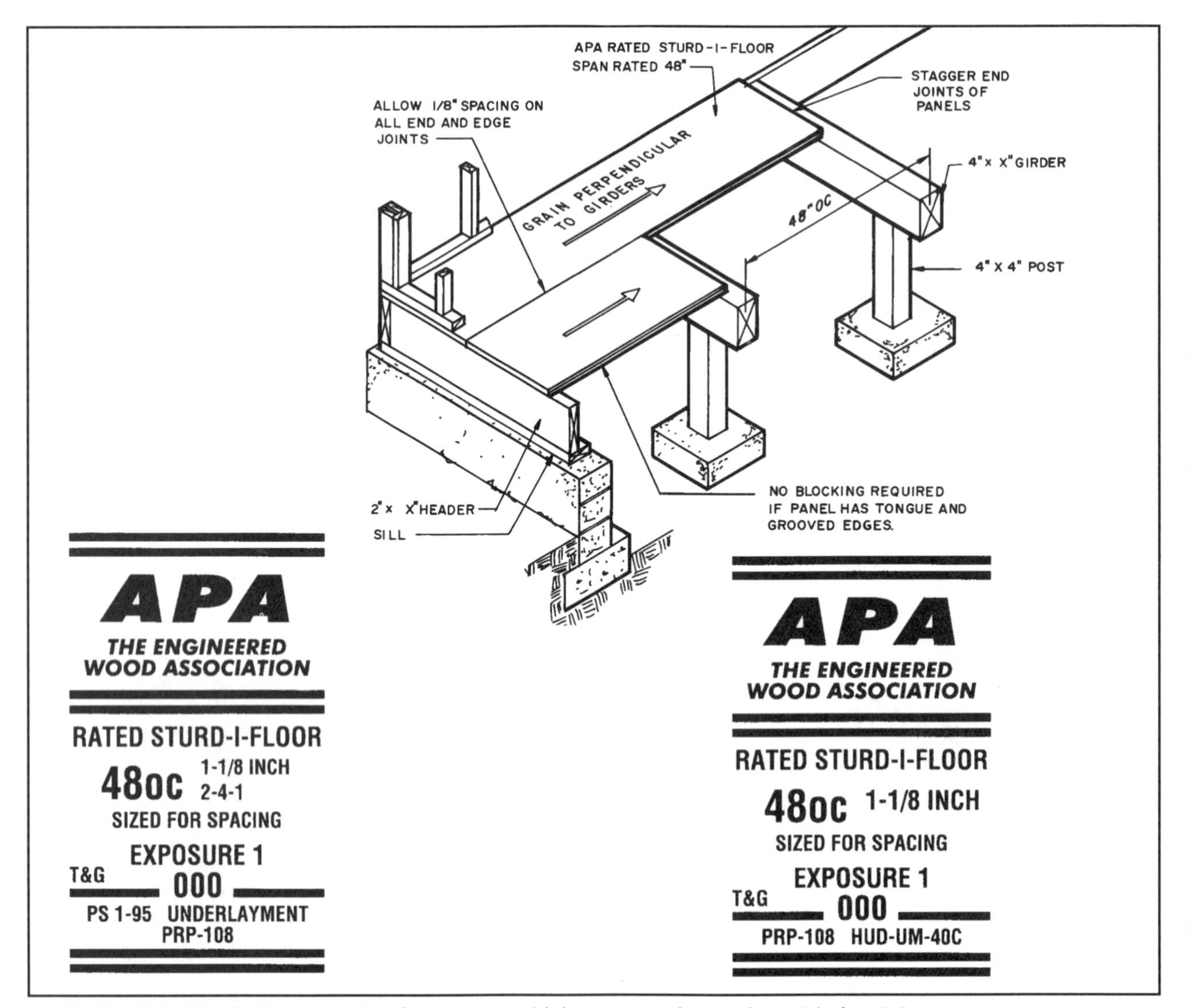

9-91 APA-Rated Sturd-I-Floor is used on floor systems with beams spaced as much as 48 inches O.C. *(Courtesy APA—The Engineered Wood Association)*

9-92 This APA-Rated Sturd-I-Floor panel is being installed on wood floor beams spaced four feet O.C. *(Courtesy APA—The Engineered Wood Association)*

9-93 Plywood subflooring is installed with the grain of the top veneer perpendicular to the floor joists. *(Courtesy APA—The Engineered Wood Association)*

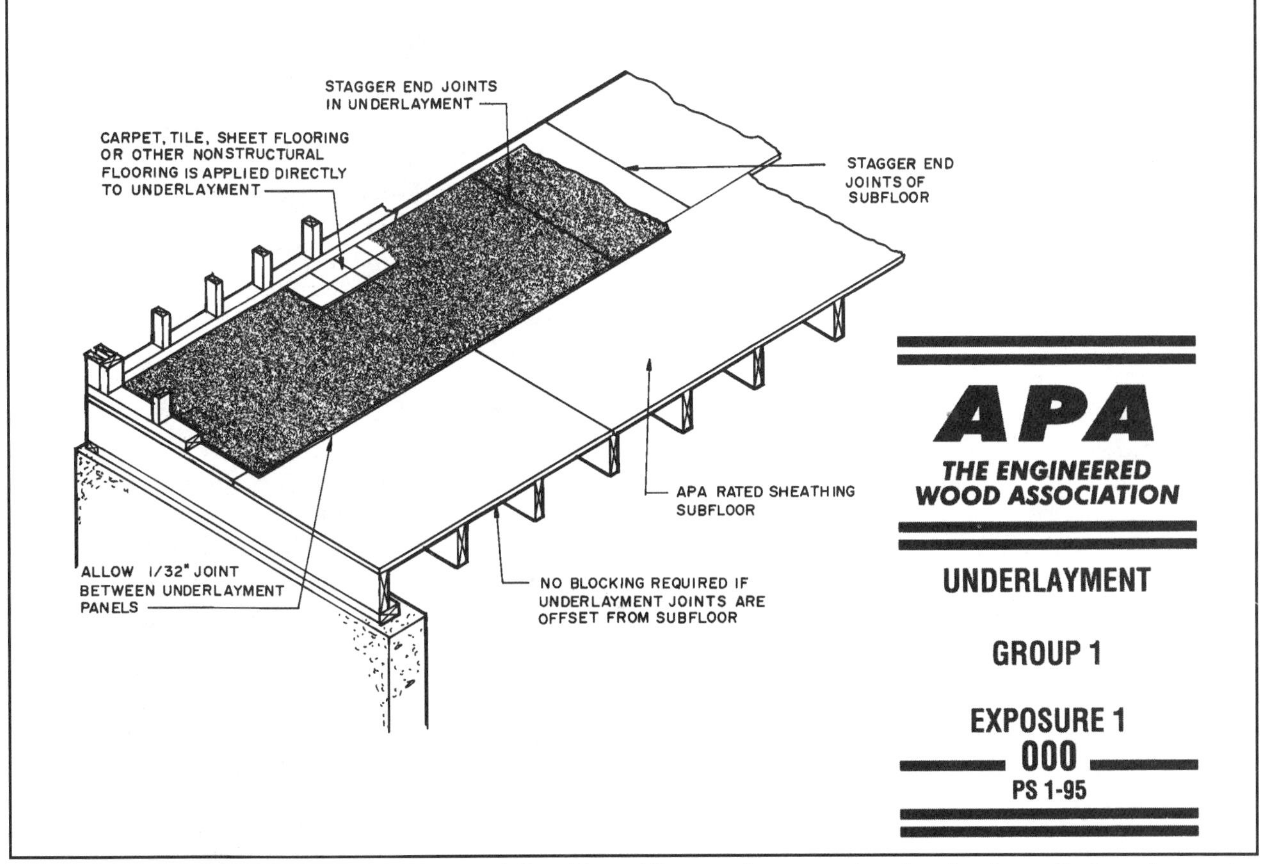

9-94 Underlayment is applied over an APA-Rated Sheathing subfloor to give a smooth surface for the application of a carpet or vinyl floor covering. *(Courtesy APA—The Engineered Wood Association)*

9-95 The joists receive a ribbon of adhesive to help bond the subfloor to the floor joists and reduce floor squeaks. *(Courtesy APA—The Engineered Wood Association)*

The American Plywood Association has a thoroughly tested glued floor system using field-applied construction adhesives to permanently secure wood-based subfloor panels to wood joists. The glue bond joins the joists and subfloor so that they behave like integral T-beam units. This increases the stiffness of the floor and is especially effective when tongue-and-groove panels are used **(see 9-95)**. The tongue-and-groove joint should also be glued. If square-edged panels are used, 2 × 4 blocking must be placed below unsupported joints. Gluing not only produces a stiffer floor but helps eliminate squeaks, bounce, and nail-popping.

A ¼-inch-diameter bead of adhesive is applied to the sill and joists. When panel ends butt, apply two beads of adhesive. As soon as a panel is set in place, it should be nailed using 6d ring- or screw-shank nails for panels up to ¾ inch thick and 8d ring- or screw-shank nails for thicker panels. The nails on glued panels are spaced 12 inches O.C. on the edges and intermediate joists for panels up to 23⁄32 inch thick. Thicker glued panels require edge nails every six inches on the edge and 12 inches on any intermediate supports. Subfloors are generally nailed with pneumatic nailers. Power screwdriving tools enable a carpenter to secure the subfloor with wood screws easily and rapidly. The tool holds the screws on a belt or cartridge and automatically sets one in position as needed **(9-96)**.

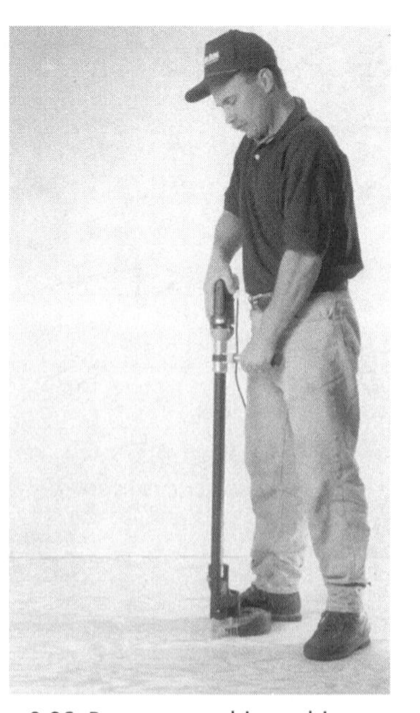

9-96 Power screwdrivers drive wood screws through the subfloor into the joist, providing a very rigid connection. *(Courtesy Quik Drive USA)*

Another method for securely bonding the subfloor to the joists uses a patented, formulated gasket that absorbs the shrinking and swelling of the floor joists. It provides a uniform coverage on the joist and helps eliminate squeaks in floors. It is used on all types of floor joists **(see 9-97)**.

After the subfloor has been finished check it for levelness with a transit. Review Chapter 5. A finished installation is in **9-98**.

9-97 This special formulated tape bonds the joist to the subfloor and provides a uniform gasket that helps eliminate floor squeaks. *(Courtesy Shadwell Company, Inc. and G. Newyear, Photographer)*

9-98 The finished floor assembly ready for the construction of the walls.

ORIENTED STRAND BOARD (OSB)

Another widely used subfloor material is oriented strand board. Information on OSB products is in Chapter 38. OSB panels, made by bonding wood wafer or strand products, have consistent strength and properties.

Sheathing grade OSB is used for subflooring. Structural 1 sheathing grade has additional cross-panel strength and can be used as subflooring when additional strength and stiffness are required. The third grade, Single Floor, is a combination subfloor and underlayment.

WALL & PARTITION FRAMING

There are two framing systems in use, platform framing and balloon framing. The emphasis in this book is on platform framing because it is the current widely used system. Balloon framing has limited use.

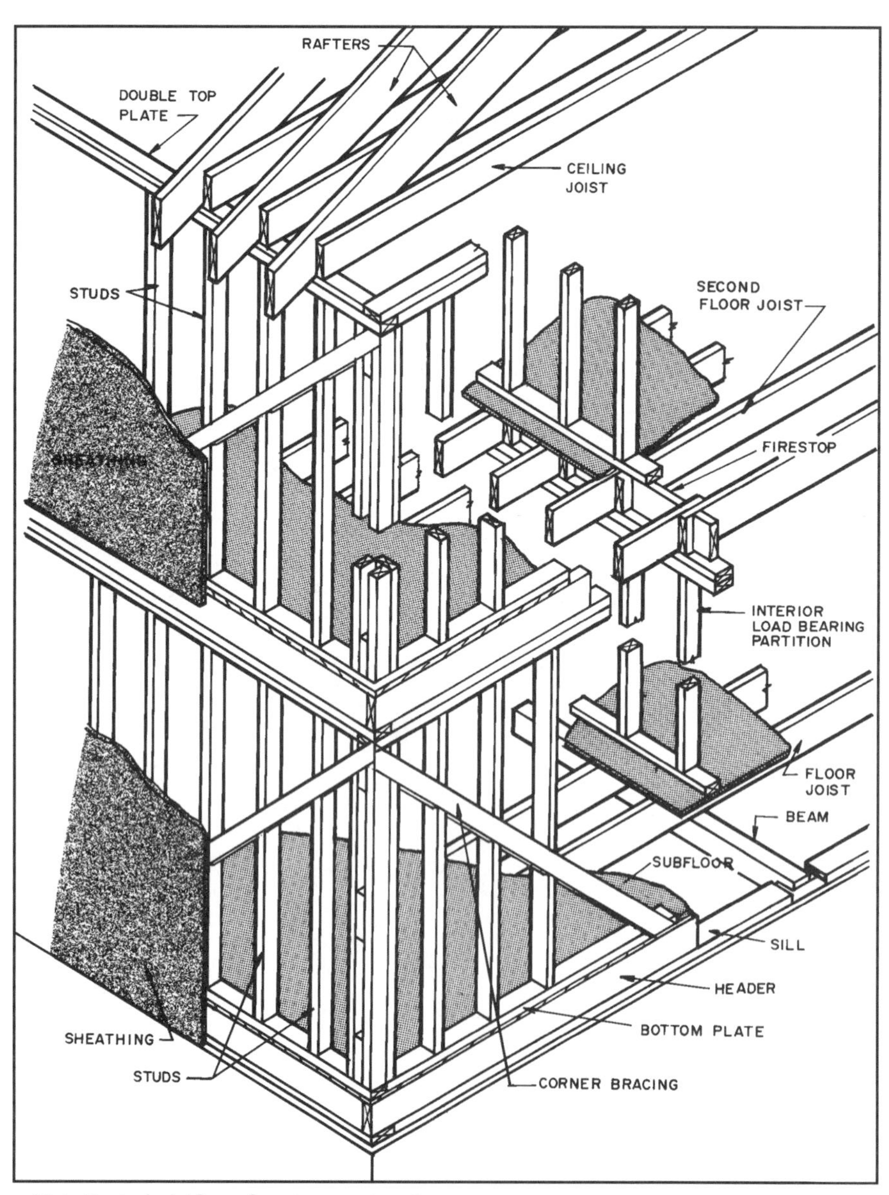

10-1 Typical platform framing construction.

PLATFORM FRAMING

Platform framing is most commonly used in residential construction. Each wall is one story high. The framing for a two-story house resembles the framing of a one-story house built directly on top of another one-story house **(see 10-1)**. Platform framing has the advantage of using shorter pieces of lumber, and at each floor level it provides a solid platform upon which the carpenter can work.

The first floor is built on the foundation, providing a flat platform upon which the interior partitions and exterior walls are erected, as shown in **10-2**. In one-story houses ceiling joists rest on the top plate of the first floor walls. These become floor joists for a two-story house. This forms a second platform upon which the second story wall and partitions are erected. The roof is framed with the rafters resting on the top plate of the second-story walls.

BALLOON FRAMING

Balloon framing uses a continuous stud running from the sill to the top plate, as shown in **10-3**. The weight of the roof and the second-floor ceiling joists is transmitted directly to the foundation through the studs. The studs rest on the sill, which should be two-inch-thick material and anchored to the foundation.

The first floor joists rest on the sill on the foundation. Notice that the second-story floor joists rest on a wood ribbon let into the studs. The studs run from the sill to the top plate of the second story. The rafters rest on the top plate of this wall in the same manner as platform framing.

PLATFORM FRAMING TECHNIQUES

The following discussion details the typical methods for assembling the structure using platform framing techniques.

Walls & Partitions

Since most light-frame buildings are constructed using platform framing, this system of framing will be emphasized in this chapter. After the floor joists are in place and the subfloor is glued and nailed to them, the framers begin constructing the exterior walls and interior partitions. The exterior wall frame serves as the base on which sheathing and siding are applied and carries most of the roof load. It also carries the weight of the ceiling, second floor if there is one, and roof. Interior **partitions** (walls) divide the space into rooms and may support a ceiling or second floor. They also support the interior wall finish material, typically gypsum wallboard. Partitions supporting an overhead load are called **load-bearing partitions** while those that simply divide up the space are **nonload-bearing partitions**. The structure of the floor must be designed to carry the loads imposed by bearing partitions as shown in Chapter 9.

10-2 The floor joists, header joists, and stringer joists covered with subflooring form a platform upon which the walls, partitions, ceiling, and roof are built.

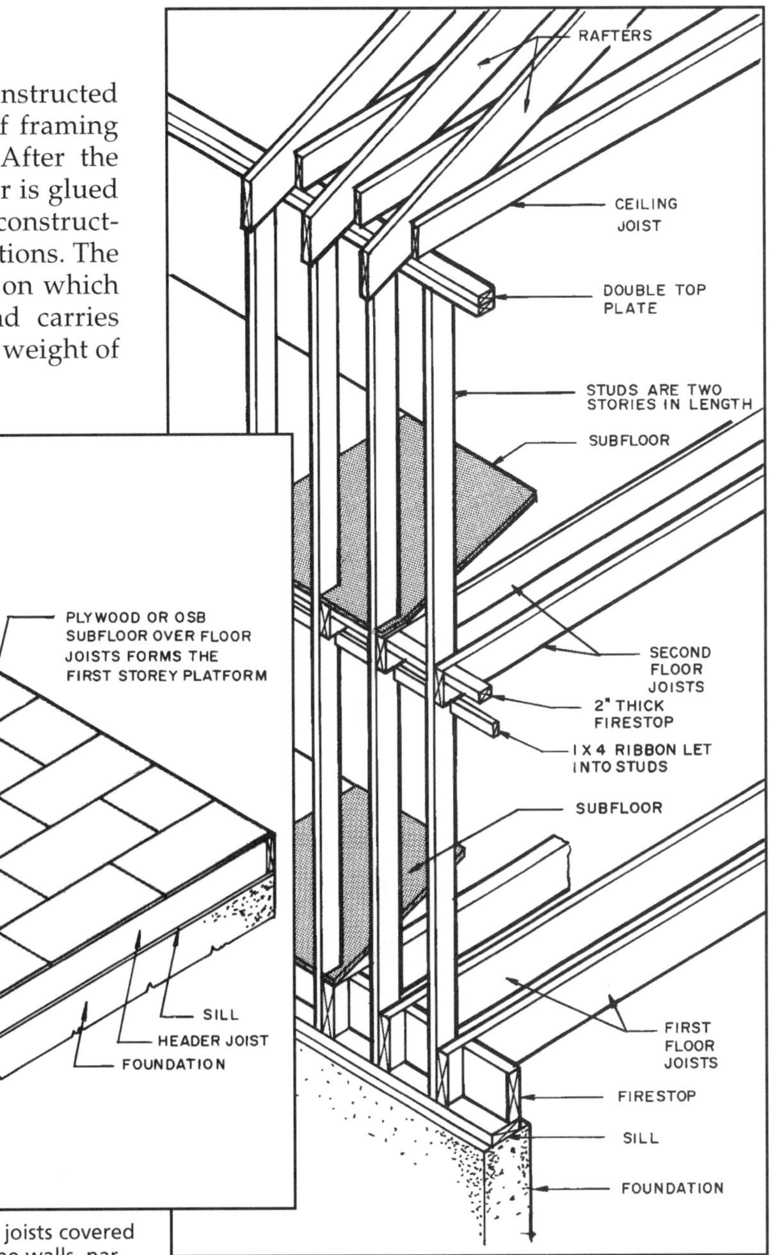

10-3 Typical balloon framing construction.

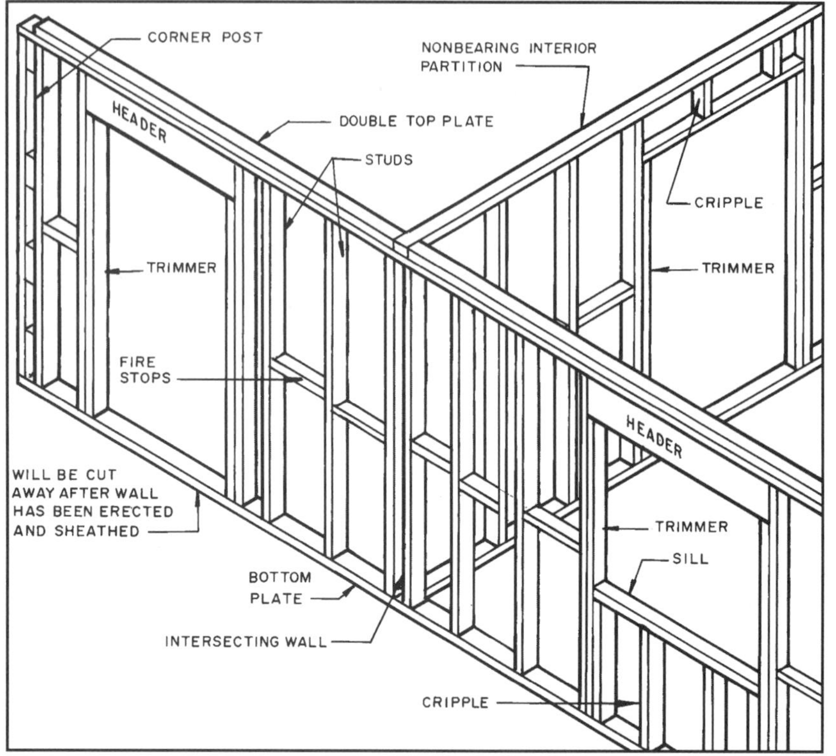

10-4 Typical wood framed wall construction.

Framing the Exterior Wall

A typical exterior wall for platform framing consists of a bottom or sole plate, studs, double top plate, sills, corner stud assemblies, trimmers, fire stops, some type of bracing, and sheathing **(see 10-4)**.

Typical residential construction uses 2 × 4-inch studs. However, 2 × 6-inch studs are used when a thicker wall is wanted so extra insulation can be installed. The 2 × 4-inch studs are generally spaced 16 inches O.C. In some cases 24 inches O.C. might be specified. The 2 × 6-inch studs are spaced 24 inches O.C. When laying out the wall or partition there must be a stud at every 16-inch or 24-inch location even though additional studs might fall in between **(see 10-5)**. These are needed to provide a nailing surface for the interior wall finish material and sheathing. This spacing must be very accurate **(see 10-6)**.

Framing lumber is available in two-foot lengths, such as 10 feet and 12 feet. However, precut studs, 92⅝ inches long, are available for framing walls when an 8-foot ceiling is desired. Using precut studs saves the framer a lot of time, because no studs have to be cut to length **(see 10-7)**. This gives a rough interior height from the subfloor to the ceiling joist of 8′-1⅛″. When floor and ceiling finish materials are installed the finish height is close to 8′-0″.

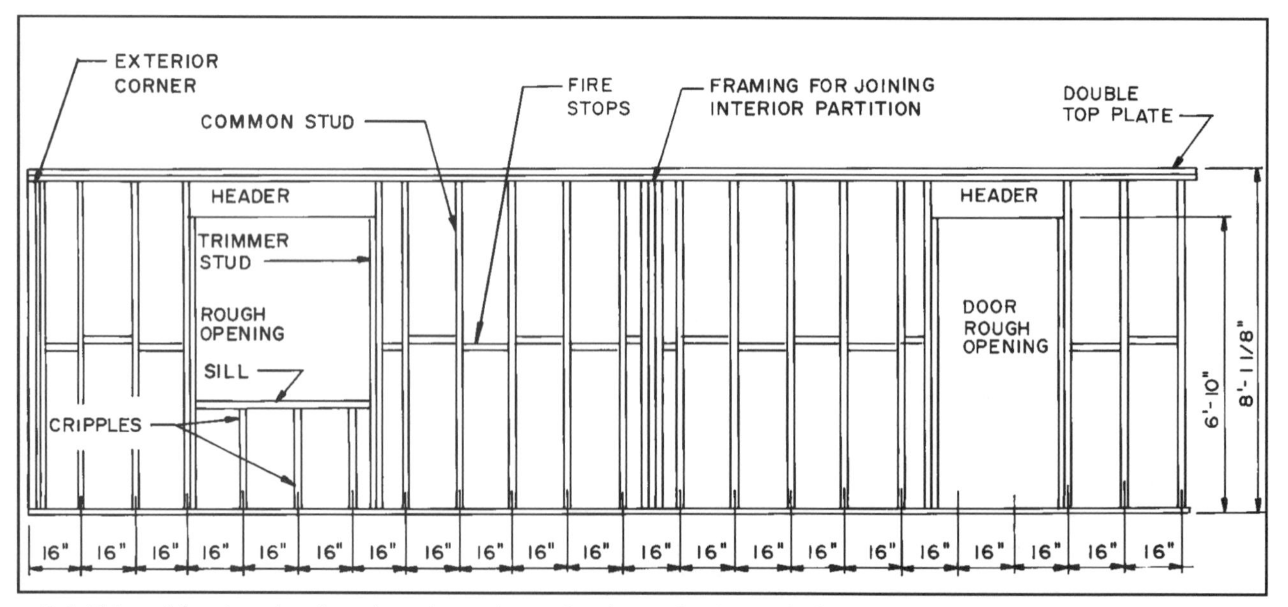

10-5 This wall framing plan shows how the studs are placed to maintain a 16-inch O.C. spacing.

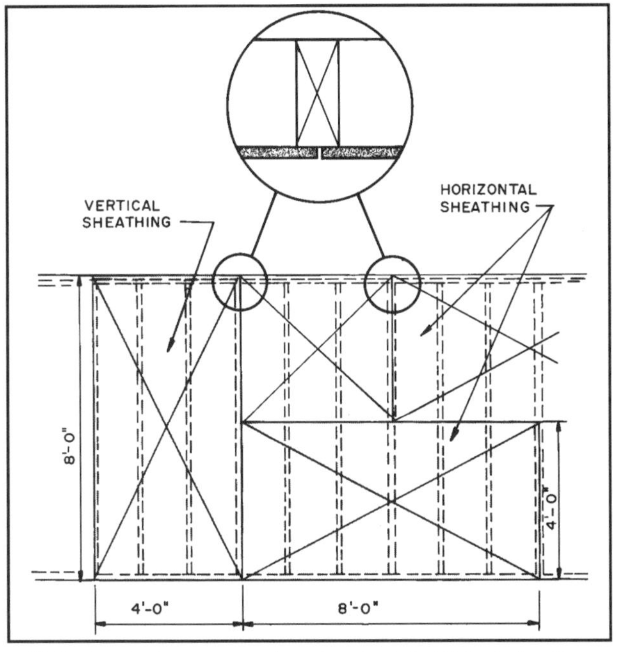

10-6 The studs are spaced so standard 4 × 8-foot sheathing panels butt on a stud making all edges nailed to a load-bearing surface.

SOLID & ENGINEERED STUDS

Most new residences are built using wood stud walls. While the only available wood studs have until recently been solid wood, now engineered-wood studs are finding increasing use.

Solid wood studs have allowances for natural defects. For example, a 2 × 4-inch stud can have knots up to 2½ inches in diameter and holes up to 1½ inches in diameter, and can contain splits, checks, bows, cups, twists, shakes, and wane. A 2 × 6-inch stud can have even larger knots and holes than the 2 × 4-inch size. These defects make it very difficult for framers to build straight, flat walls. If the studs warp after the wall is framed they may have to be straightened or replaced. They are straightened by cutting a kerf and inserting a wood wedge. When the stud is straight, wood splice plates are nailed on each side **(see 10-8)**. Green studs add an additional problem because of shrinkage and warping as they dry. This causes nails to pop as the wood dries and straight walls to become out of plumb. Never use green studs.

When preparing to frame the walls lay the studs out on the subfloor. Remove those not acceptable and save them to be cut up into blocking and backing. The good studs are laid flat on the deck with the crowns up. In this way all the studs in a wall will have the crown on the same side. After the wall is assembled and erected those needing straightening can be corrected.

Straight studs are especially important in walls, such as in the kitchen and bath, where cabinets must be fit against them and finish materials, such as plastic-laminate countertops, are required to fit snugly against the finished wall.

One type of engineered stud is a finger-jointed product **(see 10-9)**. **Finger-jointed studs** can reduce the problems caused by solid wood studs. They are made by gluing up short, dry lengths of end-jointed lumber. These short lengths have the natural defects reduced considerably, and since the lumber has been thoroughly dried it is straight with little bowing, twisting, or warping.

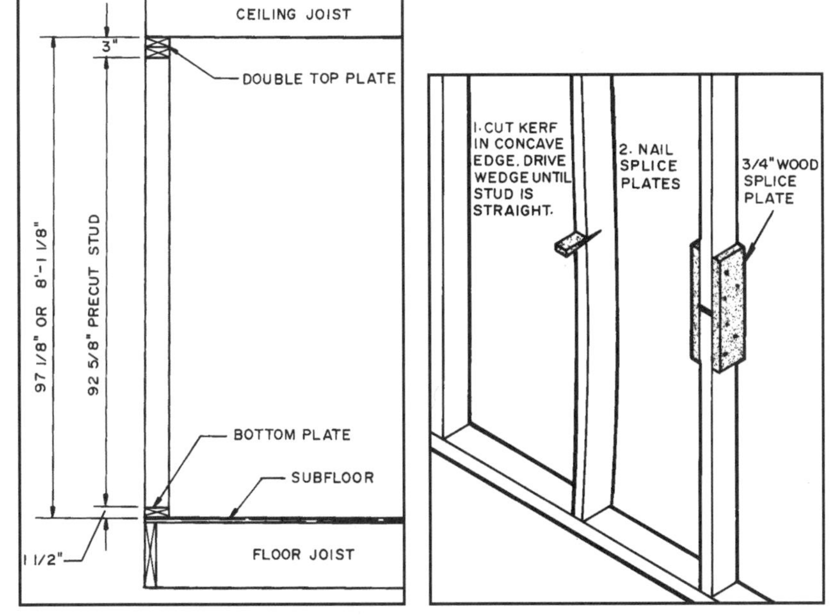

10-7 The standard wall height produced when using studs precut to a standard length.

10-8 Studs with a slight bow can be straightened as shown. If the bow is too great replace the stud.

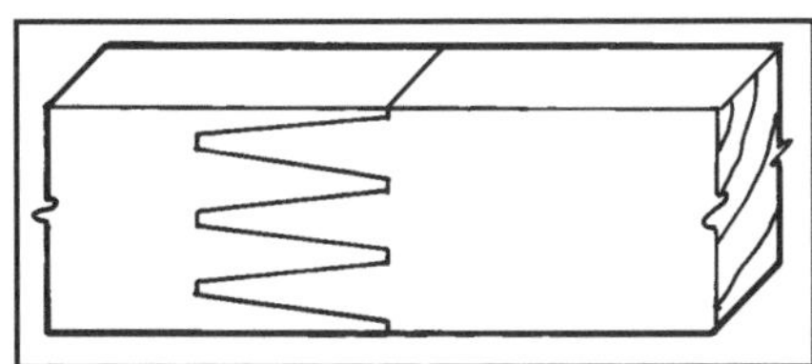

10-9 A finger joint used to manufacture studs by bonding together short lengths of clear wood.

10-10 These window openings in an exterior wall have solid wood headers.

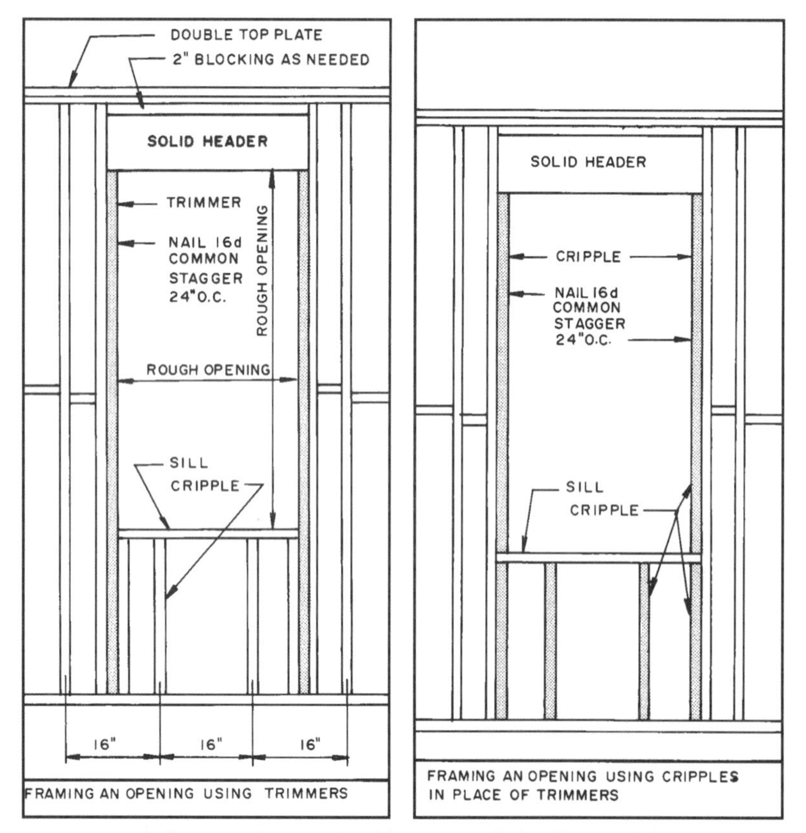

10-11 A window opening in a load-bearing wall that has a solid header. Notice the trimmer stud nailing requirements.

Finger-jointed studs are accepted by codes as a substitute for solid wood studs. These tend to be the same cost or slightly more expensive than solid wood studs.

Another engineered stud, TimberStrand® (LSL), is a laminated strand lumber. Long strands of wood are treated with a polyurethane resin and compressed into billets. These billets are sawed into 2 × 4- and 2 × 6-inch lumber in lengths up to 22 feet. Longer sizes are available. They have a uniform moisture content, and no defects, wane, pitch, warp, twist, or bow. They are very dimensionally stable. They tend to cost more than solid wood and finger-jointed studs. However, they produce the straightest wall. Some contractors use them for all partitions where straight walls are essential (such as in the kitchen).

The decision as to which to use depends upon the feelings of the contractor. A comparison must be made concerning initial cost, installation cost, cost due to call-backs after the job is finished, and the desire to produce a high-quality job.

FRAMING DOOR & WINDOW OPENINGS

Openings in walls, such as those needed for doors and windows, must be spanned by headers strong enough to carry any imposed loads. Headers are required only in load-bearing walls and partitions **(see 10-10)**. Framing for a typical window opening is shown in **10-11**. The studs on each side are doubled, and the header rests on the **trimmer** stud. The header may be one that, while it is large enough to carry the imposed load, does not fill the space up to the top plate. If this design is used, **cripples** are placed as shown in **10-12**. The cripples are spaced to maintain the 16- or 24-inch O.C. spacing. A sill is placed on the bottom side of the window

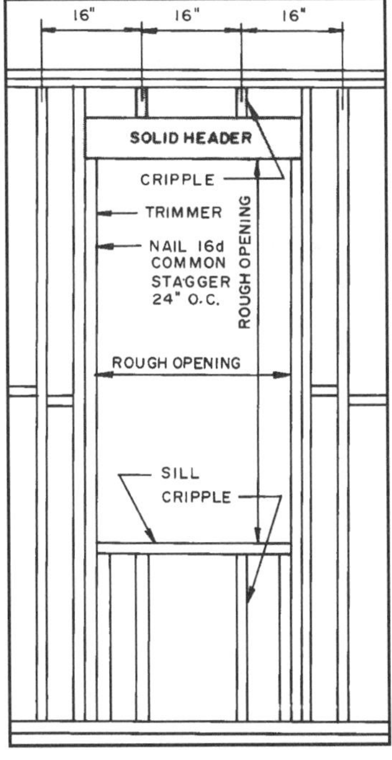

10-12 A window opening in a load-bearing wall when the minimum header is used and cripples are placed between it and the double header.

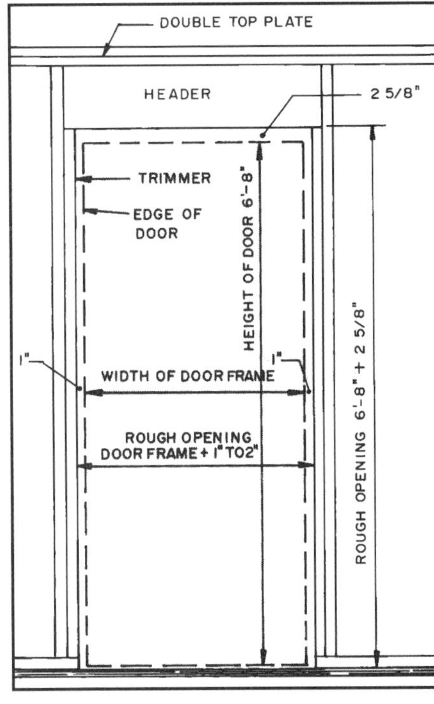

10-13 The rough framing for a door opening in a load-bearing wall or partition.

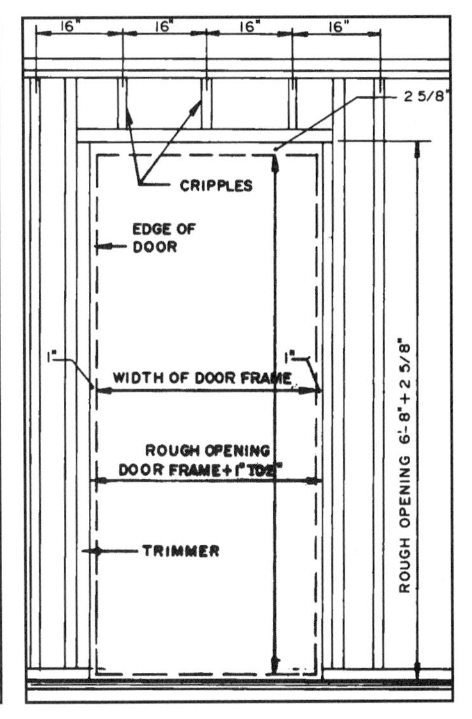

10-14 The rough framing for a door opening in a nonload-bearing partition.

opening. Some framers prefer to double the sill. Below the sill cripple studs are located, maintaining the 16- or 24-inch O.C. spacing. The rough opening for windows is established by the manufacturer and indicated on the architectural drawings on the window schedule.

Framing for a typical exterior door opening is shown in **10-13**. It is framed just like the window opening. Notice that the sole plate continues across the opening. This is needed to stiffen the wall frame until it is raised, braced, and sheathed. Then it is cut away. The size of the rough opening is determined by the size of the door to be used. While framers vary in the allowances they use, it is typical to make the rough opening one to two inches wider and 2⅝ inches higher than the width of the door frame. For interior doors the height of the opening will vary some depending on the thickness of the flooring to be used. For a door opening that is in a nonload-bearing partition, a larger header is not required, as shown in **10-14**. Commonly used door framing dimensions are given in **Table 10-1**.

Table 10-1 Standard framing dimensions for rooms with an 8'-0" finished ceiling.

Wall height with precut stud (92⅝") (subfloor to top of double plate)	**97⅛"**
Type of door	Rough opening size
Rough opening of exterior doors	Door width plus 2" Door height plus 2⅝"
Rough opening of sliding doors	Door width equals size of unit Height for 6'-8" door 80⅝"
Rough opening for interior doors	Door width plus 2" Door height plus 2⅝"
Rough opening for pocket doors	Door width actual size of door Door height plus 2⅝"
Rough opening for bypass doors	Door width actual size of door Door height plus 2⅝"

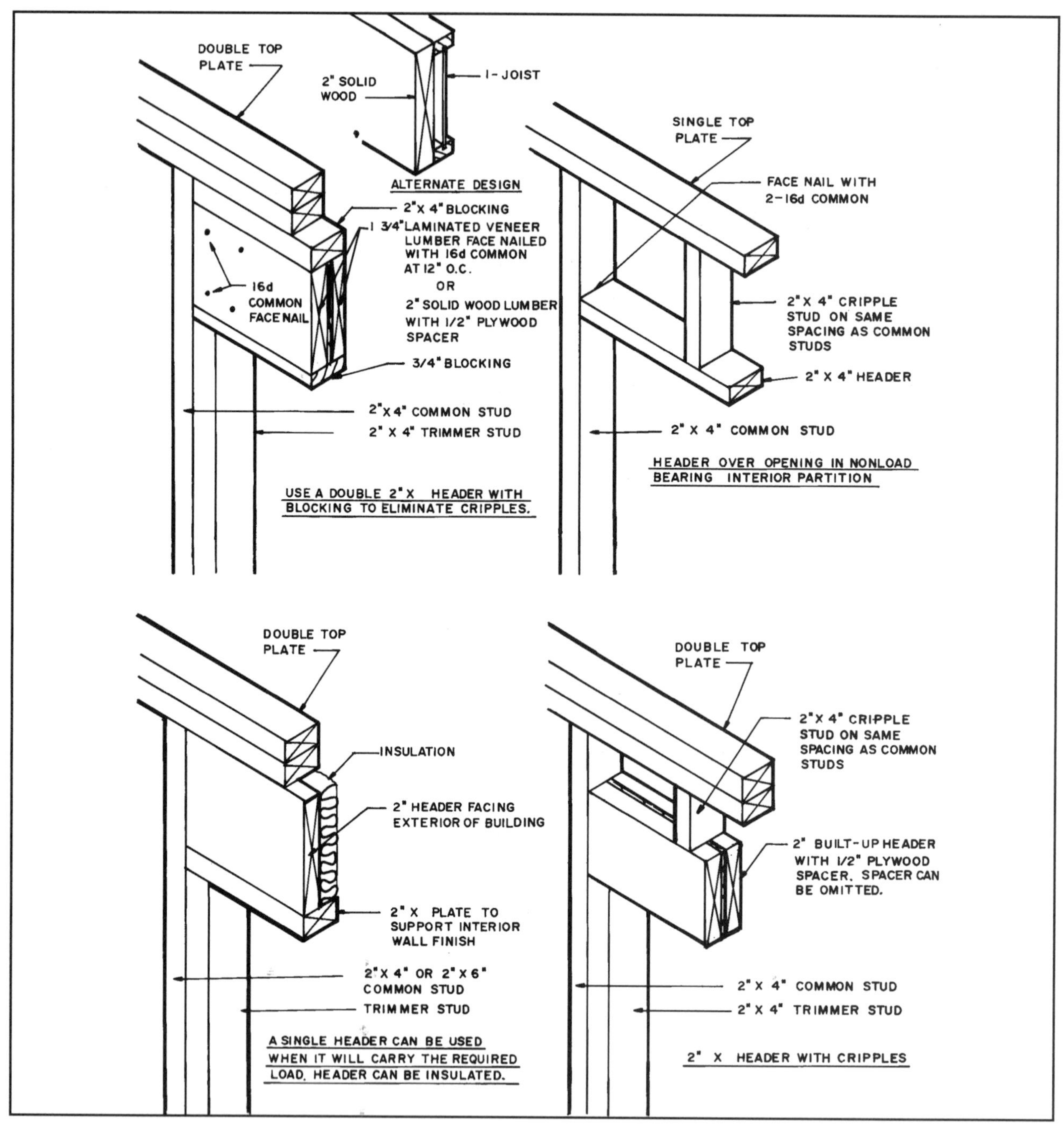

10-15 (Above and opposite) Commonly used types of headers.

HEADERS

Some of the commonly used headers are shown in **10-15** and **10-16**. One commonly used type is built using an assembly of 2-inch-thick stock with ½-inch plywood spacer strips or a solid sheet of plywood between them. This assembly when nailed together gives a 3½-inch-thick header which is the width of a 2 × 4-inch stud.

Several carpenter-built insulated headers are shown. Notice the headers made from solid wood beams or from manufactured materials such as glulam and Parallam® beams. When load conditions permit, such as for an interior wall that has no imposed load, the header can have a 2 × 4 horizontal member and cripples.

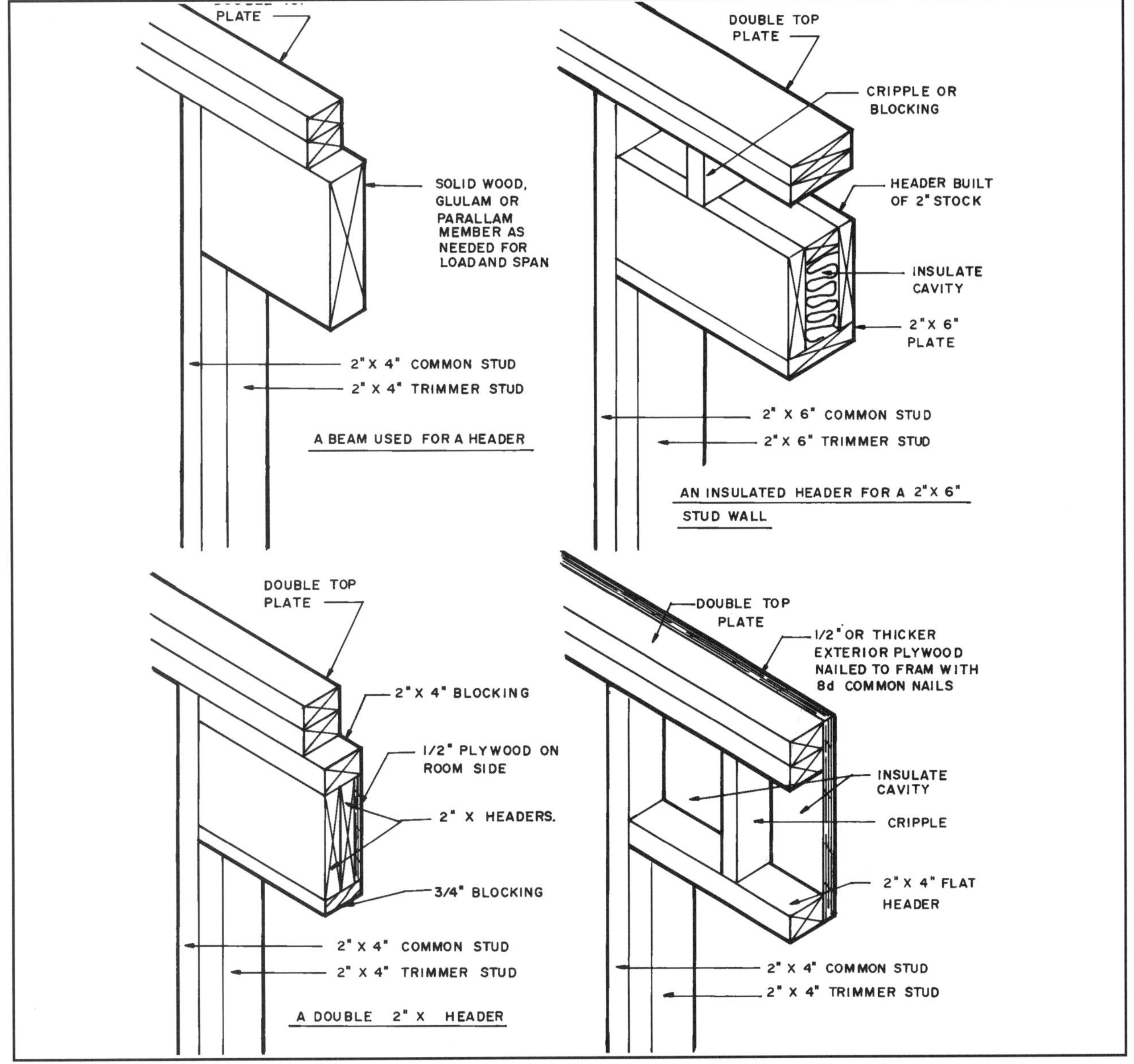

10-15 (Continued) Commonly used types of headers.

10-16 These headers show one made with solid lumber and one using Parallam® lumber.

Table 10-2 Commonly used sizes for headers.*

Header size (inches)	Supporting roof** and ceiling only	Supporting second** floor, roof, ceiling
two 2 × 4	3'-6" max.	
two 2 × 6	4'-0" to 6'-0"	4'-0" max.
two 2 × 8	6'-0" to 8'-0"	4'-0" to 6'-0"
two 2 × 10	8'-0" to 10'-0"	6'-0" to 8'-0"
two 2 × 12	10'-0" to 12'-0"	8'-0" to 10'-0"

* Spans vary depending upon strength of lumber. These data assume a minimum required bending stress of 1000.

** Based on live and dead floor load of 150 psf and roof live and dead load of 30 psf.

10-17 This is a close-up view of a plywood box beam used as a header on a wide opening in a load-bearing exterior wall.

Typical header sizes for built-up headers using two-inch material are shown in **Table 10-2**.

Plywood box beams can also be used as headers. They are light in weight, and the cavity can always be filled with insulation **(see 10-17)**.

Headers for long spans, such as a garage door, require special design considerations. A manufactured beam such as a glulam or Parallam® will span the distance, as will a plywood box beam. A suggested framing plan for a garage door or other large opening is shown in **10-18**. Headers for spans 9 feet to 12 feet are supported by double trimmer studs on each side. Triple trimmers are used on openings over 12 feet. In addition to the use of manufactured beams, other types of engineered headers are used where loads exceed the normal roof and floor loads. Some examples of designs using steel channels or plates are shown in **10-19**.

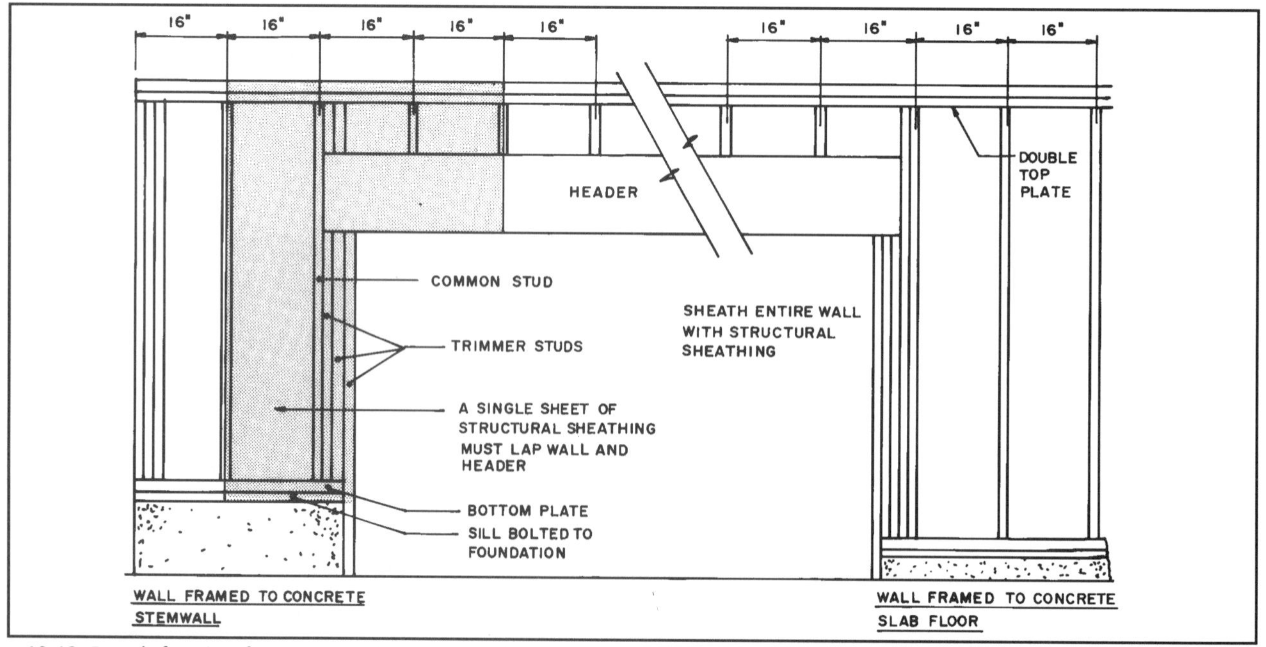

10-18 Rough framing for a garage door or other large opening.

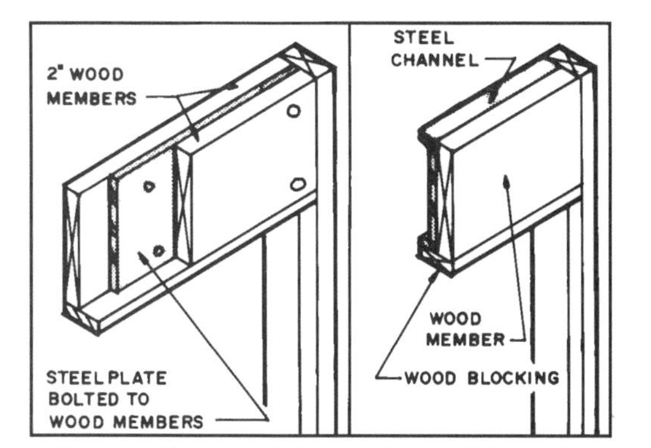

10-19 These headers are reinforced with steel channel or plate.

When framing a rough opening, the height of the header above the floor is usually 6′-10″ to 6′-11″. Normally the height of the rough opening for windows is the same as for doors. The header is placed on trimmer studs to provide the opening **(see 10-20)**. There are many ways to frame this and several are illustrated. Frequently the framer will use oversized headers to avoid the trouble of installing cripples.

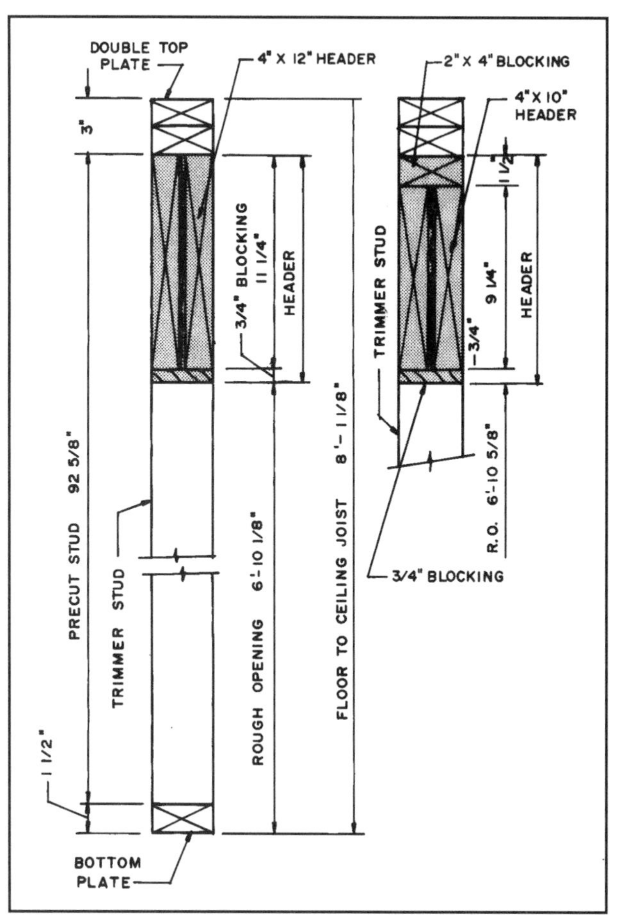

10-20 Typical lumber header installation details.

FIRE BLOCKS

Fire blocks are horizontal 2 × 4 blocking members installed between wall studs about in the middle of their length. They are staggered so they can be end-nailed. The carpenter runs a chalkline mark along the wall and nails one block above it and one below it. Examples can be seen earlier in **10-4** and other figures. They also provide a nailer for the gypsum wallboard at the middle of the wall.

FRAMING EXTERIOR CORNERS

There are several ways to frame the corners where exterior walls meet. It is important to provide some means for securing the interior wall finish material. Several ways to frame the corner when using platform construction are shown in **10-21**. The use of wallboard

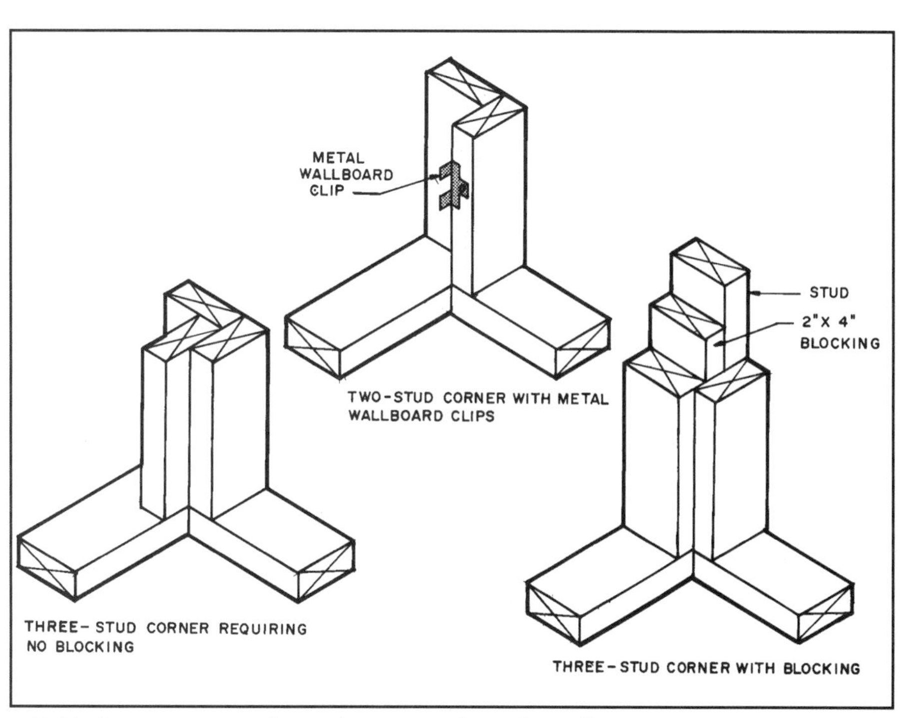

10-21 Common ways to frame the corners of exterior walls.

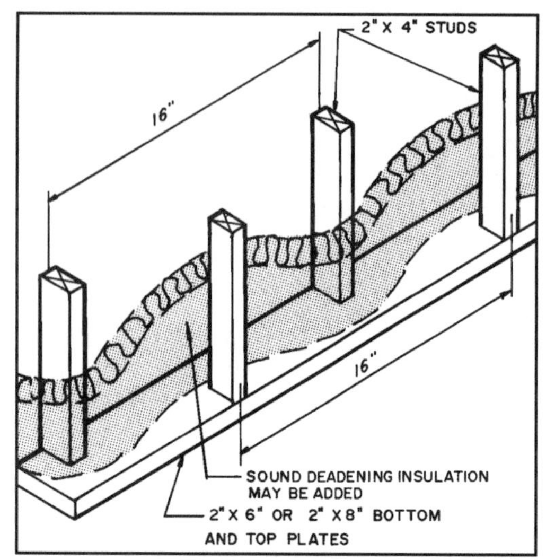

10-22 A wall framed with staggered studs is used to reduce the transmission of sound into the next room.

FRAMING INTERIOR PARTITIONS

Some interior partitions are load-bearing and carry the weight of the ceiling structure, and of a second floor if one is required. Most partitions are nonload-bearing and support only the wall finish material such as gypsum board.

Interior load-bearing partitions are framed in the same manner as exterior walls, including the use of headers. Nonload-bearing partitions use the same framing, except that large headers over door openings are not required. Generally 2 × 4 studs are used for partition construction. When extra soundproofing is needed, 2 × 6 or 2 × 8 studs can be used. Staggered studs are also used to improve soundproofing. In most cases, a 2 × 6 top and bottom plate and 2 × 4 studs are used. Sound-deadening insulation can be placed between the studs **(see 10-22)**. As a space-saving technique, 2 × 3 studs or 2 × 4 studs placed flatwise can be used. A typical example is in the wall separating closets. Typical load-bearing and nonload-bearing partitions are shown in **10-23**.

Partitions that run perpendicular to the ceiling joists can be nailed to the joists **(see 10-24)**. Notice the use of two-inch blocking on the top plate to provide a nailing surface for ceiling finish material.

Wider partitions are often necessary when the partition will contain plumbing. Generally a 2 × 6 stud partition will be adequate. If cast-iron soil pipes are used, a 2 × 8 stud partition is needed. If pipes are to be run parallel to the partition, the studs can be turned flatwise **(see 10-25)**.

Partitions running parallel to the ceiling joists must be anchored to 2 × 4 nailers **(see 10-26)**. Notice

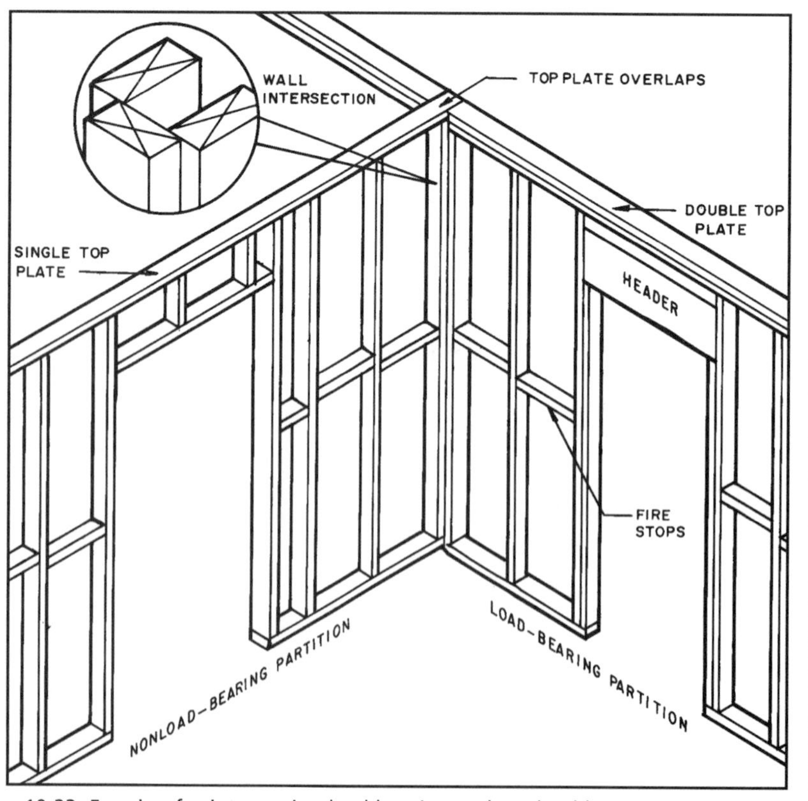

10-23 Framing for intersecting load-bearing and nonload-bearing interior partitions.

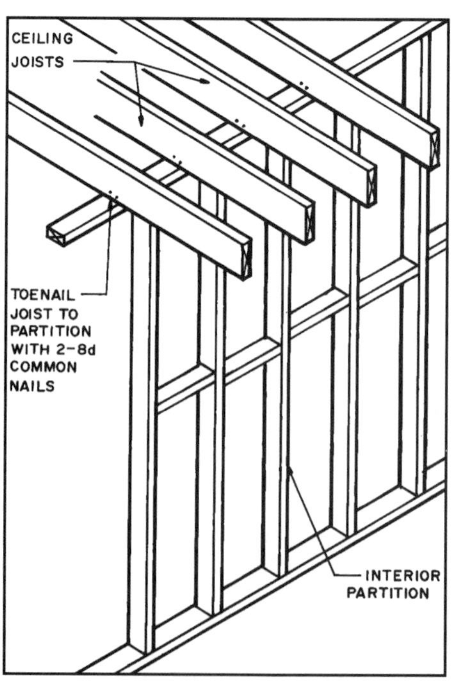

10-24 Ceiling joists are toenailed to partitions that run perpendicular to them.

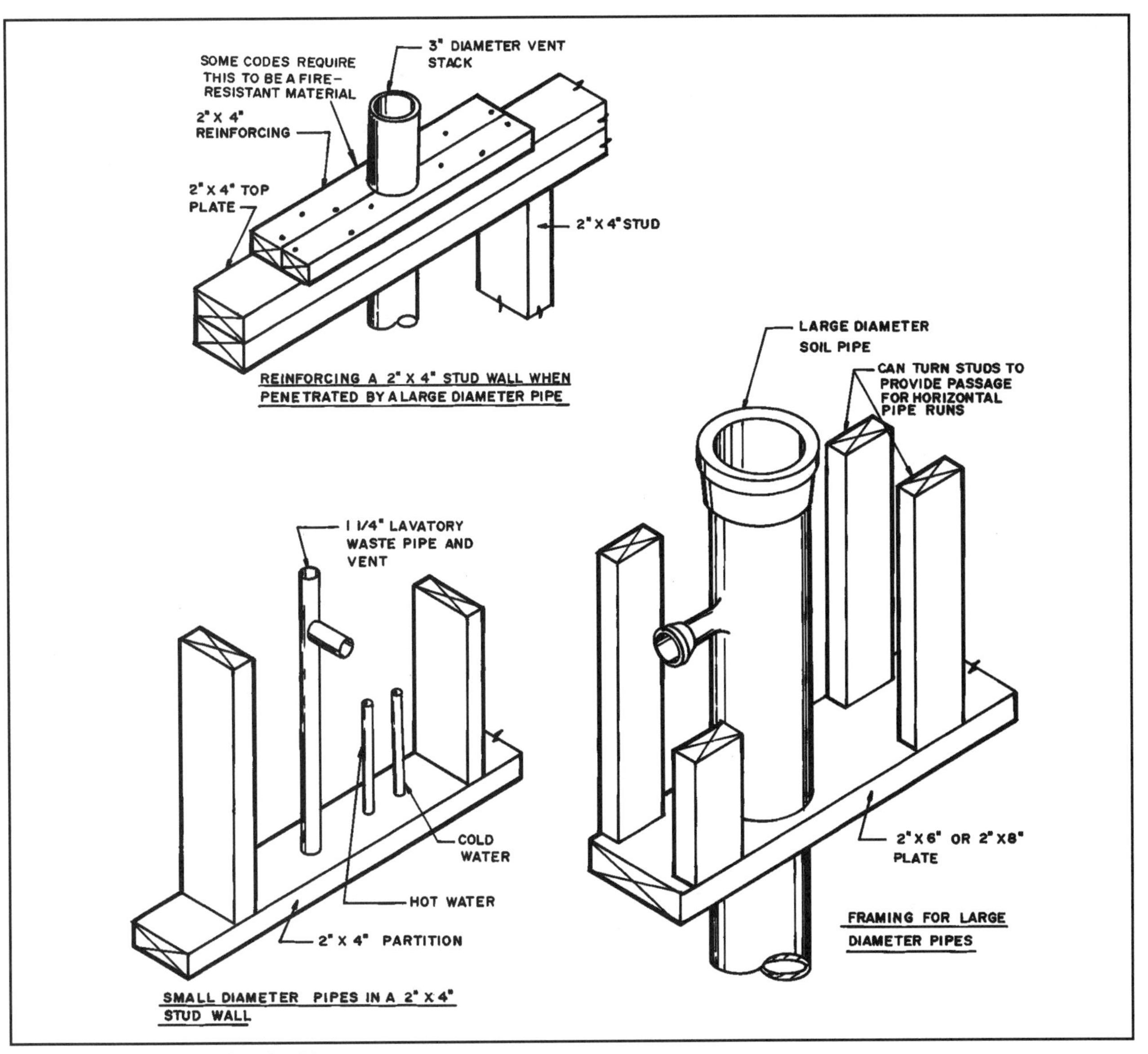

10-25 Typical framing for plumbing.

clips reduces the need for an extra stud.

the one-inch-thick boards on top of the partitions that provide a nailing surface for the finished ceiling.

Nonload-bearing partitions are adequately supported by the subfloor even if they run between, and parallel to, the floor joists, if the subfloor is ⅝ inch or thicker. They are nailed through the bottom plate into the subfloor. Floor joists should be doubled under load-bearing partitions that run parallel with the joists. Loads greater than normal may require that a beam be placed below the partition.

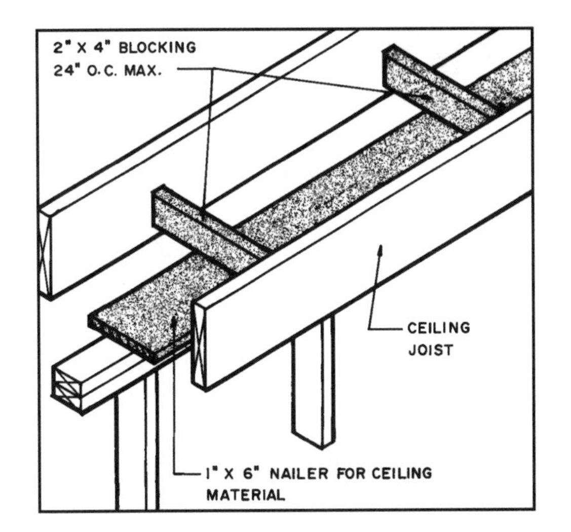

10-26 (Right) How to secure interior partitions to the ceiling when they run parallel with the ceiling joists.

INTERSECTING PARTITIONS

Partitions intersect each other and the exterior wall. The union must provide a surface to which the interior wall finish material may be installed. There are several ways this is done. One way uses 2 × 4 blocking between studs. Usually one piece at the midpoint of the wall is adequate. Metal gypsum-board backup clips are nailed to the stud.

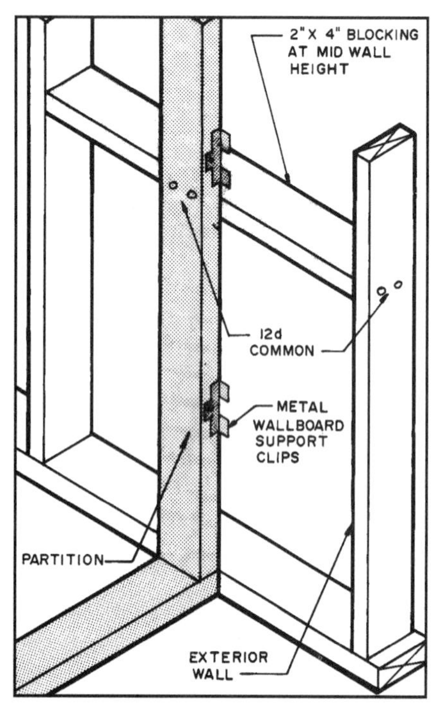

10-27 One way to secure the intersection of a partition to another wall or partition.

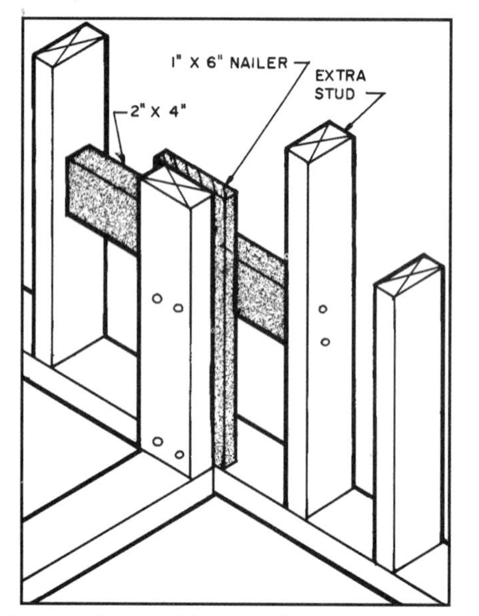

10-28 Intersecting partitions can be secured using horizontal blocking and a 1 × 6-inch nailer.

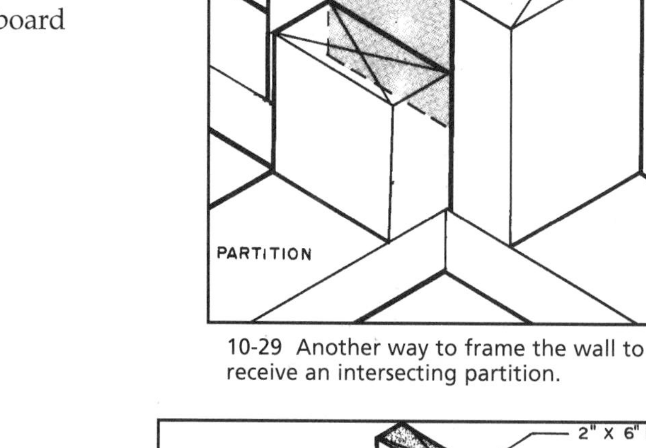

10-29 Another way to frame the wall to receive an intersecting partition.

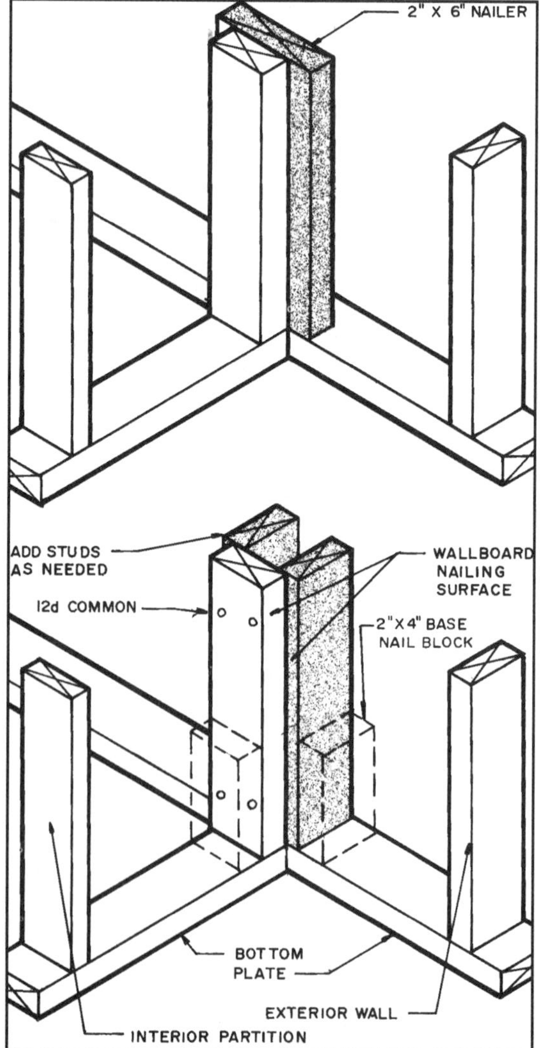

10-30 Two additional frequently used ways to secure intersecting walls and partitions.

The clips provide a nonrigid joint which tends to minimize cracking **(see 10-27)**. A variation of this is shown in **10-28**. An extra stud is inserted and a 2 × 4 block used to support a 1 × 6 nailer. The nailer provides the needed supporting surface for the interior wallboard. The use of 2 × 4 blocking is also used. This spaces the studs in the intersected wall 3½ inches apart **(see 10-29)**.

Possibly the most commonly used framing techniques are shown in **10-30**. Extra studs are used to provide the needed supporting surface. They are not needed for structural purposes. A variation of these uses a 2 × 6 nailer placed in the wall. This is a quick way to provide needed supporting surfaces. Backup and nailing surfaces for wall finish materials such as plywood paneling should be built as recommended by the manufacturer.

The top plate of the interior partition overlaps the lower top plate of the exterior partition. This provides a strong tie and structural support **(see 10-31)**.

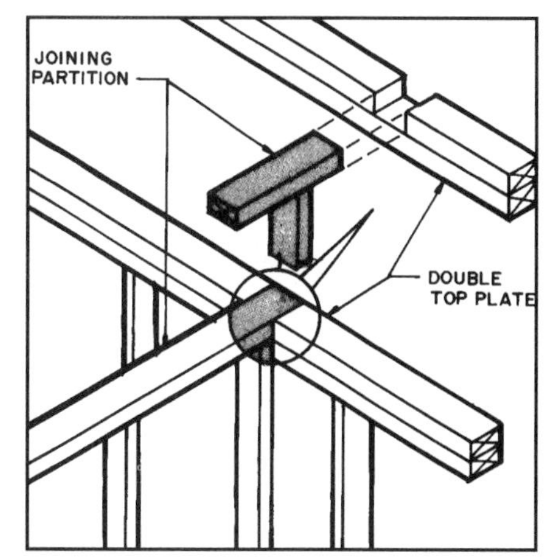

10-31 The top plate of intersecting walls and partitions overlap.

Wall Layout on the Floor

After the subfloor is in place the walls and partitions are located on it by snapping chalklines giving their exact location. These marks are used to set the assembled wall sections in place. These measurements are found on the architectural drawings. The carpenter should be aware that many architects locate partitions by their centerlines while the floor layout gives the outside edges of the bottom plate. Some carpenters prefer to locate one side of the partition and mark the side where the partition is to go with an X. Since this can lead to error many prefer to snap a line showing both sides of the bottom plate. Examples can be seen in **10-32**.

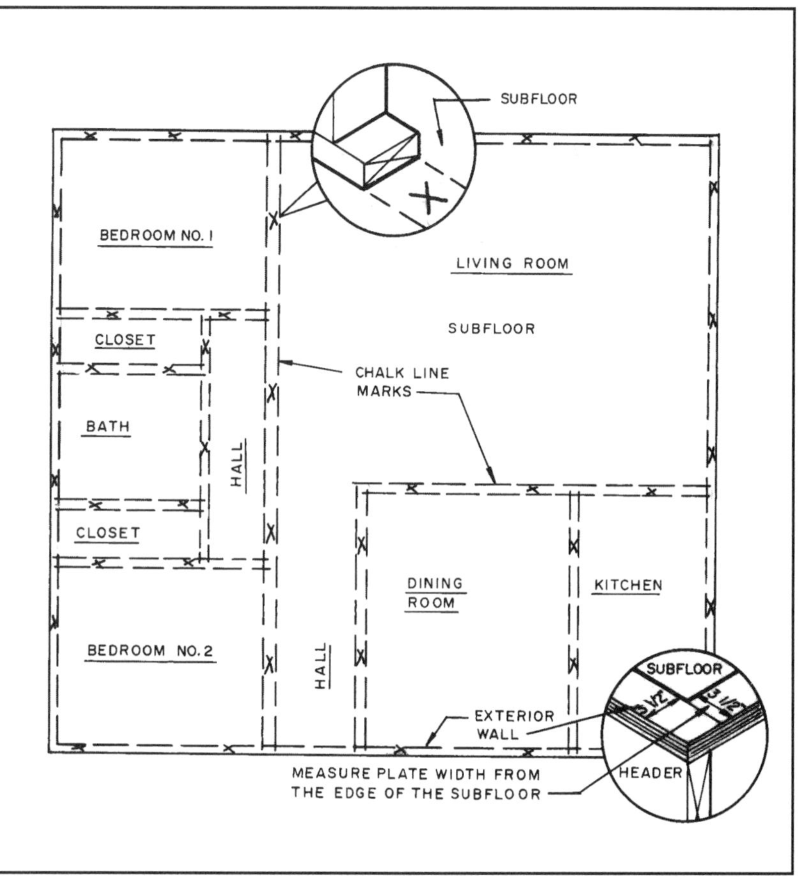

10-32 This shows the walls and partitions marked on the subfloor with a chalkline.

LAYING OUT THE PLATE

Wall and partition construction is begun by locating the studs, rough openings, trimmers, and cripples on the top and bottom plates. Select straight stock. Since it often takes several pieces to run the length of a wall, place the joint so that it will be below a stud. When possible locate the studs so that they are directly above the floor joists and so that the rafters are directly above them.

To lay out the plates, place them along the chalkline mark on the subfloor and locate the partition backers, stud trimmers, and cripples for any openings. Then locate the studs **(see 10-33)**. Symbols frequently used to mark the plates are shown in **10-34**. Other symbols may be used in various localities.

The wall layout starts at a corner of the front wall. The location of the studs should be such that when the stock 4′-0″ sheathing panels are set flush with the outside face of the end stud, they hit the center of a stud down the wall. This means the first stud beyond the corner framing is set 15¼ inches from the outside face of the end stud **(see 10-35)**. In order to maintain the 16-inch O.C. spacing on the end wall, the first stud beyond the corner stud is located 11¾ inches from the outside face of the end stud.

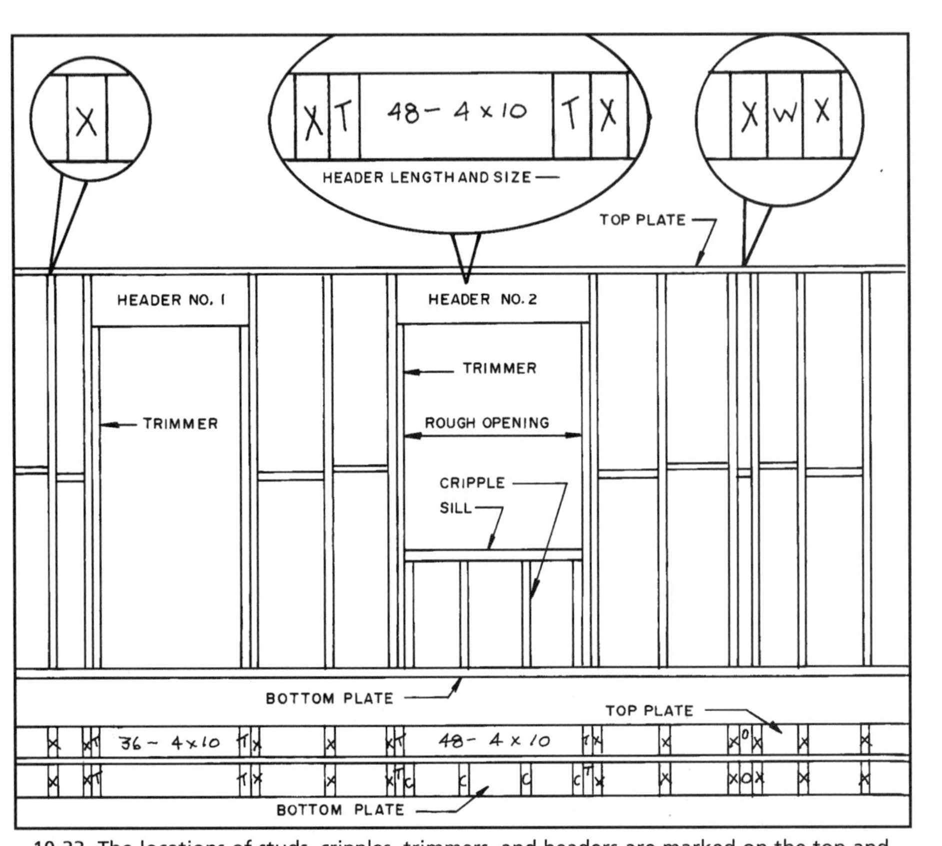

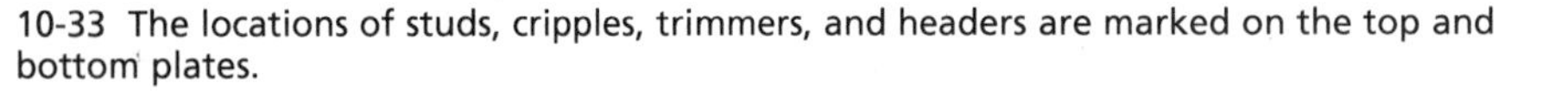

10-33 The locations of studs, cripples, trimmers, and headers are marked on the top and bottom plates.

BUILDING THE WALLS

There are many ways to assemble the parts that make up the wall framing. Most framers develop a system that suits their needs. The following is a general plan.

Usually the long exterior walls are framed first. In most cases these are the front and rear walls. The exact procedure depends on the way the framer prefers to do the job. After these long walls are assembled and raised into place, the end

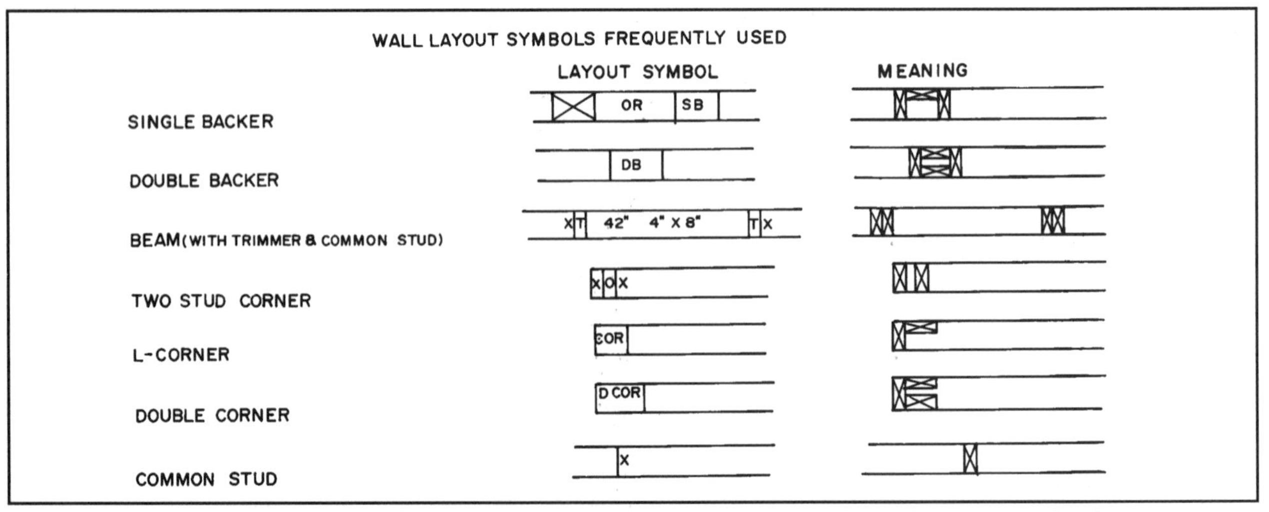

10-34 (Above and opposite) Symbols and their meanings that are frequently used for wall and partition plate layout.

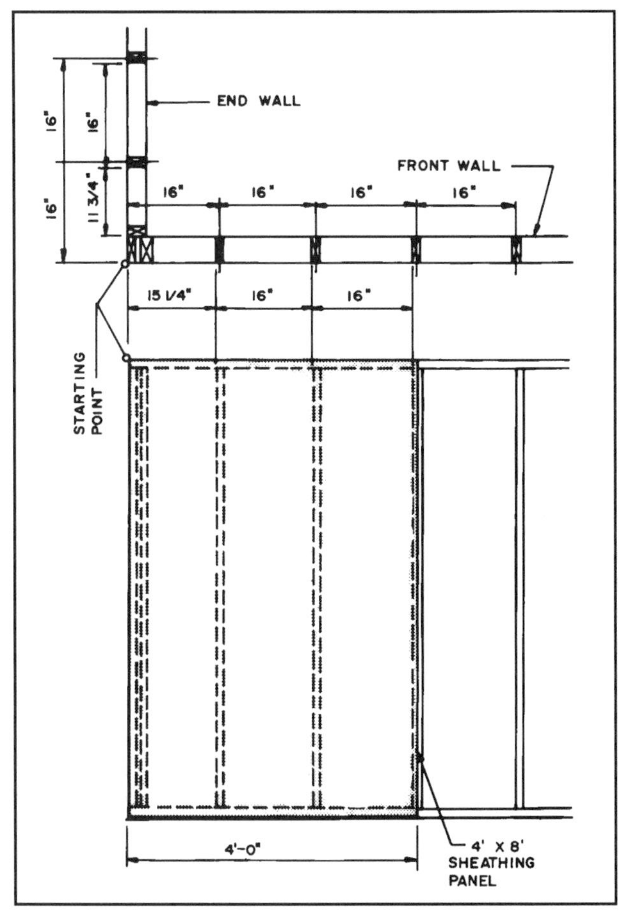

10-35 Studs must be located so the sheathing panels need not be cut. This shows a corner with the studs set for the first sheathing panel.

walls are assembled and set in between the long walls. Interior partitions are generally erected after the exterior walls are in place **(see 10-36)**.

10-36 The long exterior walls are framed and erected first. Then the partitions are framed and erected. *(Courtesy Western Wood Products Association)*

All the parts of the wall are cut before assembly begins. The studs are cut to length (unless precut studs are used), the cripple and trimmer studs are cut, and the headers are assembled and marked indicating their size and location. The top and bottom plates are laid out on the subfloor by the various walls and marked as explained earlier. Now the assembly of the walls and partitions can begin.

Symbol meaning	Plate mark
CRIPPLE STUD	C
STAGGERED STUDS	U D U D
TRIMMER (BESIDE A COMMON STUD)	X T
NONBEARING FLAT HEADER (WITH STUDS)	42" F
NONBEARING L-HEADER (WITH STUDS)	40"L
CRIPPLE HEADER (WITH STUDS & TRIMMER)	48"C
SOLID HEADER (WITH STUDS & TRIMMER)	36"– 4" X 8"
INTERSECTING WALL (WITH STUDS)	X W X

10-34 (Continued) Symbols and their meanings that are frequently used for wall and partition plate layout.

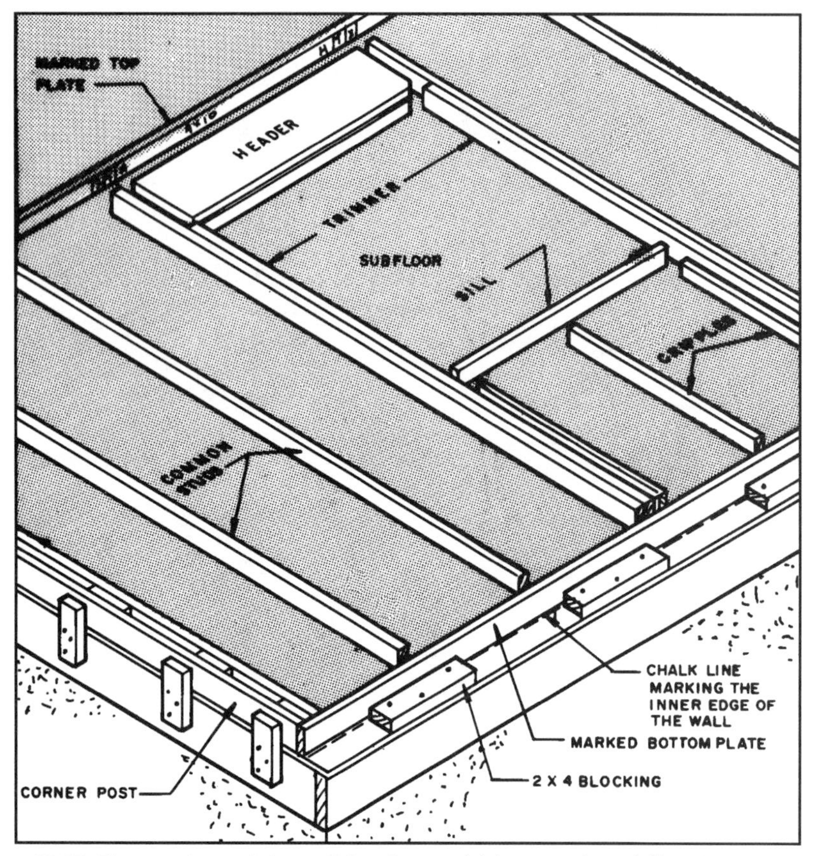

10-37 The members of the wall framing are laid out on the subfloor in preparation for assembly.

It should be noted that the **nailing requirements** mentioned are typical of those found in various codes. The carpenter should always be familiar with the local code requirements.

Assembling the Walls & Partitions

Begin the exterior-wall assembly by laying the top and bottom plates on the subfloor about 8 feet apart. Lay the precut studs, trimmers, cripples, and headers in their approximate locations **(see 10-37)**. Beginning at one end of the wall, nail the top and bottom plates to each full-length stud with two 12d or 16d common nails. Some prefer to use box nails because they are less likely to split the wood. Next, nail the trimmers to the common studs with 12d or 16d common nails **(see 10-38)**. Nail these assemblies to the plates and set the headers on the trimmer studs. Nail through the common stud into the header with 12d or 16d common nails **(see 10-39)**. Some prefer to assemble the header, trimmers, and adjacent full-length studs before placing them in the wall assembly **(see 10-40)**. Install any sills, and then nail the cripples in place. Carpenters are laying out the studs and plates for another section of the exterior wall in **10-41**.

Some carpenters prefer to add bracing and sheathing to the frame while it is still on the subfloor. Before doing this make certain the frame is square by measuring the diagonals **(see 10-42)**. If the diagonals are the same length, the frame is square. Bracing and sheathing can then be added.

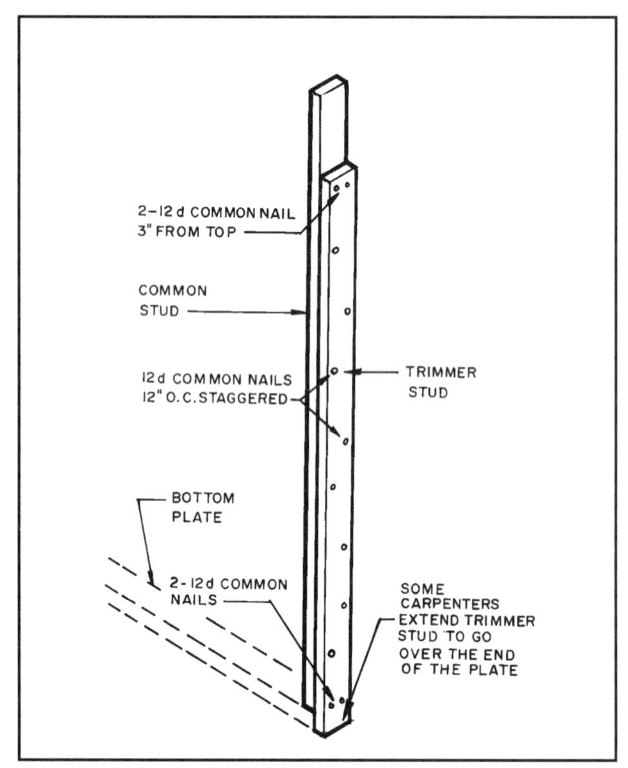

10-38 The trimmer stud is face-nailed to a common stud.

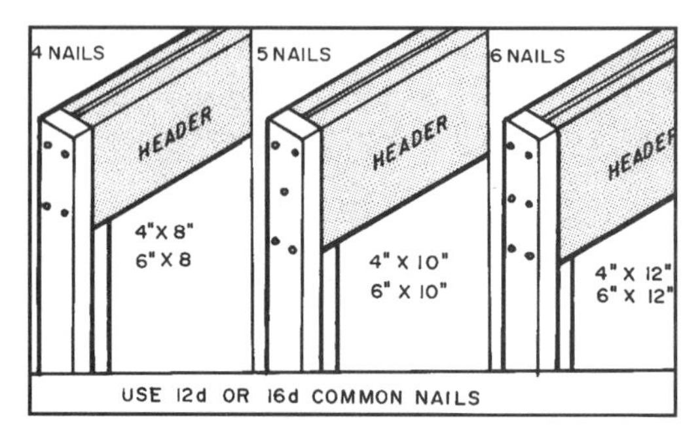

10-39 Typical header nailing requirements.

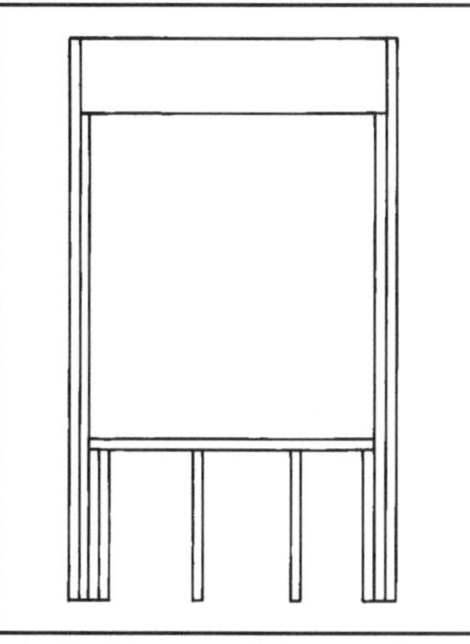

10-40 Window and door rough opening assemblies are often assembled before they are placed in the wall framing.

10-41 These carpenters are laying out the studs in preparation for nailing together a wall section.

Some framers prefer to erect the frame and apply the bracing and sheathing after it is raised into place **(see 10-43)**. Others prefer not only to sheath the frame but to install the windows before raising it.

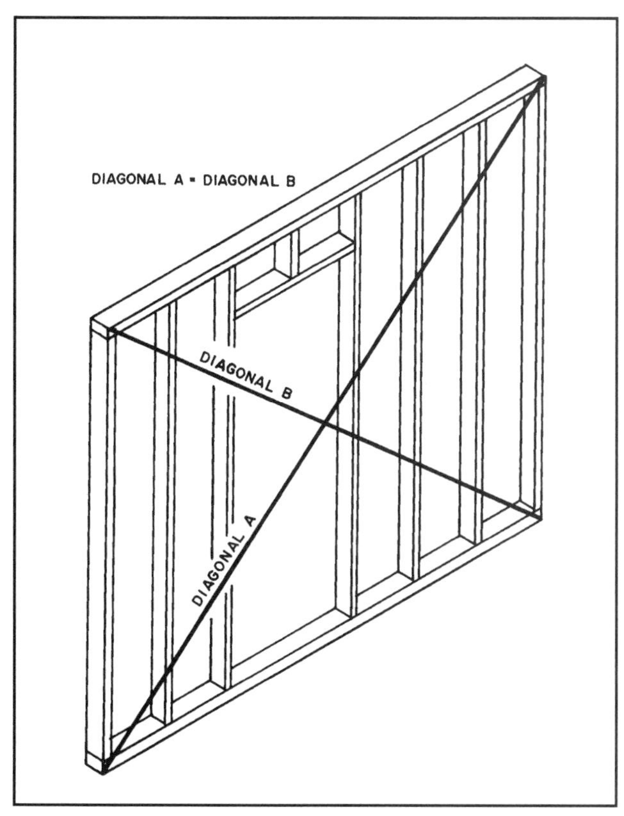

10-42 A wall is square if diagonals A and B are the same length.

Partitions are framed and erected in the same way as exterior walls.

Bracing the Exterior Walls

Building codes require that the exterior wall framing be braced to provide resistance to lateral loads such as wind. The four common ways this is done is with let-in 1 × 4-inch wood members, metal bracing straps, diagonal wood sheathing, or structural panels such as plywood or OSB.

10-43 This rigid foam aluminum foil covered sheathing is being installed after the walls have been erected. Notice the window openings have been sheathed over and cut open after the wall has been sheathed. *(Courtesy The Celotex Corporation)*

Let-in wood bracing requires the studs be notched to receive the 1 × 4-inch wood brace **(see 10-44)**. They are set on a 45-degree angle and run from the top plate to the bottom plate. The brace is nailed to each stud and plate with two 8d common nails or two 1¾-inch staples. If a window interferes, two shorter braces can be used, as you can see in **10-45**. These braces require the time-consuming work of marking and notching the studs and plates and do not provide the strength against lateral forces that is provided by plywood and OSB sheathing. When the assembled frame is square and still flat on the subfloor, cut the notches for the brace. Lay the brace across the studs and mark the location on each. Set the portable circular saw to cut ¾ inch deep and make multiple cuts in the area to be removed. Cut these away with a wood chisel **(10-46)**.

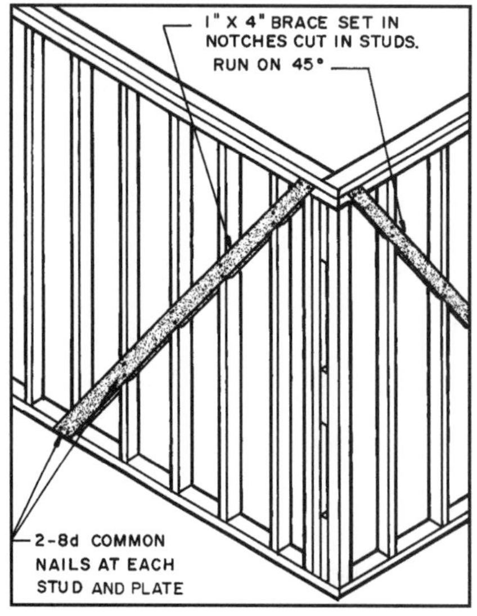

10-44 Let-in wood bracing is placed at the corners and other places along the wall as needed to provide bracing.

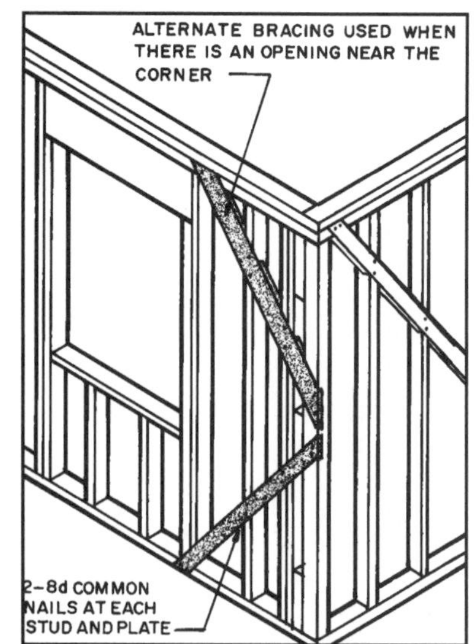

10-45 This is an alternate way to brace a wall with let-in 1 × 4-inch bracing when a wall opening is near the corner.

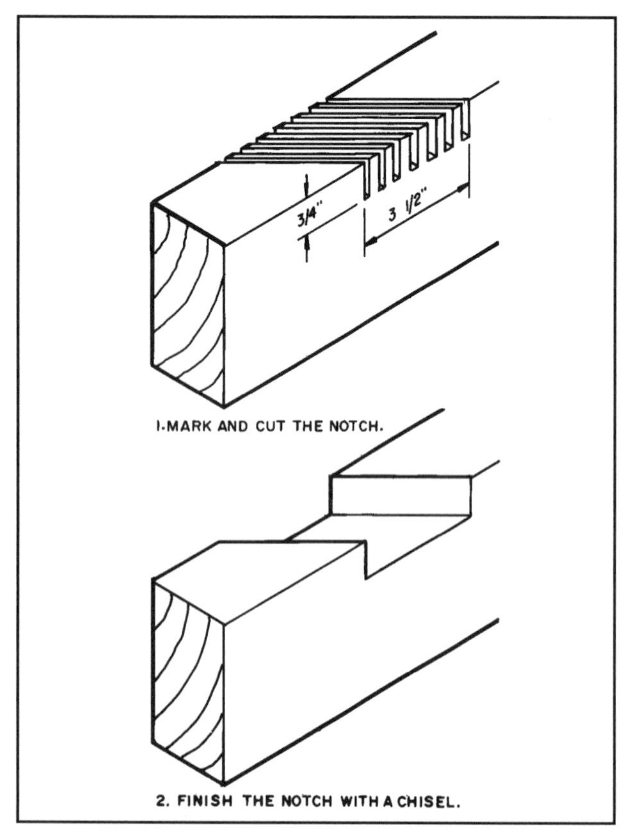

10-46 Lay out and cut the notch with a saw and a chisel.

Metal straps are nailed to the outer face of the studs on a 45-degree angle as described for wood bracing. Wood and metal strap bracing generally provide the minimum bracing required. In areas where lateral forces are greater than normal, diagonal wood or structural panels are required since they provide many times more resistance to lateral and vertical shear forces.

Diagonal wood bracing sheaths the wall with one-inch-thick solid wood boards placed on a 45-degree angle. This provides a strong wall but is very time consuming. Boards 1 × 6-inches require two 8d common or three 8d box nails into each stud **(see 10-47)**.

Plywood and OSB structural panels used as bracing are nailed on the end of each wall and in the center of walls over 25 feet long, as shown in **10-48** and **10-49**. When the entire wall is sheathed with plywood or OSB the wall has greater resistance to lateral forces than diagonal wood sheathed walls.

Plywood and OSB panels applied vertically produce a wall with the greatest lateral load capacity. Panels applied horizontally without blocking on the unsupported edges have about 75 percent of the capacity of the vertical panels. If they have two-inch blocking on the horizontal edges they have the same lateral load capacity. Walls braced with panels on each end have about 40 percent of the lateral load capacity, and those using let-in diagonal bracing have only about 10

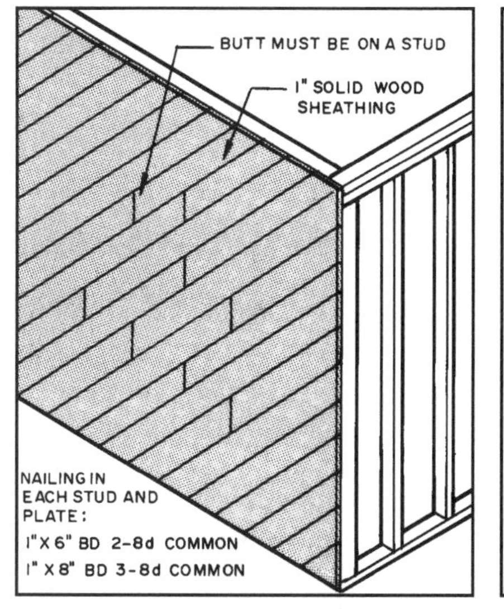

10-47 Diagonal wood siding produces a strong, rigid wall. It takes longer to install than panel sheathing products.

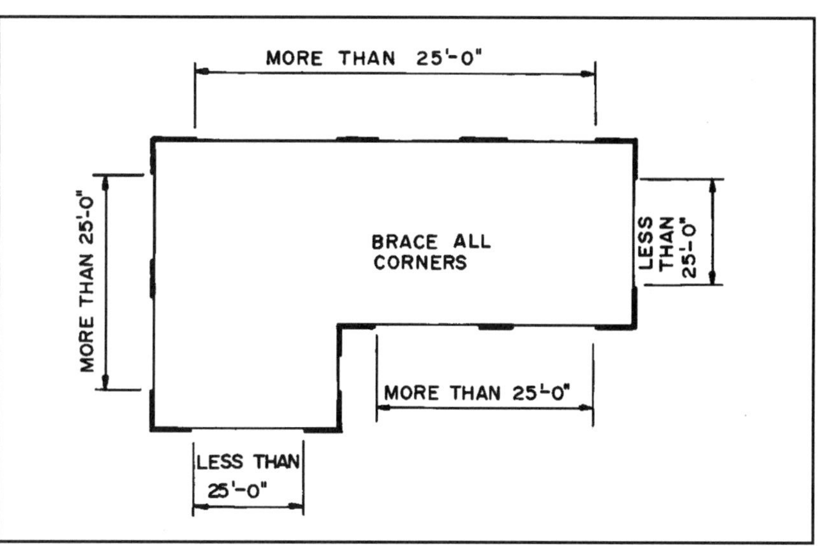

10-48 Typical examples of placing 4 × 8-foot plywood or OSB structural panels for use as bracing. This provides adequate bracing but is not as strong as a wall completely sheathed with these panels.

percent of the capacity of vertical plywood or OSB panels **(see 10-50)**.

If a frame building is more than two stories, wall bracing must be done with structural sheathing panels. Special considerations are necessary if building on a seismic zone or in an area subject to high winds. The designer and framer must consult local codes before proceeding with construction.

Additional information related to sheathing, including nailing requirements, is in this chapter.

Erecting the Assembled Walls

As mentioned earlier, the wall assembly may have the sheathing and bracing applied while it is still flat on the subfloor. This will hold the wall square as it is raised into position.

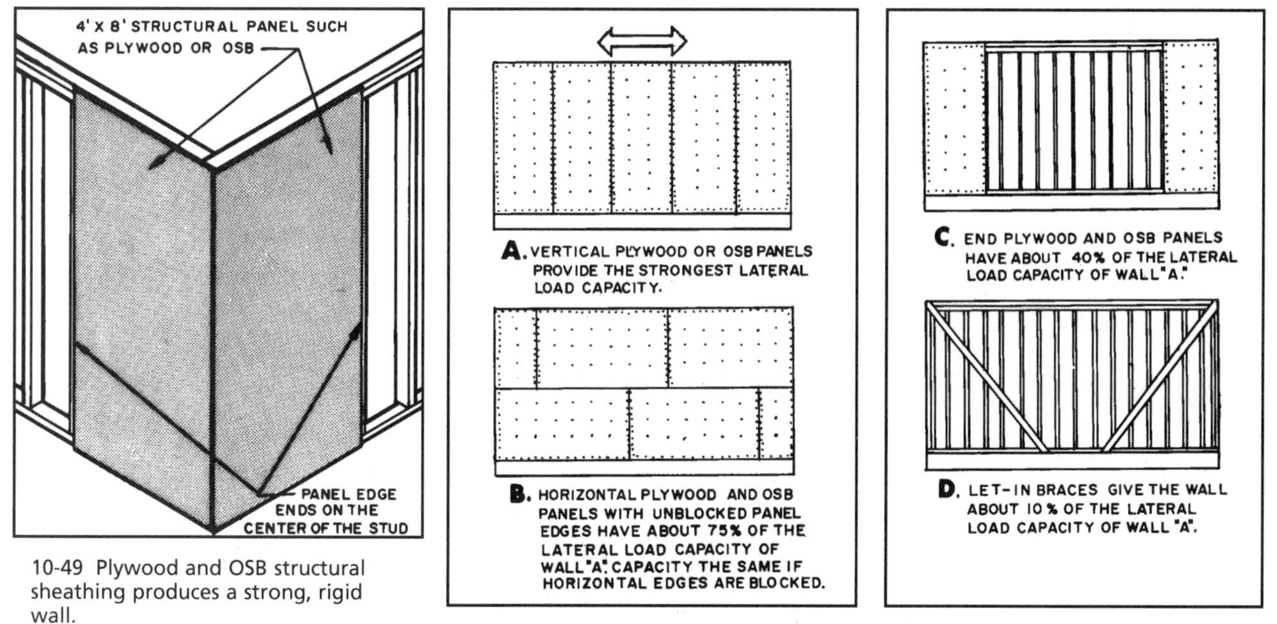

10-49 Plywood and OSB structural sheathing produces a strong, rigid wall.

10-50 The type of bracing and the method of installation greatly influence its bracing strength.

The assembled walls are raised to a vertical position by several carpenters **(see 10-51 and 10-52)**. The wall is set carefully on the edge of the subfloor and temporarily nailed to the header or stringer joist. It may be necessary to adjust it so final nailing is delayed. The raised wall is braced with 2 × 4-inch braces to blocks nailed through the subfloor into a floor joist **(see 10-53)**. Braces generally have to be loosened and readjusted as the total wall is straightened and plumbed. Be certain the braces are placed so they are clear of any interior walls to be erected. It may be necessary to tap the wall section with a sledgehammer to move it into the final desired position on the subfloor.

Install each wall section along the wall. When all are in place and plumb, nail through the bottom plate into the header or stringer joist and available floor joists with 16d common nails spaced 16 inches O.C. **(see 10-54)**.

Wall sections with large openings, as shown in **10-55**, require temporary bracing until the entire wall is erected and permanently braced.

Once all of the exterior walls are in place, plumbed, and braced, they must be checked to make sure that they are square. This is done by measuring the diagonals. If one diagonal is longer, loosen the braces involved and use a winch to pull the long diagonal shorter. After consistent repeated checks of the diagonals reveal that they are the same, then renail the braces **(see 10-56)**.

10-51 The assembled wall frame is lifted into position. This frame does not have the sheathing installed. *(Courtesy APA-The Engineered Wood Association)*

10-52 This wall has been sheathed with rigid foam plastic sheathing. Notice the sheathing extends beyond the bottom plate so the header and stringer joists are covered and insulated. *(Courtesy The Celotex Corporation)*

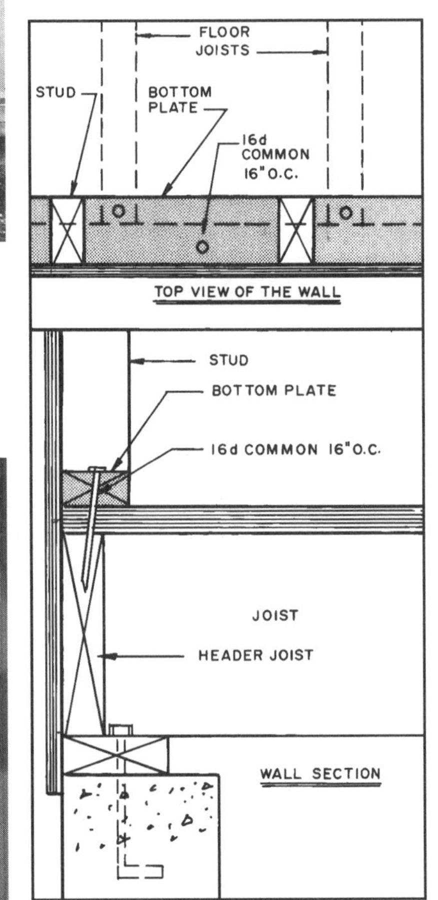

10-54 After the exterior walls are set, leveled, straightened, and plumb, they are nailed through the bottom plate into the header or stringer joists and any floor joists available.

10-53 After the walls are raised they must be plumbed and securely braced.

Straightening a Wall

The wall must also be checked for straightness. This can be done by running a chalkline around the building that is spaced out with ¾-inch blocks. If the wall is ¾ inch away from the chalkline along the length of the wall, it is straight **(see 10-57)**. If not, it will be necessary to readjust some of the braces to push the top of the wall in or out **(see 10-58)**.

10-55 Large openings in the wall must be braced before raising the wall.

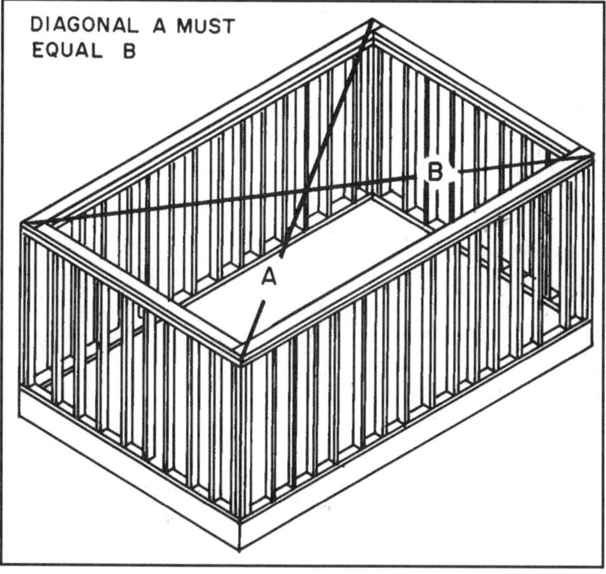

10-56 After the exterior walls have been erected, check the diagonals of the building and adjust the walls until they are equal.

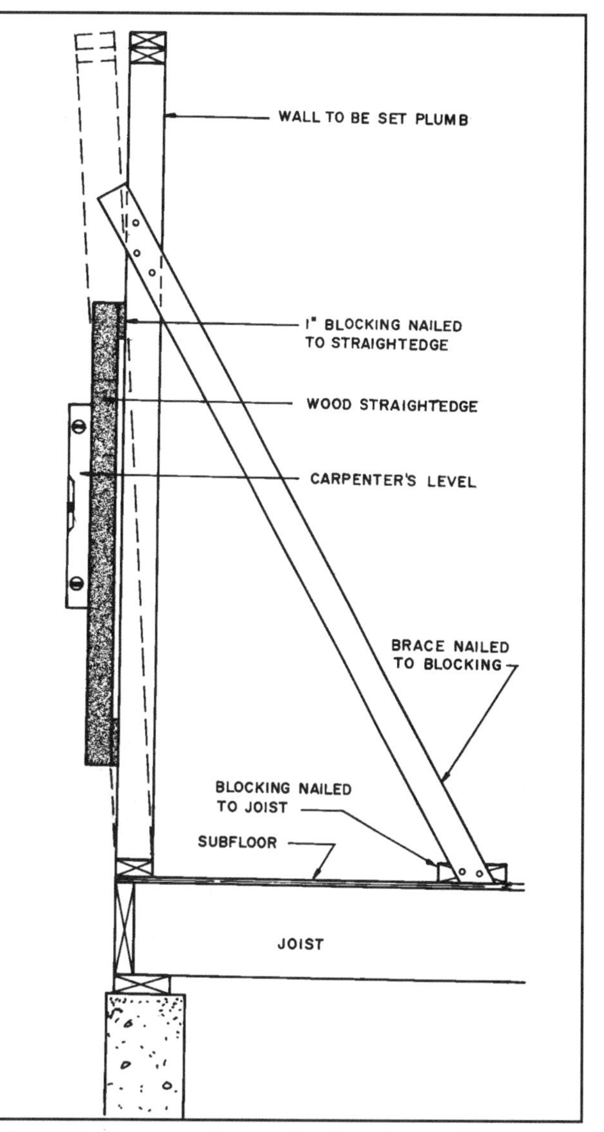

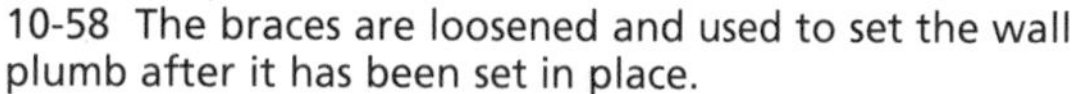

10-58 The braces are loosened and used to set the wall plumb after it has been set in place.

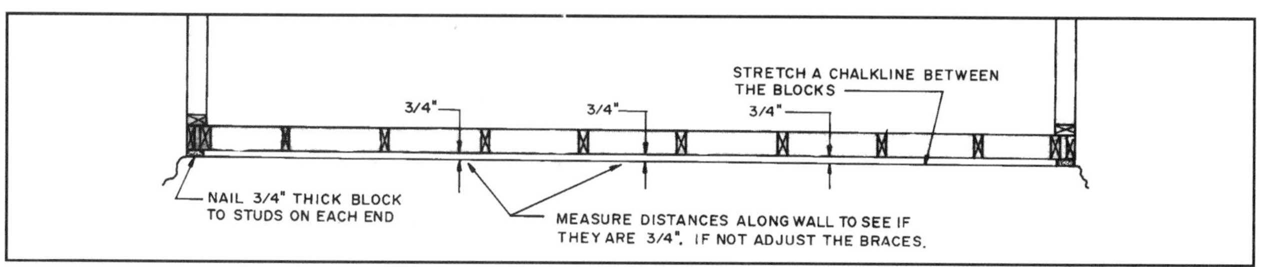

10-57 A chalkline can be used to check a wall for straightness.

The top of the wall should be checked for straightness with a chalkline. If any section is below level it may be necessary to shim a low end **(see 10-59)**.

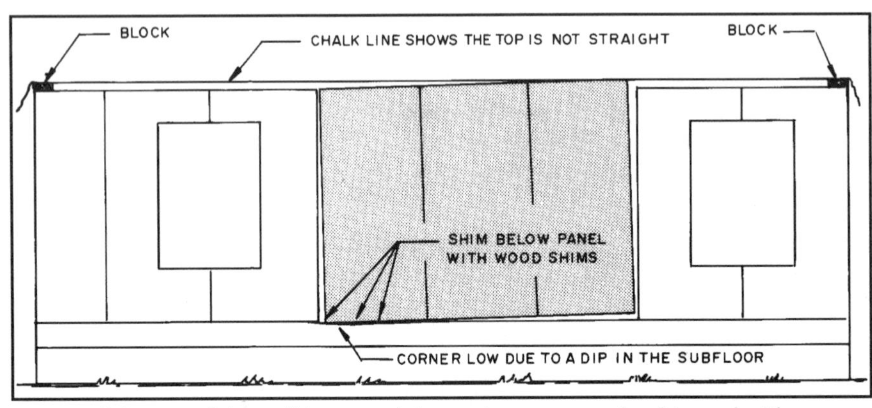

10-59 If the top of the wall is not straight any low areas can be shimmed with wedge-shaped wood shims.

Wall Jacks

A commercially available wall jack makes it possible for just a few carpenters to raise a long, heavy wall. The device shown in **10-60** is secured to the top plate with a metal bracket and to the floor with a metal foot plate. A cable winch raises the wall. The wall jack then serves as a brace after the wall has been raised. The basic steps are shown in **10-61**.

Interior Partitions

Interior partitions are often installed before the roof is framed. They support the ceiling joists and brace the exterior walls. The partitions help the exterior walls resist the forces produced by rafters, which tend to push the outside walls out of plumb. If roof trusses are to be used they are sometimes set in place before the partitions are raised. If this is done the exterior walls must have adequate bracing.

Interior partitions are nailed to the floor joists. Again, the carpenter must consult local codes for the necessary nailing requirements.

10-60 Large heavy wall sections can be raised by two carpenters using a wall jack. The jacks also serve as braces. *(Courtesy Proctor Products Co., Inc.)*

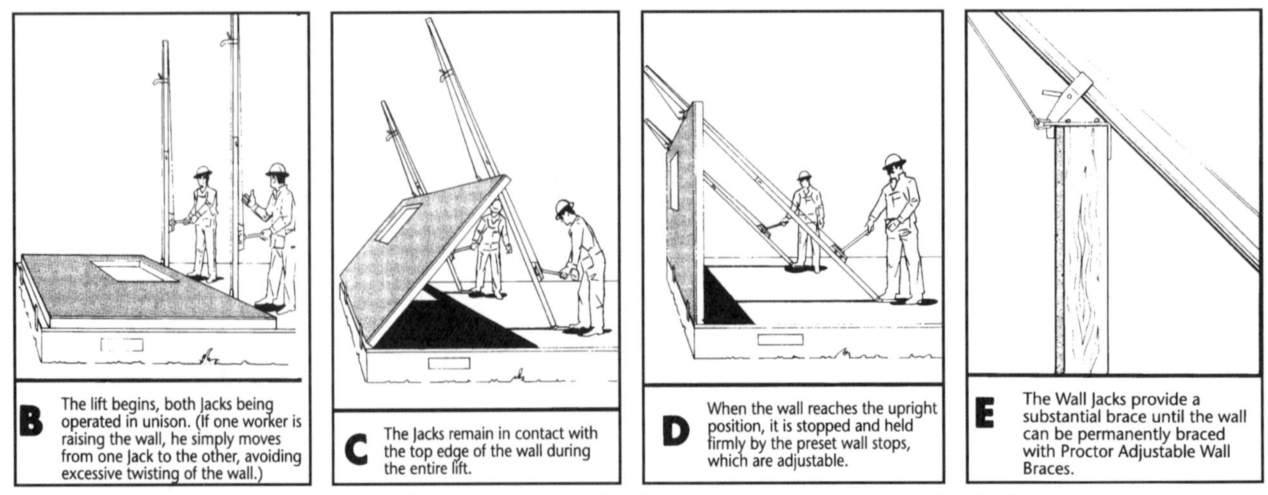

10-61 The basic steps (A to E) for installing and using a wall jack. *(Courtesy Proctor Products Co., Inc.)*

FRAMING WALLS ON CONCRETE SLABS

Concrete slab floor construction is widely used. The bottom plate of a wood-framed wall is secured to the floor with bolts or powder-driven or hammer-driven casehardened pins made for use in concrete. Bolts may be placed in the concrete when the slab is poured. The bottom plate has holes bored in it at each bolt location. After the wall is set over the bolts a washer and a nut are installed. Tighten the nut just enough to firm up the connection. Do not over tighten because it may cause the bolt to crack the concrete **(see 10-62)**.

Powder-driven or hammer-driven pins are not approved by all codes. OSHA has safety regulations pertaining to their use. The manufacturer's instructions should be observed.

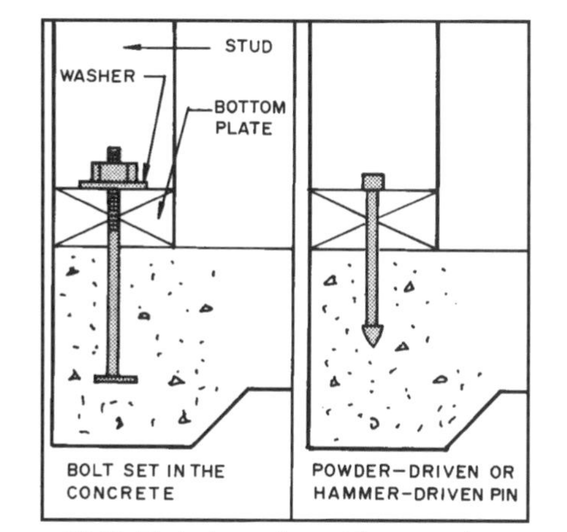

10-62 Techniques for setting wood-framed walls to concrete floors.

THE TOP PLATE

After the exterior walls and interior partitions are erected, plumbed, and braced, the top plate is nailed in place, tying them together. This plate provides resistance to forces that tend to push the walls outward. Wherever walls meet, the top member of the double top plate overlaps the intersecting wall and is nailed to it **(see 10-63)**. The top plate can be nailed onto the wall unit before it is erected, but provisions for the overlaps must be made.

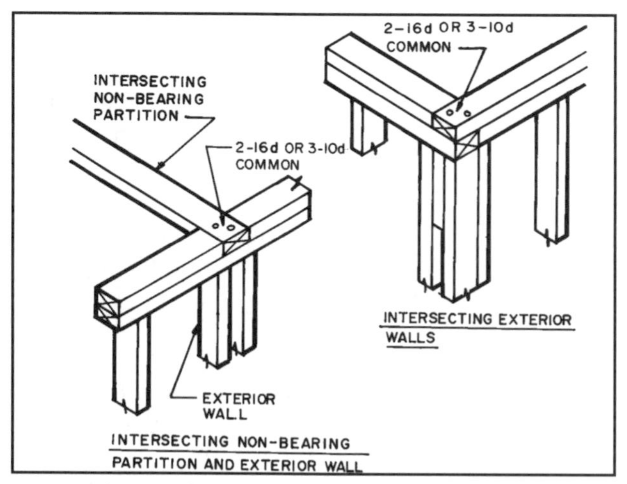

10-63 The top plates of intersecting walls overlap, tying the walls together.

SPECIAL FRAMING CONSIDERATIONS

As partitions are framed, there are many special framing applications. If plumbing fixtures are to be hung from the partition, blocking is installed **(see 10-64)**. It is important to get the correct height and make the blocking large enough to support the fixture. Generally, by the time the plumber hangs the fixture, the blocking is hidden by the wall finish material.

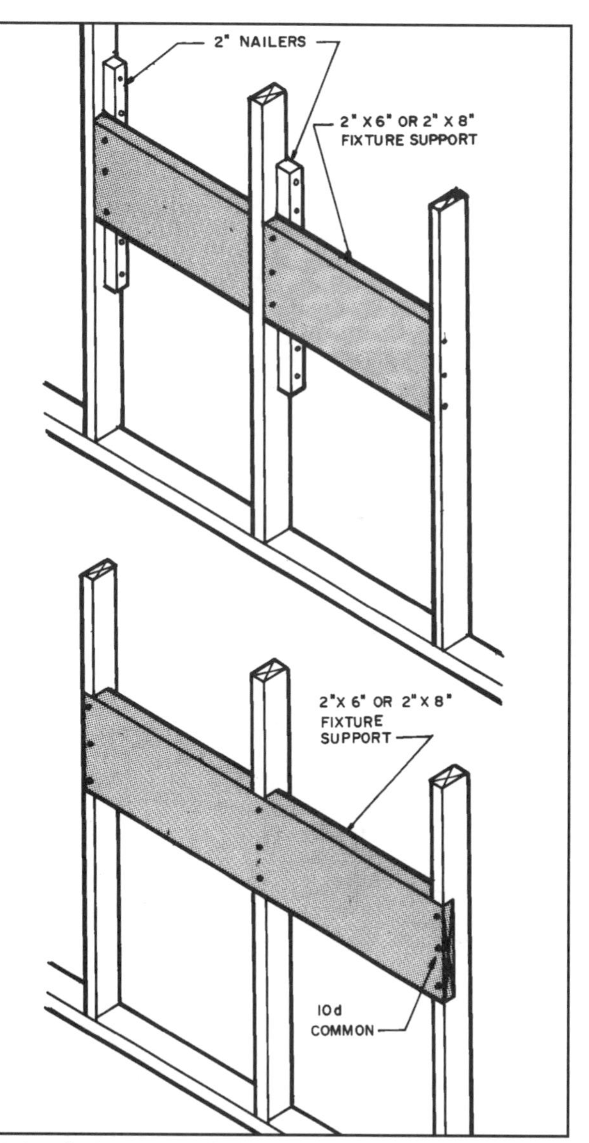

10-64 Blocking is installed to give a means of support for installing wall-hung plumbing fixtures or other items needing support.

Openings for medicine cabinets are also framed so that the cabinet can be recessed into the wall. This may require cutting a stud. The framer must know the size of the rough opening **(see 10-65)**. If safety bars, such as those used by the handicapped, are to be installed, blocking is required at each location. Bathtubs and whirlpools require supporting material **(see 10-66)**. Consult the recommendations of the manufacturer for specific instructions.

Wall-hung cabinets are easier to hang if blocking is nailed between the studs as shown in **10-67**. They can be hung from the studs, but hitting a stud is difficult. Base cabinets also are screwed to the wall along a top rail on the rear.

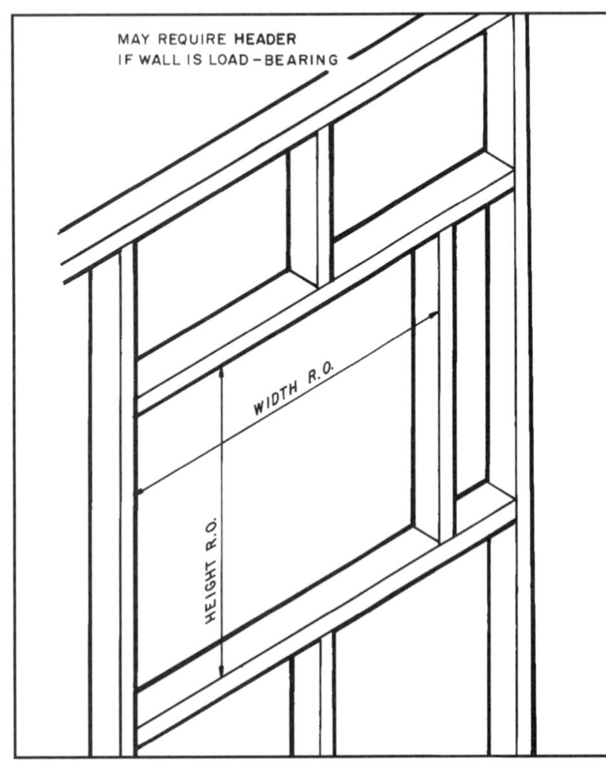

10-65 A rough opening must be framed in the wall when a recessed cabinet is to be installed.

SHEATHING THE EXTERIOR WALLS

Sheathing is secured to the studs to enclose the building. It is then covered with some type of exterior finish material such as wood siding or brick. Some types of sheathing are strong enough to brace the wall also, as shown earlier in **10-44** through **10-50**.

The most frequently used sheathing materials include plywood, oriented strand board, gypsum board, fiberboard, and various types of rigid plastic foam panels. Their basic characteristics are shown in **Table 10-3**.

PLYWOOD & OSB SHEATHING

APA-Rated plywood panels and OSB panels have a grade mark stamped on them certifying they have the required construction and structural properties. The panels are frequently applied at the corners of the building to serve as wall bracing, and the rest of the wall is sheathed with a plastic foam sheathing or fiberboard sheathing. A stronger wall is had if the entire wall is sheathed with plywood or OSB panels applied vertically or horizontally. Horizontal application requires two-inch wood blocking be nailed between the studs behind the horizontal panel joint. The panels can be nailed to this

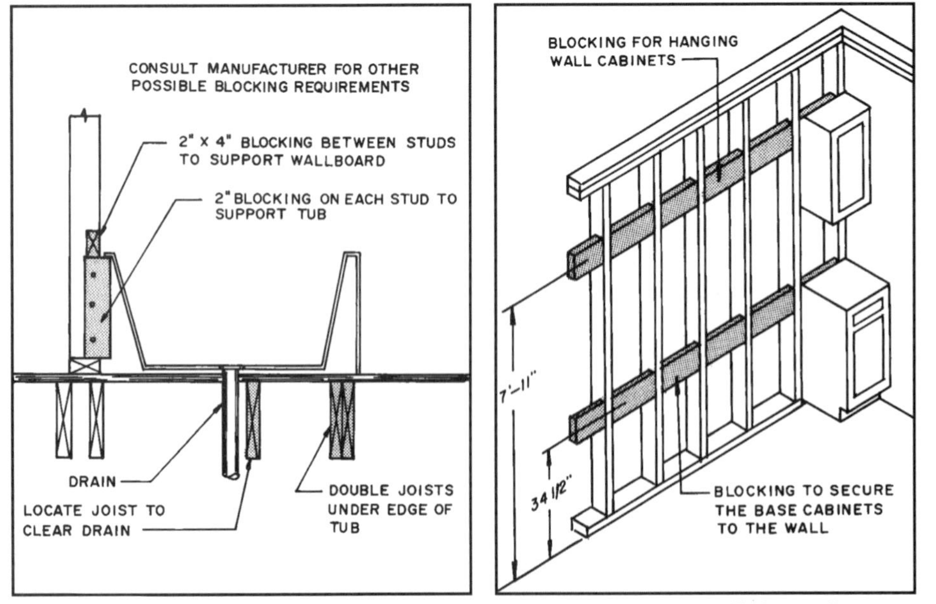

10-66 Typical blocking to support a bathtub.

10-67 Blocking is installed in walls to support base and wall-hung cabinets.

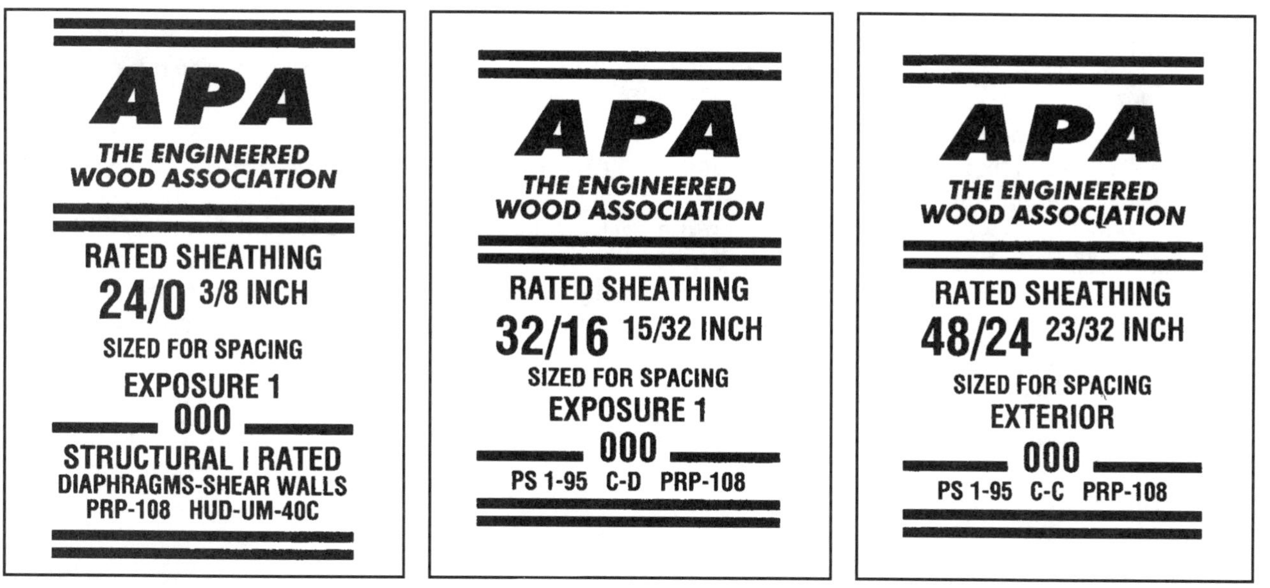

10-68 These are several of the grade trademark stamps appearing on plywood sheathing panels. *(Courtesy APA—The Engineered Wood Association)*

blocking, greatly strengthening the wall. Refer to the section in this chapter on bracing the wall for more information.

Grade marks for several plywood wall sheathings are shown in **10-68**. These grade trademarks indicate the thickness of the panel, the type of adhesive (Exposure 1 or exterior), the span rating (such as 34/16), and the number of the mill manufacturing the plywood (000 in this example). The span rating, such as 32/16, means the sheathing can be used on roofs with rafters 32 inches O.C. and on floors with joists spaced 16 inches O.C.

Table 10-3 Characteristics of wall-sheathing material.

	Gypsum	Rigid plastic foam	Fiberboard	APA Structural Panels (plywood, oriented strand board, Com-Ply®)
Panel sizes (ft)	2 × 8, 4 × 8, 4 × 9, 4 × 12	2 × 4, 2 × 8, 4 × 8, 4 × 9	2 × 8, 4 × 8, 4 × 9, 4 × 10, 4 × 12	4 × 8, 4 × 9, 4 × 10
Thickness (in)	½, ⅝	¾ to 4	½, 25/32	5/16, ⅜, 7/16, 15/32, ½, 19/32, ⅝, 23/32, ¾
Hold nails	no	no	no	yes
Serve as a vapor barrier	no*	yes	only if asphalt-impregnated	no
Serve as structural wall bracing	only under special conditions	no	only high-density type	yes
R-Value per ½ in thickness	0.5	4.00 and higher**	1.3	0.5 to 1.2

* Covered with water-repellant paper.

** Some types have aluminum foil reflective insulation bonded to the plastic foam panel.

The approved stud spacings for various plywood sheathing wall panels are in **Table 10-4**. The nailing requirements are given in **Table 10-5**.

Nailing requirements specify 6d common nails for panels under ½ inch thick and 8d common nails for those that are thicker. Nail spacing is indicated in **10-69**. A ⅓₂-inch gap is left between the long edge joints of the panels and a ⅟₁₆-inch gap is left on the end joints. Staple lengths range from 1¼ inch to 1½ inches.

APA-Rated siding can be nailed directly to the studs and serve as sheathing and the finished exterior material.

Oriented Strand Board (OSB)

Plywood and oriented strand board (OSB) basi cally have the same structural performance. They have the same performance standards and span ratings. They are both used on walls, roofs, and

Table 10-4 Recommended stud and nail spacing.

APA Panel Wall Sheathing [(a)]
(APA-rated sheathing panels continuous over two or more spans)

Panel Span Rating	Maximum Stud Spacing (in)	Nail Size [(b) (c)]	Maximum Nail Spacing (in)	
			Supported Panel Edges	Intermediate Supports
12/0, 16/0, 20/0 or wall—16 O.C.	16	6d for panels ½" thick or less; 8d for thicker panels	6	12
24/0, 24/16, 32/16 or wall—24 O.C.	24			

(a) See requirements for nailable panel sheathing when exterior covering is to be nailed to sheathing.
(b) Common, smooth, annular, spiral-thread, or galvanized box.
(c) Other code-approved fasteners may be used.
Note: Shaded construction meets Code Plus recommendations.
(Courtesy APA—The Engineered Wood Association)

Table 10-5 Recommended minimum stapling schedule for plywood & OSB wall sheathing without diagonal bracing

Plywood thickness (inch)	Staple leg length (inches)	Spacing around perimeter of sheet (inches)	Spacing at intermediate members (inches)
5/16	1¼	4	8
⅜	1⅜	4	8
½	1½	4	8

All values are for 16-gauge galvanized wire staples having a minimum crown width of ⅜ inch.

floors and have the same installation requirements. Plywood and oriented strand board also are about the same when it comes to their ability to hold nails **(see 10-70)**.

When OSB gets wet, such as from rain during construction, it expands faster around the perimeter of the panel than in the center. Plywood gets saturated faster than OSB but does not swell on the edges and dries out faster than OSB. OSB should be protected during construction to reduce swelling. Since it takes a long time to dry out it is more likely to rot than plywood. This is especially important when there is a leak in a wall or roof providing long-term moisture contact.

OSB is stronger than plywood in shear and is used as the web in I-joists for this reason.

Both are excellent wall sheathing materials.

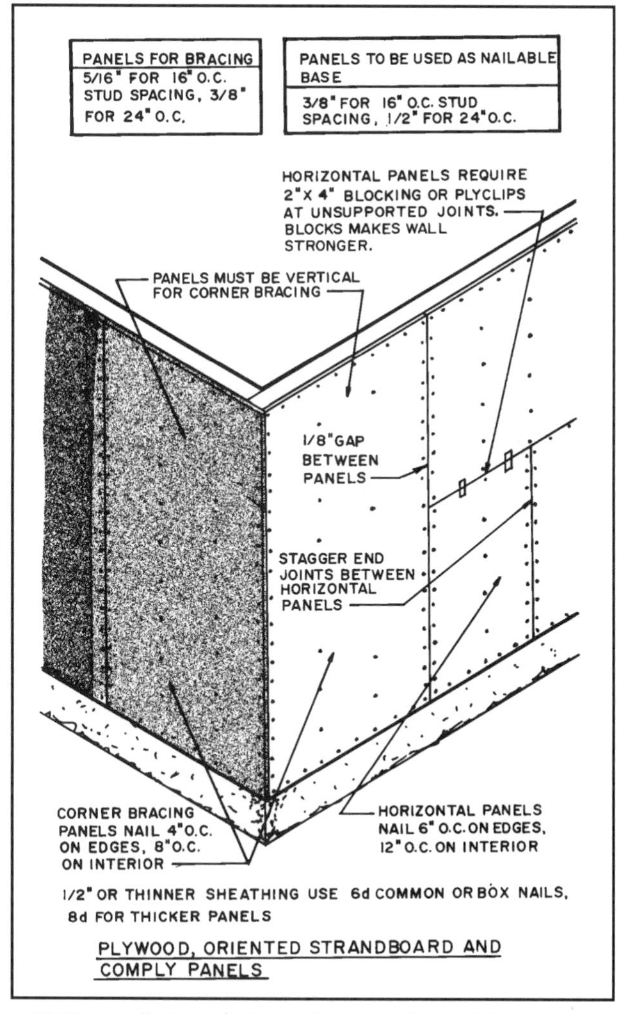

10-69 Installation of plywood, oriented strand board, and Com-Ply® sheathing.

Other Sheathing Materials

Several types of **gypsum sheathing** panels are available. Basically they are clad in a water-repellent paper on the face, back, and long edges and have a fire-resistant core. The maximum stud spacing is 24 inches O.C. and all edges must be nailed to the framing. They should be covered with finish material as soon as possible but not for more than 30 days. A typical installation is in **10-71**.

10-70 This is oriented strand board sheathing used on the subfloor, roof, and exterior walls.

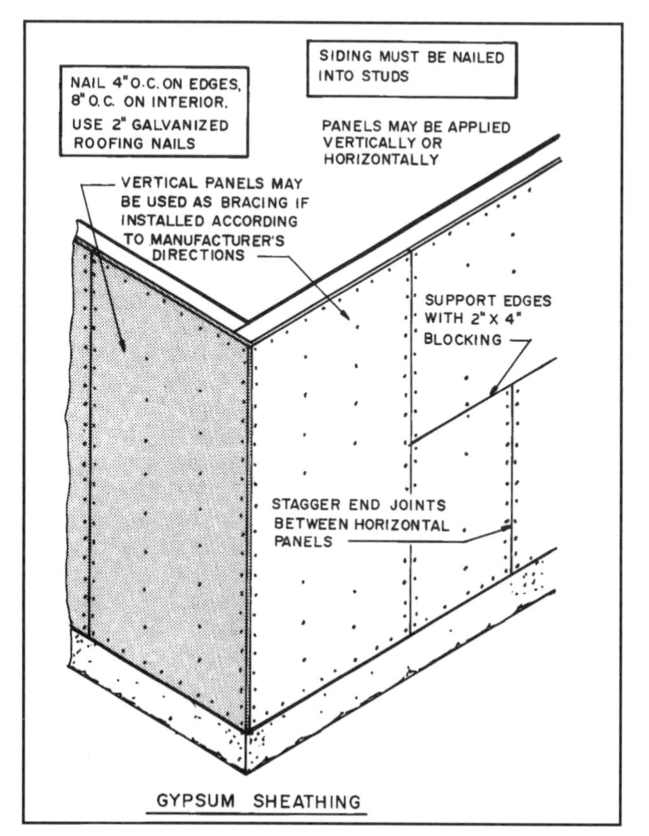

10-71 Installation of gypsum sheathing.

Fiberboard sheathing is made from wood fibers typically impregnated with asphalt. It provides insulation for the wall as well as a degree of bracing. It is available in ½-inch and 25⁄32-inch thicknesses and 4 × 8-foot and 4 × 9-foot panels. Typical installation details are in **10-72**. It is commonly used as wall sheathing when plywood or OSB is used for bracing **(see 10-73)**.

There are a variety of **rigid foam plastic sheathing** panels. Various materials are available. One product uses an extruded polystyrene panel while another uses a polisocyanurate foam core. Some have an aluminum foil facing on both sides. This provides a radiant heat barrier as well as the insulation value of the plastic core **(10-74)**. Typical installation details are in **10-75**. These panels offer no bracing properties so the wall must use plywood, OSB, or diagonal bracing as discussed earlier. The panels must be nailed to the framing on all edges.

Another type of rigid plastic foam wall sheathing is shown in **10-76**. Typically this product is a form of polystyrene foam with a transparent plastic protective film on both sides. The edges are interlocking to reduce air penetration. This sheathing may be installed with roofing nails or 16-gauge staples.

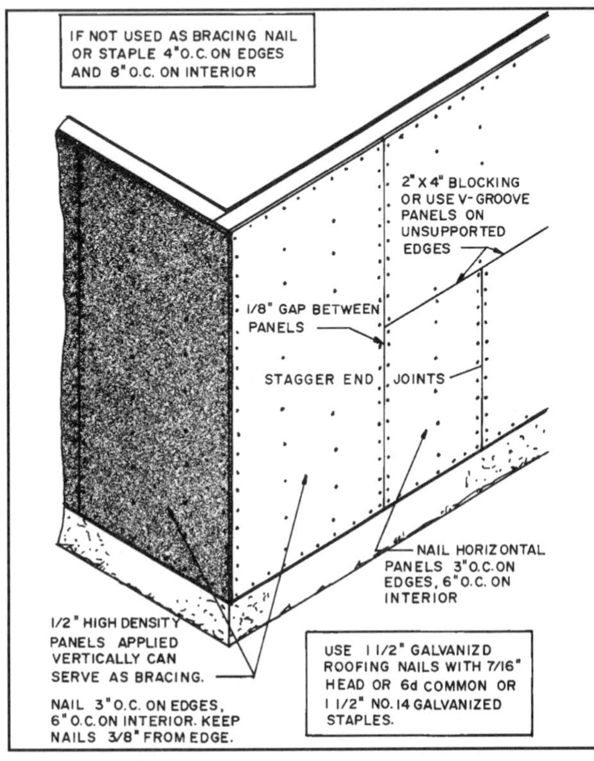

10-72 Installation of fiberboard sheathing.

10-73 Fiberboard wall sheathing is often used to cover the studs when plywood or OSB is used as the corner bracing material.

10-74 This rigid foam plastic sheathing is covered with an aluminum foil providing radiant insulation and a vapor barrier. *(Courtesy Celotex Corporation)*

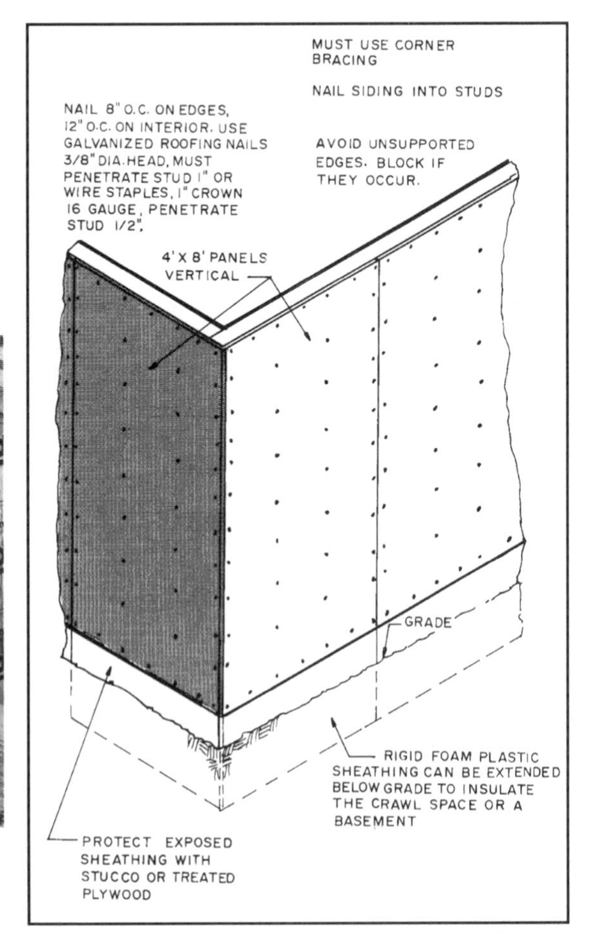

10-75 Installation recommendations for rigid foam plastic sheathing.

HOUSEWRAP

Various air-infiltration retarders are available. They are a form of spunbound olefin formed into very fine high-density polyethylene fibers. The sheets cover the entire sheathed wall **(see 10-77)**. They enhance the thermal energy efficiency of exterior walls by blocking the infiltration of air through the walls and into wall cavities but must be installed properly to be totally effective. They are permeable, therefore water vapor that may be in the wall can pass through the housewrap.

Proper Installation

Housewrap can be installed over the sheathing before the walls are raised or after they are in place. When covering walls still on the subfloor make certain the sheathing is free of debris that could be trapped below the wallwrap. Various methods can be used for this installation but flaps are usually left below the bottom plate to cover the header and stringer joists. A one-foot flap is left on the end of the wall so that after the wall is up the extra material can be wrapped around the corner.

When the housewrap is installed after the walls are up, begin by installing the first horizontal layer so the bottom edge covers the header and stringer joists. Be certain to allow extra material to wrap at least one foot around the corner. Overlap the edges of each row four inches and nail or staple through both layers. Space staples 16 inches O.C. on the edges and 24 inches O.C. on the inside of the sheet.

Wrap the material around the sides of wall openings by cutting an X-shape across the opening. Then wrap the triangular flap into the opening and staple it to the studs. Cut off the excess (the point of the triangular flap) material **(see 10-78)**.

Since the purpose of using housewrap is to stop air infiltration, all joints should be covered with a special seam tape designed for this purpose. All openings in the wrap, such as round window openings, should be taped. For maximum effectiveness taping should aggressively seal all possible points of leakage **(see 10-79)**.

10-76 This is a rigid foam wall sheathing that contributes to the R-value of the exterior wall.

10-77 The entire exterior wall is covered with housewrap to help reduce air infiltration.

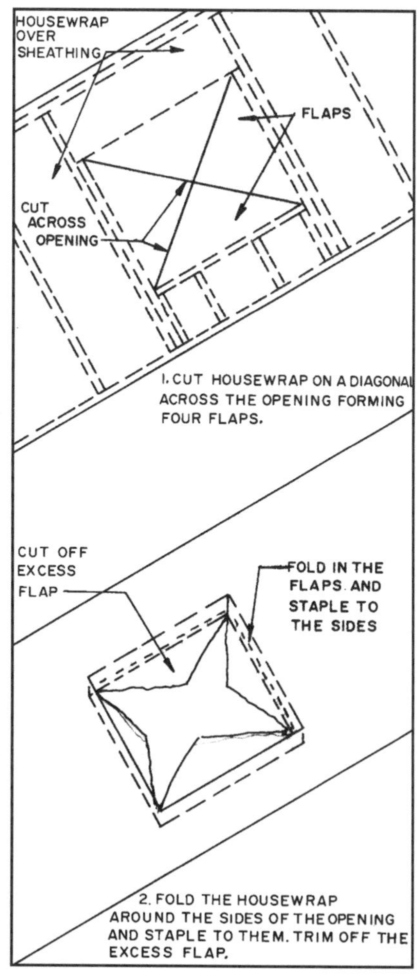

10-78 Wrap the housewrap around the edges of openings and staple in place.

10-79 The overlapping joints and other edges of the housewrap are sealed with a tape designed for that purpose.

Taping the Sheathing

When homes are fully sheathed with any material that is carefully installed, the basic sheathing is itself impervious to air infiltration. The infiltration occurs at the joints between the panels and around door and window installations. A special four-inch-wide tape designed to seal joints sticks very well to plastic and aluminum foil covered sheathing. In most cases it will stick to plywood but has some difficulty bonding to OSB. No doubt in time this will be solved with a different adhesive.

This method of reducing air infiltration is cheaper than housewrap because the tape costs less and less labor is required. Housewrap should be used in all cases where the tape will not stick to the sheathing.

Another material used to fill cracks and small openings in the sheathing, around pipes and wires penetrating sheathing, and around door and window openings is a plastic foam available in aerosol cans. When the button is pushed the foam extrudes out of a nozzle much like shaving cream. It fills the opening and expands several times as it hardens.

METAL CONNECTORS

Metal connectors are used on wood-framed walls to increase their resistance to forces tending to separate adjoining members. They are required in areas subject to strong winds, earthquakes, and other natural forces. In **10-80** they are used to secure studs to the plates. Straps are also used to tie studs to the top and bottom plates and to the header and stringer joists **(see 10-81)**. Tie plate connectors are used to tie together members that have been butted **(see 10-82)**. In areas with high winds ties are required to secure the wall to the foundation. One type secures the wall to the header and stringer joists. Another ties the wall directly to the foundation, as you can see in **10-83**. The second floor wall can be tied to the first floor wall with long metal straps **(see 10-84)**.

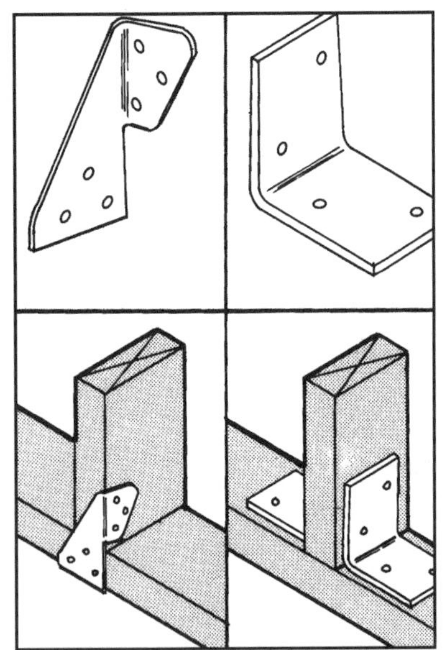

10-80 These are typical metal connectors used to strengthen the tie between studs and the plate.

STUD TO TOP PLATE

STUD TO HEADER OR STRINGER JOIST

10-81 Strap connectors can be used to connect the studs to top and bottom plates and the header and stringer studs.

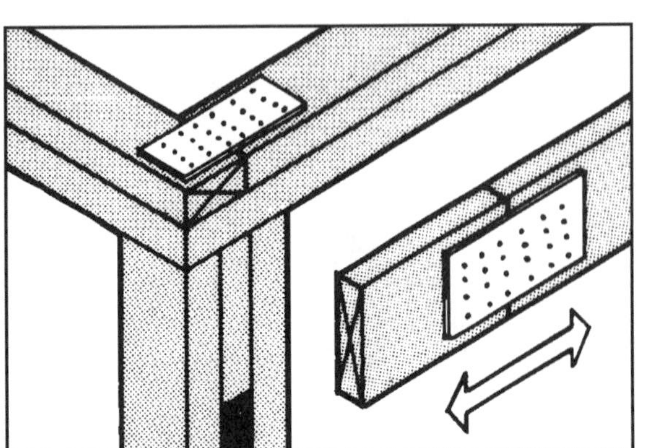

10-82 Metal tie plates strengthen members that are butted.

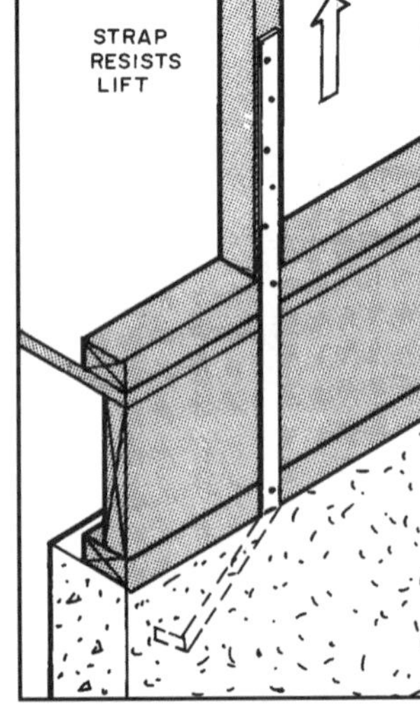

10-83 This strap connector ties the wall framing to the foundation.

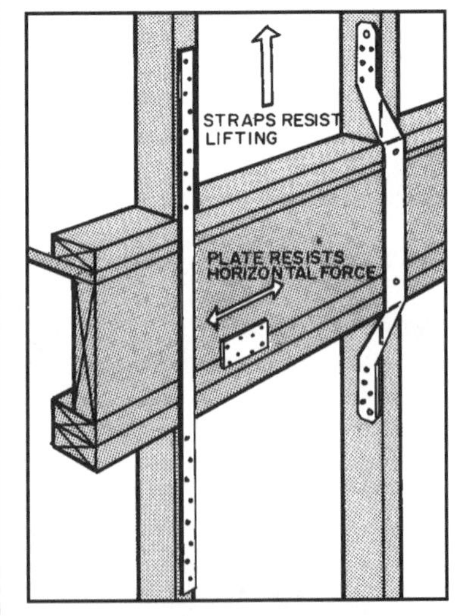

10-84 Metal strap connectors resist lifting forces and secure the second story floor and wall framing to the studs on the first floor.

MANUFACTURED WALL PANELS

A number of wall panels assembled in a factory under controlled conditions are available. The carpenter must use the installation procedures established by the manufacturer. Following is a partial explanation of products produced by AFM Corporation, Shorewood, MN 55331 **(see 10-85)**.

The AFM R-Control® panels are available in sizes from 4 × 8 feet to 8 × 24 feet. They may be used for roof, wall, and floor sections within their structural specifications. The panels consist of Exposure 1 oriented strand board skins and expanded polystyrene (EPS) core **(see 10-86)**. The panels are joined using nails or staples and a wood-to-wood bonding adhesive. One method for joining panels is with splines as shown in **10-87**. A special sealing material is placed the full length of the panel and augments the overall tightness of the structure. Another design uses panels with an overlapping top plate as shown in **10-88**.

10-85 The walls of this building are being constructed using R-Control® panels. *(Courtesy AFM Corporation. R-Control ® is a registered trademark of AFM Corporation)*

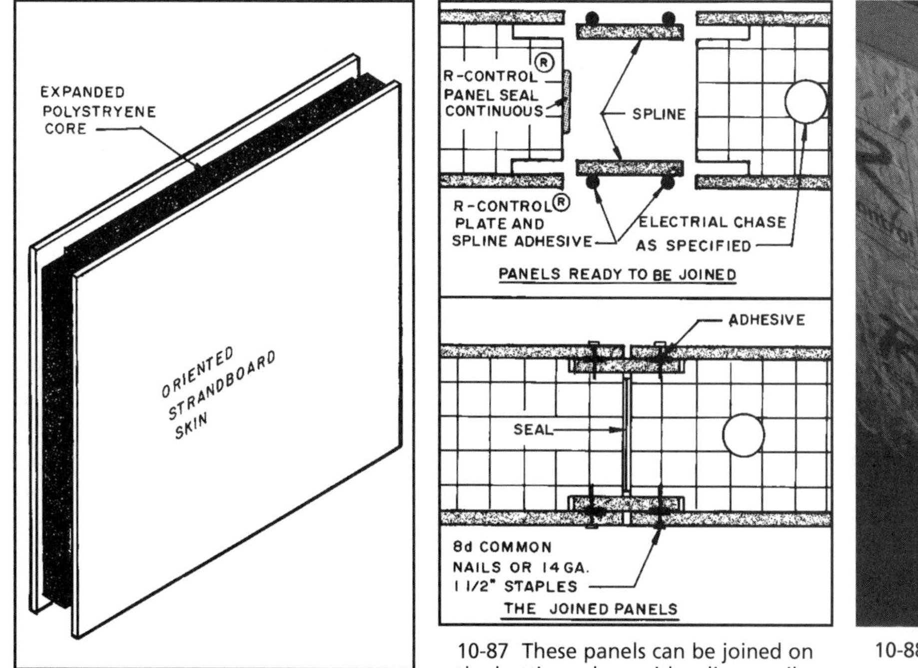

10-86 A typical AFM R-Control® panel. *(Courtesy AFM Corporation)*

10-87 These panels can be joined on the butting edges with splines, nails, and a special adhesive. *(Courtesy AFM Corporation)*

10-88 These R-Control® wall panels are made with an overlapping top plate that ties them together. *(Courtesy AFM Corporation)*

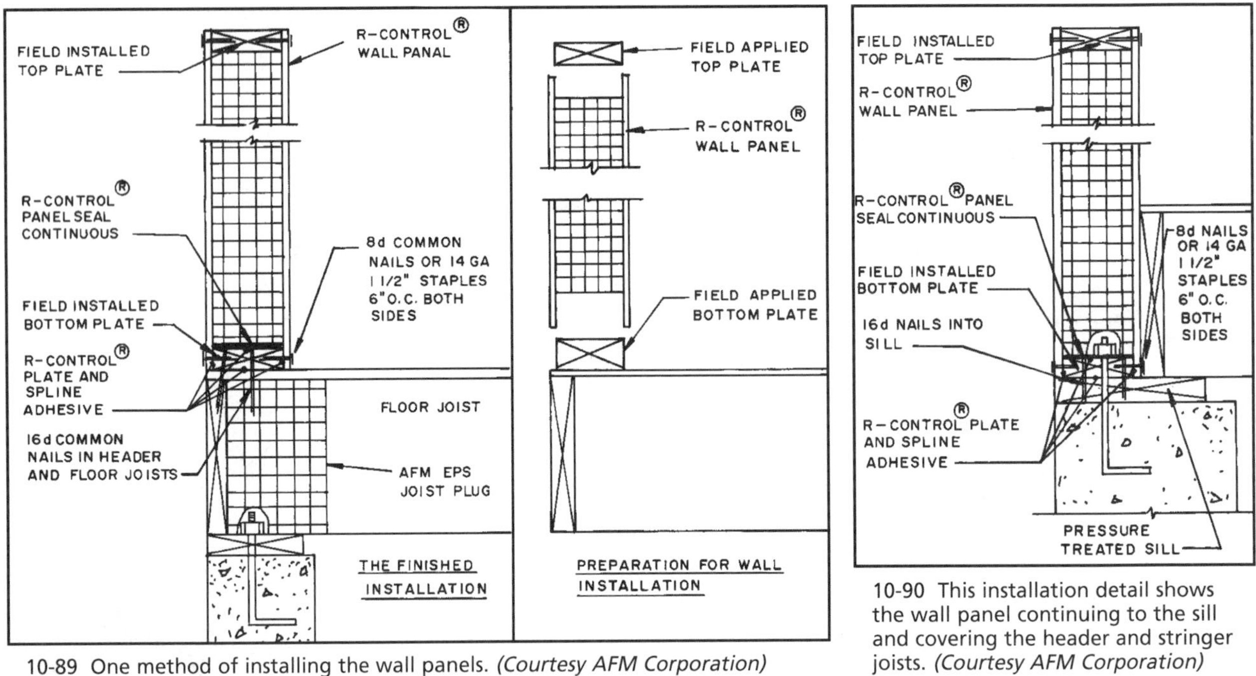

10-89 One method of installing the wall panels. *(Courtesy AFM Corporation)*

10-90 This installation detail shows the wall panel continuing to the sill and covering the header and stringer joists. *(Courtesy AFM Corporation)*

The panels are bonded and nailed to bottom plates secured on the subfloor and header and stringer joists **(see 10-89 and 10-90)**. They are tied together on the top with a top plate.

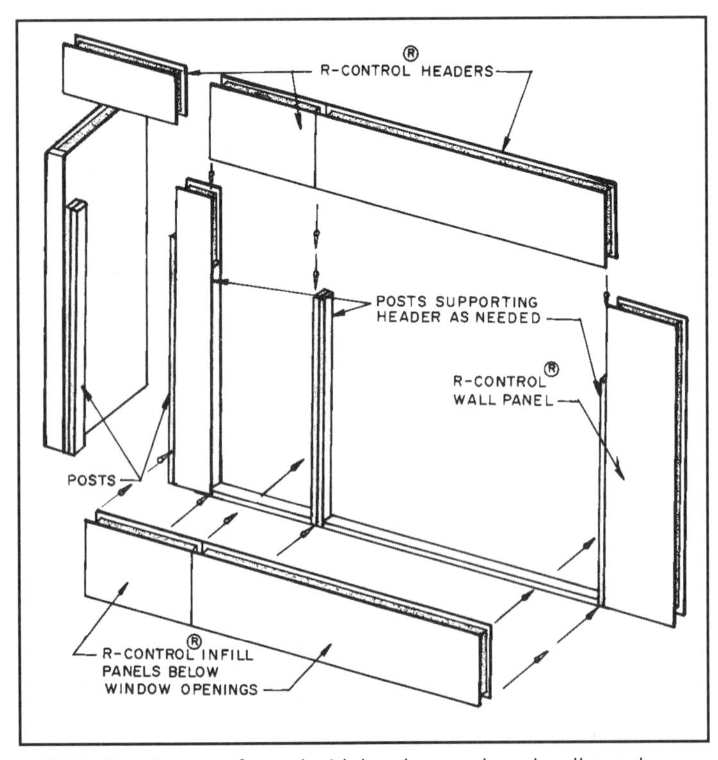

10-91 Openings are framed with header panels and wall panels. Headers are supported with solid wood posts. *(Courtesy AFM Corporation)*

Openings are built and framed with header panels and wall panels, as shown in **10-91**.

PANELIZED CONSTRUCTION

Panelized construction involves assembling floor, wall, and roof members into large sections in a factory, moving them to the site, and setting them in place with a crane. This enables the contractor to do most of the work in protected conditions. The lumber is never wetted by rain or snow and the cutting and nailing is more accurate and secure.

Before the floor panels arrive on the site the foundation is in place with the required sills, girders, piers, and columns in place. The floor panels are set with a crane as shown in **10-92**. The plywood subfloor overhangs the end joist so it overlaps and can be nailed to the end joist of the butting panel.

After the deck is set the exterior wall panels arrive. Their location has been marked on the subfloor with a chalkline. The panels are set with a crane, nailed to the header and floor joists, and braced as necessary **(see 10-93)**. The interior partitions' locations are marked on the subfloor and they are set and nailed in place **(see 10-94)**.

If there is to be a second floor the factory-assembled floor panels are set much as described for the first floor. All needed load-bearing partitions, headers, and beams are set before the floor panels are raised into place.

The roof is typically built using factory-assembled gable-ends and roof trusses **(see 10-95)**. The exterior is finished in the normal manner but the building is up and weather-tight in a very short time. The other trades can now begin work.

10-92 Factory-assembled floor panels are set with a crane. The entire floor can be set in about an hour. *(Courtesy* The Journal of Light Construction, *Richmond, VT)*

10-93 This entire wall assembly complete with sheathing, doors, and windows is assembled in a factory, moved to the site, and set in place with a crane. *(Courtesy The Journal of Light Construction, Richmond, VT)*

10-94 Once the exterior walls are set and braced, the interior partitions are hand-carried to the locations and secured in place. *(Courtesy The Journal of Light Construction, Richmond, VT)*

10-95 This two-piece factory-assembled gable-end is set in place with a crane. The remainder of the roof was framed with trusses. *(Courtesy The Journal of Light Construction, Richmond, VT)*

CHAPTER 11

FRAMING THE CEILING

After the exterior walls and interior partitions are set in place, secured to the floor framing, plumbed, and braced, the ceiling joists can be installed. Ceiling framing usually consists of joists that are supported by the exterior wall and interior load-bearing partitions as shown in **11-1**.

Typically the ceiling joists support the interior finish material (such as gypsum wallboard) and insulation. They may have a plywood subfloor placed over them, providing limited lightweight storage. If mechanical units are placed on them (such as an heating, ventilation, air-conditioning unit) they must be designed to carry the extra load. In a two-story building the ceiling joists become the floor joists for the second floor and are sized to carry the design loads expected. If a flat roof is to be used the joists serve to carry the interior ceiling as well as the roof loads **(see 11-2)**.

In addition to supporting loads, ceiling joists help tie the wall structure together and resist the outward forces produced by the rafters of pitched roofs **(see 11-3)**.

If a building uses roof trusses, the bottom chord of the truss serves as the ceiling joist.

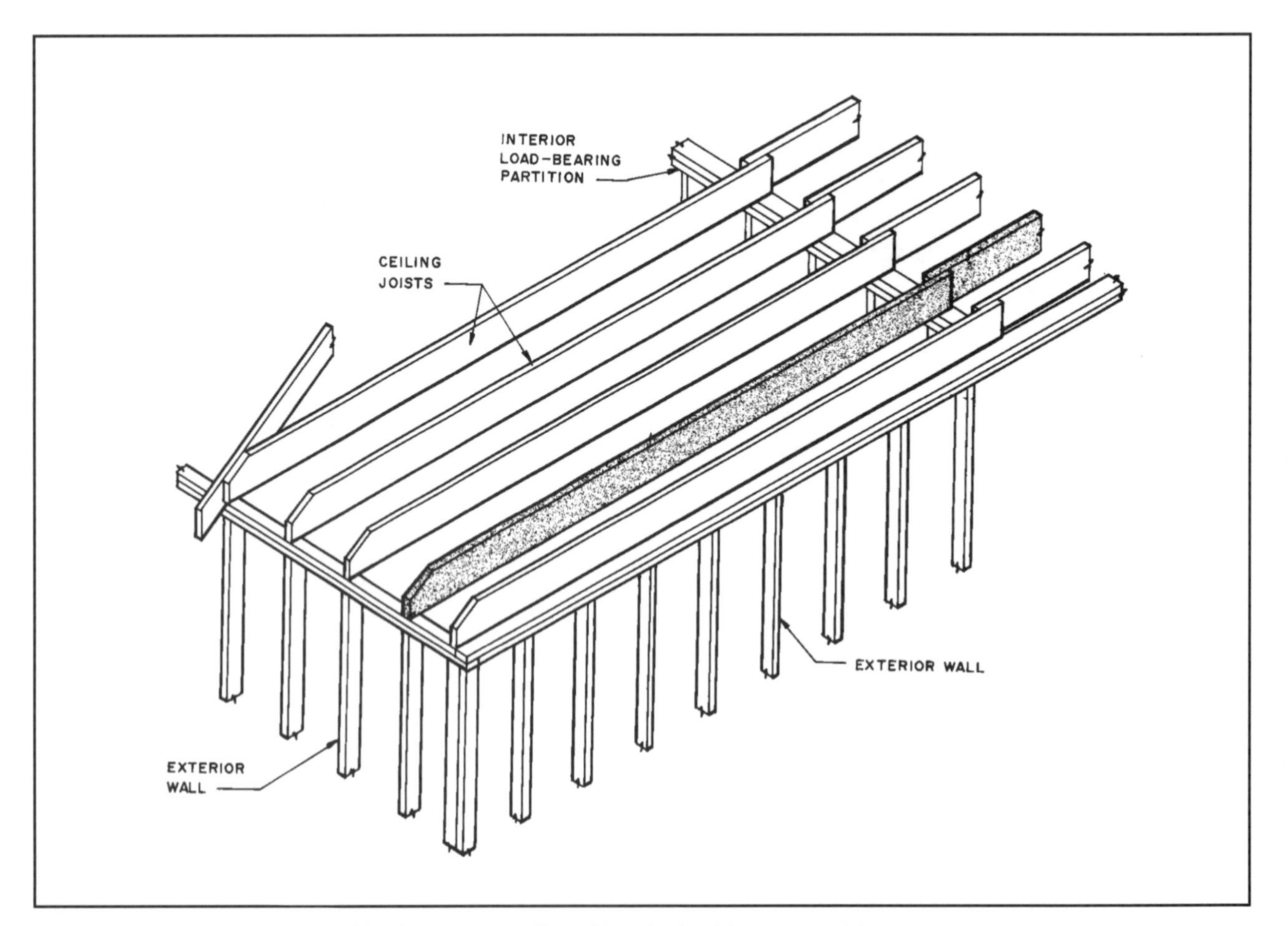

11-1 Ceiling joists are supported by the exterior walls and interior load-bearing partitions.

SELECTING CEILING JOIST SIZES

Ceiling joists are usually spaced 16 inches O.C., but in some cases they may be 24 inches O.C. The attic load to be carried, the type of finished ceiling, and the unsupported span must be known. Typically a 10-pound-per-square-foot live load is used for attics with no anticipated storage. If the attic will be used for storage but not be a living area, a 20-pound-per-square-foot live load is used. If it is to serve as a living area, 30-pound-per-square-foot live loads are used for bedrooms and 40-pound-per-square-foot live loads for uses such as a living room, dining room, or kitchen. For the 30- and 40-pound-per-square-foot live loads, floor joist tables would be used to select the joist size (see Chapter 9).

The designer must take into consideration the location and support of unusual items on ceiling joists, such as the placement of a water heater or furnace in an attic. The section of the ceiling that will support these units may require double joists or joists spaced closer together.

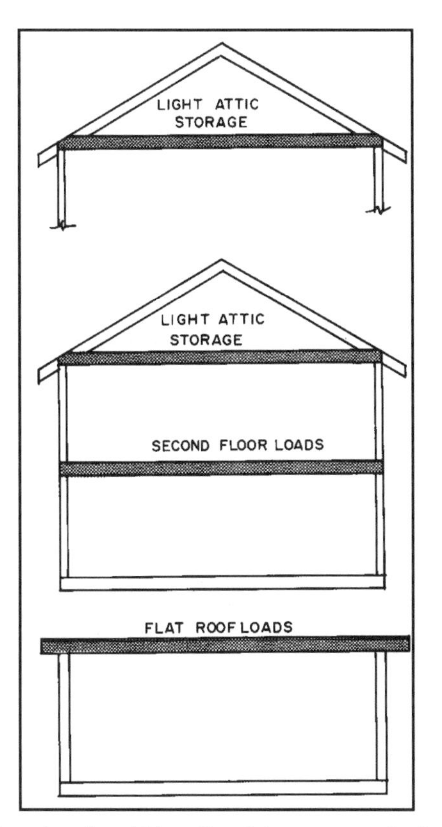

11-2 Situations in which ceiling joists support the interior finish material and additional loads.

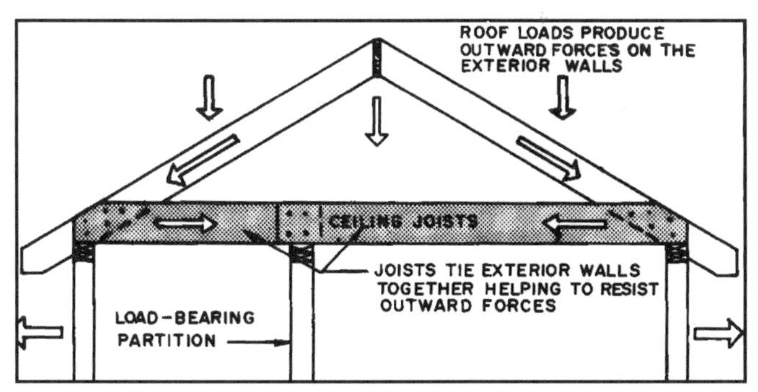

11-3 Ceiling joists help control the down forces from the roof that tend to push the exterior walls outward.

Note on Span Tables

Span tables for ceiling joists for one specie of lumber are found in Appendices E, F, K, and N. The left column gives the spacing of the joists. These tables assume a drywall ceiling. Additional tables are available for plaster ceilings. The grade of lumber is located across the top of the table, and the unsupported span is in the table body.

RUNNING THE CEILING JOISTS

Ceiling joists are generally run across the width of the building parallel with the rafters. The rafters are generally set directly above a wall stud and the ceiling joist placed next to it and nailed to the rafter and toenailed to the top plate **(see 11-4)**.

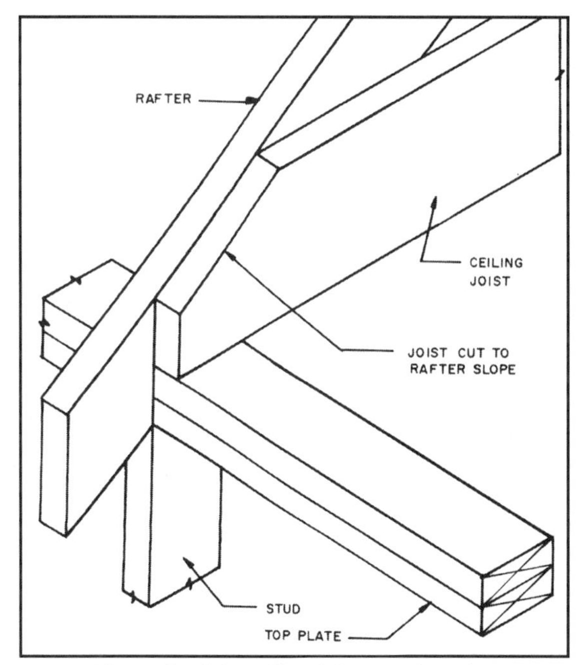

11-4 The ends of the ceiling joists are sloped to match the slope of the rafter.

Where ceiling joists run parallel to the edge of a roof and the roof has a steep slope, such as occurs with a hip roof, the last joist can be omitted and 2 × 4 blocking nailed on the top plate to provide a nailer for the interior ceiling material **(see 11-5)**.

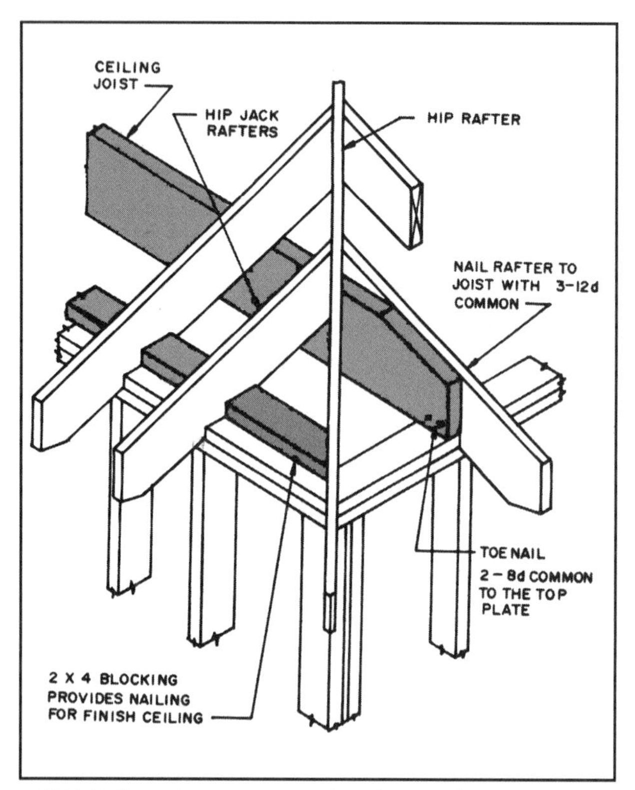

11-5 Nailing plates are secured to the top plate to provide a nailing surface for the finish ceiling material.

When the roof has a low pitch, short joists can be run perpendicular to the ceiling joists, as shown in **11-6**. Blocking is nailed between them to support the finished ceiling material.

When the ceiling joists do not run parallel with the rafters they are not providing the resistance needed to help keep the roof loads from pushing the exterior walls outward. To provide the needed resistance, strongbacks are nailed perpendicular to the joists and to the rafter as shown in **11-7**. Nail to the ceiling joists with two 16d common nails and toenail the end to the rafter. Check local codes to verify the required spacing of ties.

Sloping the Ceiling Joist Ends

On many roofs the end of the ceiling joist that rests on the exterior wall will stick above the top of the

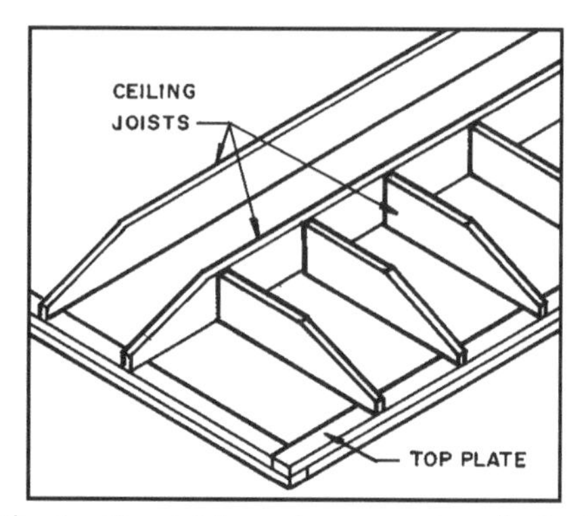

11-6 Short ceiling joists can be run perpendicular to the regular joists when low-pitched roofs make it not possible to use joists at the exterior walls.

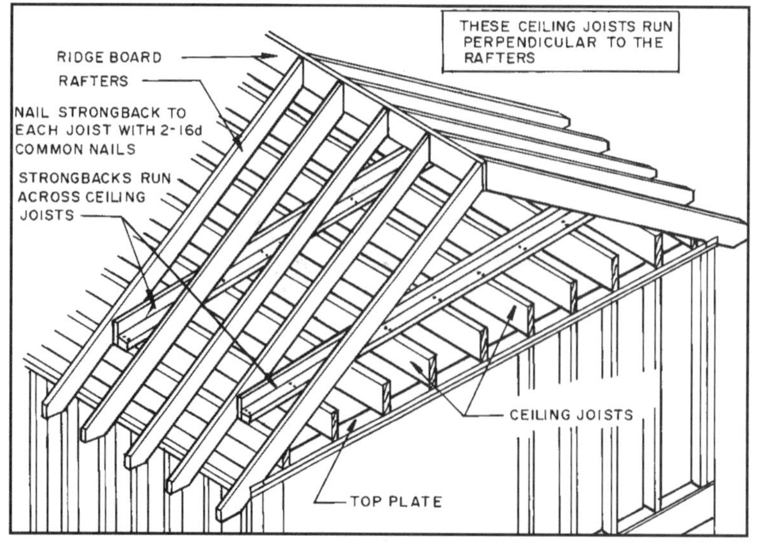

11-7 Cross ties are used to provide resistance to the thrust from the roof rafters which tend to push the exterior wall outward.

rafter and must be cut to the slope of the rafter and be slightly below it **(see 11-8)**. The cut is laid out using the distance from the bird's-mouth on the rafter to the top of the rafter and the slope of the roof.

The slope is set by locating the rise and run of the rafters on a framing square, as shown in **11-9**. Cut the joist so the sloped surface is a little below the top of the rafter.

Installing Ceiling Joists

The end joist is set next to the top plate and provides a nailing surface for the ceiling material **(see 11-10)** and the other joists are placed next to each rafter **(refer to 11-4)**. Since the rafter may not have been installed at this point, the marks on the top plate locating the rafter and ceiling joist provide the location. The top plate is marked with each rafter location as is shown later in **11-15**.

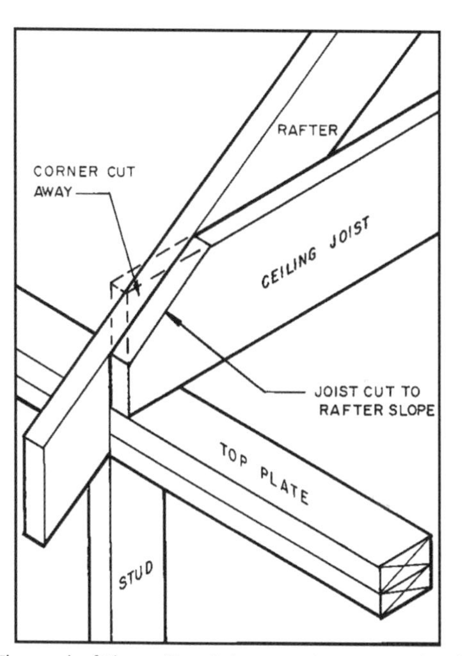

11-8 The end of the ceiling joist must be cut on the slope of the roof and be slightly below the top of the rafter.

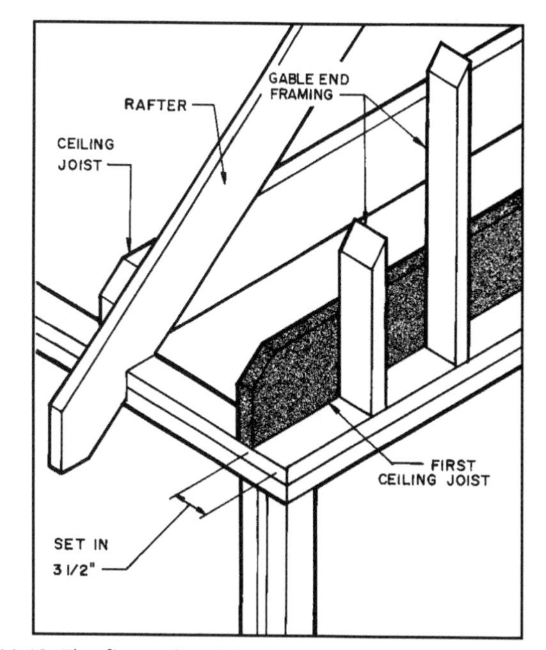

11-10 The first ceiling joist is set next to and touching the inner face of the top plate.

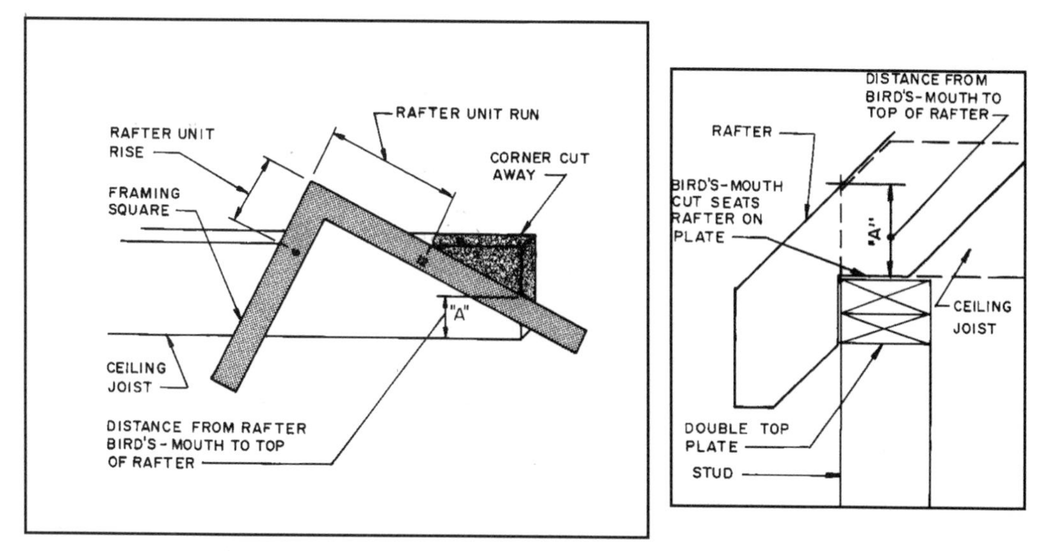

11-9 These measurements are used to lay out the slope on the end of ceiling joists.

Ceiling joists rest on interior load-bearing walls, and they may be butted and secured with splice plates or lapped and nailed as shown in **11-11** and **11-12**. Notice in **11-11** and **11-12** a 2″ × 6″ nailing plate is used. If this is not used nailing surfaces for ceiling finish material can be made using blocking between the joists. When butted, they are joined with ¾-inch plywood splice plates 24 inches long, or with metal splice plates. When lapped, they are end-nailed on each side with three 10d nails and toenailed to the top plate with two 8d nails. Remember to set joists with the crown side up **(see 11-13)**.

A large room may have a span that is beyond the span distance of normal ceiling joists. The ceiling is framed by running a beam across the room

11-11 Butted ceiling joists can be joined with metal or plywood splice plates. Notice the use of a two-inch-thick nailing plate. If this is done the studs must be shortened 1½ inches.

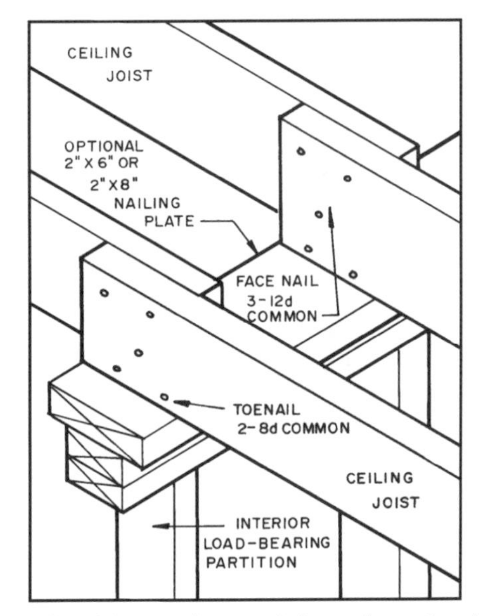

11-12 Ceiling joists are frequently lapped over interior load-bearing partitions. Notice the use of a two-inch-thick nailing plate. If this is done the studs must be shortened 1½ inches.

11-13 The carpenter should always check floor and ceiling joists for warp or crook. Notice the two-story platform framing in the building. (Courtesy Southern Forest Products Association)

and butting the joists to it **(see 11-14)**. They can be secured using metal joist hangers or a ledger in the same manner as for floor joists.

Before the ceiling joists are cut to length they are placed on top of the wall and partition plates. If the ends must be sloped they are cut before the joists are raised upon the top plates. Remember to place the crown side up as the sloped ends are cut. The joists are laid out along the top plate near the marks indicating their location. On the interior load-bearing partitions the location of the ceiling joists for those to be overlapped is shown in **11-15**. Each wall is laid out depending upon how the joists are to be lapped. If the joists are butted the layout on both walls will be the same.

Toenail to the top plates with three 8d common nails, two on one side and one on the other **(see 11-16)**. Joists that are beside rafters are face-nailed to them with three 10d common nails.

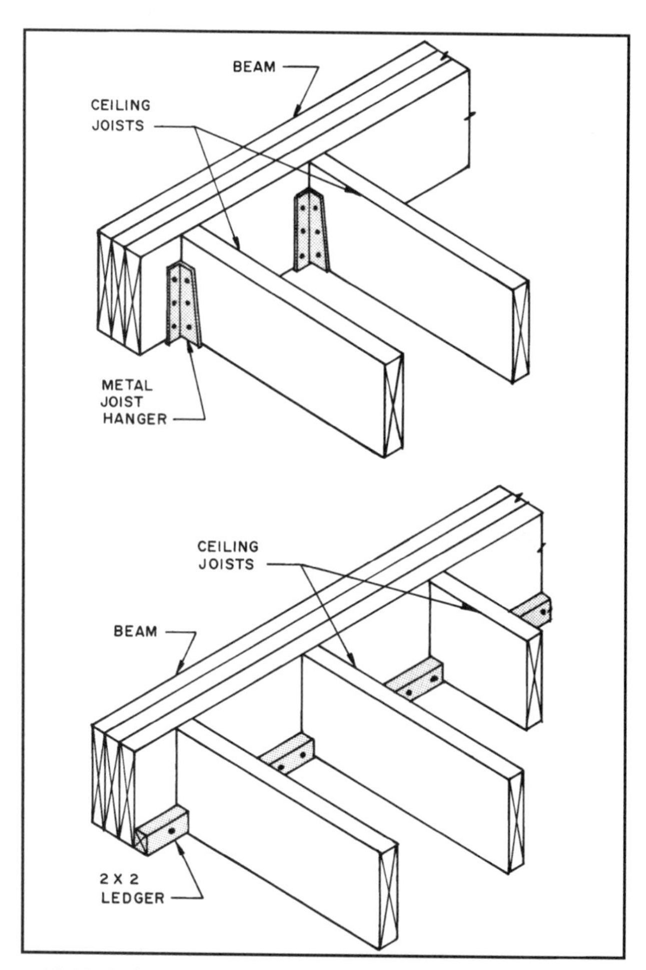

11-14 Ceiling joists can be framed to a beam with metal joist hangers or a wood ledger in the same manner as for floor joists.

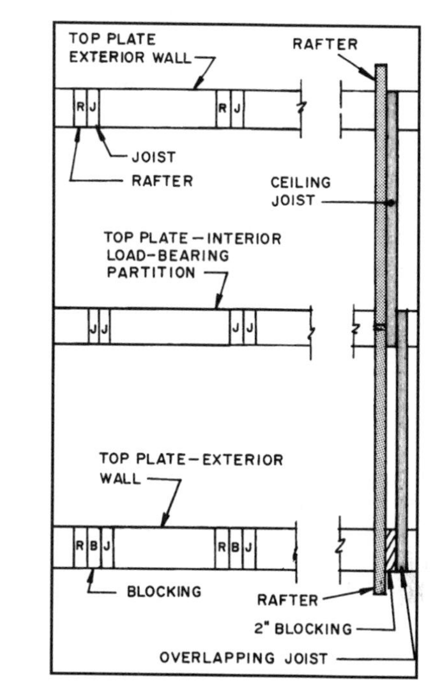

11-15 Ceiling joist locations are marked on the top plate along with those locating the rafters.

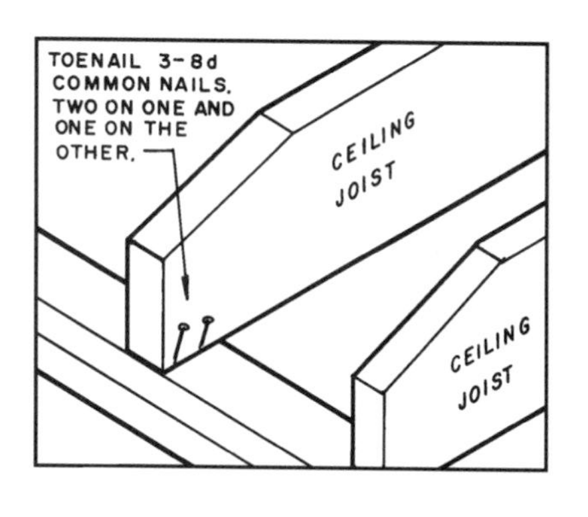

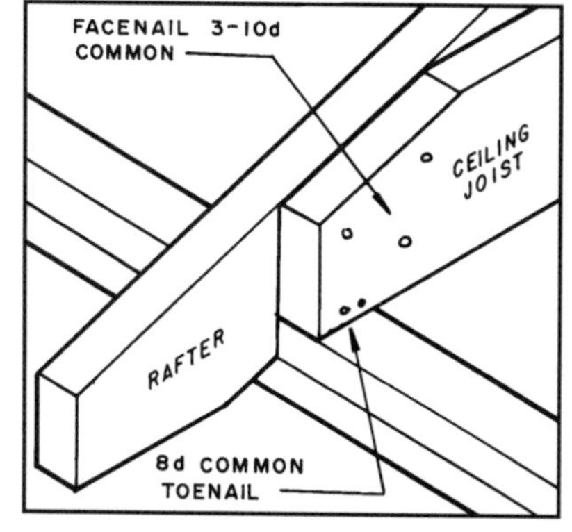

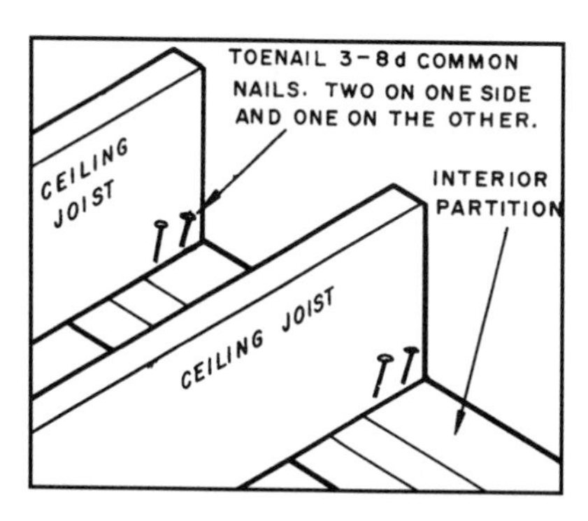

11-16 Ceiling joists are toenailed to the top plate and face-nailed to adjoining rafters.

Run the joists down one side of the building and then run the other side, lapping or butting as required. The second run of joists, if overlapped on an interior load-bearing wall, will not fall next to a rafter. A piece of two-inch blocking is nailed between the joist and rafter as shown in **11-17**. Now install the strongback or rib band and check the bottoms of the joists with a chalkline to be certain they are all in the same plane. Correct any that are below or above the plane of the ceiling. After the rafters are set install the nailing blocks on the top plate.

STRONGBACKS

Strongbacks are installed in the center of the span of a series of rafters **(see 11-18)**. They stiffen the assembly, help keep the joists in line, and reduce possible warping or bowing of the joists. This helps keep the ceiling from popping nails through the gypsum wallboards or other finish ceiling material.

The strongback is typically a 2 × 6- or 2 × 8-inch vertical member nailed to a 2 × 4-inch member forming an L-shaped structural beam. Nail the 2 × 4 to the ceiling joist with two 16d common nails **(see 11-18)**. It requires a block be installed at the exterior wall to tie in the end of the strongback.

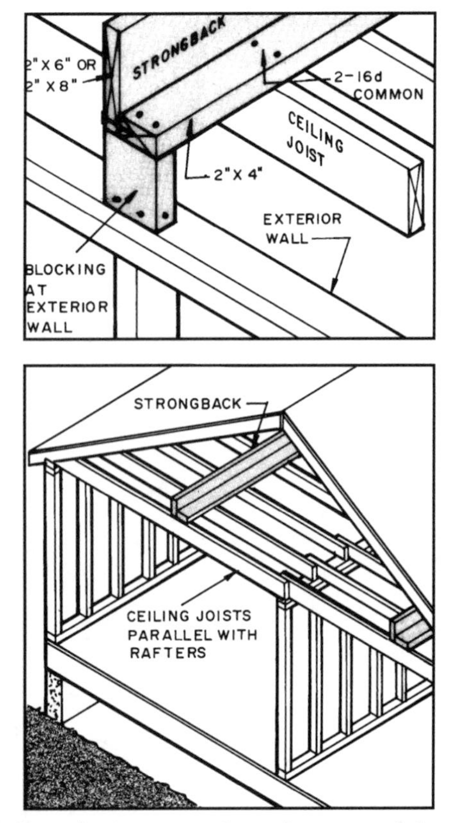

11-18 Strongbacks are run down the center of the ceiling joist span to help hold them in a flat plane and resist bow and twist. Notice that blocking is used to tie the strongback to the top plate on the exterior wall.

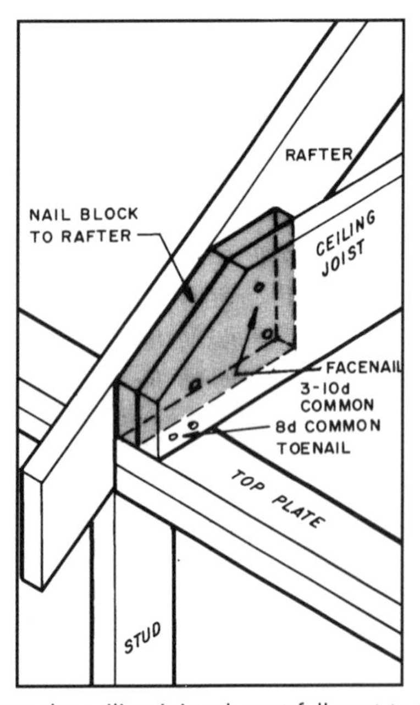

11-17 When the ceiling joists do not fall next to the rafter because of the overlap on the interior load-bearing wall, blocking is placed between the ceiling joists and the rafter.

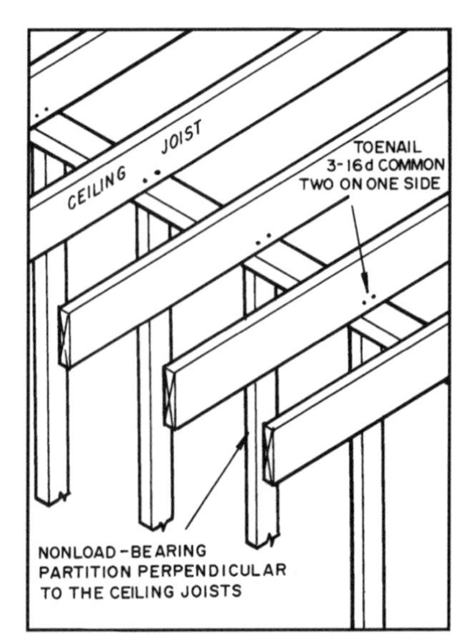

11-19 Nonload-bearing partitions that run perpendicular to ceiling joists are nailed to the joists. Notice the nailing plate blocks which provide a nailing surface for the finish ceiling material.

NONLOAD-BEARING PARTITIONS

Nonload-bearing partitions that run perpendicular to the ceiling joists are nailed to the joists through the top plate **(see 11-19)**. When they run parallel and between ceiling joists, 2 × 4-inch blocking is nailed between the joists, and a nailing plate is nailed to the top of the partition. This gives a nailing surface for ceiling material **(see 11-20)**.

OPENINGS IN THE CEILING

Openings in the ceiling occur when a pull-down stair is to be installed, an attic access opening is needed, or a fireplace penetrates the ceiling. They are framed in the same manner as described for floor construction. Single headers are used for openings less than three feet square **(see 11-21)**.

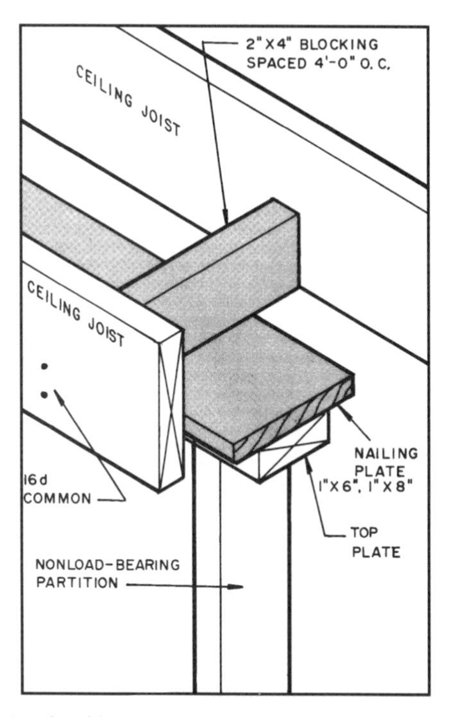

11-20 Nonload-bearing partitions that run parallel with, and between, ceiling joists are secured by blocking nailed to the top plate. Notice the nailing plate which provides a nailing surface for the finish ceiling material.

FLAT-ROOF CEILINGS

The construction of flat roofs is discussed in Chapter 16. A typical illustration is in **11-22**. Basically the roof rafters also serve as the ceiling joists. Nailing plates are placed between the rafters and on interior partitions to provide a nailing surface for the ceiling finish material around the edges of each room.

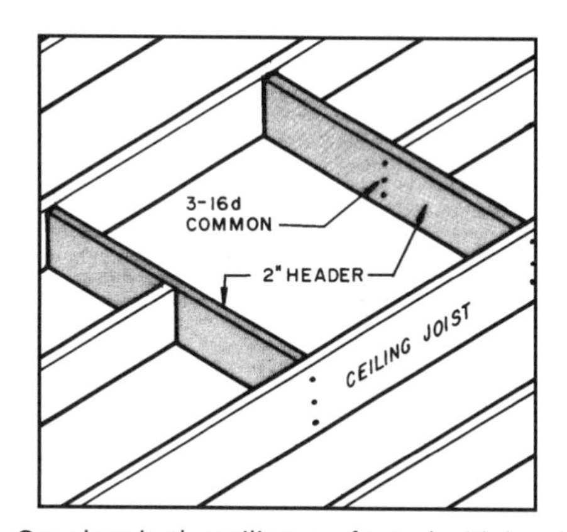

11-21 Openings in the ceiling are framed with headers in the same manner as used for floor joists. Single headers are used for openings smaller than 4'-0".

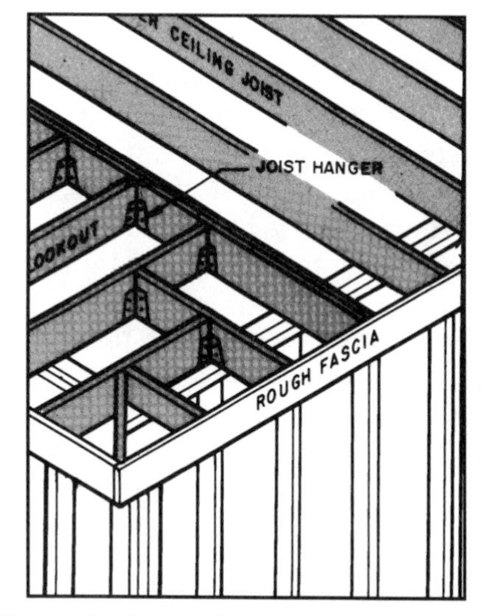

11-22 The roof rafter on flat roofs also serves as the ceiling joist.

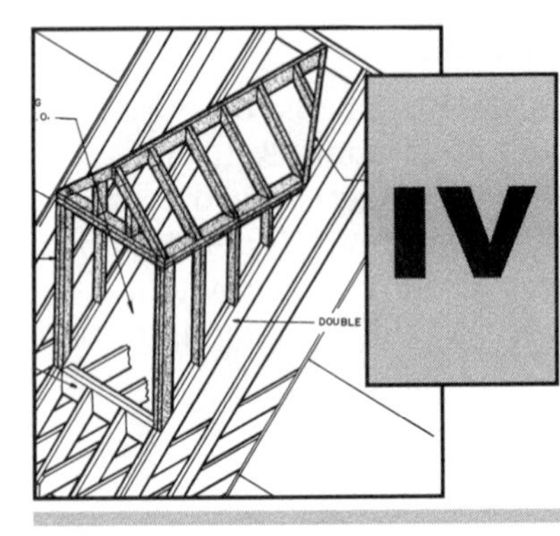

IV FRAMING THE ROOF & DORMERS

CHAPTER 12

Roof Types & Design

Framing a roof is one of the more difficult parts of constructing a building. It involves careful calculations of linear distances and angles as well as precise layout on the rafter stock. Much preplanning is required before layout and cutting is begun.

This chapter gives an introduction to the types of roofs and the technical aspects of their design. Additional chapters explain how to lay out and cut rafters for commonly used roof designs.

Types of Roof

The flat and shed roofs are much the same and are shown in **12-1**. The flat roof is built either absolutely flat or with a slight slope. It is the type most likely to leak eventually. The rafters serve as both ceiling joists and roof joists, and therefore they are often larger than those used in other types of roofs.

The shed roof is much like the flat roof except it has sufficient slope to enable water to drain. It is easy to build and is low cost. In addition to serving as a main roof it is widely used on dormers, porches, and small additions off a building.

The monitor shed roof with a clerestory typically has two shed roofs and a band of windows between them providing a source of light and ventilation.

The most frequently used roof is the gable shown in **12-2**. It is easy to frame and sheds water rapidly. A number of variations can be built, such as those with a Tudor peak on the end of the roof.

The hip roof is a popular style and is required on houses with certain architectural styles, such as French country houses. While more difficult to frame than the gable, it eliminates the gable-ends, which reduces the outside wall area requiring maintenance. One variation, the Dutch hip, shown in **12-3**, provides louvers to ventilate the attic.

Several other types of roofs are shown in **12-4**. Notice that they are an assemblage of various forms of the gable and hip roofs. The pyramid roof slopes from a center point to each side of the building. If the building is large the peak is very high and requires long rafters.

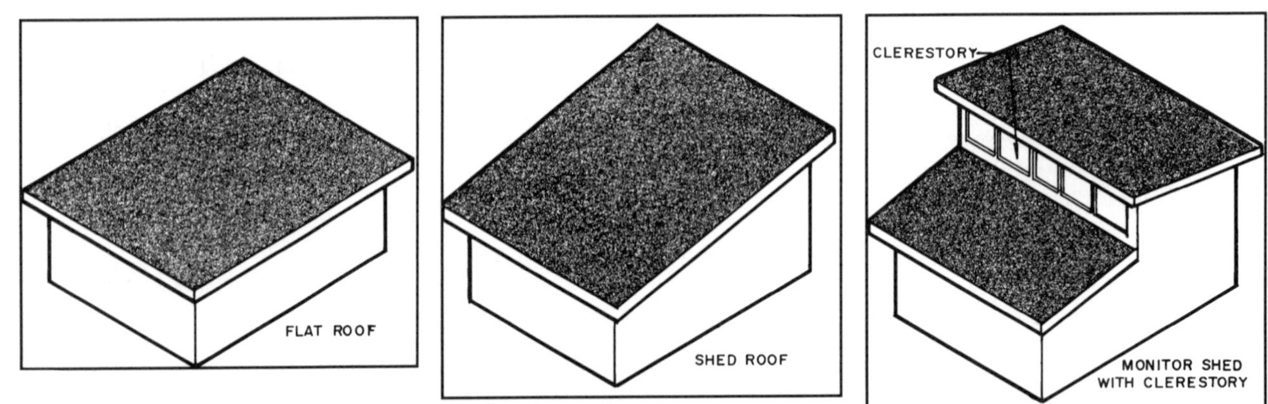

12-1 Flat and shed roofs are easy to frame and the lowest in cost.

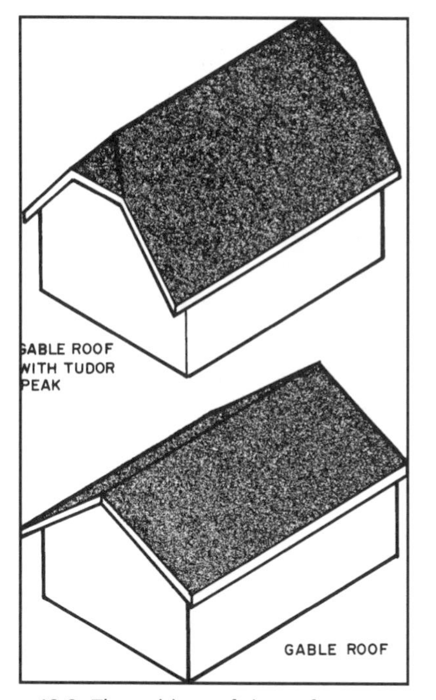

12-2 The gable roof slopes from a center ridge to two exterior walls parallel with the ridge.

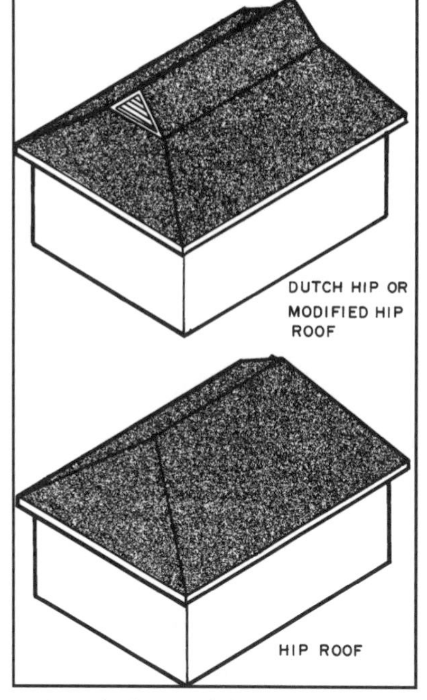

12-3 The hip roof slopes from a center ridge to all four sides of the building.

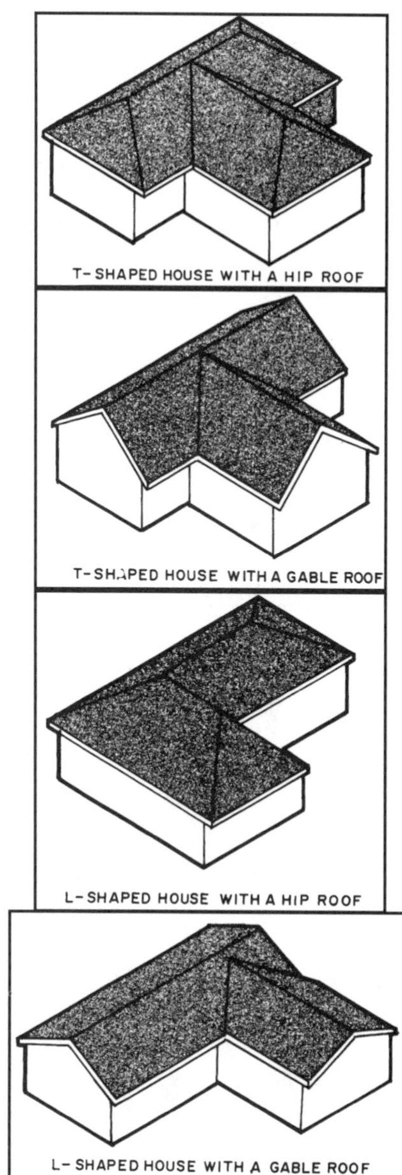

12-5 Buildings with shapes other than a rectangle require the roofs over each section to intersect, forming a valley.

The mansard roof resembles two hip roofs, one on top of the other. It provides considerable living space on a second floor and generally has a series of dormers to provide light and natural ventilation.

The gambrel roof has two sloped surfaces, each with different pitches. Like the mansard, it provides considerable living space on a second floor and generally has a series of dormers.

The examples just shown all refer to rectangular buildings. However, many buildings are built in some other shape, such as a T or an L. Examples of hip and gable roofs enclosing buildings of this type are in **12-5**. The line where the two roofs intersect is called a valley.

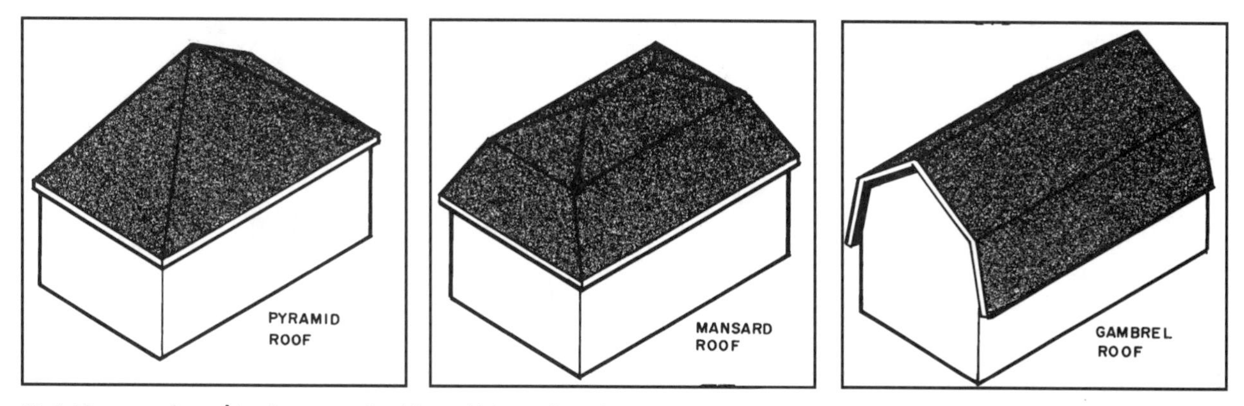

12-4 These roofs use the features of gable and hip roofs and combine them, providing double slopes.

Types of Rafter

A **rafter** is the structural member used to frame a roof and carry the sheathing and shingles. It must carry their load plus loads from snow and wind. The common types of rafters are shown in **12-6**.

Common rafters run from the top plate to the ridge at right angles to both.

Hip rafters run on a 45-degree angle from the top plate to the ridge, forming the line of intersection for the two surfaces of a hip roof.

Valley rafters connect the ridge to the wall top plate along the line where two intersecting roofs meet.

Hip jack rafters run from the top plate to a hip rafter.

Valley jack rafters run from the top plate to a valley rafter.

Cripple jack rafters run between valley jack and hip jack rafters.

PARTS OF A RAFTER

The rafter has a notch called a **bird's-mouth** which sits on top of the top plate. The rafter meets the ridgeboard with an angle called the **ridge line**. The **tail** is the overhang beyond the supporting wall. The outer end of the rafter is the **fascia surface**, and the cut on the bottom of the tail that establishes the length of the fascia surface is the **plancher (see 12-7)**. If there is no overhang the end of the rafter can have a **seat** cut as shown in **12-8**.

VENTILATION OF ATTIC

It is important that the attic have adequate ventilation. All attic spaces that are enclosed by a ceiling below must have cross ventilation. Vents must be covered with corrosion-resistant mesh with openings from ¼ to ½ inch and must protect the interior against the entrance of rain and snow. Local codes should be consulted for the required amount of net free venting. Typically this is equal to 1/150 of the area of the space to be vented. If a

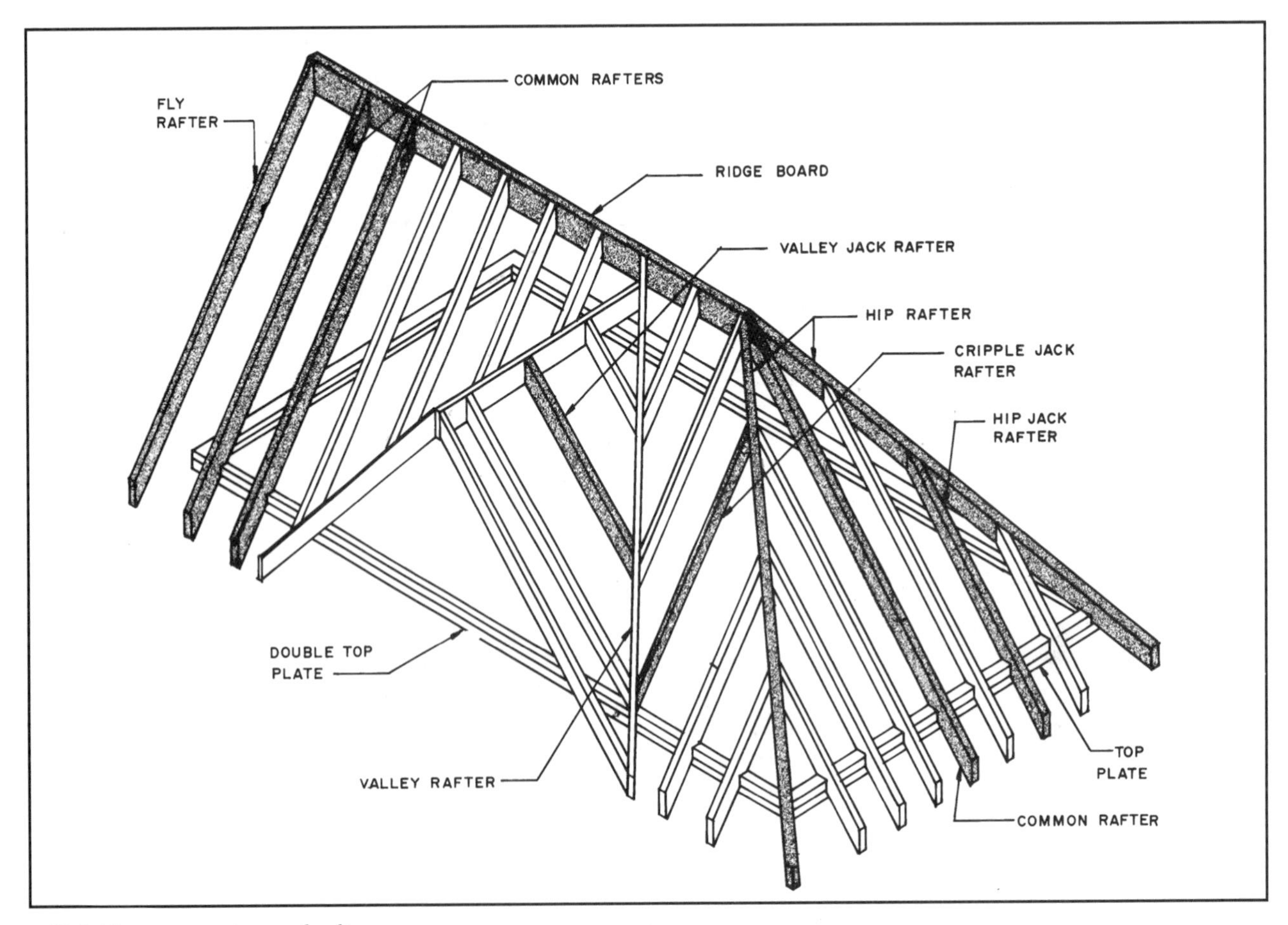

12-6 The common types of rafters.

vapor barrier is installed on the warm side of the ceiling a net free vent area of 1/300 of the floor area is commonly allowed. One way the roof can be built to allow natural venting is shown in **12-9**. The overhang of the eave will contain adequate vents that permit air to flow up and out the ridge vents. Vents on the gable-ends can be used instead of ridge vents.

Flat roofs must also be vented to clear the air between the insulation and the roof sheathing **(see 12-10)**. On large roofs metal vents are inserted as needed to provide additional venting area. Mechanical venting is also widely used. This uses an electric fan which is placed over a roof **(see 12-11)**. It is operated by a th turns it on whenever the air tempera attic reaches a preset level such as 160 d (72 degrees C).

In the following chapters the various meth for framing the roof for venting are shown.

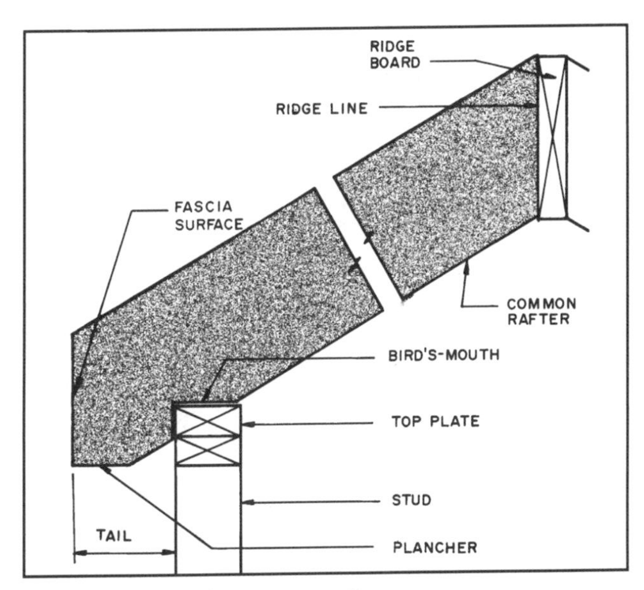

12-7 The parts of a common rafter.

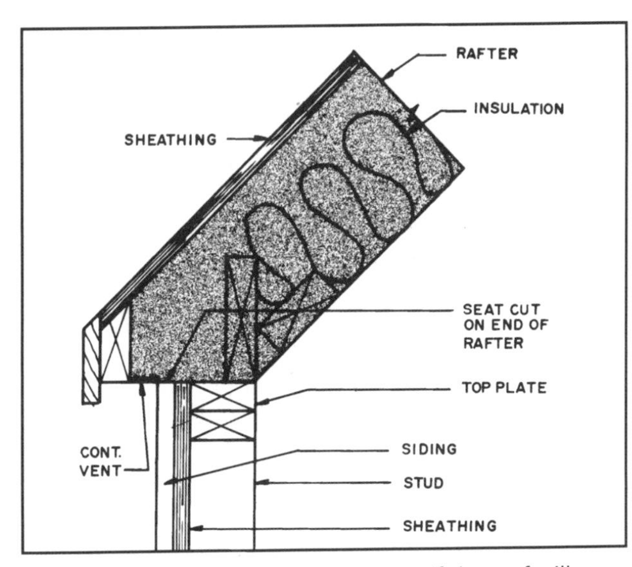

12-8 The rafter can end with a seat cut if the roof will have no overhang.

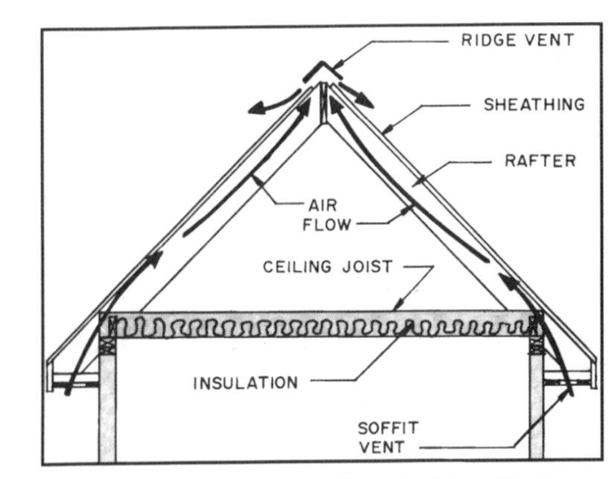

12-9 Roofs with attics are ventilated with soffit vents and some form of ridge vent or louver.

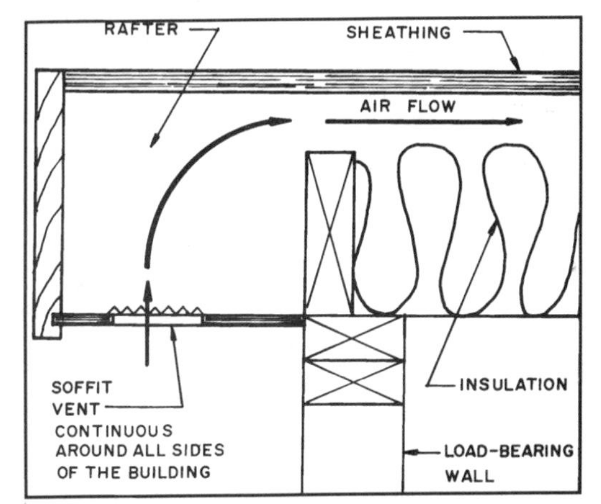

12-10 Flat roofs require ventilation between the sheathing and insulation.

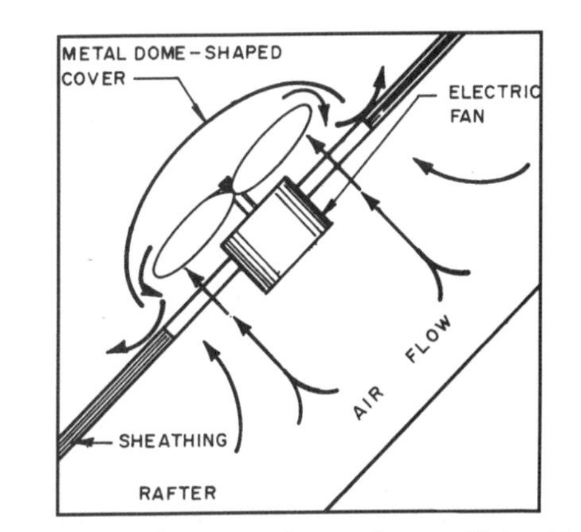

12-11 Power roof vents can be used to ventilate attics.

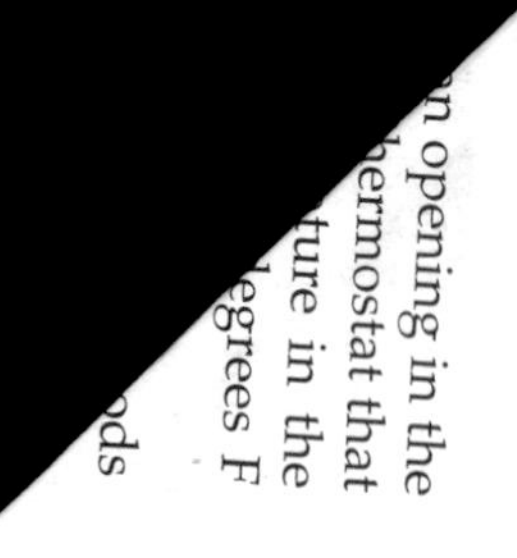

rs will prepare a roof ıg plan shows the loca- :). If the working draw- he framer will have to h the short time it takes ıtarting construction of w how many rafters are ıible framing problems ore the actual framing develop a framing plan is to lay a sheet of tracing paper over the floor plan and locate each member. Thumbnail plans can be drawn as shown in **12-13**.

DESIGN CONSIDERATIONS

As the architect designs the roof framing the live and dead loads on the members are considered, as is the run of the rafter, the use of collar beams and purlins, and the method for securing the rafters to the exterior wall top plate.

Live and Dead Loads

Live loads are loads upon the roof that occur occasionally but are not constantly present. These include snow and wind. Codes specify the local design requirements. For example, snow loads in northern states are much greater than in southern states. Sloped roofs will have different requirements than flat or shed roofs.

Dead loads consist of the total weight of the materials used to build the roof. This typically will include finished roofing, sheathing, and rafters. These loads are constantly present.

Collar Beams

Collar beams are used to tie rafters on opposite sides of the roof **(12-14)**. They are placed below the ridgeboard on every third pair of rafters or 48 inches O.C. Their purpose is to provide bracing and tie the ridgeboard and rafters into a rigid triangular configuration. They may be 1 × 6-inch or heavier material.

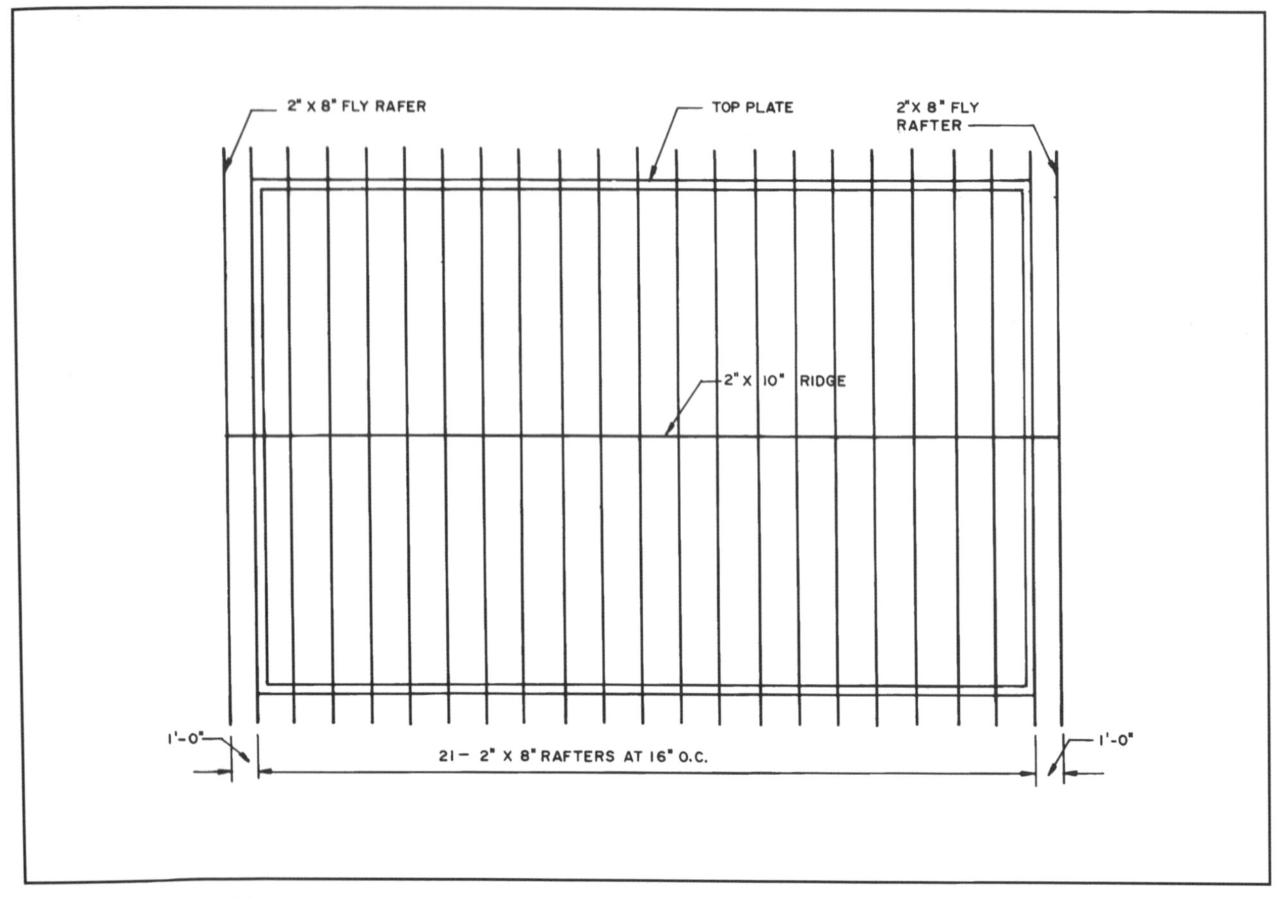

12-12 A typical roof framing plan for a simple gable roof.

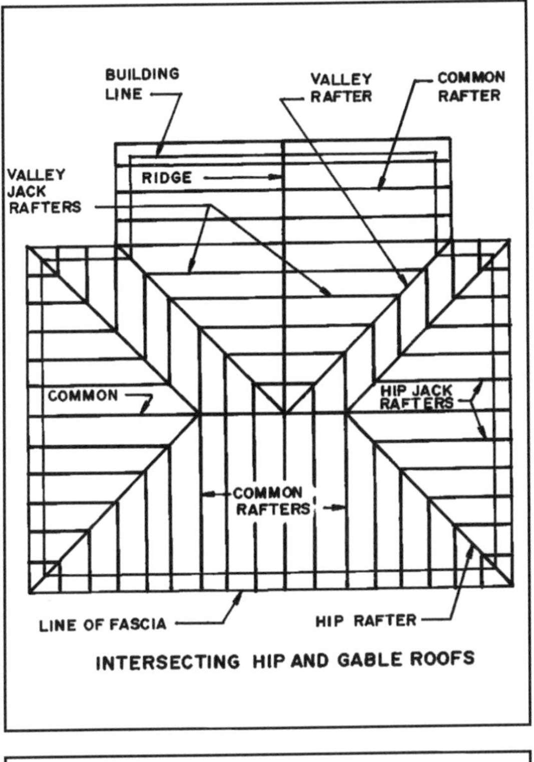

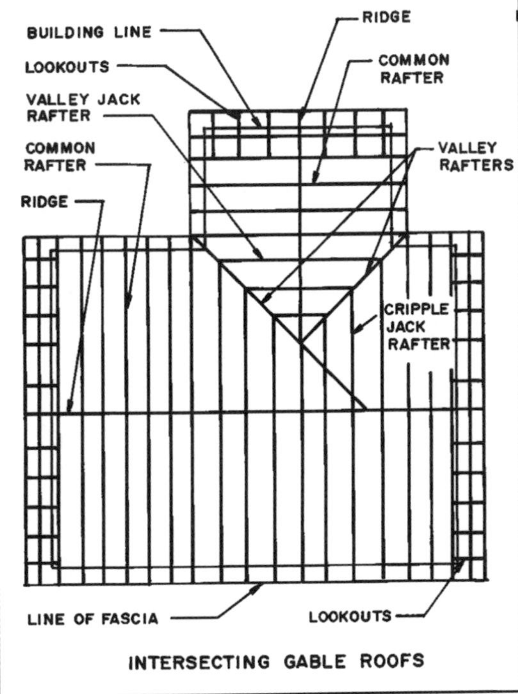

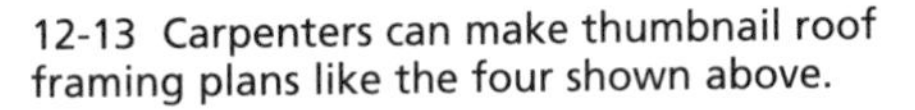
12-13 Carpenters can make thumbnail roof framing plans like the four shown above.

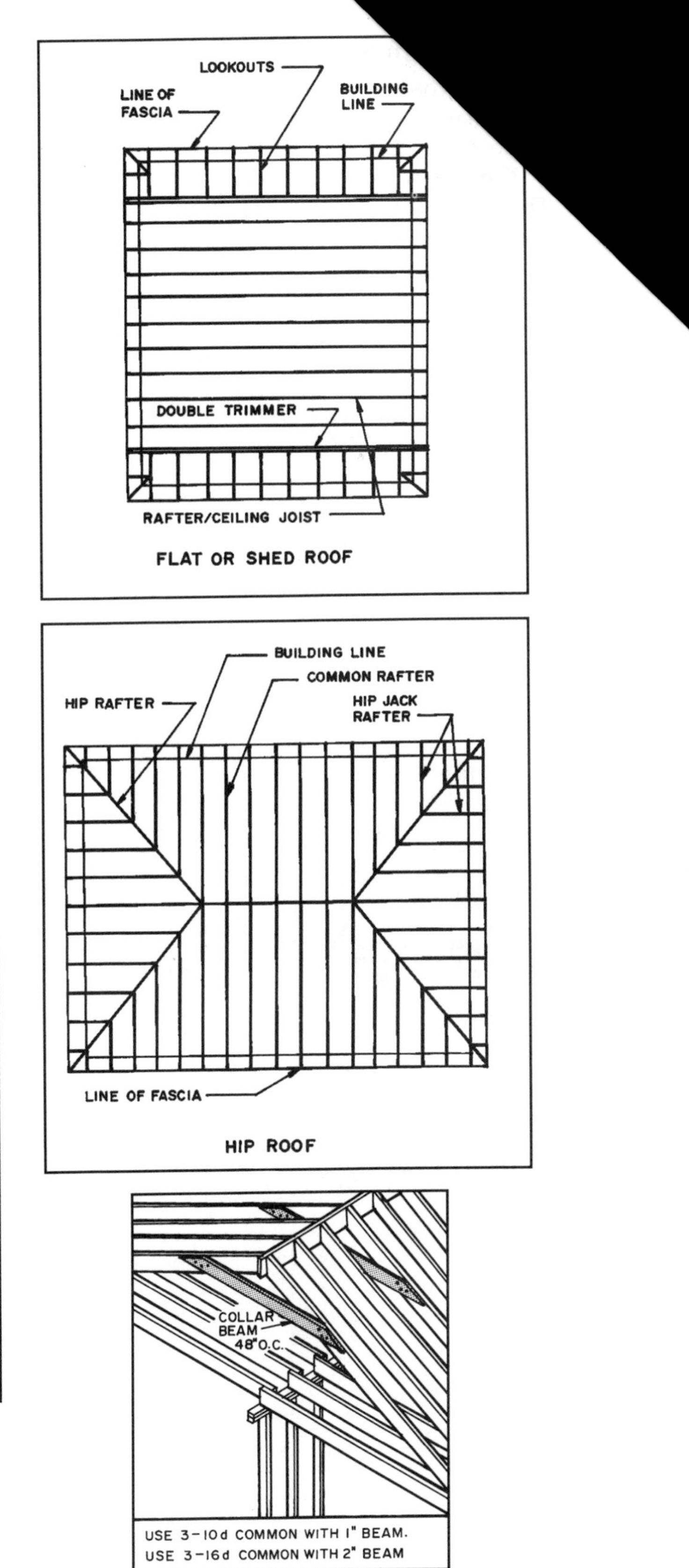

12-14 Collar beams tie opposite rafters providing additional bracing and stiffen the roof framing.

ral members that run
:s, and each is located
enter of its length **(see**
× 6- or 2 × 8-inch mate-
2 × 4-inch **purlin stud**
m, or purlin plate. The
ry depending upon the
l not be less than 45
struction supports the
ncy for it to sag, and can
be used to support rafters whose length may exceed the allowable distance.

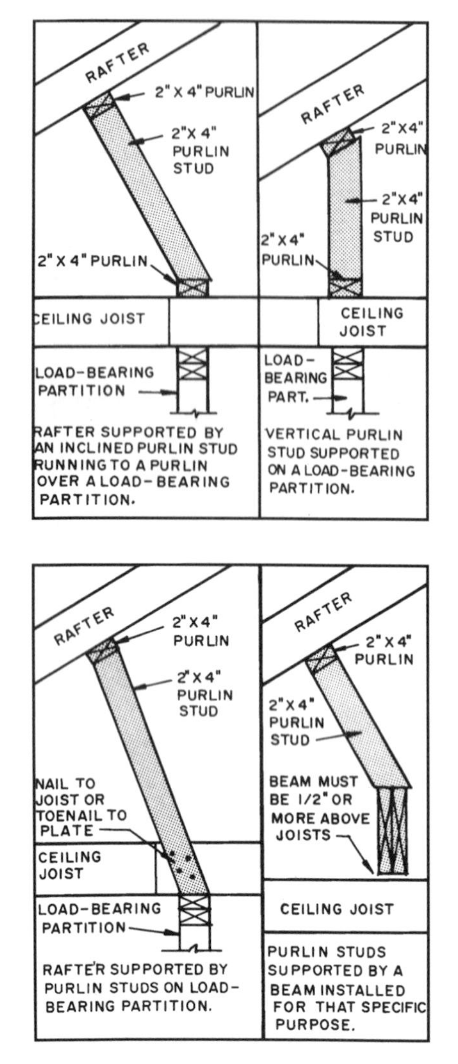

12-15 Purlins run below the rafters and with a brace to a load-bearing wall support the center of the rafter.

Roof Anchorage

Rafters are toenailed to the top plate as specified by the local code. Typically three 8d common nails are used. This connection ties the rafters to the exterior wall and resists the outward thrust on the ends of the rafters.

In some areas additional ties are required to resist seismic and wind load forces. Typically these tie the rafter to the top plate or to the studs below the top plate **(12-16)**. A wide variety of metal ties are manufactured.

TECHNICAL TERMS

Following are the technical terms used in the layout and cutting of rafters **(see 12-17)**.

Pitch is the angle the surface of the roof makes with the horizontal. It is a ratio of the unit of rise to the unit of span. A roof with a unit rise of 12 inches and a unit span of 24 inches has a pitch of 12/24 or ½. It is referred to as having a ½ pitch **(see 12-18)**.

Rise is the vertical distance from the top of the plate to the top of the ridgeboard **(see 12-17)**.

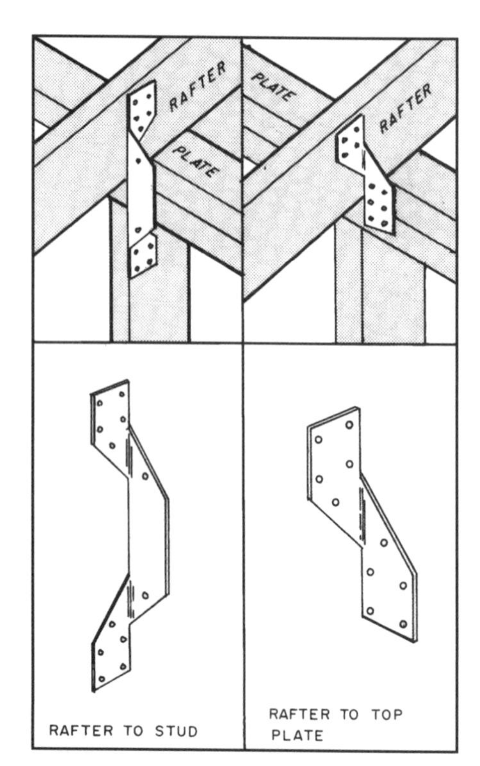

12-16 Codes in some areas require the use of metal rafter anchors to resist seismic and wind forces.

Run is the horizontal distance from the outside face of the top plate to the centerline of the ridgeboard **(see 12-17)**.

Span is the distance from the outside faces of the double top plates across the width of the building **(see 12-17)**.

Slope is the incline of the roof given in units of rise to units of run. A roof with a unit rise of 10 inches and a run of 12 inches has a slope of 10 in 12 **(see 12-19)**. The unit of run is always 12 inches.

The unit of rise is the number of inches a roof rises per foot of run **(see 12-19)**.

The unit of run is a 12-inch horizontal distance used to describe the slope **(see 12-19)**.

The unit of span is a 24-inch horizontal distance used to describe pitch **(see 12-19)**.

DETERMINING RAFTER SI

The size of a rafter depends on the must span, the roof slope, the spac rafters, the weight of the roofing mater load, snow load, and specie and grade of th ber used. Wind and snow loads vary a great across the country. Rafter span tables refer to dea loads and live loads. Dead loads are the weight of materials, such as sheathing and shingles. Live loads are those which are movable, such as snow, but do not include wind loads. Tables giving selected examples of data available are shown in Appendices G, L, M, N, and O. Detailed information is available from various organizations, such as the Southern Forest Products Association, the National Association of Homebuilders Research Foundation, the National Forest Products Association, and the Western Wood Products Association.

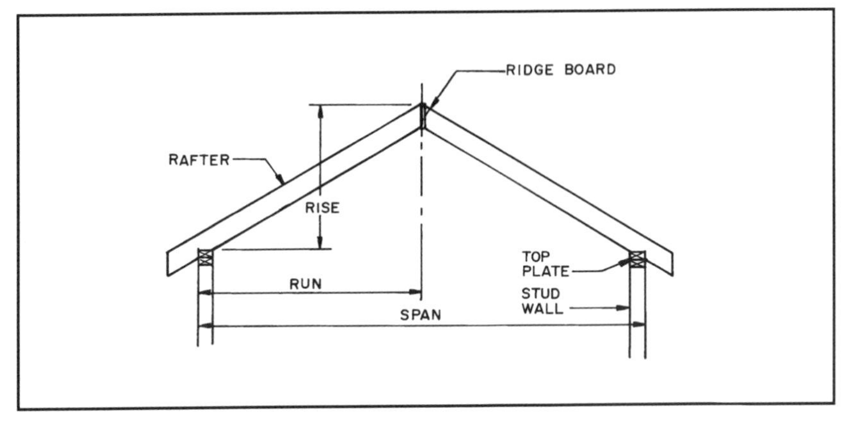

12-17 Terms used to specify roof slope and pitch.

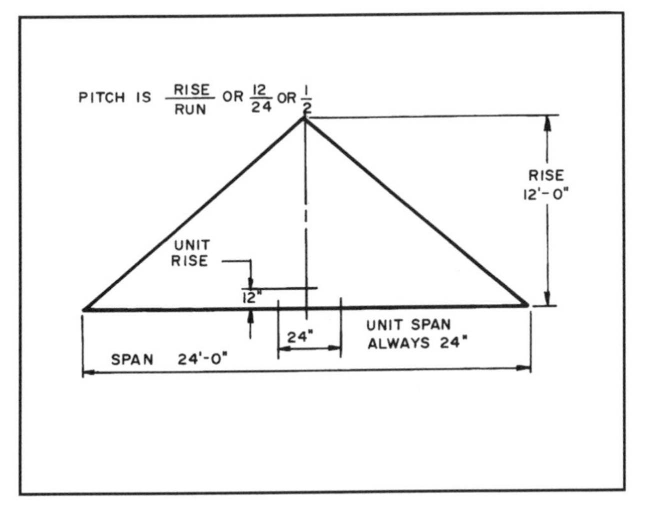

12-18 Pitch is the proportion of unit rise over unit span.

12-19 Slope is the proportion of unit rise over unit run.

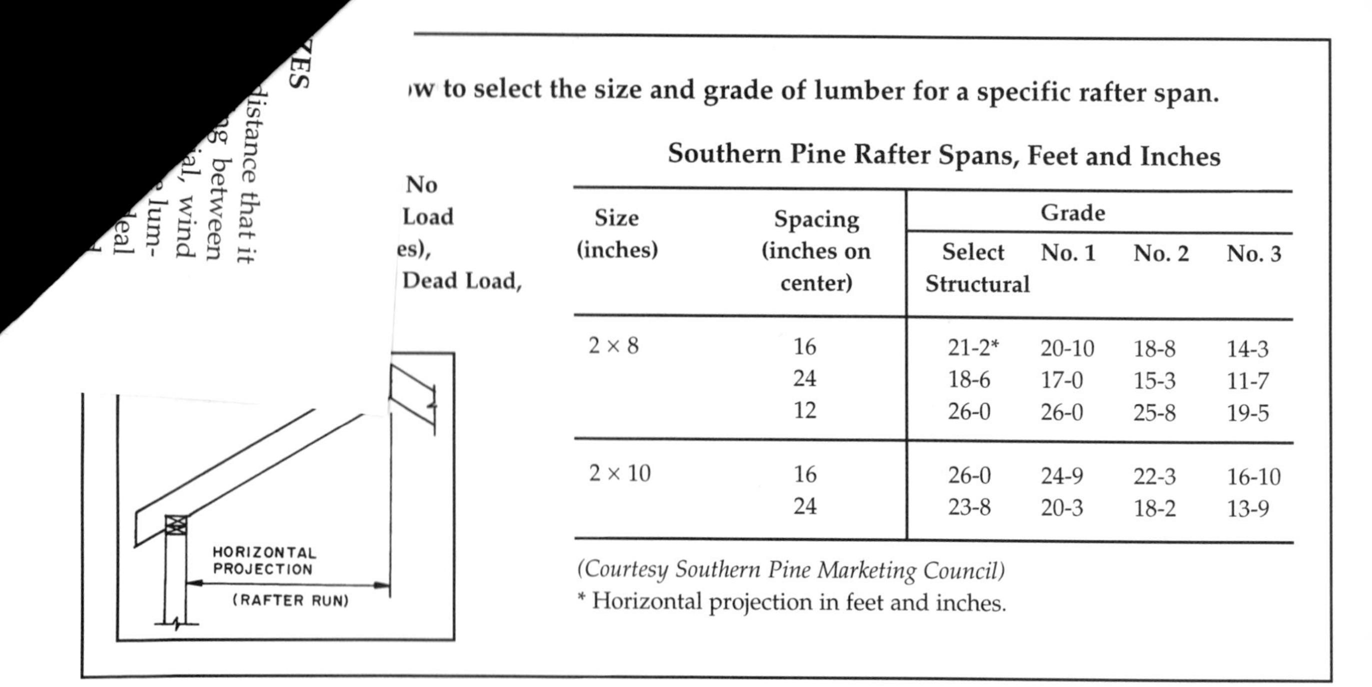

…w to select the size and grade of lumber for a specific rafter span.

No … Load … es), … Dead Load,

Southern Pine Rafter Spans, Feet and Inches

Size (inches)	Spacing (inches on center)	Grade: Select Structural	No. 1	No. 2	No. 3
2 × 8	16	21-2*	20-10	18-8	14-3
	24	18-6	17-0	15-3	11-7
	12	26-0	26-0	25-8	19-5
2 × 10	16	26-0	24-9	22-3	16-10
	24	23-8	20-3	18-2	13-9

(Courtesy Southern Pine Marketing Council)
* Horizontal projection in feet and inches.

For example, examine the partial rafter span table in **12-20**. The wood is southern pine, the loads are 20 psf live load (snow) and 10 psf dead load (weight of the materials in the roof). The allowable deflection is 180 and the load duration factor 161.15. These are explained in Section Five of this book. The unsupported distance a rafter can run is measured along the horizontal projection. For example, the 2 × 8-inch rafter as specified in **12-12**, if spaced 16 inches O.C., will run from 14′-3″ to 21′-2″ depending upon the lumber grade. If you had to run 24′-0″ you would have to use a 2 × 10-inch rafter, No. 1 grade. Extensive tables are in the appendix.

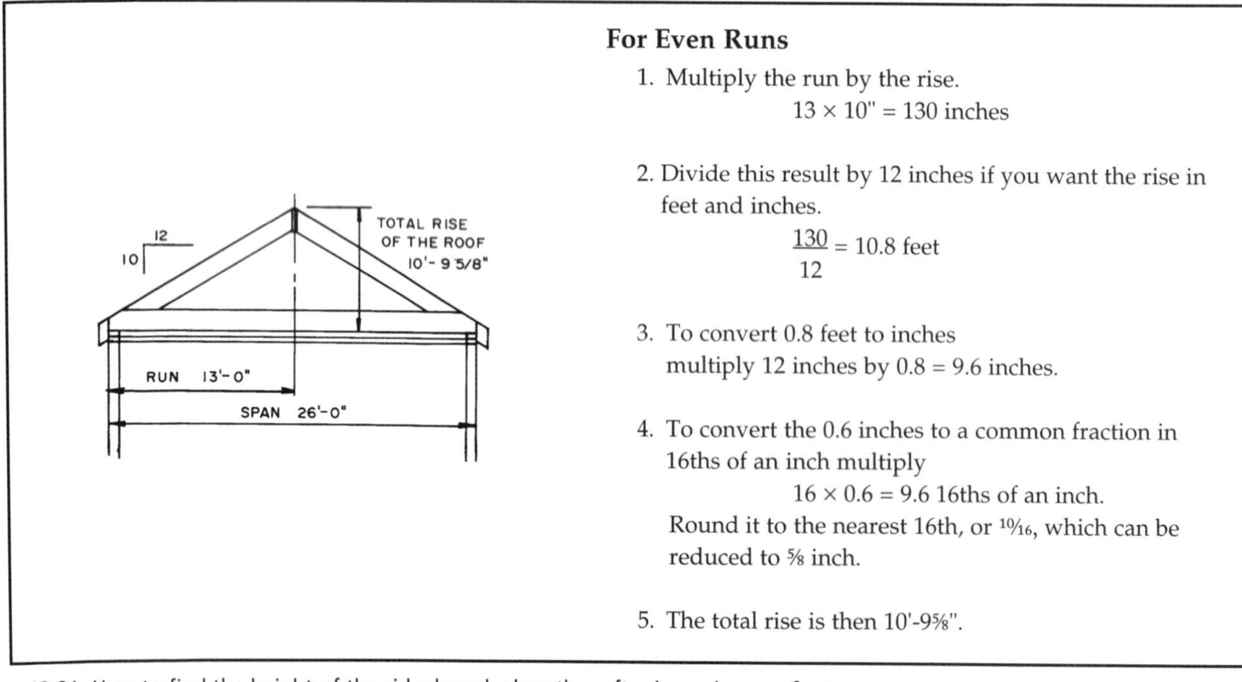

For Even Runs

1. Multiply the run by the rise.
   $13 \times 10'' = 130$ inches
2. Divide this result by 12 inches if you want the rise in feet and inches.
   $\frac{130}{12} = 10.8$ feet
3. To convert 0.8 feet to inches multiply 12 inches by $0.8 = 9.6$ inches.
4. To convert the 0.6 inches to a common fraction in 16ths of an inch multiply
   $16 \times 0.6 = 9.6$ 16ths of an inch.
   Round it to the nearest 16th, or 10⁄16, which can be reduced to ⅝ inch.
5. The total rise is then 10'-9⅝".

12-21 How to find the height of the ridgeboard when the rafter is run in even feet.

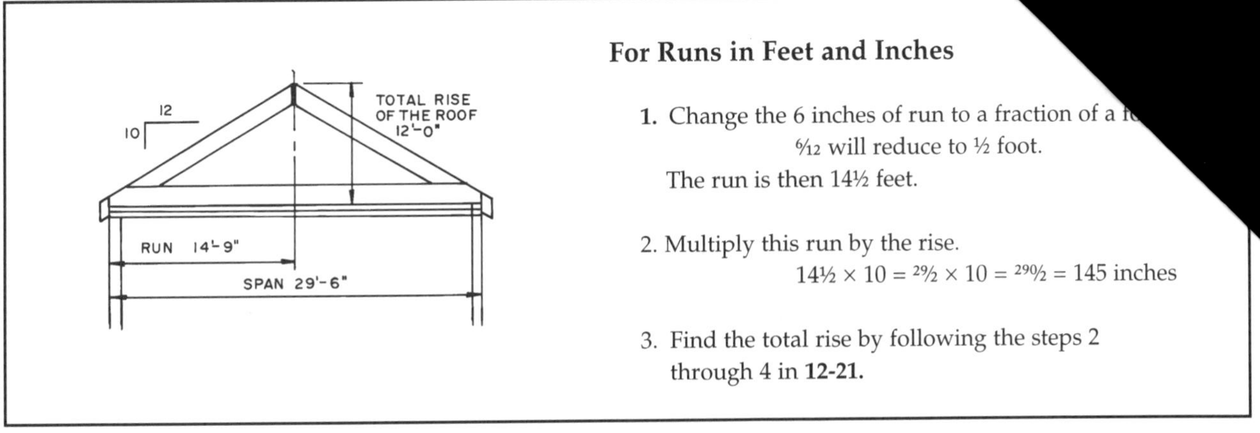

12-22 How to find the height of the ridge board when the rafter is run in feet and inches.

FINDING THE TOTAL RISE

The ridgeboard is set to the required height with temporary bracing before the rafters are installed. The carpenter needs to calculate the total rise of the roof so the ridgeboard can be set. This distance is from the top of the double wall plate to the top of the ridgeboard. If the run of the rafters is in even feet, the height of the ridgeboard can be found by multiplying the number of feet in the run by the rise as shown in **12-21**.

If the run is in feet and inches, convert the inches to a decimal and multiply the run by the rise as shown in **12-22**.

CHAPTER 13

CONSTRUCTING A GABLE ROOF

… individual rafters are … built. They are rapidly … y-assembled roof trusses (see Chapter 16). The following discussion tells how to lay out **common rafters** and shows how they are assembled to form the roof structure for a **gable roof**. Common rafters are also used on other types of roofs, such as a hip roof.

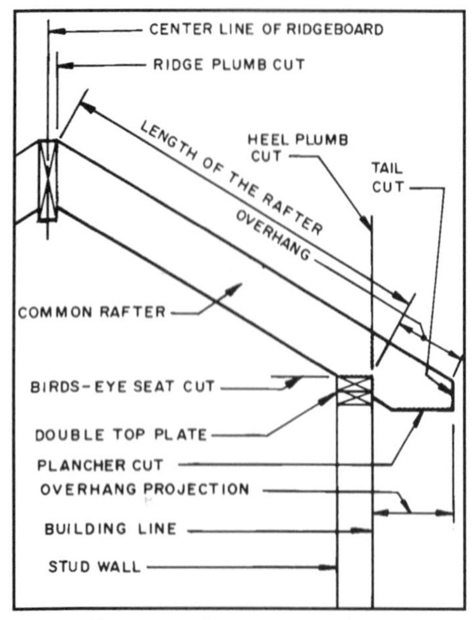

13-1 The parts of a common rafter.

THE COMMON RAFTER

The parts of a common rafter are shown in **13-1**. They typically have some overhang beyond the face of the stud. The amount of overhang and the size of the fascia cut are shown on the architect's drawings. The end of the rafter, fascia, and heel are plumb cuts. Since the rafter sits on top of the double plate, a **seat cut** (called a bird's-mouth) is made in the rafter **(see 13-2)**. It is a notch in which one side is the heel cut, and the other is perpendicular to it.

The Gable Roof Frame

The frame of a typical gable roof consists of a ridgeboard upon which the top end of the rafters are nailed. The common rafters forming the roof structure are seated on the exterior wall double top plate. The gable-end is framed and will support a exterior finished wall material. Sheathing is nailed over the rafters, making the structure a strong unified unit and also providing a nailing base for the finished roofing material **(see 13-3)**.

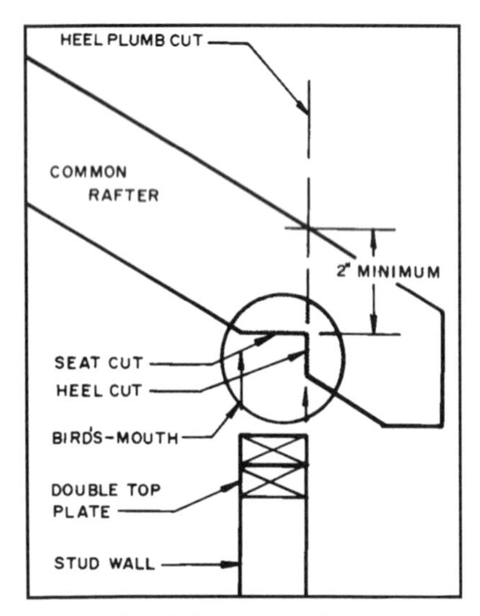

13-2 The bird's-mouth is formed with a heel plumb cut and a seat cut.

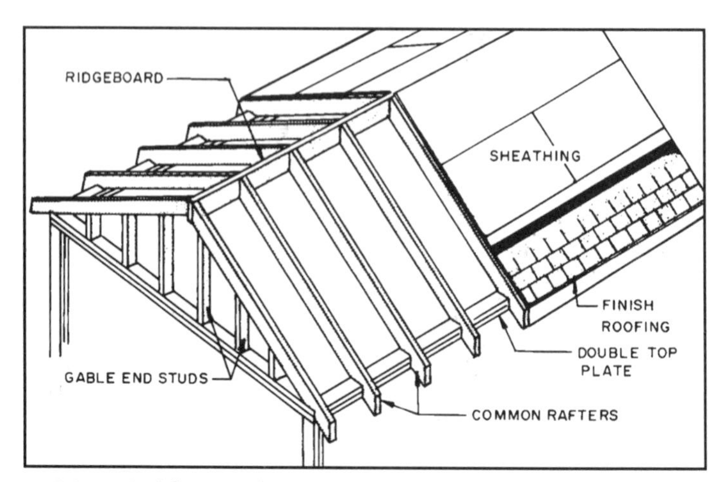

13-3 Typical framing for a gable roof.

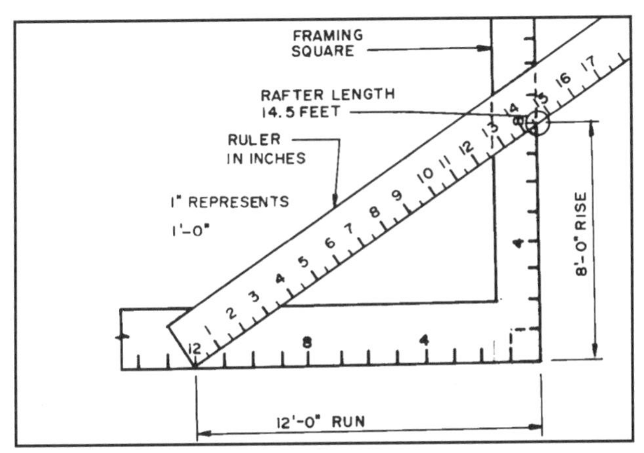

13-4 A framing square and a ruler can be used to find the approximate length of a rafter.

GETTING APPROXIMATE RAFTER LENGTH

The approximate length of a rafter can b locating the total rise in feet and total ru on the body and tongue of the framing squa **13-4)**. Each inch represents one foot. In **13-4** total rise is eight feet and the total run is 12 feet. T find the **approximate rafter length**, measure the hypotenuse and add the length of the overhang. In this case the length of the hypotenuse is 14.5 feet, and a 6-inch overhang was added, so that the rough length is 15′-0″. The framer will know to order 16′-0″ long rafter material.

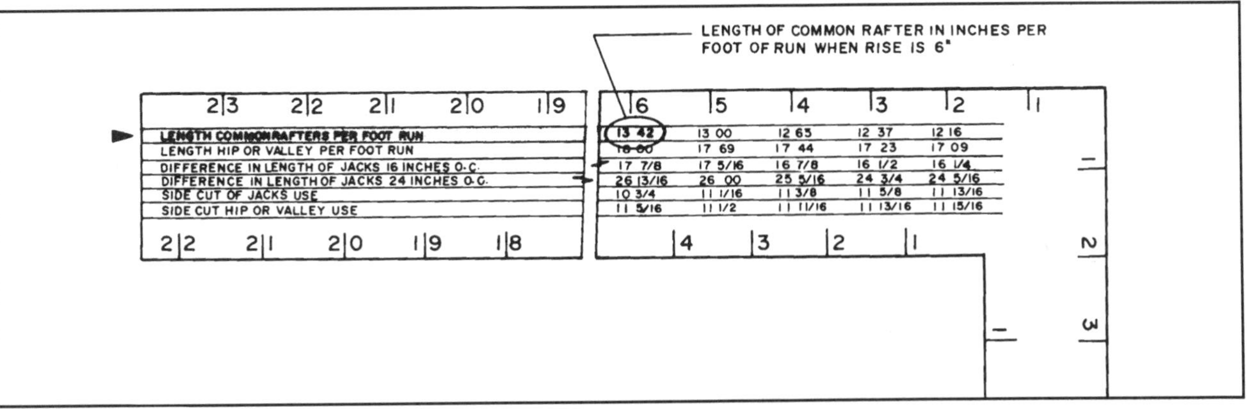

13-5 Rafter length for common rafters with rises in even inches can be found using the rafter tables on a framing square.

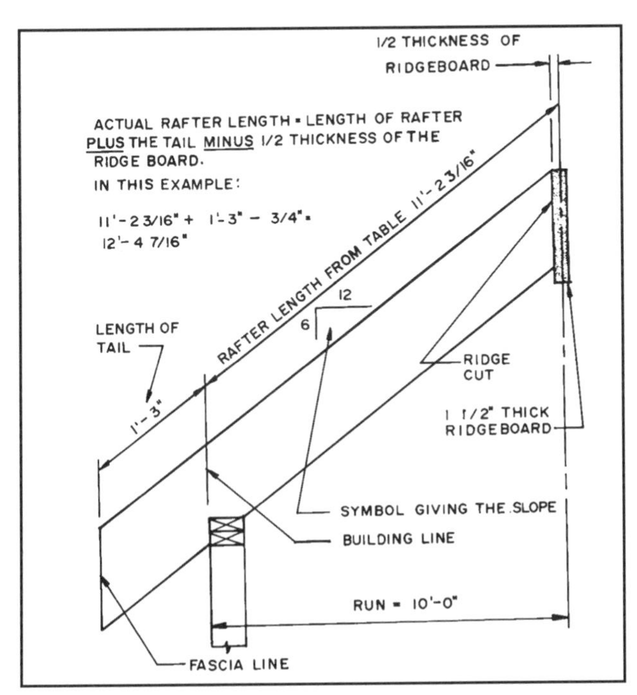

13-6 How to figure the actual rafter length using the rafter table on the framing square.

LAYING OUT A RAFTER

The two commonly used methods for finding rafter length are the use of rafter tables and the step-off method.

Rafter Tables—Framing Square

Steel framing squares have rafter tables stamped on the body. These show rafter lengths for several types of rafters. Examine the square in **13-5**. The first row of numbers gives the lengths of common rafters that have rises in even inches. For example, if a rafter has a six-inch unit rise, find the 6 on the square. Below it on the "common rafter" line is the number 13.42. This means that the length of the rafter is 13.42 for each foot of run. If the run were 10 feet the rafter would be 13.42 times 10 or 134.2 inches, or 11.183 feet or 11 feet 2³⁄₁₆ inches long.

The final total length of the rafter will include the addition of the length of the tail and the subtraction of one-half the thickness of the ridgeboard **(see 13-6)**.

ommon rafter lengths—9 foot-0 inch run.

	9'-0"			9'-0¼"			9'-0½"			9'-0¾"		
	ft	in	16th"	ft	in	16th"	ft	in	16th"	ft	in	16th"
	10'	5"	1	10'	5"	5	10'	5"	10	10'	5"	14
	10'	9"	13	10'	10"	2	10'	10"	6	10'	10"	11
	11'	3"	0	11'	3"	5	11'	3"	10	11'	3"	15
	11'	8"	9	11'	8"	15	11'	9"	4	11'	9"	9
	12'	2"	8	12'	2"	14	12'	3"	3	12'	3"	8
12 in 12	12'	8"	12	12'	9"	1	12'	9"	7	12'	9"	13
13 in 12	13'	3"	4	13'	3"	10	13'	3"	15	13'	4"	5
14 in 12	13'	9"	15	13'	10"	5	13'	10"	12	13'	11"	2
15 in 12	14'	4"	14	14'	5"	5	14'	5"	11	14'	6"	1

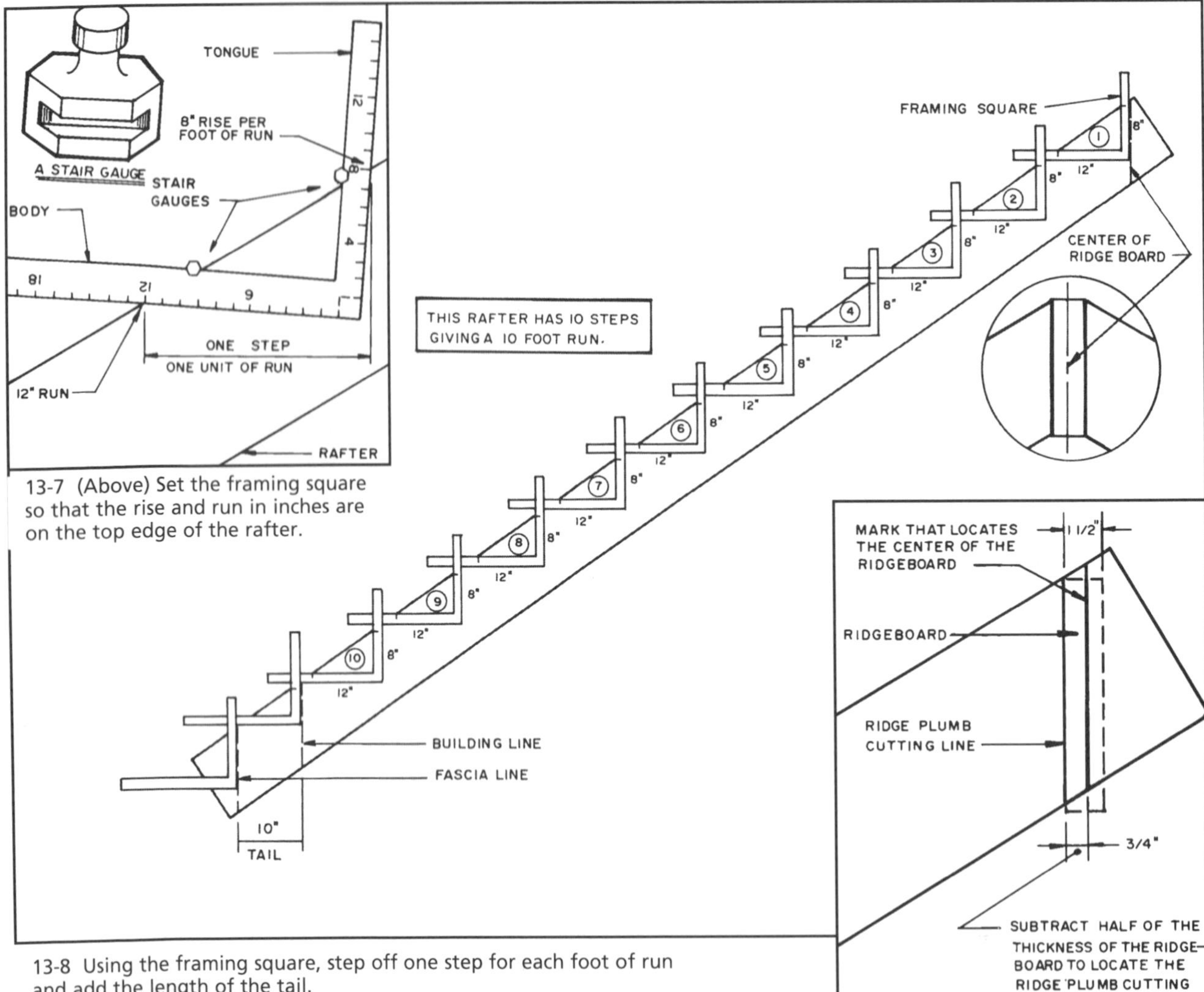

13-7 (Above) Set the framing square so that the rise and run in inches are on the top edge of the rafter.

13-8 Using the framing square, step off one step for each foot of run and add the length of the tail.

13-9 Adjust the mark at the ridge to allow for half the thickness of the ridgeboard.

Book of Rafter Tables

Common rafter lengths can be found using published manuals of rafter tables. A portion of a typical table is given in **Table 13-1**. This does not include any overhang.

Rafter Step-Off When Run Is an Even Number

The following instructions are for laying out a common rafter for a building having a run in an even number of feet. For this example, the span is 20 feet, which gives a 10-foot run. The roof has a slope of 8 inches of rise per 12 inches of run, and a 10-inch overhang is required.

1. Set the framing square so that the 8-inch and 12-inch marks are on the top edge of the rafter. Fasten stair gauges to the square so that it retains this position **(see 13-7)**. A **stair gauge** is a metal clip that is fastened to the square with a set screw. Check the rafter for straightness. If it has a slight crown, place the crown side up. If it has much warp or twist do not use it.
2. Draw the line representing the c[illegible] ridgeboard near the end of the raf[illegible] material has splits in the end, cut it ba[illegible] remove them **(see 13-8)**.
3. Step off one 12-inch step for each unit of ru[illegible] In this example it will be 10 steps. Mark this last step. This is the outside face of the top plate (the building line) **(see 13-8)**.
4. Measure an additional 10 inches, and mark the fascia line. This provides the overhang **(see 13-8)**.
5. Return to the ridgeboard centerline. Mark off half the thickness of the ridgeboard. This is the actual ridge cutting line **(see 13-9)**. Now mark the bird's-mouth as shown in **13-10**. It should be noted that codes usually require at least two inches of solid wood remain above the bird's-mouth. In **13-11** the seat can be the full 3½-inch width of the top plate because a 2 × 6-inch rafter was used. This would not be the case if a 2 × 4-inch rafter were used.

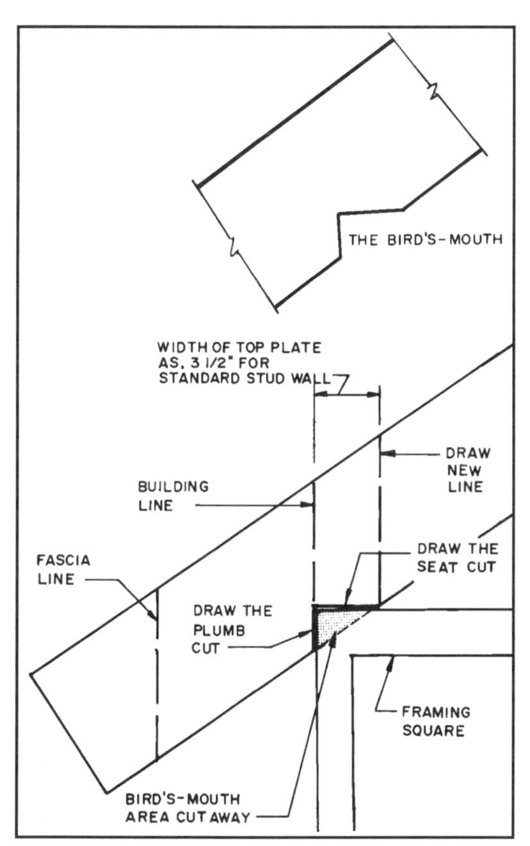

13-10 Lay out the bird's-mouth perpendicular to the building line and make the seat the width of the top plate.

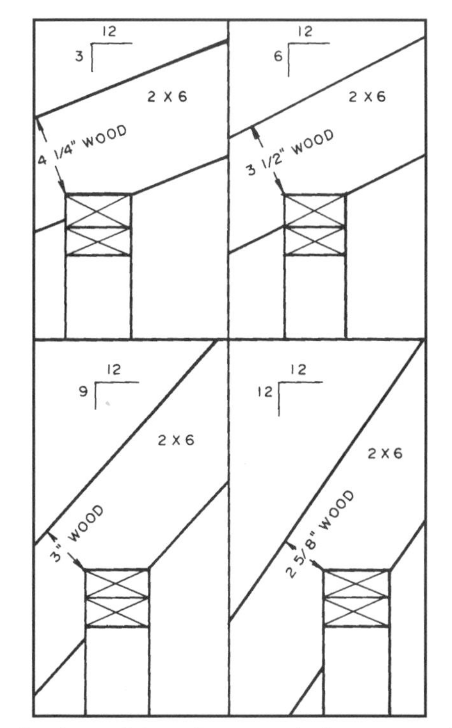

13-11 The slope of the rafter controls the amount of solid wood remaining above the bird's-mouth.

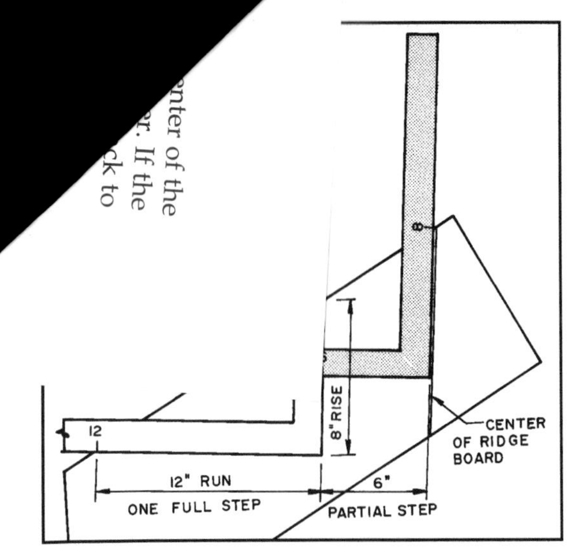

13-12 How to lay out a rafter when the run is not an even number of feet, such as 10′-6″.

Table 13-2 Length of rafter per foot run (in inches).

Rise (inches)	Common rafters	Hip valley rafters
1	12.04	17.03
2	12.16	17.09
3	12.37	17.23
4	12.65	17.44
5	13.00	17.69
6	13.42	18.00
7	13.89	18.36
8	14.42	18.76
9	15.00	19.21
10	15.62	19.70
11	16.28	20.22
12	16.97	20.78
13	17.69	21.38
14	18.44	22.00
15	19.21	22.65
16	20.00	23.32
17	20.81	24.02
18	21.63	24.74
19	22.47	25.47
20	23.32	26.23
21	24.19	27.00
22	25.06	27.78
23	25.94	28.58
24	26.83	29.39

Rafter Step-off When Span Is an Odd Number

Many times the run is not an even number of feet but will be somewhere between, such as 10 feet 6 inches. The procedure for this layout is the same as described for the even foot span except for a partial step, since an additional six inches, in this case, are needed **(see 13-12)**.

Marking the Bird's-Mouth

Return to the mark locating the building line. Measure a distance equal to the width of the top plate. Place a framing square on the building line and mark a line perpendicular to it that runs through the point where the plate width line meets the bottom edge of the rafter. This is the seat cut for the bird's-mouth **(refer to 13-7)**.

A summary of the steps to lay out a common rafter are shown in **13-13**. From marking the center of the ridgeboard through marking the tail cut and bird's-mouth, the carpenter must measure carefully to get an accurate rafter. The first rafter laid out and cut usually forms the pattern for marking and cutting all the other rafters.

Laying Out a Common Rafter Using Rafter Tables

Rafter tables give the lengths of rafters per foot of run for various slopes. They are stamped on the body of some framing squares or published in table form as shown in **Table 13-2**.

The numbers given in **Table 13-2** are the length of the rafter per foot of run for a given rise. For example, assume a roof with a run of 15 feet and a rise of 10 inches per foot of run for a common rafter. The length opposite the rise of 10 is 15.62 inches. Multiplying this by 15 feet (15.62 × 15) gives a rafter length of 234.30 inches or 19.525 feet, or 19′-6.3″ or 19′-5⁄16″. To convert 0.525 feet to inches multiply it by 12, which gives 6.3 inches. To get this to a 16th of an inch multiply 0.3 inches by 16, giving 4⁸⁄16—then round to 5⁄16 inch.

When using rafter tables, the length calculated is to the center of the ridgeboard. Therefore half the thickness of the ridgeboard must be removed from the length of the rafter, as shown in **13-14**. The tail is added as explained in the previous process.

COMMON RAFTER LAYOUT USING A SWANSON SPEED® SQUARE

The Swanson Speed® square is shown in **13-15**. It is designed with scales for laying out common hip, jack, and valley rafters using the unit of slope, such as a 5/12 slope, or in degrees.

To lay out a common rafter the proce[illegible] same as shown in **13-14**. The difference is[illegible] ridge, building line, and fascia lines are l[illegible] with the Swanson Speed® square.

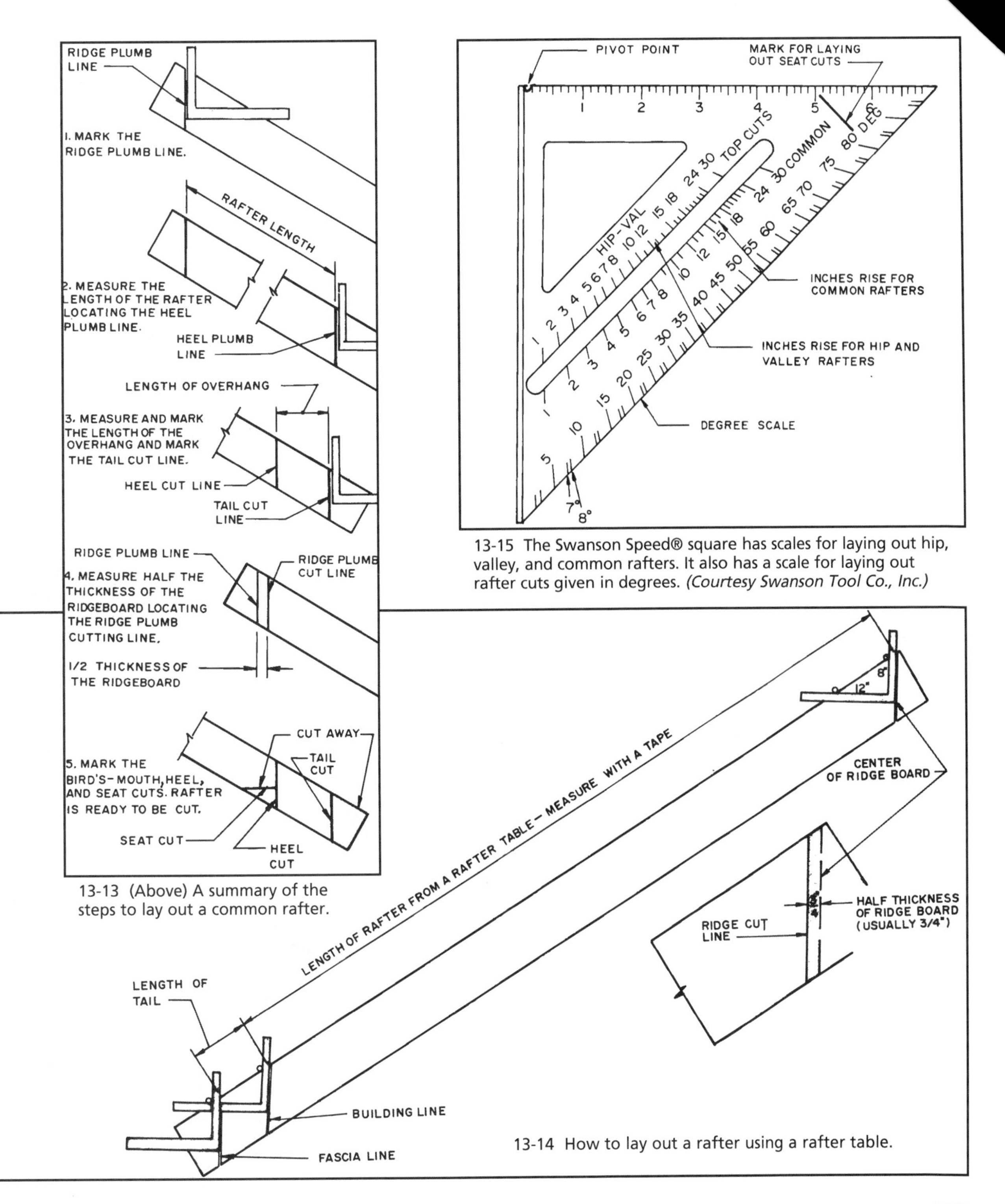

13-13 (Above) A summary of the steps to lay out a common rafter.

13-15 The Swanson Speed® square has scales for laying out hip, valley, and common rafters. It also has a scale for laying out rafter cuts given in degrees. *(Courtesy Swanson Tool Co., Inc.)*

13-14 How to lay out a rafter using a rafter table.

nson Speed® square is
n the 90-degree corner
edge of the stock. The
number representing
e of the stock. The line
) is drawn along the

ited with the Swanson
uilding line as shown in

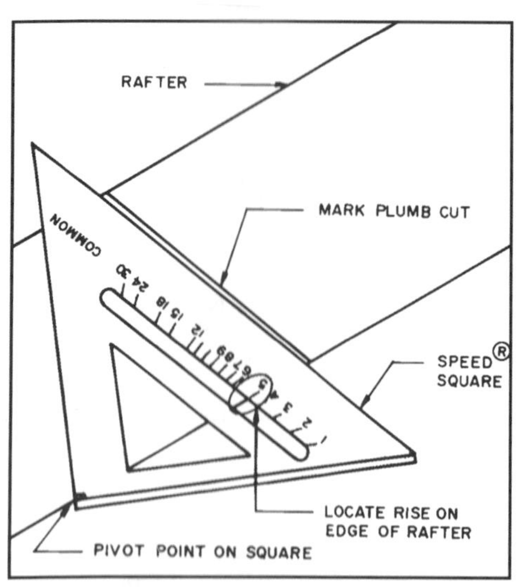

13-16 How to lay out the plumb cut at the ridge of a common rafter using the Swanson Speed® square. *(Courtesy Swanson Tool Co., Inc.)*

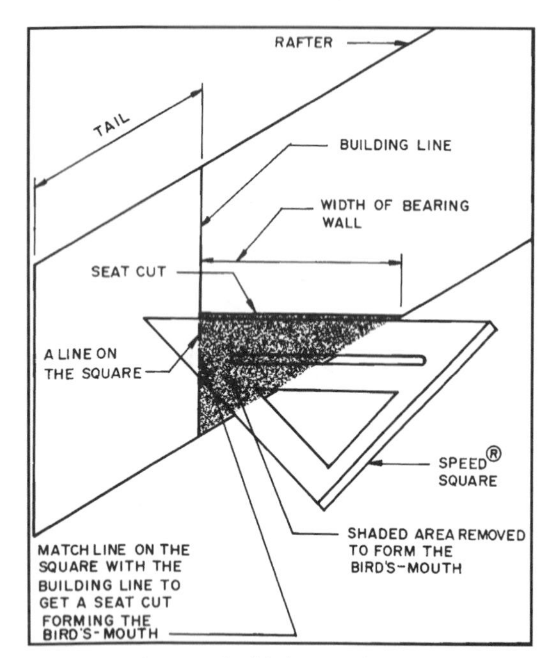

13-17 How to lay out the bird's-mouth with the Swanson Speed® square. *(Courtesy Swanson Tool Co., Inc.)*

LOCATING THE PLANCHER LINE

The end of the rafter tail usually requires a plancher cut to reduce it to the size required for the fascia. If the fascia board is as wide or wider than the fascia plumb cut, no plancher is needed **(see 13-18)**.

The size of the fascia determines the location of the plancher cut. This will depend upon how the fascia is to be framed. Three commonly used ways are shown in **13-19**. The length of the fascia line will be determined by the framing method. Once the length is determined, draw the plancher line perpendicular to the fascia line **(see 13-20)**.

Duplicating & Cutting Rafters

After the first rafter is laid out, cut it to size, and check to make certain it is correct. When certain it meets the ridgeboard and sits on the top plate properly, use it as a pattern to mark the remainder of the rafters of that size. The rafters can be cut one at a time with a handsaw or a portable circular saw.

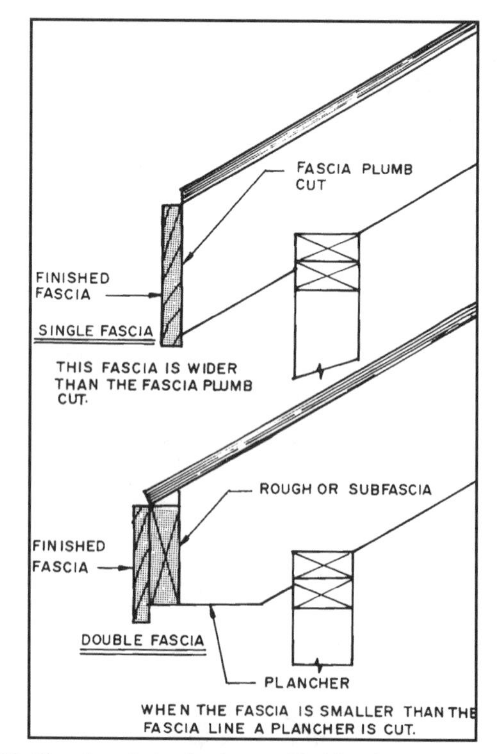

13-18 The size of the fascia specified by the architect determines if a plancher is needed.

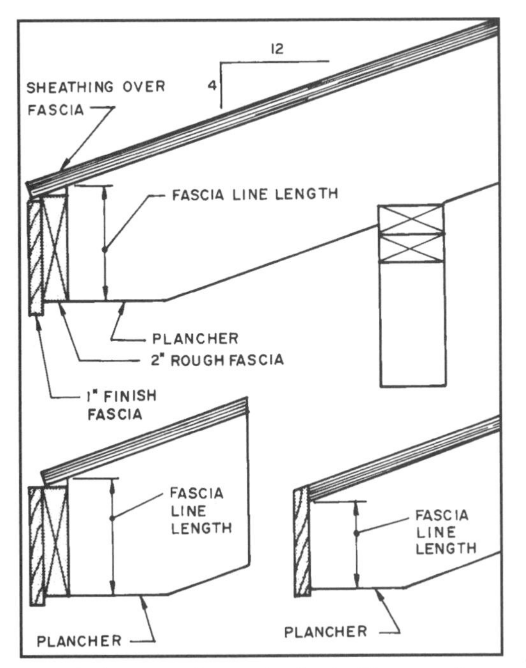

13-19 Typical ways to frame the fascia. The use of a rough fascia helps tie together the ends of the rafters, keeps the finished fascia from warping, and provides a solid nailer for the installation of gutters.

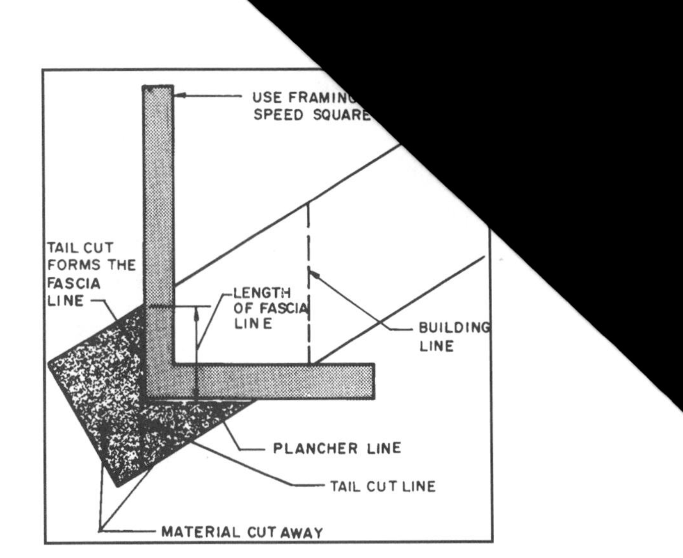

13-20 When the size of the fascia line is known it is laid out along the tail cut line.

It is quicker to set up stops on the table of a radial arm saw and set the saw to cut on the angle required. The chapter on power tools gives additional information. A basic setup is shown in **13-21**. The first rafter that was cut is used to set up the saw. The end-stop is nailed to the wood saw table. It is spaced a distance from the blade equal to the rafter length plus the length of the tail minus half the thickness of the ridgeboard. The saw is set to cut on the required angle. This setup requires that a long table be built for the saw.

When cutting the rafters place the crown, if any, away from the saw fence. If a piece has too much crown or warp do not use it.

There are several power tools that permit rafters to be gang-cut.

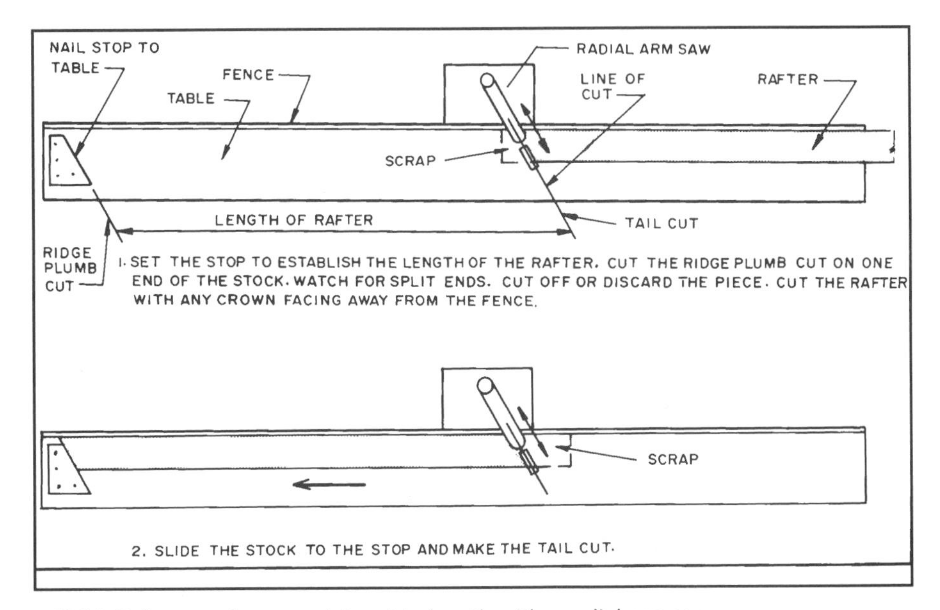

13-21 Rafters can be accurately cut to length with a radial-arm saw.

13-22 This large-diameter, portable circular saw will gang-cut most rafters and beams in one pass. It takes two persons to operate it. *(Courtesy Mafell North America, Inc.)*

One power tool that permits rafters to be gang-cut is a portable circular beamsaw with a large diameter blade. This provides sufficient depth to cut 2 × 4-inch and 2 × 6-inch rafters on low slopes. Steeper slopes may require a second cut to finish the job unless a large-diameter saw is available **(see 13-22)**. Another tool used for this purpose is a chainsaw-type beam cutter **(see 13-23)**. The bar is adjustable to angles up to 45 degrees.

To gang-cut rafters, place them on edge on sturdy sawhorses. Nail blocking to the horses to hold the rafters upright. Adjust them until they are all aligned squarely on the end to have the ridge cut. Run a chalkline along this end, which is the location of the short end of the ridge cut **(see 13-24)**. You can then measure from this line the length of the rafter to the heel cut of the bird's-mouth and snap a chalkline along all the rafters. This is a reference line for marking and making the heel cut.

The tail cut can also be marked and gang-cut with this setup. However, some carpenters prefer to install the rafters and mark and cut the tail cut after installation. To do this measure the overhang on each end and snap a chalkline. Cut each tail plumb individually with a portable circular saw **(see 13-25)**.

13-23 This chainsaw beam cutter will cut most rafters and beams in a single cut. *(Courtesy Mafell North America, Inc.)*

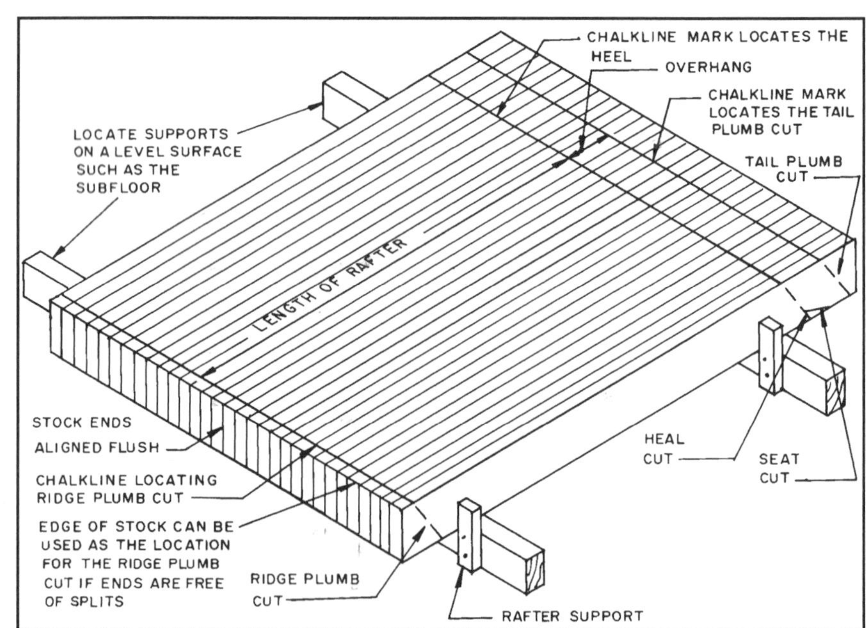

13-24 Typical preparation for assembling holding and marking rafters so they may be gang-cut with a portable circular beamsaw or a chainsaw-type beamsaw. The same setup can be used to mark rafters that are to be individually cut with a portable circular saw.

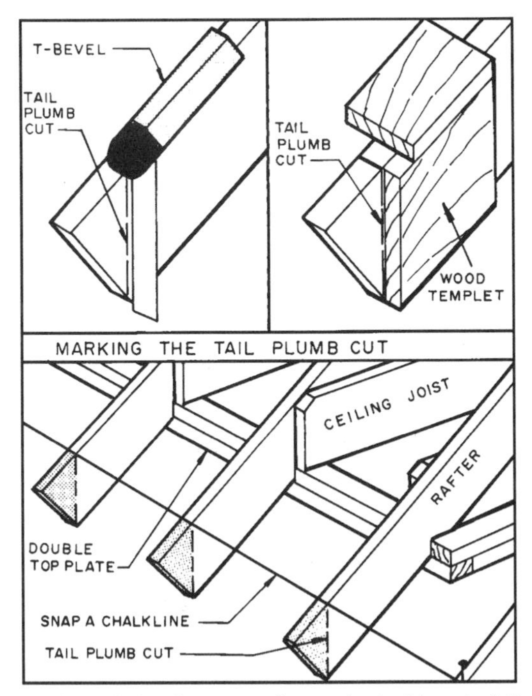

13-25 The tail plumb cuts can be marked with a chalkline and cut after the rafters are installed.

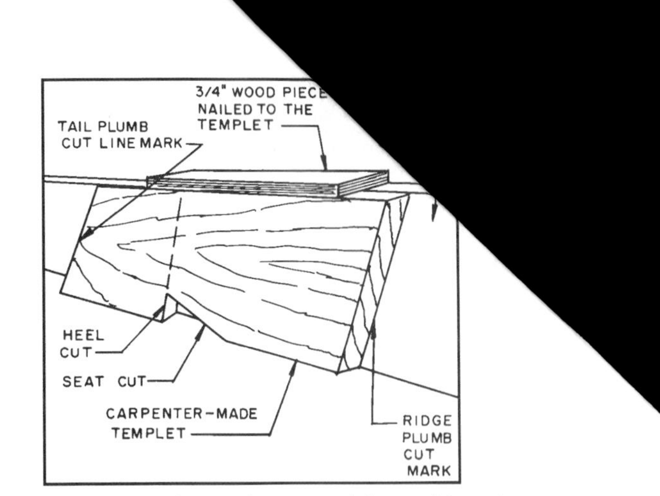

13-26 Carpenter-made templates speed the marking of the plumb cuts and the bird's-mouth.

The setup in **13-24** can also be used to mark all the rafters which can then be separated, marked, and individually cut with a portable circular saw. One way to mark rafters to be cut individually is to make a template for the plumb cuts and bird's-mouth cut **(see 13-26)**. This is used to mark each rafter.

ERECTING COMMON RAFTERS

1. Lay plywood or OSB sheets on the ceiling joists down the center of the building below the location of the ridgeboard **(see 13-27)**. This gives a working platform needed for the erection of the ridgeboard and rafters. If the ridgeboard is above the reach of the carpenter, scaffolding can be erected on the platform **(see 13-28)**.
2. Mark the rafter locations on the plate. Usually the ceiling joists are in place, and the rafter is located next to them and directly above a stud so no layout is necessary. The first rafter is set flush with the outside face of the top plate. It is often notched to receive lookouts which support the rake fascia. (This is discussed later in this chapter.)
3. Next mark the location of each rafter on the ridgeboard using the marks on the top plate **(see 13-29)**. The ridgeboard is usually two-inch-thick stock and wider than the ridge plumb cut.

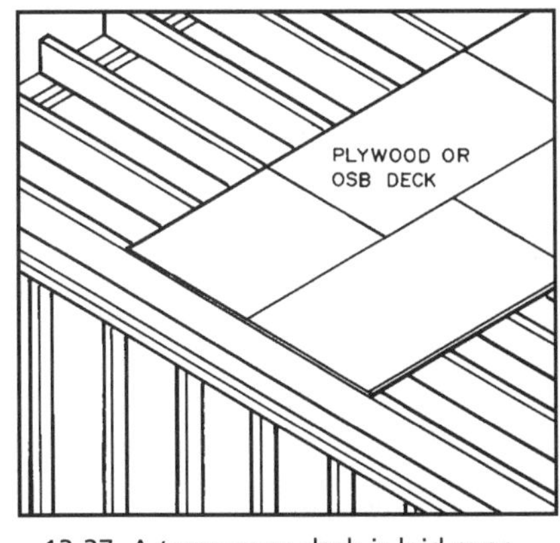

13-27 A temporary deck is laid over the ceiling joists to provide a walkway for carpenters erecting the rafters.

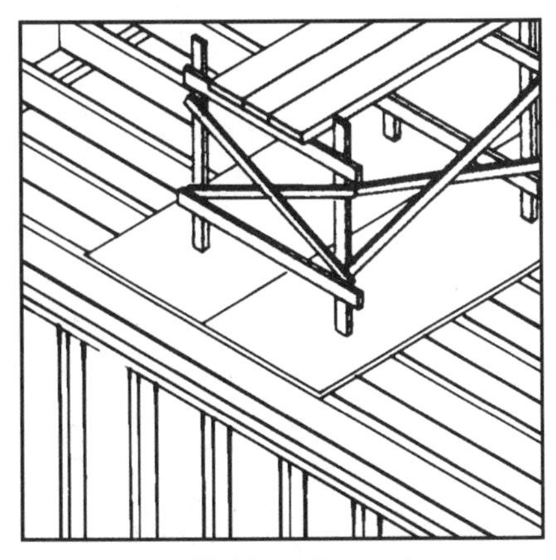

13-28 A scaffold can be used to enable the carpenters to reach the ridgeboard when installing rafters.

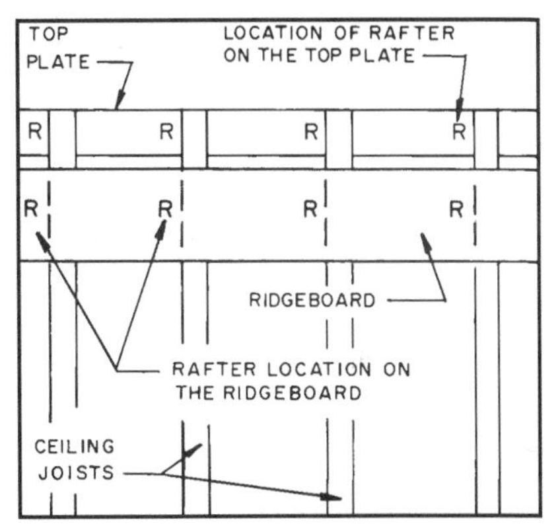

13-29 Lay the ridgeboard over the ceiling joists and mark the locations of the rafters to correspond with their locations on the top plate.

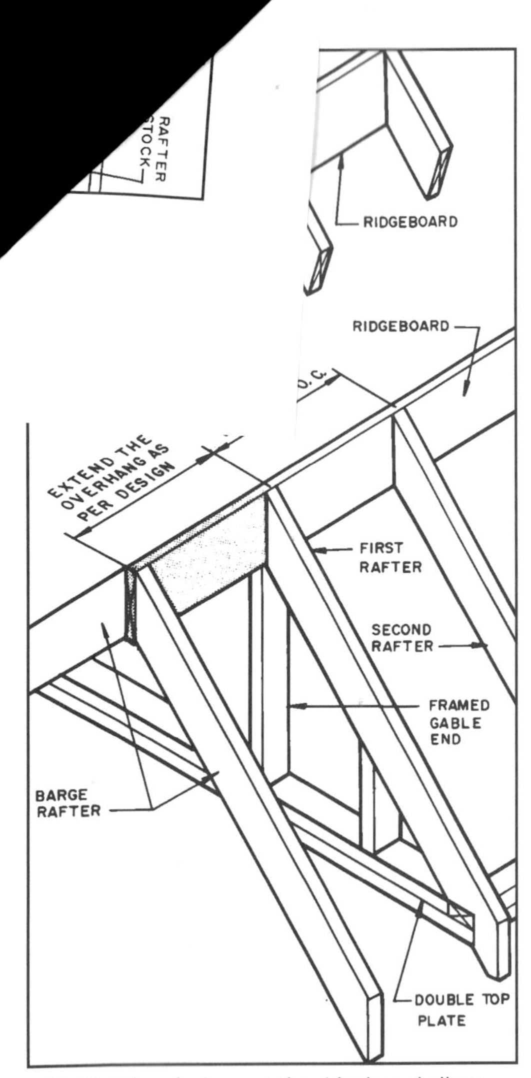

13-30 When laying out the ridgeboard allow for the required overhang needed to support the barge rafters.

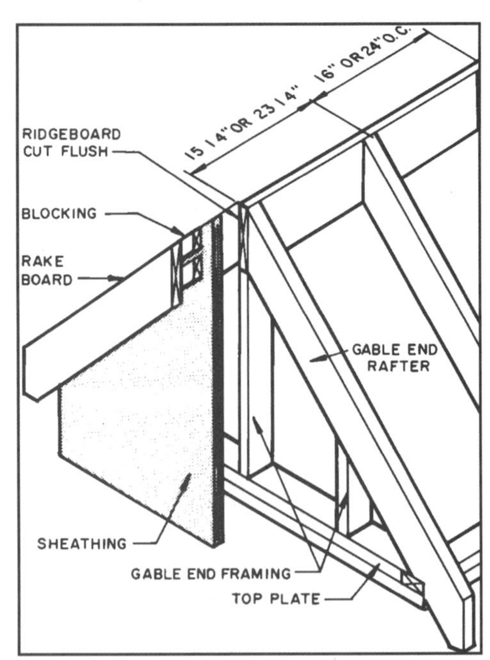

13-31 This detail shows the ridgeboard framing when there is to be no overhang at the gable-end.

The ridge must extend beyond the exterior wall an amount equal to the overhang on the gable-end with an allowance for the barge rafter **(see 13-30)**. If there is no overhang on the gable-end, the ridgeboard is cut flush with the outside surface of the plate **(see 13-31)**. Sheathing is applied and one-inch-thick blocking is added to allow for the siding. Some prefer to cut the ridgeboard overhang to length after it has been installed.

4. Next install the ridgeboard. It is supported with braced, temporary 2 × 4-inch posts. Be certain it is level and in the center of the area to be roofed (unless an irregular roof is being framed) **(see 13-32)**. Generally the ridge is made from two or more long pieces of two-inch stock. The joints should be joined between rafters with splice plates **(see 13-33)**.

   Now lay the finished rafters against the exterior wall and pull them up onto the temporary deck so they are readily available.

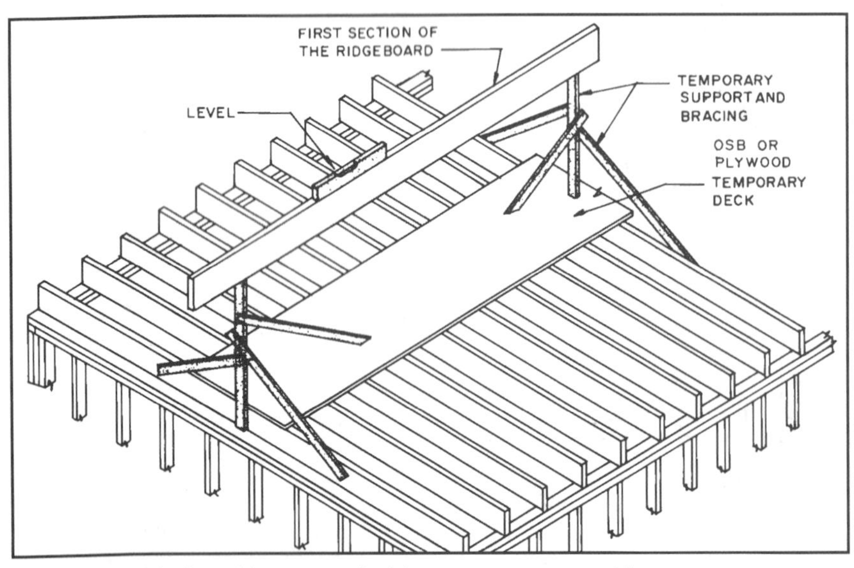

13-32 The ridgeboard is supported with temporary posts and braces.

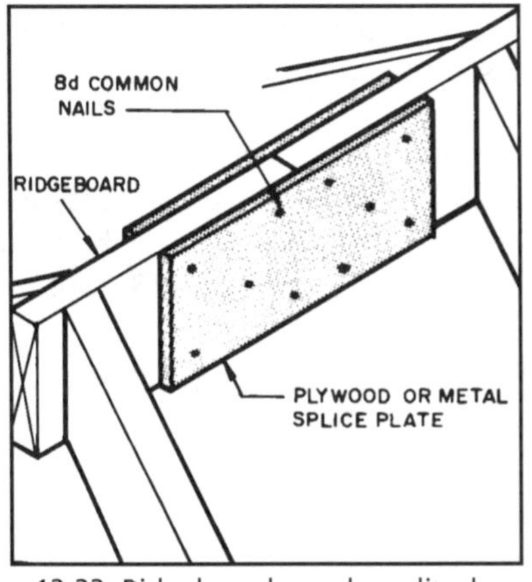

13-33 Ridgeboards can be spliced using splice plates between rafters.

13-34 After erecting and bracing the ridgeboard, install the rafters beginning with a pair at each end and in the center. Then fill in with rafters in pairs.

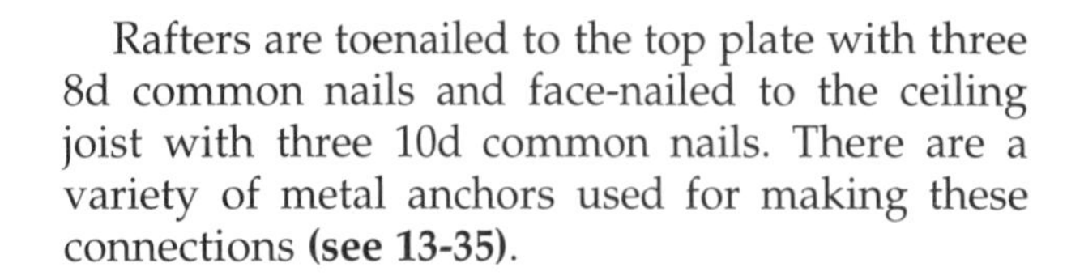

5. Nail the rafters to the ridgeboard by end-nailing the first rafter with two 16d common nails and toenailing the matching rafter on the other side with two 16d common nails. Some framers angle a 10d nail through the top of the rafter into the ridgeboard. Always erect rafters in pairs. Place a pair of rafters on each end of the section of ridgeboard and one pair in the center **(see 13-34)**. Then fill in the remainder in pairs.

Rafters are toenailed to the top plate with three 8d common nails and face-nailed to the ceiling joist with three 10d common nails. There are a variety of metal anchors used for making these connections **(see 13-35)**.

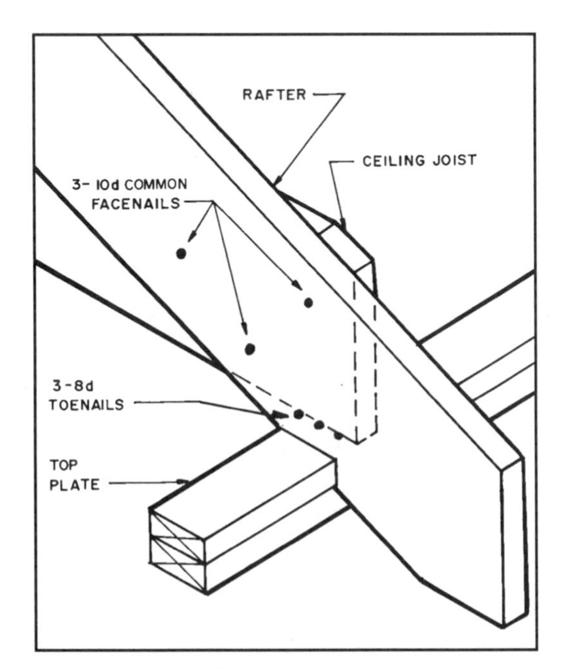

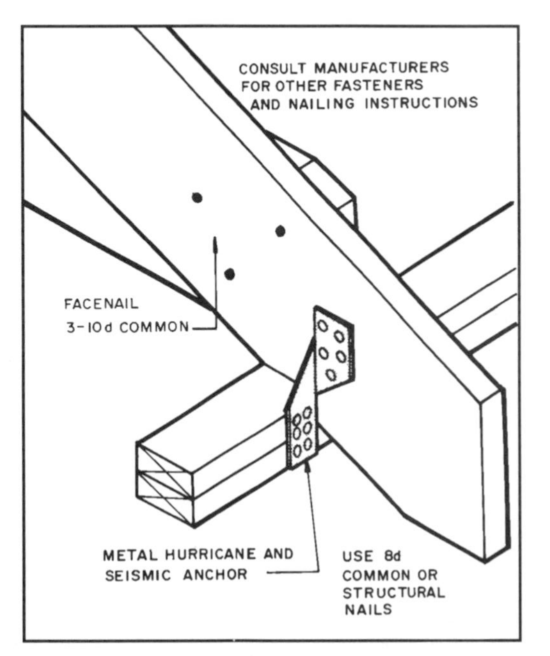

13-35 Joining rafters to the top plate.

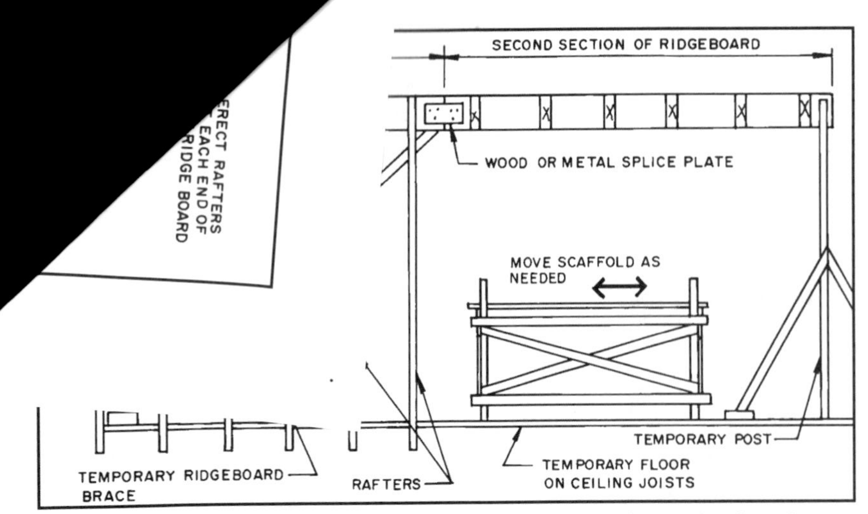

13-36 The ridgeboard is installed in sections and tied together with splice plates between rafters.

After the first section of rafters has been installed, set the next section of ridgeboard in place. Brace one end, nail the splice plates on both sides, move the scaffolding, and repeat the installation process **(see 13-36)**.

Installing Collar Beams & Sway Braces

After the rafters are in place, collar beams and sway bars are installed. Collar beams (also called collar ties) are used to help resist the tendency of the roof to push the exterior walls outward. The ceiling joists also contribute a great deal to resisting this force. Normally collar beams are 1 × 6 or 2 × 4 material nailed to every third rafter, about one-third of the distance down the rafter from the ridge **(see 13-37)**. Roofs under greater than normal loads may require collar beams on every pair of rafters. Collar beams can also serve as ceiling joists for rooms on the second floor. In this case, they are usually 2 × 6 stock.

Sway braces are used to stabilize the ridgeboard. They can be 2 × 4 or 2 × 6 material and are on about a 45-degree angle. They are nailed to the ridge and the top plate of an interior wall or 2 × 6 blocking nailed between ceiling joists **(see 13-38)**. They are spaced about every six feet.

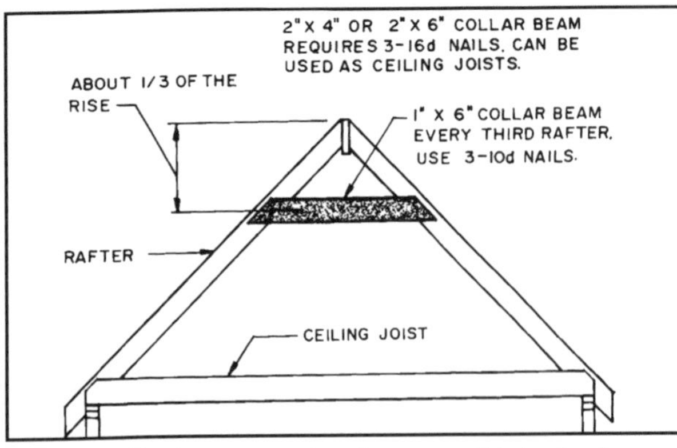

13-37 Collar beams are placed on every third pair of rafters in normal roof construction.

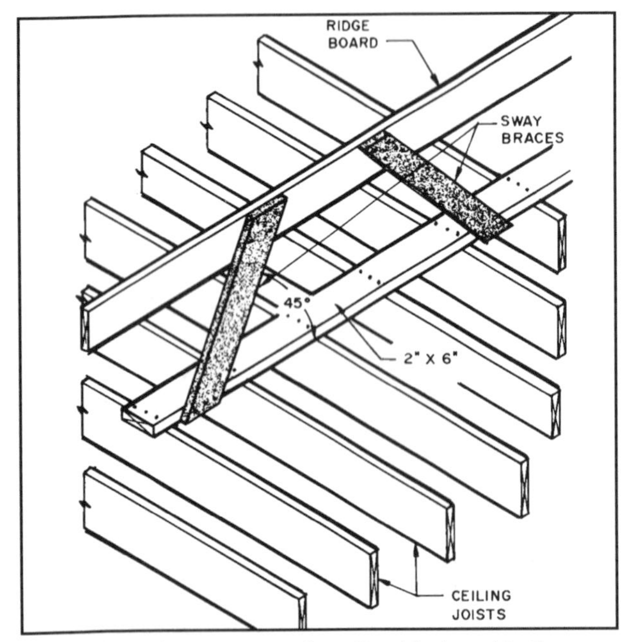

13-38 Sway braces are run from the ridgeboard to the ceiling joists.

PURLINS & PURLIN STUDS

The structural designer may specify additional bracing for rafters. In areas of high winds, a truss-type bracing may be specified. This has the effect of reducing the span of the rafters, thus increasing their ability to carry a wind load **(see 13-39)**. This bracing can also provide needed support when the span equals the maximum allowed for a rafter or exceeds it somewhat. In all cases the load must be transferred to a beam that clears the ceiling joists or to an interior bearing wall. It must never be transferred to the ceiling joists unless they have been sized to carry this extra load. Normal-size ceiling joists are inadequate for this purpose.

BUILDING THE GABLE-END PROJECTION

There are several ways to frame the roof projection over the gable-end (also referred to as the

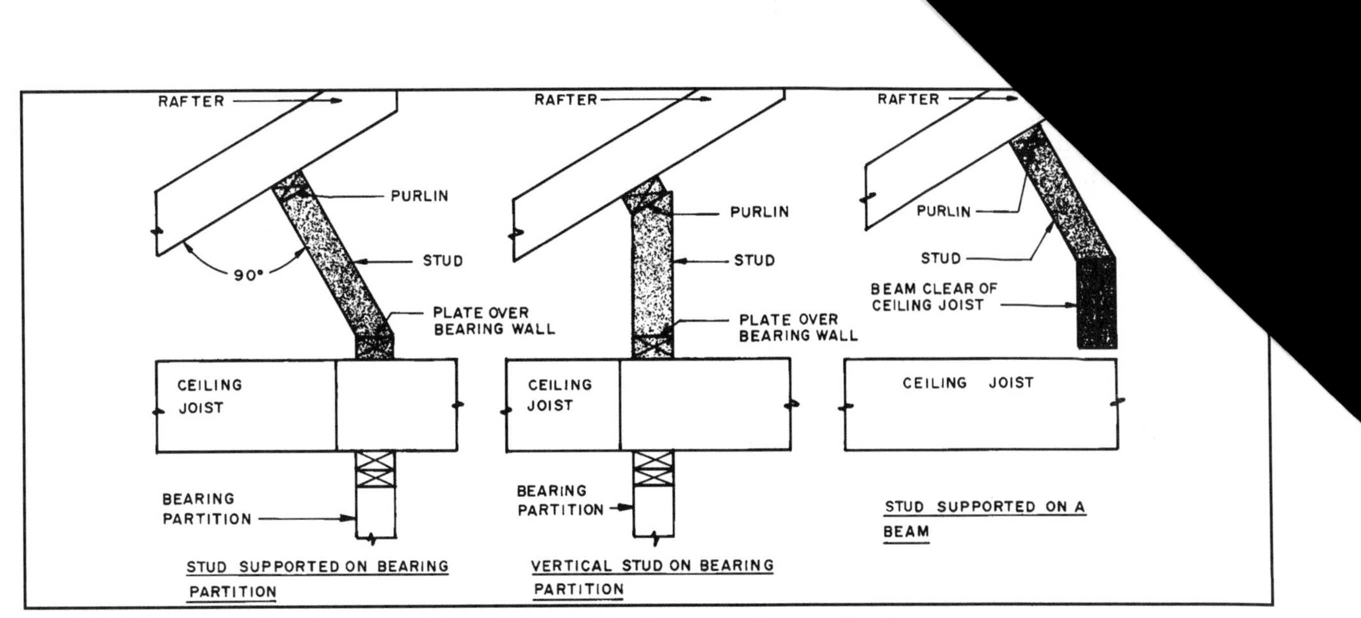

13-39 Various types of bracing are used to enable the rafters to carry an increased load.

overhanging rake). When the overhang is in the vicinity of 12 inches or less, the overhang is framed on the ground as a ladder-type unit and nailed to the end rafter **(see 13-40)**. The blocking is spaced 24 inches O.C. to support the soffit on the bottom. The finished fascia is nailed over the fly rafter (rough fascia) after the ladder unit is in place.

Larger overhangs built when a rafter is located on the end of the building are framed as shown in **13-41**. The end rafter is notched to receive 2 × 4 lookouts which run to the second rafter and can be end-nailed to it. Usually the lookouts are extended into the roof a distance equal to the amount of the overhang. The lookouts are spaced 24 inches O.C. The finished fascia is nailed over the fly rafter.

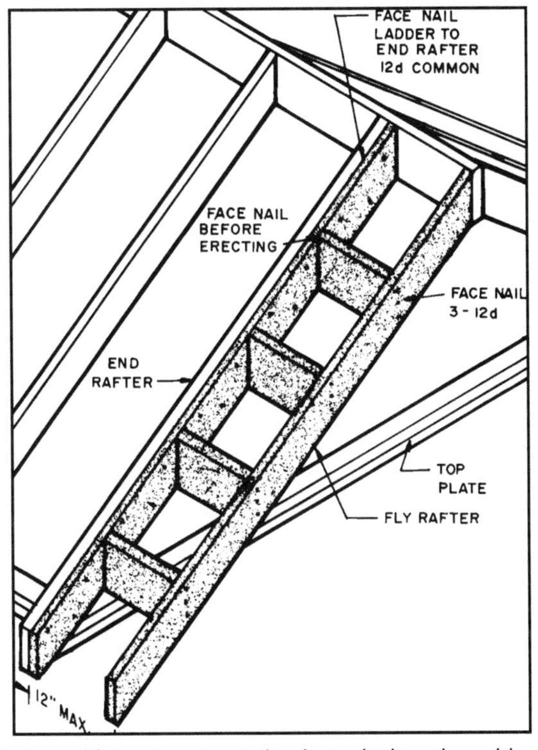

13-40 Ladder-type construction is used when the gable-end overhang is small.

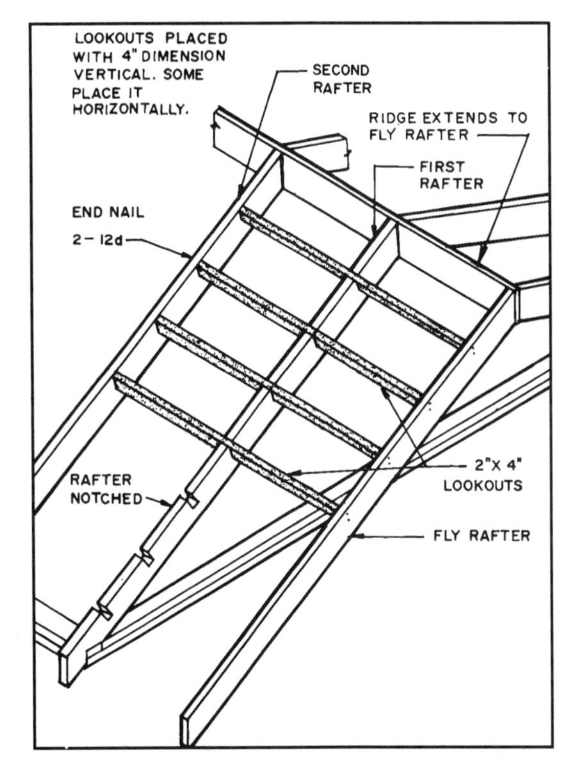

13-41 Gable-end roof overhangs of more than 12 inches are constructed using lookouts.

ıe is to omit the end
he gable-end with the
tom of the rafter. The
nd are nailed to it as

rhang at the rake and
own in **13-43**.

nmonly used to frame
sideration is whether it will have a louver. The opening for the louver is framed so that the louver can be installed later **(see 13-44)**. When the roof is framed with an end rafter, the gable-end studs can be notched and placed 16 inches O.C. as shown in **13-45**. If a rafter is not used, the gable-end framing forms the support for the lookouts on the roof overhang as shown in **13-42**. Gable-end studs are spaced 16 inches or 24 inches on center.

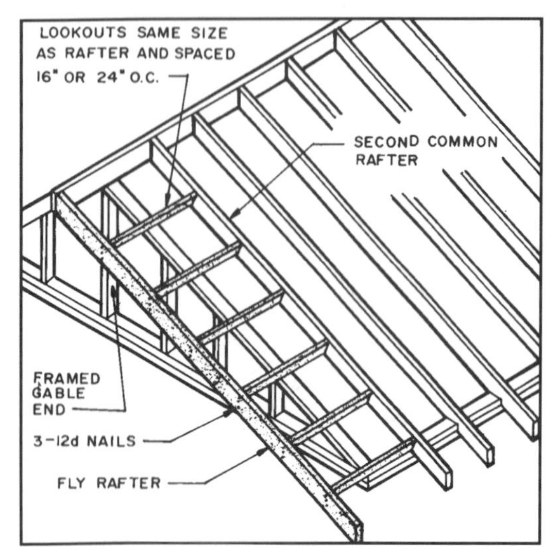

13-42 Large overhangs on the gable-end can be framed with the lookouts resting on top of the gable-end top plate framing.

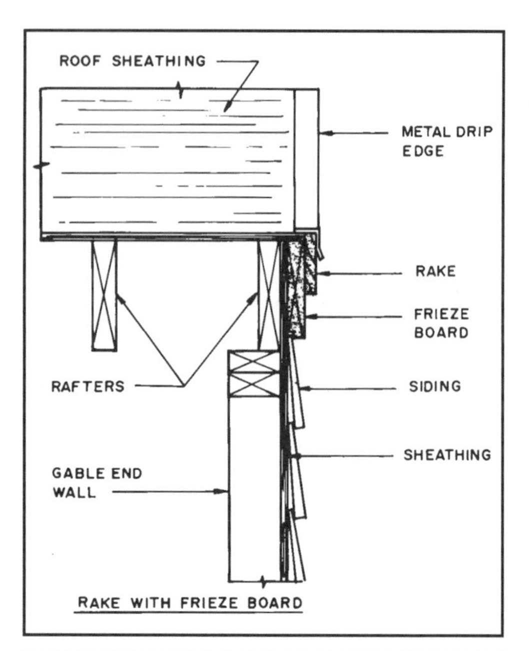

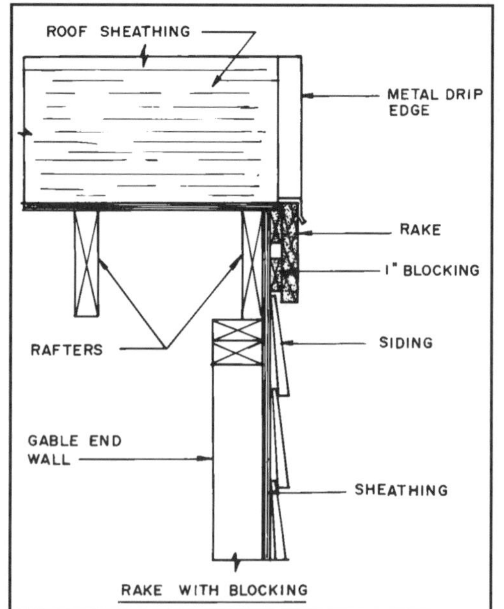

13-43 Typical ways to frame the rake on the gable-end when there is to be no overhang.

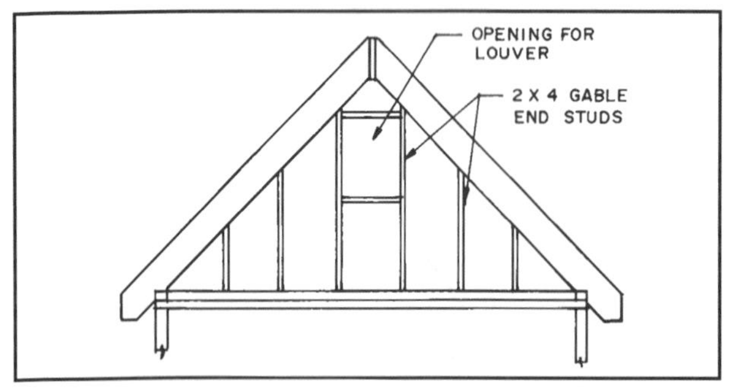

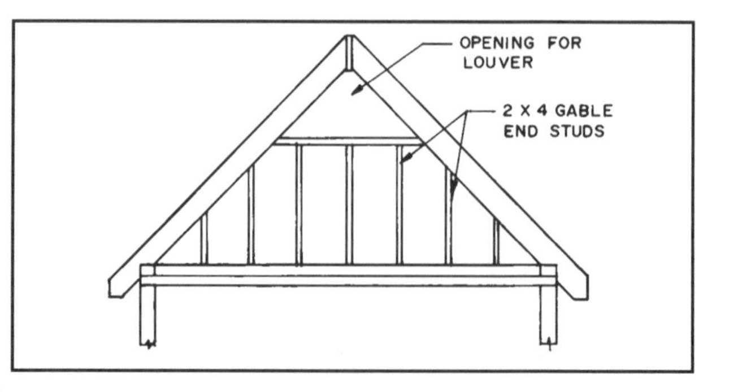

13-44 Louver openings are framed as required in the gable-end.

The gable-end studs must be cut to length and angled to butt the slope of the rafter. To find the length use the **common difference** between the studs **(see 13-46)**. Following are two methods this can be found.

The difference can be found by locating a stud, marking it along the bottom of the rafter, repeating this for the next stud, and measuring the difference in their lengths. All of the remaining studs will be this amount different in length **(see 13-47)**.

Another method uses this formula:

Common Difference = On-Center Distance (inches) × Unit Rise (inches) ÷ Unit Run (inches)

For example, a roof with a unit rise of 6 inches with studs spaced 16 inches O.C. would work out as C.D. = $16/12 \times 6 = 8.0$ inches common difference. This means that each stud is eight inches shorter than the one next to it.

The common difference can also be laid out using the framing square as shown in **13-48**. The slope of the cut can be marked on the stud as it is held against the rafter.

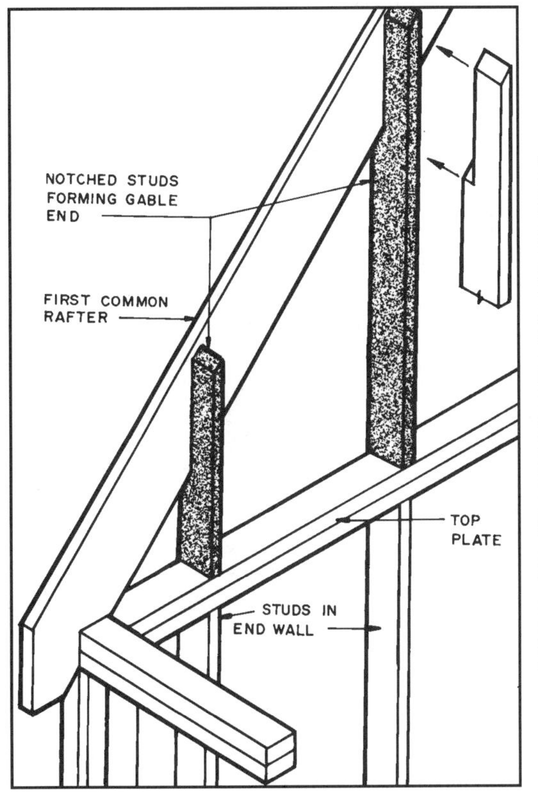

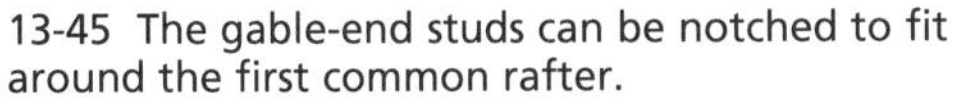

13-45 The gable-end studs can be notched to fit around the first common rafter.

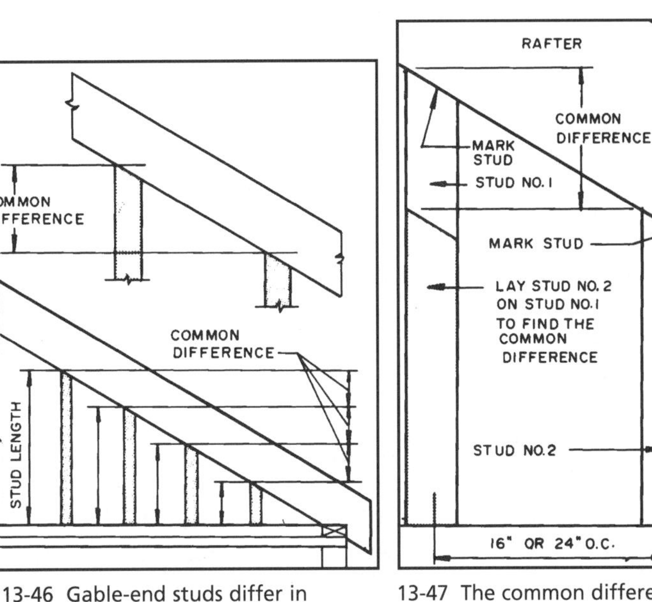

13-46 Gable-end studs differ in length toward the exterior wall by one common difference from the stud next to them.

13-47 The common difference can be found by locating two adjacent studs, marking their lengths on the bottom of the rafter, and measuring the difference in length between the two studs.

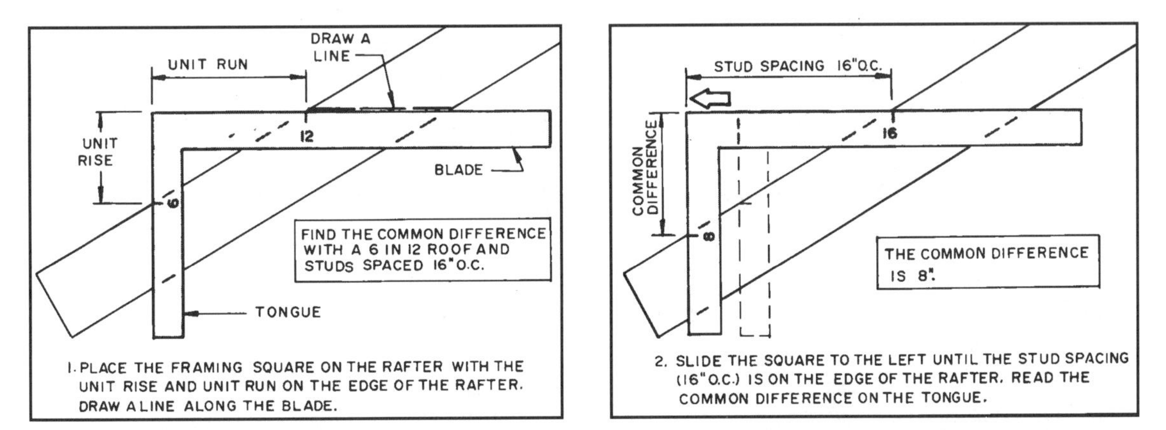

13-48 The common difference in the length of the gable-end studs can be found using the framing square.

This slope can be set on the saw for cuts on the other studs. It can also be laid out with the framing square as shown in **13-49**.

FRAMING OPENINGS IN THE ROOF

Openings in the roof are framed the same as described for floor openings. The most common opening is for a fireplace chimney. The rafters on each side and the headers between the rafters are doubled. The wood framing must clear the chimney a distance specified by local codes. A clearance of one to two inches is generally required by local code **(see 13-50)**.

FRAMING THE CORNICE

The cornice is the overhang of a pitched roof at the eave line. It usually consists of a fascia board, soffit, and sometimes molding. There are several ways a cornice can be framed. It may be an **exposed cornice (see 13-51)**, allowing the roof sheathing to be seen; **soffited (see 13-52)**; **boxed (13-53)**, using the rafter to hold the soffit material; or a **narrow box cornice (see 13-54)** giving little or no overhang. The details desired will be shown on the architect's working

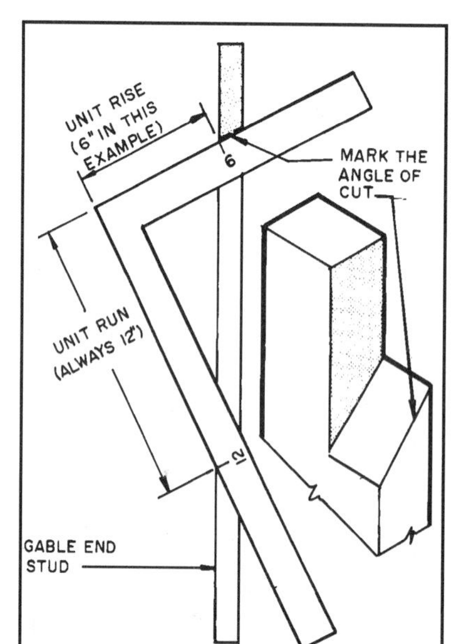

13-49 A framing square can be used to mark the angle of the cut on the top of the gable-end studs.

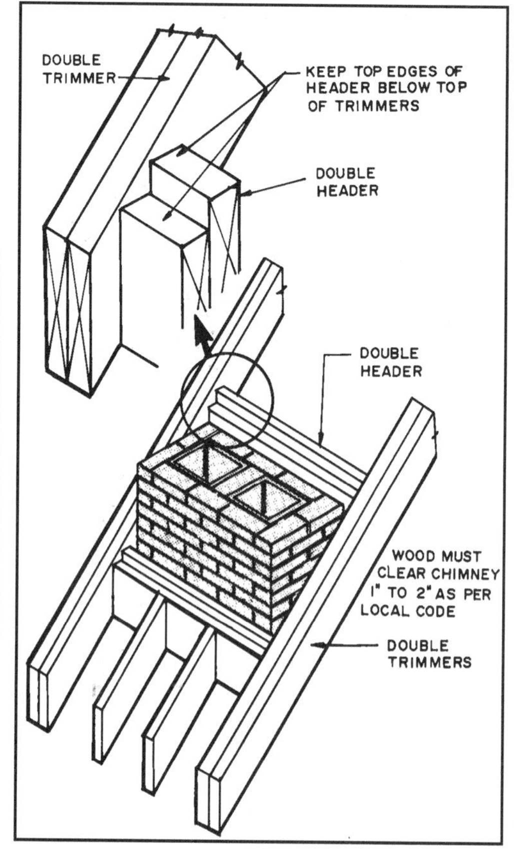

13-50 Openings in the roof are framed with headers in the same manner as openings in floors.

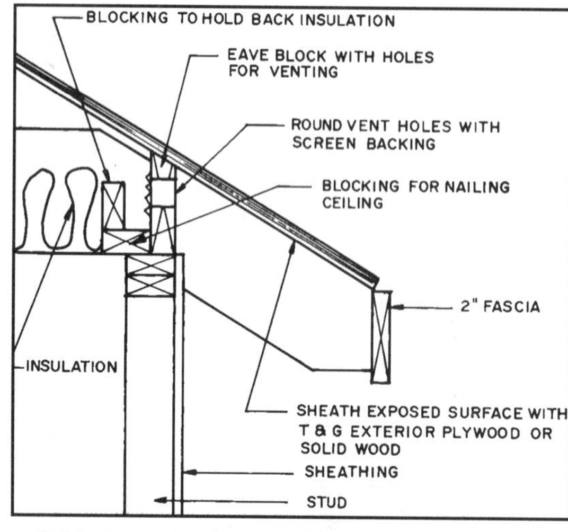

13-51 An exposed cornice has no soffit and exposes the roof sheathing and rafter tails.

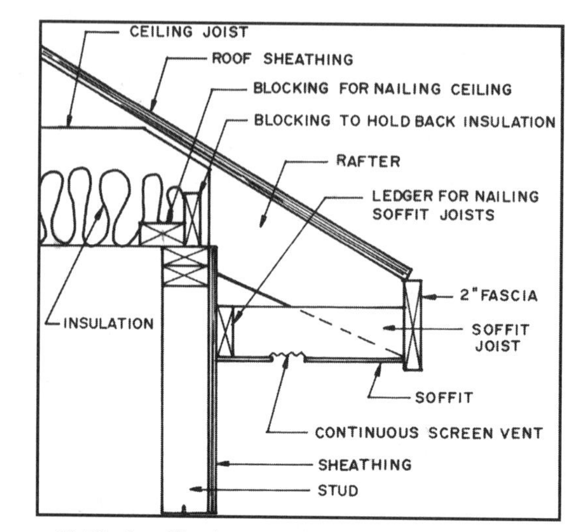

13-52 A soffited cornice has a horizontal soffit that usually has vents.

drawings. Details for framing with a cathedral ceiling are shown in **13-55**. Other methods of construction than those shown are also used.

Some framers prefer to install a rough fascia of usually two-inch-thick stock, and then install a one-inch-thick finish fascia board over it. This usually gives a better looking fascia, and the two-inch rough fascia is needed to hold gutters. A one-inch single fascia applied with no backup will often warp.

Soffit material can be hardboard, plywood, gypsum, vinyl, or aluminum. Vents installed in the soffit may be individual units or continuous strip vents **(see 13-56)**.

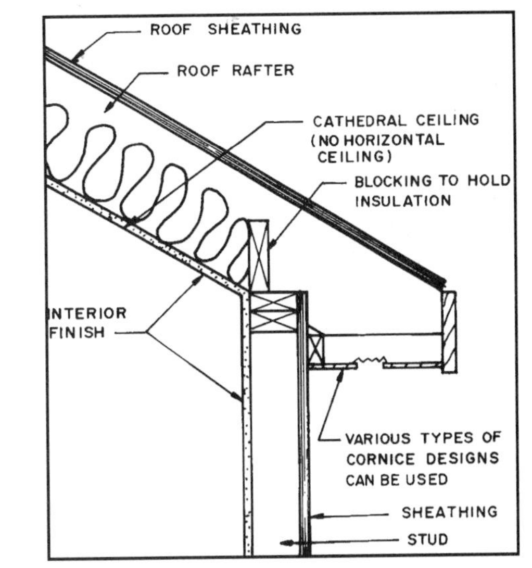

13-55 Typical cornice framing for a cathedral ceiling construction.

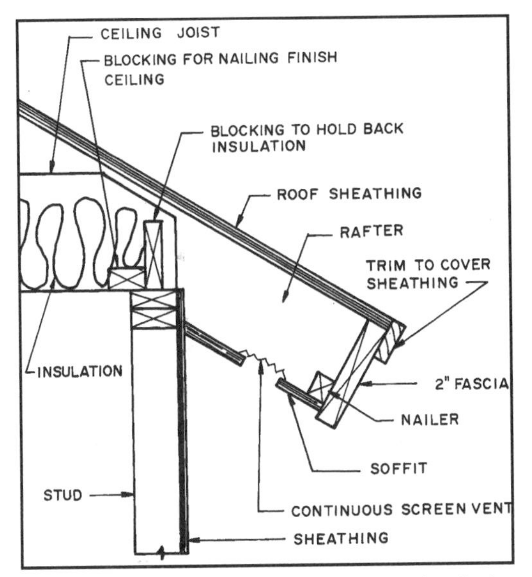

13-53 A boxed cornice has the soffit material nailed to the bottom edge of the rafter.

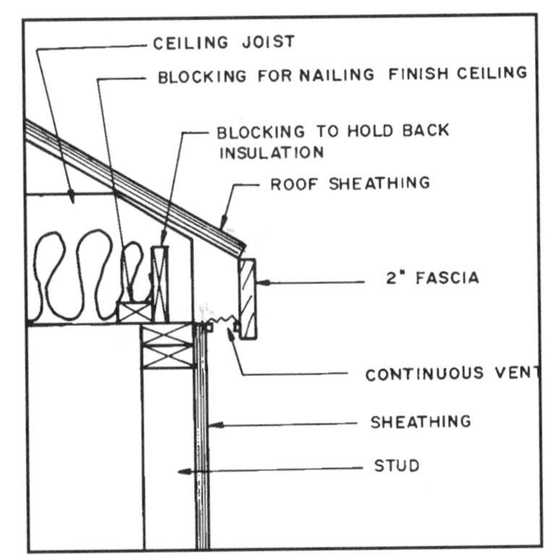

13-54 A narrow box cornice provides minimum ventilation and no protective overhang.

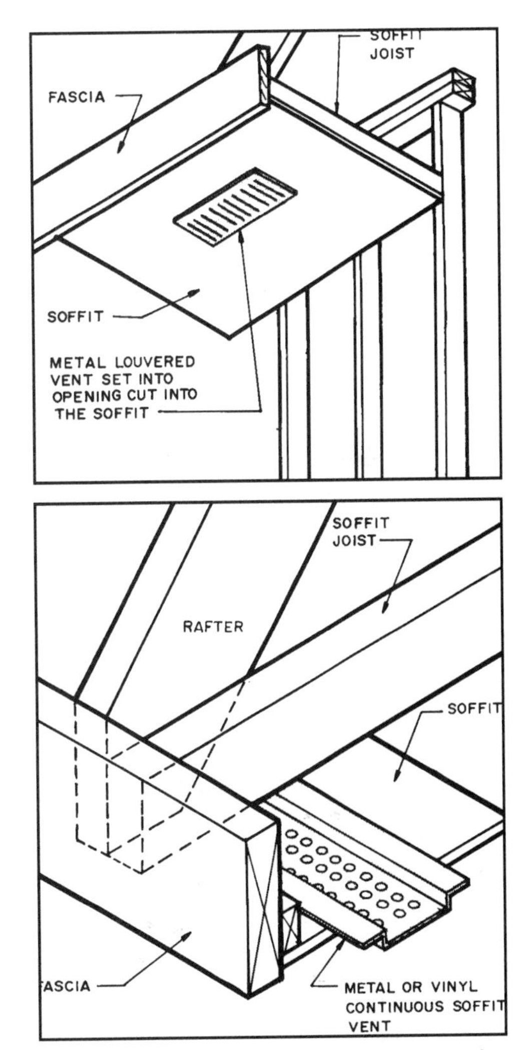

13-56 Continuous or individual vents are set into the soffit to provide attic ventilation.

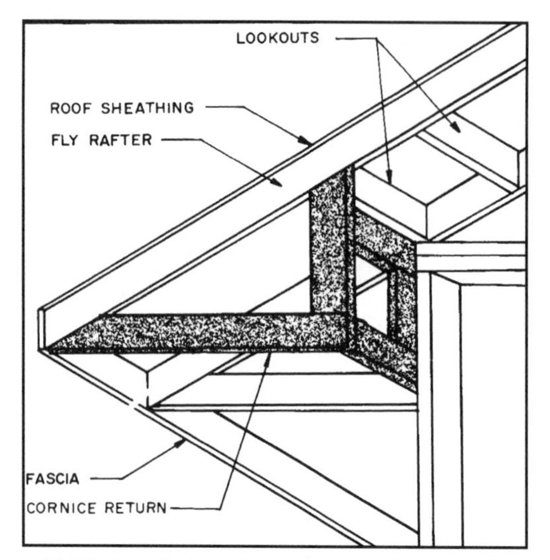

13-57 A typical cornice-return framing technique.

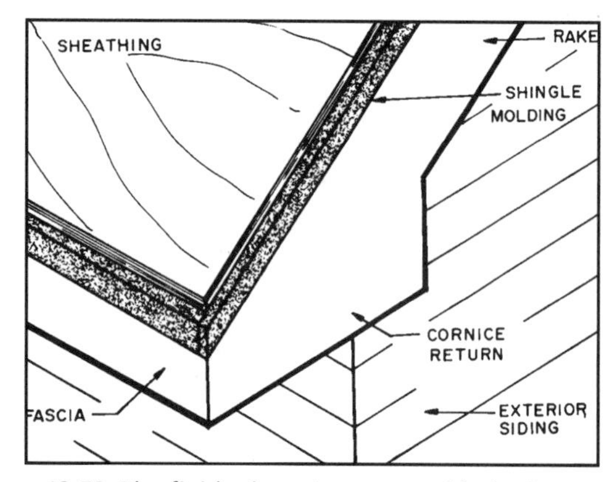

13-58 The finished cornice return with the fascia boards in place.

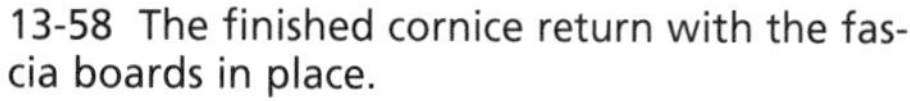

Framing a Cornice Return

The cornice return is the underside of the cornice at the corner of the building where the gable-end roofline meets the wall. It serves no structural purpose but provides a neat, decorative transition from the horizontal eave line to the sloped roofline of the gable. Framing details are shown in **13-57**.

The fly rafter is connected to cornice return nailers extending out from the wall. The rake fascia and cornice fascia meet. Shingle molding, if used, is carried around the corner and up the slope of the rake board to the ridge **(3-58)**.

Framing a Cricket

A **cricket** (also called a saddle) is a double-sloped structure built on the high side of a chimney to divert water away from the chimney **(see 13-59)**. It is usually framed with 2 × 4 members and covered with copper or aluminum.

Roof Framing with I-Joists

Manufactured I-joists are strong and lightweight and also suited for use as rafters. (They are described and their use in floor construction is illustrated in Chapter 9.) They can be joined to each other with metal straps and hangers. They are cut in the same manner as wood rafters. Since they are thin in cross section, they do not have as

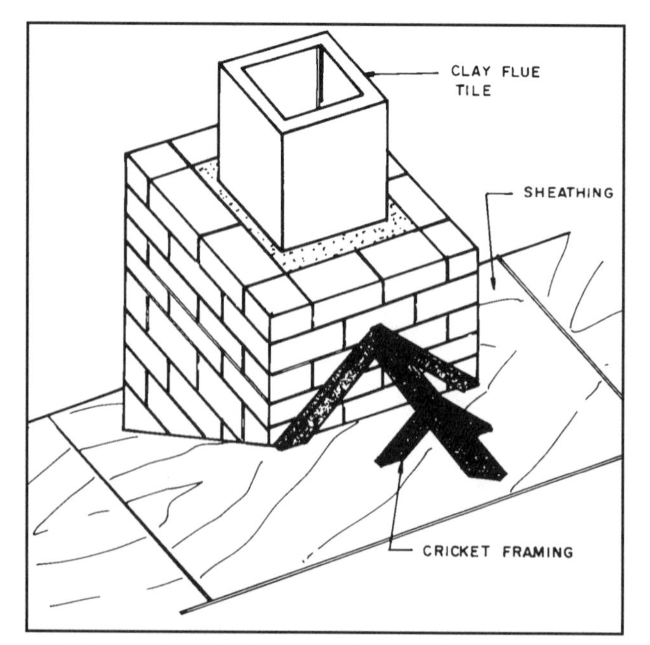

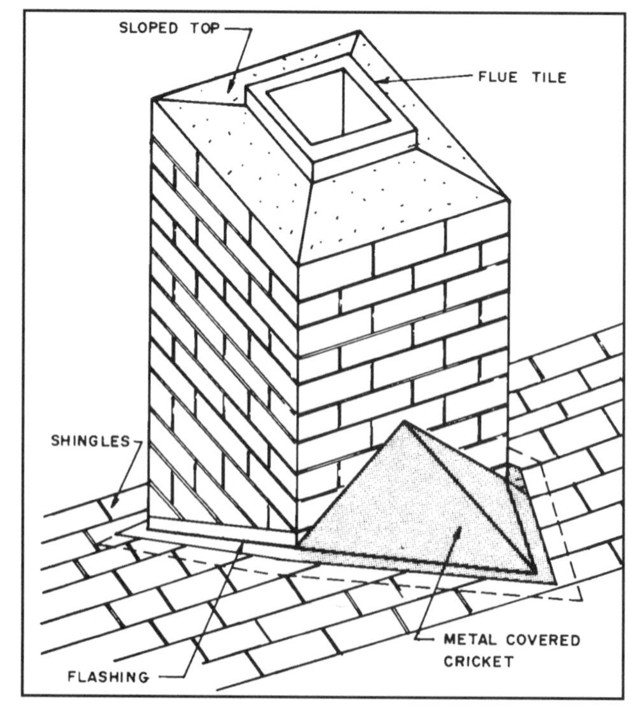

13-59 A cricket is built to shed the water behind a chimney.

much strength in compression as solid wood rafters. Therefore, wood stiffeners are added between the flanges. The same principles used for framing with solid wood rafters apply to I-joist rafters. However, manufacturers' specifications and assembly instructions must be carefully followed. Typical construction details for a few applications are shown in **13-60**.

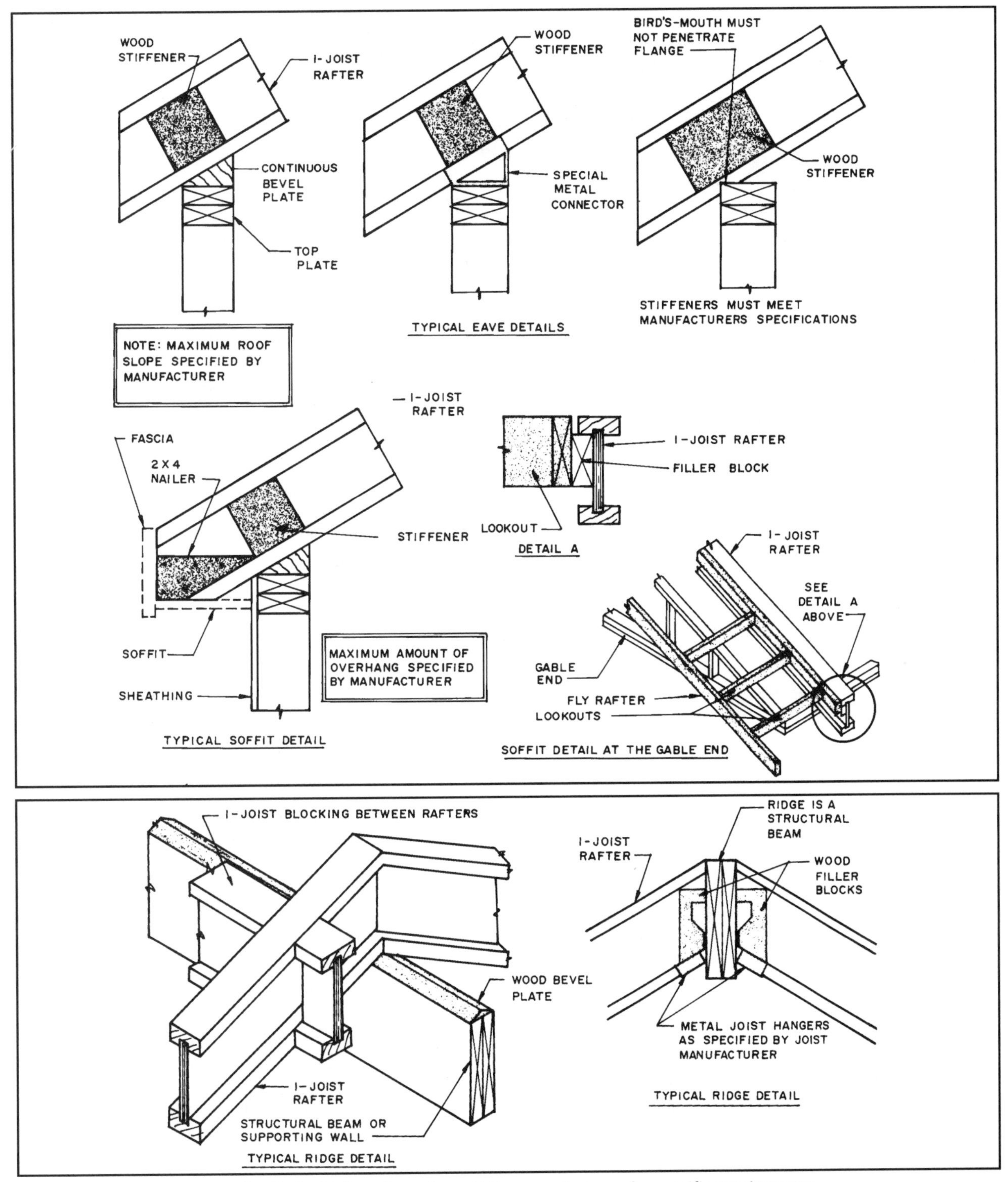

13-60 Typical construction details for I-joist rafters. Consult the manufacturer for specific requirements.

13-61 I-joist rafters can span long distances and are lightweight and easily handled by the carpenters. *(Courtesy Boise Cascade Corporation)*

13-62 These I-joist rafters butt the ridgeboard. Notice the solid lumber rafter tail extending beyond the exterior wall. *(Courtesy Boise Cascade Corporation)*

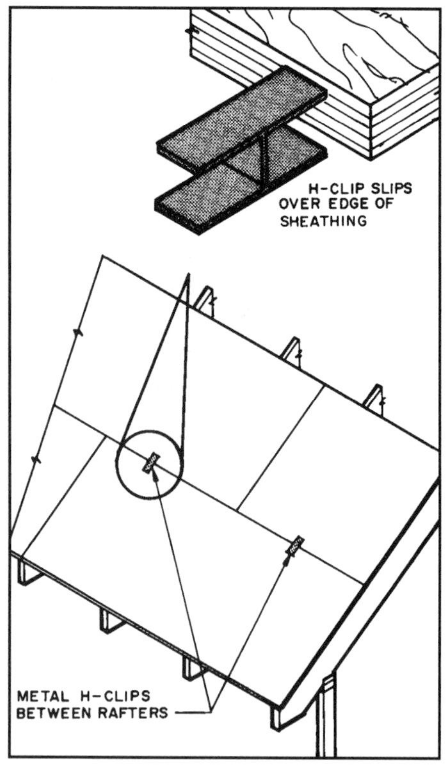

13-63 Aluminum H-clips are used on the unsupported edges of roof sheathing panels. This eliminates the need for wood blocking.

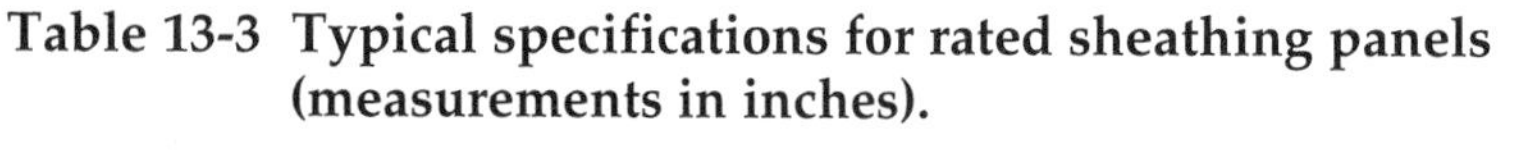

Table 13-3 Typical specifications for rated sheathing panels (measurements in inches).

Rafter O.C.	Thickness	Maximum unsupported edge length
12	5/16	12
16	5/6, 3/8	16
24	3/8, 1/2	20 for 3/8 24 for 1/2

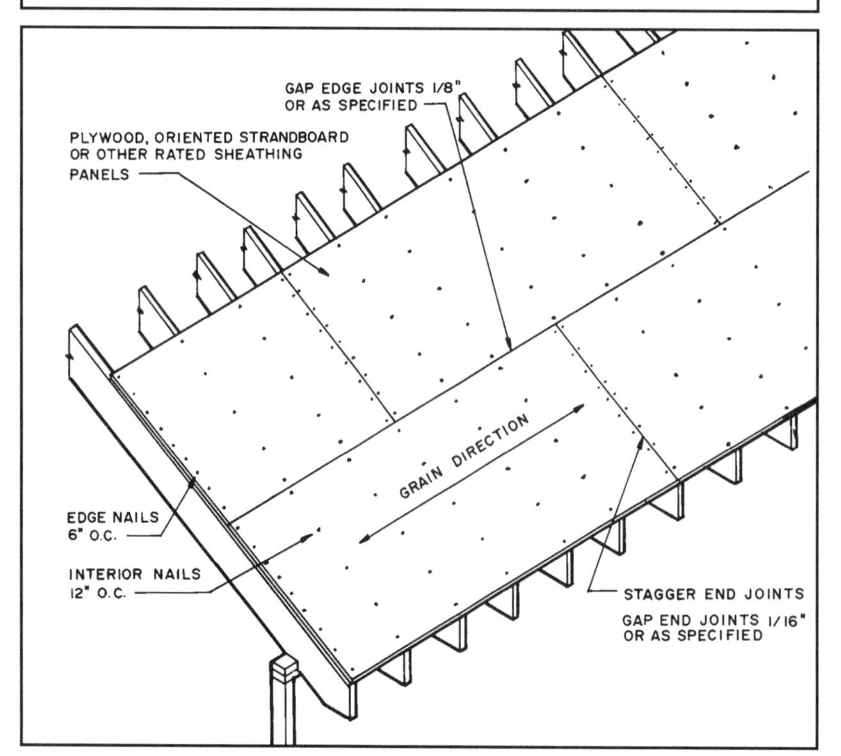

13-64 Sheathing panels are applied perpendicular to the rafters.

In **13-61** and **13-62** are views of the installation of I-joist rafters. Notice they are very long and lightweight, enabling two carpenters to install them. Notice in **13-62** that solid lumber rafter tails have been added, making the fascia board much narrower than it would be if the I-joist rafter had been extended for the overhang.

SHEATHING THE ROOF

After the framing is complete the sheathing is applied. It is normally 4 × 8-foot sheets of plywood or oriented strand board. The sheathing greatly stiffens and increases the strength of the roof. While manufacturers' recommendations should be observed when selecting sheathing, specifications typical for rated sheathing panels are given in **Table 13-3**.

If the unsupported-edge span exceeds the specified distance, aluminum H-clips are inserted in the center of the span **(see 13-63)**. Panel ends are spaced 1⁄16 of an inch apart and edges ⅛ inch apart. Panels ½ inch or less in thickness require 6d common smooth, ring-shank, or spiral-thread nails. Panels ⅝ to 1 inch use 8d common nails. The nails are spaced 6 inches apart on the edges of the panel and 12 inches apart on intermediate supports.

The sheathing is applied with the long dimension (8 feet) perpendicular to the rafters. The joints between panels should be staggered **(see 13-64)**. On roofs with overhangs at the gable-end, the sheathing should extend at least two rafter or truss spaces into the roof to provide adequate support for the overhang. Sheathing not made with waterproof adhesive must have any exposed edges protected from the weather by trim or a metal drip edge. Sometimes spaced boards are used for roofs to be covered with wood shakes, concrete, or clay tile. Normally the roof boards are 1 × 4, square-edged, and spaced the same distance on center as the shingles are laid to the weather **(see 13-65)**. Consult the manufacturer of the wood shakes, clay, and concrete roof tile for sheathing recommendations.

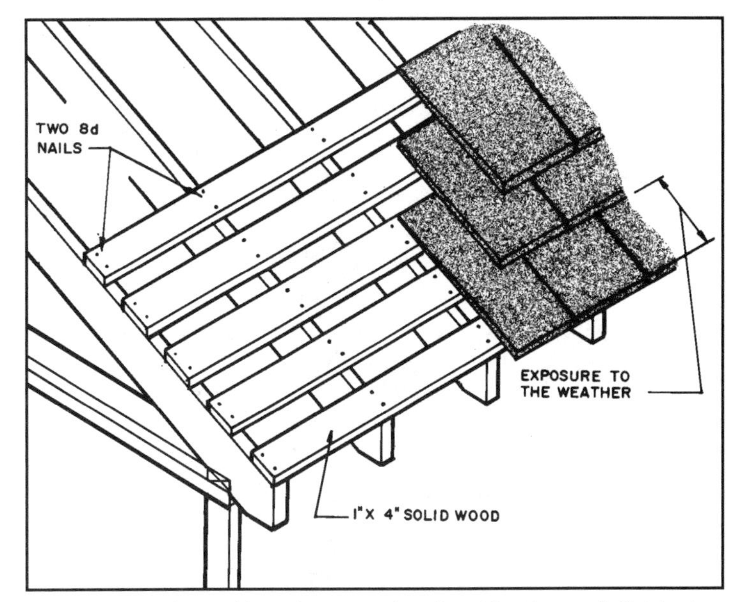

13-65 Spaced wood boards are used as sheathing for roofs finished with wood shakes.

VALLEYS

When two gable roofs intersect, a **valley** is formed. The line of intersection between the two roofs is located with a valley rafter. Valley jack rafters run between the ridge and the valley rafter **(see 13-66)**. Details for laying out valley and valley jack rafters are in Chapter 15, which follows.

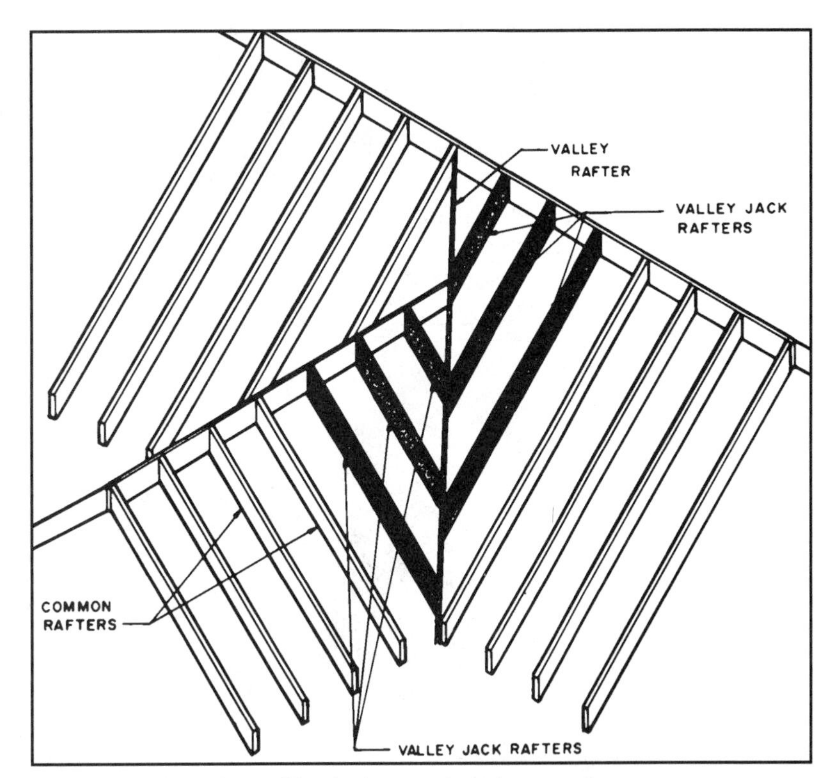

13-66 A valley is formed by the intersection of two roofs.

CHAPTER 14

Hip Roof Construction

A hip roof has slopes from all four sides toward a ridgeboard. It requires the use of a hip rafter at each corner. The roof requires the use of common, hip, and hip jack rafters **(see 14-1)**. If it is intersected by another roof, valley rafters are required. These are discussed in Chapter 15. The use of common rafters is discussed in Chapter 13.

HIP RAFTERS

Hip rafters run from the end of the ridgeboard to the corner of the double top plate, forming an angle of 45 degrees with the adjoining common rafter. Actually it is the diagonal of a square **(see 14-2)**. The unit run of common rafters is 12 inches. The unit of run for a hip rafter is 16.97 inches (use 17 inches) because the diagonal of a 12-inch square is 17 inches **(see 14-3)**. A hip rafter with all cuts made is in **14-4**. Notice the tail and ridge plumb have angled side cuts.

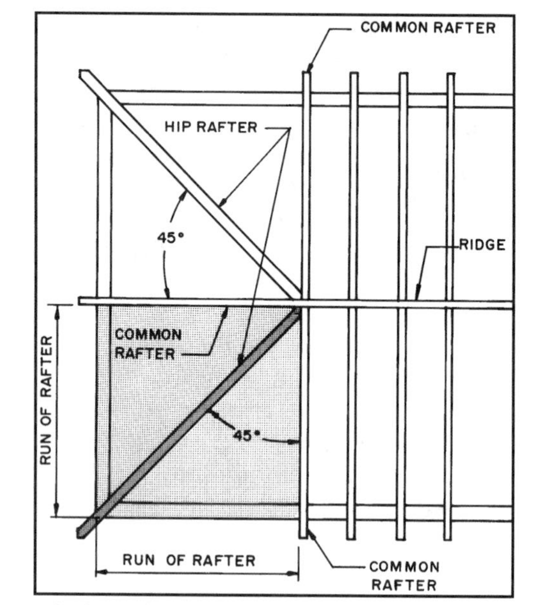

14-2 The hip rafter is the diagonal of a square formed by common rafters and the double top plate.

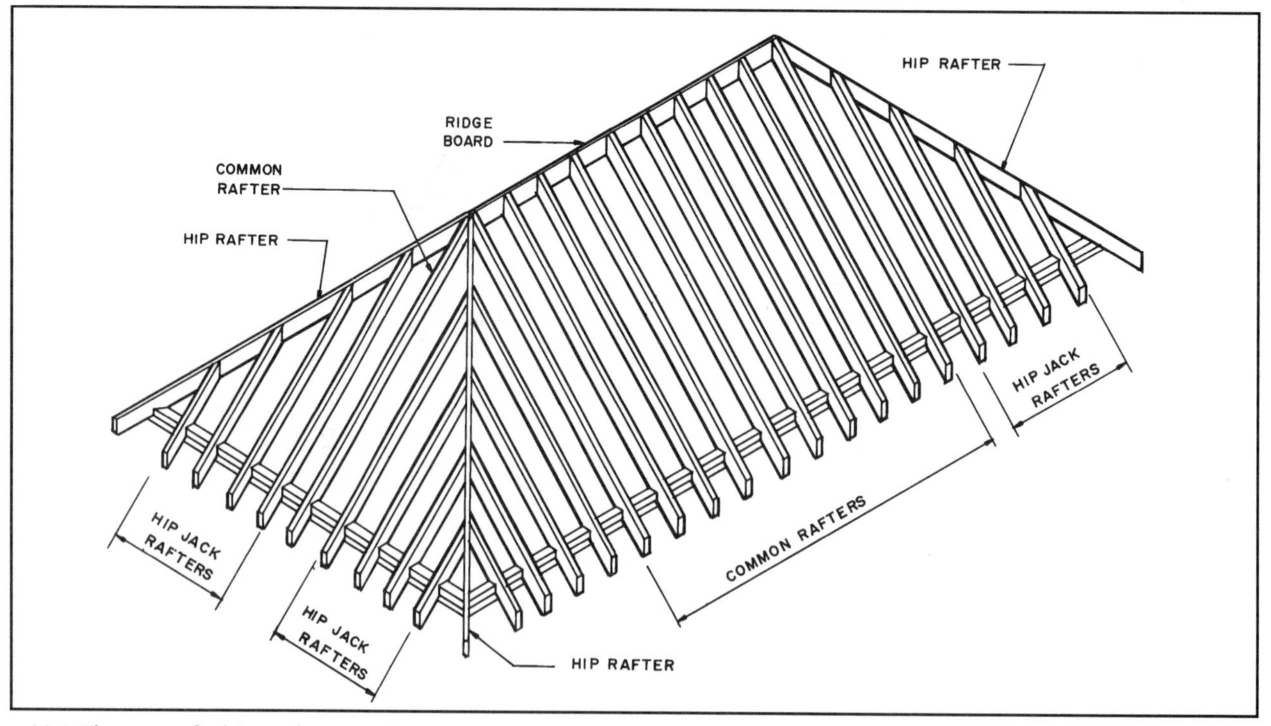

14-1 The parts of a hip roof structural system.

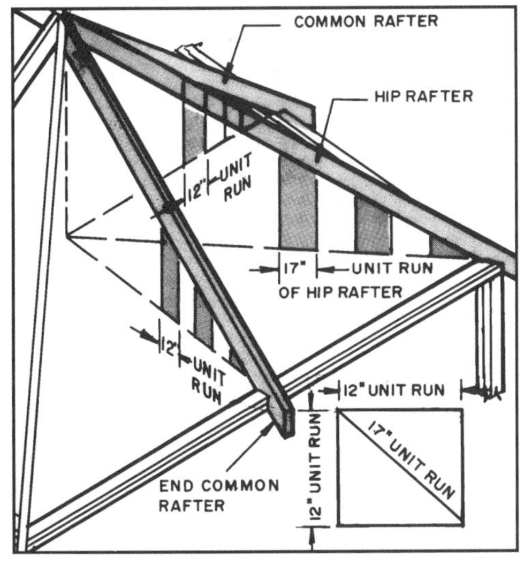

14-3 The unit run of the hip rafter is the diagonal of a square with the unit run (12 inches) of the common rafters as sides.

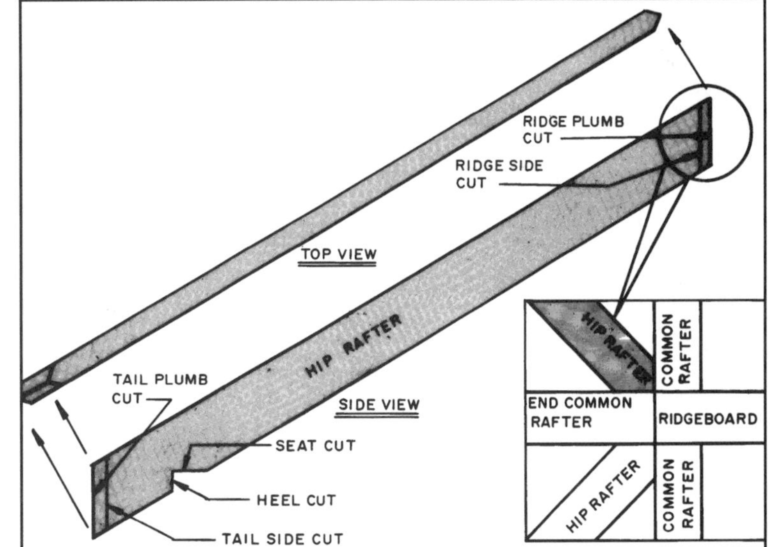

14-4 A typical hip rafter.

DETERMINING HIP RAFTER LENGTH

The hip rafter is laid out in the same way as a common rafter (see Chapter 13) except that the unit run is 17 inches. The methods that can be used to determine the rafter length are the **step-off method**, using the tables on the **steel framing square**, or standard rafter tables.

Step-off Method

The procedure for using the step-off method is the same as that described for common rafters in Chapter 13 except the unit run is 17 inches **(see 14-5)**. Generally there are a few inches left over after laying out the 17-inch units of run. These are called **odd units**. To lay out odd units, locate the distance left over on the body and tongue of the square **(see 14-6)**. Measure the diagonal. This becomes the unit of run for the odd units.

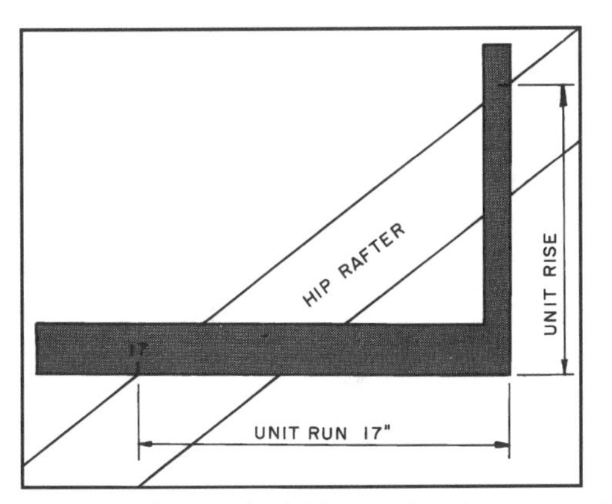

14-5 The hip rafter can be laid out using the step method as explained for the common rafter except the unit run is 17 inches.

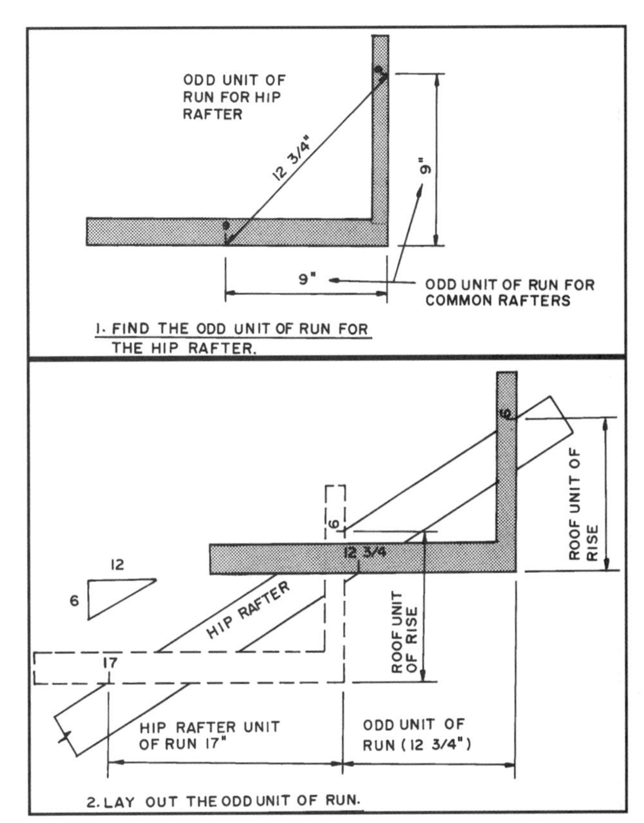

14-6 How to lay out the step for the odd unit of run for a hip rafter.

For example, if a span were 12′-9″, then the nine-inch run would be left over. Lay this out on the square and measure the diagonal. This dimension, 12¾ inches, is the odd unit of run for the hip rafter. Locate it on the rafter as shown in **14-6** on the previous page.

The hip rafter must now be shortened a distance equal to one-half the 45-degree thickness of the ridge. This measurement is shown in **14-7**.

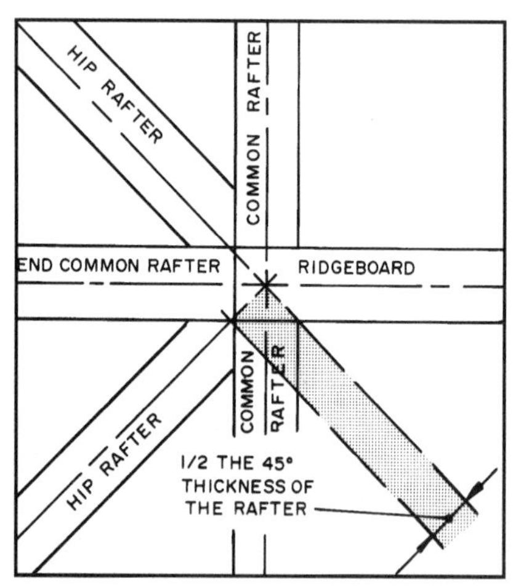

14-7 The hip rafter is shortened at the ridge by one-half of the 45-degree thickness of the rafter.

Steel Square Method

The length of the hip or valley rafter can also be found on the table on the **framing square**. The second line on the table is marked **"length hip or valley per foot run."** Find the rise on the inch scale on the outside edge of the square. Follow this down into the table. Where it meets the hip or valley row is the length of the rafter per foot of run. For example, if the roof rise is 6 inches, the length of the hip or valley rafter is 18 inches per foot of run **(see 14-8)**.

To get the length multiply the number of feet of total rafter run by the length of hip per foot of run. In the example above if the total rafter run were 15 feet the rafter length would be 15 feet × 18 inches per foot, or 270 inches or 22′-6″. This does not include the length needed for any overhang. The hip rafter must now be shortened one-half the 45-degree thickness of the ridgeboard.

Book of Rafter Tables

Hip and valley rafter lengths can be found using published manuals of rafter tables. A portion of a typical table is in **Table 14-1**. This does not include any overhang. The hip rafter must now be shortened one-half the 45-degree thickness of the ridgeboard.

Table 14-1 Hip or valley rafter lengths — 9 foot 0-inch run.

Run	9'-0"			9'-0¼"			9'-0½"			9'-0¾"		
SLOPE	Ft	In	16th"	Ft	In	16th"	Ft	In	16th"	Ft	In	16th"
7 in 12	13'	9"	3	13'	9"	10	13'	10"	0	13'	10"	6
8 in 12	14'	0"	14	14'	1"	4	14'	1"	10	14'	2"	0
9 in 12	14'	4"	14	14'	5"	5	14'	5"	11	14'	6"	1
10 in 12	14'	9"	4	14'	9"	11	14'	10"	2	14'	10"	8
11 in 12	15'	2"	0	15'	2"	7	15'	2"	14	15'	3"	4
12 in 12	15'	7"	1	15'	7"	8	15'	7"	15	15'	8"	6
13 in 12	16'	0"	6	16'	0"	13	16'	1"	5	16'	1"	12
14 in 12	16'	6"	0	16'	6"	7	16'	6"	15	16'	7"	6
15 in 12	16'	11"	14	17'	0"	5	17'	0"	13	17'	1"	4

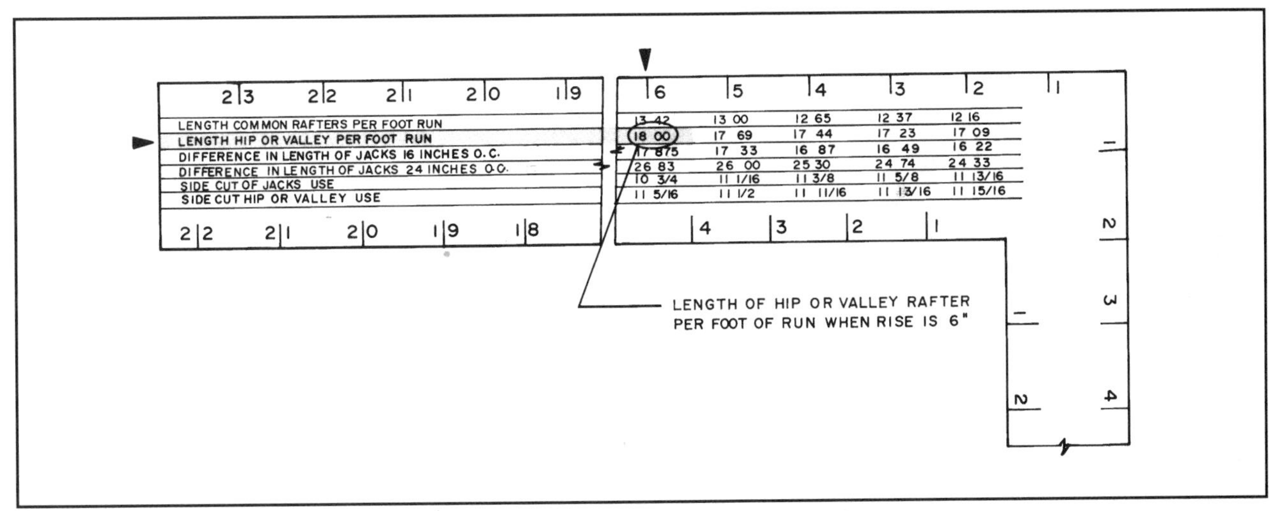

14-8 The length of the hip rafter can be found using the appropriate scale on the framing square.

HOW TO FIGURE ONE-HALF OF THE 45-DEGREE THICKNESS OF A MEMBER

The following explanation uses 1½-inch-thick lumber. The procedure is the same regardless of the thickness.

Stated simply, one-half of the 45-degree thickness of a member is equal to 1.414 times the thickness of the material, and this sum is divided by two.

See the following illustration:

1.414 inches × 1.5 inches = 2.12 inches

2.12 inches ÷ 2 = 1.06 inches

The 0.06 inches is converted to a common fraction in 16ths by multiplying

16 × 0.06 = 0.96/16 inch, or the nearest figure, 1/16 inch.

This is illustrated by the example shown in **14-9**. The length of the diagonal of a 1-inch square is 1.414 inches. The length of the diagonal of a 1½-inch square is 2.12 inches or

1.414 × 1.5 = 2.12 inches.

Half of the length of this diagonal is 1.06 inches or 1 1/16 inch.

An application of this to finding the shortening of the length of a hip rafter is shown in **14-10**.

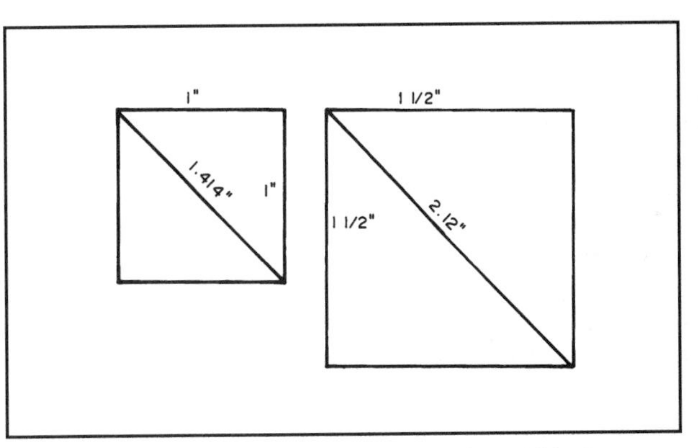

14-9 The diagonal of a one-inch square is 1.414 inches.

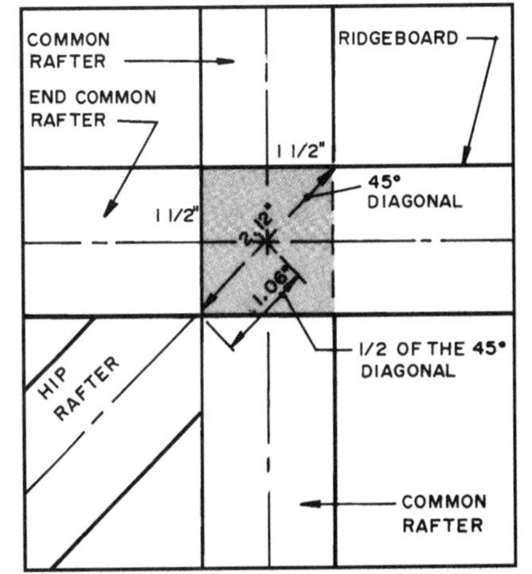

14-10 This shows how the shortening of the hip rafter is figured using the diagonal of a square based on the thickness of the ridgeboard.

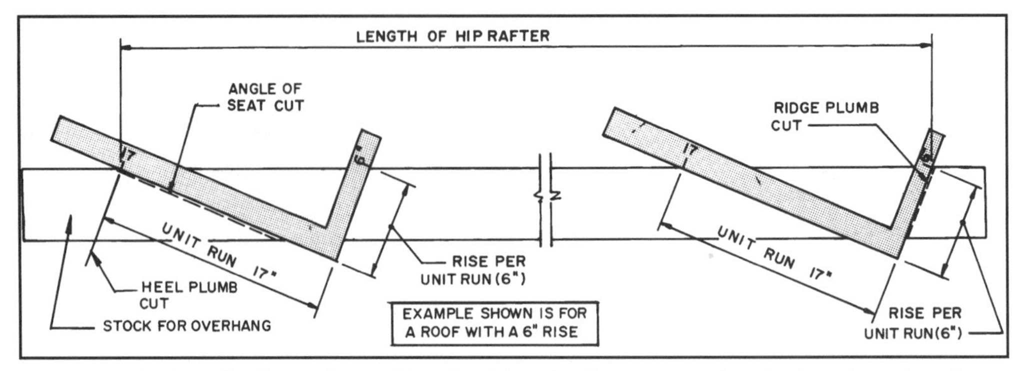

14-11 Begin the rafter layout by marking the ridge plumb cut, measuring the length, and marking the heel plumb cut.

LAYING OUT A HIP RAFTER

Begin by marking the ridge plumb and seat cuts as shown **(14-11)**. The unit rise is that specified for the roof, and the unit run is 17 inches. The plumb cut must be an angle cut because it butts the ridgeboard and two common rafters. These angled surfaces are called **cheek cuts**. To get the required angle use the table on the framing square marked **"side cut hip or valley use" (14-12)**. Use the figure under the column identified by a number representing the unit rise. Earlier in **14-8** a rise of six inches was used. The run shown on the square is 11⁵⁄₁₆ inches. Place the framing square on the top of the rafter and mark the cheek cuts as shown in **14-13**. The measurements are located from the centerline of the rafter.

After the angles are located, draw the plumb cuts. Place the square with the unit rise six inches, and the run 17 inches, on the rafter as shown in **14-14**. The tail cuts on the ends of the rafters are drawn at the same angle.

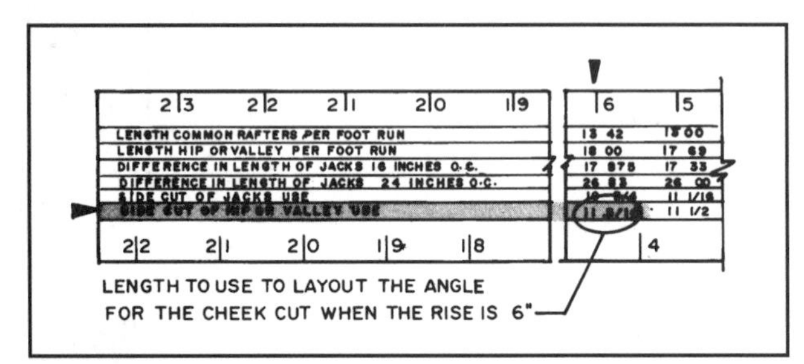

14-12 The distance for marking the angle for the cheek cuts is found on the framing square.

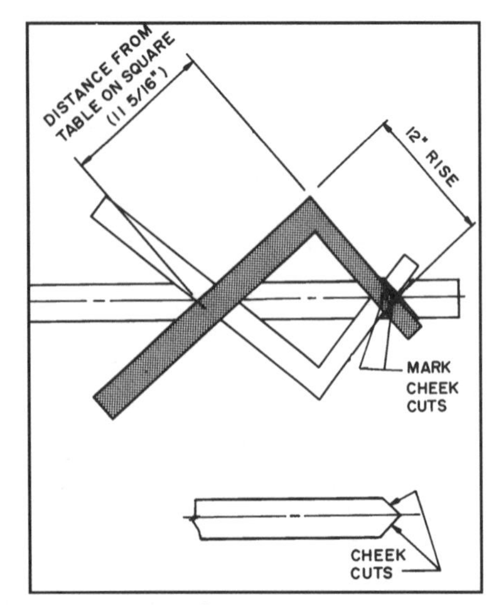

14-13 Mark the angles for the cheek cuts. See 14-11 to find the layout distance.

Marking the Overhang

Ordinarily, the hip rafter will have a tail extending beyond the seat cut. The length of the tail is found by drawing a square with sides equal to the

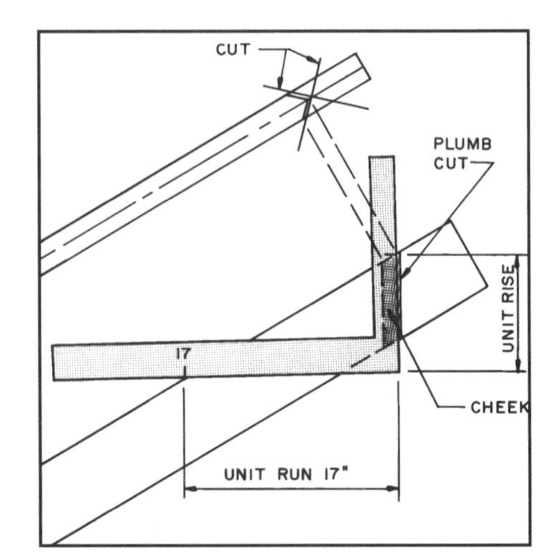

14-14 Mark the plumb cut on each side of the rafter.

desired overhang. The length of the diagonal is the run of the tail of the hip rafter **(see 14-15)**. The length of the tail of the rafter is the hypotenuse of a right triangle having sides equal to the run of the common rafter (12 inches).

The length of the overhang can be found by stepping off the length with the framing square as shown earlier in **14-5**. It can also be found by using the unit length of a hip rafter shown on the second scale on the framing square. Multiply the unit length by the number of feet of run. For example, for a roof with a six-inch rise, the unit length would be 18.00. If the overhang were to be 18 inches (1½ feet), the length of the rafter tail would be 18 × 1.5, or 27 inches.

The end of the hip rafter tail has cheek cuts on the same angle as the ridge plumb cut. The plancher is laid out and cut as described for common rafters, as is the bird's-mouth. A summary of these steps is shown in **14-16**.

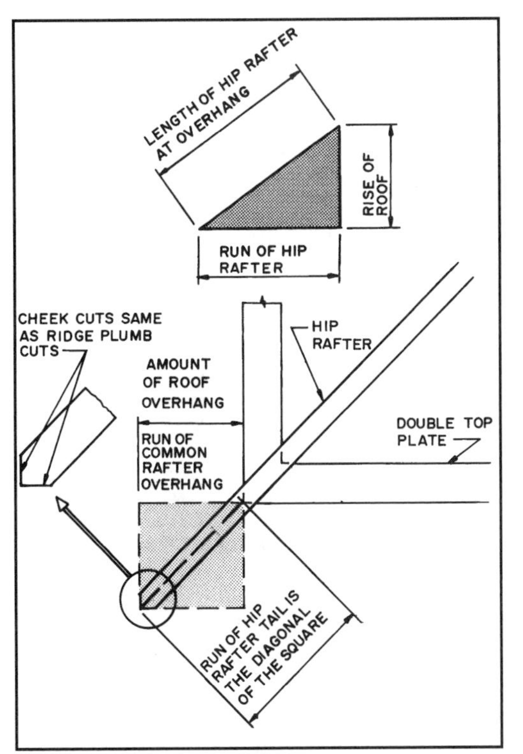

14-15 The hip rafter tail length is found after finding the run of the tail.

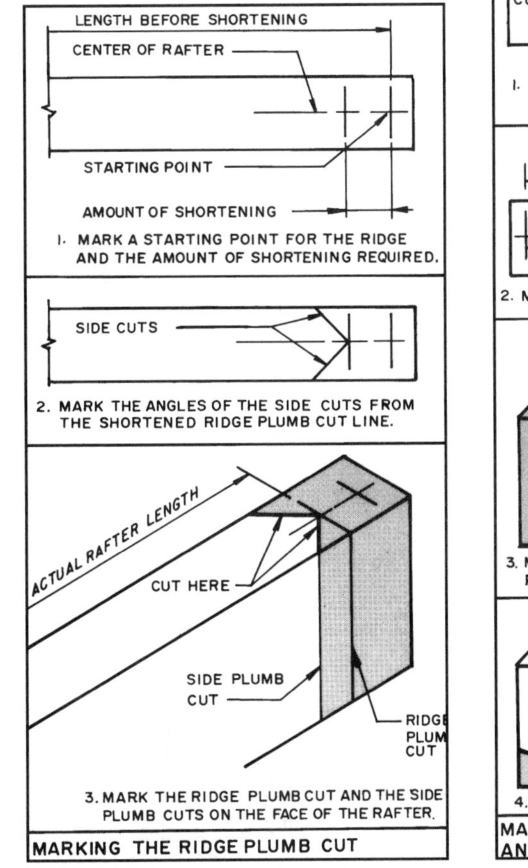

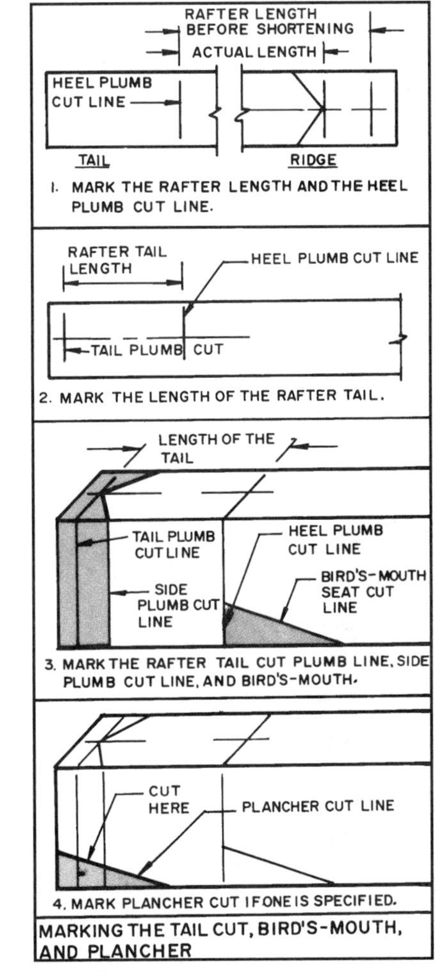

14-16 A summary of the steps for laying out a hip rafter.

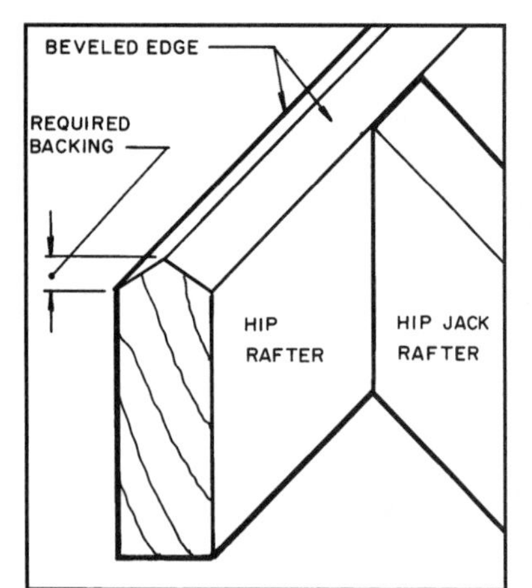

14-17 The hip rafter can be beveled to allow the sheathing to fit smoothly to it.

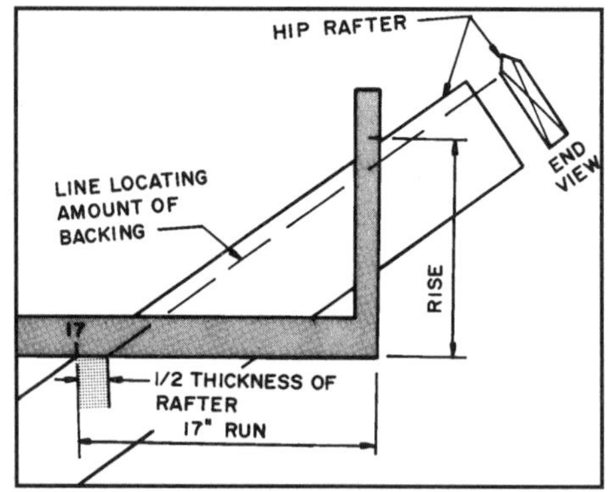

14-18 To find the amount of backing, set the square on the rafter with the rise and run on the edge. Measure half the thickness of the rafter from the 17-inch run mark. Where it meets the square locates the line of the backing.

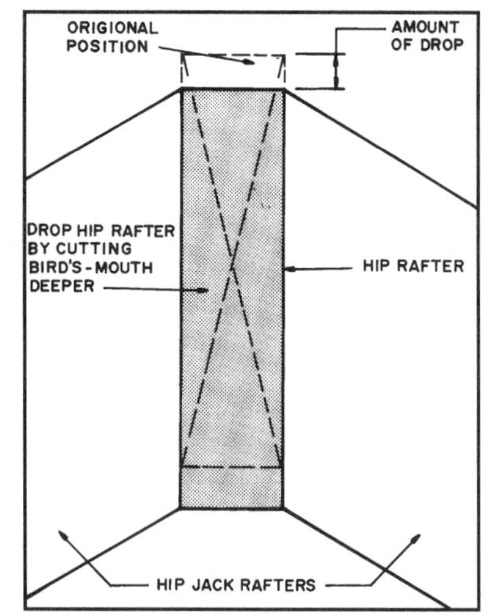

14-19 The hip rafter can be dropped to get the hip jack rafters flush with the top by cutting the bird's-mouth deeper.

BACKING OR DROPPING THE HIP RAFTER

Since the hip rafter will sit above the height of the jack rafters, an adjustment must be made. The edges of the hip rafter could be beveled **(see 14-17)**. To find the amount of bevel (also called backing), place the square on the rafter as shown in **14-18**. From the unit run, 17 inches, measure horizontally one-half the thickness of the rafter. A line drawn through this point parallel with the edge of the rafter gives the amount of bevel or backing.

Another technique is to cut the bird's-mouth deeper, dropping the height of the rafter **(see 14-19)**. To find the amount of drop, position a square on the rafter, placing the rise and run as shown in **14-20**. Measure horizontally one-half the thickness of the rafter. The vertical distance between the square and this mark on the top of the rafter is the amount of drop.

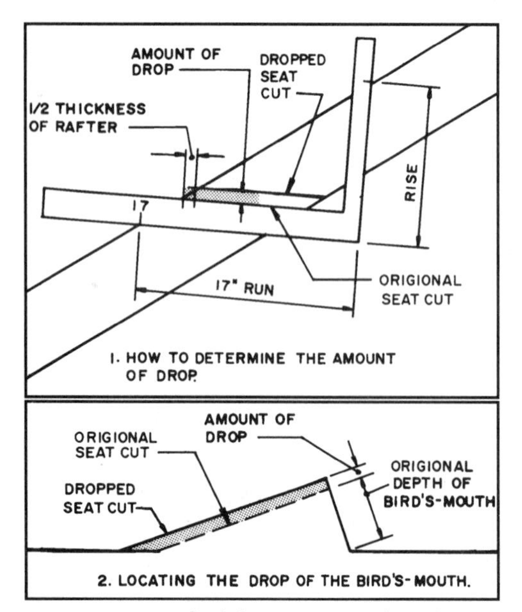

14-20 How to find the amount to drop the hip rafter.

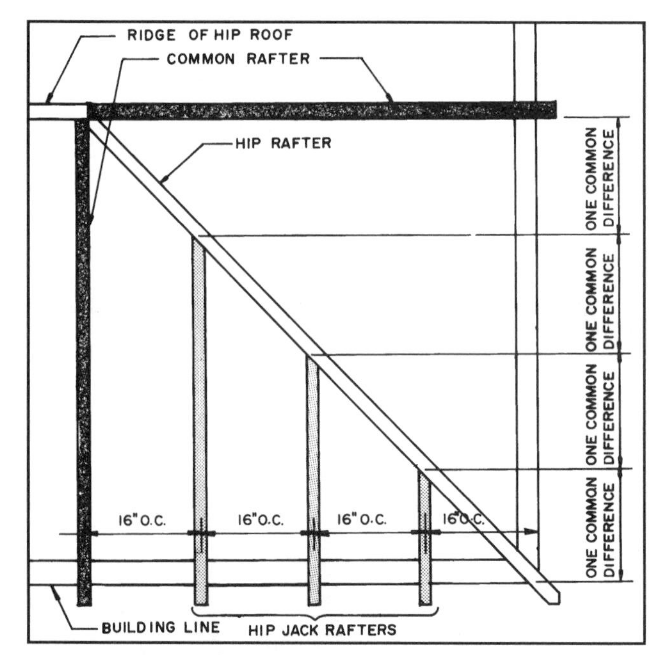

14-21 Hip jack rafters differ in length by one common difference.

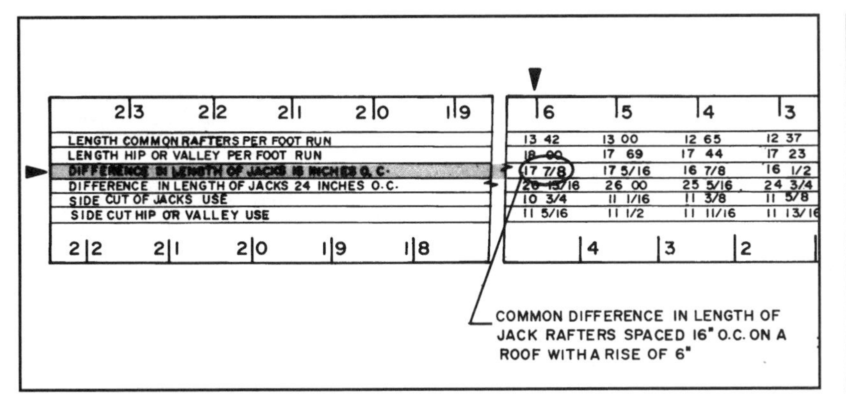

14-22 The common difference in the length of jack rafters can be found using the tables on the framing square.

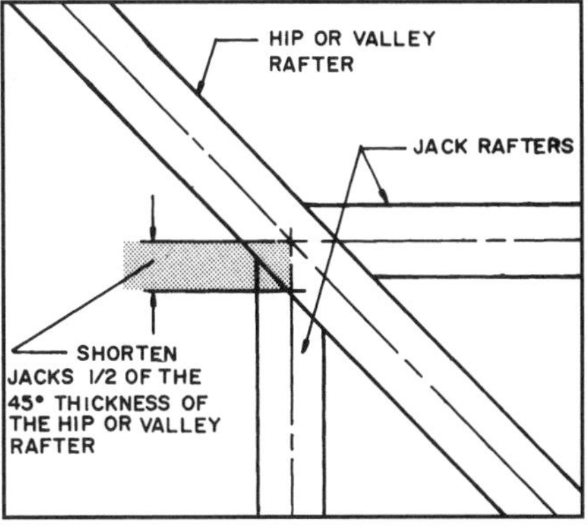

14-23 Jack rafters are shortened by half the 45-degree thickness of the hip rafter.

HIP JACK RAFTERS

Hip jack rafters run from the hip rafter to the top plate. They are spaced the same distance apart as the common rafters and have the same rise and run. Reexamine **14-1** to see how these form part of the hip roof.

Common Difference of Hip Jack Rafters

Hip jack rafters are usually spaced 16 inches or 24 inches O.C. When they are equally spaced each will have a **common difference** in length **(see 14-21)**.

The longest jack rafter will be one common difference shorter than the common rafter. The second-longest jack rafter will be two common differences shorter than the common rafter. This continues until the shortest jack rafter, which is equal to one common difference, is laid out.

The common difference can be found on the third and fourth lines of the framing square table. Line three is for rafters 16 inches O.C., and line four is for those 24 inches O.C. **(see 14-22)**. To find the common difference, locate the rise of the roof on the inch line of the square.

For example, assume a rise of 6 inches. On the third line of the table below the 6-inch mark, a common difference of 17.875 inches for rafters 16 inches O.C. is found. The shortest jack rafter is 17.875 inches. The longest is the length of the common rafter minus 17.875 inches. In addition it is necessary to subtract one-half of the 45-degree thickness of the hip or valley rafter **(see 14-23)**.

Laying Out the Hip Jack Rafter

The layout is determined by where the first rafter is located. One method is to place a common rafter at the end of the ridge and locate the jack rafters from it as shown in **14-24**.

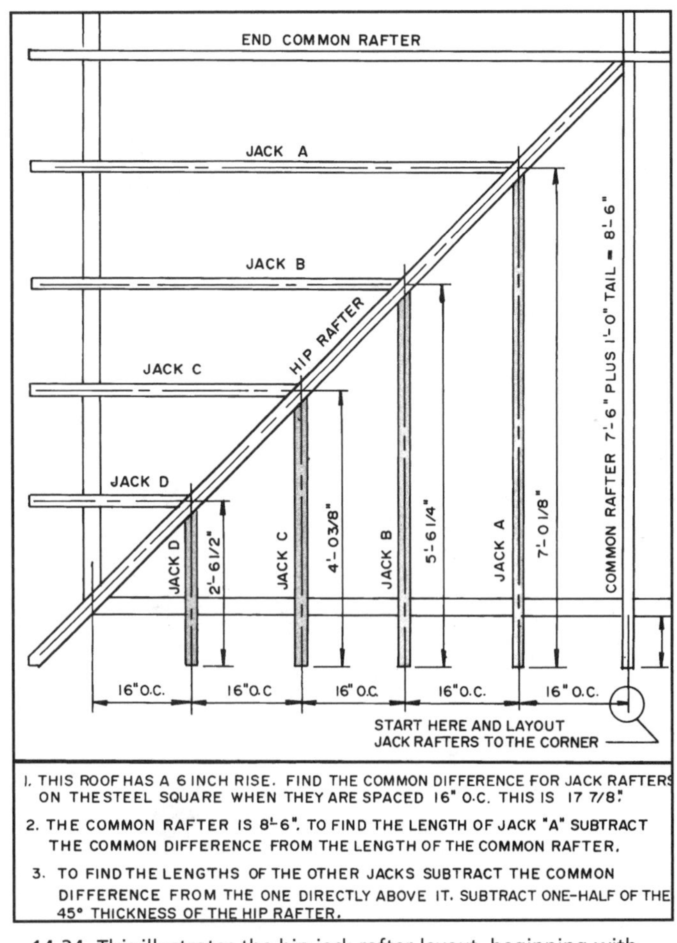

14-24 This illustrates the hip jack rafter layout, beginning with the common rafter and moving to the corner of the building.

The common difference is subtracted from the length of the common rafter. Another technique is to locate the first jack rafter in from the corner of the building a distance equal to the on-center rafter spacing and locate all other jack rafters from it as shown in **14-25**. The length of the first jack rafter is equal to the common difference. Each rafter beyond that is one common difference longer than the one below it.

A summary of the layout of a hip jack rafter is shown in **14-26**.

Jack Rafter Pattern

Patterns for all the jack rafters can be developed by first laying out the longest jack rafter and subtracting the allowance for shortening it as shown earlier in **14-23**. Then mark one common difference for each of the smaller hip jack rafters as shown in **14-27**. The bird's-mouth and tail cut for each are the same as laid out on the common rafter. This pattern can be used to lay out the cuts on each hip jack rafter.

Remember, since the rafters are installed in pairs on opposite sides of the hip rafter, the cuts must be made in opposite directions **(see 14-28)**.

FINDING THE LENGTH OF THE RIDGEBOARD

The length of the ridgeboard is found by locating the centerlines of the first common rafters from each end of the building. This is a distance equal to the run of the common rafter. The distance between these centerlines is the **theoretical length** as shown in **14-29**.

The **actual length** of the ridgeboard is the theoretical length plus half the thickness of each end common rafter. See the enlarged detail shown in **14-30**.

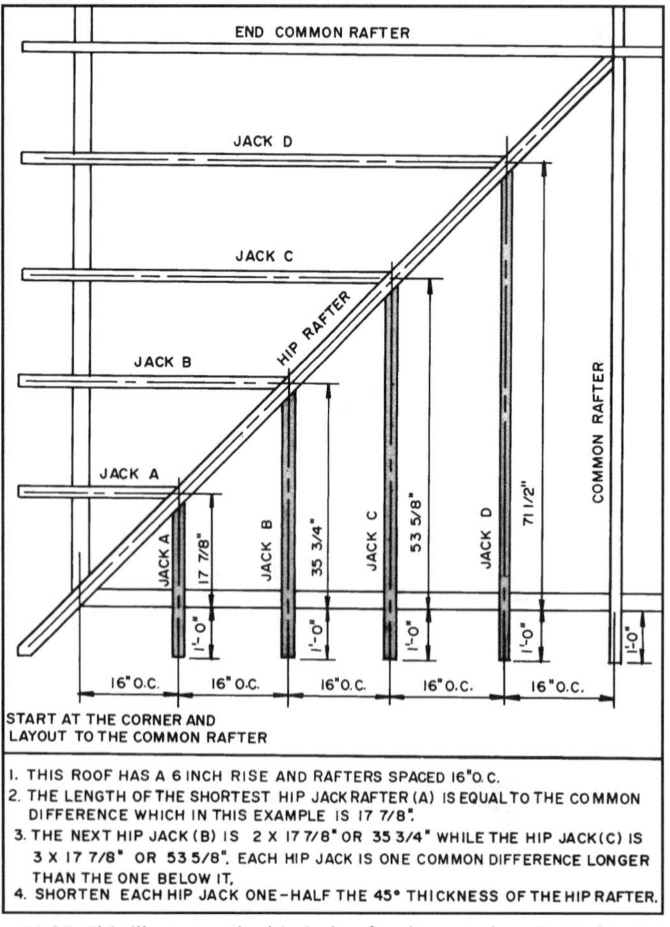

14-25 This illustrates the hip jack rafter layout when the jacks are started at the corner of the building.

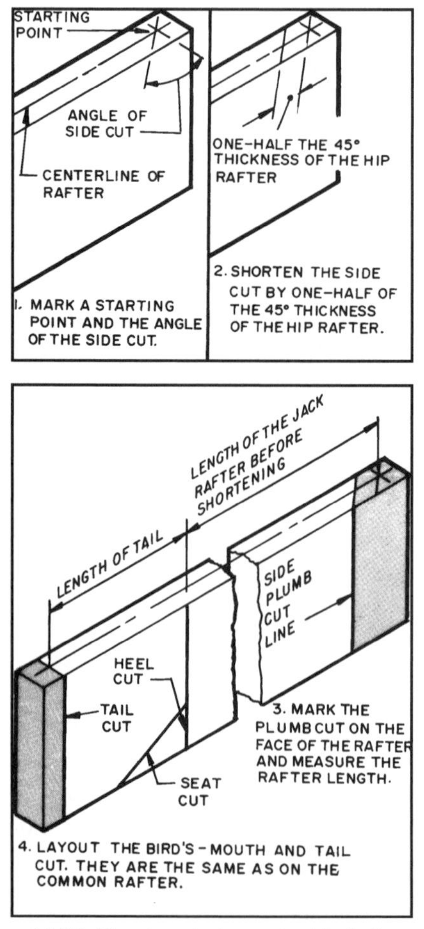

14-26 The steps to lay out a hip jack rafter.

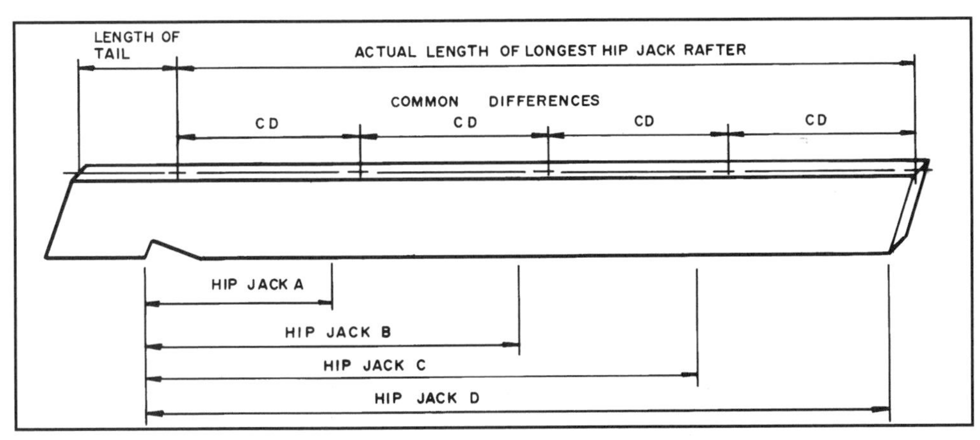

14-27 The longest hip jack rafter can serve as a pattern for all the shorter rafters.

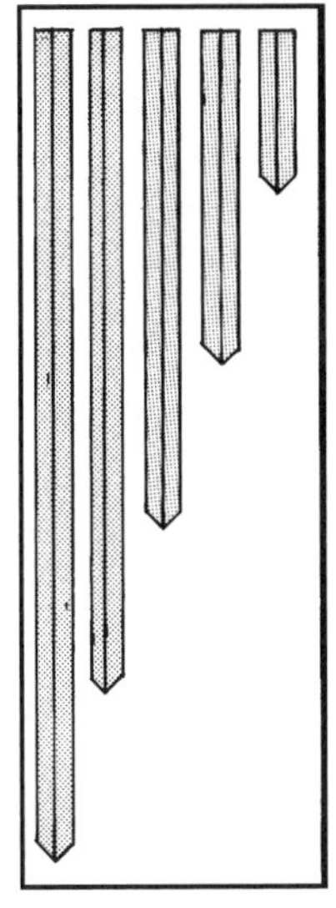

14-28 Remember to cut the hip jack rafters in pairs with the angle of cuts in opposite directions.

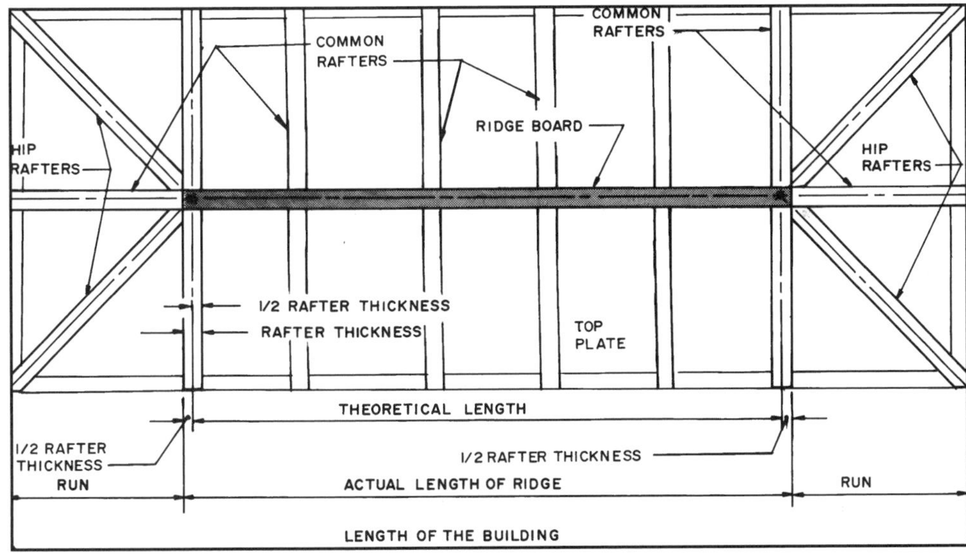

14-29 The actual length of the ridgeboard is the length of the building minus two times the run of the rafters plus the thickness of one rafter.

THE HIP GABLE

The hip gable is an interesting architectural feature used to replace the peak commonly found on the gable-ends of gable roofs. The framing is the same as that for the hip roof.

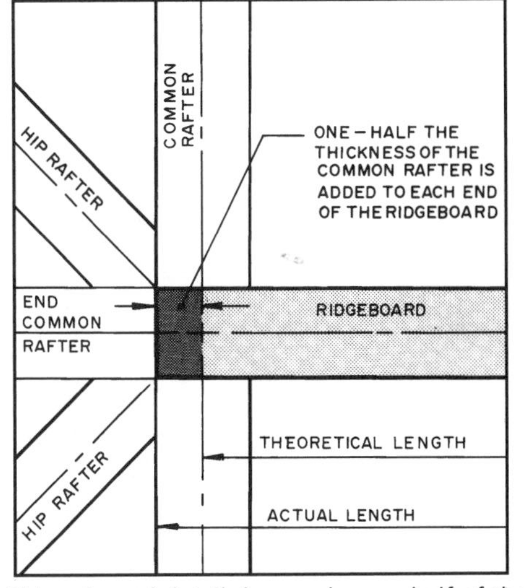

14-30 This enlarged detail shows why one-half of the thickness of the common rafter must be added to each end of the ridgeboard.

RAISING THE HIP ROOF

Carpenters have various ways they prefer to erect a hip roof. One method erects the ridgeboard with common rafters on each end set in from the end of the ridgeboard **(see 14-31, step 1, on the next page)**. The steps to erect a ridgeboard are in Chapter 13. Then nail a common rafter on the end of the ridgeboard **(see 14-32, next page)**. Next, set the hip rafter in place over the corner of the building. Toenail it to the top plate with two 16d common nails on one side and one on the other. Nail the hip rafter to the end common rafter with 16d common nails. After both hip rafters are in place nail the end common rafter to the top plate and ridgeboard **(see 14-31, step 2, next page)**. Now nail the hip jack rafters to the top plate with three 16d common nails and to the hip rafter with three 16d common nails. Install the common rafters to the length of the roof.

Another method involves erecting the ridgeboard and all common rafters and then installing the hip rafters and finally the hip jack rafters **(see 14-33)**. The ridgeboard is erected as described in Chapter 13 for gable roofs.

Common rafters are installed on each end and sometimes one pair is placed in the center of the ridgeboard. Then all the common rafters are installed. If additional sections of ridgeboard are required they are installed with common rafters. Then the hip rafters are set and the hip jacks installed **(see 14-34)**. A finished framed gable-end hip roof is in **14-35**.

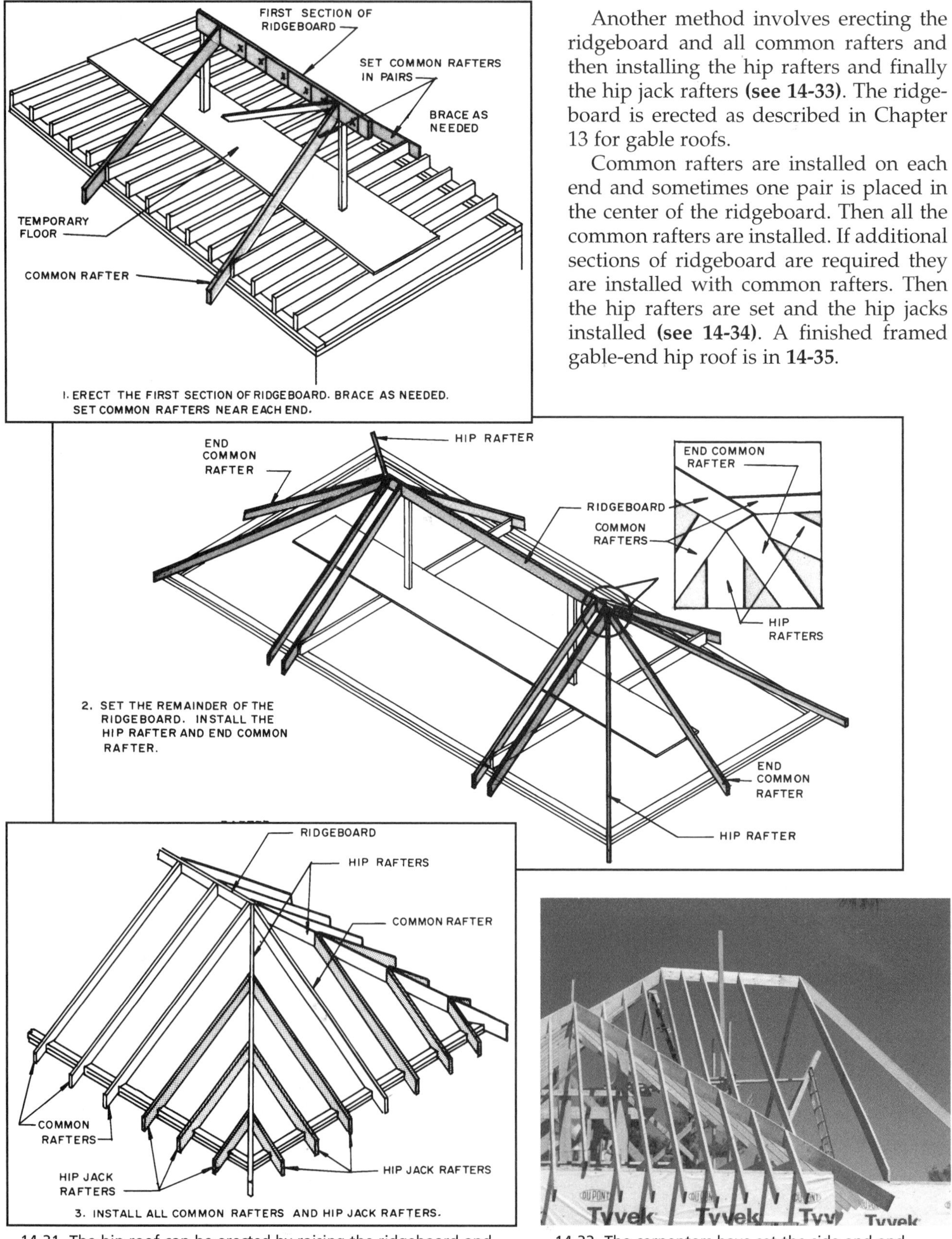

14-31 The hip roof can be erected by raising the ridgeboard and the end common rafters, installing the hip and hip jack rafters on each end, and then erecting the common rafters.

14-32 The carpenters have set the side and end common rafters. They will next install the hip rafters.

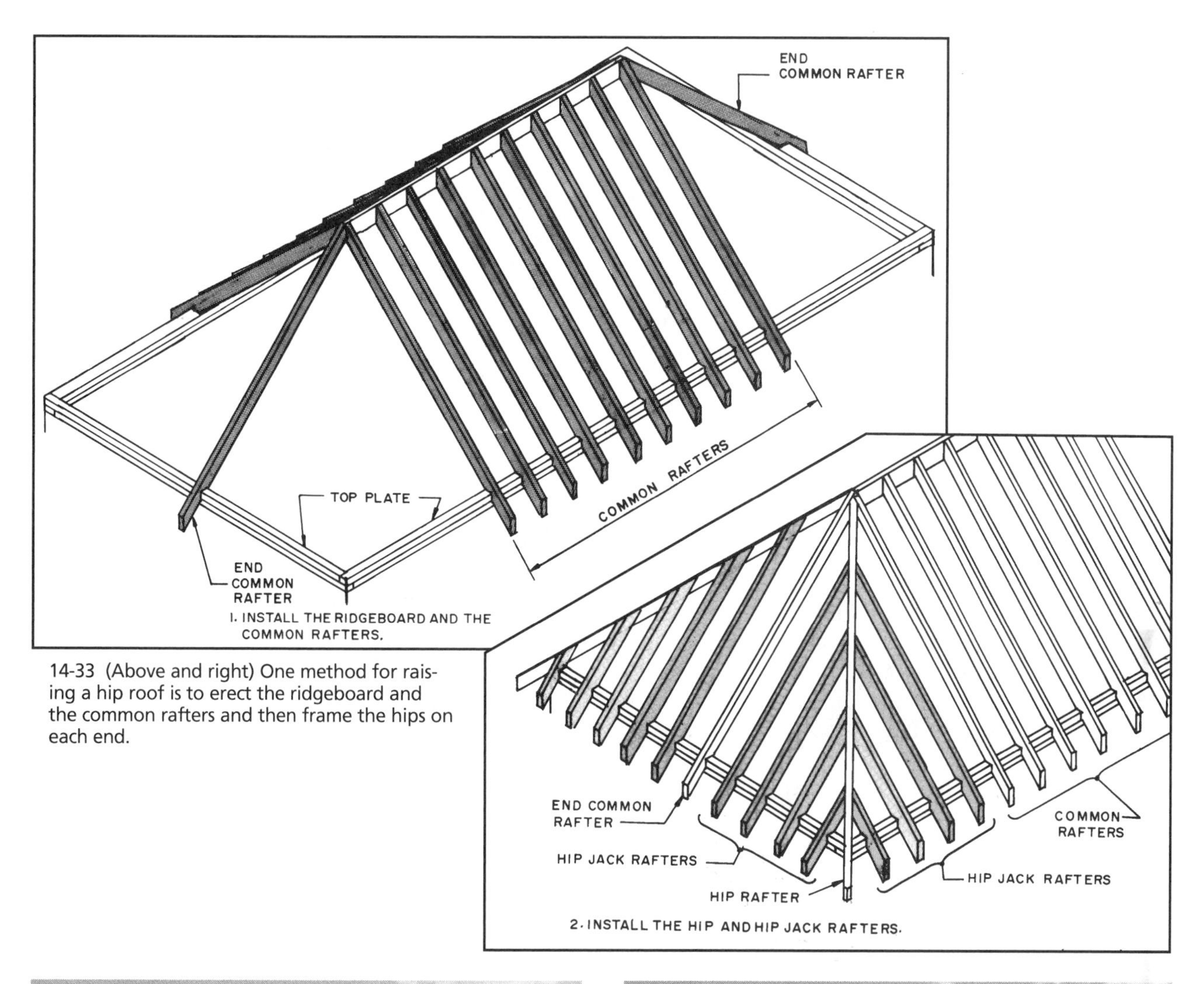

14-33 (Above and right) One method for raising a hip roof is to erect the ridgeboard and the common rafters and then frame the hips on each end.

14-34 After the hip rafter is installed the hip jacks are set in place.

14-35 This hipped gable end is ready for sheathing.

CHAPTER 15

INTERSECTING ROOFS

When the roofs of two joining sections join at a right angle, the planes of the roof intersect, forming a valley. A valley rafter is located at the line of intersection and jack rafters are used to frame the space between the valley rafter and the ridgeboards of the two roofs.

The intersecting roofs may have equal or unequal spans **(see 15-1)**. A gable roof may intersect a hip roof with jack rafters running between the valley and a hip rafter.

Basically, valley construction is similar to that of an inverted hip roof. The length of the hip and valley rafters are the same for identical spans. Valley rafters have the same 17-inch run used for hip rafters and they must be shortened in the same way. Review Chapter 14 for additional details on these procedures.

Valley rafters run from the ridgeboard to an inside corner and do not require lowering as do hip rafters. The tail plumb cut is a single side cut parallel with the ridge plumb cut.

When the roofs have equal spans, the ridges on each of the intersecting roofs are on the same level. When the spans are not equal the ridge on the small span intersects the larger roof below its ridgeboard, as shown in **15-2**.

Review Chapters 13 and 14 before continuing with this chapter.

INTERSECTING ROOFS HAVING EQUAL SPANS

Intersecting roofsthat have the same span and rise will have the ridgeboard at the same level. The construction for a T-shaped building is shown in **15-3**.

When both roofs are the gable type and have the same span, the **valley jack rafters** run from the ridgeboard to the valley rafter.

If one of the intersecting roofs is a hip type, the hip rafter and valley rafter are connected with **hip-valley jack rafters** as shown in **15-4**.

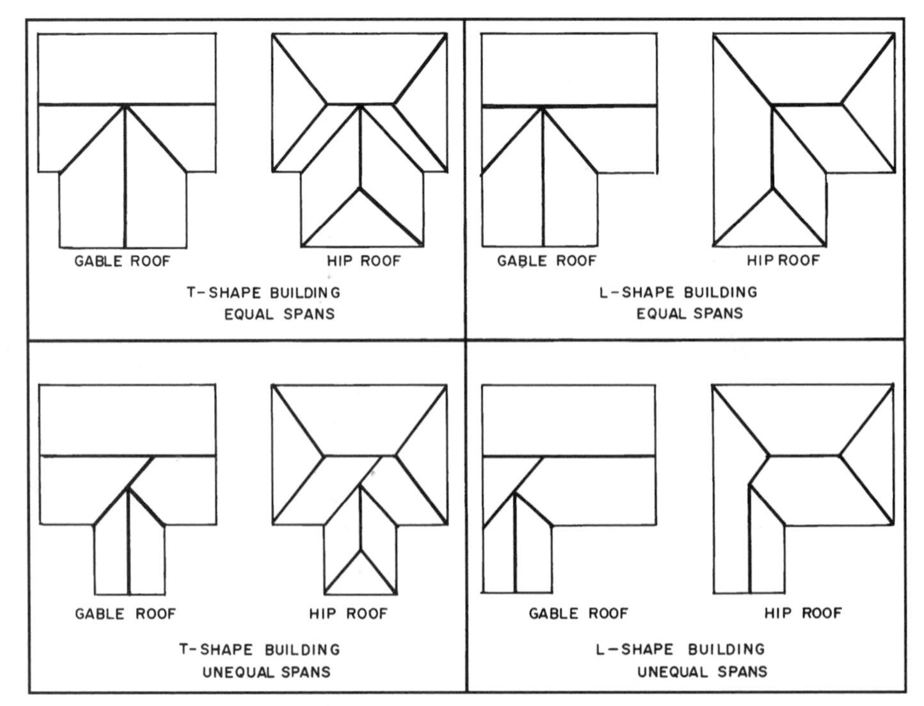

15-1 Typical intersecting roof configurations.

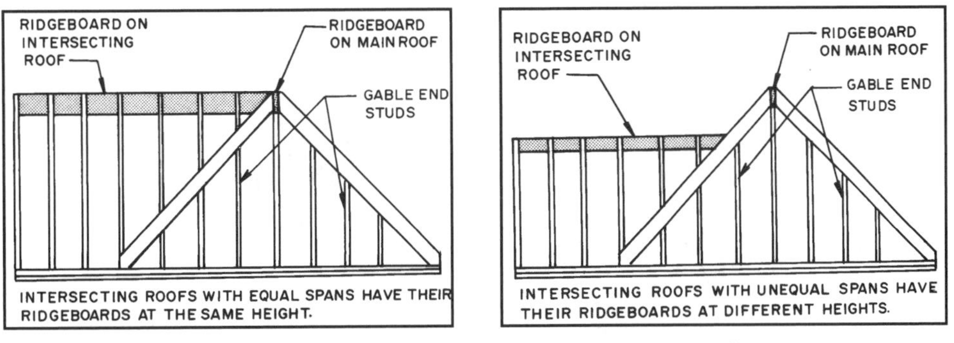

15-2 The span of intersecting roofs influences where the ridgeboards will meet.

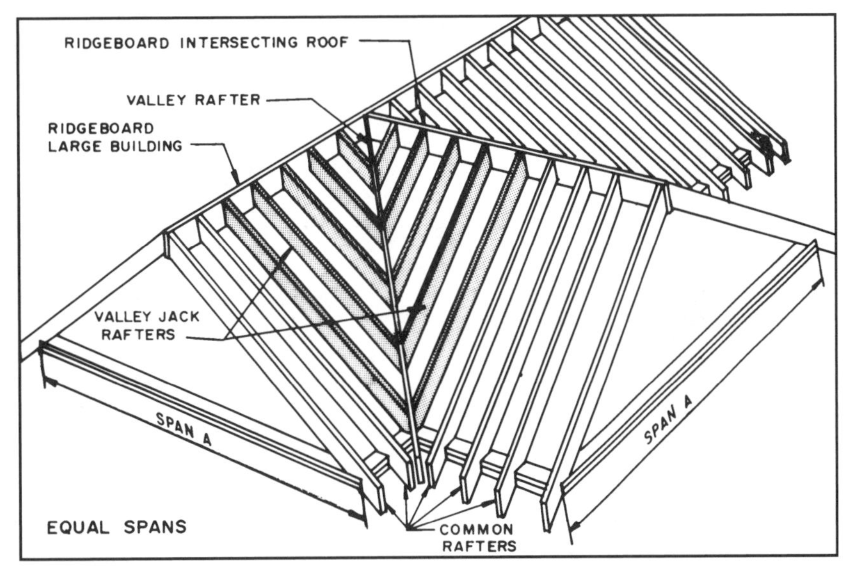

15-3 Typical framing for intersecting gable roofs for a T-shaped building when both roofs have the same span.

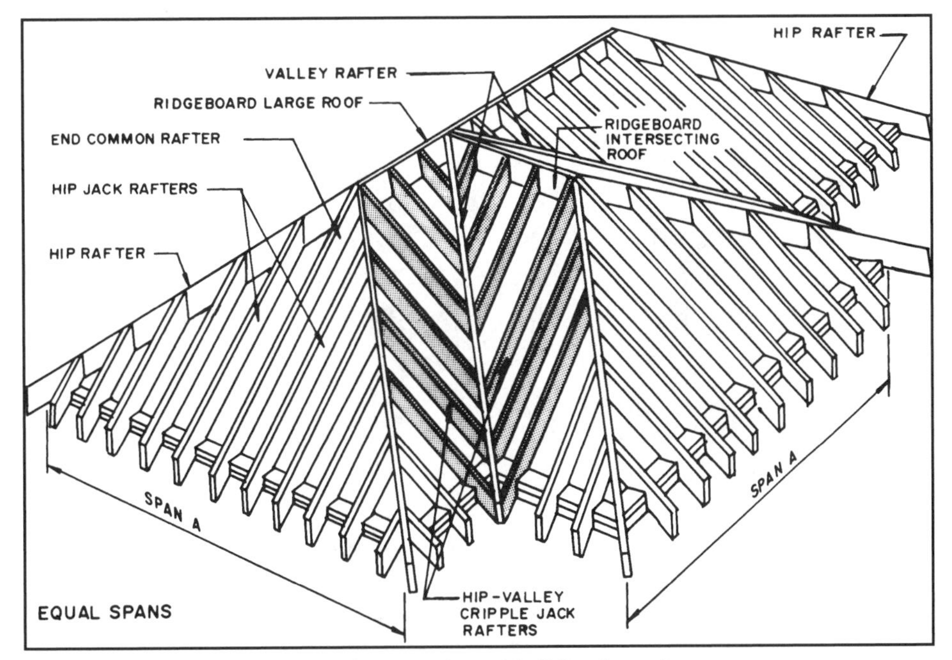

15-4 These intersecting hip roofs on a T-shaped building have the same span.

If the intersecting wing of an L-shaped building with a hip roof has the same span as the main building, the framing will appear as shown in **15-5**.

The ridgeboards meet, forming a corner, and a valley rafter runs from this intersection to the top plate on one side of the intersection.

The framing of an L-shaped building with gable roofs is shown in **15-6**.

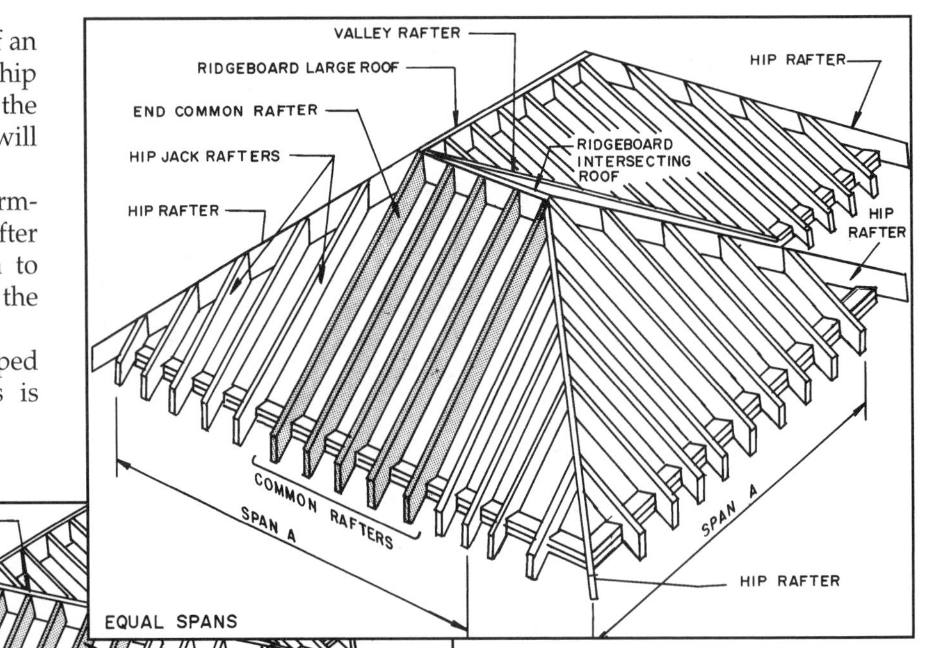

15-5 (Above) Typical framing for intersecting hip roofs for an L-shaped building when the roofs have the same span.

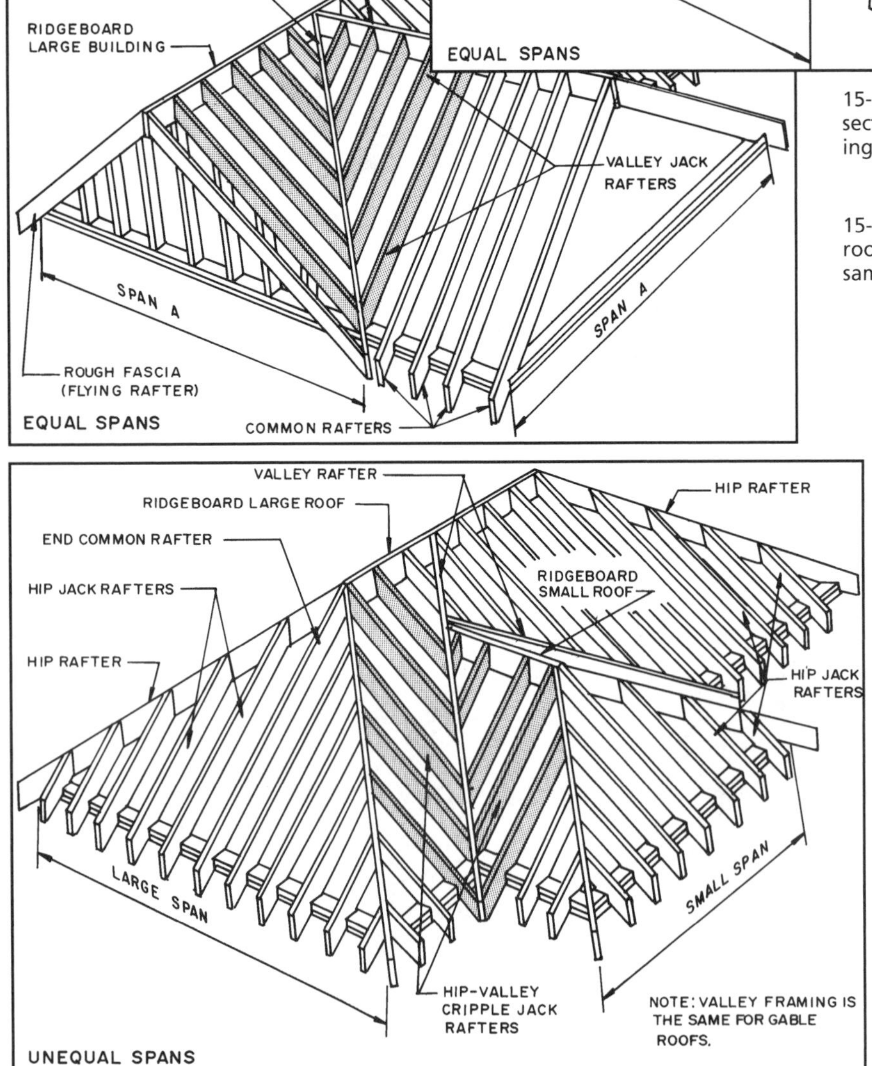

15-6 (Left) These intersecting gable roofs on an L-shaped building have the same span.

15-7 These intersecting hip roofs on a T-shaped building have unequal spans. Framing for intersecting gable roofs would be similar except there would be a gable-end instead of a hip.

INTERSECTING ROOFS HAVING UNEQUAL SPANS

Intersecting roofs having unequal spans can fall into two types. One is where the smaller intersecting roof meets the main roof somewhere along its length, forming a T as shown in **15-7**. This hip roof is framed using a continuous valley rafter running from the ridgeboard to the top plate. The other valley rafter runs from the continuous valley rafter to the top plate. The ridgeboard on the smaller roof is run to the continuous valley rafter. The roof is then framed with common and valley jack rafters. A gable roof would be constructed in the same manner.

The other situation occurs when the house to be roofed with a hip roof is L-shaped **(see 15-8)**. A partial hip rafter can be run until it intersects the ridgeboard on the smaller roof. A valley rafter is run on the one side from the intersection of the partial hip rafter and the ridgeboard. The common rafters and valley jack rafters are installed to complete the roof. Framing for intersecting gable roofs with unequal spans is in **15-9**.

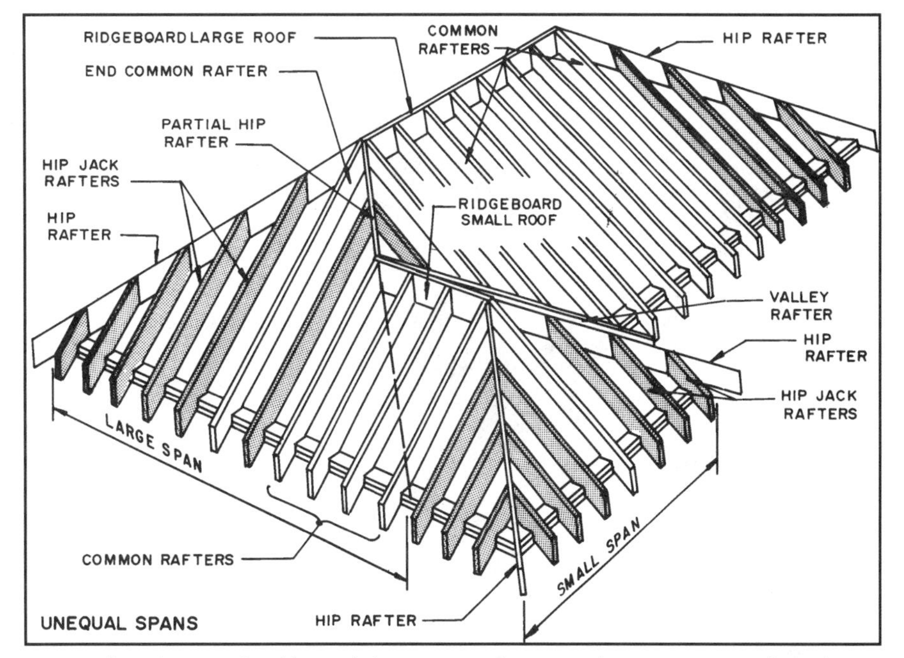

15-8 These intersecting hip roofs have unequal spans and appear on an L-shaped building.

BLIND VALLEY CONSTRUCTION

Blind valley construction is another way to build the valley intersection between two roofs. This involves first framing and sheathing the roof on the main part of the building. Then frame the roof on the intersecting structure, letting this roof run over the sheathing on the main roof. This is typically done when additions are added to an existing building. However it can also be used for new construction. This eliminates the need for the valley rafter. This technique is illustrated in **15-10**. A blind valley is laid on top of the sheathing, and the valley jack rafters are run from the ridgeboard to it.

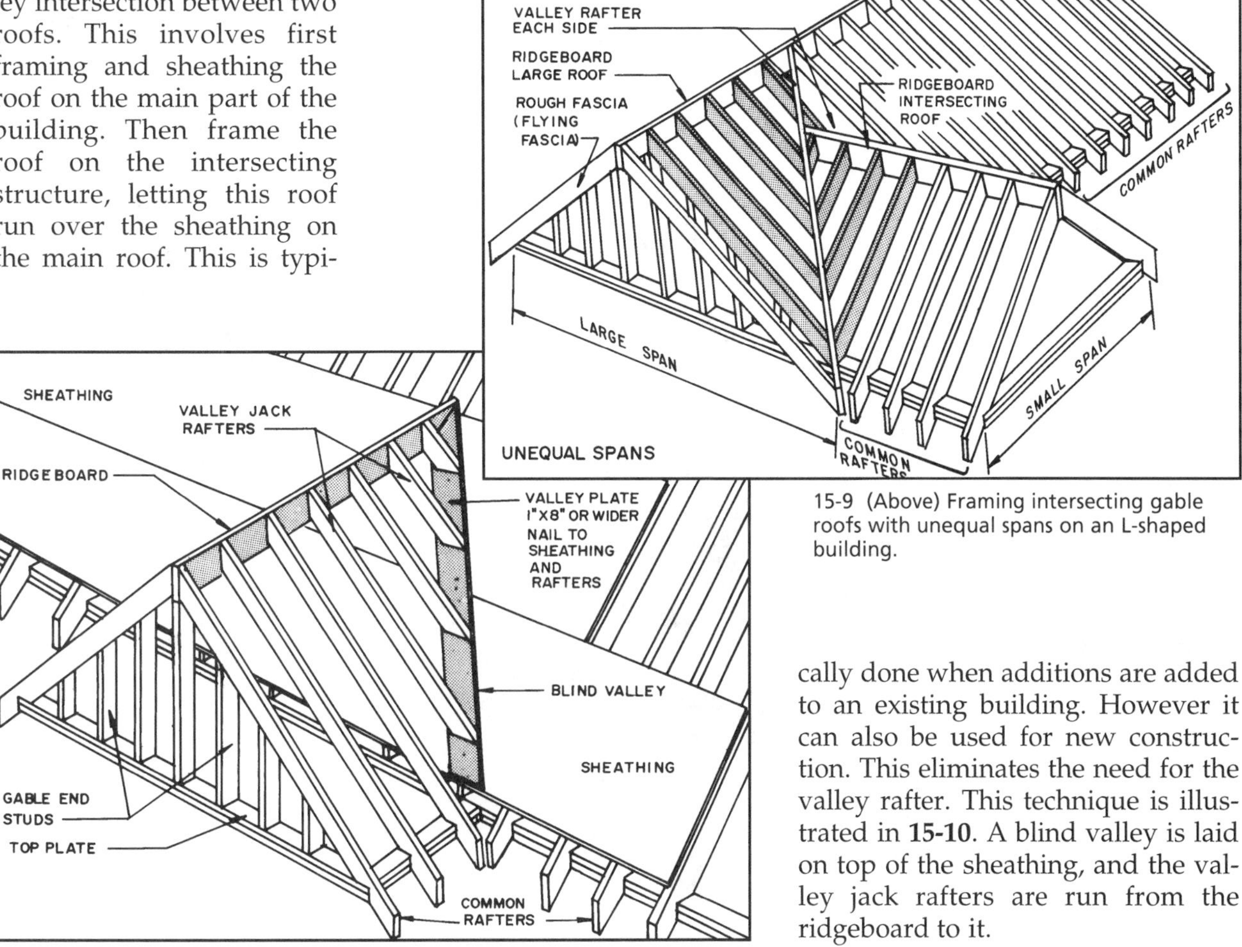

15-9 (Above) Framing intersecting gable roofs with unequal spans on an L-shaped building.

15-10 Typical blind valley construction.

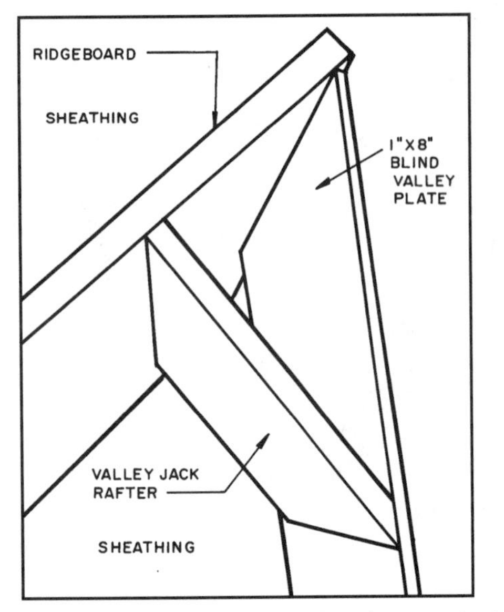

15-11 The valley jack rafters have a bevel cut on the bottom end so they can sit flat on the blind valley plate.

The valley jack rafter has a bevel cut on the end that rests on the blind valley rafter **(see 15-11)**. To get the angle where the ridgeboard meets the sheathed roof, set a leveled board, mark the angle as shown in **15-12**, cut the angle, and then measure and cut the ridgeboard to length.

The valley is located by a chalkline from the corner of the ridgeboard to the inside corner formed by the top plate. Along the chalkline nail one or more 1 × 8-inch solid wood or plywood valley plates. The width should be enough to support the bottom of the valley jack level cut. Therefore it may be necessary to use an 8- to 10-inch width. The plate should be nailed into the rafters below the sheathing.

If the overhang is large, a partial tail forming the valley can be installed as shown in **15-13**.

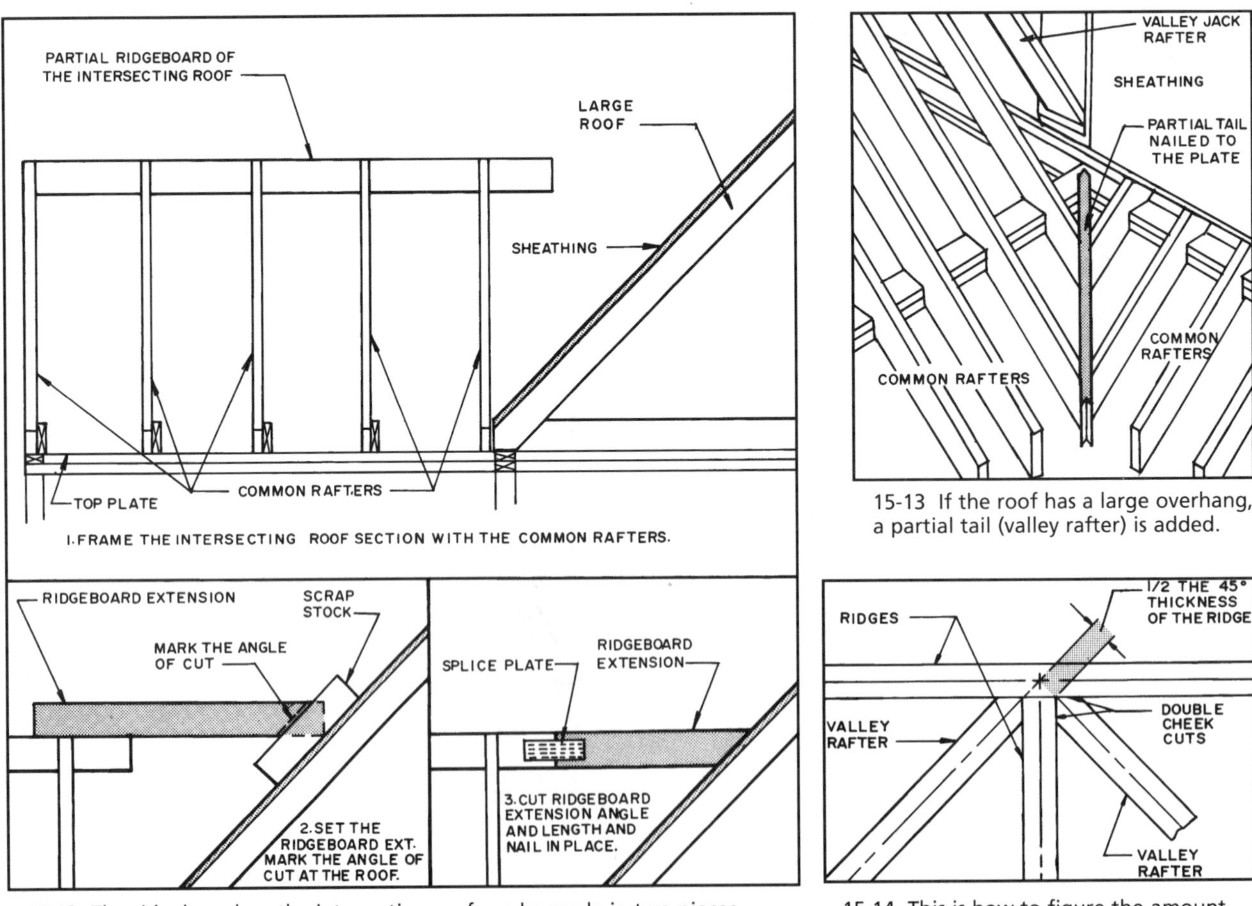

15-12 The ridgeboard on the intersecting roof can be made in two pieces, simplifying the erection process.

15-13 If the roof has a large overhang, a partial tail (valley rafter) is added.

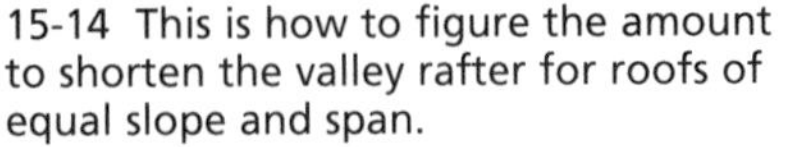

15-14 This is how to figure the amount to shorten the valley rafter for roofs of equal slope and span.

LAYING OUT VALLEY RAFTERS FOR ROOFS WITH EQUAL SPANS

At the ridgeboard the valley rafter is shortened one-half the length of the 45-degree line as shown in **15-14**. This is the same as that used on the hip rafter.

The bird's-mouth is laid out the same as described for the hip rafter. Generally the plumb cut is on an angle producing an angled section that fits against the corner of the top plate of the exterior wall **(see 15-15)**. Some framers prefer to make the seat cut square. The tail plumb cut is laid out in the same manner as the hip rafter except it has angle cuts enabling it to support the fascia **(see 15-16)**.

A summary of these steps is shown in **15-17.**

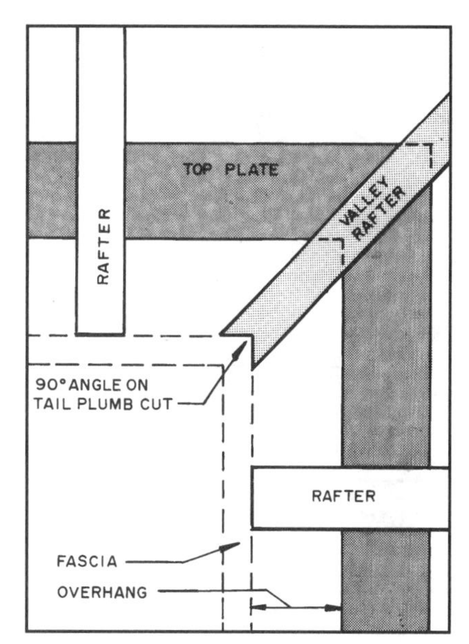

15-16 The plumb tail cut on the valley rafter has a 90-degree angle to receive the fascia.

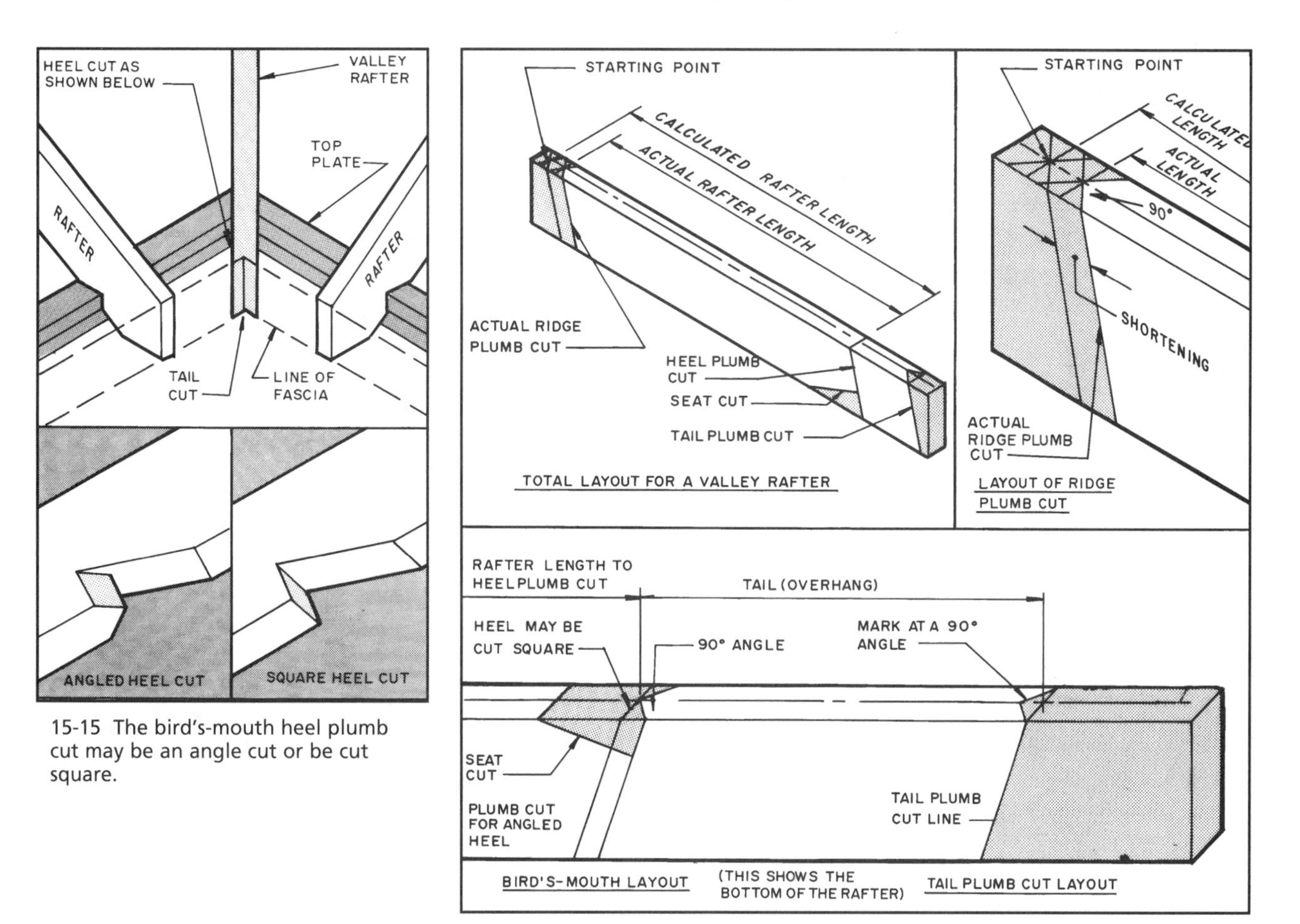

15-15 The bird's-mouth heel plumb cut may be an angle cut or be cut square.

15-17 The steps involved in a valley rafter layout.

LAYING OUT VALLEY RAFTERS FOR ROOFS WITH UNEQUAL SPANS

As shown earlier in **15-9**, roofs with unequal spans have a long valley rafter running from the ridgeboard of the large roof to the exterior wall. The short valley rafter on the other side of the smaller intersecting roof runs from long valley rafter to the exterior wall.

The length of the **long valley rafter** is figured from the center of the ridgeboard and is shortened one-half the 45-degree thickness of the ridge **(see 15-18)**. The layout is the same as a regular valley rafter. Notice it has a single angle plumb cut at the ridgeboard.

The **short valley rafter** length is figured from the point it meets the long valley rafter to the exterior wall plate. It is shortened by one-half the thickness of the long valley rafter as shown in **15-18**. The span of this rafter is from the exterior wall to a point directly below the intersection with the long valley rafter.

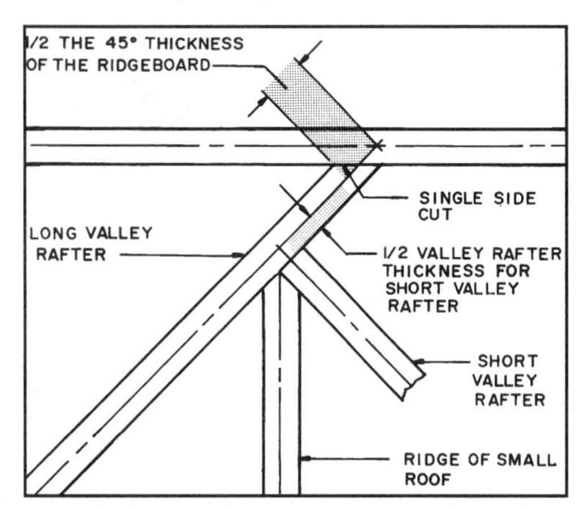

15-18 The long valley used on intersecting roofs with unequal spans is shortened one-half the 45-degree thickness of the ridgeboard. The short valley is shortened one-half the thickness of the long valley rafter.

JACK RAFTERS

The three types of jack rafters are hip jack, cripple jack, and valley jack.

Hip jack rafters run from the ridgeboard to the valley rafter. Hip jacks have a cheek cut on one end and a bird's-mouth and a plumb fascia cut on the other.

There are two types of **cripple jack rafters**, hip-valley cripple jacks and valley cripple jacks. Cripple jacks have cheek cuts on both of the ends.

Hip-valley cripple jacks run between hip and valley rafters.

Valley cripple jacks run between two valley rafters **(see 15-19)**.

Valley jack rafters have a cheek cut on one end and a plumb ridge cut on the other.

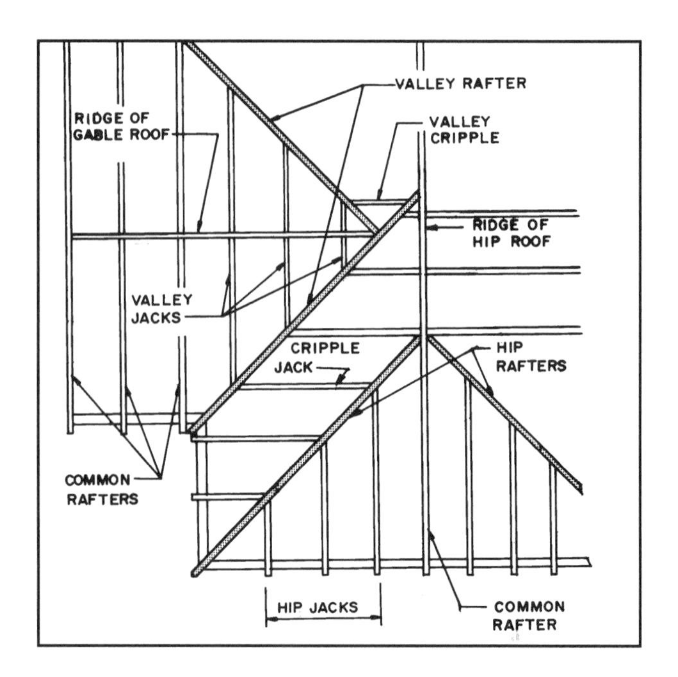

15-19 The types of jack rafters used in roof construction.

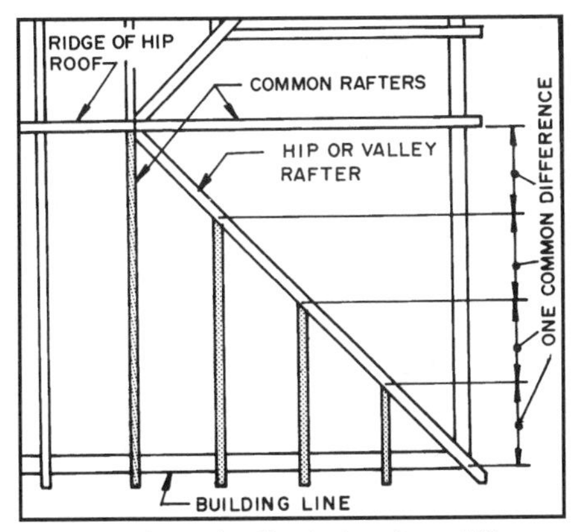

15-20 Hip and valley jack rafters differ in length by one common difference.

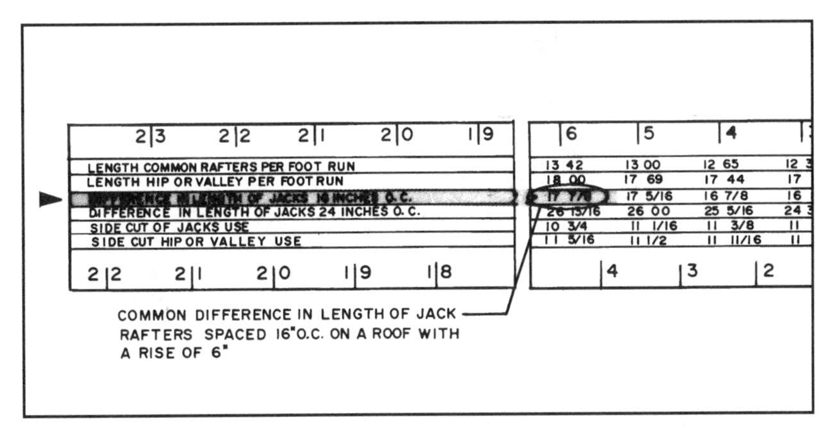

15-21 The common difference in the length of valley and hip jack rafters can be found using the tables on the framing square.

Common Difference of Hip & Valley Jack Rafters

Jack rafters are usually spaced the same, 16 or 24 inches O.C., as are common rafters. When they are equally spaced each will have a **common difference** in length **(see 15-20)**. The longest jack rafter will be one common difference shorter than the common rafter. The second-longest jack rafter will be two common differences shorter than the common rafter. This continues until the shortest jack rafter, which is equal to one common difference, is laid out.

The common difference can be found on the third and fourth lines of the framing square table. Line three is for rafters 16 inches O.C., and line four is for those 24 inches O.C. **(see 15-21)**. To find the common difference, locate the rise of the roof on the inch line of the square. For example, assume a rise of 6 inches. On the third line of the table below the 6-inch mark, a common difference of 17⅞ inches for rafters 16 inches O.C. is found. The shortest jack rafter is 17⅞ inches. The longest is the length of the common rafter minus 17⅞ inches. As well it is necessary to subtract one-half of the 45-degree thickness of the hip or valley rafter **(15-22)**. Refer to Chapter 14 for more information about the common difference.

HIP JACK RAFTERS

The procedures for laying out hip jack rafters is discussed in Chapter 14.

VALLEY JACK RAFTERS

The valley jack rafter runs from the ridgeboard to the valley rafter as shown in **15-19**. The end butting the ridgeboard is cut square and the end butting the valley rafter is cut on a bevel **(see 15-23)**.

The valley jack rafter is laid out in the same manner as the hip rafter. The common difference is the same and is found on the tables on the framing square. The allowances for shortening and cheek cuts are the same as those used on hip-valley rafters.

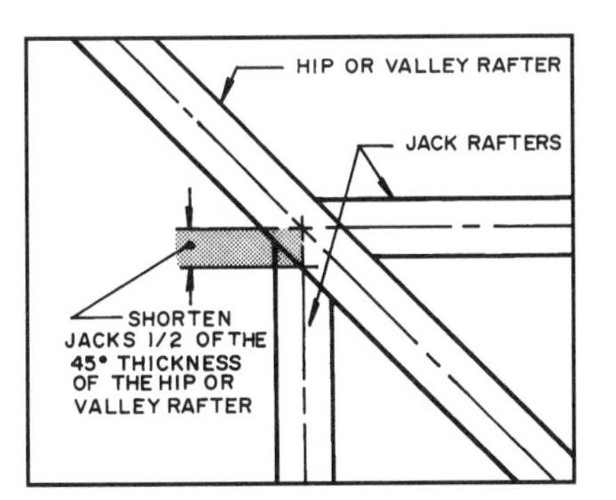

15-22 Valley and hip jack rafters are shortened by one-half the 45-degree thickness of the valley or hip rafter.

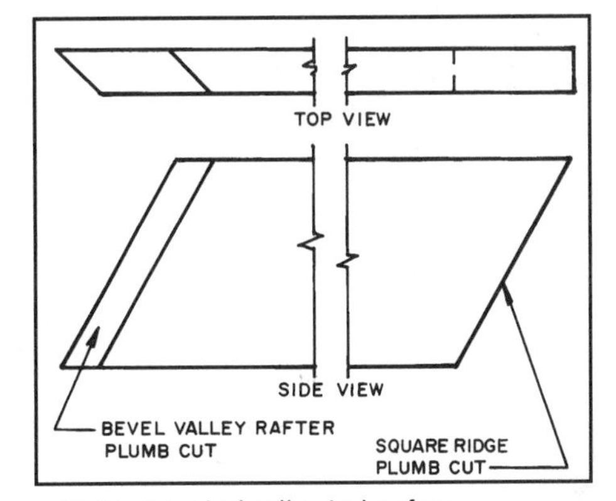

15-23 A typical valley jack rafter.

HIP-VALLEY CRIPPLE JACKS

Hip-valley cripple jacks run between hip and valley rafters as shown earlier in **15-7**. Since these are parallel, all cripple jacks are the same length. The **run** of the cripple jacks is found by measuring from the center of the hip rafter to the center of the valley rafter along the top plate. The length is then found as explained for common rafters **(see 15-24)**. The shortening allowance is one-half the 45-degree thickness of the hip rafter on one end and the 45-degree thickness of the valley rafter on the other end.

The length of the cripple jack is marked on the centerline of the rafter, and then each end is shortened by one-half the 45-degree thickness of the hip and valley rafters. The cheek cuts are located in the same manner as for hip jacks **(15-25)**.

VALLEY CRIPPLE JACK RAFTERS

Valley cripple jack rafters are found on intersecting roofs with unequal spans. The **run** of the valley cripple jack rafter is always **twice** the run of the valley jack rafter it meets at the shortened valley rafter **(see 15-26)**.

The angles for the plumb cuts and side cuts on valley cripple jack rafters are found using the steel square method as described for other jack rafters. A typical layout is shown in **15-27**.

Installing Jack Rafters

First mark the location of the rafters on the plate and hip and valley rafters. Install the rafters in pairs to help keep the hip or valley rafter straight. The first pair is usually installed near the center of the hip or valley rafter. Toenail each jack rafter with three 10d common nails. It is important that the hip and valley rafters remain straight. This may require temporary bracing.

When installing valley jacks, some framers raise them slightly above the valley rafter to help the sheathing meet in a smooth, tight corner **(see 15-28)**.

This amount is the same as the amount of drop if it were a hip rafter. The rise must be enough to allow the sheathing to meet at the line of the center of the valley rafter.

INTERSECTING RIDGEBOARD LENGTHS

A procedure for calculating the length of the ridgeboard for intersecting gable roofs with the same span is shown in **15-29**.

The length includes the run of the larger roof, the length of the building below the intersecting roof, plus any overhang on the gable end minus one-half the thickness of the ridgeboard.

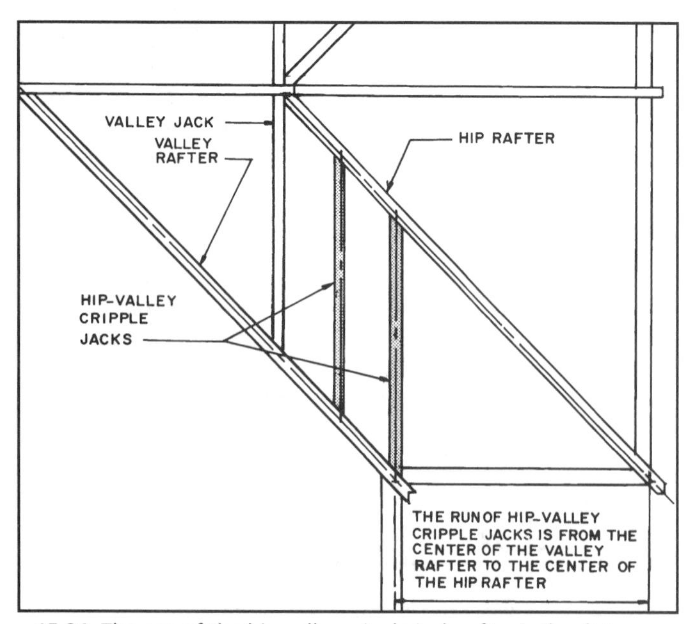

15-24 The run of the hip-valley cripple jack rafter is the distance from the center of the hip rafter to the center of the valley rafter. Shortening at each rafter is necessary.

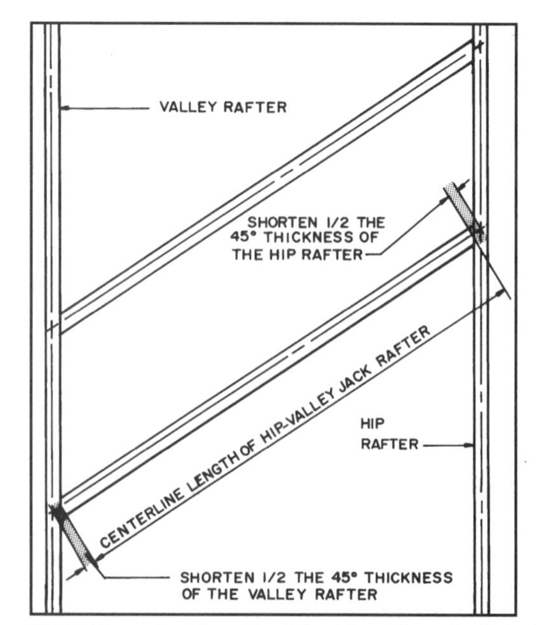

15-25 The actual length of the hip-valley cripple jack requires each end be shortened by one-half the 45-degree thickness of the hip or valley rafter.

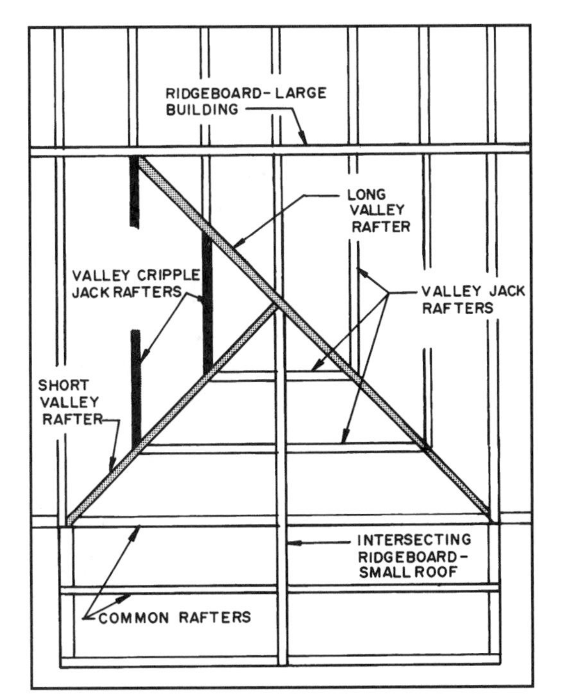

15-26 Valley cripple jack rafters run between two valley rafters.

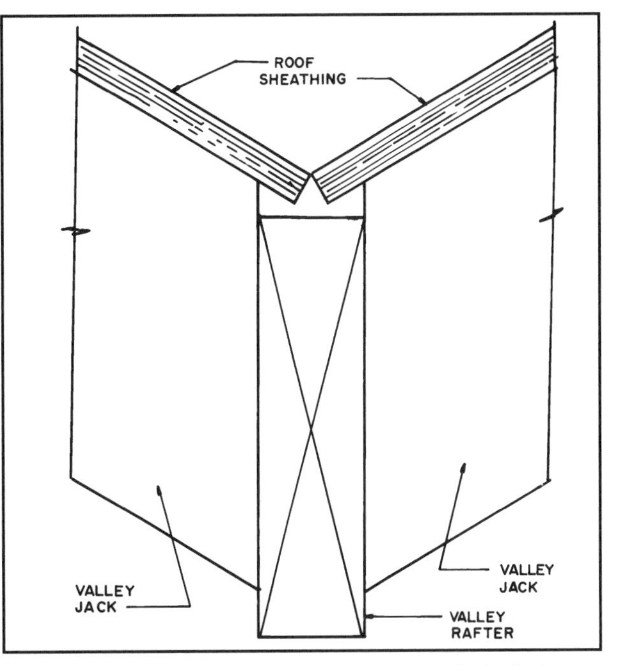

15-28 Valley jack rafters are sometimes installed a little above the valley rafter so the sheathing can form a solid valley.

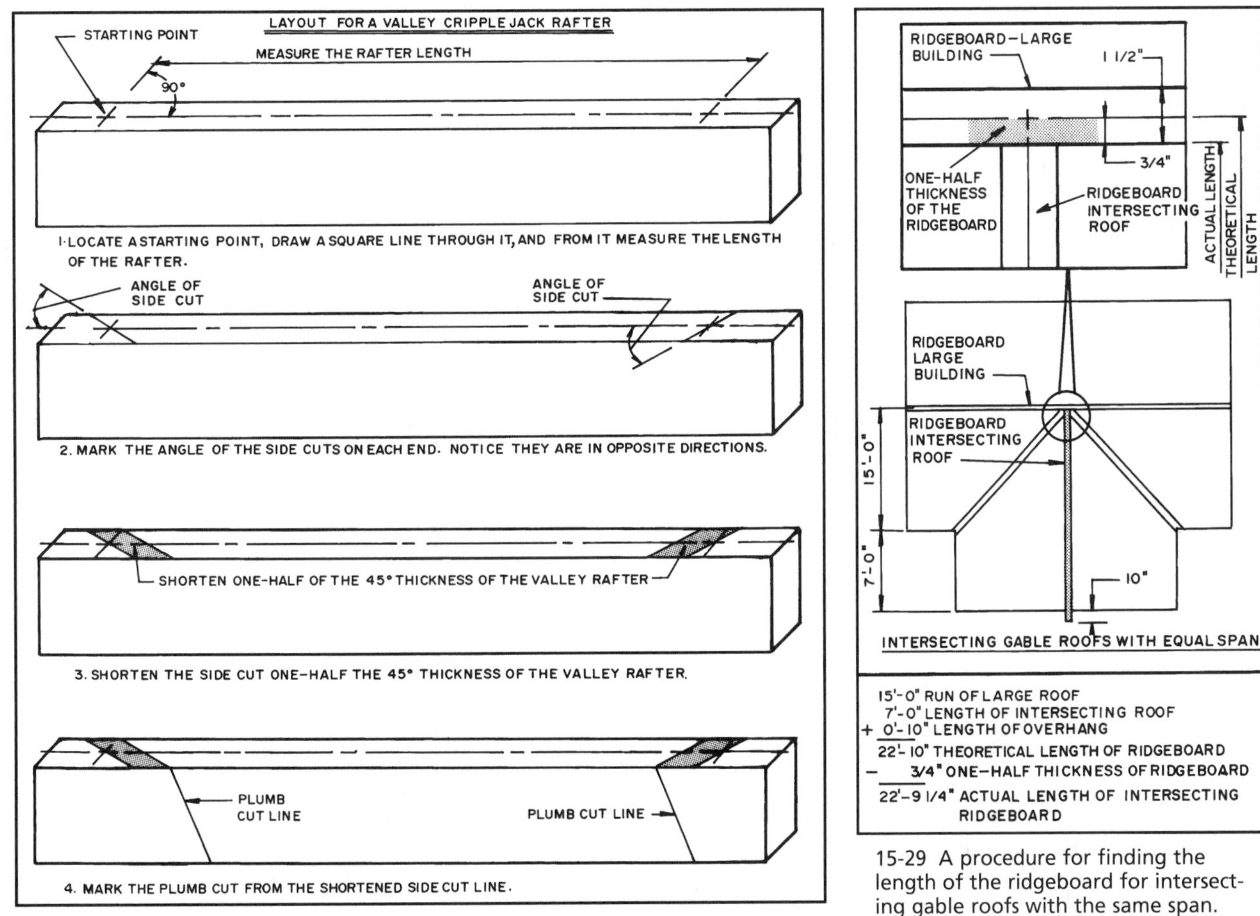

15-27 A typical layout for valley cripple jack rafters.

15-29 A procedure for finding the length of the ridgeboard for intersecting gable roofs with the same span.

The procedure for intersecting roofs with unequal spans is shown in **15-30**.

The procedures for finding the ridgeboard length on an intersecting hip roof with the same span as the larger roof is shown in **15-31**. It involves adding the run of the roof on the large building and the length of the wall of the addition and subtracting the run of the intersecting roof on the addition. If the thicknesses of the ridgeboard on the large roof and the common rafters used are the same, then no correction for them is needed. If they differ in thickness subtract this difference from the ridgeboard length.

A hip roof is shown in **15-32** intersecting a roof that has a larger span. The length of the intersecting roof ridgeboard is found by adding the run of the hip roof to the difference between the length of the wall of the addition and the run of the hip roof. It is also necessary to subtract one-half of the 45-degree thickness of the long valley rafter.

ERECTION PROCEDURE

Framers have various sequences for erecting the frames of intersecting roofs. The important factor is to provide adequate bracing and keep the ridgeboards straight. A typical sequence for a gable roof is shown in **15-33**. When the roofs have equal spans the ridge-

ONE-HALF THE 45° THICKNESS
LONG VALLEY RAFTER
RIDGEBOARD
ACTUAL LENGTH
THEORETICAL LENGTH
LONG VALLEY RAFTER
INTERSECTING ROOF RIDGEBOARD
10'-0"
15'-0"
10"
10'-0"

INTERSECTING GABLE ROOFS WITH UNEQUAL SPANS

10'-0" RUN OF INTERSECTING ROOF
15'-0" LENGTH OF INTERSECTING ROOF
+ 0'-10" OVERHANG
25'-10" THEORETICAL LENGTH OF RIDGEBOARD
− 0'- 1 1/16" ONE-HALF THE 45° THICKNESS OF THE VALLEY RAFTER
25'-8 15/16" ACTUAL LENGTH

15-30 A procedure for finding the length of the ridgeboard for intersecting gable roofs with unequal spans.

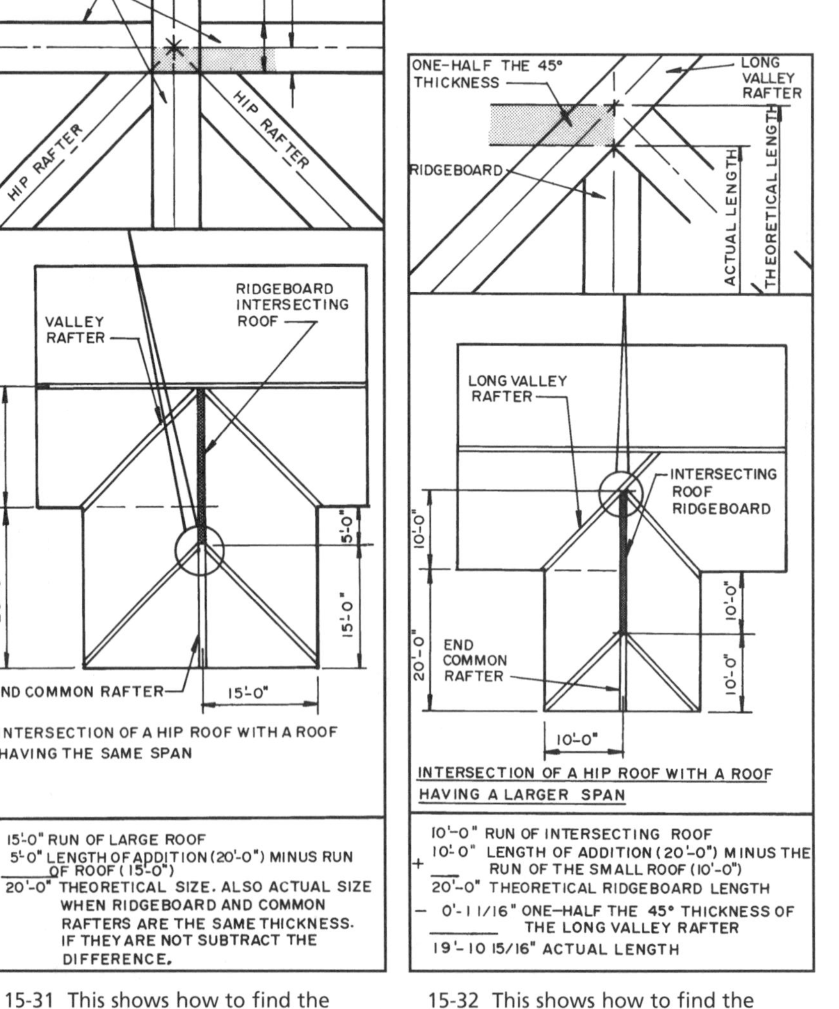

15-31 This shows how to find the length of the ridgeboard for a hip roof that intersects a roof with the same span.

15-32 This shows how to find the length of the ridgeboard for a hip roof that intersects a roof with a larger span.

boards and several pairs of common rafters are installed before the valley rafters and valley jack rafters. When roofs of unequal spans are erected the ridgeboard of the larger roof is erected first, the long valley rafter next, and finally the ridgeboard of the smaller roof. Common rafters and valley jack rafters then fill in the frame **(see 15-34)**.

Remember to install all rafters in pairs so the pressure on the ridgeboards and valley rafters is balanced.

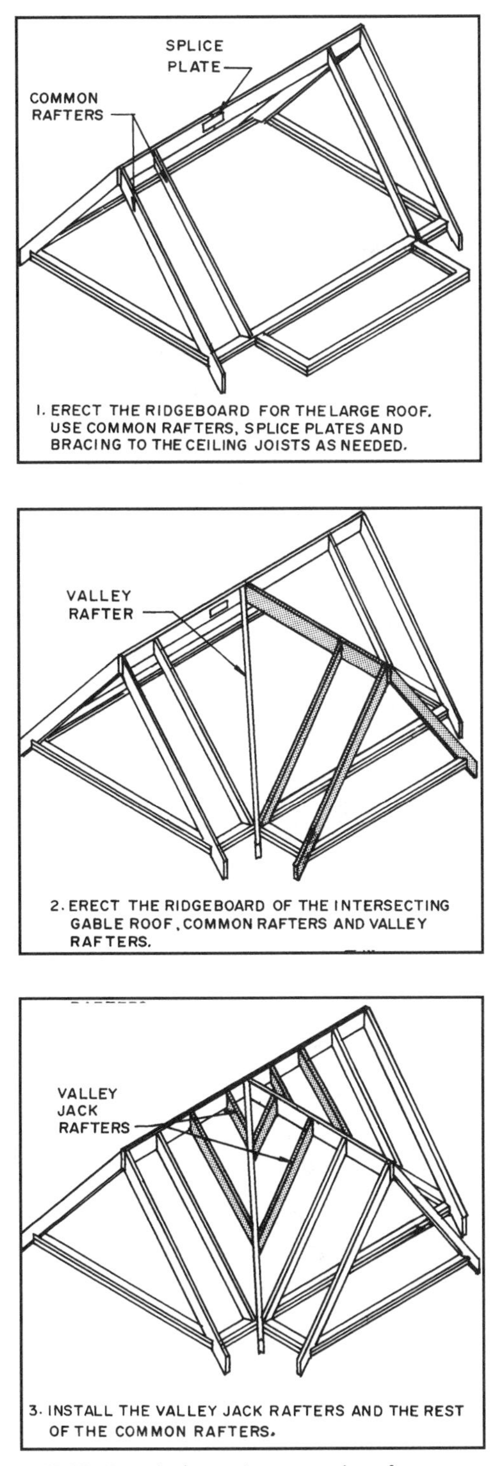

15-33 A typical erection procedure for intersecting gable roofs having the same span. Bracing (not shown) is used as needed to support the ridgeboard and keep it level.

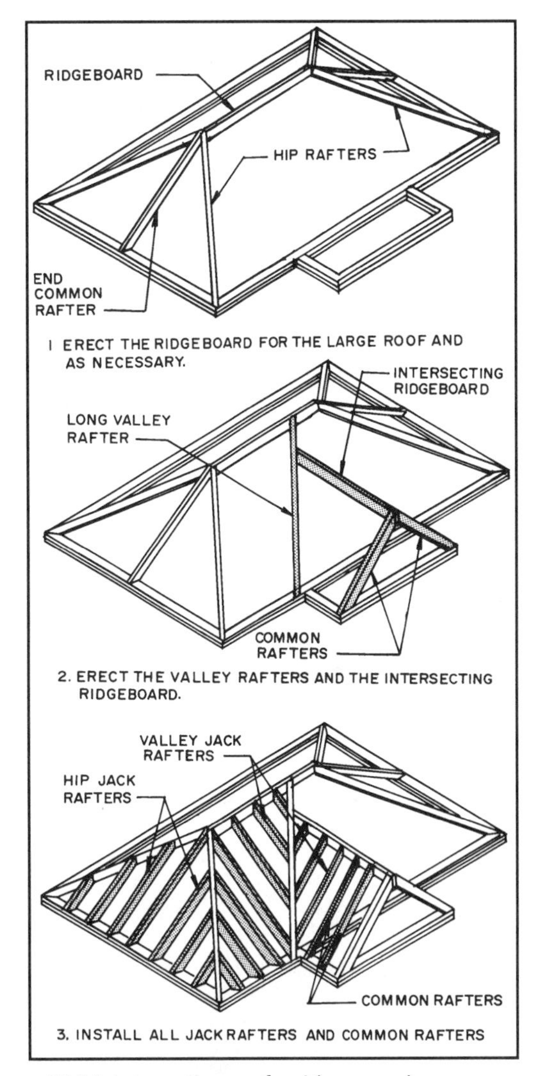

15-34 Intersecting roofs with unequal spans can be framed by erecting the ridgeboard of the large roof first and then erecting the smaller roof frame.

CHAPTER 16

FRAMING FLAT, SHED, GAMBREL & MANSARD ROOFS

While these roof types are not as widely used as the gable roof, each has certain advantages. The flat and shed roofs are easiest to build and have the lowest cost. The gambrel and mansard roofs provide more living space on the second floor than the gable but cost more.

Another factor is that each of these is typical of roofs used on certain architectural styles and must be used to reflect that style.

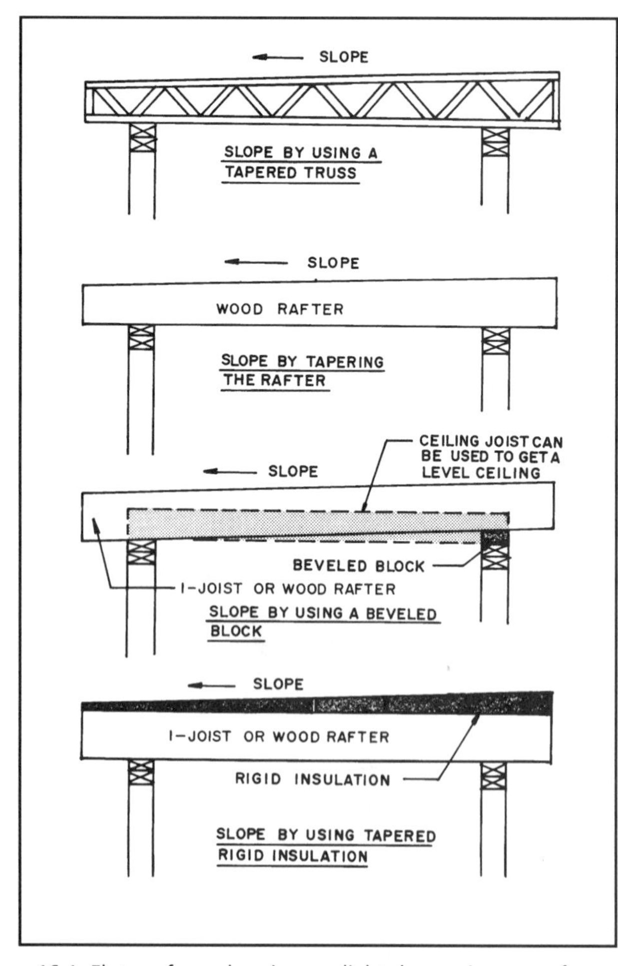

16-1 Flat roofs can be given a slight slope using one of these constructions.

FLAT ROOF CONSTRUCTION

Framing a flat roof is much the same as framing a floor. The rafters carry the roof load and the finished interior ceiling. The rafters must be sized to carry this extra load. Usually a flat roof is built with a slight slope to provide some drainage. Normally a slope of ¼ inch per foot is used. Slope can be produced by using oversized joists and cutting them on a taper, adding a beveled wood strip on top of the high plate, using trusses built with a taper, or building the roof flat and adding tapered rigid foam insulation on the sheathing. Sloped rafters will produce a sloped ceiling. If this is not desired, level ceiling joists can be installed alongside the rafters **(see 16-1)**.

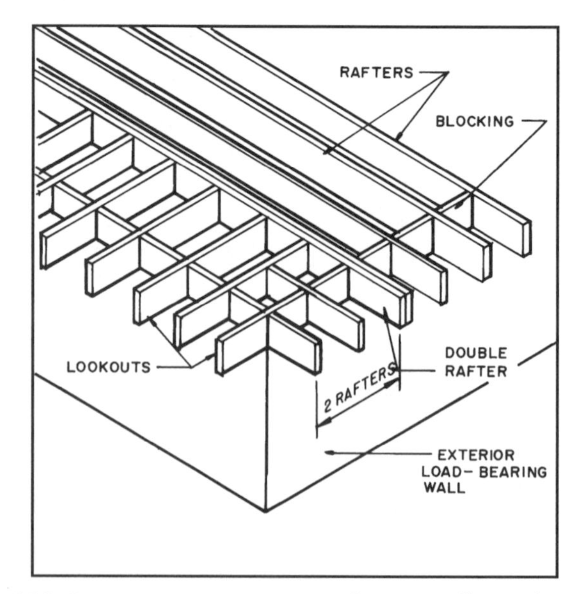

16-2 One way to construct an overhang on a flat roof using lookouts.

Typical framing details are shown in **16-2, 16-3**, and **16-4**. When an overhang is required it can be built as shown by these illustrations. The size of the joists and double joists serving as a header must be engineered to carry the loads.

It should be noted that often the building on which a flat roof is to be constructed is too wide for a single rafter to span the distance between outside walls. This will require that a beam or interior load-bearing wall be used to support the interior ends of the rafters **(see 16-5)**. Trussed rafters and I-joists can span longer distances than solid lumber and could eliminate the need for an interior support.

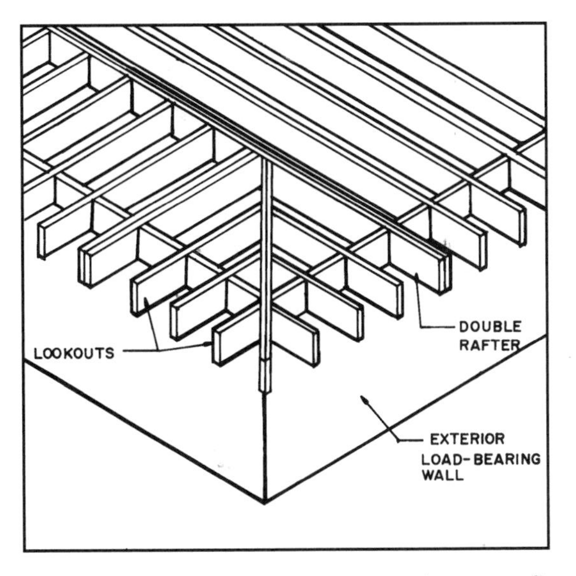

16-3 An alternate way to construct an overhang on a flat roof using lookouts.

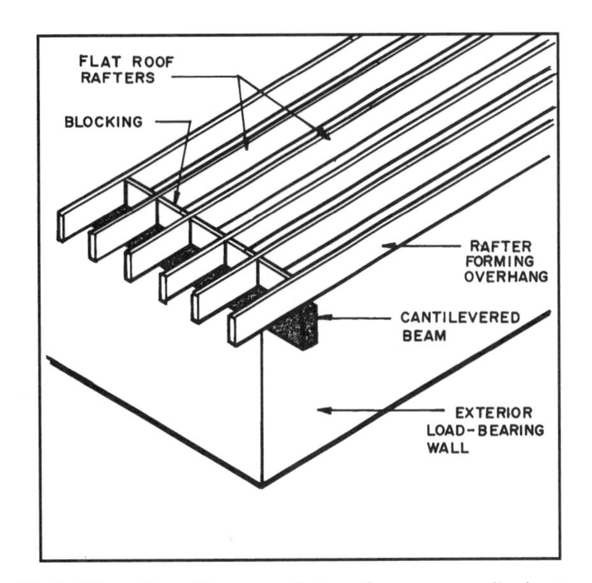

16-4 When the rafters on a flat roof run perpendicular to a wall they can extend beyond the wall. A cantilevered beam can be used to support rafters forming an overhang.

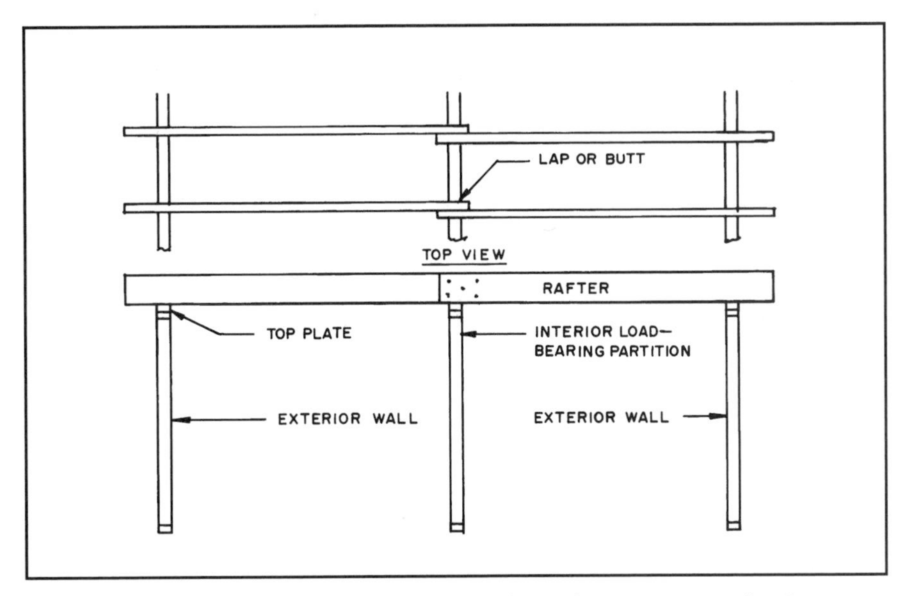

16-5 Flat roofs with long spans will need some form of interior support for the rafters.

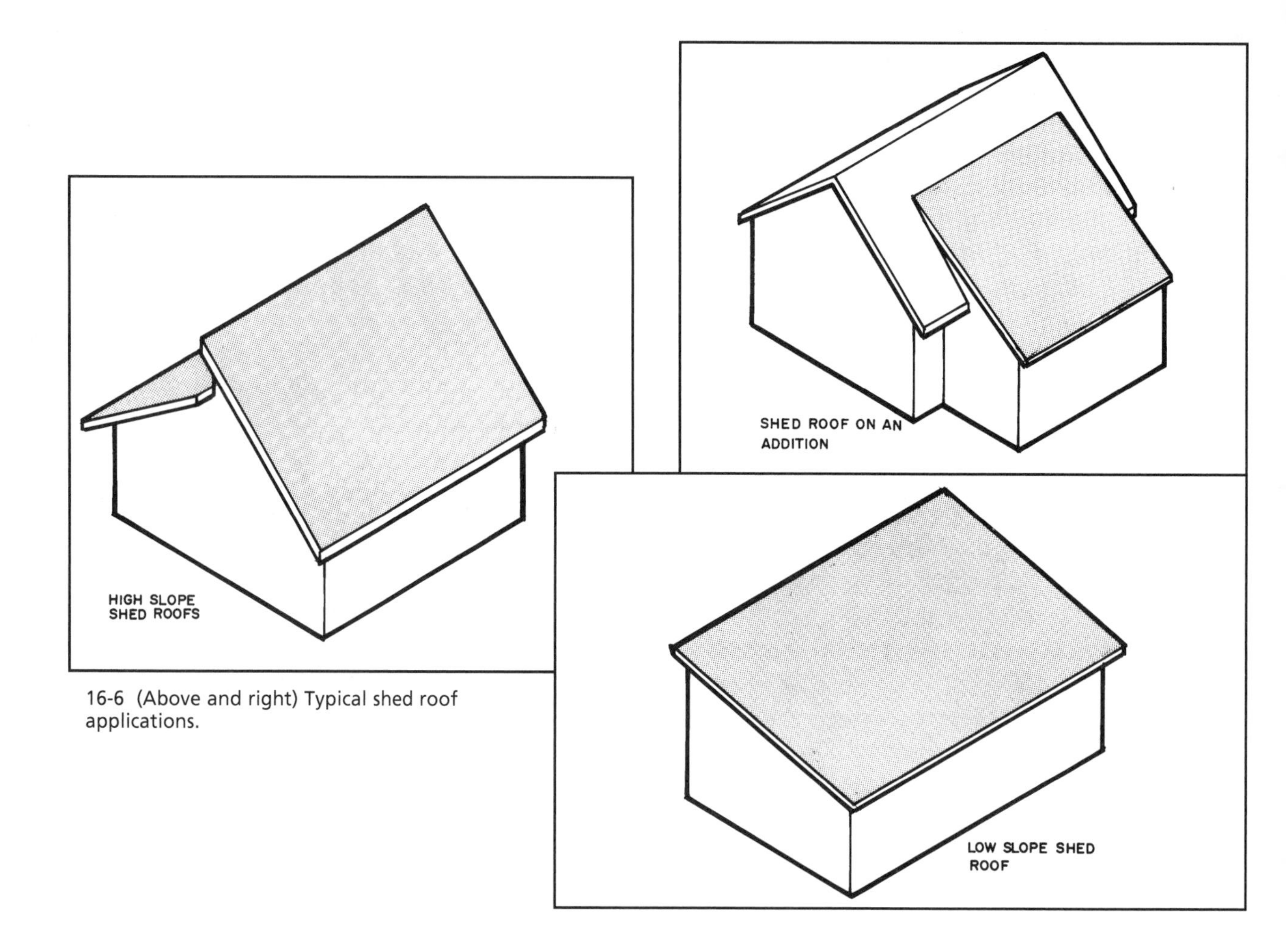

16-6 (Above and right) Typical shed roof applications.

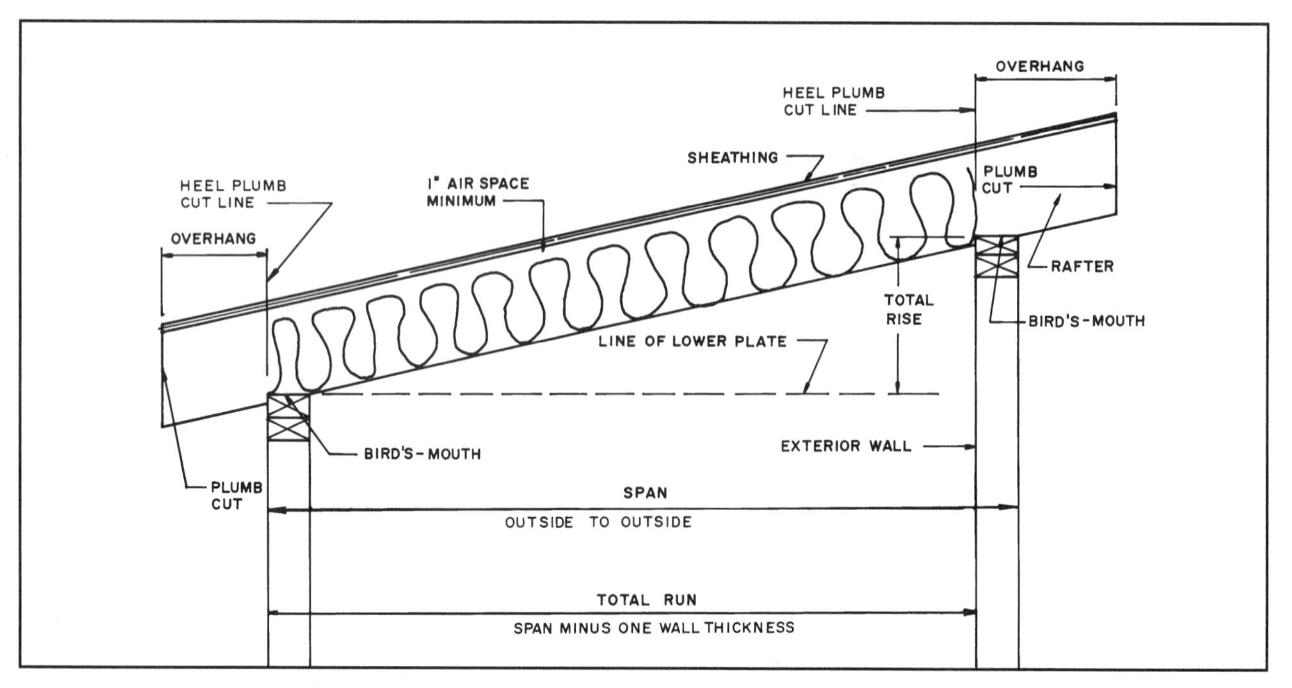

16-7 A typical rafter layout for a shed roof. Notice this single span rafter has a bird's-mouth on each end.

FRAMING A SHED ROOF

A shed roof is actually half a gable and has the slope in one direction. While some designs using shed roofs have very low slope, other styles require steeply sloped roofs **(see 16-6)**. Shed roofs are also used on porches, shed dormers, and room additions.

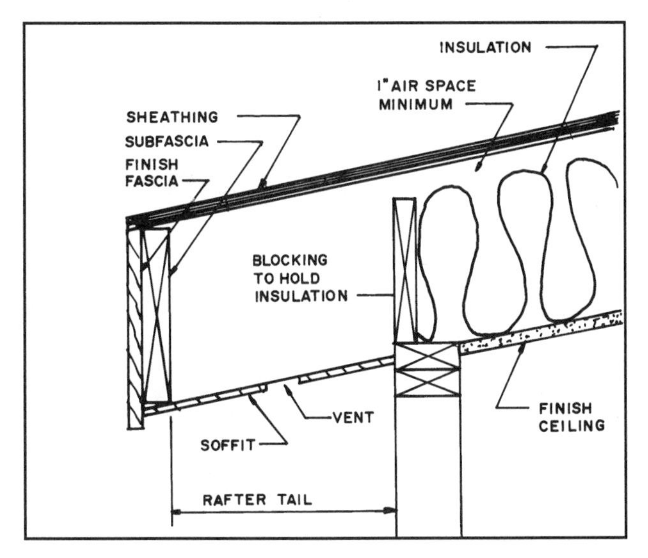

16-8 Typical finish details for the overhang on a shed roof.

The rafters are much like common rafters. On small spans they have a bird's-mouth and plumb cut on each end **(see 16-7)**. Each end has a rafter tail providing overhang.

On wide buildings the rafters cannot span the distance so one end will rest on a beam or interior load-bearing wall as shown for flat roofs in **16-5**. They may lap or butt as shown in Chapter 9 for floor construction.

The rafter length is laid out in the same manner as for a common rafter. The run of the rafter is taken from the inside face of the top plate on the higher wall to the outside face of the plate on the lower wall. When figuring the overhang at the higher wall, include the thickness of the stud wall. The rise is the vertical distance from the top of the lower plate to the top of the higher plate **(refer to 16-7)**. The overhang is constructed in the same manner as that of flat roof construction **(see 16-8)**. The rafters also usually support the finished ceiling.

A typical layout for a common shed roof rafter is shown in **16-9**. The length of the shed rafter is based on the unit rise and the total run of the roof. Review the layout procedures for common rafters for laying out the bird's-mouths.

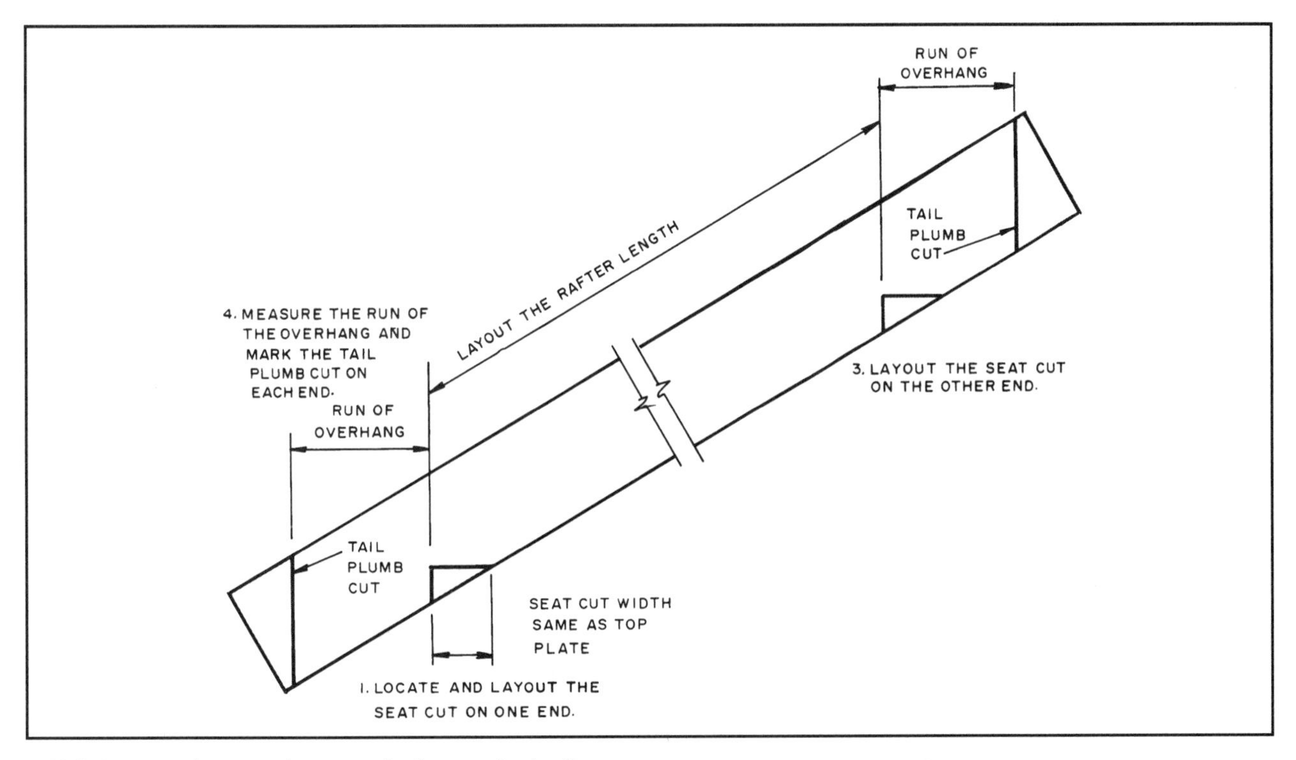

16-9 Suggested steps to lay out a single span shed rafter.

GAMBREL ROOFS

A gambrel roof consists of two gable roofs, each on a different slope. Typically, it is used to provide a large living area on the second level. The roof replaces the wall of a second floor. Normally the upper roof is on a slope of about 20 degrees, or approximately four or five inches of rise to 12 inches of run. The lower roof is very steep and is normally on an angle of about 70 to 75 degrees, or a slope of about 30 to 35 inches of rise to 12 inches of run **(see 16-10)**.

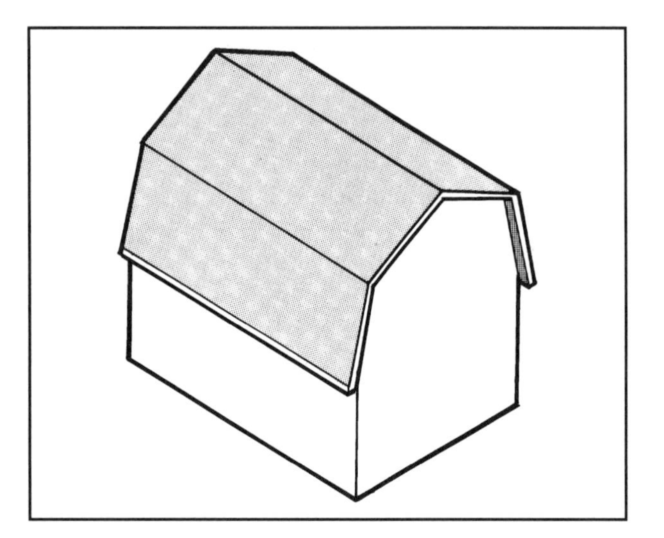

16-10 A gambrel roof.

Each rafter in the upper roof butts the ridgeboard and has a bird's-mouth that sits on a purlin. The top end of the rafter of the lower roof is notched to fit on the purlin and has a bird's-mouth that sits on the top plate of the exterior wall.

A typical layout is shown in **16-11**. The two important factors to consider are the downward thrust of the roof and resistance to wind loads. A good way to provide support for the gambrel roof is to construct a bearing wall below the purlins or double plates. While this provides less usable space than framing without the bearing wall, it does give the rooms vertical walls. The sloped area can be used for closets. The rafters are laid out as explained for common rafters.

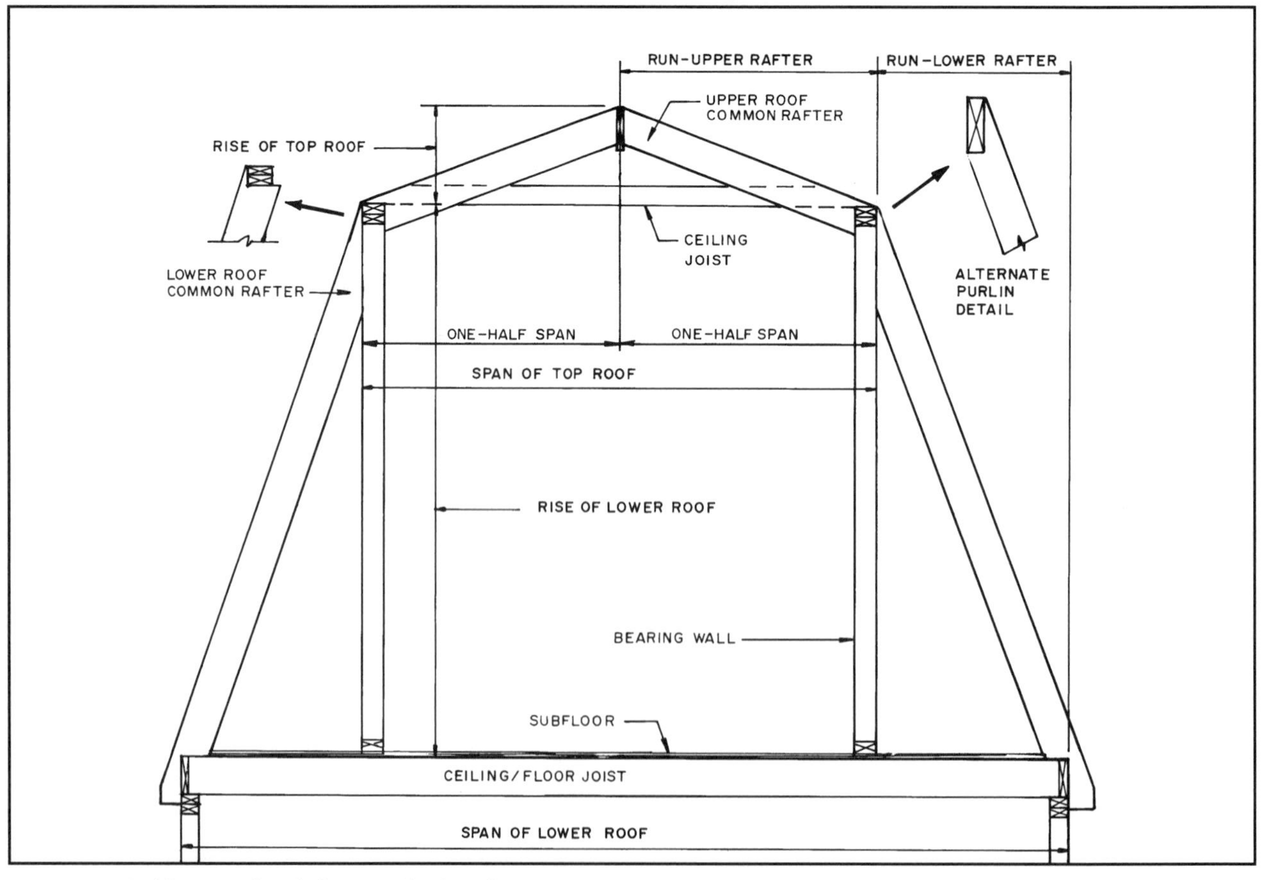

16-11 Typical framing details for a gambrel roof.

If you want to have sloping interior walls it would be advisable to consider using post and beam construction. In **16-12** the framing for the roof and gable-end is shown. Notice that a window opening has been framed in the gable-end wall. Dormers are used to place windows in the sloped gambrel roof as shown in **16-13**. They can also have recessed **(see 16-14)** or roof windows.

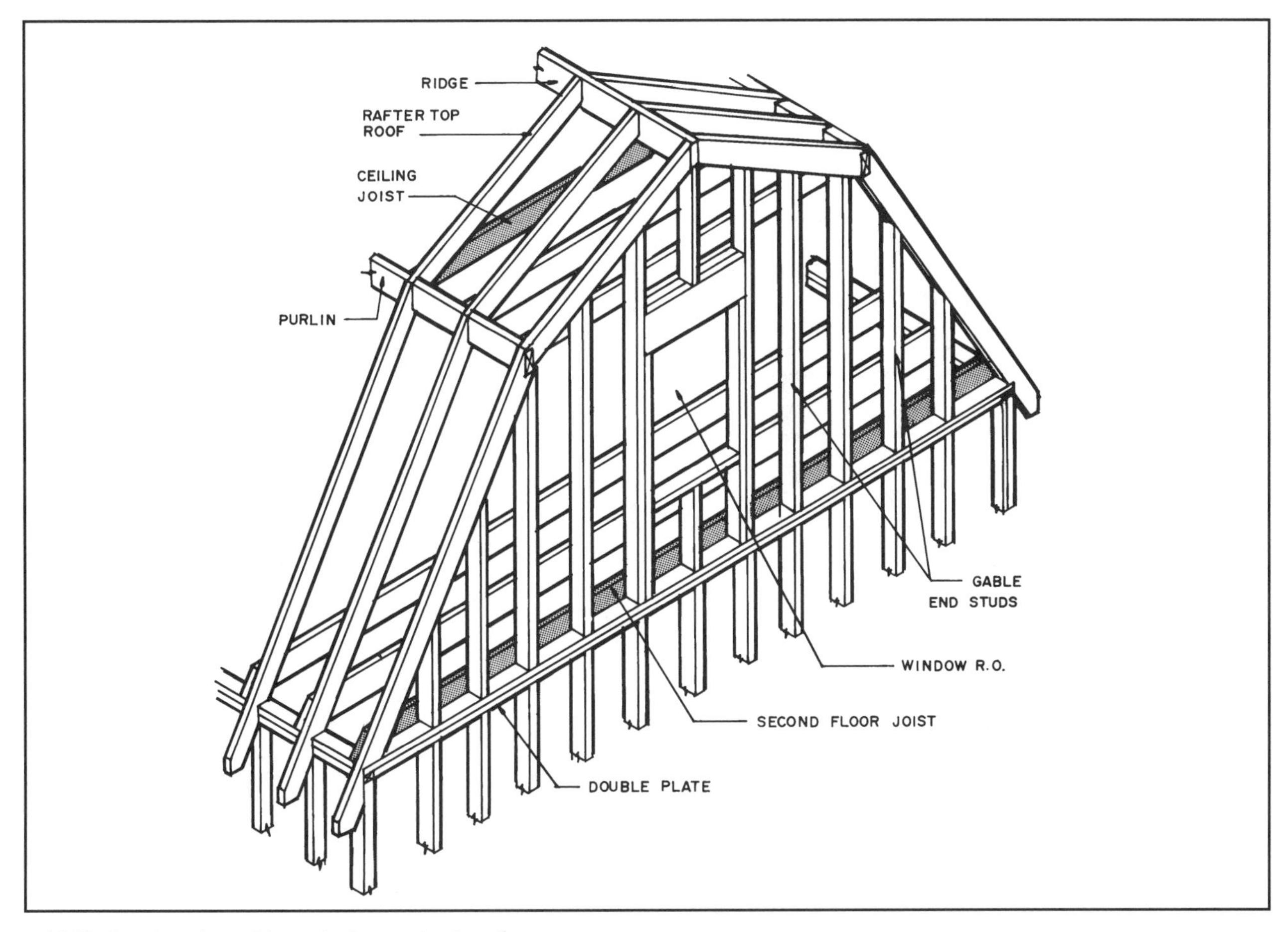

16-12 Framing the gable-end of a gambrel roof.

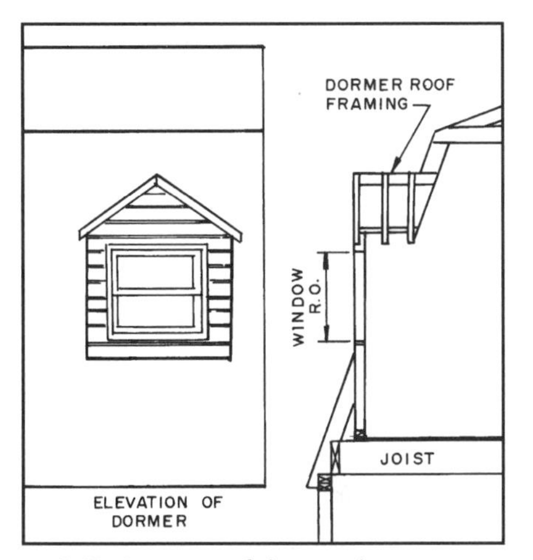

16-13 One type of dormer that may be used on a gambrel or mansard roof.

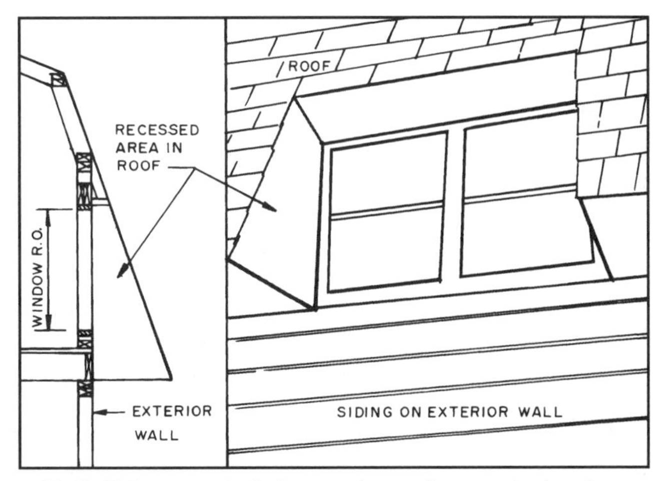

16-14 This recessed window can be used on gambrel and mansard roofs.

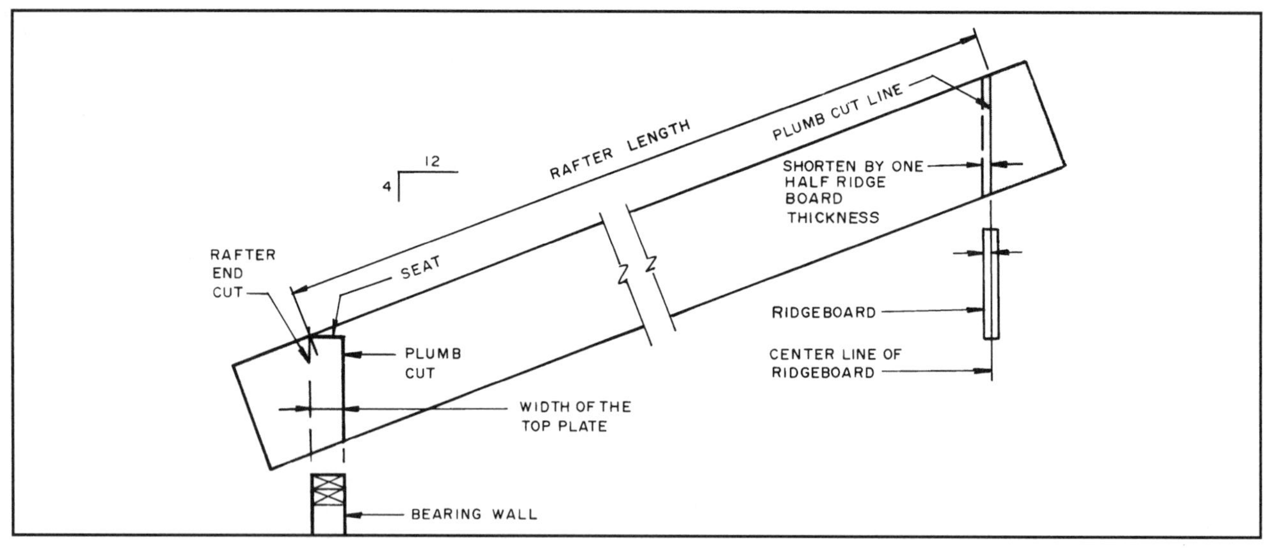

16-15 Layout for the upper rafter on a gambrel roof.

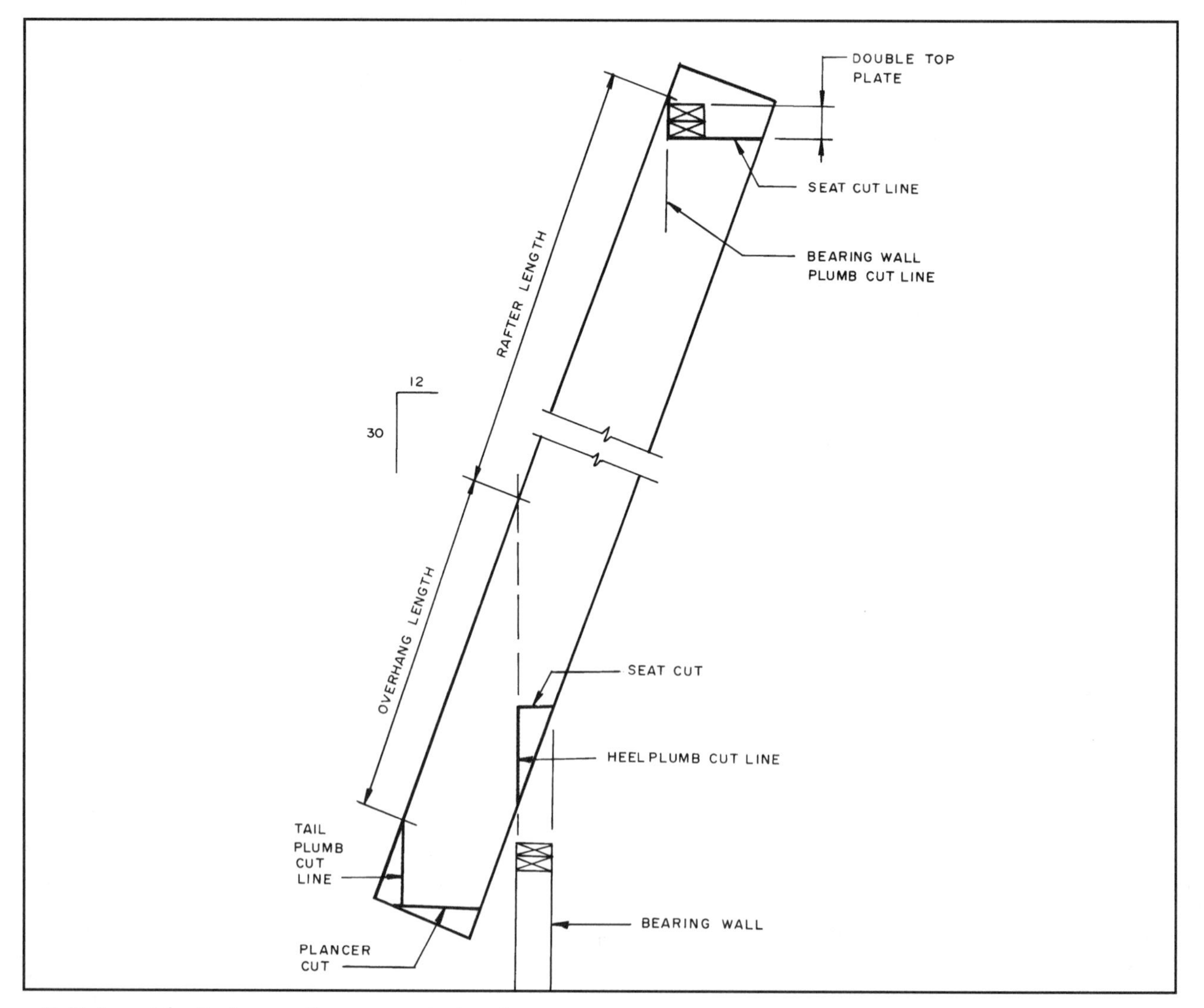

16-16 Layout for the lower rafter on a gambrel roof.

Laying Out Gambrel Rafters

The lengths of the upper and lower common rafters are based on the run of each as shown in **16-11**. A layout of the upper roof common rafter is in **16-15**. It uses the procedure described for laying out common rafters in Chapter 13. A layout for a lower roof common rafter is in **16-16**. Notice the steepness of the heel plumb cut which is due to the almost vertical position of the rafter.

Erection Procedures

Typically the load-bearing partitions are erected and braced to the subfloor before the roof construction is started. Then the ridgeboard is set in place with temporary supports and the upper roof common rafters and ceiling joists are installed in the same manner as described for gable roofs. Finally the lower roof rafters which brace the load-bearing partitions are installed **(see 16-17)**.

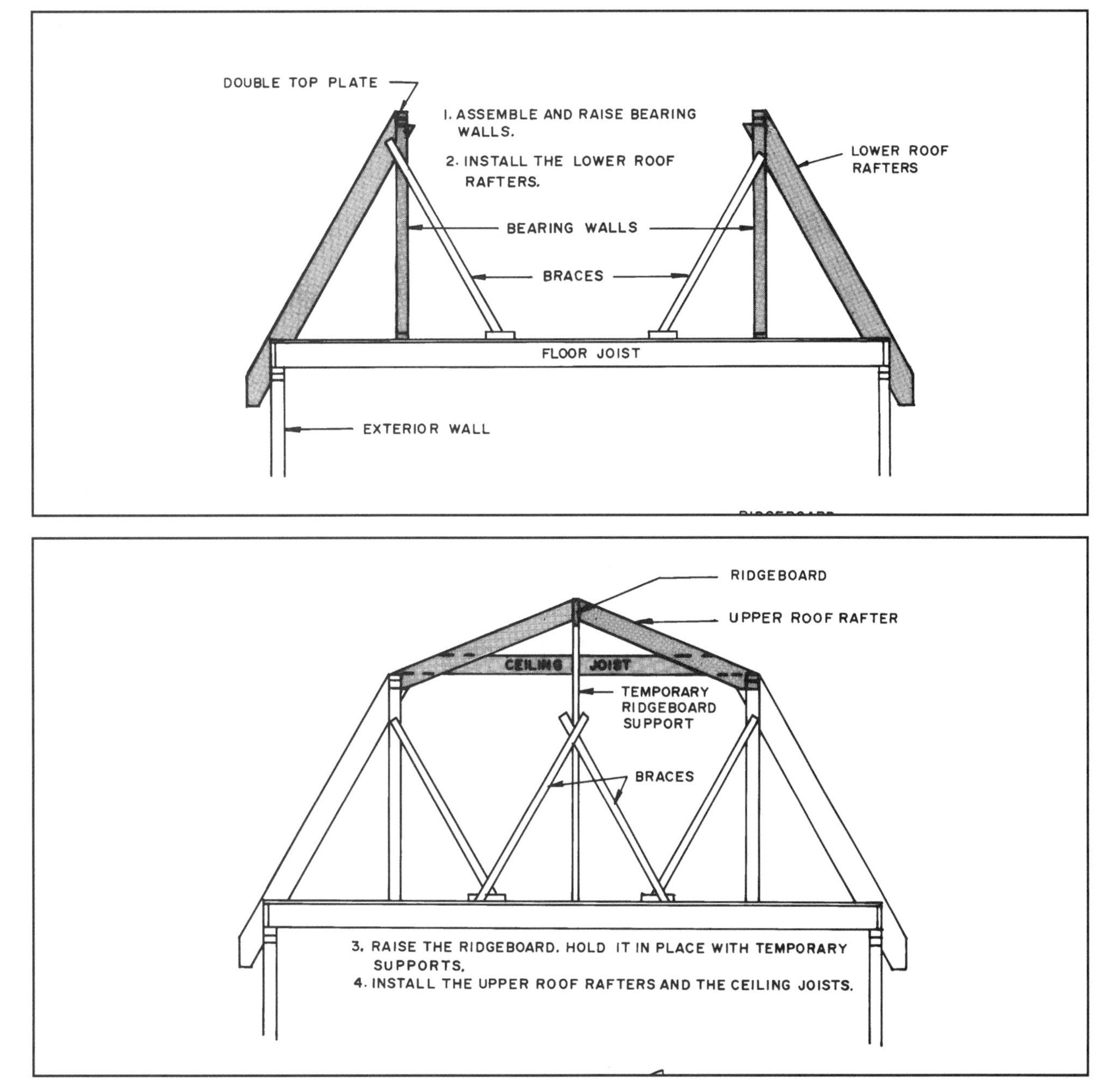

16-17 Suggested steps to erect a gambrel roof.

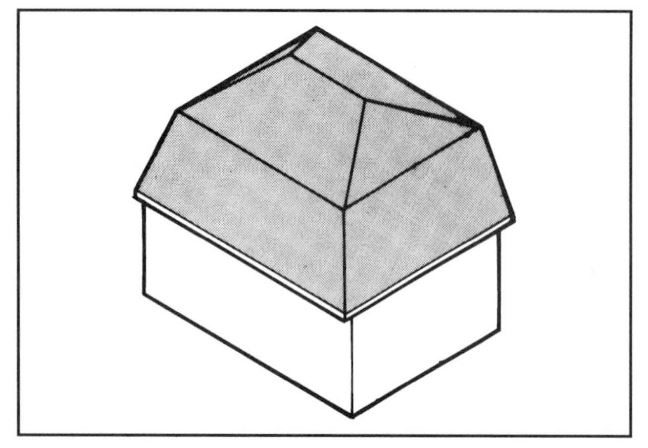

16-18 A mansard roof.

MANSARD ROOFS

A mansard roof provides full ceiling height. It is made of two roof surfaces. The top roof is a hip roof. The lower roof is very steep and serves to protect the interior wall **(see 16-18)**. It is on all sides of the building, eliminating the gable-end.

Typical construction details are shown in **16-19**. In one design the floor joists are cantilevered beyond the exterior wall. The rafters on the lower roof run from the edge of the floor to rafters on the top roof. The top hip roof is supported by the framing of the second floor wall. The hip rafters are laid out as described for hip roof construction **(see 16-20)**.

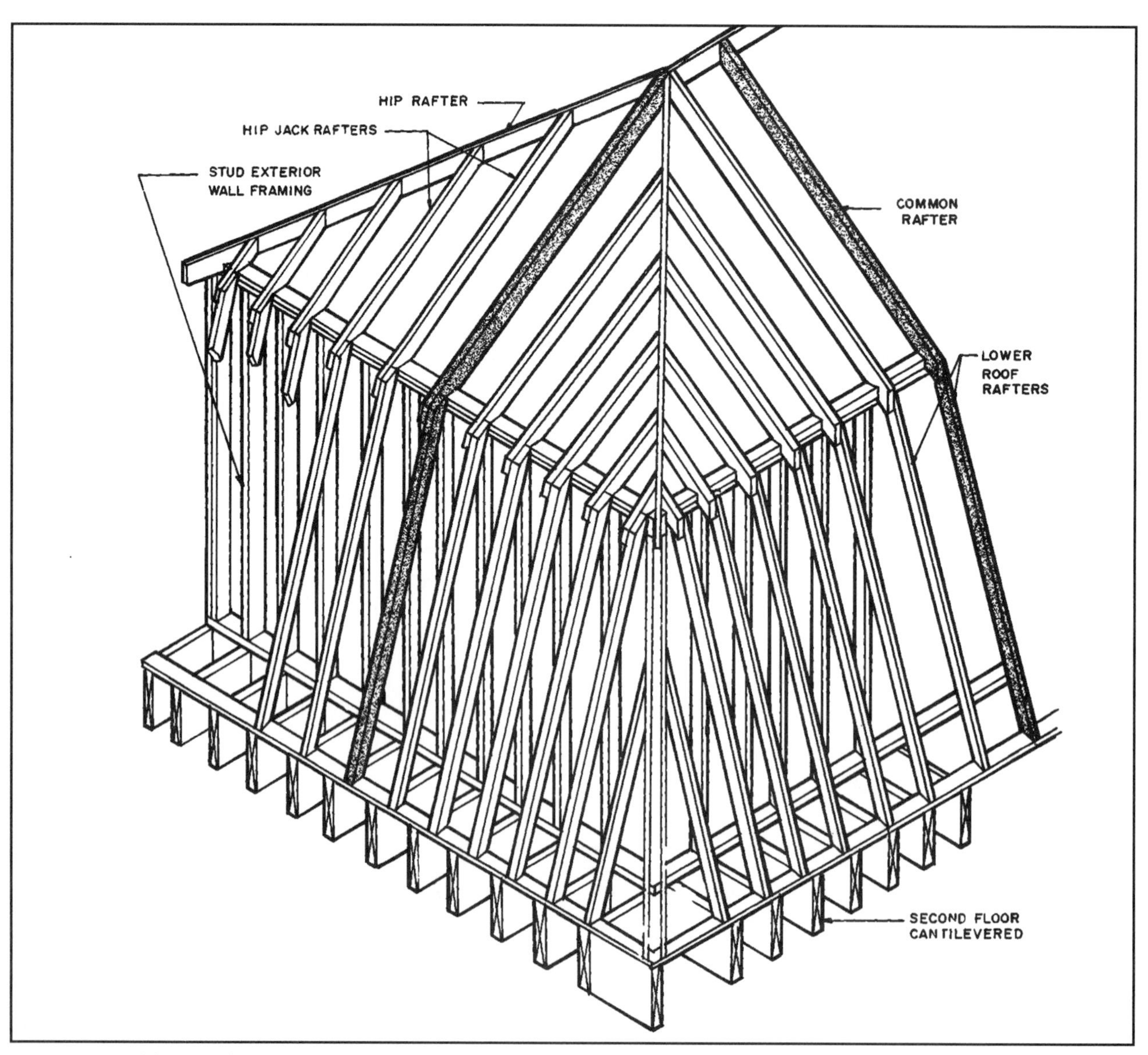

16-19 Typical framing for a mansard roof.

Another method for framing a mansard roof has the lower roof rafters sitting on the top plate of the first floor exterior wall as shown in **16-21**.

Windows are installed using dormers, recessed windows, or roof windows, as described for gambrel roofs **(see 16-13 and 16-14)**.

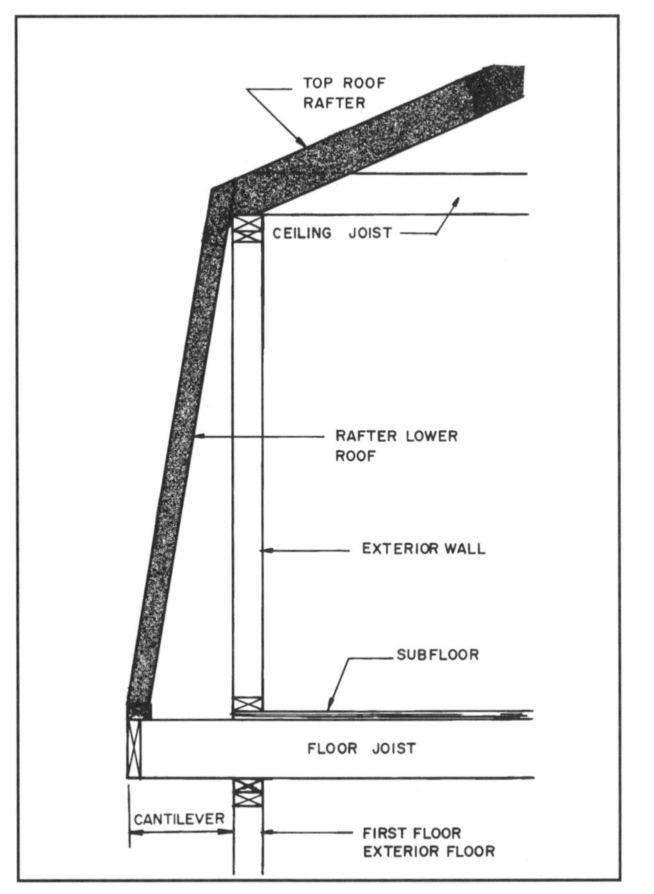

16-20 A section through a mansard roof when the floor joists are cantilevered, producing an overhang.

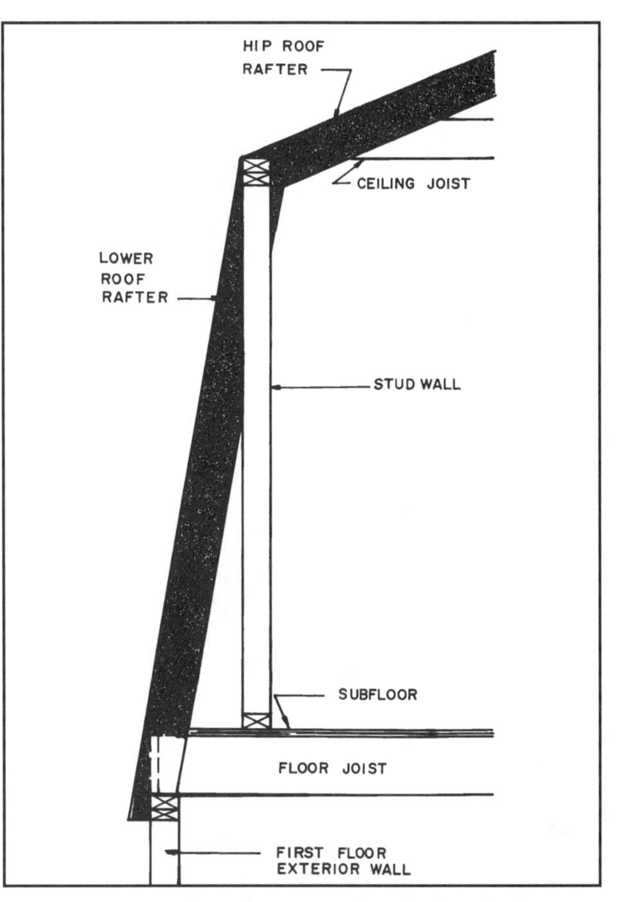

16-21 A section through a mansard roof when the lower roof rafter rests on the exterior wall.

CHAPTER 17

Dormers

A dormer is a framed structure projecting from a sloping roof to house a window or ventilating louver. This admits light and ventilation into the area under the roof (the attic). Dormers make it possible to have satisfactory living space in the attic area.

The style of dormer depends on the style of the house. The common types of dormer are the shed, gable, and hip **(see 17-1).** Shed dormers can be made very long and run almost the entire length of the house with roof overhang or no overhang.

SHED DORMERS

The roof of shed dormers may run to the ridge (see **17-2)** or fall short of the ridge **(see 17-3).**

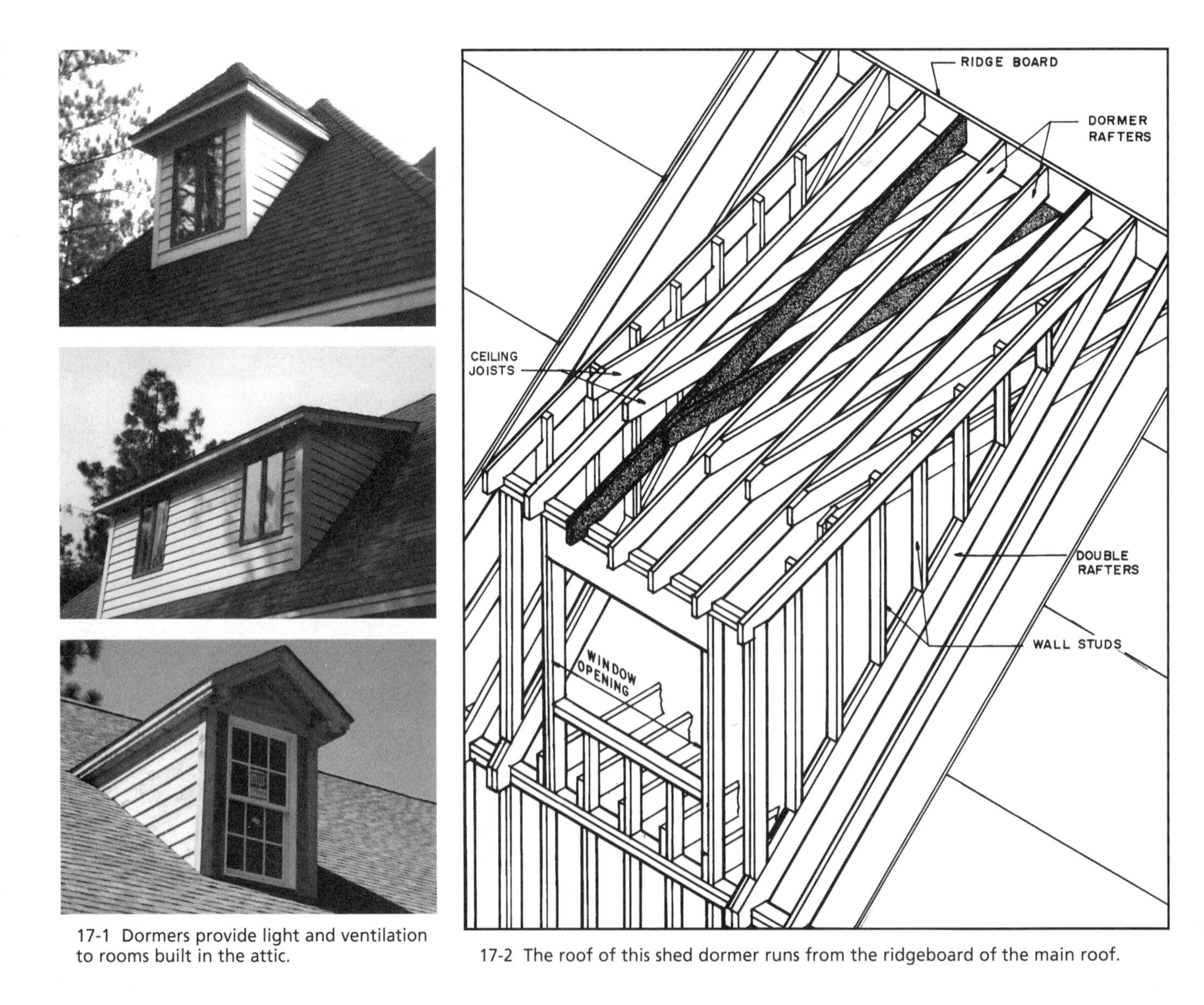

17-1 Dormers provide light and ventilation to rooms built in the attic.

17-2 The roof of this shed dormer runs from the ridgeboard of the main roof.

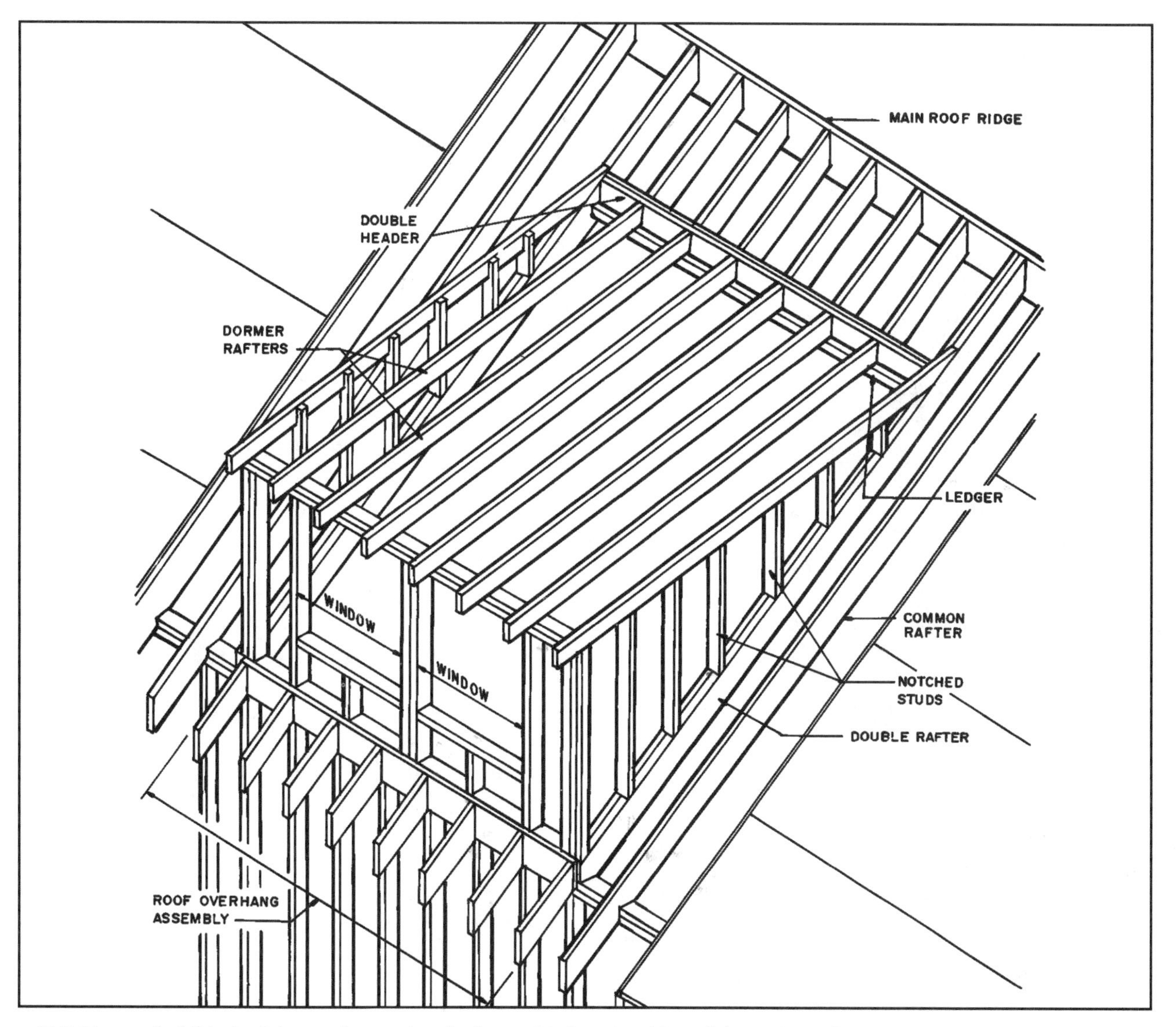

17-3 The roof of this shed dormer butts a header located below the ridge of the main roof.

The rafters are laid out and cut as described for common rafters in Chapter 13. The side walls are framed the same as for the gable-ends of gable roofs. The common rafter on which the side wall rests must be doubled. Ceiling joists are face-nailed to the dormer roof rafters and extend across the attic to the common rafters on the other side **(see 17-4)**.

The end wall usually rests on the double top plate of the exterior wall below. If this is not the case, the attic floor joists must be sized large enough to carry this extra load or doubled in the area to carry the roof load. The window openings are framed in the normal manner.

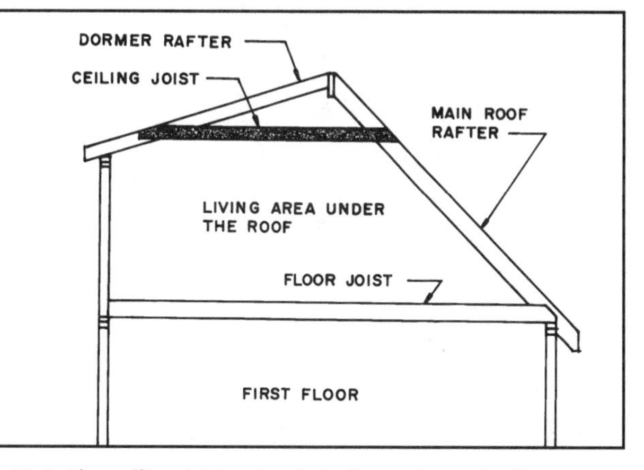

17-4 The ceiling joist extends to the main roof rafter on the other side of the roof. It also serves as a collar beam tying the roof together.

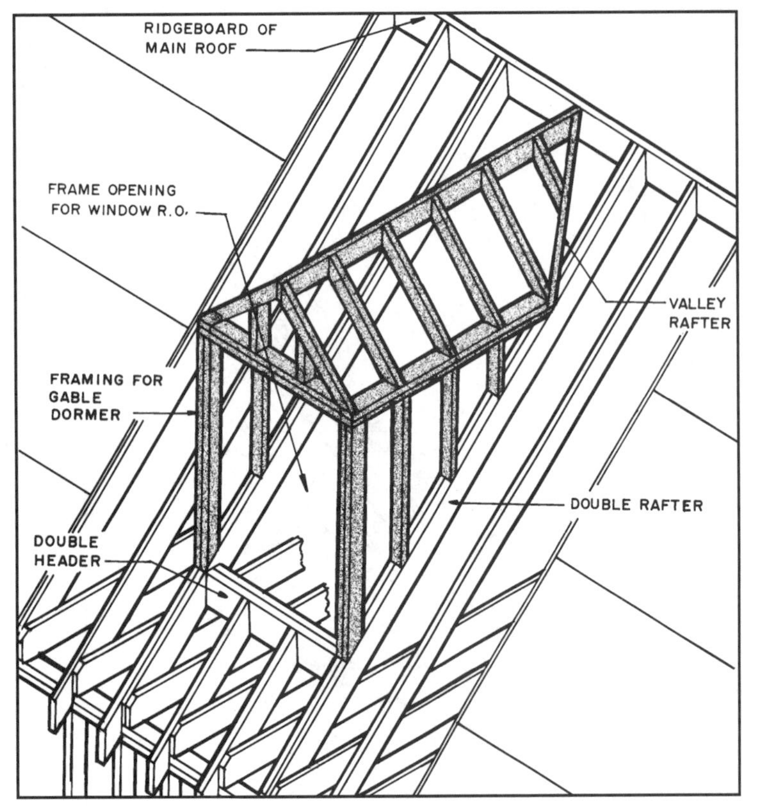

17-5 The ridge of the gable roof dormer extends to the ridgeboard of the main roof.

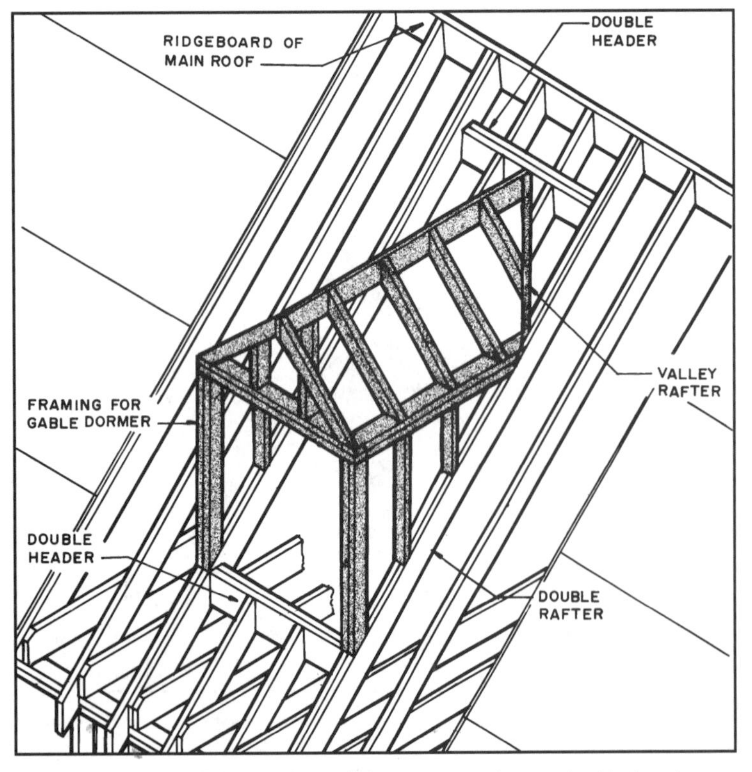

17-6 The ridge of this gable roof dormer extends to a double header located below the ridgeboard of the main roof.

GABLE DORMERS

Gable dormers are smaller than shed dormers and are usually designed to contain one window, although more could be used if the main house roof has sufficient rise to accept the higher dormer roof needed by the wider dormer **(refer to 17-1)**. The dormer roof is built as described in Chapter 13 for gable roofs. It has a ridge which may run to the ridge of the main roof **(see 17-5)** or fall below the ridge **(see 17-6)**. Notice that the gable dormer roof has a ridgeboard and common valley jack, and valley rafters. Often the bird's-mouth is not cut. If the dormer ridge does not extend to the ridge of the main roof, a double header must be installed.

A typical framing plan for the header, valley rafter, and common rafter is shown in **17-7**. Notice that an allowance must be made for the double side rafters when shortening the valley rafter. Usually a double top plate is used on the front wall, because it must be framed for a window.

Dormers are also frequently used to provide an attractive way to vent an attic. While almost any dormer design can be used, the one in **17-8** is a good solution. The opening can be designed to fit a standard louver, or custom louvers can be made.

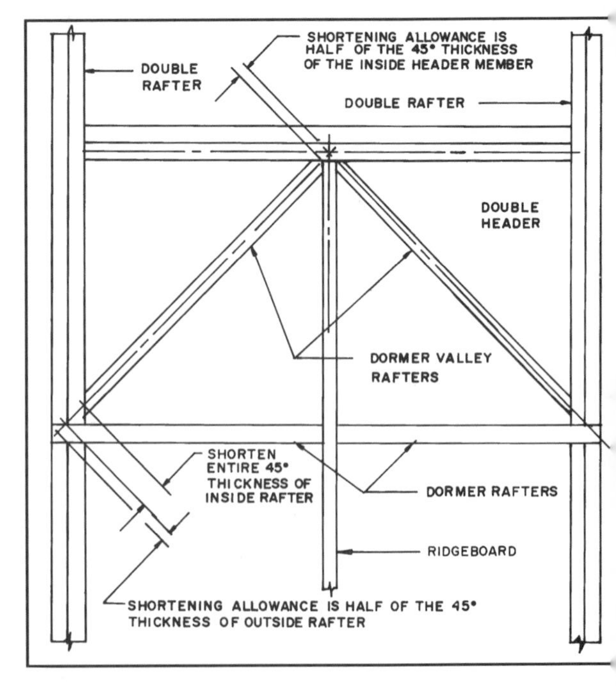

17-7 Framing details for the ridge and valley rafters for a gable or hip roof dormer.

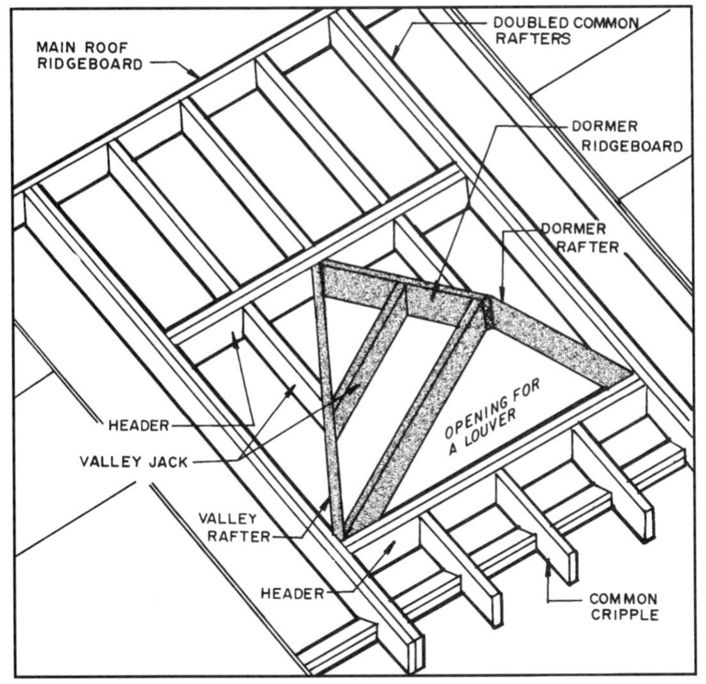

17-8 Dormers can be used to ventilate the attic by supporting some form of a louver.

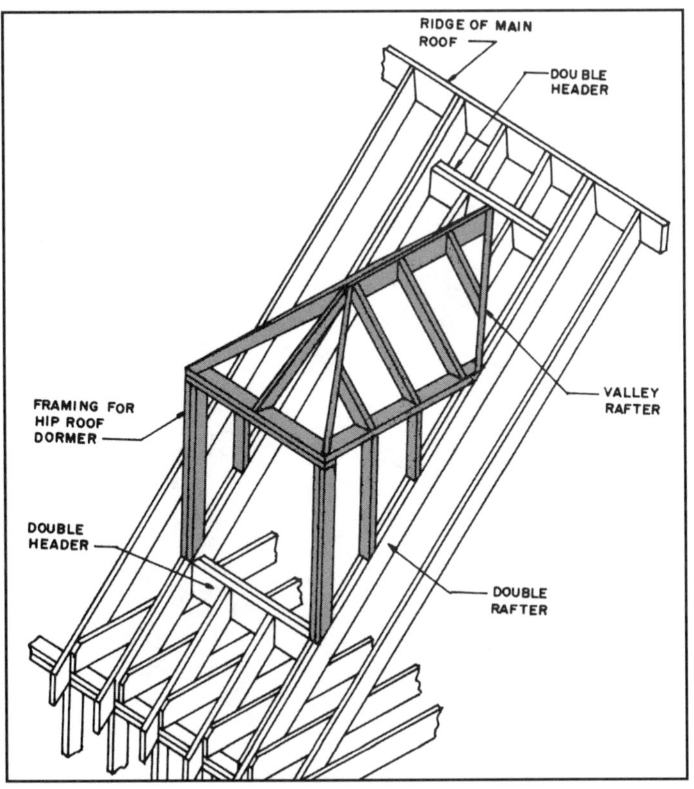

17-9 Framing details for a dormer with a hip roof.

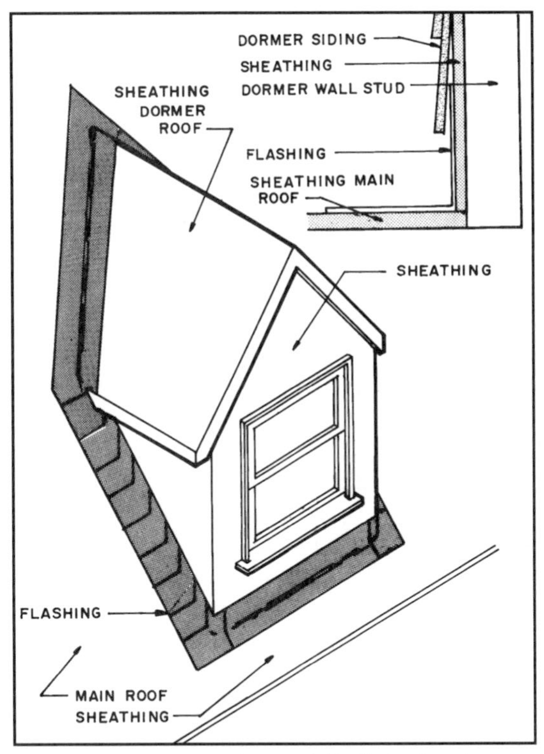

17-10 Dormers must be carefully flashed as the roof shingles are installed.

HIP DORMERS

Hip dormers are framed in the same manner as gable dormers, just described, except that the end of the roof has a hipped surface on the same slope as the hipped surface of the main roof **(see 17-9)**. Details for laying out the hip rafters are in Chapter 14.

OTHER APPLICATIONS

Dormers are also used on gambrel and mansard roofs. Gambrel and mansard roofs are illustrated in Chapter 16.

FINISHING THE DORMER

The dormer is sheathed in the same manner as the exterior of other framed walls. A major factor is the installation of flashing, which is usually done by the roofers. Flashing details are shown in **17-10**. After the flashing is in place the framers nail the siding so that it overlaps the flashing.

CHAPTER 18

TRUSSED ROOF CONSTRUCTION

A **truss** is a structural unit made up of an assembly of members usually in a triangular arrangement. It forms a rigid framework that will carry a load over a distance between two supports. The triangular shapes enable the various parts of the truss to resist the tension and compression forces produced by the load **(see 18-1)**.

A **roof truss** is a structural unit designed to frame a roof **(see 18-2)** and to support the roof material, interior ceiling, insulation, and forces caused by snow, rain, and wind. Roof trusses are supported by the exterior walls and span the width of the building.

ADVANTAGES

Roof trusses save on-site costs because they are rapidly erected, and the building is made weathertight in a minimum of time. Since they span the width of the building, no interior partitions are required for support. This allows complete freedom for room arrangement. Factory-made trusses are carefully engineered for the job and built under controlled conditions in a plant. The quality is more consistent than on-site cut-and-assembled, conventionally framed roofs. The members of a truss are usually considerably smaller than those used in conventional joist and rafter framing, reducing weight and cost.

DISADVANTAGES

One distinct disadvantage of roof trusses is that they have a series of supporting members that limit the use of attic space. Large trusses require that a crane be used to lift them in place. Since they are generally spaced 24 inches O.C., roof sheathing and interior ceiling finish material must be thicker than for conventional 16-inch O.C. spaced members.

18-1 The scissors-type roof truss is designed to carry the roof loads from one exterior wall to the other and produce a cathedral ceiling in the room below.

TYPES OF TRUSSES

There are many types of trusses available. Each type has specific design considerations and advantages and disadvantages. Many special-purpose trusses are designed by engineers working for truss manufacturers. Some of the frequently used trusses are shown in **18-3** on the following page.

The **king-post truss** is the simplest form used in light-frame construction. It consists of upper and lower chords and a center vertical post called a king post. Its spans are less than those of a W-truss when the same size members are used.

The **W- or Fink truss** is possibly the most commonly used truss for light construction. Since it uses more web members than the king-post truss, it can be made with lower-grade lumber and span greater distances for the same member size.

The **scissors truss** is designed to provide a sloping, sometimes called cathedral, ceiling. It provides solid roof construction and space for considerable ceiling insulation and for running heat ducts and electrical wiring **(refer to 18-1)**.

The **Howe truss** is often called an **M-truss** because of the design of the web members. It is

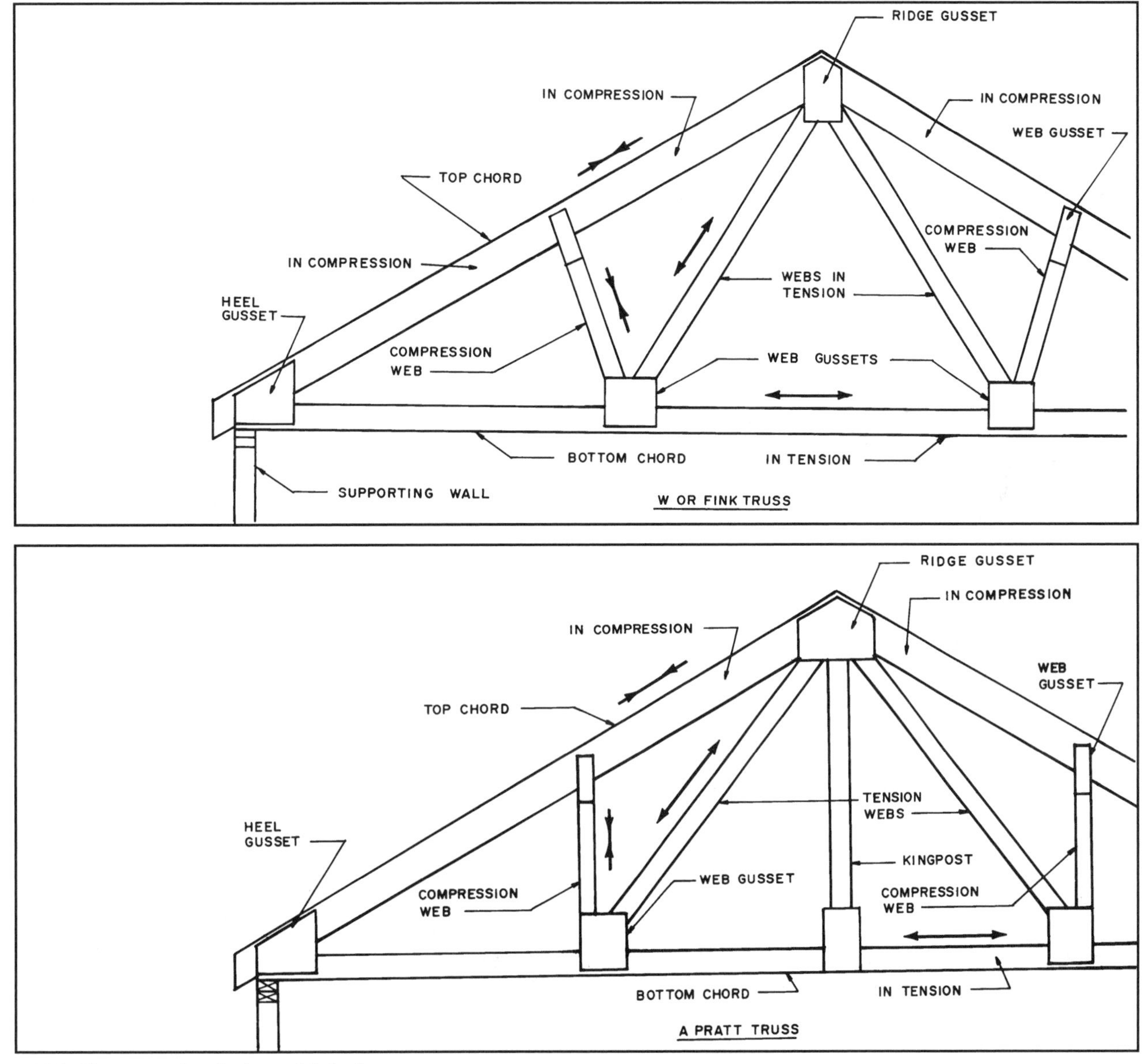

18-2 The parts of typical wood light frame trusses.

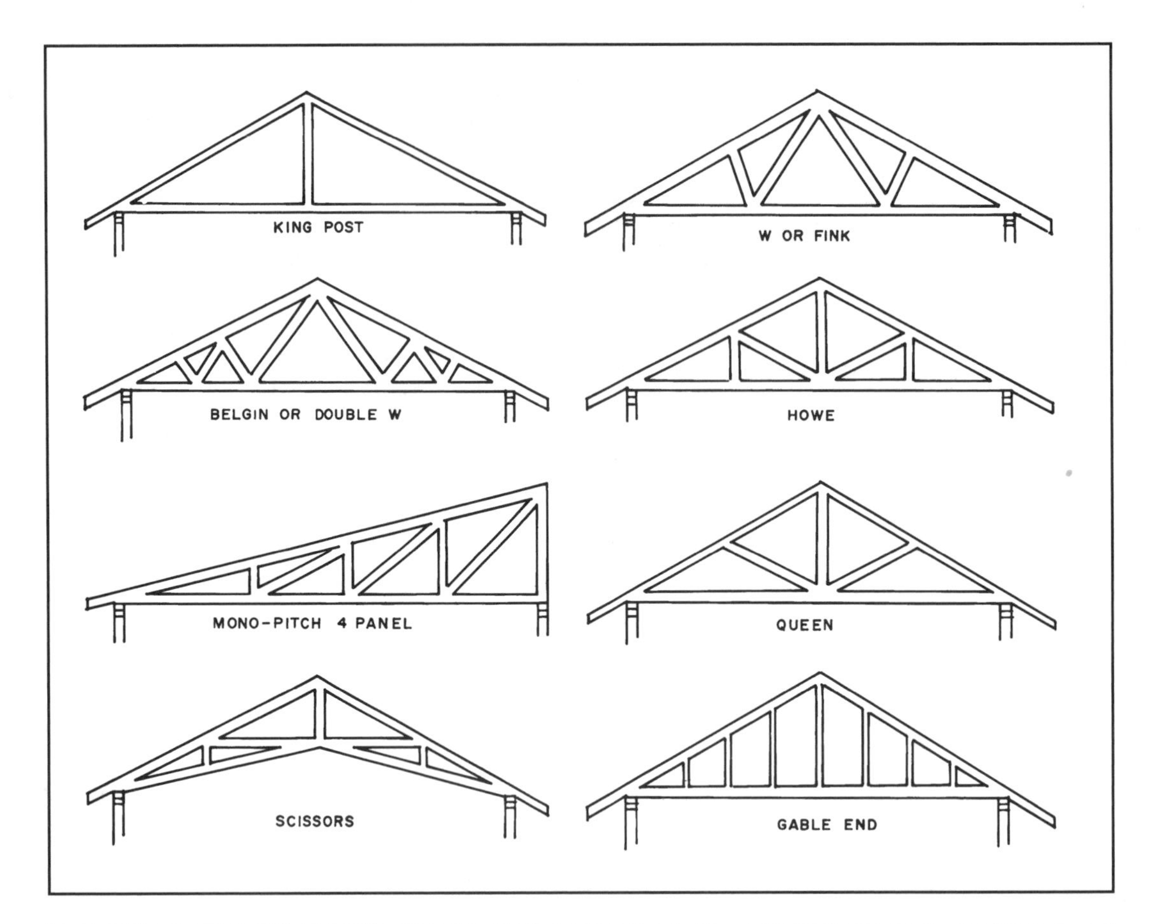

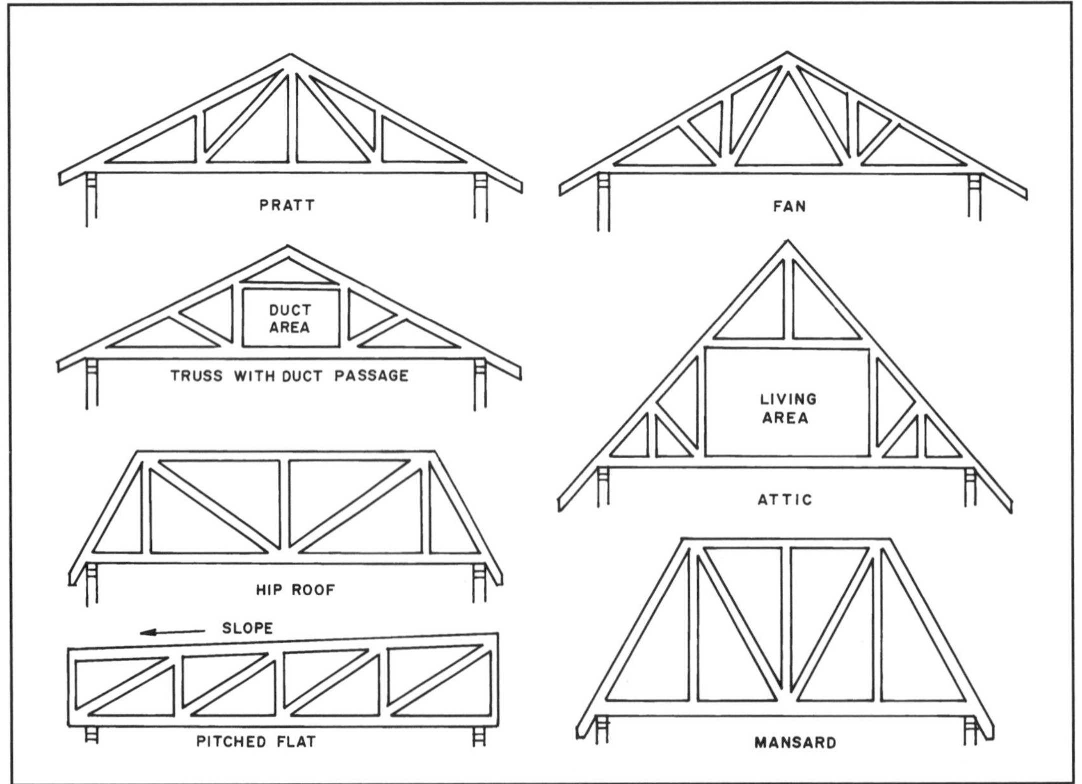

18-3 Some of the wood-framed trusses commonly available.

a king-post truss with web members running from the bottom of the center vertical member to the center of the top chord and dropping another vertical member from there to the bottom chord, which gives three vertical supports to the bottom chord. It will carry a heavier ceiling load than the W-truss, if it is designed using the same size material.

A **hip truss** is used to frame a hip roof. Hip trusses are trapezoids with equal slopes on the side members, which connect the horizontal top and bottom chords. Each hip roof uses a series of trusses each smaller than the one before.

A **mono-pitch truss** is used when framing shed roofs.

An **attic truss** is used to provide space for storage or a living area in the attic. To get headroom the roof must have a steep slope. A variation of this for low-sloped roofs is used to provide a space to run air ducts.

A **gable-end truss** is built without gusset plates. However, it is not strong enough to support a roof load, if the building length is increased at some future time. If an addition is built, add gusset plates to the truss. It is usually sheathed with structural sheathing.

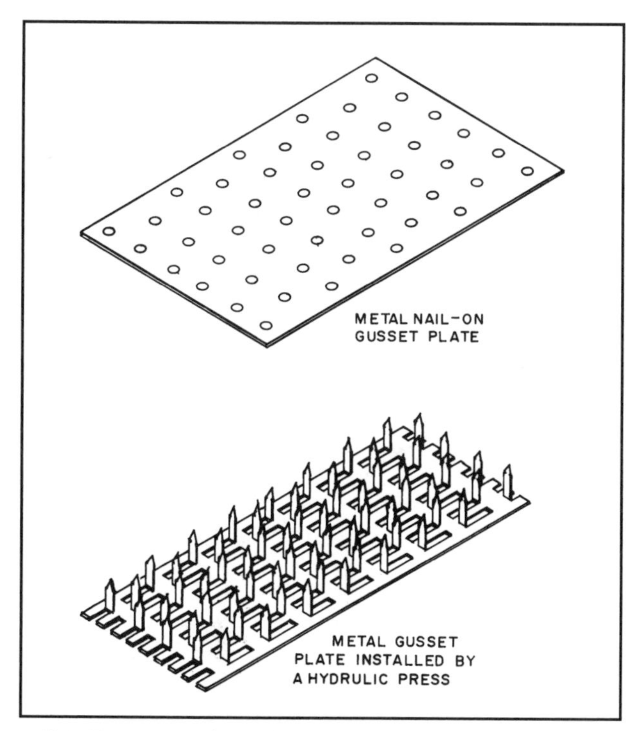

18-4 Two types of metal gusset plates used for truss assembly.

DESIGN & FABRICATION

Trusses must be designed by an engineer. Factors to be considered include the type of truss to be used, the structural properties of the wood, the imposed loads both live and dead, the span, the roof slope, and the spacing of the trusses. Imposed loads include dead loads such as the weight of the truss, sheathing, roofing, insulation, and interior ceiling. Live loads include snow and wind. Members for trusses used in light-frame construction are usually 2 × 4 and 2 × 6 stock. Notice that in the truss shown in **18-2** some web members are in compression and others are in tension. The engineer must specify the wood properties and size to carry the calculated loads.

Most trusses are manufactured in factories. The manufacturer can design a wide range of special-purpose trusses. The trusses are assembled from precision-cut parts that are held in jigs. The metal gusset plates are either power-nailed or pressed into place. Machine stress-rated lumber is used.

Two types of metal gussets are used on factory-manufactured trusses. One is a flat metal plate with punched holes for nails, whereas the other is a flat metal plate with prongs **(see 18-4)**. The plate is placed over the union of truss members and pressed in place with a hydraulic press **(see 18-5)**.

While it is best to buy trusses from a reliable truss manufacturer, small trusses can be built on the site. They must follow an approved engineering design. The proper grade of lumber must be used, it must be kiln-dried to 19 percent moisture, and it must be free of any form of warp or twist.

18-5 This gusset plate has metal prongs that have been pressed into the wood members.

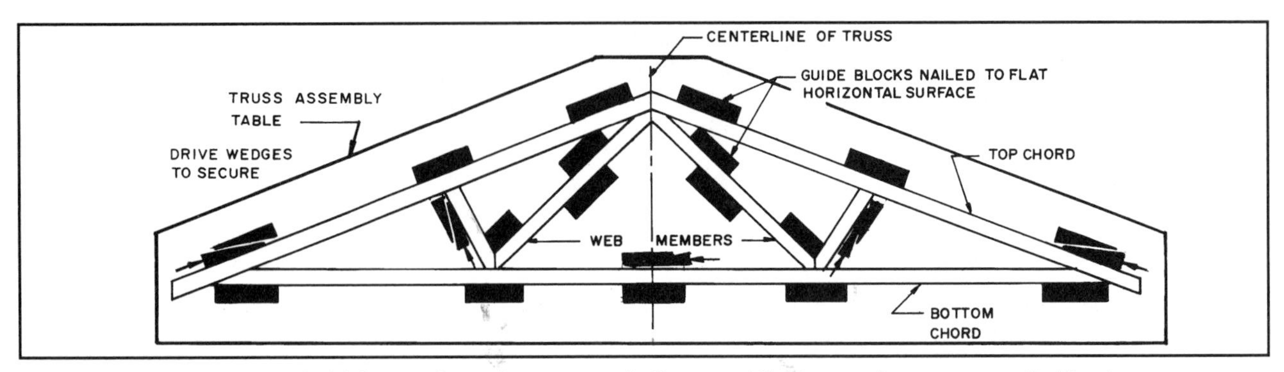

18-6 A typical jig set up to hold the members of a carpenter-built truss while the metal gussets are nailed in place.

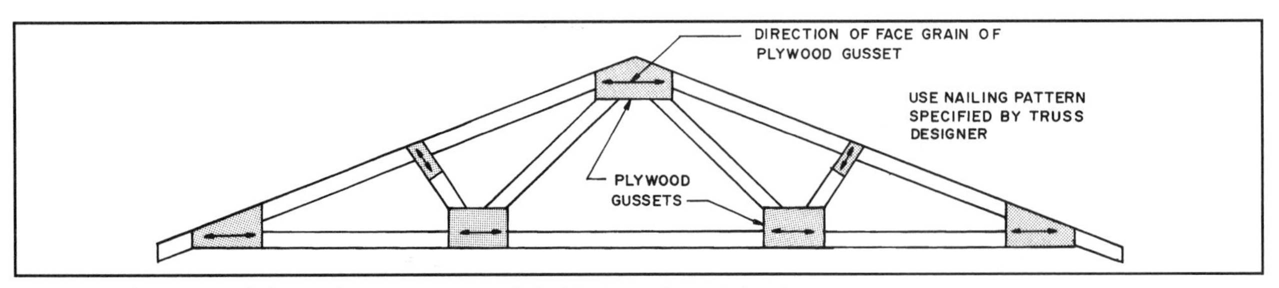

18-7 The face grain of plywood gussets runs parallel with the webs and chords.

Begin by carefully drawing the truss on a large flat surface, such as the subfloor. Cut each piece to fit, and then place it on the pattern and lightly nail in place. Check to make certain the joints fit and that the left and right sides of the truss are the same. Then nail two-inch wood blocks to the floor around the truss, forming a jig. If many trusses are to be made build the jig on legs, providing a raised work surface **(see 18-6)**. The nails in the metal gussets may be hand or power driven.

Now cut the number of pieces required for the series of trusses. Use the parts of the original truss as a pattern. A radial-arm saw will produce more accurate cuts than a portable circular saw. If plywood gussets are to be used, cut them to size. The face grain of the gussets and the heel gussets should be parallel with the chords and webs **(see 18-7)**.

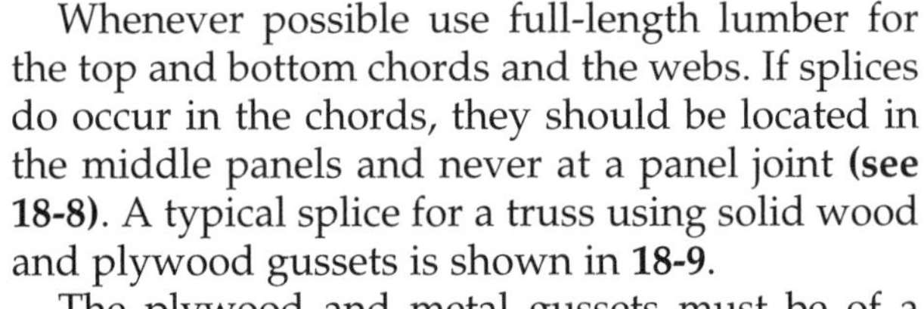

Whenever possible use full-length lumber for the top and bottom chords and the webs. If splices do occur in the chords, they should be located in the middle panels and never at a panel joint **(see 18-8)**. A typical splice for a truss using solid wood and plywood gussets is shown in **18-9**.

The plywood and metal gussets must be of a type and size specified by a certified truss designer.

When plywood gussets are used, apply glue to the truss and the gusset. The glue used is a casein type, but if there is danger of exposure to moisture, use a waterproof glue such as a resorcinol type. Almost all roof trusses have possible moisture exposure at the soffit. Spread the glue carefully over the entire surface with a brush.

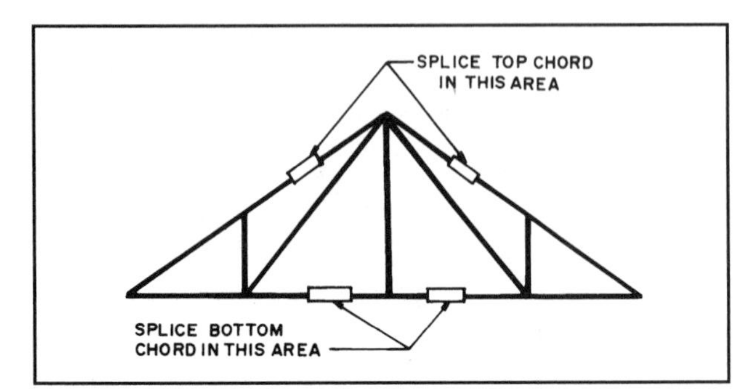

18-8 Chords are spliced near the center of panels.

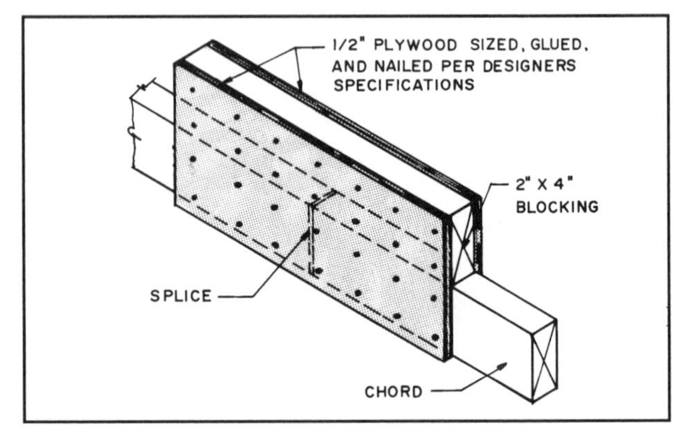

18-9 A typical plywood chord splice design.

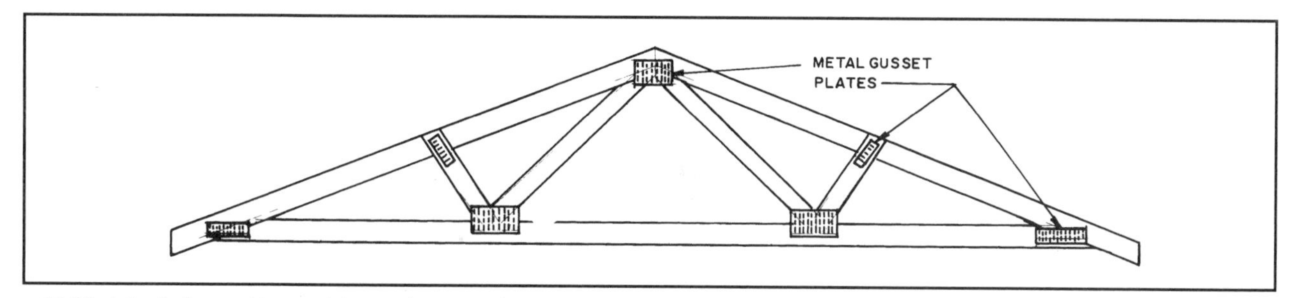

18-10 A typical wood truss with metal gusset plates.

Plywood gussets used on small trusses are usually ⅜ inch or ½ inch thick, as specified on the design drawing. Use 4d galvanized or cement-coated nails for ⅜-inch-thick gussets and 6d for those ½ inch thick. Space the nails three inches apart on ⅜-inch gussets and four inches on ½-inch gussets. When nailing into four-inch-wide members, use two rows of nails ¾ inch from each edge of the gusset. Use three rows of nails when the truss member is six inches wide.

After the gussets on one side are in place, carefully raise the truss by the peak. Have several workers help turn it over and lay it flat on the floor. Then glue and nail the gussets on the other side. Stock the trusses absolutely flat, and let the glue cure 24 hours.

A typical wood truss with metal gusset plates is in **18-10**. The plates are nailed following the truss designer's specifications. Pressed plates have a continuous array of points and are placed properly over the joints and pressed in place with a hydraulic press. The width of points must be perpendicular to the grain of the member.

HANDLING TRUSSES

When unloading trusses from the delivery truck avoid any bending of the trusses. Bending can cause joint and lumber damage.

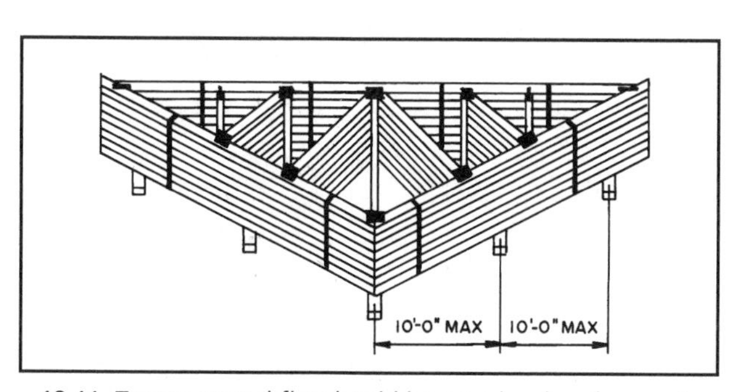

18-11 Trusses stored flat should be on a level surface and rest on solid wood blocking.

It is essential to have a crane to unload the trusses. The crane should have a spreader bar as discussed in the next section of this book. Never lift the bundles of trusses by their strapping. The cables should go below the top chords of the truss bundle. Never lift by the web members. Never lift an unbalanced load. If the trusses are not level, lower the load and adjust the cables so the load is balanced.

STORING TRUSSES

Whenever possible unload and store trusses on level ground. If they are to be stored flat on the ground, select a level place and lay blocking as shown in **18-11**. If they are to be stored vertically they should be firmly blocked and braced to prevent leaning or bowing, as shown in **18-12**. Pitched trusses should be stored with the peak up. Scissor trusses should not be stored with the peak up.

Protect the trusses from rain by covering them with sheet plastic. Arrange so some ventilation can occur.

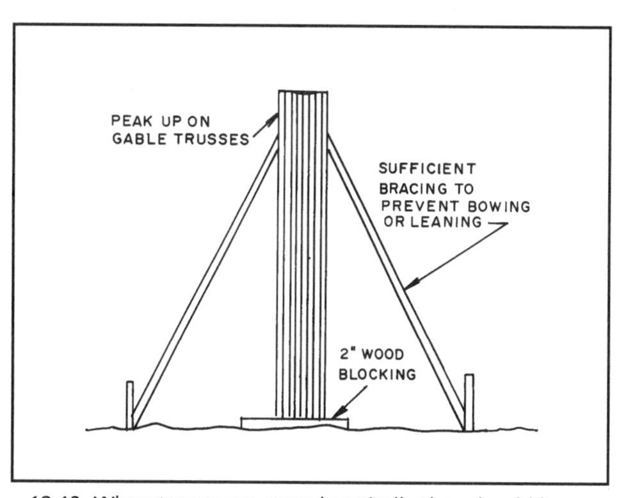

18-12 When trusses are stored vertically they should have adequate bracing to keep them straight and prevent leaning or bowing.

ERECTING THE TRUSSES

Small trusses can be manually lifted into place by several workers. One person can raise trusses 20 feet long while two persons are needed to lift trusses 20 to 30 feet long **(see 18-13)**. Workers are in position on each end of the truss and they position and nail it to the top plate.

Large trusses are set in place with a crane **(see 18-14)**. Control lines are tied to each end of the truss so that workers on the ground can keep it from swinging and can guide it into the proper position on the top plate. As above, workers are on each wall to nail it to the top plate, and one or more workers nail the needed bracing. Trusses under 30 feet in length can be lifted as shown in **18-15**. Peaked trusses with spans of 30 to 60 feet require the use of a spreader beam and three cables as shown in **18-16**.

Begin by installing and bracing one gable-end truss. It must be plumb and securely braced as shown in **18-17**. Then set each truss in place, locating

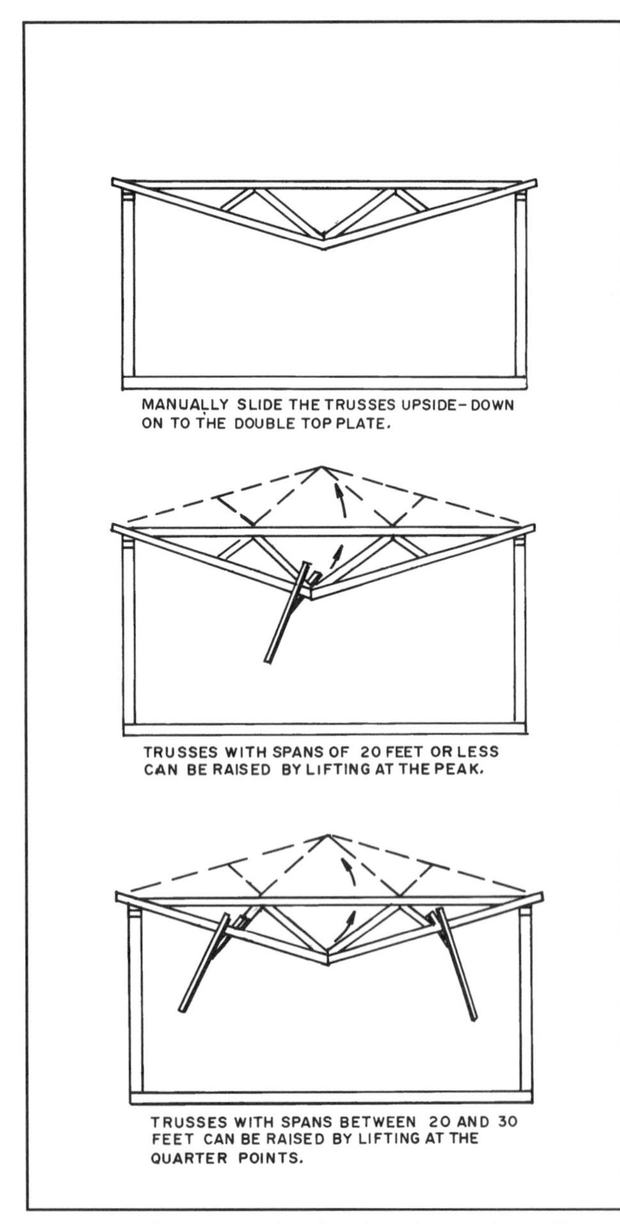

18-13 Small trusses can be placed on the exterior walls with the peak down. They are raised by workers using poles.

18-14 Large trusses are set in place with a crane.

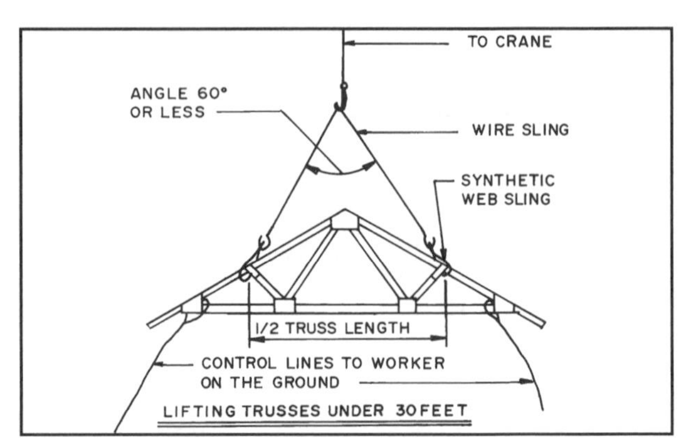

18-15 Recommended rigging for lifting trusses under 30 feet long.

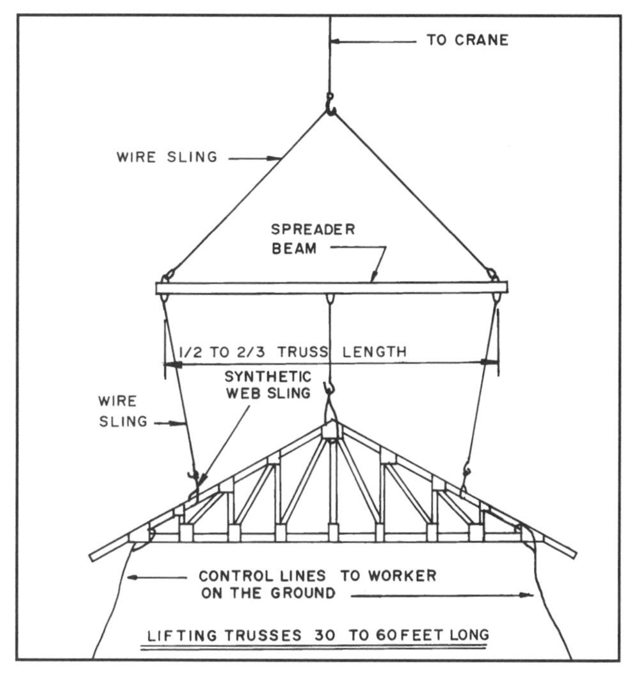

18-16 Recommended rigging for lifting peaked trusses 30 to 60 feet long.

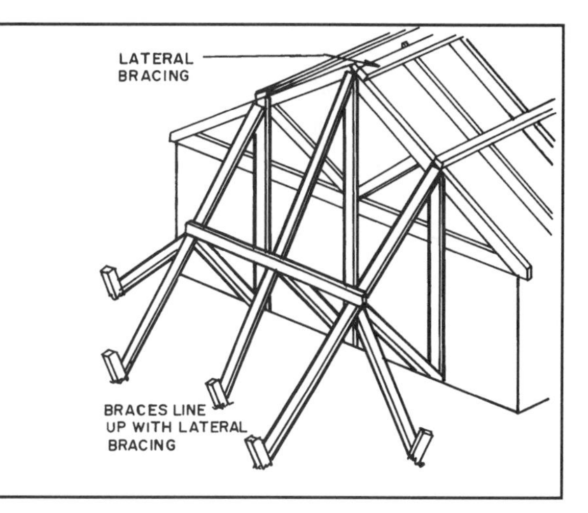

18-17 Erect and firmly brace the gable-end truss.

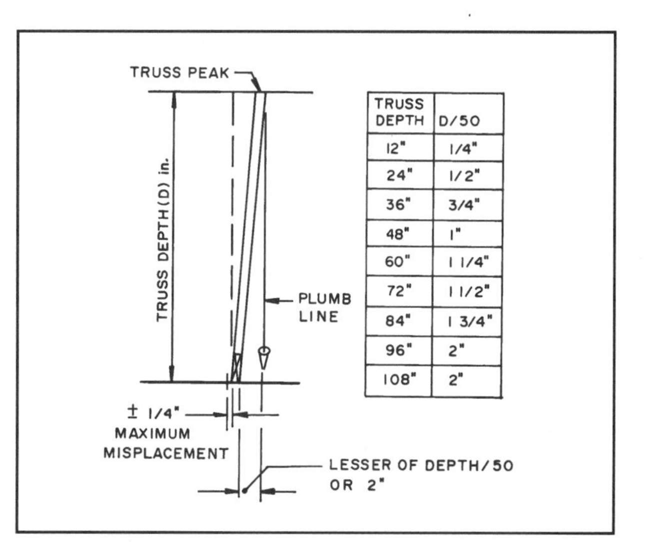

TRUSS DEPTH	D/50
12"	1/4"
24"	1/2"
36"	3/4"
48"	1"
60"	1 1/4"
72"	1 1/2"
84"	1 3/4"
96"	2"
108"	2"

18-18 Recommendations for location and plumb tolerances for truss erection. *(Courtesy Truss Plate Institute)*

it by marks on the top plate. They should not be more than ¼ inch out of line. The truss should not be out of plumb more than 1/50 of the depth of the truss **(see 18-18)** or two inches maximum for the largest truss.

As the trusses are installed temporary bracing is installed. It holds the trusses true until the permanent sheathing is installed. Continuous lateral bracing is installed on the top chord **(see 18-19)**. It runs perpendicular to the trusses and must be at least a 2 × 4-inch member. It should be spaced at least six trusses. The next piece should overlap two trusses as shown in **18-19**. The spacing between these runs of lateral bracing varies with the span of the truss. Typically spans up to 28 feet require seven feet maximum between runs of temporary bracing. Larger spans up to 42 feet can be spaced six feet, while those up to 60 feet must be spaced five feet. If there is any doubt or unusual condition, the truss design engineer should be consulted.

In addition, diagonal braces are installed on at least six trusses on each end of the roof. Braces are often nailed with duplex nails so they can be easily removed.

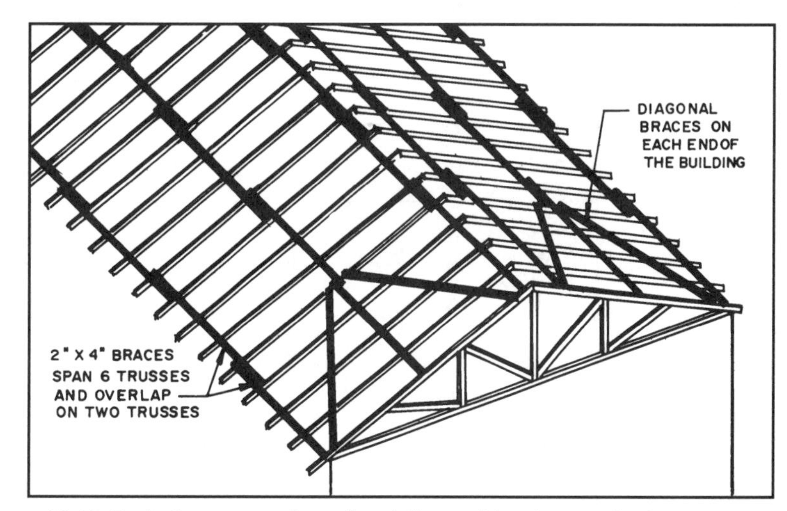

18-19 Typical temporary lateral and diagonal bracing required as trusses are erected.

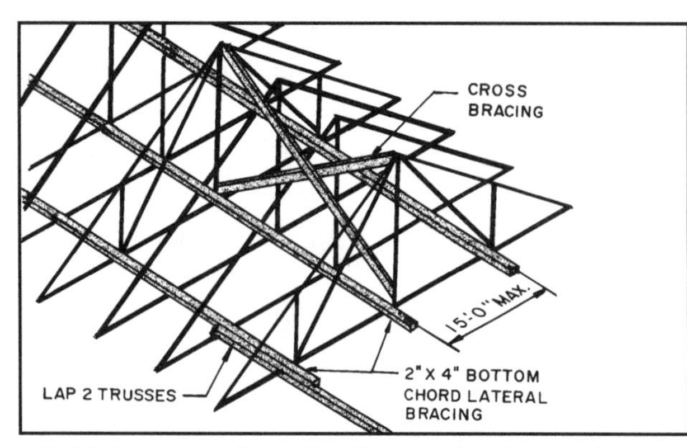

18-20 Permanent lateral bracing includes the use of 2 × 4-inch members run on the bottom chord perpendicular to the trusses.

Permanent Truss Bracing

The truss designer will indicate how the trusses are to be permanently braced. Bottom chord lateral bracing runs perpendicular to the trusses and its braces are spaced no more than 15 feet apart **(see 18-20)**. It should overlap two trusses. In addition web cross bracing is installed to opposite sides of the same group of similar web members, forming an X. It is usually installed in the first group of five or six trusses on each end of the roof **(see 18-21)** and along the length of the building at 20 feet or at smaller intervals. All bracing is nailed with two 16d common nails at each connection. No bracing is removed, including end-bracing to the ground, until the roof is completely sheathed.

The truss is toenailed to the double top plate with two 12d or 16d nails. In areas where high winds occur, metal tiedowns are required **(see 18-22)**. It is also difficult to toenail some trusses that have large metal gussets. In these cases metal tiedowns are also used.

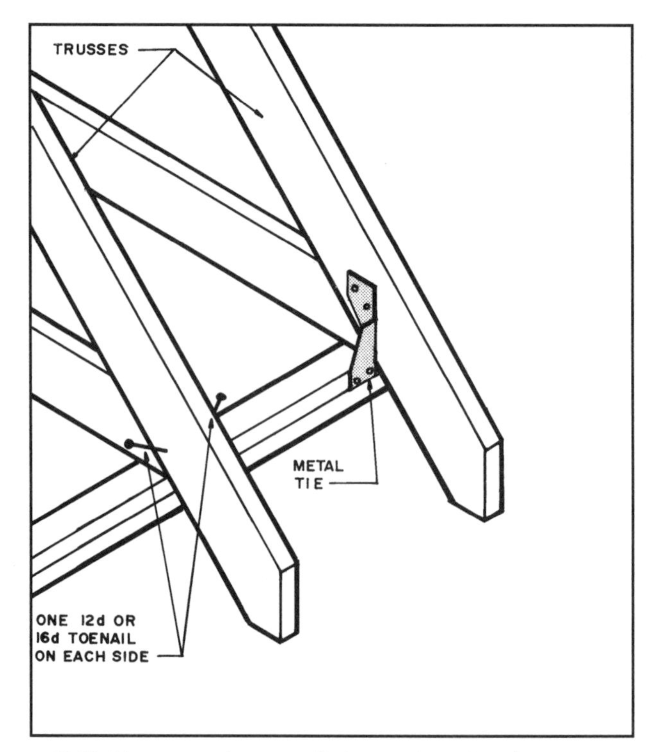

18-22 Trusses can be toenailed or anchored to the top plate with metal ties.

PIGGYBACK GABLE TRUSSES

Trusses that are too high for delivery to the job site can be made in two or more sections and assembled at the job site. These are referred to as piggyback trusses **(see 18-23)**. The truss designer prepares drawings showing the required connection details and permanent bracing to be used. The lower trusses are installed and permanently braced before the small top truss section. Sometimes the lower trusses are sheathed before installing the small top trusses.

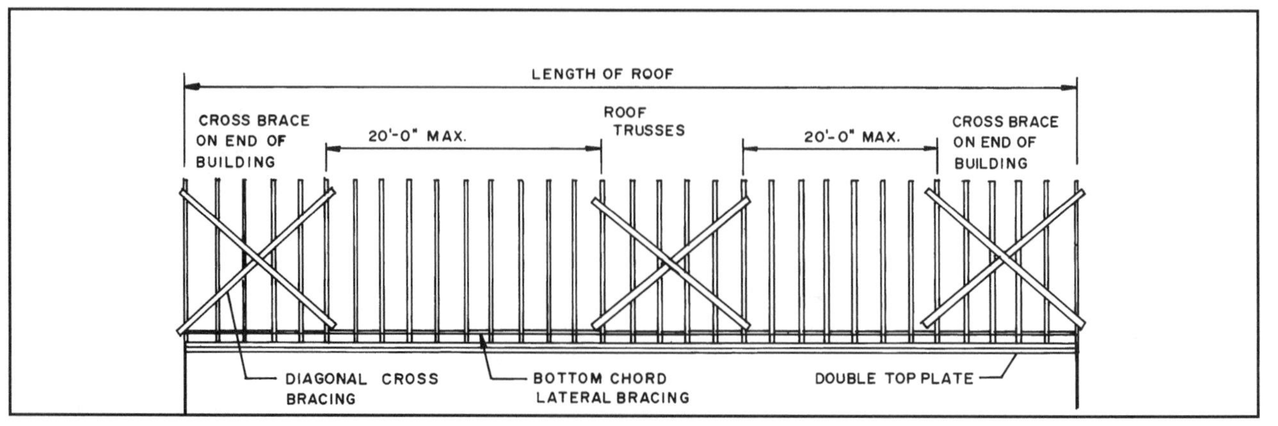

18-21 Cross bracing on trusses uses triangulation to stabilize the installation.

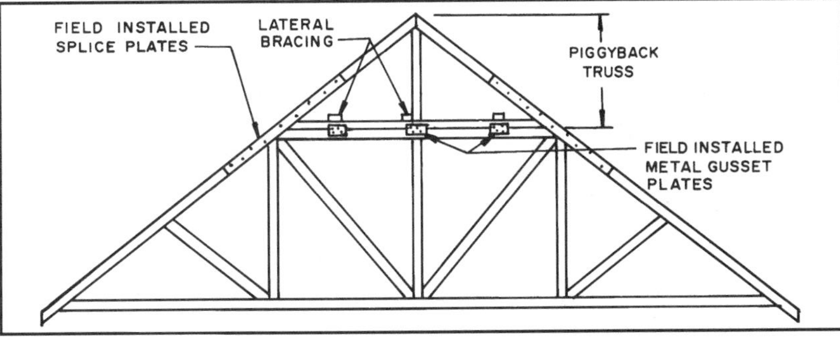

18-23 Typical connection details for a piggyback truss. Follow the instructions of the truss designer.

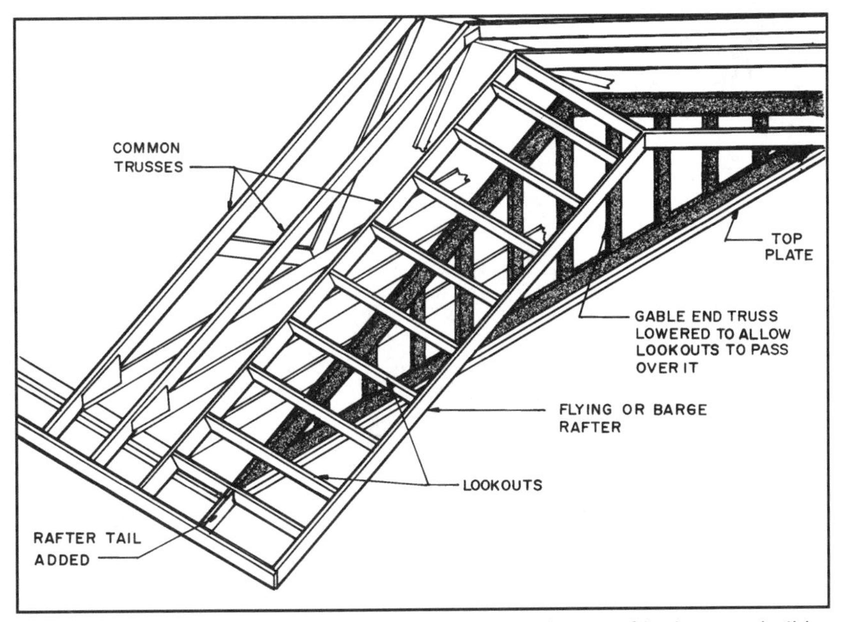

18-24 The gable-end truss can be lowered to allow the use of lookouts to build the gable-end overhang.

FRAMING A GABLE-END OVERHANG

A gable-end overhang can be framed by ordering a gable-end truss that has been dropped to permit lookouts to rest on top of it and run to the top chord of the next truss **(see 18-24)**.

FRAMING A HIP ROOF

Hip roofs are framed with roof trusses using a step-down system. This system uses a series of intermediate trusses. It uses a girder truss to carry the load of a series of common jack trusses.

A **hip jack truss** is used to form the run from the hip girder truss to the corner of the building.

Common jack trusses run from the girder truss to the end wall of the building **(see 18-25)**.

After the common trusses are in place, the hip can be constructed. Framers have various methods they prefer. Following is just one approach.

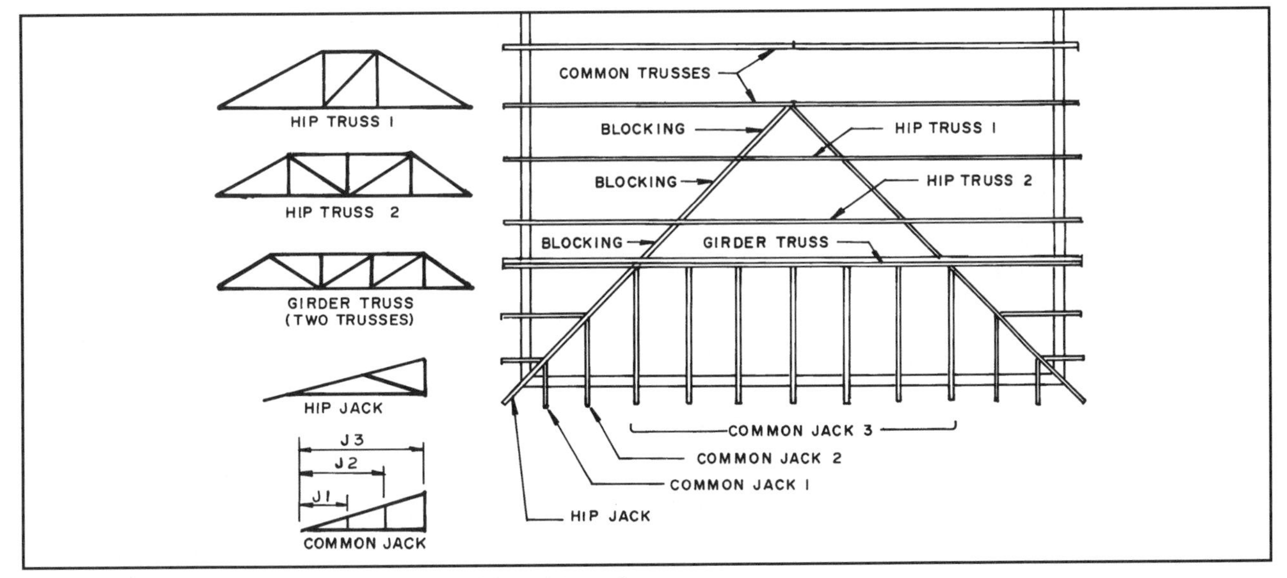

18-25 A typical layout and types of truss used for a hip roof.

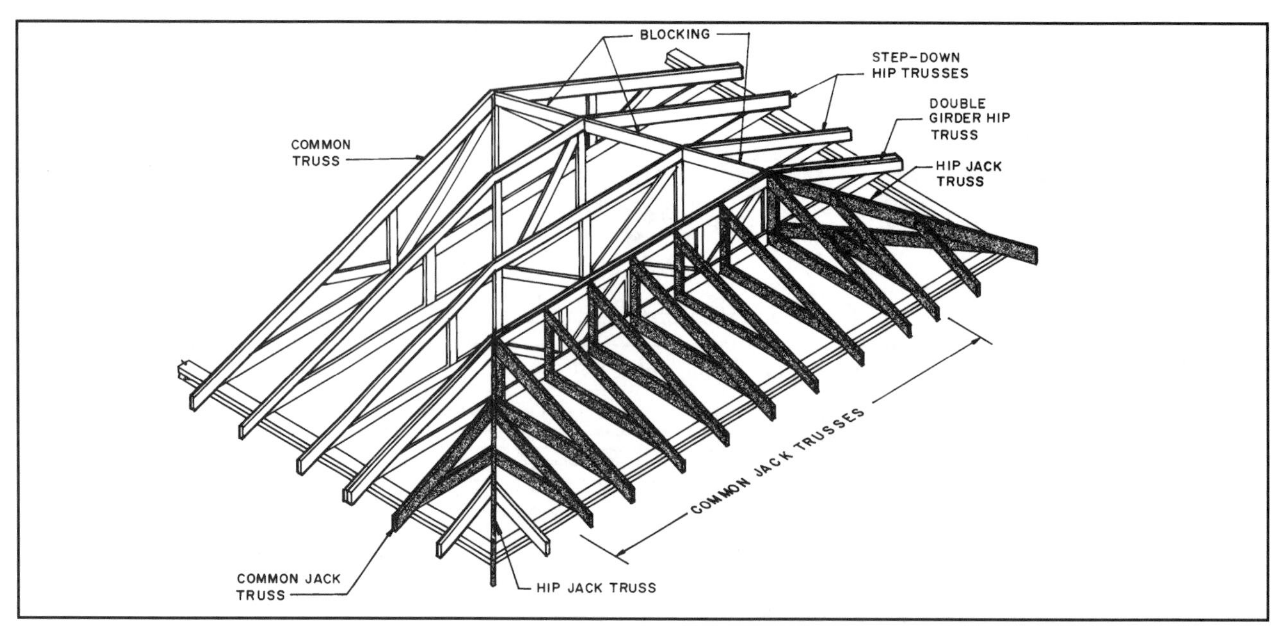

18-26 All the trusses for a hip roof have been installed and the assembly is ready for the step-down blocking.

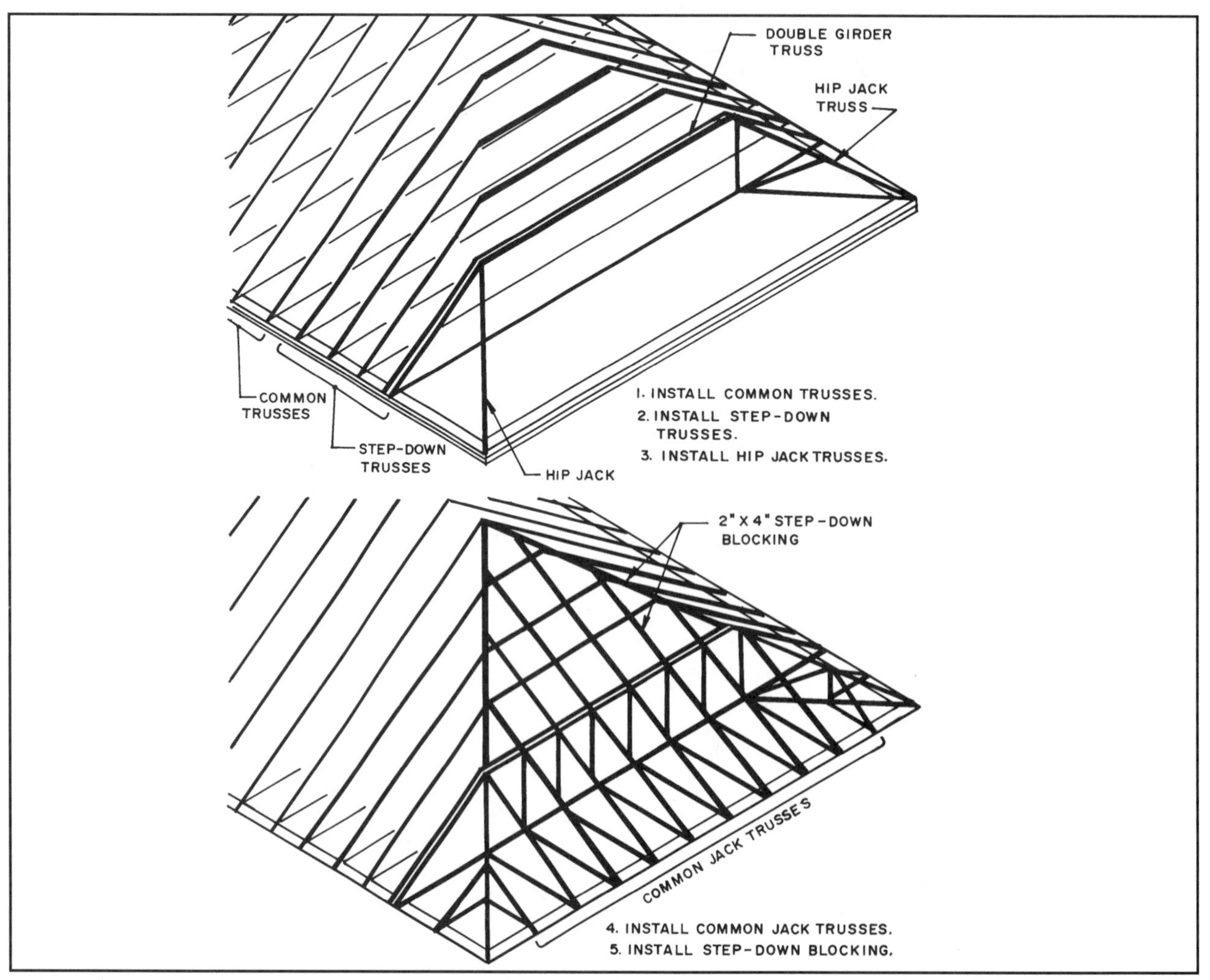

18-27 The completed assembly with the step-down blocking installed over the step-down trusses.

Install and brace the step-down trusses and girder truss. Then set the common jack trusses and the hip jack truss. Brace as specified by the truss manufacturer. The complete truss assembly before the step-down blocking is installed as shown in **18-26**.

The step-down blocking ties the top of the step-down trusses together and provides a nailing surface for the sheathing **(see 18-27)**. Two commonly used methods to install the step-down blocking are in **18-28**.

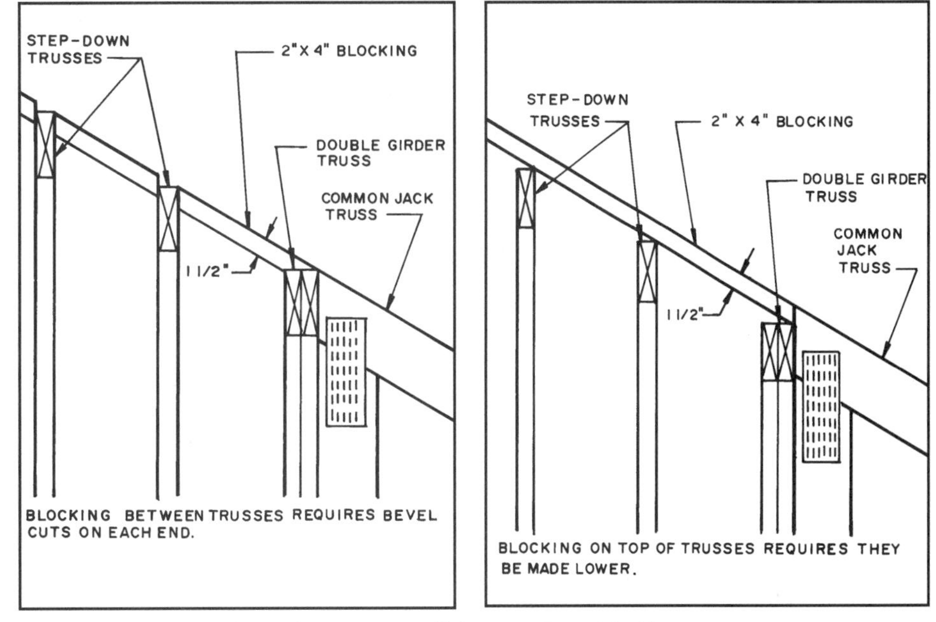

18-28 Two commonly used ways to install the step-down blocking.

FRAMING A GABLE ROOF ON AN L-SHAPED BUILDING

An L-shaped roof is produced by the intersection of two gable roofs. This forms valleys at the lines of intersection. It requires the use of a series of special trusses called valley jack trusses. Each gets progressively smaller than the previous one **(see 18-29)**. These special trusses are secured to the top chord of the common trusses in the adjoining roof. The girder truss is placed first and strongly braced. Metal truss hangers are used to secure the

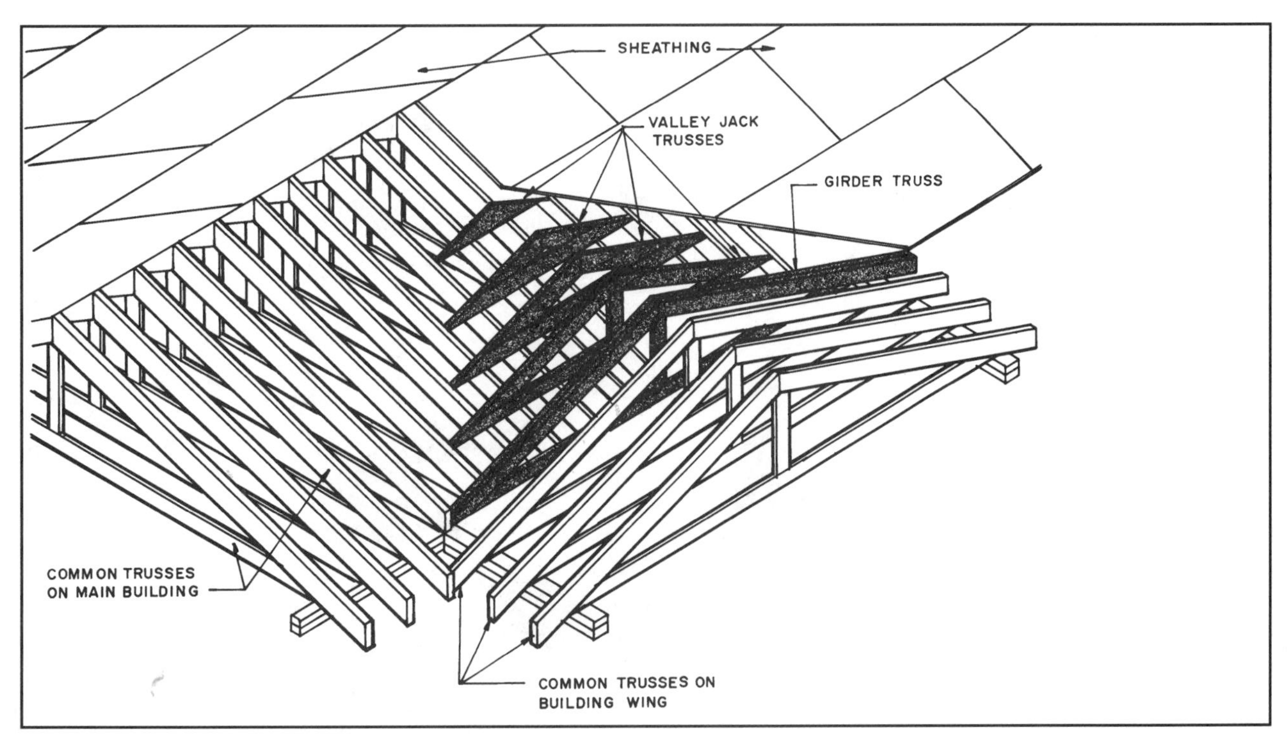

18-29 A typical truss layout for a gable L-shaped roof.

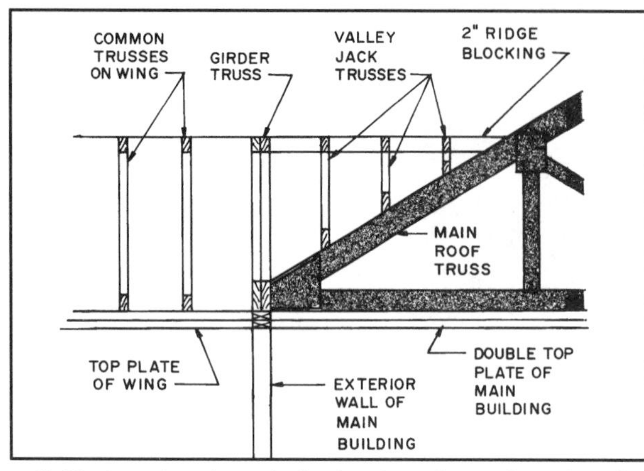

18-30 A section through the framing of an L-shaped roof using wood trusses.

bottom chord of the special trusses to the top chord of the common truss. Blocking is placed at the ridge between the trusses and is usually held with metal pocket-type hangers. This gives a nailing surface for the sheathing **(see 18-30)**.

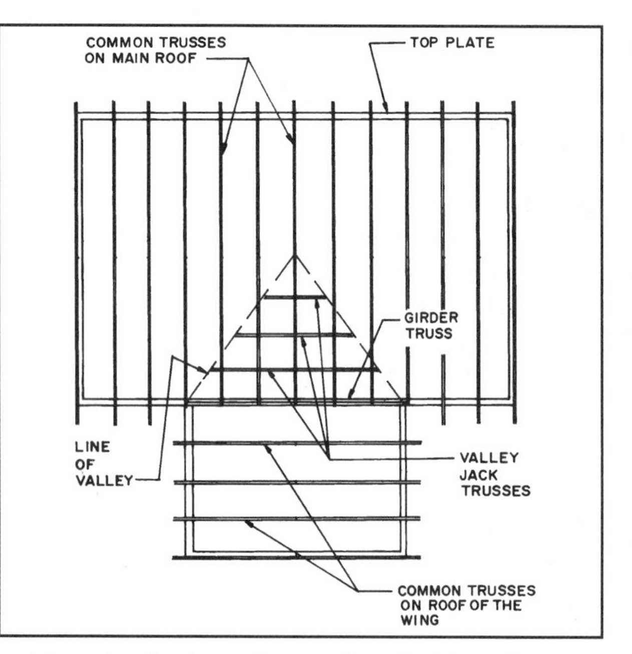

18-31 A framing layout for an L-shaped gable roof using wood trusses.

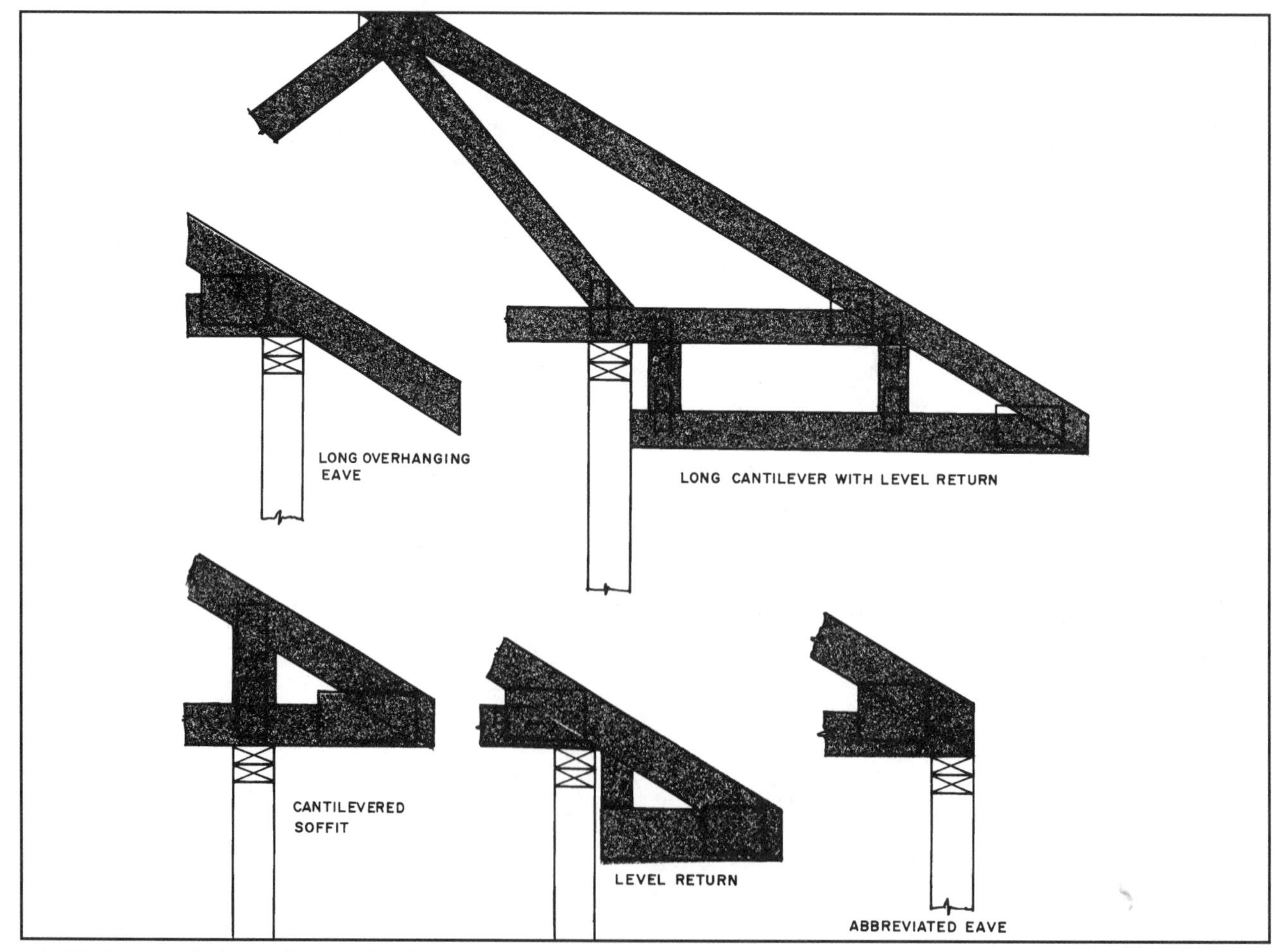

18-32 Some typical eave framing details for pitched gable roof wood trusses.

The girder truss is placed across the opening where the two parts of the house intersect. It is installed first and braced. The valley jack trusses are placed on the intersecting roof and the common trusses placed over the adjoining wing of the building **(see 18-31)**.

OTHER DETAILS

Trusses can be manufactured with a wide range of soffited eave construction possibilities. Some of these are shown in **18-32**.

The installation of a two-inch-thick nailer to support the finish ceiling material is shown in **18-33**.

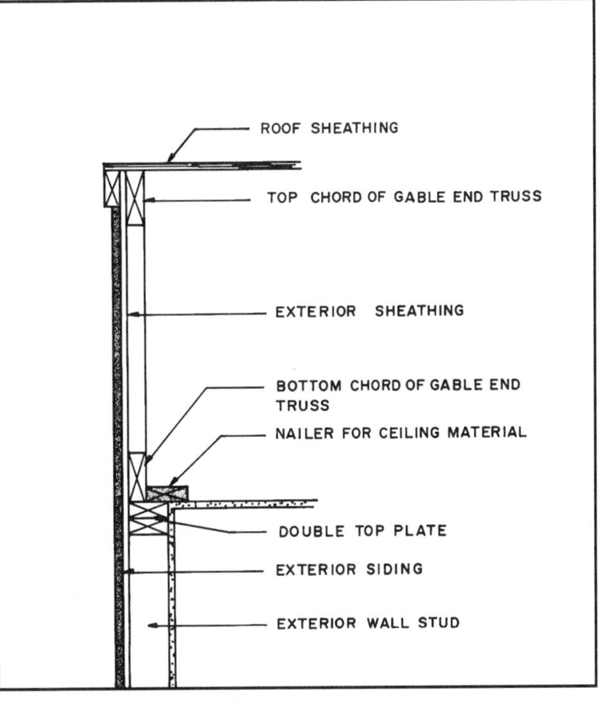

18-33 A nailer is installed on the top plate to provide a nailing surface for the ceiling materials.

ADDITIONAL INFORMATION

Wood Truss Council of America
5937 Meadowwood Drive, Suite 14
Madison, WI 53711

Truss Plate Institute
583 D'Onofrio Drive, Suite 200
Madison, WI 53719

CHAPTER 19

CORNICE CONSTRUCTION

The exterior cornice is the trim on the projection of the roof where it meets the exterior wall. On gable roofs it occurs on the sides where the roof meets the wall. Hip roofs have a cornice on all sides. Flat roofs have the cornice constructed by extending the rafters/ceiling joists beyond the exterior wall. This will occur on all sides of the building.

The three basic types of cornice are box or soffit, open, and close. The **box cornice** is the most frequently used type because it provides an attractive appearance and protects the exterior walls and windows from rain and, in some cases, exposure to the sun. The **open cornice** is much like the box because it provides overhang, but it gives quite a different appearance. It is used where a rustic appearance is desired, and is also less costly to build. The **close cornice** finishes off the ends of the rafters but has no overhang; it therefore provides no protection.

BOX OR SOFFIT CORNICES

The width of the box cornice will be specified on the working design drawings. The width can vary considerably and depends in part on the slope of the roof. The soffit return must meet the exterior wall above the head of the windows. Several roof slopes are drawn in **19-1** showing how they influence the amount of overhang. Low-sloped roofs can have wider soffits than more steeply sloped roofs.

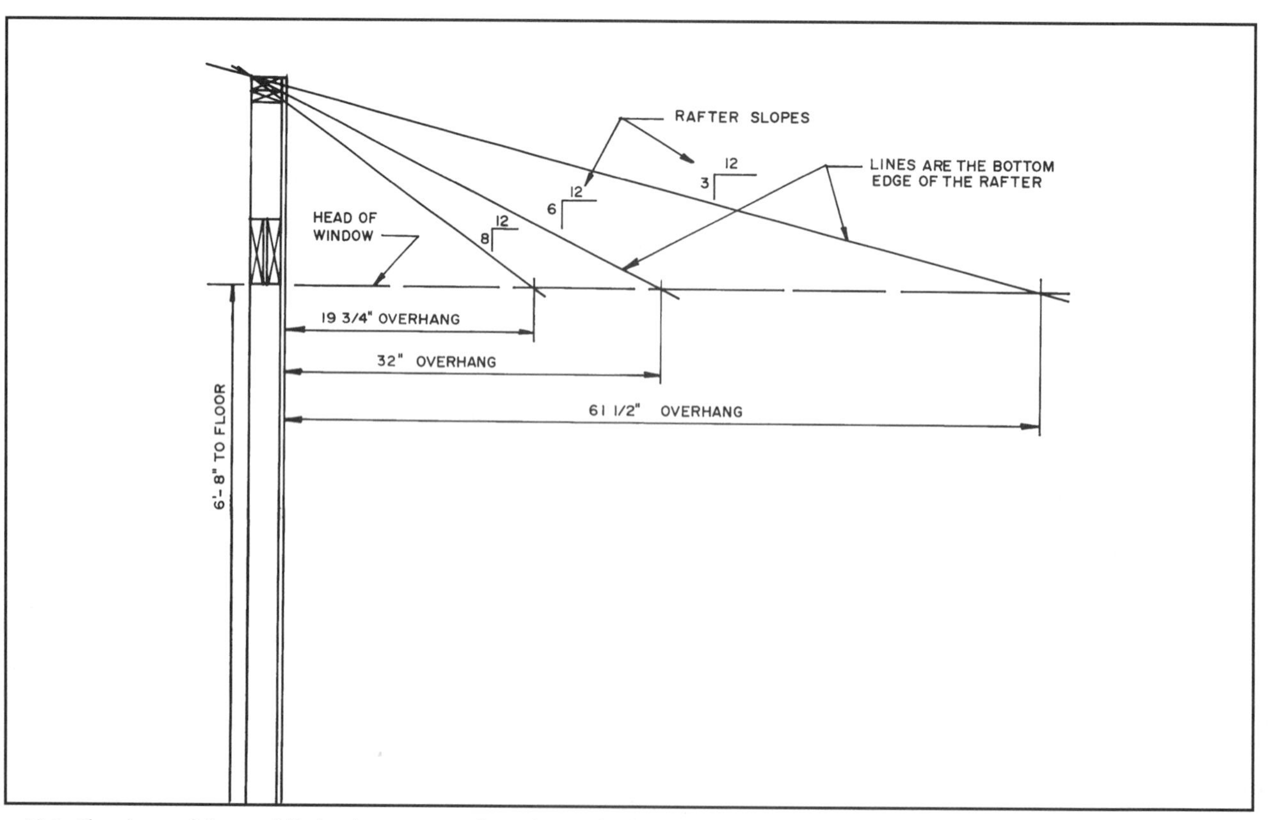

19-1 The slope of the roof limits the amount of cornice projection.

A narrow box cornice has the rafter projection cut so that it serves to hold the fascia and soffit **(see 19-2)**. A frieze board is used to cover the gap between the soffit and the siding. Often a molding is added to complete the joint. A narrow box cornice for a flat roof is shown in **19-3**. Notice the inclusion of vents to clear the air above the insulation.

A wide box cornice is used when more overhang is wanted. It uses lookouts nailed to the end of the rafter and to a ledger that is nailed to the wall studs **(see 19-4)**. The soffit can be plywood, oriented strand board, hardboard, gypsum, vinyl, or metal. There are a number of new sheet products on the market that can also be used as soffit material. The thickness required depends on the spacing of the lookouts. Typical spacing is 16 inches (4060mm) and 24 inches (6100mm) O.C. Vents are located in the soffit.

The rough fascia is two-inch (500mm)- thick material and the finished fascia is one-inch (250mm) thick. The rough fascia is nailed to the ends of the rafters, and the finish fascia is nailed to it.

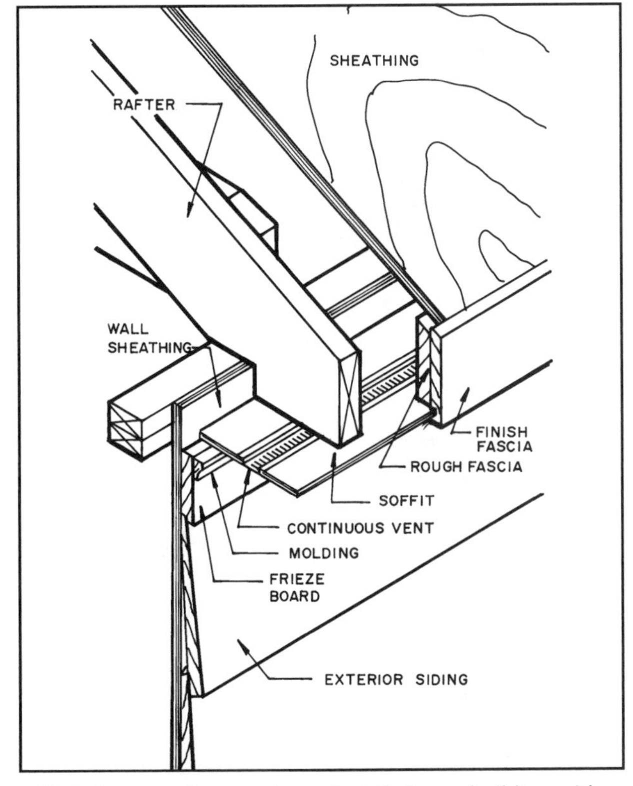

19-2 A narrow box cornice as installed on a building with wood siding.

A frieze board is commonly used with wide box cornices. It is usually a one-inch (250mm)- thick board rabbeted to receive the wood siding.

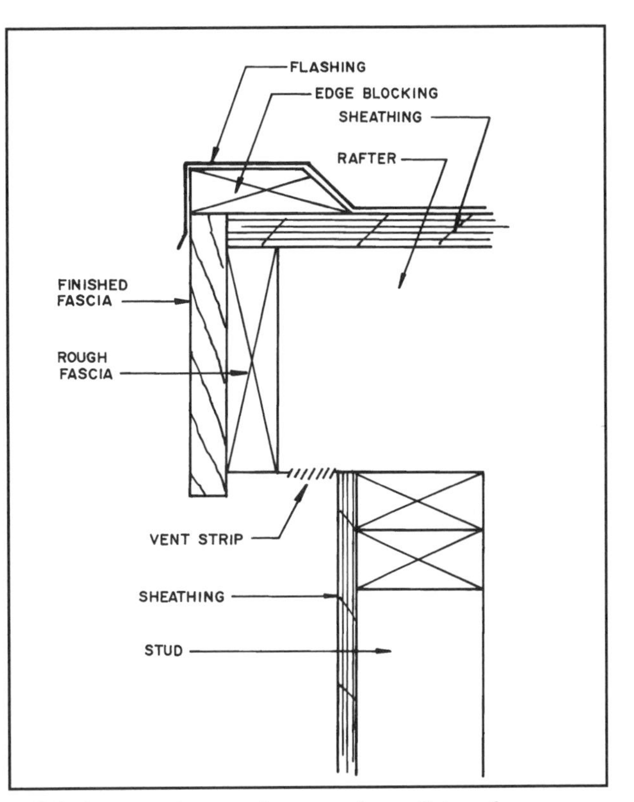

19-3 A narrow box cornice as used on a flat roof.

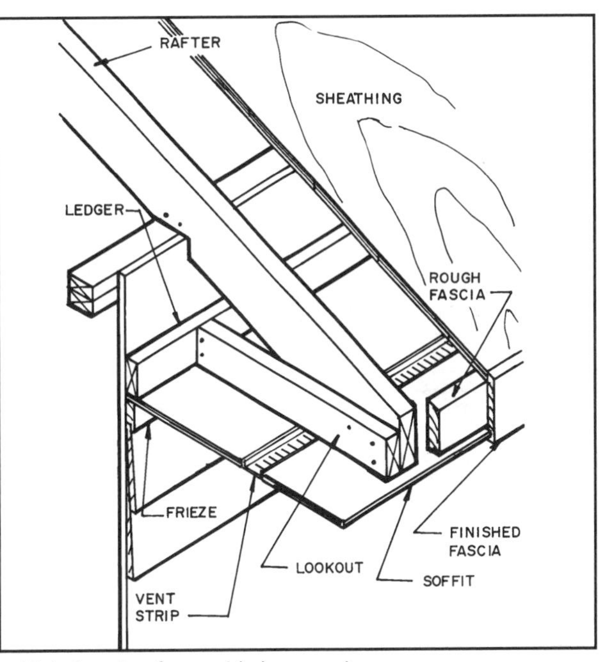

19-4 Framing for a wide box cornice.

Details for constructing other box cornices are shown in **19-5**, **19-6**, **19-7**, and **19-8**. These include truss-framed roofs, I-joist rafters, flat roofs, and brick veneer construction.

A box cornice without lookouts has a sloped soffit that is nailed to the bottom of the rafters as shown in **19-9**. The construction is the same as described for other box cornices.

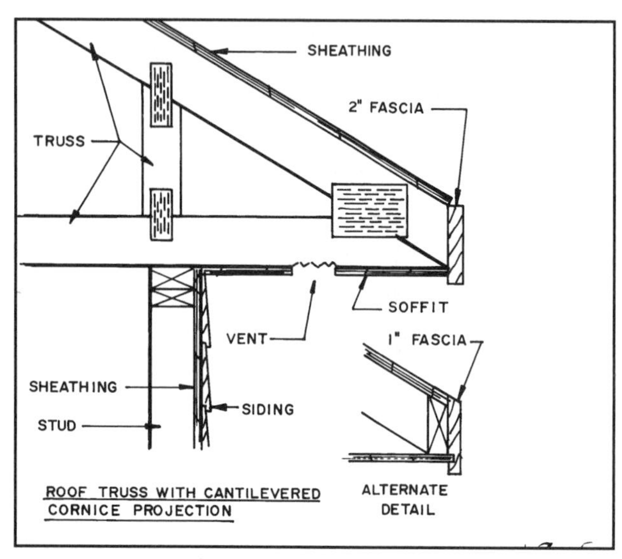

19-5 Examples of wide box cornice construction on a roof framed with wood trusses.

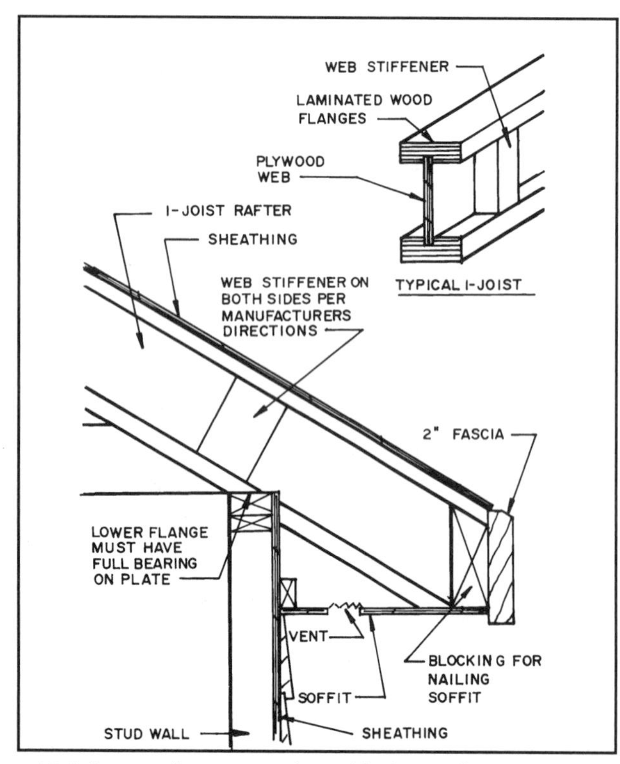

19-6 Box cornice construction with I-joist rafters.

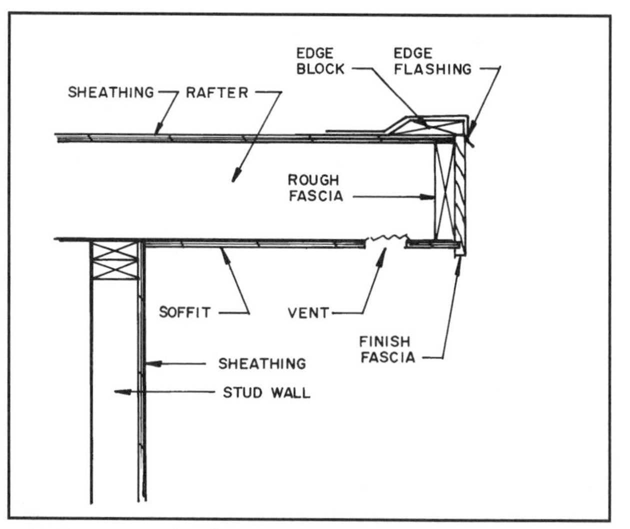

19-7 Construction for a wide cornice projection on a flat roof.

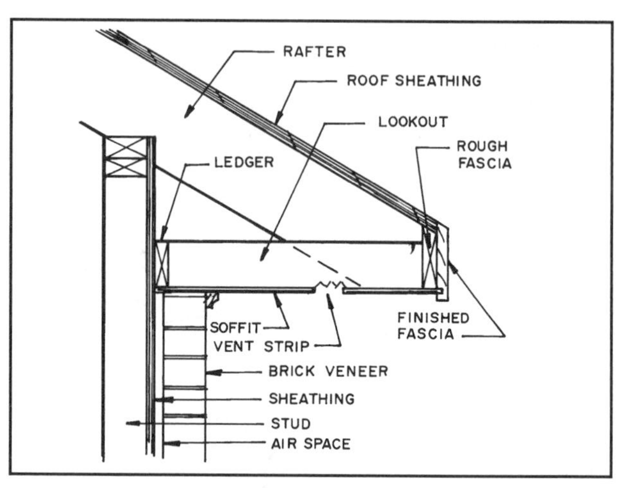

19-8 Construction details for a box cornice when brick siding is used.

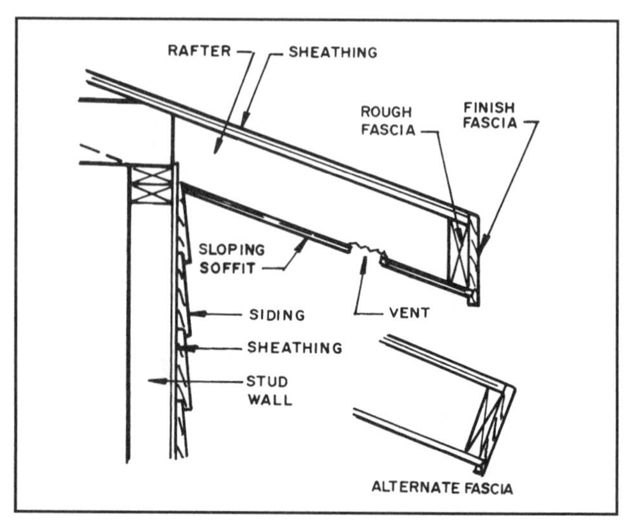

19-9 A box cornice with a sloping soffit.

OPEN CORNICES

Open cornices are built in the same manner as wide box cornices without lookouts, except that the soffit is omitted. Instead a frieze block is installed between the rafters, and vents are placed in it **(see 19-10)**. Open cornice construction with I-joist rafters is shown in **19-11**. Open cornice construction when trusses are used is shown in **19-12**.

CLOSE CORNICES

A close cornice has the rafter cut off flush with the outside of the top plate. The sheathing extends up over the ends of the rafters. A frieze board butts the bottom of the shingles and is rabbeted to receive the wood siding. Usually a molding is installed to close the joint and provide support for the shingles as they extend over the exterior wall **(see 19-13)**.

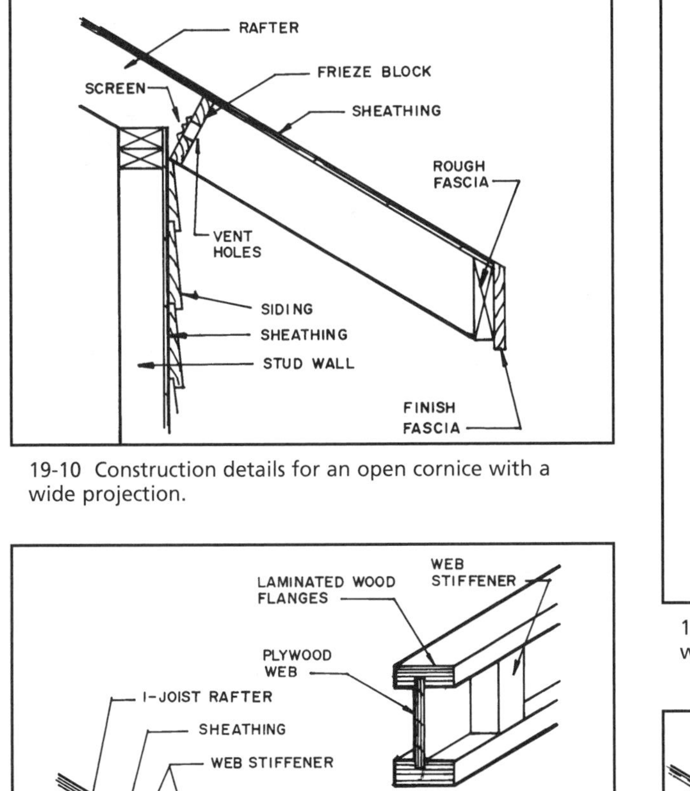

19-10 Construction details for an open cornice with a wide projection.

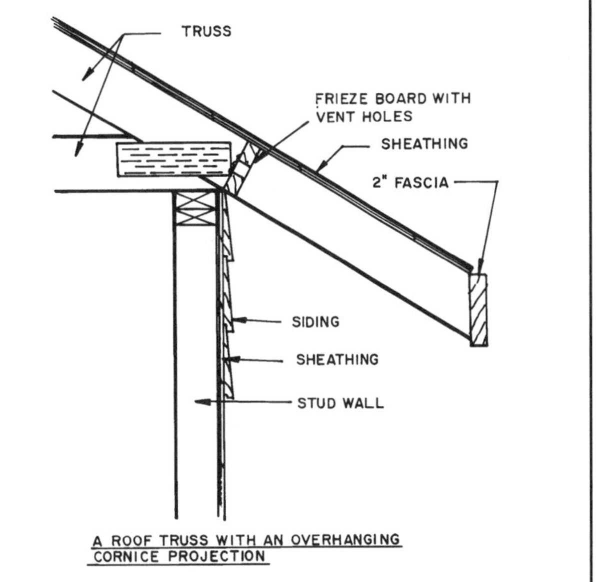

19-12 Open cornice construction when framing the roof with wood trusses.

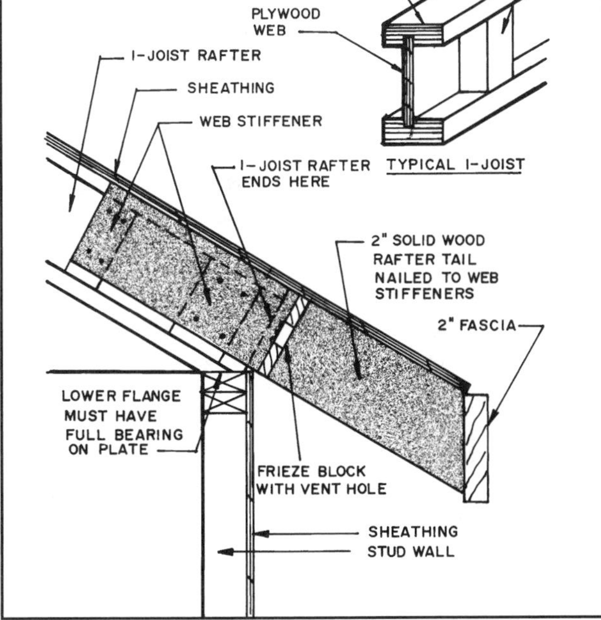

19-11 Typical open-joist cornice construction when framing the roof with I-joist rafters.

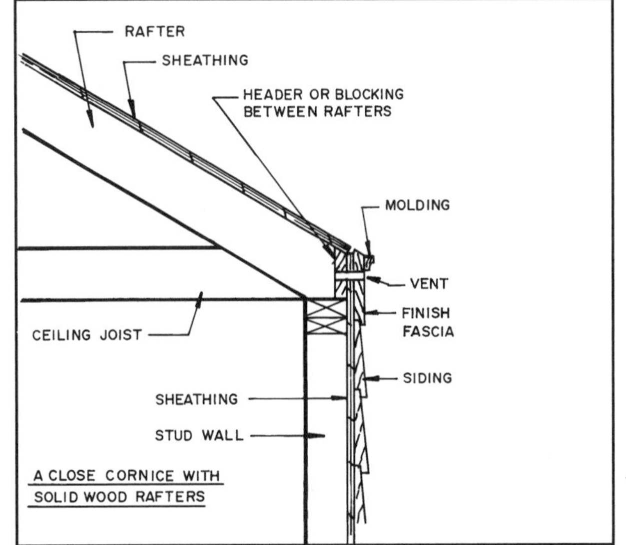

19-13 Construction details for a close, or snub, cornice with wood exterior siding.

Close cornices for roofs using I-joists, trusses, and flat roof construction are shown in **19-14**, **19-15**, and **19-16**.

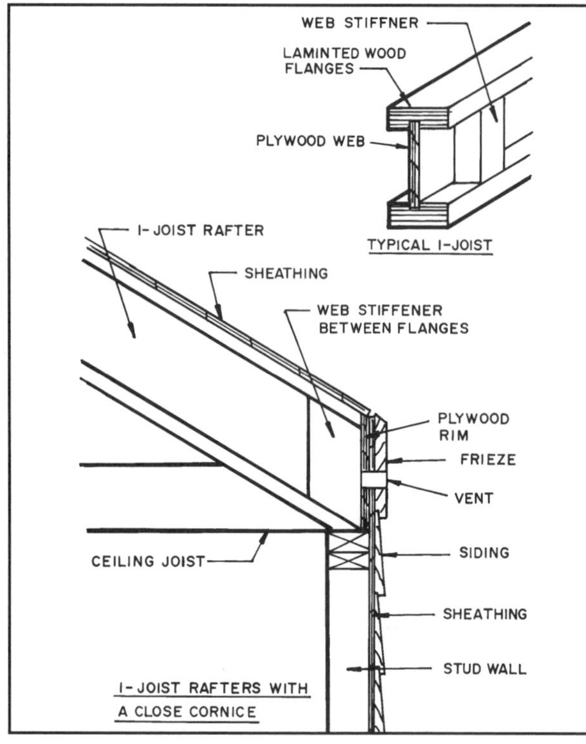

19-14 Typical close cornice construction when using I-joist rafters.

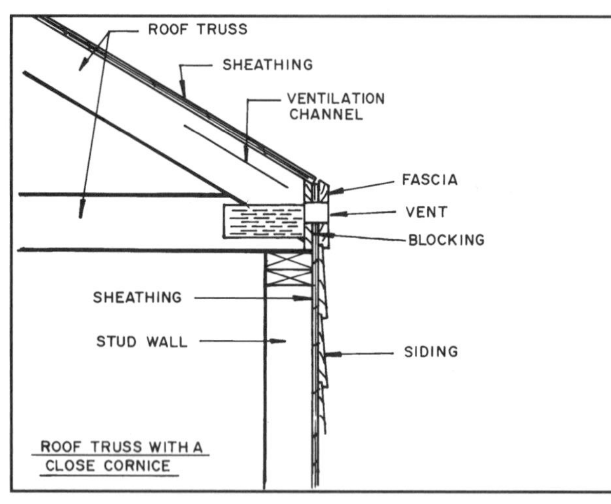

19-15 A close cornice with wood roof-truss construction.

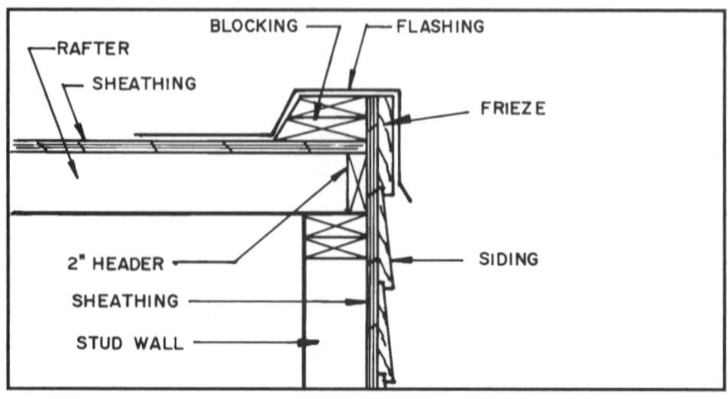

19-16 Close cornice construction with a flat roof.

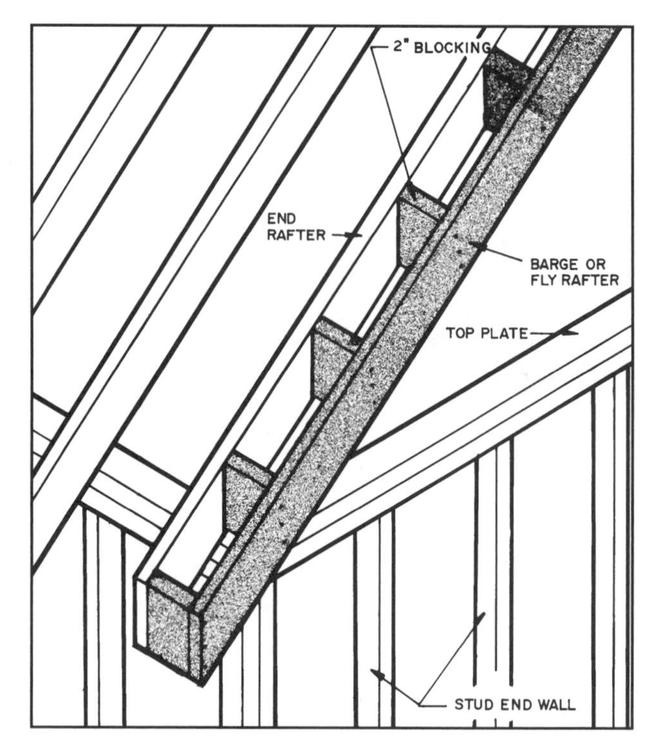

19-17 Framing for a short overhanging rake.

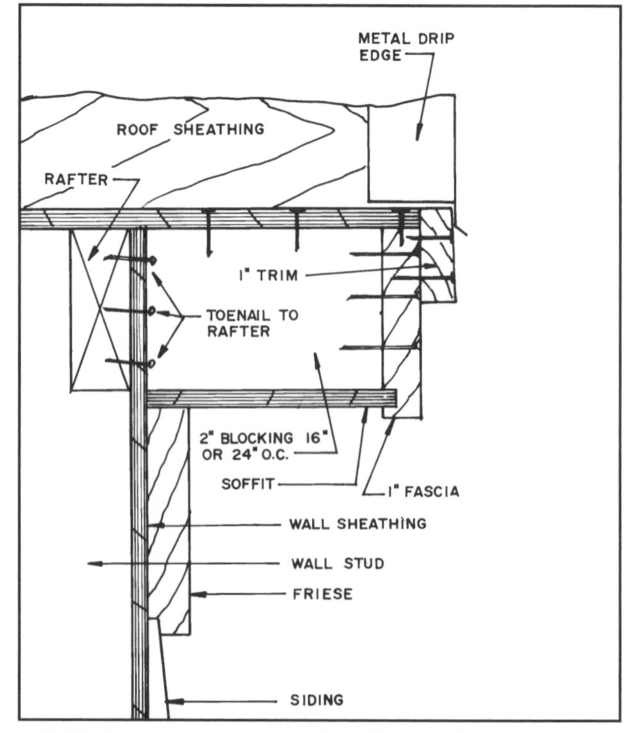

19-18 A section through a rake with a small overhang.

FINISHING THE RAKE OR GABLE-END

The rake is the finished edge of the roof on the gable-end. It may be a boxed rake, open rake, or close rake. The boxed and open rake project beyond the gable-end. When the rake projection is small, such as 4 to 8 inches (1000 to 2030mm), the fascia and soffit are nailed to short lookout blocks spaced 16 inches (4060mm) O.C. The lookouts are toenailed to the end rafter. The fascia is nailed to the lookouts and through the roof sheathing **(see 19-17)**. A frieze board is placed butting the soffit, and a molding can be added if desired **(see 19-18)**.

If the extension is in the range of 8 to 18 inches (2030 to 4570mm), it is necessary to build a "ladder" extension that is nailed through the sheathing into the end rafter. A two-inch (500mm) fly or barge rafter is nailed to the other end of the lookouts **(see 19-19)**. The lookouts are spaced 16 to 24 inches (4060 to 6100mm) O.C. The ladder is built on the ground, hoisted in place, and nailed to the end rafter. After the assembly is nailed to the end rafter, the finished fascia is nailed over the fly rafter, and a soffit is nailed to the bottom of the lookouts. A frieze board and molding can be used below the soffit.

Moderate cornice projections are sometimes framed as shown in **19-20**. The fly rafter is nailed to the end of the ridgeboard and is mitered and nailed to a two-inch (510mm)-thick fascia board. Blocking can be added to support the soffit. The sheathing is nailed to the fly rafter and adds stiffness to the assembly.

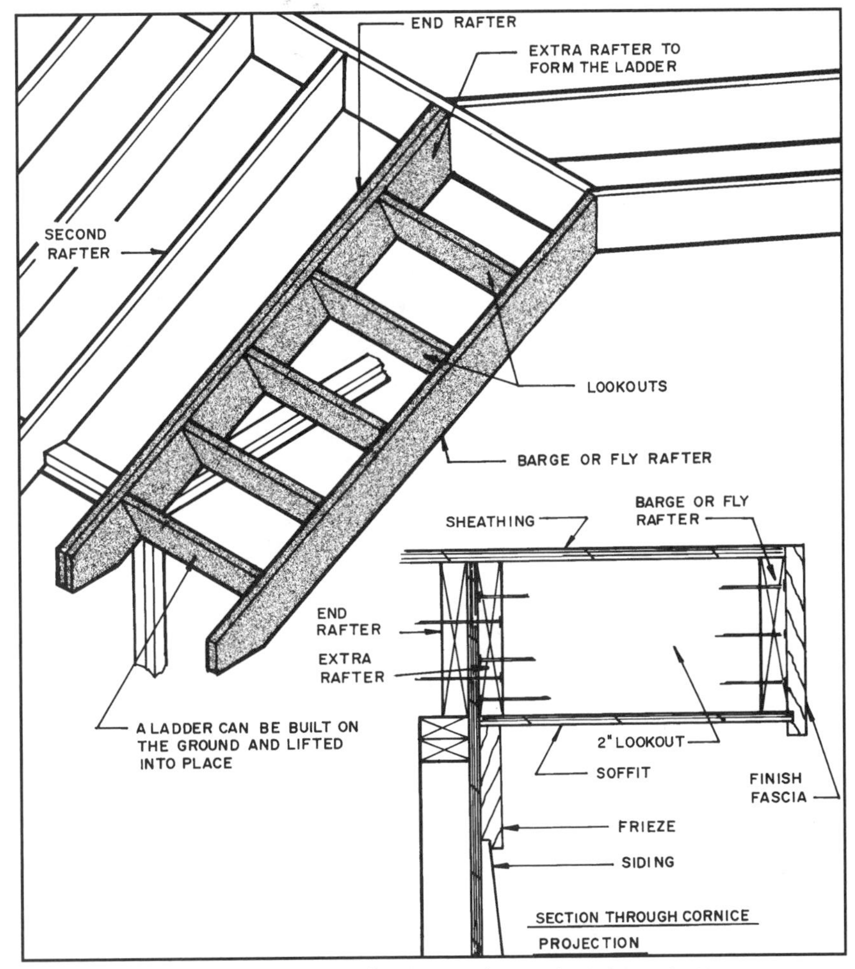

19-19 A "ladder" can be used to frame a moderate-size rake projection.

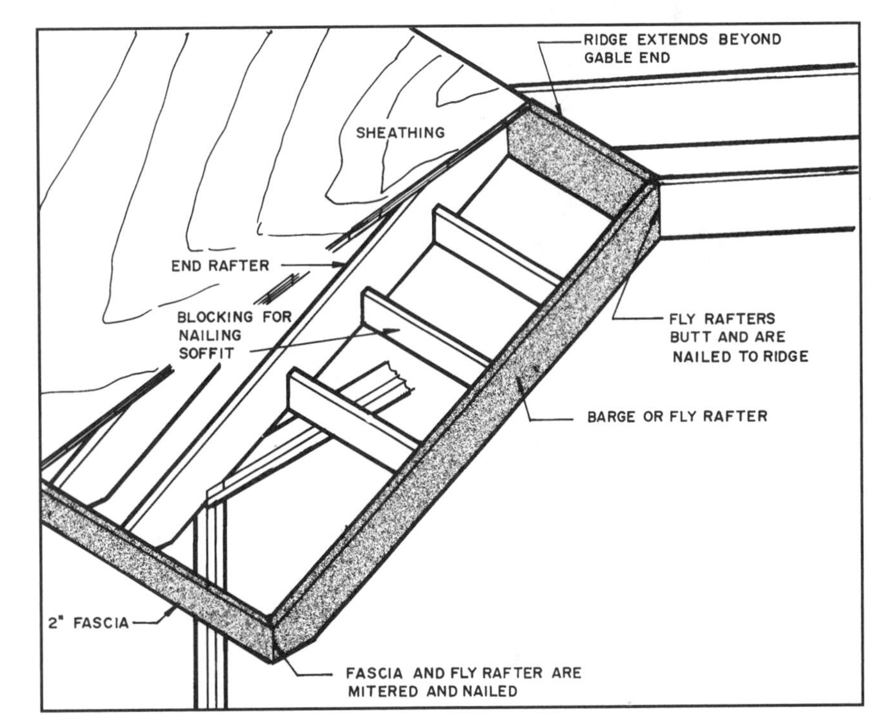

19-20 (Right) The fly rafter may be supported by the ridge board, fascia extension, blocking, and sheathing.

A rake with a wide—usually considered to be 18 inches (4570mm) or more—projection requires a rigid frame to resist snow and wind loads. A series of lookouts is used. In one case the lookouts rest on the top plate of the gable-end framing and extend to the next rafter, which is doubled. The lookouts are secured to the rafter with metal joist hangers **(see 19-21)**. Another procedure uses 2 × 4 lookouts laid flat and notched into two end rafters and face-nailed to the third rafter **(see 19-22)**. Notice in the illustrations that the ridgeboard extends beyond the end rafter or gable-end and provides support for the fly rafter and fascia. These ladder-type frames plus the roof sheathing provide adequate support for the fly rafter and the finished fascia.

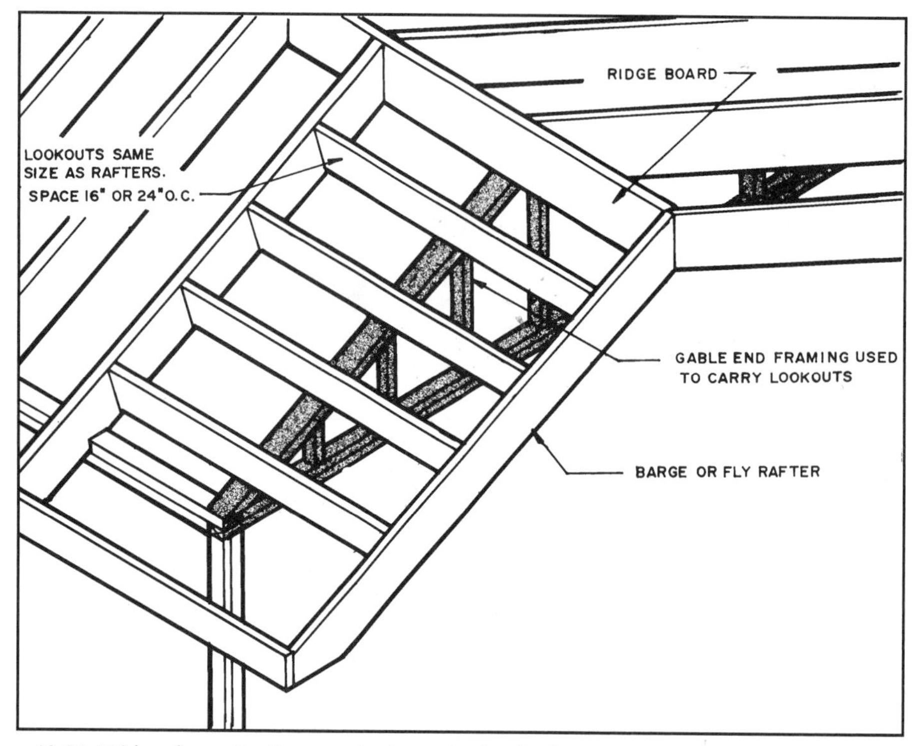

19-21 Wide rake projections can be framed using lookouts supported by the top plate of the gable-end framing.

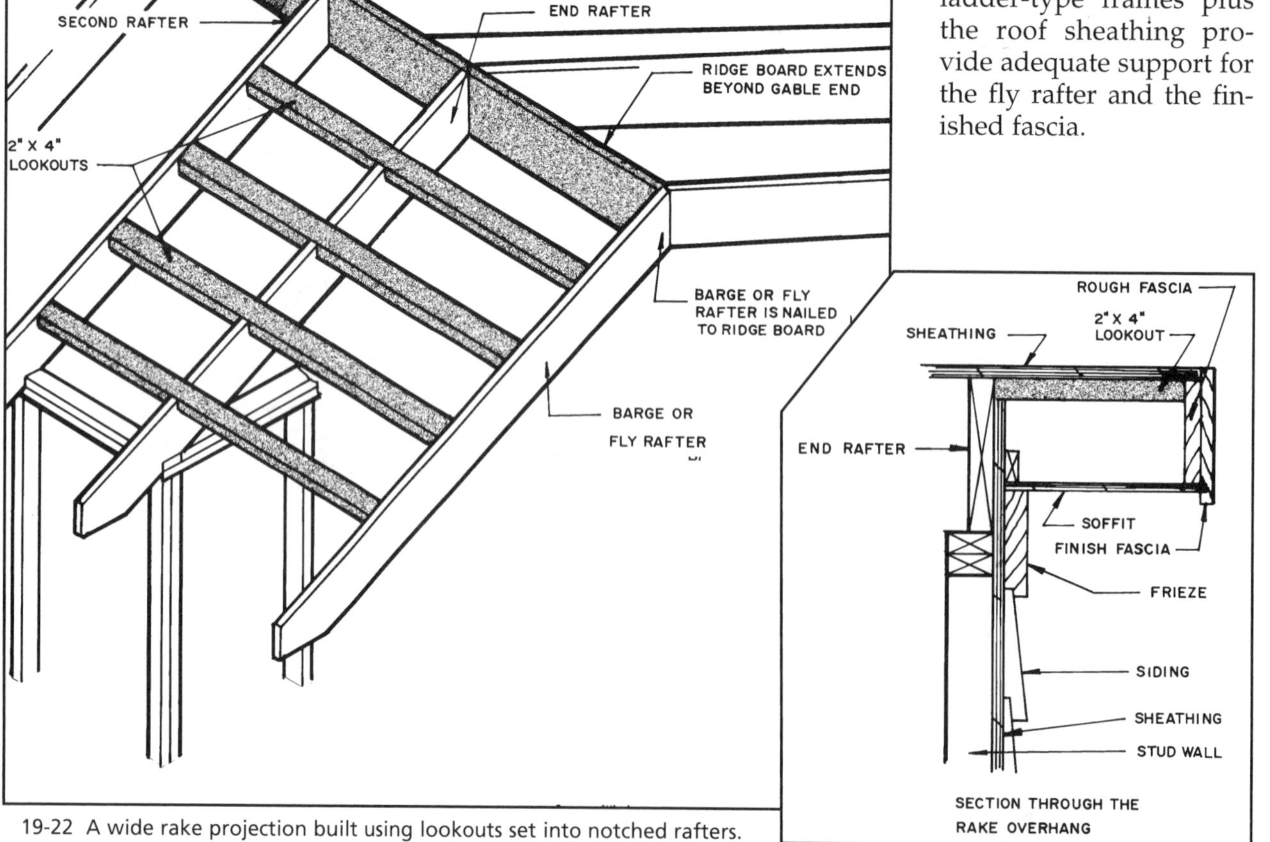

19-22 A wide rake projection built using lookouts set into notched rafters.

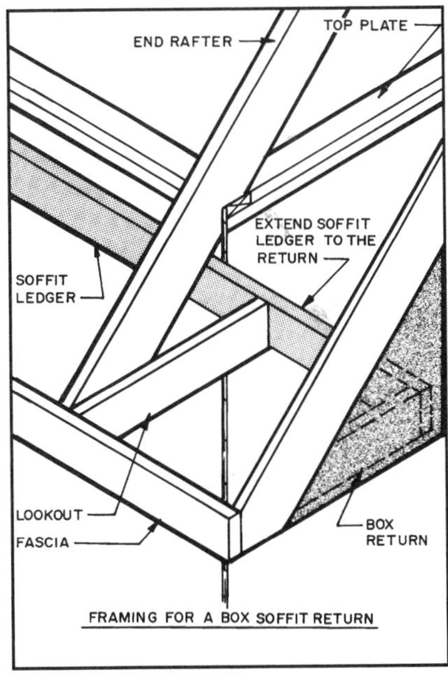

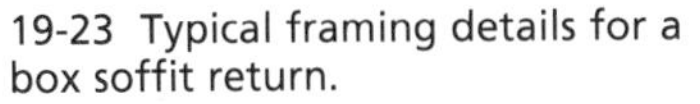
19-23 Typical framing details for a box soffit return.

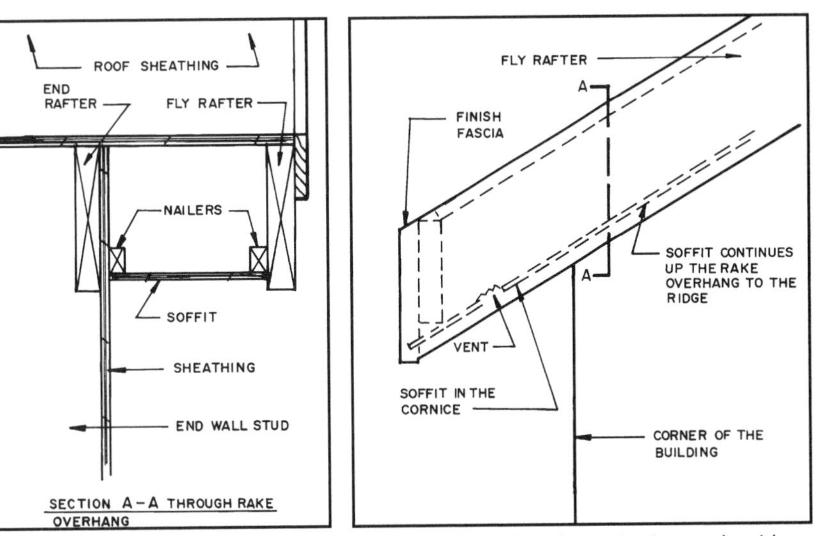

19-24 The soffit on a sloped box cornice extends up the rake projection to the ridge.

CORNICE RETURNS

The cornice return provides a finished end on the cornice as it runs from the fascia to the wall of the building. Hip and flat roof buildings have the cornice running continuously around the building. Gable roof buildings require that a finished return be constructed. The type of return required will be shown on the working design drawings. Two of the most frequently used returns are the soffit return and the Greek return. A narrow cornice can be finished using a boxed soffit return, framed as shown in **19-23**. When the cornice is a box type without horizontal lookouts, the soffit continues up the projected overhang, as shown in **19-24**.

The Greek return provides the most attractive cornice return. This is actually a little hip-type frame. Typical details of a Greek cornice return are shown in the two drawings and the photograph in **19-25**.

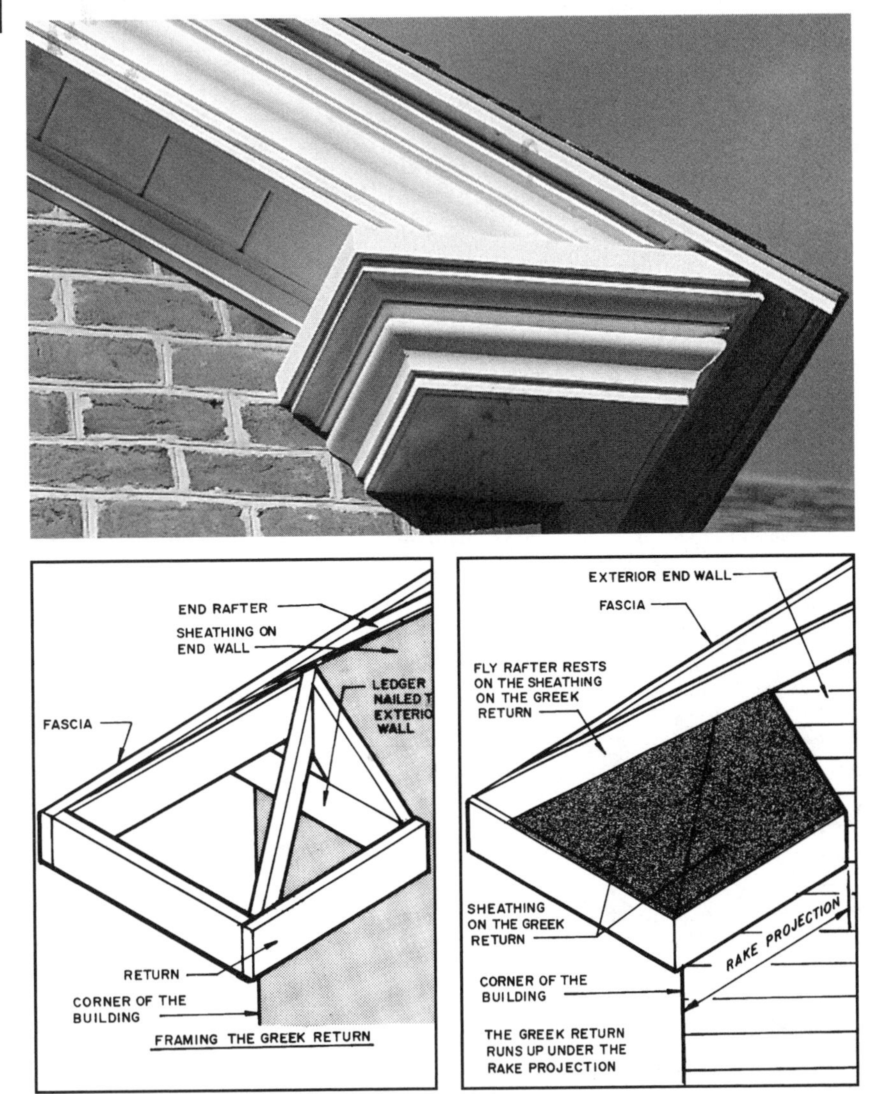

19-25 Framing a Greek cornice return.

There are situations where a fascia butts another roof. In a case such as this the butting end of the fascia is cut at the slope of the abutting roof. Keep the sloped end of the fascia at least one inch (250mm) above the surface of the shingles on the abutting roof **(see 19-26)**.

ALUMINUM & VINYL SOFFIT & FASCIA SYSTEMS

A typical detail for installing aluminum and vinyl soffits and fascias is shown in **19-27**. Exact details will vary depending on the manufacturer of the product. Installation directions are available from the manufacturer. The cornice has the rough fascia installed. It supports the fascia and soffit. The soffit panel is perforated so that it also serves as the attic vent.

TRUSS CORNICES

Cornice construction when trusses are used is part of the design of the truss. Several designs are shown in Chapter 18. The simplest design has the bottom chord overhang the exterior wall where it supports the soffit panel **(see 19-28)**. Notice the two-inch-thick rough fascia which stabilizes the trusses and provides a nailing surface for the finished fascia.

ATTIC VENTILATION

The two major problems occurring in an attic are the collection of moisture and the development of high temperatures. Even though the ceiling has a vapor barrier, moisture does tend to gravitate to the attic space. In the winter this warm moisture hits the cold roof of sheathing, condenses, and drips onto the insulation, reducing its effectiveness.

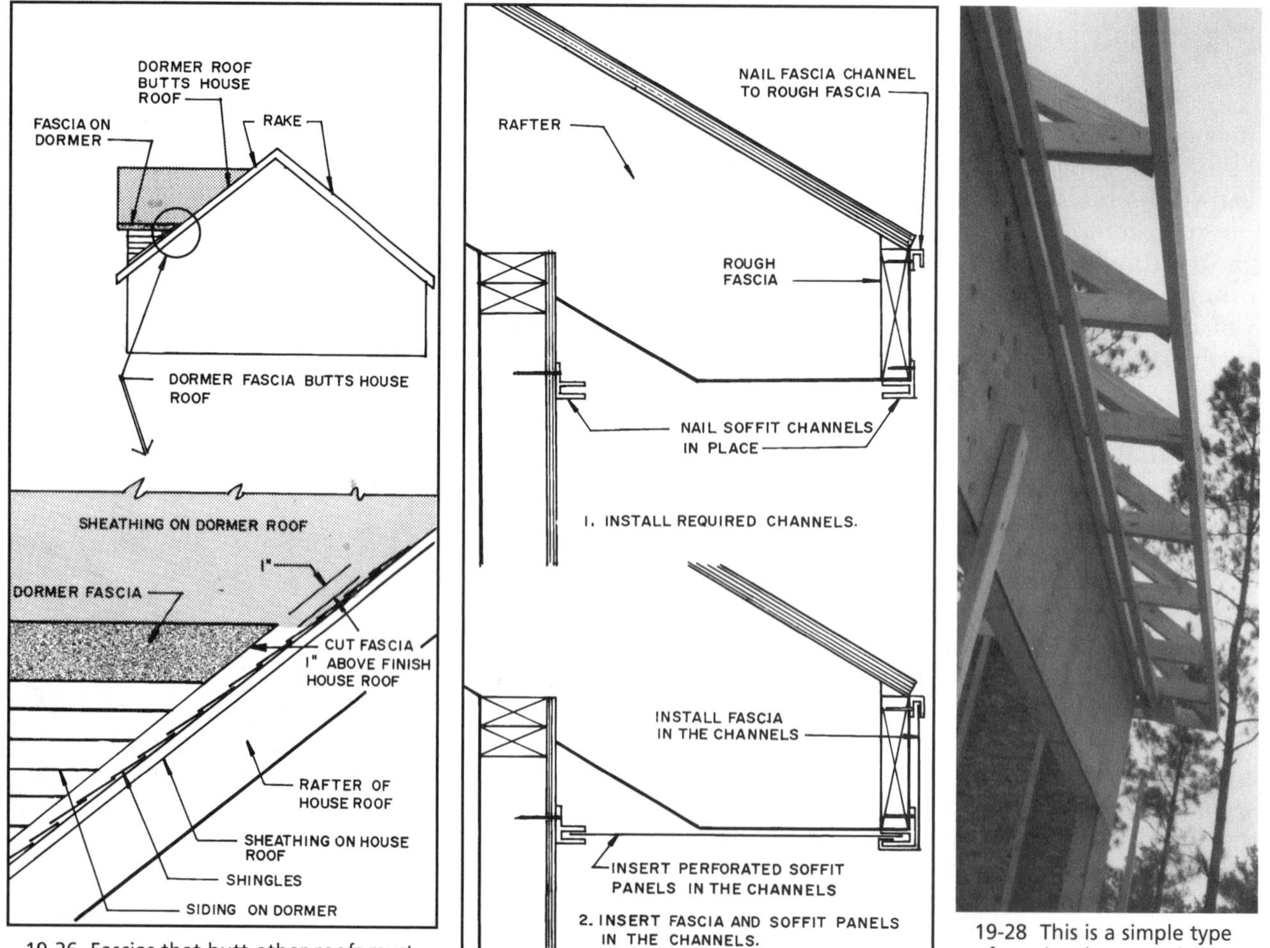

19-26 Fascias that butt other roofs must be cut so that they clear the roof by at least one inch.

19-27 Typical installation details for aluminum and vinyl soffits and fascias.

19-28 This is a simple type of cornice that uses the bottom chord to support the soffit.

In very cold climates the moisture freezes on the sheathing and eventually melts as the weather gets warmer. Attics need adequate ventilation to remove this moisture.

In the summer the air in the attic gets very hot. Even though the ceiling is well insulated, air-conditioning costs are increased by the transfer of some of this heat into the living area below. Ventilation in the summer is vital to reducing the air temperature in the attic.

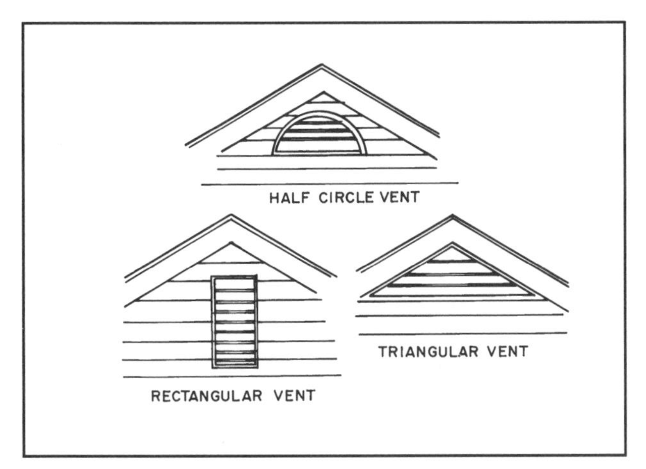

19-29 Typical gable-end attic vents.

The amount of ventilator area needed is established by the local building code. Typically the net free area of the vents should not be less than 1/150 of the area (square feet) of the space ventilated. If the area is ventilated by vents in the upper portion of the attic space, the required net area of the vents can be 1/300 of the area (square feet) to be ventilated. All vents should be covered with some form of mesh. The net area of a vent is the amount of open unrestricted air space. Since vents are covered with wire mesh the actual size of the vent unit should be double the required net area.

The air movement in the attic begins with vents in the soffit, and it exits the attic with some type of vent mounted high on the roof. Commonly used roof vents include gable-end vents, ridge vents, metal ventilators near the ridge, dormer-ventilators, cupolas, and power fans **(see 19-29, 19-30, and 19-31)**.

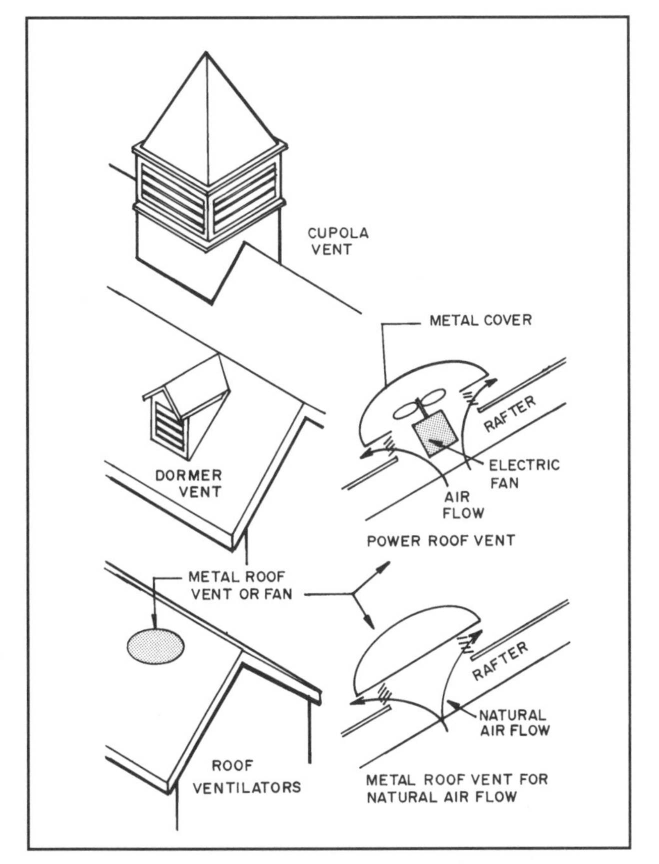

19-30 The most frequently used methods for venting the attic.

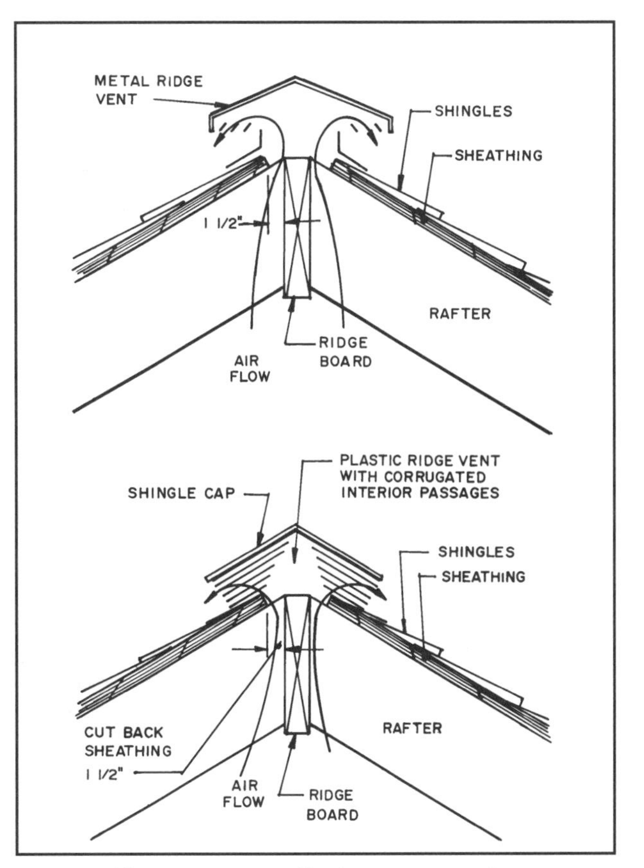

19-31 Ridge vents are an effective way to ventilate an attic.

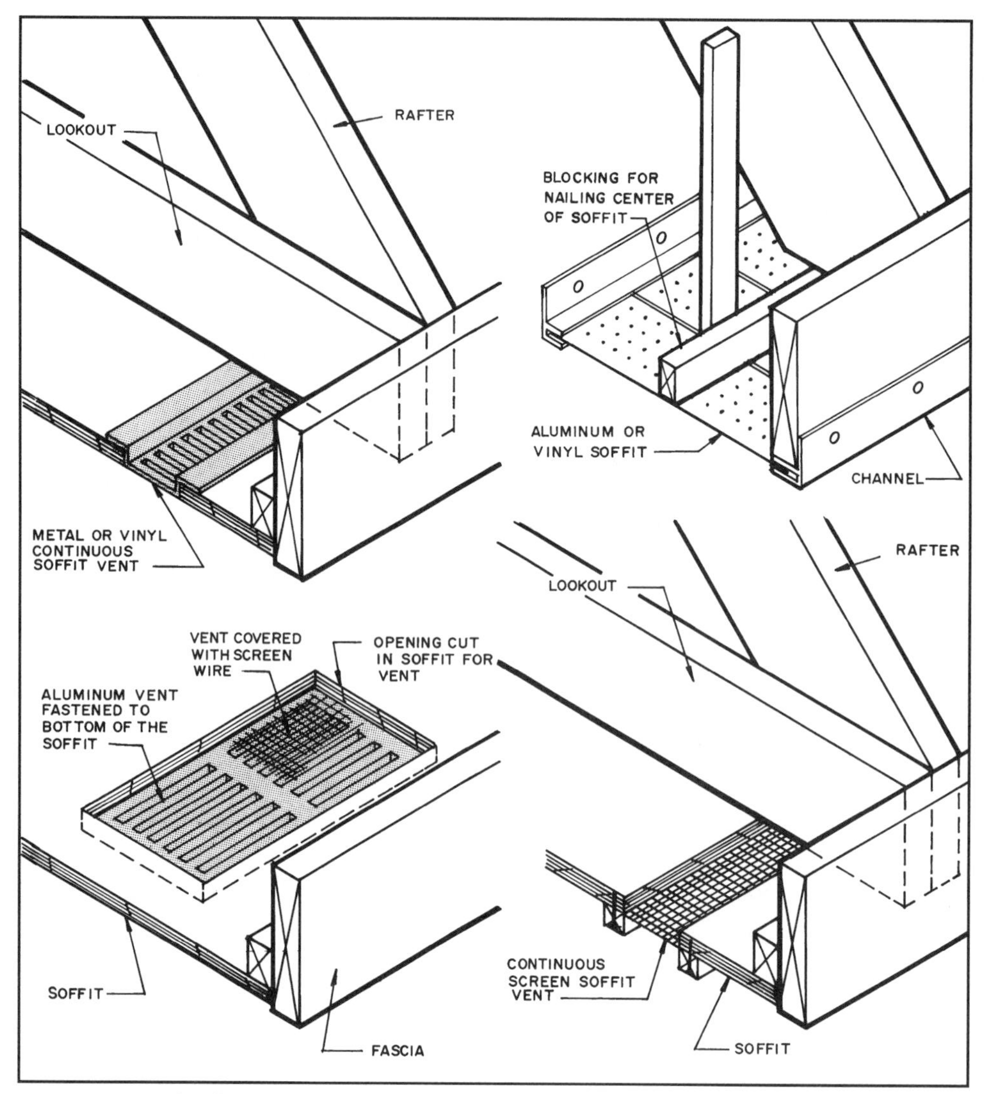

19-32 Types of soffit vent for box cornices.

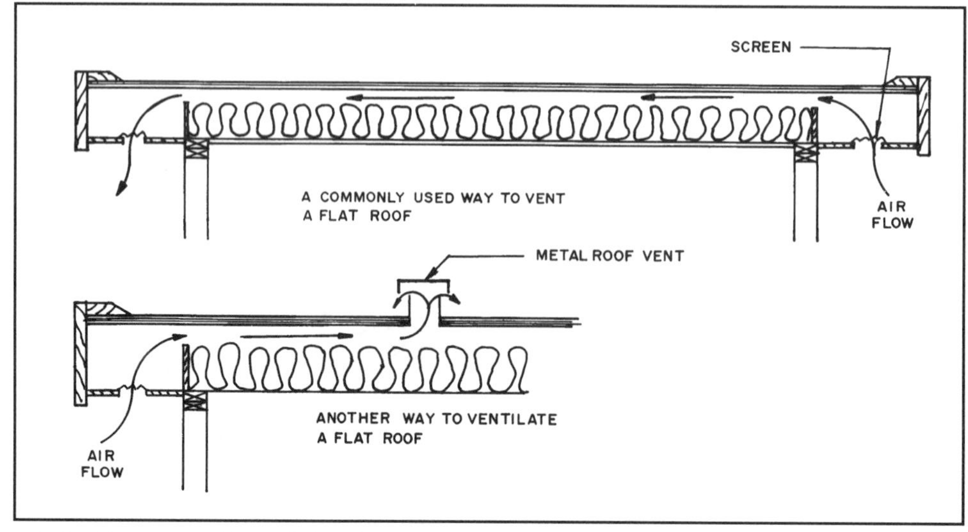

19-33 Ways to ventilate a flat roof.

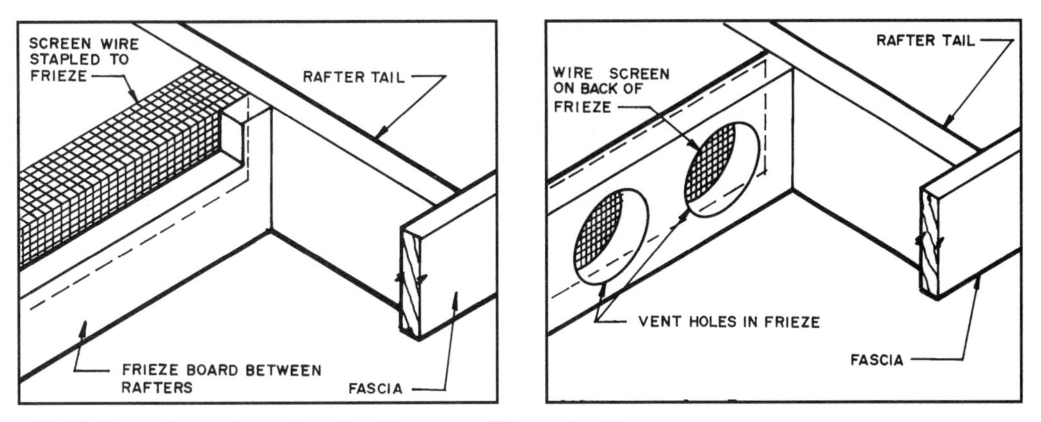

19-34 Several ways to vent roofs with open-soffit construction.

The vents in the soffit may be individual vents spaced evenly along the soffit, continuous vents, or aluminum or vinyl soffits that are perforated so that the entire soffit becomes a vent **(see 19-32)**.

Flat roofs require each rafter space be vented. The insulation must allow an air space between the top of the insulation and the roof sheathing. The most common way to ventilate a flat roof is with continuous vents in the projecting soffit **(see 19-33)**. Flat roof ventilators and parapet wall ventilators also help vent flat roofs.

Open soffit construction requires that the vents be placed in the frieze block that closes the opening above the exterior wall. Variations for venting an open soffit are shown in **19-34**.

V OTHER FRAMING METHODS

CHAPTER 20

Post, Plank & Beam Construction

Wood beams and posts have been used for centuries to construct almost any type of building. Very early buildings used wood roof timbers on masonry walls. From A.D. 1300 to 1400 wood posts and beams were assembled, forming the structural frame. This technique came to the United States when British carpenters emigrated to North America in the seventeenth and eighteenth centuries. This type of construction had hand-hewn wood timbers that were joined by wooden joints.

Currently there are a number of companies manufacturing these structural members. They help design the structural framework to your floor plan. The finished structure permits the heavy beams, rafters, and wood planking to be seen and become part of the beauty of the finished interior **(see 20-1)**.

Large commercial buildings are also built using timber framing. These members are typically larger than those for residential construction. This is referred to as **heavy timber construction (see 20-2)**.

The basic system currently in use for residential buildings uses beams widely spaced to form the floor and roof structural system. The roof beams are supported by posts located directly below them. The posts can transmit the load to piers or to a conventional foundation. The space between the posts is framed with stud walls which contain the doors and windows **(see 20-3)**.

20-1 These redwood beams and rafters form the structural system of the building and establish the interior decor. *(Courtesy California Redwood Association)*

20-2 A large post and beam wood-framed building. Notice the wood roof decking stacked on top, ready for installation. *(Courtesy Timber Structures, Inc.)*

ADVANTAGES

The **advantage** of the post, plank, and beam system is that it has fewer members, which can reduce the labor costs compared to standard framing methods. Many people like the exposed-wood structural system and the wood ceiling formed by the roof decking. The thick floor and roof decking provide good fire resistance.

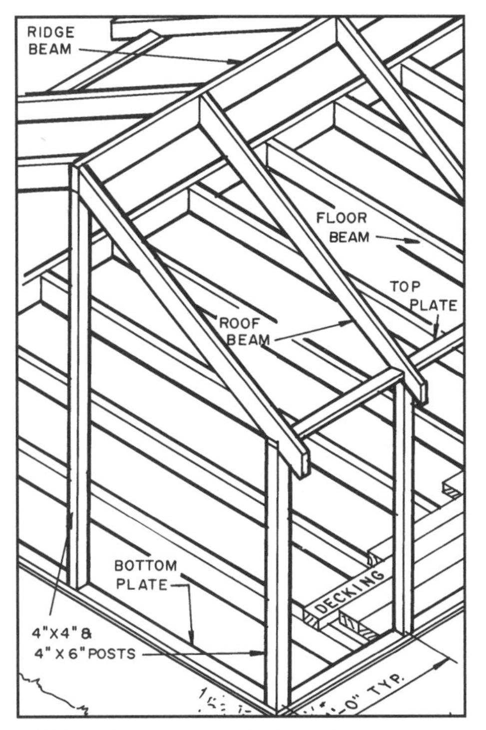

20-3 A typical post, plank, and beam structural frame.

LIMITATIONS

The **limitations** to this system include difficulty in finding concealed places to run wiring, piping, and ducts. Sometimes a beam is made up of two wood members with spacers forming a cavity for wires and pipes. The plank floors span the wide distance between the floor beams. They will not usually carry concentrated loads such as a bearing partition or a refrigerator, so extra framing is required. It is more difficult to insulate the roof because you want the wood decking exposed. Insulation will have to go on the outside of the roof.

FLOOR & ROOF BEAMS

Beams may be solid wood, built-up, laminated veneer lumber (LVL), glued laminated (glulam), parallel strand lumber (PSL), or panel-lumber beams **(see 20-4)**. (These are discussed in detail in Chapter 39.) Since roof beams are often left exposed to the inside of the building, the choice of which material to use depends not only on the strength but on the quality of the exposed surface. Exposed beams are stained or sealed with a clear, natural finish. Some beams are made using two-inch stock and spacers to permit the passage of wires and plumbing. Built-up beams are often cased with high-quality finished lumber **(see 20-5)**. The size of the beam depends on the type used, the loads, the specie of wood, and the span. These decisions must be made by an engineer.

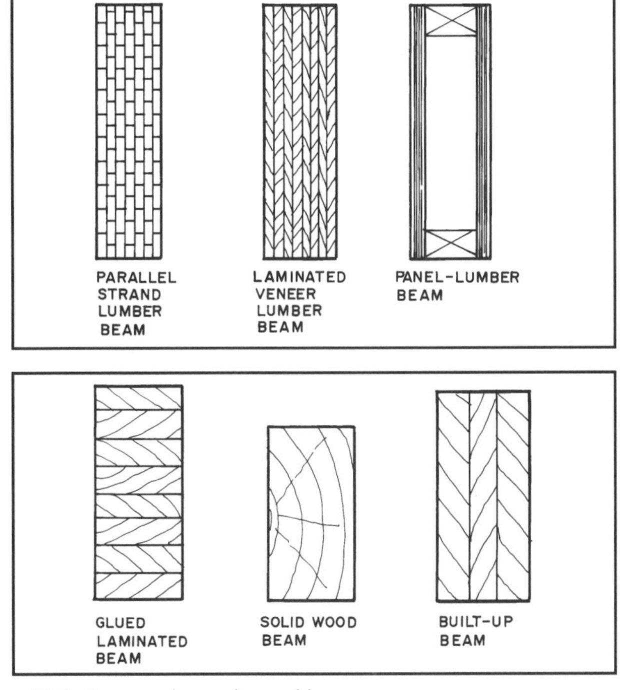

20-4 Commonly used wood beams.

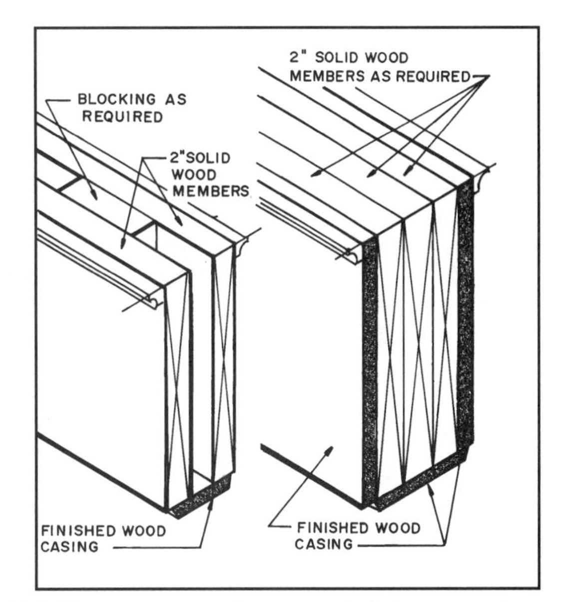

20-5 The appearance of carpenter-built beams can be improved by casing them with quality wood and moldings.

POSTS

Posts may be solid wood, built-up, glued laminated, or parallel strand lumber. They may be exposed to the interior, in which case the quality of the exposed surface is important. The size depends on the type of post, the loads it will carry, the specie of wood, the length of the post, and the spacing between posts. The longer the post, the greater it must be in cross-sectional area.

As the engineer does the calculations to select the post size, one factor of great importance is the **slenderness ratio (see Table 20-1)**. The slenderness ratio is the number found by dividing the post length in inches by the actual size of the smallest dimension of the cross section of the post. It is represented by the ratio 1/d.

For example, a 6 × 6 wood post 10 feet long is 120 inches long and 5½ inches in smallest cross section. The slenderness ratio is 120 divided by 5½, which equals 21.8. The **smaller** the slenderness ratio, the **greater** the load that can be carried.

Posts should provide an adequate bearing surface for the beams. Typically a 4 × 4-inch nominal post is adequate to carry normal roof loads. However if beams butt on a post a 4 × 6-inch post is required to give a three-inch nominal minimum bearing surface **(see 20-6)**.

PLANKS & DECKING

The words **"plank"** and **"decking"** are used interchangeably to describe wood members over one-inch thick. They are used to cover the roof beams and support the finished roofing, and they are also used to cover the floor beams and carry the live loads imposed on the floor. Commonly used wood decking is 2, 3, or 4 inches thick. Two-inch-thick decking usually has a single tongue-and-groove. Thicker types will have two tongue-and-grooves **(see 20-7)**.

The distance decking can span depends on the specie and grade of wood and the loads to be carried. A general example is given in **Table 20-2**. Consult the manufacturer's data for specific capacities.

Decking is laid up in several ways as shown in **20-8**. The **single span** has minimum stiffness and requires excessive labor. **Double span** is about twice as stiff as single span. Notice the end joints are staggered. The **random-controlled lay-up** method produces stiff floors and requires less labor because there are fewer pieces and fewer end joints. When using random decking lengths the decking must span three or more supports. The distance between end joints in adjacent rows of decking should be at least two feet and between

Table 20-1 Examples of column slenderness ratios.

Nominal size (inches)	Actual size (inches)	Slenderness ratio for 8'-0" post	Slenderness ratio for 10'-0" post	Slenderness ratio for 12'-0" post
4 × 4	3½ × 3½	27.4	34.3	41.1
4 × 6	3½ × 5½	27.4	34.3	41.1
6 × 6	5½ × 5½	17.4	21.8	26.2
8 × 8	7¼ × 7¼	13.2	16.5	19.8

Table 20-2 Total allowable uniformly distributed roof loads for 3-inch nominal (2½-inch actual) laminated wood decking.*

Span (deflection 1/240)	Inland red cedar E 1.2	Southern pine E 1.8	Douglas fir/larch E 1.8
	(pounds per square foot, psf, of roof area)		
8'-0"	54	80	80
10'-0"	27	44	41
12'-0"	16	25	24

*Simple span with end joint falling on supports.

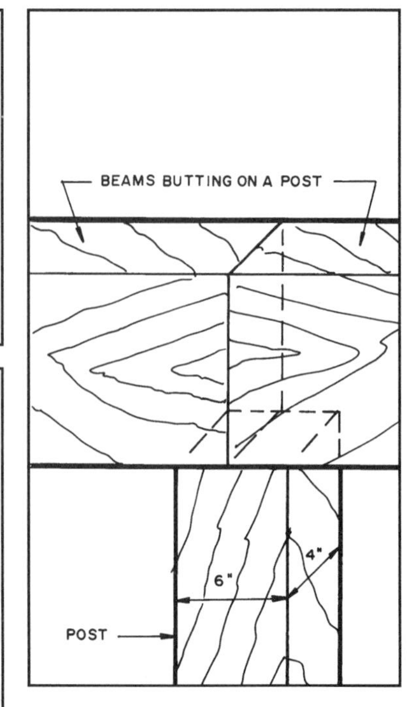

20-6 A 4" × 6" post typically provides adequate bearing for butting beams.

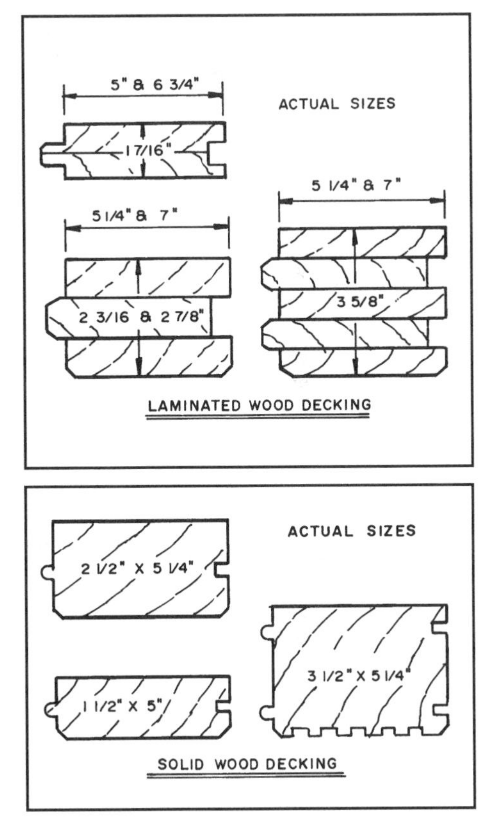

20-7 A few of the types and sizes of solid and laminated wood decking.

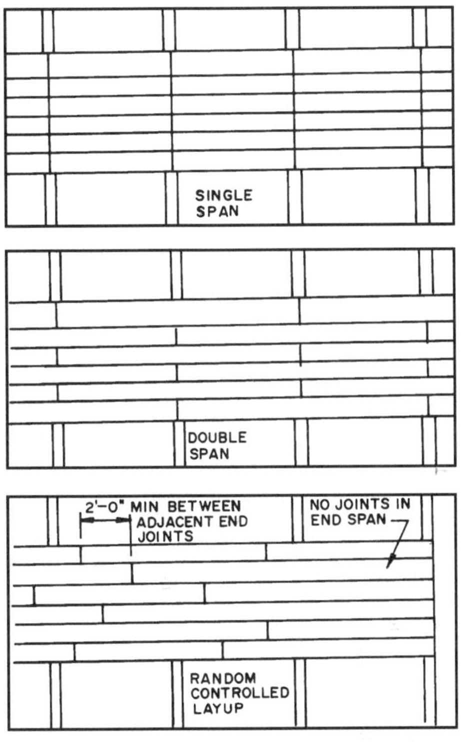

20-8 Commonly used methods for installing wood decking.

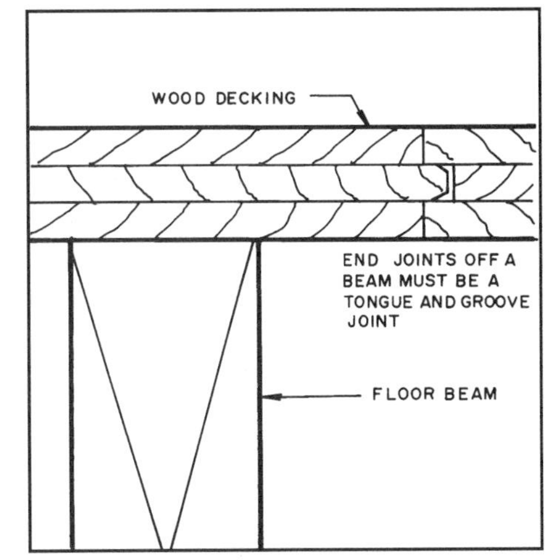

20-9 End joints in random controlled lay-ups not falling on a beam must have a tongue-and-groove joint.

decking separated by one row at least one foot. The decking should have tongue-and-groove end joints **(see 20-9)**. When decking sloped roofs, the edge with the tongue should face up the slope. Spans will vary for floor and roof construction, and the manufacturer's instructions should be observed.

Since the roof decking also serves as the exposed finished ceiling, appearance is important. Roof decking is available with a variety of joints as shown in **20-10**.

Observe the manufacturer's directions for nailing solid-wood and laminated-wood decking. The following is a typical example. Face-nail the decking to the beam with two nails. A 20d nail is used

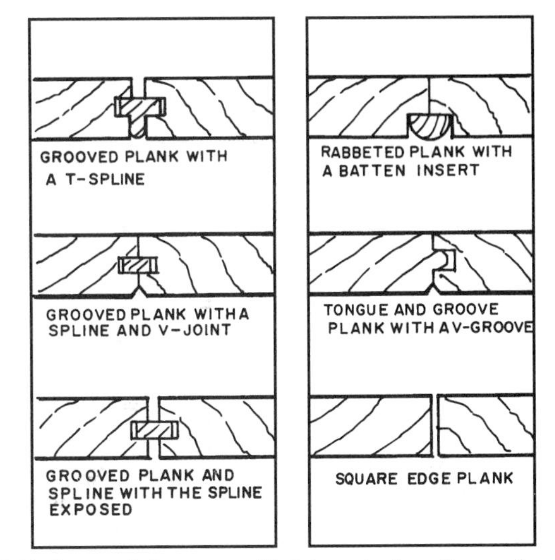

20-10 Some of the types of joint used when the roof decking is exposed, forming the finished ceiling.

for three-inch decking. Then toenail each row together through the tongue. Space these 30 inches apart and toenail within 12 inches of any end joints **(see 20-11)**.

In addition to wood decking, other products used include wood fiber roof planks, plywood, and wood panel and composite stressed-skin panels. **Wood fiber planks** are made by bonding long wood fiber strands. They are available in 2-, 2½-, and 3-inch thicknesses, 48 to 96 inches long, and as square-edged or tongue-and-groove-edged panels. They are used for roof decking. **Plywood** sheets 1⅛ inch thick with tongue-and-groove edges can span up to four feet.

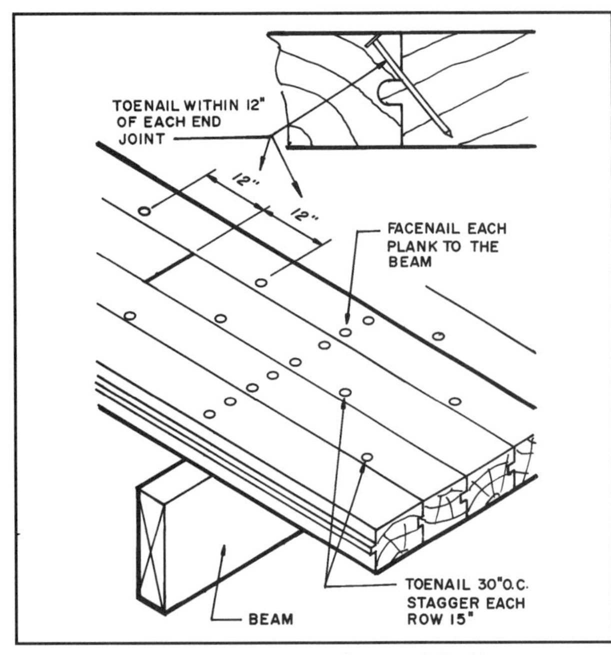

20-11 A typical nailing pattern for wood decking.

Lumber-panel stressed-skin panels use solid-lumber stringers and plywood skins to produce panels that can span long distances **(see 20-12)**. They can be fabricated on the ground, and their use speeds up the erection of the building. They must be designed following pre-engineered specifications. The panels may have a panel skin on one or two sides. Lumber blocking is required wherever there is a joint in the skin. Stressed-skin panels should be built following the specifications of APA—The Engineered Wood Association.

Composite stressed-skin panels are made in a variety of ways. One type has an isocyanurate foam insulation core with a nail base board bonded on top and a structural board bonded on the bottom **(see 20-13)**. Install following the manufacturer's nailing instructions.

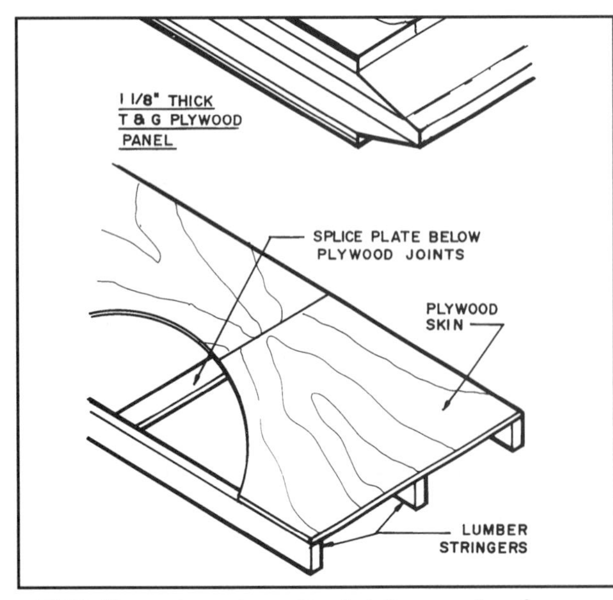

20-12 Plywood panels used to deck floors and roofs.

FOUNDATIONS

Foundations for post, plank, and beam framing may be continuous foundation walls or piers. Typically posts are spaced 8 feet apart, so a pier foundation is well adapted to this type of construction. Chapter 8 has additional information on foundations. Examples in this chapter can be seen in **20-14, 20-15,** and **20-17**.

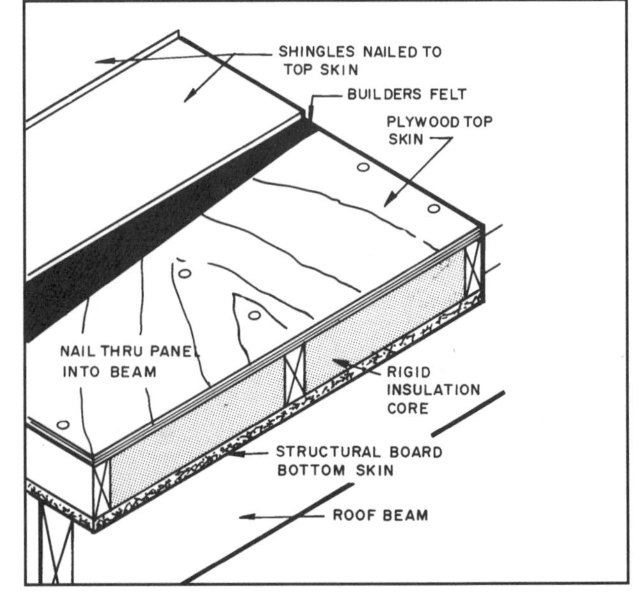

20-13 A typical composite stressed-skin roof panel having an insulating core.

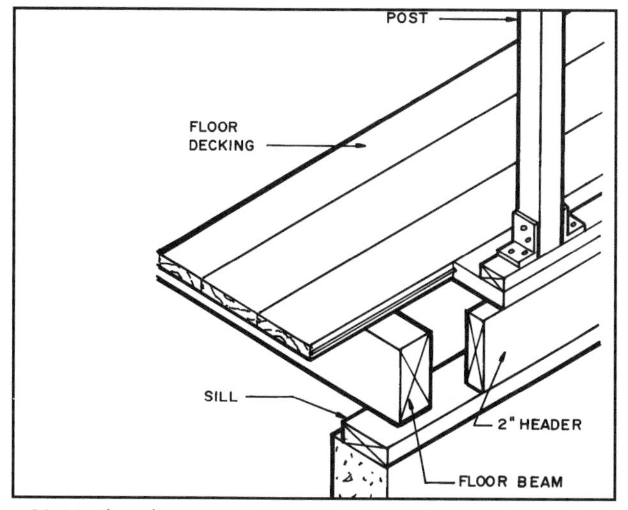

20-14 Floor beams can rest on a wood sill.

FLOOR CONSTRUCTION

The floor framing system is built several ways. One way is to rest the floor beam on a wood sill **(see 20-14)**. A header is nailed across the ends of the beam, as is done in conventional framing with floor joists. If it is desired to lower the building, the beams can be set in pockets. The pockets should allow a ½-inch air space on all sides. Some prefer to place a ¼-inch-thick steel plate between the bottom of the beam and the concrete pocket. The beam should bear at least four inches on the foundation **(see 20-15)**. Beams can be supported in the interior of the building by columns **(see 20-16)** or piers **(see 20-17)**. Several types of post anchors are available.

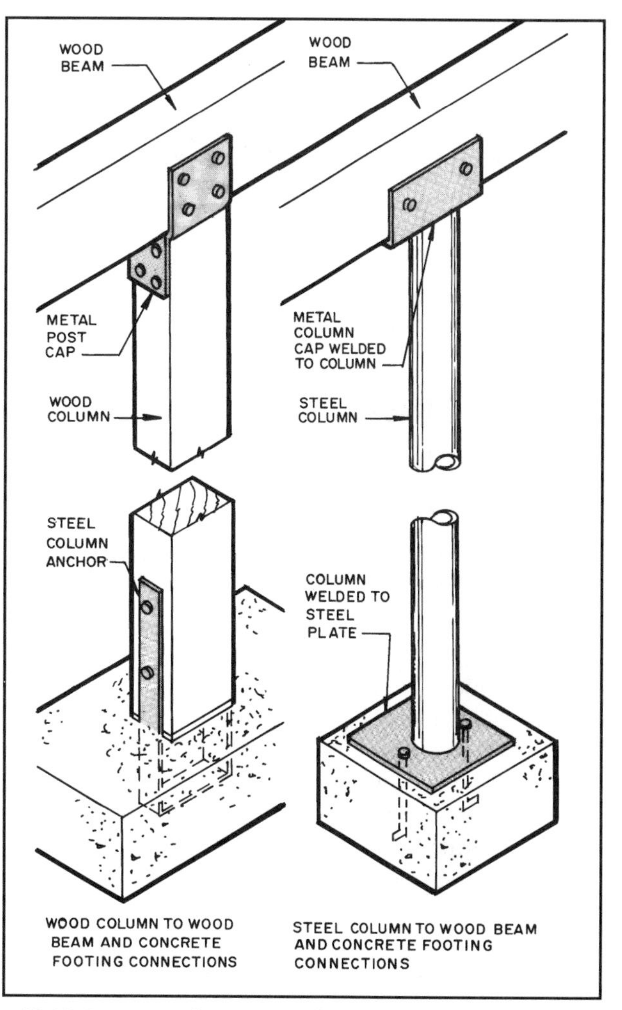

20-16 Beams can be supported within the building by columns resting on footings.

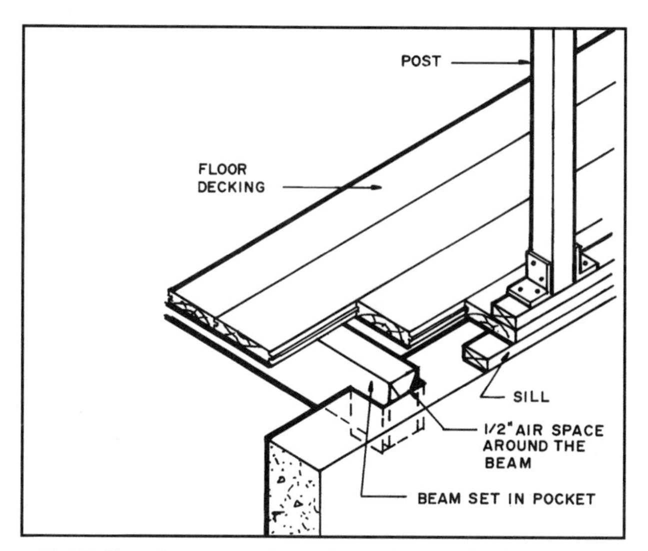

20-15 Floor beams can be set in pockets in the foundation.

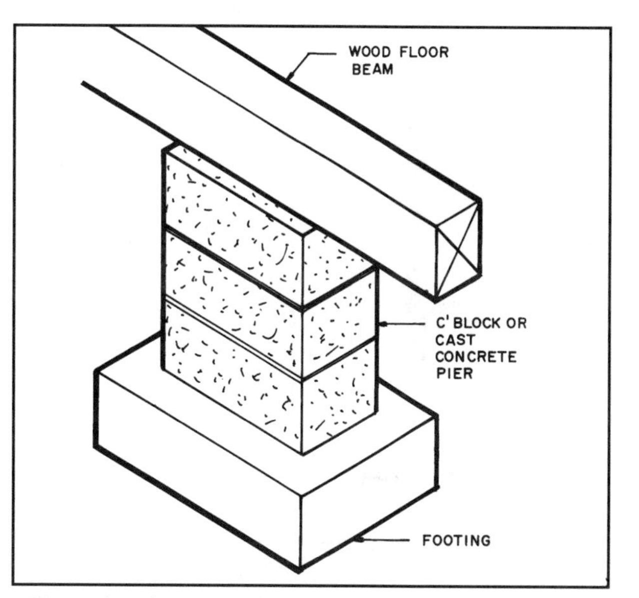

20-17 Floor beams can be supported with cast concrete or concrete block piers.

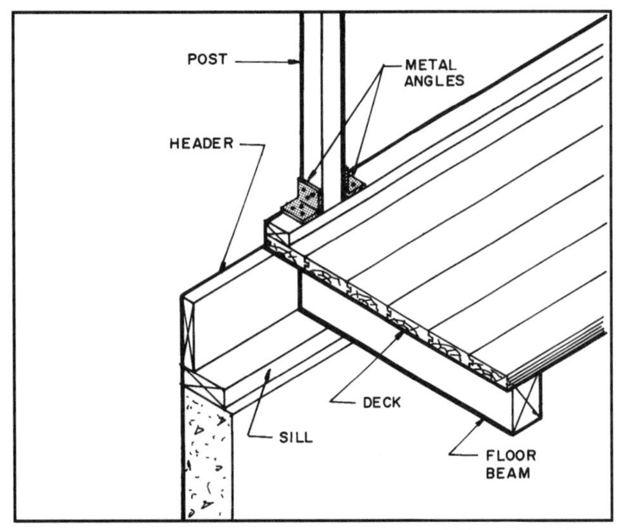

20-18 Typical wall framing with the posts falling directly above a floor beam.

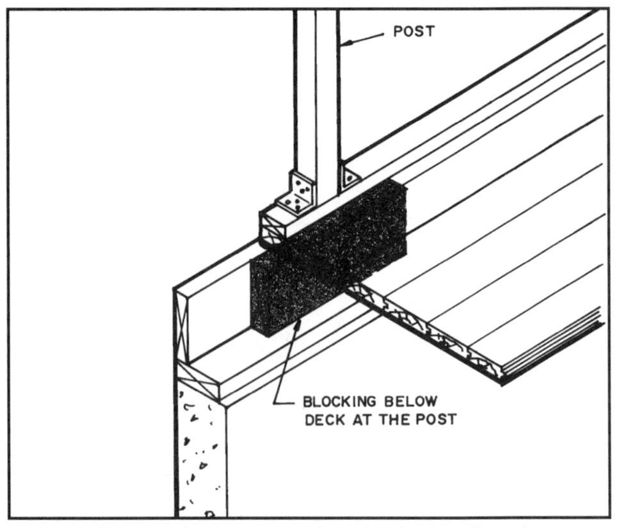

20-19 If a post does not rest directly above a floor beam, blocking below it is required.

If the floor is subject to concentrated loads, such as a heavy whirlpool, or if a nonload-bearing interior partition runs on the decking between and parallel with the floor beams, supporting framing is required above or below the decking. No extra support is required for nonload-bearing interior partitions running perpendicular to the floor beams.

EXTERIOR WALL CONSTRUCTION

The exterior structural wall posts are placed directly below a roof beam. The posts usually rest on a bottom plate directly above a floor beam **(see 20-18)**. If the posts are between floor beams, blocking is placed below them **(see 20-19)**. A wood top plate ties the tops of the posts together.

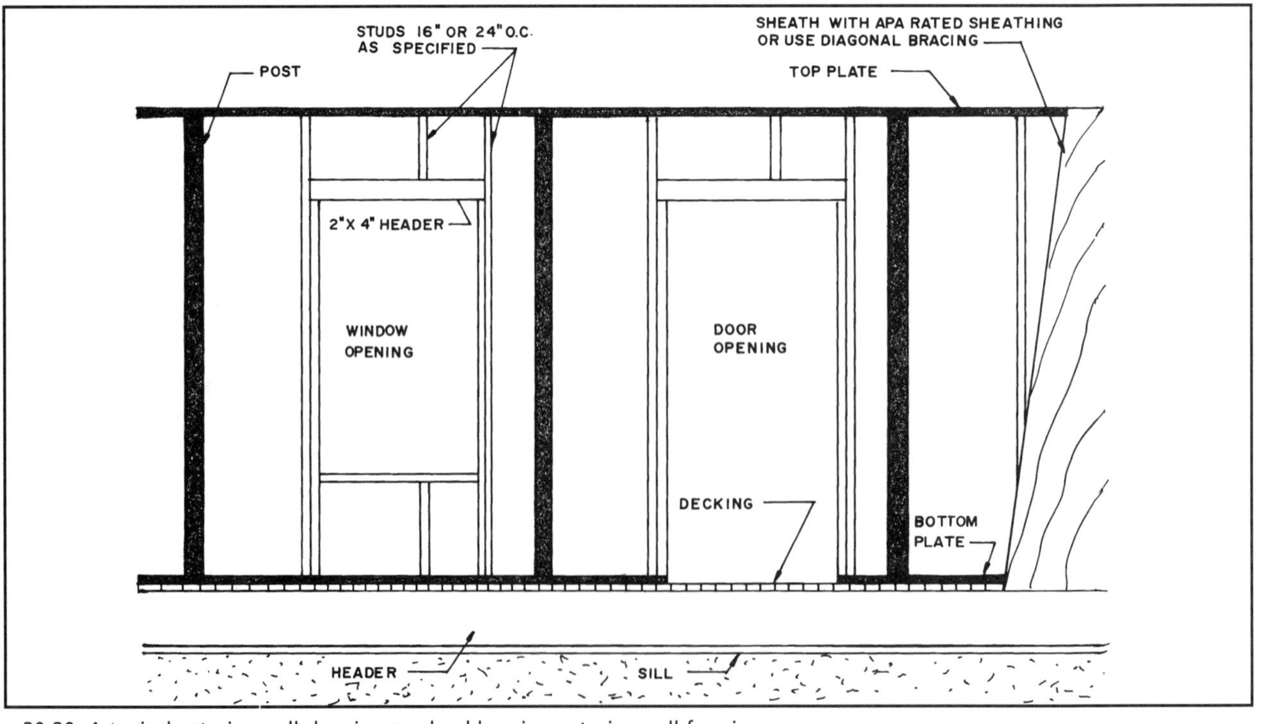

20-20 A typical exterior wall showing nonload-bearing exterior wall framing.

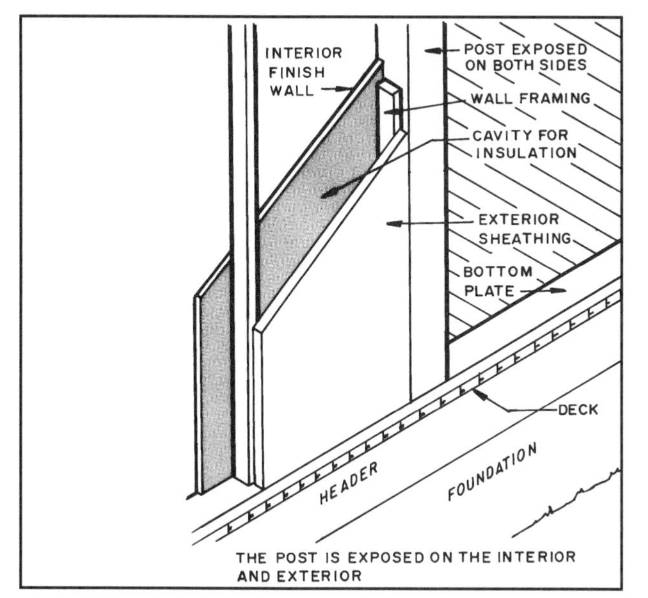

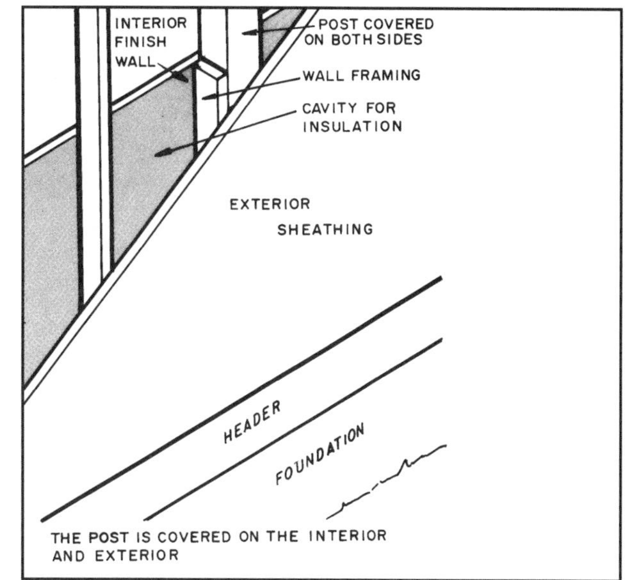

20-21 Two ways to frame the wall openings between the posts.

When possible, posts and beams are spaced at standard distances (48, 72, or 96 inches) used in conventional construction, because construction materials are manufactured to fit these distances. The sizes of the posts and beams used depend on the engineering design of the building, which includes the distance between these members.

The spaces between the posts are filled with stud-framed walls, which may have door and window openings. Since these are nonload-bearing walls, large headers are not needed. The exterior walls not only enclose the structure but also provide lateral stability for the entire wall. The posts alone will not resist lateral forces' causing racking. APA-rated sheathing provides the strength needed. It is advisable to have some sections between posts with no openings. These are needed to provide the required wall bracing **(see 20-20)**.

When framing the spaces between the posts the posts can be exposed to the interior or hidden by the constructions shown in **20-21**. Many like to expose the posts because they contribute to the overall appearance of the post and beam structure. The framing can also expose the post on the exterior wall. However, the post could be covered on the interior and exterior if desired.

A post may be toenailed to the bottom plate and through the top plate as shown in **20-22**. In areas with high winds, stud plate ties are recommended **(see 20-23)**. Some prefer to toenail the post directly to the subfloor rather than to a bottom plate.

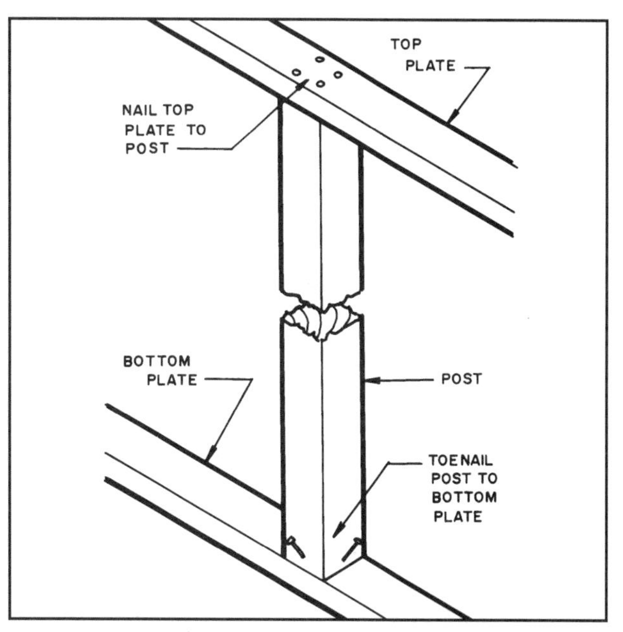

20-22 Posts can be nailed to the top and bottom plates.

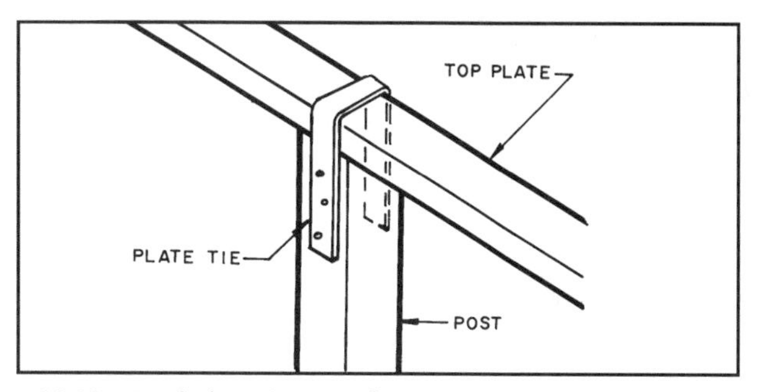

20-23 Metal plate ties are often used, especially in areas with high winds.

Typical post-to-beam connections are shown in **20-24**. There are a variety of metal connections available. The designer will specify the type you are to use. Special considerations are necessary in areas having high winds or earthquakes.

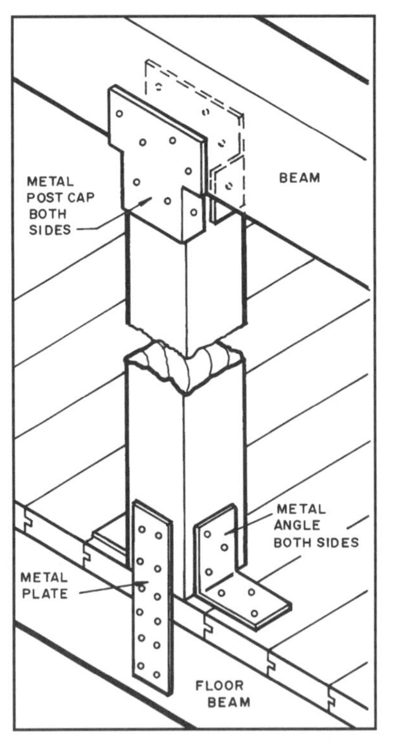

20-24 Metal post-to-beam and floor connectors provide strong connections.

Typical framing details for a two-story post, plank, and beam structure are shown in **20-25**. The posts are placed directly above each other and must rest on a solid, secure base such as a beam. This is much the same as conventional platform construction. The second floor is built on a platform formed on top of the first-floor framing. Insulation is placed between the exterior wall studs, as in conventional construction.

Interior partitions are framed in the same manner as explained for conventional construction. They generally are nonload-bearing. They provide rigidity and lateral stability to the structure. When they run parallel with the floor beams, the floor must have additional framing **(refer to 20-18)**.

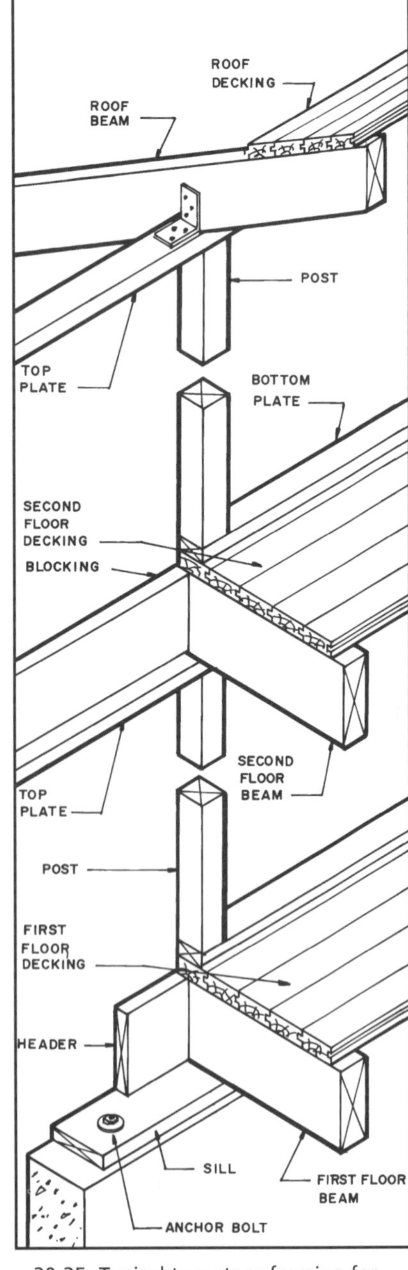

20-25 Typical two-story framing for post, plank, and beam construction.

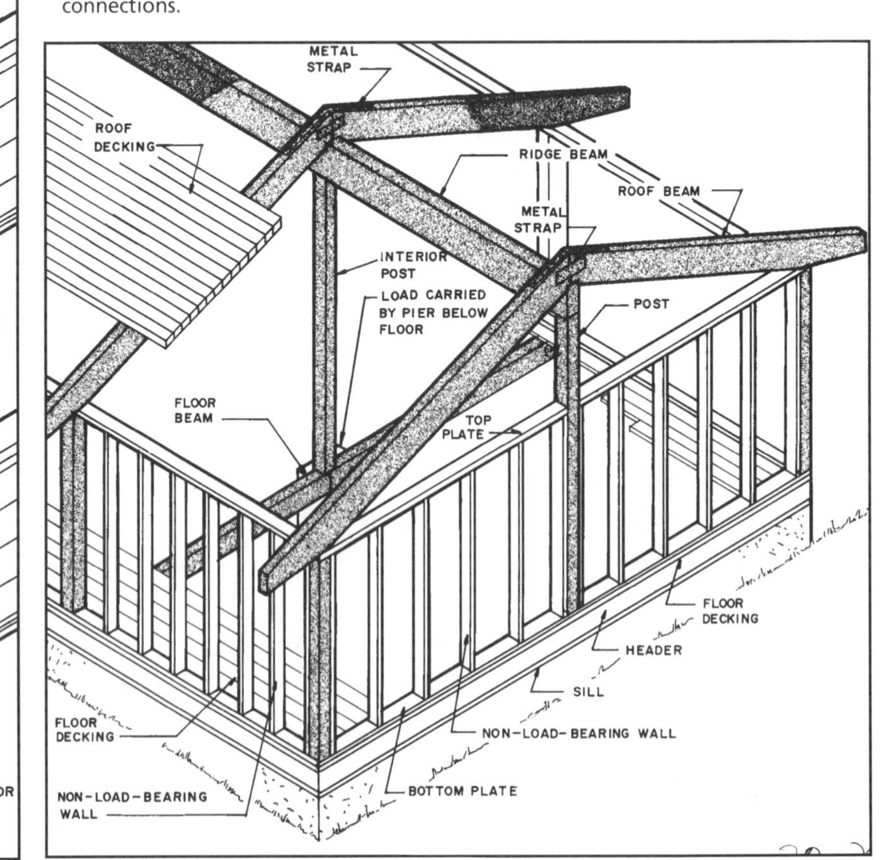

20-26 Post, plank, and beam framing can have the floor and roof beams running the width of the building.

ROOF CONSTRUCTION

The roof in post, plank, and beam construction may be built in two ways. The roof beams may run from a ridge to the exterior wall, similar to rafter-framed roofs, or they may run parallel (longitudinally) with the exterior side walls and the ridge beam **(see 20-26 and 20-27)**. The longitudinal beams are usually larger than the beams running from the ridge to the exterior wall, because they generally have to span longer distances. They can be reduced in size if supported by posts within the building.

An alternative for a parallel or longitudinal structure is shown in **20-28**. Rafters span the longitudinal beams and are sheathed with plywood. This provides a space for insulation in the roof and does not require expensive, heavy decking. The interior ceiling will have to be covered with gypsum or some other material.

The typical post, plank, and beam framing system does not provide for running electrical or plumbing service. If this is necessary in the roof, spaced beams can be used instead of solid beams **(see 20-29)**. This influences the spacing of beams and posts, because the load-carrying capacities will be different from those of solid beams.

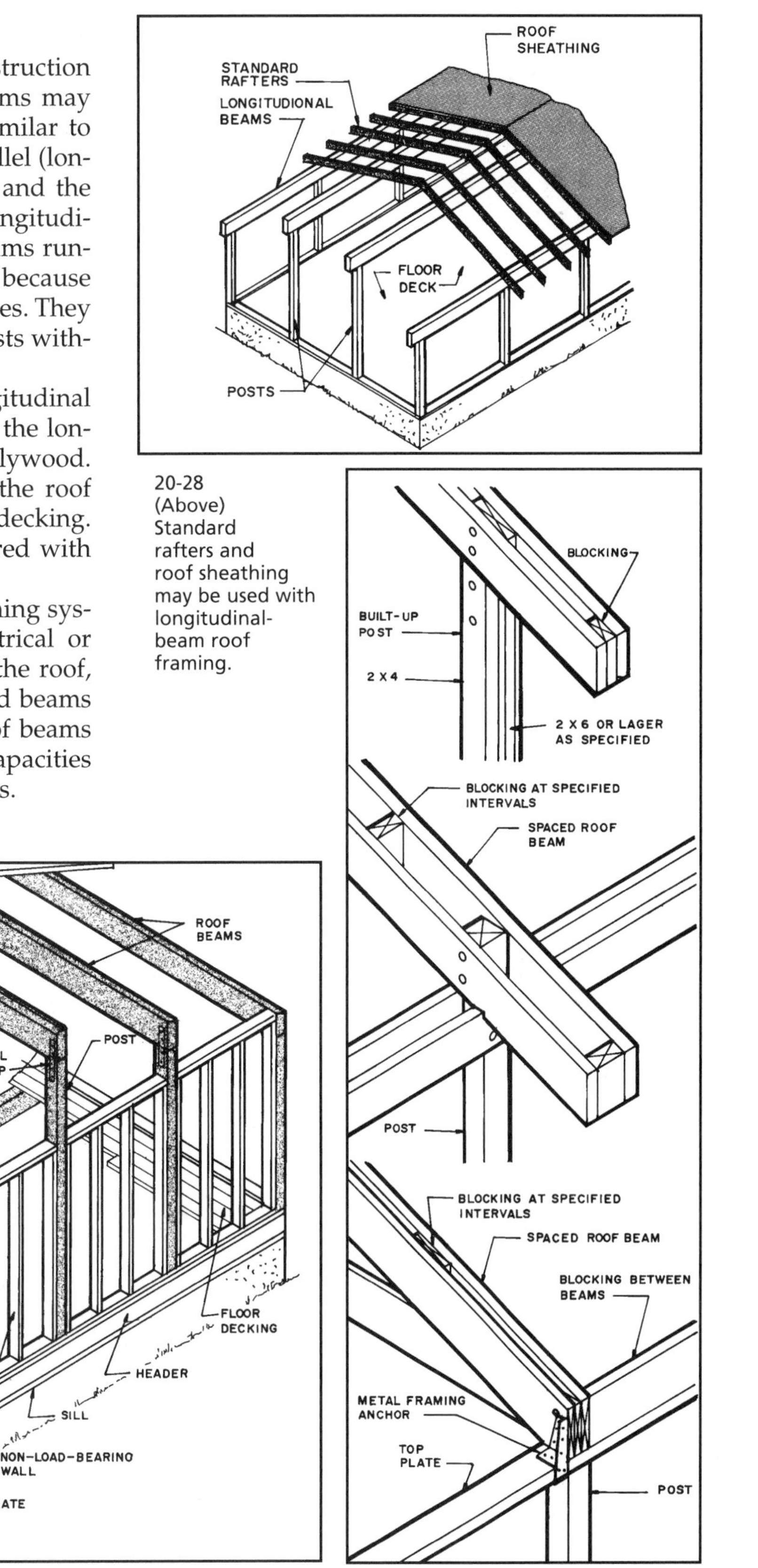

20-28 (Above) Standard rafters and roof sheathing may be used with longitudinal-beam roof framing.

20-27 Post, plank, and beam framing can be run the length (longitudinally) of the building.

20-29 Typical spaced roof beams and their column connections.

Generally, the bottom side of the roof decking is left exposed, forming the finished ceiling. The roof beams and bottom of the decking are stained or finished in an attractive manner **(see 20-30)**.

Various methods for joining roof beams to the ridge beam and supporting posts are used. Beams joining the ridge may butt it **(see 20-31)** and be secured with metal hangers, or they may rest on top of it and be secured with metal gussett plates or a tie strap over the top **(see 20-32)**. Typical rafter-to-post connections are shown in **20-33**. The choice of connectors is important, because they are an important part of the structural system.

Decking is secured to roof beams in the same manner as described for floors. While wood is a good insulating material, the roof will require additional insulation. Since it is desired for the roof beams and wood decking to be visible from inside the building, the insulation is placed on the top side of the decking **(see 20-34)**. The rigid sheets of insulation are secured to the decking. A vapor barrier is placed between the decking and the insulation. Frequently a built-up roof is laid over the insulation **(see 20-35)**.

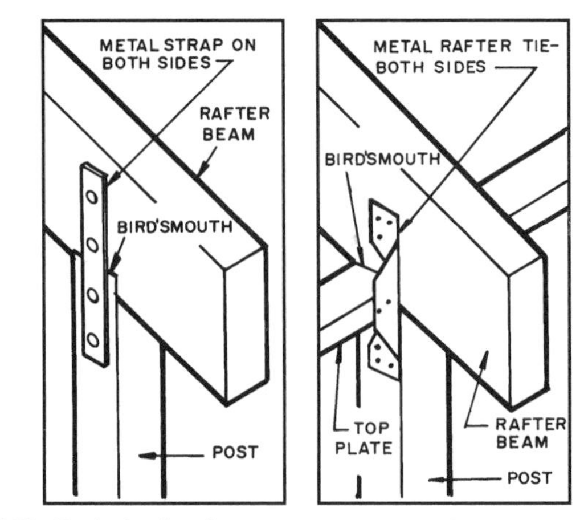

20-33 Typical rafter beam-to-post connections.

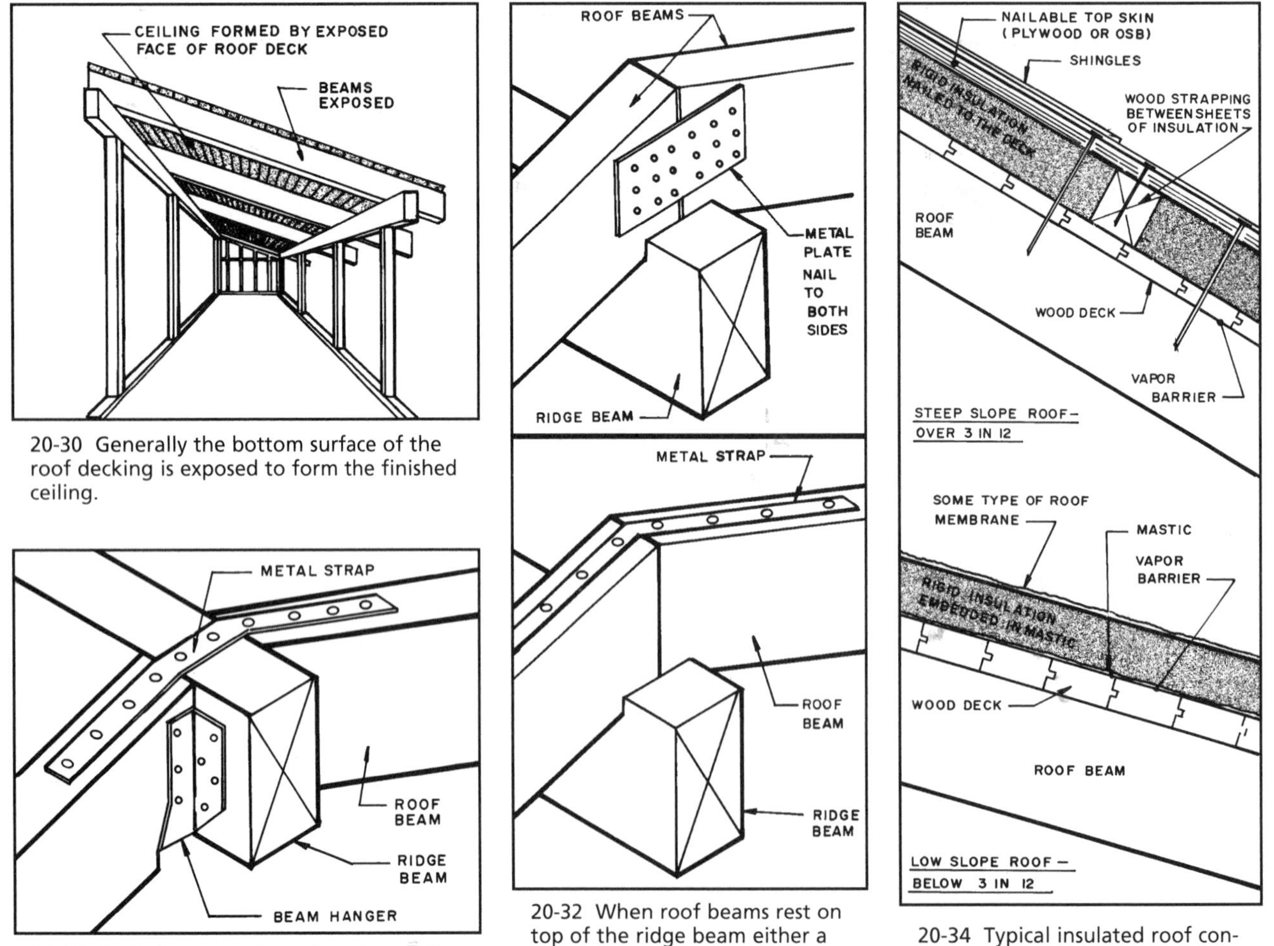

20-30 Generally the bottom surface of the roof decking is exposed to form the finished ceiling.

20-31 Typical construction when the roof beams butt the ridge beam.

20-32 When roof beams rest on top of the ridge beam either a metal strap or a tie plate can be used.

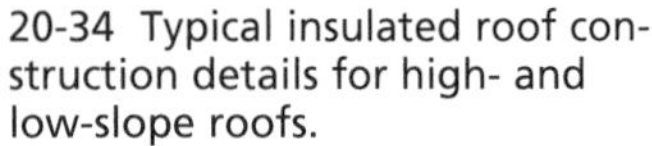

20-34 Typical insulated roof construction details for high- and low-slope roofs.

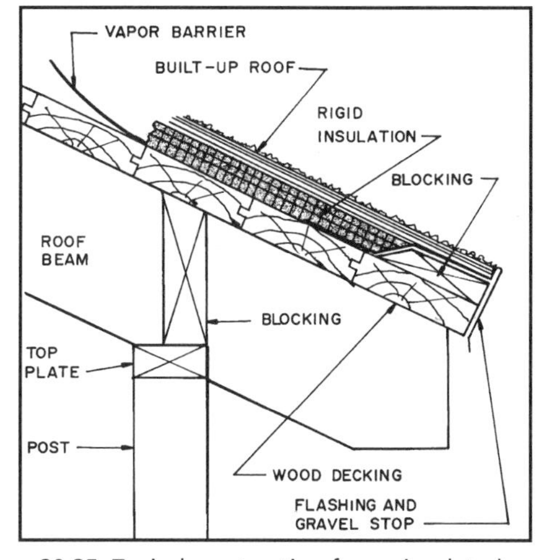

20-35 Typical construction for an insulated roof with a built-up roof membrane. Notice the blocking at the end of the roof backing the insulation.

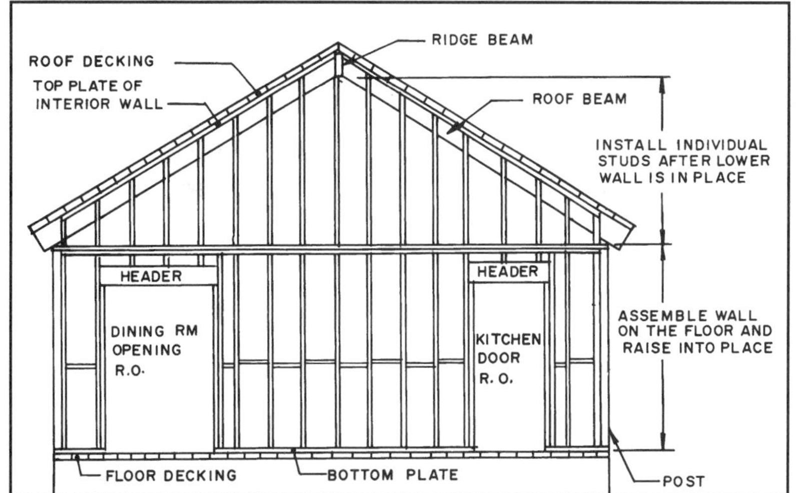

20-37 Assemble the lower wall on the floor and erect as described for conventional interior wall framing. Then install the individual studs above this wall to the roof deck.

If shingles are to be used, nailable panels are laid over the insulation and nailed through the insulation into the decking. The shingles are nailed to these panels in the usual way. If composite panels are used, they are nailed through the panel into the roof beam **(see 20-36)**.

CONSTRUCTING INTERIOR PARTITIONS

After the structural framework has been erected, the walls braced and sheathed, and the roof decking installed, the interior partitions can be installed. They are assembled on the floor as described in earlier chapters for conventional frame construction. For rooms where freedom from noise is desired the partition must be framed all the way to the bottom of the roof deck. One method is to erect a traditional eight-foot-high partition with door openings as required. Then frame in the triangular area above as shown in **20-37**. Since it is now possible to get long wood studs made from bonded wood fibers, the studs could be continuous from the floor to the roof decking as shown in **20-38**. Remember, metal studs could also be used.

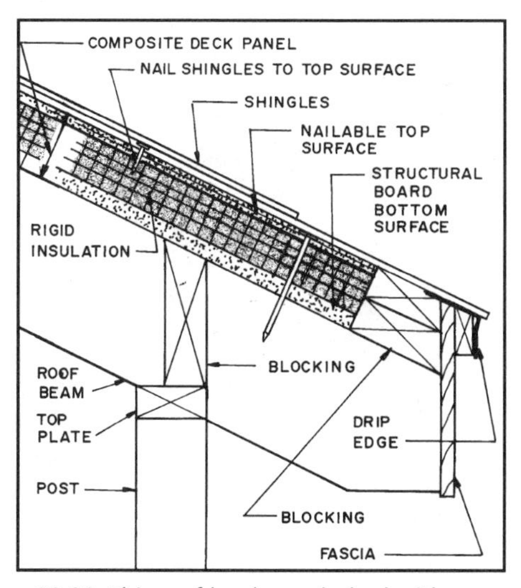

20-36 This roof has been decked with a factory-manufactured composite roof panel made with structural board skins over a rigid insulation panel.

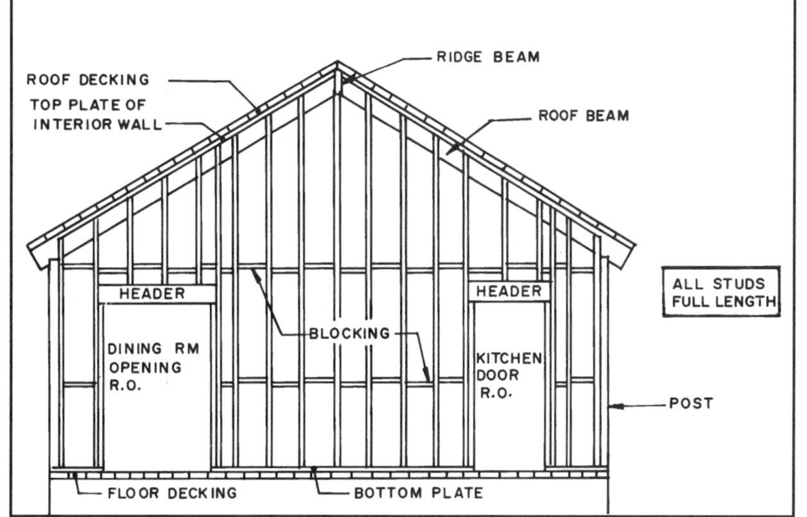

20-38 Frame the wall using full-length studs from the floor deck to the roof deck. Install headers and fireblocks in the normal manner.

ASSEMBLY PROCEDURE

The actual sequence of events for erecting a small residential post, plank, and beam structure may vary depending upon the design of the building. For a typical building made with individual posts and beams the following procedure is typical **(see 20-39)**.

First construct the floor platform on the foundation. To set, plumb and brace the corner posts. Set the bottom plate on the decking and nail in place. Mark the location of each post on it. Secure each post to the plate and temporarily brace it with the adjacent posts and the corner posts. Now nail the top plate to each post. At this point some like to frame in the exterior walls between the posts and apply the plywood or oriented strand board sheathing to get the wall rigid.

Others leave the temporary bracing in place and set the ridge beam and rafters, after which they install permanent bracing, decking, and sheathing.

In the construction of factory-manufactured buildings, often pre-assembled units of posts and roof beams are set in place with a crane. They are braced temporarily as each is erected until permanent bracing can be

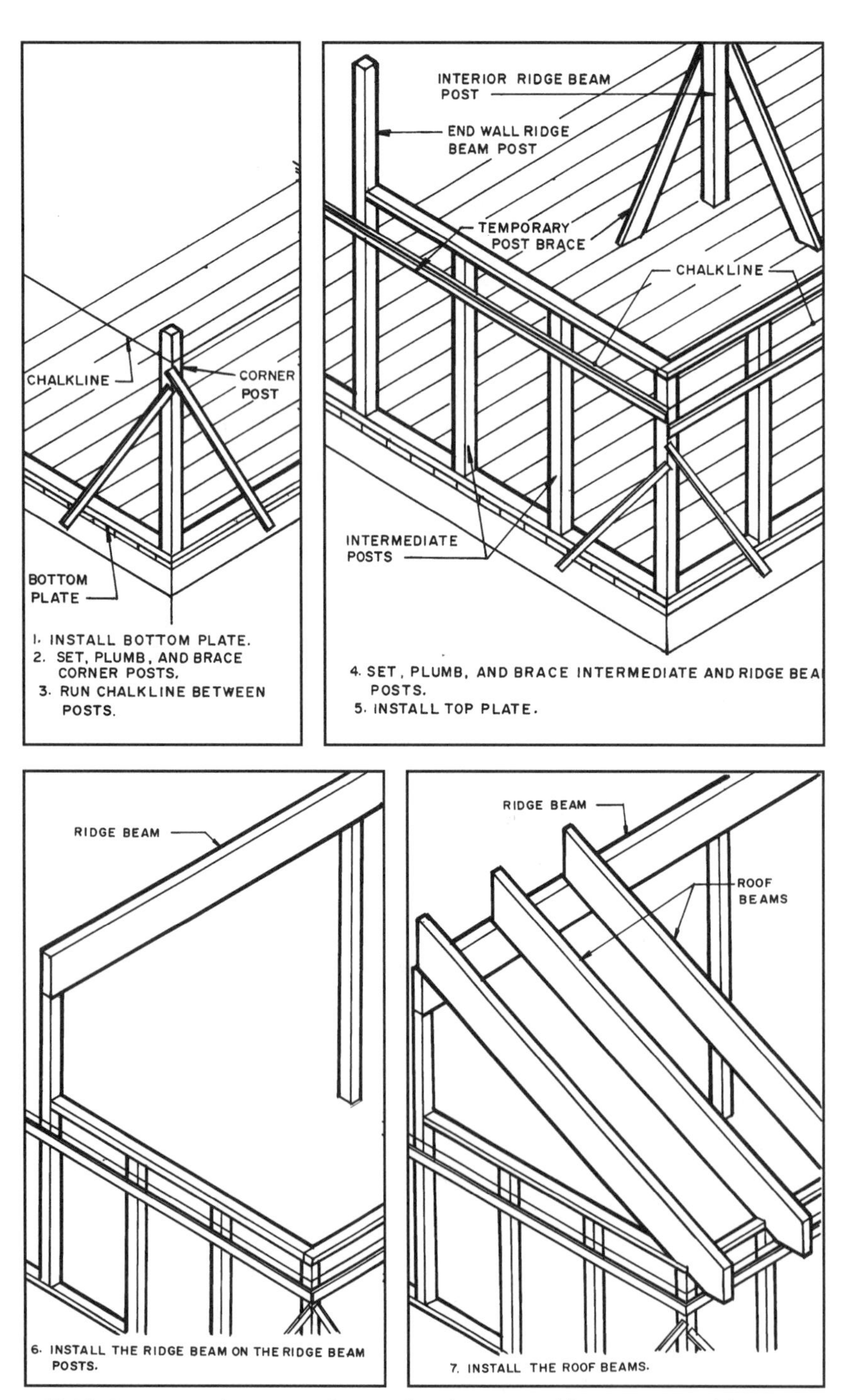

20-39 A suggested sequence for erecting the post and beam structural frame.

20-40 Factory-manufactured timber bents being erected over the subfloor. This is the second bent of the framework. *(Courtesy Dreaming Creek Timber Frame Homes, Inc.)*

applied. They are sheathed and the roof decking is installed **(see 20-40)**. The framing for the roof is shown in **20-41** from the inside of the building and the structure for a second floor balcony and living area. Manufacturers supply other related products that enhance the wood appearance. In **20-42** is a custom made door using white ash, walnut, heart pine, and zebra wood.

A typical heavy timber bent is detailed in **20-43**. Heavy timber bents are custom designed to the floor plan.

Various other buildings are made using custom-made heavy timber framing. In **20-44** you can see the framing for a small two-story house. The posts are set on concrete footings. The roof was framed with conventional trusses.

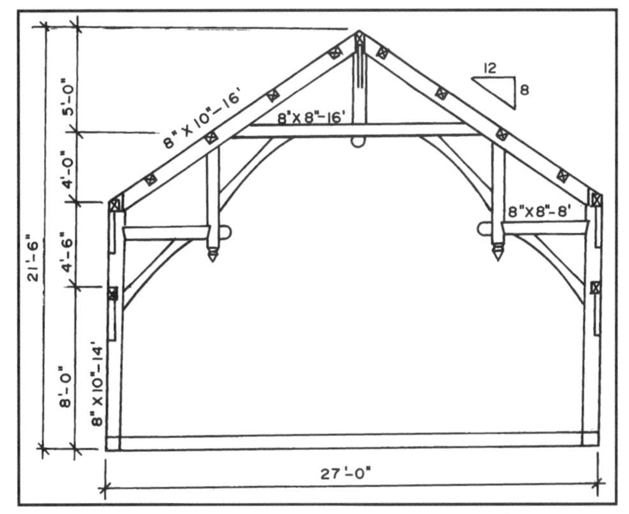

20-43 This is a custom-designed timber bent as shown in 20-40. *(Courtesy Dreaming Creek Timber Frame Homes, Inc.)*

20-41 This shows the framing for the roof and second floor balcony and living area. Oak, walnut, and cherry woods were used with a natural finish to provide an attractive, warm interior. (*Courtesy Dreaming Creek Timber Frame Homes, Inc.)*

20-42 This custom-made door blends in with the appearance provided by the timber framing. *(Courtesy Dreaming Creek Timber Frame Homes, Inc.)*

20-44 This building uses post and beam wall and floor framing and conventional wood roof trusses. *(Courtesy Strickly Barns, Inc., Rural Builder Magazine, and Jeff Gerde, photographer)*

CHAPTER 21

Heavy Timber

Heavy timber construction has been used in the United States for many years. It is one of the oldest types still used. Originally it was used for industrial and storage buildings and used large sized solid wood timbers. Today, with the availability of reconstituted wood members such as glued-laminated timbers (glulams) and stress-graded solid timbers, it has found wide use in many other types of buildings such as churches, shopping centers, schools, and auditoriums. Architects are utilizing the natural beauty of the wood frame and decking as the finished interior surfaces. While many such buildings are one-story, multistory buildings can be efficiently built **(see 21-1)**.

The advantage to using glulam beams and columns is that they are available in larger sizes than solid wood members. They are also of higher quality and can span longer distances than solid wood. For example, glued laminated structural

21-1 This attractive room relies on the beauty of the heavy timber structural members to set the color and style of the interior. *(Courtesy American Institute of Timber Construction)*

members are used for structures other than buildings. Many attractive bridges, piers, and docks are built using glulams **(see 21-2)**.

21-2 These glued laminated arches and vertical timbers form the structure of this bridge. *(Courtesy American Institute of Timber Construction)*

FLAME RESISTANCE

It should be noted that heavy timber structural members have a high degree of flame resistance. While they will burn, the burn is slow and a char forms on the outside of the member. This provides a degree of fire protection. Wood members resist failure much longer than steel framing. Steel exposed to fire rapidly loses its strength and collapses. Building codes recognize this as they establish the minimum sizes of wood structural members and the thickness and type of materials used for construction of wood walls, floors, and roofs. The types of fasteners used and construction details are also in the code for fire resistance.

FRAMING MEMBERS

As decisions are made concerning the structure of the building, the wood framing members chosen must meet the specifications of the local building code and be able to carry the loads to be imposed.

Columns

Columns should be continuous or superimposed throughout all stories and be connected by approved steel or iron caps having pintles (pins) and baseplates or a precast concrete column cap. In some cases metal splice plates can be used.

Wood columns can be glued-laminated or solid. Those used for carrying floor loads are typically at least eight inches nominal on their shortest side while those supporting ceilings or roofs are usually 6 × 8 inches nominal.

Beams & Girders

Beams and girders can be made from sawed or glued-laminated stock. The smallest member used is typically 6 × 10 inches nominal. Framed timber trusses supporting floor loads are required to have members not less than eight inches nominal in both dimensions.

Roof Framing

Glued-laminated or framed arches resting on top of the exterior wall are generally required to be 4 × 6 inches nominal. Spaced beam arches must have a tight wood cover plate over the open space. Splice plates can be used to join various sections. The sizes of these members are determined by the designer and minimum sizes are specified by the codes.

Floor Decking

Floor decking can be sawed lumber or glued-laminated members. The planks are either splined or have a tongue-and-groove. See examples in Chapter 20. Typically, tongue-and-grooved decking is not less than three inches nominal, and square-edged solid wood planks four inches nominal. The ends of all planks must rest on supporting members. It is expected that plank decking will be covered with ½-inch-thick plywood underlayment for carpet or tile or one-inch nominal tongue-and-grooved finish flooring.

Roof Decking

Roof decks can be made from sawed lumber or glued-laminated members. Typically tongue-and-grooved glued laminated planks are two inches nominal in thickness while solid wood planks are three inches nominal. Other types of roof decking such as composite stressed-skin panels can be used (see Chapter 20 for examples).

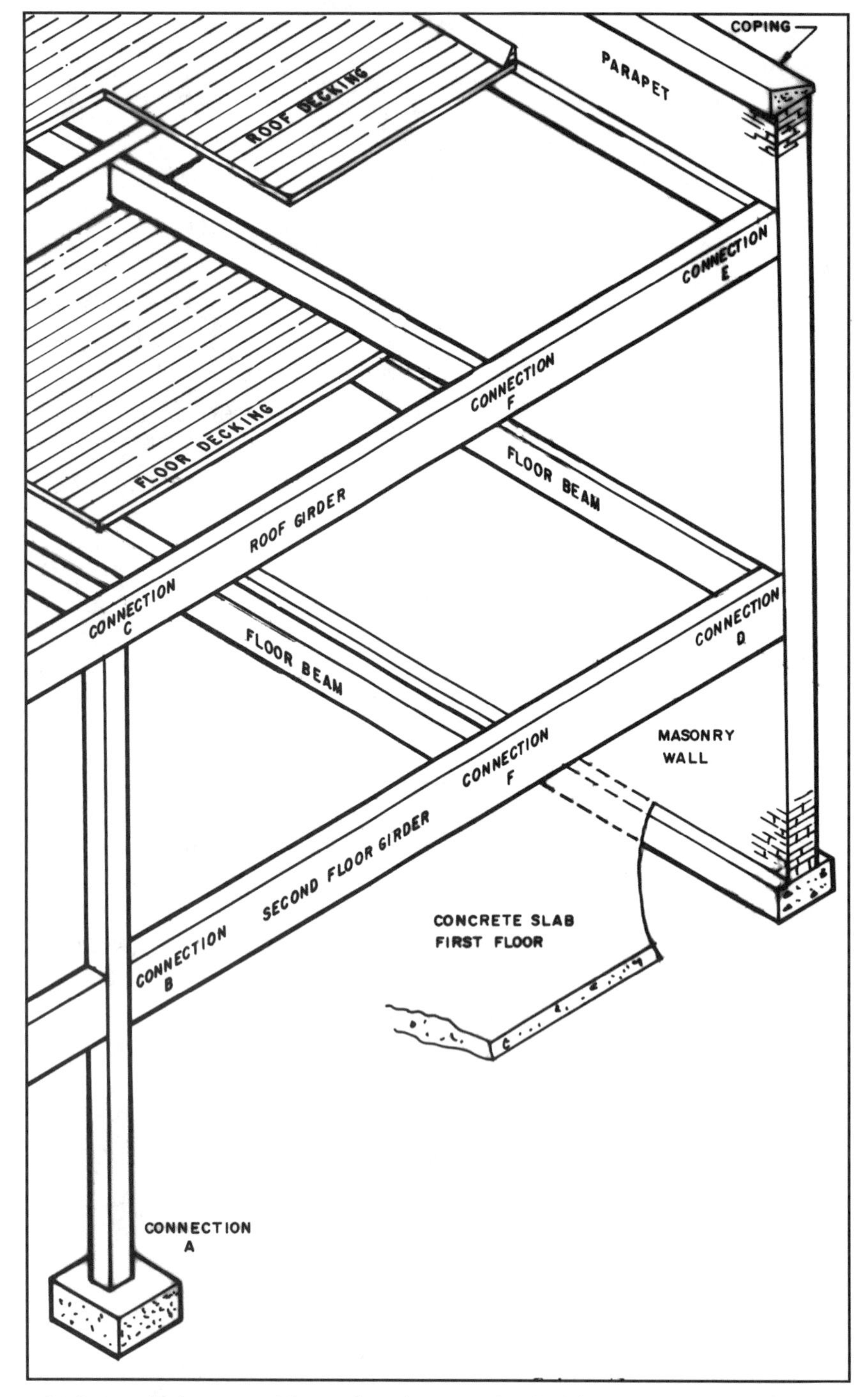

21-3 A typical structural frame for a heavy timber building. The connections noted are illustrated in the detail drawings following in this chapter. *(Courtesy American Forest and Paper Association)*

Bearing Walls

Codes require that any part of an interior or exterior bearing wall have a fire resistance rating of at least four hours. This can be had by using masonry walls or wood timbers having a fire-retardant treatment. Various codes may have additional requirements.

Nonload-bearing Walls

Nonload-bearing sections of exterior walls are required by many codes to be made using a noncombustible material or a fire-retardant treated wood. Consult your codes for additional specifications.

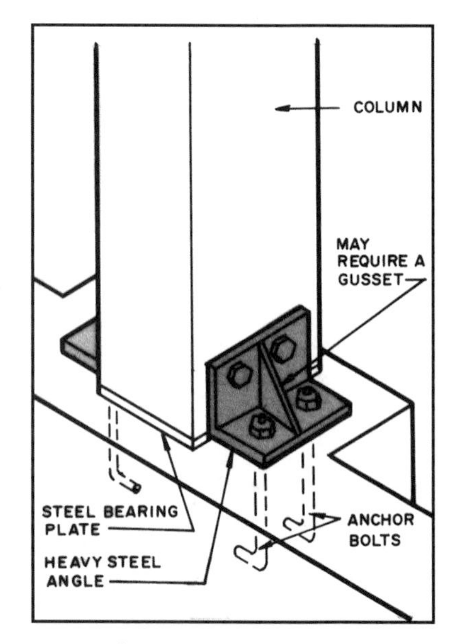

21-4 This post is connected to the concrete foundation with steel angles. This is connection A in 21-3. *(Courtesy American Forest and Paper Association)*

CONSTRUCTION DETAILS

The structural frame of a heavy timber building requires that the services of a structural engineer be engaged and that there is close consultation with the local building codes. The following examples are only a few of the typical constructions that are available.

A typical framing structure is shown in **21-3**. The connections indicated on this drawing are shown in the detail drawings which follow. These are only examples of what might be used. There are other approved ways to make these connections.

The post is connected to the foundation with some type of metal connector. The use of heavy steel anchors is shown in **21-4**. The anchor bolts are set into the concrete before it hardens. Notice the steel bearing plate below the column. This is connection **A** in **21-3**. There are many other types of metal connectors available.

Beams and girders connect with posts, forming the floor and roof structure. A metal column cap used to carry a beam or girder on top of a post is shown in **21-5**. This is the type of connection used in the structural frame for the one-story commercial building in **21-6**. Also in **21-5** is a connection using wood-bearing plates when the beam or girder butts the post which continues up above the beam. These are connections **B** and **C** in **21-3**.

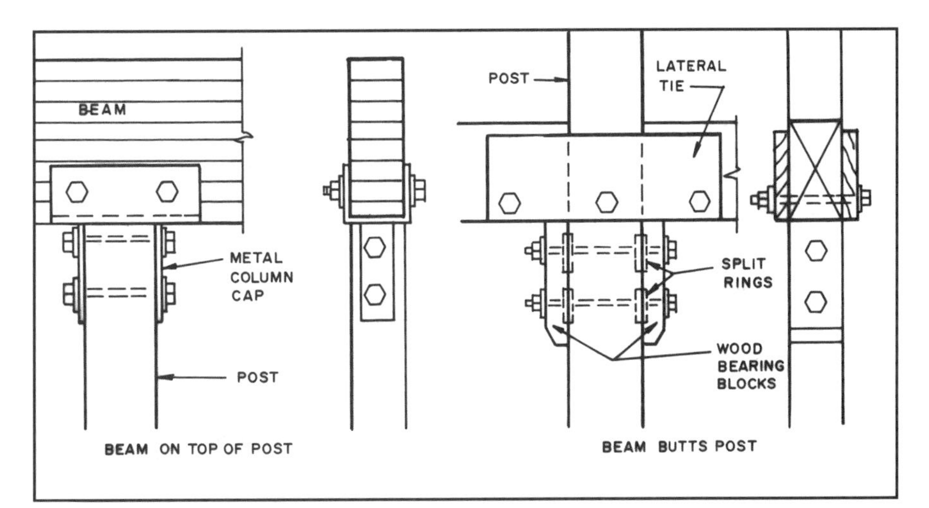

21-5 Two types of beam- and girder-to-column connections. These are connections B and C in 21-2. *(Courtesy American Forest and Paper Association)*

21-6 Heavy timber framing for a one-story commercial building.

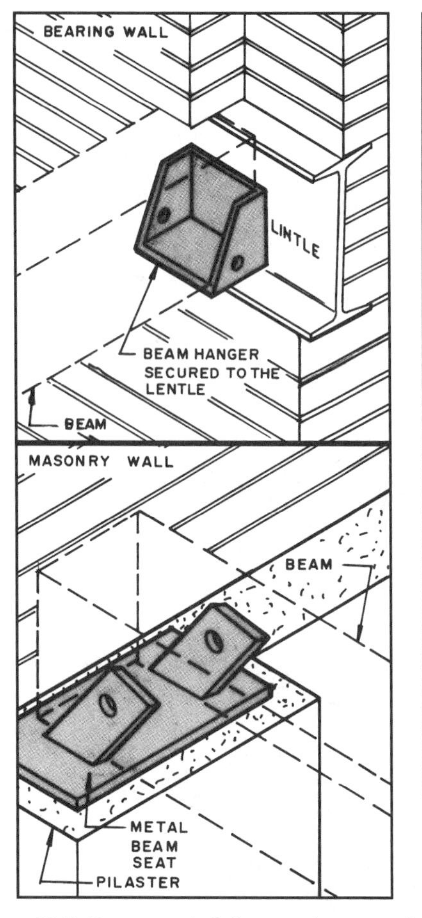

21-7 Beams and girders can be secured to load-bearing masonry walls with various metal hangers, or can rest on a pilaster. These are connections D and E in 21-3. *(Courtesy American Forest and Paper Association)*

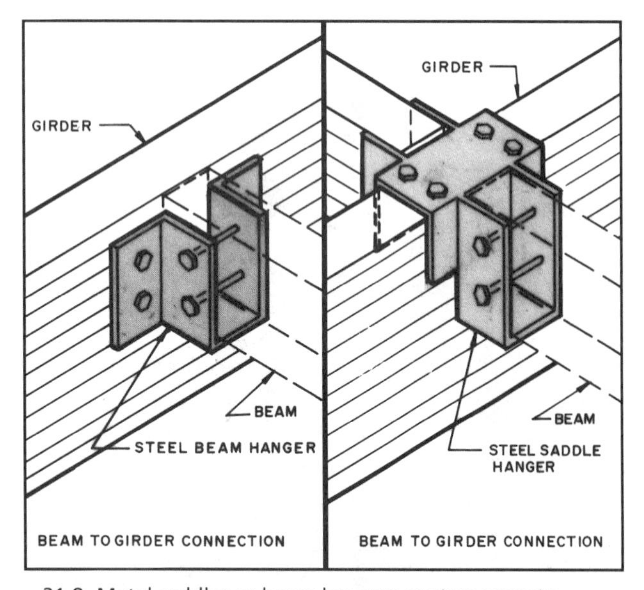

21-8 Metal saddles or beam hangers are two ways to connect butting beams and girders. This is connection F in 21-3. *(Courtesy American Forest and Paper Association)*

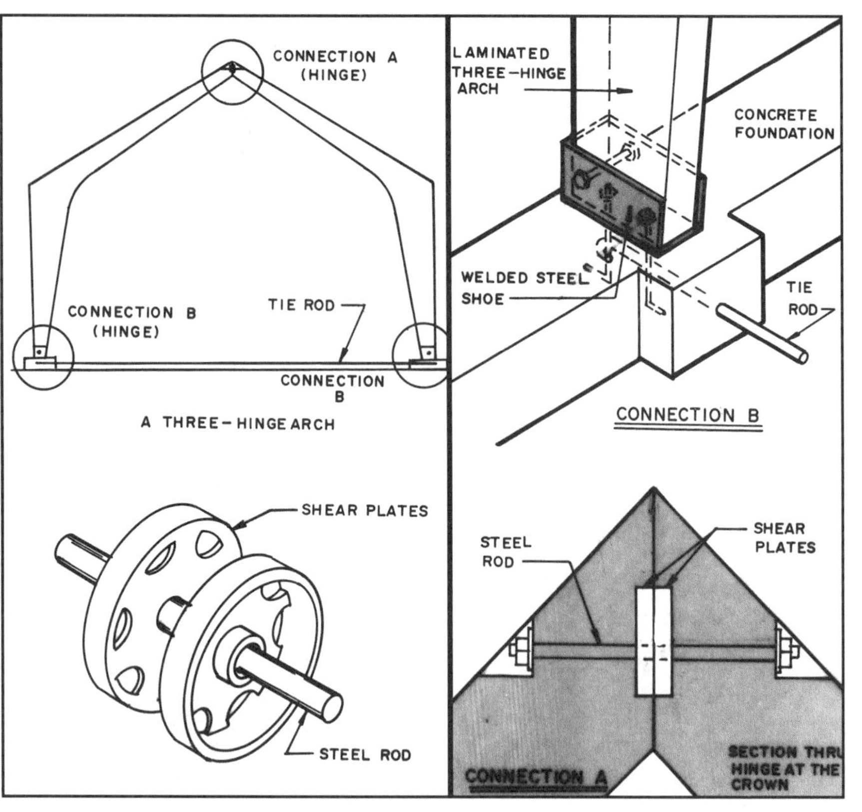

21-9 A three-hinged arch hinges at the two bearing supports and at the crown.

A floor or roof beam can be set on a pilaster or set into a metal beam hanger bolted to a steel lintel, as shown in **21-7**. The fire-resistant masonry wall is built around these beam supports.

Floor and roof beams and purlins butt and are secured to larger beams and girders with some type of metal hanger, as shown in **21-8**. The two types of metal hangers are a saddle hanger lapping over the top of the girder and a beam hanger, which is a metal pocket into which the beam is set and bolted. The hanger is bolted to the girder.

OTHER HEAVY TIMBER STRUCTURAL MEMBERS

Two commonly used structural members are the three-hinged glued-laminated arches and the two-hinged barrel arch. A three-hinged arch is one that hinges at two supports and at the crown **(see 21-9)**. Notice the arch is pinned at each foundation and with pins and has shear plates at the crown. The shear plates are recessed into grooves cut in the wood. They spread any load from the rod over a wider surface of the wood. This helps avoid splitting or crushing the wood fibers. The three-hinged arches of laminated wood in **21-10**

21-10 These three-hinged arches provide the wall and roof structural frame. Notice the purlins running between the arches to help support the roof deck. *(Courtesy American Institute of Timber Construction)*

21-11 This type of three-hinged arch is widely used for church construction. *(Courtesy American Institute of Timber Construction)*

provide the wall and roof structure. Large purlins run between arches and support the roof decking. A different type of three-hinged arch is shown in **21-11**. Notice the beauty of the exposed wood ceiling, purlins, and the arches themselves.

A two-hinged barrel arch is shown in **21-12**. It is supported at each end by steel plates mounted on a concrete base. The plates on each end ride on a pin, thus forming a hinge at each end.

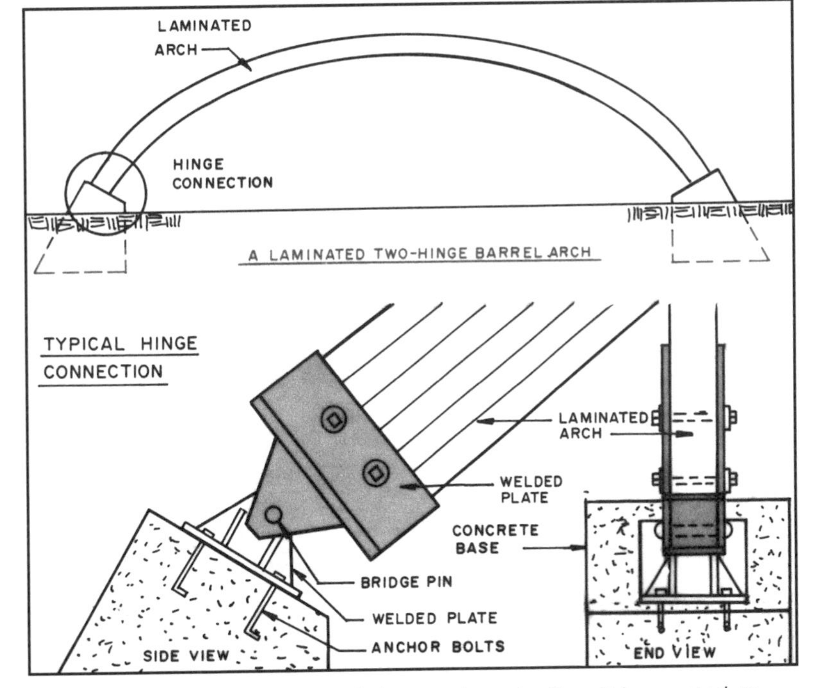

21-12 The two-hinged barrel arch forms a domed ceiling. It is mounted on substantial concrete piers.

ADDITIONAL INFORMATION

American Forest and Paper Association
1111 19th St. NW Suite 800
Washington, DC 20036

American Institute of Timber Construction
7012 South Revere Parkway, Suite 140
Englewood, CO 80112

Canadian Wood Council
350-1730 St. Laurent Boulevard
Ottawa, ON
Canada KIG 5L1

CHAPTER 22

POLE CONSTRUCTION

Evidence of the use of pole construction extends back to prehistoric times. The poles were set into holes in the earth. The floors could be elevated, giving the occupants a degree of protection from animals and floods.

Within our times this type of construction was widely used for construction of farm buildings. The problem with this construction was that the poles eventually rotted. This has largely been overcome by the use of pressure treatment of the poles by chemicals that retard their rotting. Pole construction is finding use in the building of single-family residences, apartments, beach houses, and a variety of commercial buildings **(see 22-1)**.

DESIGN

Pole buildings must be designed by civil or structural engineers to assure the structure meets the building codes and any special local requirements. The potential effects of wind forces, earthquakes, floods, connections, preservative treatment, bearing capacity of soils, depth of embedment of the poles, and loads to be placed on the structural frame are items to be considered. Following is a general discussion of the basic construction techniques.

TYPES OF FRAMING SYSTEMS

The typical types of framing include platform without embedding the poles, platform with embedded poles, and a pole frame. The platform-type construction is generally used on sites where masonry or concrete foundations are not practical, such as on a steeply sloped site or one with wet or unstable soil such as at the beach or in wetlands. A conventional framed wood building is built upon the platform. The pole frame utilizes the poles to support the roof and floors.

22-1 A typical small pole-frame residence.

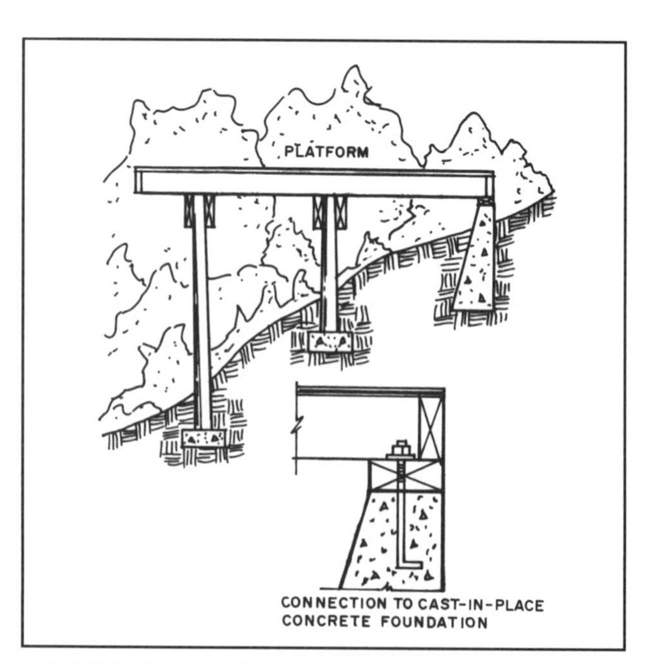

22-2 This shows platform framing without embedding. The poles are not set deeply into the earth and a concrete foundation is required to resist horizontal forces.

22-3 The soil permits these poles to be embedded the required depth to stabilize the structure and resist horizontal forces.

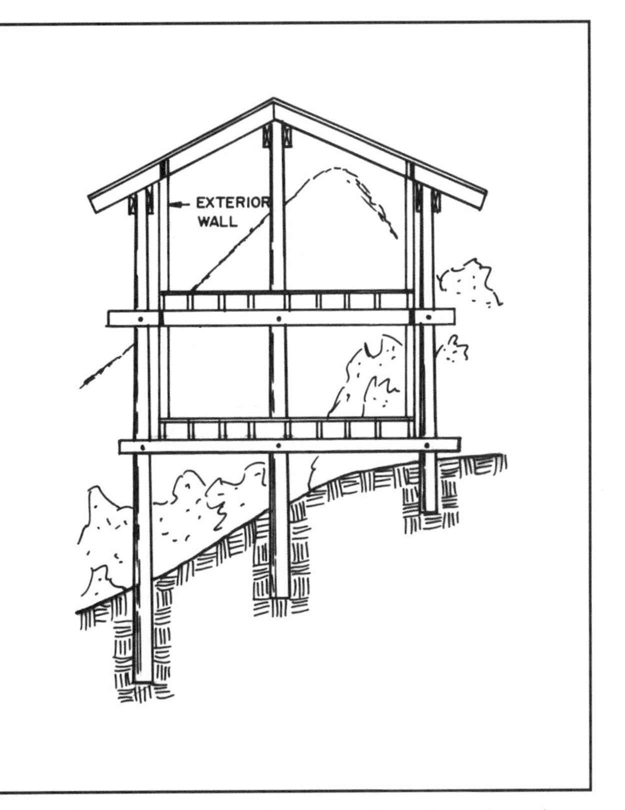

22-4 This is a typical pole-frame building where the poles extend from the ground to the roof.

A platform without embedding poles is shown in **22-2**. Lack of embedding means the poles are set in the ground but cannot be set deep enough to provide the required resistance to horizontal forces. On some sites the soil is so rocky it is not possible to get the depth of embedment required. A key wall of concrete is set into the soil to tie the structure to the hill and resist horizontal forces.

Constructing a platform with embedded poles is an excellent way to build on a sloped lot if the poles can be embedded deeply enough to resist the horizontal forces. Proper embedment plus adequate bracing will provide a stable platform. A conventionally framed building is then built on top of this platform **(see 22-3)**.

Pole-frame construction is shown in **22-4**. The poles must be embedded to resist the horizontal forces of the floor platform and the roof. The poles run from the ground all the way to the roof. Since the poles are very long this helps to brace the structure. Such construction can be used for two-story structures. Since there are only a few poles on the interior of the building there is great flexibility in placing interior partitions. An overall framing plan is shown in **22-5**.

22-5 A typical framing plan for a small pole-frame construction building.

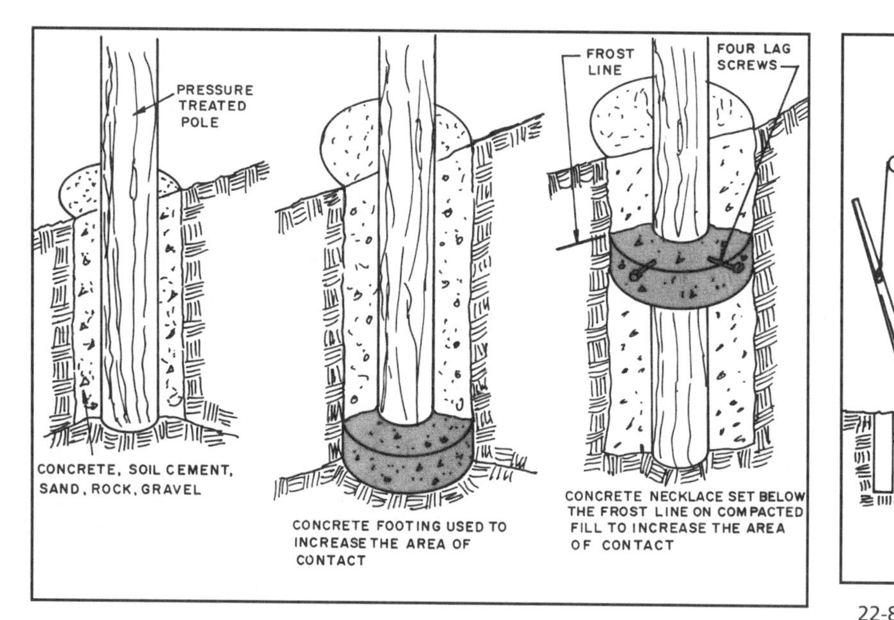

22-6 These are commonly used methods for embedding poles.

22-8 Poles are set in the holes in the earth with a crane.

22-7 This auger mounts on a track or backhoe and uses a reversible hydraulic motor. It carries augers up to six feet long. *(Courtesy Denuser Machine Company, Inc.)*

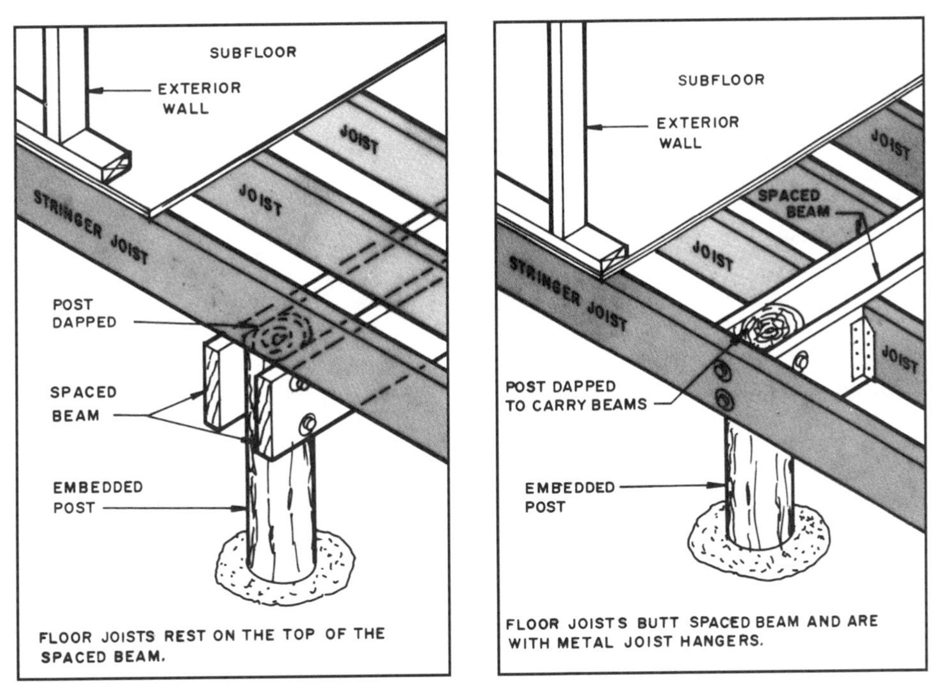

22-9 Two ways used to frame for pole platform construction.

EMBEDMENT

Frequently used methods for embedding the poles are shown in **22-6**. Poles may be embedded with soil cement, concrete, sand, gravel, rock, or the natural earth. Sometimes a concrete footing pad is used to provide additional support, or a concrete necklace used to increase the contact of the pole with the soil. When earth is used it must be compacted. Sand will compact if flooded with water. Soil cement is a mixture of five parts of natural soil and one part Portland cement. It is mixed dry and compacted around the pole. Sometimes a little water is added as it is compacted. The pole hole should be about eight inches or larger than the pole diameter. Holes are dug with a power auger **(see 22-7)** and the poles are placed with a crane **(see 22-8)**.

POLE PLATFORM FRAMING

Details for typical pole platform framing are in **22-9**. The poles extend out of the earth to the first floor of the building and serve as the foundation. The stringer joists and spaced beams are bolted to the posts, forming the structural frame. Floor joists are hung on the header joists as is done in normal floor framing. The subfloor is secured to the floor joists and the platform is ready for the assembly of the walls, ceiling, and roof. An alternate plan is to place the floor joists on top of the spaced beams, again as is done in conventional floor framing.

The band and header joists can be bolted to the round surface of the pole. Many prefer to dap (notch) the pole so they fit against a flat surface **(see 22-10)**.

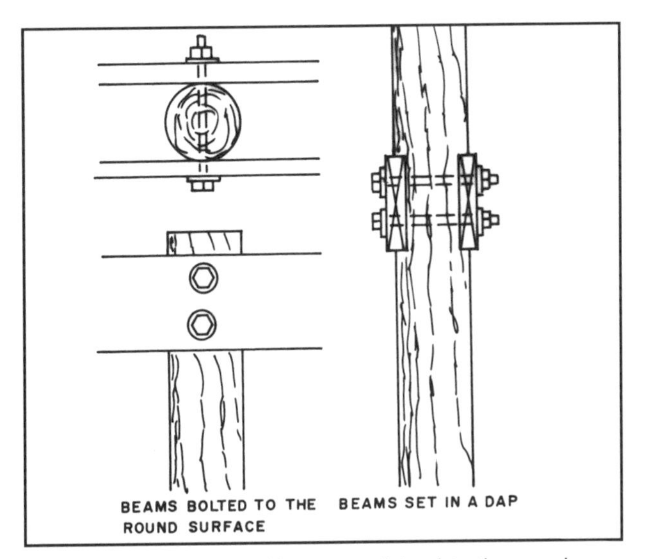

22-10 Two ways spaced beams are joined to the wood posts.

POLE-FRAME CONSTRUCTION

Details for typical pole-frame construction are shown in **22-11**. This method of framing places the poles outside of the exterior walls of the building. The poles support the floor and walls and continue up to support the roof. The poles are embedded as discussed on the previous page.

FLOORS

The floors can be framed the same as for a conventional house with the typical 16 inches O.C. (on center) spacing, as shown in **22-9** and **22-11**. However, if the posts are widely spaced and floor beams are used, the subfloor could be 1⅛-inch plywood decking which spans 4'-0". See Chapter 20 for additional information on the use of floor beams instead of joists.

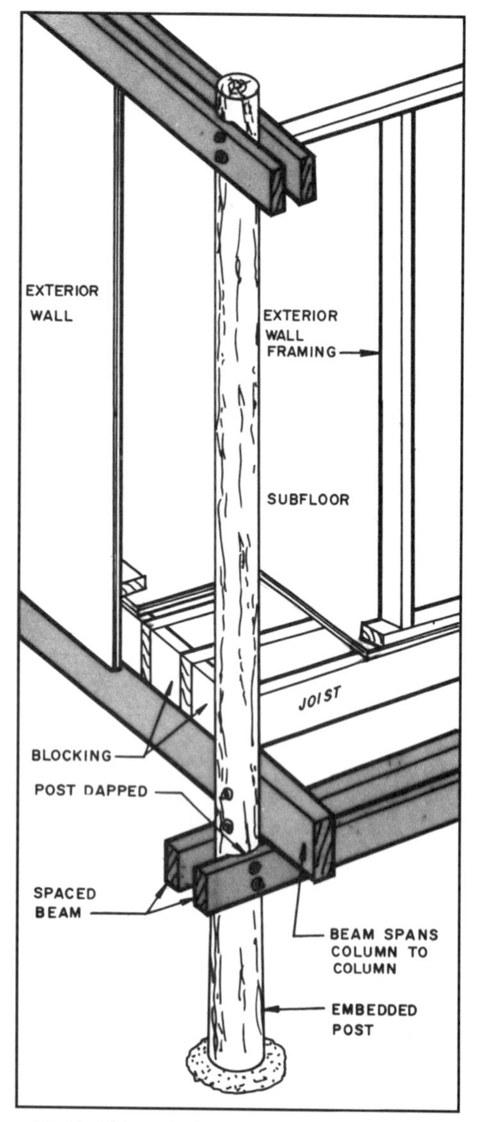

22-11 This pole framing construction places the poles on the outside of the exterior walls of the building.

BRACING

Properly embedded poles can have additional bracing using wood members. A typical knee brace is shown in **22-12**. Braces are usually 2 × 6-inch stock.

While they may be nailed in place, bolting gives a better connection. Bracing is determined by the engineer and should be installed exactly as shown on the drawings.

Steel rods can be used as bracing on the understructure **(see 22-13)**. The rods are usually ⅝ to ¾ inch in diameter and are threaded on the ends. They are fitted through holes in wood members. A cast beveled washer is placed between the wood and the nut. A single diagonal brace or cross-bracing are specified by the needs of the particular situation.

All bracing is specified by the engineer designing the structure.

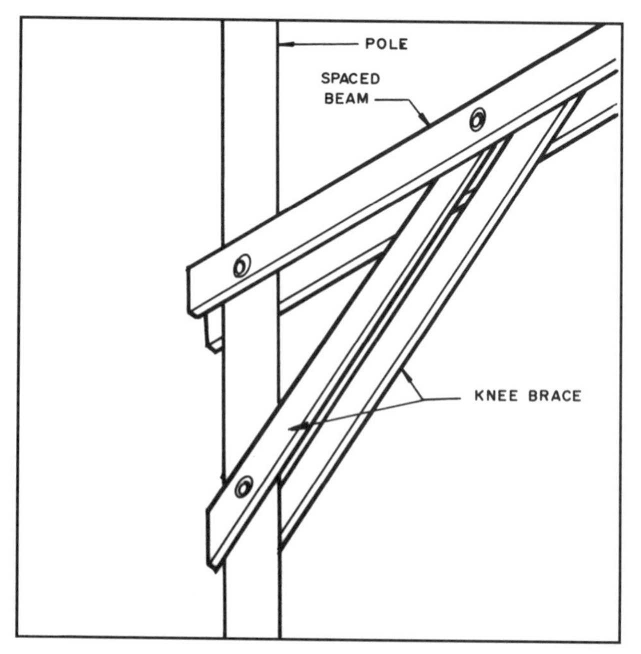

22-12 Wood knee braces stiffen the structure and sometimes reduce the required depth of embedment.

INSULATION

Since the bottom of the floor is completely exposed it presents a special problem in cold climates. One technique is to insulate below the floor and seal it to protect the insulation. The entire underside of the house could be enclosed and used for storage.

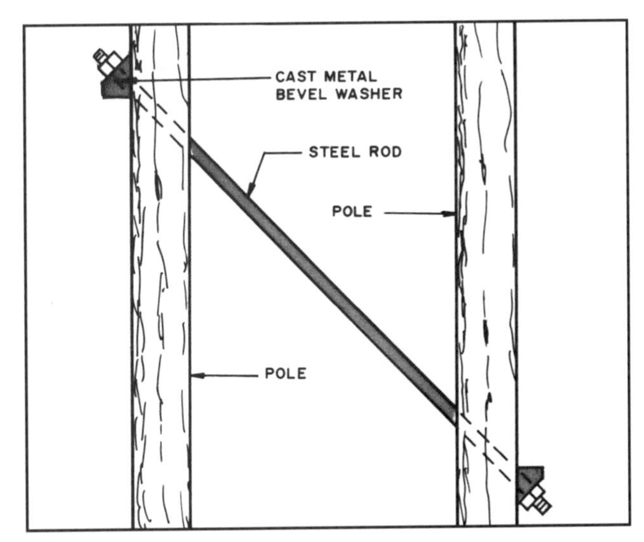

22-13 Steel rods on a diagonal between posts are frequently used for bracing.

ADDITIONAL INFORMATION

Forest Products Laboratory
USDA—Forest Service
1 Gifford Pinchot Dr.
Madison, WI 53705

American Wood Preserves Association
1945 Old Gallows Rd. Suite 150
Vienna, VA 22182-3931

American Institute of Timber Construction
7012 S, Revere Pkwy
Suite 140
Englewood, CO 80112

U.S. Department of Housing and
Urban Development
451 7th St. SW
Washington, DC 20410

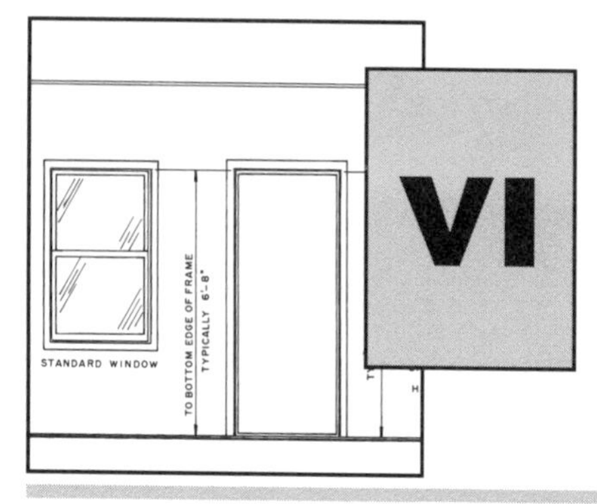

DOORS & WINDOWS

CHAPTER 23

INSTALLING & TRIMMING DOORS

Exterior and interior doors and their frames are usually assembled by the manufacturer. The exterior doors are installed after the building has been enclosed with sheathing and the roof is in place. This is the same time the windows are installed. The interior doors are installed later when the partitions are in place.

Information about the types and sizes of door is shown on the architectural drawings in a door schedule. Often elevations of the doors are drawn. Schedules are shown in Chapter 4.

DOORS

There are many styles and types of **interior door** manufactured. Examples of those most commonly used are shown in **23-1**. The solid wood door with wood **panels** is a popular choice **(see 23-2)**.

23-1 A few of the many interior doors available.

Various panel configurations are available that can be used for interior or exterior doors. They are made in softwoods for painting and in hardwoods for stained or natural finishes. The wood panels can be replaced with glass or louvers. Materials that are molded from wood fibers—such as a solid particleboard core covered with a wood fiber skin—can be used to make a door with a similar appearance **(see 23-3)**. The surface is embossed to create the appearance of a wood grain. These molded doors have the advantage of being dimensionally stable, thus reducing the possibility of warping. The panel design door comes factory primed and ready for the finish coat of paint.

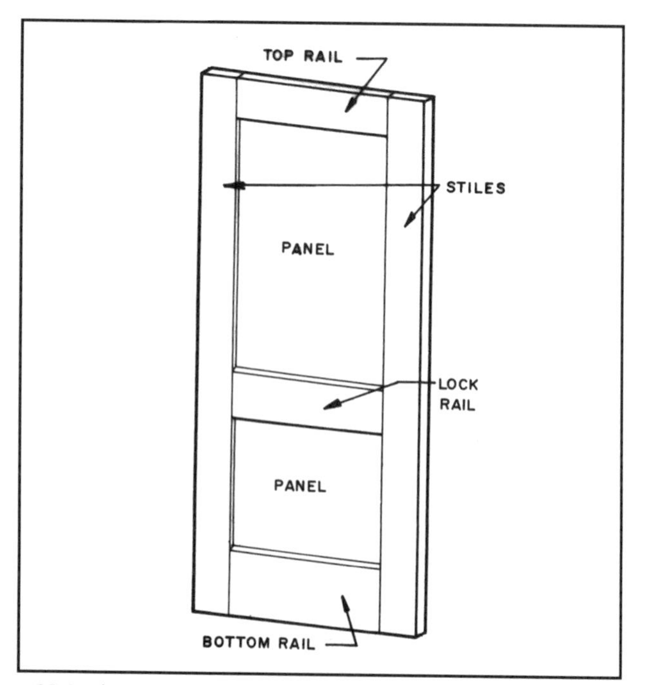

23-2 The parts of a typical solid wood panel door.

Another popular style of door is the **flush door**. The hollow-core flush door is built with a wood frame and an internal bracing of cardboard or wood strips that form a honeycomb interior **(see 23-4)**. Over this internal structure plywood veneers or plastic laminate is glued, providing a smooth, flush door face. Flush doors are available in a wide variety of hardwood veneers. The flush door is also available with a solid core **(see 23-5)**. The solid-core flush door is heavier and stronger; it is generally used as an exterior door, because it provides great resistance to damage from fire or potential intruders. The same style is made from molded hardboard. The exterior surface has embossed wood grain and is available in a variety of natural color finishes, such as oak and walnut.

23-3 This beautiful interior door is made with a particleboard core overlaid with a wood fiber skin and bonded with adhesive at high pressure. *(Courtesy Jeld-Wen® Inc. Elite® Doors)*

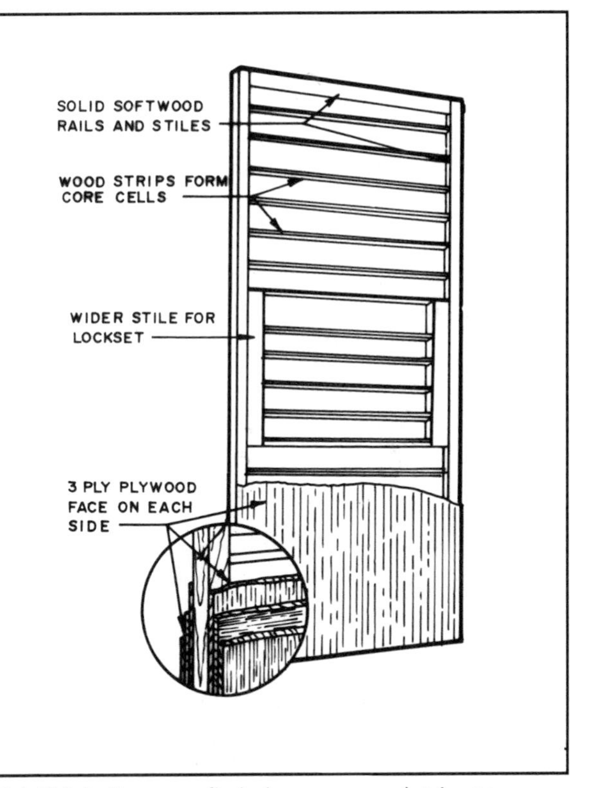

23-4 This hollow-core flush door uses wood strips as dividers on the interior. Other types use a heavy cardboard honeycomb.

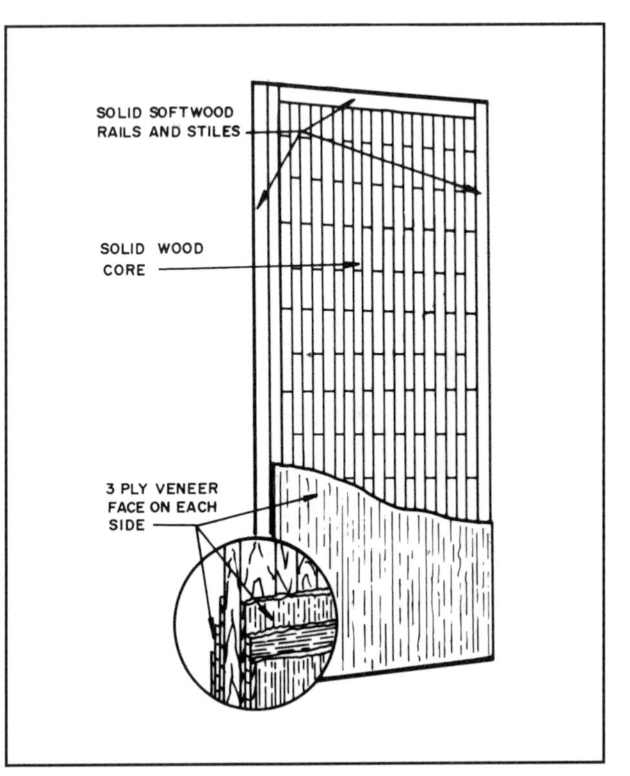

23-5 A solid-core flush door uses solid wood blocking to fill the interior core.

23-6 A few of the many exterior door designs available. *(Courtesy Stanley Door Systems)*

Many **exterior doors** in use today have a **steel** or **fiberglass** outer surface over a wood framed interior structure. The voids inside the door are filled with polyurethane foam, providing better insulating qualities than a solid wood door. Steel-sheathed doors are primed, ready for paint.

23-7 Storm doors protect the front entrance door and provide additional security for the occupants of the house. *(Courtesy Weather Shield Windows and Doors)*

Fiberglass doors are embossed with a wood grain pattern, and can be stained to the color of various woods and finished with a clear urethane top protective coating. They can also be painted. Steel doors have the disadvantage of denting if struck. Fiberglass doors resist such denting. Both are highly weather resistant and do not absorb moisture or warp or twist. Some of the door designs available are in **23-6**.

Storm doors are typically made from galvanized steel or aluminum **(see 23-7)**. They protect the expensive exterior door, add to the energy efficiency of the door, and possibly most important provide a measure of additional security. Some typical designs are in **23-8**. Some have sliding glass panels permitting air to enter when ventilation is wanted. They must use tempered glass to protect against breakage.

The stock sizes of interior and exterior doors can be found in **23-9**.

23-8 Some storm door designs available from one manufacturer. *(Courtesy Weather Shield Windows and Doors)*

Handling Doors on the Job

Doors are very expensive, and the property owner expects each one to be installed free of damage. When handling and storing doors on the job, observe the following:

1. Store doors flat in a clean, dry place. Place cardboard between the stacked doors to protect them from damage.
2. Protect doors from exposure to excessive heat, moisture, dryness, and direct sunlight.
3. When handling doors, wear gloves or be certain your hands are clean.
4. When moving doors, lift them clear of the floor and carry them. Do not drag them across one another or across the floor.
5. Store doors in the area in which they are to be installed for a few days, so that they can adjust to the relative humidity of the area before they are hung.
6. If wood doors are to be stored for any length of time on the job site, seal all edges and ends with an effective wood sealer to prevent absorption of moisture.

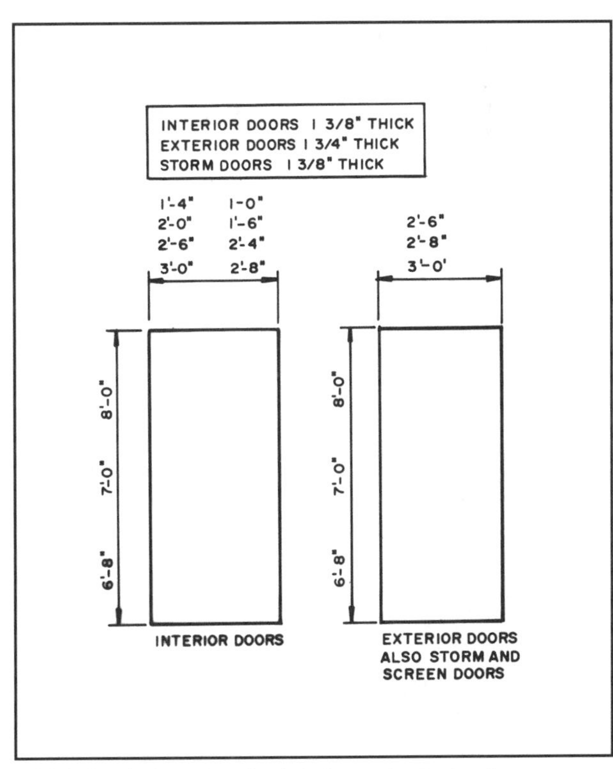

23-9 Stock door sizes for interior and exterior doors.

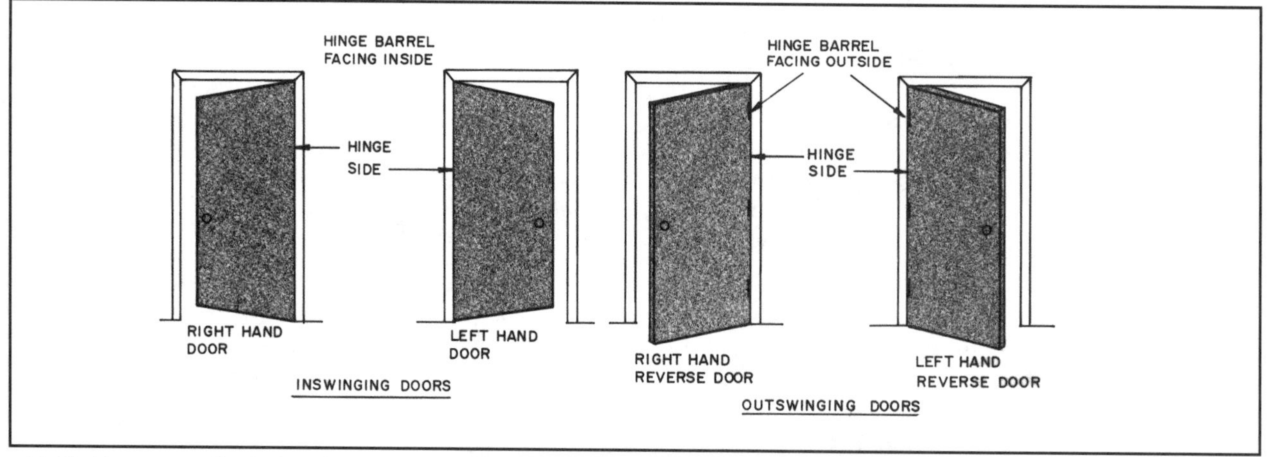

23-10 How to specify the hand of a door.

Door Hands

Before the door frames can be hung, the finish carpenter must ascertain the **swing**, or **hand, of the door**. The hand of a door refers to which side of the jamb will have the hinges and which the lockset. This relates to the way the door will swing. An examination of the floor plan shows which way the designer wants the door to swing.

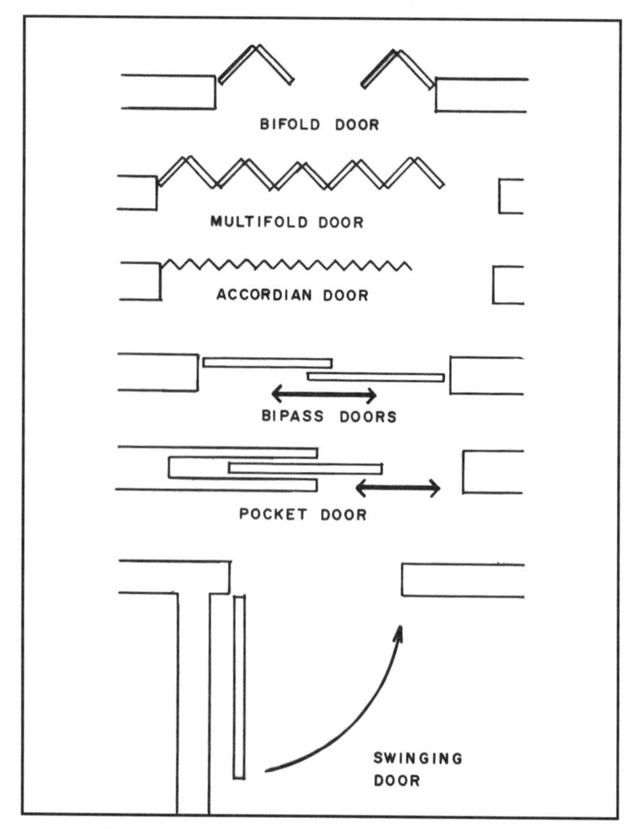

23-11 Types of interior door operations.

The carpenter must install the frame to produce this result.

While there is not universal agreement on determining the hand of a door, the following is a typical procedure.

The hand of a door is determined by looking at it from the "outside." For example, you view a bedroom door from the hall to determine the hand. You view an exterior door from outside the building. The various possible hand designations are shown in **23-10**.

When you view the door from the **outside**, in the standard designation of handedness—right hand, left hand—the door will **swing away** from you. If the hinges are on the right side, then in this case the door is a **right-hand door**. If, however, you view the door from the outside, but the door swings **towards** you, e.g., into the hall, then this is designated as a reverse door. Thus, in this instance, if the hinges are on the right side, the door is then a **right-hand reverse door**. Note that the lockset opening is on the side opposite the hinges.

When ordering locksets, it is necessary to specify the hand of the door. Some locksets are designed so that they can be changed to fit either hand.

DOOR OPERATION

Exterior doors are usually the swinging type, though sliding doors are popular for access to porches and patios. Interior doors can be of these types or accordion, multifold, bifold, bipass, or pocket types **(see 23-11)**. The type is shown on the architectural drawings and detailed in the door schedule.

CHECKING THE ROUGH OPENING

The framing carpenters who built the exterior walls and framed the interior partitions are responsible for framing door openings to hold the doors and their frames, and to leave room for making adjustments to set the frames plumb. It is advisable to check the size of the rough openings to see that they are correct. If they are too large, additional material may be added to the trimmer stud to reduce the size. If an opening is too small, the studs will have to be removed, repositioned, and a longer head built. Commonly used rough opening data are given in **23-12**. Also check to see that the gypsum drywall was installed properly so that it can be covered by the door jamb.

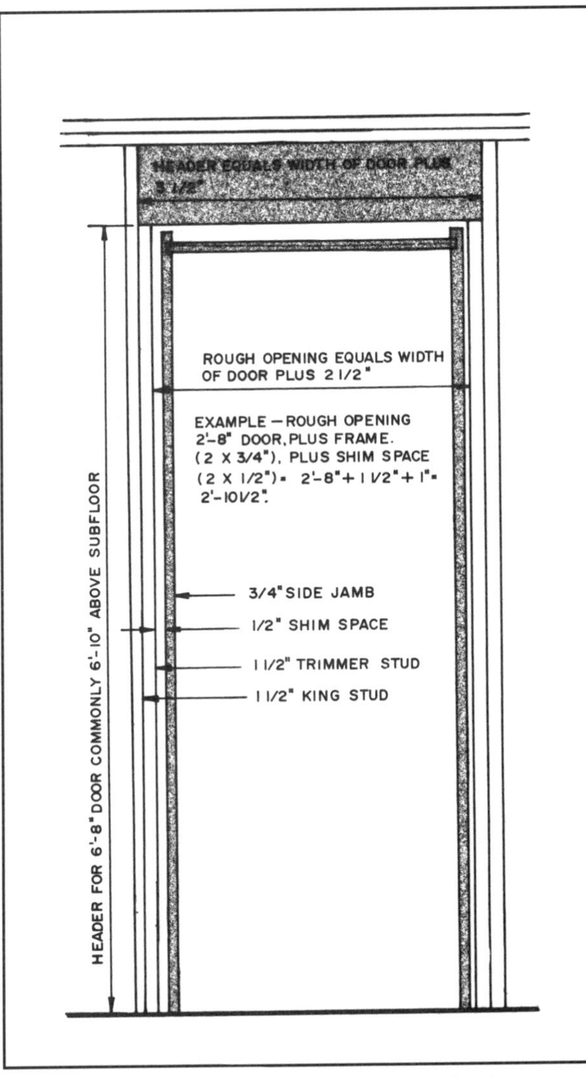

23-12 Commonly used rough opening sizes for doors.

WEATHERSTRIPPING

Factory-manufactured exterior frame and door assemblies have the weatherstripping installed as part of the unit. Two ways this is done on the sides and top of the door frame are shown in **23-13**. At the sill a gasket is usually placed along the inside and the door presses against it **(see 23-14)**. Other systems are also in use.

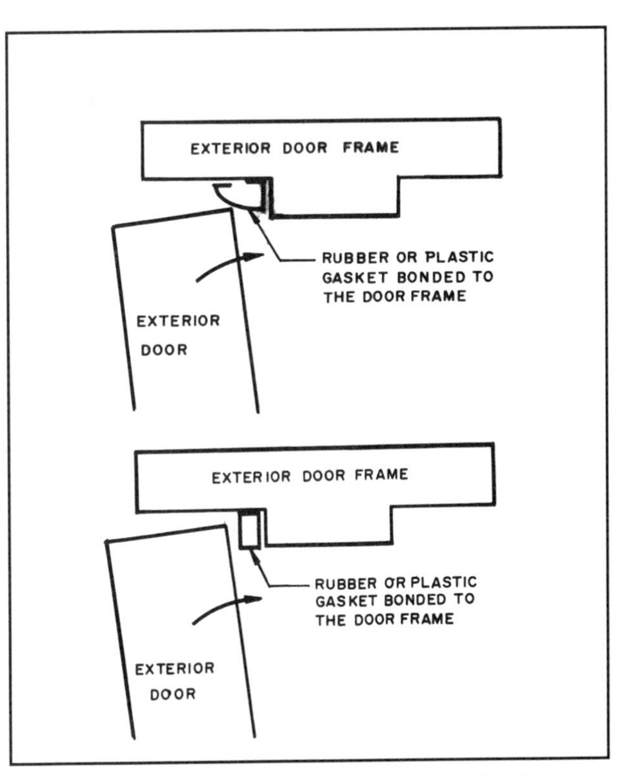

23-13 Two types of weatherstripping used on the door frame of factory-manufactured and assembled exterior frame and door assemblies.

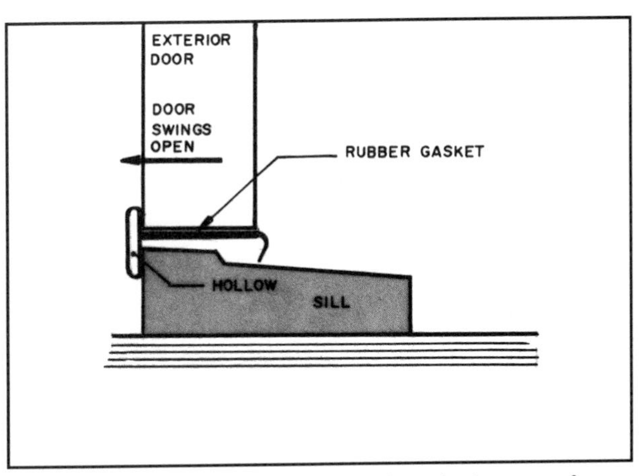

23-14 Typical weatherstripping used on the bottom of an exterior door.

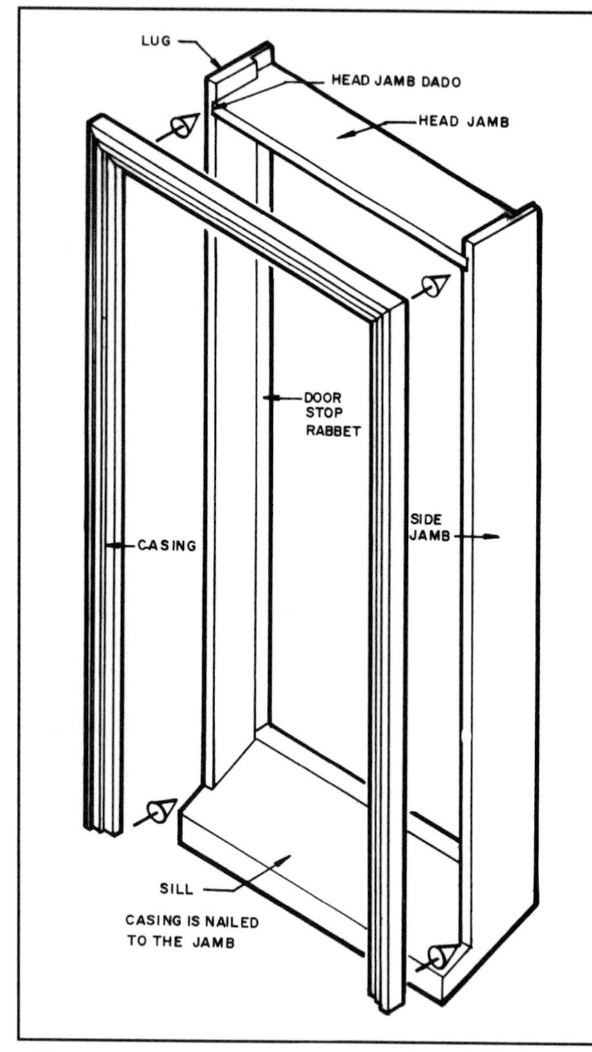

23-15 An exterior door frame and casing.

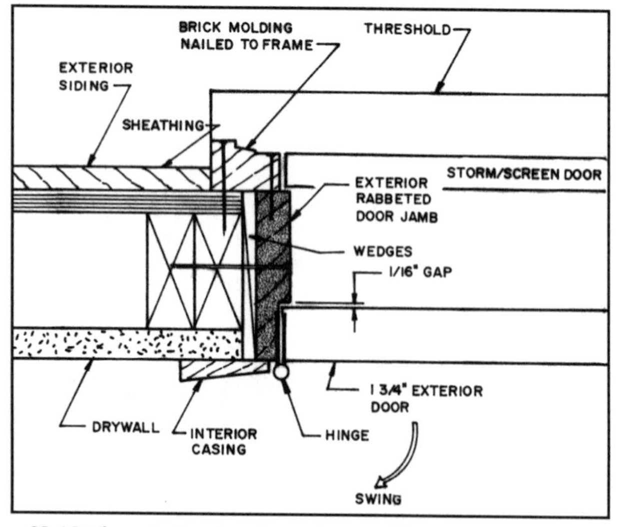

23-16 The exterior door frame has a rabbet cut to serve as a door stop for the exterior door. The storm or screen door butts the casing.

WOOD EXTERIOR DOOR FRAMES

Wood **door frames** are manufactured in large quantities. They are available either in knocked-down or assembled form. Since all the joints are cut, it is easy to assemble them on the job **(see 23-15)**. Some door frames are shipped with the door already hung. Exterior door frames have a wood sill. In masonry buildings the sill is often stone or concrete rather than wood.

The exterior door frame has a ½-inch-deep rabbet on the inside edge to receive the door. It also has a rabbet on the outside edge, formed by the casing. This is for a screen or storm door **(see 23-16)**. Exterior doors swing into the house, so this rabbeted edge must face in. The screen or storm door swings out.

INSTALLING PREHUNG EXTERIOR DOORS

The exterior door and frame are usually delivered to the site completely assembled, weather-stripped, and ready for installation. The frame is often braced, and the door held centered in the frame with wedges. Do not remove these until the frame is securely in place in the wall.

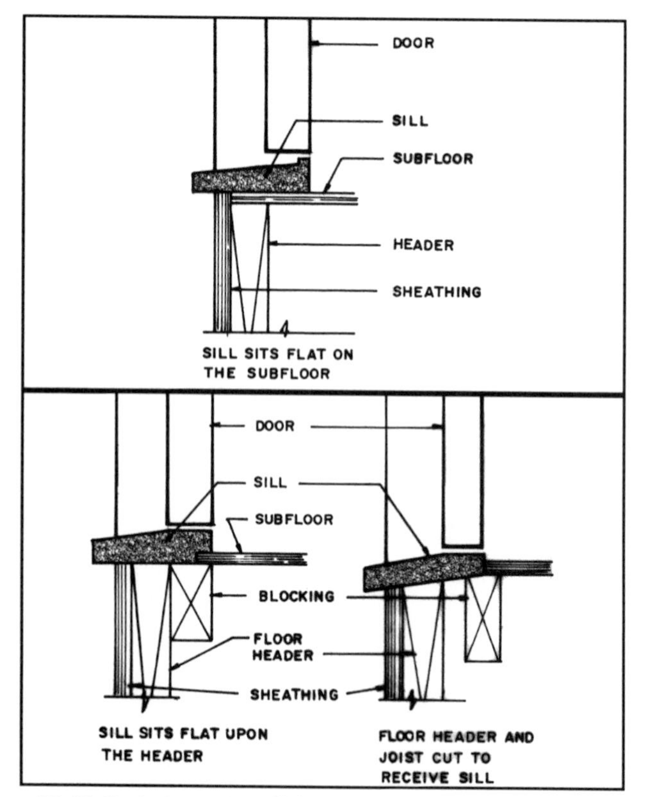

23-17 Typical exterior sill details.

23-18 This exterior door frame includes a prehung door and sidelights. *(Courtesy Weather Shield Windows and Doors)*

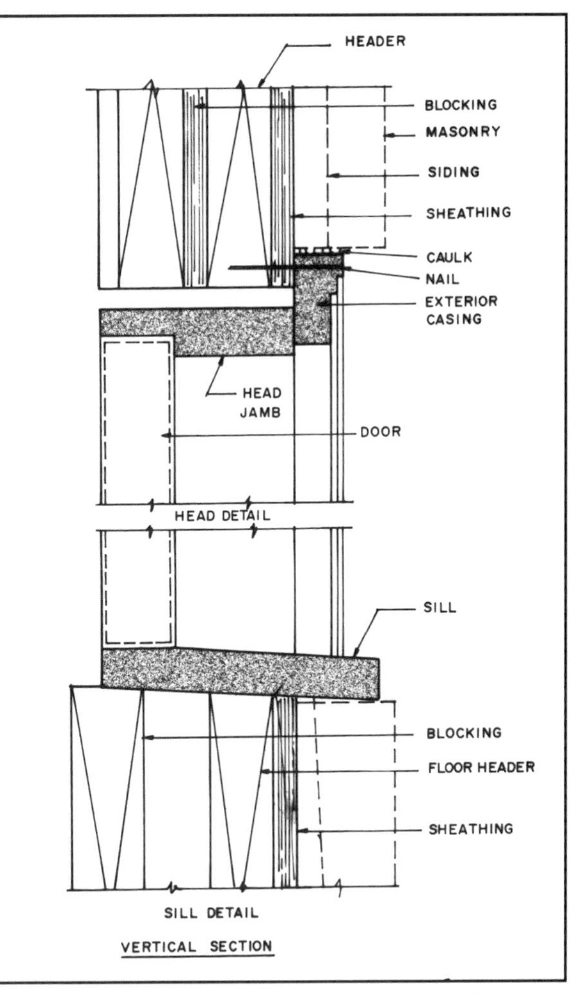

23-20 A vertical section through an exterior door frame showing the head and sill.

There are several types of frames and sills manufactured. Examples of the commonly used sill details are shown in **23-17**. One type requires the header and joist to be cut away a little to allow for the sloped sill. Another is designed to sit flat on the subfloor, while another sits on the header and the subfloor butts against it. Some exterior doors are installed with metal sills.

Prehung exterior door units are available as single-door units, double-door units, and units with sidelights. They should be installed following the directions of the manufacturer. Usually a printed instruction sheet accompanies each unit **(see 23-18)**.

Horizontal and vertical sections through typical exterior door frames are shown in **23-19** and **23-20**. The exterior casing fits flush against the sheathing. The siding or masonry veneer butts against the exterior casing.

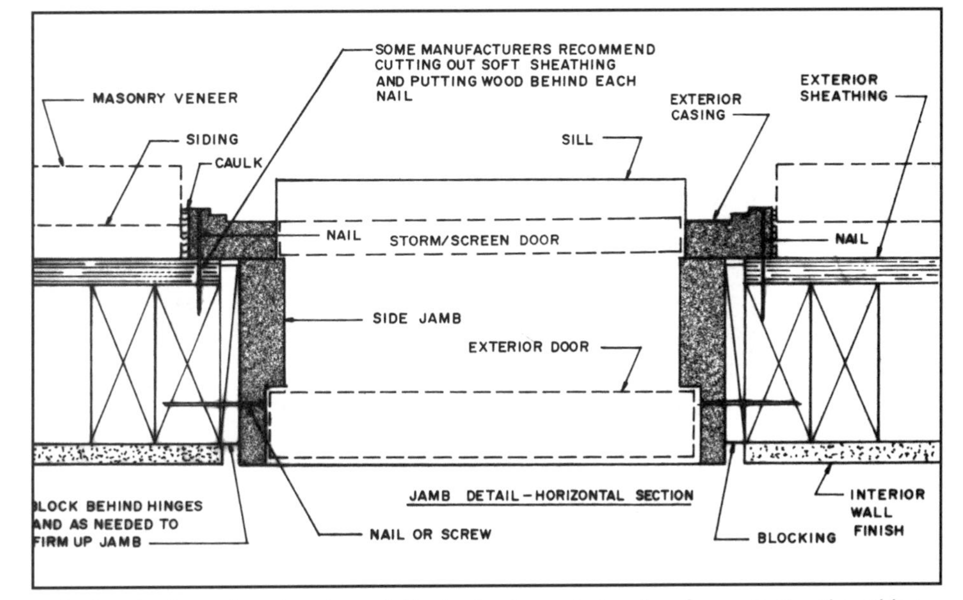

23-19 A horizontal section through the jamb of an exterior door frame. Notice the rabbet for the door and the location of the screen or storm door.

Following is a typical procedure for installing prehung exterior doors.

1. Be certain the opening has been covered with building paper and the sill flashed with aluminum.
2. Set the door into the rough opening from the exterior of the building.
3. Check to be certain the sill is level. If it is not, shim it at the floor. Some framers place several rows of caulking below the sill to get a tight seal. Some manufacturers recommend placing a screw through the sill into the subfloor.
4. The door frame must be plumb in two directions. It must be plumb and square so that the door fits properly. It must also be plumb in relation to the wall so that the door does not swing open by itself, or swing closed by itself when it should stay open **(see 23-21)**.

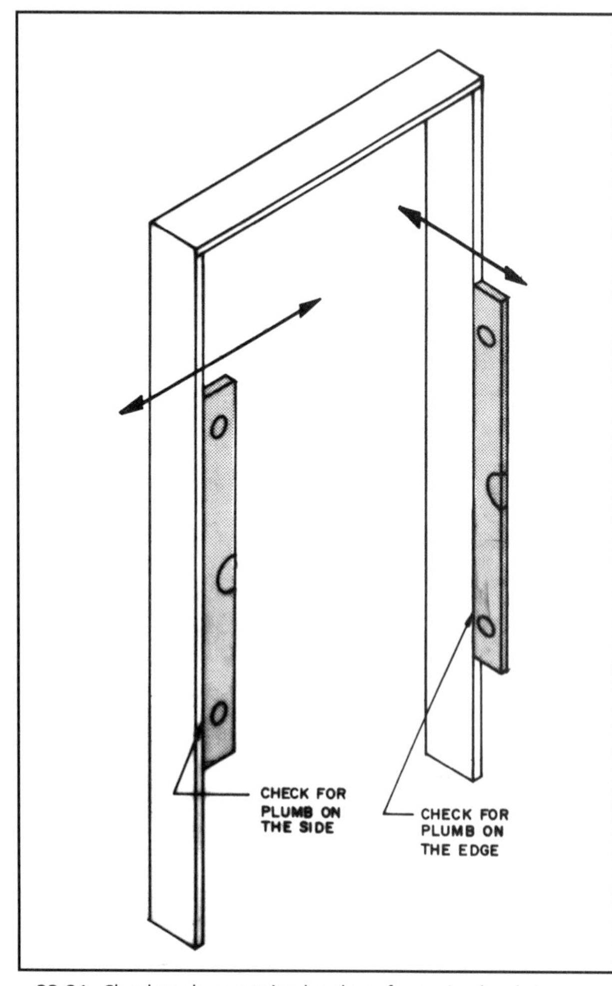

23-21 Check to be certain the door frame is plumb in both directions.

5. Place wedge-shaped blocking between the door jamb and trimmer stud to hold the frame plumb. Be certain the casing (sometimes called brick molding) is firm against the sheathing. Drive one casing nail through the casing into the studs on both sides. Do not set the nails at this time, because they may have to be removed for readjustment **(see 23-22)**.
6. Place blocking between the jamb and studs on the hinge side, and, when plumb, nail through the jamb and blocking into the stud **(see 23-23)**. Use 16d galvanized casing nails. Place blocking behind each hinge location and nail with two 16d casing nails **(23-24)**.
7. Now nail the other jamb, placing blocking at the lockset and other locations as shown in **23-24**. Place 16d galvanized nails in the lockset area, after making certain that the lockset hole in the door and the hole in the jamb are in line. The head jamb may require blocking and be nailed to the header to make certain that it is straight and level.
8. Some framers prefer to place the jamb with the hinges tight against the stud, shimming only enough to get it plumb **(see 23-25)**. The rough opening must be narrow enough to allow the exterior casing to lap over the sheathing and trimmer stud on the lockset side.

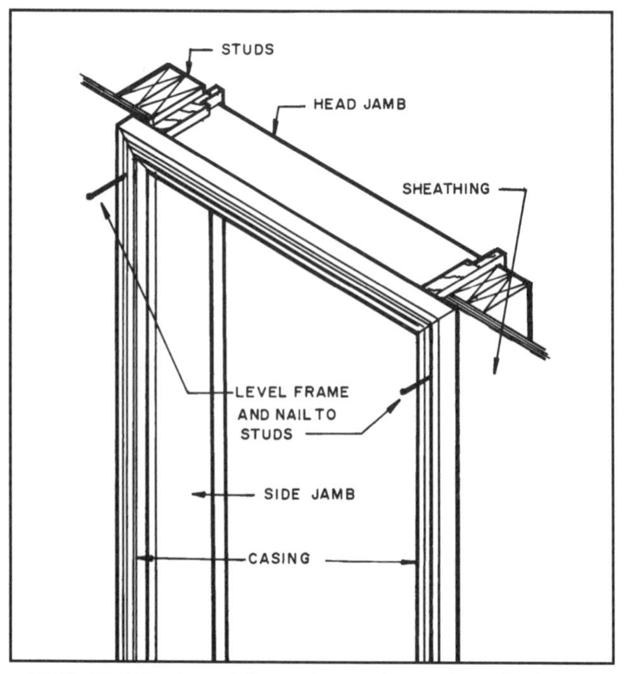

23-22 Nail the top of the casing to the studs as the frame is leveled.

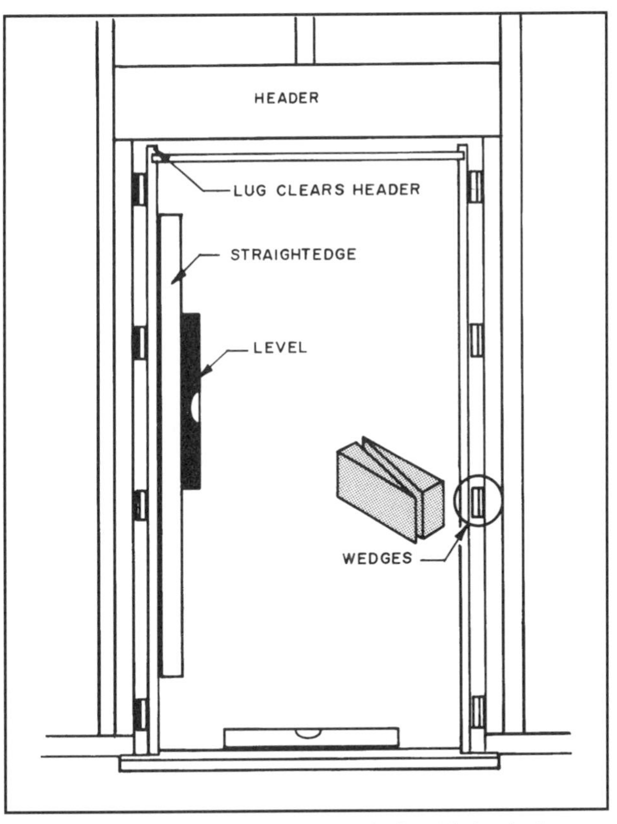

23-23 Once the frame is level, plumb the side jambs by inserting wedges between the frame and stud and nail through the wedges. Finish nailing the casing.

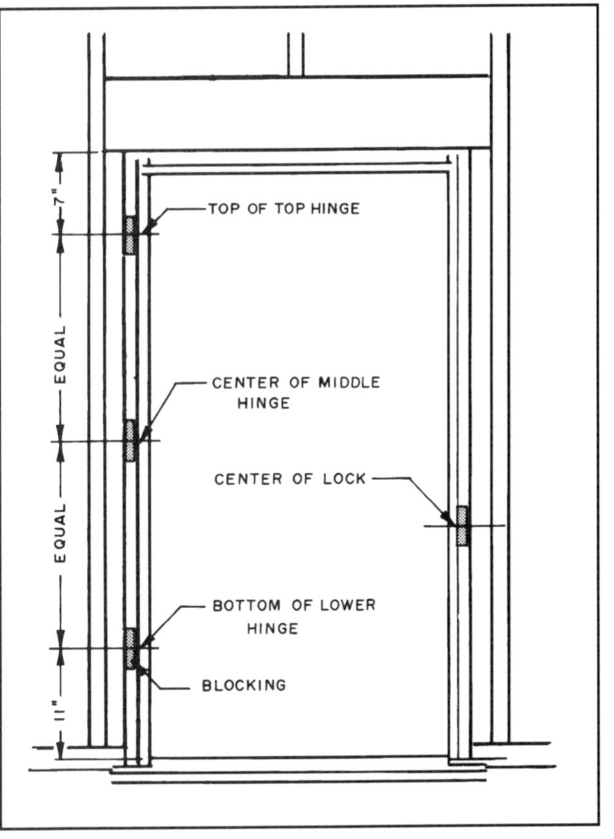

23-24 Install blocking at the location of each hinge on the lockset.

9. Check to be certain that when the door closes, the opening around the edges is uniform, and that it does not rub on the jamb. Also check to be certain the door hits the door stop uniformly on all sides.

10. If all is plumb and the door fits the opening correctly, set the nails and cut off any shims sticking beyond the frame.

11. Then nail through the casing into the studs on all sides of the frame, and set the heads.

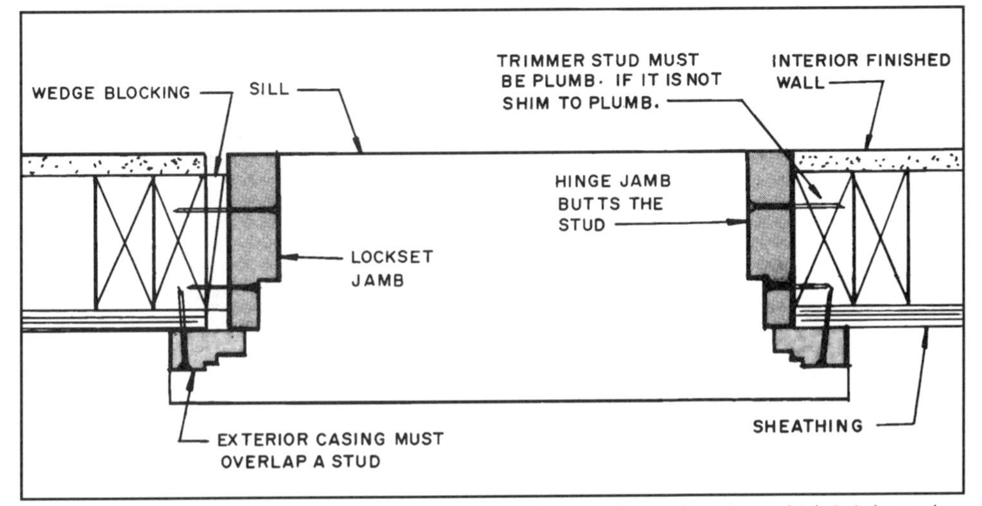

23-25 An alternate way to install an exterior door frame. The stud against which it is butted must be plumb. If not, shim as necessary.

SETTING A DOOR FRAME IN A SOLID MASONRY WALL

In solid masonry construction the door frames are set in place after the subfloor is laid. The masons lay the wall to the frame.

The head of the door must be the proper height above the finished floor. The side jambs might have to be cut shorter to meet this height.

Set the door frame on the masonry sill. The frame should have diagonal braces and a spacer at the bottom. These keep it square and the correct size. Brace the frame to the subfloor. Make certain that it is level and plumb and securely nailed in place **(see 23-26)**.

The inside edge of the jamb must line up with the inside surface of the finished wall material. The plan must show a typical wall section. This tells the carpenter the amount of furring to be used inside and the thickness of the finished wall **(see 23-27)**.

The bricklayer lays the bricks up next to the casing on the door frame. The frame is anchored to the brick wall as it is laid. The crack between them is later caulked **(see 23-28)**.

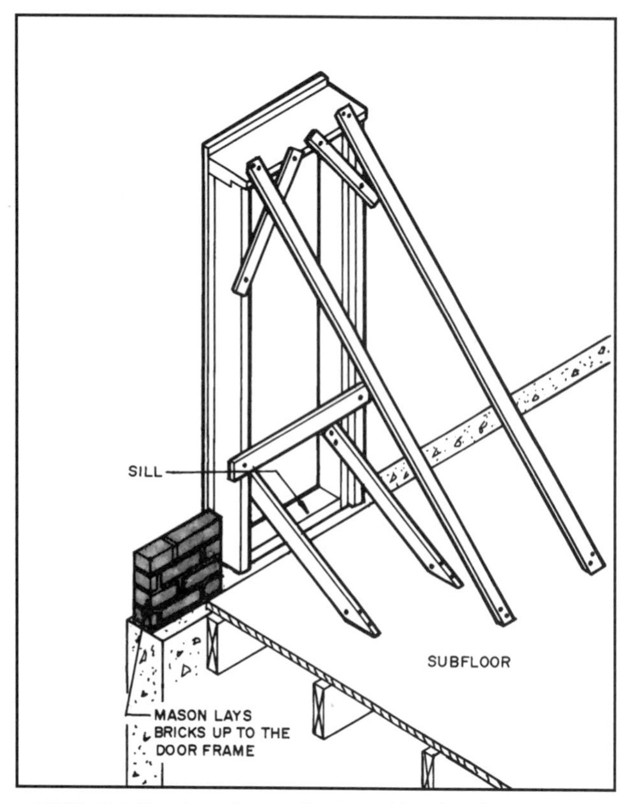

23-26 Set the door frame plumb and level and brace it. The mason will lay the brick exterior wall to the casing on the door frame.

WOOD INTERIOR DOOR FRAMES

Interior door frames are flat. They do not have a rabbet or a sill **(see 23-29)**. The door butts against a door stop that is installed after the frame and door are set in the wall opening. Most commonly used widths are 4½ inches for walls with ½-inch drywall and 5¼ inches for walls with plaster. The frames are cut to finished size and sanded in a factory. They are shipped knocked down for assembly on the job. Since the dado is cut for the head jamb, assembly is easy. Use three 8d casing or finishing nails.

Also available are adjustable split jambs. They can be adjusted to fit different wall thicknesses **(see 23-30)**.

Many interior doors come with the frame assembled and the door installed. The frame requires no assembly but is simply installed in the wall opening.

INSTALLING INTERIOR FRAMES WITH PREHUNG DOORS

Most **interior doors** and frames are assembled in the factory and have the door prehung in the frame on hinges. The door usually has holes

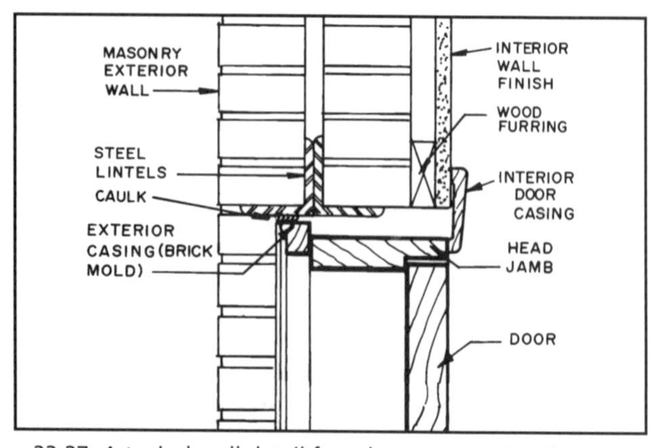

23-27 A typical wall detail found on construction drawings. It shows the construction of a solid masonry wall and how the head of an exterior door is to be installed.

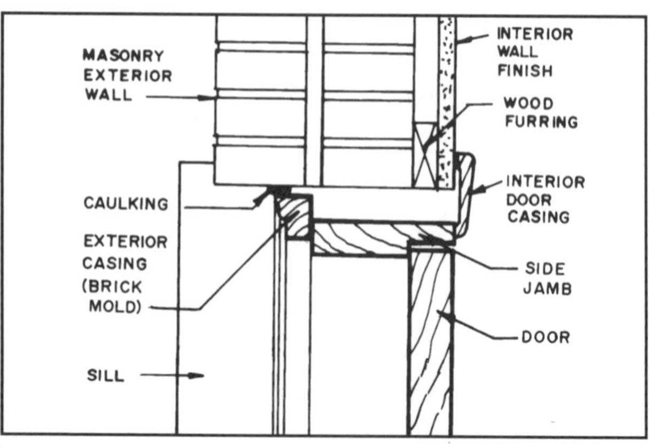

23-28 This is a horizontal section through the door jamb as it butts a solid masonry wall. The joint at the casing is caulked after the masonry units are laid.

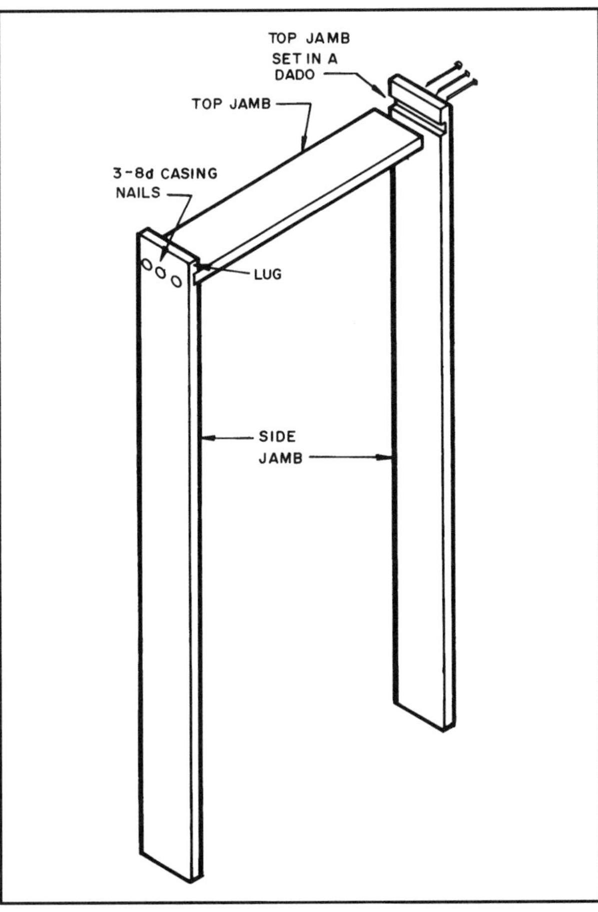

23-29 A typical interior door frame.

that are already bored for the lockset. The unit consists of a frame and door and, if ordered, a precut casing **(see 23-31)**. There are two types of frames generally available, a one-piece frame and a split frame **(see 23-32)**.

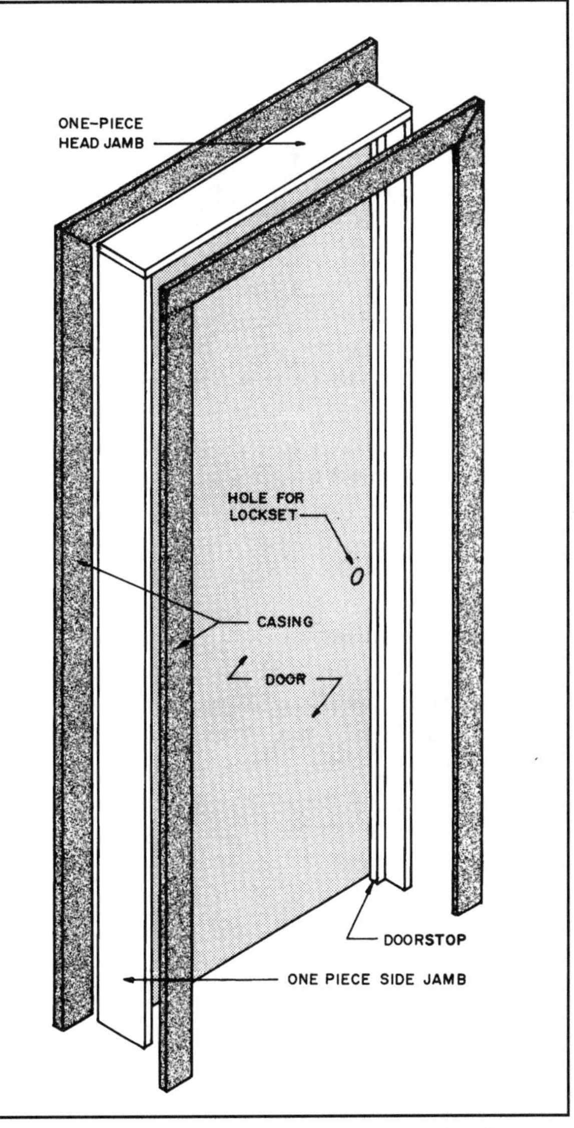

23-31 A typical interior door frame, casing, and prehung interior door.

23-30 Some interior door frames have split jambs permitting the frame to fit walls of various thicknesses.

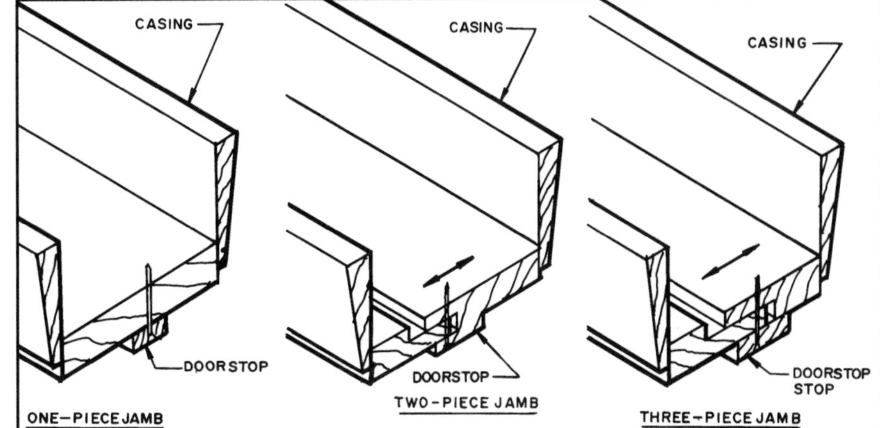

23-32 Three types of interior door frames.

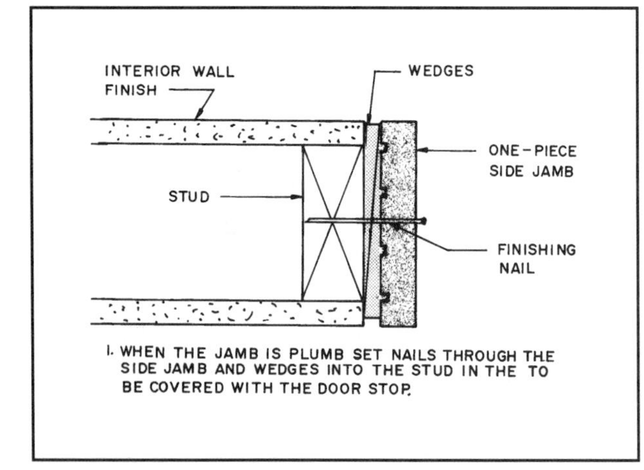

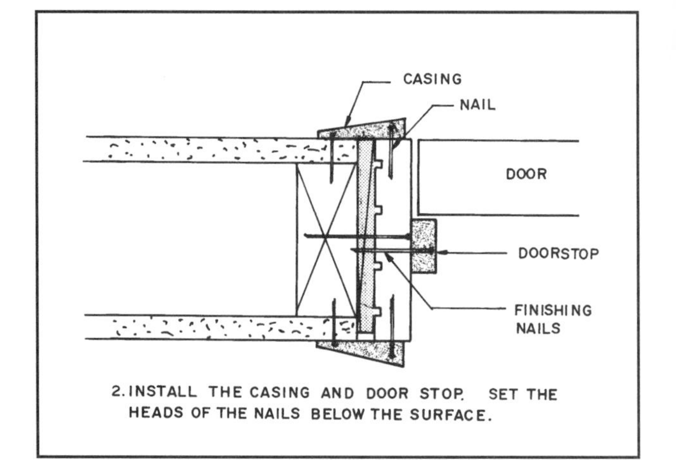

23-33 How to install a one-piece interior door frame.

One-piece jambs are installed by placing the frame in the rough opening, adjusting it so that it is plumb, installing blocking on each side, and nailing through the side jamb and blocking into the stud **(see 23-33)**. Wedge-shaped blocks are often used, because they are easy to adjust when plumbing the frame. The nails are placed in the area to be covered by the doorstop **(see 23-34)**. The blocking is placed behind each hinge and the lockset area and in other places as needed to plumb and reinforce the frame **(see 23-35)**. Once the frame is in place, the casing is installed.

Some types of interior door frames have the casing nailed on one side in the factory. After installation, the precut casing for the other side is nailed in place **(see 23-36)**. The door frame width should be the same as the width of the interior wall. This is specified when ordering. Widths wider than normal may arrive with an extender strip that is glued and nailed to the basic jamb **(see 23-37)**.

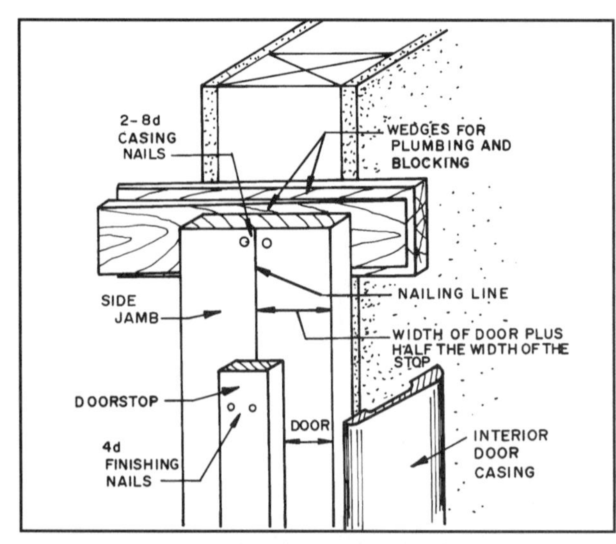

23-34 Locate the nailing line so that the door stop covers the nail heads. Nail through the jamb and wedges into the stud. Put a nail on each side of the nailing line.

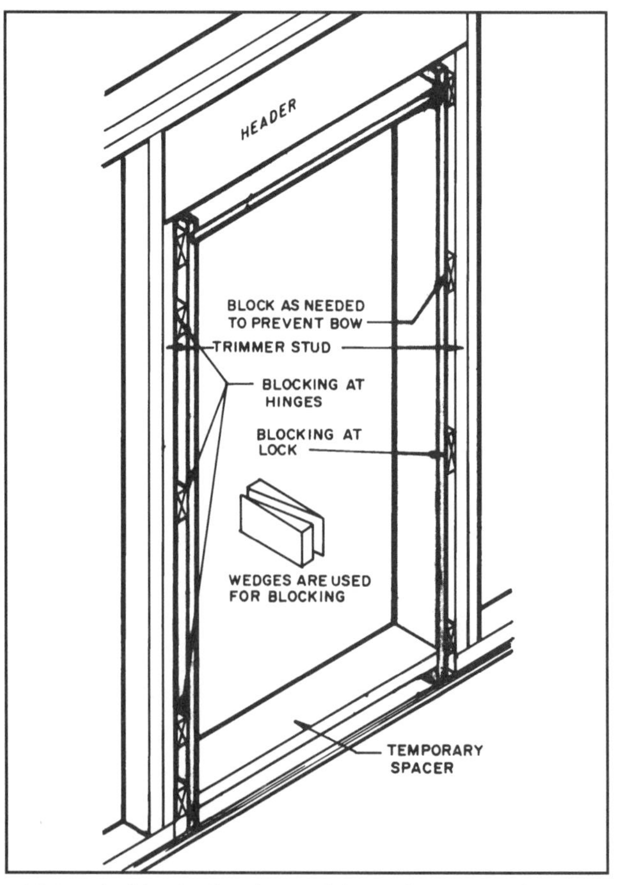

23-35 Blocking is placed at each hinge location and at the lockset.

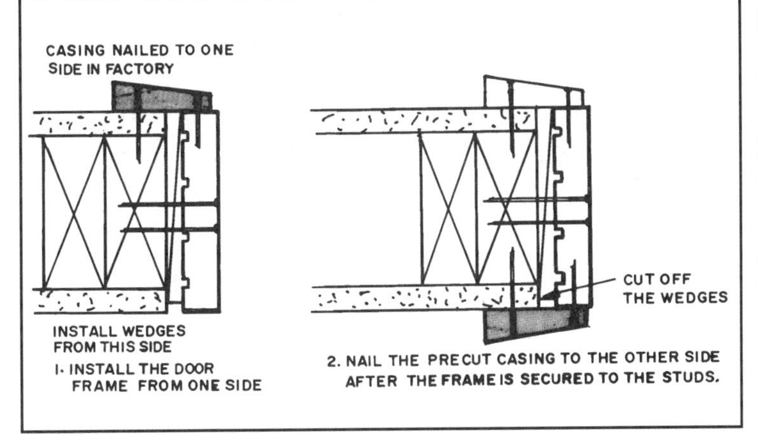

23-36 How to install a preassembled interior door frame that has the casing installed on one side at the factory.

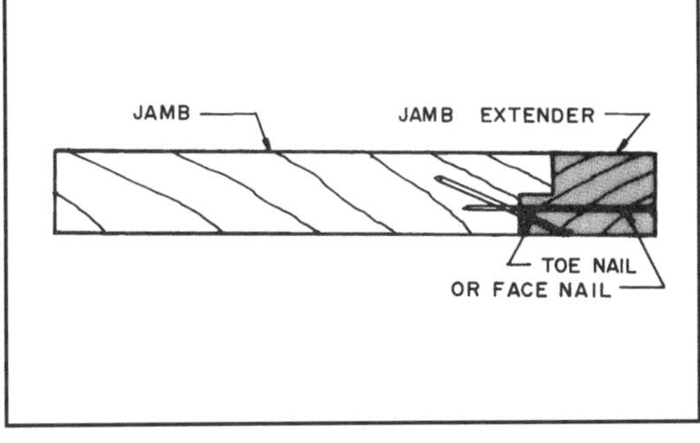

23-37 Jambs can be made wider by adding a jamb extender.

Split interior door jambs usually have the casing applied in the factory. One side is installed first, by nailing the casing to the stud. The jamb then has blocking installed as described for one-piece jambs. Then, the other half is inserted from the other side and nailed in place through the casing. The doorstop is nailed as shown in **23-38**. The tongue-and-groove slip joint enables the jamb to adjust for variances in wall thickness **(see 23-39)**.

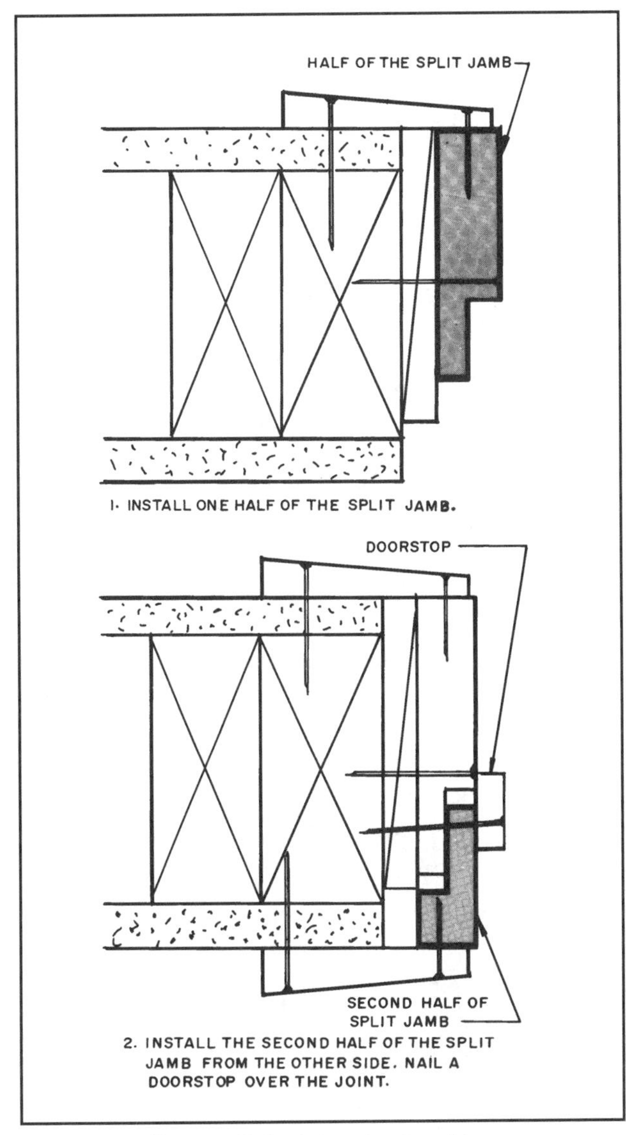

23-38 Installing a split-jamb interior door frame.

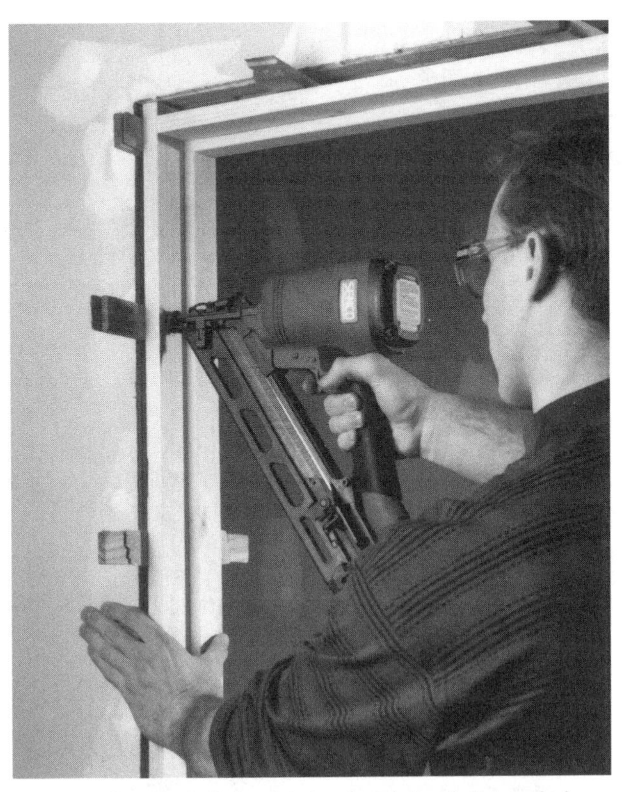

23-39 After the split jamb is level and plumb it is nailed securely to the studs. A power nailer does a fast and secure job. Notice the blocking above the head jamb. *(Courtesy Senco Fastening Systems)*

If the building is to have a hardwood floor, it is easier to install the floor first and then set the door frame on top of it. However, many carpenters prefer to set the frame first, placing it on the subfloor. The hardwood flooring has to be fitted around the base of the side jambs or the jambs may be cut off to allow the flooring to slide under them. Another technique is to raise the bottom of the jamb above the subfloor a distance equal to the thickness of the hardwood flooring, rather than trimming later.

An example of the door frame set on the subfloor is shown in **23-40**. The frame is one inch (25mm) longer than the door. This leaves a space below the door for carpet. In some cases it is necessary to trim a little off the bottom of the door to get the desired clearance above the floor.

Another consideration is locating the head of the frame above the floor. In most cases the head of the door frame and the heads of the windows are kept the same distance above the floor. When this is the case, the desired height can be marked on the studs that form the rough opening, and the door frame head lined up with this. This may require the bottom of the frame to be raised slightly above the subfloor, but this space will be hidden by the carpet.

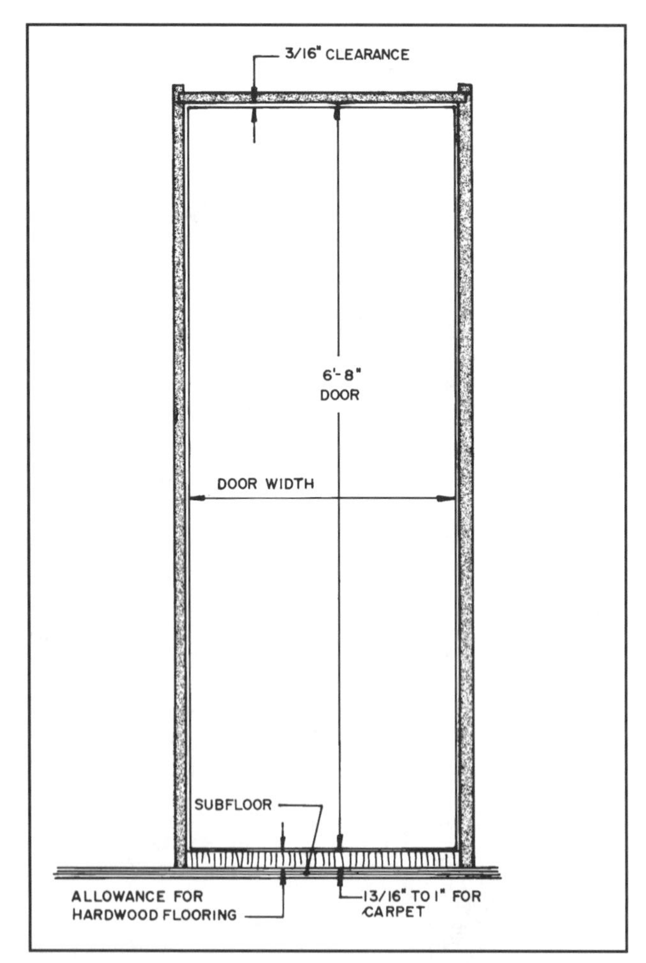

23-40 The height of the door frame head may vary depending on the type of finished flooring to be used.

Once the height and floor decisions have been made the door frame can be installed as follows:

1. Remove the hung door from the frame by pulling the hinge pins.
2. Set the frame in the rough opening. Position it so that the head is in line with the height mark. Nail the side jamb with the hinges to the trimmer stud first **(see 23-41)**. Use a six-foot level to be certain the jamb is plumb and in the same plane as the wall. Insert wedge-shaped shims between the stud and jamb to adjust the jamb to a plumb position **(see 23-42)**. It is best to use shims made for this purpose. While wood shingle scraps can be used, they tend to split and fall out when nailed or cut to length.
3. Insert shims from each side of the wall, and tap lightly until they bind. Adjust until the jamb is plumb. Place the first shims near the top and bottom of the jamb. Nail through the jamb and shims with 8d casing nails to secure the frame to the stud. Place the nails in the center of the jamb so that they are covered by the door stop. Do not set the nails until the frame installation is complete. Additional shims are placed where each hinge is to be located. Other shims can be added, wherever needed, to produce a plumb jamb. (Interior doors should have three hinges.) Cut off the shims so that they are flush with the edge of the jamb.
4. Check to be certain that the jamb is square with the partition. Use a framing square as shown in **23-43**.
5. Move the strike jamb (the side to have the lock) up and down until the head jamb is horizontal and at right angles with the hinge-side jamb. The head jamb is held in place by the strike side jamb, which is now nailed in place.
6. Place shims near the top and bottom of the strike jamb. Check to see that it is at right angles with the head jamb and is plumb. Nail through the shims into the stud with 8d casing nails. Place shims behind the jamb in the area where the lock will be located. Place other shims as needed to get the jamb straight.

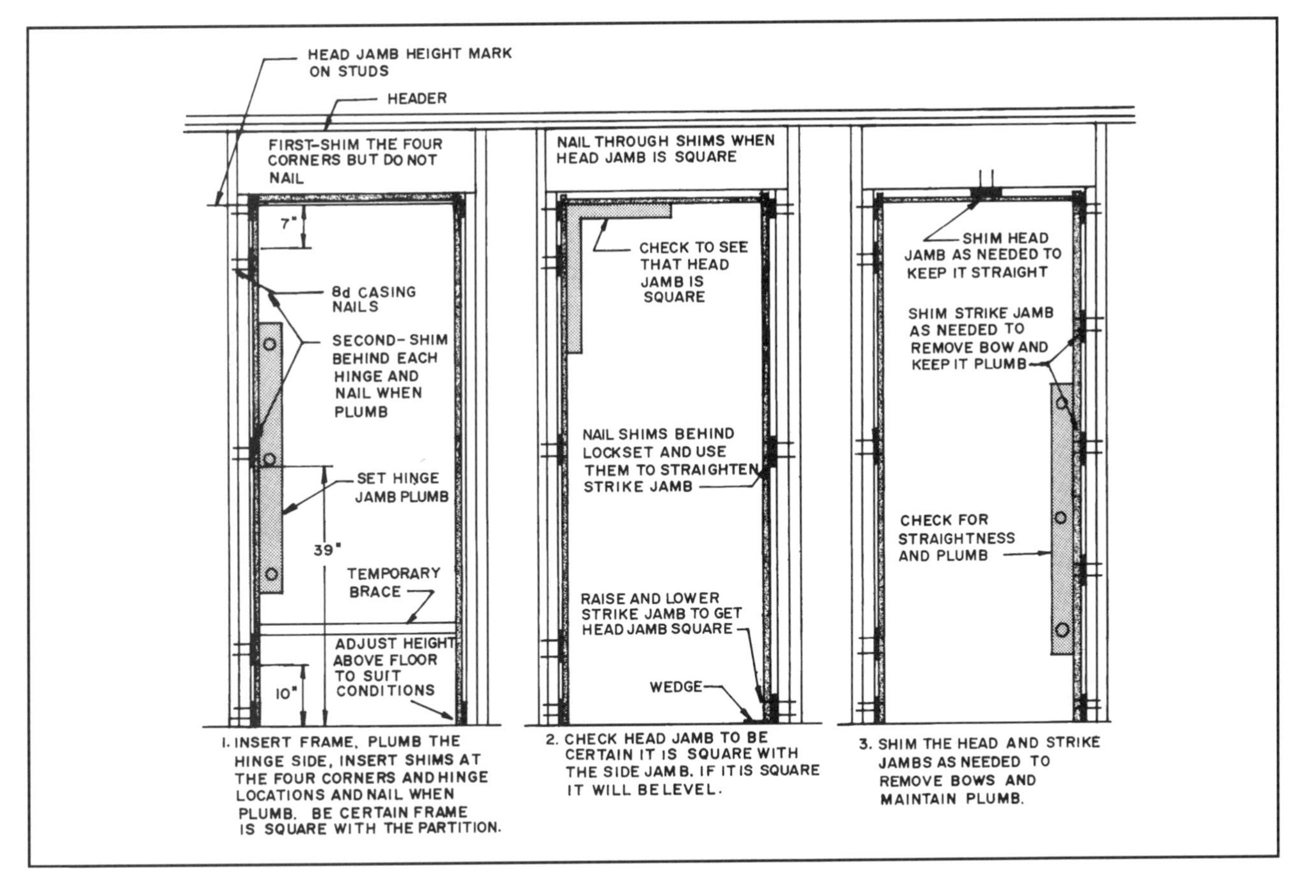

23-41 The basic steps for installing an interior door frame.

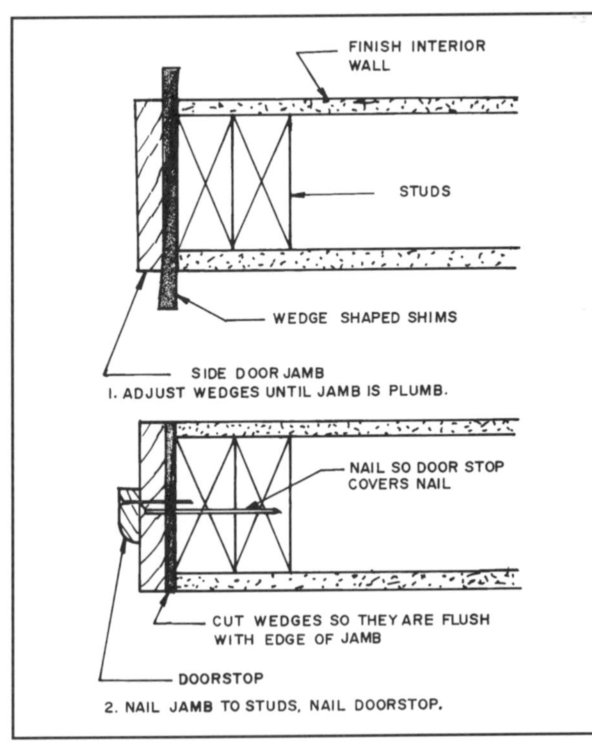

23-42 Wood wedges are used between the door jamb and trimmer stud to set the jamb plumb.

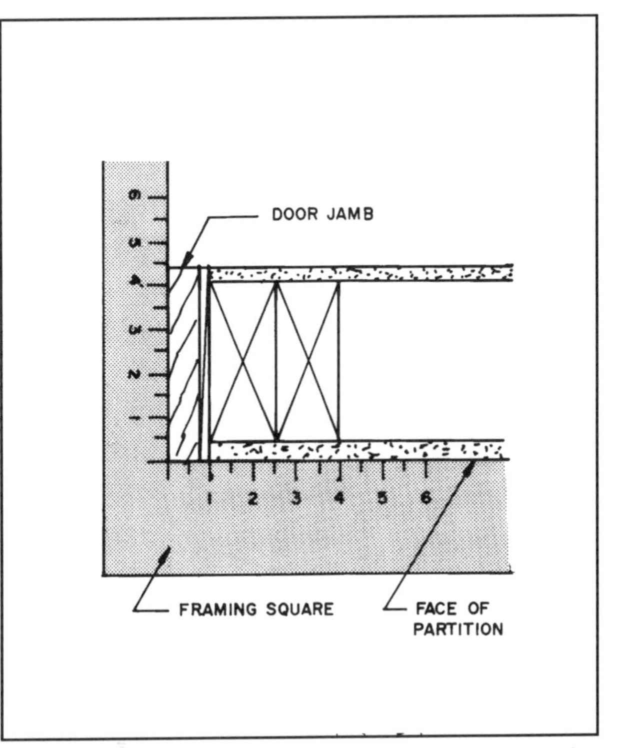

23-43 Be certain the jamb is at right angles to the face of the partition.

7. Finally shim the head jamb to prevent it from eventually bowing.
8. The frame can be checked for squareness by measuring the diagonals of the opening. They must be exactly the same length. The real test comes when the door is placed on the hinges. Check the opening around the edges to be certain it is uniform and that the door opens and closes properly. The installed frame with the door in place is in **23-44**. When the jambs are plumb and the door operates properly, set the nails.

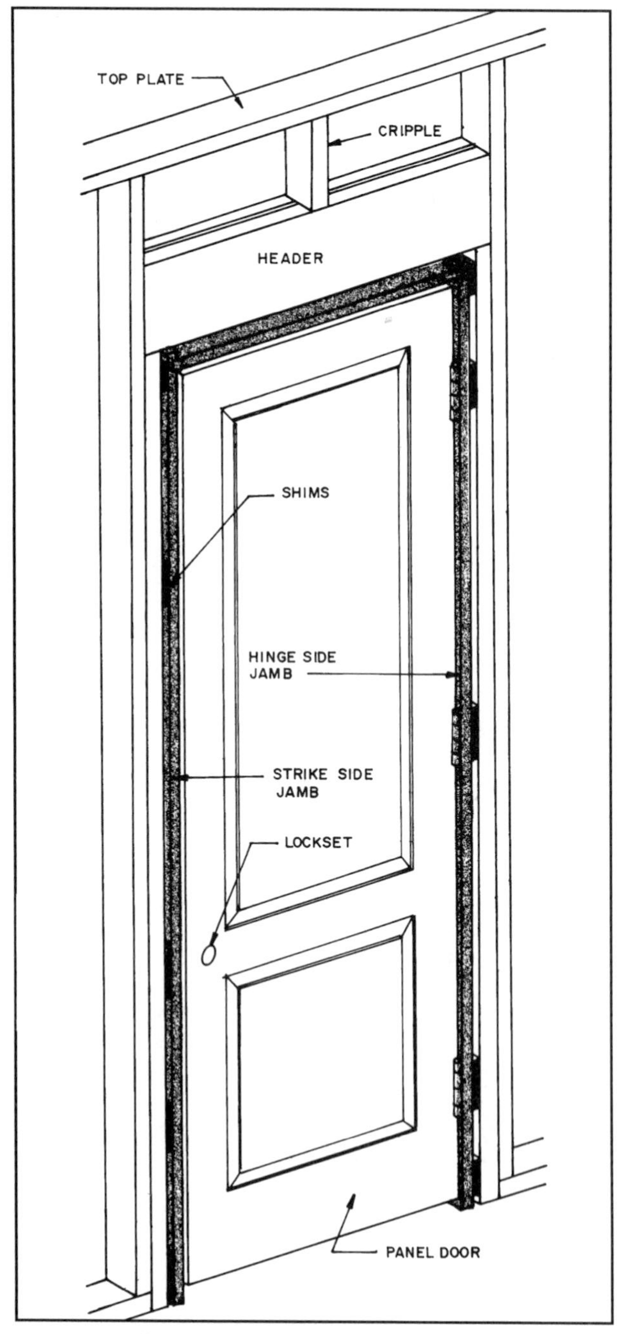

23-44 The finished installation of a prehung interior door and frame.

HANGING A DOOR

If you face a situation where prehung interior doors are not to be used, you will receive the door frame in a knocked-down condition and a supply of doors with no preparation for hinges and maybe not even the exact size for the finished opening.

Begin by selecting the size door frame specified on the drawing. Nail it together and nail a cross brace to keep the jambs apart. Install it in the wall opening as described earlier for prehung door frames.

Select the proper door specified for the opening. Notice how it is to swing, and mark the hinge side. Measure the frame opening and the size of the door. The door may require a little trimming. Usually, very little need be removed from the door.

Never cut a great deal off any one edge. This will produce a thin stile and a weak door.

Trim the door so that it clears the door frame 1⁄16 inch on all sides. Usually 5⁄8-inch clearance on the bottom will be enough. If metal weather stripping is to be used, a side clearance of 1⁄8 inch will be necessary.

Since most doors require very little trimming, the amount required is removed with a plane. A hand or power plane can be used. Stand the door on edge. The best way to support it is with a door holder. One is shown later in **23-53** . Remove some material from each side. If it must be trimmed with a power saw, use a fine-toothed blade to get a smooth cut. Cut a little large so that the edge can be planed to its final size. Once the door is its final size, plane a bevel on the edge on the lock side of the door **(see 23-45)**. This is needed so that the edge will not hit the door frame as the door opens. Usually, a bevel about 1⁄8 inch will be enough. If an aluminum threshold is used, bevel the bottom of the door so it swings open easily yet presses against the gasket as shown in **23-46**.

Now sand the planed edges and corners so that they are slightly round and smooth. Always sand with the grain of the wood. Use a fine-grit abrasive paper.

The door is now ready to fit into the frame and have the hinges and lock installed.

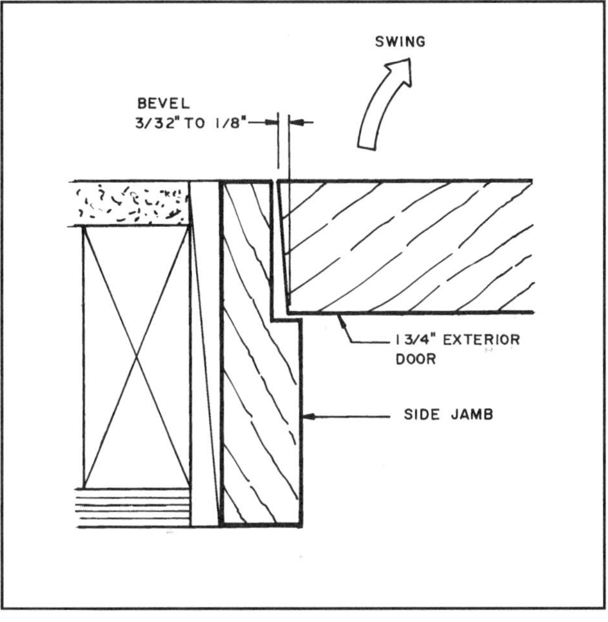

23-45 The slight bevel of the edges of the door allows a little extra clearance when the door swings open.

Installing the Hinges

When prehung doors and frames are used, the hinges are already in place. When frame kits are used or the carpenter builds the frames, the hinges must be located and mounted on the jamb and the door.

The hinges commonly used for interior doors are **loose-pin butt hinges**. Exterior doors use fixed pin hinges so thieves cannot pull the pin and remove the door. For most doors in residential construction, the residential-grade butt hinge is adequate. For heavy doors commercial-grade or ball-bearing hinges are recommended. The parts of a hinge are shown in **23-47**. Butt hinges used to hang doors are swaged; this means that the leaves are bent to provide a small space between them when the door is closed **(see 23-48)**.

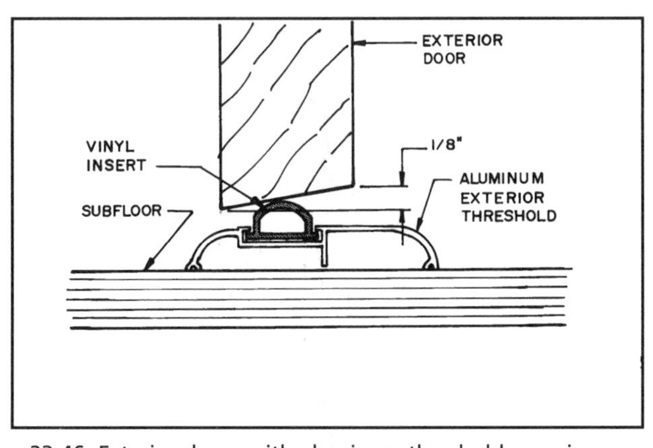

23-46 Exterior doors with aluminum thresholds require the bottom of the door be beveled to clear the gasket yet seal it against air leakage.

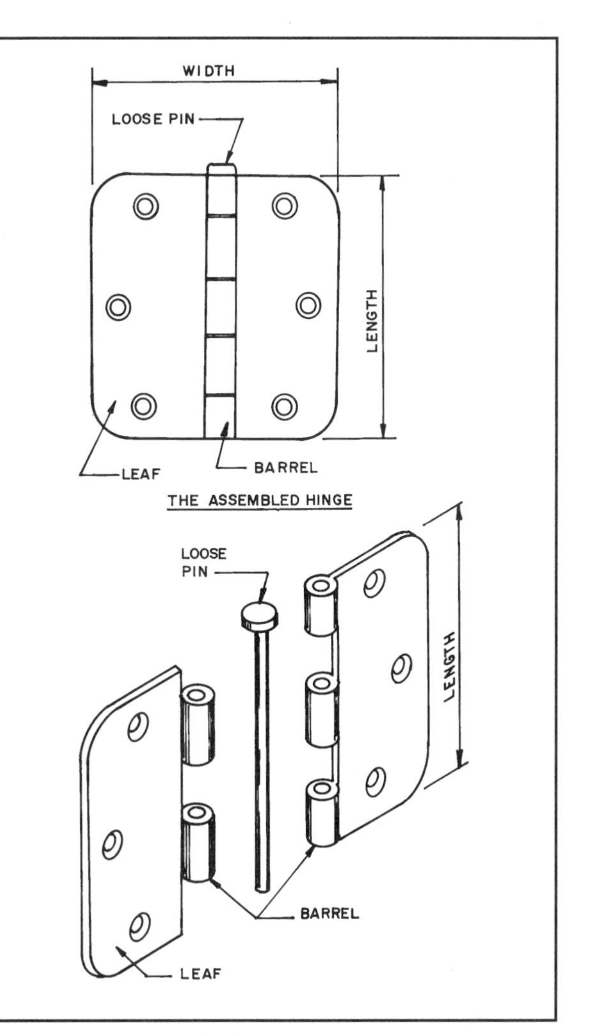

23-47 A butt hinge typically used to hang interior and exterior doors. The pin in exterior door hinges is fixed so it cannot be removed.

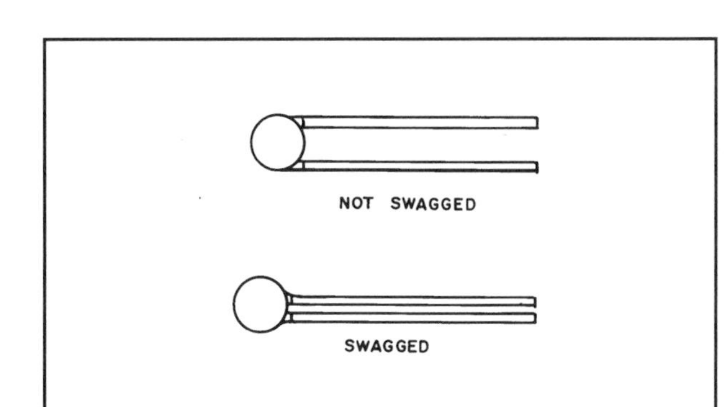

23-48 Door hinges are swaged.

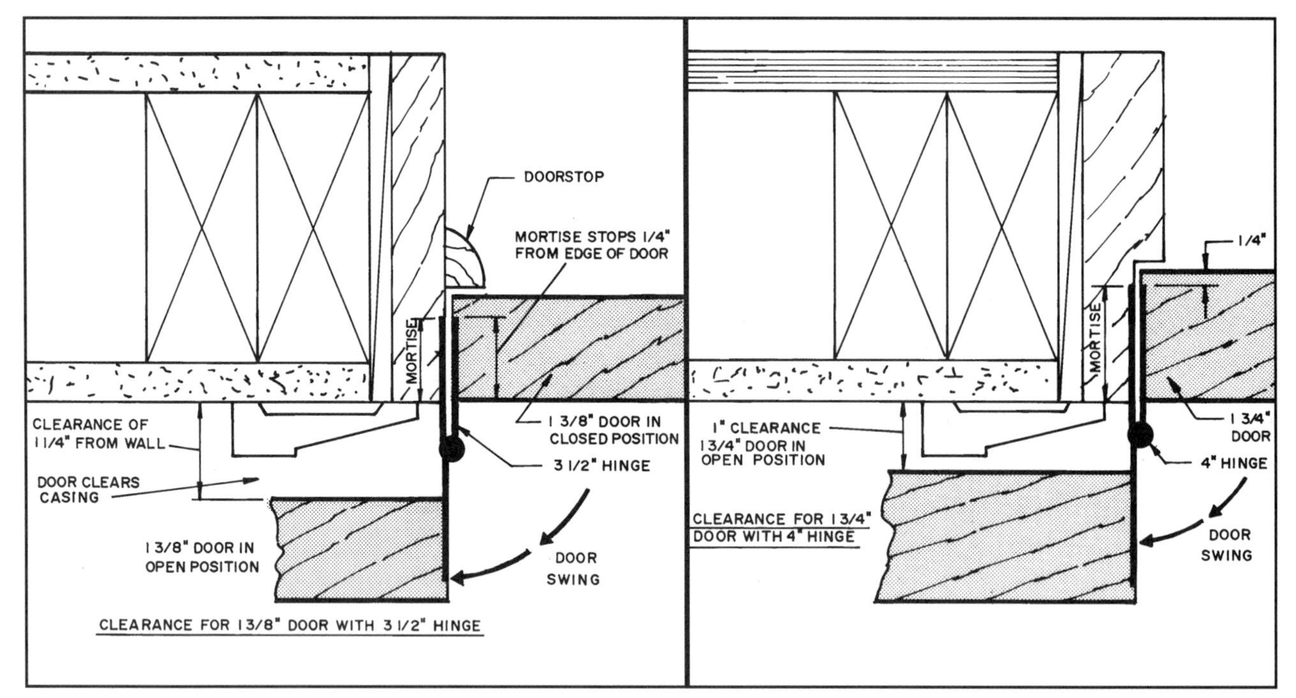

23-49 When selecting the proper hinge width, the required clearance must be considered.

Recommendations for selecting the height of a hinge are shown in **Table 23-1**. While these are not mandatory sizes, they do provide a guide. Doors that are heavier than usual or doors on openings that will have considerable use should use larger hinges of commercial quality or have additional hinges.

Hinge dimensions are specified by giving the height of the leaf (parallel with the barrel) and the width, which is measured when the hinge is open in a flat position. The width chosen relates to the thickness of the door and to the clearance between the open door and the wall that is required to clear the casing or other wall moldings **(see Table 23-2)**. The clearances provided by various width hinges for the most commonly used door thicknesses are shown in **Table 23-2**. These are based on the assumption that the mortise stops ¼ inch from the edge of the door. Examples of how the information in **Table 23-2** applies to 1⅜-inch- and 1¾-inch-thick doors is shown in **23-49**. Interior and exterior doors should have three hinges.

The location of the hinges and lockset are shown in **23-50**. These are typical locations recommended for standard doors used in residential construction.

The layout of the hinge on the jamb and door is shown in **23-51**. Hinges are available with round or square corner leaves.

Table 23-1 Recommendations for selecting hinge sizes.

Door Thickness	Door Width	Hinge Height
1⅜"	up to 32"	3½" or 4"
	33" or larger	4" or 4½"
1¾"	up to 36"	4½"*
	36" to 48"	5"*
	over 48"	6"*
2¼"	up to 42"	5" heavy**
	over 42"	6" heavy**

*Use heavy-weight hinges for heavy doors.
**Use heavy-weight ball for bearing hinges.

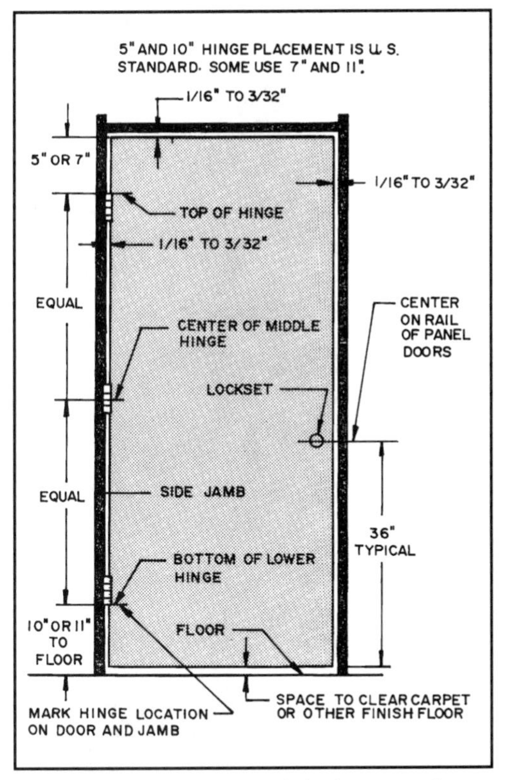

23-50 Vertical locations of the hinges and lockset.

Table 23-2 Selecting the proper width hinge.

Door Thickness (inches)	Hinge Width When Open (inches)	Clearance of Door from Wall Provided with Mortise Stopping ¼" from Edge of the Door (inches)
1⅜"	3½"	1¼"
	4"	1¾"
1¾"	4"	1"
	4½"	1½"
	5"	2"
2¼"	5"	1"
	6"	2"

The steps to manually cut the **hinge mortise** (also called a **gain**) are shown in **23-52**. This is used to produce square corner mortises. Begin by laying out the edges of the hinge, and then mark the thickness of the leaf to establish the depth. Use a wood chisel to outline the edges of the mortise, and then cut a series of crosscuts along it. Use the chisel to remove the surface to the required depth.

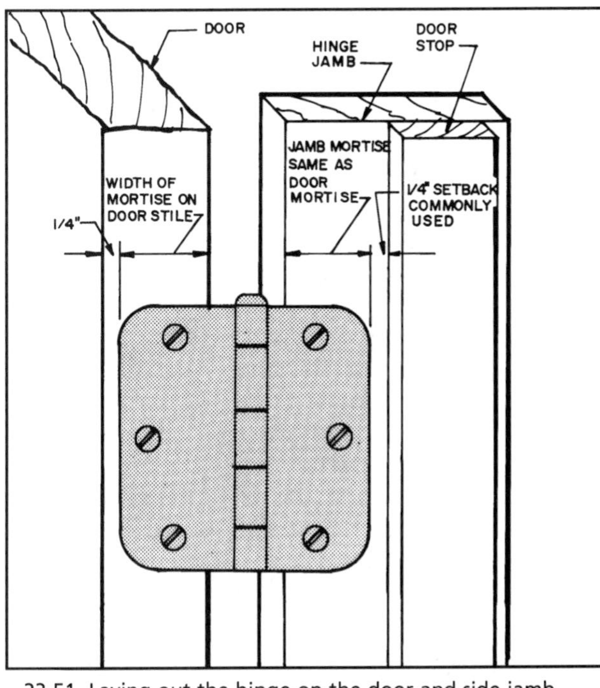

23-51 Laying out the hinge on the door and side jamb.

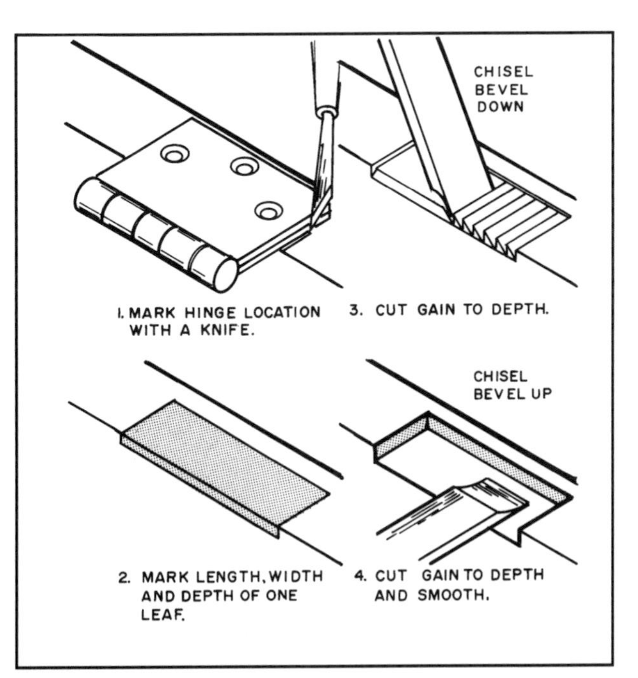

23-52 The steps to hand-cut a mortise for a square corner hinge leaf.

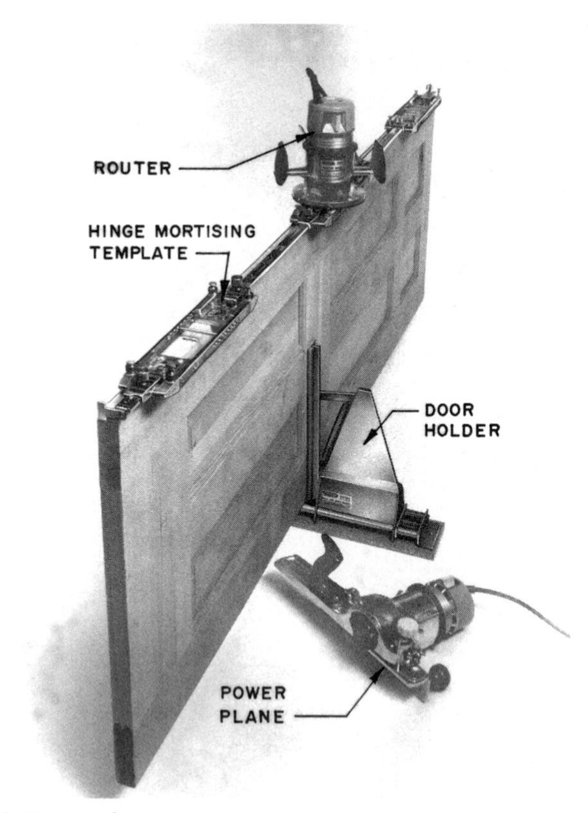

23-53 Round corner mortises in the door edges are cut using a hinge mortising template and a router. The door is held with a door holder. The edges of the door can be trimmed down with the portable power plane shown. *(Courtesy The Stanley Works)*

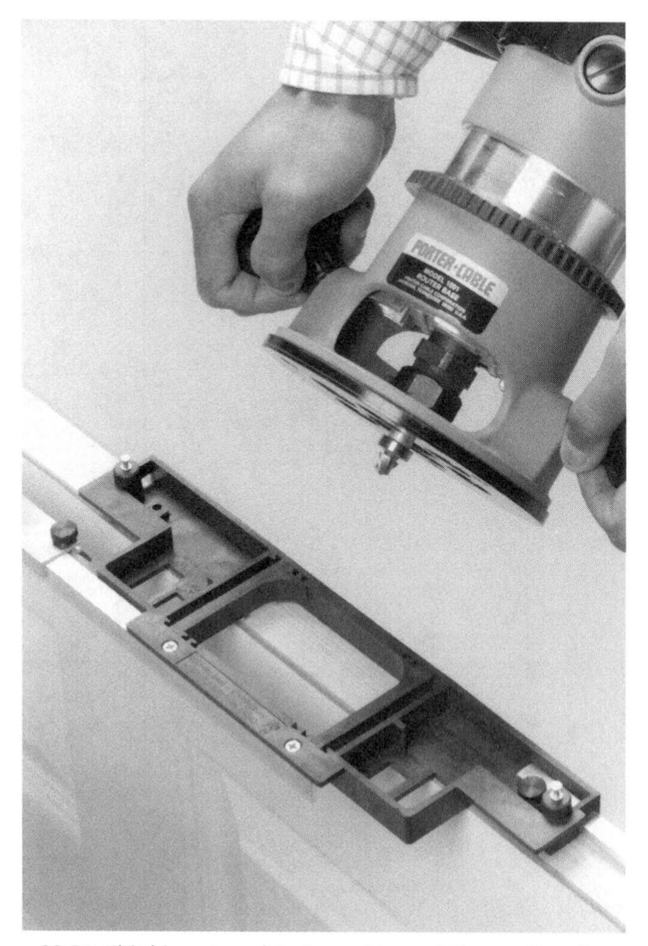

23-54 This hinge template is used to rout the mortises for hinges on doors and jambs. It can be used to rout hinges with ¼- and ⅝-inch radii. *(Courtesy Carey Template Company)*

Round corner mortises are cut with a portable router and a hinge template. The door can be held with a door holder as shown in **23-53**. The hinge template is clamped to the door or jamb and the recess is routed with a router **(see 23-54)**. The router bit will cut a flat bottom and must be set to cut the required depth. The router is moved around the template, cutting the profile, and then used to remove the material in the mortise.

23-55 This shows the hinges mounted on the door and jamb.

Once the mortise is cut the hinges can be installed. Install the hinges on the door, being certain the loose pin is located so it is removed by **pulling up**. Otherwise it will eventually fall out. Remove the pin from the hinge, and install each leaf in the mortise on the door and jamb.

Bore small-diameter holes for each screw. This provides a better seat for the screw and reduces the chance of splitting the wood. Be certain the hinge is firmly touching the back side of the mortise. Some carpenters set the screws slightly **off center** towards the back of the mortise to be certain that the hinge firmly presses against the back of the mortise **(see 23-55)**. Once the hinges are installed, set the door in place, and see if the hinge barrel can slide together. Often minor adjustments are needed at this point. Also check the clearances around the edges of the door. The space should be 1/16 to 3/32 inch (1.5 to 2.4mm) wide and uniform in width. If the space on the lockset side is too small, remove the hinges from the door and cut the mortise a little deeper. If the space at the lockset side is too large, remove the hinges and place pieces of hard, incompressible cardboard or wood veneer in the mortise under the hinge.

INSTALLING POCKET DOORS

Pocket doors are used when there is not enough room to use a swinging door. They slide inside the partition. A special metal frame with a track on the top for rollers is manufactured. The frame is shipped knocked down and is assembled and installed by the carpenter **(see 23-56 and 23-57)**. Hardware for 2 × 4- and 2 × 6-inch partitions is available. Manufacturers supply detailed installation instructions including how to adjust the door to move properly after it has been installed.

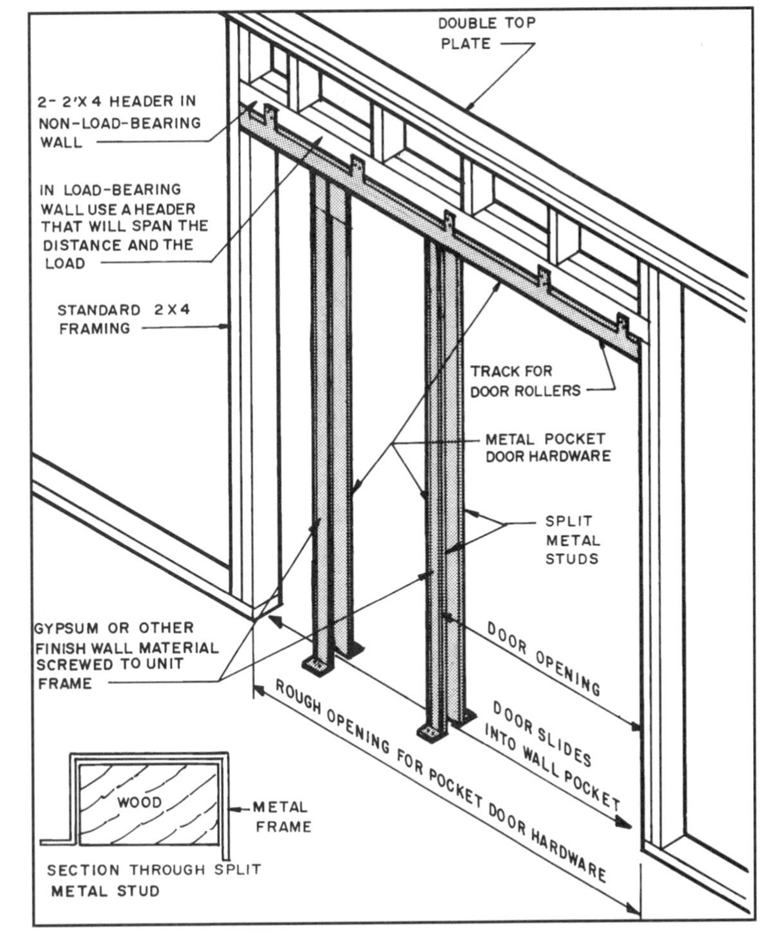

23-56 A typical installation of pocket door hardware in a frame partition.

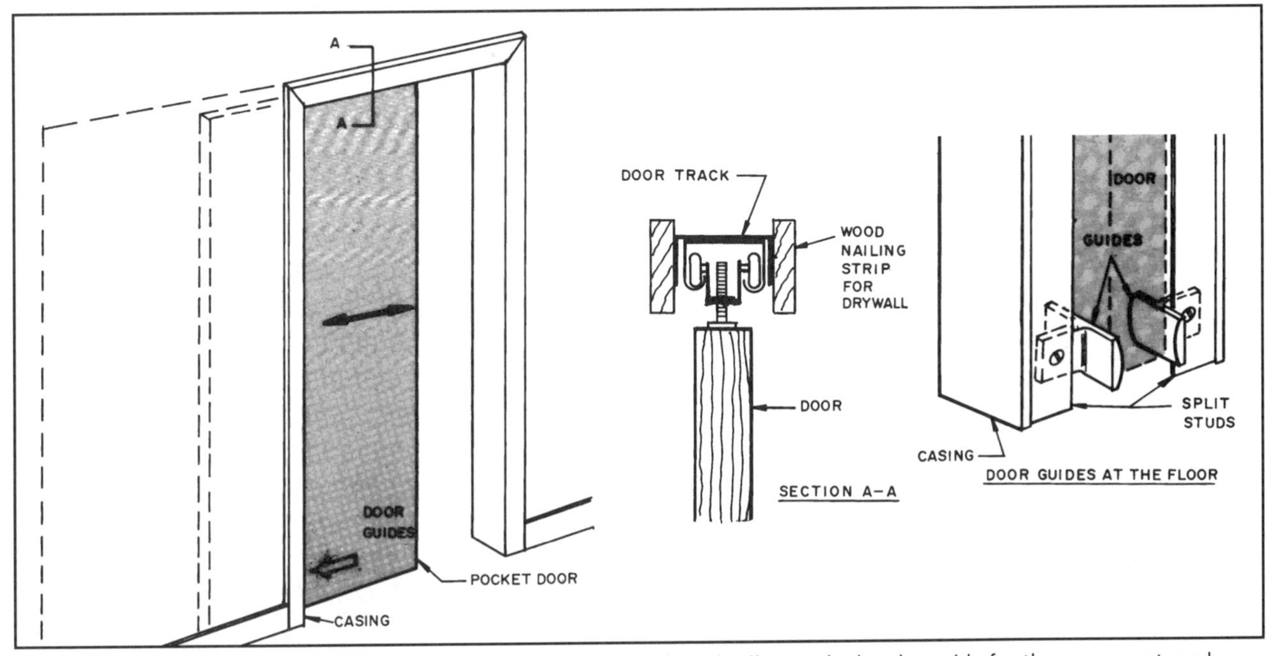

23-57 The installed pocket door slides inside the partition. A track and rollers at the head provide for the movement, and guides at the floor help keep the door in line.

INSTALLING BYPASS SLIDING DOORS

The rough opening for a **bypass sliding door** is framed as described for other doors. The door frame and trim are installed in the normal manner. One-piece frames are used, because door stops are not required.

The door slides on rollers that run in an overhead track **(see 23-58)**. The track is screwed to the top jamb. The rollers are secured to the doors. Any type door can be used. However, the track selected must accommodate the door. Most commonly 1⅜-inch-thick interior doors are used. After the doors are hung, the floor guides are screwed to the floor. Be certain to have the doors plumb before securing the floor guide.

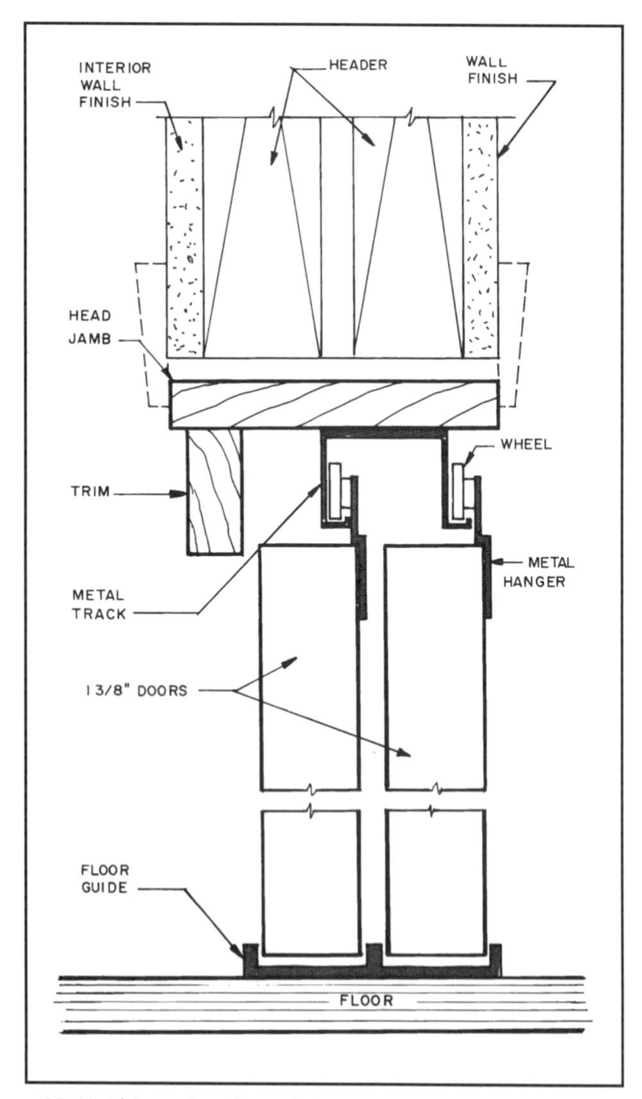

23-58 This section through bypass doors shows the metal track at the head and the floor guide that keeps the doors running in line.

INSTALLING BIFOLD DOORS

The opening for **bifold doors** is framed with a solid frame, as described for other doors. The doors are hinged, enabling them to fold as they open **(see 23-59)**. The sides of the doors next to the jamb have a metal pivot installed on the floor and top jamb **(see 23-60)**. A metal track is screwed to the top jamb and runs across the door opening. A pin is placed on the top edge of the interior door. The pin runs in the track, keeping the doors moving in a straight line.

INSTALLING SLIDING EXTERIOR GLASS DOORS

Sliding glass doors are available in a variety of sizes with wood or metal frames **(23-61)**. They are weather-stripped and glazed with safety glass. They may have single or double glazing which provides an air space between the two panes of glass.

23-59 Bifold doors open easily and do not require a lot of open space in front as is necessary for swinging doors. *(Courtesy Jeld-Wen® Elite® Doors)*

23-60 A typical bifold door installation.

23-61 Sliding exterior doors, here seen from outside, bring considerable light and ventilation to the interior as well as provide a view of the exterior scene.

The most common types of sliding glass door have two or three doors. One door moves and the others are fixed **(see 23-62)**.

The frame is often shipped disassembled. The carpenter must carefully follow the assembly instructions provided with the door. The door opening is framed in the conventional manner with a header spanning the opening **(see 23-63)**. After the sheathing has been applied the door frame can be installed.

23-62 Two styles of sliding exterior doors.

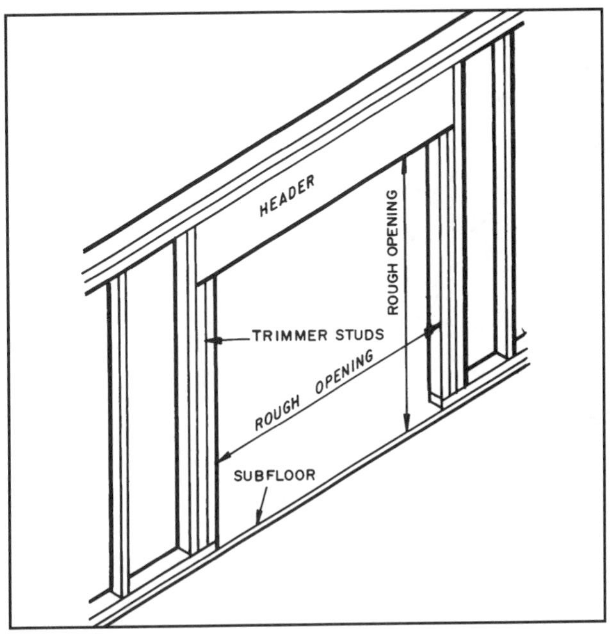

23-63 Typical framing for the rough opening for the sliding exterior door.

Follow the instructions that come with the **sliding exterior door** unit. Procedures differ with each brand of door. Following is a typical installation procedure.

1. Remove the doors and frame from the carton. If the frame is preassembled leave all shims and braces in place until the door is installed. Perform any needed assembly on the door frame.
2. Check to see that the rough opening is the correct size and that the studs are plumb and the floor level. Be certain that clearances on all sides are those recommended by the manufacturer.

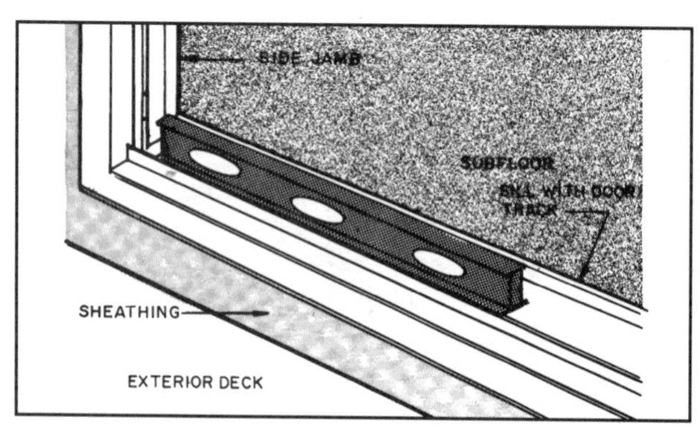

23-64 Level the sill and shim it as necessary.

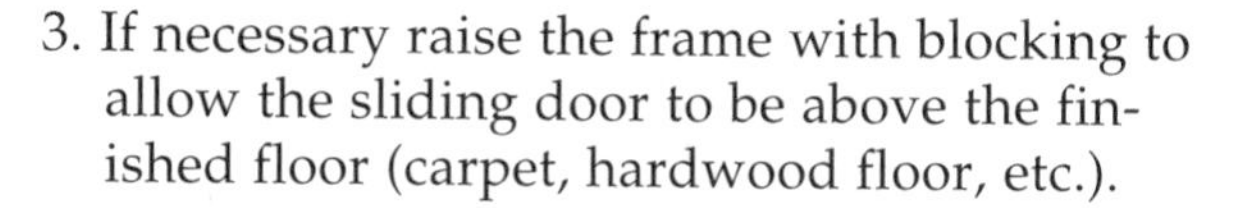

3. If necessary raise the frame with blocking to allow the sliding door to be above the finished floor (carpet, hardwood floor, etc.).
4. Clean the area around the door opening so that it is free of dust and debris. Lay three beads of caulking across the opening to seal the sill to the subfloor. Run the caulking several inches up the sides on each side.
5. Slide the door frame into the opening from the outside and center it in the rough opening.
6. Level the sill and shim it if necessary **(23-64)**. Secure it to the subfloor as directed by the manufacturer's instructions.
7. Now shim the jambs until they are plumb and secure them to the wall studs **(see 23-65)**. Be certain to use plenty of shims along each jamb and at the head. It is important to keep the side jamb and head jambs straight and solidly secured to the wall studs **(see 23-66)**. When the jamb is plumb, secure it to the studs. This might involve nailing through the casing (brick molding) or through a metal fin-type flange as used on windows **(23-67)**. Install flashing along the top of the frame if it is required. Some types of door provide additional support by driving screws through holes in the side of the frame **(see 23-68)**.

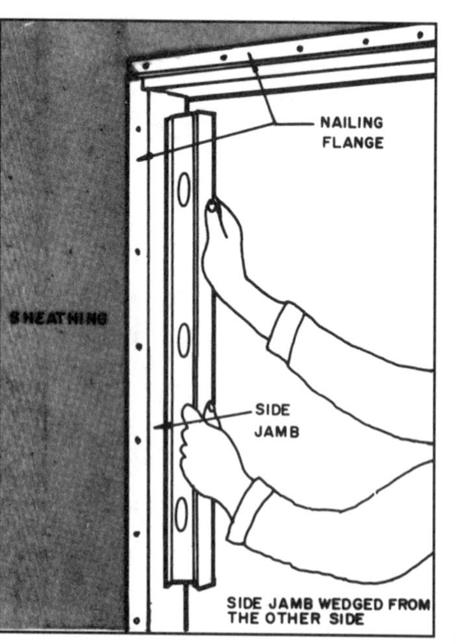

23-65 After the sill is level and secured to the subfloor, shim and plumb the side jambs. Secure to the wall as specified by the door manufacturer. Be certain the nailing flange is flush against the sheathing.

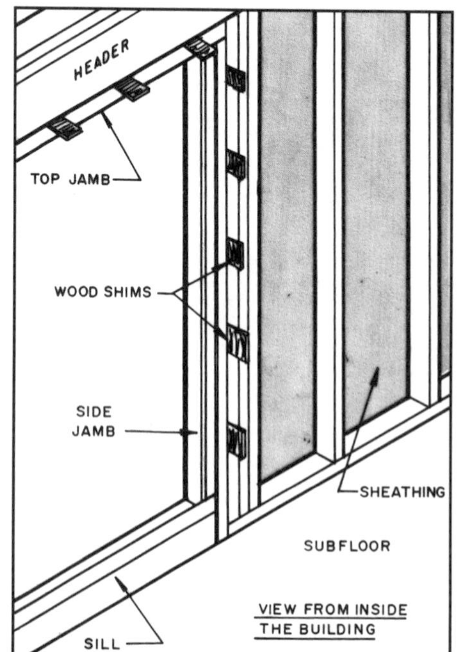

23-66 Use enough shims to keep the side and top jambs straight. Secure to the studs with screws as specified by the manufacturer.

8. Now position the fixed door in the outer track. Push it tightly against the jamb and secure it as specified by the manufacturer. Check to see if it is fitting tightly against the jamb. Some doors have screws you can turn to align the door **(see 23-69)**.
9. Position the sliding door on the track and secure as specified by the manufacturer. Now slide it back and forth to see if it runs smoothly. Check to see that it butts the

frame squarely. Most doors have adjusting screws that permit you to make some adjustment within the frame **(see 23-70)**. Check to be certain the door lock catches in the plate on the frame.

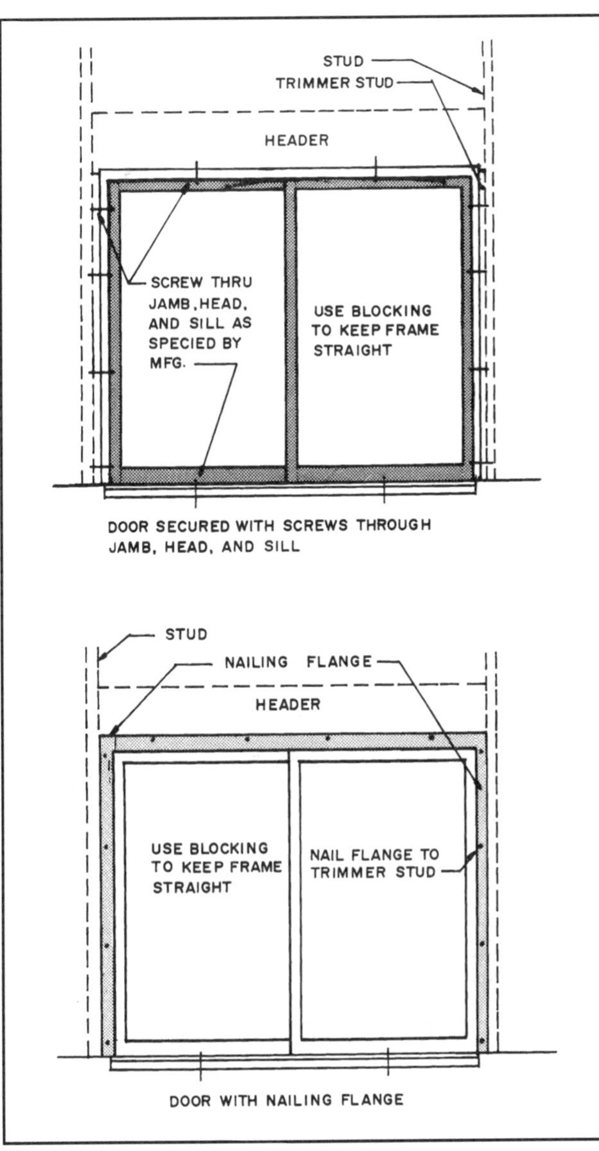

23-67 Two ways manufacturers install sliding glass exterior doors.

23-68 Some manufacturers require that the frames be screwed to the studs and header after they have been leveled and plumbed.

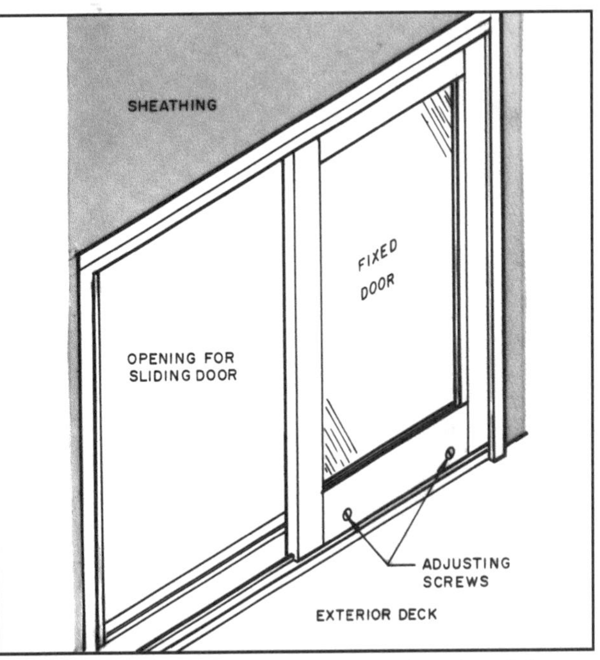

23-69 After the frame is securely set in place, install the fixed door as directed by the manufacturer.

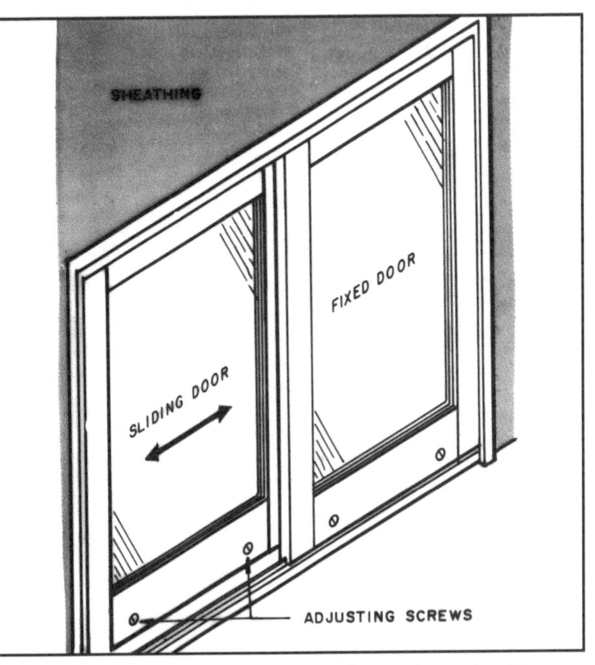

23-70 Position the sliding door on the track on the sill. Check to see that it runs smoothly. Adjust it with the screws as necessary.

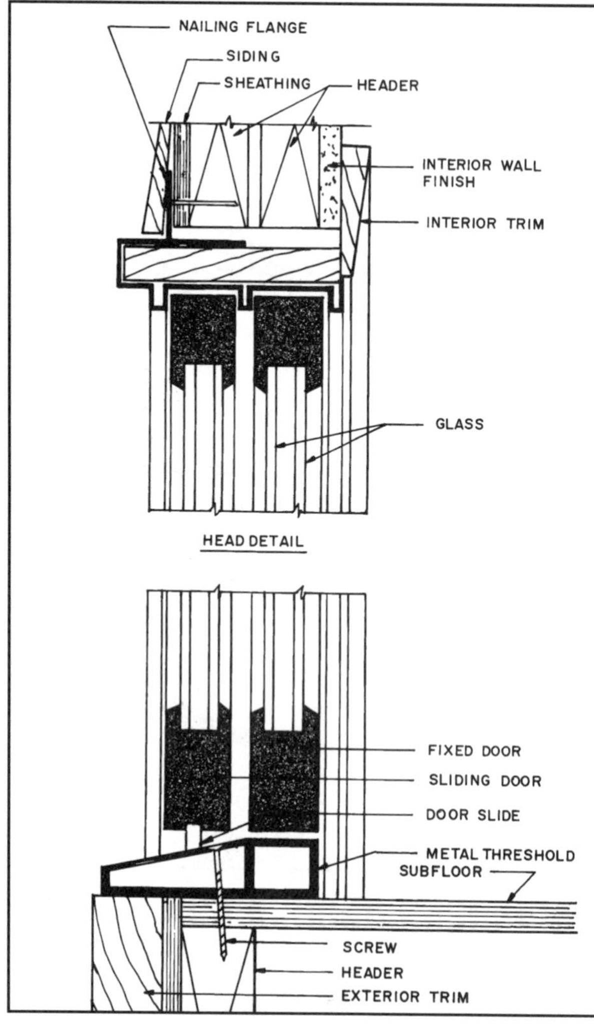

23-71 A generalized section showing details of a sliding exterior door installation.

10. When all is finished, caulk around all sides on the exterior of the door. A section through a typical installation that is installed using nailing flanges is in **23-71**.

OVERHEAD GARAGE DOORS

Roll-up garage doors made up of several hinged sections are the most widely used type for residential and commercial installations **(see 23-72)**. They are available in a wide variety of designs and can be made from wood, steel, aluminum, an aluminum frame with fiberglass panels, or polyethylene.

23-72 A wood-framed, hinged-section, roll-up door. *(Courtesy Overhead Door Corporation)*

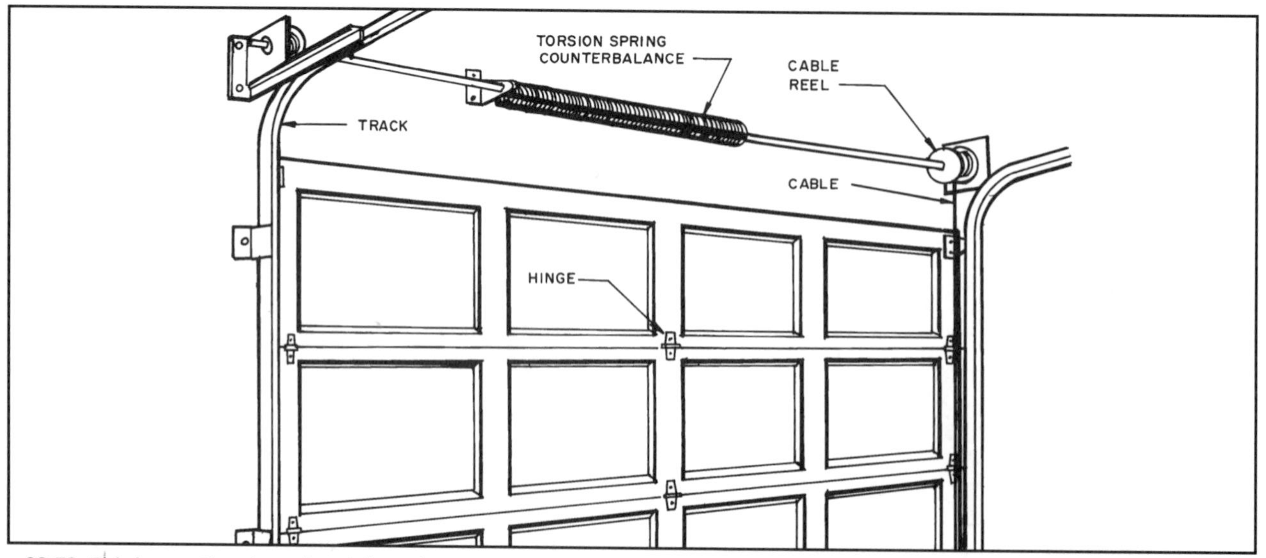

23-73 This is a sectional overhead door that moves on tracks that are bolted to the wall framing and ceiling joists. *(Courtesy Overhead Door Corporation)*

The size of the door is shown on the working drawing. A single garage door is usually 7′-0″ high and 8′-0″, 9′-0″, or 10′-0″ wide. A double door is usually 7′-0″ high by 16′-0″ or 18′-0″ wide. Other heights and widths are available.

The door with hinged sections opens using some form of track **(see 23-73)**. Wheels on the sides of the door roll inside the track **(see 23-74)**. Electric door openers also can be used. The electric door opener is activated with a remote control transmitter. The swing-up door uses several metal levers to hinge the door up against the ceiling of the garage. A section through a typical installation is shown in **23-75**.

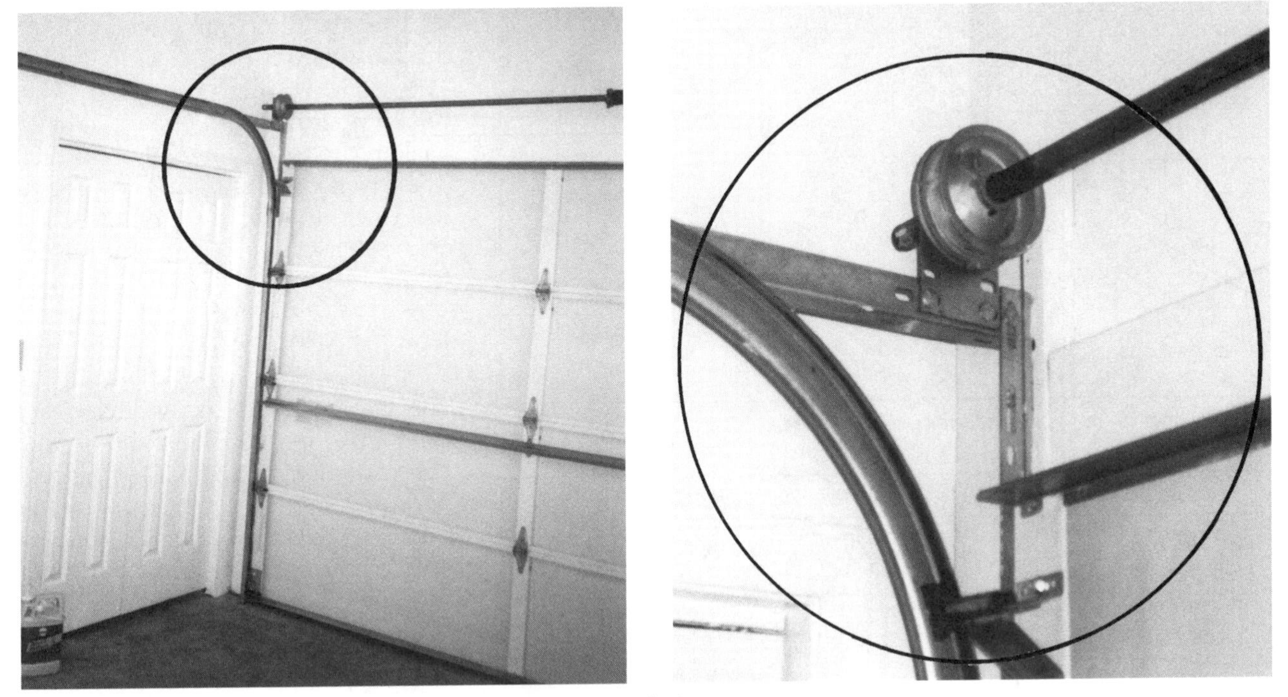

23-74 (Above and left) Typical track cable reel and roller installation.

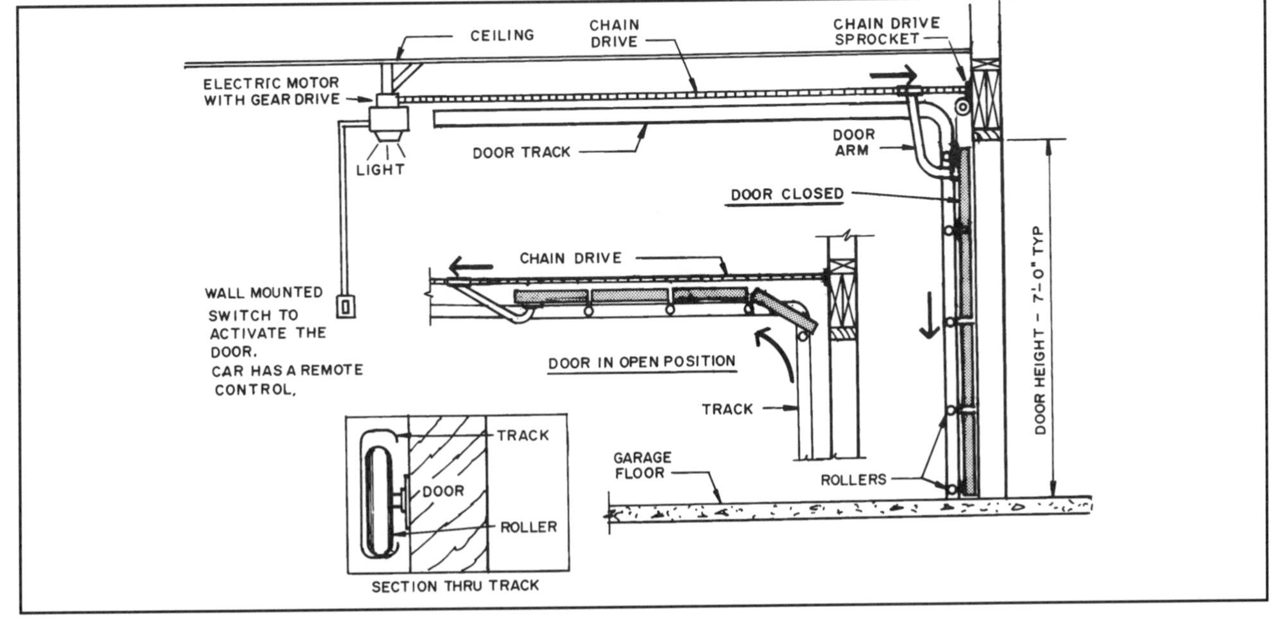

23-75 A section through a typical overhead door installation.

Garage Door Frames

Details for framing a garage door opening are shown in **23-76**. The side jamb and head board are flat stock with no rabbets. They are installed as described for door frames. Each part is set plumb and level using wedges. When set, it is nailed through the jamb and wedge into the rough framing **(see 23-77)**. The inside dimensions of the finished door frame are about ¾ inch larger than the size of the door. It is best to refer to the manufacturer's directions that come with the door.

After the door frame is installed, the hardware for the door is hung. The door is then mounted. The doorstop trim which has a rubber gasket can now be nailed to the door frame. In this way it can be set to fit against the door, sealing any space between it and the door frame.

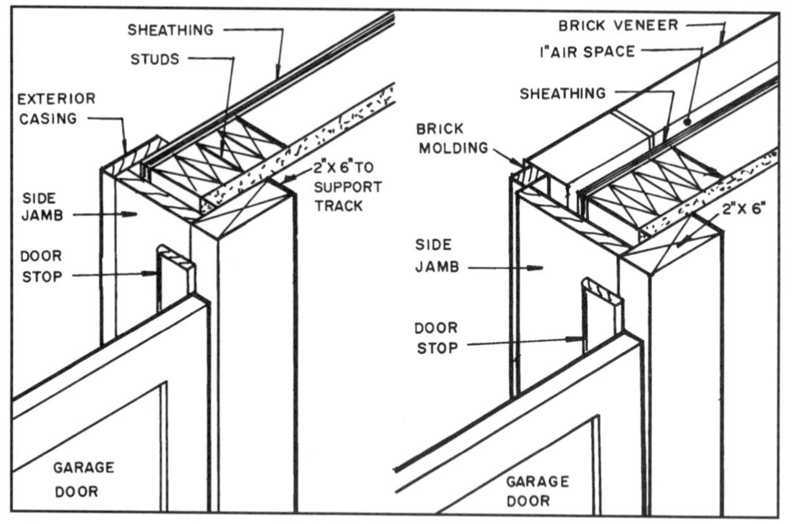

23-76 How to frame the jamb for a garage door.

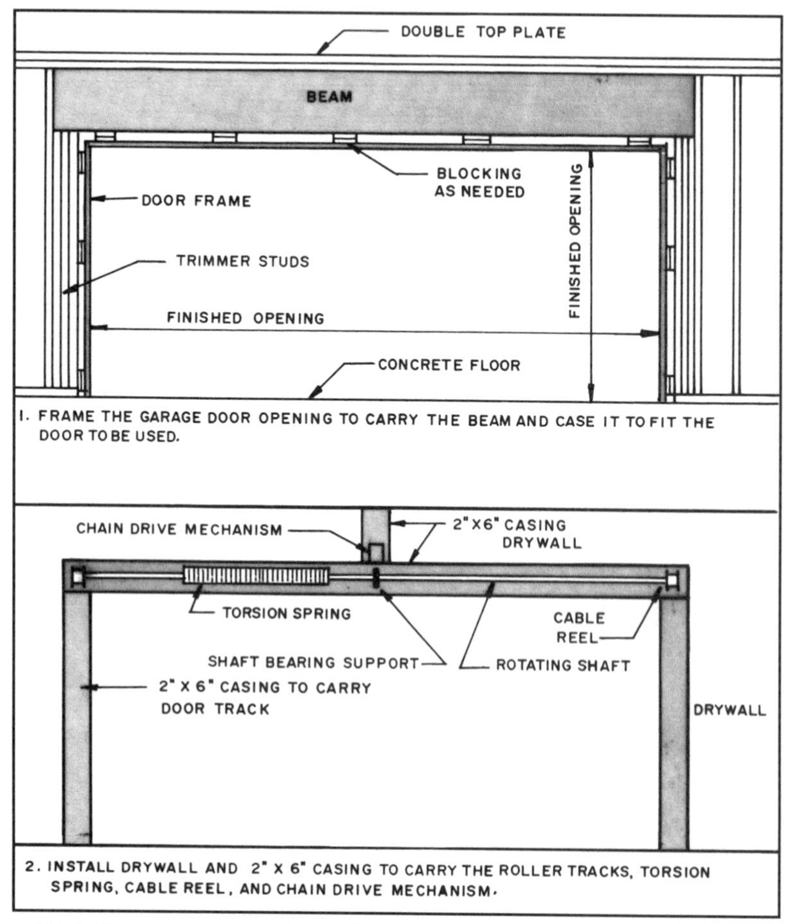

23-77 The proper way to frame the garage door opening.

Installing the Door

The garage doors are installed by companies specializing in this work. The carpenter prepares the finished door opening. The garage door contractor installs the door.

If a carpenter installs the door, the installation directions that come with it should be followed carefully.

CASING STYLES

The style of casing on interior door openings is generally the same as used on the windows. There are exceptions, such as when it is desired to have more elaborate door opening casings which are exposed to view and simpler casing on the windows which are covered with draperies. A few door casing designs are shown in **23-78**. The actual design could take many forms. The designer can draw up specifications to match a classic style or devise an original design using standard flat and profiled casings and various moldings. Custom-designed casings can also be ordered. Refer to Chapters 27 and 28 for additional information on trim and molding.

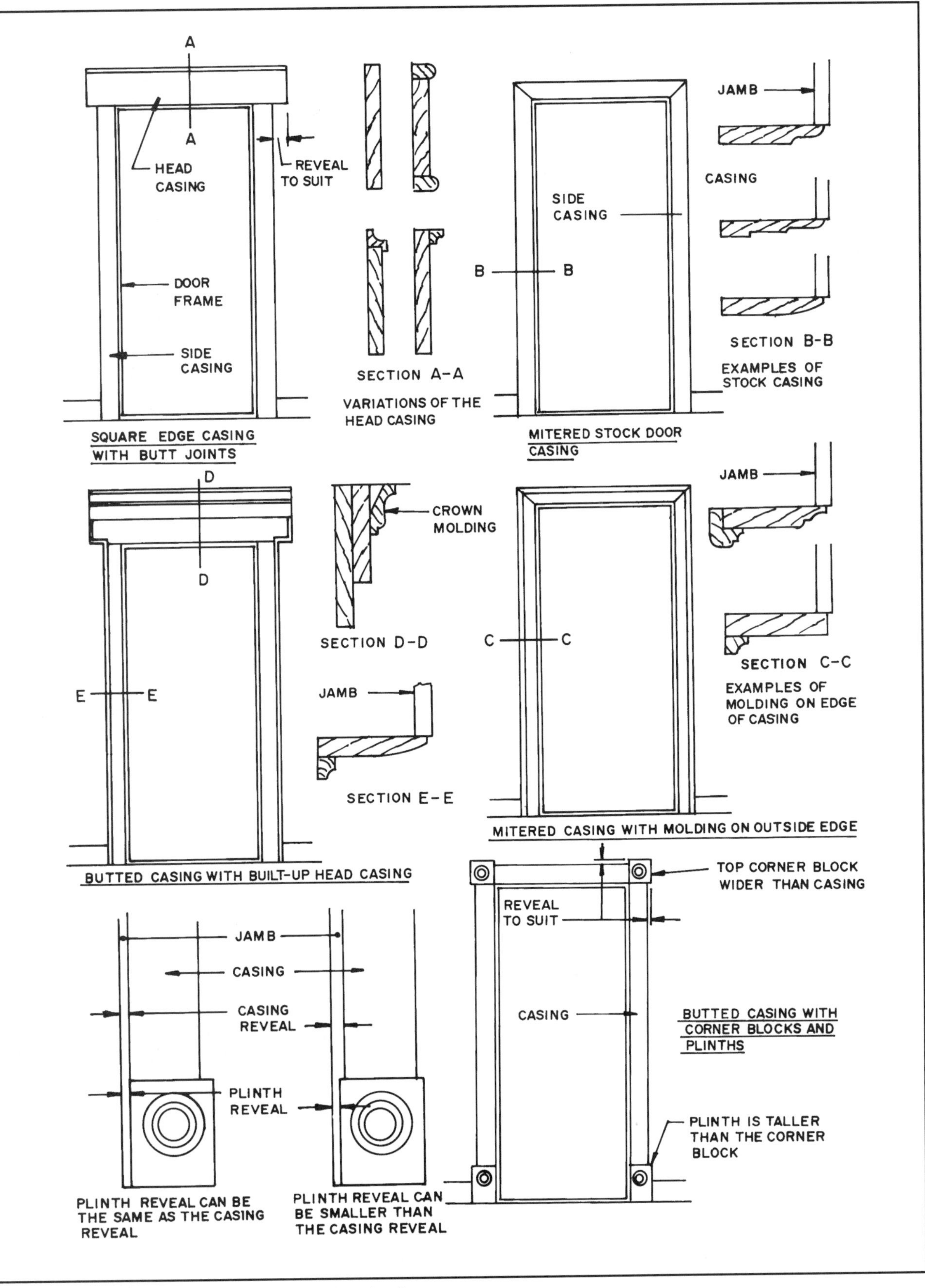

23-78 Some typical casing techniques used on interior doors.

INSTALLING DOOR CASING

The **door casing** is a decorative wood molding covering the opening between the door frame and the finished wall material. In addition to being attractive, it structurally stiffens the union between the frame and the wall. It should, therefore, be nailed not only to the frame but to the trimmer stud before forming the rough opening **(see 23-79)**. The door casing is set back from the edge of the jamb, giving a reveal of 3/16 to 1/4 inch (4.5 to 6mm); the same reveal is used when casing the windows. (The **reveal** is the amount the casing is set back from the edge of the frame.)

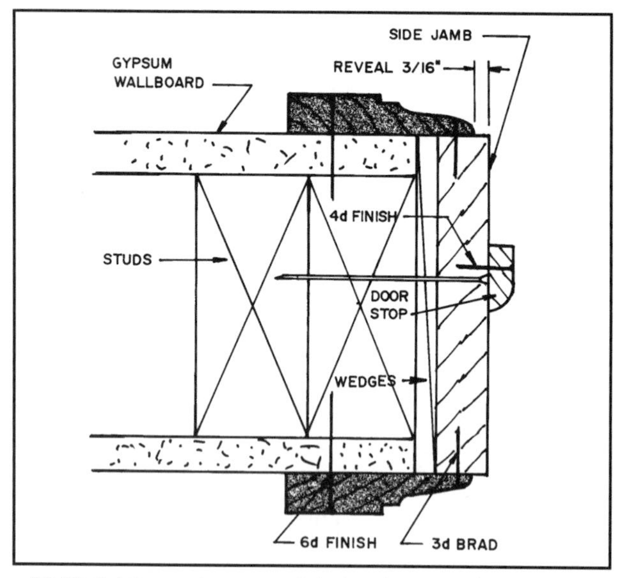

23-79 Setting casing on an interior door opening.

Possibly the mitered casing is the most commonly used style. (The technique for cutting and installing is basically the same as discussed for windows in Chapter 24.) Some finish carpenters prefer to install the mitered top casing first. Mark the reveal on each frame. Measure and cut the top casing for all doors to the same size. Nail it to the frame with 3d or 4d finishing nails spaced 8 to 10 inches (20.3 to 25.4cm) apart. Do not set the nails at this point, because it may be necessary to remove them to adjust the top casing to the side casing.

Cut the side casings to length. Some prefer to cut them about 1/4 inch (6mm) short of the floor. This allows some room to adjust the side casing as it meets the top casing. The space at the floor will be covered with the carpeting or other finished floor material. Nail the side casings to the door frame. When they are in place with the correct reveal and a tight miter, nail them to the header and studs with 6d or 8d finishing nails.

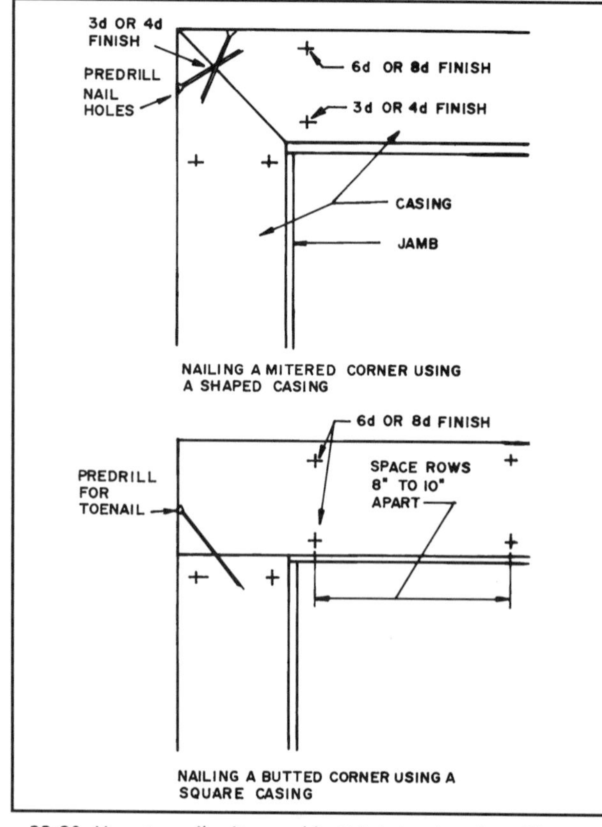

23-80 How to nail miter and butt joints when installing casing.

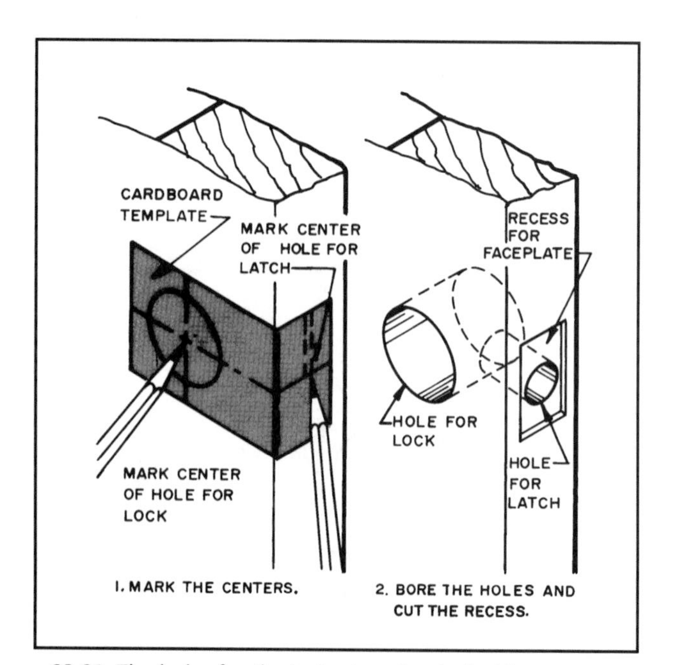

23-81 The holes for the lockset are located with a template supplied with each unit.

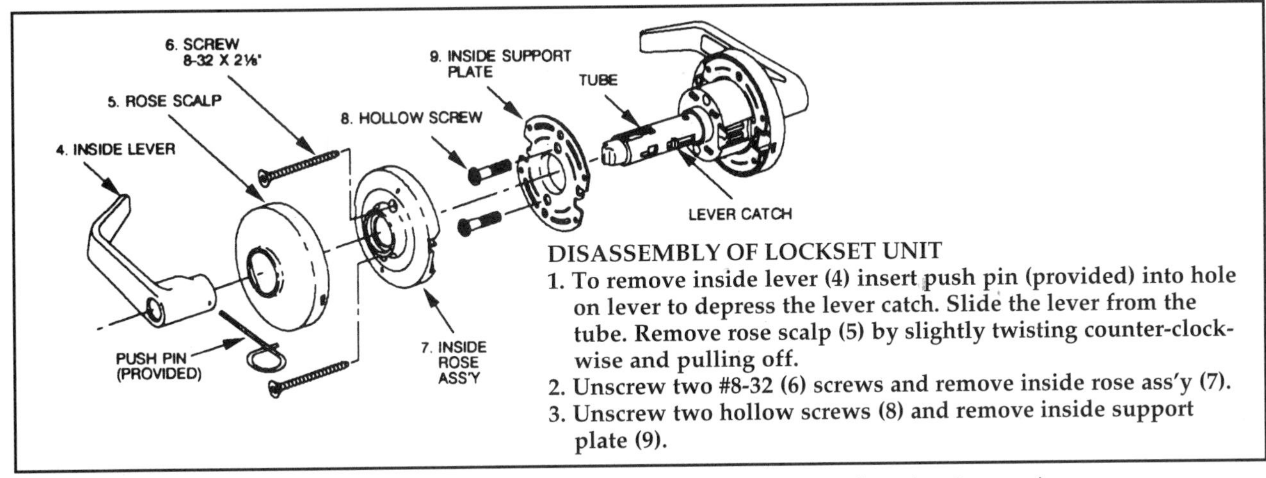

23-82 An exploded view of a disassembled lockset unit. *(Courtesy Arrow Lock Manufacturing Company)*

Do not nail too close to the miter. Instead, toenail the corner to keep it closed. If there is a danger of splitting the casing, drill small holes for each nail. A suggested nailing pattern for a butt joint casing is shown in **23-80**.

23-83 A decorative entrance handle.

INSTALLING A LOCKSET

Many prehung doors have the holes bored to receive a **lockset**. This eliminates a rather difficult task for the finish carpenter. If the doors are not pre-bored, it is necessary to locate and bore the holes for the lock. Each set of hardware comes with detailed instructions and a **template** to be used to locate the centers of the holes. It also specifies the required hole diameters. The steps to locate and bore the holes for a lockset are in **23-81**.

Generally the center of the lock is located 36 inches (914mm) above the bottom of the door. The amount the center is set back from the edge of the door varies with the type and brand of lock.

Details of a typical lock used on interior and exterior doors are shown in **23-82**. A decorative handle used on entrance doors is shown in **23-83**. A deadbolt is shown in **23-84**.

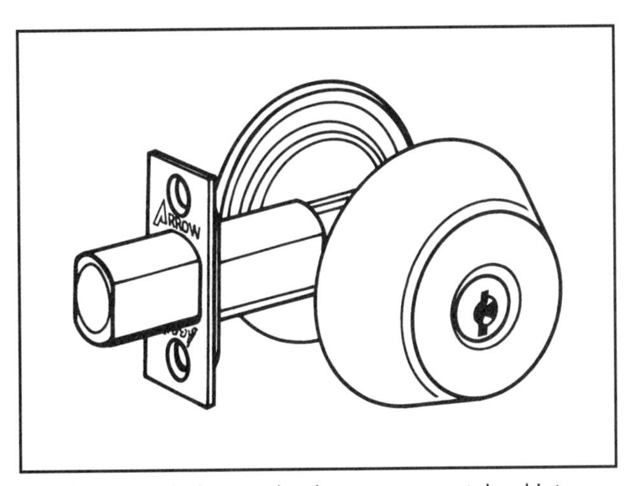

23-84 A deadbolt extends a long, strong metal rod into a metal plate screwed to the door frame. *(Courtesy Arrow Lock Manufacturing Company)*

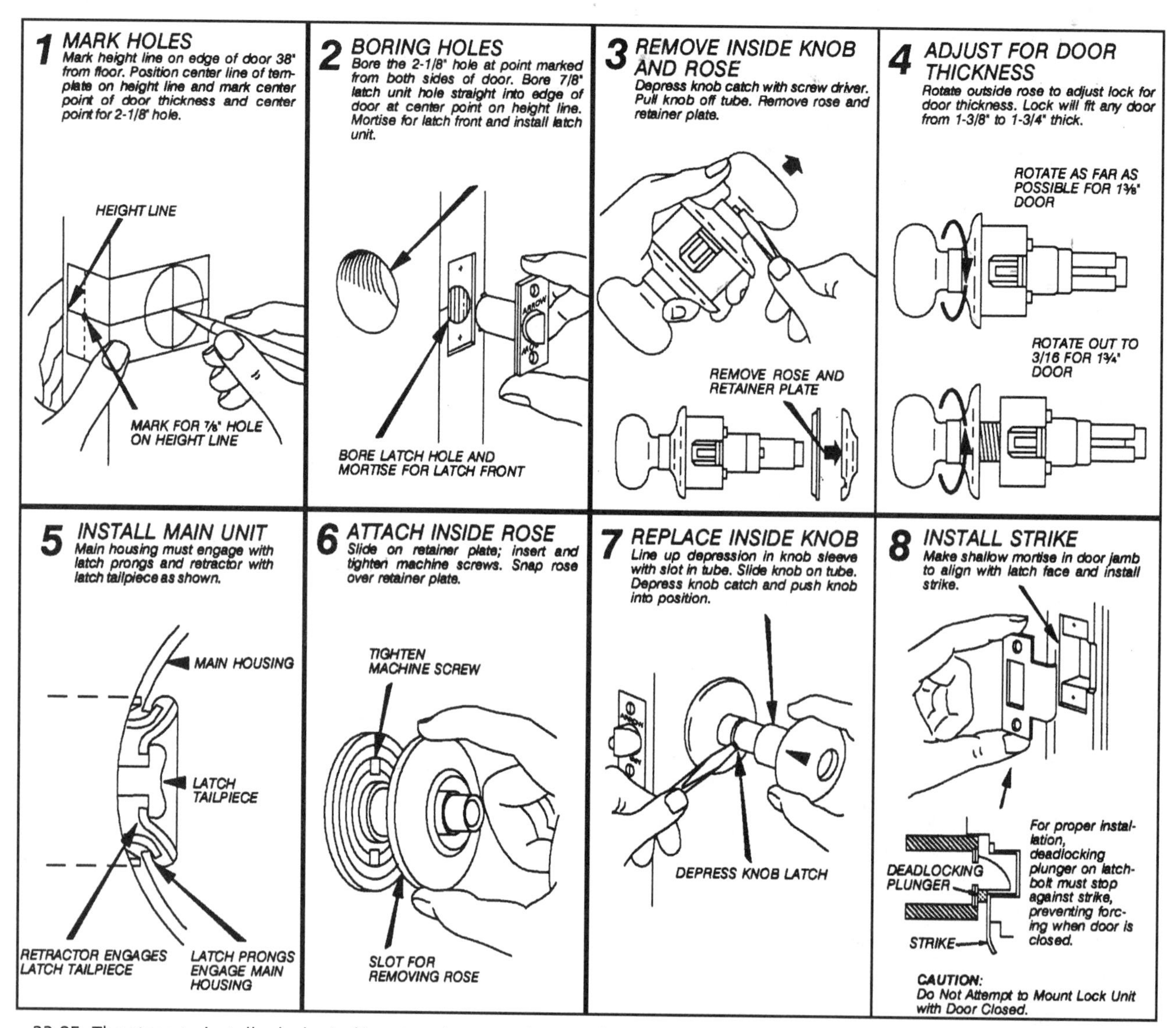

23-85 The steps to install a lockset. *(Courtesy Arrow Lock Manufacturing Company)*

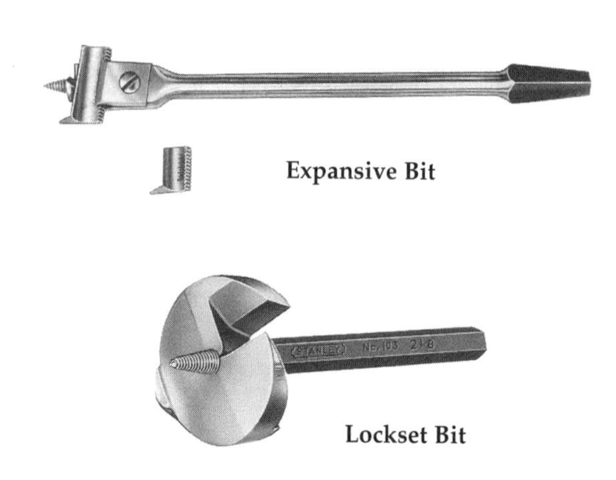

23-86 Bits used to bore large diameter holes required for installing locksets. *(Courtesy Stanley Tools © 1994)*

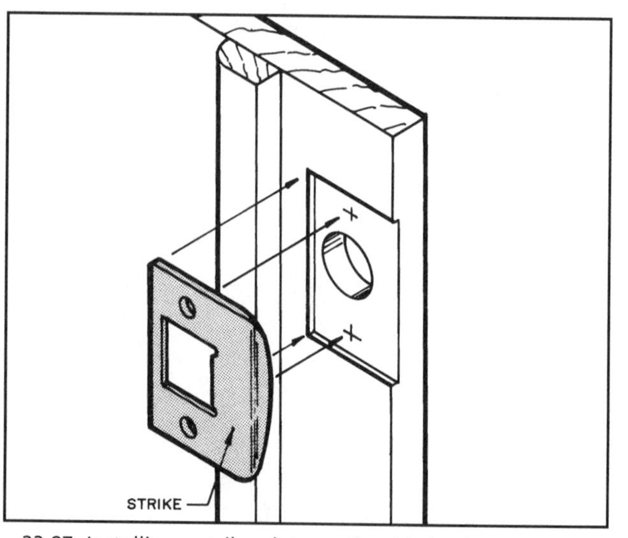

23-87 Installing a strike plate on the side jamb.

To install a lock, measure 36 inches (91.4cm) from the bottom of the door, and make a mark across the end of the door. Using this as a guide, place the centerline of the cardboard template that comes with the lock on this mark. Mark the center for the holes on the face and end of the door **(see 23-85)**. Bore the holes at the diameter specified.

The large-diameter hole can be bored with an expansive bit, lockset bit, or hole saw **(see 23-86)**. **Be careful; do not bore all the way through the door.** Bore from one side until the tip of the bit just shows through the other side. Finish boring by coming from the other side. Then bore the smaller hole from the edge of the door.

Install the lock as instructed by the manufacturer's directions.

The **strike** is installed on the door jamb; locate it opposite the faceplate on the latch **(see 23-87)**. Place the loose strike over the installed latch, and carefully close the door. Hold the strike in place, and mark the top and bottom edges of the jamb. Then draw a line along the edge of the door onto the strike.

Open the door, place the strike on the marks, and draw around it. Chisel the recess so that the strike is flush with the surface of the jamb. Hold the strike in the recess, and mark the hole for the latch to enter the jamb. Bore the hole; then install the strike with the wood screws provided **(see 23-88)**.

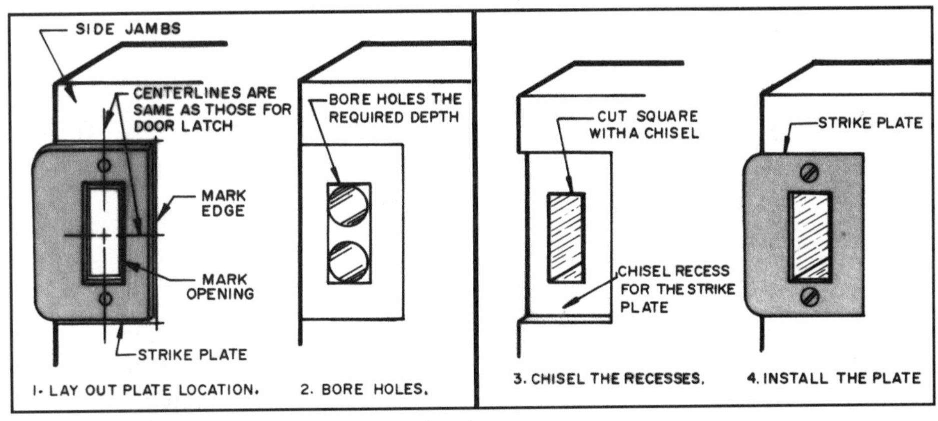

23-88 How to lay out and install the strike plate.

INSTALLING DOOR STOPS

Interior doors have **door stop molding** nailed to the frame to serve to stop the door as it is closed. Typical door stop molding is 7/16 inch (11mm) thick and from 7/8 to 1 3/8 inch (22 to 35mm) wide. Two types of door stop molding are available. One has a profiled edge and must be mitered at the corners of the frame. Another type is a rectangular, square-edged molding that can be butted at the corners.

The door stop is installed after the lockset is in place **(see 23-89)**. Close the door so that it is held in place by the lockset. Cut the top molding to length. Place it **lightly** against the closed door, and nail it to the top jamb. If there is a danger of damaging the door with the hammer, mark the location of the stop on each end, open the door, and then nail the stop to the head jamb. Do not set the nails, because it may be necessary to move the stop. Now close the door, and see if it is firm against the door. Now cut and nail the stop on the lockset side; this is usually set rather firmly against the closed door to keep the door from rattling. Install the hinge-side stop last. It can clear the door by about 1/32 inch (0.8mm). Use a piece of cardboard or wood veneer as a spacer between the door stop and the molding. Check the installation carefully before setting the nails.

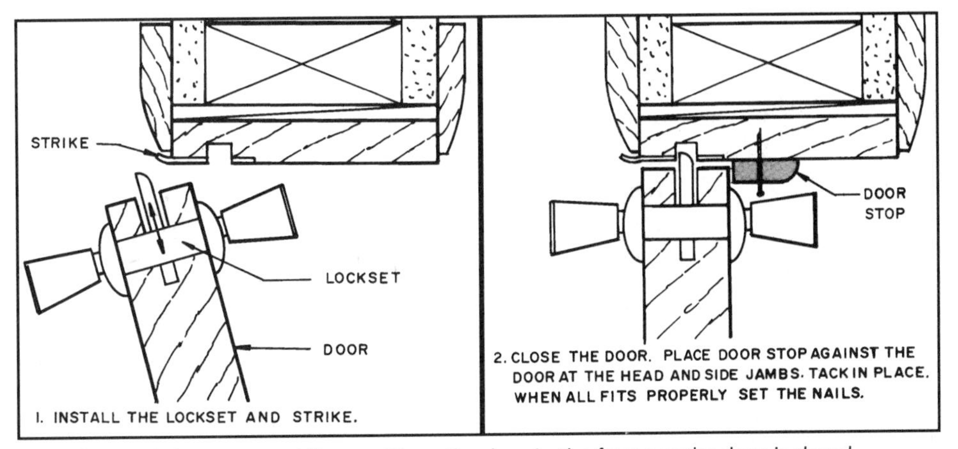

23-89 Wood door stop molding positions the door in the frame as the door is closed.

CHAPTER 24

INSTALLING & TRIMMING WINDOWS

Windows are an important part of the exterior design of a building. The architect selects the types and sizes that suit the architectural styles of the building **(see 24-1)**. The sizes must be in proportion to the exterior wall and overall mass of the building.

Windows have the function of bringing light and ventilation into the building. In addition they are a means of escape in an emergency such as a fire. Building codes place some regulations on windows.

BUILDING CODES

Codes require every sleeping room to have one operable exterior window or door. These exterior windows or doors must be able to be opened from the inside. If windows are used to meet this requirement some codes specify the maximum height of the sill above the floor, such as 44 inches (112cm). The clear opening height and width are also specified.

The glazed area of occupied rooms is specified. Typically this is eight percent of the floor area. Skylights may be used to meet natural light and ventilation requirements. Mechanical ventilation can in some cases be substituted for natural ventilation.

ENERGY-EFFICIENT WINDOWS

Windows are a source of major heat loss in the winter and gain in the summer. Energy can be saved if the leaks around the window frames can

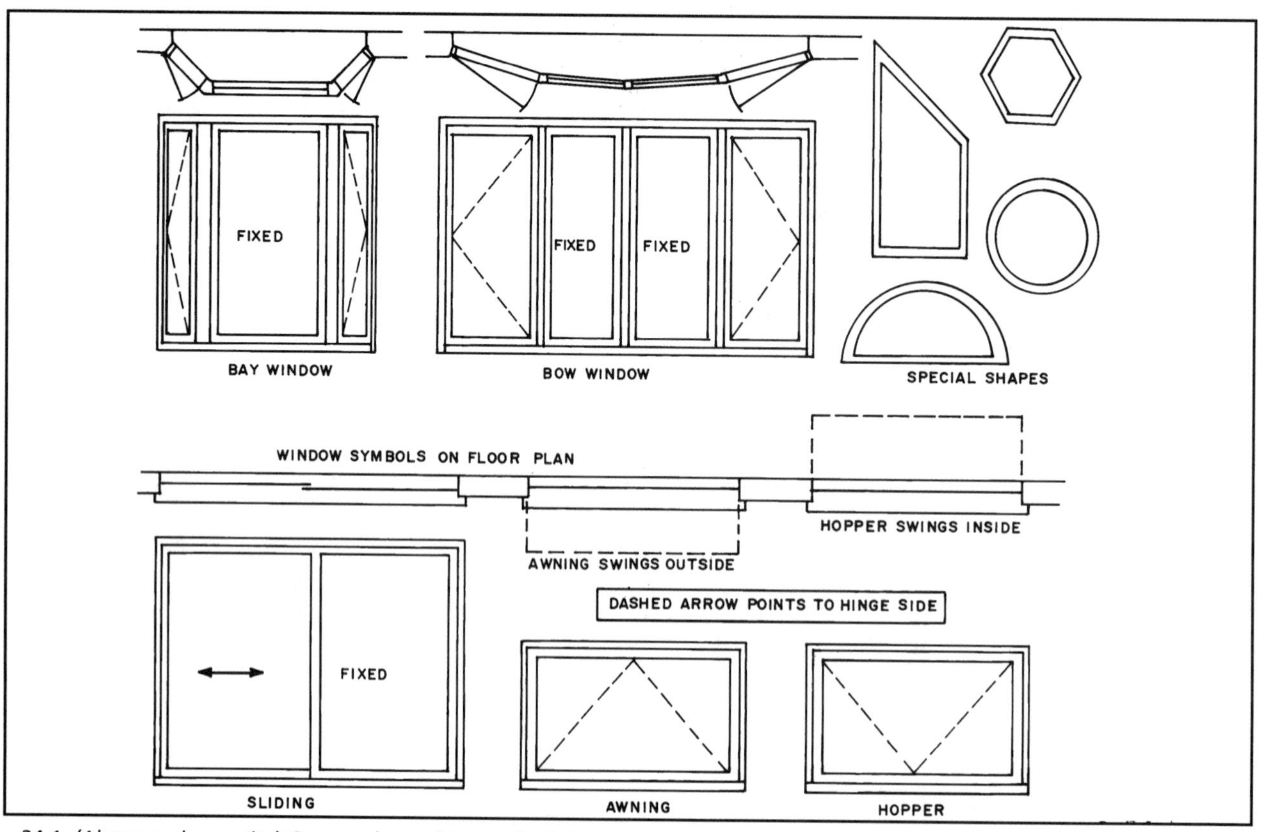

24-1 (Above and opposite) Commonly used types of windows.

be sealed by caulking and the sashes have adequate weatherstripping. The caulking is the carpenter's job. The weatherstripping is the manufacturer's responsibility.

Various types of glazing also play a role in controlling energy losses.

Low-emissivity (low-e) glass has a coating that reduces heat transfer back through a window.

Heat-absorbing glass has tints that absorb some of the incoming solar energy.

Reflective glass is coated with a reflective film which controls solar heat in the summer.

Plastic glazing materials are available. They are no better at controlling energy loss than glass.

Storm windows are installed over the regular windows and provide a dead air space, thus reducing heat loss.

Various **multiple-pane window** glazing units provide dead air spaces, reducing energy loss. Some have the spaces filled with a gas, increasing the resistance to heat flow **(see 24-2)**.

24-2 This triple-pane glazing unit is highly energy efficient. Its rated value is R-10, which is about the same as 2½ inches of glass fiber insulation. The air space is filled with a mixture of krypton and argon gas. *(Courtesy Weather Shield Windows and Doors)*

TYPES OF WINDOW

The style and type of window is indicated in the architectural drawings on the building elevations and the window schedule **(refer to 24-14 and 24-16)**. Windows are of three basic types: sliding, swinging, or fixed. Sliding windows move vertically or horizontally. Swinging windows are hinged at the top, bottom, or side.

Double-hung windows have two sashes that slide vertically. They are held in position by friction or various devices, such as springs **(see 24-3)**.

24-3 A double-hung window and two double-hung windows with a circle top.

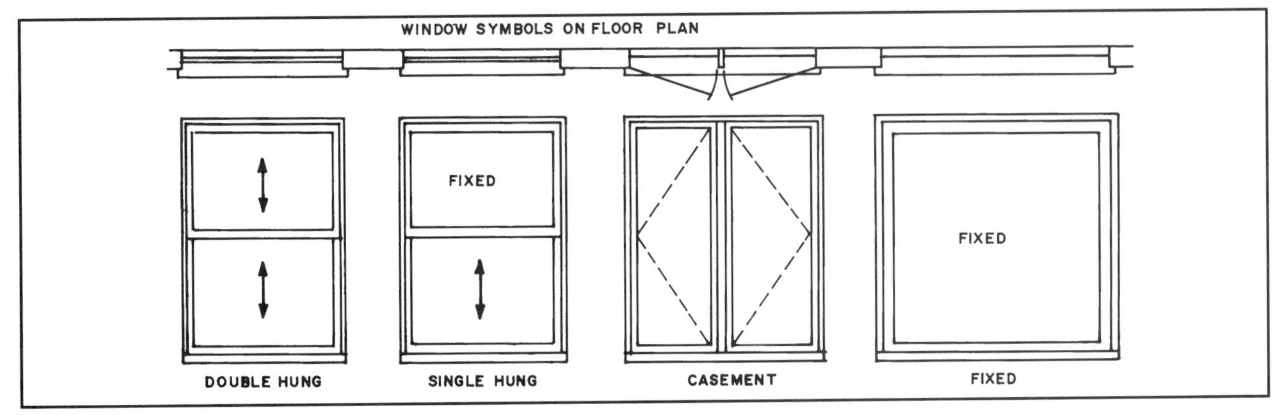

24-1 (Continued) Commonly used types of windows.

24-4 A horizontal sliding window. *(Courtesy Andersen Windows, Inc.)*

Single-hung windows have the top sash fixed, while the lower sash moves vertically.

Sliding windows move horizontally in a metal track. Generally a two-sash sliding window will have one sash fixed and one that moves **(see 24-4)**.

Casement windows are a type of swinging window that is hinged on the side. They have some form of crank arrangement to open and close the sash. Casement windows are made in a variety of arrangements, ranging from a single swinging sash to a multiple-sash unit in which some sashes are fixed **(see 24-5)**.

Awning and hopper windows are a type of swinging window that hinges from the top. They usually come as single units and can be stacked one above the other or beside the other, to produce multiple openings. They have the advantage of allowing the window to be open during a rain **(see 24-6)**. **Hopper units** have the hinge at the bottom, and the top swings out. This permits the window to be open, but blocks the direct flow of air on the people inside the area.

Fixed windows are secured to the frame so that they do not move. They allow light to enter the area and permit an unobstructed view from inside. They are often combined with moving types of windows **(see 24-7)**.

24-5 Various styles of casement windows.

24-6 These awning windows are hinged at the top. They are generally used to admit light and air yet provide a degree of privacy.

24-8 A round or circle top window.

Window manufacturers offer an extensive array of **special window** shapes which are usually fixed in the frame. These include round tops, which are generally placed on top of one of the standard types **(see 24-8)**. Other types are illustrated earlier in **24-1**.

Roof windows and skylights are installed on the roof and have a framed opening into the building, permitting light and, in some cases, ventilation. Those with moving sashes have the hinge on the top of the unit. They are opened with a crank **(see 24-9)**.

24-7 This fixed window has a transom sash on top.

24-9 Roof windows let in light and ventilation. *(Courtesy Weather Shield Windows and Doors)*

24-10 Bay windows extend outside the building. The side sashes are on a 30- or 45-degree angle. *(Courtesy Andersen Windows, Inc.)*

24-11 Bow windows typically have 5 or 7 sashes, several of which open to provide ventilation. *(Courtesy Andersen Windows, Inc.)*

Bay windows extend outside the building and have three glazed sashes. Generally one or more of these can be opened. The side sashes are available on 30-degree and 45-degree angles **(see 24-10)**.

Bow windows are much like bay windows except they are larger. They typically have five or seven sashes, of which some open. The opening sashes are typically the casement type **(see 24-11)**.

24-12 Some attractive variations of the basic window types commonly used. *(Courtesy Weather Shield Windows and Doors)*

There are many interesting variations on the design of these window types. Some of these are shown in **24-12**.

Manufacturers have a variety of **replacement windows** available. One type provides new sash and side guides which mount on the wood jamb. Another provides a complete new window and frame custom-made to fit into the frame of the old window. Replacement windows are generally made from polyvinyl chloride **(see 24-13)**.

24-13 This replacement window is a total vinyl unit that replaces a worn, old window. *(Courtesy Weather Shield Windows and Doors)*

LOCATING WINDOWS ON THE PLAN

Window types are identified on architectural drawings by a symbol and a mark. They are located by dimensions **(see 24-14 and 24-15)**.

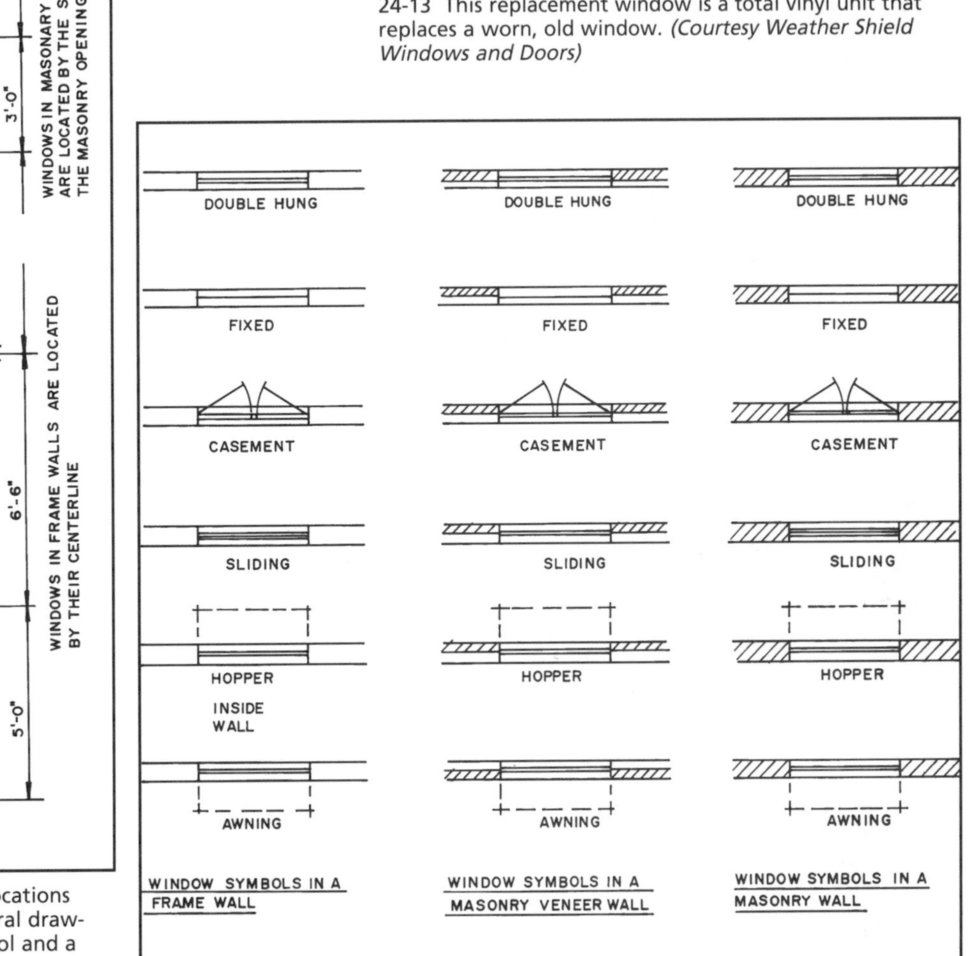

24-14 Window types and locations are identified on architectural drawings by an identifying symbol and a mark. They are located by dimensions.

24-15 Window symbols as they appear on the floor plan.

WINDOW SCHEDULE						
MARK	UNIT SIZE	ROUGH OPENING	TYPE	MATERIAL	GLAZING	REMARKS
W1	2'-5" X 2'-9"	2'-5" X 2'-10"	FIXED	WOOD	DOUBLE	PLASTIC COVERED COLOR-SAND
W2	5'-5" X 5'-1"	5'-6" X 5'-2"	D.HUNG	DO	DO	DO
W3	4'-11" X 2'-0"	5'-0" X 2'-05/8"	CASEMENT	DO	DO	DO

DO MEANS DITTO

24-16 Specific data about the windows to be used is given in an annotated list referred to as the window schedule.

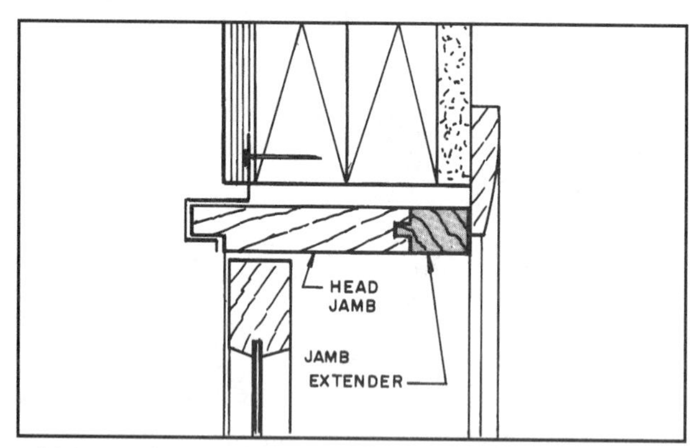

24-17 Jamb extenders are secured to the standard window frame to enable it to fit various wall thicknesses.

24-18 These double-hung and circle-top windows have plastic strips inserted on the inside of the window to represent muntins.

Specific data about the windows to be used is given on the architectural drawing in an annotated list labeled and referred to as the window schedule **(see 24-16 and 24-24)**.

FRAME & SASH MATERIALS

Window frames and sashes are available in wood, wood clad with vinyl or aluminum, vinyl, aluminum, and fiberglass.

Wood frames have good energy efficiency ratings and withstand variations in temperature. They do require regular painting to protect from moisture resulting in warping, twisting, and cracking. Many manufacturers treat all wood surfaces with a preservative and then preprime the exterior surfaces. This greatly reduces potential damage from the weather. They can be clad with vinyl.

Vinyl frames are typically made from polyvinyl chloride (PVC). They are moderately energy effective and are not damaged by the weather. They require little maintenance and can be used to produce custom-designed windows. Vinyl frames are not as strong and rigid as wood or metal, which limits the weight of glass they can handle. Under some conditions they might soften, warp, or twist.

Aluminum frames have a low energy efficiency rating (they rapidly conduct heat or cold). This conduction is reduced somewhat by placing insulating thermal breaks between the inside and outside parts of the frame and sash. Aluminum frames are strong and resist weathering, therefore requiring little maintenance.

Fiberglass frames have a high energy efficiency rating. They weather well and resist swelling, twisting, and rotting. They are made in a variety of colors and are strong enough to hold large panes of glass.

JAMB EXTENSIONS

The width of the window jamb often needs to be increased because of the wall thickness. Most window manufacturers build their units to a standard width. They supply jamb extensions to make them fit walls of various thicknesses **(24-17)**.

SCREENS

Screens are needed to keep insects from entering the house. They are available in factory-built units sized to fit the various windows. Most use a light metal frame. They are available in sizes that cover the entire window opening or just half the opening. If storm windows are installed, screens will be a part of the unit and run in one of the metal tracks.

MUNTINS

Muntins are the vertical and horizontal bars that break up a large glass pane into smaller panes **(see 24-18)**. Windows used to be made with the muntin built into the sash. The glass panes were actually small. Today, it is too expensive to make windows this way. The sash is made with one large piece of glass. Wood and plastic strips are made which clip to the frame and present the appearance of muntins. These are supplied by companies manufacturing the windows and cost extra.

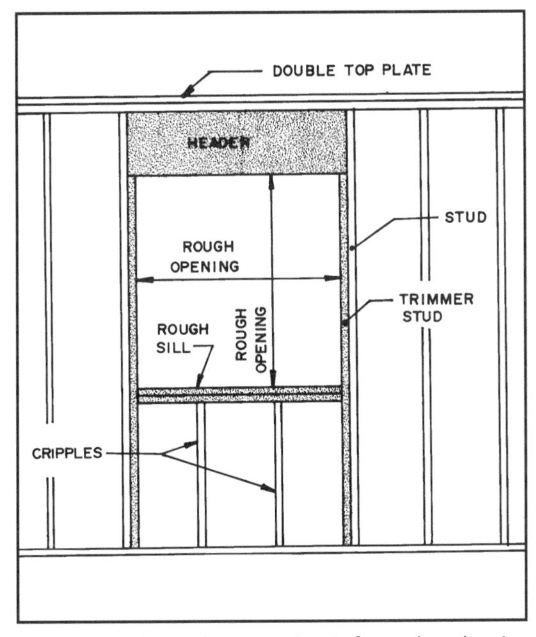

24-19 The rough window opening is framed to the size specified by the window manufacturer.

PREPARING FOR INSTALLATION

Check the rough opening to be certain it is the correct size. The opening should be framed as shown in **24-19**. The edges are wrapped with builder's felt or polyethylene plastic sheet material to prevent air infiltration and control moisture penetration **(see 24-20 and 24-21)**.

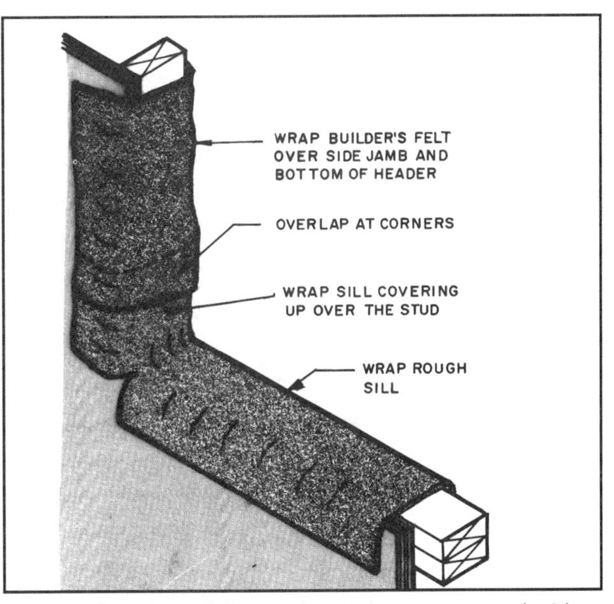

24-20 The edges of the rough opening are wrapped with builder's felt or sheet plastic before the window is installed.

24-21 A finished installation showing the wrapping material around the rough opening.

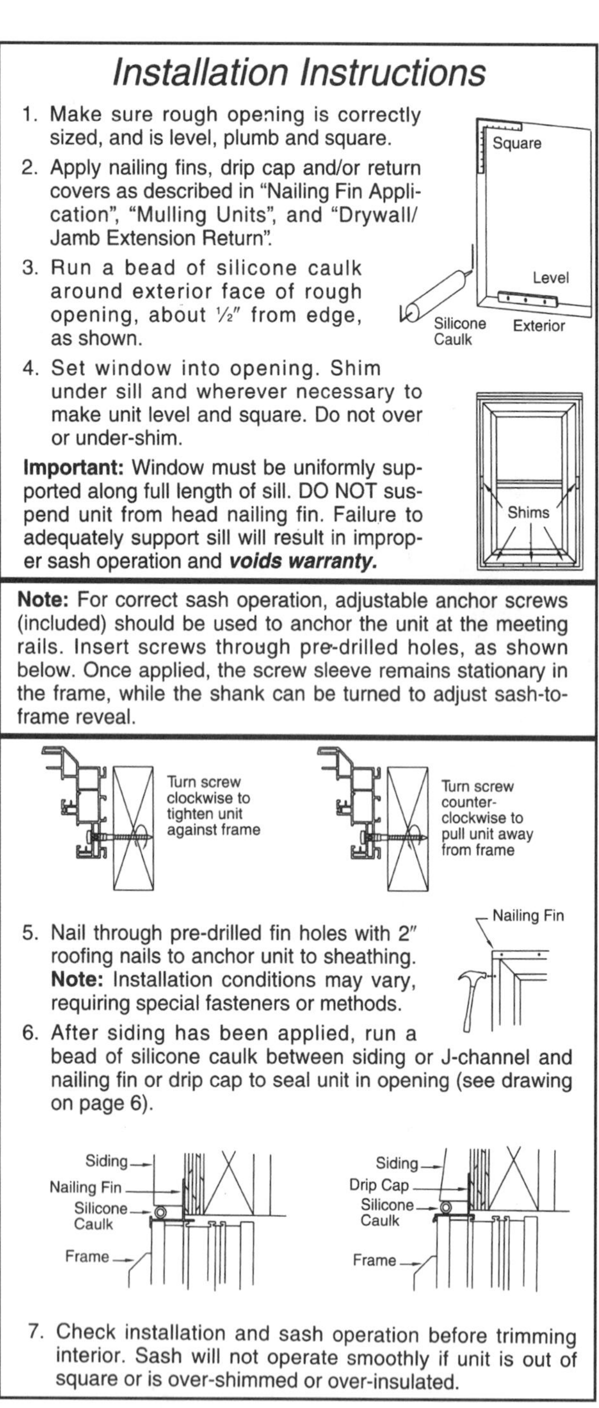

Installation Instructions

1. Make sure rough opening is correctly sized, and is level, plumb and square.
2. Apply nailing fins, drip cap and/or return covers as described in "Nailing Fin Application", "Mulling Units", and "Drywall/ Jamb Extension Return".
3. Run a bead of silicone caulk around exterior face of rough opening, about ½" from edge, as shown.
4. Set window into opening. Shim under sill and wherever necessary to make unit level and square. Do not over or under-shim.

Important: Window must be uniformly supported along full length of sill. DO NOT suspend unit from head nailing fin. Failure to adequately support sill will result in improper sash operation and ***voids warranty.***

Note: For correct sash operation, adjustable anchor screws (included) should be used to anchor the unit at the meeting rails. Insert screws through pre-drilled holes, as shown below. Once applied, the screw sleeve remains stationary in the frame, while the shank can be turned to adjust sash-to-frame reveal.

5. Nail through pre-drilled fin holes with 2" roofing nails to anchor unit to sheathing. **Note:** Installation conditions may vary, requiring special fasteners or methods.
6. After siding has been applied, run a bead of silicone caulk between siding or J-channel and nailing fin or drip cap to seal unit in opening (see drawing on page 6).
7. Check installation and sash operation before trimming interior. Sash will not operate smoothly if unit is out of square or is over-shimmed or over-insulated.

24-22 Installation instructions provided with each window unit by the manufacturer. *(Courtesy Weather Shield Windows and Doors)*

INSTALLING A WINDOW

Specific instructions are available from the manufacturer. They should be followed carefully. If they are not, the warranty that comes with the window will be void. A typical instruction sheet that is packed with each unit is shown in **24-22.**

Check to be certain the window is the one that is supposed to be placed in the opening. The architectural drawings will identify each unit. The floor plan shows the location for each window **(see 24-23).** Usually, windows are identified by a code. All windows of the same size and type use the same letter identification code.

The other details needed are in the window schedule. It identifies the windows by code, the number needed, the sash size, the rough openings in frame and masonry walls, and often the specific brand **(see 24-24 and refer to 24-16).**

Windows arrive on the job carefully packed to prevent damage. Place the units in the rooms where they will be used. Do not remove them from the packing until they are needed for installation. After removal, examine them for damage. Keep all factory-applied bracing and spacer strips on the units until they have been installed.

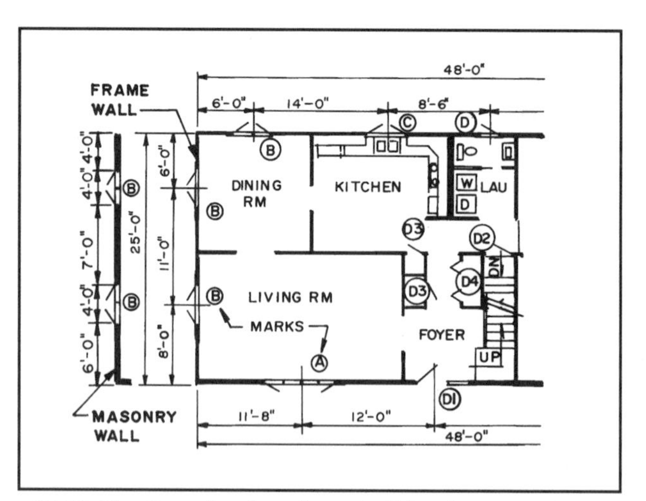

24-23 The floor plan for a frame building locates each door and window by its centerline. A solid masonry building gives the rough window opening from side to side of the masonry opening.

MARK	QUAN	NO. LTS	UNIT DIMENSION	ROUGH OPENING	DESCRIPTION
A	1	3	4'-11" X 8'-0 1/2"	4'-11 1/2" X 8'-1"	CASEMENT ANDERSEN CP25C15
B	3	2	4'-11 7/8" X 4'-0"	5'-0 3/8" X 4'-0 1/2"	CASEMENT ANDERSEN C25
C	1	2	3'-4 3/4" X 3'-4 3/4"	3'-5 3/8" X 3'-5 1/4"	CASEMENT ANDERSEN CN235
D	1	1	2'-11 15/16" X 2'-0 1/8"	3'-0 1/2" X 2'-0 5/8"	CASEMENT ANDERSEN C13

24-24 A typical window schedule as found on construction drawings.

24-25 Most of the window heads will be kept the same height as the doors. Special windows typically extend beyond this height.

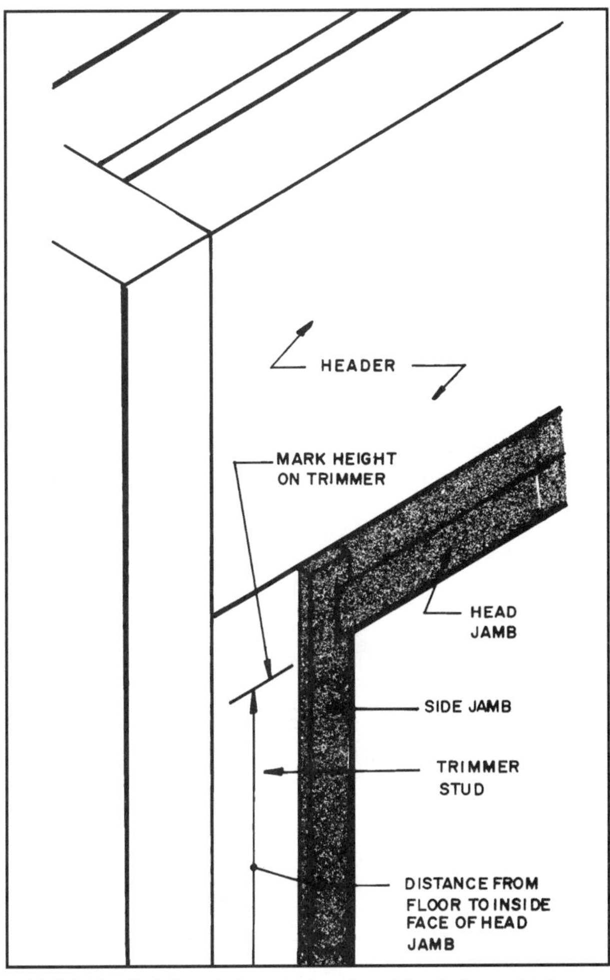

24-26 The height of the window head jamb is marked on each trimmer stud.

Windows not shipped in cartons should be wrapped in plastic and stored in a safe, dry place on the job.

In most cases the designer wants the window heads to line up with the door head jamb. This is usually 6′-8″ (2.03m) above the floor **(see 24-25)**. Mark the trimmer stud at each window with the required height **(see 24-26)**. Frequently a **story pole** is used. A story pole is a straight piece of wood with the desired height marked on it. It saves time because it locates the height without having to open up a tape rule each time a measurement is required **(see 24-27)**.

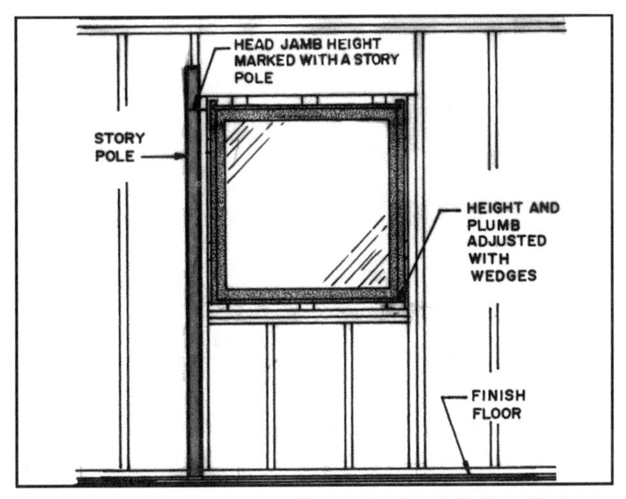

24-27 The head height can be marked rapidly for all windows using a story pole.

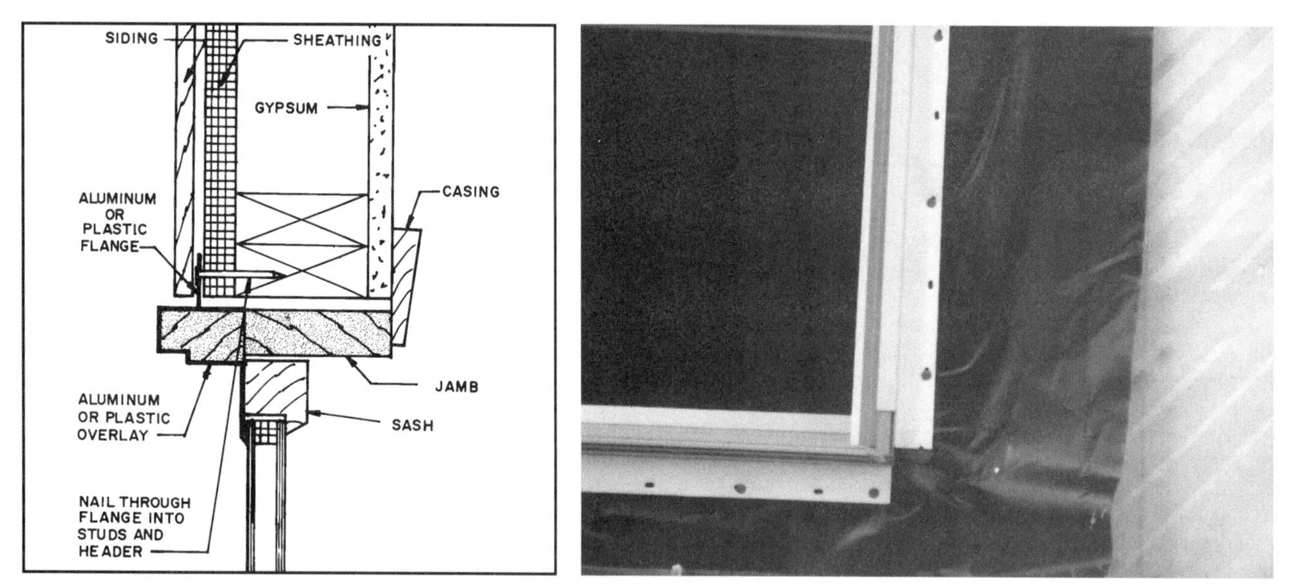

24-28 This window is installed by nailing through a flange into the studs and header.

There are two frequently used ways windows are secured to the wall. One uses an aluminum or plastic flange that runs around all sides of the window frame. It has holes punched at evenly spaced intervals for nails **(see 24-28)**. Another type has a thick wood casing nailed to the frame. The window is nailed to the wall with casing nails driven through this wood casing **(see 24-29)**.

Following are the steps frequently used to install a window. Various carpenters may use a different sequence. Always consult the manufacturer's directions because each one has very specific instructions as to how to install the unit. Some windows come knocked down and must be assembled on the job. They come with full directions and all the screws and hardware required.

1. Set the window unit into the opening. If the sashes are to remain in the frame, lock them closed.
2. Place shims under the sill to level it, and adjust the space at the head and sill so it is about equal. Check to be certain it is level **(see 24-30)**. Adjust the shims as necessary.

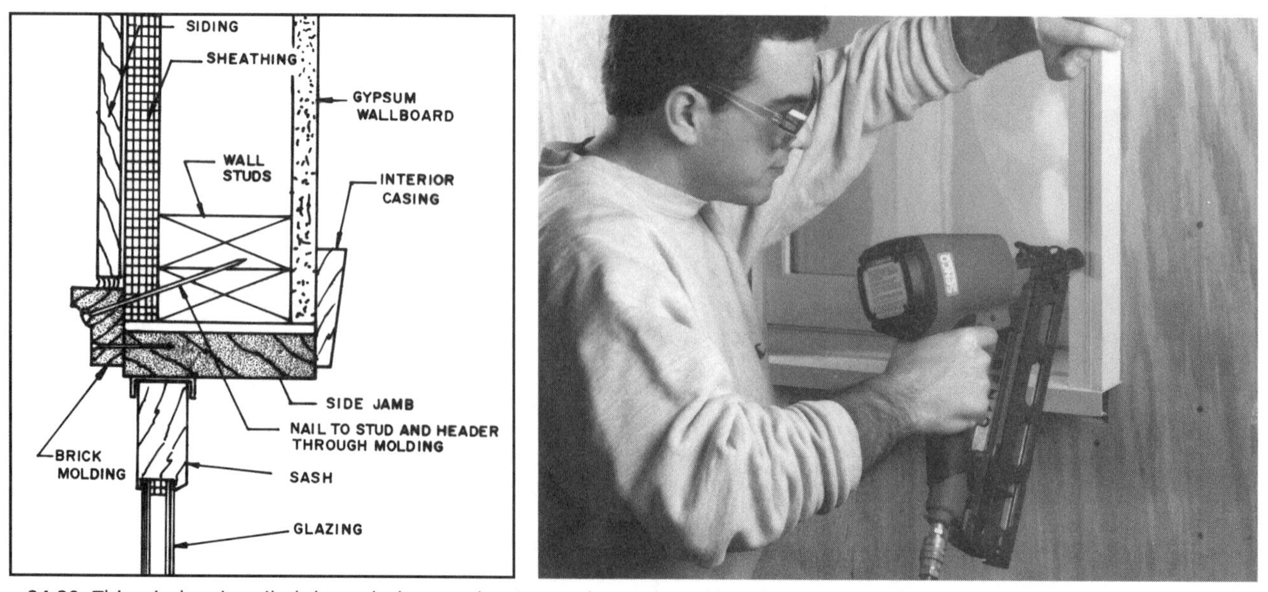

24-29 This window is nailed through the wood casing to the studs and header. *(Courtesy Senco Products, Inc.)*

3. Shim the side jambs, firming the window in place. Keep the space on each side about equal. Check to be certain they are plumb **(see 24-30)**. Adjust the shims as necessary.
4. Drive a nail at each top corner through the casing or nailing flange. Do not set the nails at this time **(see 24-31)**. Each nail must be driven through a solid material, such as plywood; therefore if rigid foam sheathing has been used, solid wood blocking must be inserted wherever there is to be a nail.
5. Temporarily tack the bottom corners to the studs **(see 24-31)**.
6. Measure across the width of the frame in several places to be certain it is square and has no bow. Check the side jamb for bow with a long level and adjust the wedges if necessary **(see 24-32)**.
7. If the sashes have remained in the frame, move them to be certain they move easily. If they have been removed to set the frame, install them now and check for movement **(see 24-33)**.
8. Nail all around the window through the casing or flange **(see 24-33)**.
9. Fill the space between the frame and studs, header, and rough sill with insulation. Do not pack it in too tightly or it may bow the frame. It also insulates better if left loose **(see 24-34)**.

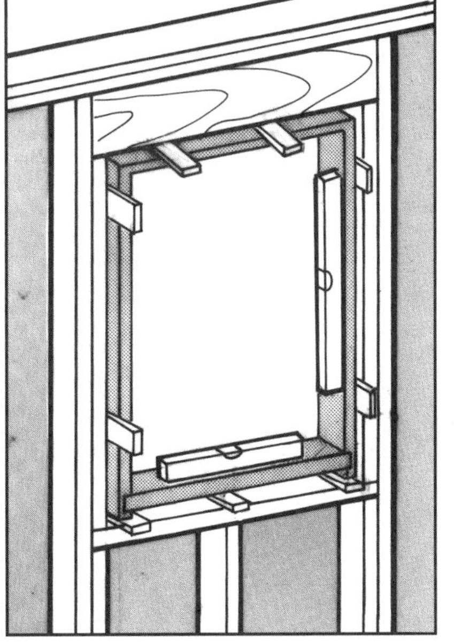

24-30 After sliding the window unit into the rough opening, level and plumb it with wood wedges.

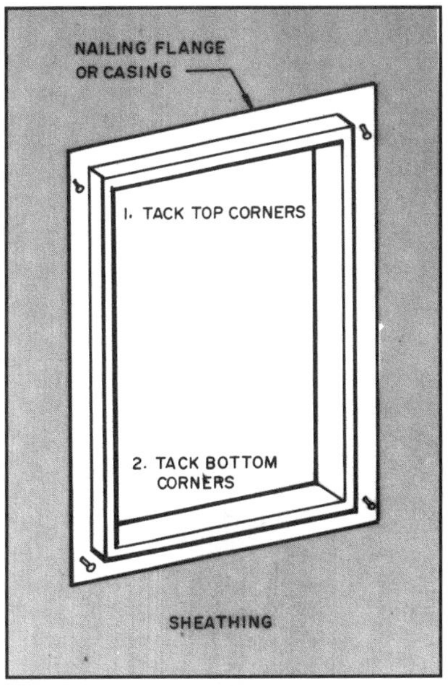

24-31 Drive a nail at each top corner. After being certain the unit is still level and plumb, tack a nail at each bottom corner.

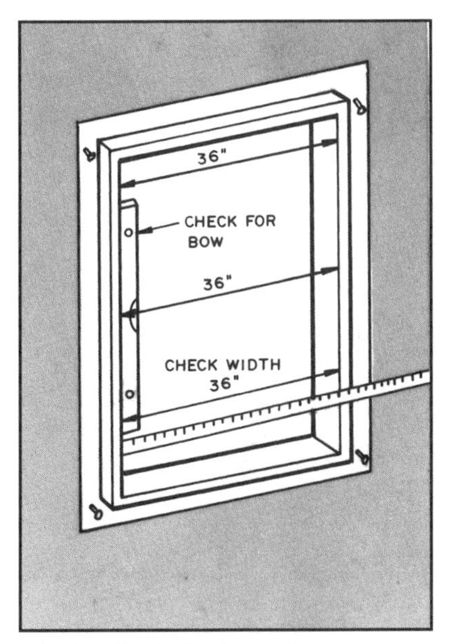

24-32 Check the frame for squareness and bow.

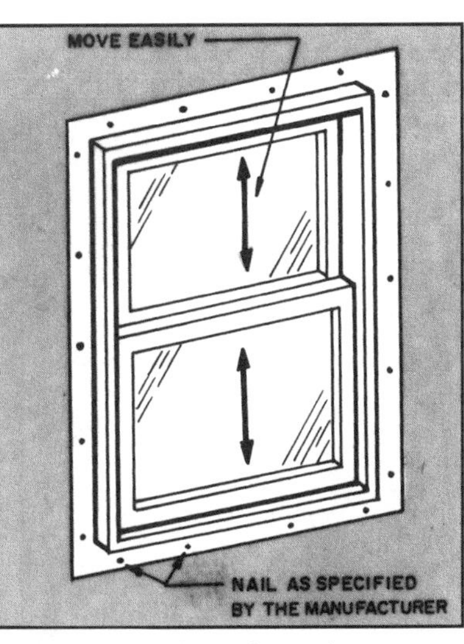

24-33 Move the sashes to be certain the frame is square and they work easily. Then nail the casing or flange to the studs, header, and rough sill.

24-34 Fill the space between the window frame and the studs, header, and rough sill with loose insulation.

Generally this completes the installation of the window. However, since the side jamb and head jamb on long windows can bow, it is advisable to place shims between the frame and trimmer stud, and nail through the window frame and shims into the trimmer stud with casing nails. The number of these used will depend on the length of the window, but a spacing of 16 to 20 inches (approx. 41 to 51cm) is often used.

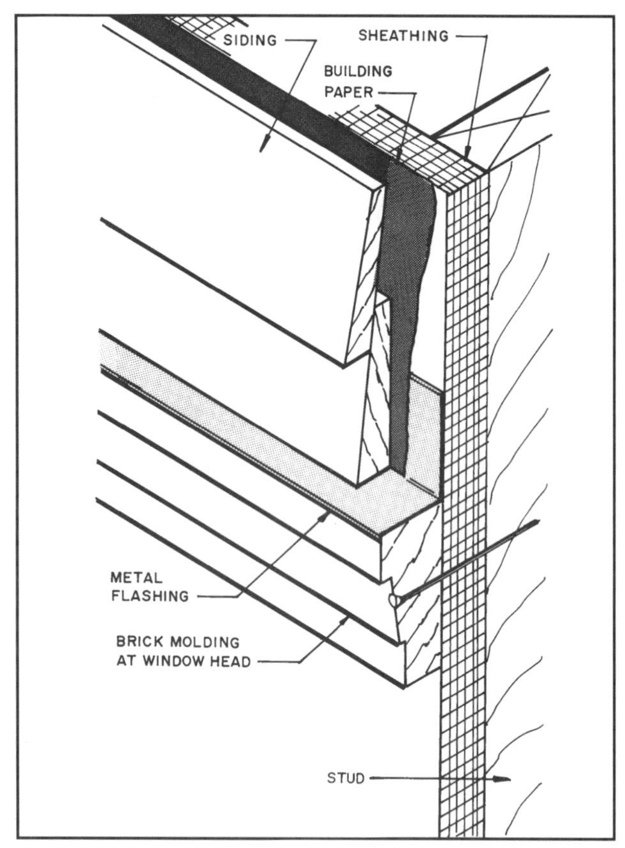

24-35 Windows secured with wood casing require flashing along the top of the casing. Side casings must be carefully caulked after the siding is installed.

The top flange on windows that use flanges for installation serves as the **flashing**. Wood-cased windows require installation of a small aluminum flashing strip as shown in **24-35**. The flashing is covered by the exterior siding material.

When nailing through a casing, use aluminum or galvanized casing nails long enough to go through the casing and sheathing, and penetrate the stud by 1½ inches (38mm). When nailing flanges, use 1¾-inch (44mm) or longer galvanized roofing nails. When nailing through soft-foam-type sheathing, it is good practice to cut away a small section where each nail will go, and replace it with wood of equal thickness. This will stiffen the window installation. If you do not do this, be certain to drive the nails until they are firm, but do not overdrive because this could cause them to break through the flange.

Installing Multiple-Window Units

Frequently window units are butted side by side or one on top of the other. Window manufacturers have prescribed ways to install these units and can supply the exterior mullion for some units. The manufacturer gives the sizes of the rough openings for multiple unit assemblies. Typical details are shown in **24-36**, **24-37**, and **24-38**. In one example the side jambs are butted and have a narrow block-

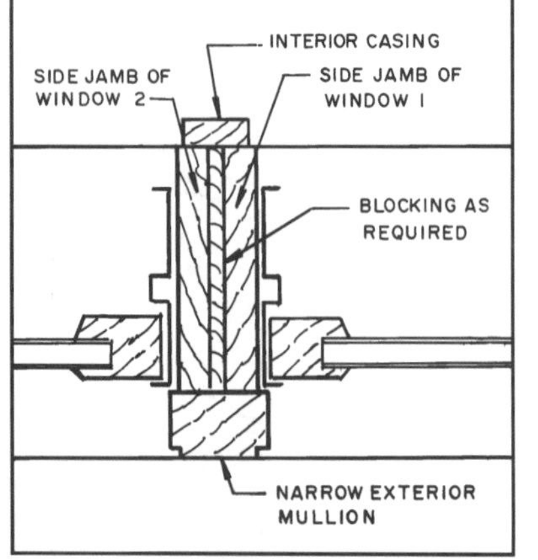

24-36 This narrow nonload-bearing mullion is formed by butting the window frames and using a little blocking if needed.

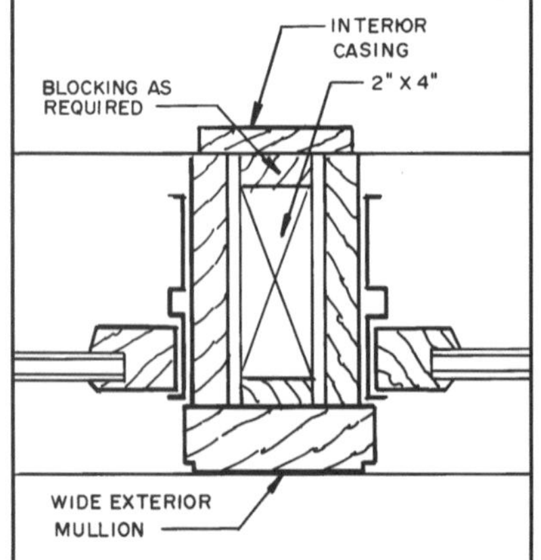

24-37 This wider mullion is made with a 2 × 4 stud spacer. It will carry some load.

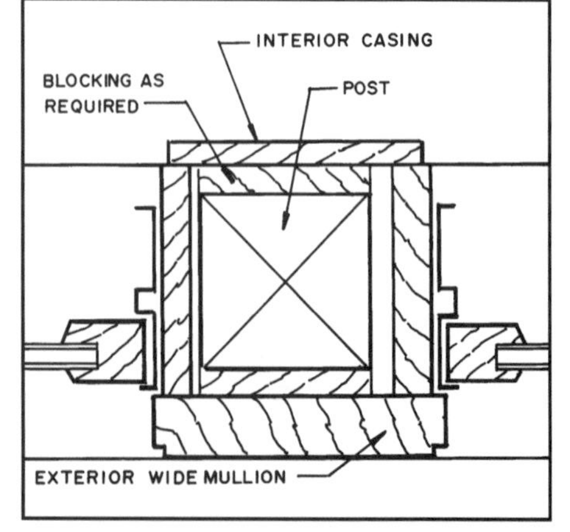

24-38 This wide mullion is made using a 4 × 4-inch post. Any size mullion can be made by varying the internal load-bearing members.

ing between them. The exterior mullion is narrow. This is done if no structural support is needed and a narrow mullion is wanted for appearance. The other examples show 2 × 4 and larger posts between the two window units. This provides structural support and a wide exterior mullion. It should be noted that the design of the casing and mullion will vary depending upon the manufacturer of the window. Those shown are generalized examples.

INSTALLING ELLIPTICAL & CIRCLE TOP WINDOWS

Elliptical and circle top windows are installed as individual units and on top of standard doors and windows. Refer back to **24-18**.

When installed on **top of a door or window** the assembly appears as shown in the section view in **24-39**. The circle head unit is fastened to the top frame of the door or window as specified by the manufacturer.

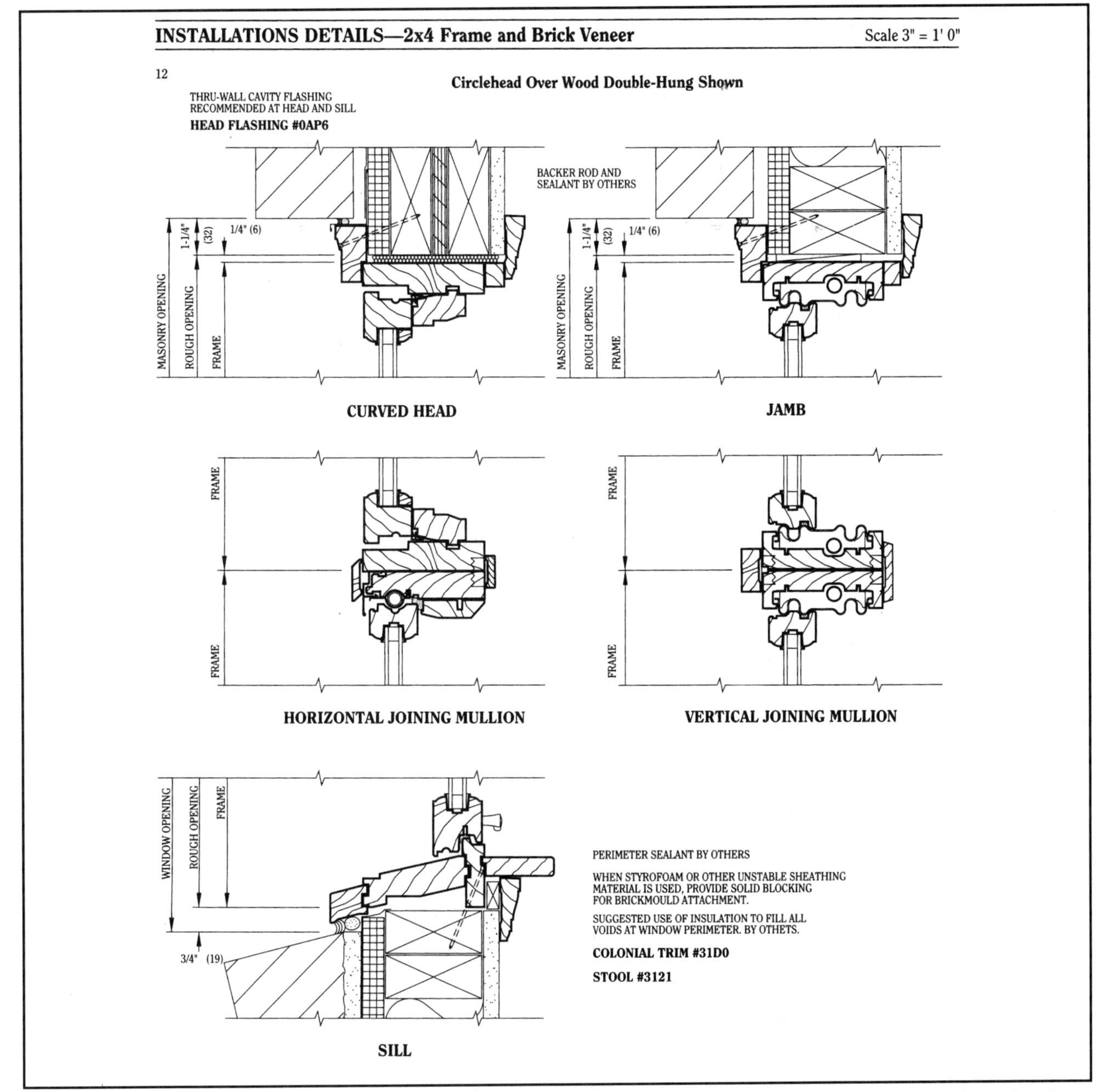

24-39 Installation details for a circlehead window over a wood double-hung window. This shows it in a wood-framed wall with a brick veneer exterior. Consult the manufacturer for additional details. *(Courtesy Pella Corporation)*

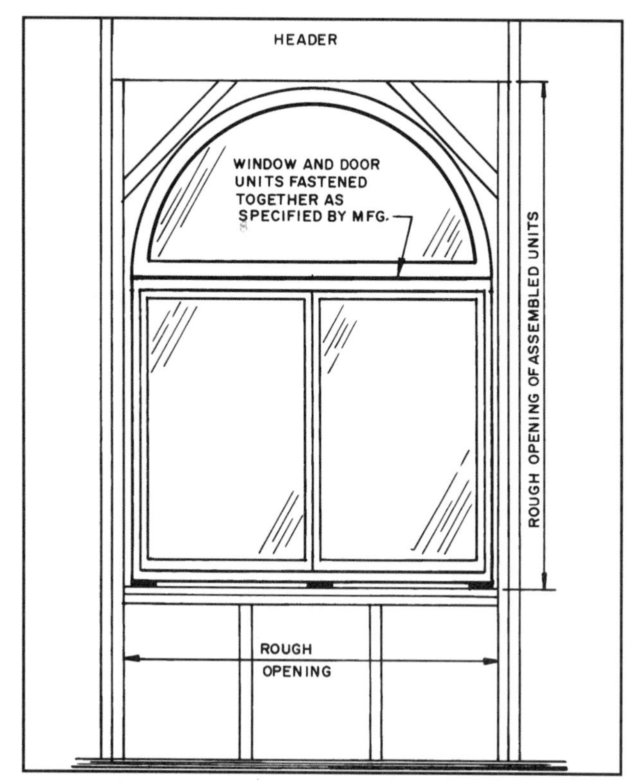

24-40 The circle-top window is fastened to the standard window as specified by the manufacturer. The assembled unit is installed as discussed for standard windows and doors.

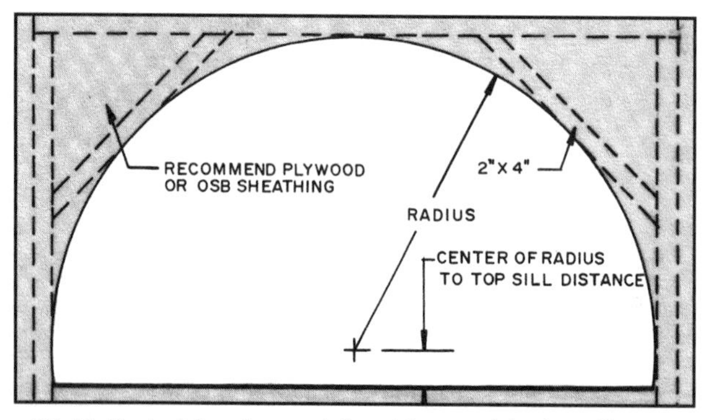

24-41 Typical framing and sheathing used for installing circle top windows.

The opening for the circle head unit and the door or window below is framed as shown in **24-40**. The rough opening required is specified by the manufacturer. The unit is leveled and plumbed as described for standard doors and windows. It is nailed through its flanges into the sheathing and wood framing. Plywood or OSB sheathing is recommended. If soft insulating sheathing is used it is best to place solid wood blocking behind the flange at each nail. If it is nailed through the soft sheathing do not set the nail. Nail it firm and quit. If set too hard the flange will bend and possibly break.

Elliptical and circle top windows are often installed as a separate, independent unit. Suggested steps for installing independent circle-top and elliptical windows follow:

1. Insert the unit in the opening prepared as shown in **24-41**. Cover the edges of the opening with builder's felt.
2. Shim the ends to get the unit level **(see 24-42)**.
3. Tack the flange in the top center hole **(see 24-43)**.
4. Recheck for levelness. Adjust shims as necessary. Now nail the flange using 1¾-inch roofing nails.
5. Caulk around the perimeter of the window after the siding is in place **(see 24-44)**.
6. From the inside of the building insulate the space between the framing and the window unit.

INSTALLING BOW & BAY WINDOWS

Bow and **bay windows** usually arrive on the site fully assembled. If they come knocked down, special instructions from the manufacturer must be followed to assure proper assembly. These units are very heavy. They require several people to move them and to lift them into the rough opening. Be certain that the unit is not allowed to twist or bend.

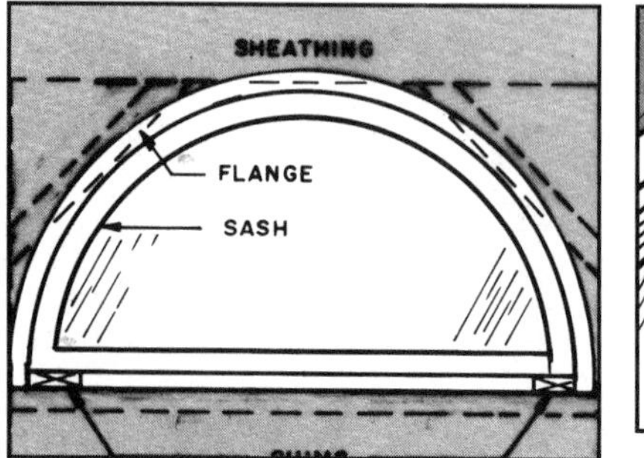

24-42 Insert the window into the opening. Level and shim along the rough sill.

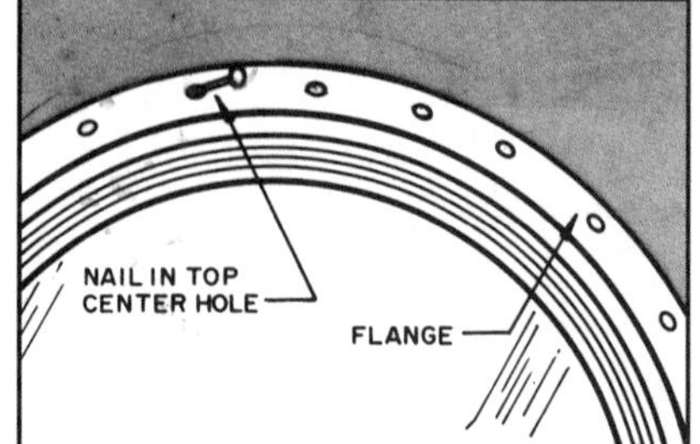

24-43 After the window is level, tack it to the wall through the top center hole. Recheck for levelness before you nail the entire flange.

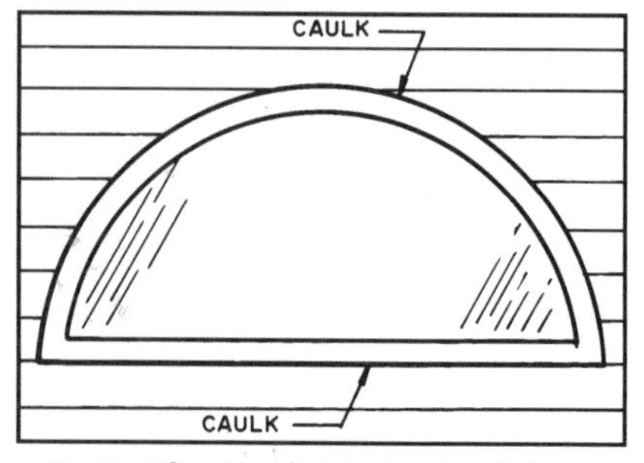

24-44 After the finished siding is in place, caulk on all sides. Install batt insulation in the space around the window on the inside of the building.

Installation instructions vary a great deal from one manufacturer to another. It is important to follow the instructions that come with the unit.

Following is a generalized description of an installation process that covers many of the instructions to be followed.

1. The rough opening must be of the size recommended by the manufacturer and be square, level, and plumb. The rough sill should be doubled. The header must be designed to carry the roof over the rough opening.

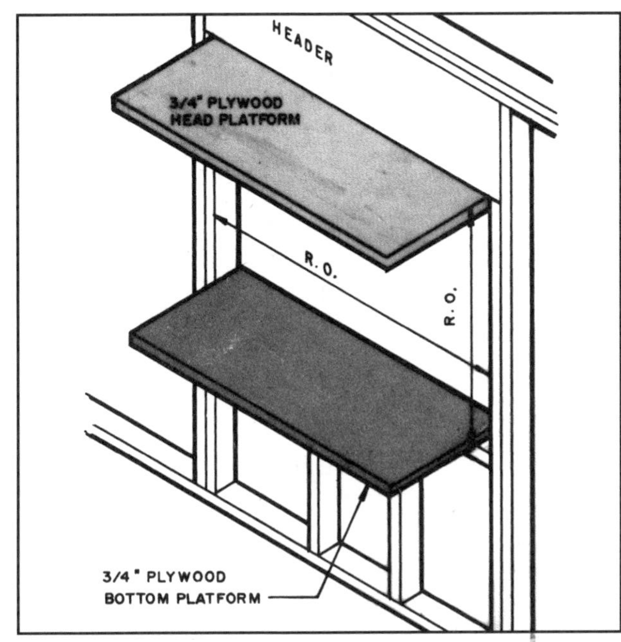

24-45 The head and bottom platforms are nailed to the wall framing.

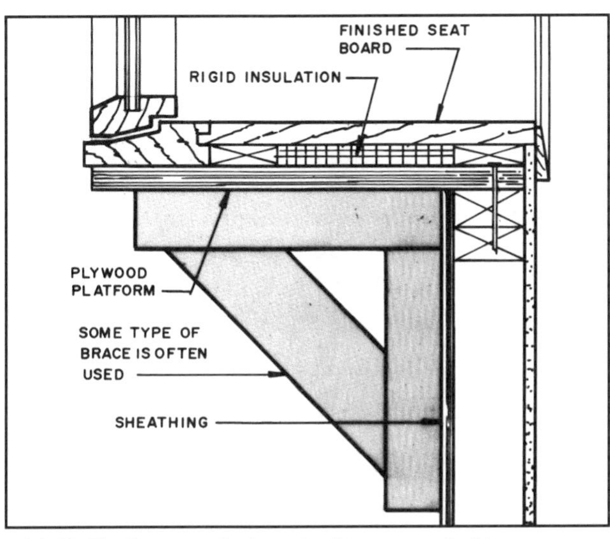

24-46 The bottom platform is often supported by some type of brace.

2. Some form of platform is required to provide support. In **24-45** the unit has a ¾-inch plywood platform that is nailed to the sill and a head platform that is nailed to the header. Another system uses a ¾-inch plywood platform that is supported by braces below it **(see 24-46)**. Follow the manufacturer's recommendation for the brace below the unit.

3. Slide the unit into the rough opening. Block and shim it until it is centered in the opening and is level and plumb. Notice that the seat board extends past the stud and covers the interior wall finish material **(see 24-47)**.

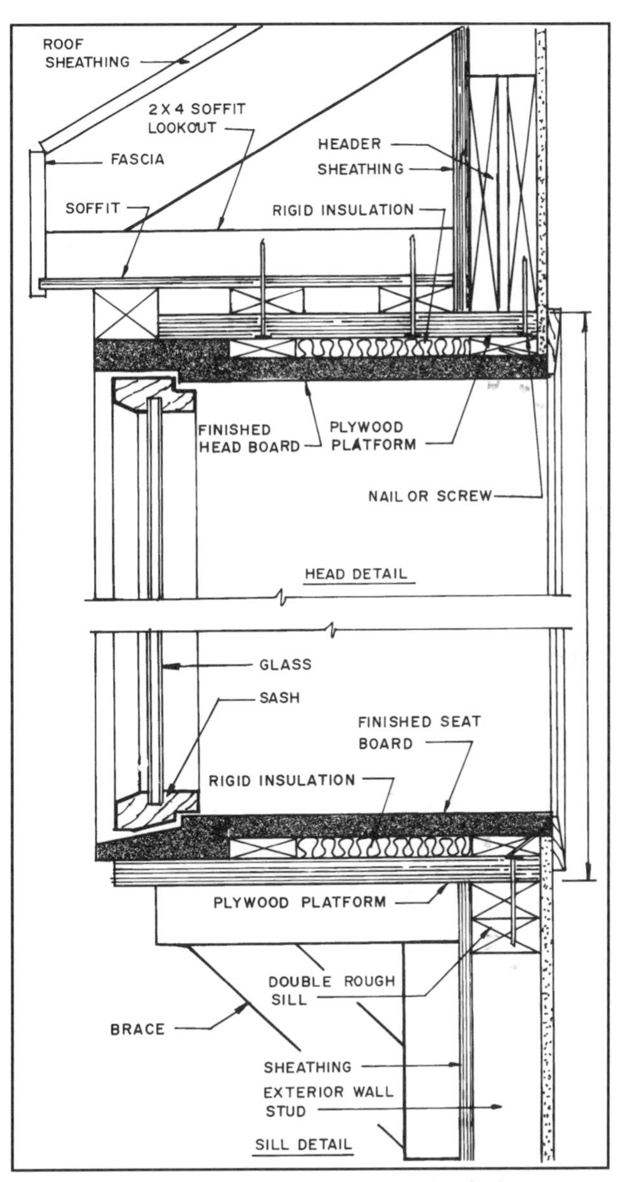

24-47 A section through a generic example of a bow or bay window set into the rough opening.

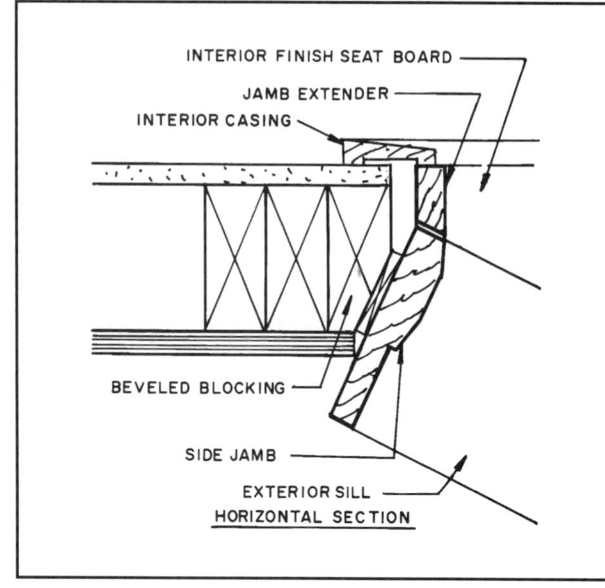

24-48 A typical horizontal section through the side jamb as it butts the studs forming the rough opening.

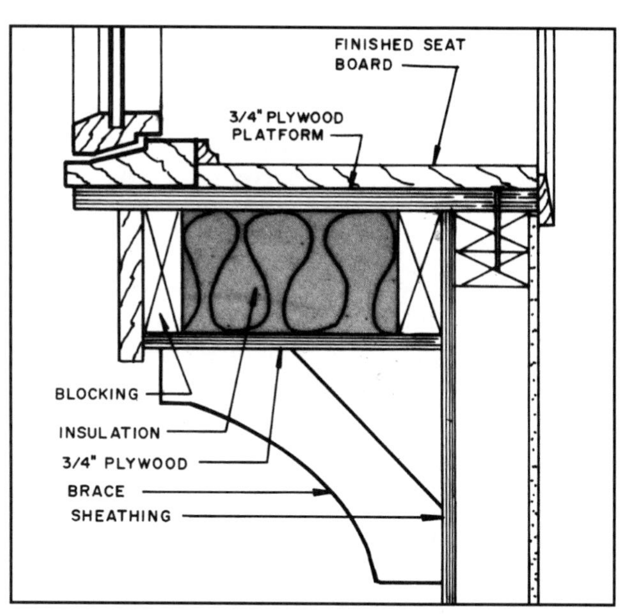

24-49 Extra insulation can be placed below the platform by framing a wider opening.

A typical horizontal section in **24-48** shows one type of construction where the unit butts the studs on the side of the unit. A jamb extender is used to finish the interior exposed frame. This is supplied by the manufacturer.

4. The platforms are nailed to the sill and header with 10d ring-shank nails spaced 8 inches apart. The top platform is also nailed into the soffit lookouts. The finished head and seat boards are then installed over a layer of rigid insulation. Additional insulation can be installed as shown in **24-49**.

5. Apply exterior trim supplied by the manufacturer.

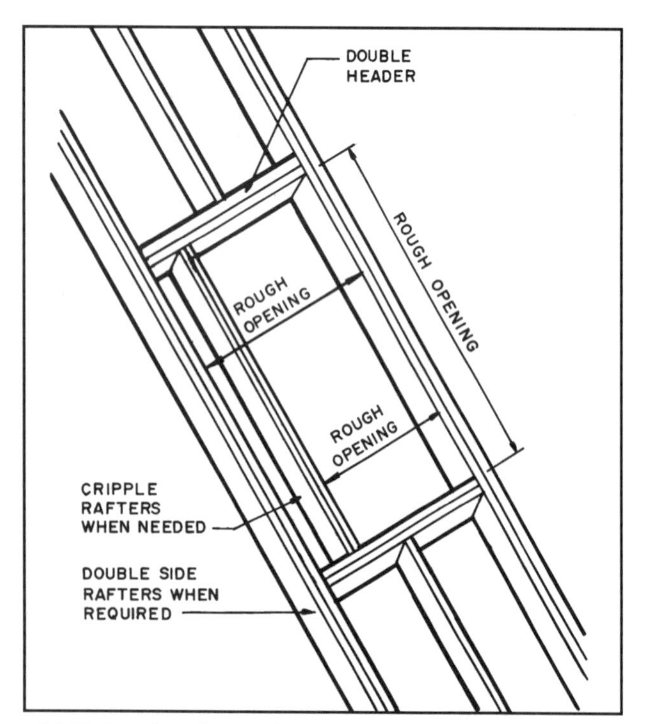

24-50 Framing the rough opening for a roof window.

INSTALLING ROOF WINDOWS

The opening in the roof must be framed to support the roof load. Some roof windows fit between rafters spaced 24 inches O.C. Others are wider and require a rafter be cut. A typical framed opening is shown in **24-50**.

When the opening requires a rafter be cut, double the rafters on each side of the opening. They should be the same size as the regular rafters and run the full span. The headers on each side of the opening must be doubled. The size of the rough opening is specified by the window manufacturer. The opening must be square and very close to the required size.

Some manufacturers size their roof windows 22½ inches to fit between rafters spaced 24 inches O.C. without cutting a rafter. A 30½-inch wide unit will fit between two 16-inch O.C. rafters, requiring one rafter to be cut and headers installed. A 46½-inch unit will fit between two rafters spaced 24 inches O.C., requiring one rafter to be cut.

The installation can take several forms, as shown in **24-51**. The one used depends on the specifications in the architect's design. The opening is framed with 2 × 4 stock and then covered with an interior finish material such as drywall gypsum board.

Following are typical directions for installing a roof window.

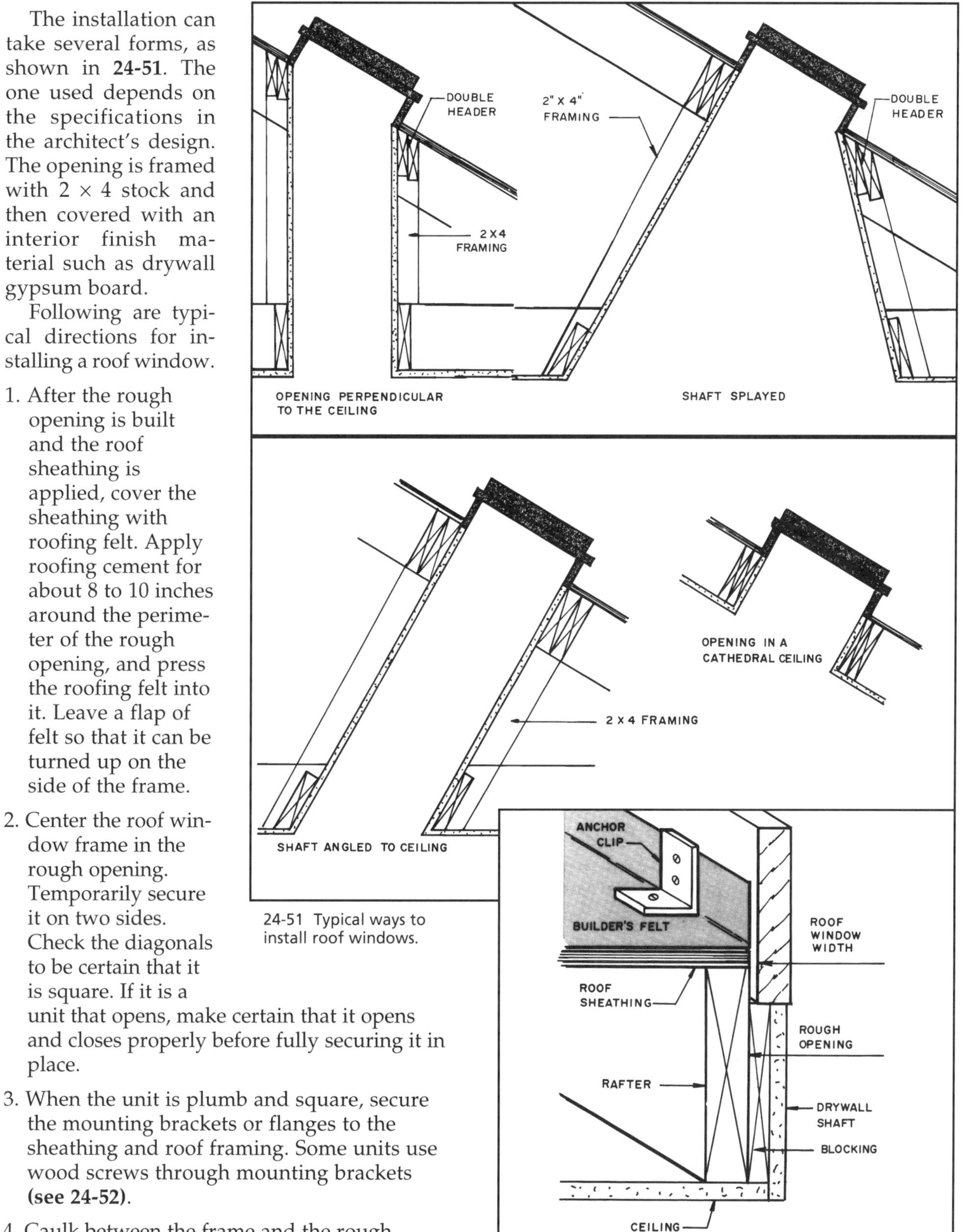

24-51 Typical ways to install roof windows.

24-52 A section through the side jamb of a roof window showing one type of metal fastener used to secure it to the roof sheathing.

1. After the rough opening is built and the roof sheathing is applied, cover the sheathing with roofing felt. Apply roofing cement for about 8 to 10 inches around the perimeter of the rough opening, and press the roofing felt into it. Leave a flap of felt so that it can be turned up on the side of the frame.
2. Center the roof window frame in the rough opening. Temporarily secure it on two sides. Check the diagonals to be certain that it is square. If it is a unit that opens, make certain that it opens and closes properly before fully securing it in place.
3. When the unit is plumb and square, secure the mounting brackets or flanges to the sheathing and roof framing. Some units use wood screws through mounting brackets **(see 24-52)**.
4. Caulk between the frame and the rough opening.
5. Run the roofing belt up the side of the frame.

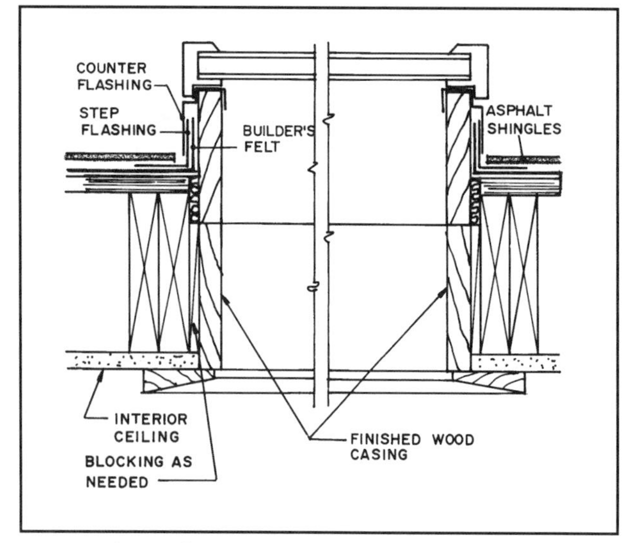

24-53 This roof window is flashed using step flashing with asphalt shingles. See the information on roofing for step flashing installation details.

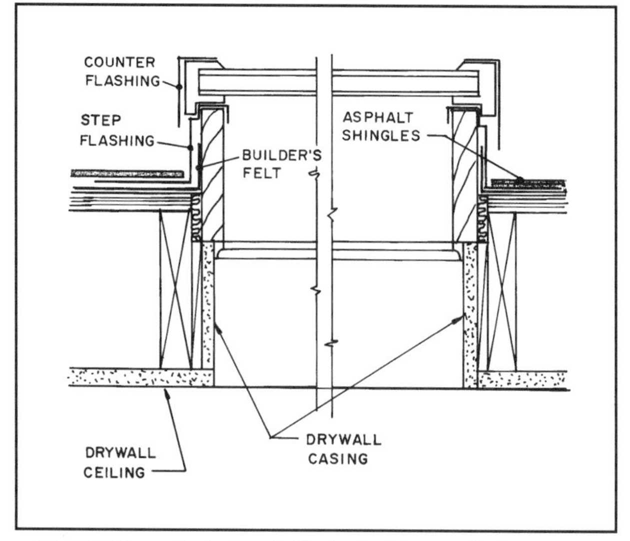

24-54 This roof window is flashed with step flashing that runs up and over the side of the window frame. A strip of counterflashing extends down from the sash.

6. Install flashing as shown on the instruction sheet that is supplied by the manufacturer. Some typical installation details are shown in **24-53** and **24-54**.

7. With the roof window in place, you can construct the light shaft. Some installers may prefer that the light shaft framing is constructed before the window has been set into the roof opening.

Begin by locating the opening in the ceiling. This is done by dropping plumb lines from the corner of the roof opening **(see 24-55)**. Frame this opening in the same manner as described for the roof opening **(see 24-56)**. Now frame the walls of the light shaft with 2 × 4 studs in the same manner as you build interior partitions. Remember to provide nailers for the installation of the drywall or other interior finish material **(see 24-57)**. Insulate between the studs as you would an exterior wall.

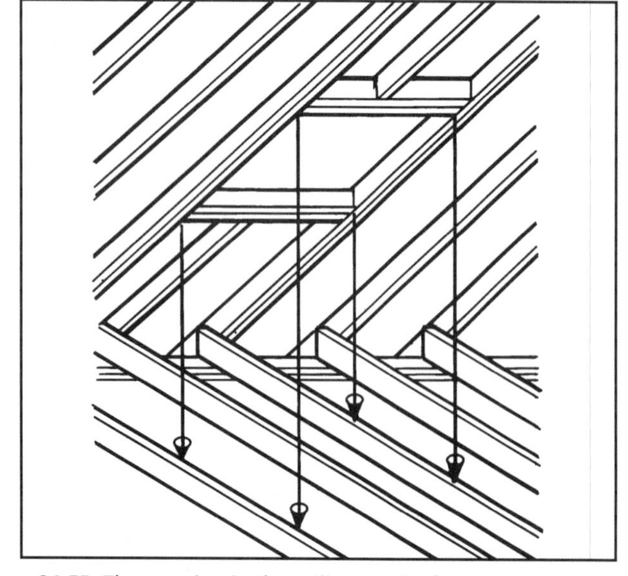

24-55 The opening in the ceiling can be found by dropping plumb lines from the opening in the roof.

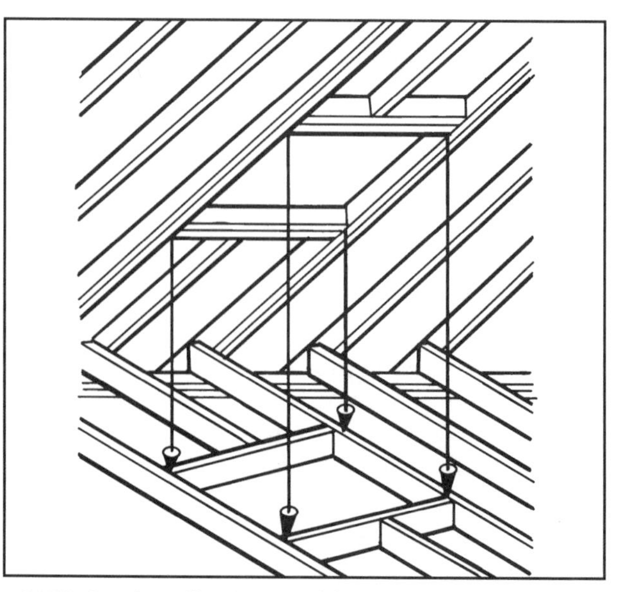

24-56 Cut the ceiling joists and frame the opening in the ceiling.

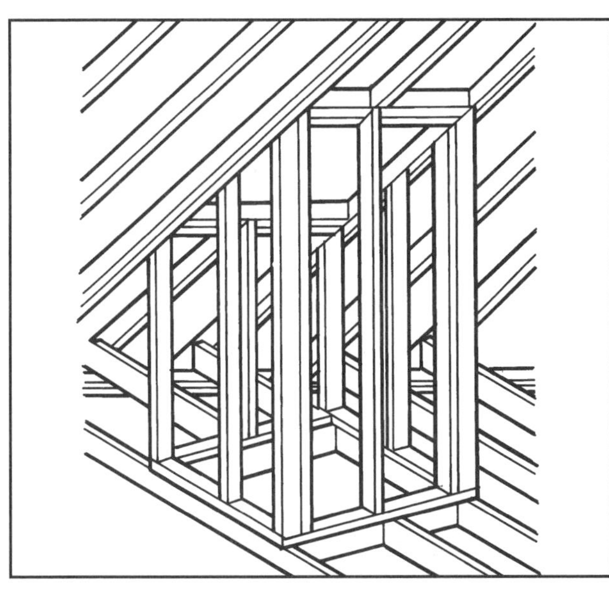

24-57 Build the partitions forming the shaft using 2 × 4 studs.

INSTALLING TUBE SKYLIGHTS

Tube skylights enable you to easily locate spots of natural light inside the building. They consist of a roof dome and a light tubing **(see 24-58)**. The unit comes complete with the exterior dome, light tube flashing at the roof, and an interior ceiling diffuser as shown in **24-59**.

24-58 This tube skylight permits natural light to be directed into dark areas of the building. *(Courtesy Solatube International, Inc.)*

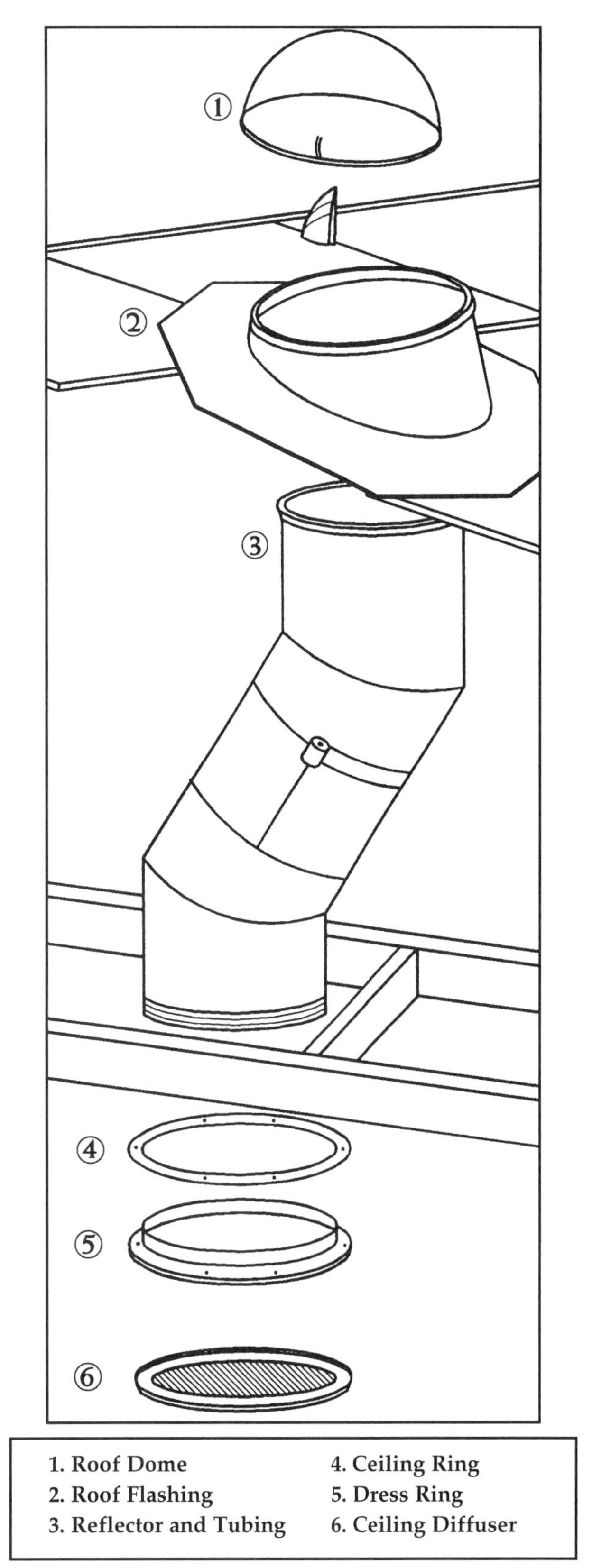

1. Roof Dome	**4. Ceiling Ring**
2. Roof Flashing	**5. Dress Ring**
3. Reflector and Tubing	**6. Ceiling Diffuser**

24-59 These are the parts that make up the complete tube skylight. Notice it includes the roof flashing. *(Courtesy Solatube International, Inc.)*

INSTALLING REPLACEMENT WINDOWS

Many wood windows in older homes have deteriorated due to poor maintenance. Manufacturers produce windows designed to replace these without tearing up the wall. They fit inside the frame of the old window. If the old frame has rotted it will have to be replaced.

Details concerning one available replacement window are shown in **24-60**. It consists of two glazed sashes and vinyl jamb liners. This is a tilt-sash unit allowing the sash to lean inwards so it is easy to clean both sides of the glass **(24-61)**. The steps to install this particular window are shown in **24-62**.

24-60 This is a replacement window kit containing new glazed sash and vinyl jamb liners secured to the wood frame cased inside the window opening. *(Courtesy Weather Shield Windows and Doors)*

1. Remove the old sash from inside. Using a pry bar or putty knife, take out the inside stop moldings from the side jambs.

2. Cut out all cords and weights and remove bottom sash. Repeat with the top sash.

24-61 This replacement window is a double-hung tilt window made from vinyl. The sash tilts inward to make cleaning easy. *(Courtesy Weather Shield Windows and Doors)*

3. Prepare the frame. Position the brackets for the vinyl jamb liners. Start about 4 inches from the top and finish about 4 inches from the bottom. Secure in place.

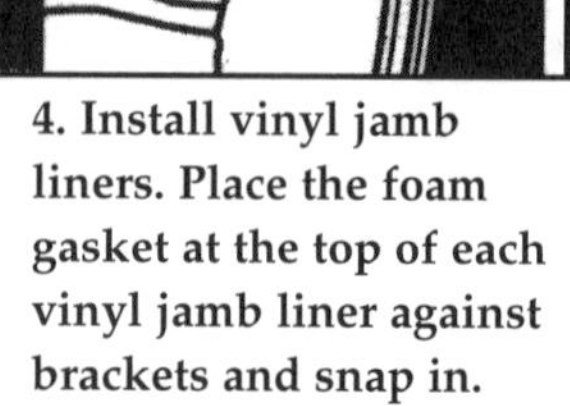

4. Install vinyl jamb liners. Place the foam gasket at the top of each vinyl jamb liner against brackets and snap in.

5. Install the sash. Hold top sash at 90 degrees, level with the cams in the vinyl jamb liner. Engage corner pins. Tilt sash upright into vinyl track while sliding down. Repeat with the bottom sash.

24-62 Here are five steps to install one type of replacement window. *(Courtesy Weather Shield Windows and Doors)*

24-63 This certification label of the American Architectural Manufacturers Association indicates the aluminum or vinyl door or window meets specified standards. *(Courtesy American Architectural Manufacturers Association)*

PRODUCT CERTIFICATION

Aluminum and vinyl prime and replacement windows and doors that conform to product and performance standards of the American National Standards Institute (ANSI) and the American Architectural Manufacturers Association (AAMA) are identified by the AAMA certification label. The AAMA serves to assure, by inspection of the manufacturer's facility, that the products or services bearing the certification label conform to the requirements of the standard. Examples of the labels are in **24-63**.

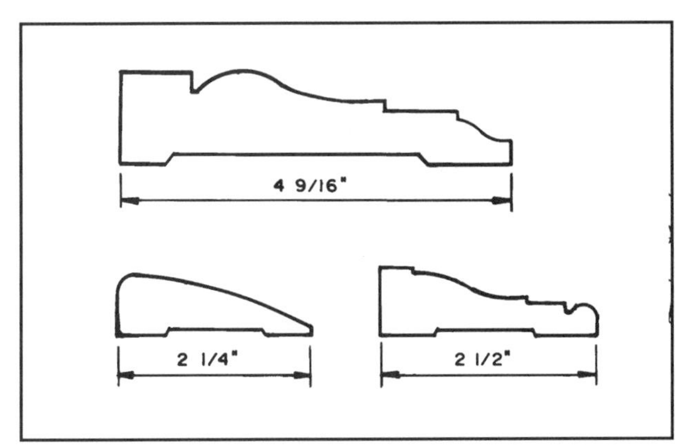

24-64 A few of the many interior casing designs available.

STYLES OF INTERIOR WINDOW TRIM

Many designs of interior casing are available **(see 24-64)**. The decision as to the style to use is made by the designer and owner.

Possibly the most commonly used style is some form of mitered frame. The window can be framed with the casing on all four sides, or on three sides with a stool and apron at the bottom **(see 24-65)**.

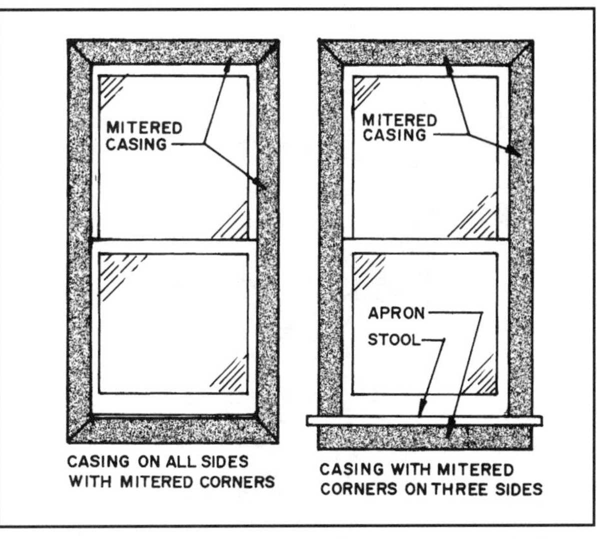

24-65 Two ways to install mitered corner interior window casings.

Another style uses a rectangular casing and joins the head in a butt joint. A variation of this is made by adding various moldings. A few examples are shown in **24-66**. One type has a flat head casing extending beyond the side casing. Another type uses some form of molding to enclose the trim. Corner blocks are typically used when the designer is reproducing a period style of traditional house. The corner blocks are available from various manufacturers. Window casing and door casing usually follow the same design.

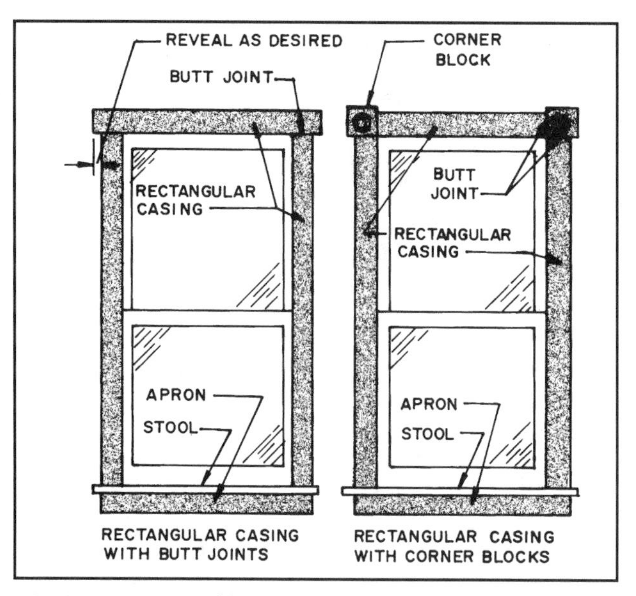

24-66 Two types of butted-corner interior window casings.

PREPARING TO INSTALL THE CASING

The interior finish carpenter will most likely not be the one to install the windows. Therefore, before starting to cut and fit trim, it is wise to check the window frame to see that it is plumb, level, and square. This is the time to make corrections. Check to see if the windows open and close smoothly.

If the exterior wall is built using 2 × 6 studs, the window frame will fall short of extending over the drywall. It is necessary to install **jamb extensions**. Jamb extensions are wood strips added to the frame to bring it flush with the drywall **(see 24-67)**. Window manufacturers will supply jamb extensions that are to be glued into grooves in the edge of the jamb. They can be firmed up by driving a few finishing nails. Jamb extensions can also be made on the job. White pine is the recommended wood, because most windows are made from that material. If the extension used is wider than the wall, the width can be marked by drawing a line where it touches the drywall **(see 24-68)**.

The jamb extension is usually glued to the window frame. Narrow extensions can be held by nailing or screwing through them into the frame. Wider extensions may require blocking between them and the trimmer stud, and they are nailed through the extension to the stud. Careful work is required to get a good joint.

Before the casing is installed, **insulation** should be inserted into the space between the frame and the studs and header. Loose insulation can be inserted, but it must be kept loose and fluffy. Expanding foam insulation (which looks like shaving cream) is available in small pressurized cans; it is good for filling small cracks. Since it expands greatly, only a very small amount is required.

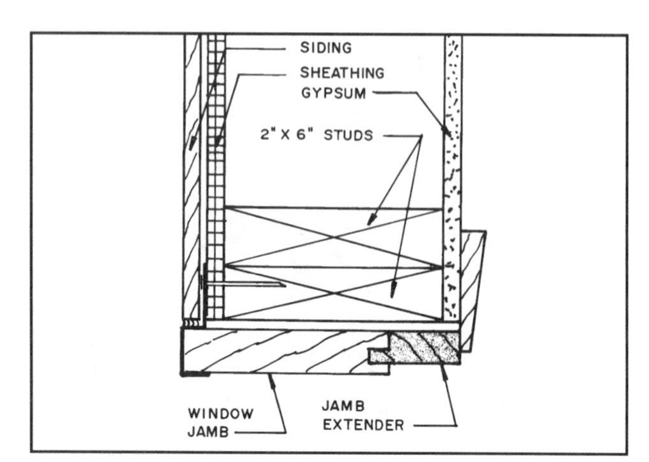

24-67 Window jambs are made wider by adding a jamb extender.

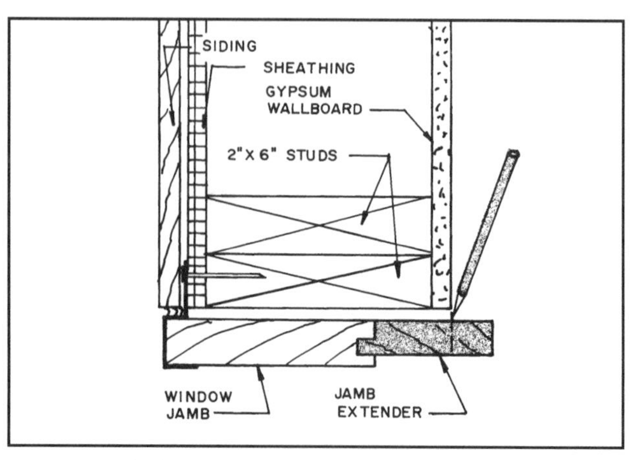

24-68 Mark the jamb extender and cut it so that it is flush with the finished interior wall.

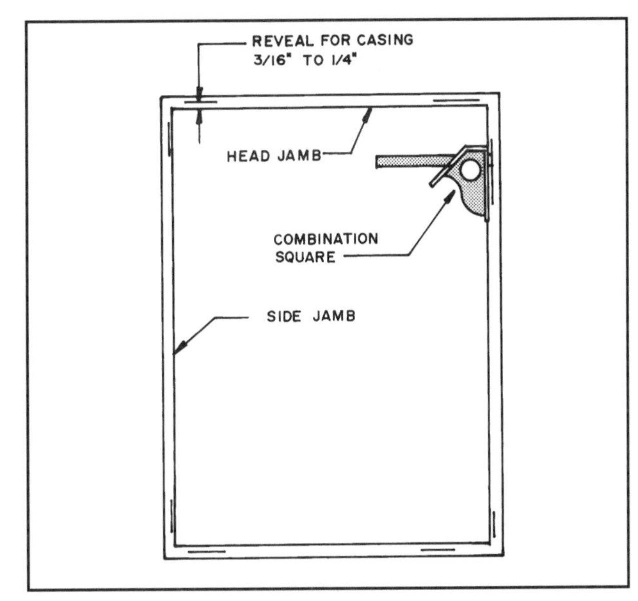

24-69 Mark the reveal on the casing. A combination square does a good job.

INSTALLING A WRAPAROUND MITERED CASING

A **reveal** is the amount the casing is set back from the edge of the frame, typically 3⁄16 or ¼ inch (5 to 6mm). Begin by marking the reveal on the corners of the casing. An easy way to do this is to set a combination square with the blade extending out the amount of the reveal. Make a pencil mark on each corner **(see 24-69)**.

A wraparound miter casing has two lengths of casing. The length of each can be measured on the short or long side of the casing. In **24-70** the lengths are taken to the short side of the casing. This includes the distance inside the jambs plus twice the reveal. When cutting the miters, place the edge with the marks next to the fence **(see 24-71)**.

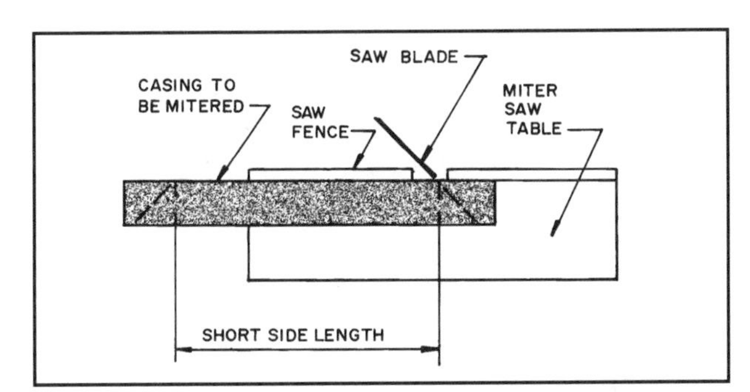

24-71 To cut the casing, place the length marks next to the miter saw fence. This shows cutting to length when the short-side measurement is used.

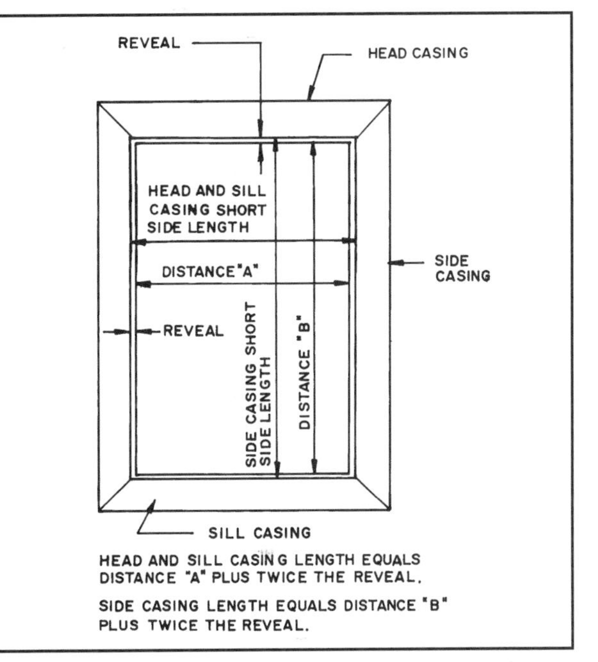

24-70 The length of the casing can be found on the short side by measuring the distance between the jambs and adding twice the size of the reveal.

If you prefer to cut to the long side of the casing, the length will be equal to the width inside the jambs, plus twice the reveal, plus twice the width of the casing material, as shown in **24-72**.

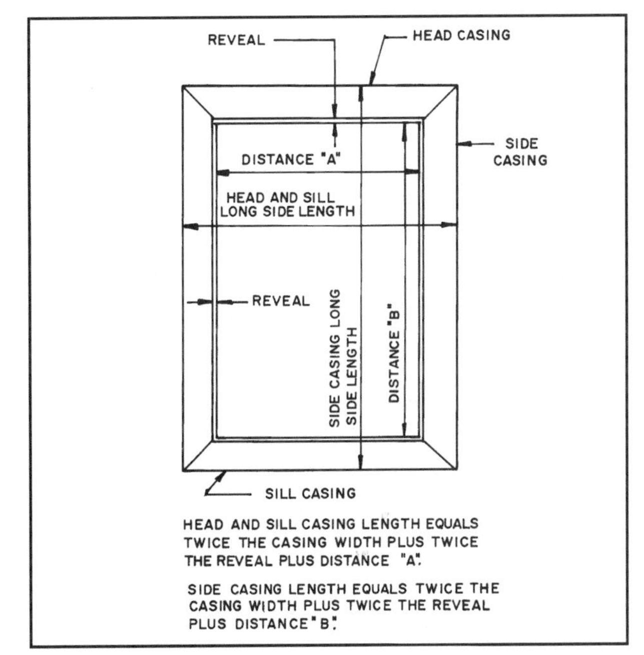

24-72 The length of the casing can be found on the long side by measuring the distance between the jambs and adding twice the width of the reveal and twice the width of the casing material.

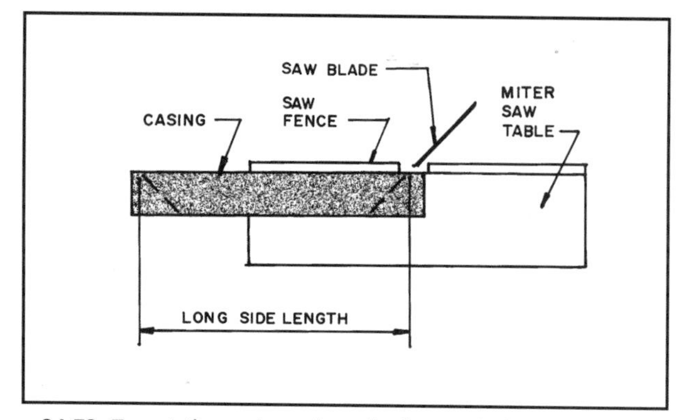

24-73 To cut the casing, place the long-side length marks next to the miter saw fence when using the long-side measurements.

The long side of the casing is placed on the miter saw as shown in **24-73**.

Interior finish carpenters approach the installation of mitered casings in different ways. Following is one suggested procedure. Start the installation by nailing the head casing to the jamb edge with 3d or 4d finishing nails. Do not set the nails firmly, because it may be necessary to adjust the casing to get a tightly closed miter. Be certain it fits flat to the surface of the drywall. If it does not, it may be necessary to file off some of the drywall or plane a little off the jamb. Be certain the casing lines up with the reveal marks; actually, it should just cover the marks **(see 24-74)**.

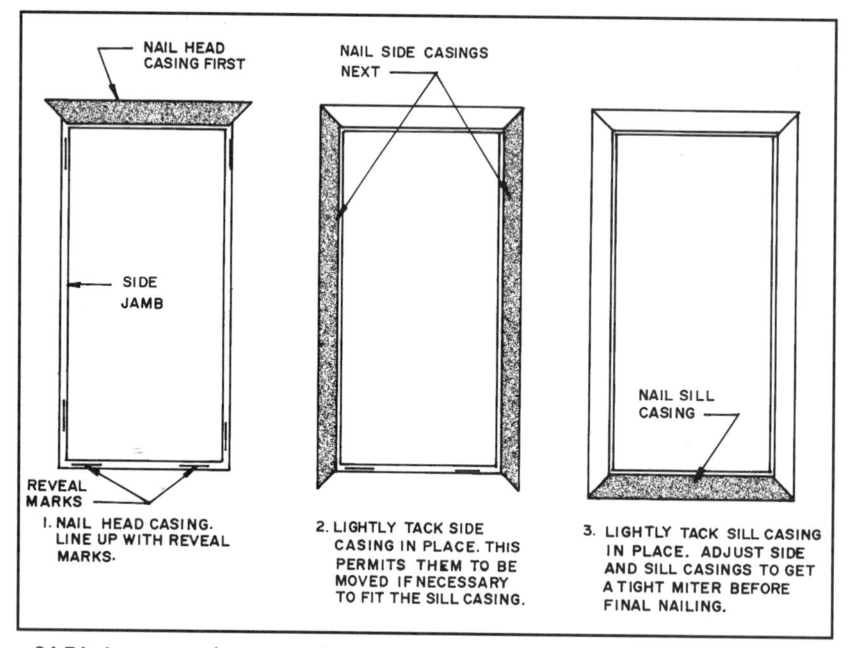

24-74 A suggested sequence for installing wraparound mitered casing.

Now install the side casings, being certain that they line up with the reveal marks and that the miters close. If this is the case, nail them to the frame. Check again to be certain the miters are closed after nailing. If the miters do not close, adjust them as described in the next paragraph. Finally, cut and install the sill casing. You can measure the actual length required, or cut a miter on one end, place the stock across the bottom, and mark the location of the other end. Cut and install the bottom casing. If all miters are closed, finish nailing by placing nails about 8 to 10 inches (approx. 20 to 25cm) apart along the edge on the frame. Then, using 6d or 8d finishing nails, nail the thick side of the casing to the header or stud. Suggested nailing patterns are shown in **24-75**.

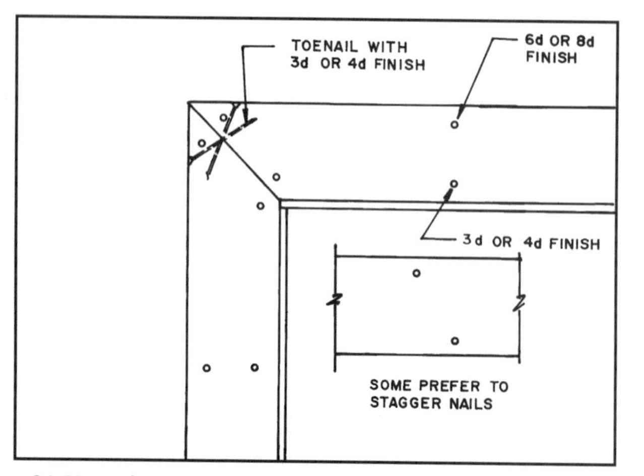

24-75 Nailing mitered casing.

The miters will not always fit tightly on the first try, and the casing will have to be removed and the miter adjusted. In **24-76** are the types of corrections that will need to be made. The top drawing shows a miter that has no gaps and is square. This is how all miters **should** appear, but, because of variances in the wall surface or the window frame, some adjustments to the miter often need to be made. The casing must always keep the reveal uniform on all sides of the window.

The second drawing in **24-76** shows a miter with the toe touching, but open at the heel. The third shows the miter touching at the heel and open at the toe. In these drawings the amount of gap is greatly exaggerated; in actual practice the

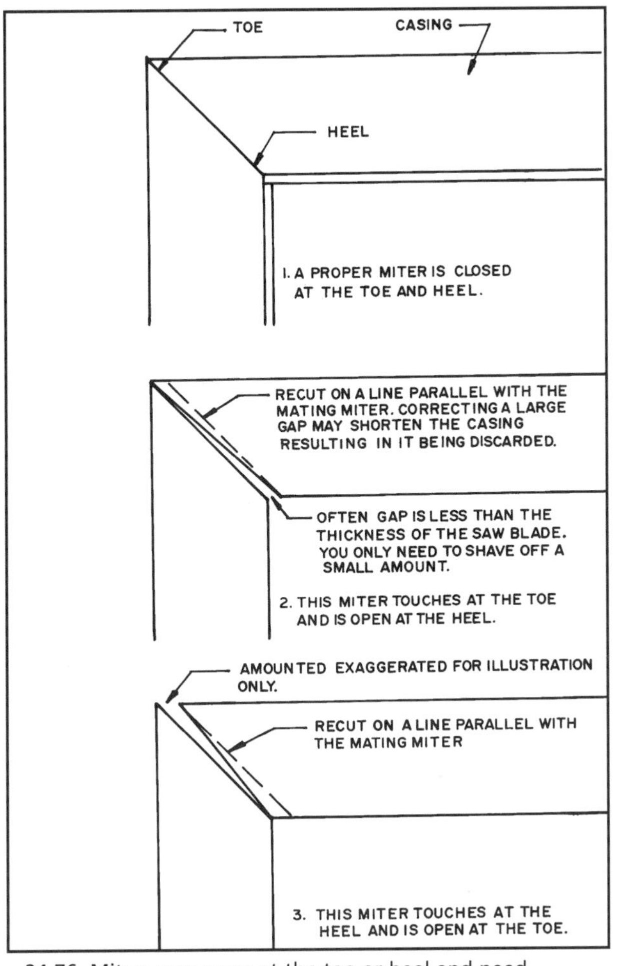

24-76 Miters may open at the toe or heel and need adjustment so they close tightly.

gap will be much smaller. Usually the gap is in the range of 1/64 to 1/32 inch (0.4 to 0.8mm).

CLOSING A GAP IN A MITER

The gap in a miter may be closed by removing wood from one side of the miter. This may be done several ways. One way is to place the casing on the miter saw, place a small wedge between the casing and the fence, and make a very fine cut across the miter. All you will get will be some sawdust off of the end to be lowered **(see 24-77)**. The amount to remove is judged by examining the joint and observing the size of the gap. Usually a thin piece of cardboard gives enough change in the angle of the miter to close the joint. Check the joint after cutting, and remove more if needed.

Another way to lightly trim a miter is to use a block plane, as shown in **24-78**. Plane "down the slope." Slant the plane a little across the surface, and move it with a slicing motion.

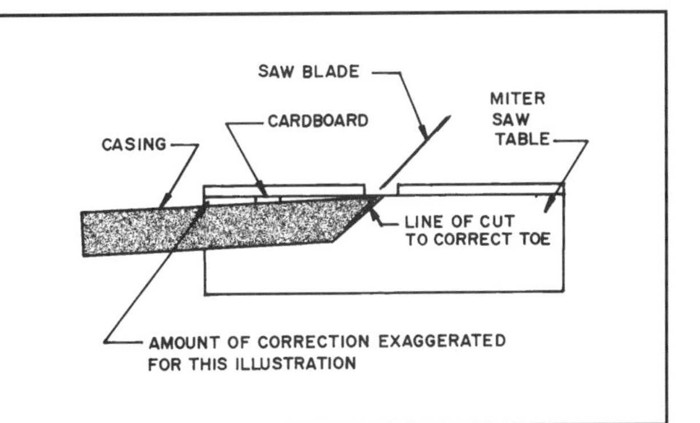

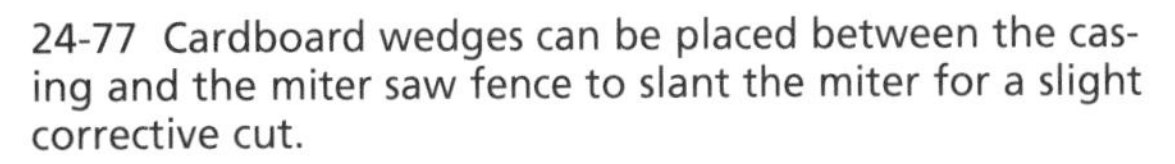
24-77 Cardboard wedges can be placed between the casing and the miter saw fence to slant the miter for a slight corrective cut.

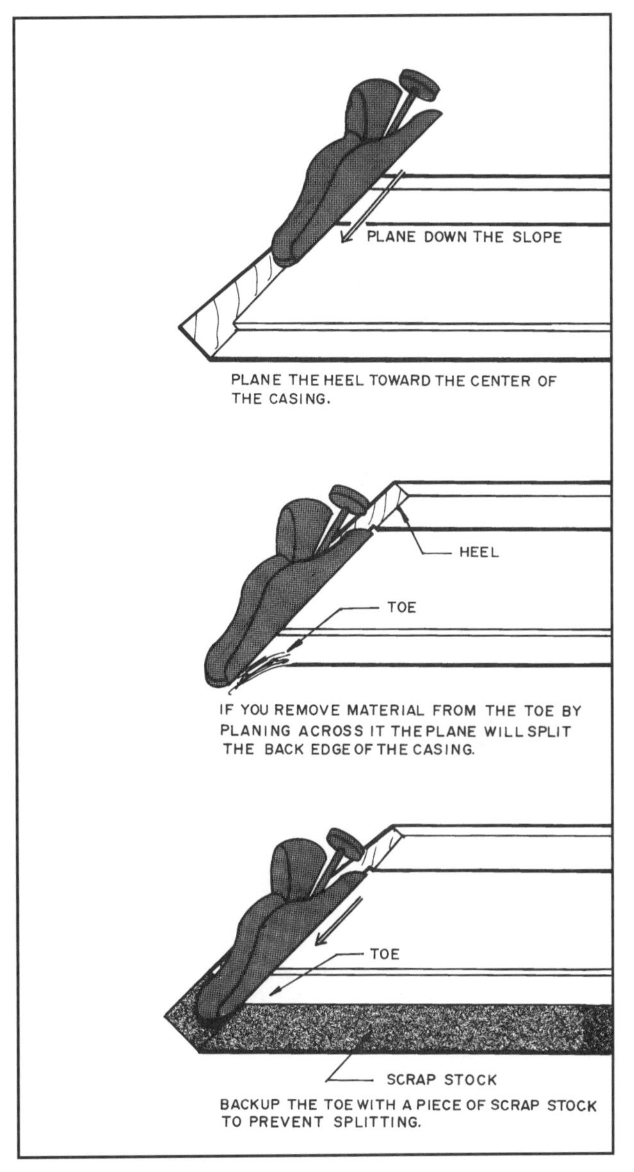

24-78 A miter can be adjusted by carefully removing stock with a block plane.

Take very light cuts, and be certain the plane is very sharp. Caution must be exercised to keep the surface square with the front of the casing. Never slant the surface to the front; however, it can be slanted a little to the back surface. This will actually help close the miter. Often a miter will have parallel edges, but will not close. This is an indication that the surfaces of the miter slant towards the front. Remove some wood off the back edge until the miter closes, as shown as **24-79**.

Some finish carpenters use files to correct miters. While files do remove material, it is difficult to keep a surface flat or straight. Filing tends to round the surface, which, if it occurs, will keep the miter from closing. Wood chisels and multiblade forming tools can also be used.

INSTALLING CASING WITH A STOOL & APRON

A butted casing with a **stool** and **apron** is shown in **24-80**. Often the head casing and side casings are the same width. In this instance, the head casing usually overhangs the side casing so that its length is equal to the length of the stool. This is typically ½ to ¾ inch (12 to 18mm). However, the head casing may be wider and have various moldings applied to it. Typical stool profiles are shown in **24-81**. They are available as stock material from building materials suppliers. In some cases the window manufacturers have stool material for their specific window units available. A stool with an angled bottom surface is used on windows made with a sloping sill **(see 24-82)**.

Fitting the Stool

First cut the stool to its finish length. This includes the width between the window jambs, plus two times the reveal, plus two times the width of the casing, plus two times the length of the horn. Check the distance between them to verify it is the same as the distance between the side jambs. Then measure the window unit to see how deep a notch is needed to form each horn, and lay this out on the stool. Cut the horn notches, and fit the stool to the window frame. Trim and adjust as necessary to get the required fit, as shown in **24-83**. Be certain the stool is level. If it is not, it may be necessary to shim it.

If the window needs a jamb extension, some carpenters prefer to install the stool first and then add the jamb extension. Others prefer to add the jamb extension and cut the stool to it **(see 24-84)**.

After you are satisfied with the fit of the stool, consider shaping the ends of the horn. They can be left square, as cut, if you sand out the saw marks. Some finish carpenters prefer to round the horn or shape a profile on it, as shown in **24-85**.

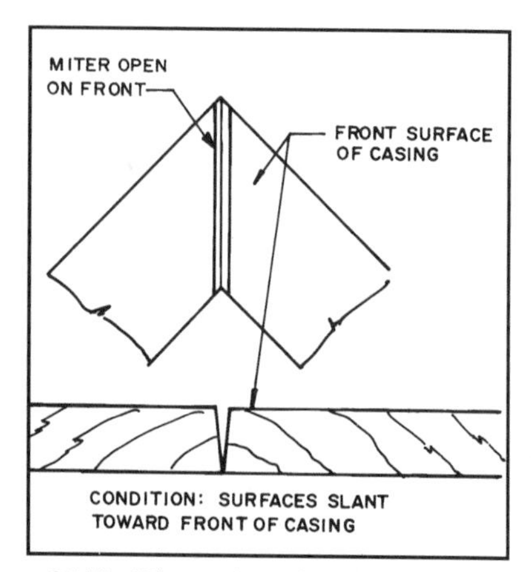

24-79 When mitered surfaces slope towards the front of the casing the miter will be open even if it is cut clearly on a 45-degree angle. Trim the back edge to close the gap.

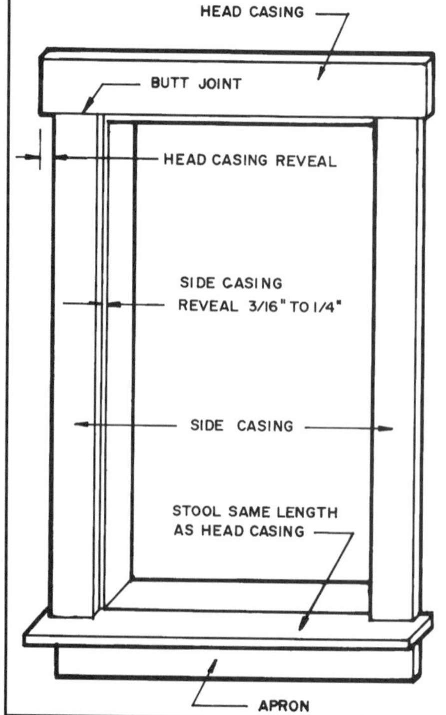

24-80 A square-edged butted casing with a stool and apron.

24-81 Typical stock stools.

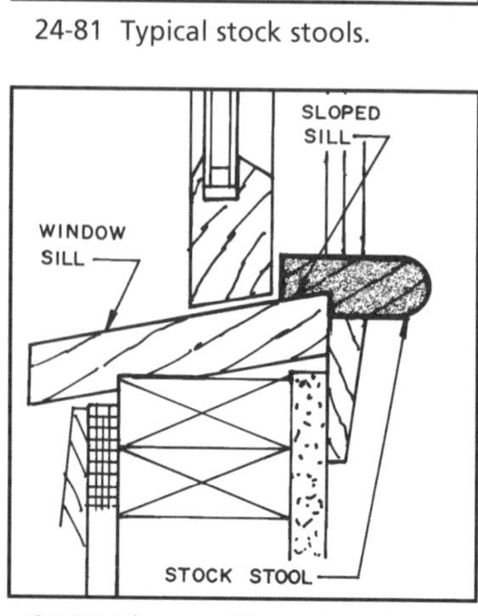

24-82 The stool is machined to rest upon the windowsill.

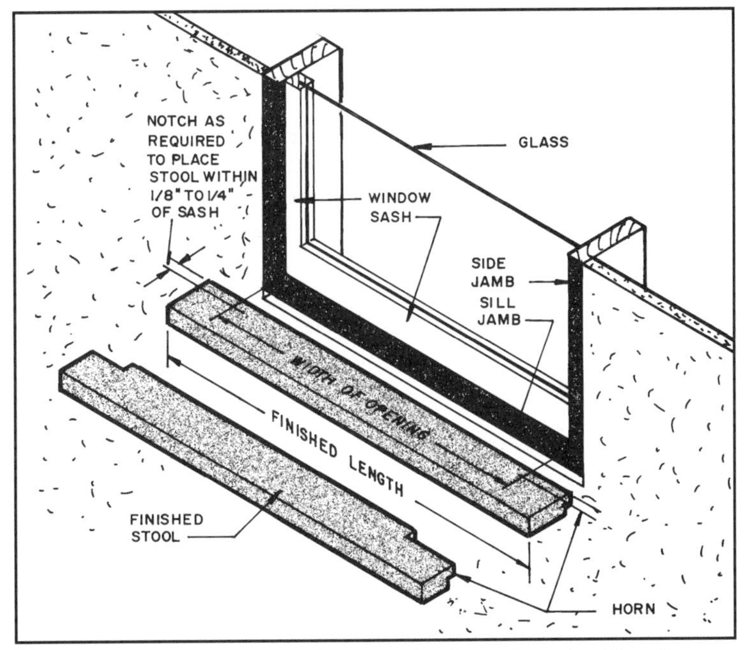

24-83 Notch the stool at the horns so that it overlaps the sill and butts the window sash.

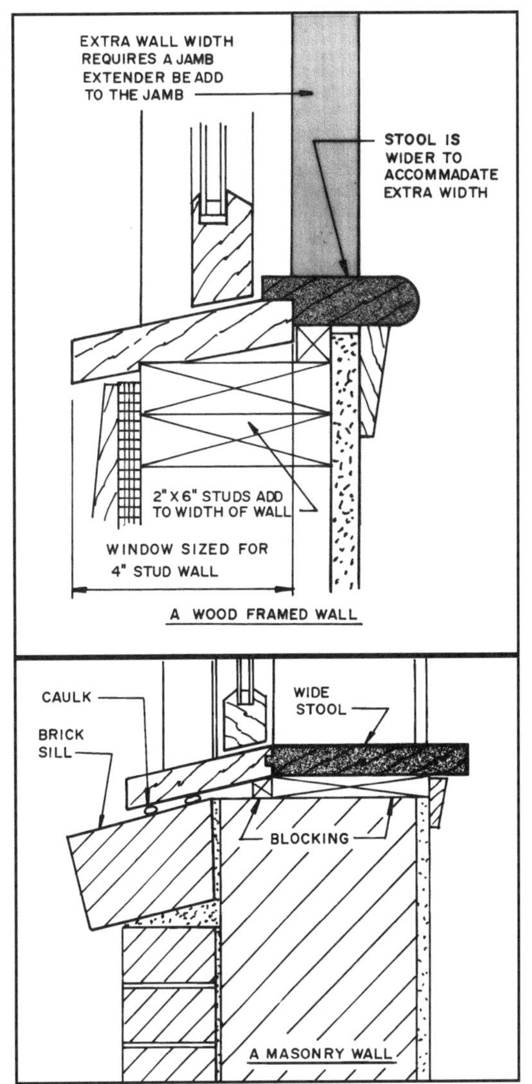

24-84 When the wall width requires a jamb extender because it is extra wide the stool must also be wider. This can occur with wood-framed and masonry exterior walls.

Now nail the stool in place. Nail through the stool into the rough sill (if possible) using 8d finishing nails. After the apron is installed, nail the stool to it with 6d finishing nails **(see 24-86)**. Since the actual design of window frames varies, the stool may have to be nailed into the sill below it. On some stock windows, it is not possible to install a stool.

Installing the Apron

The apron is usually made from the same material used to trim the window. It is placed below the stool and covers any opening between the sill and the drywall. It is placed with the thick edge up against the stool **(see 24-86)**.

The length of the apron is usually made the same as the overall width from one side of the window casing to the other. This permits the stool horn to extend beyond the apron.

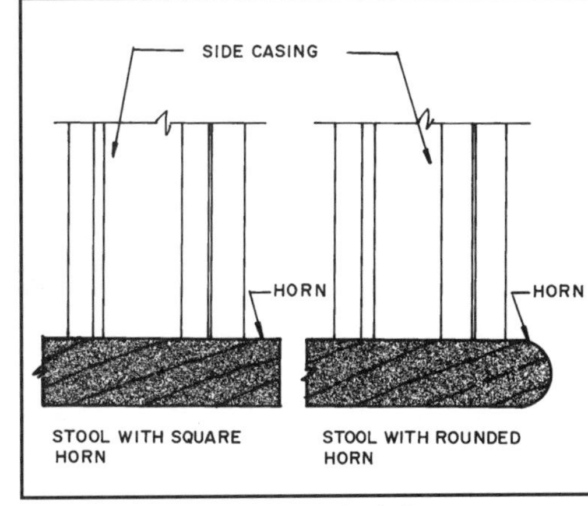

24-85 The stool horn may be left square, rounded, or shaped to another profile.

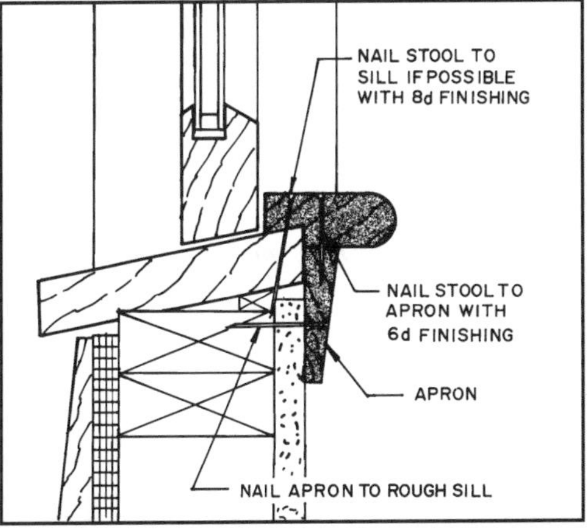

24-86 How to nail the stool and apron to the window and wall.

The ends of the apron may be finished in several ways, as shown in **24-87**. The easiest is to cut them square. Another is to cut them square and then, with a copying saw, cut the profile of the face design around the end. Another is to miter the ends and glue a return.

Installing the Casings

The casing with mitered corners is installed in the same manner described for installing a wraparound mitered casing. The only difference is that after the head casing is installed, the side casing is mitered on one end and cut square on the other, so that it butts against the stool **(see 24-88)**.

The butted window casing is installed by first cutting the side casings to length. Both ends are cut square. The top end should touch the reveal marks. Be certain that the end butting the stool has a tight, closed joint when the casing is in line with the reveal marks. If not, you will have to trim the butt end as described for correcting miters. Once the side casings are correct, nail in place temporarily. Now cut the head casing to length with square ends. It should be the same length as the stool. This provides the same overhang of the casing at the head as occurs at the stool. Locate the head casing with equal extensions beyond the casing, and nail in place. A finished installation is shown earlier in **24-80**.

Corner Blocks

Butted casings on some traditional houses use **corner blocks**. The corner blocks are available from building materials suppliers. They are typically wider than the casing, and overhang the casing the same amount as the horn on the sill.

Begin by installing the stool and apron. Then install the side casings as described for butted casings. Next tack the

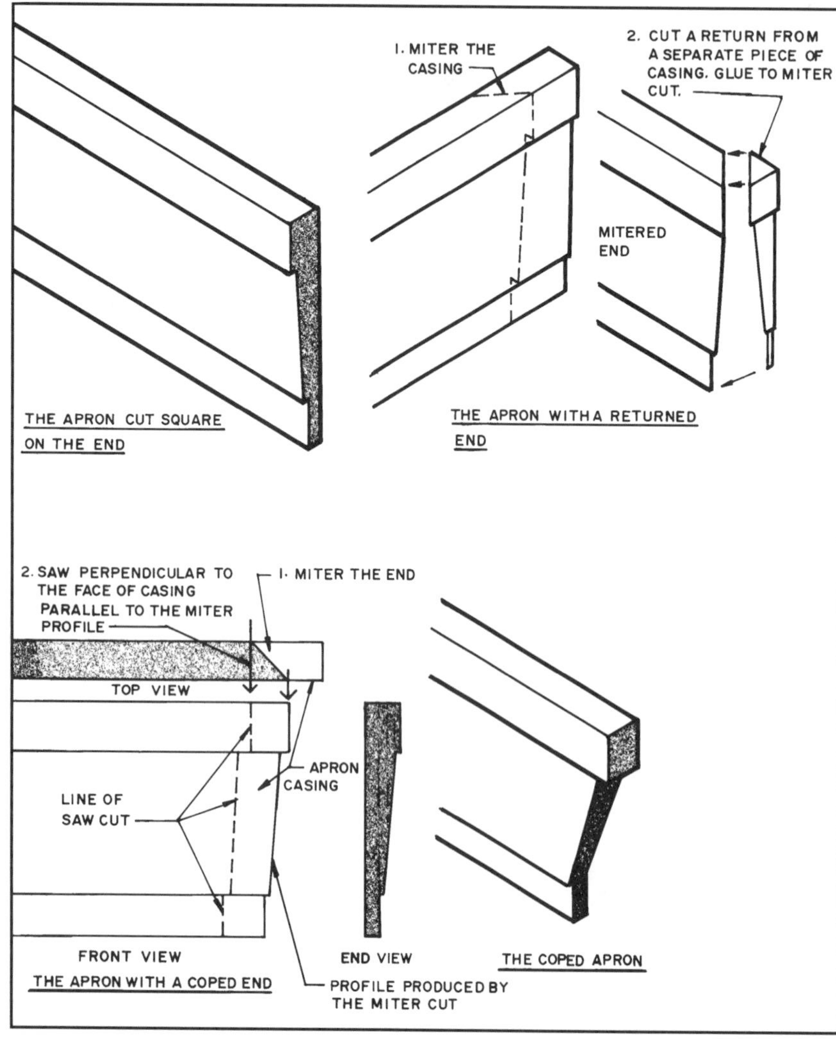

24-87 Various ways to finish the ends of the apron.

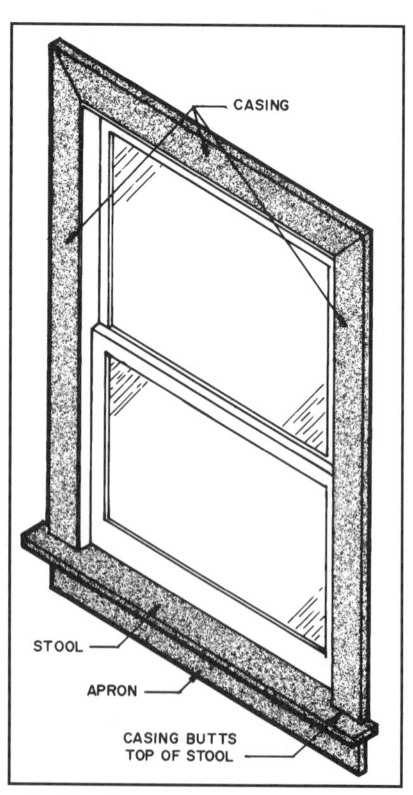

24-88 When mitered casing is installed and a stool is used, the side casings butt the stool.

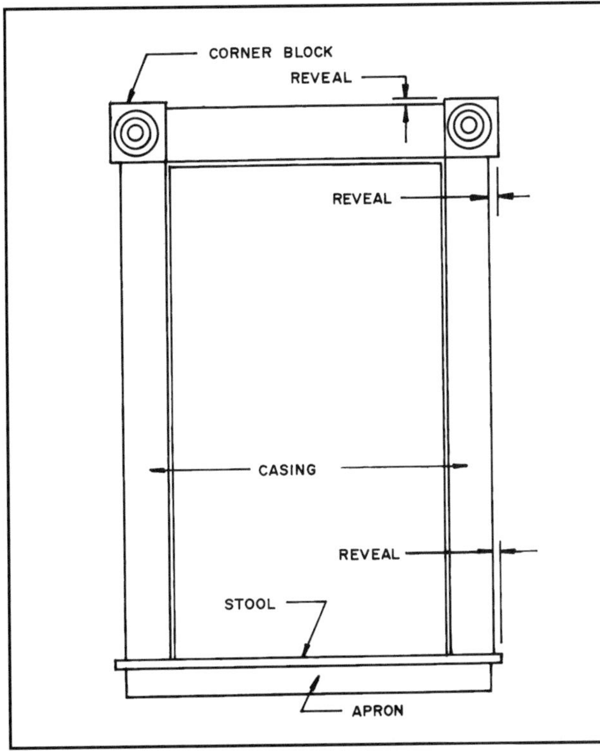

24-89 This square casing uses corner blocks.

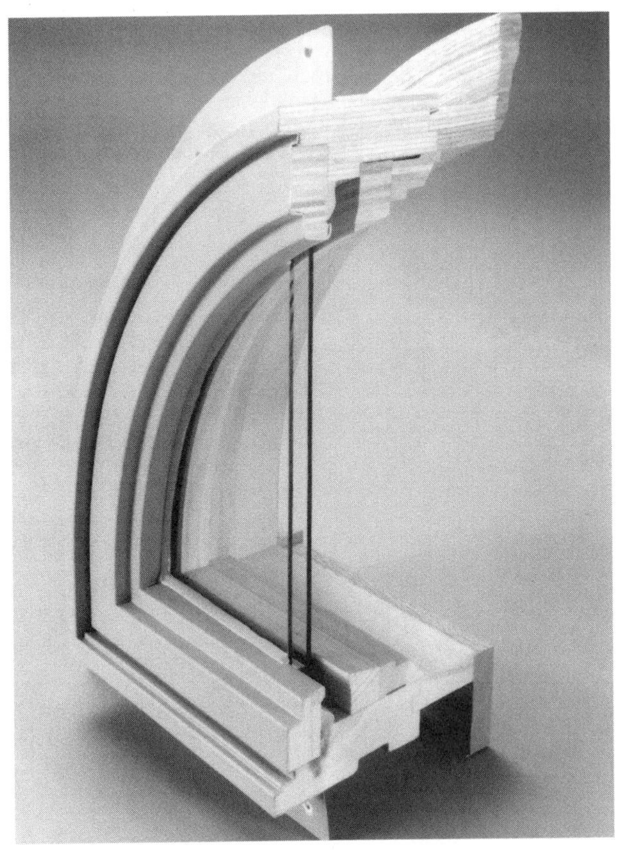

24-90 A section through a circle top window showing its construction and the curved finished interior and exterior casing provided. *(Courtesy Andersen Windows, Inc.)*

corner blocks on top of each side casing, carefully measuring the overhang, which, as mentioned, is usually the same as the horn of the stool. Measure the distance between the corner blocks and cut the head casing. It must butt tightly against each corner block. Usually some minor adjustments are needed to get the tight joints desired. In all cases keep the reveals uniform **(see 24-89)**.

CIRCULAR & ELLIPTICAL HEAD CASING

The manufacturers of circle top windows also supply the necessary casing cut to fit the units **(see 24-90)**. Usually several casing styles are available. They also supply **plinth blocks**, which the curved casing and the horizontal casing below the window join in a butt joint **(see 24-91)**.

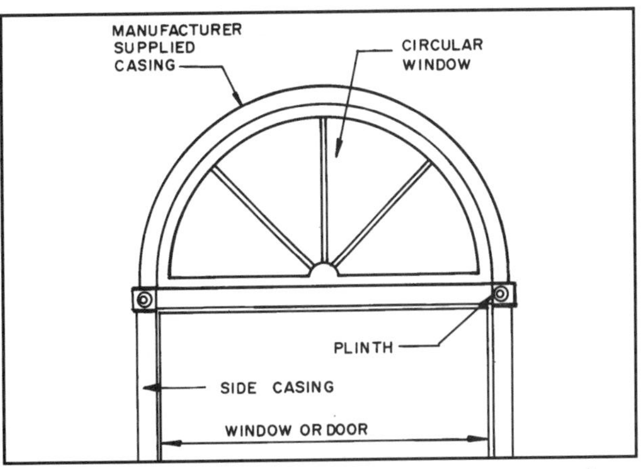

24-91 The manufacturer of the circle top window supplies the casing.

SPECIAL HEAD CASING DESIGNS

The designer may copy an ornate traditional head casing or prepare an original design. Each situation may have particular requirements so that detailed drawings are necessary to allow the finish carpenter to build exactly what is specified. Special casings could also be built at a cabinetmaking or millwork shop, and delivered to the job site assembled. Details for installation are the same as described in the earlier parts of this chapter.

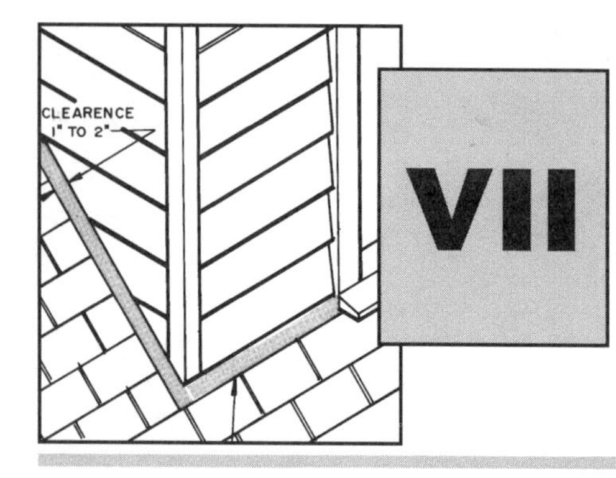

VII FINISHING THE EXTERIOR

CHAPTER 25

Finishing the Exterior Walls

The choice of exterior siding is influenced by many things, such as the style and design of the house. For example, some architectural styles traditionally use wood siding while others use brick, stone, or stucco. Materials other than wood, such as plastic or fiber-cement, are used to produce siding that looks like wood but has different properties. These are chosen when consideration of long-term maintenance is concerned. This relates to resistance to warp, rot, mold, and attack by insects, as well as the need to regularly paint the siding.

The exterior siding is installed over the wall sheathing. In addition to protecting the walls and interior from the weather it serves as the "face" of the building to the viewer. In some cases it strengthens the wall, but this is not a major consideration.

TYPES OF SIDING

There is a wide range of exterior siding materials available. In addition to the traditional **solid wood, brick, stone,** and **Portland cement stucco**, a number of specially formulated materials are used.

Aluminum and **vinyl** siding formed in long strips like boards and on panels have been used a long time. In addition, a number of other plastic and metal products in the form of individual **shingles** are available.

Another product is formed by bonding wood chips with resin, forming a **hardboard** siding.

A **wood fiber-cement** mixture is used to produce a durable siding in the form of boards.

Synthetic stucco, an acrylic-based cementitious material, is widely used instead of Portland cement stucco. It is lighter and has excellent insulation value.

Portland cement stucco is a combination of Portland cement-based cementitious materials and aggregate mixed with water. It provides a strong, durable exterior finish.

Wood siding is in the form of boards machined to a variety of shapes, individual shingles, and various types of plywood siding products.

Steel siding panels are generally used on commercial buildings, but some types find use on residential structures. They are strong and have a durable permanent finish.

GETTING READY

Before the exterior siding is installed, the doors and windows should be in place. Review Chapters 23 and 24 for information on installing doors and windows. Adequate scaffolding must be on the job and erected properly **(see 25-1)**. Review Chapter 2 for details on using scaffolding and ladders. It is common practice to cover the sheathing with 15- or 30-pound builder's felt or one of the plastic membrane housewrap materials available. This protects the sheathing in case there is a leak in the siding or around a door or window opening. It also reduces the amount of air infiltration **(see 25-1)**. All flashing around doors, windows, and other openings should be in place. The cornice is usually finished before the siding is installed. This is especially important when siding other than wood is used. See Chapter 19 for information on cornice construction **(see 25-2)**. Check to see that the proper siding has been delivered in sufficient quantity to justify starting to work. Be certain it is stored properly so it is not damaged. Wood, vinyl, and plastic siding should be stored flat and raised above the ground with supports to keep them from bowing. They should be covered with plastic sheets to protect from the weather.

NAILS USED WITH WOOD SIDING

Several kinds of nails are used with wood siding. These are finishing, casing, sinker-head, and sinker-head ring-shank nails **(see 25-3)**. These nails can be set below the surface of the siding with a nail set. After the siding is primed, the hole is filled with putty.

These nails must be of aluminum or be galvanized. Steel nails will rust even if set and puttied.

The diamond point is the most commonly used type of nail point. If there is danger of the wood splitting, the nail can be blunted. Blunt-pointed nails are available. The threaded shank nails have greater holding power.

25-1 This scaffolding is being used to install the brick siding. Notice the horizontal guardrails are missing. The sheathing has been covered with a housewrap material to protect it from moisture and reduce air infiltration.

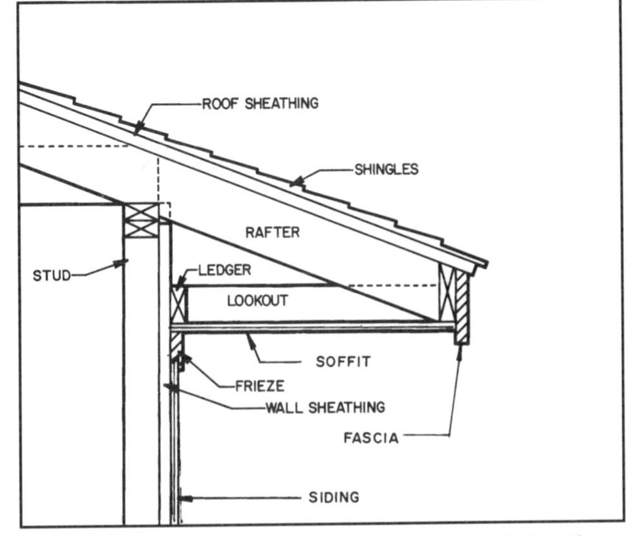

25-2 A typical horizontal cornice seals the area below the roof overhang and is butted by the wall siding.

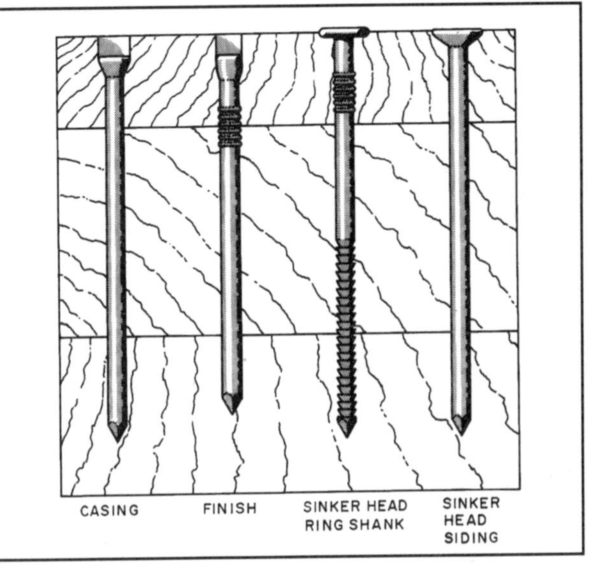

25-3 Nails used in installing exterior wood siding.

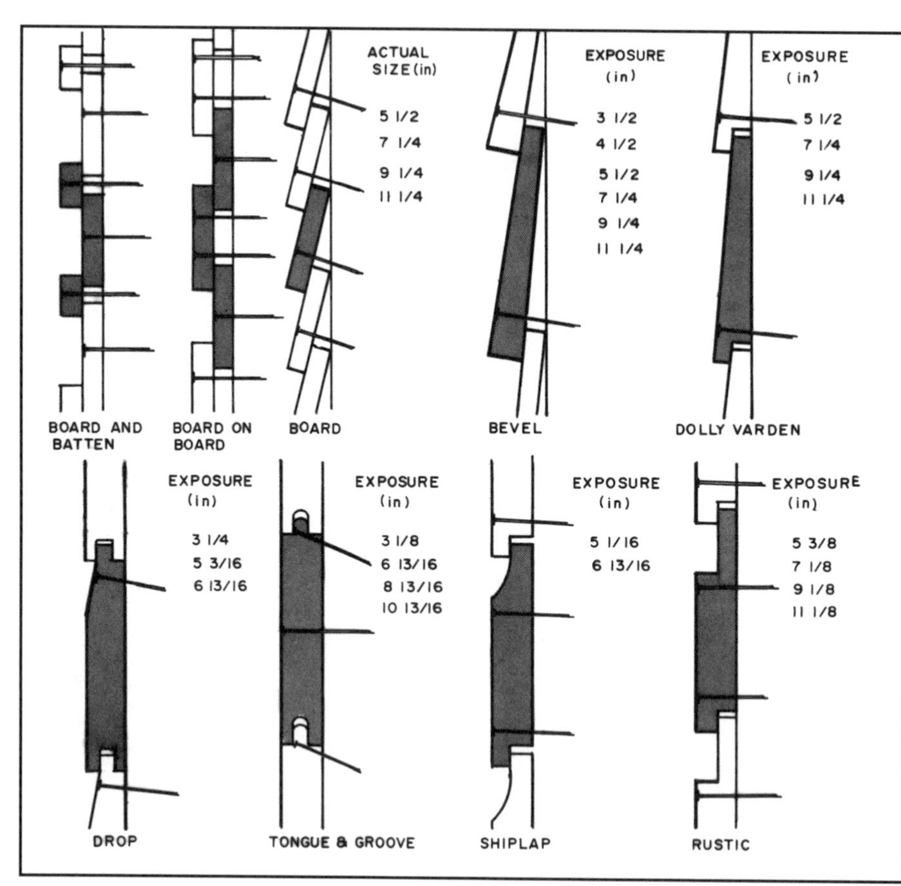

25-4 Common types of wood siding and recommended nailing techniques.

25-5 This rough-surfaced solid lap wood siding was stained to preserve the natural wood look and color.

Typical types and sizes of solid wood siding are in **25-4**. These will vary some with the different lumber manufacturing associations in various parts of the United States or Canada. Sometimes one side has a rough-sawed surface and the other a smooth-planed surface. The rough surface on bevel siding is often exposed when a stain is to be used **(see 25-5)**.

SOLID WOOD SIDING

Quality wood siding is easy to work, relatively free from warp, and holds paint well. Best-quality wood siding should be free from knots and pitch pockets, because they will tend to bleed through the paint. It should have a moisture content of 10 to 12 percent, except in dry areas, where the moisture content could be as low as eight percent. Wood siding should be stored flat and protected from the weather. It should be primed as soon as it has been installed.

The architectural drawings or the specifications should specify the specie of wood. Some species hold up better, and others are more expensive. The best species include cypress, redwood, cedar, eastern white pine, sugar pine, and western white pine. Yellow poplar, spruce, Ponderosa pine, and western hemlock are also good. Southern pine, western larch, and Douglas fir are rated as fair.

HORIZONTAL WOOD SIDING

Horizontal wood siding is available in three types—**bevel, drop,** and **boards**. It is manufactured in a variety of patterns, from solid wood, hardboard, and plywood.

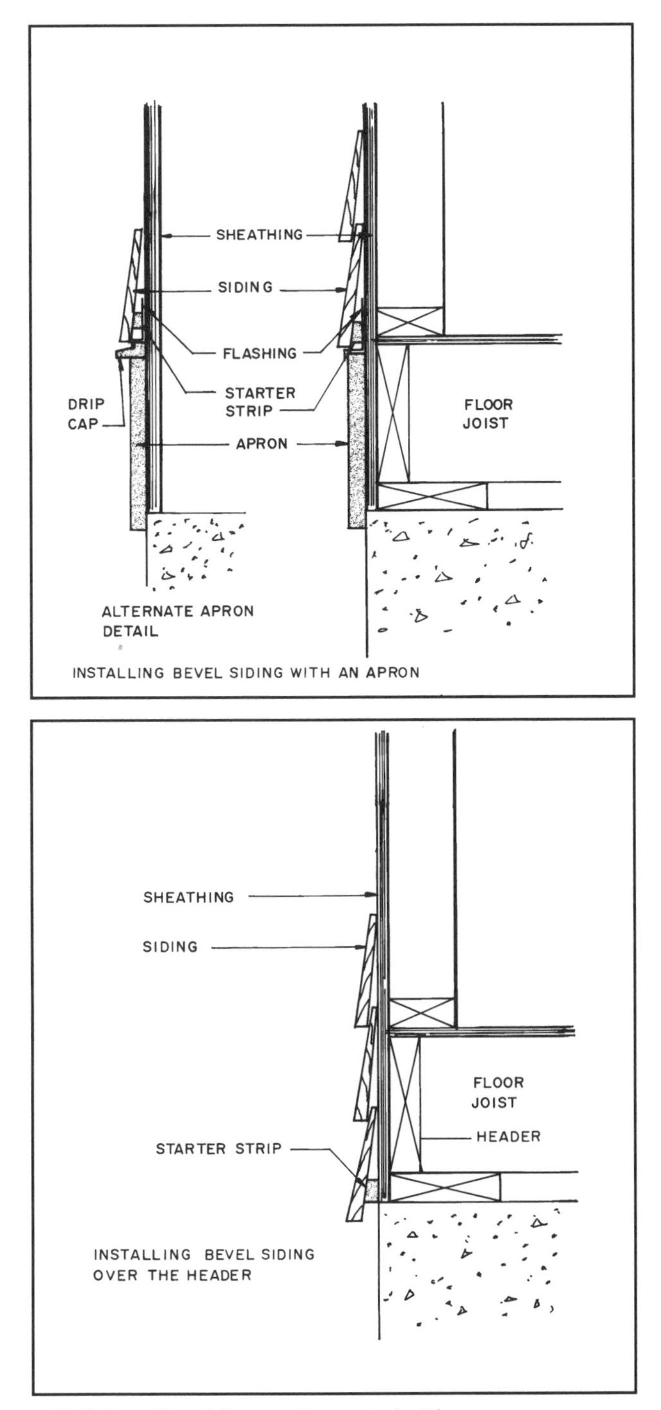

25-6 Bevel lap siding can be started with an apron or run completely over the floor joist header.

INSTALLING HORIZONTAL BEVEL SOLID WOOD SIDING

Bevel siding begins with the bottom course at the sill. There are several ways to start the bottom course **(see 25-6)**. One uses a wood **starting strip** that holds the bottom of the first course out from the foundation. Another uses an **apron board**, sometimes called **fascia**, around the sill, and sometimes has a **drip cap** on top. The choice depends on the specifications of the architect, which will appear on the elevations and wall section.

Plain bevel siding should overlap at least one inch. The exact overlap and exposed surface depend on the width of siding required and the need to space around windows. If possible the top edge of the siding should butt up against the bottom of the windowsill **(see 25-7)**. To see if this is possible, measure the distance from the windowsill to the top of the foundation, allowing some overlap of the foundation. If this distance can be divided to give one-inch overlap or more and have a course end at the bottom of the window, then this is the measurement used.

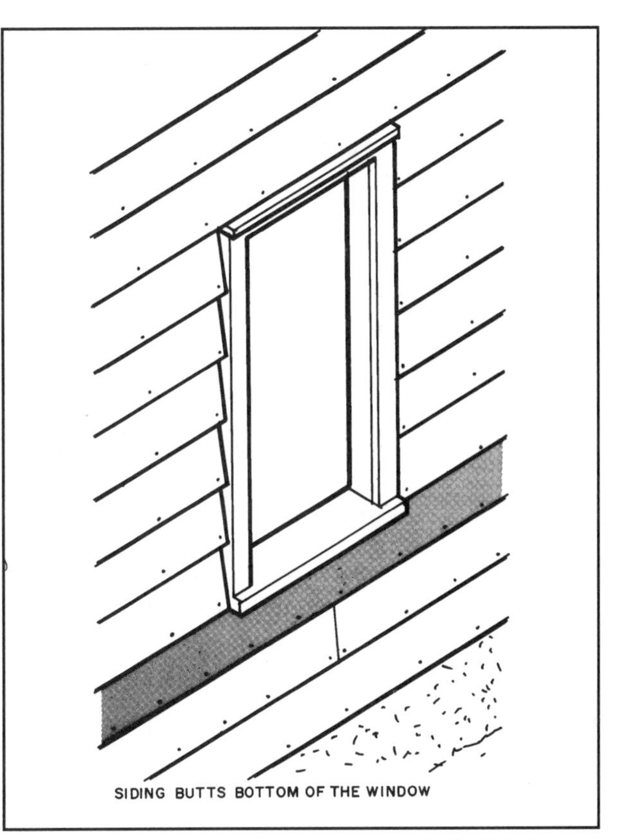

25-7 This siding layout places a course to fit flush below a window.

25-8 This siding layout requires a course be notched around the window.

25-9 A carpenter fitting notched redwood siding around the windows. *(Courtesy California Redwood Association)*

If the distance cannot be divided to give a one-inch overlap or more and it does not work out to have a course end at the bottom of the window, then the course at the window will have to be notched around the window. This should be carefully done so that a minimum of space is left **(see 25-8 and 25-9)**.

Following is an example of how siding spacing could be figured. Assume that 10¼-inch bevel siding with 9¼-inch exposure is used. The distance from one inch below the foundation to the bottom of the window is 28 inches. The length of the window from the bottom of the sill to the top of the drip cap is 48 inches. Dividing these by 9¼ inches reveals that the lower area needs three courses. The window needs five courses. The exposure on the window area is 9¹⁰⁄₁₆ inches. The exposure on the lower wall is 9⁵⁄₁₆ inches. These are both very close to the desired 9¼-inch exposure. They are different exposures, but the difference is so small that it will not be noticed. If it becomes necessary to go much below the one-inch overlap, it would be best to notch the siding **(see 25-10)**.

Many horizontal wood sidings have rabbeted edges. The amount of overlap cannot be changed much. The location of the first course below the foundation can be varied some, but this is the only possibility for adjustment.

The size of nail to use depends on the type of sheathing and the thickness of the siding. The nail must penetrate the stud 1½ inches **(see Table 25-1)**.

Table 25-1 Wood siding nailing recommendations.

Siding thickness	With plywood sheathing	With nonnail-bearing sheathing
½"	7d	2" plus thickness of sheathing
¾"	8d	2¼" plus thickness of sheathing

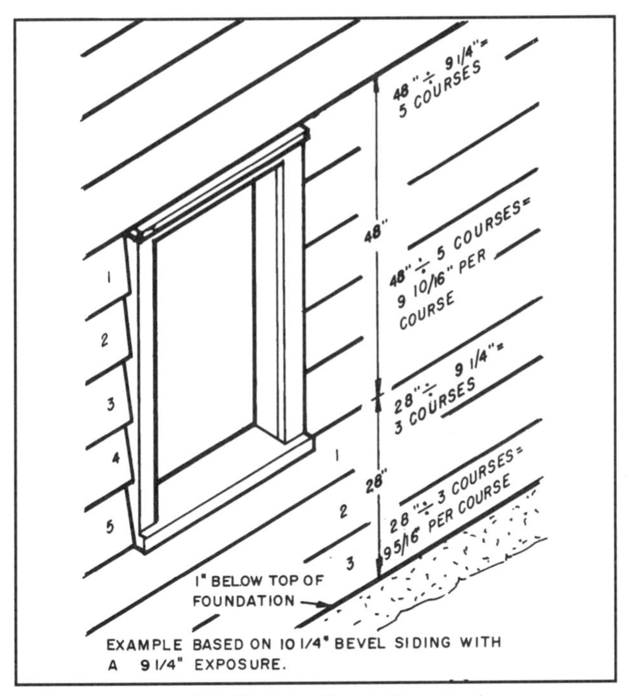

25-10 An example of how to figure the actual exposure at a window or other opening.

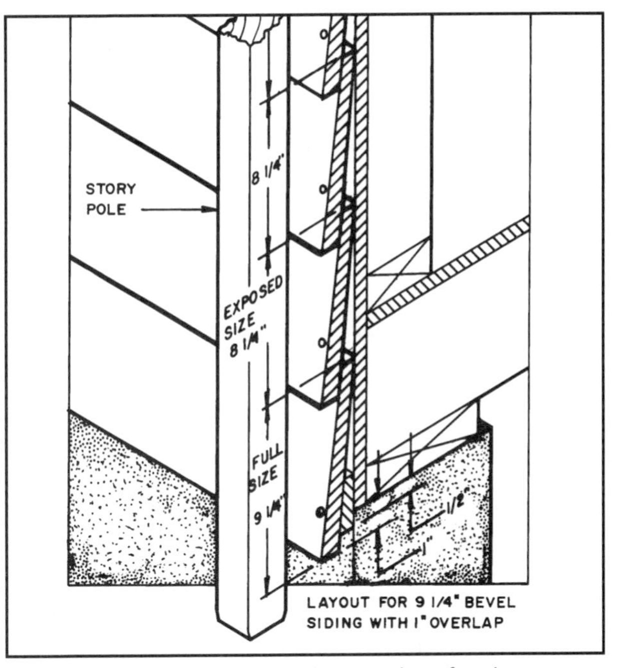

25-11 The story pole locates the top edge of each course of siding. It starts with a full course that overlaps the foundation one inch.

When nailing the siding, the nail should be placed so that it misses the bottom edge of the course below. This is necessary to allow for expansion and contraction of the siding. Examples are shown earlier in **25-4**.

After deciding upon the size of the exposed surface, mark this distance on a story pole. A **story pole** is a 1 × 2-inch wood member on which is marked the location of the top edge of each course of siding **(see 25-11)**. The first mark is the actual width of the siding. It allows a one-inch overlap on the foundation. The other marks are the exposed face of the siding. These marks are laid out on the door and window frames and the corners of the building. Check each course as it is installed **(see 25-12)**.

Now install the first course by snapping a chalkline on the sheathing to locate the top edge of the siding **(see 25-13)**. Nail the starter strip in place so it overlaps the top of the foundation about ½ inch. It should be the same thickness as the thin edge of the siding. Now nail the first course of siding to the sill.

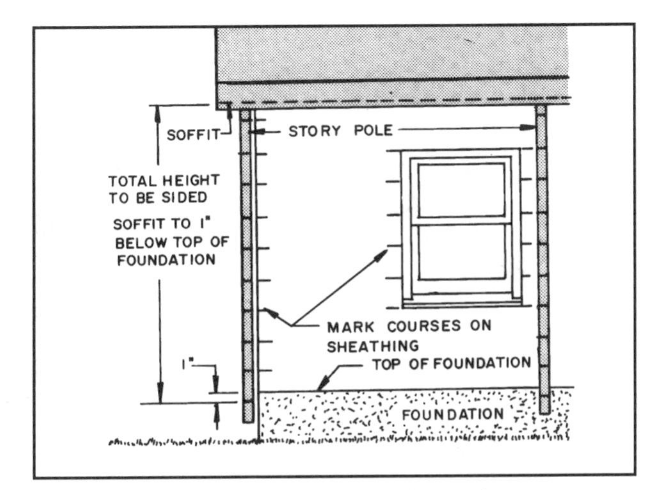

25-12 Mark the courses on the sheathing using a story pole to locate them.

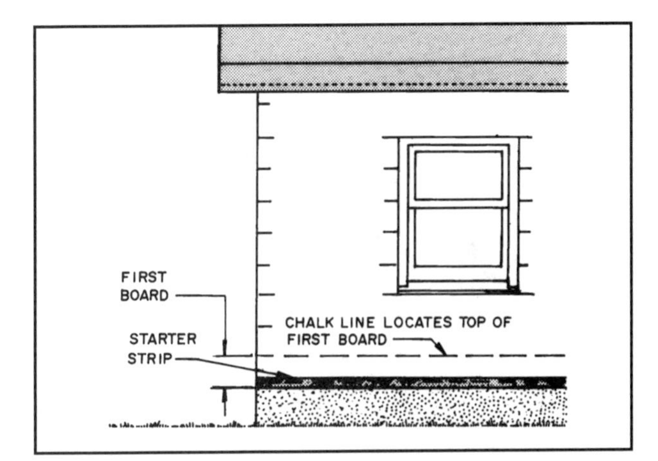

25-13 To locate the first course of siding, snap a chalkline from one mark to the other across the building. This locates the top edge of the siding.

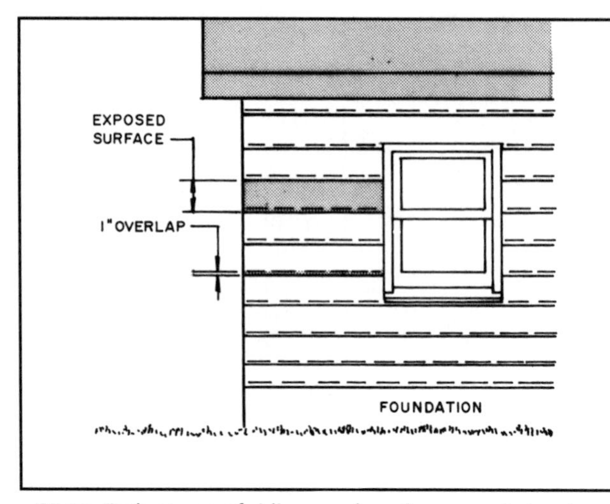

25-14 Each course of siding overlaps the course below by one inch.

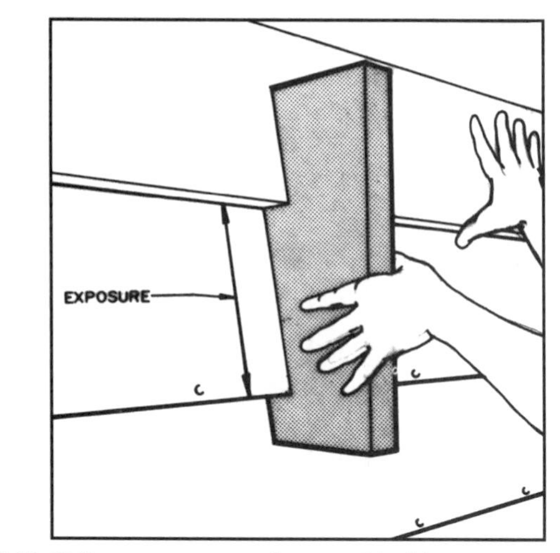

25-15 Siding exposure can be set with this carpenter-made notched device.

After the first course is in place snap a chalkline from the first mark on the story pole, locating the bottom edge of the second course **(see 25-14)**. Another way to space the courses is to make a notched board which is used to measure the exposure **(see 25-15)**.

As the courses are installed try to avoid butt joints. Use the shorter pieces between windows and between windows and corners. If a butt is necessary it should occur over a stud unless plywood or oriented strand board sheathing is used. The courses should be staggered **(see 25-16)**.

Do not fit pieces between windows or windows and corners too tightly. Leave a small space to allow for expansion on each end. This space will be caulked to prevent a leak into the wall. It is recommended that each cut end be coated with a water repellent before it is nailed in place.

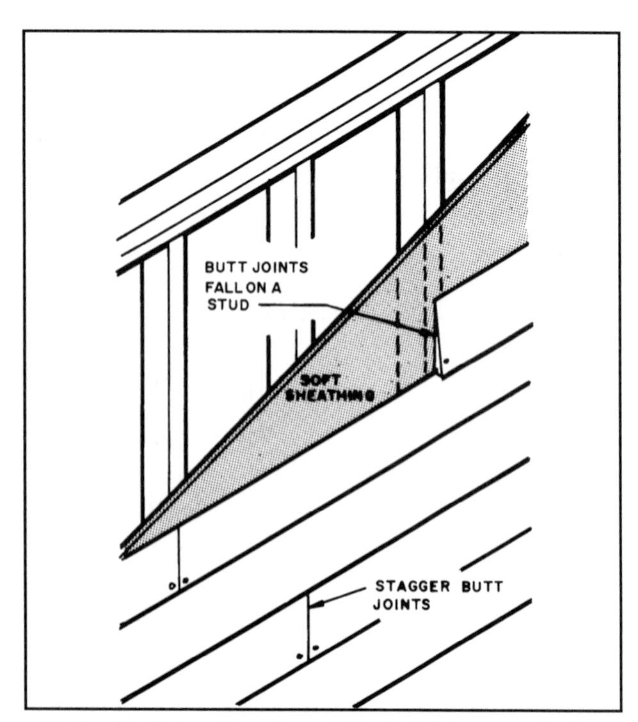

25-16 If soft sheathing is used, the butt joints must be located on a stud. All butt joints should be widely staggered.

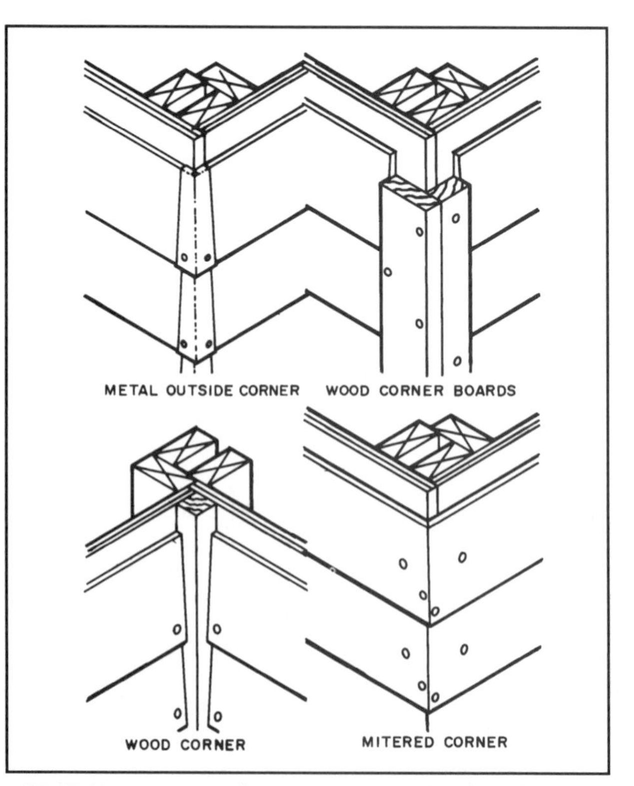

25-17 Here are several ways to construct inside and outside corners when installing wood bevel siding.

Siding that is rabbeted or tongue-and-grooved is installed in the same manner as bevel siding. The nailing patterns are shown earlier in **25-4**.

When tongue-and-groove paneling is used as horizontal siding, it is blind-nailed through the tongue. It is face-nailed in the center of each piece if it is six inches or wider.

HORIZONTAL SIDING AT THE CORNERS

In **25-17** are shown several ways to finish inside and outside corners when installing wood bevel siding. The design used depends upon the architect's design. Outside corners can have the siding mitered; however this is very difficult to cut because the siding is tapered and nailed on a slope. The mitered corners must fit tightly so moisture does not penetrate. It is advisable to coat the cut ends with a wood preservative and caulk the joint after installation.

Often, specially manufactured metal corners are used. This gives the same appearance as mitered corners. The metal corners are nailed in place as each course of siding is installed.

Corner boards are also used **(see 25-18)**. They are usually 1⅛ to 1⅜ inches thick. The width varies depending upon the instructions of the designer. Usually, a rather narrow board is used. The corner boards are butted and nailed through the sheathing into the corner stud. Notice that one board is narrower than the other. They are set in place before the siding is installed. The siding must be cut to length carefully so that it fits against the corner board. The crack between them is caulked after they are primed.

Interior corners are butted against a 1⅛- or 1⅜-inch square corner board. The board is nailed to the stud before the siding is installed. The joint between the siding and the corner board is caulked after the siding is primed.

25-18 This horizontal wood siding is butted to the corner board. Notice the short pieces were saved and used on the return to the brick wall.

When siding meets a masonry wall, a watertight joint is needed. One way to frame an exterior corner is shown in **25-19**. A corner board is nailed to the corner studs. The space between it and the masonry is caulked. It is recommended that a layer of roofing felt be nailed around the corner before the siding and masonry units are set in place.

The siding on an interior corner butts the masonry wall. The space between them is caulked **(see 25-20)**.

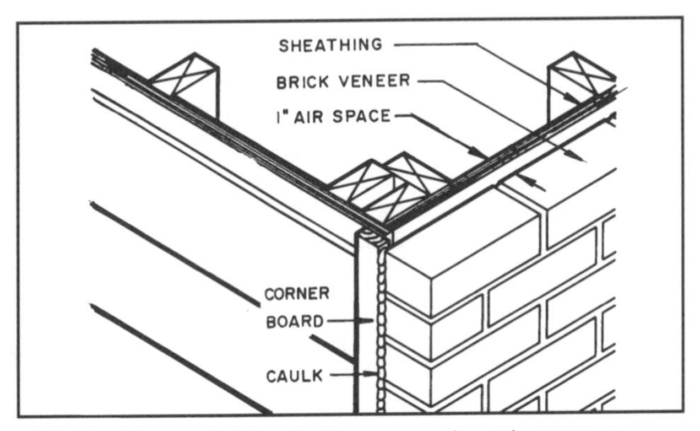

25-19 This wood siding butts a corner board next to a masonry wall. It is important to caulk the joint.

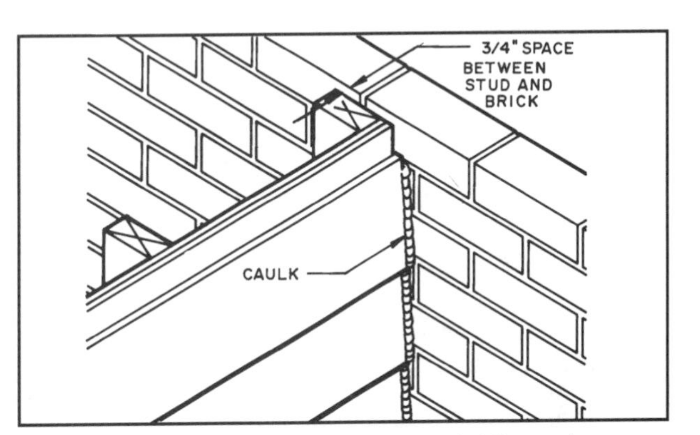

25-20 This construction allows the wood siding to butt the masonry wall. Caulking is vital to a waterproof joint.

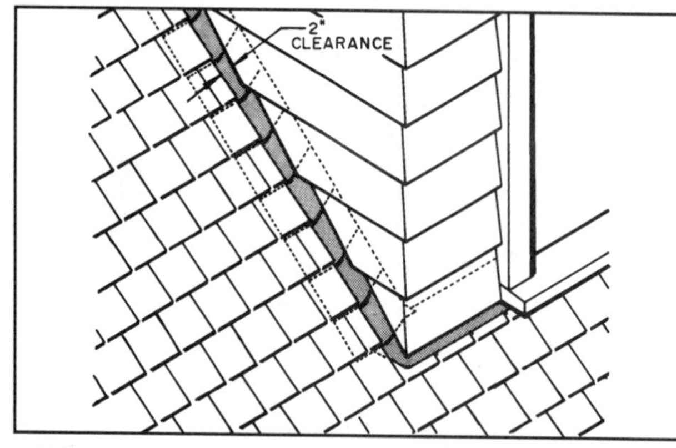

25-21 Wood siding on dormers should clear the shingles two inches. The cut end should be coated with a water repellent.

When siding ends against a roof surface, the edge is cut parallel with the angle of the roof **(see 25-21)**. It should clear the roof surface about two inches. The siding is on top of the flashing. The cut end should be coated with a water-repellent preservative.

Siding meets window and door frames as illustrated later in **25-39**.

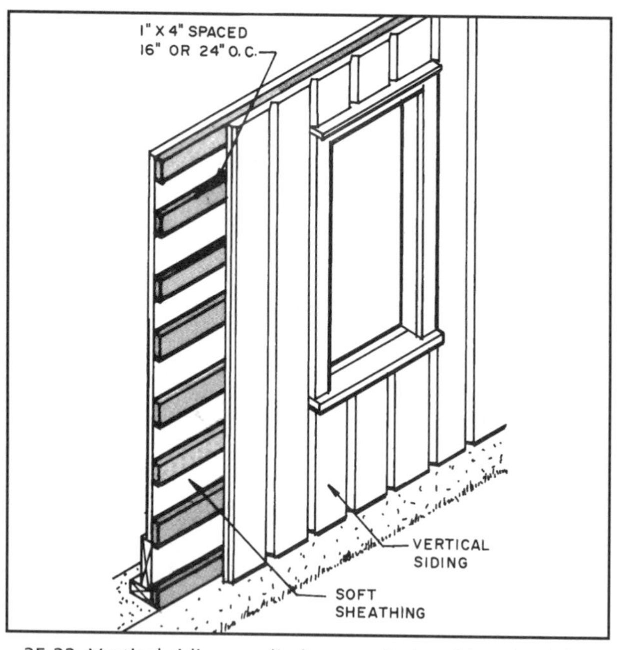

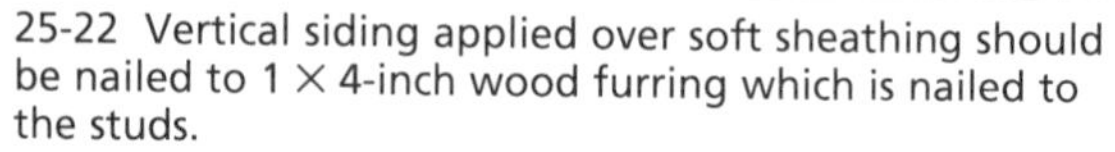
25-22 Vertical siding applied over soft sheathing should be nailed to 1 × 4-inch wood furring which is nailed to the studs.

25-23 Horizontal 2 × 4-inch blocking between studs can be used to provide a nailing surface for vertical siding installed over soft sheathing.

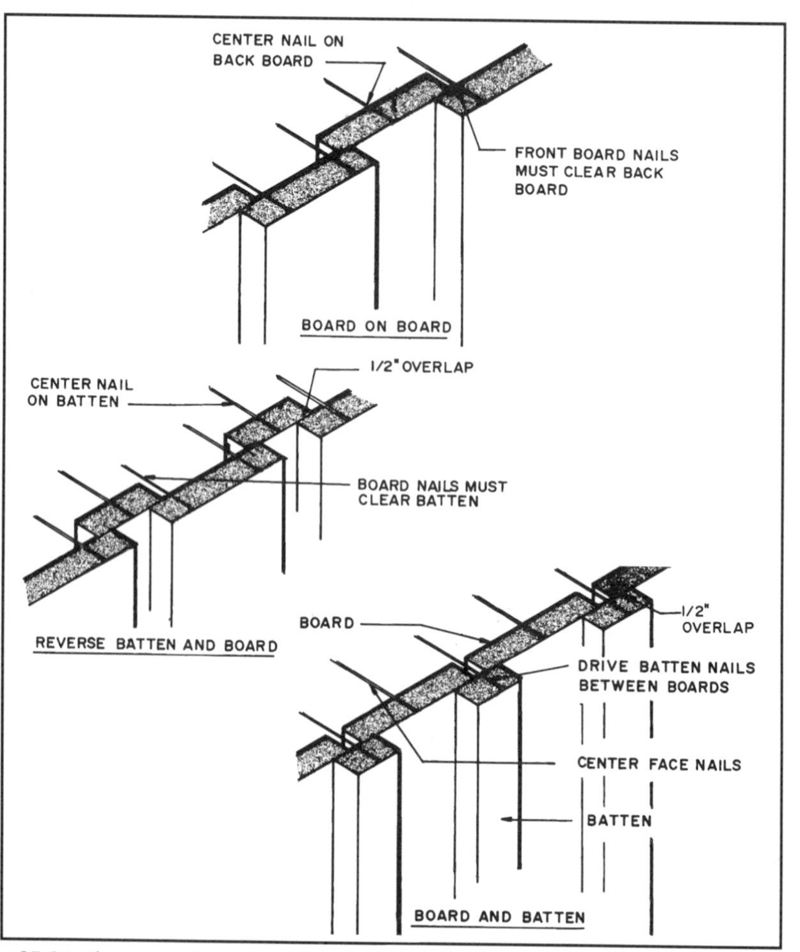

25-24 Three ways boards are used as vertical siding.

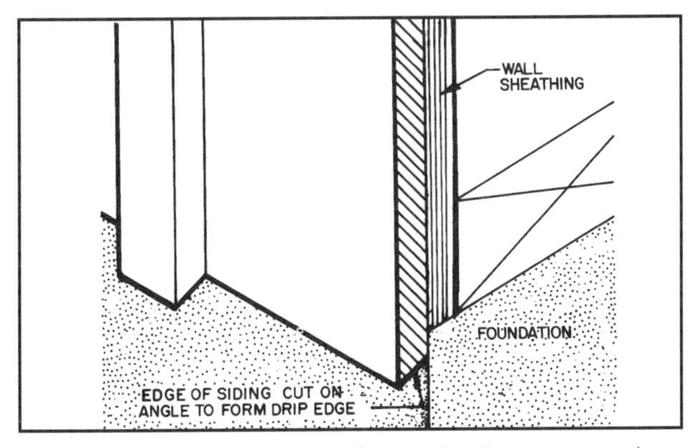

25-25 Cut the bottom edge of vertical siding on an angle to form a drip edge. Treat the cut surface with a water repellent.

INSTALLING VERTICAL WOOD SIDING

The common types of vertical wood siding include board and batten, board on board, tongue-and-groove, and rustic. These profiles and nailing patterns are shown earlier in **25-4**. Vertical wood siding is best if applied over plywood, oriented strand board, or solid wood sheathing, because it can be nailed directly to it. If fiberboard, gypsum, or plastic foam-type sheathing is used, 1 × 4-inch wood strips must be nailed horizontally along the wall. Space them 16 to 24 inches apart and nail them to the studs **(see 25-22)**. Another technique used when soft sheathing is used is to nail 2 × 4-inch blocking spaced 16 or 24 inches O.C. between the studs as shown in **25-23**.

Square-edged boards are also used as vertical siding. These are commonly referred to as board and batten, reverse board and batten, and board. They require the same sheathing preparation as just discussed. The layer of boards or battens next to the sheathing is nailed with 8d nails at each blocking strip. Wide under-boards require two nails. Battens require one nail. The nails in the top boards should miss the underboards. Use 12d nails through the top boards. The nailing patterns and board arrangements are in **25-24**. The overhanging bottom edge of vertical siding is often cut on an angle to form a drip edge **(see 25-25)**. The cut edge should be coated with a water repellent.

Typical details for finishing vertical siding at inside and outside corners and at windows and doors are in **25-26**. Installation at doors and windows is shown later in **25-39**.

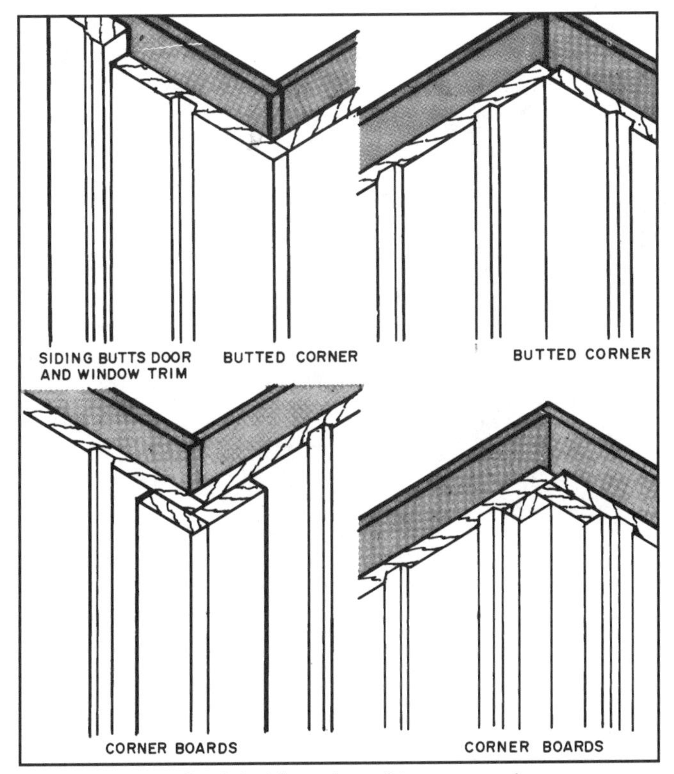

25-26 Ways to finish inside and outside corners when installing square-edged board siding.

GABLE-END TREATMENT

Often, the designer uses a different siding or treatment on the **gable-end**. This requires the carpenter to prepare the joint between these treatments so that it is neat and waterproof. One way to do this is with a **drip cap (see 25-27)**. The drip cap is nailed to the plate. It is then flashed the same as discussed for windows. The gable siding is applied over the flashing.

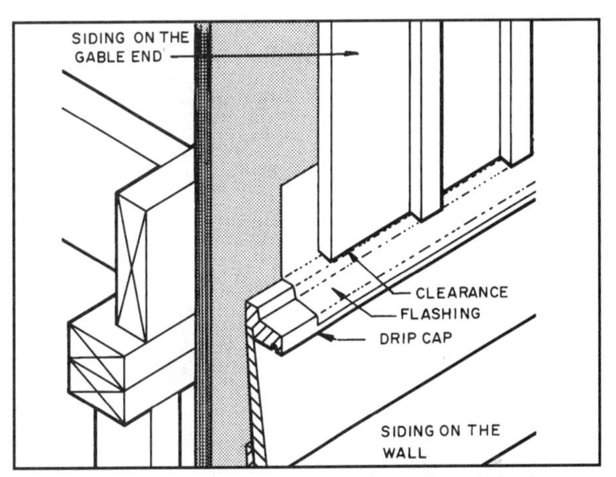

25-27 The vertical siding on a gable-end can join the siding below by using a flashed drip cap.

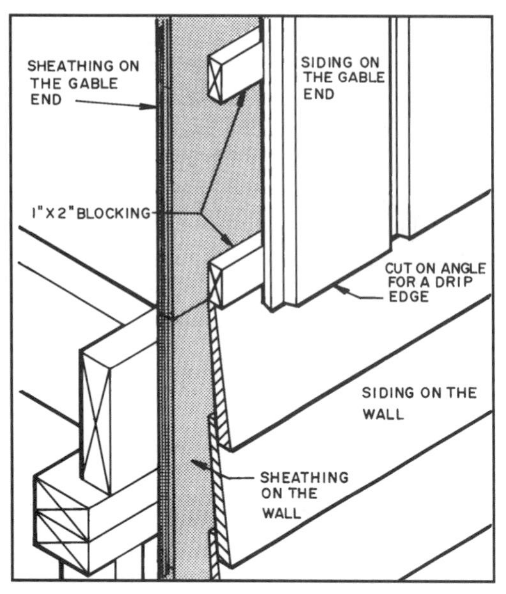

25-28 The gable-end vertical siding can be set up over the siding on the wall using 1 • 2-inch wood strips.

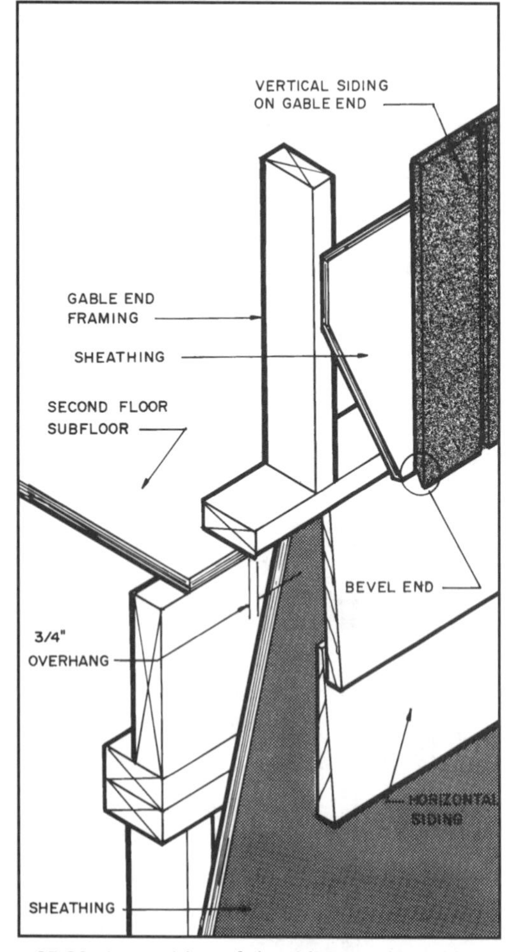

25-29 A transition of the siding on the gable-end can be made by setting out the gable-end framing, providing an overhang.

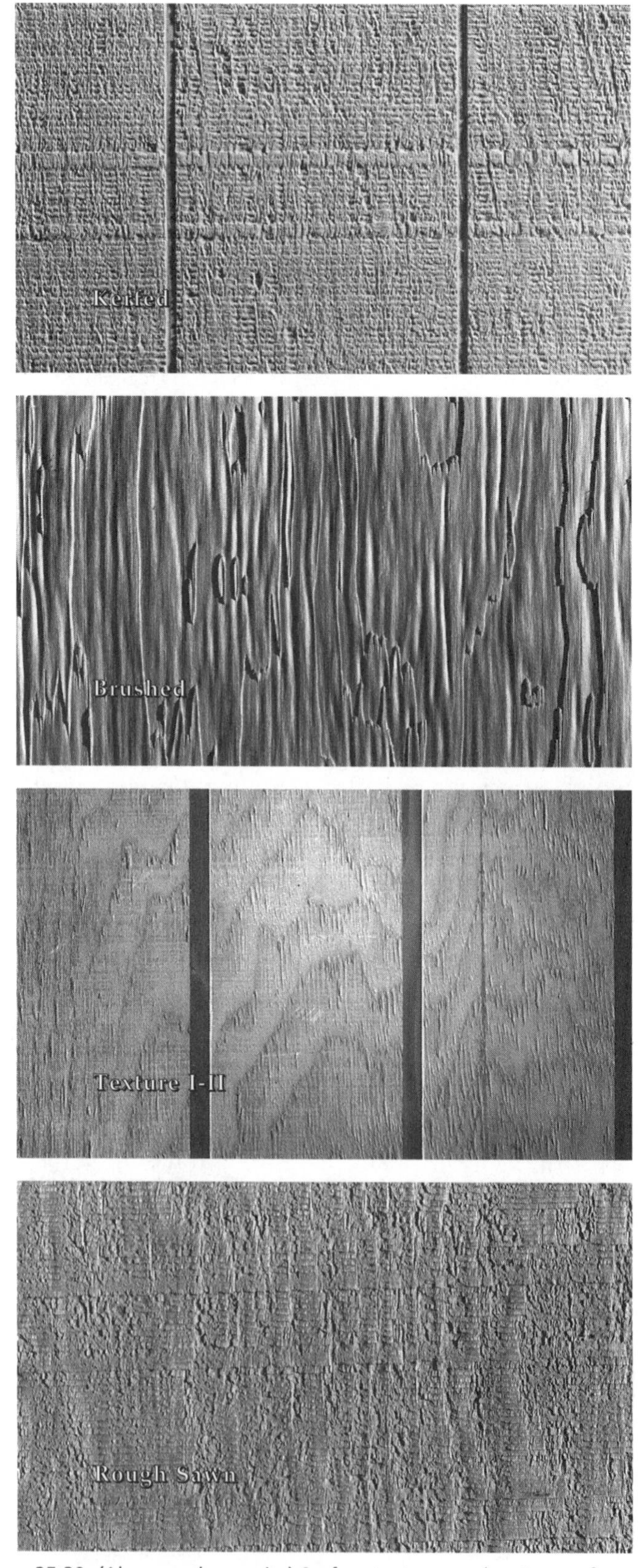

25-30 (Above and opposite) Surface textures and patterns of selected plywood siding panels. *(Courtesy APA—The Engineered Wood Association)*

Table 25-2 Typical sizes of plywood siding.*

Lap Siding		
width (in)	length (ft)	thickness (in)
6, 8, 9½, 12	12, 16	11/32, 3/8, 1/2, 19/32, 5/8

Panel Siding		
width (ft)	length (ft)	thickness (in)
4	8, 16	7/16, 1/2, 19/32, 5/8

*Consult manufacturer for specific size information.

Another technique is to apply 1 × 2-inch blocking over the sheathing on the gable-end **(see 25-28)**. This places the gable siding out over the lower siding. The edge of the gable siding is undercut to form a drip edge. Still a third technique is in **25-29**. Here the gable-end framing is set out over the top plate, forming the overhang.

PLYWOOD SIDING

Plywood siding is available in panels and lap siding. It is made by bonding wood veneers with a waterproof adhesive and is available in a variety of wood species and textures **(see 25-30)**. The most common texture is a rough-sawed top veneer. Another type, medium-density overlay (MDO), has a resin-treated fiber overlay sheet bonded to it. This provides a smooth, tough, check-free surface that holds paint well. It is available in a variety of surface patterns **(see 25-31)**.

Typical sizes of lap siding and panels are in **Table 25-2**. Product sizes will vary depending upon those available from various manufacturers.

25-30 (Continued) Surface textures and patterns of selected plywood siding panels. *(Courtesy APA—The Engineered Wood Association)*

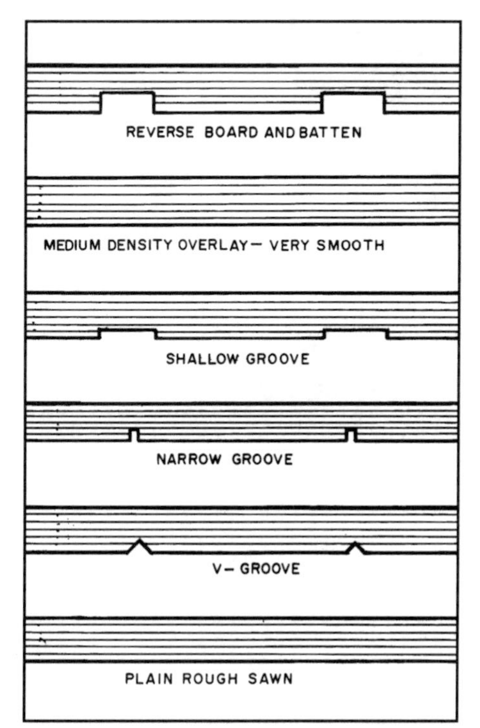

25-31 Typical plywood panel siding patterns.

Siding manufactured to the standards of the APA—The Engineered Wood Association is identified by a grade mark on each piece. Several such grade marks are shown in **25-32**.

Installing Plywood Lap Siding

Plywood lap siding is available in a range of widths up to 12 inches. It requires a minimum overlap of one-inch and ⅛-inch clearance between butted siding ends and unions with casings on doors and windows and other trim. This clearance must be caulked.

Plywood lap siding can be nailed directly to studs, if it meets the APA span rating or to nailable sheathing. The span rating is the allowable maximum spacing of the studs. The studs must be braced with diagonal bracing or a structural sheathing. Building paper is stapled to the studs before installing the siding. Plywood lap siding that is not span rated must be applied over nailable sheathing.

Keep the bottom panel at least 6 inches above grade and allow a ⅛-inch gap between the siding and door and window frames and trim.

Plywood lap siding when applied over nailable sheathing is nailed 8 inches O.C. along the bottom edge of the siding, with the nails driven to penetrate the top edge of the lower course **(see 25-32)**. Use 8d nonstaining box, casing, or siding nails for siding ¾ inch or thicker. Use 6d nails for siding ½ inch thick or thinner.

If nailed directly to the studs or over nonnailholding sheathing, nail the lap siding at the bottom edge at each stud **(see 25-33)**.

Installing Plywood Panel Siding

APA-rated siding panels can be applied directly to the studs, and no diagonal bracing is needed if the stud spacing does not exceed that shown on the grade mark **(see 25-34)**. Building paper can be omitted if the panel joints are battened or shiplapped. Square-edged joints require backing with building paper. Plywood siding panels can also be applied over any type of sheathing **(see 25-35)**.

APA-rated siding panels can be applied horizontally, if the horizontal joints are backed with 2 × 4 blocking **(see 25-35)**. Use 6d nonstaining box, casing, or siding nails for panels up to ½ inch thick and 8d for thicker panels. Space six inches O.C. on panel edges and 12 inches O.C. on interior studs.

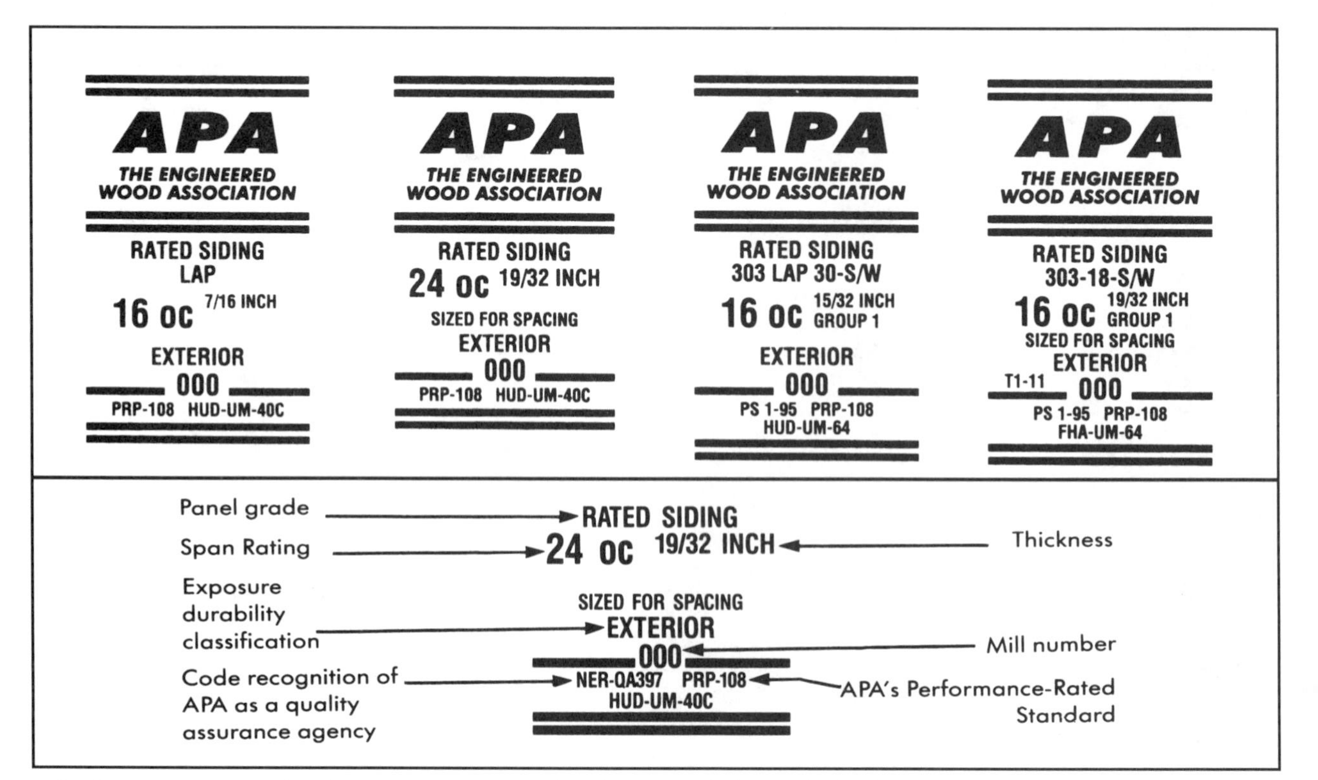

25-32 Grade marks used on APA-rated siding. *(Courtesy APA—The Engineered Wood Association)*

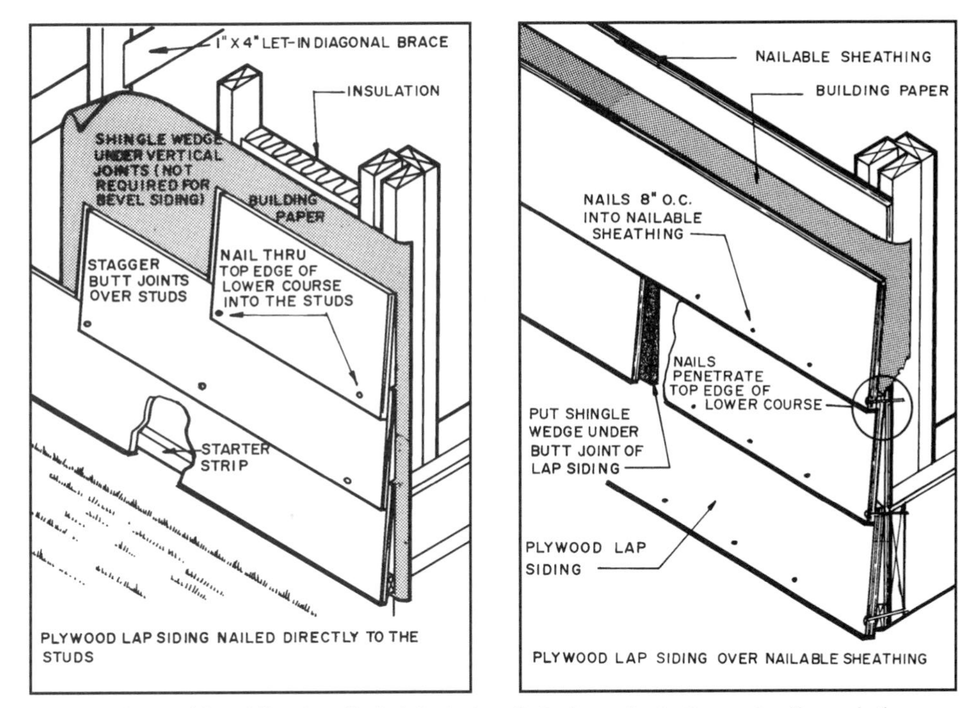

25-33 Plywood lap siding is nailed eight inches O.C. along the bottom edge through the top edge of the course below.

25-34 The plywood siding has been nailed to the studs and the soffit is being installed. *(Courtesy APA—The Engineered Wood Association)*

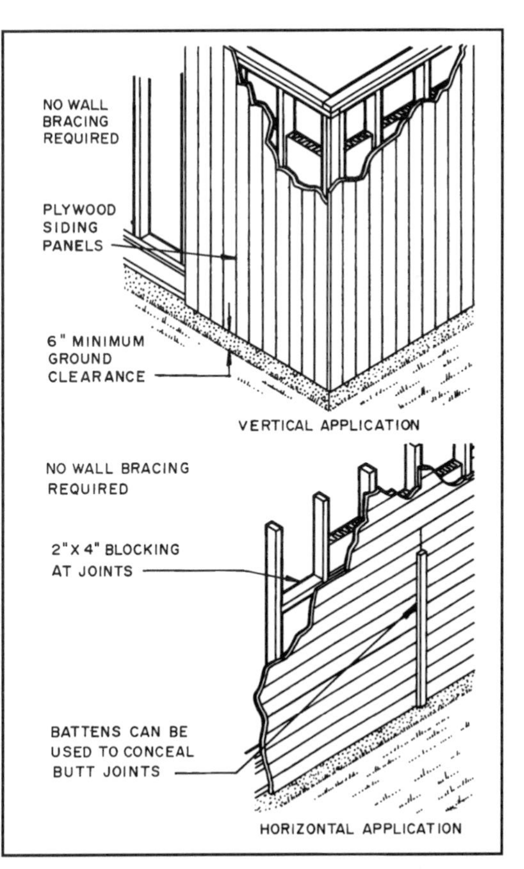

25-35 Plywood siding panels can be installed vertically or horizontally. Space nails 6 inches O.C. on panel edges and 12 inches O.C. on interior supports.

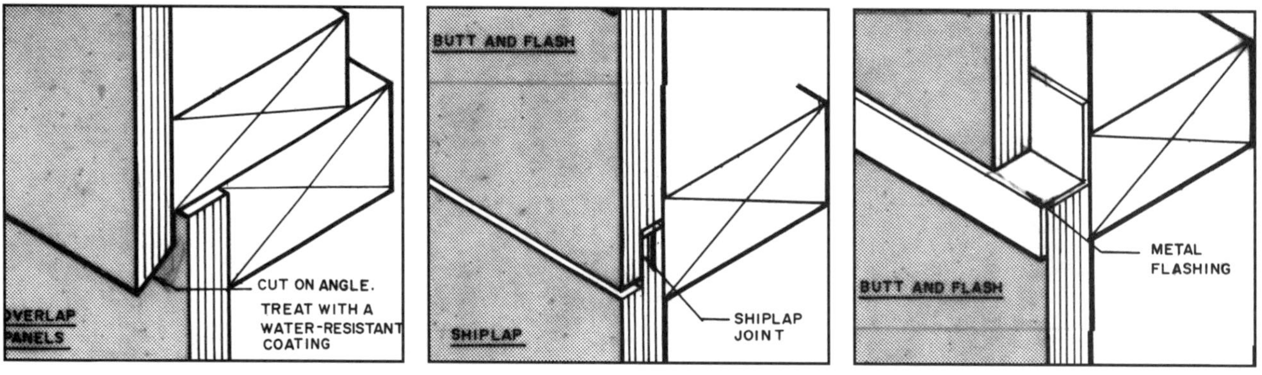

25-36 Types of horizontal joints that are used with plywood siding panels.

Horizontal joints are installed as shown in **25-36**. These include a shiplap joint, overlap joint, or butt joint with metal flashing. Seal edges of panel with water-resistant sealer.

Vertical joints between wall panels can be installed as shown in **25-37**.

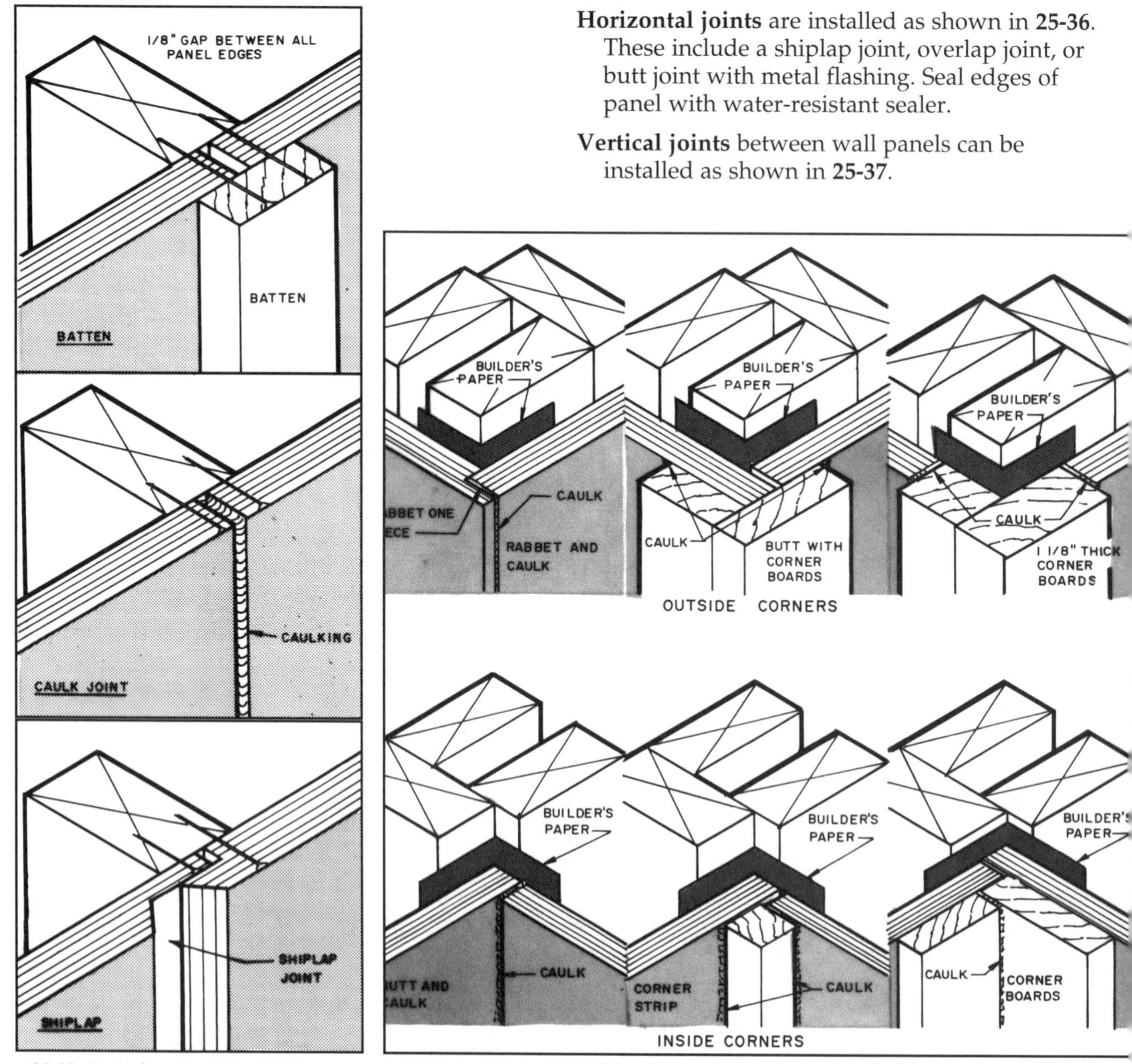

25-37 Typical vertical joints between plywood and hardboard siding panels.

25-38 Installation techniques for vertical inside and outside corners used with panel siding.

Inside and outside corners are formed as shown in **25-38**. As with all joints, caulking is essential to providing a waterproof joint.

Framing at windows and doors is much like that for solid wood. Details are in **25-39**.

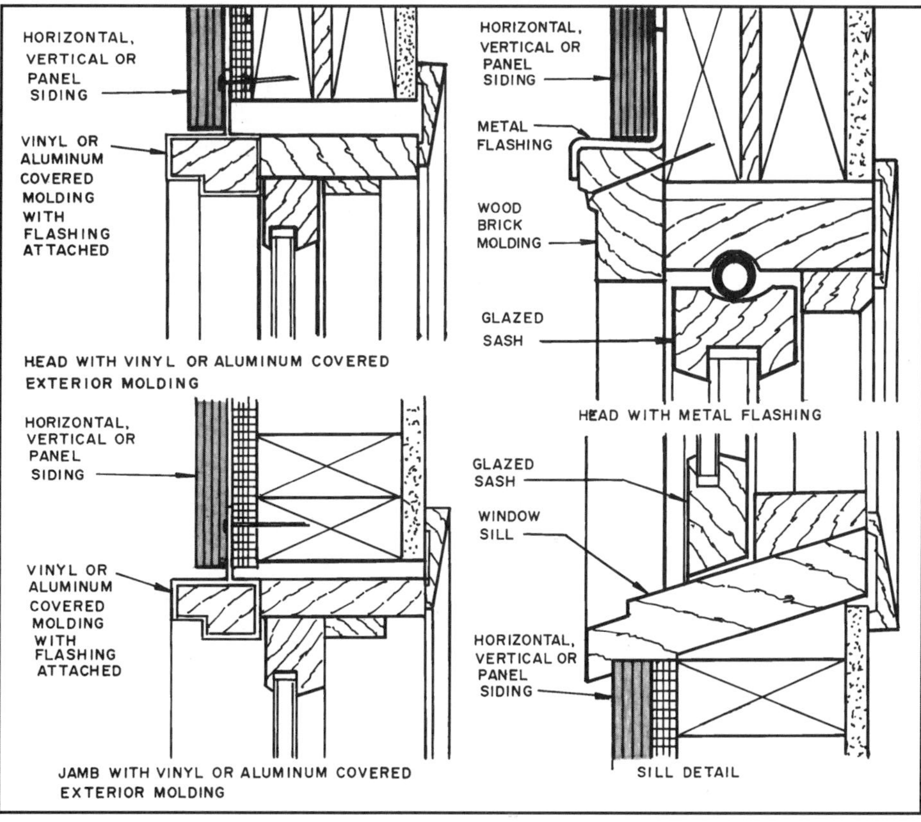

25-39 Typical installation details for vertical, horizontal, and panel siding at doors and windows.

HARDBOARD SIDING

Hardboard siding is available as lap siding and panels. Hardboard is made from wood chips converted into fibers and bonded into panels under heat and pressure. It is made following the standards of the American Hardboard Association. It is available in a variety of textures and patterns similar to those discussed for plywood. The lap siding is usually in the form of single, double, or triple laps. The panels are available in several designs **(see 25-40)**.

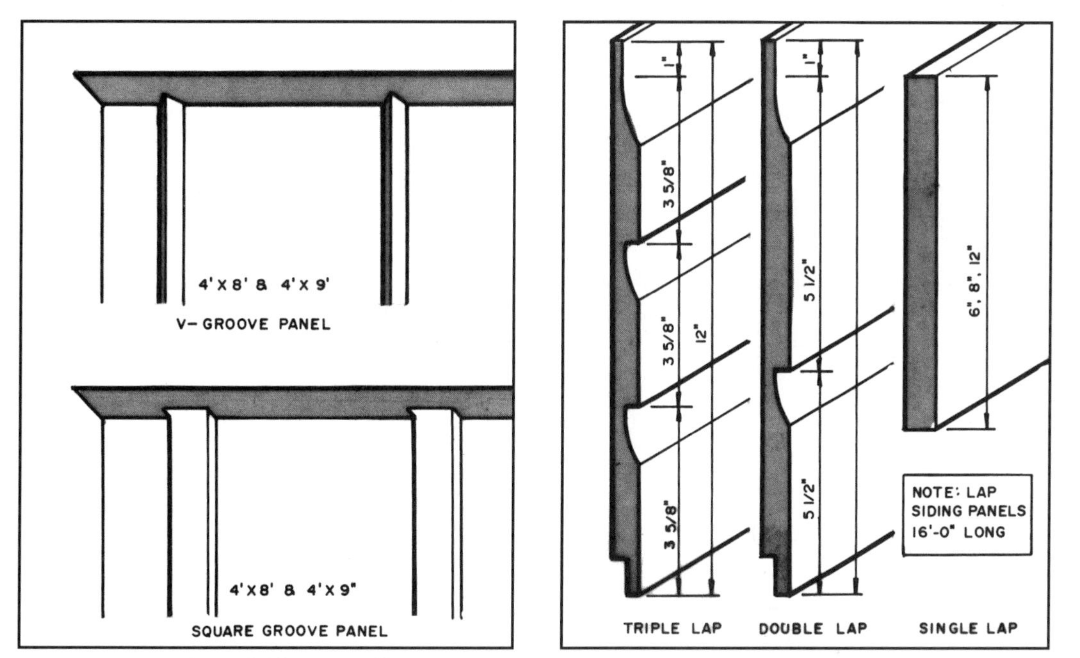

25-40 Commonly available types of hardboard lap and panel siding.

Typical sizes of hardboard siding panels are given in **Table 25-3**. Panels are available primed or unprimed and with a smooth or wood grain embossed texture. Some companies provide a vinyl-coated product. Panel products are most often installed vertically as shown in **25-41**.

25-41 Panel hardboard siding is often placed with the long dimension vertical. *(Courtesy of Georgia-Pacific Corporation)*

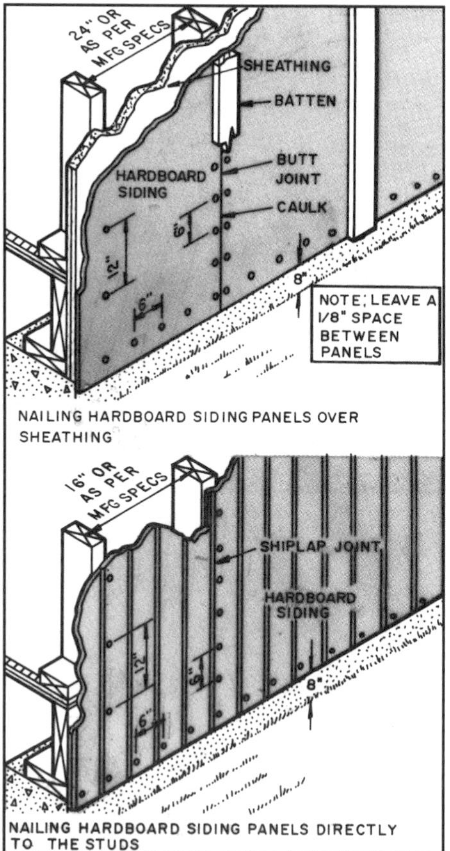

25-42 How to install hardboard siding panels.

Installing Hardboard Siding

Hardboard siding is cut with standard hand and power woodworking tools. It can be nailed directly to studs spaced 16 inches O.C. Panels used for structural purposes, such as exterior wall bracing, must be clearly identified by the mark of an approved agency. If siding is placed over a wall covered with structural sheathing the studs can be spaced 24 inches O.C. The lowest edge should be at least eight inches from the ground **(see 25-42)**. Consult the manufacturer for bracing, spacing, and nailing requirements.

The panels should be applied with 8d corrosion-resistant nails spaced six inches apart on the edges. They should be in ½ inch from the edge of the panel. Nails should be 12 inches apart on intermediate members. There should be a gap of ⅛ inch between butting panel edges.

Table 25-3 Typical sizes of hardboard siding.*

Lap Siding		
width (in)	length (ft)	thickness (in)
6, 8, 12	16	½

Panel Siding		
width (ft)	length (ft)	thickness (in)
4	7, 8, 9	7/16

*Consult manufacturer for specific size information.

Hardboard panels are installed in the same manner as described for plywood panels **(see 25-43)**. One type of horizontal joint that can be used

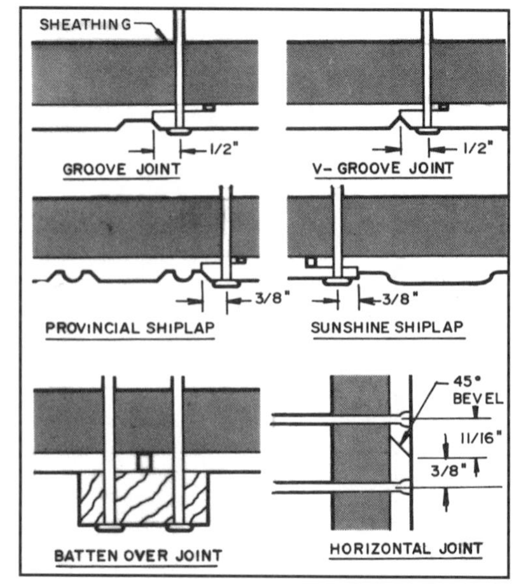

25-43 Typical surface patterns and joints available in hardboard siding panels.

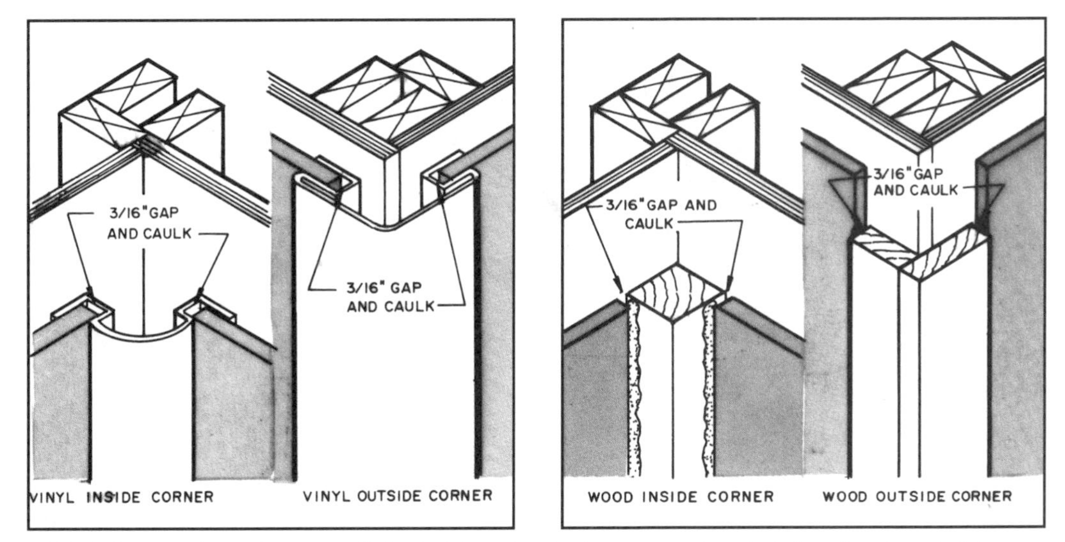

25-44 Commonly used ways to finish inside and outside corners when installing hardboard siding. *(Courtesy ABTCO Building Products)*

is a bevel cut. This is caulked before the two parts are set together. Inside and outside corners can be closed with wood corner boards. Some manufacturers supply vinyl corner covers as shown in **25-44**. They also supply vinyl J-trim to cover the ends of panels where they meet doors and windows **(see 25-45)**. Notice a gap is recommended allowing for expansion. The gap is closed with flexible, expandable caulking.

Hardboard lap siding is installed by nailing through the face of the single-lap siding and the underlying siding along the overlap as shown in **25-46**. Nails should penetrate the studs 1½ inches. For most hardboard lap siding, 8d galvanized box nails are used, spaced a maximum of 16 inches O.C. Hardboard panel siding uses 8d galvanized box nails, spaced six inches O.C. around the perimeter and 12 inches O.C. at stud locations inside the panel.

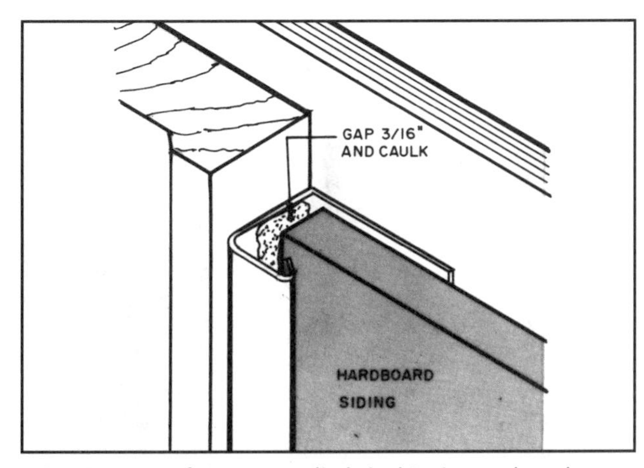

25-45 A manufacturer-supplied vinyl J-trim used on the ends of panels as they butt doors and windows. *(Courtesy ABTCO Building Products)*

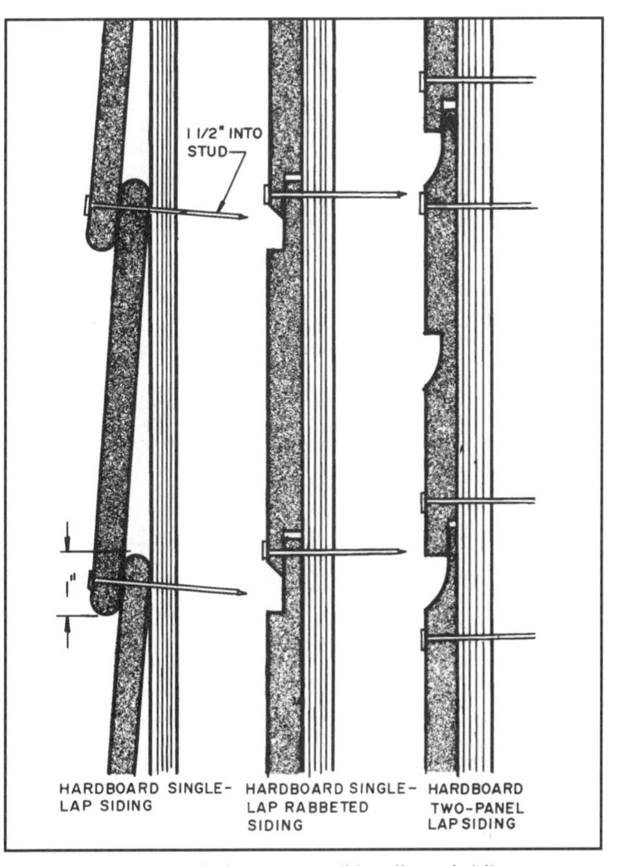

25-46 Recommended ways to nail hardboard siding.

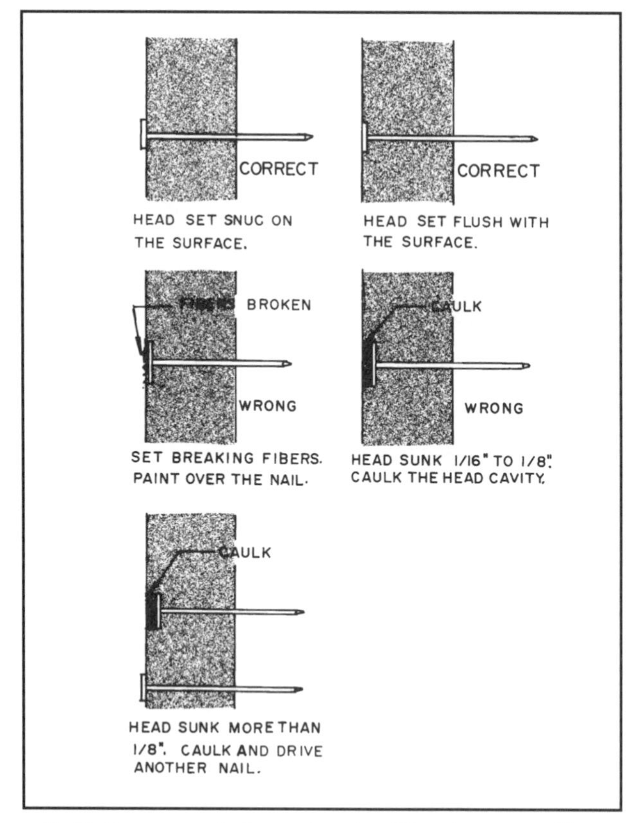

25-47 Recommended ways to set nail heads when nailing hardboard siding and remedial actions when improperly driven.

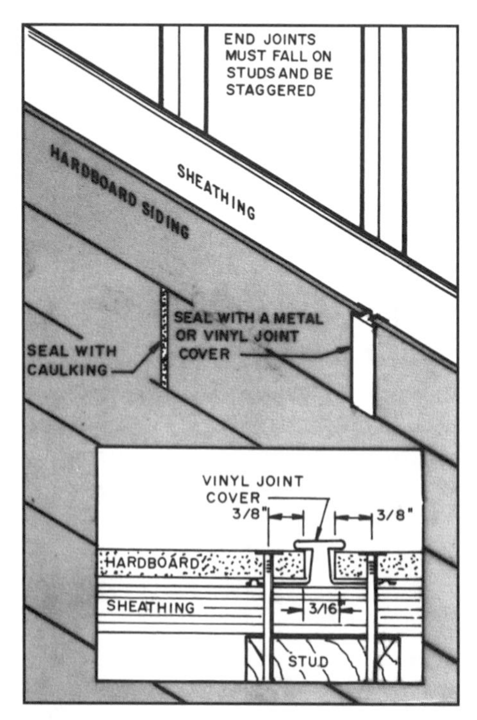

25-48 Two ways end joints in hardboard lap siding can be sealed. *(Courtesy ABTCO Building Products)*

The nail must be snug or flush with the surface of the siding. If it breaks the surface and is recessed, it must be painted or caulked. Nailing recommendations of the American Hardboard Association are shown in **25-47**. The end joints may be caulked, though some companies provide a metal or vinyl butt joint cover **(see 25-48)**.

FIBER-CEMENT SIDING

Fiber-cement siding is available as horizontal lap siding and as panels. It is made from a mixture of cellulose fiber, Portland cement, ground sand, selected additives, and water. It is hard, tough, weather and insect resistant, and noncombustible. The siding is available in various surface textures, one of which is shown in **25-49**. Typical sizes are in **Table 25-4**.

CUTTING FIBER-CEMENT SIDING

Manufacturers of fiber-cement siding make recommendations on how to cut it. Typically they recommend using blades with carbide-tipped teeth or an abrasive blade that has diamond tips. A shear is also available for cutting. Carbide-tipped blades cut rapidly but last only a few hours. However, they produce less dust than abrasive blades. The abrasive blade lasts much longer but should be used on a saw that has an efficient dust removal system. The dust produced is hazardous to your health. Wear a dust mask if a dust removal system is not on your saw. Cutting with a shear does not produce dust. One shear cuts like a paper cutter while another type is like a guillotine.

25-49 This fiber-cement lap siding has a rough wood grain. *(Courtesy James Hardie Building Products 1-800-9-Hardie)*

INSTALLING FIBER-CEMENT SIDING

Fiber-cement siding can be hand nailed or power nailed with coil nailers. It is essential that the nails not be overdriven. The nail head must sit on the surface but not break the surface. A broken surface greatly reduces the holding power of the nail and can lead to wall failure. You must use corrosion-resistant nails recommended by the manufacturer. The nails should penetrate the wood stud at least one inch (25.4 mm).

The siding can be installed directly to studs spaced up to 24 inches (607mm) O.C. Builder's felt or some other weather-resistant barrier should be placed over the studs. If installed over a soft sheathing, such as rigid foam plastic, it is necessary to nail 1 × 3-inch (25 × 75mm) wood batten strips over the sheathing at each stud.

The installation is much like that for wood siding. It starts with a wood starter strip at the foundation **(see 25-50)**.

Overlap the lap siding 1¼ inches (31mm). All end butt joints must be on studs. Some framers plan where these will occur and install two studs there to give a larger nailing surface. The siding can be blind-nailed at the top and bottom of each piece. Blind-nailing hides the nail but does not offer the strength needed in areas where high winds are common **(see 25-50)**.

The end butt joints must be carefully sealed with a latex sealant. Unsealed edges tend to permit water to wick into the panel.

Interior and exterior corners are butted to wood corner boards the same as wood siding. Leave a space of about ⅛ inch (3mm) and seal carefully.

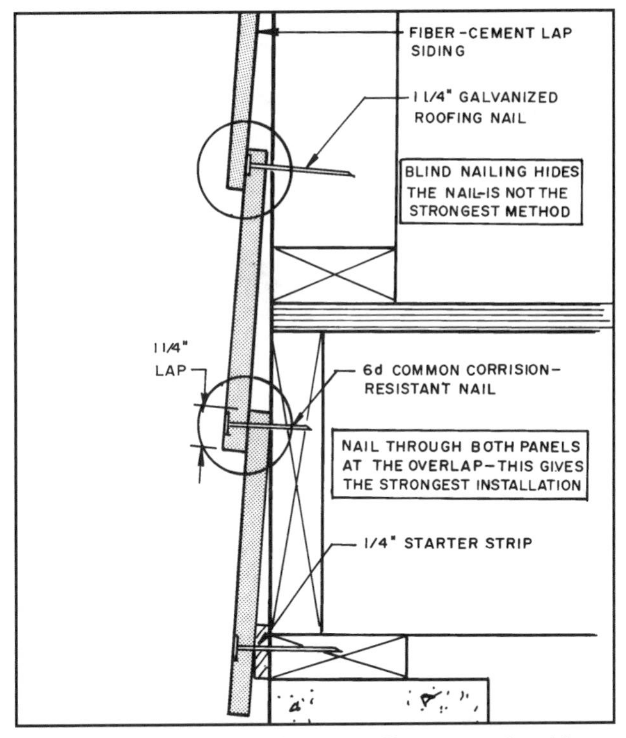

25-50 Recommendations for nailing fiber-cement lap siding. *(Courtesy James Hardie Building Products 1-800-9-Hardie)*

Cement-fiber panels are installed vertically. A typical installation is shown in **25-51**. Studs may be spaced 16 or 24 inches O.C. The spacing between nails varies with the type of panel and the manufacturer. Typical nail spaces are four, six, and eight inches.

Table 25-4 Typical sizes of cement-fiber siding.*

Lap Siding		
width (in)	**length (ft)**	**thickness (in)**
6, 7½, 9½, 12	12	5⁄16

Panel Siding		
width (ft)	**length (ft)**	**thickness (in)**
4	8, 9, 10	¼, 5⁄16

*Consult manufacturer for specific size information.

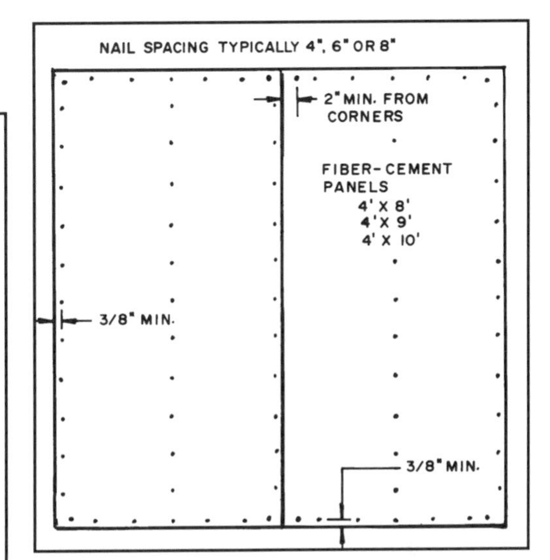

25-51 Nailing pattern of vertically applied fiber-cement panels. (*Courtesy James Hardie Building Products 1-800-9-Hardie)*

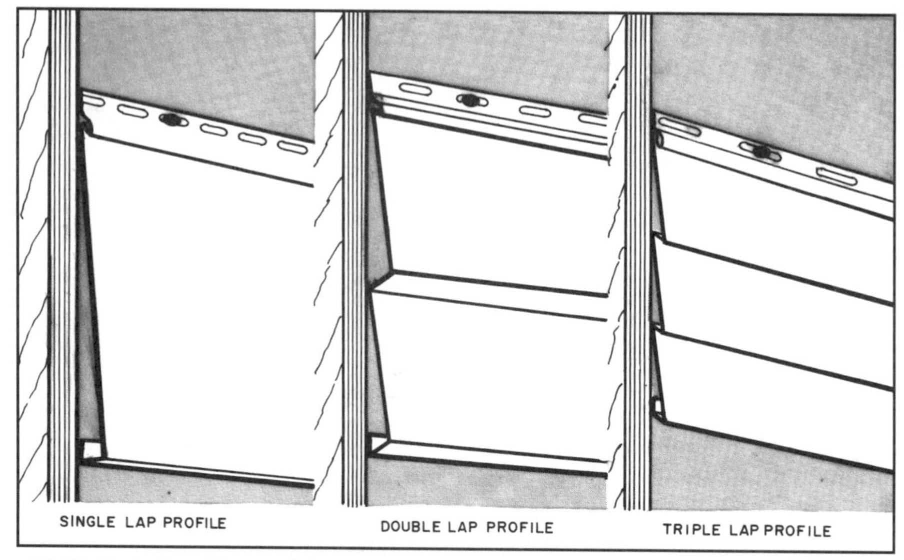

25-52 Typical profiles available in vinyl siding.

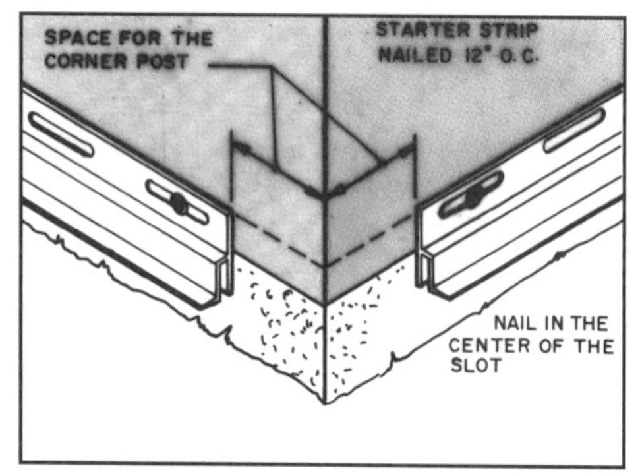

25-53 Snap a chalkline to locate the starter strip. Leave space for the corner posts. Drive the nails in the center of the slot. *(Courtesy Vinyl Siding Institute)*

Table 25-5 Typical sizes of vinyl siding.

Vinyl Lap Siding	
width (in)	length (ft)
Triple 3 Double 4, 5 Single 6.5, 8	10, 12, up to 40
12 Vertical	10

Aluminum Lap Siding	
width (in)	length (ft)
4, 5 Double	12.5
8 Single	12.5
12 Vertical	10.0

*Consult manufacturer for specific size information.

VINYL & ALUMINUM SIDING

Vinyl siding is available as lap siding. Various configurations and styles are available from the several companies manufacturing the product **(see Table 25-5)**. It is made from polyvinyl chloride and has a range of surface textures available. Typically an embossed wood-grain pattern and a smooth wood grain are available. It is made in one piece as single-, double-, and triple-formed lap siding **(see 25-52)**. Some companies make a 12-inch-wide vertical panel that may have a V-groove or a batten formed on the surface.

Aluminum siding is available in the same basic forms as described for vinyl siding.

Installing Vinyl & Aluminum Siding

Both types have no structural strength and should be installed over nailable sheathing. If soft rigid plastic sheathing has been used, 1 × 3-inch wood batten strips should be nailed over the sheathing to each stud. If installing vertical vinyl siding, nail the 1 × 3-inch strips perpendicular to the studs, spaced 12 inches O.C. Begin by installing the starter strip supplied by the manufacturer. Snap a chalkline along the bottom of the wall to locate the top of the strip. Leave the required space for the inside and outside corner posts **(see 25-53)**. Then install the inside and outside corners **(see 25-54)**. Finally install the channels that go around the doors and windows **(see 25-55)**. Be certain to leave the spacing between these as recommended by the manufacturer. Now you are ready to install the horizontal siding.

Place the first panel over the starter strip and nail every 12 to 16 inches through the slot in the top of the panel **(see 25-56)**. The nails should be centered in the slot and driven so they are 1⁄32 inch from the siding **(see 25-57)**. This allows the piece to expand and contract. If this is not done the panel will buckle. Keep the nail in the center of the slot and drive it perpendicular to the wall. Nail the corner strips 6 to 12 inches O.C. as recommended by the manufacturer.

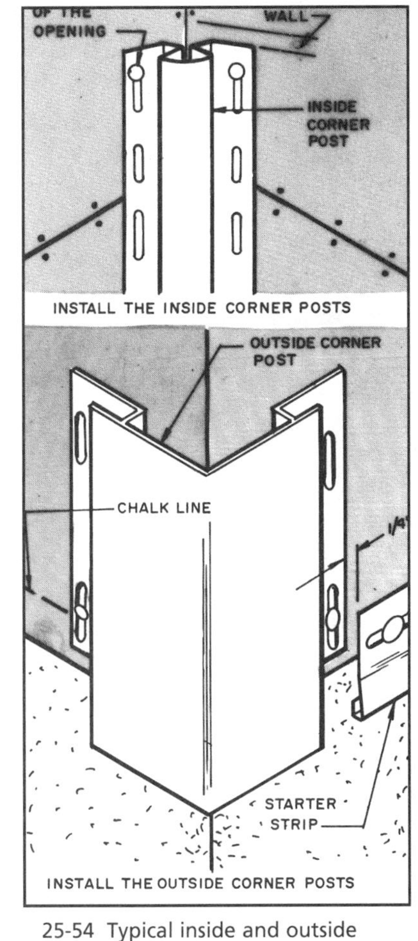

25-54 Typical inside and outside corner posts. *(Courtesy Vinyl Siding Institute)*

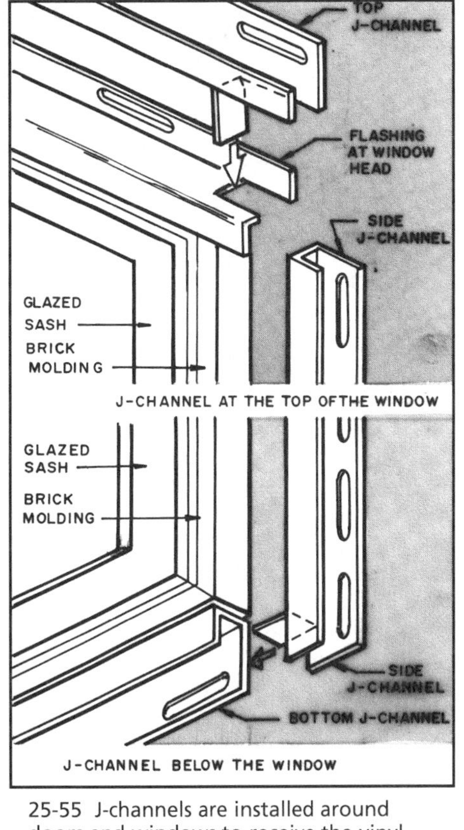

25-55 J-channels are installed around doors and windows to receive the vinyl siding. *(Courtesy Vinyl Siding Institute)*

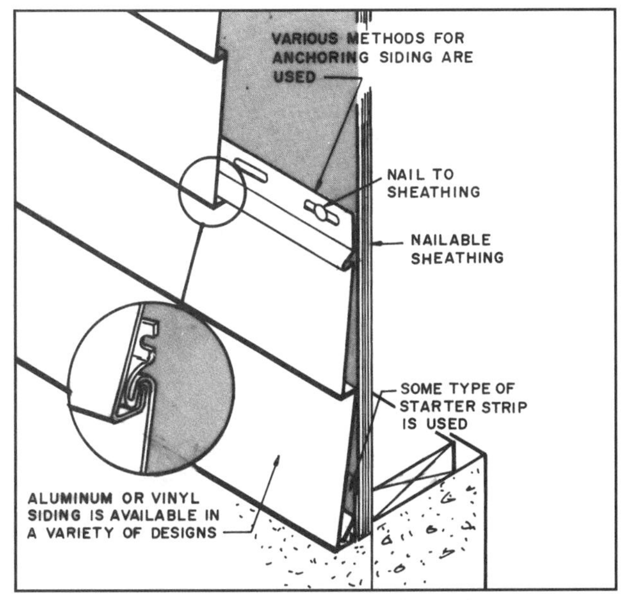

25-56 The first strip of siding is placed over the starter strip and nailed to the nailable sheathing. Do not pull the end joint tight. Let it have some movement. *(Courtesy Vinyl Siding Institute)*

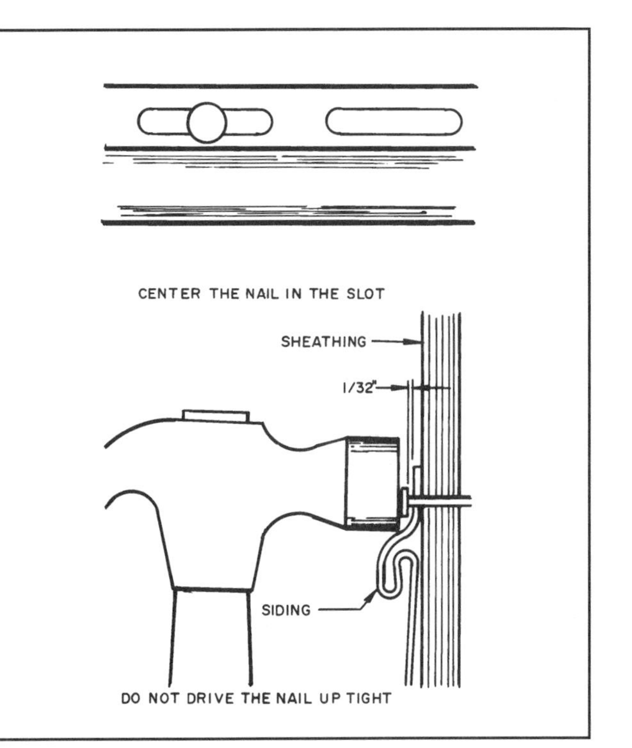

25-57 Center the nail in the slot and drive up to within 1/32 inch of the face of the siding. *(Courtesy Vinyl Siding Institute)*

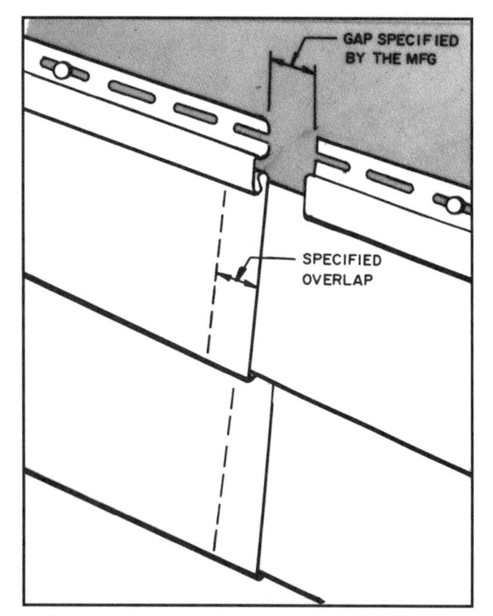

25-58 After overlapping the butting ends of the panel be certain to leave a gap at the top as recommended by the manufacturer. *(Courtesy Vinyl Siding Institute)*

The horizontal panels are leveled as explained for wood siding. Be certain they hang freely but are locked where they connect on the bottom **(refer to 25-56)**. When panels must be end-lapped be certain they have room to move freely **(see 25-58)**.

Vinyl and aluminum soffit material is widely installed using manufacturer-supplied channels **(see 25-59)**.

Nails used should be aluminum or galvanized steel for vinyl siding and aluminum for aluminum siding. They should have a head at least 5⁄16 inch in diameter and penetrate solid material about 1½ inches.

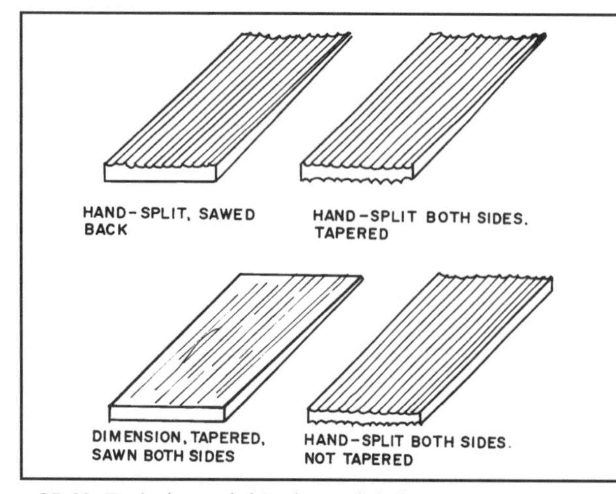

25-60 Typical wood shingles and shakes.

25-59 Vinyl and aluminum soffit panels are installed by placing them in J-channels nailed to the sheathing along the overlap. *(Courtesy Heartland Building Products)*

WOOD SHINGLES AND SHAKES

Wood shingles and shakes are made from red cedar, redwood, southern pine, and cypress and are available in a variety of sizes and patterns. **Shingles** are sawed flat on both sides and are available in what are called dimension shingles, in widths from 3 to 14 inches as well as in random widths. They come in 16-, 18-, and 24-inch lengths **(see 25-60)**. They are available in grades No. 1 (blue label), No. 2 (red label), and No. 3 (black label). Redwood shingle grades are No. 1 and No. 2. Bald cypress grades are No. 1, Best, Prime, Economy, and Clipper. No. 1 grade is best in each case.

Random-width shingles are packaged by the **square** (100 square feet) and dimension widths are packaged with 1000 shingles per bundle. Shingles are thinner than shakes.

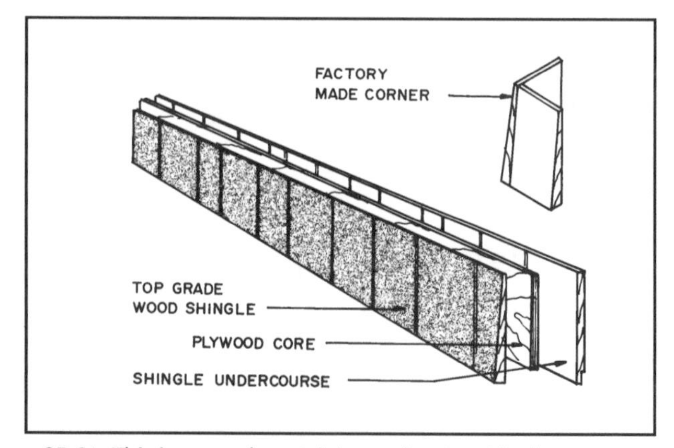

25-61 This is a panel containing red cedar shingles bonded to a plywood backer board.

Another wood shingle product is a panel product having wood shingles bonded to a wood panel backer board. The panels may be one, two, or three shingles wide and are eight feet long **(see 25-61)**.

Shakes are handsplit and have a rough front and back surface. One type has both faces sawed but is much thicker than shingles. Shakes are available in widths from 4 to 14 inches and in 18- and 24-inch lengths **(refer to 25-60)**. Shakes have a thick butt end and taper to a thin end. Red cedar shakes are available in No. 1 and premium grades. Southern pine shakes are taper-sawed, producing relatively smooth faces. They are 18 and 24 inches long and available in No. 1 (best) and No. 2 grades.

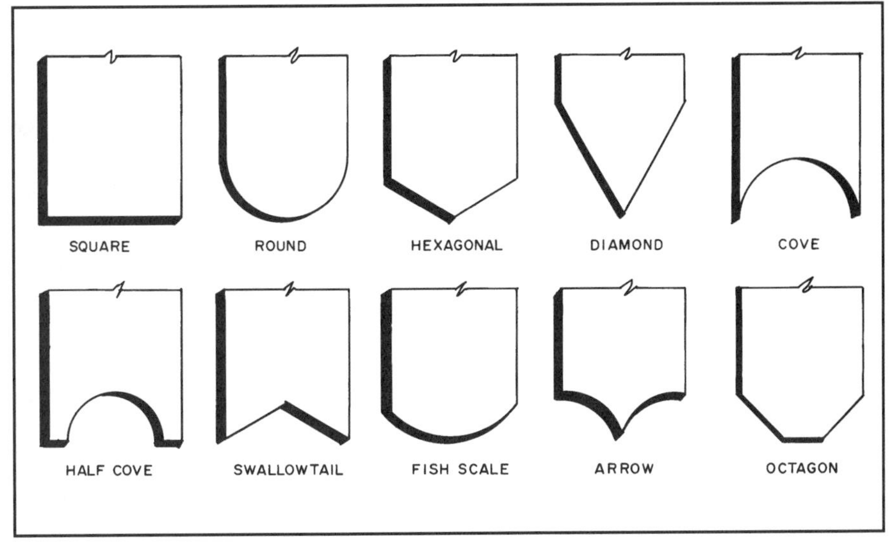

25-62 Typical patterns for decorative wood shingles.

Shingle Patterns

Wood shingles are cut with a variety of decorative edges which are used to produce the visual appearance desired by the architect. The typical patterns available are in **25-62**. These can be installed using the same pattern shingle on the entire wall, as shown in **25-63**. Occasionally several patterns are laid together, producing even more interesting visual effects **(see 25-64)**.

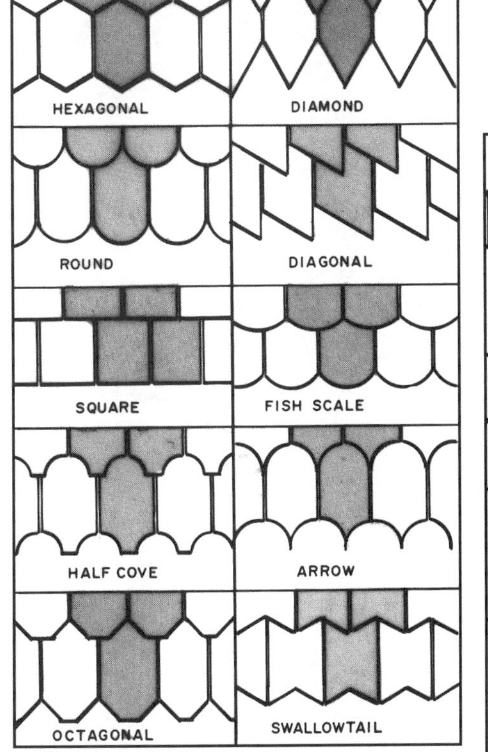

25-63 These wood shingle installations use the same pattern shingle over the entire wall.

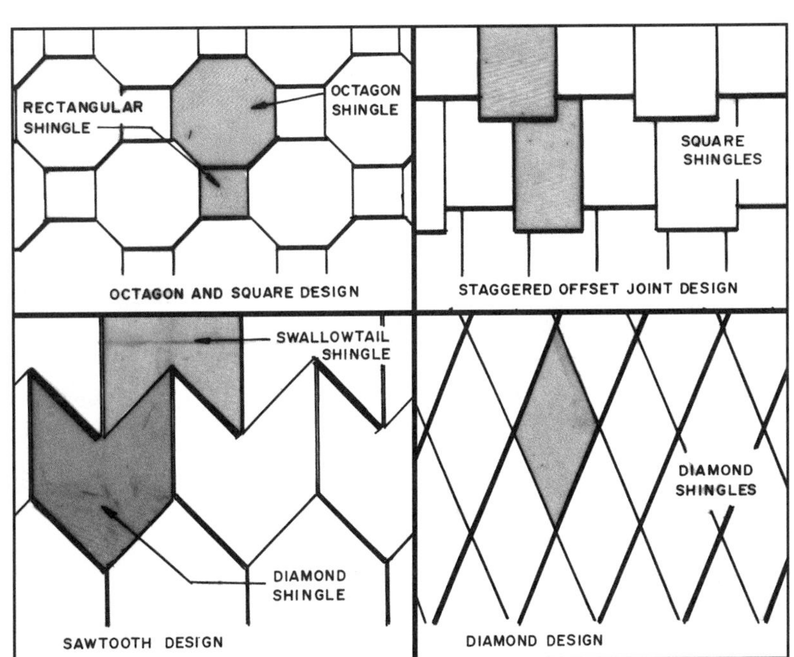

25-64 These are a few of the patterns that can be developed using several of the decorative shingles.

Installing Wood Shingles & Shakes as Siding

25-65 Wood shingles and shakes are secured to solid wood furring strips or nailable sheathing with corrosion-resistant nails or staples. *(Courtesy Senco Products, Inc.)*

Both wood shingles and shakes are installed with corrosion-resistant nails or staples **(see 25-65)**. The wall to receive the shingles must have a sheathing, such as plywood, that will hold nails. If soft sheathing, such as foam plastic, is used, 1 × 4 furring strips must be nailed over the sheathing and into the studs. The spacing between the furring strips equals the shingle exposure **(see 25-66)**. The sheathing must be covered with a building paper that permits the passage of vapor from inside the building.

The shingles are laid in uniform courses, similar to bevel siding. The amount exposed to the weather can be varied to help meet the heads and sills of windows in a proper manner.

Usually, only one-half or less of the shingle is exposed. The amount of exposure depends upon the desired appearance **(see 25-67)**. Wood shingle and shake siding exposure recommendations are given in **Table 25-6**.

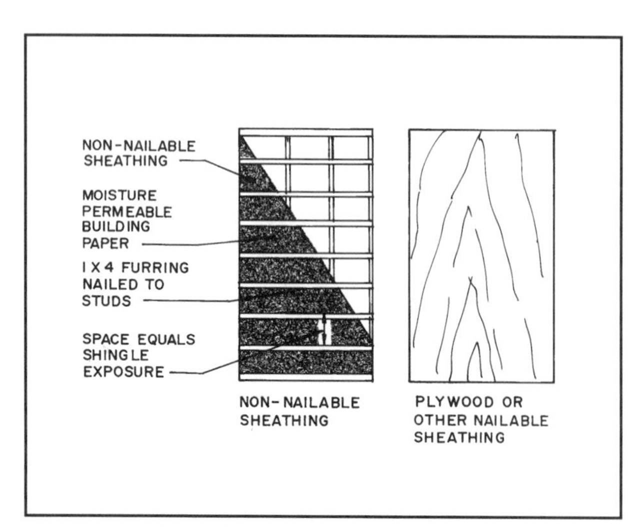

25-66 Wood shingles and shakes must be nailed or stapled to nailable sheathing or wood furring strips.

25-67 Nail with half or less of the shingle exposed. *(Courtesy Red Cedar Shingle and Handsplit Shake Bureau)*

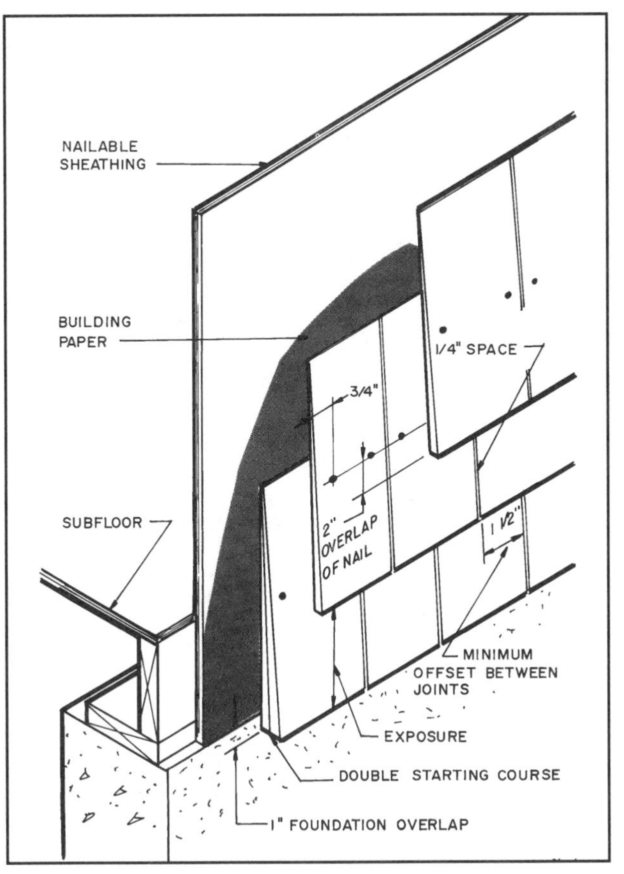

25-68 How to apply single-course wood shingles and shakes.

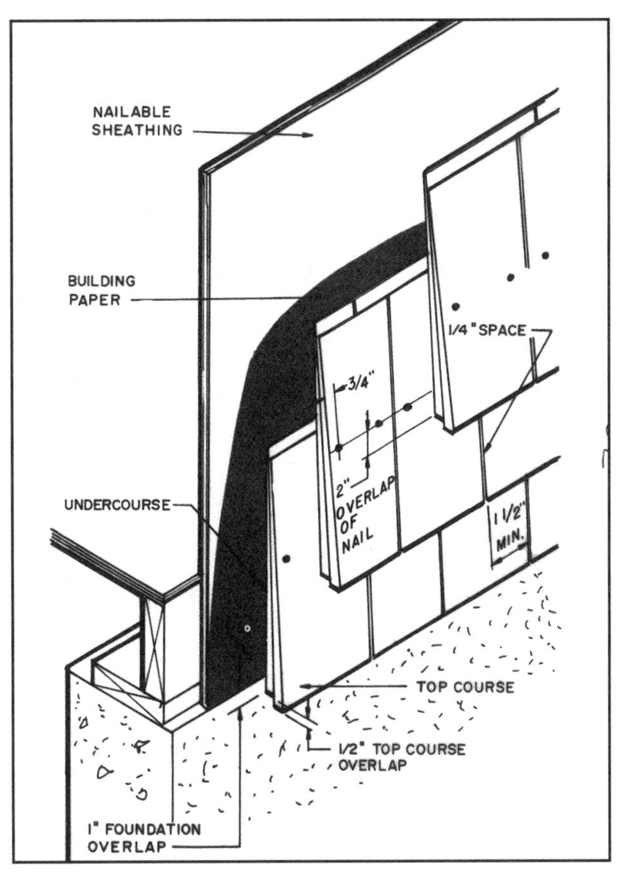

25-69 How to apply double-course wood shingles and shakes.

Wood shingles are applied as a single or double course. Single-course application is shown in **25-68**. The first course is doubled, providing a drip edge on the outside shingle. It should overlap the foundation by one inch. Each course is nailed over the one below, leaving the calculated exposure. It is nailed so that the shingle above overlaps the line of nails by two inches. A ¼-inch (6mm) space is left between shingles, and the space should be offset at least 1½ inches (38mm) from a joint above or below it.

The double-course application is shown in **25-69**. Each course is doubled. The top shingle in each course overhangs the undershingle by ½ inch. In both cases the lower-grade shingles are used on the underlayers.

Table 25-6 Wood shingle and shake siding exposure recommendations.

(inches)	Shingles			Shakes		
Length	16	18	24	18	24	32
Single Course	6 – 7½	6 – 8½	8 – 11½	8½	11½	15
Double Course	8 – 12	9 – 14	12 – 16	14	20	—

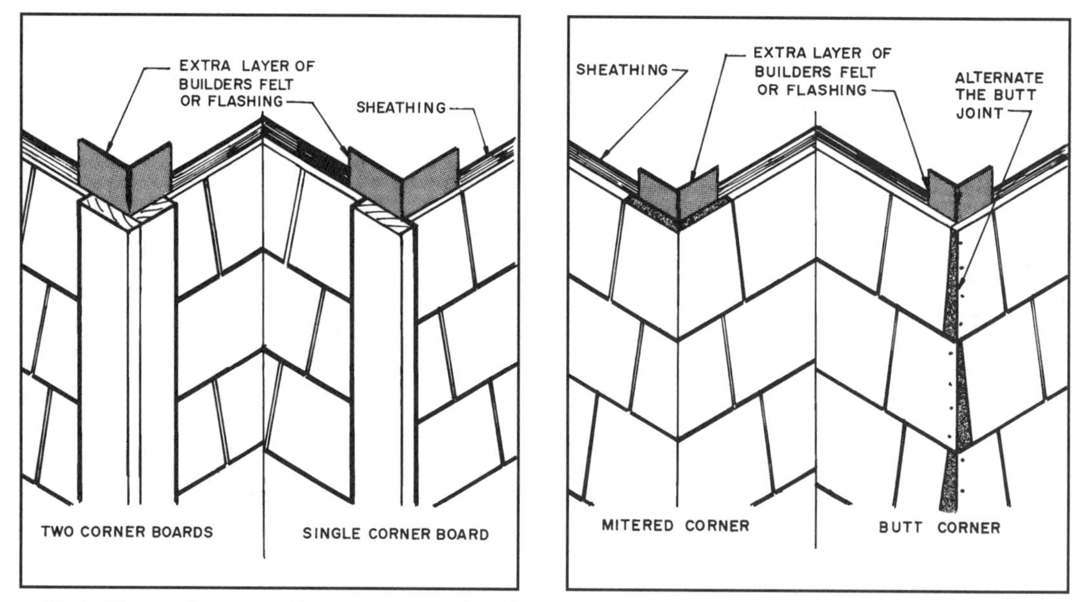

25-70 Ways to frame outside corners when installing wood shingles and shakes.

Interior and exterior corners can be formed using corner boards or by cutting and fitting the shingles **(see 25-70 and 25-71)**.

The long shingle panels available are installed by applying builder's felt over the sheathing and nailing the panels directly into the studs. If nail-holding sheathing is used, they can be nailed to it **(see 25-72)**. It is important to get the first course level because the panels have a rabbet in the bottom edge that fits onto the top of the panel below. This makes the courses self-aligning. The corners are covered with a manufacturer-supplied corner **(see 25-73)**.

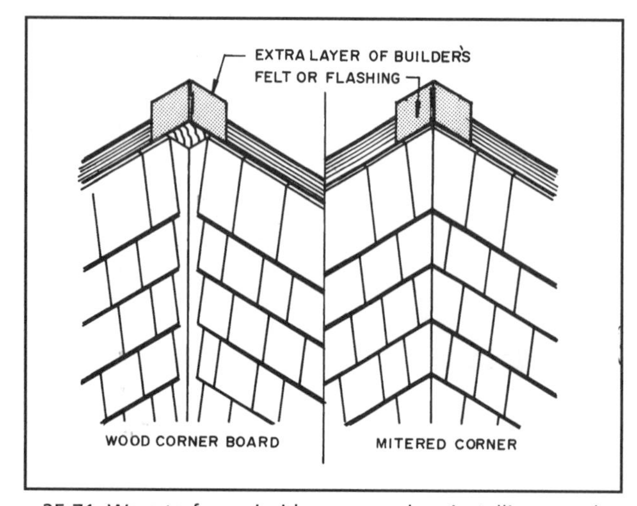

25-71 Ways to frame inside corners when installing wood shingles and shakes.

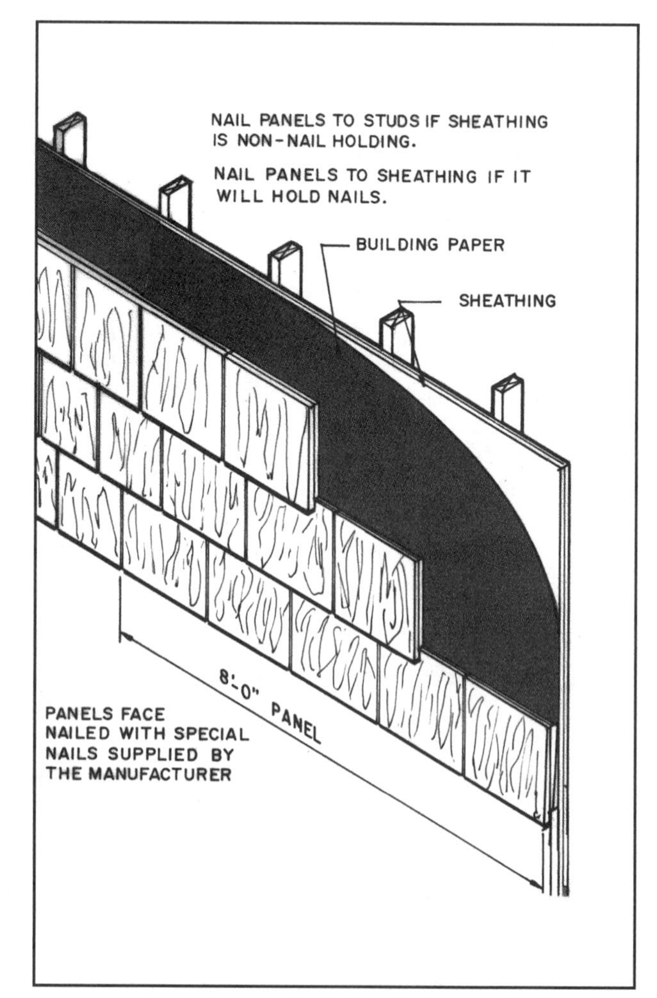

25-72 Wood shingle siding panels are nailed to the studs or to nailable sheathing.

Nails Used with Wood Shingles

Rust-resistant nails are required. The screw type are recommended. Normally, 3d nails are used on the undercourse and 5d on the overcourse. Longer nails can be used, if needed. The nails are placed about ¾ of an inch from each edge of the shingle. Shingles narrower than 8 inches require two nails. Those wider require three nails. The nails should be driven flush with the surface of the shingle but should not crush the wood below the head.

When staples are approved for use, typically 1¾- to 2-inch lengths are recommended.

OTHER SHINGLE PRODUCTS

On the market are a number of shingle products manufactured from plastics and other materials. One product is injection molded from a thermoplastic resin with selected additives. Also available are brick and stone patterns. This product is manufactured in panels 18 × 36 inches.

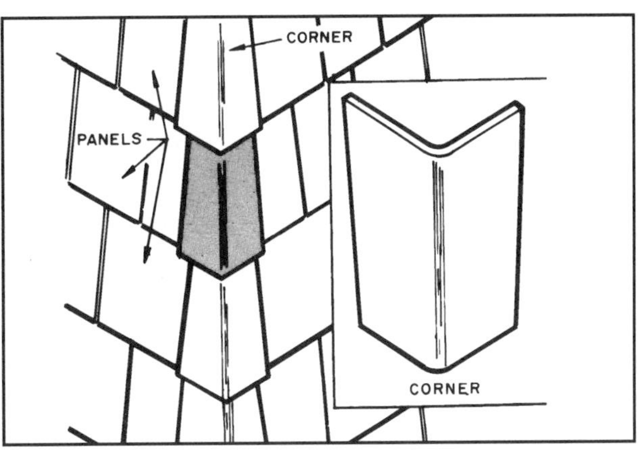

25-73 Manufactured corners are used to cover the outside corners of wood shingle panels.

These molded products are installed much like vinyl siding. The manufacturer supplies complete installation instructions and the necessary trim and channels. A residence finished with these shingles is in **25-74**.

25-74 This building has been covered with shingles made by injection molding, a thermoplastic resin formulated with special additives. *(Courtesy Nailite International)*

25-75 The walls and roof of this building are finished with a galvanized steel product that has a baked-on enamel finish. *(Courtesy Wheeling Corrugating Company, a division of Wheeling-Pittsburgh Steel Corporation)*

STEEL SIDING

Steel siding is available in panels 38 inches wide and 6 through 40 feet in length. They are zinc coated and given a factory-applied painted coating. Panels are available in many colors.

The manufacturer supplies various trim pieces such as preformed corner channels and trim to surround doors and windows into which the vertically installed steel wall panels are fitted. The roof is covered in the same manner as the walls. A finished installation is shown in **25-75**.

The panels are secured to the studs with screws supplied by the manufacturer. The edges overlap and are sealed together as the screw enters the wood. One joint design is shown in **25-76**.

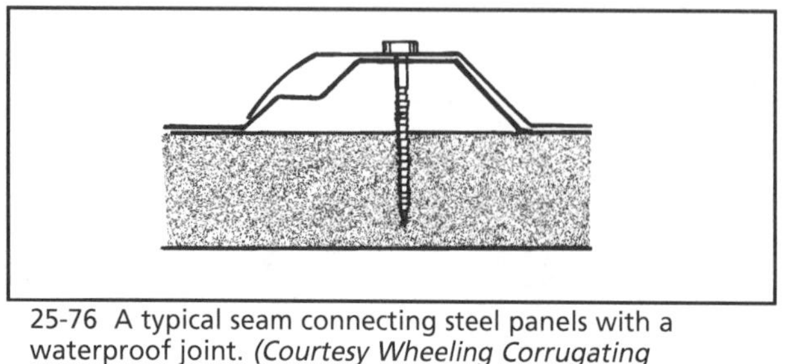

25-76 A typical seam connecting steel panels with a waterproof joint. *(Courtesy Wheeling Corrugating Company, A Division of Wheeling-Pittsburgh Steel Corporation)*

PORTLAND CEMENT PLASTER (STUCCO)

Portland cement plaster, also referred to as **stucco**, is a combination of Portland cement-based cementitious materials and aggregate mixed with water to form a plastic mass. It adheres to a surface and cures forming a hard, durable finish material. Various textures can be imposed on the surface while it is still plastic.

Most commonly the wall to be plastered is sheathed with a nailable material such as plywood or oriented strand board. The sheathing is covered with two layers of builder's felt or some other waterproof building paper. If gypsum board, insulation board, or expanded polystyrene sheathing is used, the wall must have bracing installed to stop racking (being forced out of shape or out of plumb). Metal lath, woven-wire mesh, or welded-wire lath is installed over this surface by nailing it to the studs **(see 25-77)**. This serves as the plaster base and reinforces the stucco base layer. Nails should be corrosion resistant and long enough to penetrate at least ¾ inch into the studs. A typical layered installation is shown in **25-78**.

If no sheathing is used, the wall framing must have diagonal bracing let into it. Rows of soft 18-gauge steel line wire are stretched horizontally across the studs. They are spaced six inches apart and nailed or stapled to every fourth stud. The builder's felt is nailed over the wire. It is lapped three inches on the edges. The metal lath is nailed over the felt **(see 25-79)**.

25-77 Woven-wire lath is being installed over nailable sheathing to provide base and reinforcement for the stucco-coat base layer.

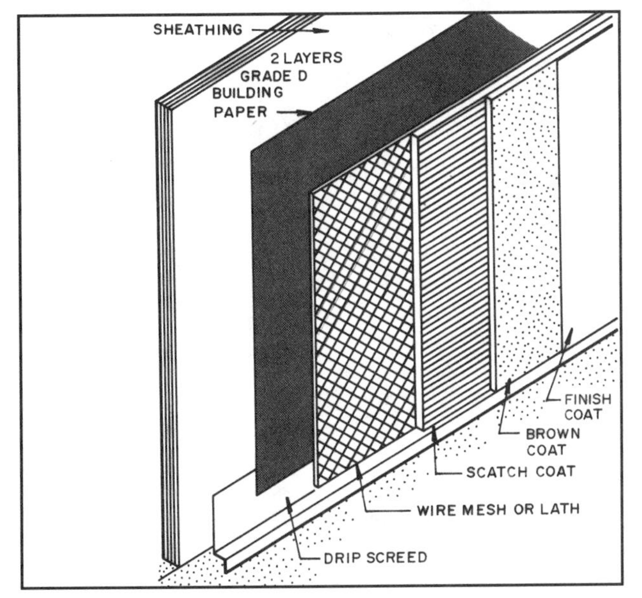

25-78 The layers forming a Portland cement stucco wall when sheathing is used.

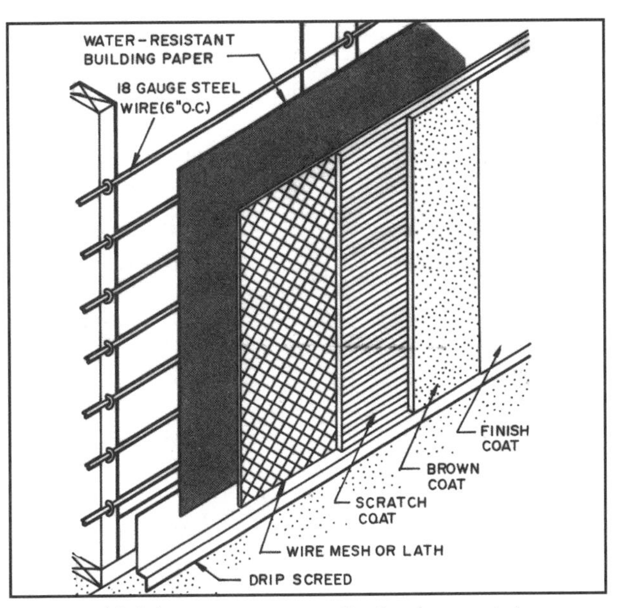

25-79 This is how to construct a Portland cement stucco exterior wall finish when sheathing is not used.

Usually, three coats of Portland cement plaster are applied. The first coat is the scratch coat. It is worked into the metal lath. The second coat is the brown coat. It builds up the thickness of the layer and smooths irregularities in the scratch coat. The final coat is the finish coat. This can vary depending upon the texture desired. It can be troweled smooth or brushed to a rough texture. Coloring can be added to the finish coat. The wall can be painted.

Usually, a metal molding is applied on the edges and around openings in the wall **(see 25-80)**. If the wall area is large, control joints are installed. These provide some movement, thus relieving stresses and reducing the possibility of cracking.

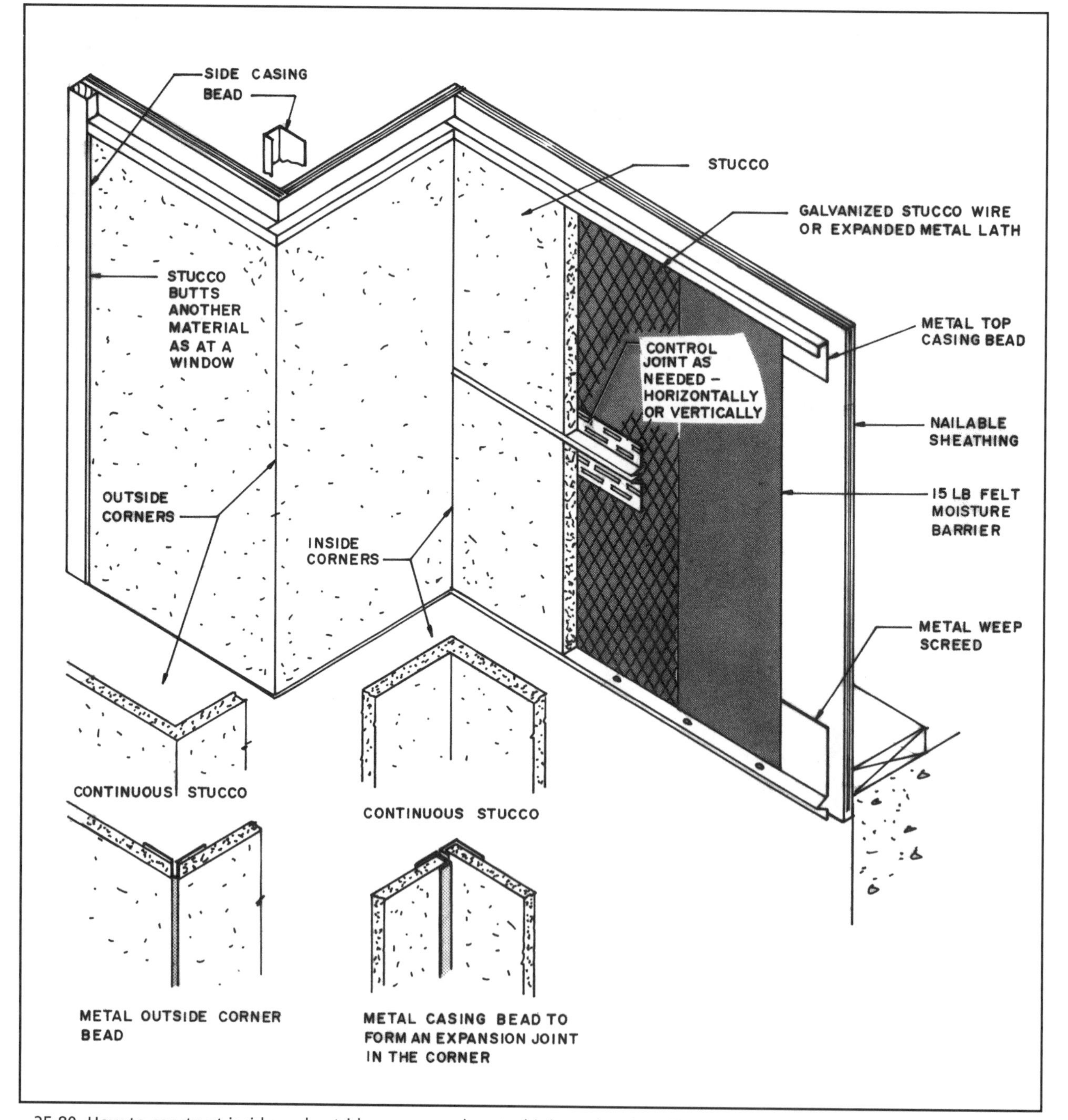

25-80 How to construct inside and outside corners and control joints when installing Portland cement plaster stucco.

EXTERIOR INSULATION & FINISH SYSTEM (EIFS)

This exterior wall finish system is often referred to as synthetic plaster or synthetic stucco. One system uses an approved substrate as sheathing, covered with a watertight, airtight membrane **(see 25-81)**. A proprietary expanded polystyrene (EPS) insulation panel with drainage channels is fastened to the sheathing with mechanical fasteners supplied by the manufacturer. This permits any moisture that may penetrate into the wall to drain to the bottom and weep at the flashing.

A base coat formulated by the manufacturer is troweled over the drainage board and a reinforcing mesh is embedded in it. Finally another base coat is troweled over the mesh. When it has hardened the synthetic plaster finish coat is troweled and textured as desired over the base coat **(see 25-82)**.

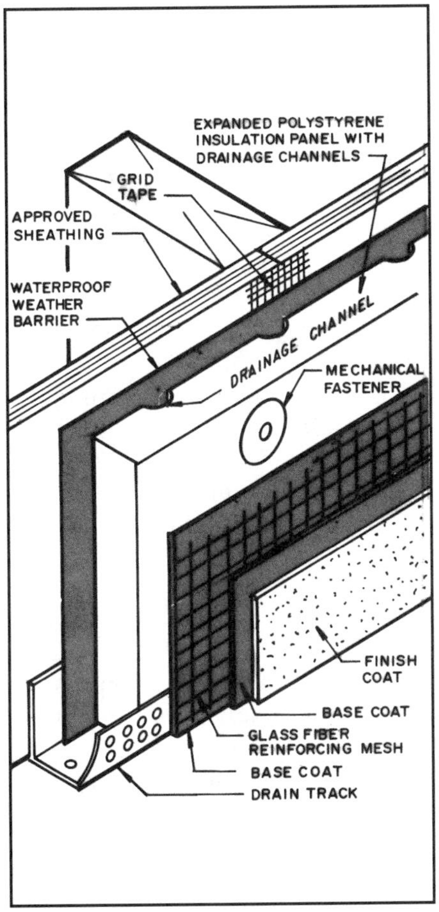

EXPANDED POLYSTYRENE INSULATION PANEL WITH DRAINAGE CHANNELS
GRID TAPE
APPROVED SHEATHING
WATERPROOF WEATHER BARRIER
DRAINAGE CHANNEL
MECHANICAL FASTENER
FINISH COAT
BASE COAT
GLASS FIBER REINFORCING MESH
BASE COAT
DRAIN TRACK

25-81 This exterior finish wall construction is an example of the exterior insulation and finish system (EIFS). It illustrates the products and design of the Dryvit® Residential MD (moisture drainage) System. *(Courtesy Dryvit Systems, Inc.)*

25-82 The finish coat of synthetic plaster is troweled over the base coat. *(Courtesy Dryvit Systems, Inc.)*

25-83 This residence has the exterior finished using the exterior insulation and finish system (EIFS).

A residence that has been finished with the exterior wall finish system is shown in **25-83**.

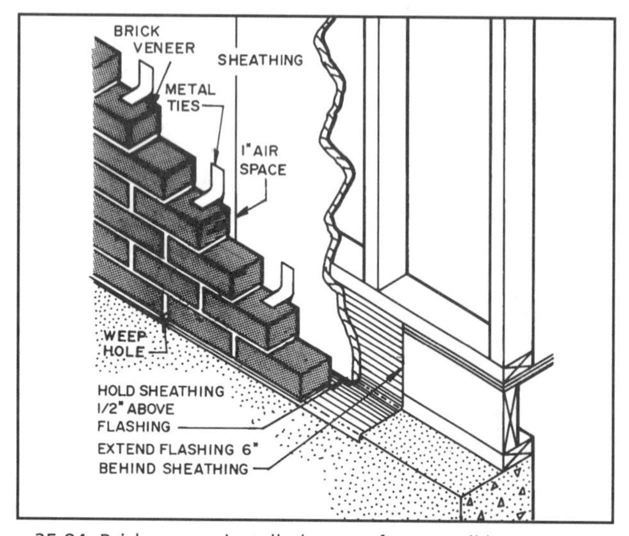

25-84 Brick veneer installed over a frame wall has a one-inch air space and weep holes to permit moisture to drain.

Be certain to consult the technical manuals of the company manufacturing the system. They give specific material specifications and detail drawings showing how to prepare inside and outside corners, how to butt doors and windows, and other details.

MASONRY VENEERS

A wood-framed wall can use brick or stone as the finish material. The foundation provides a five-inch ledge for the masonry units. A one-inch air space is left between the sheathing and masonry. Openings between the end joints in the bottom course of bricks permit moisture to weep. When using plywood sheathing or oriented strand board sheathing, cover it with builder's felt **(see 25-84)**.

Flashing at the base of the wall is needed. The flashing should extend below the bottom of the sheathing. It should extend at least six inches up on the sheathing. If the sheathing is covered with builder's felt, it should lap over the flashing.

Corrosion-resistant ties are used to join the masonry to the sheathing. They are spaced in rows 16 inches apart and spaced 32 inches apart vertically. If the sheathing will not hold nails, the ties must be nailed into the studs.

DORMERS

Dormers usually have wood siding in a horizontal or inclined application. The siding is applied as just described, except it is cut one to two inches short of meeting the roof. This cut end must be coated with water-repellent material and carefully painted to keep moisture out of the wood **(see 25-85)**. Other siding materials described in this chapter are also widely used.

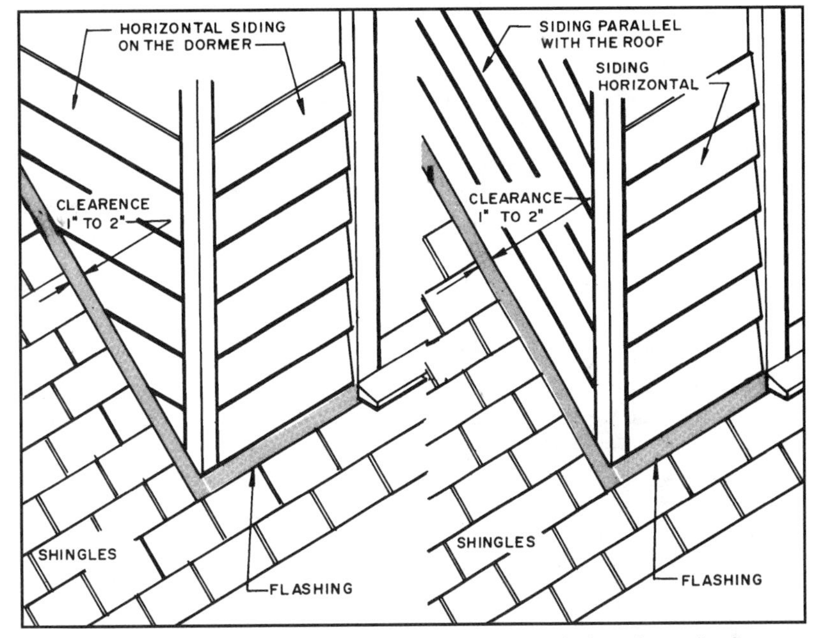

25-85 Wood siding on dormers should be kept 1 to 2 inches above the finished roof.

CHAPTER 26

FINISHING THE ROOF

The framing carpenters typically construct the cornice and sheath the roof. Cornice construction varies depending upon the type of roof and the architectural design. Various methods of cornice construction can be found in Chapter 19. Once the cornice has been finished and the roof sheathed, the finished roofing can be installed.

BUILDING CODES

Building codes will specify the physical requirements that a roofing material must meet and the method of installation. For example, in areas that have occasional high winds, the local code may specify the approved type of nail to use and ban the use of staples for installing asphalt or fiberglass shingles. Codes will also specify the level of fire resistance. The American Society for Testing and Material's standards are used to classify fire resistance into four groups.

Class A materials have the highest resistance to fire. These include materials such as slate, concrete, tile, and clay tile.

Class B materials are used where moderate resistance to fire is required. These include some types of composition shingles and metal roofing.

Class C materials are effective against light exposure to fire and include some asphalt shingles and fire-retardant-treated wood shingles.

Nonclassified materials include untreated wood shingles. The manufacturers indicate the verified rating for each of these materials.

ROOF SHEATHING

The common types of roof sheathing include lumber, plywood, oriented strand board, waferboard, and planking. These materials are nailed to the rafters. They support the finished roofing material. Plywood or solid wood sheathing is used for pitched roofs. Planking and wood fiber decking are used for homes in which the underside of the material forms the exposed ceiling.

Lumber Sheathing

Lumber sheathing can be laid closed or spaced **(see 26-1)**. The boards should be six to eight inches wide. If placed on rafters spaced 16 to 24 inches O.C., they should be ¾ inch thick. Use two 8d common nails at every rafter. If square-edged boards are used, their ends must meet on a rafter. Not more than two adjacent boards should have joints

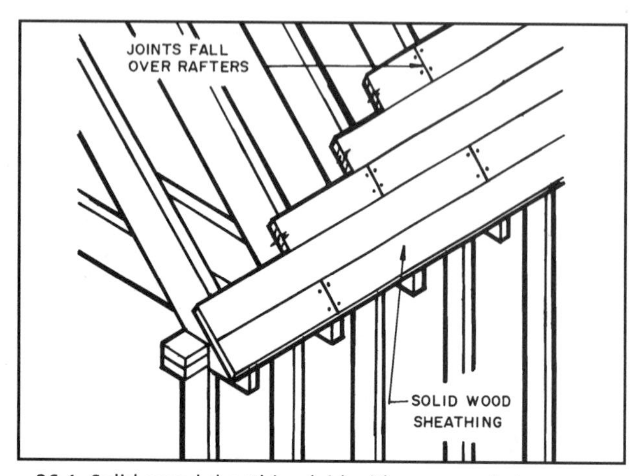

26-1 Solid wood sheathing laid with no space between the boards.

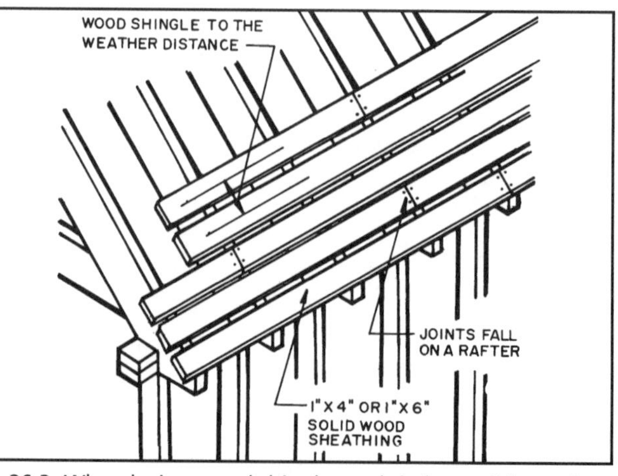

26-2 When laying wood shingles and shakes, solid wood sheathing is frequently laid with a space between the boards.

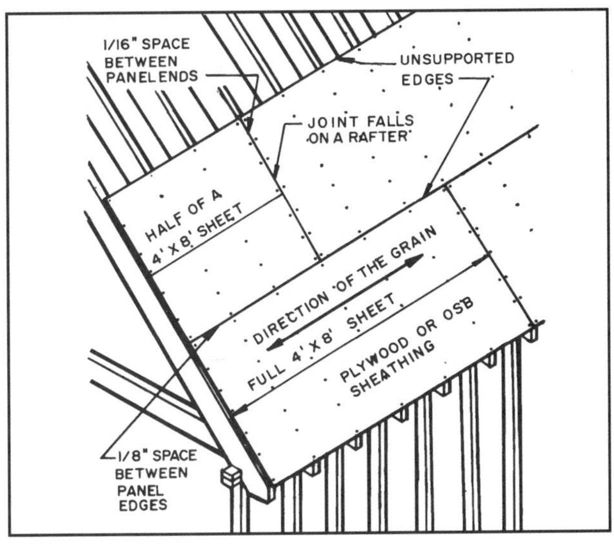

26-3 Plywood or oriented strand board sheathing is laid with the face grain or long edge perpendicular to the rafters.

on the same rafter. If end-matched tongue-and-groove boards are used, the end joints can fall between rafters. In no case should the joints of two adjoining boards be in the same rafter space. Each board must be supported by at least two rafters.

When **wood shingles** or **shakes** are used in damp climates, the boards should be spaced. This helps in drying the roof. Wood shingles require 1 × 4-inch board, and wood shakes, 1 × 6-inch. They are spaced on center the same as the shingles are laid to the weather. "Laid to the weather" means the amount of shingle actually exposed. For example, a 10-inch shingle lapped five inches by the one above it is laid five inches to the weather. For shingles laid five inches to the weather over 1 × 4-inch lumber, there will be about 1½ inches of space between the boards. Two 8d common nails should be placed in each board at each rafter **(see 26-2)**.

Plywood Sheathing

Plywood sheathing is laid with the face grain perpendicular to the rafters **(see 26-3)**. The sheathing grades of plywood include C-D Interior, C-D Interior with Exterior Glue, C-C Exterior, Structural I C-D Interior, Structural II C-D Interior, Structural I C-C Exterior, and Structural II C-C Exterior. These are all unsanded engineered grades. The C-D Interior with Exterior Glue is the most frequently used for roof and floor sheathing. Carpenters often refer to this as CDX.

The thickness of the plywood used depends upon the distance between rafters, the length of the unsupported edge of the panel, and the roof load. Design data are given in **26-4**. The panel identification index is a set of two numbers separated by a slash, such as 32/16. The first number, 32, is the recommended center-to-center spacing of rafters when the panel is used as roof sheathing. The other number, 16, is the center-to-center spacing of floor joists when the panel is used for subflooring **(see 26-5)**.

Under normal conditions, 5/16-inch plywood is used for 16-inch rafter spacing when wood or asphalt shingles are used. When the span is 24 inches, 3/8-inch plywood is used. Heavier roofs, such as slate, tile, and asbestos cement, require 1/2-inch plywood for 16-inch rafter spacing and 5/8-inch for 24-inch spacing.

If interior-type sheathing is used, no edges of the plywood panel should be exposed to the weather.

The carpenter should make a roof sheathing layout. This can be a freehand sketch to approximate scale. Draw a rectangle representing half the gable roof. One side is the length of the roof. The other is the rafter length plus overhang.

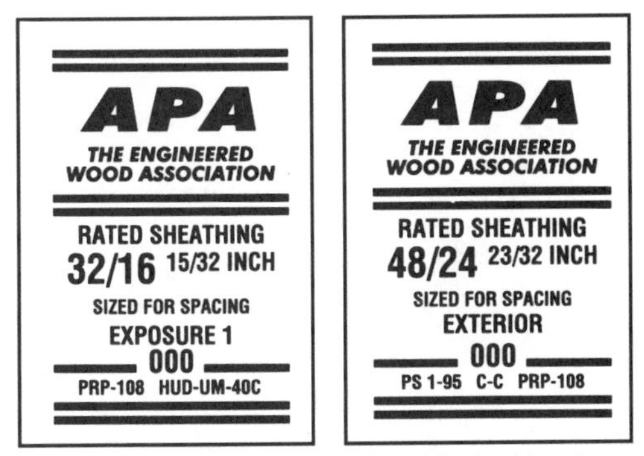

26-5 APA grade trademarks with the identification index for sheathing. *(Courtesy APA—The Engineered Wood Association)*

26-4 Roof sheathing span ratings.

Identification Index	Plywood Thickness (inches)	Oriented Strand Board (inches)	Waferboard Thickness (inches)
16/0	5/16, 3/8	3/8	3/16 no clips
24/0	3/8	7/16	3/16 with clips

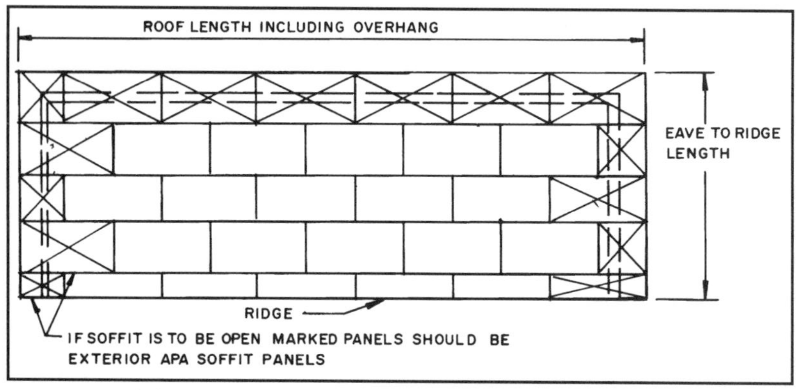

26-6 A roof sheathing layout for one side of a gable roof.

Start the layout at one corner with full 4 × 8-foot sheets. The second row of sheets starts with a half panel. In this way the joints are staggered. The top row will most likely have panels that must be cut to fit the remaining width. This layout can also be used to figure the number of sheets needed. Remember, it represents only half the roof. The number of sheets needed will have to be doubled **(see 26-6)**.

It the roof is to have an open soffit and sheathing with interior glue is used, the plywood covering the soffit will have to be panels with exterior glue. These are marked with an X on the layout. Generally, the soffit panels used will be thicker than the others. This is so that the roofing nails will not break through and be visible from below. The carpenter will have to shim up the thinner panel where the two types meet. The panels not over the soffit could be the same thickness, but this costs more.

Panels 5⁄16 to ½ inch thick are fastened with 6d common, ring shank, or spiral thread nails. Panels up to one inch thick require 8d common nails. They are spaced 6 inches O.C. on panel edges **(see 26-7)**. They are spaced 12 inches O.C. inside the panel. Always allow 1⁄16 inch between panel ends and ⅛ inch between panel edges.

Sometimes the unsupported edges of sheathing panels require blocking to increase stiffness. Instead of wood blocking, aluminum clips can be used. They are made in thicknesses from 5⁄16 to 13⁄16 inch. They are tapered to their edges. They are installed by slipping them over the edges of the plywood panels **(see 26-8)**.

A fast way to move plywood panels to the roof is with a conveyor or crane **(see 26-9)**.

On low-pitched roofs, the sheets can be stacked on the rafters. Steep-pitched roofs require that a few pieces of bracing be nailed to the rafters to keep the sheet from sliding off the roof.

On small jobs a carpenter on the ground can raise the sheets to a person on the roof. This person pulls them onto the roof and stacks them. It helps if they can be moved from the delivery truck directly onto the roof.

26-7 Plywood roof sheathing is nailed to the rafters. A ⅛-inch space is left between the long sides of the panels.

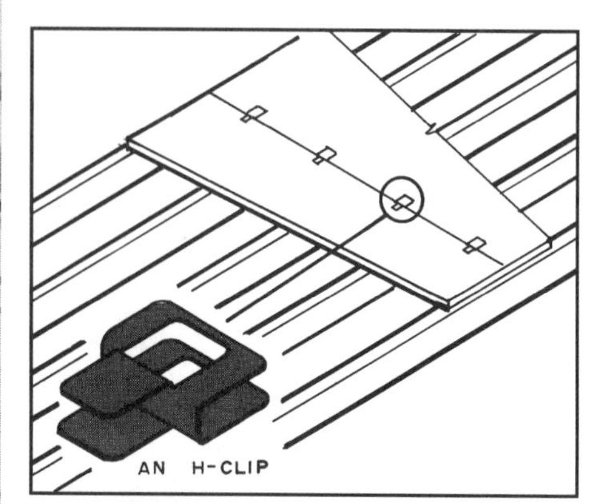

26-8 When unsupported sheathing edges require support, 2 × 4 blocking or aluminum H-clips can be used. The H-clips are much faster to install.

26-9 Conveyors can be used to move sheathing to the roof. *(Courtesy APA—The Engineered Wood Association)*

Oriented Strand Board & Waferboard

Oriented strand board (OSB) is strand-like wood particles arranged in perpendicular layers and bonded with a phenolic resin. **Waferboard** is made by compressing wafer-like wood flakes and bonding them with phenolic resin. They are marked with the same product identification index as described for plywood. Normally, oriented strand board used for roof sheathing is 7⁄16 inch thick for rafters 16 inches O.C. and 3⁄8 inch thick if rafters are 16 inches O.C. with no edge clips and will go 24 inches O.C. if edge clips are used.

Plank Decking

Plank decking is tongue-and-groove wood two inches and thicker. It is used in flat and low-pitched roofs and in post-and-beam construction. It will span distances of 6 to 8 feet between supports. Consult the manufacturer for span information. Examples can be seen in Chapters 20 and 21.

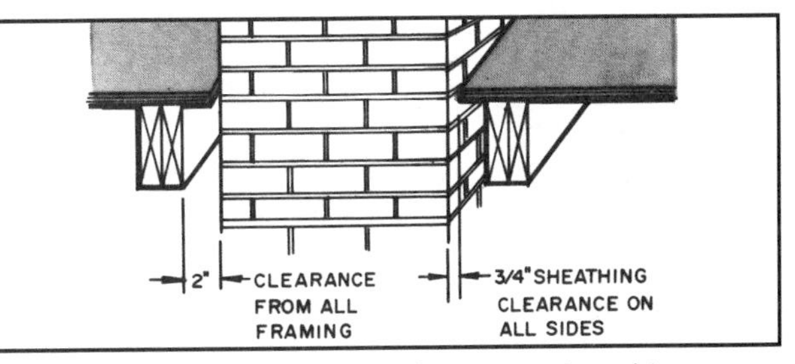

26-10 Wood framing and sheathing must clear chimneys as specified by local codes.

SHEATHING AT CHIMNEY OPENINGS

Rafters and headers should clear chimneys at least two inches. Sheathing should overhang these but clear the chimney ¾ inch **(see 26-10)**. Check your local building code to be certain you have the required clearances.

SHEATHING AT THE GABLE-END

When gable-ends have little or no projection, the sheathing is set flush with the outside of the wall sheathing. If it extends beyond the end wall, the sheathing should span at least three rafter spaces. This provides necessary strength and prevents sagging. Projections beyond 16 inches require special ladder framing as shown in Chapter 13.

UNDERLAYMENT

The sheathing is covered with an **underlayment** to provide protection for the sheathing until the finish roofing is installed. It also protects if a leak occurs or snow backs water up under the shingles.

Asphalt-saturated felt is used for the underlayment. It is required below fiberglass and asphalt shingles as well as various metal and clay roofing tiles. Sometimes it is used below wood shingles.

The layers are edge-lapped two inches and end-lapped four inches **(see 26-11)** and lapped six inches over both sides of all hips and ridges. Plastic disks can be nailed across the underlayment to help hold it in place until the finished roofing material is applied.

26-11 Underlayment is lapped at the ends and edges of sheets. It is placed over the drip edge at the eave and below it on the rake.

DRIP EDGE

The edge and rake of the roof should have a metal **drip edge** applied. It is usually galvanized steel. It is designed to protect the edges of the sheathing and prevent leaks. At the eave the underlayment goes over the drip edge. At the rake of the roof it goes under the drip edge **(see 26-12)**.

ROOF FLASHING

Flashing is metal or some other material placed where special protection is needed to prevent water from entering. Anything that pierces the roof requires flashing. Examples include pipes, chimneys, and dormers. When two roofs intersect, as at a valley, flashing is necessary.

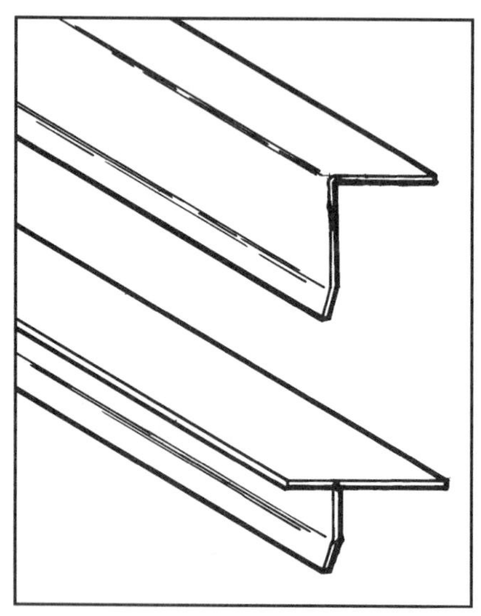

26-12 Two types of metal drip edges.

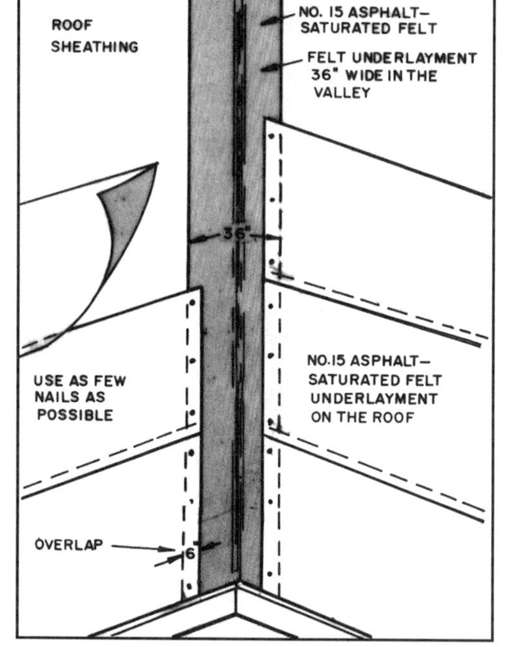

26-14 When installing the underlayment, cover valleys with 36-inch-wide pieces of No. 15 asphalt-saturated felt.

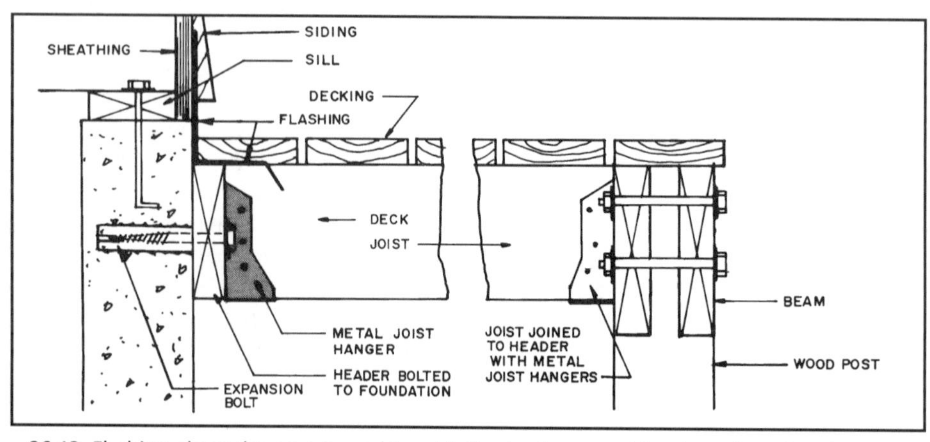

26-13 Flashing along the eave is used in cold climates to prevent ice dams from forming.

Flashing is generally galvanized metal, aluminum, copper, or stainless steel. The nails used to hold the flashing in place should be of the same material as the flashing. Mineral-surfaced roll roofing is also used.

Eave Flashing

The eaves must be flashed in areas where the outside design temperature is zero degrees F (–18 degrees C) or colder or where there is the possibility of ice forming along the eaves. This keeps the moisture backed up from penetrating the shingles and underlayment. The exact procedure varies with the type of finish roofing material. Typically it involves adding an additional layer of 50-pound asphalt-saturated builder's felt or mineral roll roofing from the edge of the roof up over the first layer of underlayment **(see 26-13)**.

This flashing should extend two feet over the warm interior of the building. This helps prevent snow on the roof from melting and refreezing, forming an ice dam. An ice dam will permit water to run up under the shingles and possibly leak to the interior of the building.

Valley Flashing

As the underlayment is laid, a 36-inch-wide strip is placed over the valley as shown in **26-14**. The valley flashing is placed over this underlayment. The three types of valleys are open, woven, and closed cut.

The **open valley** can be flashed with two layers of mineral-surfaced roll roofing applied as shown in **26-15**. Notice that the shingles are cut back, leaving the valley flashing exposed. The flashing is nailed only along the edges. The roof shingles then are cemented to the flashing.

The line of the shingles on open valley flashing is located with a chalkline. Snap a chalkline from the ridge to the eave. Space six inches apart at the ridge and slope about ⅛ inch per foot of valley to the eave. The shingles are cut to the chalkline mark, forming a straight edge **(see 26-16)**.

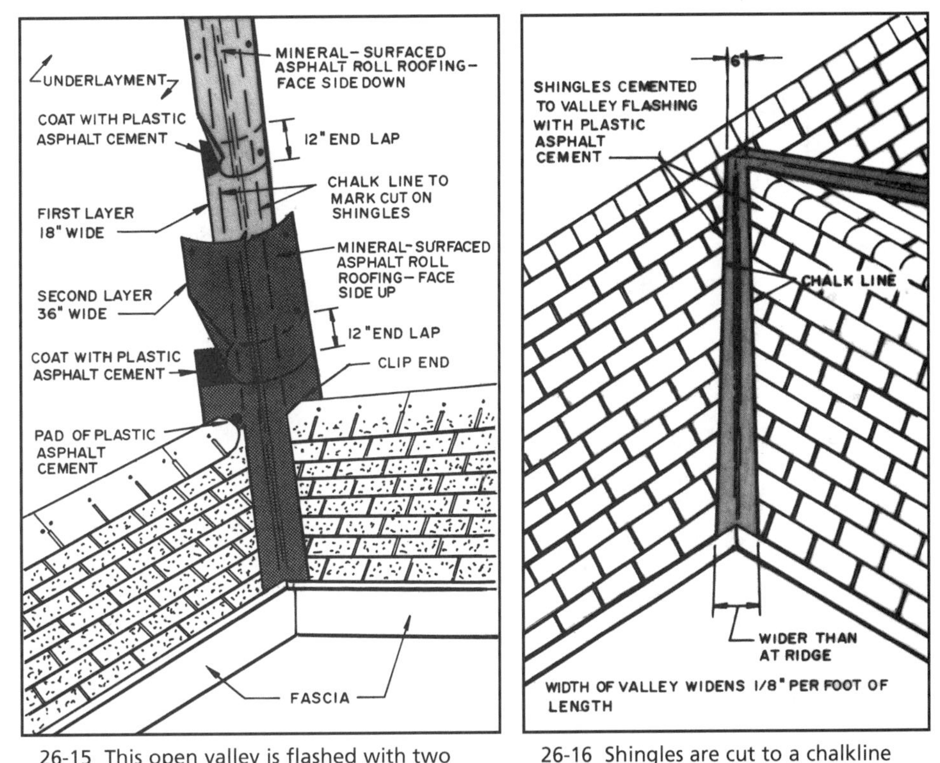

26-15 This open valley is flashed with two layers of mineral-surfaced roll roofing.

26-16 Shingles are cut to a chalkline along the valley flashing. The valley widens as it nears the eave.

As the shingles are laid to the chalkline, their upper corner is cut on about a 45-degree angle. This helps prevent water from running under the shingles. The shingles are cemented to the valley lining with asphalt cement. No nails should be exposed along the valley.

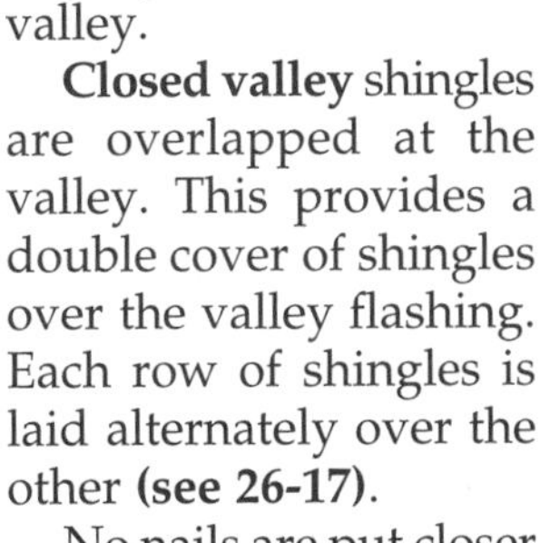

Closed valley shingles are overlapped at the valley. This provides a double cover of shingles over the valley flashing. Each row of shingles is laid alternately over the other **(see 26-17)**.

No nails are put closer than six inches from the center of the valley. Use two nails at the end of each shingle.

A **closed-cut valley** is shown in **26-18**. The valley is covered with 36-inch-wide roll roofing felt.

The shingles on one side lap over the valley and are nailed to the sheathing. The shingles from the other side are cut on a line parallel with the valley. They are cut about two inches short of the valley.

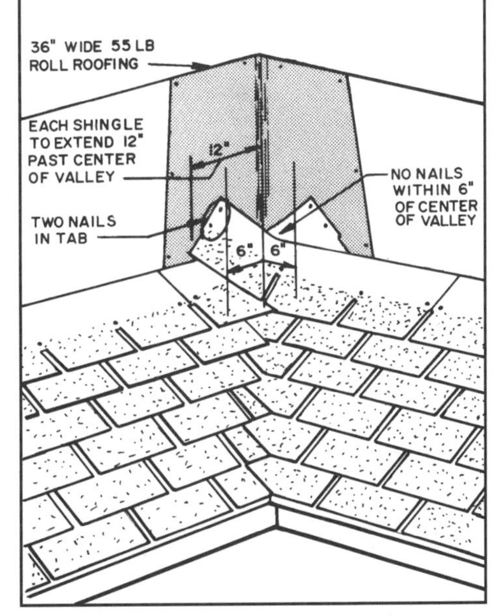

26-17 A closed valley is flashed with a layer of 55-lb. roll roofing felt.

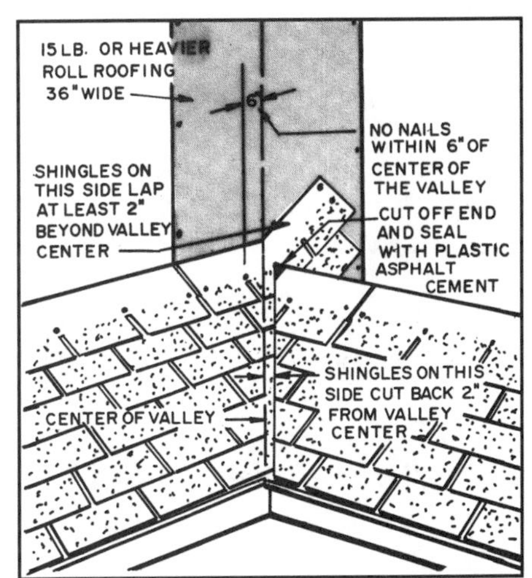

26-18 Details for a closed-cut valley. The cut side is set back about two inches beyond the center of the valley.

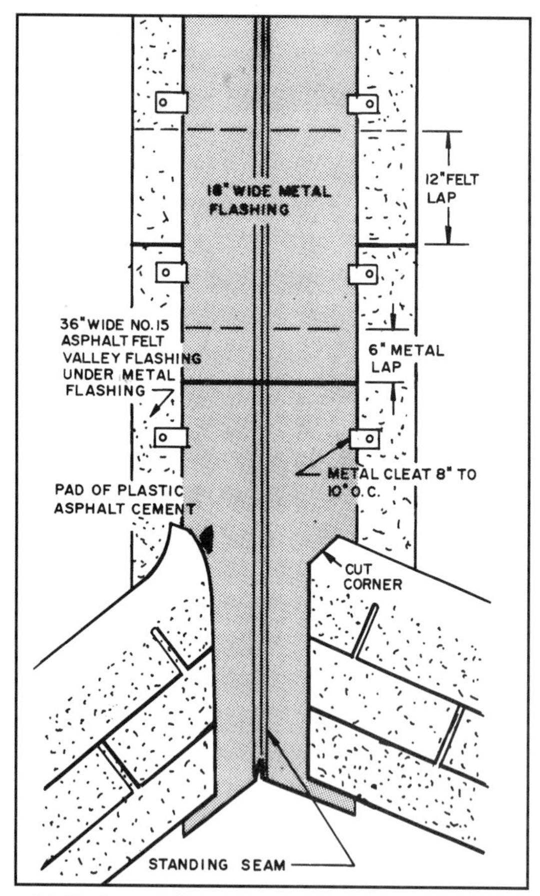

26-19 Open valley metal flashing is underlain with 36-inch-wide asphalt felt flashing.

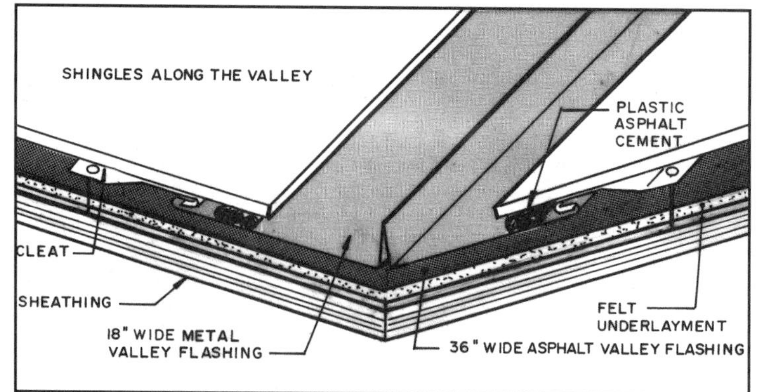

26-20 Metal flashing is held to the sheathing with metal cleats.

Valleys can also be flashed with metal flashing. This is usually aluminum or galvanized steel. It helps if the metal has a seam folded in the center **(see 26-19)**. This helps keep the water from one roof from running under the shingles. The flashing is secured to the sheathing with metal clips **(see 26-20)**. A layer of asphalt builder's felt is placed below the metal flashing.

Vertical Wall Flashing

When the roof meets a wall, the intersection must be flashed. Generally, this is done with stepped metal flashing applied over the end of each course of shingles **(see 26-21)**. The flashing is two inches wider than the exposed face of the shingle. It is laid so that each piece overlaps the two inches. It is bent so that four inches are against the wall and two inches extend on to the roof. Each flashing piece is placed slightly up the roof from the exposed edge of the shingle. In this way it will not be visible. Nail each metal piece to the wall sheathing with one nail in the top corner. After the flashing is in place the siding is nailed over it. The siding should clear the roof one to two inches **(see 26-22)**. Notice that the asphalt felt underlayment is turned up about four inches along the vertical wall.

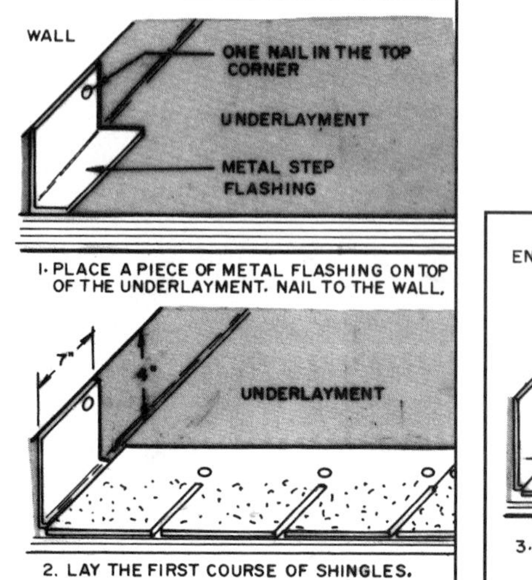

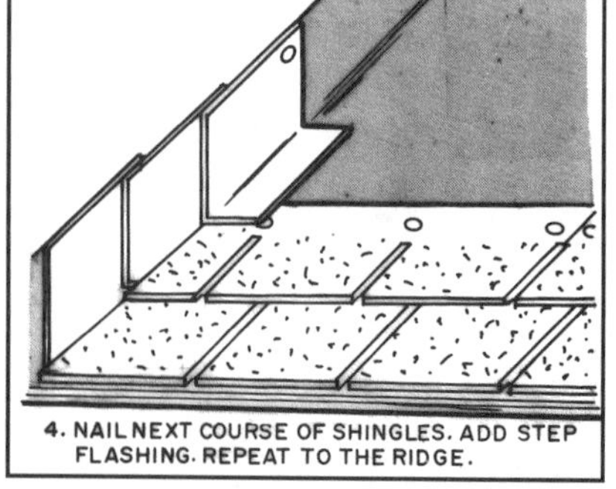

26-21 The intersection of a roof and a wall is flashed with metal stepped flashing.

FLASHING A CHIMNEY

The chimney is built on a footing separate from the rest of the foundation. This permits it to move slightly due to possible settling. The flashing between the chimney and roof must have a base flashing that will allow movement. This is covered with counterflashing which is connected to the chimney but not the sheathing.

Begin by running the shingles up to the front of the chimney and build a cricket on the back side **(see 26-23)**. The cricket sheds the water to the sides of the chimney.

Install base flashing on the front first and then on the sides. It should lie out at least four inches from the chimney on top of the roof and 12 inches on the face. Bond it to the chimney with plastic asphalt cement **(see 26-24)**. Base flashing is typically galvanized sheet steel or copper. On the front it will go over the shingles. Nail it through the shingles into the sheathing. Next flash the cricket.

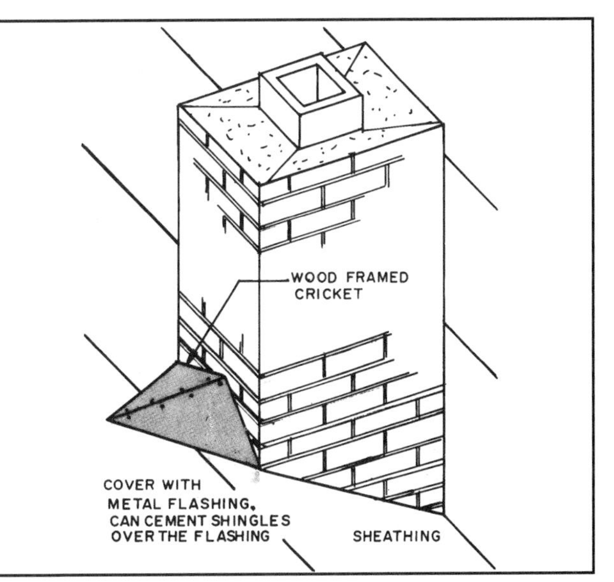

26-23 A cricket is built on the back of the chimney to divert the water to the sides.

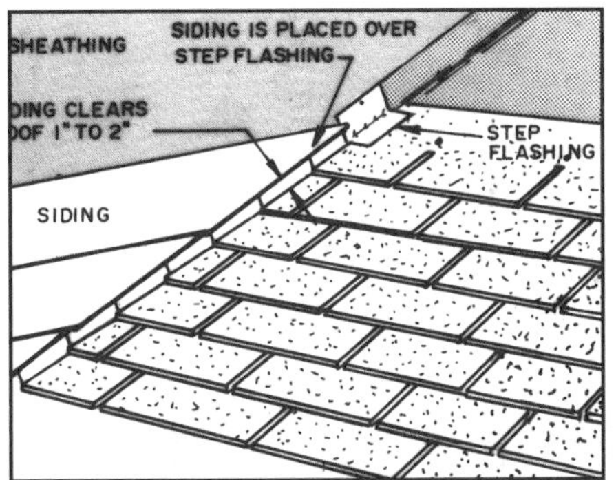

26-22 Siding is placed over the step flashing and kept 1 to 2 inches above the surface of the roof.

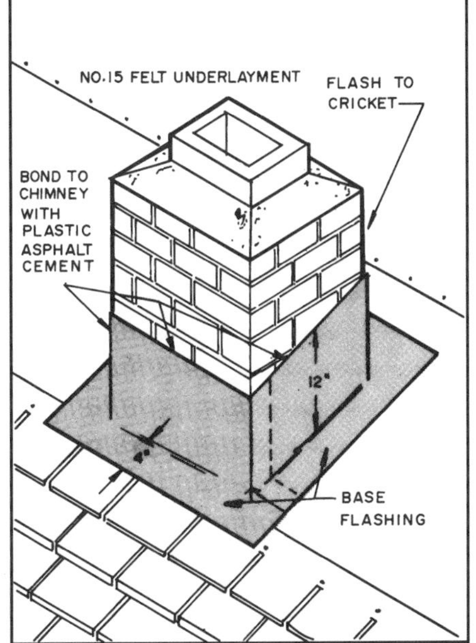

26-24 To flash a chimney, begin by installing the base flashing.

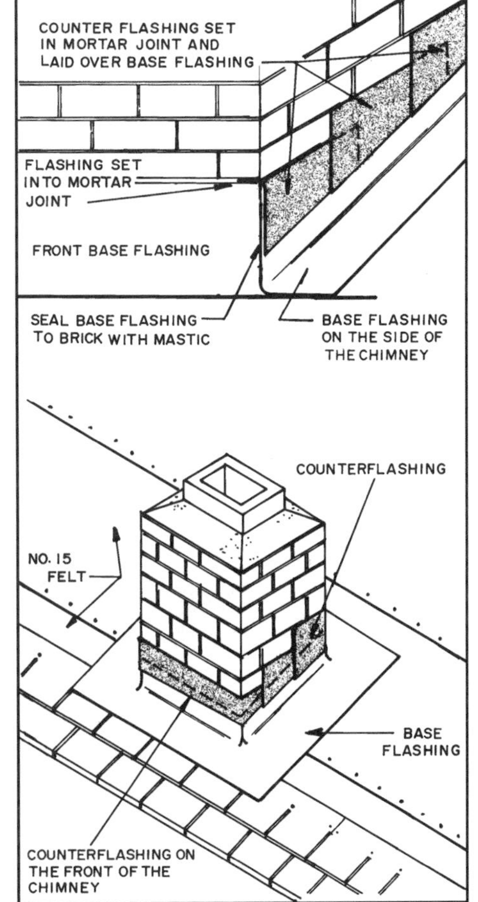

26-25 Chimney flashing is completed by installing counterflashing over the base flashing.

Next install the counterflashing. Step flashing is used on the side of the chimney and one solid piece is on the front. They are set into the mortar joint which is then filled with mortar. When it sets, bend the pieces down over the base flashing as shown in **26-25**. The stepped counterflashing should overlap the base flashing three inches and each other two inches. Now lay the shingles up to the chimney over the flange of the base flashing. Cement the shingles to the flashing.

Stack Flashing

Vent pipes from the waste disposal system project through the roof. These are flashed as shown in **26-26**. Begin by laying the shingles up to the pipe. Cut a hole in the shingle that falls over the pipe. Several types of flashing units are available. The one shown has a metal base with a rubber sleeve.

The pipe slides through the sleeve, fitting tightly against it, sealing out rain. Nail the metal base to the roof in the area above the shingles. Continue shingling the roof and cut the shingles to fit around the pipe. The shingles go on top of the metal base. Apply plastic asphalt cement between the base and the shingles to bond them together.

Another type of vent pipe flashing unit is shown in **26-27**. It consists of a metal unit that fits over the vent pipe. The connection between the flashing base and pipe is flexible, allowing it to be adjusted for a roof with different slopes. The top of the flashing pipe is folded over the top of the vent pipe. The shingles are fitted around the flashing as described in **26-26**.

ROOFING MATERIAL CHOICES

On steep-sloped roofs (a slope of 3 in 12—written 3⁄12—or greater) the following materials are used:

1. Fiberglass asphalt and organic asphalt shingles.
2. Wood shakes and shingles.
3. Natural slate.
4. Clay tile and concrete tile.
5. Metal sheet and tile roofing.

On flat and low-sloped roofs (usually a slope of 3 in 12—3⁄12—or less) the following materials are used:

1. A built-up membrane.
2. Certain types of metal sheet roofing.

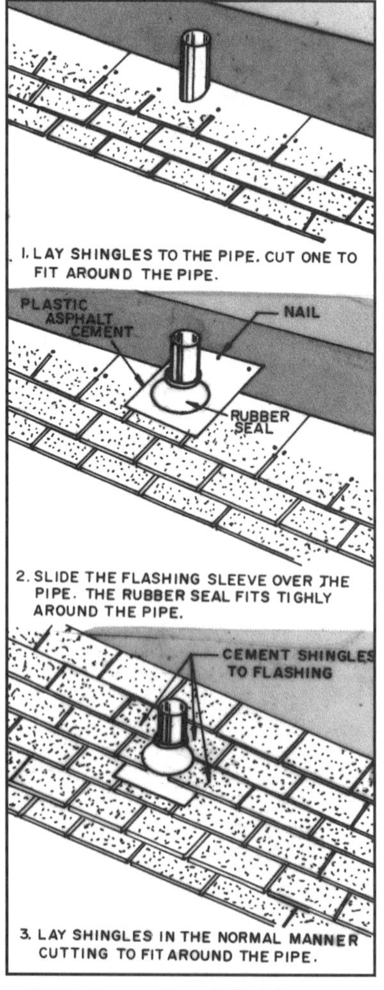

26-26 How to install flashing sleeves on pipes that penetrate the roof.

ROOFING MATERIALS

Roofing materials shield the interior of the building from the elements, provide fire protection, and enhance the exterior appearance. The finished roof is often the dominant feature seen when viewing a house.

The type of finished roofing material to be used is shown on the working design drawings and detailed in the specifications. The local building code also has a bearing on the type of material that can be used. A wide choice of roofing materials is available to the architectural designer and homeowner. Things considered include the appearance, architectural style, cost, weight, availability, and fire rating. Following are those materials in general use.

Asphalt shingles are made using a felt layer that is constructed of organic materials such as cellulose fibers and is saturated with asphalt and has a finished, exposed surface of granular material, such as marble chips, slate, or granite **(see 26-28)**.

Fiberglass asphalt shingles have a similar construction to asphalt shingles, except that they are made on a fiberglass base. Some of the types of shingles are in **26-29** and **26-30**. The adhesive strip softens when the sun heats the shingle and bonds it to the shingle below.

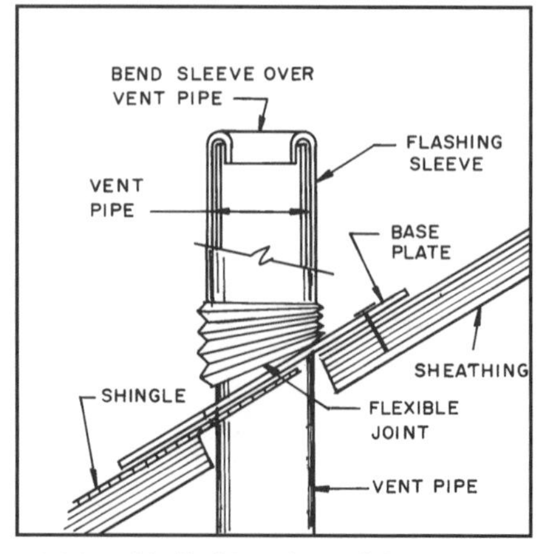

26-27 This flashing sleeve folds over the top of the vent pipe, providing a positive seal.

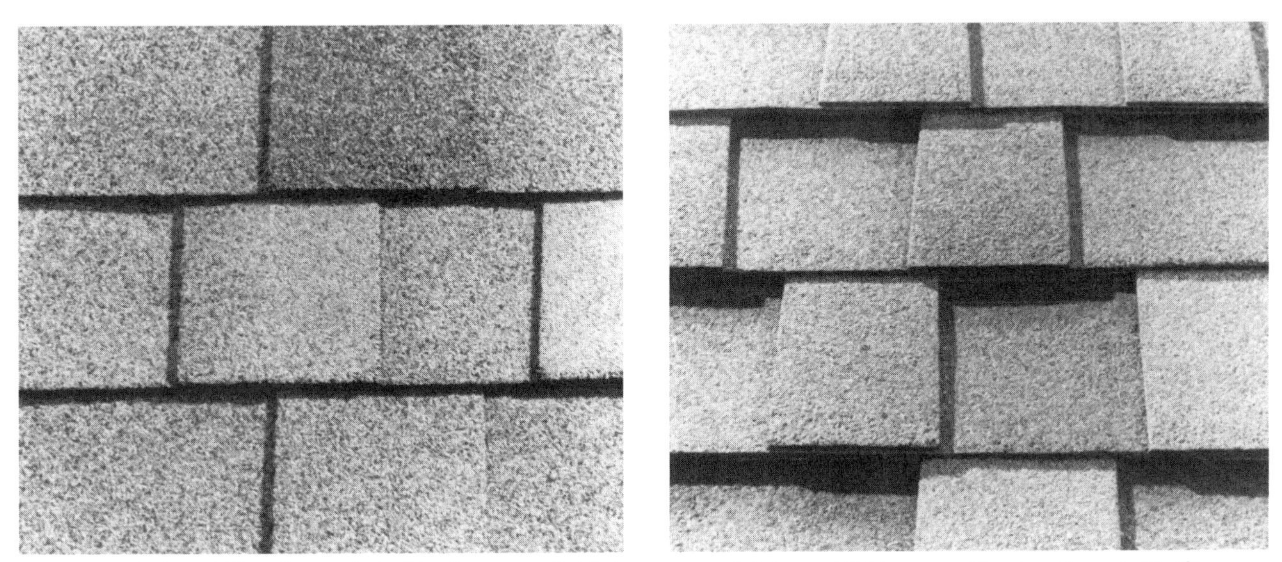

26-28 Asphalt and fiberglass asphalt shingles are available in a range of styles and thicknesses. *(Courtesy Georgia-Pacific Corporation)*

Wood shingles are sawed from logs of various species including western red cedar, eastern white cedar, Tidewater cypress, California redwood, and southern pine. The grading differs for the various species. For example, western red cedar shingles are graded as No. 1 (the best), and No. 2, No. 3, and No. 4 (used for undercourses). They are available in 16-, 18-, and 24-inch (406, 457, and 609mm) lengths.

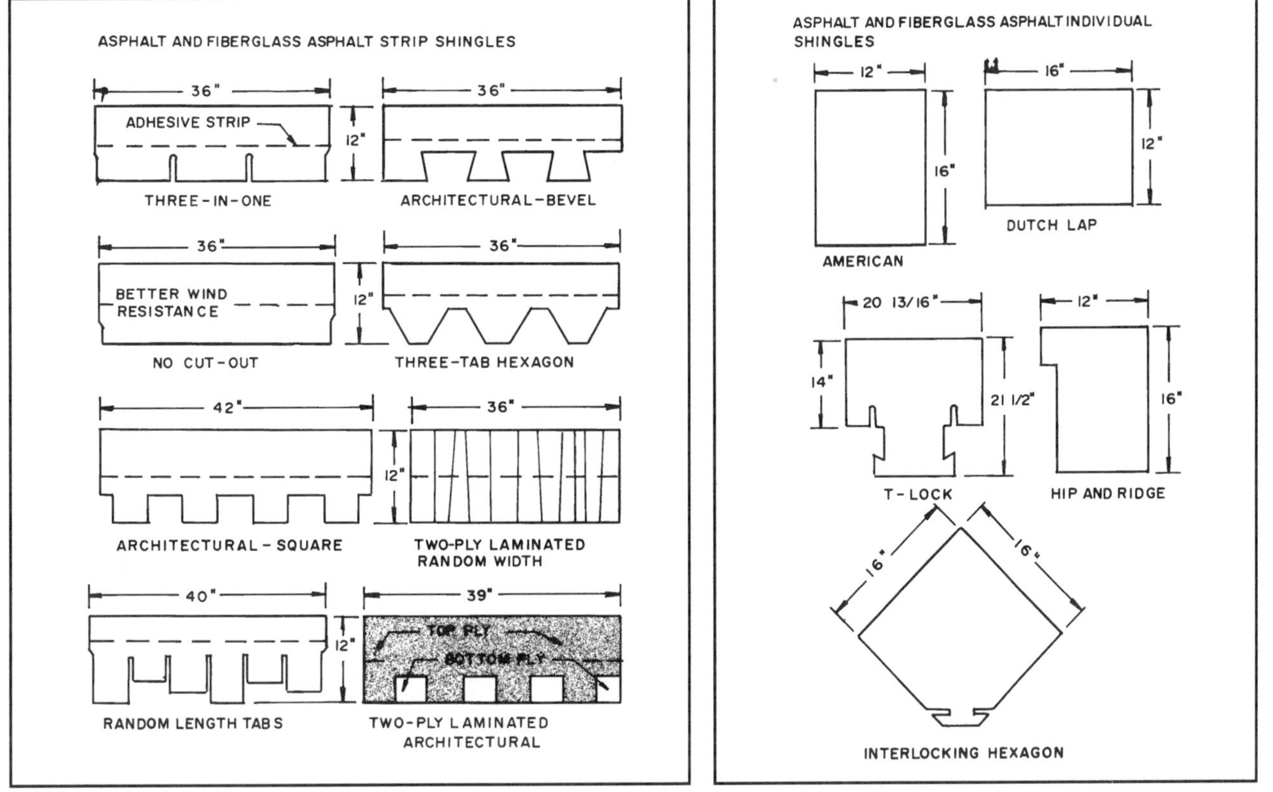

26-29 Some of the styles of asphalt and fiberglass asphalt strip shingles.

26-30 Typical asphalt and fiberglass asphalt individual shingles.

26-31 Wood shakes are handsplit, producing a rough surface.

26-32 Slate shingles are available in a range of colors and hues. *(Courtesy Evergreen Slate Co., Inc.)*

Wood shakes are much like wood shingles, except that they are split from the logs rather than sawed. Some are split and resawed, producing a shake with a rather smooth side and a rough split side, whereas others have the rough split surface on both sides. They are available in various thicknesses ranging from ½ to ⅝ inch (12 to 16mm) and lengths of 18 and 24 inches (457 and 609mm) as shown in **26-31**.

Slate is a natural rock that is mined, cleaved along natural planes of weakness, and then cut to standard sizes. It has holes drilled in it for nailing to the sheathing. Roofing slate is available in a variety of colors including red, green, purple, blue-gray, gray, black, and blue-black. It is very heavy, and the roof framing must be designed to carry this extra load **(see 26-32)**.

Clay tile roofing is made from mined clays much the same as those used for brick. It is available in a variety of stock sizes and colors. The finished tile may be glazed or unglazed. Like slate, it is heavy, and the roof structure must be designed to carry it **(see 26-33)**.

Concrete tiles are extruded and steam-cured, producing a strong, dense, stain-resistant roofing material. They are sealed so that they are waterproof and have a semigloss finish. Concrete tiles are available in standard sizes similar to those used for clay roofing tiles. They are also very heavy **(see 26-34)**.

26-33 Clay roof tile is heavy, durable, and fire resistant. *(Courtesy Gladding McBean Co.)*

26-34 Concrete tile is made in several styles and a wide range of colors. *(Courtesy Monier Lifetile)*

Metal roofing that is most commonly used is made from steel or aluminum sheet. Steel is usually galvanized and has a colored finish coat. Aluminum roofing is also made with a wide range of colors as a finish coating. Such roofing is manufactured in stock sizes and has a variety of systems for securing it to the sheathing and for locking the panels together to get a waterproof joint.

Metal roof tiles that look like clay tiles are also available **(see 26-35)**.

Built-up roofing is made by laying alternate layers of organic or fiberglass roofing felt and hot bitumen coating. Generally three or more layers of felt with a bitumen coating over each are used. The final top coat of bitumen has some type of aggregate spread over it forming the finished roof surface.

26-35 Metal roofing is available in sheets and individual tiles. *(Courtesy Berridge Manufacturing Co.)*

SAFETY

The installation of finished roofing is a hazardous occupation because the danger of falls is always present. Study all of the safety suggestions in Chapters 1 and 2. Specifically observe the following suggestions.

1. Use scaffolding and ladders properly. They should be strong and also properly set in place.
2. Wear shoes that are comfortable and have soles providing traction, such as rubber soles.
3. Do not work on wet roofs.
4. Beware of any overhead items, such as tree branches or power lines, that may cause an accident.
5. Keep the working surface free of debris but be cautious when discarding unwanted materials off the roof.
6. When required by OSHA regulations use life lines.
7. Use temporary railings around openings and roof edges where a fall is possible.

INSTALLING ASPHALT & FIBERGLASS ASPHALT STRIP SHINGLES

Remember, do not walk on hot shingles. The granular surface is easily marred and the damaged shingle will have to be replaced. Likewise, installing shingles when the temperature is near freezing may cause their surfaces to check or crack.

Before starting to work check the **sheathing** to be certain it is properly nailed, no nails are sticking above the surface, and sufficient end and edge spaces have been left **(refer to 26-3)**. Install the underlayment if it has not already been set in place **(see 26-11)**.

Check to see that all flashing has been completed and that anything that must pierce the roof, such as a sewer vent stack, has been installed and the flashing for it is available **(refer to 26-13 through 26-27)**.

If drip edges are specified be certain they are in place **(see 26-11)**. Finally, sweep the underlayment so it is free of any nails, wood blocks, or other objects.

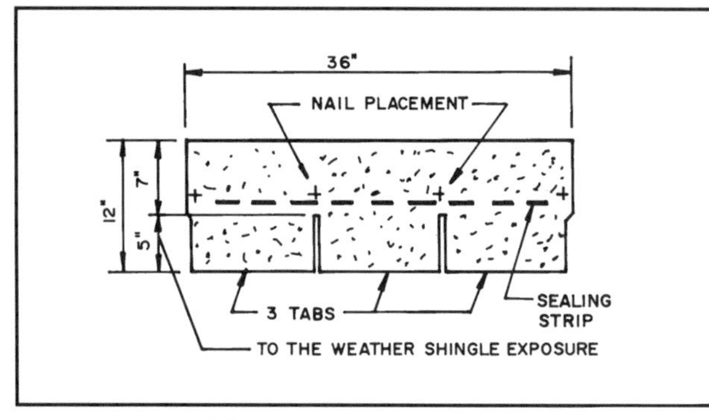

26-36 A typical three-tab organic asphalt or fiberglass asphalt shingle.

Laying the Starter Strip & the First Course

While there are a variety of these types of shingles, this discussion will center on the one most commonly used, the 12 × 36-inch (approx. 30 × 91cm) square butt, three-tab shingle **(see 26-36)**.

Begin by installing the starter strip at the eave. It can be cut from a roll of roofing that matches the shingles or it can be a row of actual shingles turned upside down so that the space between the tabs is up **(see 26-37)**. This should overhang the fascia by about ½ inch (13mm). This gives coverage below the slots in the first row of shingles.

Lay the first course on top of the starter strip. Start it with a full shingle as shown in **26-38**.

Laying Additional Courses

The second row is begun one-half of a tab off the first. This allows the space between the tabs to line up vertically **(see 26-39)**. The third course is started with a full shingle and then the procedure repeats itself **(see 26-40)**.

Before actually nailing the first row, check it out across the roof to the other side. You do not want to end up with a very narrow tab on the other edge; these usually break off over time. If this configuration occurs, move the first row over to allow a wider end tab on the other side.

Place one nail behind the slot between each tab, as shown in **26-38**. As additional courses are laid, the lower edges of the shingles should line up with the tops of the slots on the course below them. This gives the exposure required by the shingles being used. The dark asphalt strips that are above the exposure area are used to seal

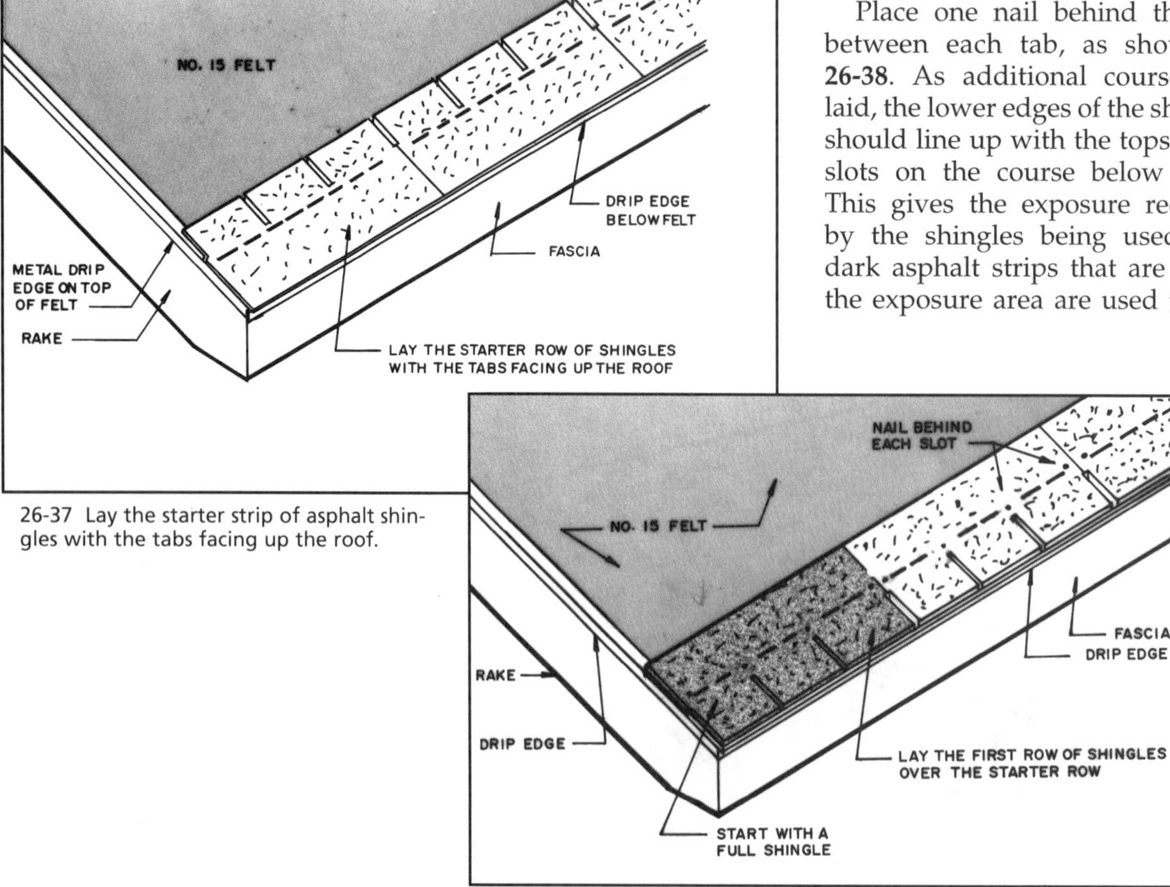

26-37 Lay the starter strip of asphalt shingles with the tabs facing up the roof.

26-38 Start the first row of shingles with a full shingle.

26-39 Start the second course of shingles with one-half of one of the tabs cut away.

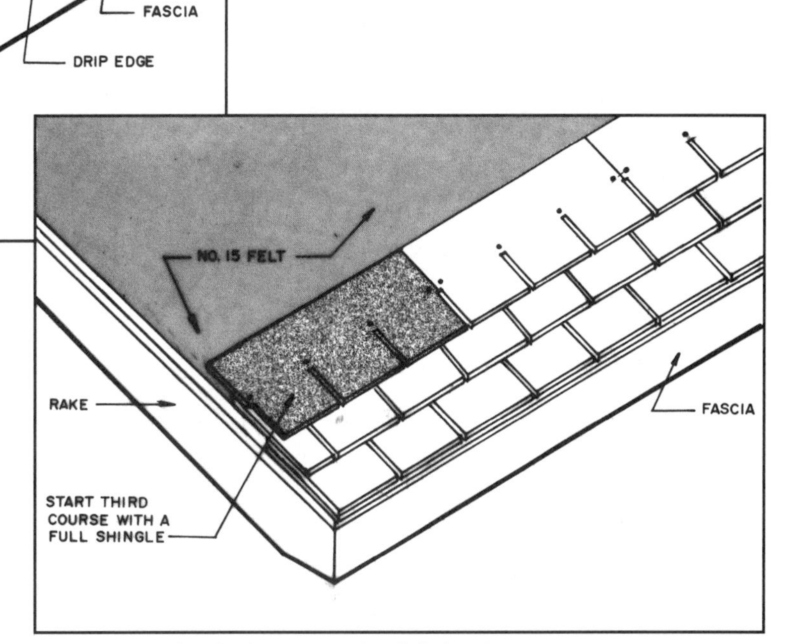

26-40 Start the third course with a full shingle.

the shingles together; as the sun warms up the shingles, this dark strip softens, and bonds the shingles together. This process is essential if the shingles are to resist high winds.

When the shingles reach something projecting through the roof, they are cut to fit over it. Then the flashing is placed over the shingles, and the roofing continues **(refer to 26-26)**.

When the shingles reach flashing, such as on a chimney, lay the shingles up to the chimney. Then install the flashing so that it is over the shingles on the front of the chimney. If the flashing is already in place, slip the shingles under it. Then proceed to install the shingles around the chimney, letting them overlap the flashing as shown earlier in **26-24** and **26-25**. Notice that the flashing is in two parts. The base flashing rests on the roof sheathing, laps up the chimney, and is bonded to it with mastic. The counterflashing is set in a mortar joint and overlaps the base flashing.

When the shingles butt against a vertical wall, such as a dormer wall, the joint is flashed using metal flashing shingles that are applied over each course of shingles **(refer back to 26-21 and 26-22)**.

Asphalt shingles can be nailed or stapled. In some areas, as mentioned, building codes regulate what is used because of a need to resist high winds. In **26-41** asphalt shingles are shown being installed with a roofing coil-nailer. The nails are coiled in the cylinder below the handle and are driven by compressed air.

26-41 Asphalt and fiberglass-asphalt shingles can be rapidly laid with a roofing coil-nailer operated by compressed air. *(Courtesy Senco Products, Inc.)*

26-42 The shingles are lapped over the ridge and will be covered with ridge shingles. *(Courtesy CertainTeed Corporation)*

26-43 The hip is covered by the hip shingles. *(Courtesy CertainTeed Corporation)*

Finishing the Ridge & Hips

When the shingles reach the ridge and hip, lap them over it as shown in **26-42**. Then the ridge and hips on a hip roof are covered with ridge shingles that are available from the manufacturer or can be made by cutting the butt shingles into 9 × 12-inch (approx. 23 × 30cm) pieces **(see 26-43)**. They are laid over the ridge or hip and nailed to the sheathing. The same exposure is used for the roof **(see 26-44)**. The nails are placed behind the exposure, and one nail is placed on each side.

Shingles at Valleys

Valleys are flashed as shown earlier in **26-14** through **26-17**. The line of the shingles on open valleys is marked using a chalkline. Appearance suffers if the shingles are not cleanly and neatly cut.

Installing Individual Lock-type Shingles

Individual lock-type shingles have interlocking notches and tabs that tie them together. This helps hold them down during high winds **(see 26-45)**.

Ridge Vents

There are various ways to ventilate the area below the roof. It is important to change the air in the summer because the temperature can reach well over 100 degrees F (38 degrees C), which damages the sheathing and finished roof. In the winter warm moisture from inside the house may penetrate into the attic. If it does it strikes the cold sheathing, condenses, and freezes, forming a frost-like coating. When things warm up this melts, soaking the insulation, rafters, and ceiling drywall, causing great damage.

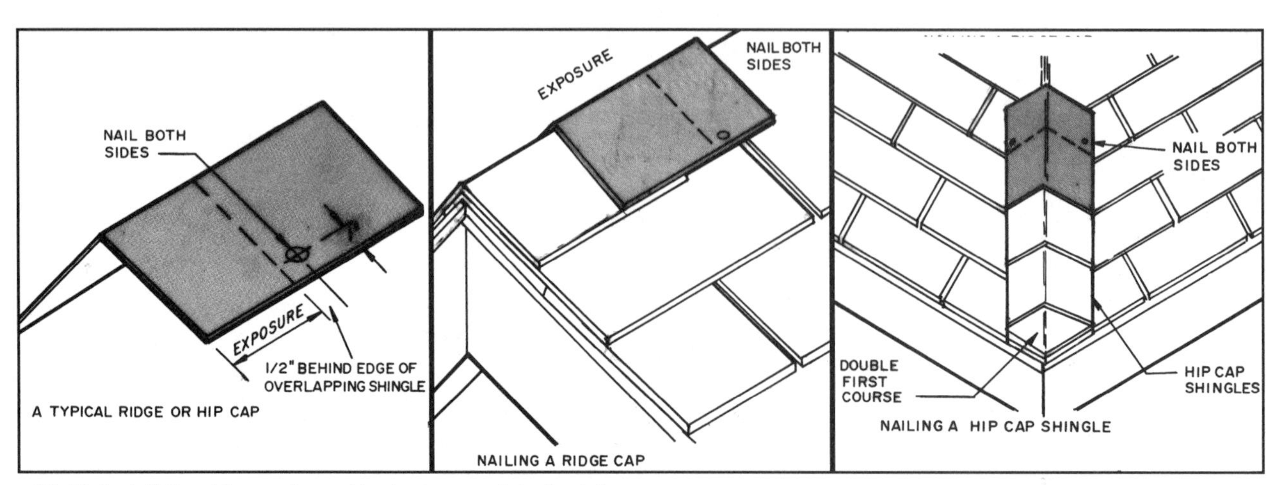

26-44 Install the ridge and cap shingles to complete the job.

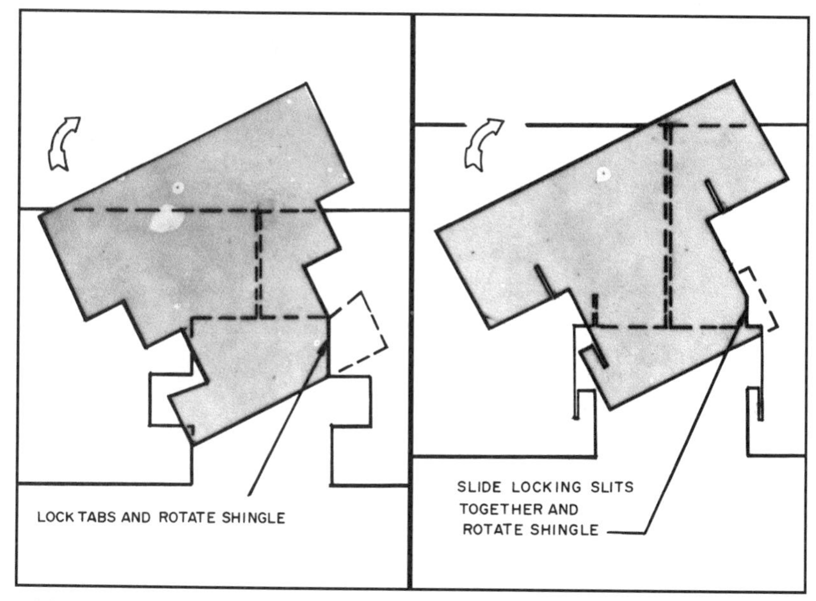

26-45 Installation details for a couple of lock-type shingles.

26-46 This ridge vent is available in roll form and is made from a tough corrugated plastic. It will withstand hot and cold exterior temperatures. *(Courtesy Trimline Roof Ventilation Systems)*

One type of vent that is installed as part of the finished roofing system is a ridge vent **(see 26-46)**. The sheathing at the ridge is cut back, leaving a ventilation opening.

The ridge vent in **26-46** is in roll form and is rolled over the ridge and nailed to the sheathing. Since it has hollow cells the air can escape through them. The standard ridge shingles are nailed over the ridge vent. Vents are installed in the soffit, permitting air to flow up through the attic and out the ridge vent. Various other venting methods are shown in Chapter 19.

INSTALLING WOOD SHINGLES & SHAKES

The minimum roof slope for which wood shingles are recommended is 3/12 and for wood shakes it is 4/12. Lower-sloped roofs can use these materials if special methods of application are used.

The sheathing may be plywood, oriented strand board, or solid wood. If solid wood is used, it is spaced so that air can circulate between it and the wood shingles. There are two ways to install spaced wood sheathing for use with wood shingles. If 1 × 4-inch lumber is used, it is spaced from the front edge of one board to the front edge of the next by a distance equal to the exposure to be used. If 1 × 6-inch lumber is used, the distance is equal to two exposures **(see 26-47)**.

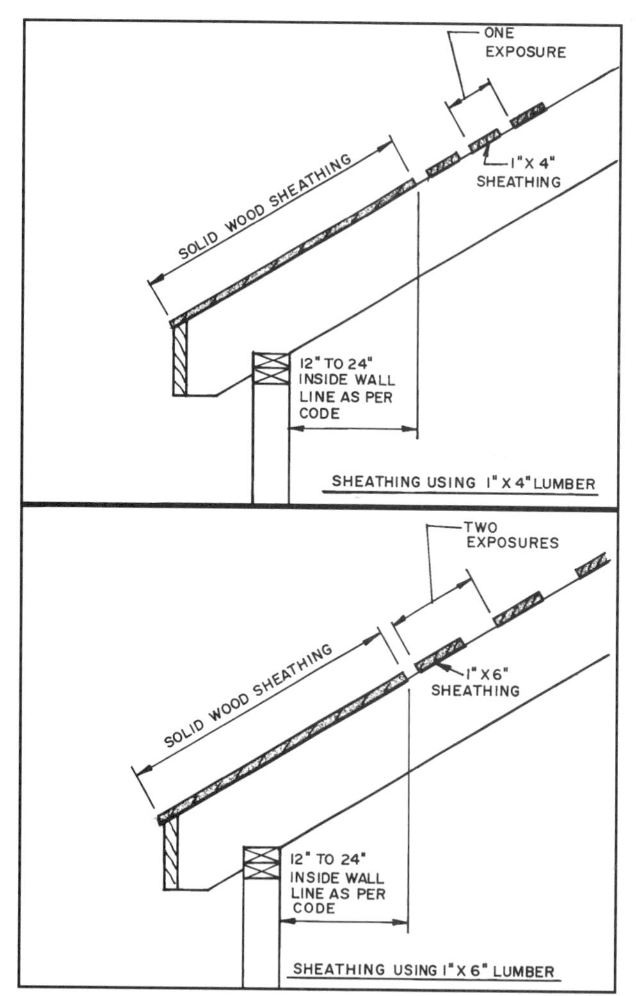

26-47 Spacing wood sheathing for the installation of wood shingles.

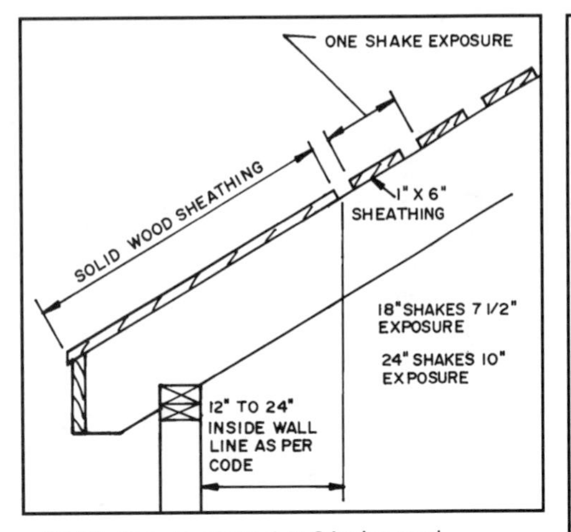

26-48 How to space 1 × 6-inch wood sheathing when wood shakes are to be installed.

26-49 Recommended exposure for western cedar shingles and shakes.

Length	Shingles: Slope 3/12 to 4/12	Shingles: 4/12 and steeper
16"	3¾"	5"
18"	4¼"	5½"
24"	5¾"	7½"

Length	Shakes: Slope 3/12	Shakes: 4/12 and steeper
18"	Cannot use on this slope	7½"
24"	Cannot use on this slope	10"

(Courtesy Cedar Shake and Shingle Bureau)

There are exceptions to this for some exposures. When using wood shakes, the sheathing should be 1 × 6-inch stock with the spacing equal to the exposure, but never more than 7½ inches (19.1cm) for 18-inch (45.7cm) shakes and 10 inches (25.4cm) for 24-inch (61cm) shakes **(see 26-48)**. Solid sheathing should be installed from the edge of the rafters to a point 12 to 24 inches (30.5 to 61cm) inside the exterior wall. Solid sheathing is recommended on the entire roof in areas where wind-driven snow is expected. A 36-inch- (91.4cm) wide layer of No. 15 felt is installed at the eave and extends up the roof over the exterior wall.

The allowable exposure depends on the grade of wood shingle or shake and the slope of the roof. Recommended data for No. 1 western cedar shingles and shakes are given in **26-49**.

The nails used should be corrosion-resistant aluminum, zinc coated, or stainless steel. Recommended sizes are shown in **26-50**. They should penetrate at least ½ inch (13mm) into the sheathing; deeper penetrations are required by some codes. They must be flush with the surface of the shingle or shake. Two nails must be used with every shingle or shake. Units over eight inches (20.3cm) wide should be split to form two shingles or shakes.

If codes permit the use of power-driven staples, they should be aluminum or stainless steel, 16 gauge, with 7/16-inch (11mm) minimum crowns. They should penetrate at least ½ inch (13mm) into the sheathing and be driven flush with the surface of the shingle. Two staples must be driven in each shingle or shake in the same location as specified for nails. Nails or staples should be placed ¾ to one inch (19 to 25mm) from the side edges of the shingles or shakes and 1½ to two inches (38 to 51mm) above the butt line of the following course.

26-50 Recommended nail type and length for use on cedar shingles and shakes.

Shingles—New Roof	Nail Type	Nail Length
16" and 18"	Box	3d (1¼")
24"	Box	4d (1½")

Shakes—New Roof	Nail Type	Nail Length
18" straight split	Box	5d (1¾")
18" and 24" split and resawn	Box	6d (2")
24" Tapersplit	Box	5d (1¾")
18" and 24" tapersawn	Box	6d (2")

(Courtesy Cedar Shake and Shingle Bureau)

Installing Wood Shingles

Begin the installation of wood shingles by laying a course of starter shingles with the butt edges overhanging the finished fascia 1½ inches (38mm). Then go over this with a course of standard-size shingles, making certain the joints between shingles are no closer than 1½ inches (38mm) as shown in **26-51**. Space the edges of the shingles at least ¼ inch (6mm) apart, but no more than ⅜ inch (10mm).

Start the second course with the butt above the butt of the first course a distance equal to the required exposure, to maintain the desired exposure. Set the pin on the head of the shingling hatchet so that the exposure is the distance from the pin to the face of the hatchet **(see 26-52)**. Another way is to run a chalkline or wood straightedge across the roof. Continue each course, observing the required 1½-inch (38mm) spacing between edge joints.

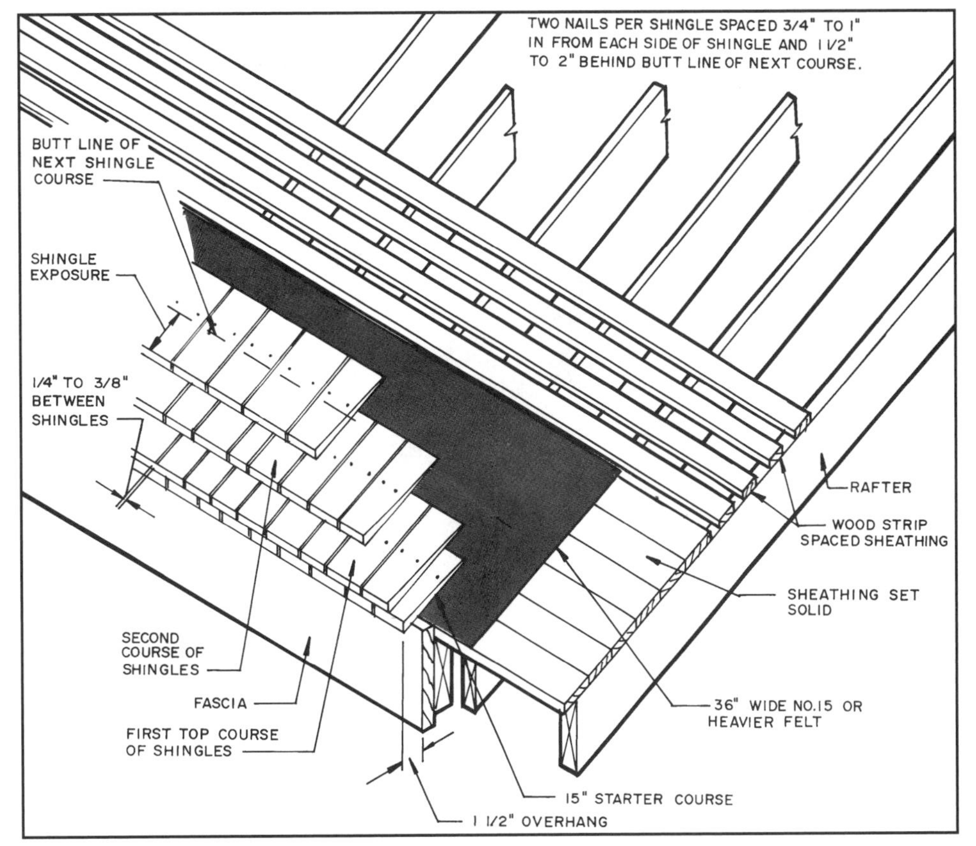

26-51 How to install wood shingles.

Usually a felt interlay is not required when using wood shingles. It is usually required, however, with wood shakes; consult the local codes on this requirement. In both cases, a felt layer is installed over the solid wood sheathing at the eave. This provides protection from possible water damage to the inside of the house by ice dams that may form on the cornice. These occur when snow on the roof is melted by heat lost to the attic; the water drains to the cornice and freezes there, forming an ice dam. The water then begins to back up under the shingles and leaks into the attic. In areas where this is a likely problem, additional felt may be required over the cornice and on the roof sheathing over the interior of the building **(see 26-53)**. Most codes require this flashing to run two feet over the warm interior of the building. Refer back to **26-13**.

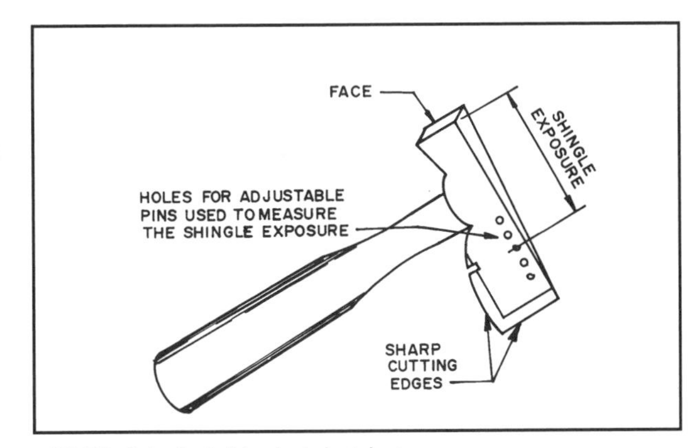

26-52 A typical shingler's hatchet.

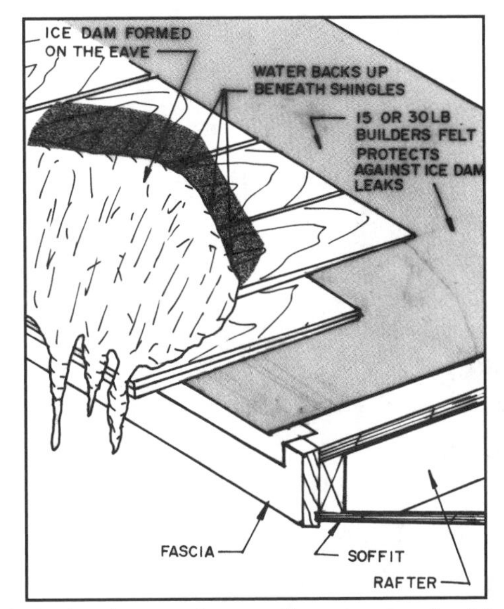

26-53 An ice dam can form on the eave when heat lost to the attic melts some of the snow on the roof. Builder's felt flashing must be installed over the soffit and two feet over the building interior.

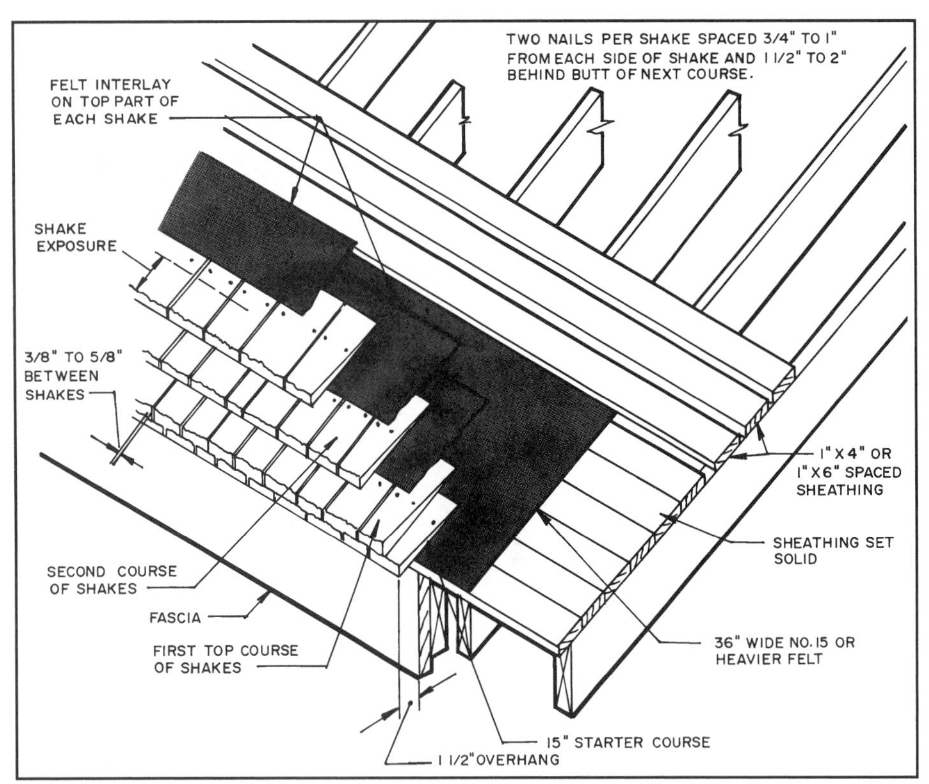

26-54 A roofing felt interlay is placed between each course of shingles.

Installing Wood Shakes

A 36-inch- (91.4cm) wide, No. 15 felt is installed at the eave. The first course uses a 15-inch (38.1cm) starter shake which is covered with a second course of standard-length shakes. The butts of the first course overhang the fascia by 1½ inches (38mm). The joints between shakes should be spaced at least ⅜ inch (10mm), but no wider than ⅝ inch (16mm).

An 18-inch- (45.7cm) wide, No. 30 felt interlay is placed between each course of shakes **(see 26-54)**. The bottom edge of the felt is located above the butt a distance equal to twice the exposure **(see 26-55)**. There are some cases where this felt can be omitted; consult local codes, as always, for specific requirements.

Other Installation Details

Hips and ridges are capped with preassembled hip and ridge units. These can be factory-made or site-made caps. When installed, the overlap joint must alternate sides of the ridge or hip **(see 26-56)**. The exposure to the weather is the same as that used for the roof. It requires longer nails because they penetrate both the cap and the roof shingle. Most codes require that the nails penetrate the sheathing ½ inch (13mm) or go completely through it. The end cap is always doubled **(see 26-57)**.

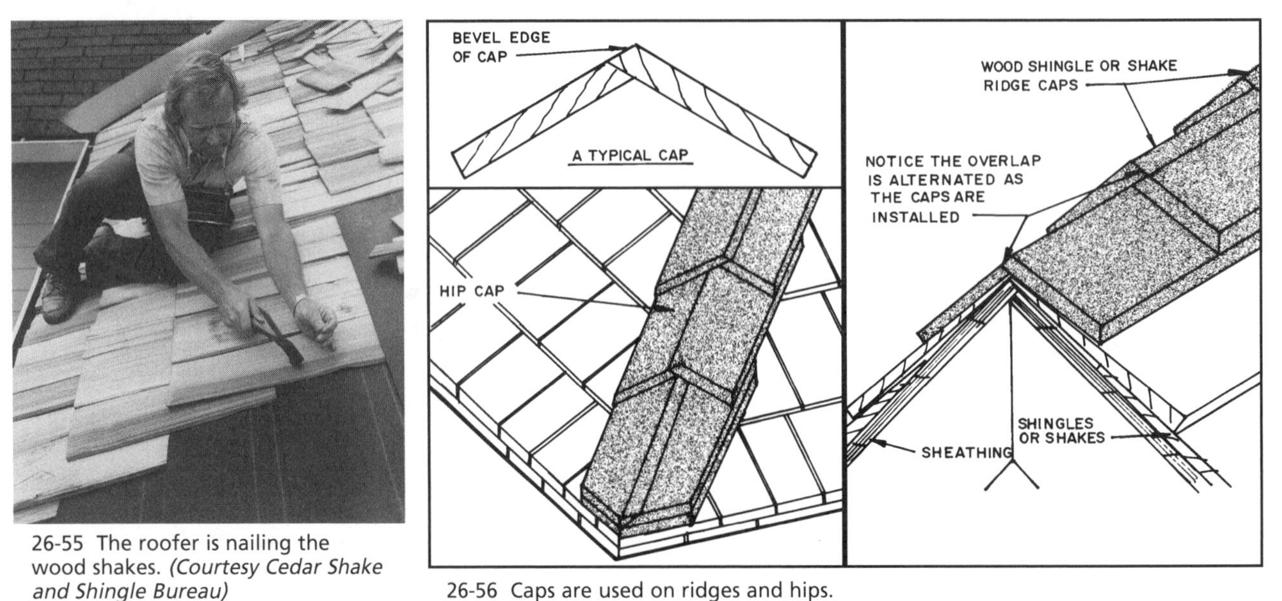

26-55 The roofer is nailing the wood shakes. *(Courtesy Cedar Shake and Shingle Bureau)*

26-56 Caps are used on ridges and hips.

Valleys are flashed by first installing a layer of No. 15 or heavier felt up the valley. Metal flashing is installed over this. It should be painted galvanized steel (paint both sides) or aluminum. The metal flashing should extend over the fascia to the butt of the first course of shingles **(see 26-58)**. The grain of the shingles should not be parallel with the centerline of the valley. Joints between the shingles should not break into the valley.

With valleys for wood-shingle roofs over a slope of 12:12, the flashing should extend not less than 7 inches (17.8cm) on each side of the centerline. For roof slopes less than 12:12 this distance should be increased to 10 inches (25.4cm) as shown in **26-59**.

26-57 Woodshake caps close the roof along the hip. *(Courtesy Cedar Shake and Shingle Bureau)*

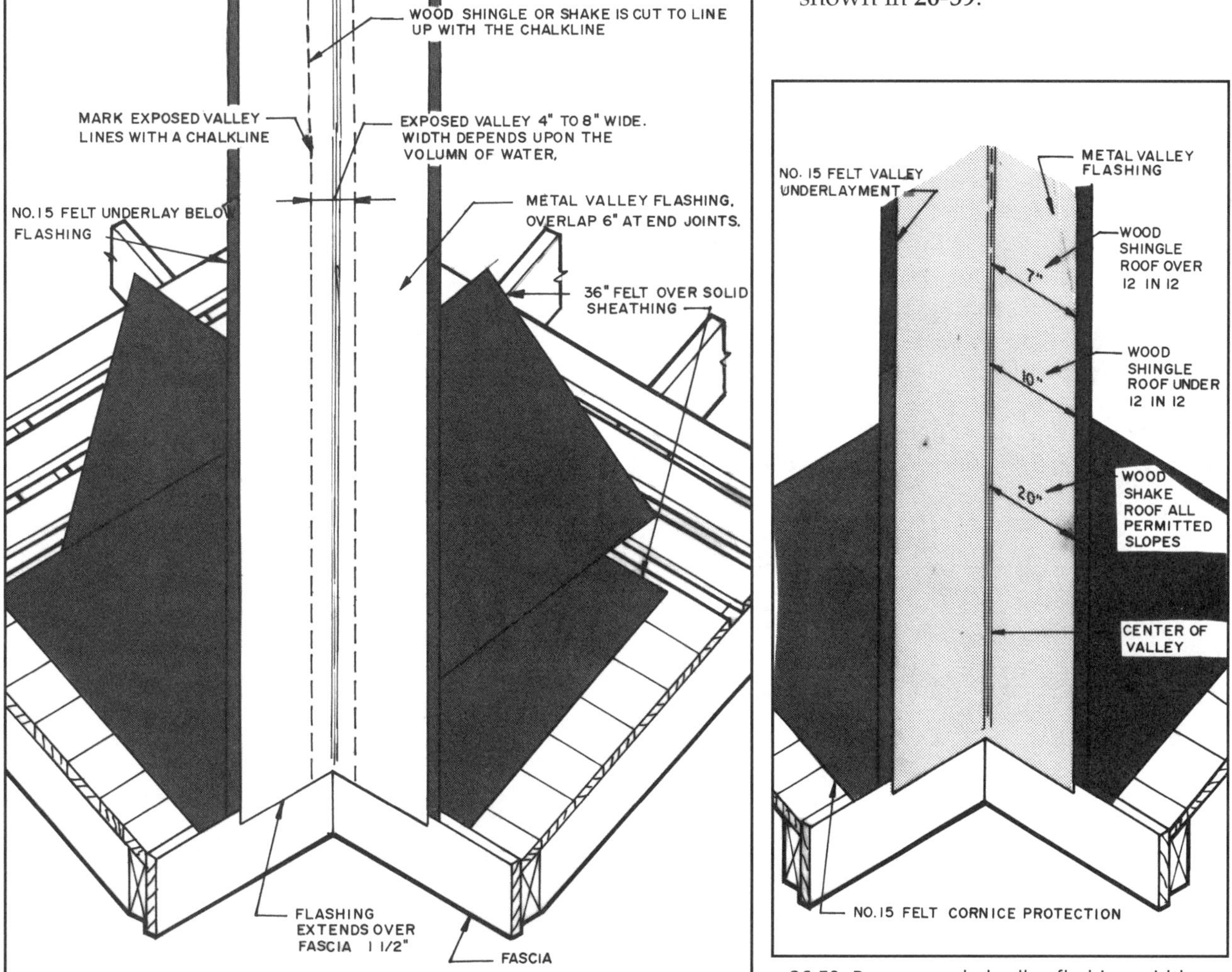

26-58 How to flash a valley when installing wood shingles or shakes.

26-59 Recommended valley flashing width for wood shingles and shakes.

26-60 A metal valley is used with wood shingles and shakes. *(Courtesy Cedar Shake and Shingle Bureau)*

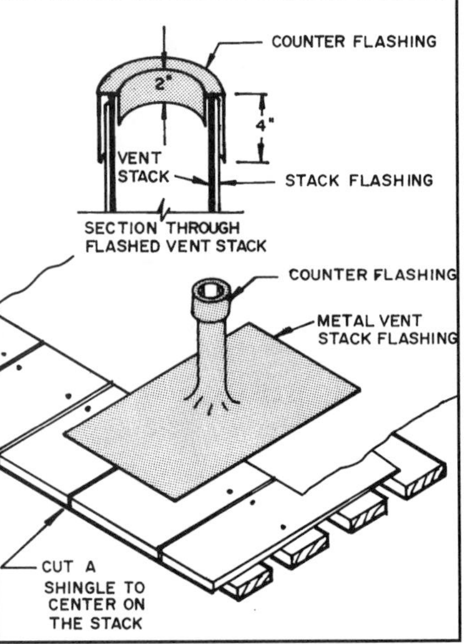

26-61 Typical vent stack flashing when wood shingles or shakes are used.

Valleys for wood shakes should have a minimum width of 20 inches (50.8cm) from the centerline. Codes in some areas require wider spacing. The grain of shakes should not be parallel with the centerline of the valley. Joints between shakes should not break into the valley **(see 26-60)**.

Projections through the roof, such as a vent stack, should be flashed and counter-flashed about the same as for asphalt shingles **(see 26-61)**. The flashing should extend three inches (76mm) under the sheathing paper. It should be long enough to cover the shingle or shake course just below the pipe and extend up under the straight course above the pipe. The positioning of shingles and shakes around the projection is shown in **26-62**.

Details for flashing and installing shingles or shakes when the roof butts a wall or chimney are shown in **26-63** and **26-64**. Nails should not penetrate the flashing.

Additional technical information is available from the Cedar Shake and Shingle Bureau, 515 116th Ave. NE, Suite 275, Bellevue, WA 98004-5294.

CLAY & CONCRETE TILE

Roofing tile is available as a clay or concrete product. The shapes and sizes are about the same for each. The manufacturers of these products can supply information on the sizes they produce and their recommended installation details **(see 26-65)**.

26-62 Placing wood shingles and shakes around a flashed vent pipe.

Clay roofing tile may be glazed or unglazed. Interlocking-type tiles are used on roofs with slopes of 3⁄12 or higher. Flat shingle-type tiles are used on roofs with slopes of 5⁄12 or more. The roofing is heavy, durable, and fire-resistant.

Concrete tiles are manufactured in a wide range of colors and tend to be less heavy than clay tiles. Following are examples of typical installation recommendations.

Flat interlocking shingle tiles are installed over solid sheathing that is covered with roofing felt. The installation at the eave begins with a flat under-eave tile or metal drip edging. The gable rakes are covered with right- and left-side rake tiles or metal drip edging. Special V-shaped tiles are used on the ridge and hips. A typical installation is shown in **26-66**.

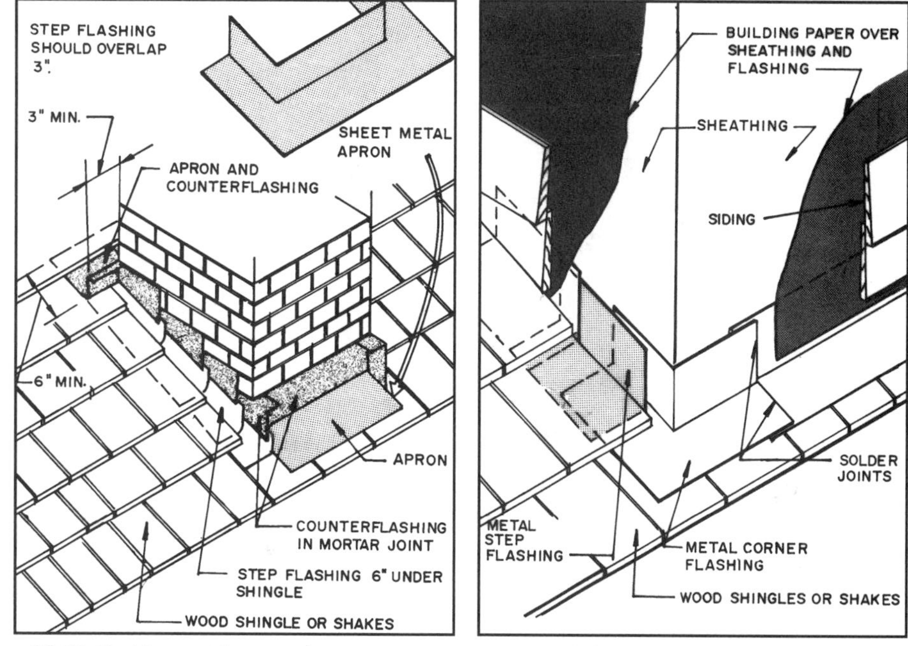

26-63 Flashing a chimney when installing wood shingles or shakes.

26-64 Flashing a roof when the wood shingles or shakes butt a wall.

26-65 These curved concrete roofing tiles are nailed to the sheathing. *(Courtesy Monier Lifetile)*

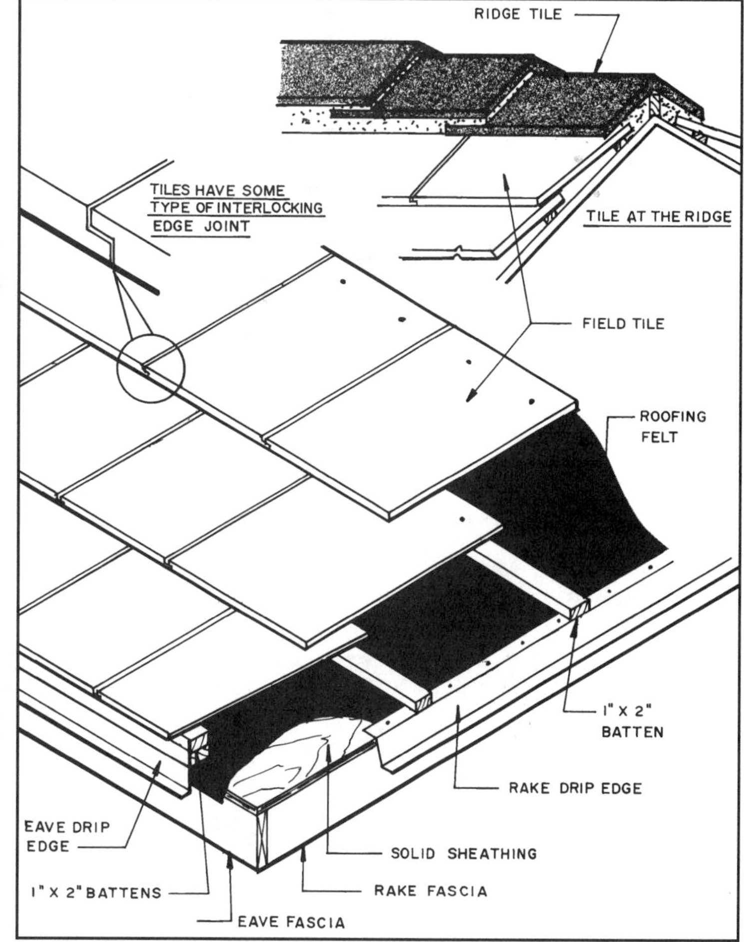

26-66 Typical installation details for flat concrete and clay roofing tile.

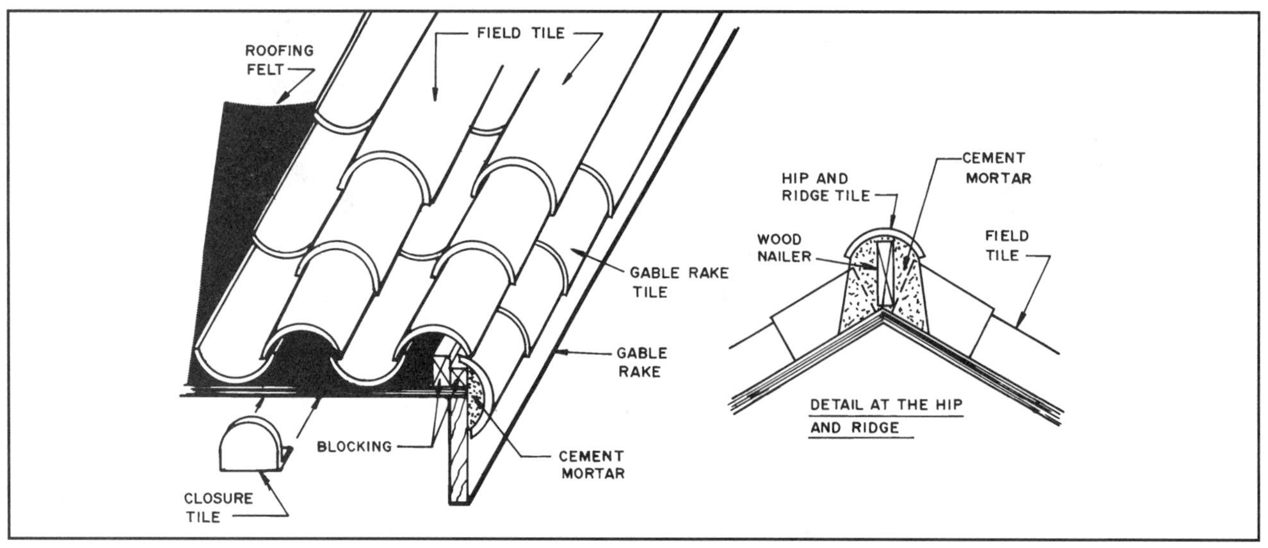

26-67 Installation details for one type of curved clay and concrete roofing tile.

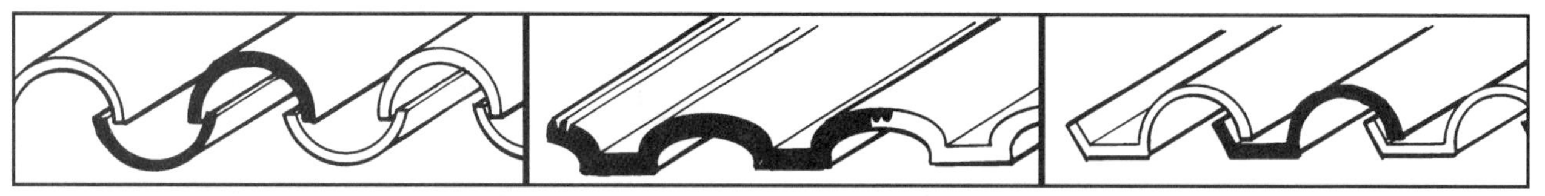

26-68 Some of the curved tiles available.

26-69 The concrete ridge caps finish the roof. (*Courtesy Monier Lifetile*)

26-70 The metal standing-seam roof panels run from the ridge to the eave as a single panel. *(Courtesy ATAS International, Inc.)*

Curved clay and concrete tiles are made in several ways. Installation details for one type are shown in **26-67**. Several types are shown in **26-68**. The hips and ridges are covered with curved tiles that are set in a cement mortar and nailed in place **(see 26-69)**. Both types are flashed in the same manner as other roofing materials.

METAL ROOFING

Two types of metal roofing in common use are panels and tiles. The panels can cover a long section of roof **(see 26-70)**. They are joined at the edges with watertight joints. The design of the joints depends on the manufacturer of the roofing. On steep-sloped roofs, batten, capped, and standing seams are used **(see 26-71)**. Some panels must be installed over solid sheathing, whereas others have sufficient strength to span unsupported distances between purlins.

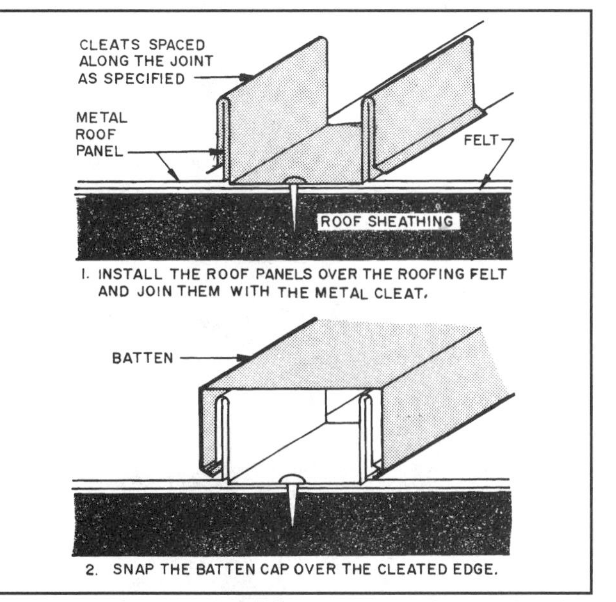

26-71 Construction details for a metal panel steep-sloped roof with batten seams.

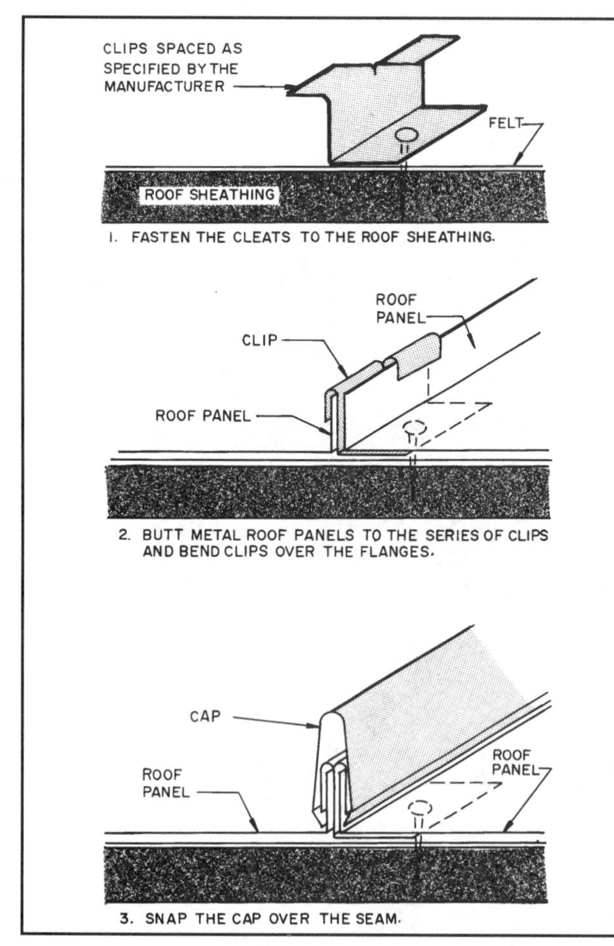

26-72 Construction details showing the use of a capped seam on a steep-sloped metal panel roof system.

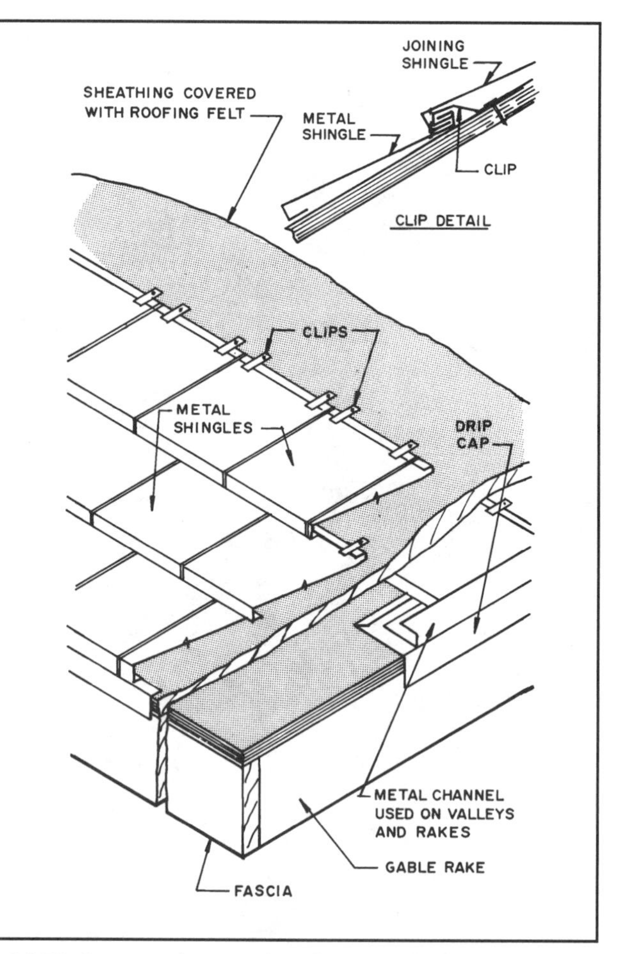

26-74 One way that metal roof tiles are fastened to the roof sheathing.

Details for other standing-seam assemblies for steep-slope metal panel roof installations are shown in **26-72 and 26-73**.

Metal shakes and shingles are lightweight, yet have the appearance of heavier roofing materials such as tile and slate. They are fastened to the roof sheathing with metal clips. One system is shown in **26-74**. The actual system used depends on the design provided by the tile manufacturer.

The surface is embossed in a range of textures, patterns, and colors, including metal tiles that look as if they are clay **(see 26-75)**.

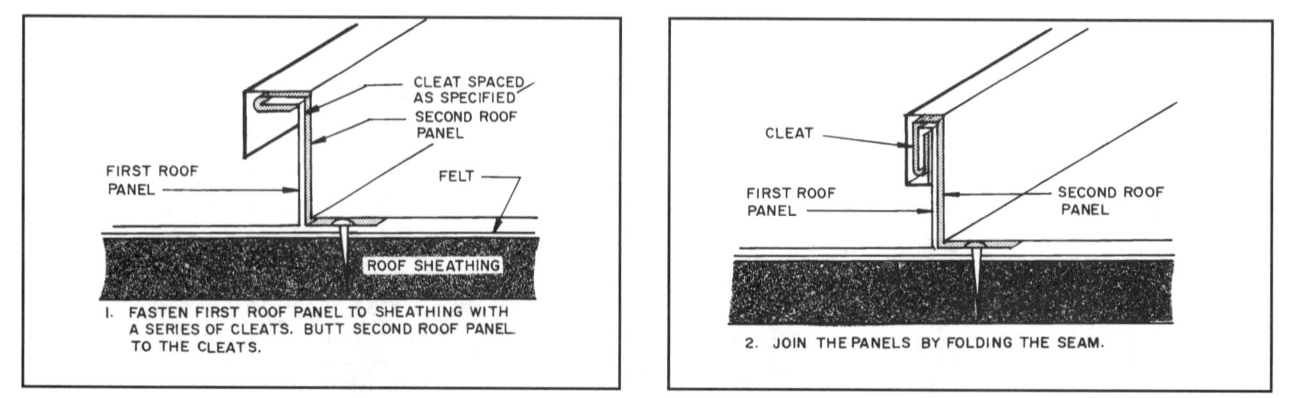

26-73 A typical detail for a standing-seam assembly for a steep-slope metal panel roof installation.

26-75 These metal tiles look like clay tiles but are much lighter. *(Courtesy ATAS International, Inc.)*

CHAPTER 27

Decks & Porches

Wood-framed porches are an important part of the design of home styles of traditional homes and find wide use. They usually have some form of a roof. They typically provide access to the entrances into a house **(see 27-1)**. Decks are usually larger, generally on the rear of the house, and provide a private area for relaxation. Both have similar construction requirements. A most important consideration is that they are exposed to the elements and therefore must resist rot, decay, expansion, contraction, and checking **(see 27-2)**.

JOISTS

Joists for decks and porches are the same as those used for residential floor framing. They are typically 2 × 6, 2 × 8, 2 × 10 or 2 × 12, depending upon the span and expected floor load. Typically design loads of 40 lbs/sqft and 60 lbs/sqft are used. Use standard joist span tables. Pressure-treated lumber is typically used.

DECKING

Decking is typically pressure-treated wood or a specie, such as redwood or cypress, that naturally resists decay when exposed to the elements. The decking is frequently two-inch-thick planks. However 5⁄4 (1¼-inch) radius edge decking (RED) is widely used. RED has a rounded edge which reduces the possibility of splintering along the edge of the piece. Typically 5⁄4 RED decking spans joists spaced 16 inches O.C. while two-inch decking will span 24 inches O.C.

Porch floors typically are tongue-and-grooved decking providing a smooth-closed floor. Water tends to get trapped between the boards so proper finishing is needed. Spaced square-edged or RED decking is also used. Porch floors, such as those used on a second floor porch or balcony, must be waterproofed to keep the water from dripping onto the area below. They are built, flashed, and waterproofed much like a flat roof. The decking is typically waterproof plywood.

27-1 This front porch is enclosed with a vinyl railing. *(Courtesy Thermal Industries)*

27-2 Decks provide an attractive outdoor space to relax or entertain. This decking is a product made from recycled redwood fibers and polyethylene plastic. *(Courtesy Choice Dek by A.E.R.T., Inc.)*

POSTS

Posts that are to be placed in the ground should be of ground-contact grade, treated with preservative. Those set on footings above grade can be preservative-treated or a naturally decay-resistant species. Naturally decay-resistant posts should not come in contact with the soil. Post sizes generally used include 4 × 4, 4 × 6, and 6 × 6 inches. The size depends upon the load and the required length. The longer the post, the larger it must be for the same load. Some prefer 6 × 6-inch posts because smaller posts tend to appear too slender.

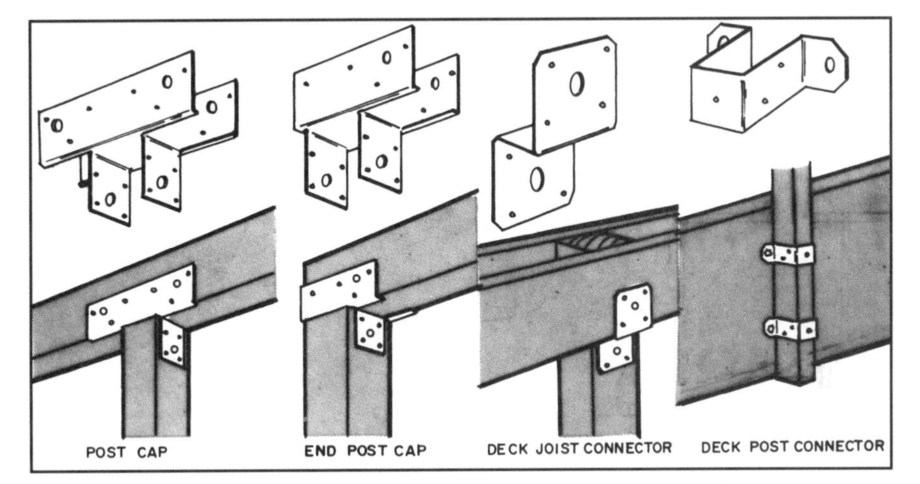

27-3 Post and deck connectors. *(Courtesy Simpson Strong-Tie Co., Inc.)*

CONNECTORS

Metal connectors, such as joist hangers, angles, and rail-to-post connectors, must be heavily galvanized and designed for exterior use **(see 27-3 and 27-4)**.

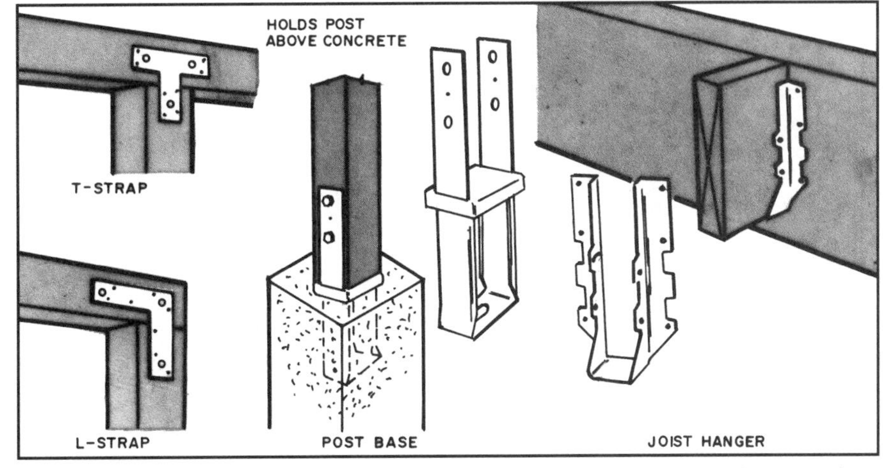

27-4 Straps, post, base, and joist connectors. *(Courtesy Simpson Strong-Tie Co., Inc.)*

Nails must be galvanized or aluminum. Since the wood is exposed to the weather, it is subject to expansion and contraction. This tends to loosen nails and weaken joints. Nails should be spiral shank or annular grooved. Smooth common nails will work loose and should not be used **(see 27-5)**.

Screws should be of the type designed for this purpose. The thread is different from standard woodscrews. It is designed to drive rapidly and has good holding power. They should be hot-dipped galvanized or stainless steel. Square recess and Phillips heads are available **(see 27-6)**. They are not to be used to secure joint hangers or other metal connectors. Use standard thread wood screws for these.

Lag screws, bolts, and **expansion bolts** are used to connect pieces of two-inch stock or members to the wall or foundation of the building. Lag screws must have pilot holes drilled before installing. They should be long enough so that half of the screw penetrates the wood. Washers must be used under the heads and nuts of all these fasteners. Since carriage bolts cannot accept a washer they are not recommended. These fasteners must be hot-dipped galvanized or stainless steel **(see 27-6)**.

Other **deck-to-joist connectors** are available. Several are shown in **27-7**. These produce a deck surface free of nail or screw heads.

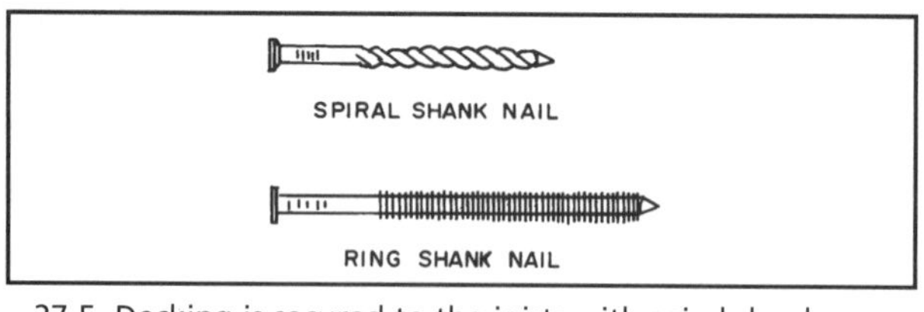

27-5 Decking is secured to the joists with spiral shank or annular grooved nails.

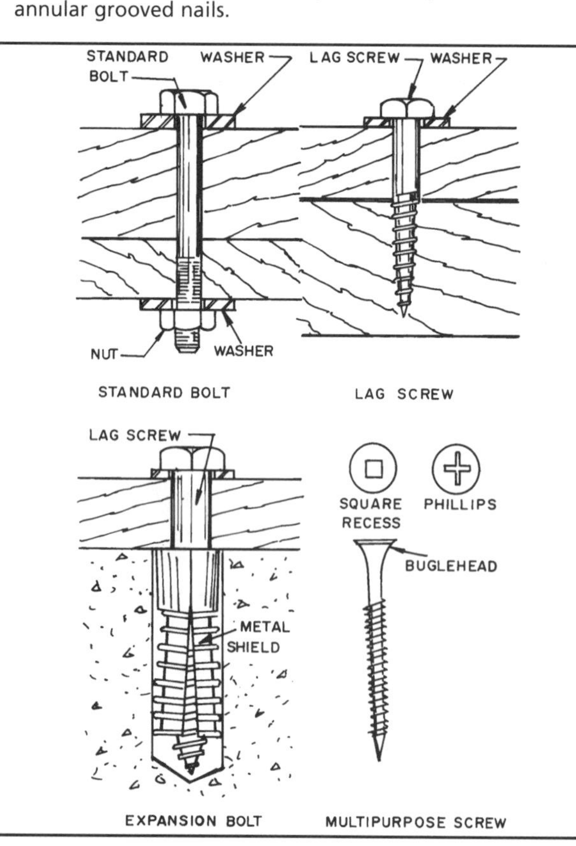

27-6 Screws and bolts recommended for use in deck construction.

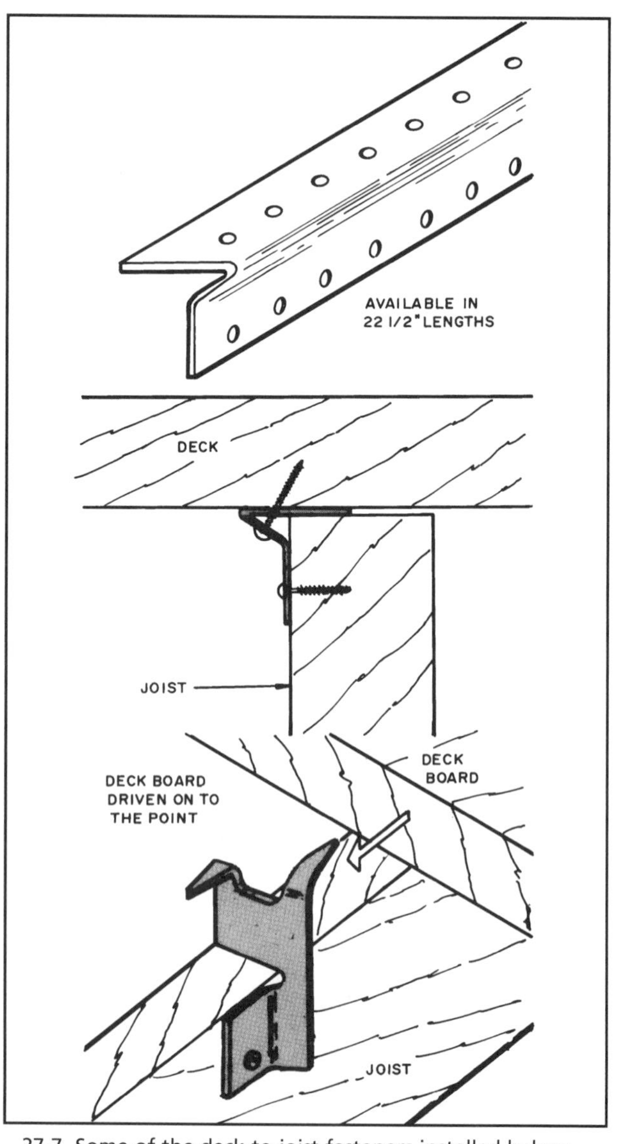

27-7 Some of the deck-to-joist fasteners installed below the decking.

FINISHES

Decks are frequently left natural and treated with a penetrating-type finish. This may be a clear water repellent or water preservative, or a semi-transparent stain. The difference between the water repellent and the water preservative is that the water preservative, while also water repellent, includes a mildewcide to help resist the growth of mildew on the deck. The semitransparent stain has a slight pigment coloring the wood as it penetrates yet allowing the grain to be visible. Both types are used on wood treated with a preservative and on naturally decay-resistant woods.

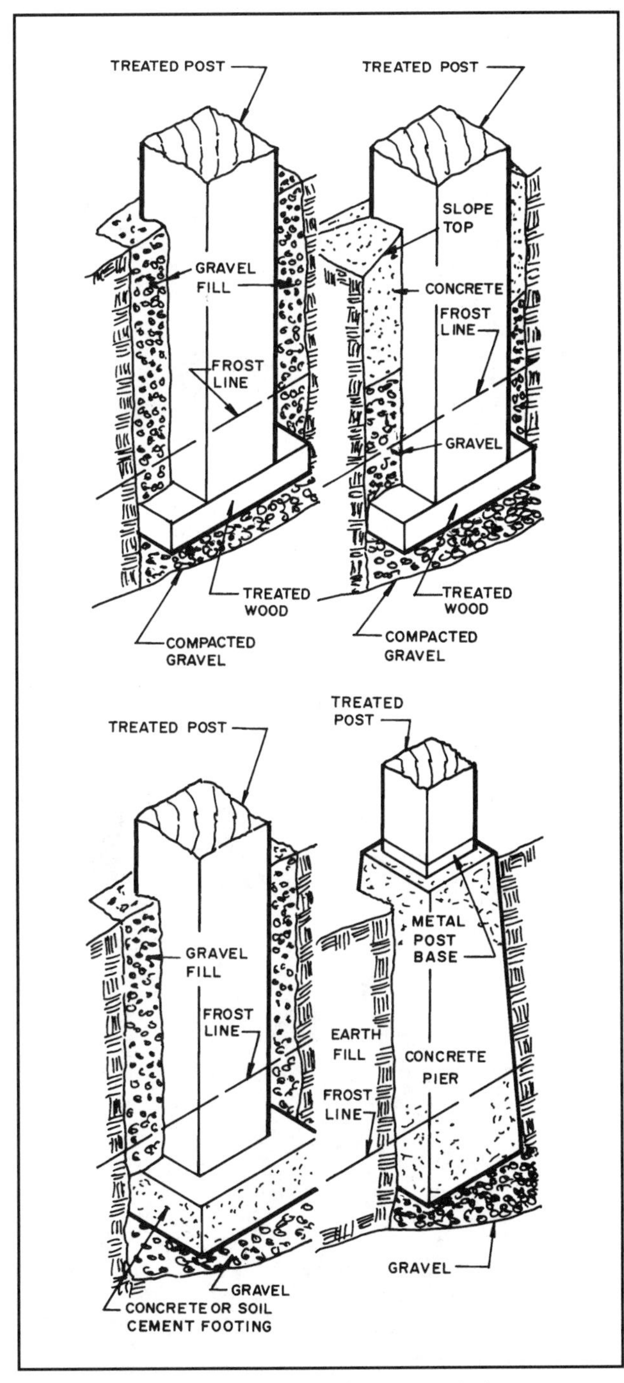

27-8 Commonly used methods for setting deck posts.

FOUNDATIONS

The posts supporting the deck may rest on gravel or concrete footings set below the surface of the soil or on concrete footings that extend above the soil, keeping the ends of the posts above it **(see 27-8)**.

The design depends upon local conditions. The footing must extend below the frost line. When posts rest on top of a footing above the soil a metal post anchor should be used. This holds the bottom of the post off the concrete and allows it to dry, thus retarding checking and preventing damage that might occur were it resting directly on the concrete.

DECK DESIGN

The design of the deck can vary as needed to suit the conditions and desires of the owner. Decks can take any shape required, as shown in **27-9**.

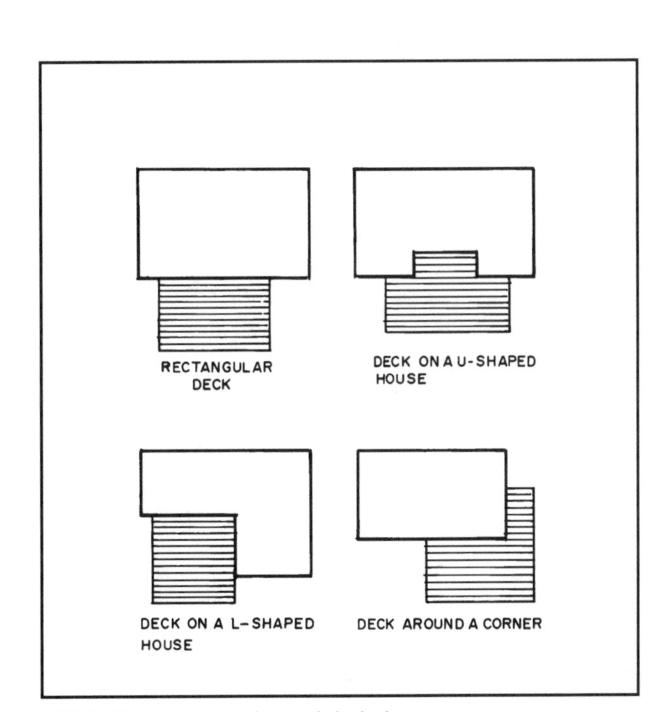

27-9 Some commonly used deck shapes.

27-10 Decks can be on several levels.

Decks can also be built on several levels. This is especially useful when the lot slopes steeply and the deck would have long spindly support posts exposed **(see 27-10)**.

The unattractive exposed area below decks can be covered with some form of lattice or screening **(see 27-11)**. In some developments this is required by the covenants.

27-11 The area below a deck should be screened using some form of lattice work and shrubs.

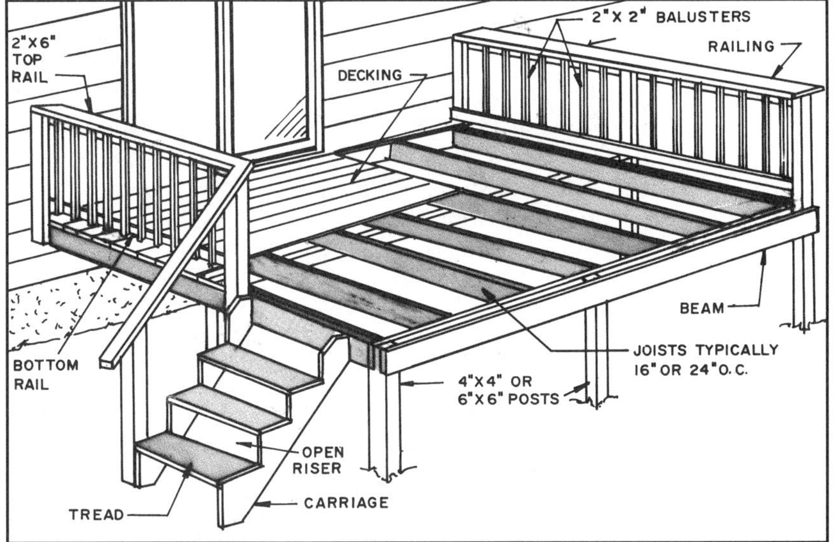

27-12 Parts of a typical small deck.

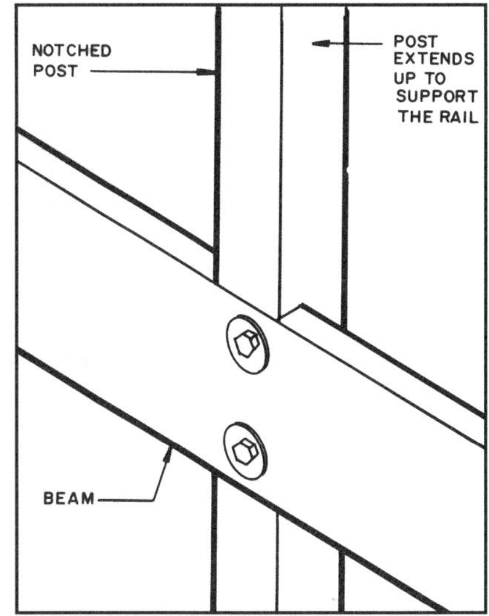

27-13 Posts can continue above the deck and support the railing.

BUILDING WOOD-FRAMED PORCHES & DECKS

The parts of a typical deck are shown in **27-12**. Notice that it is framed much like that for wood-framed floors.

The following illustrations show commonly used construction details. For an open deck that joins the house at the foundation, a wood pressure-treated header is bolted to the foundation. The deck joists are secured to this with metal joist hangers in the same manner as used in floor construction. The wood posts are bolted to the beams on the outer side of the deck, and joists are joined to the beams with metal joist hangers. The floor of the deck can be placed level with the floor of the house or dropped below it. In areas of heavy snow it is important to drop the floor of the deck somewhat, so that the door can be opened after a snow. The posts can be extended above the deck to form a support for a railing **(27-13)**.

Typical details for an open deck that joins the house against horizontal wood or plywood siding are shown in **27-14**. The drawing sets out the edge of the decking from the wall to allow rain to pass down below the deck. Notice the use of metal flashing to protect the sheathing.

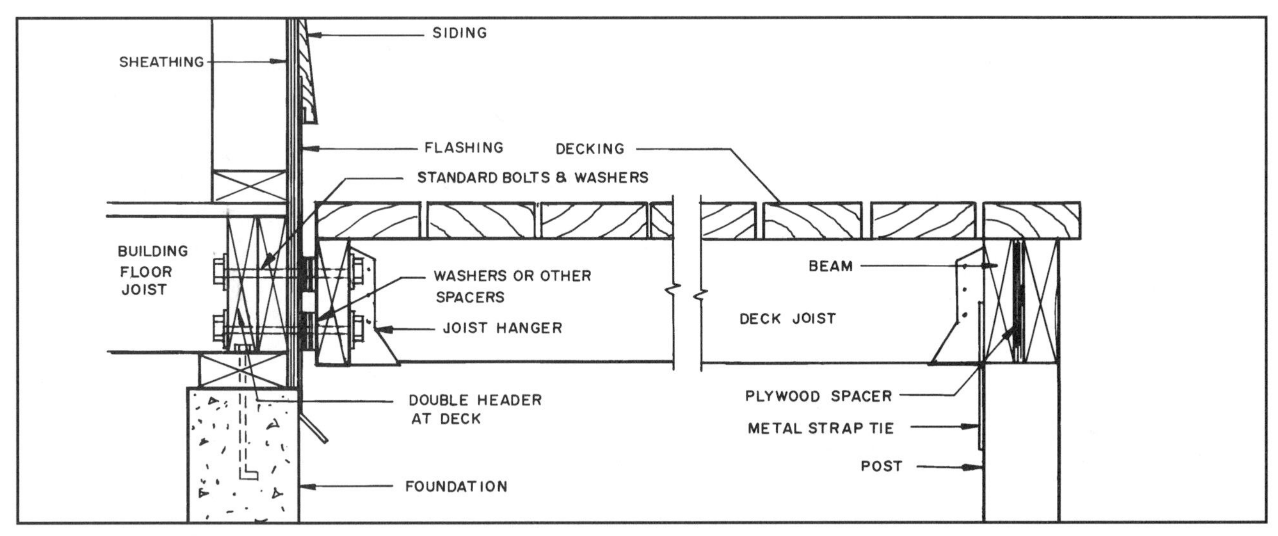

27-14 One way to join the deck to the wood-framed wall and allow space for air circulation.

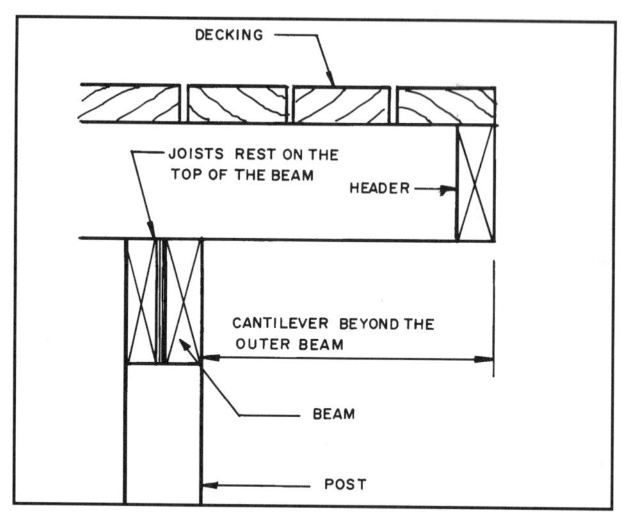

27-15 Decks can be cantilevered beyond the outer beams.

Decks can be cantilevered beyond the row of posts or piers. A beam is placed on the posts, and the joists rest on top of it. The sizes of the beam and joists depend on the span. Careful planning should occur so that the cantilever does not exceed safe distances. A typical detail is shown in **27-15**. A rule of thumb that some use is to limit the amount of overhang to one-fourth of the span of the outer joists.

Post-to-Beam Connections

It is best to have the beam rest on top of the post and tie them together with one of the many types of metal connectors **(refer to 27-3 and 27-4)**. The post can be notched when two two-inch-thick members form the beam **(see 27-16)**.

Joist-to-Beam Connections

Joists can be joined to beams by hanging them with metal joist hangers or resting them on top of the beam as shown in **27-17**.

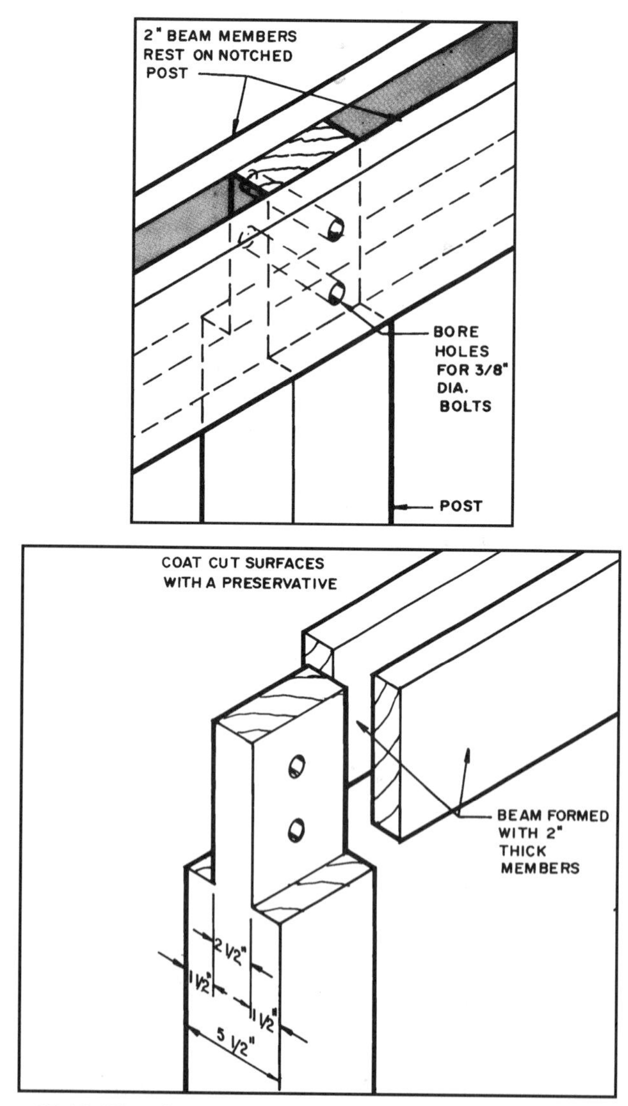

27-16 When beams are made using two 2-inch members, the supporting post is notched.

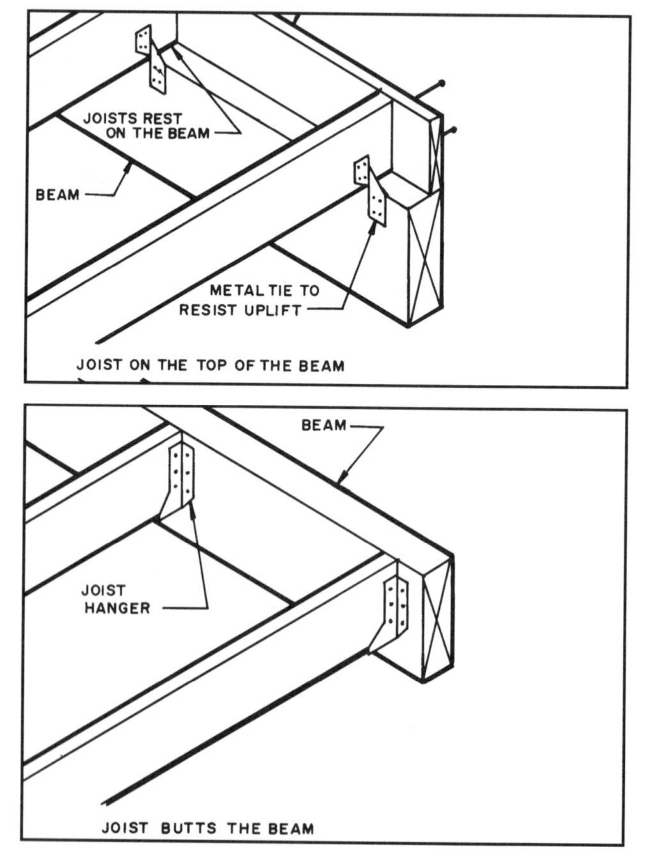

27-17 Joists can be joined to beams with metal joist hangers or by resting them on top of the beam.

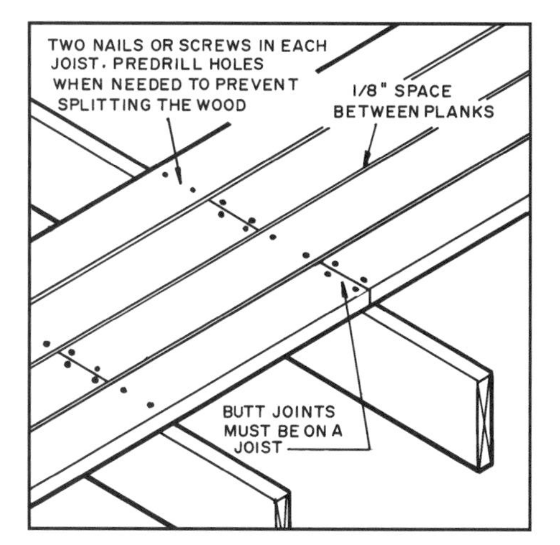

27-18 Butt joints in the decking must fall on a joist.

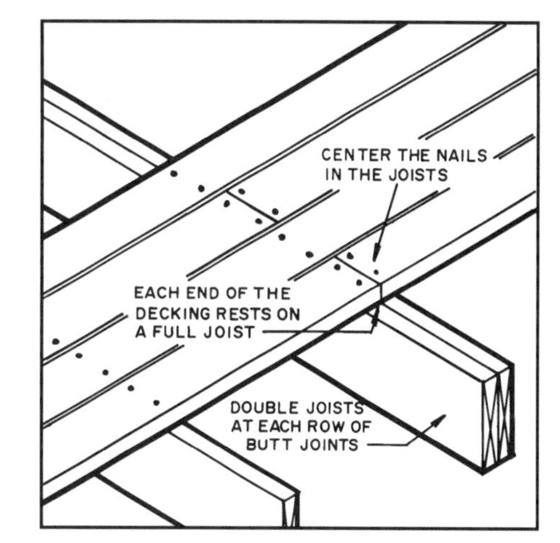

27-19 The best practice is to double the joists below each row of butt joints.

Installing the Decking

The decking planks must be installed so that air can circulate between them and promote drying. The space should be about ⅛ inch apart. Some place a 16d nail between them as a spacer.

Hot-dipped galvanized or stainless steel nails or screws are used to secure the decking to the joists. The decking should have 2 nails at each joist. When pieces butt this must occur over a joist. It is best to predrill the nail holes to prevent splitting **(see 27-18)**. An alternate and better technique is to double the joist under each row of butted decking, as shown in **27-19**.

When installing the decking, sometimes a piece has some bow and requires force to bring it straight before it is nailed in place. Various methods are used to pull the piece in place. One tool designed to do this easily and rapidly is shown in **27-20**. Some carpenters nail a scrap wood block to a joist and use a jack to straighten the piece.

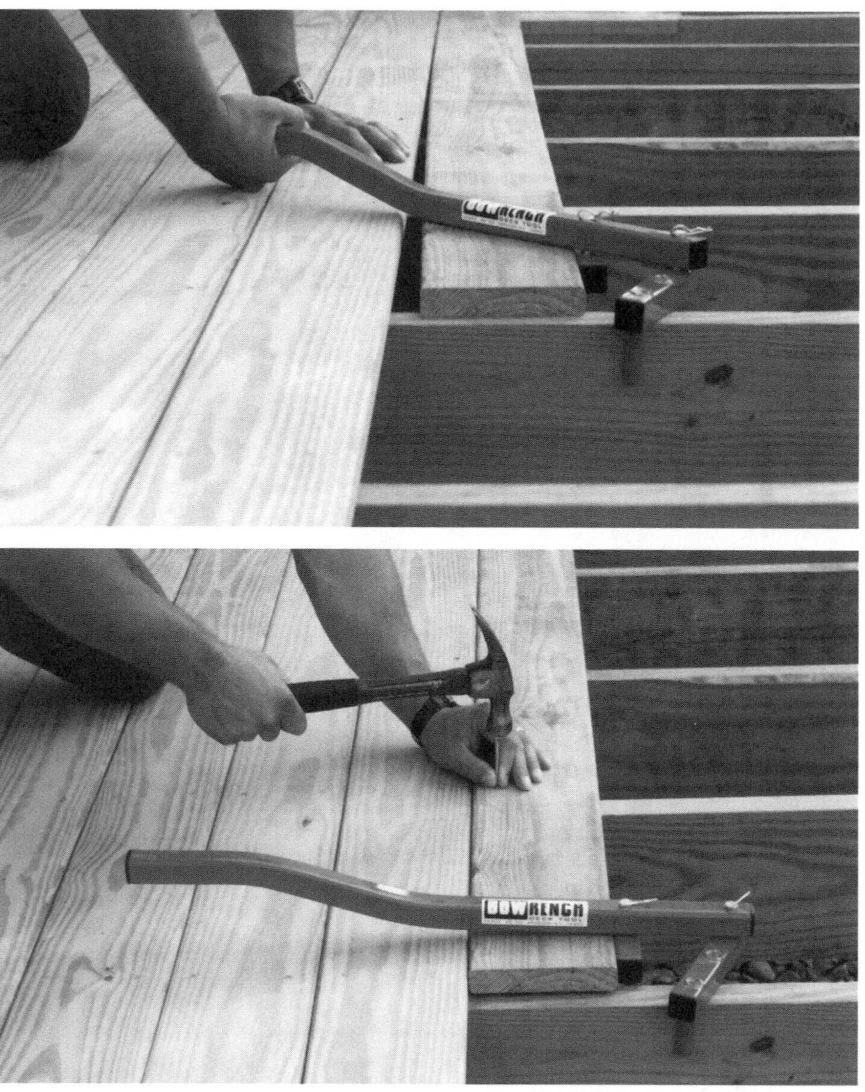

27-20 This tool pulls bowed decking into line and holds it there as it is nailed or screwed to the joist. *(Courtesy Quik Jack Flooring Jack, Cepco Tool Company)*

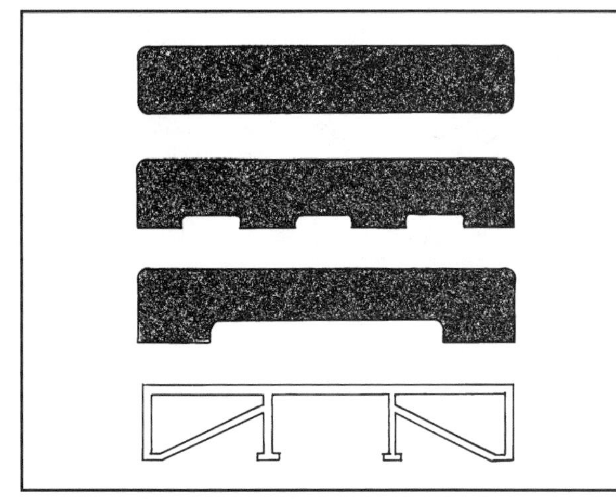

27-21 Profiles of some of the synthetic decking available.

There are a number of products available that can be used to secure the deck planks to the joists from below, eliminating all nails on the surface. Typical examples are shown earlier in **27-7**.

27-22 This is a wood-polymer lumber accepted by three major code agencies. It is available in various hues, from grays to browns. *(Courtesy Trex® Easy Care Decking)*

The various **synthetic decking** products are installed in the same manner. Most are nailed or screwed to the joists the same as solid wood decking. Consult the manufacturer for specific requirements.

SYNTHETIC DECKING

The national interest in recycling has led a number of manufacturers to produce a variety of synthetic decking from recycled wood and plastics.

Other types are made entirely from plastics, such as vinyl. For example one type is made from recycled high-density polyethylene while another is a composite made from recycled wood and plastics. These products are produced in a number of profiles as shown in **27-21**.

One such product, Trex® Easy Care Decking, is made from recycled plastics and wood **(see 27-22)**. It does not need protective sealants, will not crack, and resists termites and other insects. It is designed for use as decking and not for posts, beams, or joists. Trex® Easy Care Decking should be cut with carbide-tipped saw blades.

27-23 These exterior stairs have wood-polymer lumber treads and risers. *(Courtesy Trex® Easy Care Decking)*

The decking is secured with hot-dipped galvanized or stainless steel nails or screws. The planks should be gapped width-to-width at least ⅛ inch. End-to-end gapping should be $\frac{1}{16}$ inch for every 20-degree-F (11-degree-C) difference between the hottest temperature expected. Instructions for gapping are stapled on the end of every board. Trex® is available in three colors—Winchester gray, natural, and brown.

Trex® $\frac{5}{4} \times 6$-inch decking used on residential installations will span joists 16 inches O.C., and 2×6-inch will span 20 inches O.C. Trex® can also be used for constructing exterior steps **(see 27-23)** and railings.

Another product, Dream Deck, is made from 100 percent polyvinyl compounds **(see 27-24)**. It requires no waterproofing or other maintenance. The joists are framed as for all decks. They are spaced 16 inches O.C. when the decking is installed perpendicular to them. The planks are secured to the joists with manufacturer-supplied tracks. The tracks are screwed to the joists and the planks placed over them and pushed down and snapped into place. Dream Deck is also used for docks, stairs, and other exterior applications **(see 27-25)**.

Another product is made from redwood fibers and polyethylene plastic. It is cut and installed in the same manner as solid wood decking. This product, Choice Dek, has a uniform cedar color which weathers to a natural driftwood color. It is available in $\frac{5}{4} \times 6$-inch and 2×6-inch planks.

Synthetic decking holds up extremely well under all weather conditions and is very suitable for decks and docks **(see 27-26)**.

27-24 This is a decking made from polyvinyl compounds. It requires no waterproofing or other maintenance. *(Courtesy Thermal Industries, Inc.)*

27-25 The deck, stairs, and rail were built using polyvinyl materials. *(Courtesy Thermal Industries, Inc.)*

27-26 This dock decking is made from redwood fibers and polyethylene plastic. It is also used on decks and exterior stairs. *(Courtesy Choice Deck by A.E.R.T.)*

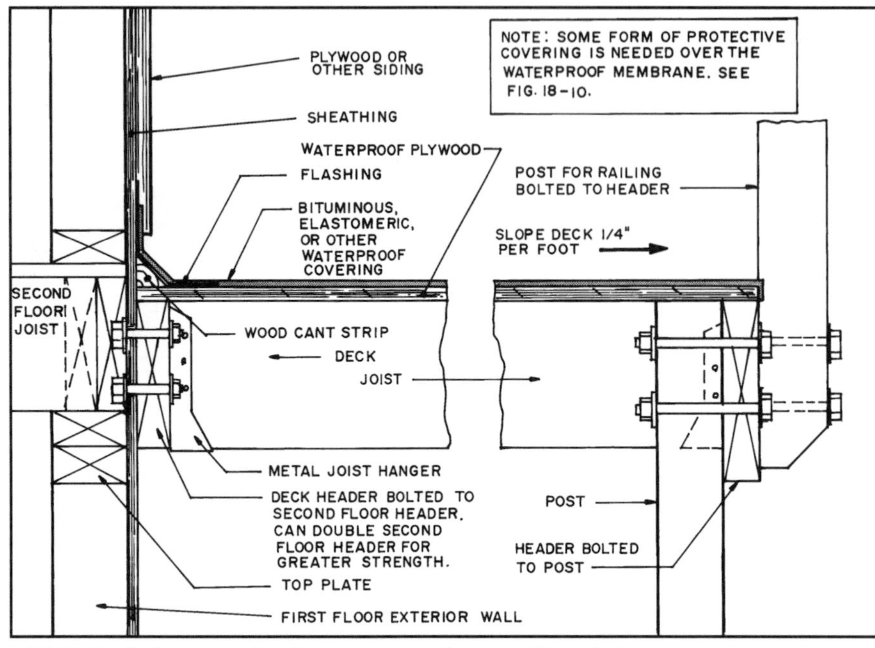

27-27 Typical construction for a waterproof second floor deck.

height of the railing. It is desirable to have the floor waterproof so water does not drip down on the area below. There are many ways to build this floor. One is shown in **27-27**. This uses waterproof plywood decking covered with some form of waterproof membrane which could be bituminous, elastomeric, or metal. The deck must slope to the outer edge. To protect the membrane some type of protective surface is required, such as the wood grating shown in **27-28**. This is built and laid loose over the membrane.

SECOND FLOOR DECKS & PORCHES

These installations are typically built over a first floor deck or porch. The structure is built as described for decks, with close attention to the size and spacing of structural members. Local code will specify requirements for loads and for the

BUILDING RAILINGS

Railings are built in many ways. The height of the railing is specified by the building code. Railings are required for porches and decks with the floor above the ground a specified distance, such as 30 inches. The minimum height of the railing is spec-

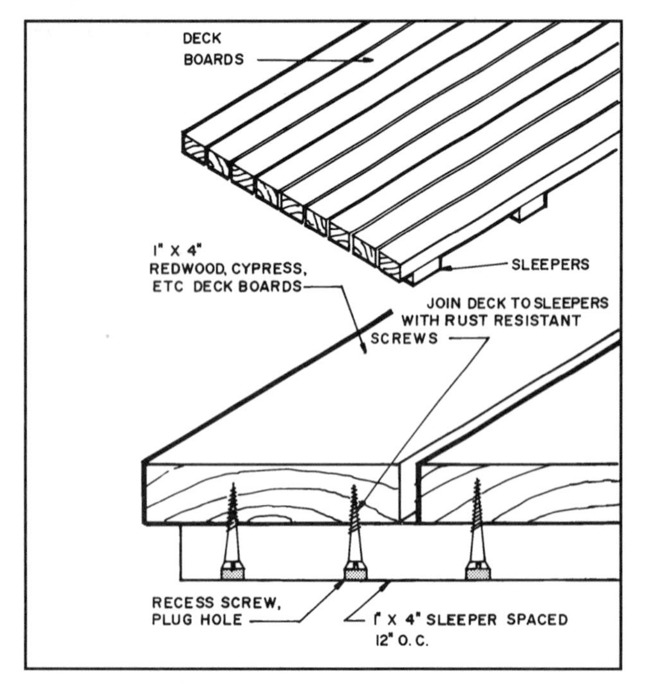

27-28 A wood grating can be placed over the second floor deck to protect the waterproof membrane.

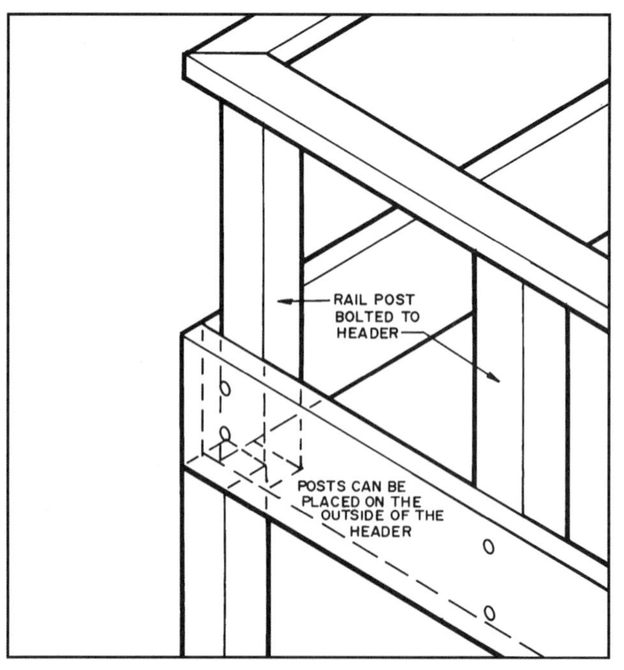

27-29 Rail posts can be bolted to the end joists and the header.

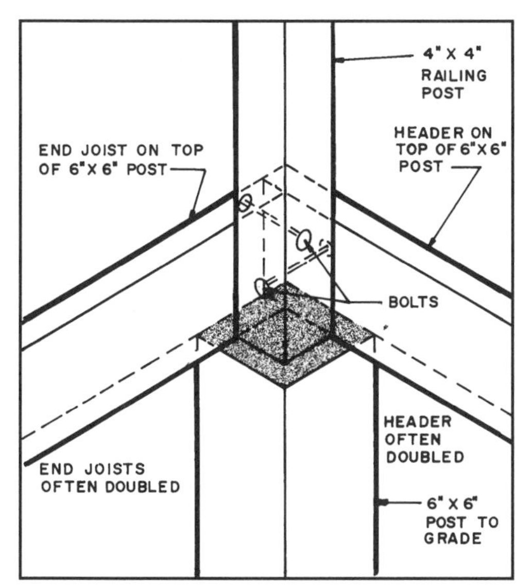

27-30 The corner is made up with the header and end joist on top of the post, and the rail post rests on top of the post and is bolted to the header and joist.

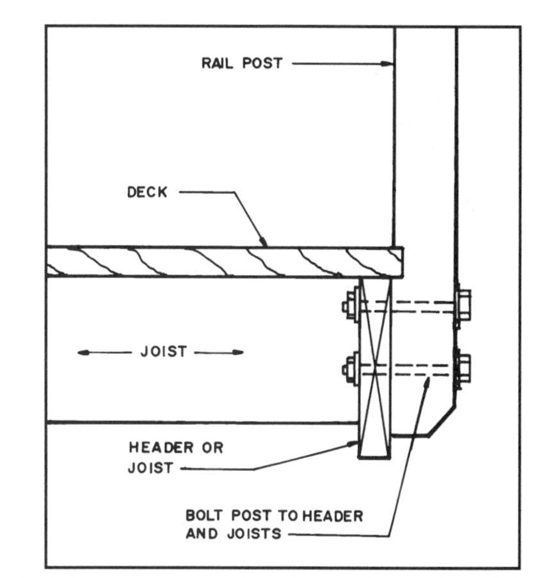

27-31 The rail post can be bolted to the outside of the end joists and header.

ified, such as 36 inches. It is necessary for the railing to be rigid and strong. Railing posts can be bolted to the floor header and a joist, as shown in **27-29** and **27-30**. The floor decking is cut to fit around the post. The posts could extend above the floor and form the vertical rail support, as shown in **27-16**. The railing posts could be bolted to the outside of the header and end joists, as shown in **27-31**.

Rails are also manufactured from vinyl **(refer to 27-25)**. They are weather resistant and require no maintenance. They are installed following the recommendations of the manufacturer. Wood posts are used to support the rails and balusters. Vinyl covers are placed over the posts.

The design of the railing can vary considerably. One of the many design possibilities is shown in **27-32**; more are on the following page.

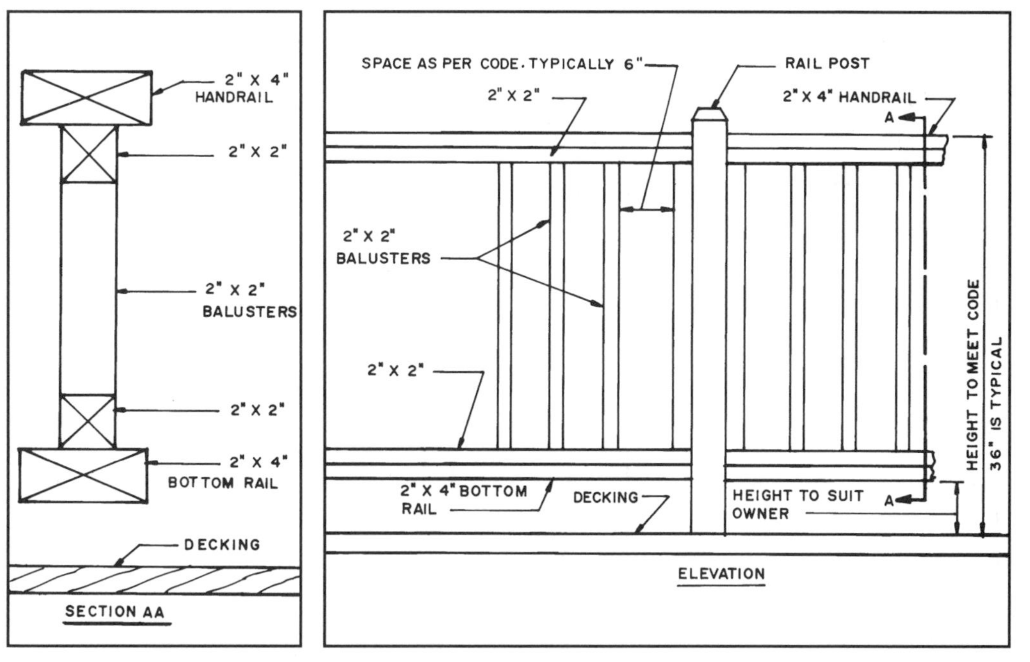

27-32 This railing design has the balusters butt the top and bottom rails.

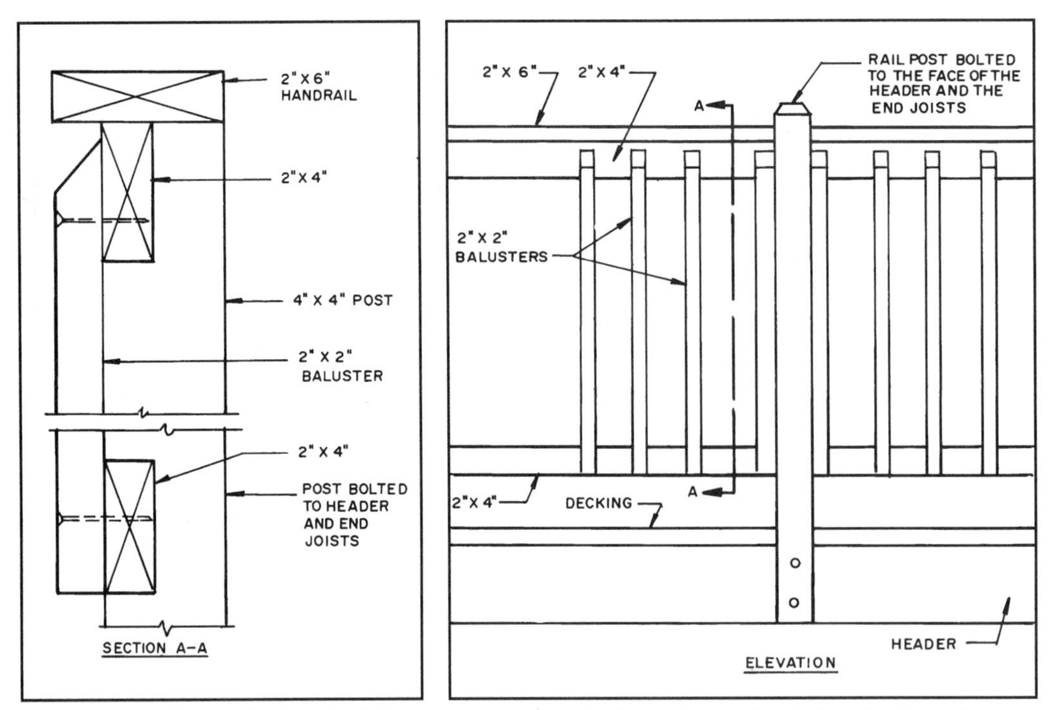

27-33 This railing design has the balusters mounted on the face of a member below the top rail.

A few more of the many railing design possibilities are shown in **27-33**. If the deck or porch is close to the ground and does not require a railing, some people like to build some form of seat around the edge **(see 27-34)**. Another seat design is shown in **27-35**.

STEPS

Some type of wood steps is required to get from the porch or deck to grade **(see 27-36)**. Pressure-treated lumber should be used. The design of the stair must meet local building codes. The following is typical of such codes. Stairs with four treads or less do not require a handrail. Handrails must be 30 to 40 inch-

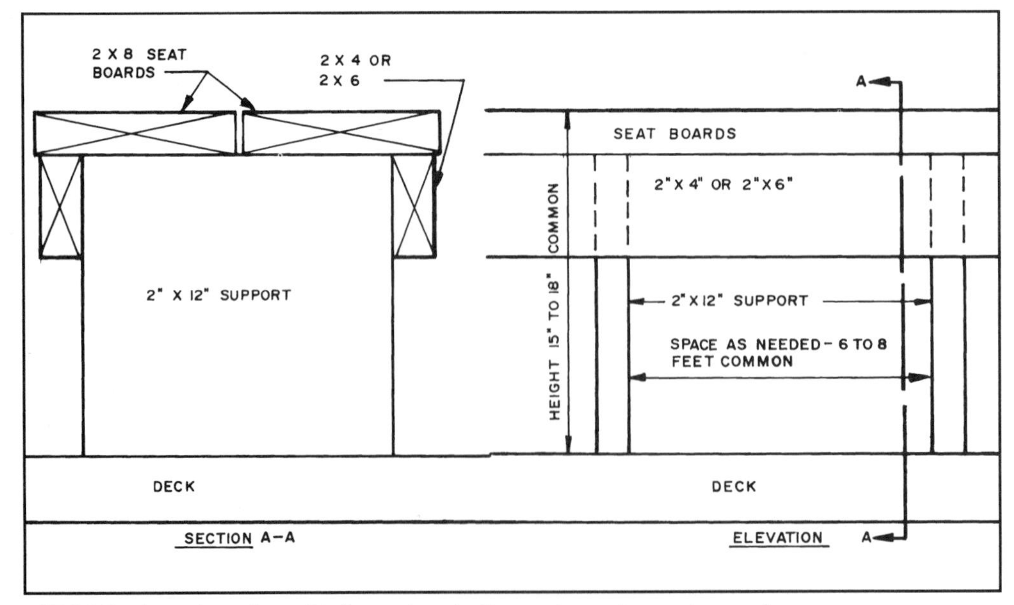

27-34 Decks and porches with floors close to the grade can have a low seat built around the edge instead of a railing.

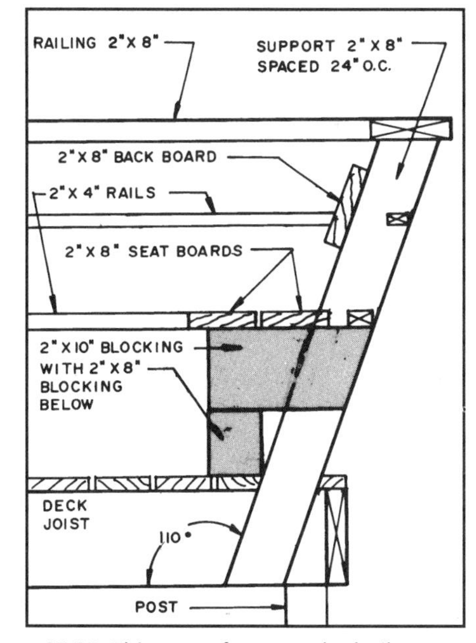

27-35 This type of seat can be built as part of the railing.

27-36 These stairs to the deck use notched stringers and have open risers.

es above the stair tread. The stair rise must be at least four inches and not more than eight inches. The minimum tread width is nine inches, and the minimum stair width is 30 inches.

The stair stringers can be nailed to the header and rest on a 2 × 2 ledger or joined with a metal joist hanger. The end of the stringers on the ground can rest on four-inch concrete blocks set into the soil **(see 27-37)**.

Generally the treads are nailed to an open stringer, as shown in **27-37**. Metal staircase hangers are available as shown in **27-38**.

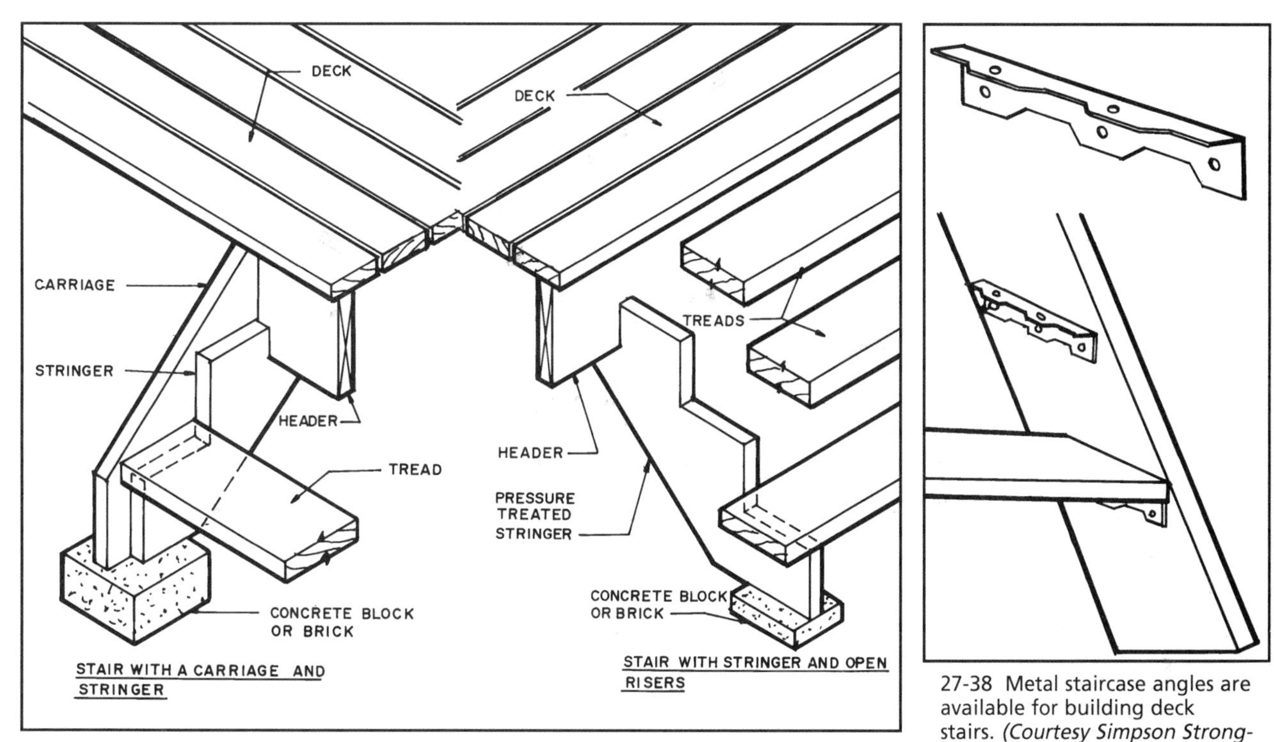

27-37 Typical details for porch and deck stairs.

27-38 Metal staircase angles are available for building deck stairs. *(Courtesy Simpson Strong-Tie Company, Inc.)*

The handrail posts can be bolted to the stringer and the handrail design is usually the same as the deck railing **(see 27-39)**.

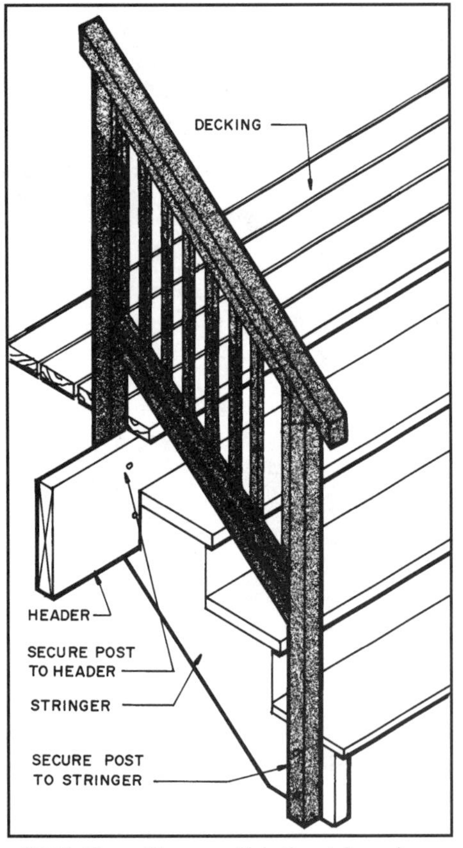

27-39 The railing parallels the stair and usually uses the same design as the deck rail.

PORCH ROOF CONSTRUCTION

The porch roof is framed in the same manner as discussed for roof construction (refer to Section 4, "Framing the Roof & Dormers,"and specifically to Chapter 16 on "Framing Flat, Shed, Gambrel & Mansard Roofs").

Most often a porch roof has a single sloped surface, although a gable roof can be used. The posts extend up to the headers, which carry the rafters. The size of these members must be selected so that they carry the required loads. The ridge is secured to the wall of the house. As the shingles are put on, flashing must close the joint between the roof and the wall. Typical details for porch roof construction are shown in **27-40** and **27-41**.

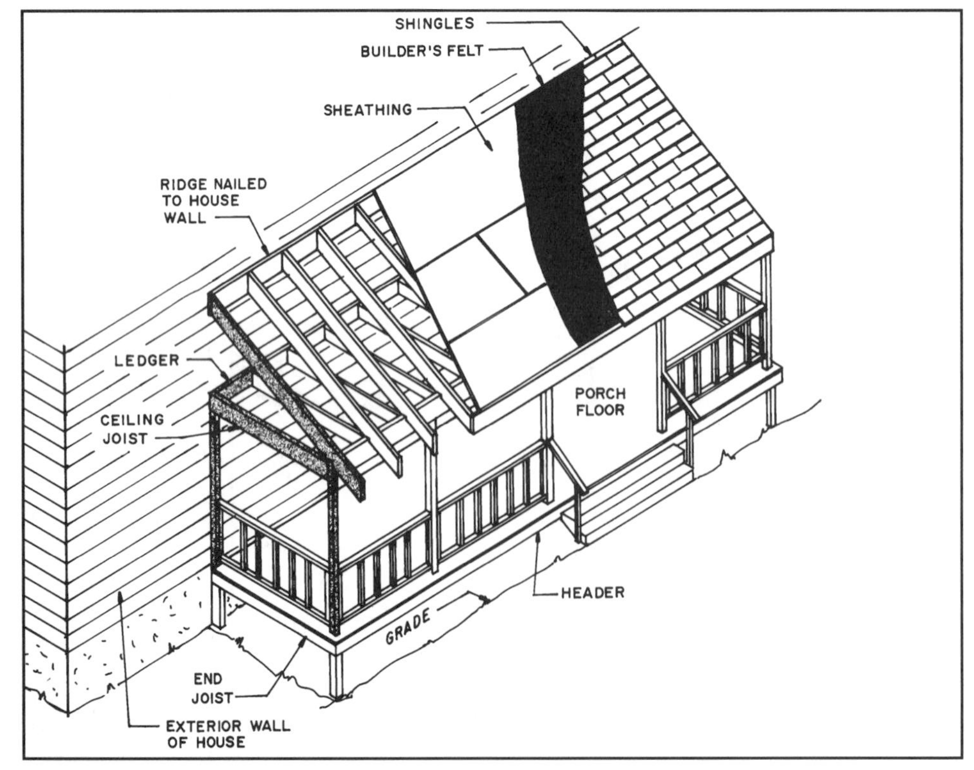

27-40 One way to frame a porch.

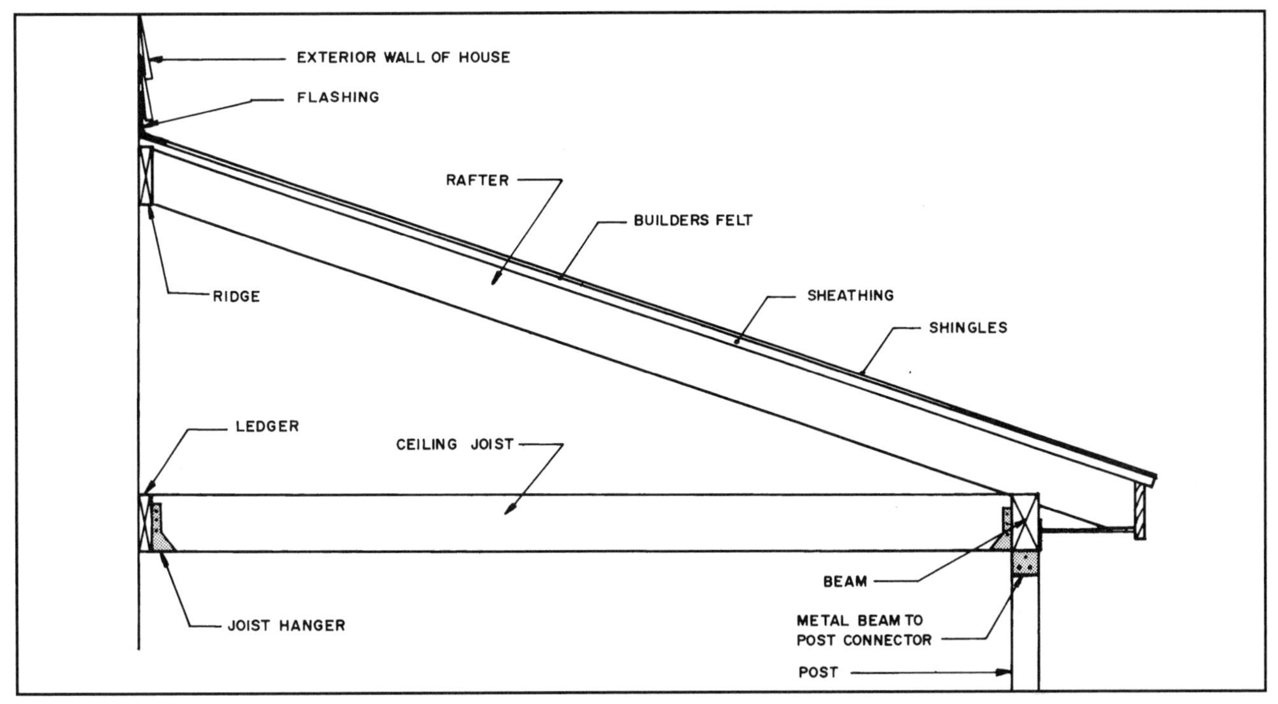

27-41 Typical porch roof construction.

ADDITIONAL INFORMATION

Wood Decks, Materials, Construction, and Finishing

U.S. Department of Agriculture
Forest Service, Forest Products Laboratory
One Gifford Pinchot Drive
Madison, WI
53705-2398

PANELING & MOLDING

CHAPTER 28

MOLDINGS

The use of moldings on the interior and exterior of a house greatly enhances the appearance. Their choice and location are part of the detailing decisions made by the architect and the homeowner. The choice is often related to the style of house.

Interior moldings, such as are used where the wall and ceiling meet, add a great deal to the atmosphere of the room **(see 28-1)**. Exterior use, such as at the intersection of the exterior wall and the soffit, gives the building a classic, finished appearance **(see 28-2)**.

The choice of the moldings to be used in new construction is made by the architectural designer in consultation with the owner of the building under construction. With old construction being remodeled or renovated, an interior designer may be involved in such specifications. The carpenter must be certain the specified moldings are used and should not make a substitution without written approval from the designer or owner.

A wide variety of molding profiles are available for interior and exterior use. Satisfaction with the finished job depends on the selection of the molding profile, the use of moldings made from quality materials, and the installation by a skilled finish carpenter. Moldings are available from three sources: stock moldings, special moldings made to order by a molding manufacturer, and carpenter-made moldings. These are available in a variety of wood species and plastic polymers.

28-1 This attractive interior molding is a major design feature of the room. *(Courtesy Focal Point)*

28-2 Moldings are used on the exterior as part of the detailing features. *(Courtesy Fypon, Ltd)*

STOCK MOLDINGS

Stock moldings are those manufactured in large quantities and sold by local building materials suppliers. The actual designs and sizes available vary from region to region. It is less costly to use stock moldings than have specially designed moldings made. However, the choice of profiles is limited. They can be bought from a local building supply dealer and from a number of mail-order supply houses. They have catalogs showing the stock moldings available, and orders are placed by phone, fax, or mail. The molding is shipped to the construction site.

The quality of wood moldings varies depending on the species of wood used and the method of cutting them. While the production of stock moldings is a mass-production process, the cutting tools must be kept sharp and the profile of the cutters consistent so that smooth, accurately cut profiles are produced.

Some companies supply wood moldings that are fireproofed and prefinished. Prefinished moldings must be handled carefully so that they are not damaged during shipping, storage, cutting, and installation.

CUSTOM-DESIGNED MOLDINGS

Sometimes it becomes necessary to match a molding in a very old house that is being restored. The architect can send a sample of the molding to a specialty shop. The architect may want to produce original molding designs for new construction. Carefully dimensioned drawings are made and submitted to the specialty shop. Here they shape cutters to the profile desired and run the molding. Usually a complexly shaped molding will require several different sets of cutters and several passes through the molding machine **(see 28-3)**. The process is similar to that for carpenter-made moldings explained later in this chapter. In the carpenter-made process, finish carpenters can buy quality wood stock and use a router or shaper to produce moldings that are different from the stock types available.

28-3 This beautifully designed and assembled hardwood wall opening illustrates the complexity and detail that can be provided by molding manufacturers. *(Courtesy White River Hardwoods-Woodworks, Inc.)*

STOCK WOOD MOLDINGS

Examples of the basic molding shapes are in **28-4**. Examples of some of the commonly available wood moldings are shown in **28-5**.

Baseboards cover the lower edge of the wall where it meets the floor.

Base caps are attached to the top of some types of baseboard to give an attractive profile.

Base shoes are placed on the floor where the baseboard meets the floor.

Casing is used to trim the interior edges of door and window frames; it covers the space between the frame and the interior wall finish material.

Chair rails are applied to the wall in rooms, such as a dining room, where chairs may rub against the wall.

Corner moldings are used to protect wall corners from damage.

Cove moldings find a wide range of applications, including use on inside wall corners.

Crown moldings (see 28-6) are used where the wall and ceiling meet and are also used on exterior cornice construction.

Door stops are nailed inside the door frame so the door closes against them when it is closed.

Panel molding is used to provide a profile on the top of wood paneling and to the outer edge of casing.

Picture rails are applied on the wall near the ceiling and are shaped to hold a metal hook used to hang pictures.

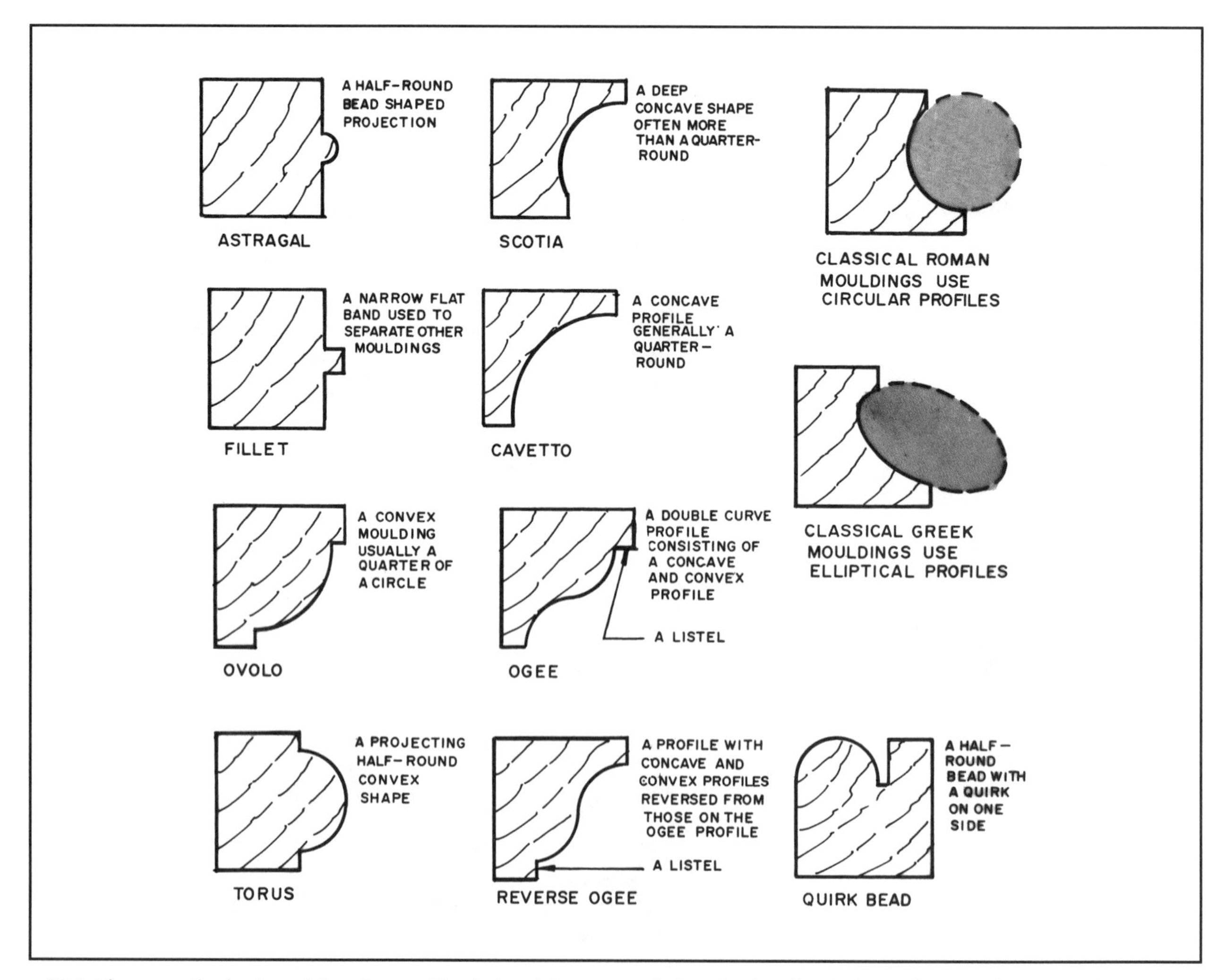

28-4 These are the basic molding shapes. Classical moldings were designed using the circle or ellipse as the design element.

28-6 This molded polymer crown molding is used to enhance the junction of the wall and ceiling. *(Courtesy Fypon, Ltd)*

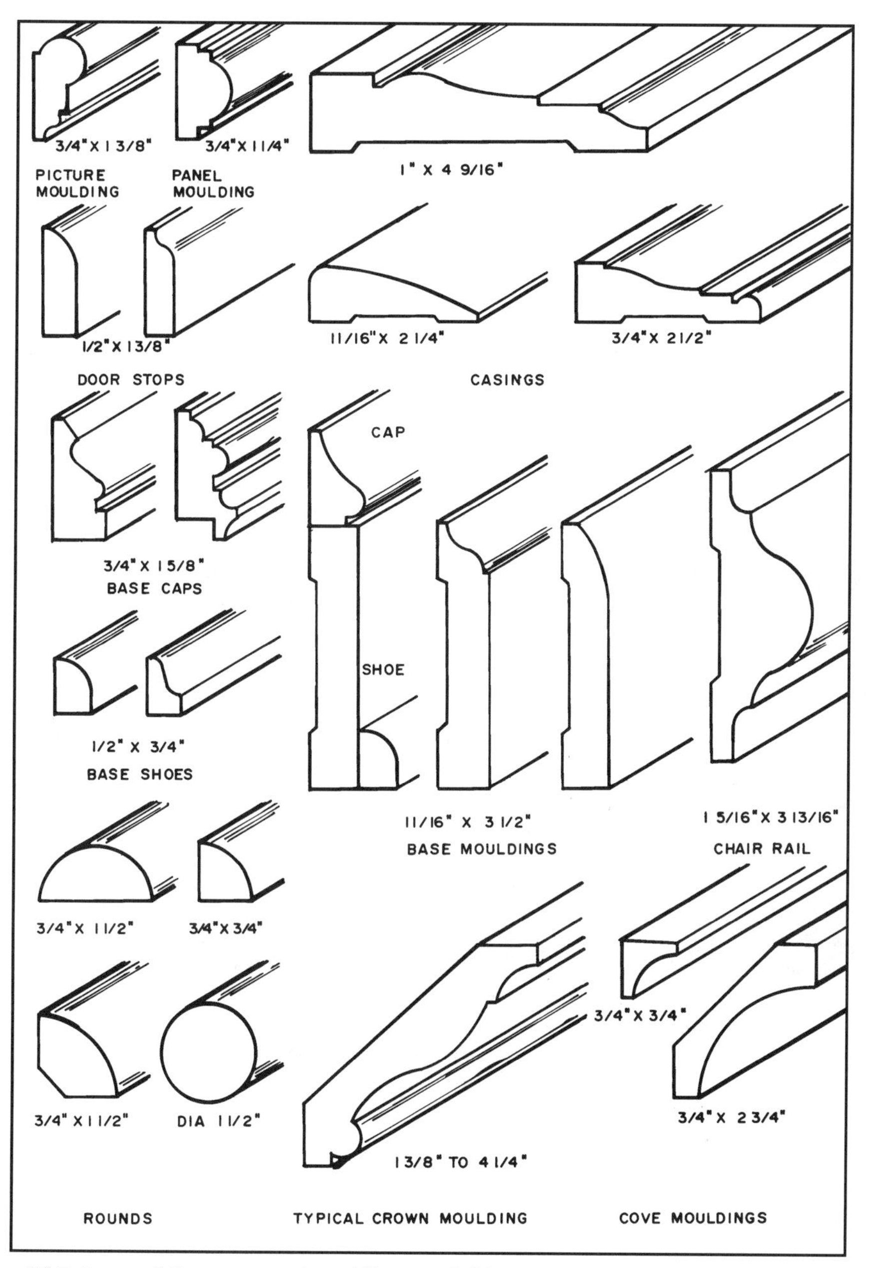

28-5 Some of the many stock moldings available.

POLYMER MOLDINGS

Polymer moldings are manufactured from a high-density polyurethane foam. The foam is cast in a silicone-lined mold and as it cures a smooth, hard outer surface develops. It has an impact resistance similar to softwood, but dents can be easily repaired. It is much lighter than wood moldings and can be used on the interior or exterior of the building. It is available in many of the stock molding profiles used with wood, plus a wide range of other products such as ceiling medallions, decorative wall panels, and door and window pediments **(see 28-7)**.

Polymer moldings do not rot and will not swell and shrink with changes in humidity. They also save some labor costs because complex moldings are made as a single piece while wood of the same profile would require several separate pieces be cut and nailed.

28-7 This molded polymer arch surround with arch pilasters provides a decorative, weather-resistant exterior design feature. *(Courtesy Fypon, Ltd)*

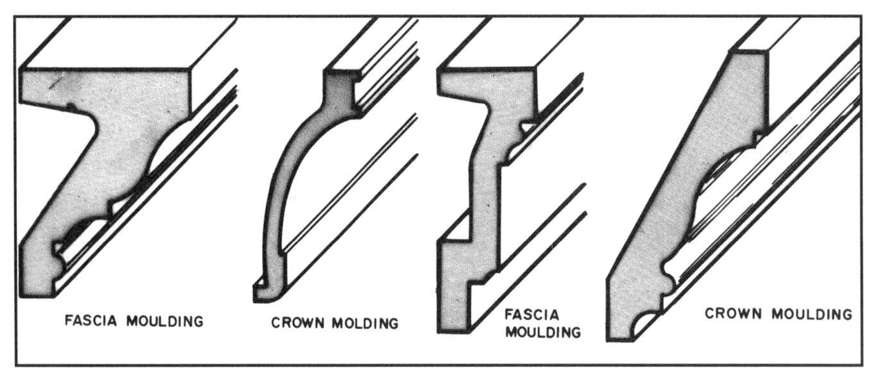

28-8 A few of the many complex, single-piece polymer moldings manufactured.

Some building supply dealers stock polymer moldings, or moldings can be ordered directly from one of the companies manufacturing the product. These moldings are available in 8-, 10-, 12-, and 16-foot lengths. Examples of some of the complex profiles are in **28-8**. Some manufacturers will make custom-designed moldings.

Polymer moldings can be painted or stained. Some are very flexible and can be bent to the desired contours **(see 28-9)**. They can be cut with regular woodworking tools. A fine-toothed saw or sharp utility knife is typically used.

28-9 Polymer architectural moldings are flexible and can be bent to fit on curved applications. *(Courtesy Flex Trim, Inc.)*

WOOD MOLDING GRADES & SPECIES

The selection of the grade and species of wood molding influences the cost and quality of the finished job. Softwood moldings cost less than hardwoods but will dent upon impact much easier. The products can be divided into two major types: **paint grades** and **stain** or **natural finish grades**.

Paint Grades

Paint grades are usually softwoods because these have a smooth, tight grain and are less expensive. Commonly used species include various types of pine, such as white, Idaho, northern, and Ponderosa, as well as sugar and yellow hemlock, poplar, fir, and redwood, and occasionally a hardwood such as maple, beech, or birch. Since they will be covered with an opaque coating the painter can fill any poorly fitted joints, dents, and defects, and set and cover nails. The surface is then sanded smooth and painted.

Stain or Natural Finish Grades

Stain-grade moldings are commonly some type of hardwood such as oak, walnut, birch, mahogany, or cherry. The choice of grade and specie is made by the designer and owner. Stain-grade moldings are not usually stocked by local building materials suppliers, so an order for these moldings must be placed well in advance of when they are to be needed so that they can be manufactured.

Stain-grade molding requires more careful cutting of joints because poor-fitting joints cannot be hidden by filling them. The job appears better if long lengths are used, thus reducing the number of end joints, which are never attractive. This requires the highest level of skilled joinery and installation.

Hardwood moldings are more difficult to nail and frequently the holes have to be drilled for the nails. Obviously, hardwood moldings will withstand bumps and abrasions better than softwoods. If for some reason moisture is a possible problem, molding made from rot-resistant wood, such as redwood, cedar, or cypress, can be used.

28-10 A wood shaper used to cut profiles on wood parts. *(Courtesy Delta International Machinery Corporation)*

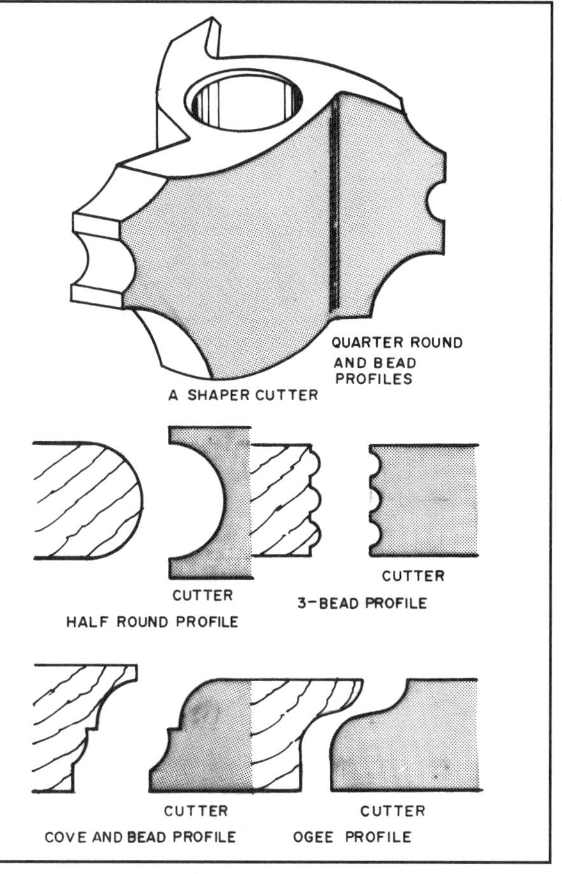

28-12 These are just a few of the profiles available on stock shaper cutters.

Detailed information on wood moldings of various designs and species is available from the Architectural Woodwork Institute, 2310 S. Walter Reed Drive, Arlington, VA 22206-1199.

CARPENTER-MADE MOLDINGS

The finish carpenter can produce special moldings using a woodworking shaper, circular saw, and a portable router. Specific details for operating each of these machines are available in woodworking books. A shaper is shown in **28-10**. It uses a three-lip cutter having the profile ground in the edges of each lip **(see 28-11)**. A wide variety of cutters with various profiles are available; some are shown in **28-12**. The shaper cutter extends through a round opening in the shaper table. The cutter is raised and lowered until the desired profile is exposed. The wood is placed against a fence and moved into the rapidly rotating cutter **(see 28-13)**.

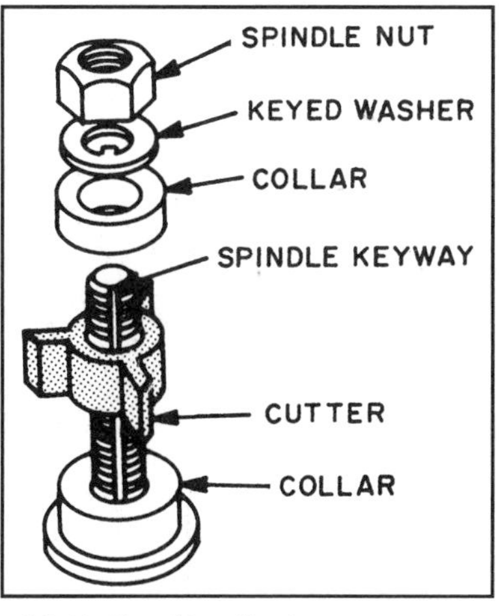

28-11 The three-lip shaper cutter is mounted on the shaper spindle.

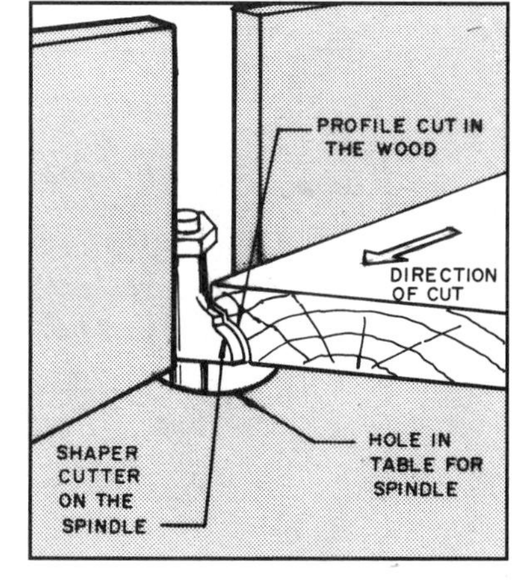

28-13 Place the stock against the shaper fence and move it into the rotating cutter.

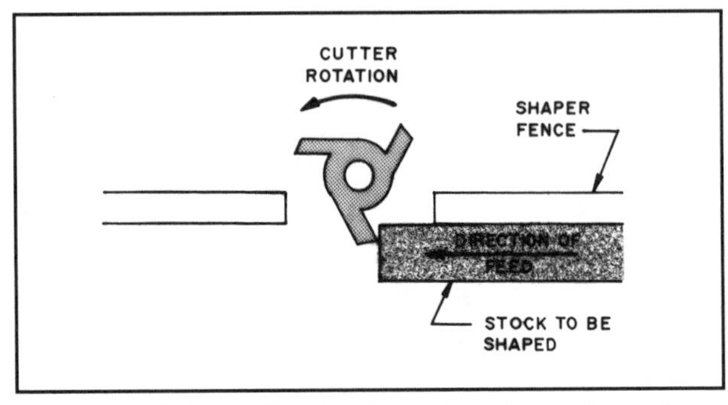

28-14 Feed the stock to be shaped into the rotation of the shaper cutter.

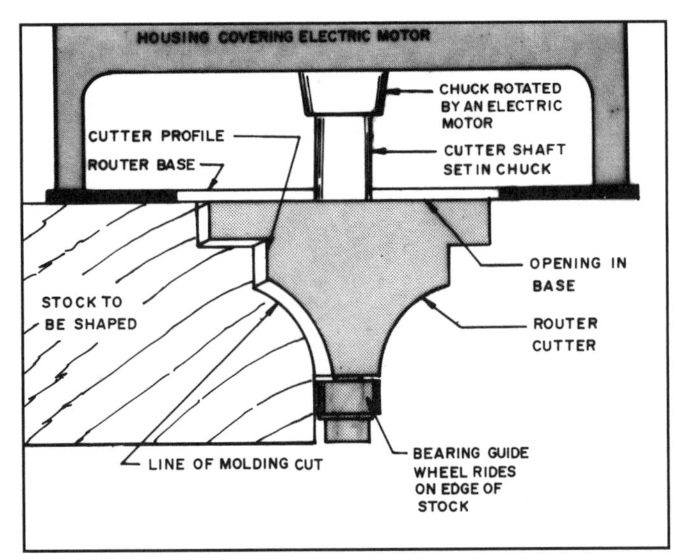

28-15 The router bit has a bearing guide wheel to help control the depth of cut.

Always feed the stock against the direction of rotation of the cutter **(see 28-14)**. Remember that the direction of rotation of the cutter on a shaper can be reversed by a simple switch, so it is important to feed the stock from the proper direction.

Small moldings can be cut with a portable router. (The router is discussed in Chapter 42.) It is advisable to use a sufficiently powerful router such as one that has one or more horsepower. A variety of bits are available in various profiles. The stock to be shaped will have a square edge against which the router-bit guide rides. The cutter removes the wood forming the profile **(see 28-15)**. The depth of cut used influences the profile cut. This is adjusted by raising and lowering the motor **(see 28-16)**. Notice in **28-17** how this adjustment changes the profile of the molding.

Moldings produced with a router can be made more easily and more accurately if the router is mounted on a router **shaper** table **(see 28-18)**. The router is mounted upside down below the table. A fence is used to guide the stock past the cutter. This is much the same operation as using a shaper.

The stationary circular saw (discussed in Chapter 42) can be used to cut a limited variety of moldings using the circular saw blade. It can cut bevels and rabbets using the rip fence as a guide. To cut a bevel, the blade is tilted to the desired angle, and the stock is placed

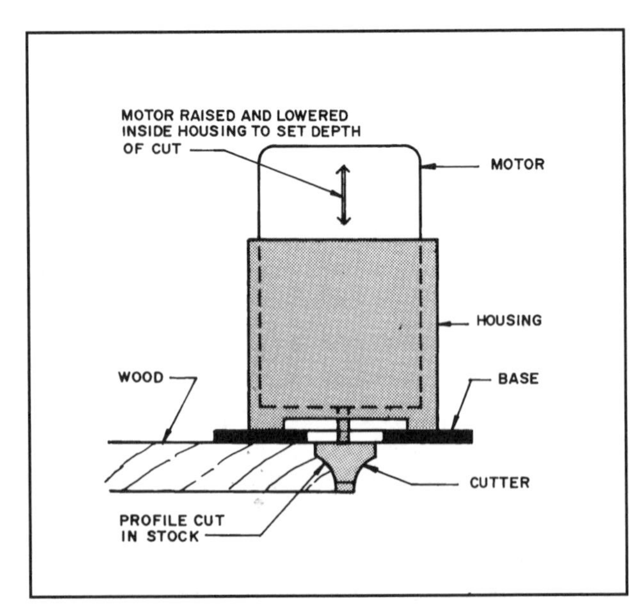

28-16 The base of the router rests on the surface of the wood, and the motor is raised and lowered to adjust the cut.

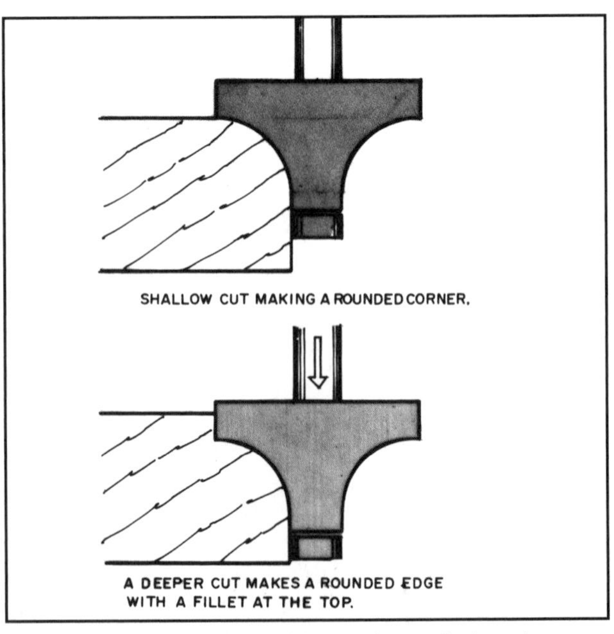

28-17 The shape of the molding is changed when the router motor is raised or lowered.

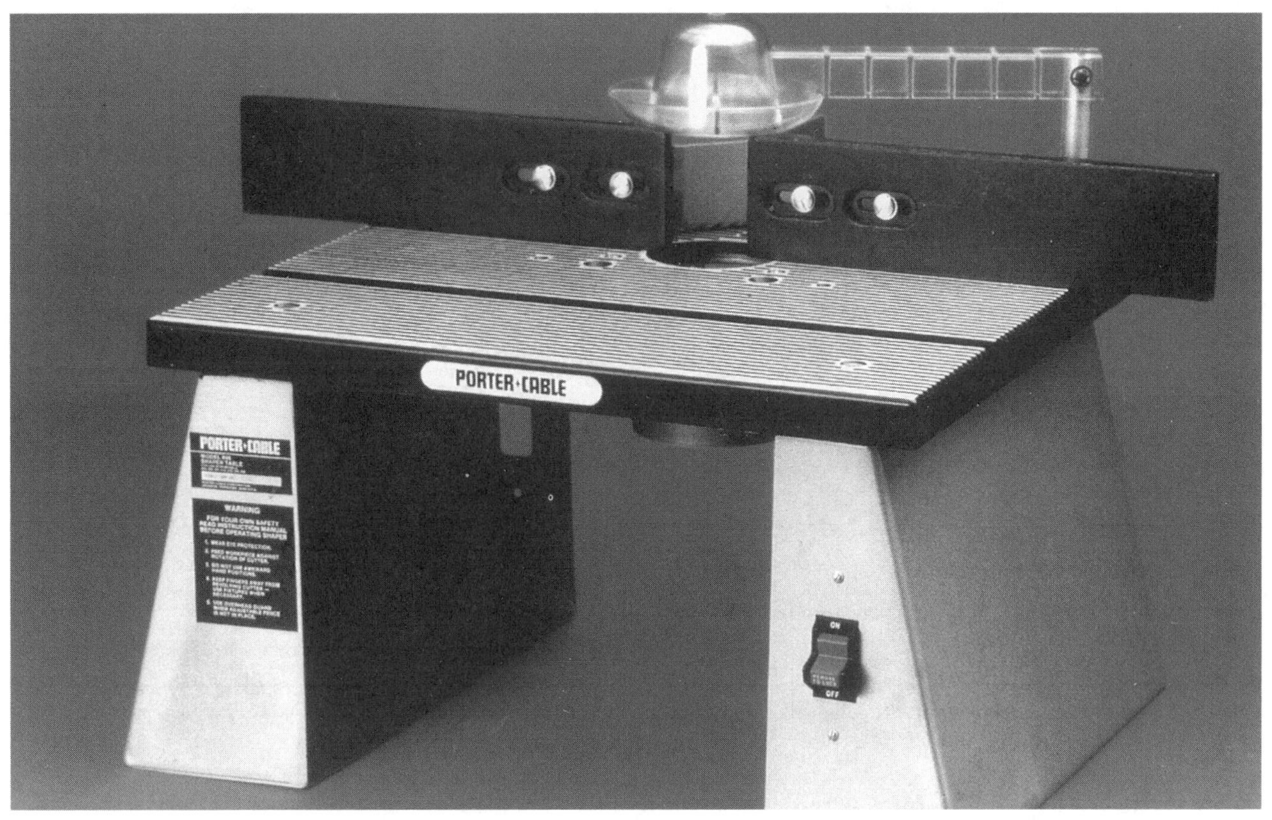

28-18 The router is mounted on a router table, allowing it to operate much like a shaper. *(Courtesy Porter-Cable Corporation)*

against the rip fence and moved past the blade **(see 28-19)**. To cut a rabbet, the saw is set to the desired depth of cut, and the width of the rabbet is set between the blade and the fence and this is cut. Then the stock is turned over, the fence is reset, and the second cut is made as shown in **28-20**.

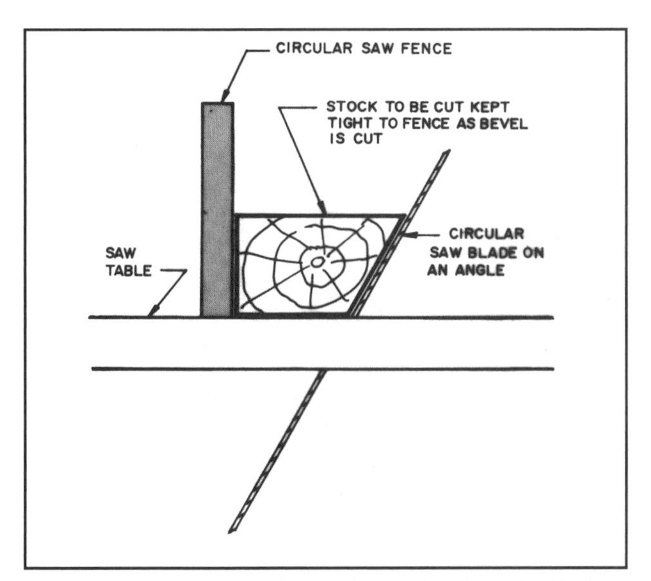

28-19 Bevels are cut on the circular saw by tilting the blade.

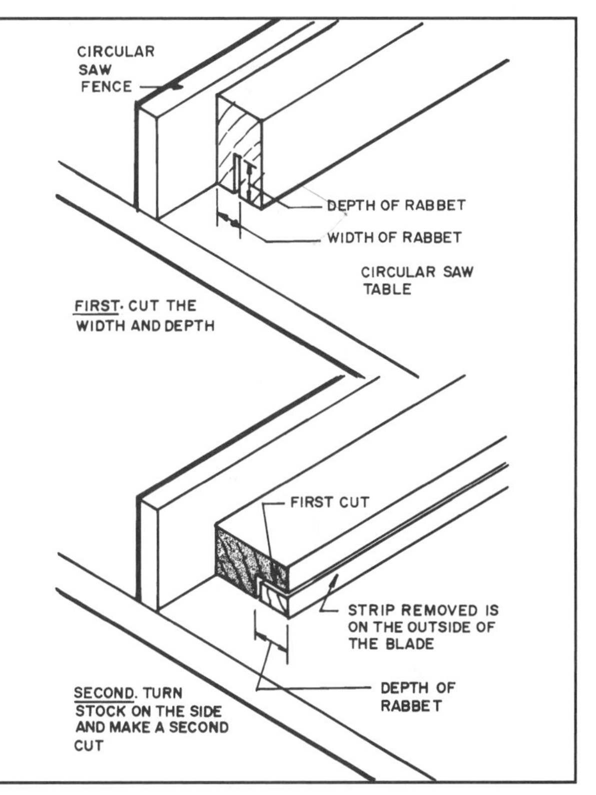

28-20 A rabbet can be cut on the circular saw by making two cuts.

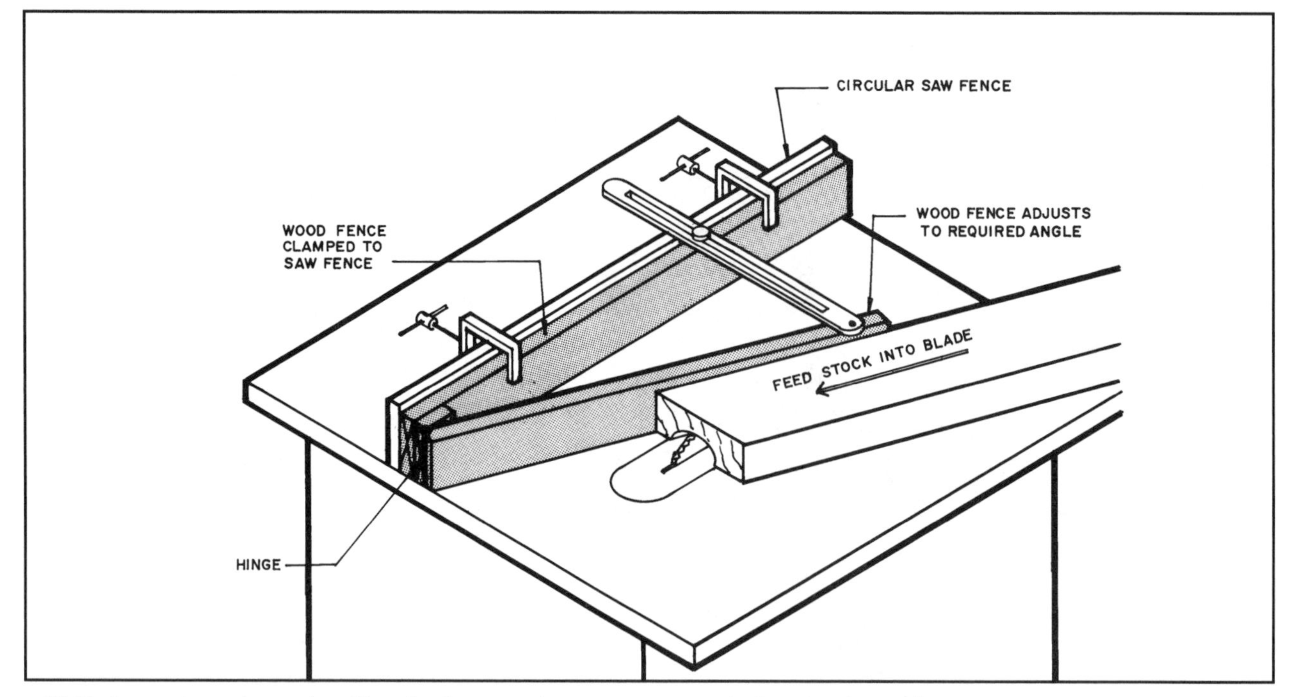

28-21 Cove cuts can be made with a circular saw using a carpenter-made fence as the guide.

Cove cuts can be made by feeding the stock across the saw on an angle. An adjustable fence can be made as shown in **28-21**. It is clamped to the rip fence and adjusted to the desired angle. The size and shape of the cover is varied by adjusting the angle. Make some trial cuts in waste stock until the desired cove shape is produced. Tightly lock the adjustable fence at this angle. Raise the blade so that it protrudes ¼ inch (6mm) or less above the surface of the table. Place the stock against the temporary fence, and push the wood across the saw. Then raise the saw blade ⅛ to 3⁄16 inch (3 to 4.5mm) higher and repeat the cut. Continue until the depth of the cove required is reached. Cove moldings of various profiles can be made by combing the cove cut with other cuts on a shaper or molding head, or with a router, as shown in **28-22**.

There are a number of molding heads available for use on the circular saw. They replace the saw blade and have cutter blades much like those used on shapers and routers **(see 28-23)**. The stock to be

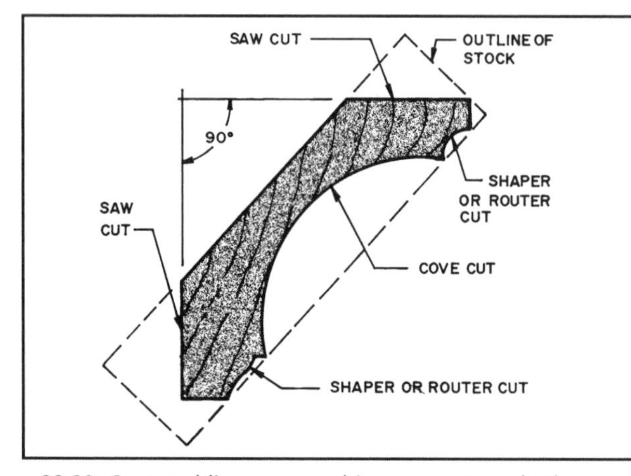

28-22 Cove moldings can combine cove cuts and other profiles.

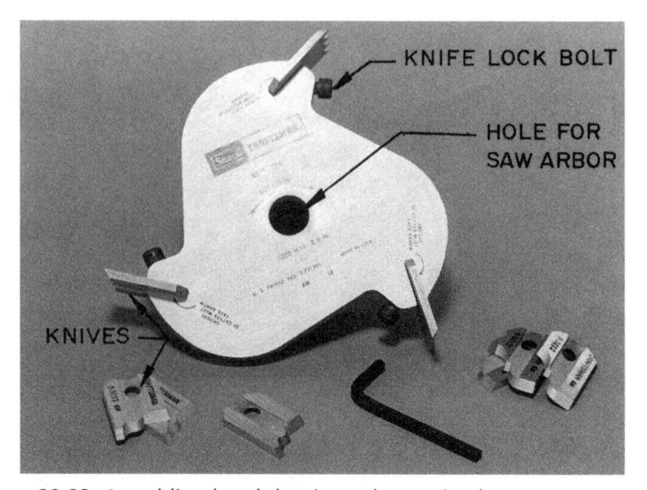

28-23 A molding head that is used on a circular saw.

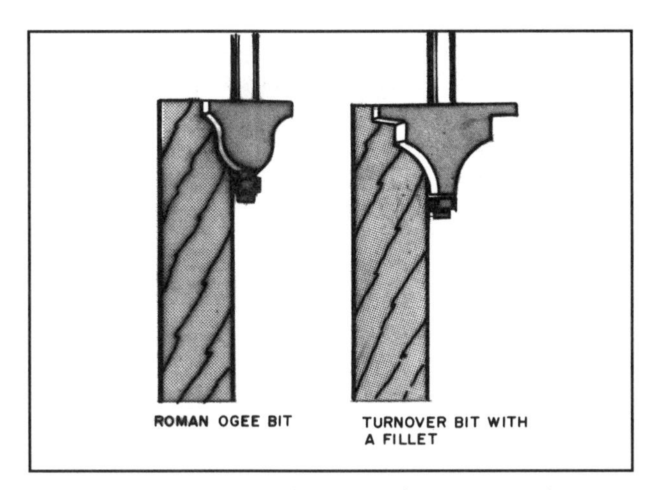

28-24 Carpenters can make base and casing using the shaper or router.

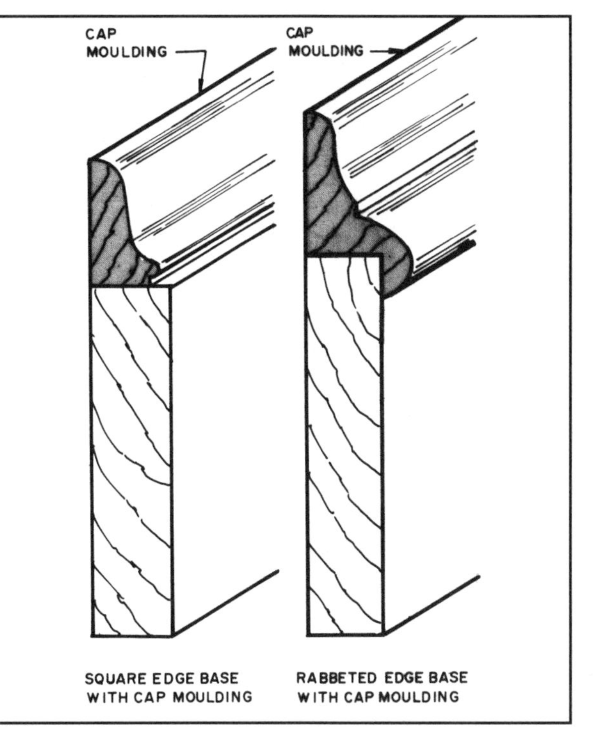

28-25 The base can have a separate profiled cap molding installed on top of the square-edged base material.

cut is placed against the rip fence and moved over the rotating molding head. Manufacturers' instructions should be carefully studied and observed when using molding heads on circular saws.

The process of fashioning carpenter-made molding can take many forms. A knowledge of the capabilities of the power tools and ingenuity in applying these to molding design give the carpenter quite a range of possible designs.

Start with good-quality, kiln-dried, straight stock free of knots, warps, and other defects. In some cases, square-edged stock becomes the baseboard or door and window casing. It can have various profiles cut, if a more decorative design is desired. Selected examples are shown in **28-24**. If a baseboard with a finished cap is desired, the cap could be purchased ready-made or made by the carpenter, as shown in **28-25**. The square-edged base may require a rabbet in the top edge. The base cap can be machined on the edge of a wider board and cut to width **(see 28-26)**.

When the carpenter proceeds to make the molding, it is best to make trial cuts on scrap stock before cutting into the more expensive material. Moldings to be painted can be cut from white pine or other softwoods. These can be purchased cut to standard widths and planed to standard thicknesses. If hardwoods are to be used, they are usually sold in random widths and lengths and are not surfaced. It will be necessary to have hardwoods planed to thickness and cut and squared to the desired widths. It is more difficult to get long pieces of hardwoods, so more joints will occur as the molding is installed than with softwoods.

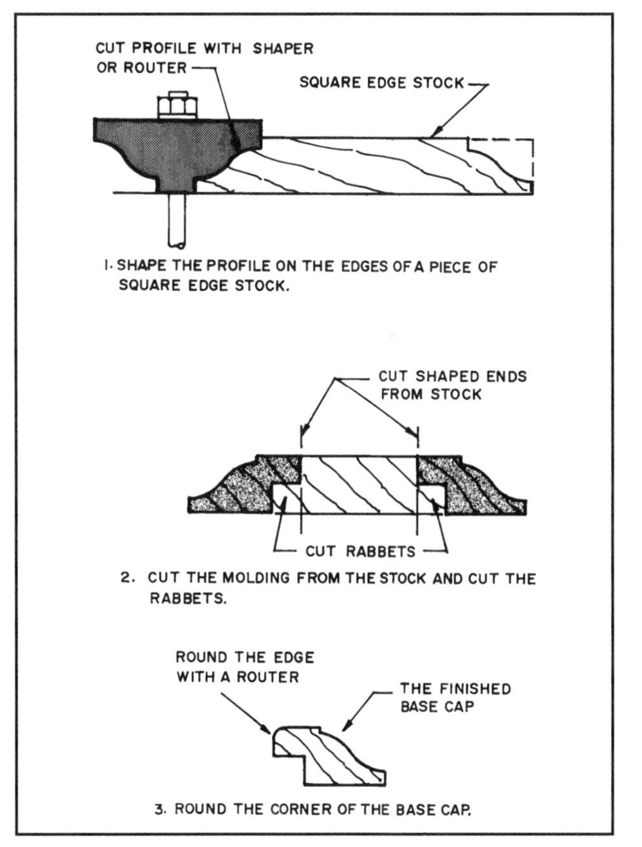

28-26 Carpenter-made base caps can be shaped and cut from flat stock.

ESTIMATING QUANTITIES OF MOLDING

Trim and molding may be ordered by **specifying lengths** for certain installations or by giving the **total number of lineal feet** required. Often lengths are specified for casing doors and windows. Quantities of this type are ordered by giving descriptive information such as the width, profile, and length of each piece.

For example, to case an interior door, each side would take two pieces long enough to run the 6′-8″ length plus the width of the molding required for a miter. In most cases, a 7′-0″ piece would be long enough. However, without detailed specifications, casing is sold in two-foot lengths, starting with six feet; so four eight-foot pieces would be required for the trim on both sides of the wall.

The head casing will vary by door width, but if it is a 2′-8″-wide door, it will have to be that long plus twice the molding thickness for a miter on each end, plus up to about ½ inch for the setback from the door frame. Therefore, a 2′-8″ door plus ½ inch plus 5½ inches for twice the 2¾-inch-wide casing gives a finished length of 3′-2″ for the head casing. Two pieces could be cut from one 8-foot length with 1′-8″ waste. Three pieces could be cut from a 10-foot piece with 6 inches of waste. Some door and window manufacturers supply casing in required lengths for the units **(see 28-27 and 28-28)**.

Most moldings are sold by the lineal foot and are installed using the lengths received. This is frequently referred to as **running molding**. These moldings can be joined with scarf joints to form the required long pieces **(see 28-29)**. In most cases, therefore, the exact lengths received are not critical. Base, shoe, crown, cove, panel, and other moldings are usually ordered by giving the total number of lineal feet needed. This figure includes the number of feet actually required plus a percentage for waste. A waste factor of 10 to 20 percent is typically used.

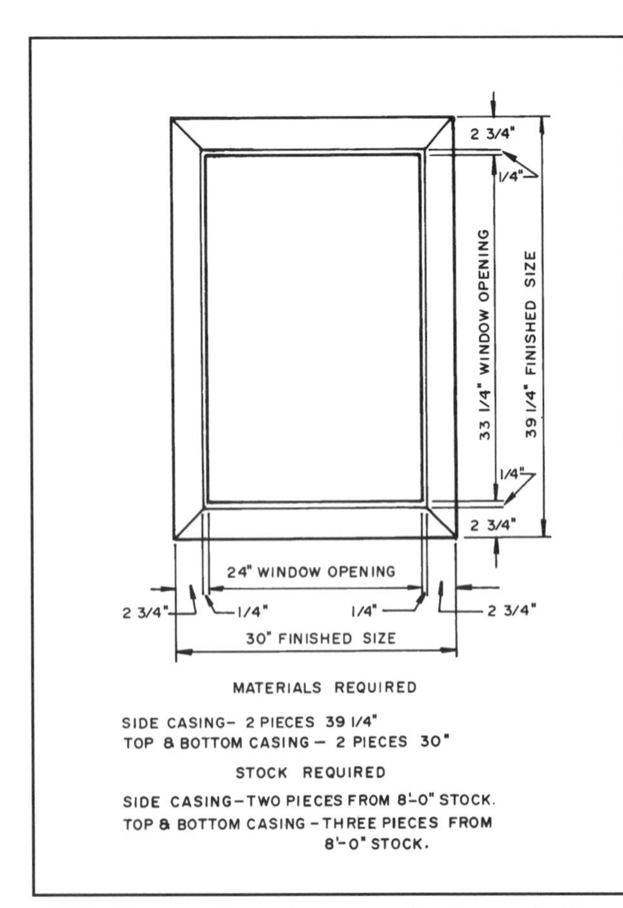

28-27 How to estimate the amount of casing needed for a window.

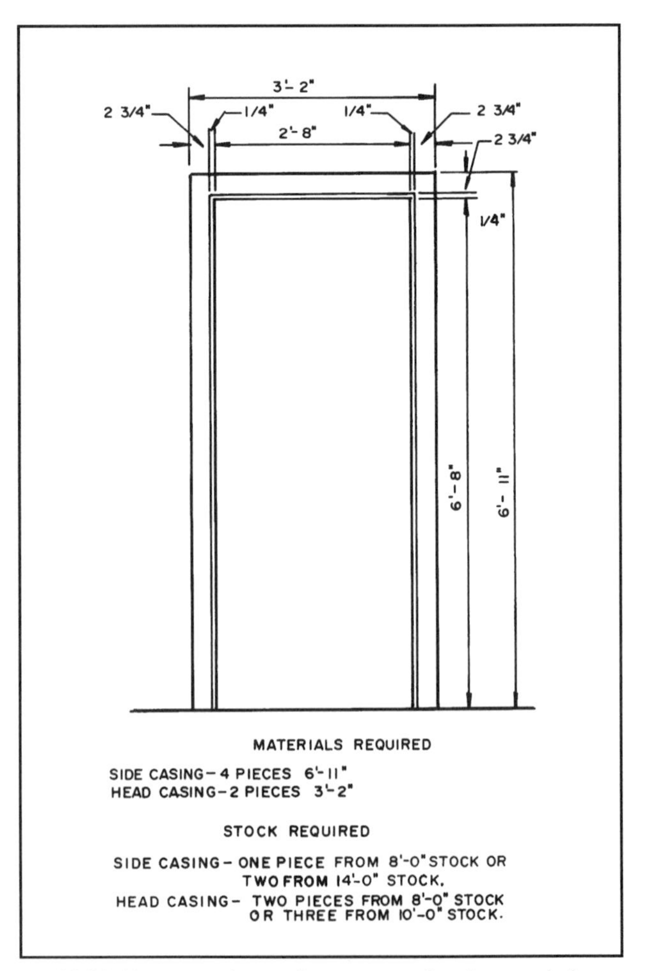

28-28 How to estimate the amount of casing needed for an interior door opening. Remember casing is used on both sides of the opening.

An example is shown in **28-30**. To figure the amount of base, the length of the perimeter was found, ignoring openings, to be 65 feet. A 10 percent waste factor was added to give a total of 72 lineal feet of base and shoe to be ordered. The quantity received may be slightly more than this, because the supplier will send the lengths on hand that add up close to the amount ordered. If you make the decision as to the sizes you need for each wall and order these specific lengths, you will be buying 68 lineal feet.

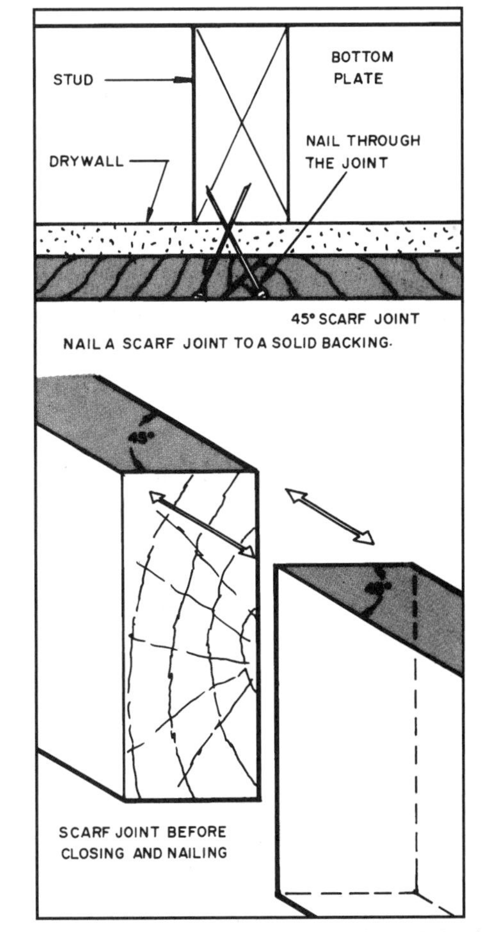

28-29 Butting molding and base materials are joined with a scarf joint.

SECURING & STORING WOOD MATERIALS

The general contractor, with the consultation of the owner and designer, purchases the finish materials in most arrangements. The species of wood, quality of all materials, and so on may be stated in the specifications and be part of the general contract. When the finish carpenter arrives on the job the materials should be there.

However, many small jobs require the finish carpenter to choose and purchase these materials. As always, the obligations of the contract between the carpenter and contractor must be observed. As discussed, moldings are available in various grades and species, and these choices influence the cost of the materials.

When the molding is delivered to the job, it must be **stored flat on a dry surface**. Storage inside the building on the dry wood subfloor is best. If the subfloor is wet or if the material is stored on a concrete garage floor, plastic sheeting must be placed below it. The material must also be covered with plastic to protect it from moisture. All drywall plaster work must be finished and completely dry before storing the stock in the building. The building must be ventilated to reduce the moisture in the air. The wood molding is kiln dried for interior use and will absorb moisture from the air. It is recommended that the wood products be allowed to be in the building a few days before they are used so that they can equalize their moisture content with that in the air.

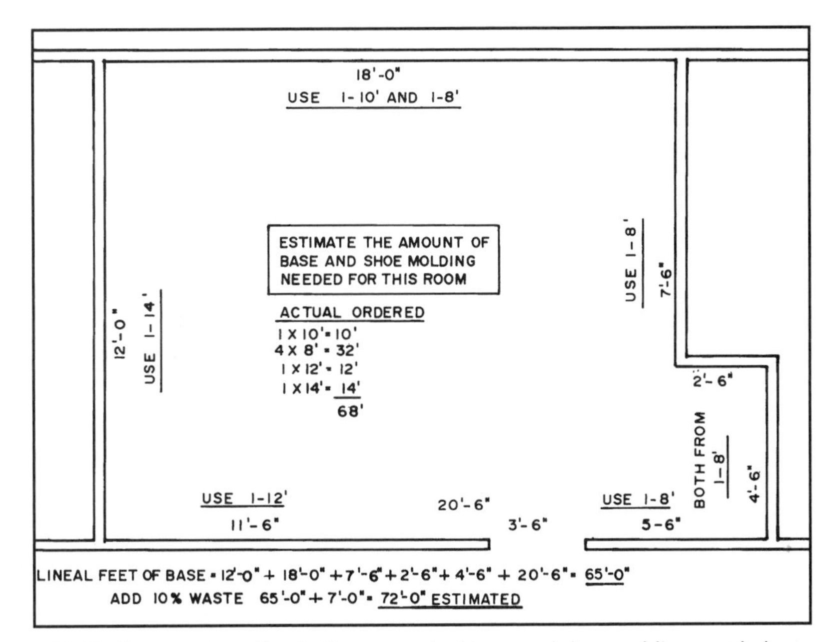

28-30 One way to estimate the amount of base and shoe molding needed for a room.

CHAPTER 29

INSTALLING BASE, CROWN & OTHER MOLDINGS

Interior base and other moldings should not be installed until after the walls are finished. The drywall should be taped and thoroughly dry. If the walls are of plaster, it must be completely dry and the rooms free of moisture. The wood molding will absorb moisture and warp if installed in an area of high humidity. Actually the molding should not be brought into the building until the atmosphere is at normal humidity levels for that region. Plastic molding is not affected by moisture.

BUILDING CODES

Under most building codes used for residential construction, interior trim such as baseboard, shoe molding, window casing, picture molding, chair rails, and ceiling molding can be wood. This will not violate restrictions pertaining to fire regulations. Foam plastic interior trim may be used if it meets a series of specifications in the code and does not cover more than a specified percentage of the wall or ceiling area. A 10 percent maximum coverage is typically specified.

PRELIMINARY WORK

Before installing the baseboard, you should check to see that the installation conditions are favorable. See that the walls are plumb and straight; see that the floors are level. If the floors have an irregular surface or any other conditions are unfavorable, it will be difficult to do a good job of installing the base. Some corrective action may be called for before you start to cut and install the base.

In some cases the corners of a room will not be 90 degrees or will have excess drywall compound that changes the angle. This occurs on both inside and outside corners.

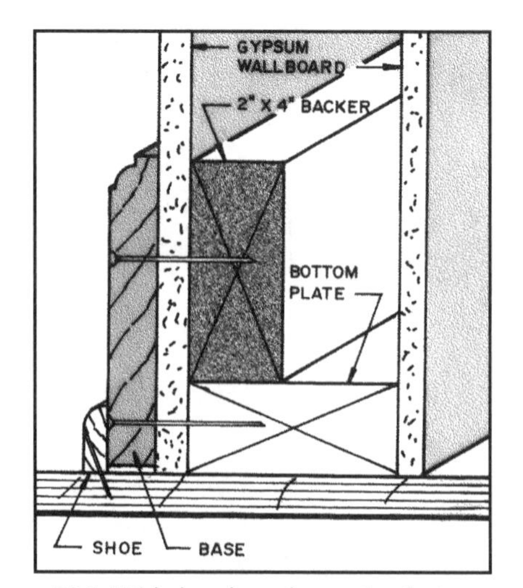

29-1 Wide baseboards require that a backer board be nailed between the studs so they can be secured to the wall.

Verify that proper backing for nailing wide baseboards was installed. Baseboards under five inches (127mm) wide can be nailed to the studs and bottom plate. If baseboards wider than five inches are used, the framers should have installed backing boards as shown in **29-1**. Correcting these and other defects is discussed later in this chapter.

Baseboards are installed after the door and window casings are in place and after the wainscoting, paneling, and drywall are installed. Check the plans to be certain the interior is finished and ready for the base.

Find out what type of finished floor is to be used and whether a shoe molding is specified. Some carpenters install the base ¼ inch (6mm) above the subfloor. This helps avoid most irregularities in the floor. If vinyl floor covering is used, a shoe molding covers the space at the floor. When carpet is used, shoe molding is generally omitted, and the carpet butts the base. If hardwood flooring is specified, it is best if it can be installed before the base. If this is not the case, use pieces of the flooring below the base as spacers to allow the clearance needed for the floor to be installed. The hardwood floor may go under the base, not butt it. Since wood expands when exposed to moisture, the floor could buckle if expansion space is not left around the wall. The shoe covers this space at the base **(see 29-2)**.

Finally, check the molding received. Select the best pieces for use on the walls that are most visible. Less desirable pieces may be used where short runs are needed and defects can be cut away, or they can be used in closets and other areas where they are seldom seen.

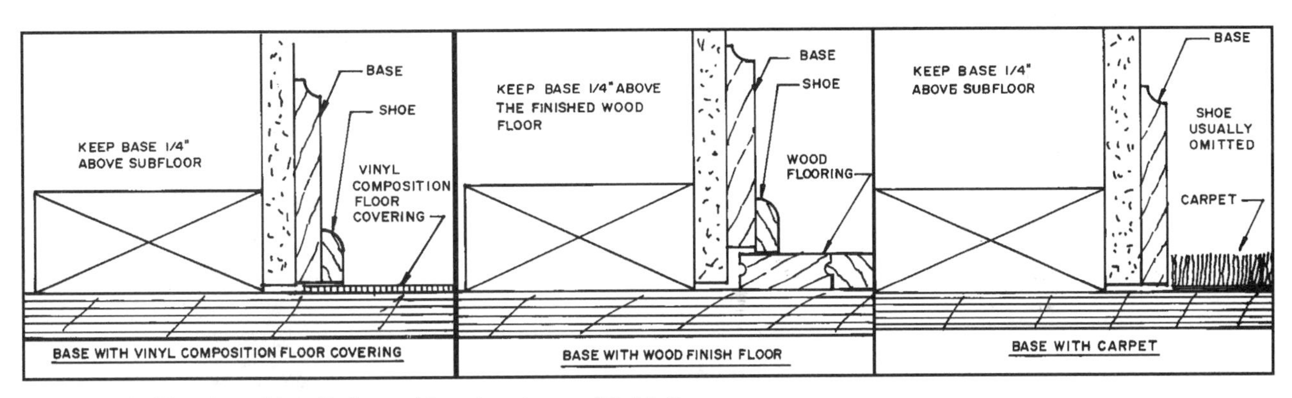

29-2 Typical baseboard installations with various types of finish floors.

CORRECTING DEFECTS IN FLOORS & WALLS

Sometimes the floor will have a slight bulge causing the base to rock on it and making it difficult to get it level or form good joints. If this defect is small, it can be compensated for by planing a small amount off the subfloor and scribing the bulge on the base and planing off some of it. One way to scribe the bulge onto the base is shown in **29-3**. Plane the base until you get near the line, then place the molding on the floor to check it. Continue removing small amounts, until the fit is satisfactory. Generally it is helpful to remove some from both the floor and the base rather than trying to take it all from one or the other. A small gap at the floor causes no problem because it will be covered with carpet or a shoe molding.

Another problem arises when the base butts a wall that is not plumb. Before cutting the base to its finished length, place one end against the wall, and scribe the slope on the base as shown in **29-4**. Cut to this scribed line. A flat carpenter's pencil is an excellent tool to use to scribe the line.

If a corner is not square, it may be possible to cut a miter that is not 90 degrees. To do this, measure the actual angle of the corner with a T-bevel **(see 29-5)**. Find the number of degrees between the handle and blade with a protractor. Divide this angle in two and use it to set the miter saw. You can also cut the corner on a 45-degree angle and then, by trial and error, lightly trim the mitered surfaces until a good fit is achieved. This will require you to leave a little extra length to allow for the additional trimming.

Sometimes an inside or outside corner is out of square because extra drywall compound was applied on the corner. If this is only a small amount, the excess compound can be removed with a Surfoam® tool, wide wood chisel, or a wooded block wrapped with coarse sandpaper.

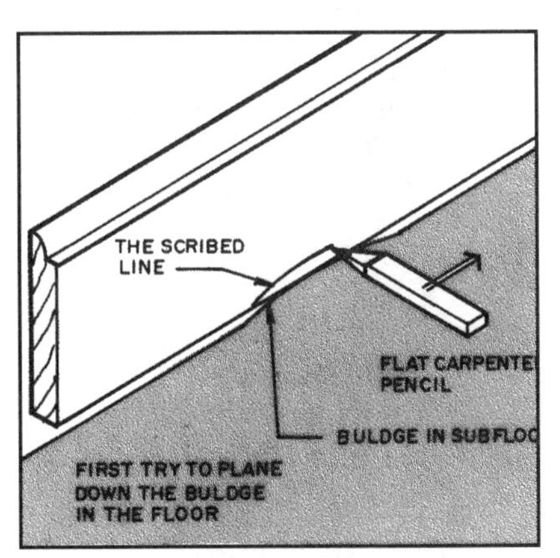

29-3 To scribe a floor bulge on the base, hold the base level while sliding a pencil along the floor, marking the bulge on the base.

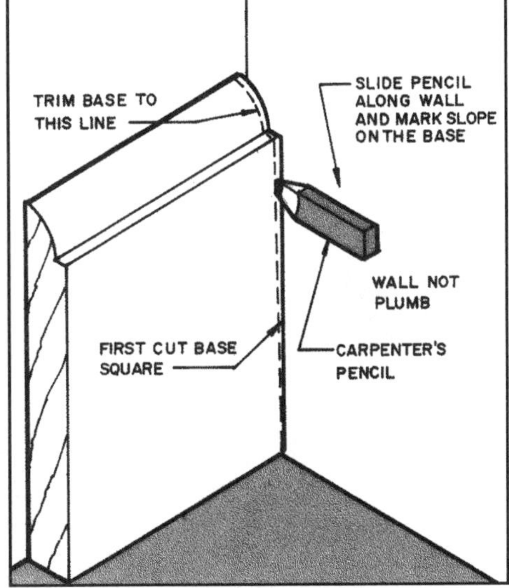

29-4 The base must be scribed to the slope of a wall that is out of plumb.

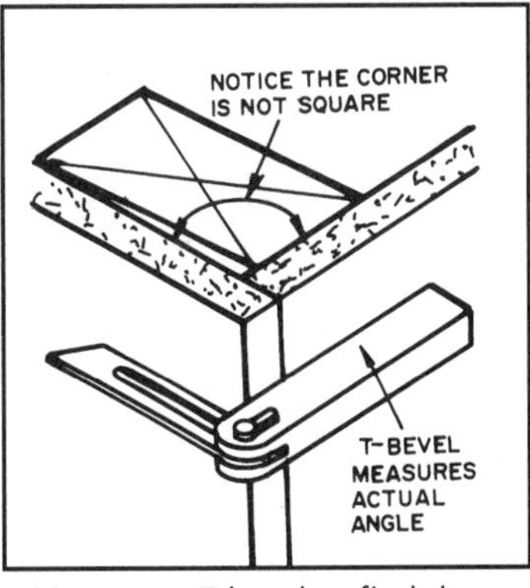

29-5 Use a T-bevel to find the angle of corners that are other than 90 degrees.

CORNER JOINTS

The base meets at inside corners and outside corners. Outside corners are formed by cutting a 45-degree miter. Inside corners can be formed by cutting a miter or coping the molding. Coping is preferred because, if the joint does open up over time, it will not be noticeable. A square base can use a butt joint at inside corners.

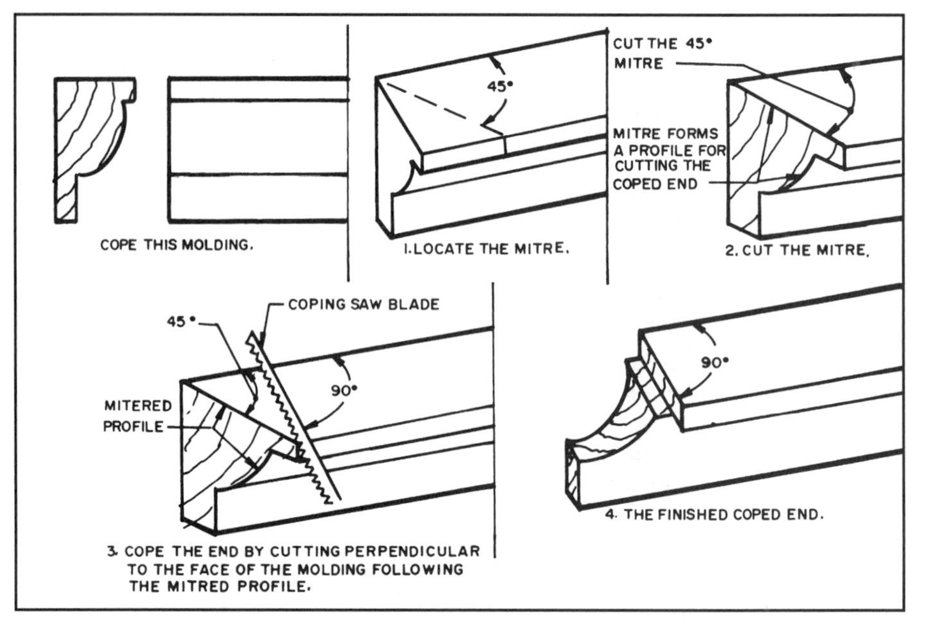

29-6 The steps to cope the end of a molding.

CUTTING A COPED JOINT

The **coped joint** is used on **base, shoe, crown,** and other moldings where they form inside corners. It is more time consuming to make than a miter, but it makes openings in the joint that occur as the molding shrinks and swells with changes in humidity less noticeable. The cope is cut on one of the two moldings forming the corner. The coped profile butts against the other base, forming the inside corner.

To cope a molding **(see 29-6)**, follow these steps:

1. Miter the end of the molding so that the length of the molding is slightly longer than required. This provides some wood so that the saw can get started and cut on the waste side of the line forming the profile. The miter should slope in the same direction as it would if an inside miter were cut.
2. Cut the profile using a coping saw or a portable power jigsaw **(see 29-7)**. The edge formed by the miter is the line of the profile to cut, forming the coped end.
3. Keep the saw perpendicular to the front of the molding and saw following the curved profile formed by the miter.
4. Test the joint by placing the coped end against the butting molding. Use a file or small power sander to smooth the cope and alter the shape until a close-fitting joint is developed.
5. Some carpenters will angle the cope so that the back edge is slightly behind the front. This backcut leaves the front edge firmly touching the butting molding.
6. The coped joint is not glued. Since it contains end grain the glue would not give any special strength or help hold it closed.

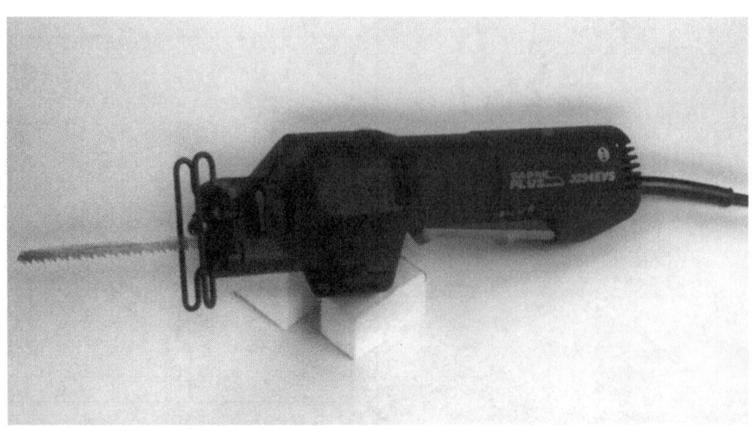

29-7 This portable power jigsaw speeds the cutting of the cope. *(Courtesy Bosch Power Tools)*

CUTTING MITERS

Miters can be cut with a hand or power miter saw. Both produce a straight, smooth cut. The piece to have the miter may have the other end coped or cut for a butt joint. In any case it is necessary to mark the exact length from that end to the corner of the wall where the outside miter will occur. The steps to locate the miter on an **outside corner**

are shown in **29-8**. Begin by butting the square end of the base against the wall at the inside corner. At the outside corner mark the length and direction of the miter. Cut the miter and lightly tack the base in place. Place the second piece against the wall and mark the length and direction of the miter. Cut it and nail it in place. Before nailing permanently in place check to see if the miter fits tightly. Trim as necessary to get a close fit.

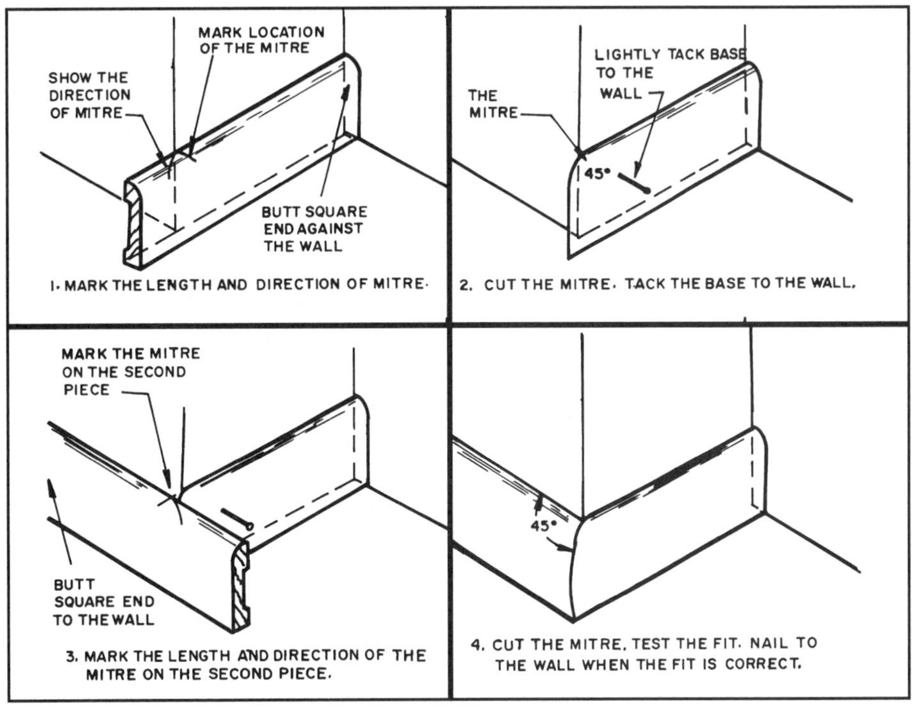

29-8 The steps to measure, mark, cut, and install base at an outside corner.

If the miter opens at the top, it will have to be trimmed a little at the bottom **(see 29-9)**. If it opens at the bottom, the top will have to be trimmed. Many carpenters do this by placing a thin piece of cardboard on the table of the miter saw and then taking a light cut from the edge to be lowered. The cut should remove just a fine layer of sawdust. Try the pieces at the corner until the joint closes. This same technique can be done with a sharp plane. It is advisable to remove a little from each piece rather than trying to take it all from one piece.

The installation of base at an inside corner when the butting pieces are to be coped is shown earlier in **29-6**.

Trim carpenters have various ways they use to measure base to inside and outside corners. The distance to the face of the base to be coped can be found by placing a loose piece of base on one wall and scribing a pencil mark on the adjoining wall. Then measure to this mark as shown in **29-10**.

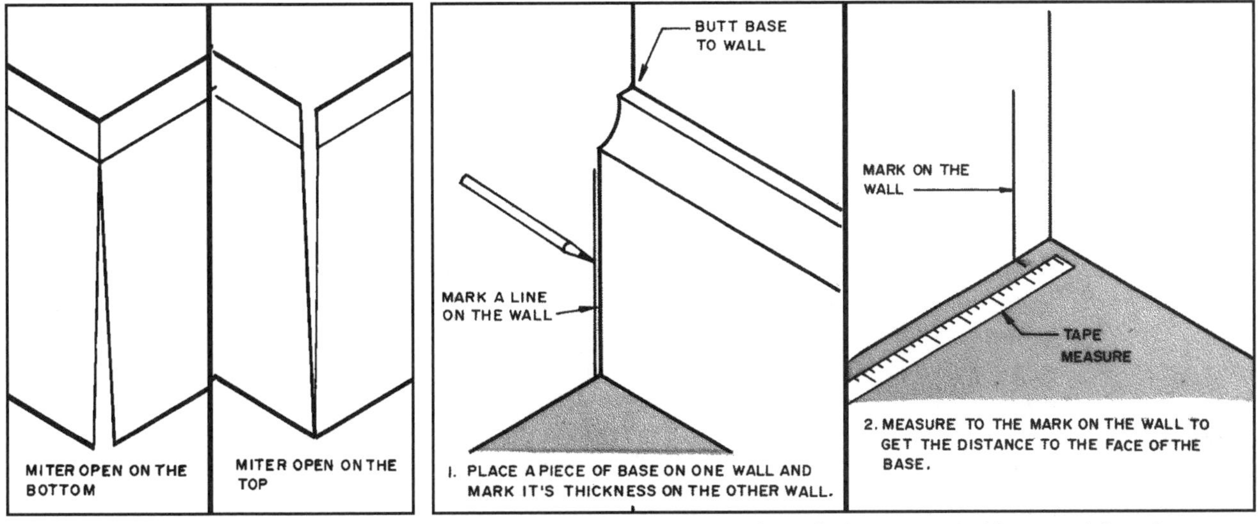

29-9 Miters that open on the top or bottom need to be trimmed to close.

29-10 When measuring to the face of a base on an inside corner it is easier to mark the face on the wall, remove the piece of base, and measure to the line.

Outside corner measurements can be laid out by marking the width of the base on the subfloor. The length of the miter can also be located as shown in **29-11**. You can also use the marks shown in **29-11** to mark the miter on the piece of base **(see 29-12)**.

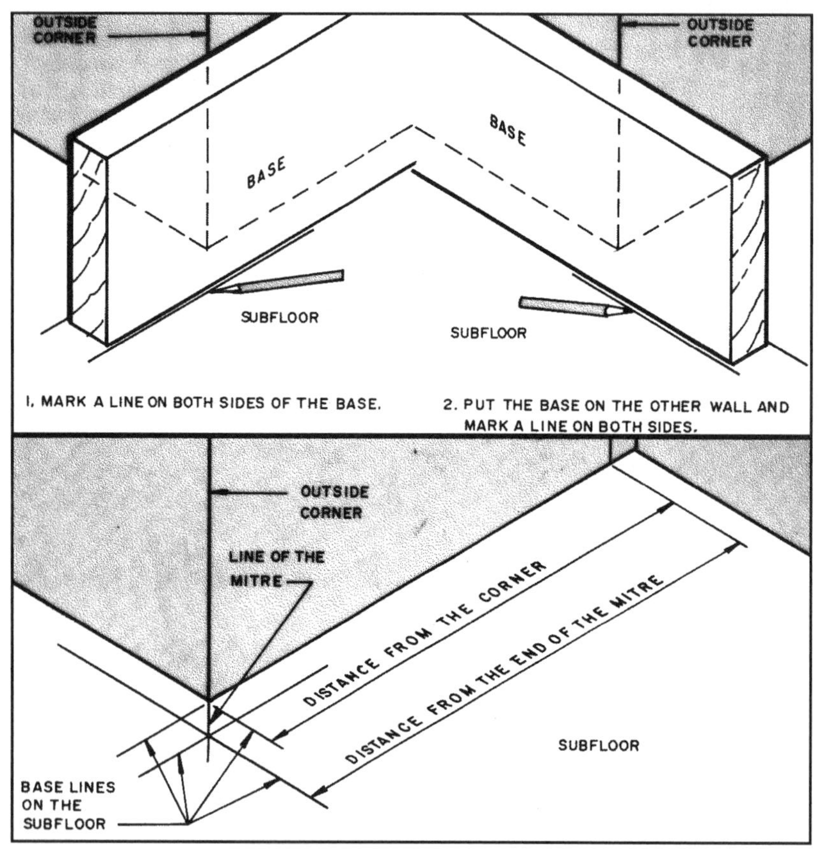

29-11 The base can be drawn on the subfloor and used to measure the lengths of base required.

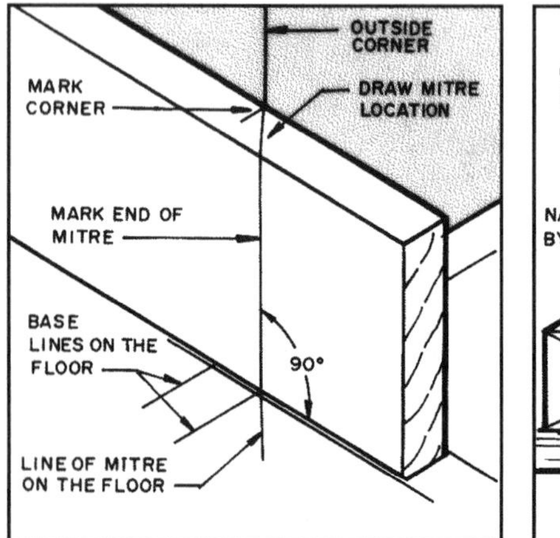

29-12 The base lines on the subfloor are used to measure the lengths of base required.

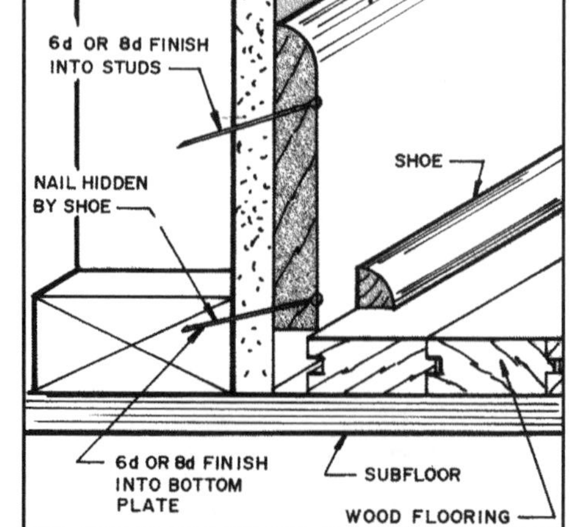

29-13 Narrow base is nailed to the bottom plate and to each stud.

NAILING THE BASE, SHOE & CAP MOLDING

Softwood base can be installed by nailing through the finish wall material into the studs with two finish nails long enough to give good penetration into the stud and plate. Usually 6d or 8d finishing nails are adequate. Place one nail about ½ inch (12mm) from the bottom into the bottom wall plate **(see 29-13)**. This will be covered with the shoe molding or carpet. The other nails will have to be set, and the holes filled. Nail near the ends of each piece to help stabilize the joint. If splitting is a problem, drill small holes for the nails. The procedure for nailing wide base is shown earlier in **29-1**. Hardwoods often require that the nails have predrilled holes. This helps prevent splitting and bent nails.

When nailing, there are several things to consider. One is that the base and other moldings can be pulled tight against a slightly wavy wall surface. This can be accomplished when there is adequate backing to nail into. Another thing to consider is what is the best way to nail shoe and cap moldings. Some prefer to nail the shoe molding to the floor **(see 29-14)**. When this is done the base can move vertically without exposing a gap at the floor; the shoe molding keeps any gaps covered. However, when this movement occurs the line of paint between the base and shoe will break, and unfinished base will be exposed. If the shoe is nailed to the base, they will move vertically together, but will leave a gap at the floor. Carpenters will generally use the procedure that they believe has caused them the least trouble.

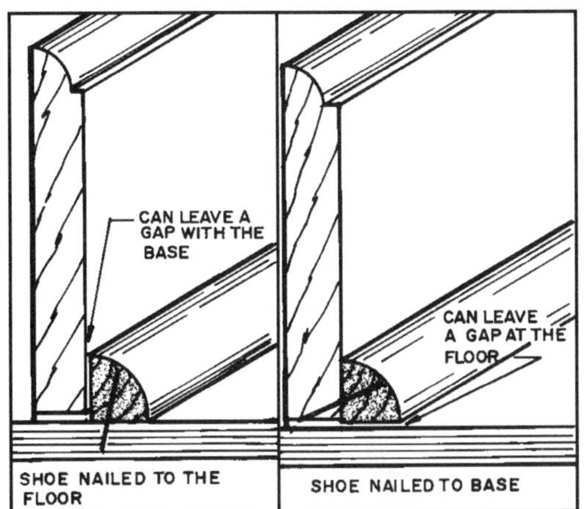

29-14 Two ways to nail the shoe molding.

29-15 Power nailers are widely used to install interior trim. *(Courtesy Hitachi Power Tools)*

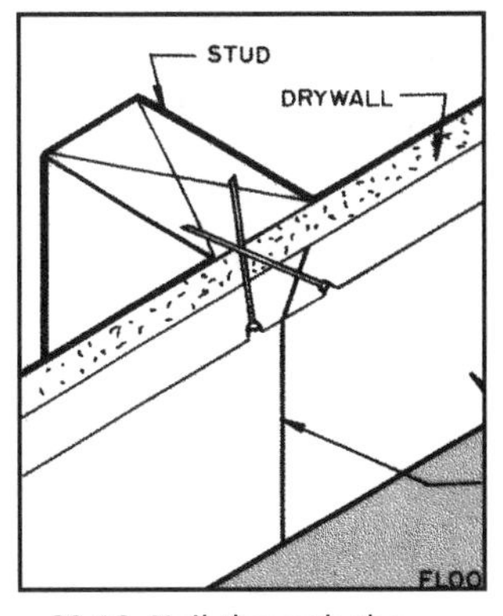

29-16 Nail through the scarf joint into a solid backing. Some glue as well as nail the joint.

If the shoe molding is placed on top of wood flooring that has not been sanded and finished, lightly tack it in place. The floor finishers will want to remove it so that they may sand closer to the baseboard. The shoe molding is nailed in place **after** the floor is finished. When prefinished wood flooring has been installed, the shoe can be permanently nailed as it is cut and fit.

The cap molding is usually nailed into the top of the base and moves with it.

Many finish carpenters prefer to use power nailers to install moldings. These power nailers are fast and do a good job. Power nailers have thinner nails available for trim work and therefore tend not to split the wood as often **(see 29-15)**.

SPLICING BASE & SHOE MOLDING

When base and shoe molding has to be spliced to span a long wall, it is cut on a 45-degree angle to form a **scarf joint**. This joint should be located at a stud so that there is adequate nailing surface to close and hold the joint. A typical situation that arises when base is being installed is shown in **29-16**. The one side is nailed to the stud. The other is nailed through the joint into the studs. Use two nails in each piece. Some carpenters glue the joint.

INSTALLING THE BASE

Finish carpenters develop their own favorite procedures for installing wood base. The following are commonly used techniques. The procedures include installing single-piece square base, single-piece base with a shaped top edge, and multipiece base.

Begin by planning the order of installation of each piece in the room. Generally the first piece installed is the one on the longest wall **(see 29-17)**. This piece can be cut with square ends and butted to each wall. While some prefer to cut wood base about ⅛ inch (3mm) longer than the wall length to get a tight fit, others prefer to cut it to the exact length or even a bit short. If a base fits too tightly and the moisture content increases, it will warp and buckle away from the wall.

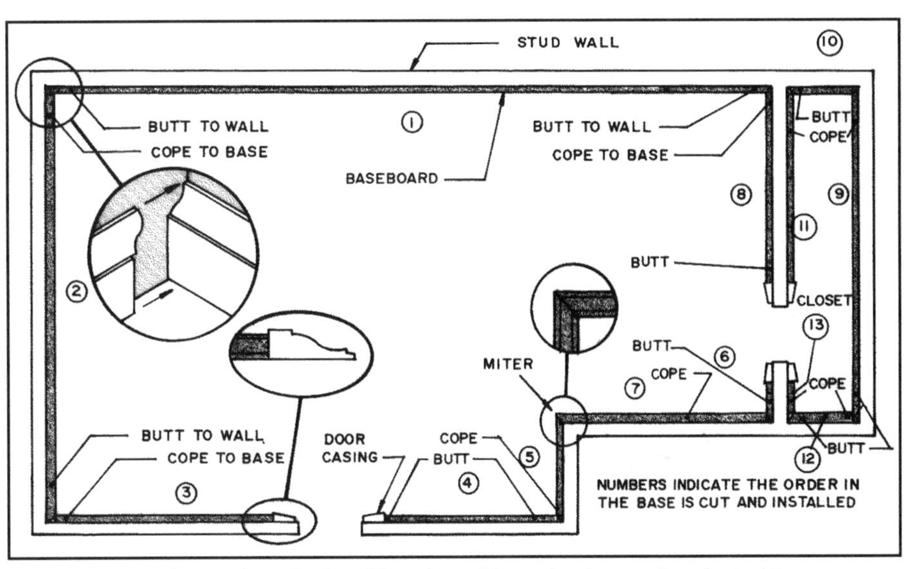

29-17 A typical procedure for installing shaped base that is coped at the inside corners.

Some carpenters prefer to cut the next pieces following the walls to their left, and others go to the right. The direction to proceed is simply a matter of habit or personal preference. In **29-17** it was decided to work to the left. The longest piece, No. 1, was butted wall to wall. The adjoining pieces are coped and mitered as shown. Notice that the base butts the door casing, is coped at inside corners, and is mitered at outside corners. It can be mitered at inside corners instead of being coped, but, since walls are seldom at right angles, miters tend to open. Coping tends to hide imperfections in the corners.

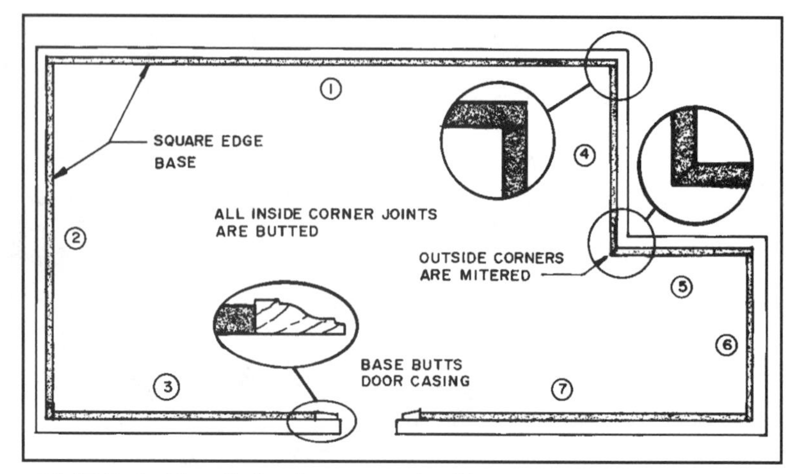

29-18 Typical installation for square-edged base.

INSTALLING SQUARE-EDGED BASE

Square-edged base is installed as shown in **29-18**. The exterior corners are mitered. The interior corners are often butted, but can be mitered. If shoe and base cap molding are to be installed, they will be coped.

INSTALLING SINGLE-PIECE SHAPED BASE

Single-piece shaped base has the exterior corners mitered and the interior corners either mitered or, better still, coped. Refer to **29-17** for specific details.

INSTALLING MULTIPIECE BASE

Multipiece baseboards are usually built around square-edged base stock. They are assembled using the base caps such as those shown in Chapter 28. Several examples are in **29-19**.

Multipiece base is installed as described for square-edged base. Then the shoe and base cap moldings are cut and nailed in place. These two moldings are usually coped at inside corners **(29-20)**.

MITERLESS BASE INSTALLATION

This wood product eliminates the need for miters by supplying internal and corner pieces. After these

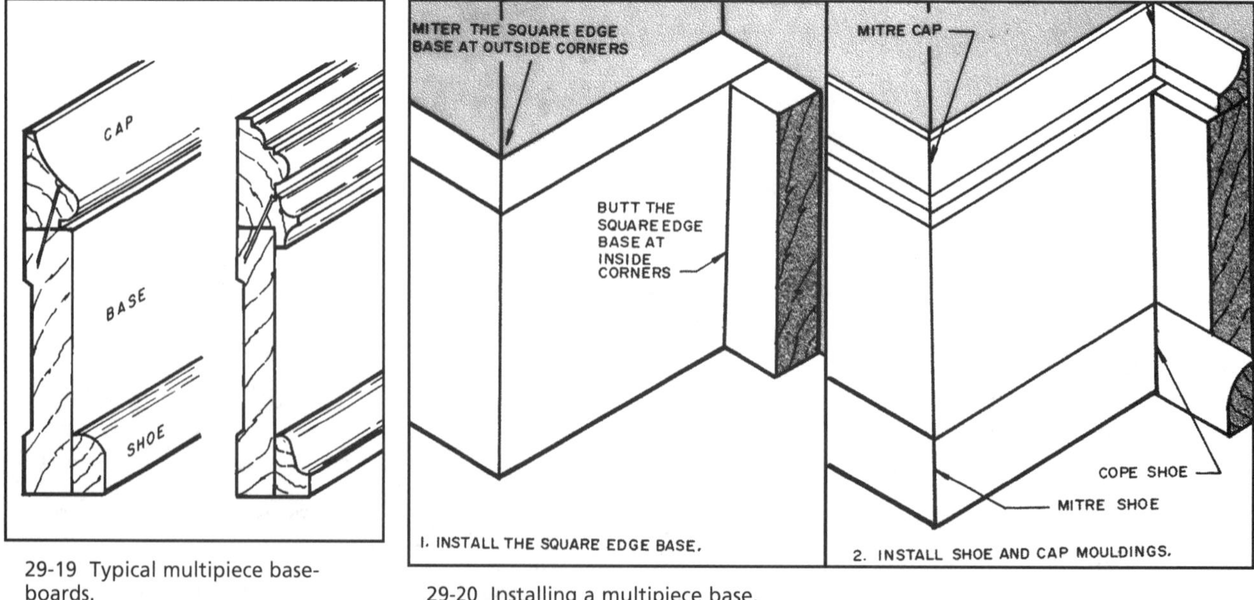

29-19 Typical multipiece baseboards.

29-20 Installing a multipiece base.

are installed by nailing and gluing to the wall, the base is cut to length at 90 degrees on each end and fitted between the end pieces. The nails are countersunk and filled. The base can be stained, painted, or varnished **(see 29-21)**.

29-21 The inside and outside corner pieces are installed. Then the base is cut to length and butted to each corner piece. *(Courtesy Ornamental® Moldings)*

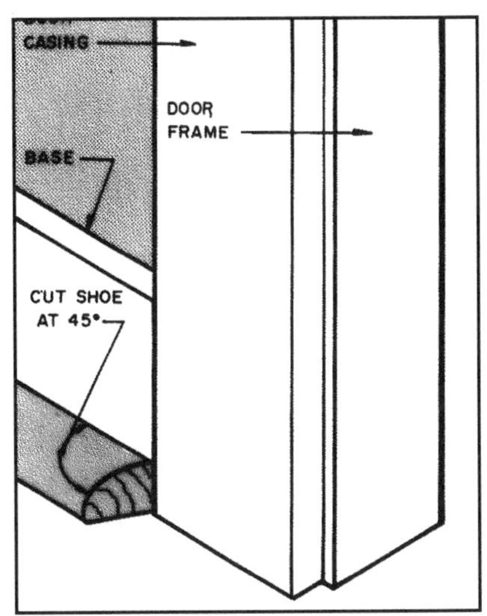

29-22 The shoe molding is mitered when it meets the door casing.

SPECIAL SITUATIONS

Often when a shoe molding is used, it will stick out beyond the casing at the doors. In this case the shoe is cut on a 45-degree bevel **(see 29-22)**.

Some traditional interiors use **plinth blocks** and **rosettes** on the door openings. The casing and base butt against these blocks. Plinth blocks and rosettes of various designs are available from companies that manufacturer moldings. While there is no hard and fast rule, the block is often wider than the casing and higher than the base **(see 29-23)**. It should be thicker than both so that it has a reveal on the sides where molding butts it. If a base has a molding applied to the top that is wider than the base, it will stick beyond the casing. Where it meets the casing, the molding is notched around the casing and smoothed to produce a finished return **(see 29-24)**.

If the base ends along a wall in a location where it does not terminate in a casing or butt the wall, the exposed end should be finished by coping it or mitering and cutting a return. A coped end is shown in **29-25**.

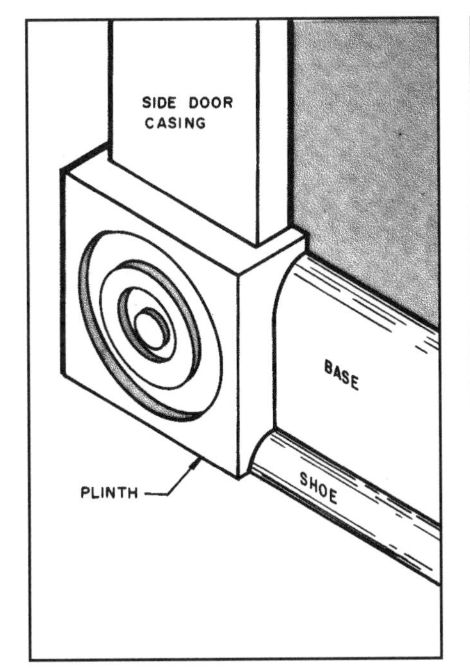

29-23 The base and shoe moldings butt against the plinth.

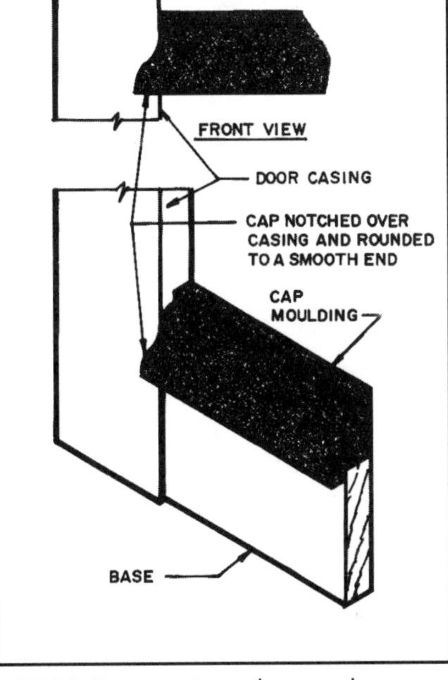

29-24 One way to end an overhanging base cap molding as it meets a door casing.

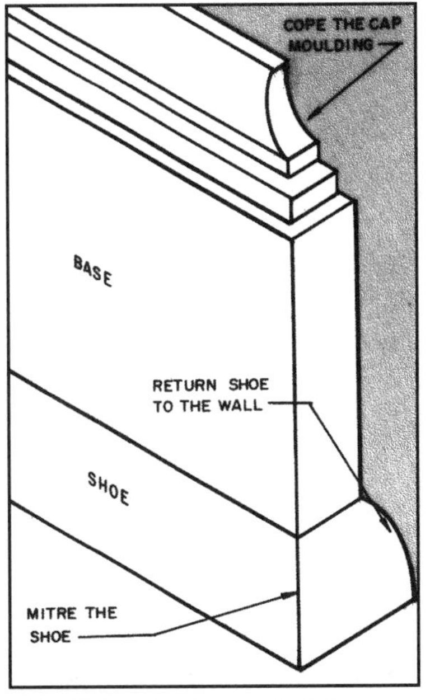

29-25 The exposed end of the base cap can be coped and the shoe mitered and returned to the base to produce a finished termination to the base.

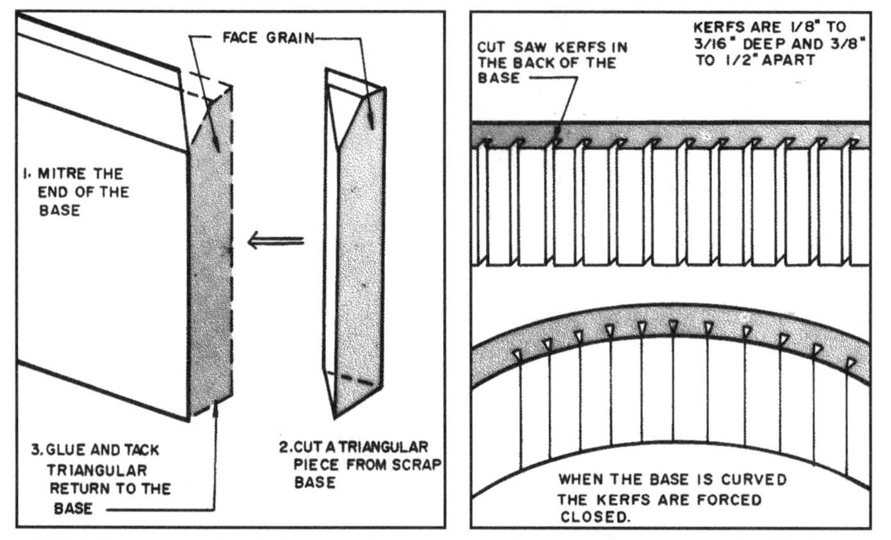

29-26 The exposed end of a one-piece base can have the end grain hidden by covering it with a mitered return piece.

29-27 Wood base can be installed on walls with curves having a large radius by cutting saw kerfs in the back side.

A mitered end is formed as shown in **29-26**. The end is cut on a 45-degree angle. Cut the return piece from the end of another piece of molding. Glue the return to the base. If an attempt is made to nail it, the danger of splitting is so great that the nail hole should be predrilled. Use a smaller-diameter wire brad.

If wood base must be installed along a curved wall, it will have to have saw kerfs cut in the back. Cut the kerfs about 1/8 to 3/16 inch (3 to 4.5mm) deep and 3/8 to 1/2 inch (10 to 13mm) apart. Use good quality, straight, grained stock. The kerfs will be visible on the top of the base **(see 29-27)**. If it is to be painted, they can be filled and the top sanded smooth. A better solution is to plan to use a base having a small, separate molding nailed to the top. This will bend around the curve easily to cover the saw kerfs. Another solution is to use molded plastic base for the job. It is easily made to follow curved walls without the need for notching.

Sometimes an electric outlet box has been located in the area to be covered by the base. This is especially true in traditional-style homes having very wide baseboards. The location of the outlet box must be marked on the base. One way to do this is to cut and fit the base to the wall. Then rub chalk on the edges of the outlet box. Place the base against the wall and press against the box to produce a chalk outline on the back of the base. Drill holes through the base at each corner, and then cut from hole to hole with a saber saw **(see 29-28)**.

While most heat registers used with a central air system are now placed in the floor, some are still installed to open through the wall at the floor. They are usually higher than the baseboard. Therefore the baseboard can be butted to the sides of the register. If a more pleasing appearance is desired, run molding from the top of the base and miter around the register, as shown in **29-29**. If the molding is separate from the

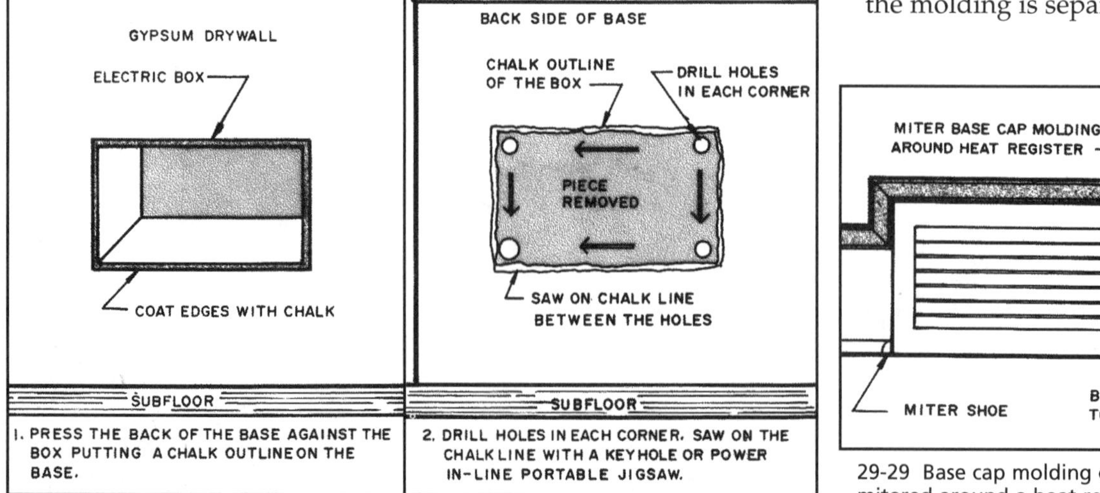

29-28 Openings are sometimes needed in the base, such as for an electrical outlet.

29-29 Base cap molding can be mitered around a heat register or other wall protrusions.

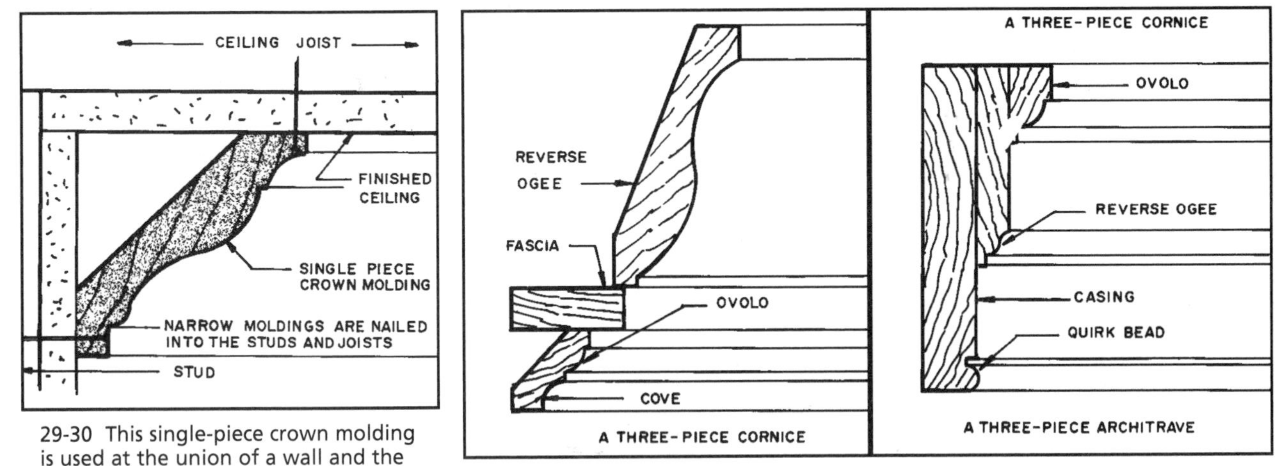

29-30 This single-piece crown molding is used at the union of a wall and the ceiling.

29-31 Two typical multipiece corner moldings.

base, it is simply a matter of cutting and fitting it around the register. If it is a one-piece shaped base, the top shaped portion can be cut from a piece of base and mitered around the register.

INSTALLING CROWN MOLDINGS

Crown molding is available in wood or plastic. Much that has been described earlier in this chapter also applies to the installation of crown molding. Crown molding is installed to cover or decorate the joint formed where the interior wall and ceiling meet. While the crown molding is a single shaped molding **(see 29-30)**, it is often combined with other moldings to form interior and exterior cornices. An ornamental molding located at the junction of an interior wall and ceiling or an exterior wall and the roof is called a **cornice**. In **29-31** are shown a multipiece cornice and a three-piece architrave. An **architrave** is an ornamental molding that is placed around the face of a jamb, lintel, or other wall opening.

If the ceiling molding has flat members, they are installed in the same manner as baseboards. Be certain they are level. Any gaps between the ceiling and flat molding will be covered with the crown molding **(see 29-32)**.

Crown moldings up to about four inches (102mm) wide may be nailed to the studs and ceiling joists. Nail through the flats on the top and bottom edges. Wider moldings or wider multipieced ceiling moldings require blocking to be installed by the framing carpenter between the studs and possibly between ceiling joists **(see 29-33)**.

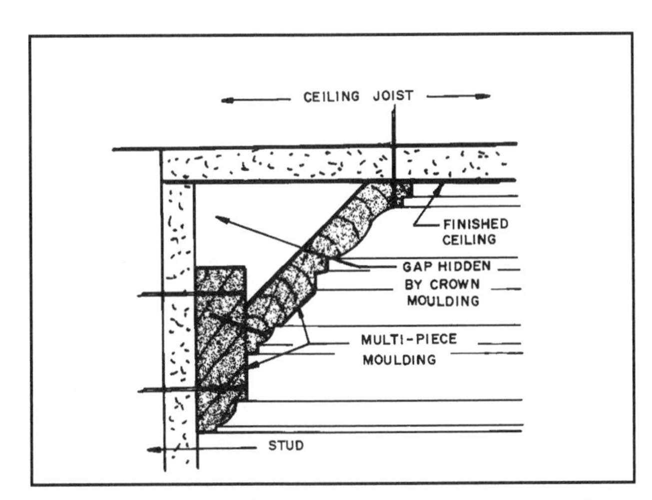

29-32 Gaps at the ceiling are covered by the crown molding.

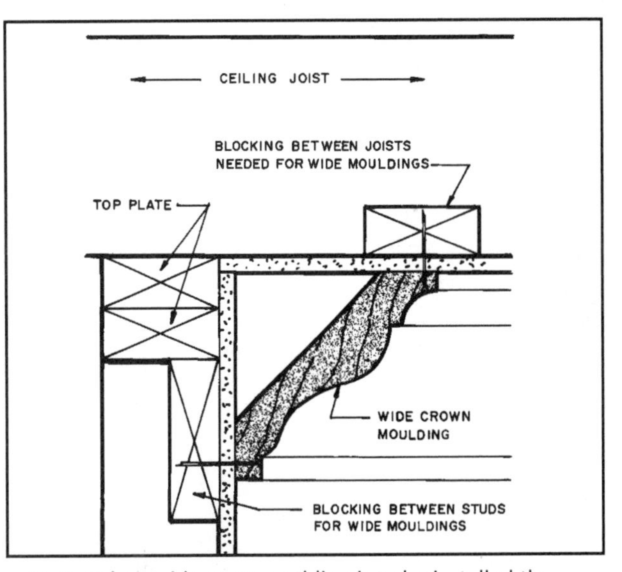

29-33 When wide crown molding is to be installed the carpenter should install two-inch-thick blocking between the studs.

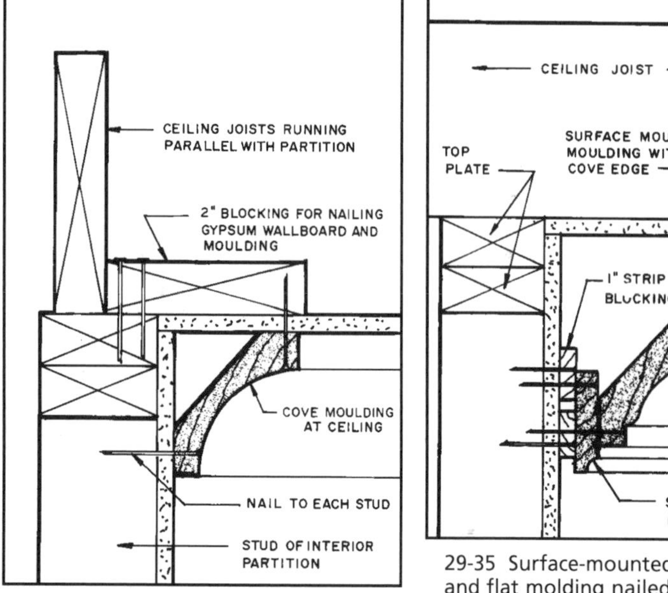

29-34 Blocking to support wide crown molding can be nailed on top of the double top plate on walls running parallel with the ceiling joists.

29-35 Surface-mounted wood strips and flat molding nailed to the ceiling joists and wall studs can be used to mount crown molding when blocking has not been placed between the joists.

On walls running parallel with ceiling joists, blocking should be nailed to the top plate and cantilevered out beyond the top plate. This is done to provide nailing for the drywall. It would have to be wider for holding crown molding **(see 29-34)**. If this blocking has not been installed, it is still possible to install a wood backing on the outside of the drywall, and then nail the ceiling molding to this. The molding must be designed to cover this backing board, as shown in **29-35**. Notice in **29-36** that the crown molding is nailed to a surface-mounted molding that provides the nailing base.

JOINERY

Crown molding is mitered at outside corners and can be mitered or coped at inside corners. Since most walls do not meet at exactly right angles, the miter usually has to have considerable adjustment, as explained for mitering base molding. Also, if there is any movement in the wall, the miter will open, producing a gap that is plainly visible. It is best to cope the joints at inside corners, as explained for base molding.

The big difference in coping crown molding is that the miter cut must be made with the molding placed on the saw table **upside down and backwards** from the way it will rest on the wall. It will have one inside face flat against the fence and the other flat against the saw table **(see 29-37)**. This produces the required miter cut needed to cope the end of the molding. The miter on the end to be coped must slant towards the wall. It is difficult to hold the crown molding firmly in place when cutting the miter. To help hold it, some carpenters clamp a wood strip to the saw table to serve as a stop.

After the miter has been cut, cope to the line produced by the cut. Cut perpendicular to the face of the molding. Trim and fit to the butting molding using a file **(see 29-38)**.

29-36 This crown molding is being power-nailed to surface-mounted molding. *(Courtesy Senco Fastening Systems)*

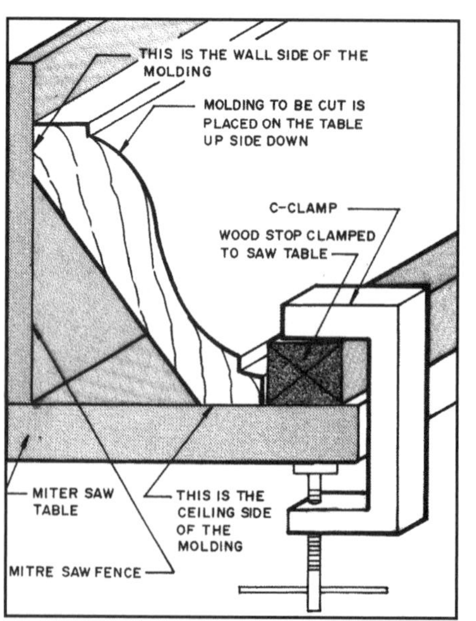

29-37 How to position crown molding on the miter saw when cutting the miter.

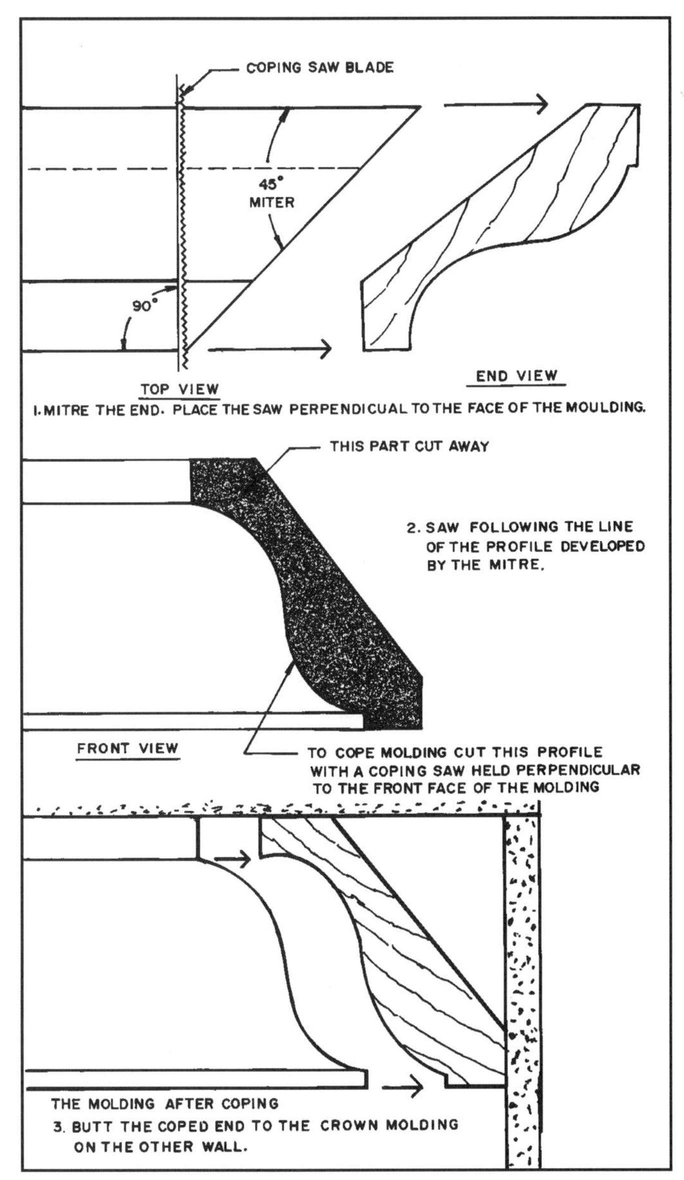

29-38 How to cope crown molding.

END-JOINING CROWN MOLDING

When crown molding on long walls must be joined, a scarf joint is used as described for base molding **(see 29-39)**. The molding is placed on the miter saw as shown earlier in **29-37**.

Begin by cutting the miter on the underlying molding first. Locate the end on the wall and measure from it to the end wall to get the required length of the second piece. Some cut the piece a bit long and trim it as needed to get a tight fit. Place glue on the faces of the scarf joint before nailing them in place. If the pieces do not line up perfectly after nailing them in place, work them down with sandpaper.

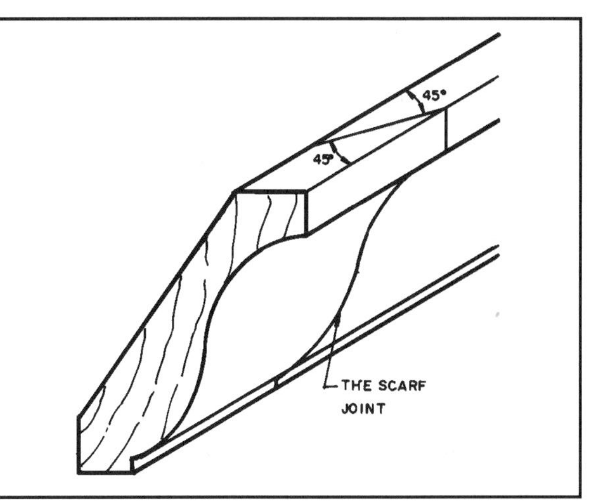

29-39 Crown moldings are joined with a scarf joint.

INSTALLING CROWN MOLDING

Finish carpenters have their own techniques and procedures for accomplishing this difficult task. There are any number of ways that the various problems and procedures may be handled. Following are some that are used.

1. Find and mark the studs and ceiling joists. Mark with a light, soft-leaded pencil. Do not use ballpoint pens, because it is very difficult to cover the ink marks with paint. To find the studs and joists, use a stud finder or look for the rows of nails covered by drywall compound. To get the exact location, nail through the drywall and hit each member. Do this in an area where it will be covered by the molding.
2. Check to see whether the ceiling is flat and level. This can be done a number of ways. One is to measure a fixed distance from the floor, such as eight feet, in each corner and run a chalkline. The variance in the ceiling can then be marked. If a ceiling is not flat, it is most likely to bow slightly in the center. Mark the low spot, if one exists, as shown on the following page in **29-40**. Then measure down from the low point a distance equal to the height of the molding in its installed position. Run a level line through this point from wall to wall. When installing the crown molding, place its bottom edge on this line. If things are going well, the gap between the ceiling and the molding will be small. The top edge of the

molding is sometimes relieved (sloped a little to the back). This helps disguise a small gap. If the gap is too large to let go, the drywall contractor can feather a layer of compound to reduce the size of the gap **(see 29-41)**.

3. Plan the layout of the crown moldings. As was recommended for the installation of base, run one piece on the longest wall and butt it to each wall. Doors and windows are not shown, because they do not interrupt the run of the molding.

4. Cut and install the first piece. Cut it to the exact length or a bit short. If cut too long and the molding absorbs moisture from a period of high humidity, it may buckle. Any tiny gap at the wall is covered by the adjoining molding. Align the bottom of the molding with the mark, and nail through the flats at the top and bottom of the molding. The nails should be long enough to give good penetration into the studs and joists. Sometimes the molding can be pulled to the wall a little. However, too much pull will produce a wavy result. If the wall or ceiling are out of plane quite a bit, it may be necessary to place wood shims between the wallboard and molding to keep the molding flat and straight.

5. Continue to cut, fit, cope, and butt each piece as you proceed around the room. You can work to the left or right, as desired.

MITERING OUTSIDE CORNERS

It is difficult to make tight outside miter joints with crown molding. After the miters are cut, it usually takes some adjusting to get the joints to close. Whenever possible the molding containing an outside miter on one end should butt the wall on the other end **(refer to 29-40)**. As the miter is adjusted, the butted end may clear the wall a little, but this is covered by the coped molding that will butt it. The length of the molding can be measured by finding the distance from the wall to the edge of the outside corner. If a molding has to have a coped end and a mitered end, some carpenters prefer to cut the coped end first, then mark the length to the bottom corner of the miter and cut the miter. Cut the miter a bit long so that it can be

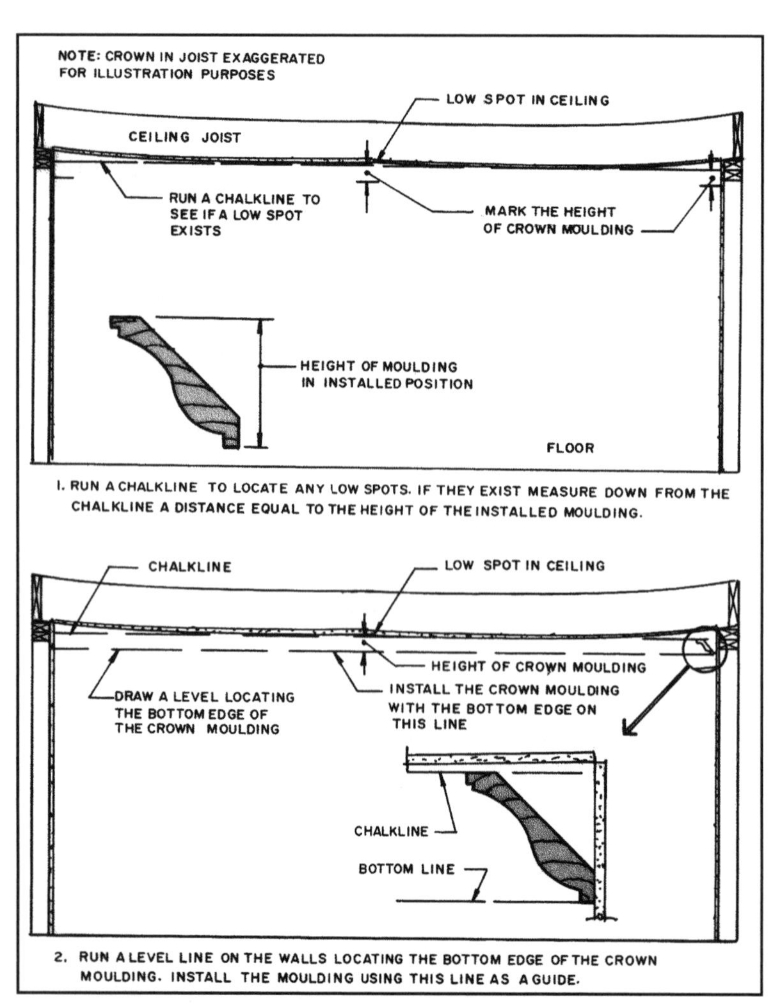

29-40 One way to position the crown molding when the ceiling has a bow.

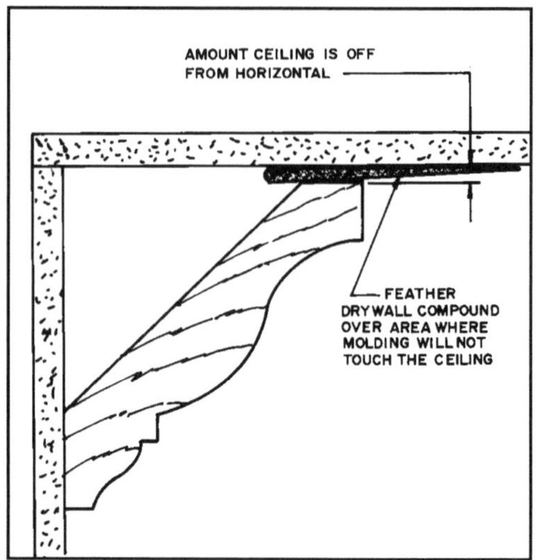

29-41 Small gaps between the crown molding and the ceiling can be filled by feathering out a layer of drywall compound in the low areas.

trimmed to fit the butting miter. Nail the molding in place, but drive the nails only partially in so that the molding can be removed if the miter needs additional adjustment. Do not set any nails until the molding has a perfect fit on both ends.

Remember, when cutting miters, that the point must face away from the wall on outside corners. This will help keep you from cutting the miter in the wrong direction. Inside corner miters point towards the wall.

Outside miters are glued and nailed. It is important that the nails do not break out the front or back of the molding. The continued closure of the joint depends on a good nailing job.

Probably the most commonly occurring problem with an outside miter joint is when it has a parallel open outside gap. When this occurs, change the angle of the cut by a few degrees to remove wood from the back side of the miter. If there is a gap and it tapers, remove a fine amount from the high side of the face of the miter.

MITERLESS CROWN MOLDING

This wood product eliminates the need for miters by supplying internal and external corner blocks. The crown molding is cut to length with square ends and butts these corner pieces, as shown in **29-42**. The molding and corners are glued and nailed to the wall.

29-42 This wood crown molding butts factory-made inside and outside corners, eliminating the need to cut miters and cope joints. *(Courtesy Ornamental®*

Estimating Trim Quantities

Base, shoe molding, ceiling molding, and chair rail quantities are determined by finding the lineal feet in the perimeter of the room plus any related areas, such as closets. Make no allowances for doors; this will give a little extra for waste. Some add 10 percent for waste.

Panel molding quantities are determined by calculating the lineal feet around the perimeter for each panel and adding 10 percent for waste. In order to avoid splicing the molding, some carpenters determine what they can get from a standard length.

Quantities of door and window casing can be figured by calculating the lineal feet around each opening and then adding 20 percent for waste. Many manufacturers supply these in precut lengths as sets. You can order a set for each opening.

INSTALLING CHAIR RAIL MOLDINGS

The design and height of the **chair rail molding** above the floor will be specified on the working drawings. If these are not available, it is necessary to contact the owner or designer. Normally chair rail is placed 32 to 36 inches (81.3 to 91.4cm) above the floor. The chair rail serves to keep furniture from scraping the wall, and originally it was used in dining rooms. It can be used almost anywhere, and now serves as a divider between two different wall finishes. For example, the area below the chair rail, called the **dado**, may have paneling and the area above, called the **frieze**, may be covered with wallpaper **(see 29-43)**. Chair rail can be a one-piece molding or built up from several pieces.

Chair rail is installed as described for base. The corners are mitered and coped in the same manner. For best results, it is recommended that a 1 × 4 wood member be let in the studs to provide a solid backing for nailing the chair rail or 2 × 4 blocking be nailed between the studs.

The rail should fit tightly against the wall. It can usually be pulled some when nailing. If the wall is too wavy, the drywall contractor may have to come back and add additional compound. Some prefer to glue it and nail it in place. Generally 8d finish nails are used.

29-43 This beautifully cased door opening is butted by a double chair rail. Notice the difference in wall finish above and below the chair rails. *(Courtesy Ornamental® Moldings)*

INSTALLING PICTURE MOLDINGS

Picture moldings are shaped to hold metal clips which hold long cords secured to the rear of picture frames. They are installed 84 to 96 inches (2.1 to 2.4m) above the floor. At the 96-inch (2.4m) level they form a ceiling molding of rooms with 8-foot (2.4m) ceilings. Installation is the same as described for chair rails. Since they may have to support heavy pictures, they need to be securely nailed or screwed into the studs and backing strip. Screws with oval heads may be left exposed. Flathead screws should be set below the surface and plugged.

29-44 These panels were made using hardwood panel moldings mounted over wood paneling. Notice the hardwood cornice which has large dentil blocks below the cornice moldings. *(Courtesy White River Hardwoods-Woodworks)*

INSTALLING PANEL MOLDING

Panel moldings are used to produce decorative panels **(see 29-44)**. Often the area inside the panel created has a different finish from the rest of the wall. The design of panel moldings can include single-piece stock, special machine material, or can be made up by combining several moldings. The size and location is determined by the designer and should appear on the working drawings.

The key to success is careful layout of the exact location as specified on the plan. The sides must be vertical and the top and bottom members horizontal. These can be leveled with a long carpenter's level, and marked on the wall with a soft lead pencil. Prior to installing the finish wall material, such as gypsum drywall, provision should be made to install blocking to which the molding can be nailed. Joints are mitered.

INSTALLING MOLDED POLYMER MOLDINGS

Polymer moldings are cut with a fine-toothed handsaw or power miter saw. They are fastened with pneumatic nailers or can be hand nailed. Joints can be trimmed and fitted with a hand plane or power sander.

Following is a recommended way to install molded polymer moldings.

1. Begin the installation from the most conspicuous outside corner, and work around the room. Plan to end in the most inconspicuous corner.
2. Hold the molding in place on the wall, and mark the top and bottom projections. A line can be drawn from one end to the other to give an accurate mark for installation. Also mark the location of each stud below the bottom edge mark.
3. Measure and cut as described previously for wood molding, using a hand or power miter saw. Cut the pieces slightly longer than required. Allow about ⅛ inch (3mm) per five feet (approx. 1.5m) of length. Miter outside corners and cope inside corners.
4. Before applying the adhesive, fit the pieces to the wall and check the joint for thickness. Shave off the back until the faces of the butting pieces are flush. If a piece is too low, it can be shimmed. The face surfaces can be lightly sanded, if necessary.
5. Apply the adhesive recommended by the manufacturer to the top and bottom bedding edges of the moldings.
6. Press the molding in place on the wall. Since it was cut slightly longer it will bow and have to be forced against the wall. This is necessary to get a tight end joint. When the piece of molding does not reach the full length between two walls, nail an end block to the wall so that the molding can be sprung into place. Such an end block is simply some form of carpenter-made blocking. This is needed to get good corner and butt joints on a long wall.
7. Drive a finishing nail into each stud or backer board at least every 16 inches (approx. 41cm). Narrow moldings can be nailed into each stud.

Wide moldings should be nailed to a backer board installed behind the gypsum drywall. However, remember the adhesive is what provides the long term connection to the wall and between butt and miter joints.

8. Remove any adhesive that squeezes out of the closed joints and wipe the surface with mineral spirits.
9. Some carpenters prefer to miter both inside and outside corners. If the pieces are cut about ⅛ inch long, a good miter has been made, and adequate adhesive used, then a satisfactory corner joint can be produced.
10. Countersink the nails, and cover with spackling. Apply a manufacturer-supplied vinyl patching compound over the filled nail holes and any sanded areas. Gaps at the ceiling can be filled with caulking. Some use spackling compound or drywall compound.
11. Allow the installation to set for 48 hours before applying the paint or stain finish coats.

FIREPLACE MANTELS & SURROUNDS

Some fireplaces are designed so they protrude into the room and the wall butts them on each side **(see 29-45)**. This type does not use a surround but may have a mantel attached to the brick facing. This mantel may be designed by the architect and shown on the working drawings. Usually mantels are made with stock moldings.

29-45 Some fireplace designs do not use a surround or a mantel.

Many fireplaces are set so the face masonry is flush with the drywall finish on the butting wall. This leaves a gap to be covered. Usually this is covered by some type of **surround (see 29-46)**. The surround may be fabricated on the site by the carpenter using various flat stock and moldings or be purchased from a wood products manufacturer.

BUILDING CODES

When designing or selecting the size of a stock surround and mantel, be certain to observe your local building code. Basically codes require that combustible material be at least six inches from the fireplace opening and any combustible material above the opening (the mantel) that projects more than 1½ inches be at least 12 inches from the edge of the opening.

STOCK MANTELS & SURROUNDS

Stock mantels and surrounds are available in a range of sizes. Some companies custom-manufacture them to the architect's design.

When ordering, select a size that will meet the building code and cover the wall area required. Basically you will need to specify the information. If a return is needed be certain to specify the amount needed.

29-46 Wood mantels and surrounds are manufactured in a wide variety of sizes and designs.

CHAPTER 30

PANELING & WAINSCOTING

Wainscoting is a decorative and protective facing applied to the lower part, the dado, of an interior wall. Typically it is solid wood, plywood, hardboard, or some other bump-resistant facing material **(see 30-1)**. The top edge is covered with some type of wainscot cap. It can be used in almost any room and in halls. It is typically 36 to 40 inches (914 to 1016mm) high.

Paneling uses the same materials described for wainscoting. However, it runs from the floor to the ceiling **(see 30-2)**. One difference is that wainscoting has the top edge covered with a wainscot cap and paneling usually has some form of molding, called a cornice, at the ceiling. The ceiling molding can be crown, bed, cove, or some form of flat molding (see Chapters 28 and 29).

BUILDING CODES

Interior wall and ceiling finish materials are regulated by local building codes. The finish carpenter and the designer must consider these code specifications in the selection and installation of materials.

Solid-wood and plywood wainscoting and paneling are acceptable for most residential applications. Exposed foam-plastic surface materials are generally not acceptable. The code also specifies the smoke-development rating and the flame-spread rating required of the paneling materials to be used.

Typically codes require plywood or hardboard paneling under ¼ inch (6mm) thick to be installed over a fire-resistant backing, such as gypsum drywall. The drywall is applied to the studs and the paneling applied over it. Gypsum drywall can also be used under solid-wood paneling, when a fire-resistant backing is required behind it.

30-1 Wainscoting is a decorative and protective facing applied to the lower part of the wall. Notice the extensive use of molding on the wainscoting and upper wall. *(Courtesy Ornamental® Moldings)*

30-2 Paneling can be used to cover the wall from the floor to the ceiling. This paneling is flush-joint western hemlock board paneling. Notice the walls and ceiling are paneled. *(Courtesy Western Wood Products Association)*

INTERIOR WALL PANELING

Interior wood paneling products provide the beauty of the color and the grain of the species of wood used. Products available include plywood, hardboard, and solid-wood paneling.

Plywood Paneling

Plywood paneling is made by bonding three or five layers of wood veneer with their grains at 90 degrees to the bonding veneer. The highest-quality panels are made with the top facing veneer being a quality soft- or hardwood such as oak, birch, southern pine, cherry, fir, mahogany, or redwood. The surface usually has some type of groove cut running the length of the panel. Most panels are prefinished. They are available with a wide range of finishes such as a clear, natural coating, a stain to enhance the grain and richen the color, or a translucent coating, such as a white pigment, that lets the grain show. Some panels have the face covered with a durable simulated woodgrain paper overlay. These are less expensive than those with a quality wood face veneer.

Panels are available in thicknesses such as 5.0mm, ¼ inch (6.4mm), and ⅜ inch (9.5mm). The typical sheet size is 4 × 8 feet (1220 × 2440mm) but some companies can make larger sheets available.

Since the panels are prefinished at the factory great care must be taken on the site to protect the surface.

The panels have various edge joints prepared for connecting the panels. Some have a rabbeted edge that butts a rabbet on the adjoining panel, forming a recessed groove, while another type uses a beveled edge **(see 30-3)**.

Hardboard Paneling

Hardboard is made from wood chips converted into fibers which are then bonded into panels with the application of heat and pressure. Panels are available in thicknesses from ⅛ inch (3.2mm) to ¼ inch (6.4mm). The ⅛-inch (3.2mm) thickness must be applied over a solid backing such as gypsum wallboard or plywood. The ¼-inch- (6.4mm) thick panel can be applied over a solid subsurface or over furring strips spaced not over 12 inches (305mm) O.C. Panels are available in 4 × 8-foot sheets.

The panels are overlaid with a simulated woodgrain paper, a plastic laminate, or a baked-on melamine plastic finish.

Solid Wood Plank Paneling

Solid wood paneling is manufactured as 5⁄16-inch (7.6mm) to ¾-inch (19mm) boards 96 inches (2430mm) long and ranging from 4 inches (100mm) to 12 inches (305mm) wide. Softwood species such as southern pine, fir, redwood, and cedar are widely used because they are less expensive and more widely available than hardwoods. The face surface may be smooth or have decorative grooves machined into it **(see 30-4)**. It may have square, tongue-and-groove, or rabbeted edges. It can be applied vertically, diagonally, or horizontally.

ON-SITE STORAGE

Paneling should not be brought into a building until all operations that might create high humidity have been finished and the humidity level returns to normal. This includes letting the drywall compound thoroughly cure and plaster set and cure. This allows the paneling to be conditioned to the room temperature and humidity level. Allow it to sit several days before installing.

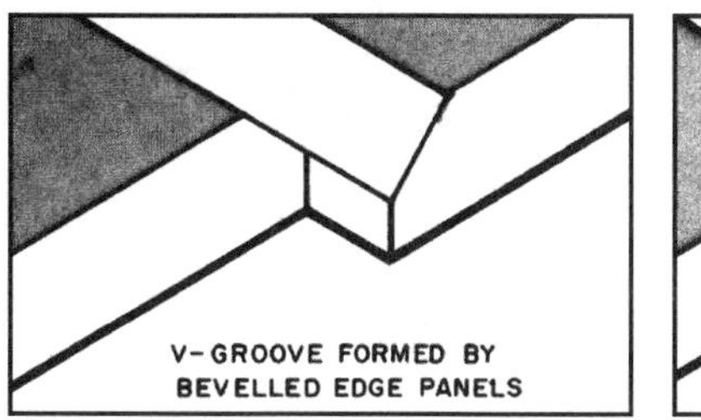

30-3 Sheet wall panels typically have a beveled edge or a rabbet on their long edges.

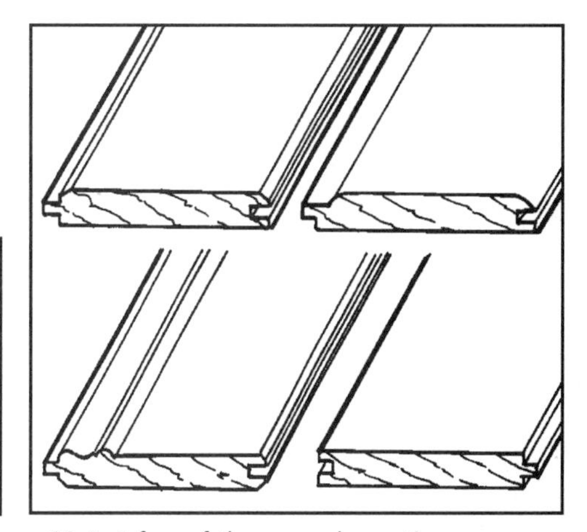

30-4 A few of the many decorative grooves cut into solid wood plank paneling.

Plywood and **hardboard** panels are stored flat with one-inch (25mm) wood spacers between panels to allow air to circulate between them. Be careful not to damage the finished faces. If space is limited they may be stored on their edges and spaced with sticks to allow air to circulate.

Wood plank paneling is stored flat with wood spacers **(see 30-5)**. Use enough spacing sticks to keep the panels or planks from bowing.

30-5 Paneling can be stored flat or in a vertical position. It must be separated by wood spacers to allow air to plow between the panels to condition them to the temperature and humidity in the room.

SOLID-WOOD PLANK PANELING

Various types of solid-wood paneling are available. Perhaps the most commonly used is some form of **tongue-and-groove paneling**. It is available in various widths and profiles **(refer to 30-4)**. **Square-edged paneling** is also used. Often rough-sawed boards are used as well for these types of installations.

Paneling should be ⅜ inch (9mm) thick for 16-inch (406mm) O.C. framing, ½ inch (12.7mm) thick for 20-inch (508mm) O.C. framing, and ⅝-inch (16mm) thick for 24-inch (610mm) O.C. framing. It must have been kiln dried to eight percent moisture.

Wood paneling is usually manufactured in four- to eight-inch (102 to 203mm) widths. Wider stock tends to cup and may require extra blocking and nailing in the center of the panel.

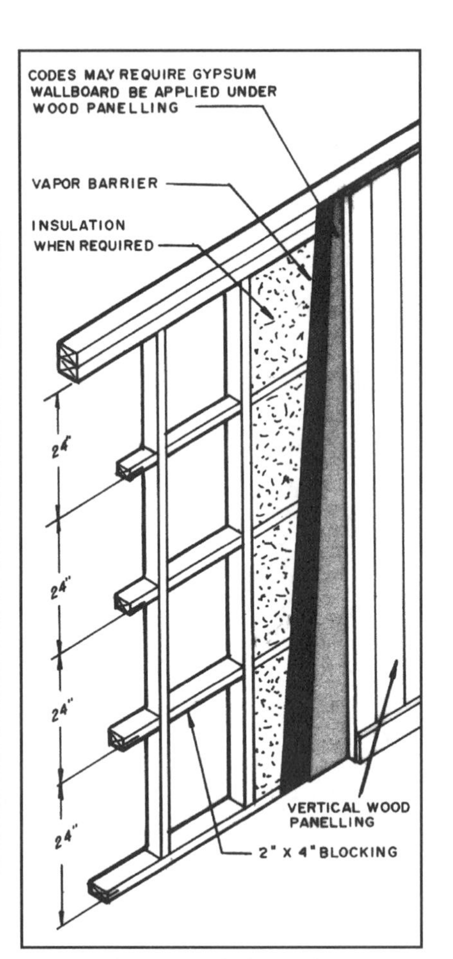

30-6 Solid wood plank paneling requires blocking be installed between the studs.

INSTALLING VERTICAL SOLID-WOOD PANELING

Solid-wood paneling may be installed horizontally, diagonally, or vertically. The same basic techniques apply to each. Usually blocking is installed between the studs when it is applied vertically **(see 30-6)**. When installed on an exterior wall, the insulation is in place

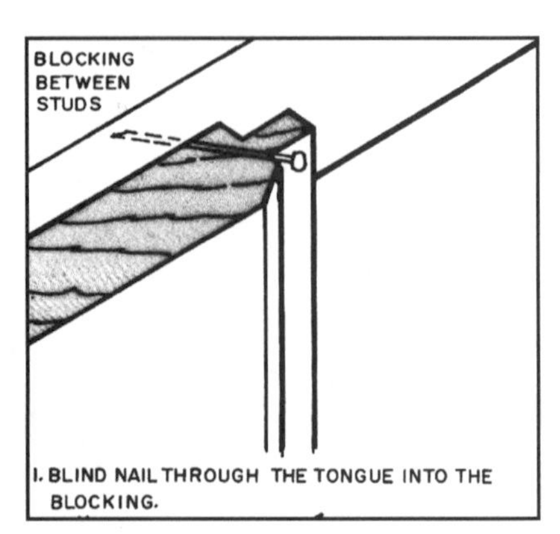

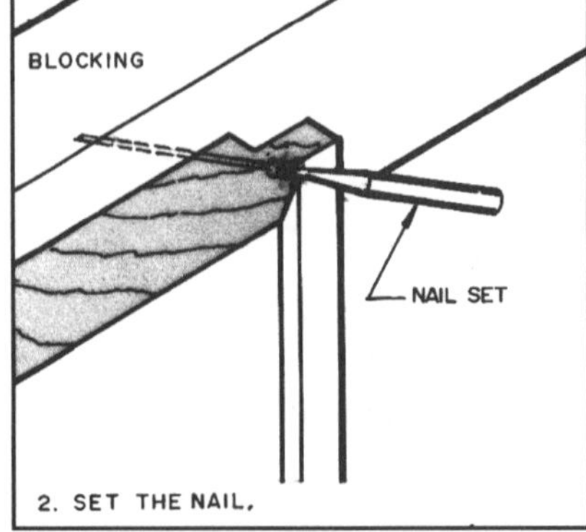

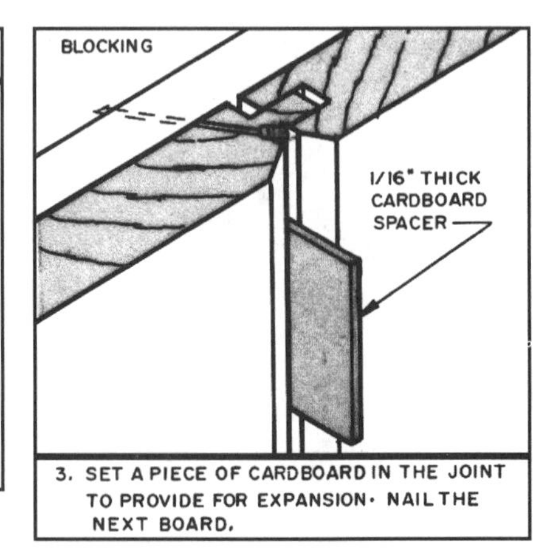

30-7 The steps to nail solid wood paneling that has a tongue-and-groove edge.

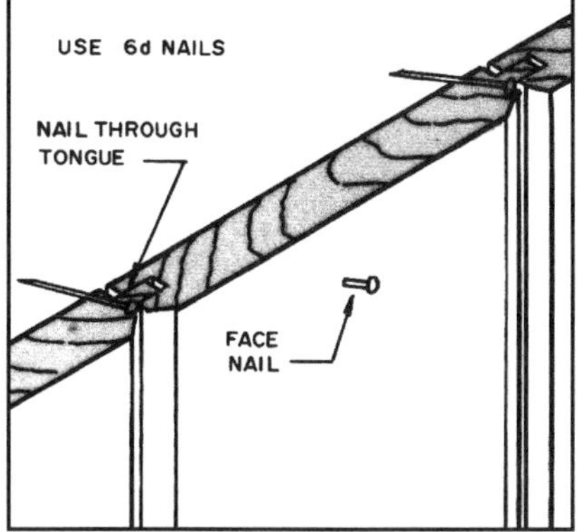

30-8 Tongue-and-groove paneling is nailed through the tongue. Wider planks are also face nailed.

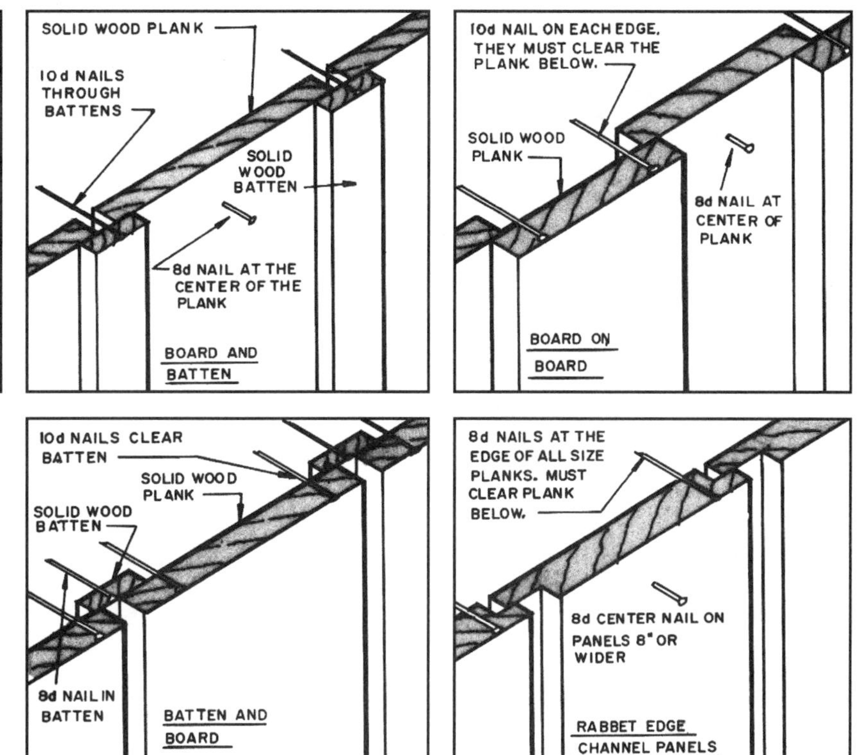

30-9 Various ways to install and nail square-edged and rabbeted-edged channel paneling.

and a vapor barrier is stapled over it. Usually 6d finishing nails are used, and the panels are nailed to the blocks and studs.

Tongue-and-groove paneling 4 to 6 inches wide is often nailed with one 6d finishing nail through the tongue. Some prefer to add one face nail on six-inch planks. Eight-inch planks must have a nail in the tongue and one face nail. The steps to set a nail at the tongue are in **30-7**. Nailing recommendations for tongue-and-groove plank panels are shown in **30-8**. Nailing square-edged and rabbet-edged channel paneling is shown in **30-9**.

Plan ahead how you will proceed **(see 30-10)**. Some carpenters prefer to start at the outside corners and work towards the inside corners. One reason to start at the outside corner is that it is more visible; beginning there will give full-width boards until you get to the inside corner. The final board may have to be narrower; this is less noticeable at the inside corner. The outside corner can be mitered or butted **(see 30-11)**. Since it is a long joint, it is very difficult to miter.

If there are no outside corners, then, of course, begin at an inside corner, and work towards doors and windows and other inside corners.

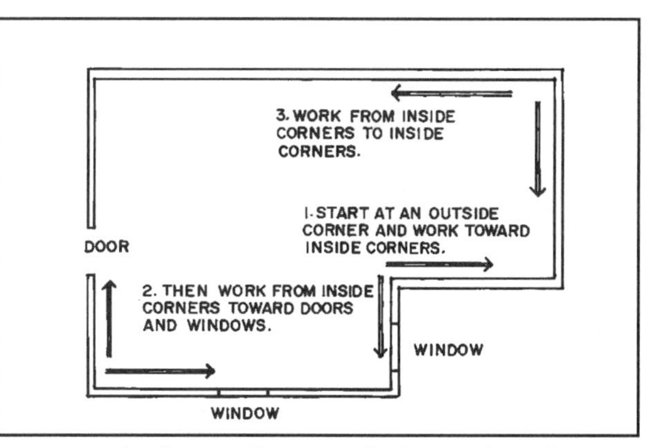

30-10 A plan for installing vertical solid wood paneling.

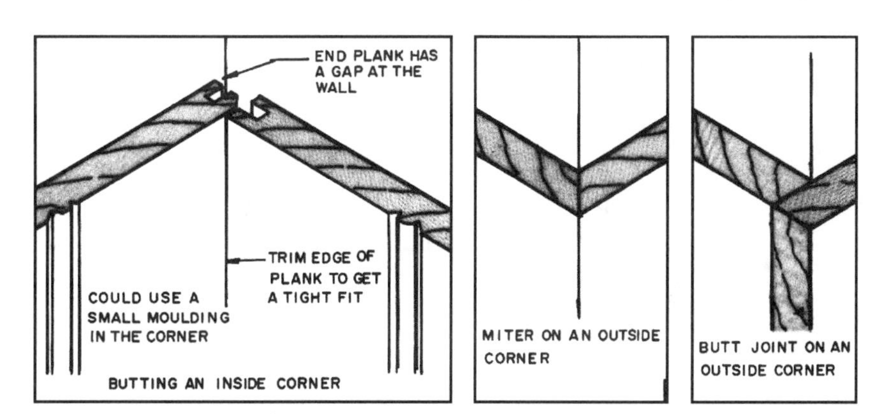

30-11 Techniques for finishing inside and outside corners.

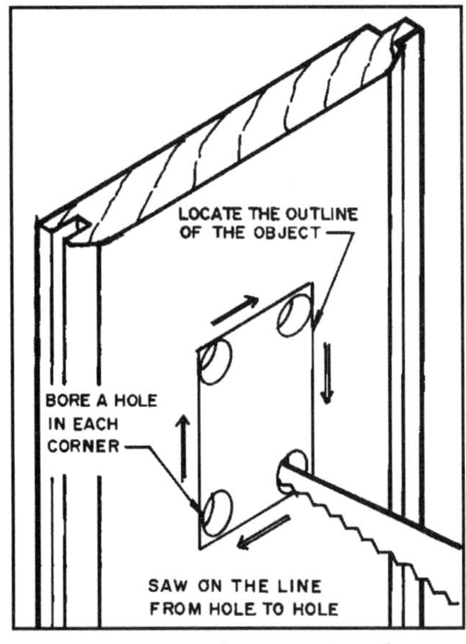

30-12 Locate and cut openings for items penetrating the paneling.

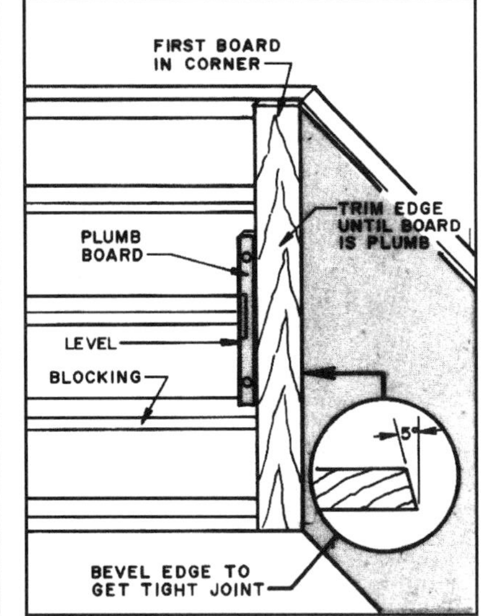

30-13 Start nailing solid wood paneling in a corner. Be certain the first piece is plumb.

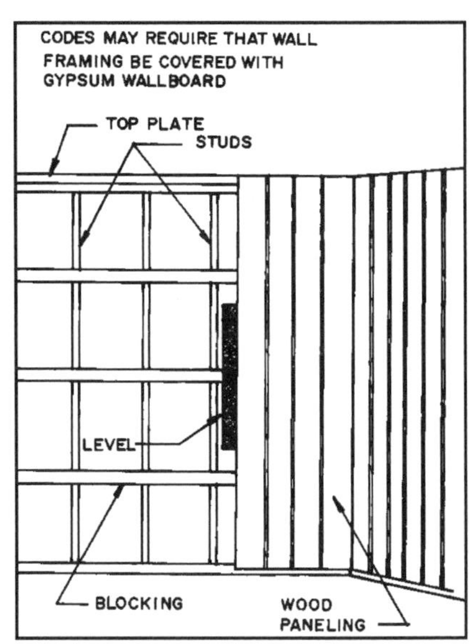

30-14 Check each piece of paneling for plumb. Adjust as necessary to keep it plumb.

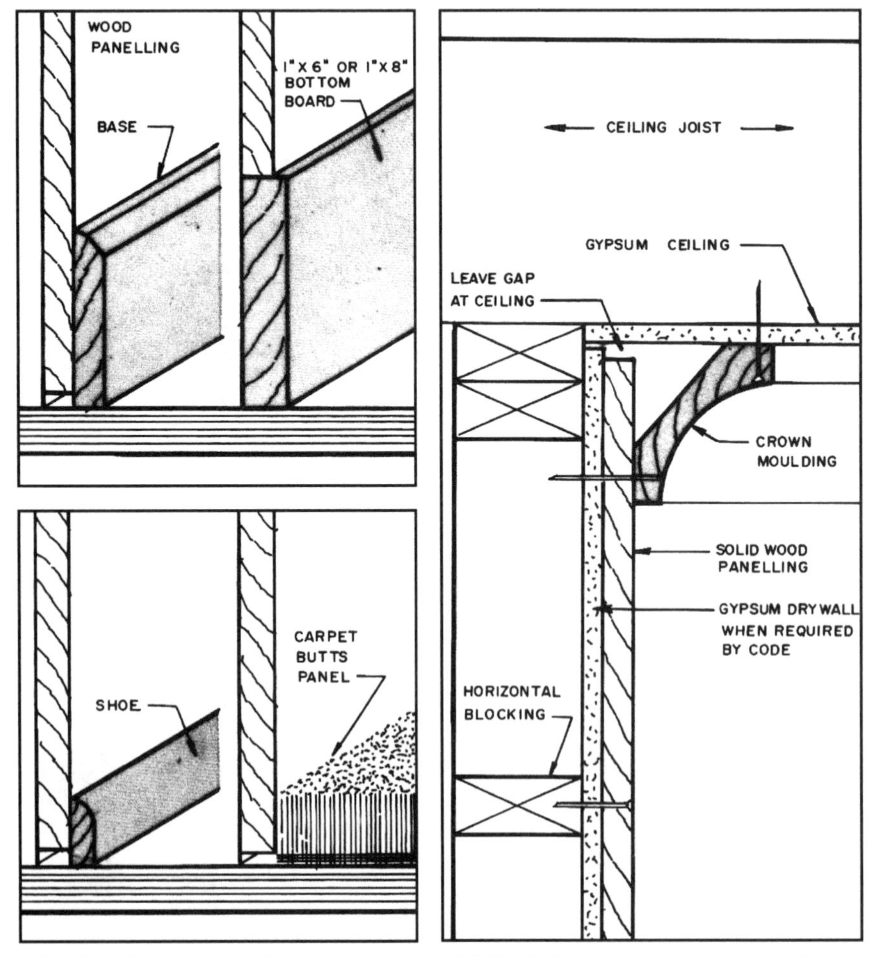

30-15 Solid wood paneling and wainscoting can be finished at the floor with any of these methods.

30-16 Solid wood paneling is usually finished at the ceiling with some type of molding.

Before beginning to work, line up the planks along the wall and attempt to match up the grain so the color flows smoothly along the wall. Also remember to locate and cut openings for electrical outlets, pipes, and other things that have to penetrate the panel **(see 30-12)**. See Chapter 29 for additional information.

Begin installation by scribing the first piece to the butting wall, setting it plumb, and nailing it in place **(see 30-13)**. As additional pieces are installed, check each for plumb. When necessary, adjustments are made to keep the boards plumb **(see 30-14)**. The last piece is scribed to the end wall and nailed to the backing.

The baseboard is installed on top of the paneling unless the paneling is butted to a wider board at the floor **(see 30-15)**. Ceiling molding is used to conceal the joint between the paneling and ceiling **(see 30-16)**. Refer to the previous chapter, Chapter 29, for more information.

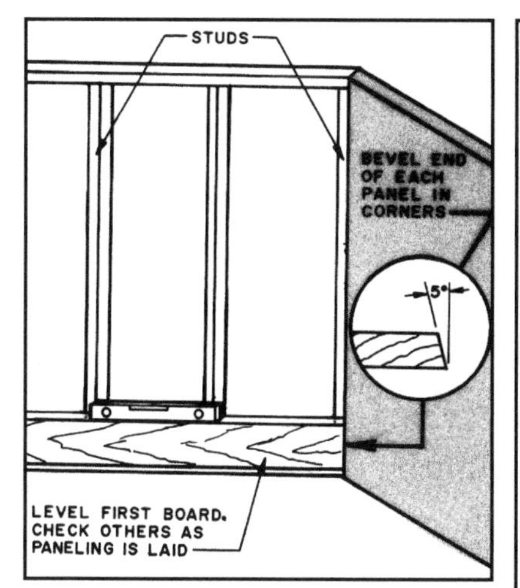

30-17 Begin installing horizontal solid wood paneling by leveling the first piece on the floor.

Estimating Solid-Wood Paneling

The following is based on the plan shown in **30-18.**

1. Calculate the area of the walls to be covered. This is the length times the height of each wall.

   20 × 8 × 2 walls = 320 sqft
   14 × 8 × 2 walls = 126 sqft
   446 sqft total

2. Subtract door and window areas. Assume the following:

   1 door, 3 × 7 = 21 sqft
   2 windows, 3 × 4 × 2 = 24 sqft
   45 sqft

   Other wall-covering objects, such as fireplaces and cabinets, can also be subtracted.

3. Total wall area is 446 – 45 = 401 sqft

4. Multiply the number of square feet by the area factor for the type of paneling. The area factor is found by dividing the nominal width by the face size, as given in **30-19**. If you plan to use 1 × 6 tongue-and-groove paneling, the area factor is 1.19. Therefore we need 401 × 1.19 = 478 board feet of paneling.

5. Some add a 5 percent or 10 percent waste factor. At 5 percent, the total would be 478 + 24 = 502 board feet of paneling.

INSTALLING HORIZONTAL SOLID-WOOD PANELING

Horizontal wood paneling is nailed directly to the studs. Begin by installing the first row at the floor. Leave a small gap between the board and subfloor. Be certain the first row is level **(see 30-17)**. As additional layers of boards are installed, constantly check to be certain they remain level. If it is tongue-and-groove paneling, leave a 1/16-inch (1.5mm) open space between the end of the tongue and the bottom of the groove. Outside corners are mitered, and inside corners are butted. Tongue-and-groove paneling is blind-nailed, as explained for vertical paneling. Square-edged paneling is face-nailed. Ceiling molding is used, and baseboard is installed in the usual manner.

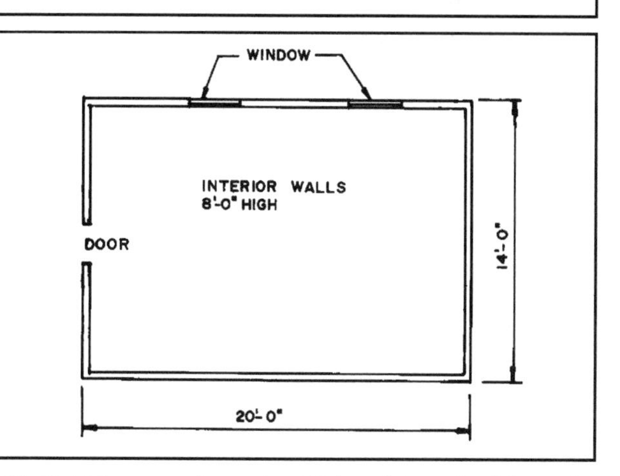

30-18 This room will require 502 board feet of solid wood tongue-and-groove paneling to cover the walls from the floor to the ceiling.

30-19 Coverage factors for solid-wood paneling (inches).

	Paneling Width			
	Nominal size	Dress size	Face size	Area Factor
Tongue-and-Groove Paneling	1 × 6	5 7/16"	5 1/16"	1.19
	1 × 8	7 1/8"	6 13/16"	1.18
	1 × 10	9 1/8"	8 13/16"	1.14
Square-edge Paneling	1 × 6	5 1/4"	5 1/4"	1.14
	1 × 8	7 1/4"	7 1/4"	1.10
	1 × 10	9 1/4"	9 1/4"	1.08

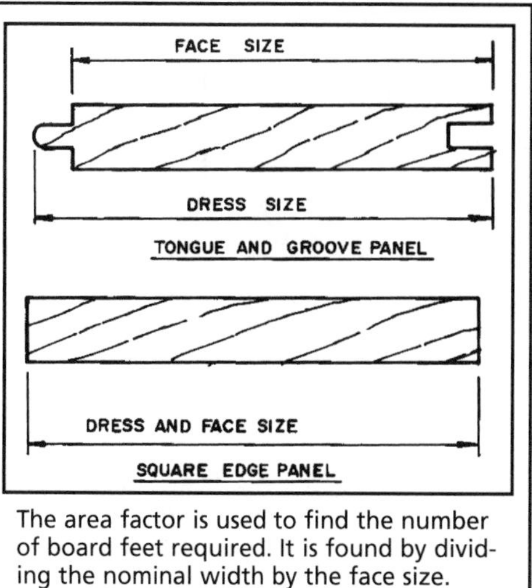

The area factor is used to find the number of board feet required. It is found by dividing the nominal width by the face size.

30-20 This dining room has plywood paneled walls, giving the room a warm, pleasant atmosphere. *(Courtesy Georgia-Pacific Corp.)*

PLYWOOD PANELING

Plywood paneling is available in a large variety of wood veneers. Typical thicknesses are 5⁄32, ¼, 5⁄16, and 7⁄16 inch (3.9, 6.4, 7.9, and 11mm). The most commonly used panel is 4 × 8 feet (1.2 × 2.4m). Most paneling is prefinished, so it must be handled with care **(see 30-20)**. The thickness of plywood paneling to use depends on the spacing of the studs **(see 30-21)**.

The humidity in the room must be at a normal level. All drywall compound and plaster must be thoroughly cured. Panels should be stored in the room 48 hours before installing so that their moisture content equalizes with the surrounding area.

Before installing plywood panels, arrange them side-by-side along the wall and observe the differences in grain and color. Rearrange them so you have the best flow of grain and color.

30-21 Minimum plywood paneling thicknesses for various stud spacings.

Stud Spacing	Minimum Thickness
16" O.C.	¼"
20" O.C.	⅜"
24" O.C.	⅛"

Number the panels on the back in the order in which they should be installed **(see 30-22)**.

Generally panels over ¼ inch (6.4mm) thick can be applied to wood studs or furring. Panels ¼ inch or thinner usually must be applied over a fire resistant backing such as gypsum wallboard.

PREPARING THE WALLS TO RECEIVE THE PANELING

If the paneling is to be applied over **wood stud wall framing,** check to be certain the studs are set to fall on four foot centers. If they are not, you will have to install additional studs **(see 30-23)**. Another solution is to nail horizontal furring strips spaced 16 in O.C. across the studs **(see 30-24)** and then nail to these furring strips.

Many carpenters prefer to install horizontal blocking to provide an additional base for the panels **(see 30-25)**.

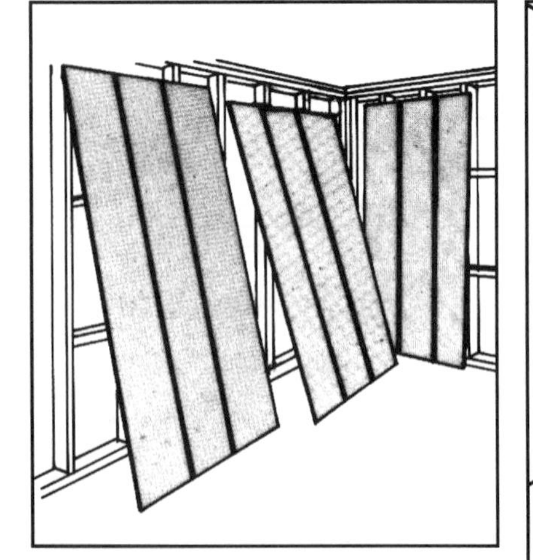

30-22 Line up the panels along the wall and arrange so the grain and color produce the most pleasing appearance.

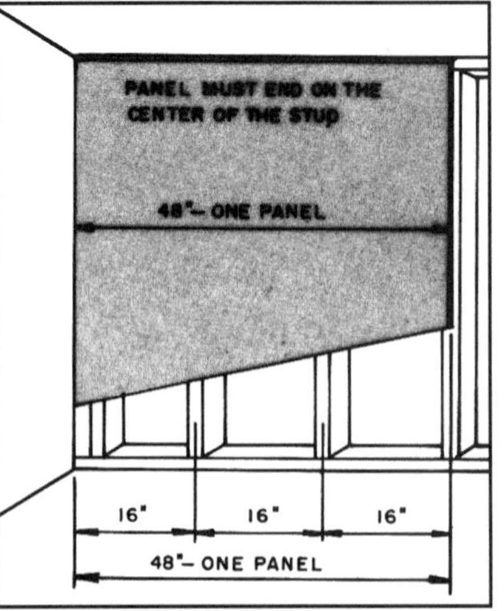

30-23 Studs must be spaced so the edges of the panel fall on the center of the studs.

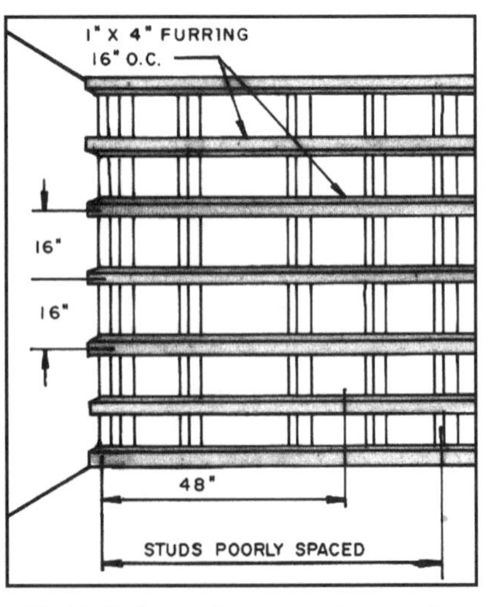

30-24 If the studs are poorly spaced, nail 1 × 4-inch furring strips horizontally and nail the paneling to them.

30-25 Horizontal blocking is often installed between the studs to firm up the paneling.

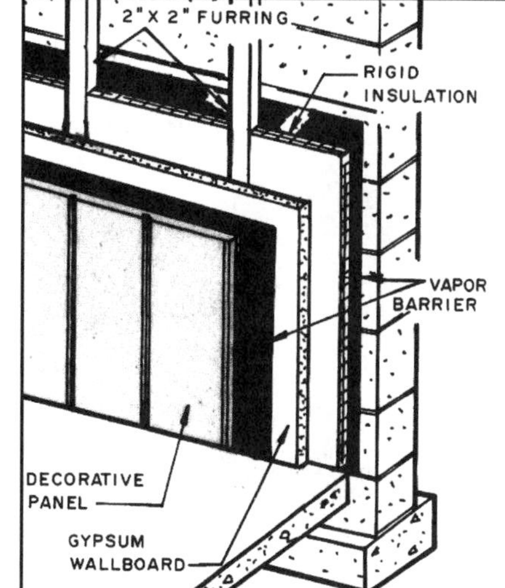

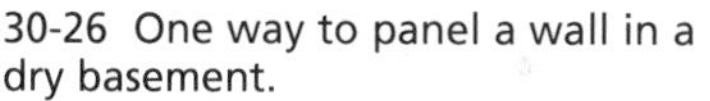

30-26 One way to panel a wall in a dry basement.

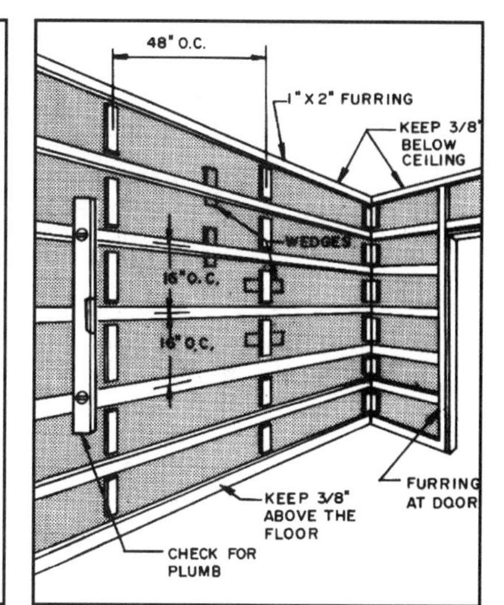

30-27 Masonry walls can be furred out with 1 × 2-inch furring secured to the wall. Wedges are used as necessary to get the strips plumb.

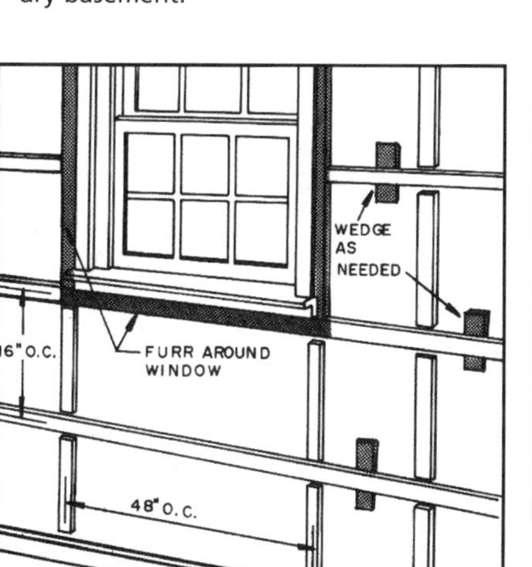

30-28 Furring is placed around windows and doors to provide a nailing base for the paneling.

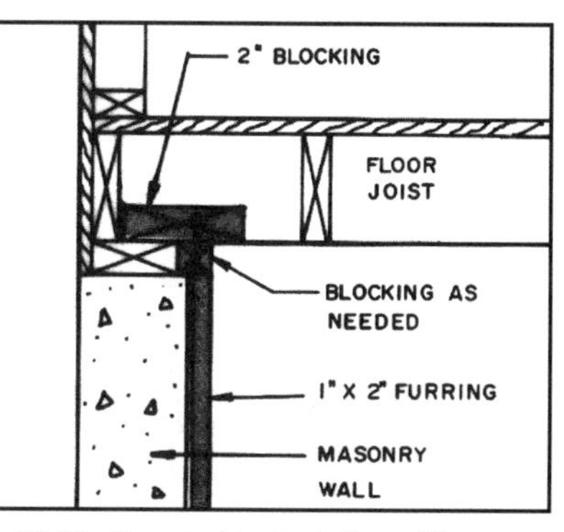

30-29 How to block at the ceiling when paneling masonry walls, such as in a basement.

Check the walls for straightness with a chalkline. Adjust as needed to secure a flat, vertical plane surface.

If a **masonry wall**, such as in a basement, is to be paneled it should be dry and have no seepage of moisture from the exterior. In dry basements apply a vapor barrier over the wall and then install 2 × 2-inch (25 × 25mm) furring vertically to the wall with masonry nails or some type of mechanical fastener. Place rigid foam insulation sheets between these strips **(see 30-26)**. Nail ½-inch (12.7mm) gypsum wallboard over the insulation and staple a vapor barrier over this. Then nail the paneling through the gypsum into the wood furring. Interior masonry partitions do not require insulation or vapor barriers. Remember to check the installation so the strips form a flat plane. This will usually require that the furring be adjusted with wood wedges **(see 30-27)**. Install vertical furring between these at 48-inch centers. Furring must also be placed around doors and windows **(see 30-28)**.

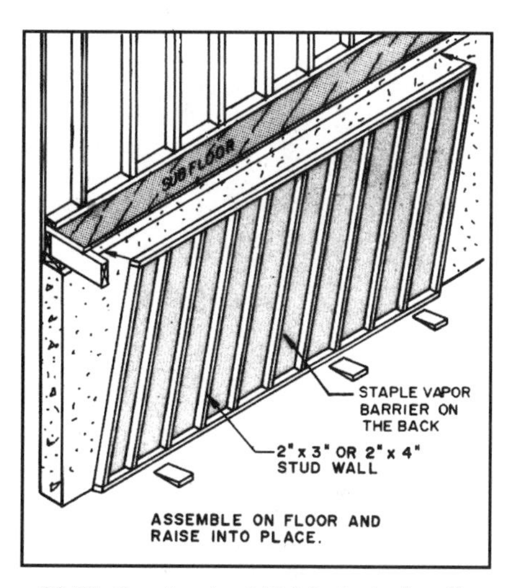

30-30 Construct a 2 × 4-inch stud wall over the masonry wall.

When furring basement walls you will need to install special blocking at the ceiling where it runs parallel with the floor joists **(see 30-29)**.

Another technique for paneling masonry walls is to build a wood frame wall with 2 × 4-inch (50 × 100mm) studs as shown in **30-30**. If this is an exterior wall it will need a vapor barrier and insulation as shown above in **30-26**.

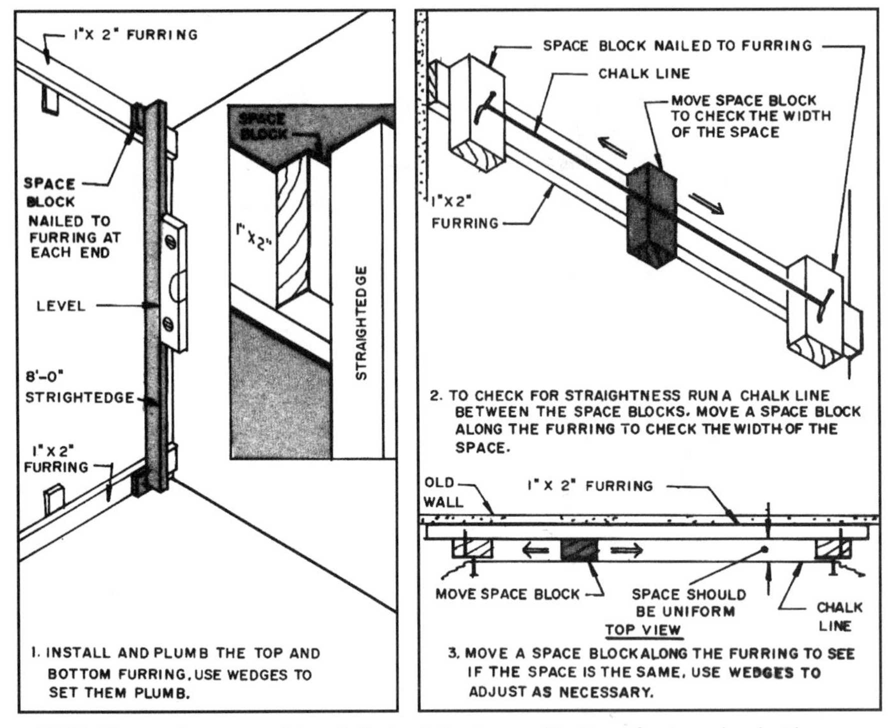

30-31 To panel over an old wall, first set the top and bottom furring plumb. Then check for straightness with a chalkline and space blocks.

It is sometimes necessary to panel over old walls which may be rough, cracked, and out of plumb. Install 1 × 2-inch (250 × 50mm) furring strips over the wall. It may be necessary to adjust the furring with wood wedges to bring them into a flat, plumb plane.

Begin by nailing the top and bottom horizontal furring strips. Keep them about 3/8 inch (9.5mm) from the ceiling and floor. Check them for plumb with a long straightedge and a level.

Check to see that they are straight (not bowed) with a chalkline **(see 30-31)**. Adjust with wedges as necessary. Finally install the rest of the furring spaced 16 inches (406mm) O.C. If the wall has an unusually high spot that interferes, it may be necessary to cut it back. Install vertical furring between the horizontal at 48 inches (1219mm) O.C. as shown earlier in **30-27**.

The furring strips over old walls with wood studs can be nailed into the studs. You will normally use 6d or 8d nails depending upon the thickness of the old wall finish.

The furring strips can be applied to masonry walls with adhesive and specially hardened masonry nails.

CUTTING PLYWOOD PANELING

Plywood paneling usually has the final finish applied in a factory. The carpenter must handle the sheets carefully so that the surface does not get scratched. Also, when cutting the panels care must be taken to prevent splintering the edges. If the panel is cut on a table, type circular saw, the finished surface should face **up**. When cutting with a portable circular saw or saber saw, cut with the finished face down. If cutting with a handsaw, use a fine-toothed saw and keep the finished face up.

LOCATING & FITTING THE PANELS

Begin by marking the location of each stud. A stud finder will help you locate them **(see 30-32)**. Mark a mark on the ceiling and floor indicating the stud locations **(see 30-33)**.

Since the panels are usually installed with a 1/16-inch (1.5mm) gap between panels to allow for expansion, some spray paint the wall or stud with black paint where each joint will occur so the lighter color of the underlaying material will not show through the joint **(see 30-34)**.

30-32 You can use a stud finder to help locate the studs behind the finished gypsum wallboard. *(Courtesy Georgia-Pacific Corporation)*

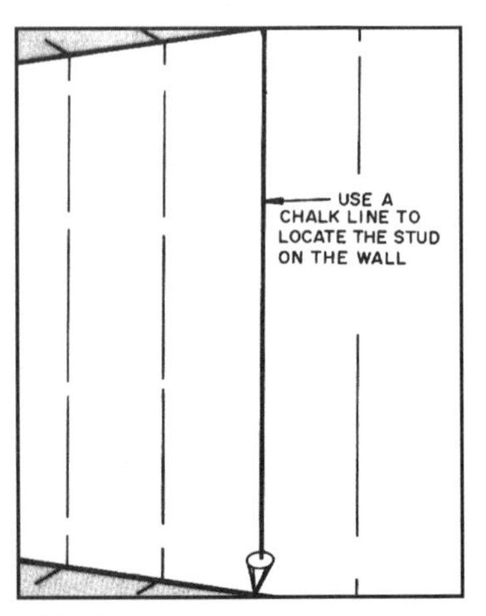

30-33 Locate each stud at the ceiling and floor and along the wall when necessary.

30-34 Spray paint the drywall black where the panels will meet so the light color will not show through the gap between the panels. *(Courtesy Georgia-Pacific Corporation)*

30-35 When the panel meets a wall out of plumb, scribe the slope on the panel and cut on the line.

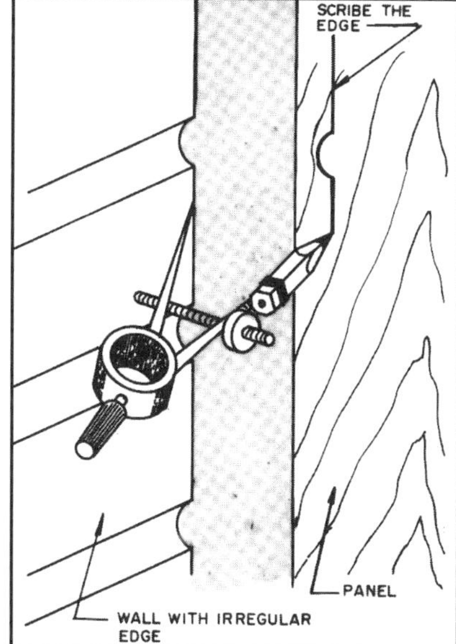

30-36 When a panel meets an irregular surface, scribe the profile on the panel and cut on the line.

It is important to get the first panel absolutely plumb. Since the wall it will butt may not be plumb, the edge of the panel must be scribed to it. Usually this is only a very small amount.

Place the panel on the wall, and push it up to the butting wall. Using a level, set it plumb. If the butting edge leaves a gap that will not be covered when the adjoining wall is paneled, scribe a line on the edge parallel with the out-of-plumb wall **(see 30-35)**. Cut or plane the edge to this line. If the butting wall is irregular, such as stone, brick, or grooved wood, scribe as shown in **30-36**. If the paneling has V-grooves 16 inches O.C., check to see that they fall on a stud. If not, adjust the width of the first panel so that this occurs **(see 30-37)**.

Cut the panels to length so they are about ½ inch (12.7mm) shorter than the floor-to-ceiling height.

Remember to locate and cut openings for electrical outlets and other items that penetrate the wall. You can locate an opening by measuring from the edge of the nearby panel and the floor as shown in **30-38**. Some prefer to cover the edges of the item with chalk and press the panel against it **(see 30-39)**.

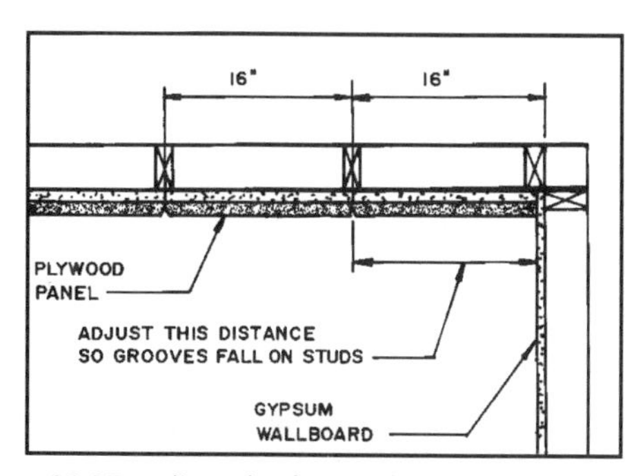

30-37 Adjust the first panel so that the grooves fall on the studs to allow the nails to be hidden in the grooves.

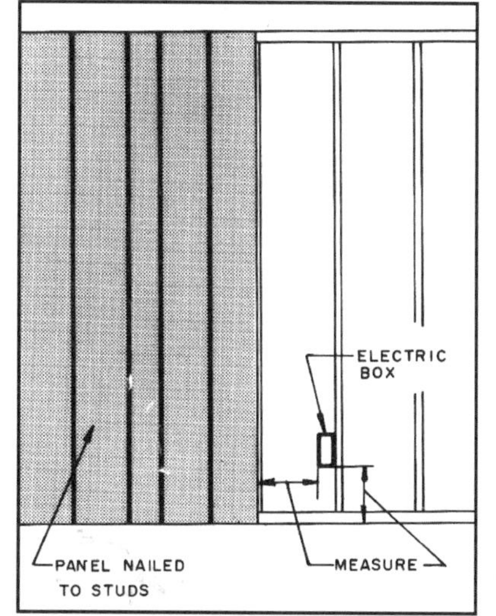

30-38 The location of items penetrating the paneling can be located by measuring the location.

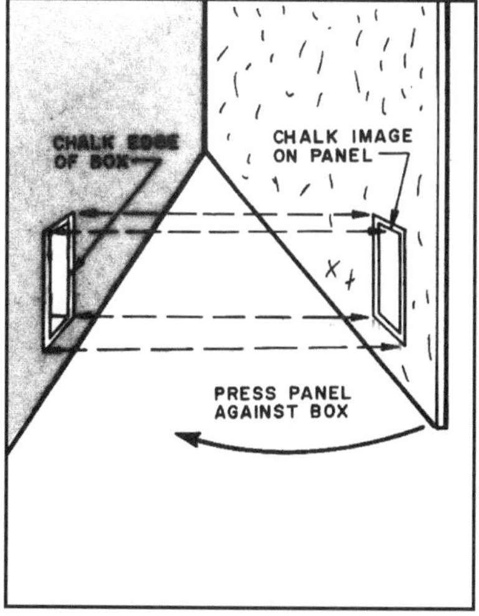

30-39 Chalk can be used to imprint on a panel the location of an outlet box or other item penetrating the panel.

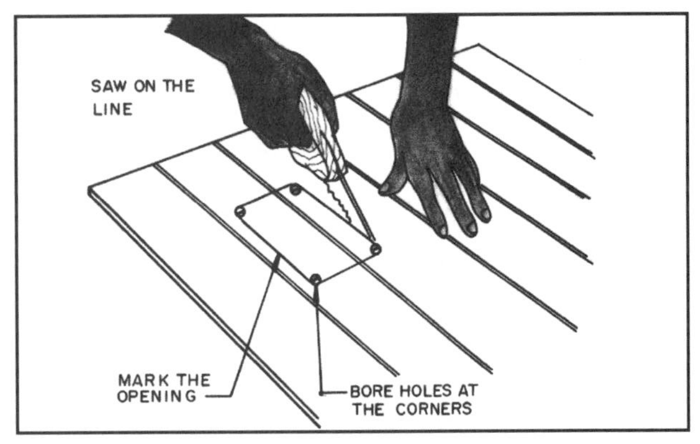

30-40 Bore holes in the corners of the opening and cut with a compass or in-line power jigsaw.

Bore holes in each corner and cut the opening with a compass saw or an in-line power jigsaw **(see 30-40)**.

Sometimes a ceiling is higher than 8′-0″. This can be handled in several ways, depending upon the distance. If the distance beyond 8′-0″ is small, the panel can be raised off the floor and a molding can be used at the ceiling. Higher ceilings require that a small panel be cut to fill the space above the 8′-0″ sheet. Usually, some type of wood molding is placed over the end joint between the panels **(see 30-41)**.

NAILING THE PANELING

Panels can be installed directly to studs with 3d nails. If installed over gypsum wallboard or some other finished wall material, use 6d nails. Nails may be set and covered with colored putty, which is available in stick form. Nails with the heads painted to match the color of the panels are also used and are not set. The nails are spaced six inches (150mm) apart on the edges and 12 inches (approx. 300mm) apart on the interior of the panel **(see 30-42)**. Do not nail closer than ⅜ inch (9.5mm) to the edge of the panel. While panels appear to have randomly spaced grooves, there is a groove every 16 inches (406mm). If the first panel is properly set, a groove will fall on each stud. Nail to the stud, putting the nails in the groove; this helps hide them.

The first panel is installed in one corner of the room. Using one or two nails, lightly tack it in place. Then use a level or plumb line and set the panel plumb **(see 30-43)**. Set the panel so it is about ⅜ inch (9.5mm) above the subfloor and ceiling. This gap will be covered by the base or other molding. It may help position the panel to insert wedges below it at the floor. They can be adjusted to get the panel plumb **(see 30-44)**. When you are certain the panel is plumb, nail it to the studs.

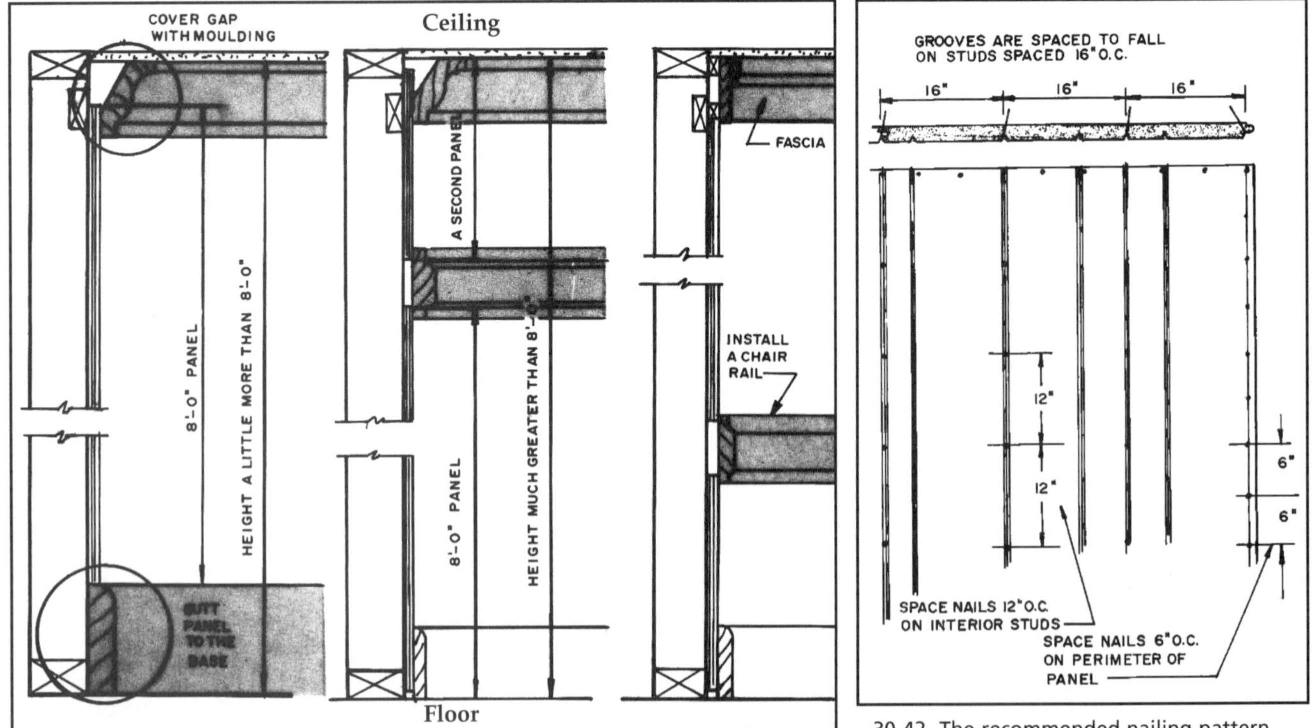

30-41 How to panel rooms with ceilings higher than the normal 8′-0″.

30-42 The recommended nailing pattern for installing plywood paneling.

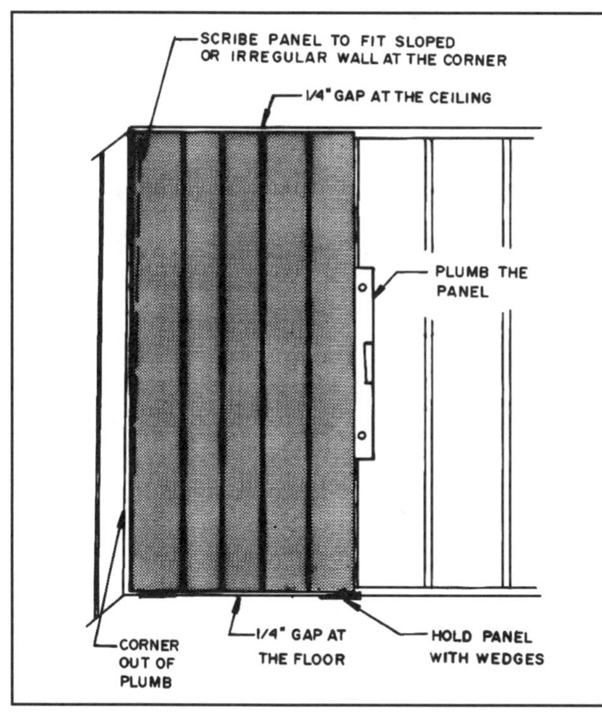

30-43 Check the panel to be certain it is plumb before nailing it to the studs.

30-44 Wedges below the panel at the floor help hold it plumb and off the floor as it is nailed.

Continue installing panels across the room. Check each for plumb. Carefully measure and cut the last piece to finish the wall.

Since plywood paneling is thin, it is difficult to get tightly butted inside and outside corners. To handle this, some type of molding, such as cove molding, is often used on inside corners, and corner molding is used on outside corners **(see 30-45)**.

The panels can butt door and window casings. To do this requires careful measuring and cutting so that a gap does not occur. Paneling is finished at the floor and ceiling as shown for solid wood paneling shown earlier in **30-15** and **30-16**. It can meet the window and door casing as shown in **30-46**.

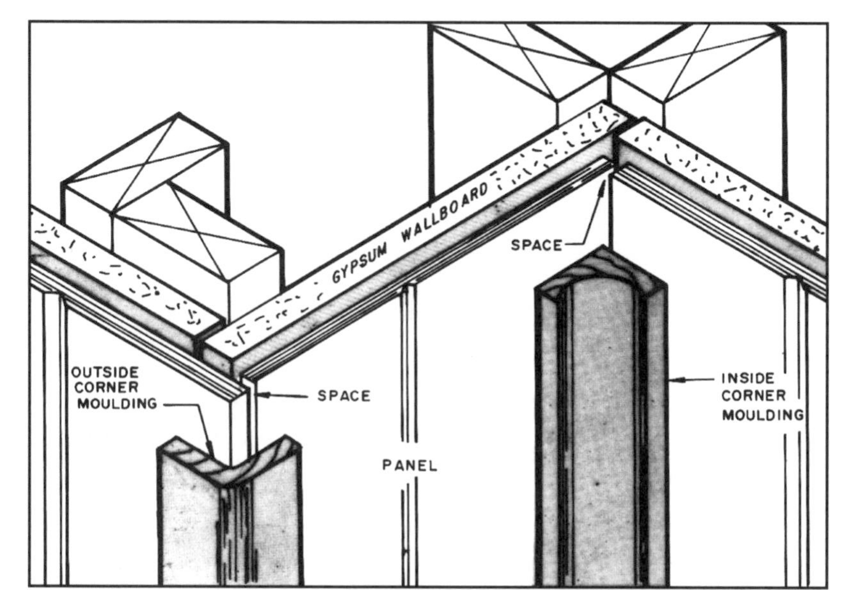

30-45 Moldings are used to form inside and outside corners with plywood paneling.

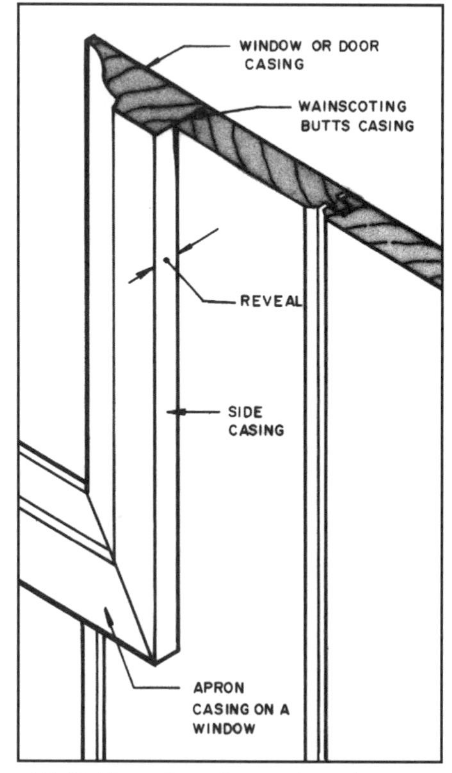

30-46 The wainscoting can butt door and window casing.

30-47 This paneling is installed horizontally, making the room appear longer. *(Courtesy Georgia-Pacific Corporation)*

30-50 Paneling applied in a herringbone pattern. *(Courtesy Georgia-Pacific Corporation)*

INSTALLING PLYWOOD PANELING HORIZONTALLY

Occasionally, plywood paneling is applied horizontally **(see 30-47)**. This makes the ceiling appear lower and the walls longer. Since all panel edges must rest on two-inch blocking, horizontal members are needed when installing the sheets horizontally **(see 30-48)**. If the plywood paneling has tongue-and-grooved edges, the horizontal blocking can be omitted.

If the studs are spaced 16 inches O.C., the paneling should be at least 1/4 inch thick if no backing is used. If spaced 24 inches O.C., 3/8-inch paneling should be used. Plywood paneling less than 1/4 inch thick must have a backing material. Often 3/8-inch gypsum board or 5/16-inch plywood sheathing is nailed to the framing and the thin plywood paneling is fastened over this **(see 30-49)**. It is best if all plywood paneling is applied over a backing material.

INSTALLING PLYWOOD PANELING TO PRODUCE A HERRINGBONE PATTERN

A herringbone pattern produces a dramatic appearance **(see 30-50)**. To produce this pattern, lay out the paneling cuts on a 45-degree angle **(see 30-51)**. Match up two sheets of paneling.

Place them so that the groove pattern is the same on the edge where they meet. After cutting, pair up the opposite cuts where panels are nailed to the wall. The finished wall should have a batten strip applied vertically at the midpoint. A butt joint could be used instead.

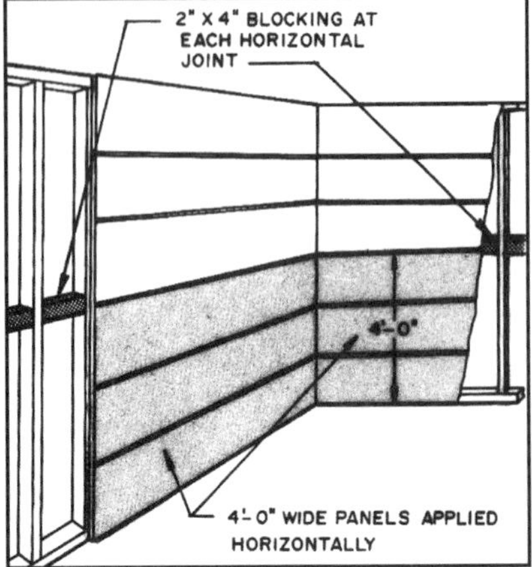

30-48 When installing plywood or hardboard panels horizontally, blocking is needed to support joints between the edges of the panel.

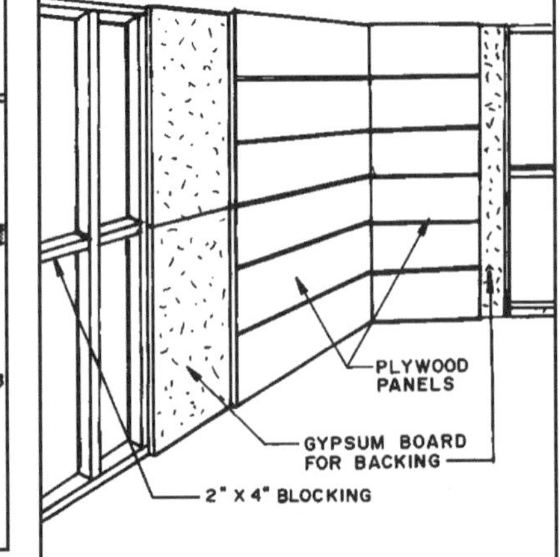

30-49 It is best to install plywood paneling over a gypsum board backing.

INSTALLING PLYWOOD PANELING WITH ADHESIVES

Plywood paneling can be bonded to studs, gypsum drywall, or any other sound wall surface with adhesives recommended by the plywood manufacturer. After the panel has been scribed and fitted, apply a solid bead of adhesive to all four edges and, on the interior studs, beads three inches (76mm) long spaced about six

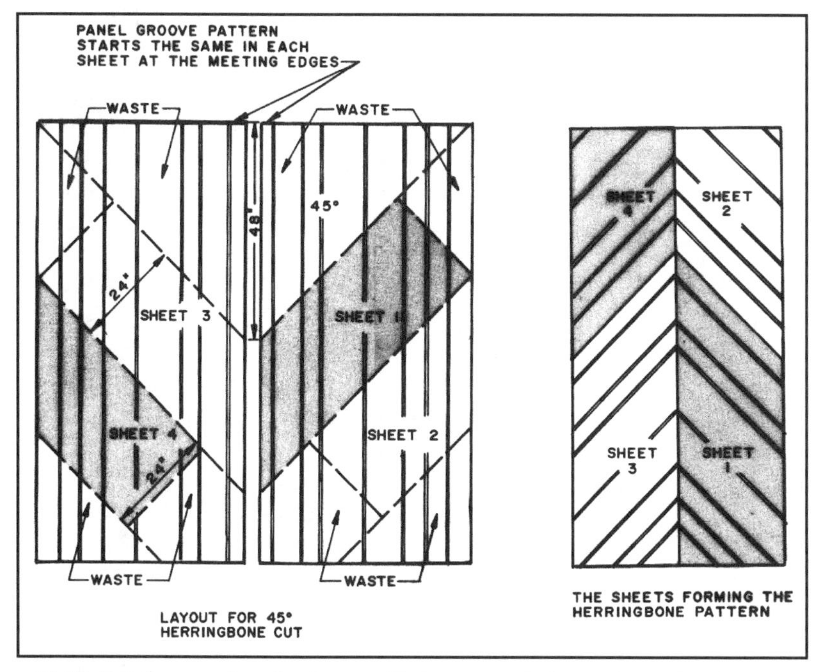

30-51 Cut the paneling sheets on a 45-degree angle to produce a herringbone pattern.

inches (152mm) apart as shown in **30-52**. Place a strip of wood on the subfloor, and rest the bottom of the panel on it. Then press the panel against the wall. Tap over the glue lines using a wood block covered with cloth until you get a good bond between the panel and the backing material. Some adhesives require that the panel be pressed against the wall, but pulled away again for a few minutes to let the solvent evaporate rapidly, and then pressed back, and tapped until it is firm **(see 30-53)**. Some recommend placing nails at the top and bottom where they will be covered by molding or base. Some place a few nails within the panel to hold it as the adhesive sets.

INSTALLING HARDBOARD PANELING

Hardboard paneling is available in a wide range of simulated wood grains and colors. Most panels are ¼ inch (6mm) thick and 4 × 8 feet (1.2 × 2.4m) in size. Other products, such as hardboard planks, are available.

Hardboard is handled in the same way as described for plywood paneling. The sheets have a finished surface that must be protected from damage. If ⅛-inch-thick hardboard is used, a backing is necessary. Sheets ¼ inch thick can be used over studs spaced 16 inches O.C. Sheets 7/16 inch thick can be used with studs spaced 24 inches O.C.

When nailed to studs or over a backing, the nails are spaced four inches apart on the outer edges and eight inches at intermediate studs. The nail should be long enough to penetrate the stud ¾ inch **(see 30-54)**.

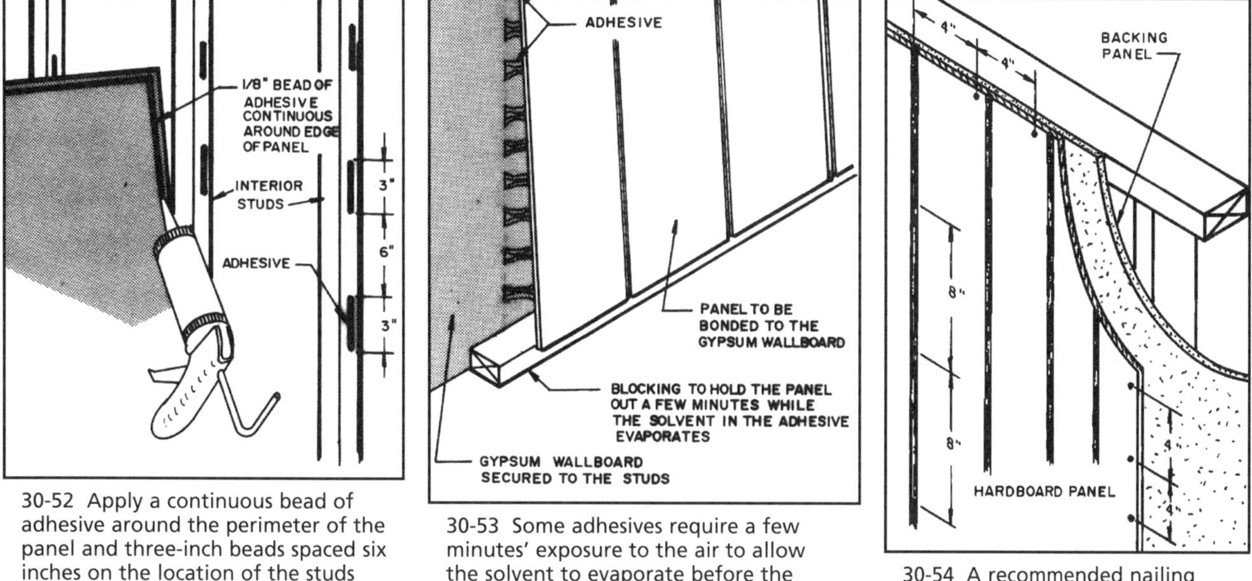

30-52 Apply a continuous bead of adhesive around the perimeter of the panel and three-inch beads spaced six inches on the location of the studs within the panel.

30-53 Some adhesives require a few minutes' exposure to the air to allow the solvent to evaporate before the panel is pressed against the wall.

30-54 A recommended nailing pattern for hardboard paneling.

Gluing Hardboard Panels

When applying hardboard panels with adhesive to studs, apply a continuous ⅛-inch bead to the stud at the outer edge of the panel. On intermittent studs apply three-inch-long strips of adhesive with six inches between each. The same application is used on panels with backing. Moldings and trim are applied as described for plywood paneling. To trim edges and form corners, prefinished aluminum moldings are available **(see 30-55)**.

INSTALLING SOLID-WOOD WAINSCOTING

Solid-wood wainscoting is installed in much the same way as solid wood paneling. Begin by making certain the walls have the required blocking. This is typically 2 × 4-inch (38 × 100mm) stock installed horizontally between the studs, providing nailing for the wainscot cap and intermediate support. Some prefer to let 1 × 4-inch (25 × 100mm) stock into the studs instead **(see 30-57)**. Usually gypsum wallboard is applied over the studs before the paneling is installed.

To begin a wainscoting, measure the desired height from the floor to one end. Run a chalkline that is perfectly horizontal to the other wall and snap a line on the wall **(see 30-58)**. Plan to allow the paneling to have a small gap at the floor.

Next, check the first wall for plumb with a level. If it is out of plumb, measure out a distance equal to the width of one board, and draw a plumb line. Measure the distance from the wall to the line at the top and bottom. Plane a little off the edge of the paneling that will butt this wall. Refer again to **30-58**.

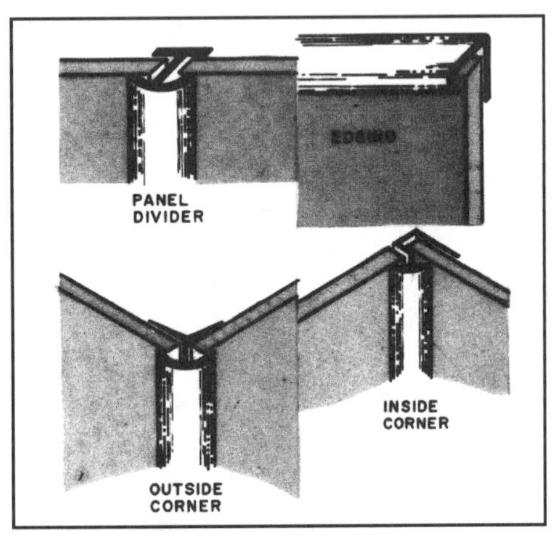

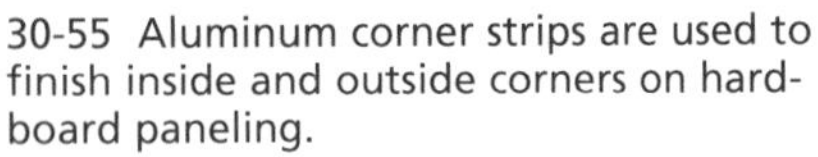

30-55 Aluminum corner strips are used to finish inside and outside corners on hardboard paneling.

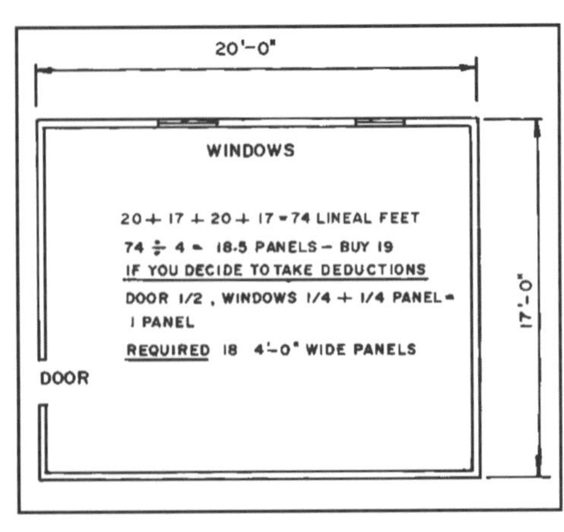

30-56 Estimating the number of 4 × 8-foot plywood or hardboard panels that are needed to panel a room.

Estimating the Number of Panels Needed

The following is based on the plan shown in **30-56**.

1. Determine the lineal feet of wall around the room. For this example, this equals 20 + 17 + 20 + 17, or 74 feet.
2. Divide by 4 (4 feet per panel), which equals 18.5 panels. Purchase 19 panels.
3. Some estimators deduct for doors and windows as follows:

   Doors— deduct ½ panel.
   Windows—deduct ¼ panel.

   You can also deduct for other things such as wall and base cabinets and fireplaces.

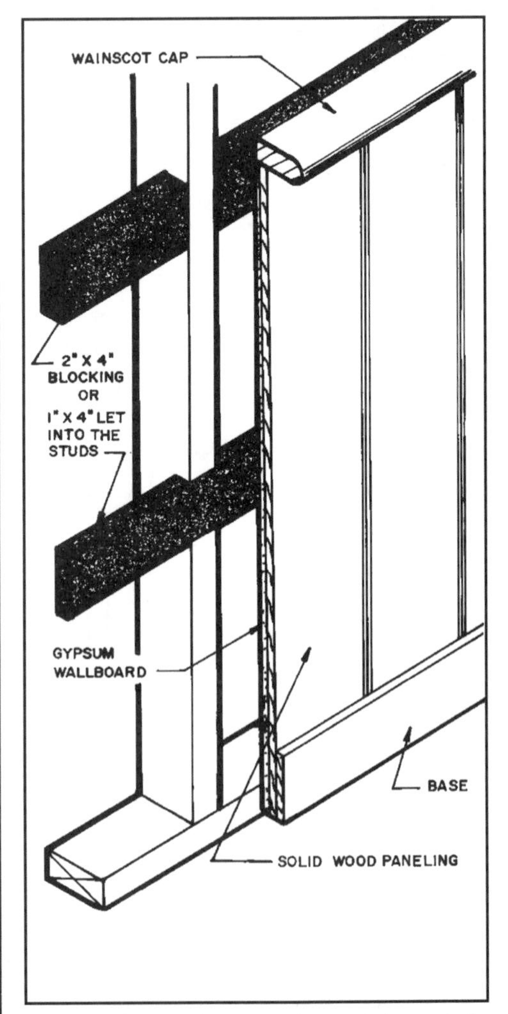

30-57 Blocking between the studs is needed to secure the wainscot and wainscot cap.

Now install pieces of paneling across the wall. Every now and then measure the distance to the other wall. If the top and bottom measurements are close to the same, continue on across the wall. If it appears that a large difference is developing, plane a very slight taper on each piece of paneling, thus reducing the size of the angle that will be needed on the last piece. This is not an easy thing to do and takes some experience. If the last piece has some slope on the butting end in an inside corner, it might be left because this is not too noticeable.

As you proceed across the wall, you need to measure the remaining distance to the corner. Since you know the width of the paneling and the distance left, you can find the exposed width of the last piece. The makeup of the corner will also influence this width.

If the space left is large and a near-full-width panel is desired on the end, it will be necessary to plane a small amount off the grooved edge. It may be necessary to re-bevel the edge to match the original designs. If the butting wall is a little out of plumb, it will be necessary to scribe the last piece and cut it on a slight taper to form a closed joint. Cut it a little too wide, and get a close fit by smoothing the edge with a hand plane, power plane, or jointer. The jointer will give the straightest edge. An exaggerated illustration is shown in **30-59**. If the work is carefully done, the actual taper will be small. The procedure for scribing the last piece is shown earlier in **30-35**.

Nail the paneling as described earlier for solid-wood paneling. This includes leaving a 1/16-inch (1.5mm) gap between panel edges and around 3/8 inch (9.5mm) at the floor. Nail through the tongue and face-nail wide boards. The last board in the corner will be totally face nailed.

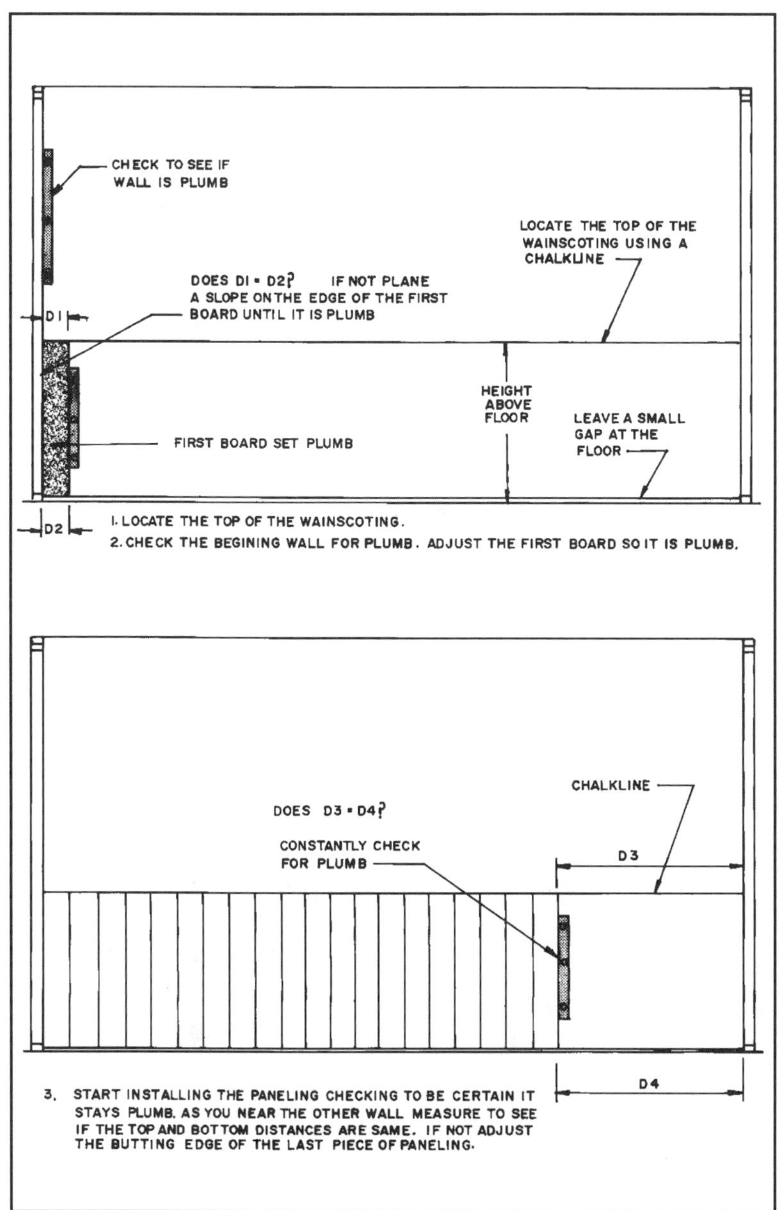

30-58 The steps to lay out and install vertical wood wainscoting.

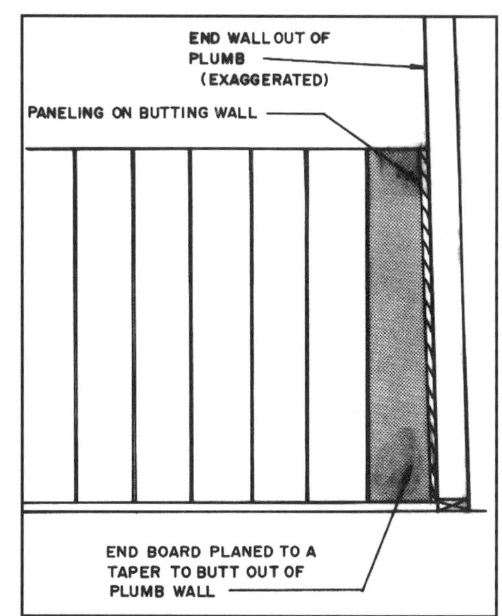

30-59 When necessary, scribe and plane the last board to a taper so that it butts the end wall.

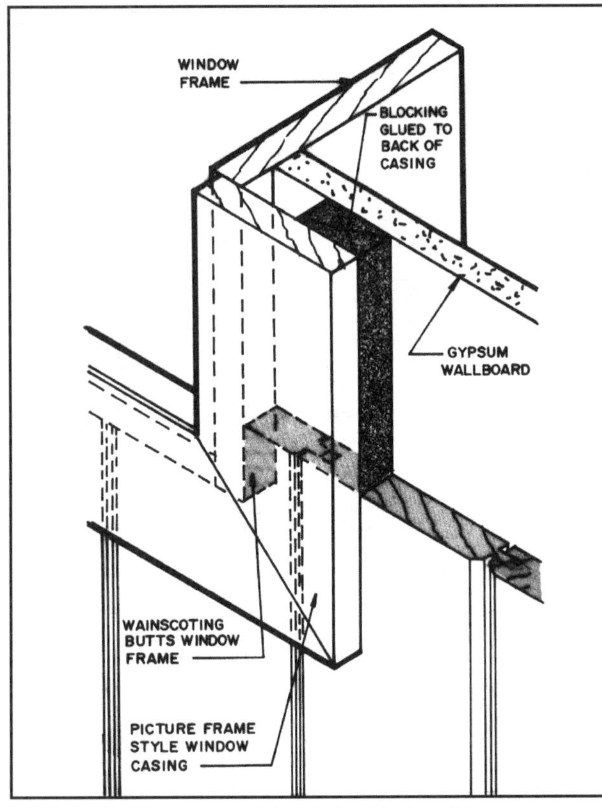

30-60 If the wainscoting butts the window frame, blocking is needed behind the casing on all sides of the window.

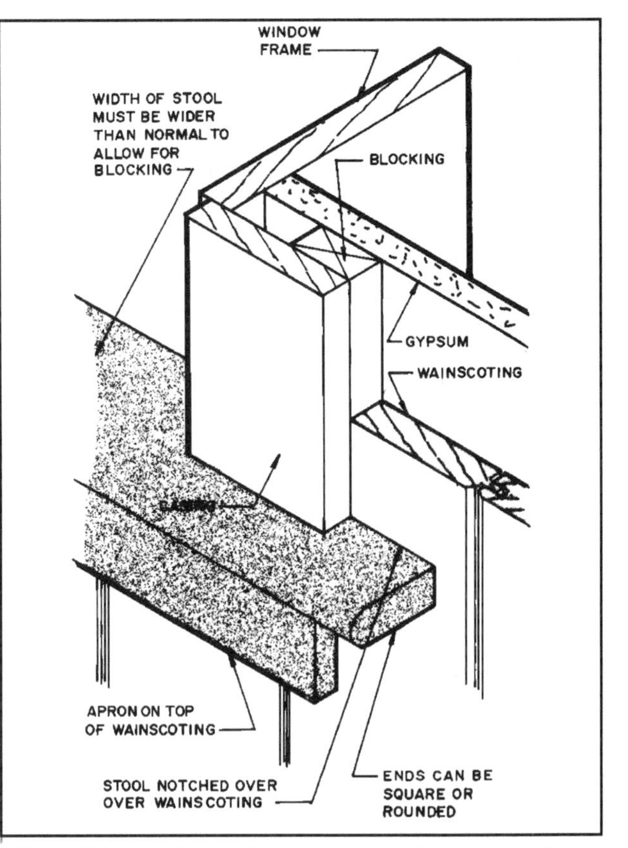

30-61 When a stool and apron are used they are installed over the wainscoting.

The inside and outside corners are made as shown earlier in **30-11**. Remember to locate and cut openings for electric outlets and other items penetrating the paneling.

At the floor the wainscoting can be covered with a base or can butt a 1 × 6-inch (250 × 150mm) or 1 × 8-inch (25 × 200mm) board which may have a shoe molding at the floor. Refer back to **30-15**.

When the wainscoting meets the door or window casing, it can butt it as shown earlier in **30-46**. The casing should be thicker than the paneling so that a slight reveal is exposed. The casing can be

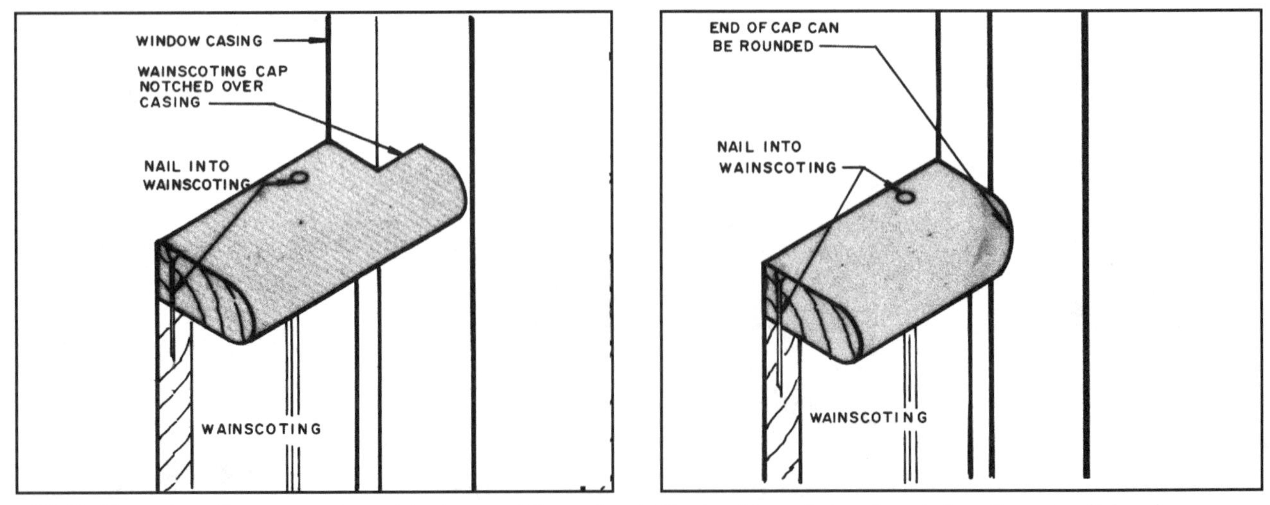

30-62 The wainscoting cap can be notched around the casing, butted and rounded, or beveled on the extended end.

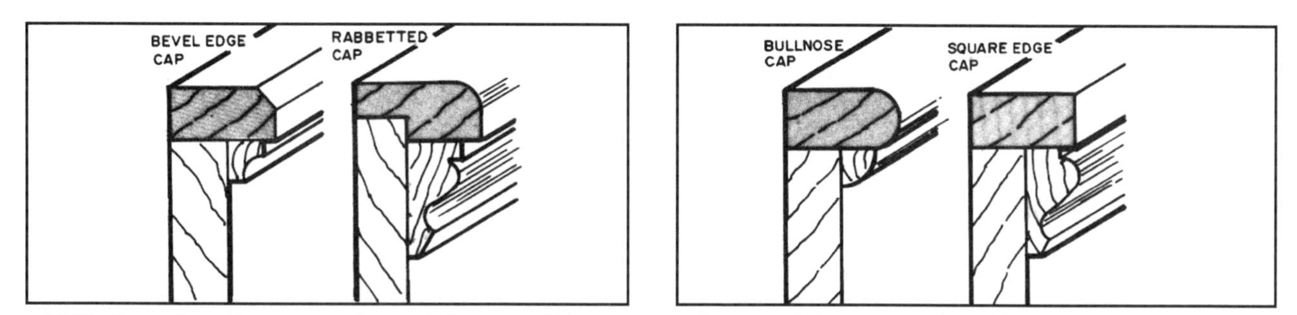

30-63 The wainscoting caps can be made in many different ways using any of the many stock moldings available.

the same thickness as the paneling, but this is not as attractive.

Another technique is to install the paneling before the casing is applied and let it overlap the window frame. The casing is then applied on top of the paneling. This requires that blocking be installed behind the casing above the paneling to fill the void that is created by the thickness of the paneling. This can be nailed to the window frame or glued and nailed to the rear of the casing. This permits the use of thinner casing **(see 30-60)**.

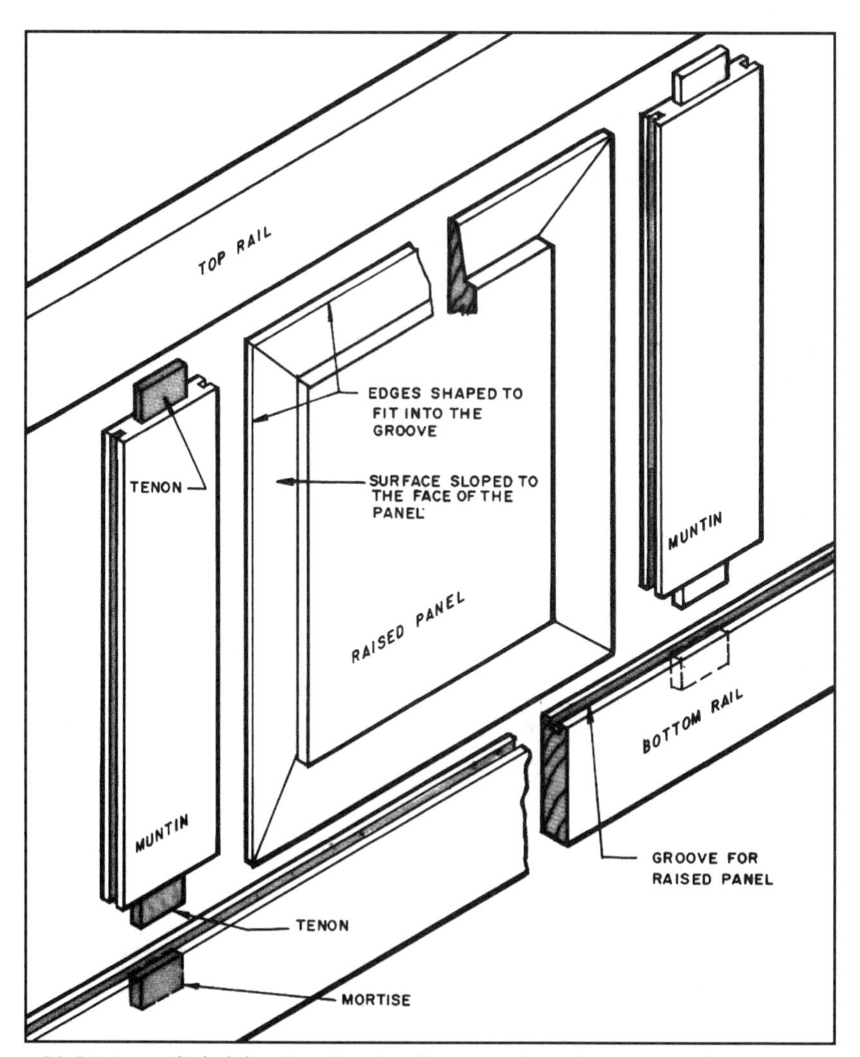

30-64 An exploded drawing showing the parts of a wainscoting made by using an assembly of raised panels and solid wood rails and muntins.

If the window has a stool, generally the wainscoting is installed before the casing is applied. Then cut and install the stool, notching it over the casing. The casing and apron are nailed over the paneling **(see 30-61)**.

The wainscoting cap is usually notched over the door or window casing, as shown in **30-62**. It could be beveled or rounded, but the notched end is most attractive. The wainscoting cap can be specially milled to the specifications of the designer or made up of stock moldings **(see 30-63)**. The wainscoting cap is nailed through the top of the cap into the top edge of the paneling. The baseboard is installed over the paneling in the normal manner.

PANELED WAINSCOTING & WALLS

Paneled wainscoting and wall panels are fabricated by machining solid-wood raised panels to fit into solid-wood frames **(see 30-64)**. Some are made by skilled cabinetmakers or finish carpenters with the highest woodworking skills and good woodworking equipment.

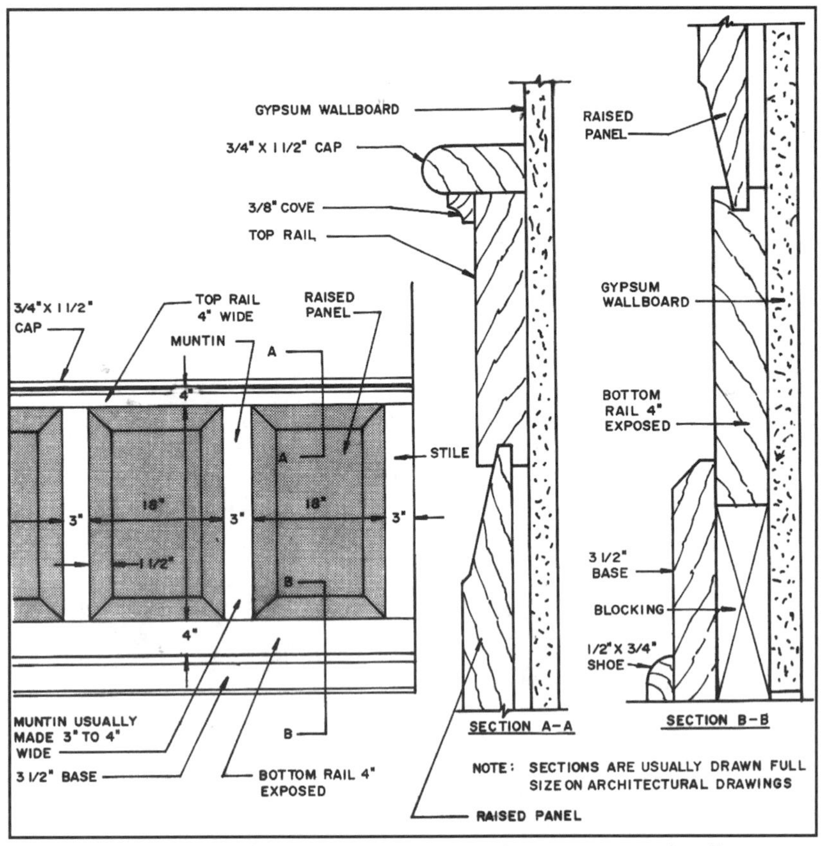

30-65 Typical construction details for raised panel wainscoting and walls.

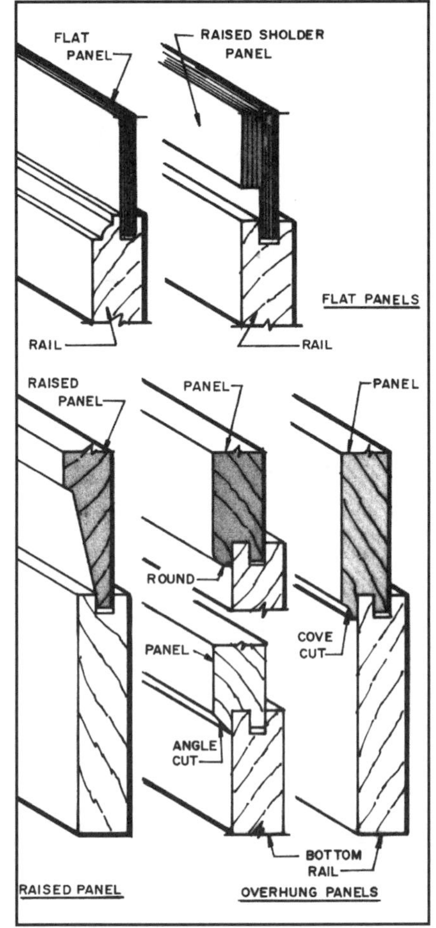

30-66 Some typical wall-panel construction.

The working drawings should show an **elevation** (a projection of the structure or an object onto a vertical plane) of each wall to be paneled and give **sections** through the panels to show the profiles required and construction details. A typical example is shown in **30-65**. Before starting to make the panels the carpenter or cabinetmaker must go to the job site and measure the actual distances the panels will cover.

The design of wall panels can take many forms. A few are shown in **30-66**. It is desirable that outside corners be mitered, since these panels are of high-quality woods, usually hardwoods, and the very best joinery is expected. A butt joint would be satisfactory if the exposed edge grain is not considered objectionable. Inside corners should be designed so that the exposed stile on the butting panels is the same width. Since the panels are designed for a particular space, the let-in panels can be adjusted in size as the unit is made to allow the **stiles** and **muntins** to be the same size. The inside corner could be a butt joint, but a better job is had if a tongue-and-groove joint is cut. Examples are shown in **30-67**.

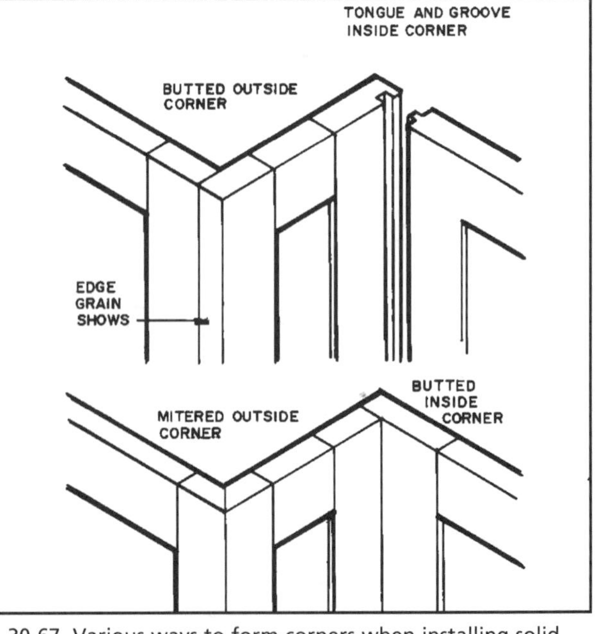

30-67 Various ways to form corners when installing solid wood panel wainscoting and wall panels.

The method of joining the stiles, muntins, and rails may be specified by the designer. This is commonly done using dowels, mortise and tenons, or oval biscuits (also called plates) installed with a biscuit joiner. These are shown in **30-68**.

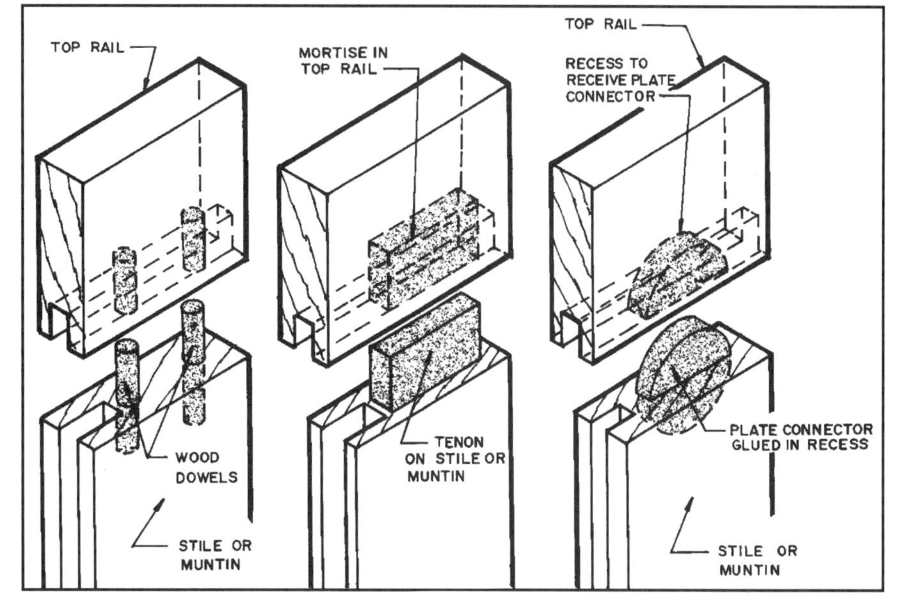

30-68 Methods used to join wood frames.

INSTALLING PANELED WAINSCOTING & WALL PANELS

This procedure is much the same as that described for wood paneling, except that the paneled units are long and heavy. Measure in one corner the desired height of the panel. Snap a chalkline across the wall, producing a level line for the top of the panel. Allow room so that the bottom of the panel will clear the floor. Set the first panel in place. Which panel will be "first" depends on how the joints at the corners were designed. The panel with the wider stiles for inside corners must be installed first.

Check the walls that the panel will butt for plumb. It may be necessary to plane a little off the stile, if the wall is out of plumb.

To set the panel, place it on the wood blocks to allow the floor gap desired. Nail through the top rail into the studs. If all goes well, these nails will be covered by wainscoting cap or ceiling molding. Nail through the bottom rail into the bottom plate about every 16 inches (approx. 410mm). These nails should be covered by the baseboard. The end stiles can be face-nailed into the corner studs. These can be set and filled for paint-grade work. If it is hardwood with a stained or natural finish, the nails can be set and covered with colored fillers as used on furniture construction. Wood screws could be used. They can be set in a counterbore and covered with wood plugs cut from the same wood used to make the panels **(see 30-69)**.

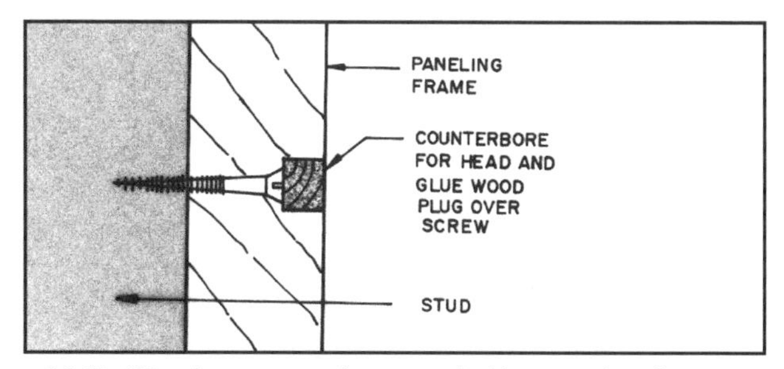

30-69 Wood screws can be concealed by covering them with a wood plug.

Once the panels are installed, the baseboard, ceiling molding, and wainscoting cap are installed, if required.

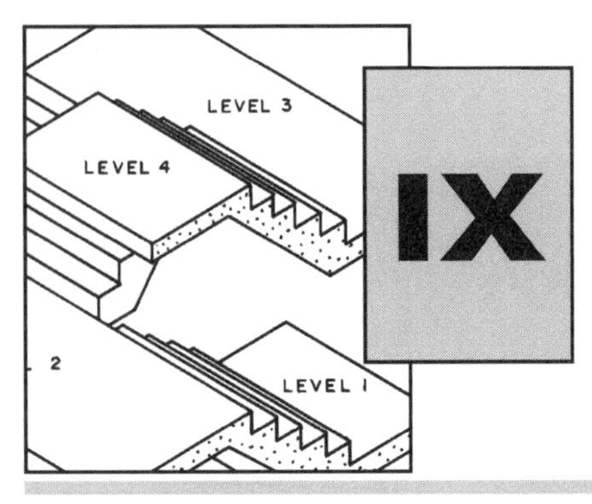

IX FLOORS & STAIRS

CHAPTER 31

Installing Wood Floors

Wood flooring is available as a solid wood or laminated wood product. Solid wood is available prefinished in the factory or unfinished—to be sanded and finished after installation. Site-finished wood floors are typically finished with polyurethane, and prefinished wood floors are finished with acrylic. Laminated flooring is prefinished.

Wood flooring is available in hardwoods and softwoods, in a variety of species. The most commonly used hardwoods are oak and maple although beech, birch, and pecan are also used. The most frequently used softwoods are yellow pine and Douglas fir **(see 31-1)**.

31-1 Wood flooring provides a warm, attractive finished floor. *(Courtesy Harris-Tarkett, Inc.)*

THE SUBFLOOR

The subfloor will usually be plywood or oriented strand board but some prefer to use solid boards. Whatever is used, check it to be certain it is even and all nails are driven in place. Uneven areas should be planed or sanded smooth. If the floor has any bounce it must be renailed, or better yet, screwed to the joists. Sometimes it gets wet as the building is being erected. It must be dry. If it has started to deteriorate because of the rain it should probably be replaced. Hopefully the subfloor was glued and nailed to the floor joists. If glue was not used it will most likely begin to squeak in a few years.

GRADES OF SOLID WOOD FLOORING

The various grades of solid wood flooring are established by various manufacturers' associations. Such grades are shown in **31-2**. The grade determines the appearance of the exposed surface. All grades are equally strong and can be used with confidence of structural integrity. Wood flooring is available unfinished or with a factory-applied finish.

STORING ON THE SITE

Solid wood flooring should not be delivered to the job site until the structure is weather tight and concrete work, drywall, plaster, and other humidity-raising installations are dry. The subfloor and framing members must have a moisture content below about 12 to 14 percent before the flooring is delivered. In the winter the building should be heated to 65 to 70 degrees F (21 degrees C). Operation of the air-conditioning unit is helpful to control temperature and moisture. Various types of laminated wood flooring have even more restrictive requirements. Consult the manufacturer's specifications for proper storage and installation temperature and humidity requirements.

It helps if the area below the floor, such as a basement or crawl space, is dry and well ventilated. The bare ground in a crawl space must be covered with a six-mil polyethylene vapor barrier.

The bundles of unfinished flooring should be broken open and spread about in the rooms in which they are to be installed. Flooring material should be laid flat on the subfloor and stored in this manner at least three days before the start of installation so that it can adjust to the temperature and humidity in the building.

Some types of prefinished flooring are shipped shrink-wrapped in plastic and are therefore protected from moisture. The shrink wrap should be kept intact until the flooring is installed; it is not necessary to have any on-site acclimation time. Consult the manufacturer's directions usually accompanying each shipment.

PROTECTION AFTER INSTALLATION

To assure years of satisfactory use of the hardwood flooring, the flooring contractor should be certain that the general contractor provides a permanent, controlled atmosphere to protect the

31-2 Flooring grades for oak, maple, and southern pine.

Hardwood Flooring Grades of the National Oak Flooring Manufacturers Association

Unfinished Oak Flooring	Beech, Birch, and Hard Maple	Pecan	Prefinished Oak Flooring
Clear Plain or Clear Quartered—best appearance	First Grade White Hard Maple—face all bright sapwood	First Grade Red—face all heartwood	Prime—excellent
Select and Better—mix of Clear and Select	First Grade Red Beech—face all red heartwood	First Grade White—face all bright sapwood	Standard and Better—mix of Standard and Prime
Select Plain or Select	First Grade—best	First Grade—excellent	Standard—variegated
Quartered—excellent appearance	Second and Better—excellent	Second Grade Red—face all heartwood	Tavern and Better—mix of Prime, Standard, and Tavern
No. 1 Common—variegated	Second—variegated	Second Grade—variegated	Tavern—rustic
No. 2 Common—rustic	Third and Better—mix of First, Second, and Third	Third Grade—rustic	

Softwood Flooring Grades of the Southern Pine Inspection Bureau	Maple Flooring Grades of the Maple Flooring Manufacturers Association
Grades	**Grades**
B and Btr—best quality	First—highest quality
C and Btr—mix of B and Btr and C	Second and Better—mix of First and Second
C—good quality, some defects	Second—good quality, some imperfections
D—good economy flooring	Third and Better—mix of First, Second, and Third
No. 2—defects but serviceable	Third—good economy flooring

flooring. The building should be built so that the basement or crawl space will remain dry. This can be assured by waterproofing the foundation, sloping the soil away from the building, installing drain tile around the inside of the foundation, and providing lots of ventilation for the area under the floor. In addition the installed flooring must be protected from excess heat generated by a furnace below the floor or by uninsulated heating ducts. One solution is to install ½-inch (12mm) or thicker insulation board between the joists. Anything installed over a furnace must, of course, be a nonflammable material.

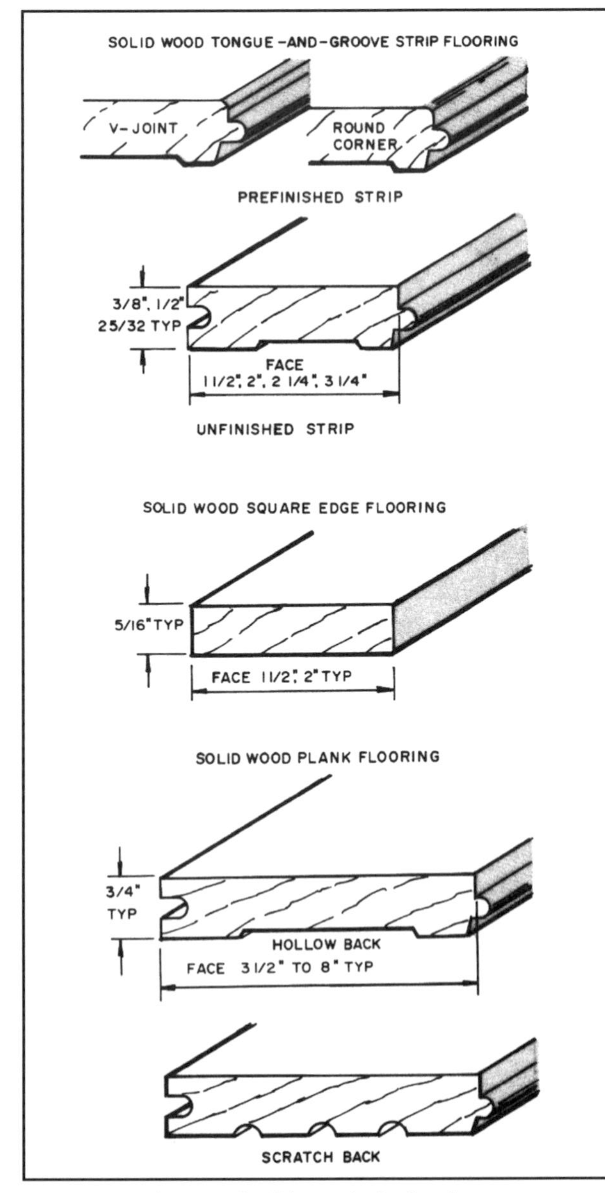

31-3 Typical types of solid woodstrip flooring.

TYPES OF WOOD FLOORING

Wood flooring is available in solid wood strips and planks, laminated strips and planks, and parquet wood blocks.

WOOD STRIP FLOORING

Wood strip flooring is available as solid wood or laminated products. **Solid wood flooring** is available as tongue-and-grooved strips or square-edged strips **(see 31-3)**. The center of the bottom surface on tongue-and-grooved strips is recessed. This provides two areas to support the strip. The center recess will prevent any minor bumps from causing the strip to rock.

Solid wood strip flooring with tongue-and-grooved edges (side-matched) also has tongue-and-grooved ends (end-matched) as shown in **31-4**. It is available prefinished or unfinished. The unfinished strips have square, tight-fitting edges while the prefinished strips have a V-joint or are rounded on the edges. Strip flooring is available in widths from 1½ to 3¼ inches (38 to 82mm). The back may be hollow back or scratch back.

Laminated Wood Flooring

Laminated wood flooring is manufactured by gluing together very accurately produced wood veneers **(see 31-5)**. It is glued up in layers like plywood. Therefore it does not expand and contract with changes in moisture and temperature as much as solid wood flooring. This means that fewer and smaller gaps will occur between the strips of flooring. It is also less likely to warp and cup.

The manufactured product looks like solid wood because it is made in the same widths and has the top veneer one of the commonly used woods. It is available in thicknesses from ⅜ inch to 9⁄16 inch. The number of plies used varies but the more plies used the more stable the flooring. Three and five plies are commonly used.

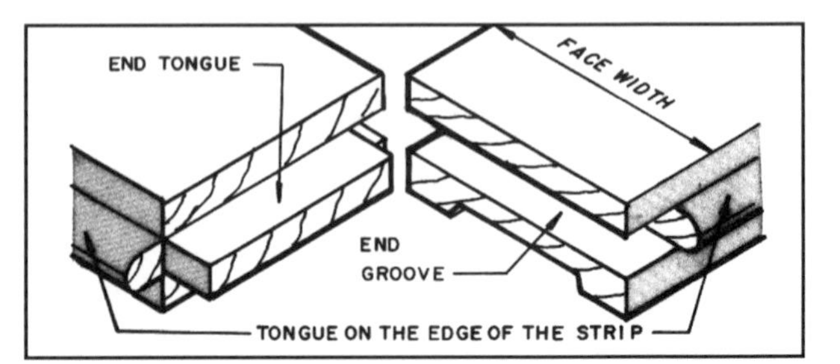

31-4 Strip and plank wood flooring are end matched.

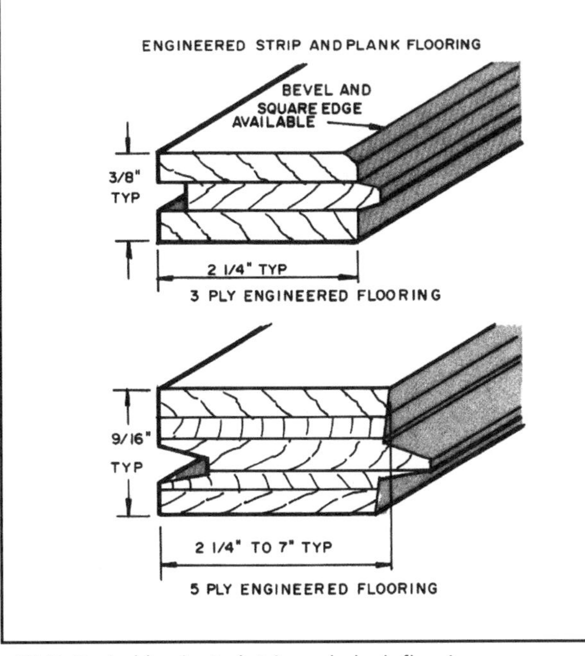

31-5 Typical laminated strip and plank flooring.

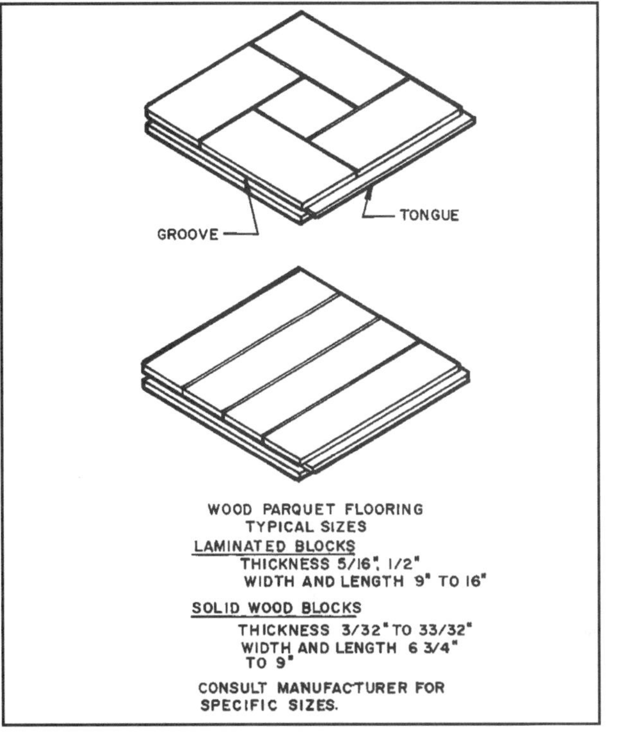

31-6 Typical wood parquet flooring blocks.

Planks

Planks are available as solid wood or laminated products. They range in width from 3½ to 10 inches (8.7 to 20.4cm). Solid wood planks are typically ¾ inch (19mm) thick while laminated planks are thinner, usually ⅜ to 7⁄16 inch (9 to 11mm) thick **(see 31-5)**. Usually planks five inches (12.7cm) wide are edge nailed to the subfloor with screws through the face. These are covered with wood plugs.

Since plank floor boards are wider and have fewer edge joints they tend to develop wider gaps between planks than narrower stripflooring as they age.

Parquet Blocks

Parquet blocks may be made from solid wood or laminated veneers, much like plywood. The solid wood blocks are made from pieces of strip flooring bonded along the edges. The exposed edges are tongue-and-grooved to facilitate installation and produce a tight joint. The laminated blocks are bonded with a moisture-resistant adhesive. Parquet blocks are available in a wide range of sizes, from 3 × 9 inches (7.6 × 23cm) to 12 × 12 inches (30.5 × 30.5cm), as shown in **31-6**. They have a variety of patterns on the exposed face **(see 31-7)**.

31-7 A wood parquet floor. *(Courtesy Harris-Tarkett, Inc.)*

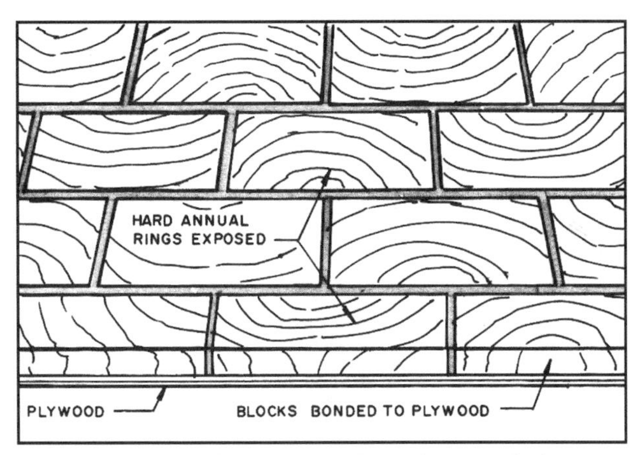

31-8 Wood block floors expose the end grain, which is very hard and makes the floor wear well.

31-9 Inlaid borders and floor panels are made by bonding various hardwoods cut in geometric patterns. *(Courtesy Kentucky Wood Floors)*

31-10 Sweep the subfloor so it is absolutely clean. *(Courtesy National Oak Flooring Manufacturers Association)*

End-Grain Flooring

End-grain flooring is made by cutting thick stock, such as 2 × 8- or 4 × 8-inch members into pieces about ¾ inch (19mm) thick. These are glued to the subfloor, and the spaces left between them are filled with a ground cork filler. End-block flooring exposes the hard annual rings, which withstand hard wear **(see 31-8)**.

Inlaid Panels

Inlaid panels are made by bonding pieces of various wood species cut in geometric shapes together to form a complex, attractive product **(see 31-9)**. Companies manufacturing these panels have a choice of stock designs but can produce a custom design when requested. They also create narrow multipiece products that are used as borders within a wood floor. When installing, follow their directions carefully. The panels are applied with a mastic.

Plastic Laminate Flooring

Plastic laminate flooring has a top surface made from a high-pressure decorative laminate. This is bonded to a core of medium-density fiberboard. It is discussed in Chapter 32.

LAYING SOLID WOOD STRIP FLOORING

After inspecting the subfloor and making certain it is satisfactory, sweep it to remove all debris **(see 31-10)**. Then cover the subfloor with 15-pound asphalt-saturated building paper, lapped four to

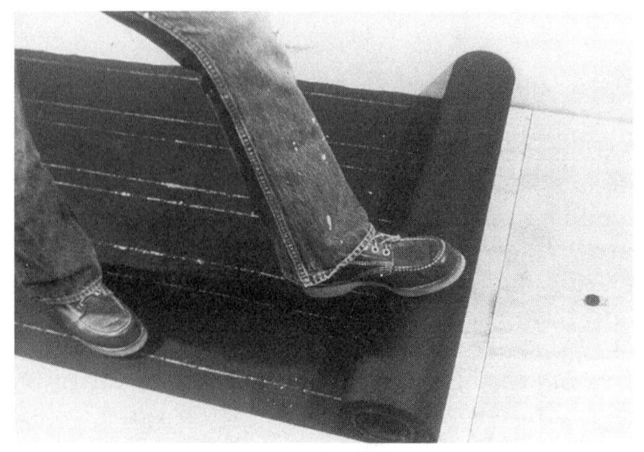

31-11 After sweeping the subfloor, cover it with 15-pound asphalt-saturated felt. *(Courtesy National Oak Flooring Manufacturers Association)*

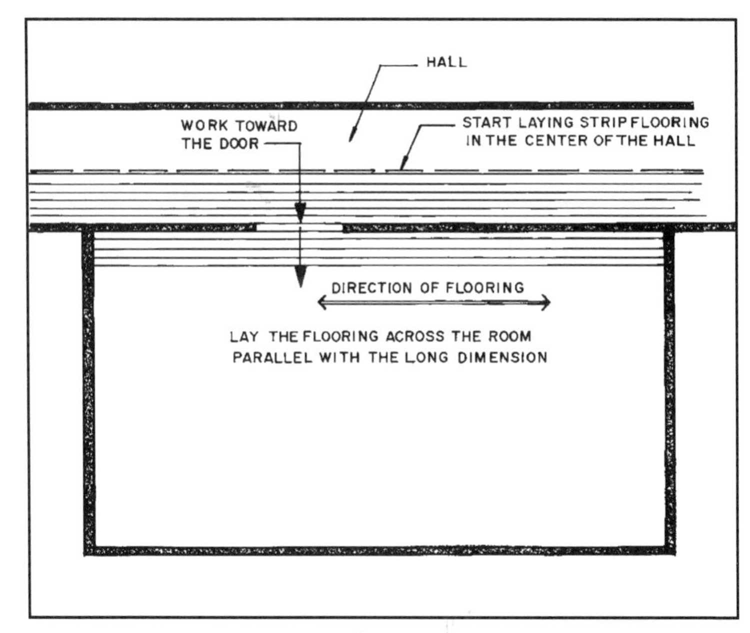

31-12 When the flooring is also in the hall, start in the center of the hall and lay the flooring into the room.

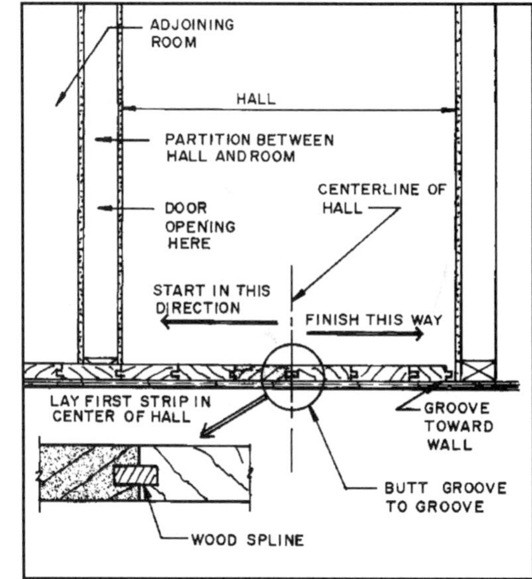

31-13 Use a wood spline in the groove of the starting strip as a tongue for working the tongue-and-groove flooring towards the other wall.

six inches (100 to 152mm) on the sides and ends. Staple it to the subfloor. This helps keep dust from penetrating the floor and serves as a moisture barrier **(see 31-11)**.

Locate and mark the position of each floor joist on the builder's felt so that the nails are certain to penetrate into them. This can be done by snapping a chalkline at each joist.

Strip flooring gives the best appearance if it runs parallel with the longest wall of the room, and it is best if it runs perpendicular to the floor joists so that it can be nailed through the subfloor into the joist. However, if the long wall runs parallel with the joists, the strip flooring can be nailed to the subfloor alone if it is ¾-inch (18mm) plywood. If the subfloor is ½-inch (12mm) plywood, the strip flooring will have to be run perpendicular to the joists even if it parallels the short wall of the room.

Also you must consider how the direction of the flooring may affect the floor in a hall or other room opening onto it. Plan for the strip flooring to work through a door and into a hall when the hall is also to have strip flooring **(see 31-12)**. If the hall parallels the long dimension of the room and the flooring is installed parallel with the long dimension, start by installing the flooring in the middle of the hall and working towards the door of the room. Snap a chalkline down the middle of the hall to get started. Place the groove in the flooring towards the centerline. Lay the flooring across the hall, through the door, and into the room. To finish the hall, insert a spline in the groove, and lay the flooring to the other wall **(see 31-13)**.

If it is desired to change the direction of the flooring in an abutting room, it can butt a wood sill made from flooring or a marble sill **(see 31-14)**. Cut off the groove or tongue so that the flooring butts a solid, square edge. Face-nail each piece to the subfloor.

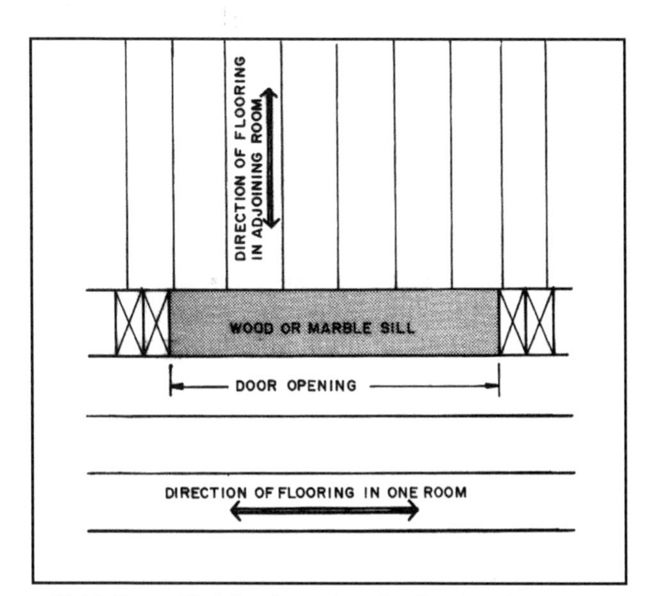

31-14 Use a sill at the door when the flooring changes directions from one adjoining room to another.

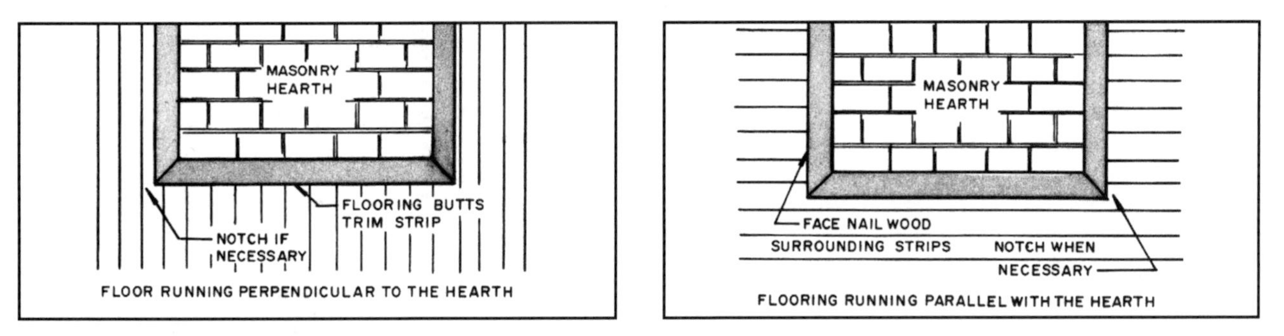

31-15 How to install wood flooring when it meets a hearth or other protrusion into the room.

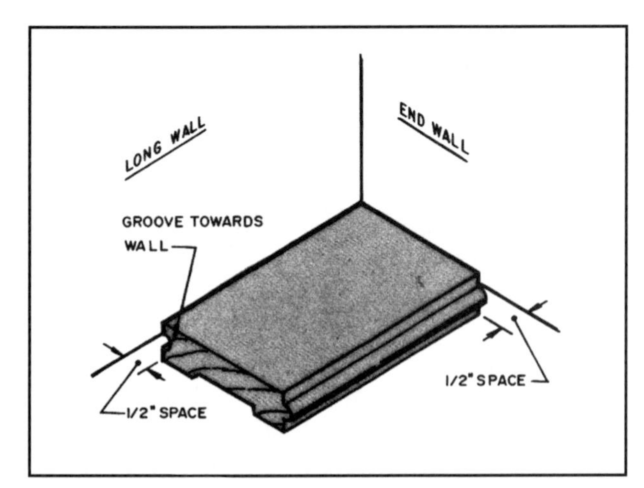

31-16 Set the first strip of wood flooring away from the wall to allow for expansion.

If the floor butts a fireplace hearth or some other feature protruding into the room install solid wood strips around the edge to provide a butting surface. A four- to six-inch (102 to 152mm) wide surrounding strip is commonly used. Miter the corners to give a finished appearance. It will usually be necessary to notch the piece of flooring that butts this trim strip **(see 31-15)**.

Starting to Lay Strip Flooring

The first strip should be placed ½ inch (12mm) from the wall, with the groove towards the wall **(see 31-16)**. The gap allows for expansion of the floor and will be covered by the base and shoe molding.

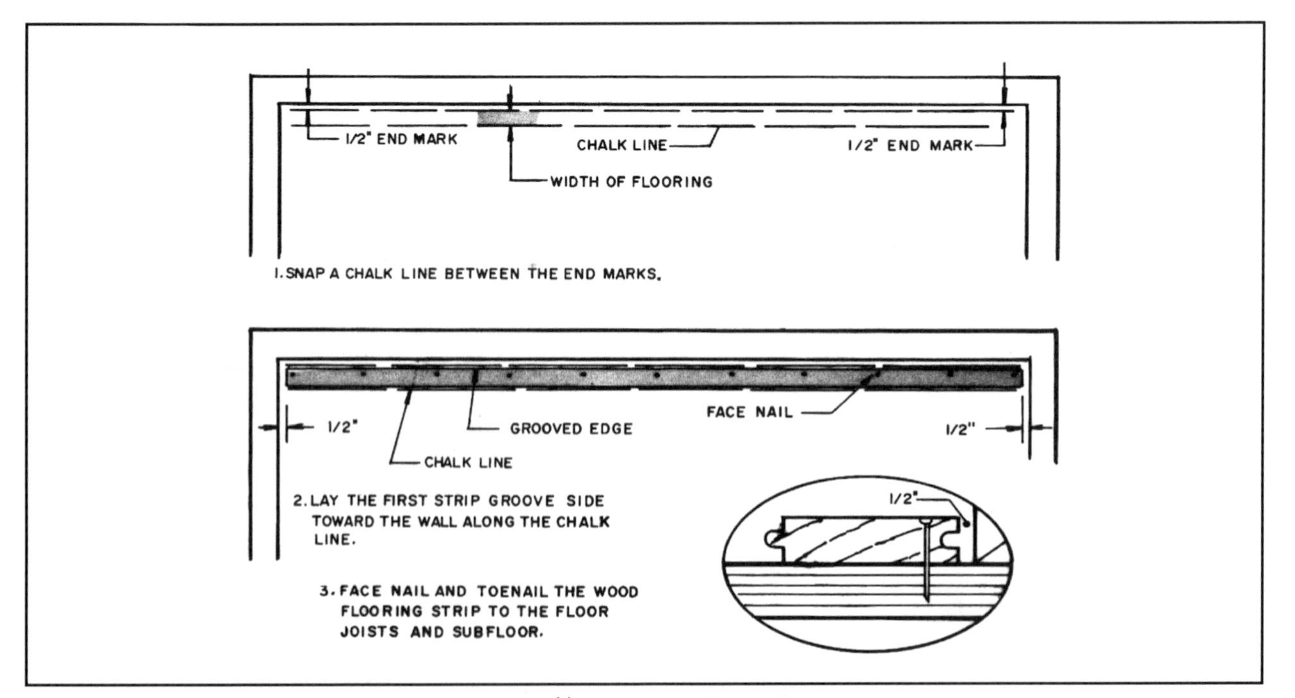

31-17 Install the first strip parallel with the wall and ½ inch (12.5mm) away from it.

1. Begin by measuring ½ inch (12mm) plus the width of the flooring from the wall on each end of the room to establish a baseline for the first strip. Snap a chalkline from one end of the room to the other **(see 31-17)**.
2. Check the room to see if the walls are parallel. If they are not the gap at one wall will most likely not be covered by the base. To allow for this find out how much the walls are out of parallel. Allow half of this on each end. For example, if the walls are one-inch (25mm) out of parallel measure out a distance equal to the gap (½ inch) and the ½ inch needed to straighten the flooring as shown in **31-18**.

This will leave a gap of one-inch (25mm) at one end of the wall. If the base and shoe or carpet will not cover this, cut the drywall ¾ inch (19mm) above the floor and set the first strip under it but ½ inch (13mm) from the bottom plate.

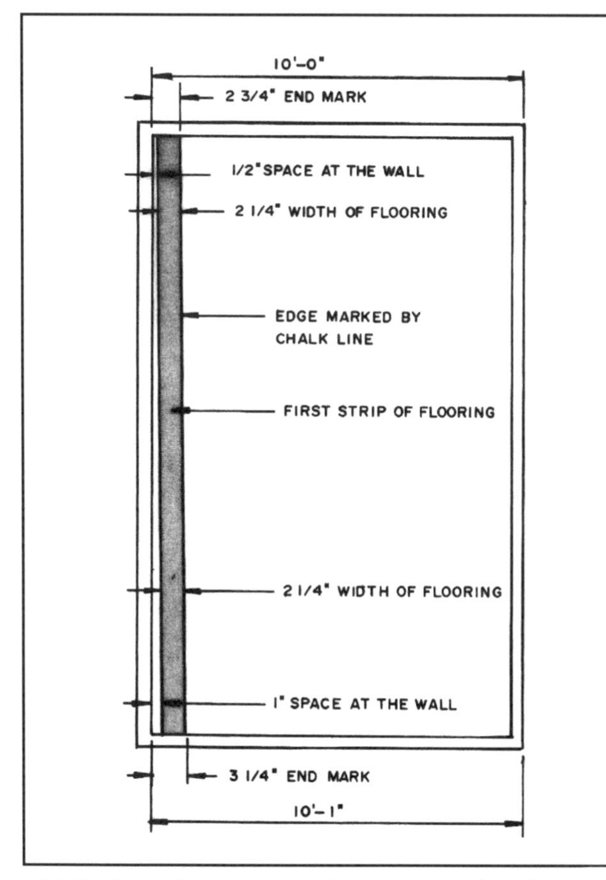

31-18 When the opposite walls are not parallel adjust the first strip of flooring by adding half the amount they are out of parallel to the space allowed (½ inch) at the wall on the wide end.

3. Place the first strip with the edge along the chalkline. Be certain it is parallel with the wall. Face nail into the floor joists, and place one nail between joists using 8d finish nails. If there is a danger of splitting the wood, predrill each hole. These nails should be close enough to the edge to be covered with the shoe molding **(see 31-19)**.
4. Blind-nail all the rest of the strips through the tongue into the floor joist, as shown in **31-20**. The nail is driven just above the tongue and on about a 45-degree angle.

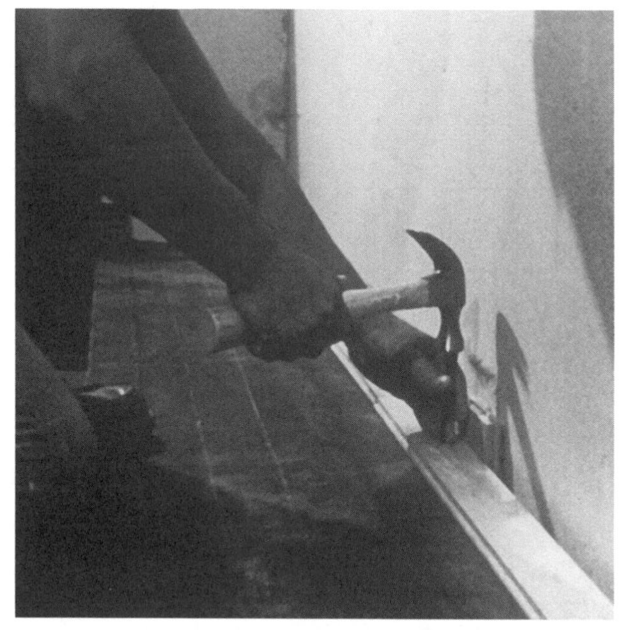

31-19 Face-nail the first strip into the floor joists and at the end of the strips. *(Courtesy National Oak Flooring Manufacturers Association)*

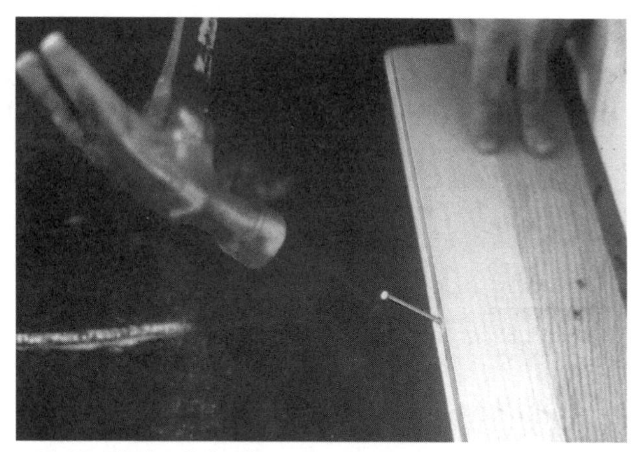

31-20 Blind-nail the flooring strips at the tongue into the floor joists and set the nail. *(Courtesy National Oak Flooring Manufacturers Association)*

Do not try to drive the nail flush with a hammer, because you will most likely damage the edge of the flooring. The nail can be set using a nail set **(see 31-21)**.

Nails can be driven using a hammer or half hatchet **(see 31-22)**. The carpenter stands with the feet on each side of the nail location and leans over to drive the nail. A nailing machine is often used. It has a lip that fits under the tongue. The carpenter drives the nail by hitting the nailing block on the end of the machine with a heavy mallet **(see 31-23)**. Another tool is a flooring stapler that is operated by compressed air. It will hold 100 staples which are driven through the top of the tongue **(see 31-24)**.

5. Select the proper nails. A suggested nailing schedule is shown in **31-25**.

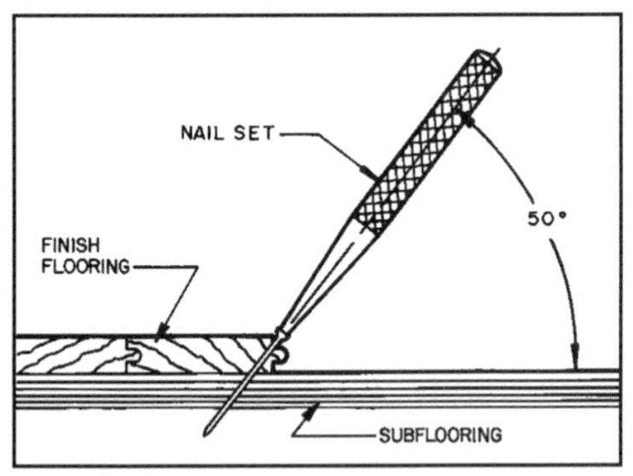

31-21 Set the nails into the strip flooring with a nail set.

31-22 To hand-drive nails in strip flooring, straddle the spot to be nailed.

31-24 This is a compressed-air-driven hardwood floor stapler. *(Courtesy Stanley-Bostitch, Inc.)*

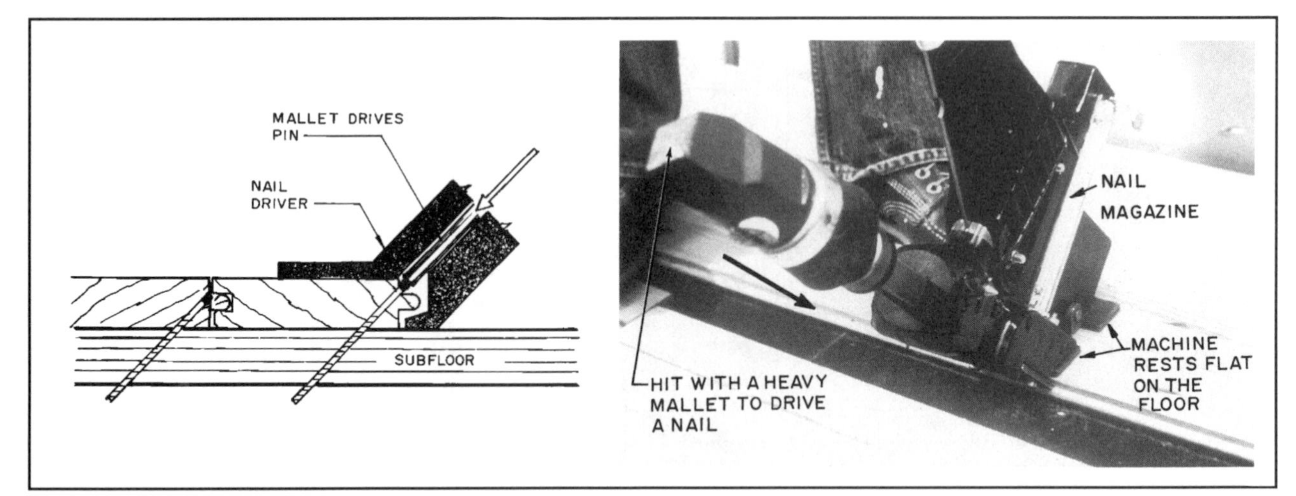

31-23 This mallet-activated strip wood floor nailing machine blind-nails through the top of the tongue. *(Courtesy National Oak Flooring Manufacturers Association)*

31-25 A suggested nailing schedule for solid-wood flooring.
(Courtesy National Oak Flooring Manufacturers' Association)

NOFMA hardwood flooring must be installed over a proper subfloor.
Tongue-and-grooved flooring MUST be blind-nailed;
Square-edged flooring must be face-nailed.

A slab with screeds 12" O.C. does not always require a subfloor.

Strip T & G Size Flooring	Size Nail to Be Used	Blind-Nail Spacing along the Length of Strips. Minimum two Nails per Piece near the Ends (1"–3")
¾" × 1½", 2¼", & 3¼"	2" serrated-edge barbed fastener, 7d or 8d screw or cut nail, 2" 15-gauge staples with ½" crown. On slab with ¾" plywood subfloor use 1½" barbed fastener.	In addition—10–12" apart—8–10" preferred For ½" plywood subfloor with joists at a maximum 16" O.C., fasten into each joist with additionalfastening between, or 8" apart
	MUST Install on a Subfloor:	
½" × 1½" & 2"	1½" serrated edge barbed fastener, 5d screw, cut steel, or wire casing nail.	10" apart ½" flooring must be installed over a MINIMUM ⅝" thick plywood subfloor.
3¼8" × 1½" & 2"	1¼" serrated edge barbed fastener 4d bright wire casing nail	8" apart
Square-edge Flooring 5⁄16" × 1½" & 2"	1" 15-gauge fully barbed flooring brad	two nails every 7"
5⁄16" × 1⅓"	1" 15-gauge fully barbed flooring brad	one nail every 5" on alternate sides of strip
Plank ¾" × 3" to 8"	2" serrated-edge barbed fastener, 7d or 8d screw, or cut nail. Use 1½" length with ¾" plywood subfloor on slab.	8" apart

Follow manufacturer's instructions for installing plank flooring.

Widths 4" and over must be installed on a subfloor of ⅝" or thicker plywood or ¾" boards. On slab use ¾" or thicker plywood.

For additional information, write to:
National Oak Flooring Manufacturers Association,
P.O. Box 3009, Memphis, TN 38173-0009

6. Next lay out several rows of flooring **(31-26)**. Select long pieces, and arrange them so that the end butt joints are at least 6 inches (152mm) apart. It is recommended that they be farther apart than this when possible **(see 31-27)**. Fit the second strip to the first, and position the end ½ inch (12mm) from the wall. Drive the piece tightly against the first strip, being certain the joint closes. To close the joint place a piece of scrap flooring next to it, and tap it with your hatchet so as to not damage the piece **(see 31-28)**. The joint can also be closed by tapping a ripping bar into the subfloor and using it as a pry **(see 31-29)**.

A tool that speeds up the installation of hardwood flooring is shown in **31-30**. It jacks boards into place and holds them while they are being nailed.

After the second piece is in place, blind-nail it through the tongue. Space the nails as shown in the nailing schedule on the previous page **(refer to 31-25)**. Generally you will put one nail into each joist and one nail in between joists into the subfloor.

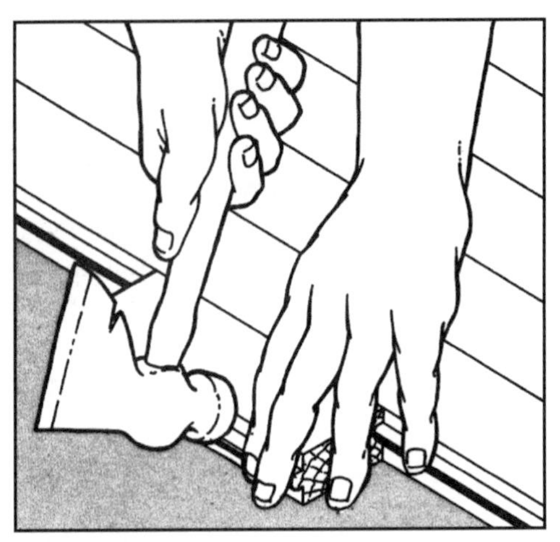

31-28 Flooring strips can be tapped into place with a scrap block and a hatchet.

31-29 A pry bar can be used to pull flooring strips into place. *(Courtesy Cepco Tool Company)*

31-26 Lay out several rows of flooring and arrange so the end joints fall 6 inches or more between adjacent runs. *(Courtesy National Oak Flooring Manufacturers Association)*

31-27 This southern pine flooring has staggered end joints. Notice it has the scratch back and tongue-and-grooved edges. *(Courtesy Southern Pine Council)*

31-30 This tool, the Quik Jack, rapidly pulls bowed flooring strips into place and holds them while they are being nailed. *(Courtesy Cepco Tool Company)*

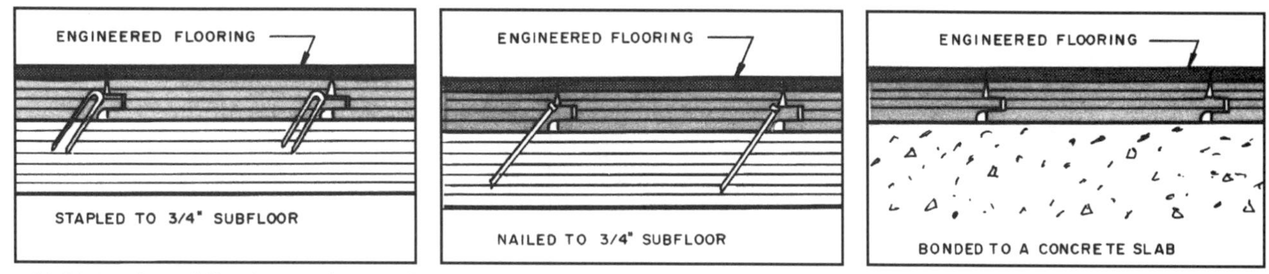

31-31 Laminated flooring can be installed to wood subfloor or a dry concrete slab.

7. Continue working across the room, being careful to space the end joints so that the overall appearance is attractive. Avoid bunching up end joints or several short pieces. Keep short pieces of flooring for use in closets. As you get to the other side of the room, remember to allow a ½-inch (13mm) gap along the wall. This may require that the last piece be ripped to width to fit the space remaining. It will be necessary to face-nail the last two or three rows.

INSTALLING LAMINATED WOOD FLOORING

Laminated wood flooring can be nailed or stapled to any subfloor suitable for wood flooring installation **(see 31-31)**. Most types can be glued directly to a concrete slab floor. This is done with an adhesive supplied by the manufacturer. It is spread with a notched trowel and the flooring is pressed into the adhesive. Details for a floating floor installation are in Chapter 32.

INSTALLING PLANK FLOORING

Plank flooring is normally wider than strip flooring, running as wide as 8 inches (20.3cm). It is installed by blind-nailing and then secured with wood screws set in counterbores and covered with wood plugs **(see 31-32 and 31-33)**. The plugs are furnished by the flooring manufacturer. It is installed in the same manner as described for strip flooring. Usually it is made up of planks of several widths. Start the first row with the narrowest boards; then use the next widest, etc. After one row of each width has been laid, repeat the pattern. Manufacturers provide instructions on how they want their product installed.

Use one-inch (25mm) screws for flooring laid over ¾-inch (19mm) plywood on a concrete slab and 1- or 1½-inch (25 or 38mm) screws for flooring laid over plywood subfloor or on screeds bonded to a concrete slab. Screws should be spaced as recommended by the manufacturer.

31-32 Plank flooring is installed by nailing. The screws provide additional fastening and are decorative when covered by wood plugs. *(Courtesy National Oak Flooring Manufacturers Association)*

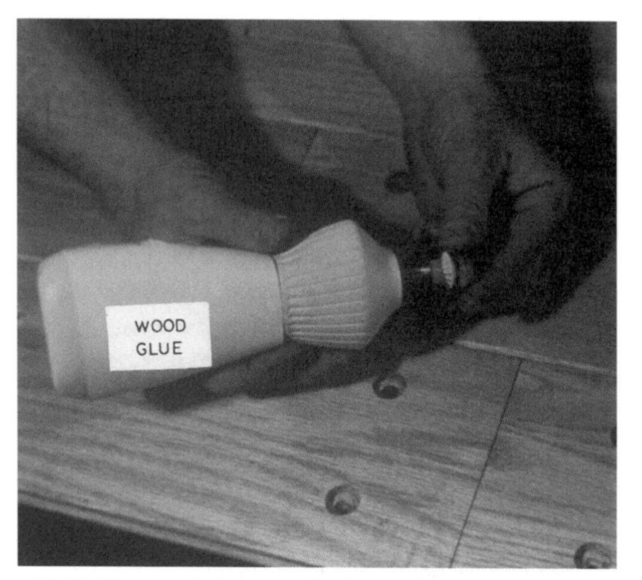

31-33 The wood plugs are glued over the counterbored screws. *(Courtesy National Oak Flooring Manufacturers Association)*

INSTALLING STRIP & PLANKS OVER A CONCRETE SLAB

It is vital that any installation over a concrete slab have a polyethylene vapor barrier on the ground below the slab and on top of the slab below the flooring. Moisture from the earth can penetrate a concrete slab, causing swelling and buckling of the wood floor.

Test the Slab for Moisture

The concrete slab must be tested for moisture before attempting to install wood flooring over it. If it fails the test, either some provision must be made to correct the moisture problem or wood flooring should not be installed over it. The following tests are recommended by the Oak Flooring Institute of the National Oak Flooring Manufacturers Association.

1. **Rubber Mat Test.** Lay a flat, noncorrugated rubber mat on the slab, place a weight on top to prevent moisture from escaping, and allow the mat to remain overnight. If there is "trapped" moisture in the concrete, the covered area will show water marks when the mat is removed. Note that this test is worthless if the slab surface is other than light in color originally.
2. **Polyethylene Film Test.** Tape a one-foot square of heavy, clear polyethylene film to the slab, sealing all edges with plastic packaging tape. If, after 24 hours, there is no "clouding" or drops of moisture on the underside of the film, the slab can be considered dry enough to install wood floors.
3. **Calcium Chloride Test.** Place one-quarter teaspoonful of dry (anhydrous) calcium chloride crystals (available at drugstores) inside a three-inch (7.6cm) diameter putty ring on the slab. Cover with a glass so that the crystals are totally sealed from the air. If the crystals dissolve within 12 hours, the slab is too wet for a hardwood flooring installation.
4. **Phenolphthalein Test.** Put several drops of three percent phenolphthalein solution in grain alcohol at various spots on the slab. If a red color develops in a few minutes, there is a moist alkaline substance present; it would be best not to install hardwood flooring. (This solution is available at drug or chemical stores.)

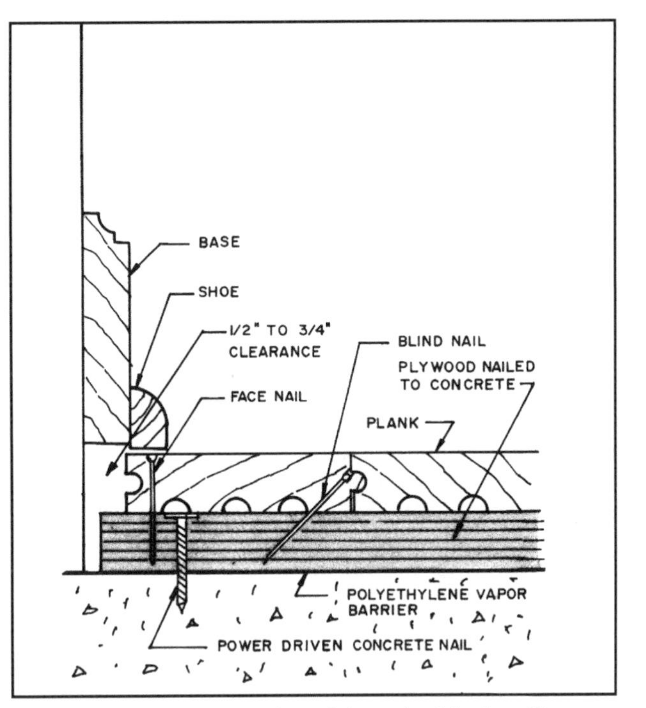

31-34 This is the plywood-on-slab method for installing wood flooring.

Preparing the Concrete Slab

Two ways to prepare the slab to receive residential wood flooring are the **plywood-on-slab** method and a method using **wood screeds**. For commercial construction there are other procedures that will not be discussed here for the installation of wood flooring.

The **plywood-on-slab** method involves placing a 6-mil polyethylene plastic film over the slab, lapping all edges four to six inches (10 to 15cm). It is laid on the slab and not bonded with a mastic. The ¾-inch (19mm) plywood subfloor is laid on top of the polyethylene film and secured to the slab with power-actuated concrete nails. Start nailing the panel in the middle and work out towards the edges. Use at least nine nails per panel. Additional nails will help assure that it will stay flat **(see 31-34)**.

Arrange the panels so that the end joints are staggered every 4 feet (1.2m). Leave a ½-inch (12.5mm) space at the walls and a ¼-inch (6mm) space between the edges and ends of the panels. At places where there will not be base or shoe, such as at a door opening, cut the plywood to fit within ⅛ inch (3mm) of the space.

The **screed method** involves coating the concrete slab with an asphalt primer and allowing it to dry. Then hot or cold asphalt mastic is applied,

31-35 The screed method for installing wood strip flooring over a concrete slab involves embedding wood screeds in a bed of hot asphalt mastic or "cold stick" troweled asphalt mastic. *(Courtesy National Oak Flooring Manufacturers Association)*

covering 100 percent of the floor area. Wood 2 × 4 (5.1 × 10.2cm) screeds 18 to 48 inches (45.7 to 122cm) long are embedded in the mastic. The ends of the screeds are lapped 4 inches (10.2cm) and the ends at the walls have a ¾-inch (1.9cm) gap. Space the screeds 12 to 16 inches (30.5 to 40.6cm) O.C. The wood screeds should be pressure treated, but creosote should not be used **(see 31-35)**.

Next spread a 6-mil polyethylene-film vapor barrier over the screeds. Lap the edges and ends at least 6 inches (152mm). Do not nail or staple the film in place, because this will put holes in it, reducing its moisture-resisting purpose **(see 31-36)**. A vapor barrier should be placed below the slab.

The strip or plank flooring is nailed to the screeds, as discussed earlier. It may be necessary to use a 5d nail, rather than the usual 6d, if there is a danger that it may hit the concrete below the screed. This obstruction would make it difficult to see the nail and to get a tight fit.

INSTALLING PARQUET FLOORING

The procedures for installing parquet flooring will vary somewhat depending on the manufacturer. Instructions accompany the flooring when it is delivered to the job. Basically the same procedures as described for strip flooring, unfinished and prefinished, are used.

An important factor in the installation of parquet flooring is the planning of the preliminary layout. Procedures will differ somewhat depending on the circumstances. The next sections on the following pages cite some of the things that must be considered.

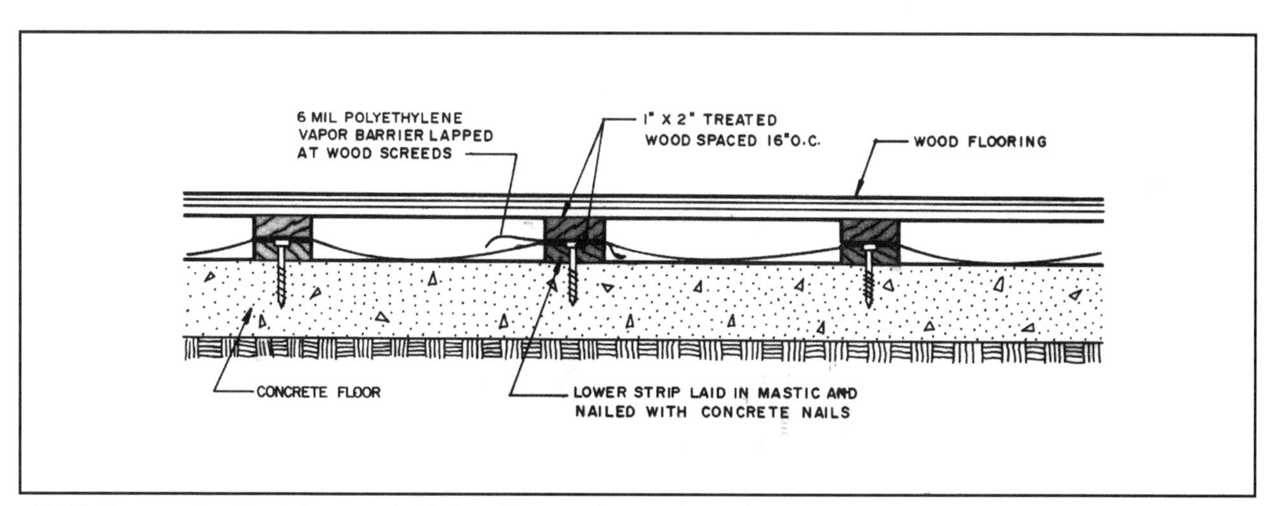

31-36 The screed method places a polyethylene film over the wood screeds.

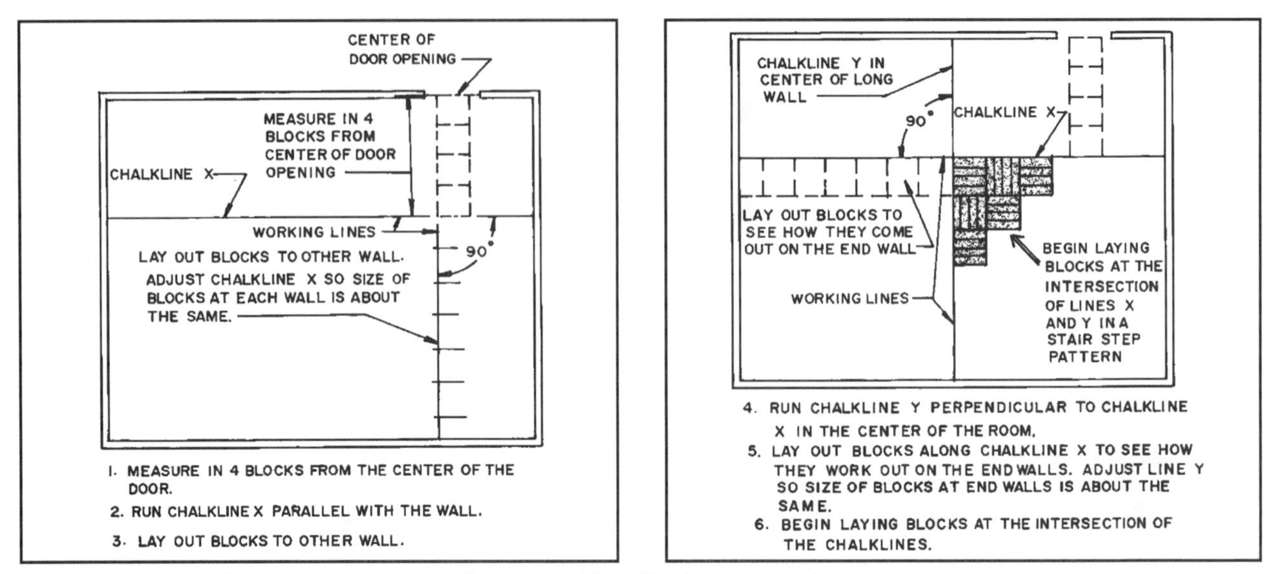

31-37 The steps to lay out a square pattern for parquet block flooring.

Lay Out a Parquet Floor on a Square Pattern

This discussion relates to laying out a parquet floor with the blocks parallel with the long wall of the room. There are two ways this is done. One method, shown in **31-37**, recommends snapping a chalkline (called the "working line") four or five blocks from the wall containing the major entrance to the room. This distance is measured from the middle of the doorway. This line is identified as line X in **31-37**. Then find the middle of the room measuring along line X. Snap a chalkline perpendicular to line X through the center point (line Y) and 90 degrees to it. Before laying the first block, check the distance to the other wall to see how the blocks will work out. If the last block is reasonably near a full block, the pattern will be satisfactory. If, however, the last block is just a narrow piece, some carpenters prefer to move line X **(see 31-37)** away from or closer to the wall to get the edge blocks closer to the same size. You will have to see how this move changes the size of the block that covers the area in the doorway and include this in your final layout decision.

Once the layout is established, lay the first block at the intersection of the two lines **(see 31-38)** and build them in a **stair-step pattern**, as shown in **31-39**.

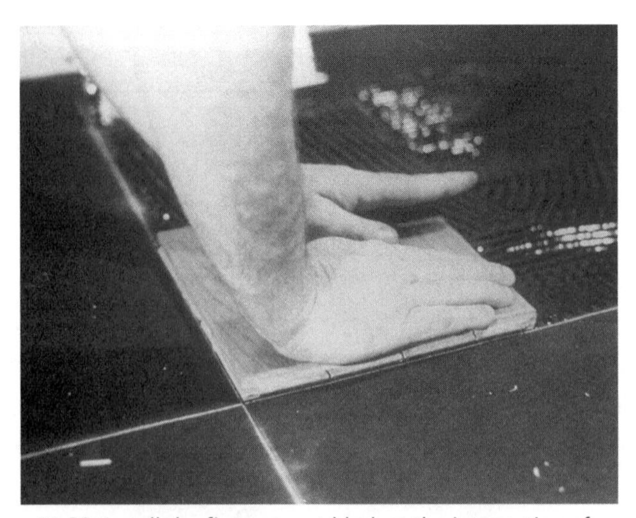

31-38 Install the first parquet block at the intersection of the working lines. *(Courtesy National Oak Flooring Manufacturers Association)*

31-39 Lay the 3/4-inch-thick tongue-and-groove parquet blocks in a stair-step fashion. *(Courtesy National Oak Flooring Manufacturers Association)*

Do not lay parquet flooring in rows. The stair-step procedure helps prevent small inaccuracies of size in the blocks from producing the appearance of misalignment. Complete the installation in one quadrant of the room, leaving any cutting and fitting at the walls until later. Then proceed to fill each quadrant using the stair-step method.

Cut the blocks to fill in the edges leaving a ¾-inch (1.9cm) gap, which is covered by the base and shoe molding. Place wood blocks around the edges in this ¾-inch gap to help keep the blocks from moving. Some recommend placing cork blocks in the gap and leaving them, because they will compress and allow the floor to expand. If wood spacers are used, remove them after the adhesive has set **(see 31-40)**.

When laying parquet with adhesive, the blocks tend to slide if sidewise pressure is applied. To help avoid moving them after they are laid, you can lay plywood panels over the laid area and use these as kneeling boards so that direct pressure is not on individual blocks.

Laying Out Parquet Flooring on a Diagonal

Lay out a diagonal line on a 45-degree angle from a corner of the room, as shown in **31-41**. One way to do this is to measure equal distances right and left from the corner. Connect the ends of these, and measure to find the center of the diagonal. Snap a chalkline from the corner of the room through this center. It will be at 90 degrees to the base line. Lay the first block at the intersection of the lines, and fill in each area using the stair-step method. This requires cutting triangular blocks along the wall. Maintain the ¾-inch gap on all walls.

Securing Parquet Flooring

Parquet flooring is secured in the same ways described for strip and plank flooring. Again, manufacturers will supply instructions for doing several different types of installation, such as blind-nailing, direct-gluing to the subfloor, and bonding to foam plastic underlayment.

INSTALLING HARDWOOD FLOORS OVER RADIANT HEATING

With the use of current building techniques and insulation, a radiant heat floor can be provided when hardwood floors are to be used. A typical hot water heating system consists of a source of heat, such as a boiler, to heat the water, and a series of pumps and controls to move the heated water through a network of tubes placed below the floor. The tubes release the heat to the floor as the water passes through and returns to the boiler to be reheated.

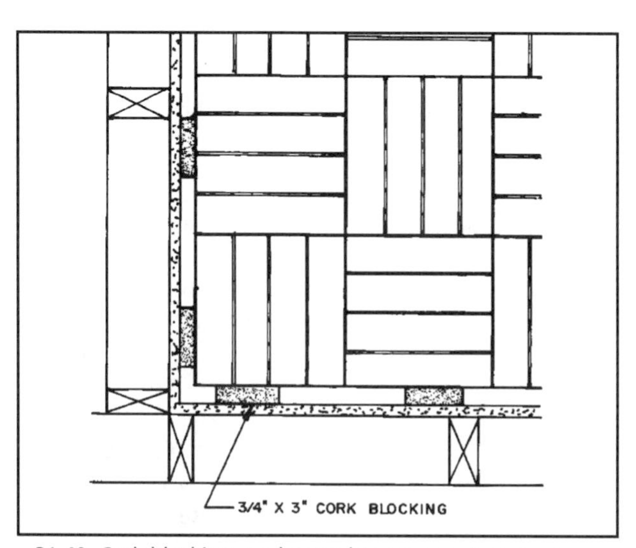

31-40 Cork blocking can be used to space out parquet blocks from the wall.

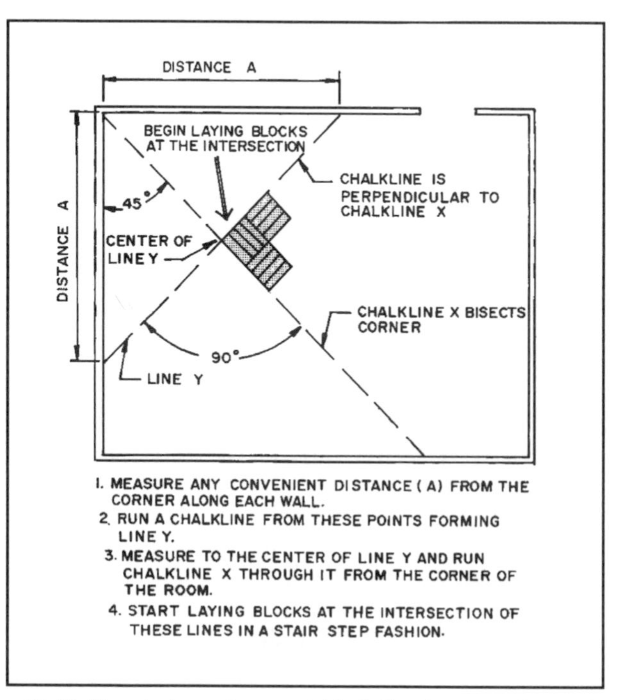

31-41 How to lay out parquet block flooring on a diagonal.

Following are some suggestions from the Hardwood Council for designing and installing such a system. A qualified heating system designer should be employed to design the system and select the subfloor option that best meets your needs. Two such designs are shown in **31-42**. The heating tubes should never touch the hardwood finish flooring.

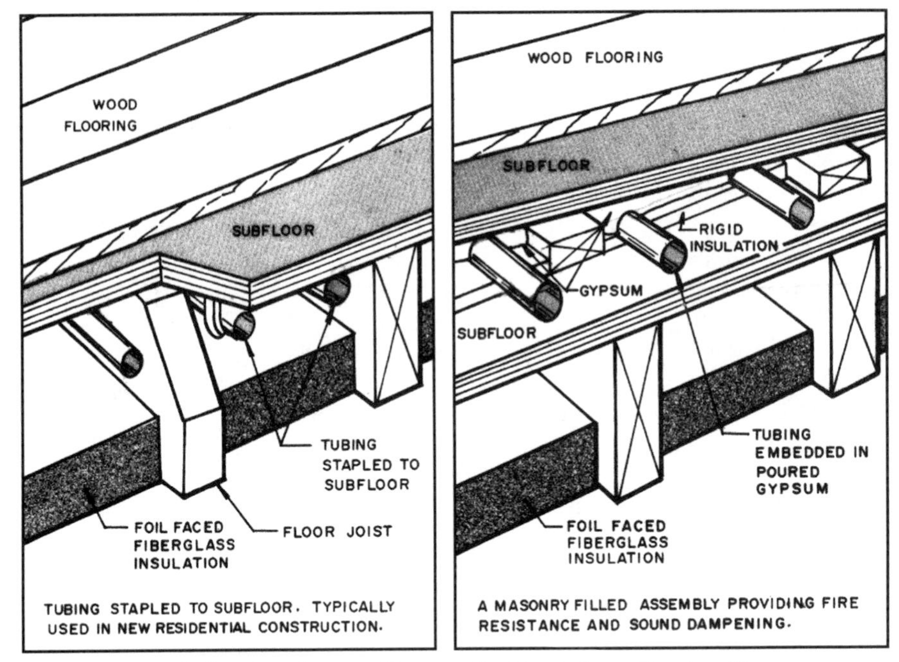

31-42 Two typical options for installing a radiant heat floor. *(Courtesy The Hardwood Council)*

The building should have the subfloor, tubing circuits, and controls for the heating system installed. The heating system must be fully operational and run for at least 72 hours to balance the content of the moisture inside the building. If the plywood or oriented strand board subflooring is wet, turn on the heating system and heat the interior until the air has reached the proper relative humidity. Particleboard subflooring is not recommended.

The moisture content of the hardwood flooring should be between six and nine percent. When the hardwood flooring is received on the job, check it with a moisture meter. If it is too wet spread it around the rooms and let it dry. If it is below six percent consult the manufacturer.

Flooring with tongue-and-grooved ends and edges is recommended. Beveled edges will show fewer cracks if cracking occurs. Do not install planks wider than three inches (76mm).

Install the flooring as described earlier. Constantly monitor the moisture content of the subfloor and the flooring strips. It must be kept within the range of six to nine percent.

To install flooring over concrete be certain the concrete has cured completely. This will vary depending upon the time of year and section of the country. It could be several weeks or even several months. Then turn on the heating system and let the hot water circulate in the pipes buried in the concrete floor. Check for moisture by taping a four-foot-square section of polyethylene plastic sheeting to the slab. Turn on the heat. If moisture gathers below the plastic continue heating the slab until no moisture appears. Install an eight-mil polyethylene vapor barrier over any slab that is in contact with soil.

The flooring is made by edge-gluing two strips of hardwood flooring having tongue-and-grooved edges. A kerf is wet on the bottom of each piece near the tongue. Special metal clips are inserted in the grooves to hold the planks together. The flooring is laid over a water-resistant membrane that has been placed over the concrete slab.

SURFACING THE FLOOR

After unfinished hardwood flooring is installed, a floor finishing crew sands it and prepares it for the application of the finish coating. However, there is usually a little more interior work to be completed after the floor is installed; so do not sand it until all the other trades have finished. Protect the floors from damage by other trades.

Before starting to sand, be certain any exposed nail heads are set below the surface. The opening produced will be filled later. Since dust is created it is best to seal off other rooms and seal heating duct openings and cold air returns. If there are cabinets in the room, cover them with plastic drop cloths, and protect surfaces that may be damaged with a covering of heavy corrugated cardboard. Finally, sweep the floor to remove all wood scraps and other debris.

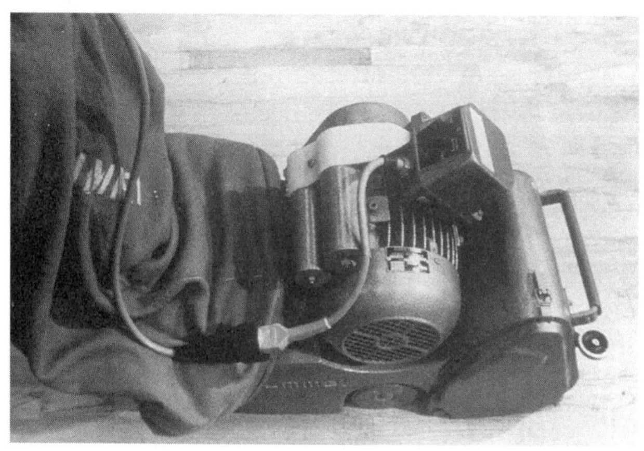

31-43 Unfinished wood flooring is sanded with a drum/belt sander. *(Courtesy National Oak Flooring Manufacturers Association)*

Using the Drum/Belt Sander

The floor is made flat and ready to finish with a drum/belt sander **(see 31-43)**. Although the sander has a dust bag which collects much of the wood dust produced, some does escape into the air. It is also a noisy machine. Therefore the operator should wear an ear protection device and some form of dust mask. The abrasive paper is on a drum at the front of the machine, or it may be a continuous belt. The abrasive action pulls the sander forward, so the operator must keep firm control.

Following are some things to observe as the sander is used.

1. Sand in the direction that the flooring is laid. In other words, sand with the grain of the wood. If there is a spot that is unusually uneven, it can be sanded **briefly** on a 45-degree angle, but leave enough wood to finish sanding with the grain.
2. Keep the sander moving whenever the drum or belt is touching the floor. If it sits still for even a second or two, it will cut a concave depression in the floor.
3. Start the drum/belt sander next to the right side wall and about two-thirds of the way from one wall. The drum or belt should be clear of the floor. This is done by tilting the sander backwards. When the drum or belt has reached full speed, slowly lower it to the floor, and move it forward to the end wall.
4. Just before it reaches the end wall, gradually raise the drum/belt off the floor. This produces a tapered or "feathered" cut at the wall. Then move the sander back towards the center of the room, lowering the drum/belt to sand the floor on the return cut.
5. Lift the drum/belt off the floor and move the sander to the left, letting it overlap the first cut by about half the width of the drum/belt. Repeat the above sweep to the wall and back to the middle of the room. Repeat these sweeps until the width of the floor on one end has been sanded. Then sand from the middle to the other end **(see 31-44)**.
6. When sanding hardwood floors, start with a coarse abrasive, such as a 36 grit, for the first sweep. Then use fine paper, such as 40 grit, followed by an 80 grit, and a final sweep with 100-grit abrasive. Finer abrasive, such as 120 grit, can be used if an even smoother finish is desired. When sanding softer woods such as pine, start with a 50-grit abrasive. Remove all wood dust by vacuuming the floor between each sanding.

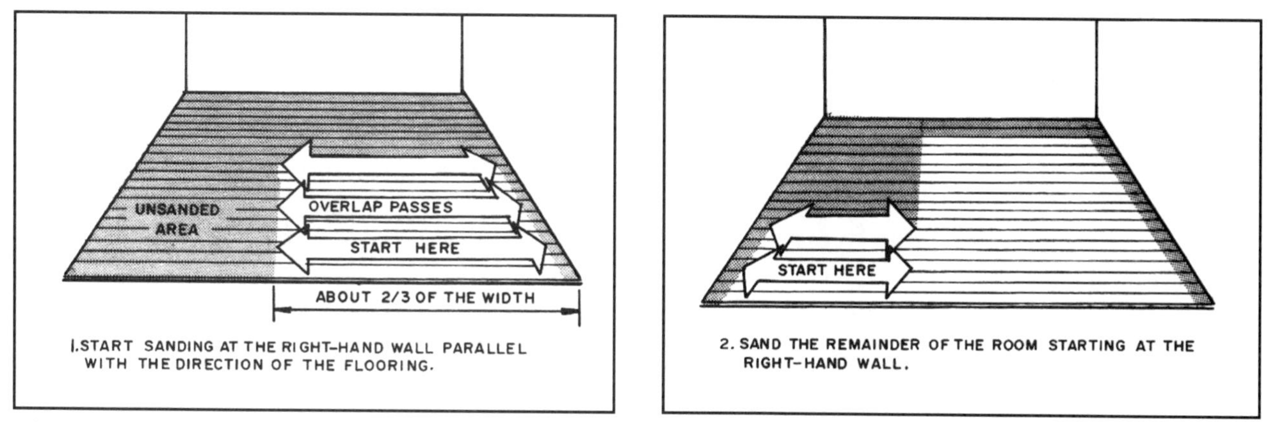

31-44 A recommended procedure for sanding wood floors with a drum/belt sander.

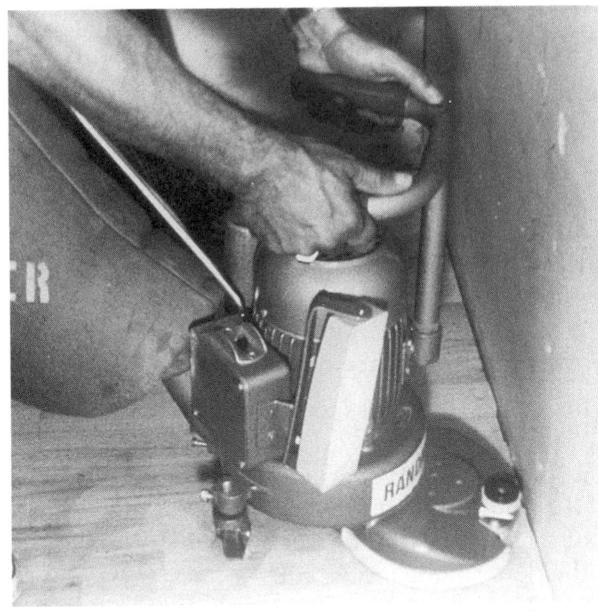

31-45 An edge sander is used to sand along the wall where the drum/belt sander cannot reach. *(Courtesy National Oak Flooring Manufacturers Association)*

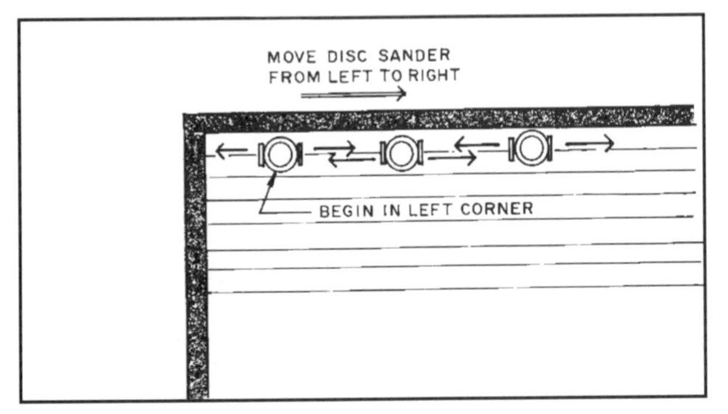

31-46 Start sanding with the disc sander in the left corner and move to the right.

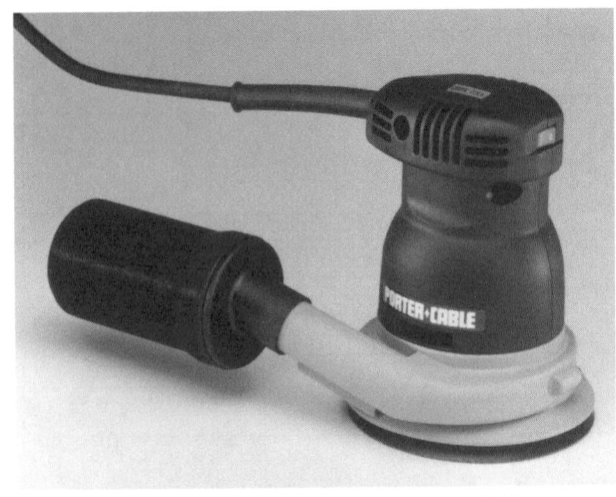

31-47 A small finish sander. *(Courtesy Porter-Cable Corporation)*

Using the Edge Sander

After the main area of the floor has been rough-sanded, the edge next to the wall is rough-sanded with a disc sander. Usually the edge sander will use an abrasive the same grit as on the drum/belt sander or one level finer. After the main floor area has been sanded with the next-finest-grit abrasive, the edges are also sanded. This continues until the finest-grit abrasive has been used **(see 31-45)**.

Following are some things to observe as the disc sander is used:

1. Place the sander on the floor. Tilt it back or adjust the rollers so that the disc is clear of the floor.
2. Hold the sander by the handles and turn the switch on.
3. When the sander reaches full speed, lower the disc to the floor and immediately begin moving it along the edge.
4. Move the sander back and forth in a slow sweeping motion of about 15 to 18 inches (approx. 38 to 46cm).
5. Begin sanding in a left hand corner as you face the wall. Move the sander along the wall to the right **(see 31-46)**. Remember to keep the sander moving while the disc is turning; this avoids creating an unwanted depression that occurs if the machine remains in one spot while sanding.
6. Allow the weight of the sander to apply pressure on the abrasive disc. Do not push down on it to try to speed up the cut.
7. When it is necessary to shut off the sander, tilt it back so that the disc is clear of the floor, and then turn off the switch. Remember to empty the dust collection bag frequently. Never leave dust in the sander bag overnight, because it can catch fire through spontaneous combustion. Always empty the dust bag before leaving the job at the end of the day.

Finishing the Sanding

The disc sander may leave some marks where it overlaps the drum/belt-sanded area. These marks may be removed using a handheld power sander having a random orbital or straight line motion **(see 31-47)**. Areas such as in a corner, where no sander can reach, may be smoothed with a hand scraper and hand sander **(see 31-48)**.

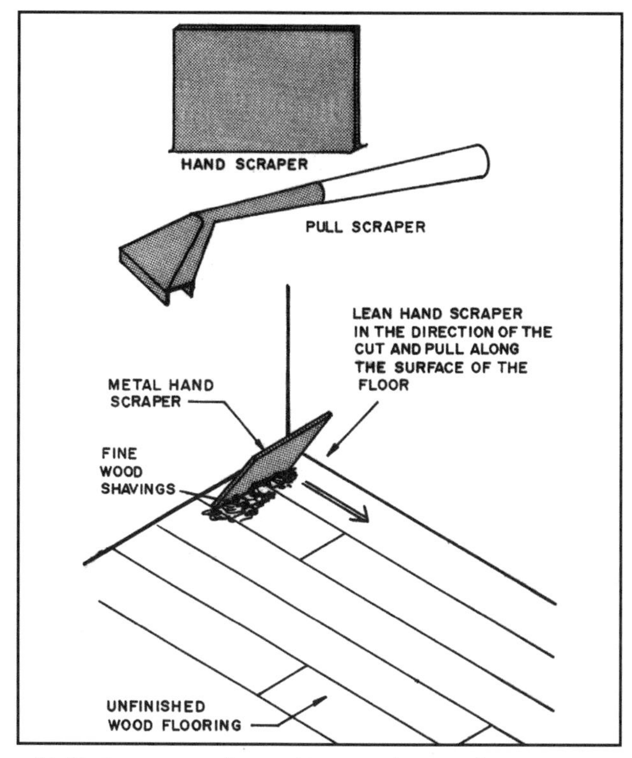

31-48 Scrapers can be used to smooth areas that cannot be reached with sanders.

FINISHING UNFINISHED WOOD FLOORS

After the floor has been sanded all dust must be removed, usually with a powerful vacuum cleaner. At this point the floor must be protected from any traffic. Therefore finishers usually proceed immediately with the finishing process. If the color is to be altered a stain is applied. After it has dried it is often covered with a sealer. Open-grain woods, such as oak, require a paste wood filler be applied next to fill the open pores in the wood. The manufacturer of the finish top coating will specify if a sealer is to be applied over the paste wood filler. The floor is now ready for the application of the finish top coating.

There are a number of frequently used top coatings. Study the manufacturer's specifications and application directions. Of special importance is the minimum drying time specified. Following are some of these top coating materials.

Polyurethane Polyurethane coatings are widely used as the top coat on wood floors. There are two types available, oil-based and water-based.

Oil-based polyurethane takes longer to dry and fills the room with noxious fumes. Oil-based coatings penetrate deeper into the wood surface than the water-based type.

Water-based polyurethane coatings dry rapidly and produce a harder finish. It tends to raise the grain of the wood so the floor has to be sanded more between coats. Both types provide good mar and abrasion resistance.

Oil and Wax Finishes These finishes produce a duller finish. The polyurethane coating gives a very glossy appearance, so if this is not desired you may consider an oil or wax finish. It is not as mar resistant as polyurethane but can be easily refinished by reapplication of the material. It will blend in with the floor surrounding the refinished area. Spills may pass through these finishes. Therefore the finishes require regular maintenance.

Varnishes Varnish has been used for many years as the major floor finish material. Modified phenolic varnishes form a hard, abrasion-resistant surface and are suitable for finishing wood floors. They may yellow over the years.

Penetrating Sealers Penetrating sealers are absorbed by the wood and do not provide a surface coating. They remain below the surface yet provide wear resistance and prevent soiling of the surface. It is easy to restore areas that have worn.

Acrylic and Urethane Acrylic and urethane top coatings are factory applied to prefinished flooring. The urethane is used for residential flooring. Acrylic is tougher and is used for residential and commercial applications.

CHAPTER 32

OTHER TYPES OF FLOORING

In addition to wood flooring a large range of other flooring products are widely used. Each has its own special merits. Carpet is soft and has sound-deadening properties. Laminate flooring provides a hard, durable surface and is installed as a floating floor. Clay tile is tough, durable, and can be used indoors or outdoors. Resilient flooring is durable, easily installed and cleaned, and available in a wide array of colors and patterns.

INSTALLING CUSHION & CARPET

Carpet is installed over concrete, plywood, oriented strand board, or other firm smooth substrates. It must be clean and smooth and any openings, such as a knot hole, must be filled with wood filler. In some cases the carpet is secured directly to the subfloor. However, generally a carpet cushion is installed next to the subfloor. The cushion used can vary, as shown in **32-1**. The cushion can be a separate pad that is then covered with the carpet. Some types of carpet have the cushion fused to the back of the carpet. A third type has a thin backing and is laid without a thicker cushion.

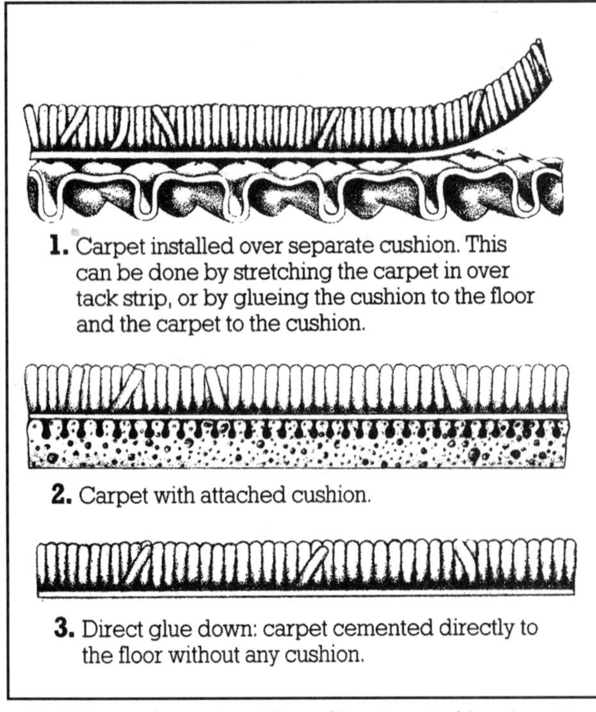

32-1 Types of carpet cushions. *(Courtesy Cushion Carpet Council)*

Using Carpet Strips

One method for installing carpet is to use wood carpet strips that have sharp metal points protruding from the surface. Metal strips with points are also available.

Nail the strips to the subfloor around the perimeter of the room. Keep them about ¼ inch from the baseboard. Cut and place the cushion inside the carpet strips and staple it to the subfloor around the edges of the room **(see 32-2)**.

Now unroll the carpet and cut it to fit inside the baseboard. Place it next to one wall and press it down on the tacks on the carpet strip. Stretch the carpet across the room to the other side. A carpet stretcher is used to pull it up. Then press it onto the tacks on the other side.

In many cases the carpet is left butting tightly against the baseboard and the shoe molding is omitted. When the carpet meets a thin finished flooring, such as vinyl, a metal strip is installed over the joint **(see 32-3)**.

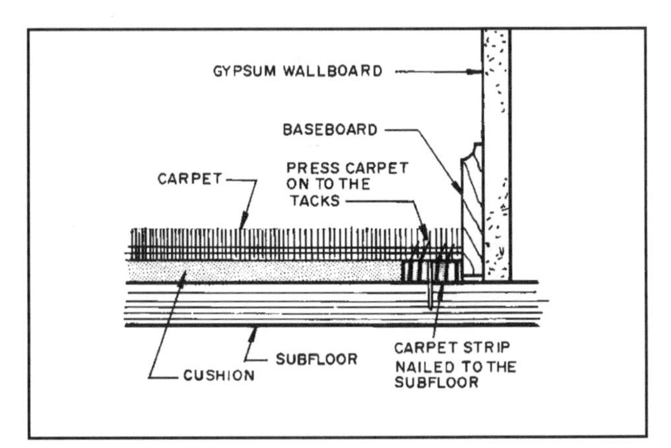

32-2 Carpet can be installed by securing it to wood carpet strips nailed around the walls of the room.

When it is necessary to join two pieces of carpet, the butting edges are cut clean and straight. A special carpet tape is applied to the back of one side with half the width extending beyond it **(see 32-4)**. The butting piece of carpet is placed against the first piece and pressed into the tape. Some tapes require the use of a special heating tool to bond the tape to the carpet. The joint should fit closely but not be forced together, because excessive force may cause the carpet to get some lumps.

Carpet can also be bonded directly to the subfloor with an adhesive, as shown in **32-5**. Typically this is done with carpets made with a thin cushion bonded to the back.

Carpets have to be cut to fit around openings in the floor. In **32-6** the carpet fits around the heat register and butts the metal frame of a sliding glass door.

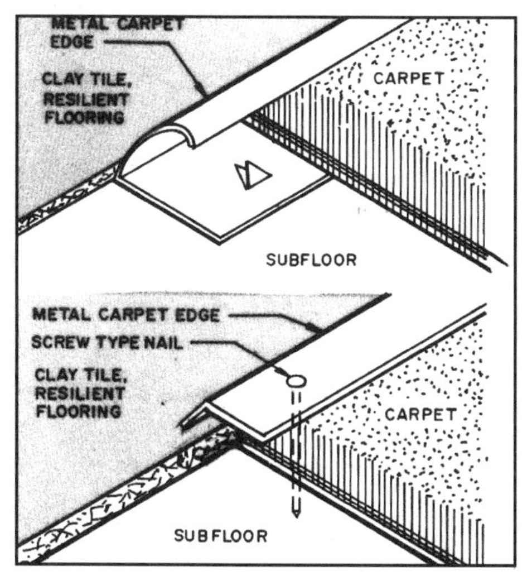

32-3 When carpet butts against another finish flooring material, metal carpet edges can be used to cover the union.

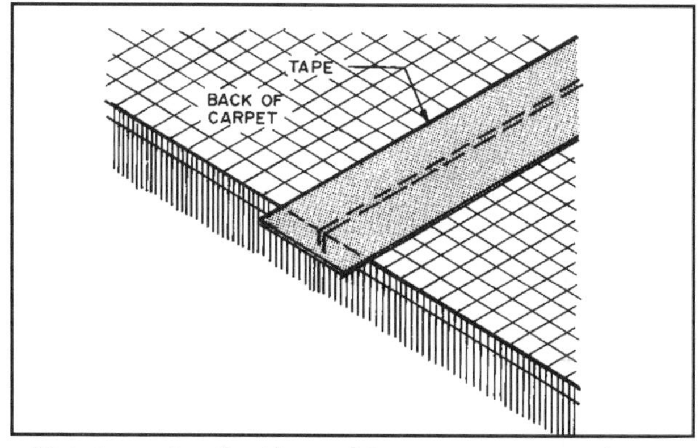

32-4 Carpet is spliced by applying a carpet tape to the back of the joint.

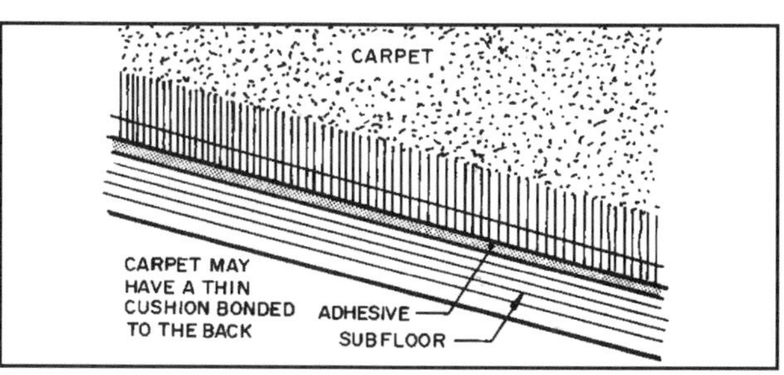

32-5 Carpet is sometimes bonded to the subfloor with an adhesive.

RESILIENT FLOORING

Resilient flooring generally available is vinyl, cork, and rubber. Vinyl composition materials are most commonly used. They are called resilient materials because they are flexible, having some "give" when pressure is applied and then returning to their original condition. They wear well, resist damage from dropped items, and are not damaged by moisture. Generally they are laid by flooring specialists. However, the typical homeowner can do a good job as well.

Before starting the installation, get the subfloor in first-class condition. Holes and dents must be filled, protruding nails set or removed, and the floor swept absolutely clean. If the subfloor is in poor condition, nail ¼-inch-thick underlayment panels over it. These are typically plywood, hardboard, or oriented strand board.

32-6 Carpet is cut to fit around floor openings. Notice the tight fit made against the metal frame of a sliding glass door.

Most resilient flooring can be cut with a large pair of scissors or sheetmetal shears. However, a utility knife may also be used. Always use the adhesive recommended by the flooring manufacturer.

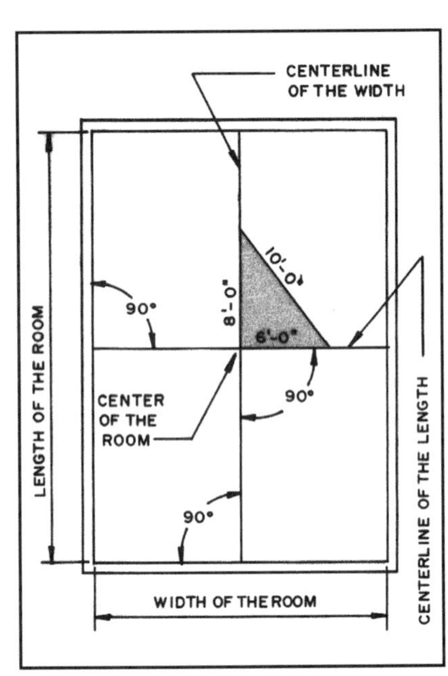

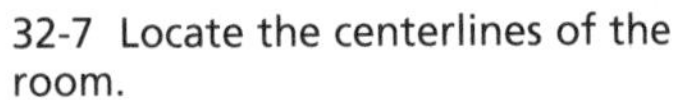
32-7 Locate the centerlines of the room.

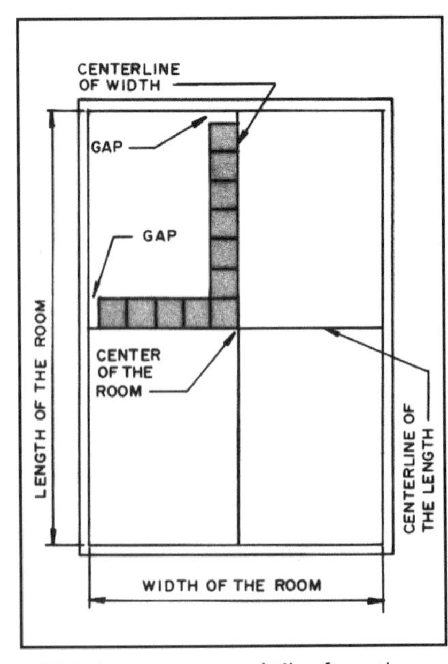

32-8 Lay uncemented tiles from the center of the room to the walls to see how much of a gap occurs at the walls.

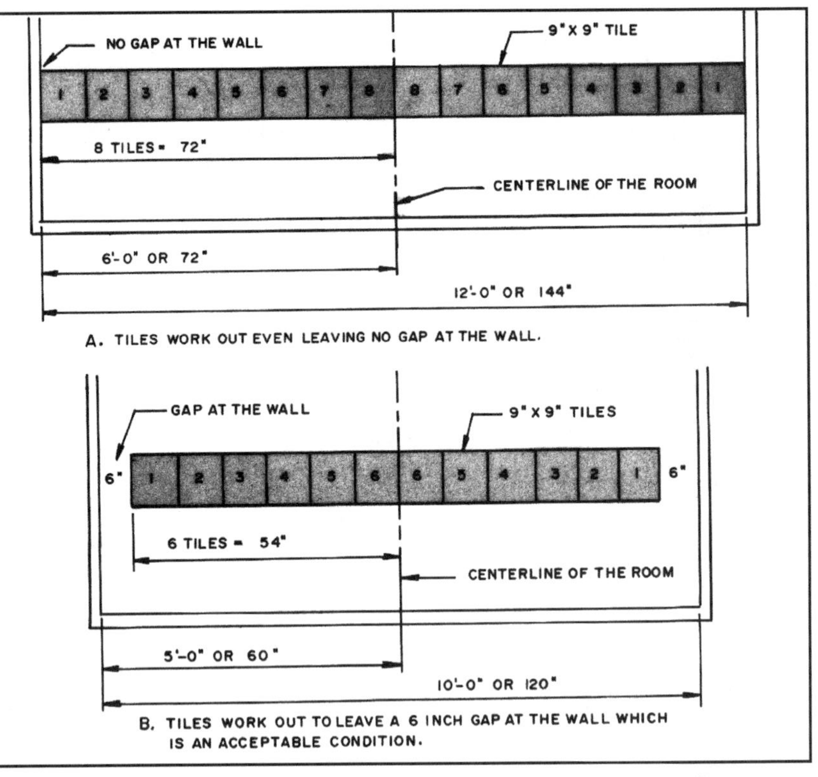

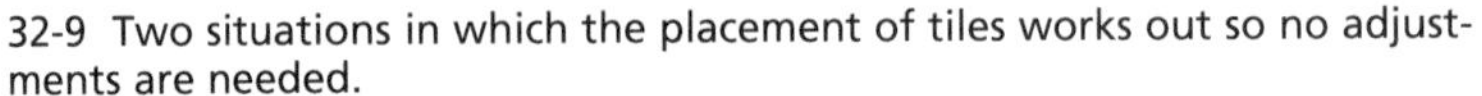
32-9 Two situations in which the placement of tiles works out so no adjustments are needed.

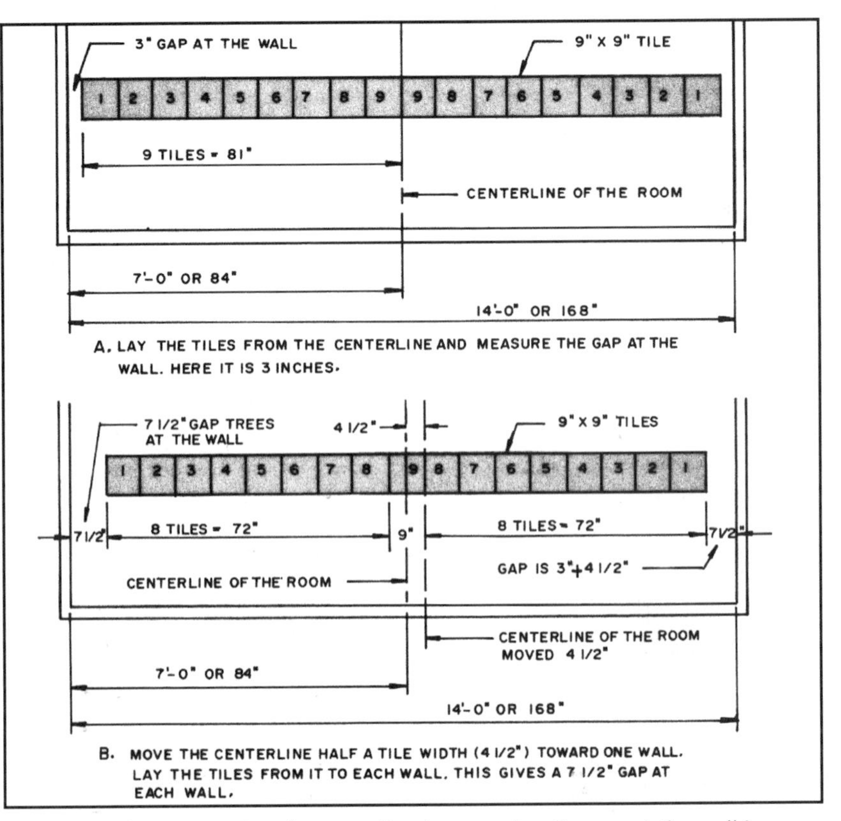

32-10 This layout requires the centerline be moved so the gap at the wall is equal to or greater than half the width of the tile.

Installing Resilient Floor Tiles

Begin by dividing the room along its centerlines. Measure the distance on each wall and snap the centerlines on the floor with a chalkline (see **32-7**). They should be 90 degrees to each other. This can be checked by using the 6, 8, 10-foot triangle shown in the figure.

Begin by laying uncemented tiles from the center of the room to each wall **(see 32-8)**. It may be necessary to shift the centerlines so that the tiles meeting each wall are not smaller than about one-half a tile. Study the examples in **32-9**. Example A shows a plan for a room 12 feet wide with 9 × 9-inch tiles. The distance from the centerline to the wall is 72 inches, which is equal to eight tiles nine inches wide, so no gap exists at the wall. In example B, a room 10 feet wide has 12 tiles 9 × 9 inches with a gap at each wall of 6 inches. Since this gap is wider than half the tile (4½ inches) no adjustment is needed.

In **32-10** is a room 14 feet wide. If the tile is laid from the centerline, as shown in example A, a three-inch gap occurs at each wall. This is not acceptable. To widen the gap, move the center tile 4½ inches toward one wall and space out the other tiles. This gives 16 tiles plus a 7½-inch gap at each wall.

In all cases allow about a ¼-inch gap at the walls for expansion. This will be covered with shoe molding or a vinyl cove base.

Begin by applying adhesive with a notched trowel to one quarter of the room. Place one tile at the center as shown in **32-11** and lay a row of tiles along the chalkline to each wall. Then fill in the area. Be certain to place each tile exactly where it belongs. Do not slide tiles after placing because they will pick up adhesive on the edges and you will not get a clean, closed joint between tiles. When postitioned press down on them to firmly set in place. Then finish another quarter.

Finally measure, cut, and place the tiles along the wall. Place an uncemented tile on top of the row next to the wall. Place a third tile on top and slide it until it touches the wall. Mark the width on the second tile as shown in **32-12**. Cut it about ¼ inch short to allow for expansion. If the wall is straight you can cut enough at this measure to do the length of the wall. If it slants you will have to repeat this every few tiles. The same procedure can be used to mark corner cuts **(see 32-13)**.

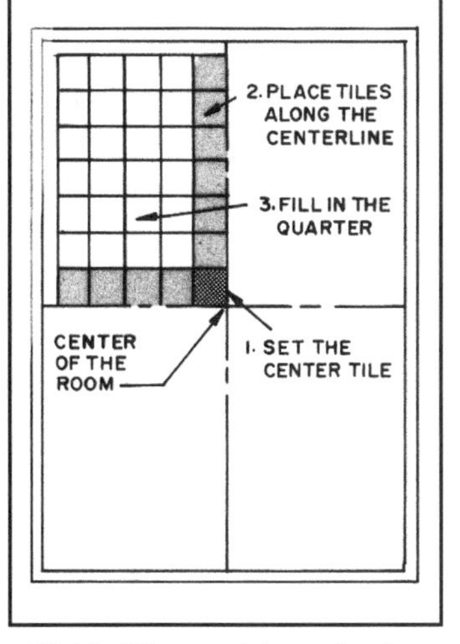

32-11 After applying adhesive to one quarter of the floor, set the center tile, lay tiles along the centerline, and then fill in the quarter.

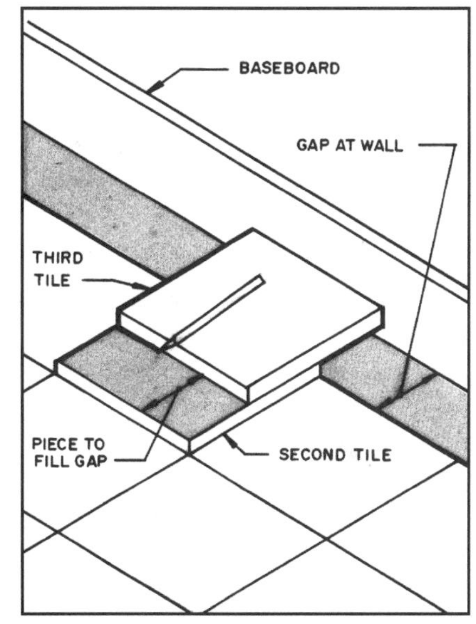

32-12 One way to mark the tiles to fill the gap at the wall.

32-13 A procedure for marking tiles to fill the gap at the wall at a corner.

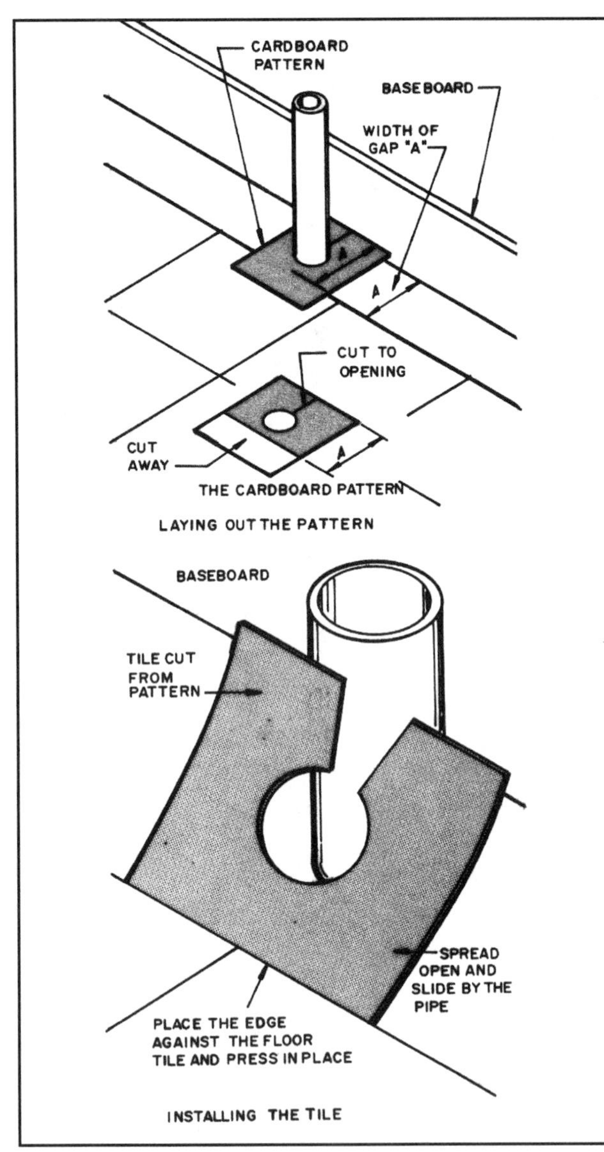

32-14 A cardboard pattern is made to get the shape of the tile that fits around a pipe or other protrusion.

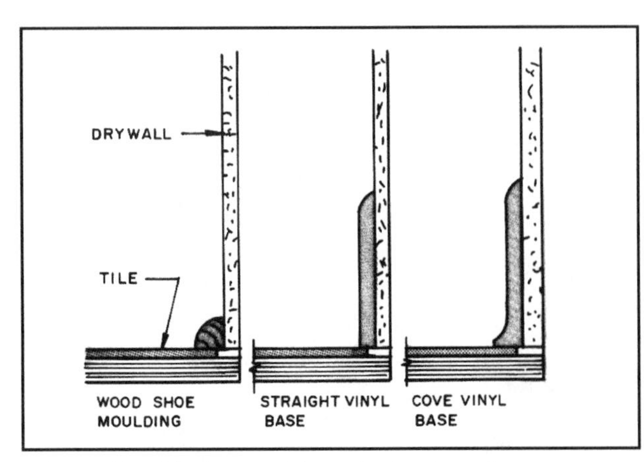

32-15 The ¼-inch space left at the baseboard for expansion can be covered with shoe molding or a vinyl base.

To cut around a pipe or other protrusion, make a cardboard pattern. You can cut and trim until you get a good fit. Draw this on a tile and cut to shape. Cut through the back side to the opening and install as shown in **32-14**.

The space around the edge or the wall can be covered with wood shoe molding or a vinyl base. Vinyl bases may be straight or coved **(see 32-15)**.

INSTALLING RESILIENT SHEET FLOORING

Begin by measuring the room and making a sketch as seen in **32-16**. Locate any corners, cabinets, or other protrusions into the floor area. Prepare the subfloor as described for resilient tile. Bring the roll of flooring into the room and unroll it. Cut it to length. Roll back about half the length of the sheet and spread the adhesive on the subfloor. Roll the sheet back over the adhesive and press into place. Then roll up the other half and repeat the process. When smoothing the sheet to the adhesive, work from its center toward the edges. A heavy roller is used for this process.

If the room requires a second sheet carefully measure the required width. Measure in several places because rooms are often not square. Cut the sheet to width. It is best if the joint can be made by butting two factory-prepared edges. Be certain the

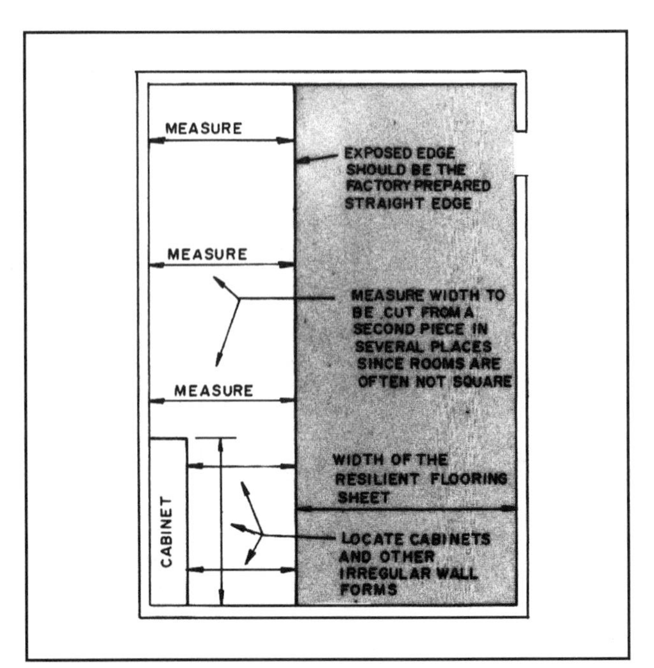

32-16 How to plan for cutting the resilient vinyl sheet flooring to size.

pattern matches at the joint. If this is not possible lap one sheet over the other and, using a long straightedge, cut through both sheets at the same time. This way even if the cut is a little off the edges will still match **(see 32-17)**.

Measure for any cutouts as required for cabinets. Some prefer to leave a little extra material, let it lap up on the cabinet base, and trim it after the sheet is in place **(see 32-18)**.

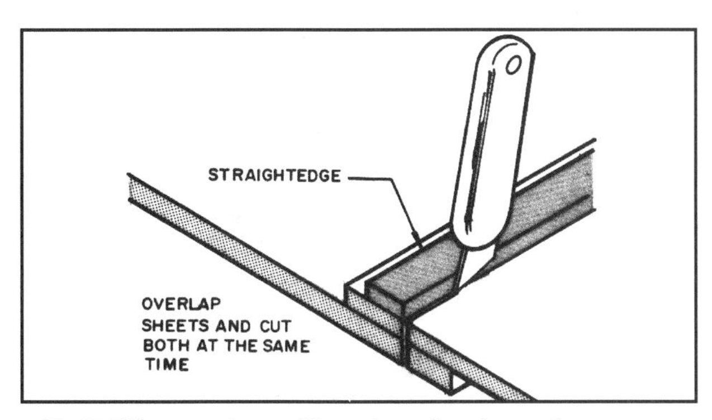

32-17 When cutting resilient sheet flooring to form a joint it is best to cut both pieces at the same time. Be certain to position them so the pattern is not broken.

INSTALLING CERAMIC TILE

Ceramic tiles are durable and moisture resistant. They are used where heavy traffic and moist conditions are expected **(see 32-19)**. They are used on walls and floors.

Quarry tile is an unglazed floor tile. Clay floor and wall tiles are available in a wide variety of shapes but square, rectangular, hexagonal, and octagonal are most frequently used.

Ceramic tiles are available as individual tiles around four inches square and larger and as mosaic tiles (small tiles about one inch square) that are glued to a fabric backing in sheets around two feet square.

Ceramic floor tile may be set in a bed consisting of a cement–plaster mixture. The mix is troweled level and smooth and when it is plastic (firm) the tiles are placed on it and pressed lightly in place. Tiles used for this method are soaked in water before placing. After the tiles are in place the spaces between them are filled with grout. This is leveled in the cracks and tooled to get a smooth finish. All of the grout is cleaned off the surfaces of the tile.

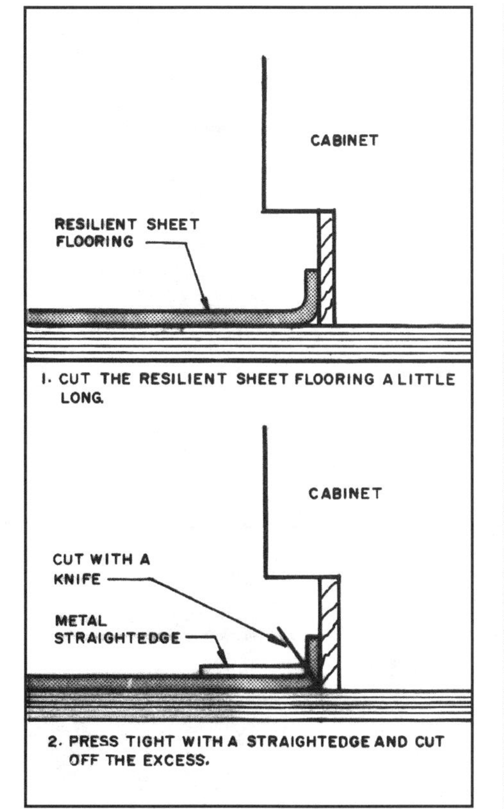

32-18 Cut the resilient sheet flooring a bit long and trim to fit after it is laid.

32-19 Ceramic tile is very durable and used where wear will be heavy, as in this entrance hall.

An easier and more commonly used installation bonds the tiles to the subfloor with an adhesive in the same manner as described for resilient tile. The spaces between them are grouted and the faces of the tiles are wiped clean **(see 33-20)**. A baseboard covers the space between the tile and the wall. Some type of molding is often used around the base of cabinets as shown in **32-21**.

LAMINATE FLOORING

High-pressure laminate flooring has the beauty of wood or stone yet is tough and durable. It can be used on any interior application but should not be used where it will be continuously exposed to high temperatures, humidity, or large quantities of water. It requires no finishing after installation.

32-20 Ceramic floor tile can be bonded to the subfloor with an adhesive. *(Courtesy American Olean Tile Company)*

32-21 The edges of the ceramic tile at the cabinet base are covered with a molding.

The composition of one brand is detailed in **32-22**. The top surface is a high-pressure decorative laminate similar to that used on countertops but ten times more durable. The core is a 50- to 55-lb. medium-density fiberboard that has tongue-and-groove joints. The bottom layer is a backer ply of high pressure laminate that balances the construction. This flooring is available in 15½-inch square tiles and four-foot-long planks 7¼ inches wide.

The tiles and planks have a patented tongue-and-groove edge design producing a tight-fitting seam **(see 32-23)**.

Installing Laminate Flooring

Laminate flooring can be installed over existing flooring such as wood, vinyl tile, ceramic tile, sheet vinyl, and some types of glue-down carpeting. If laid over a bare concrete floor a polyethylene vapor barrier must first be installed.

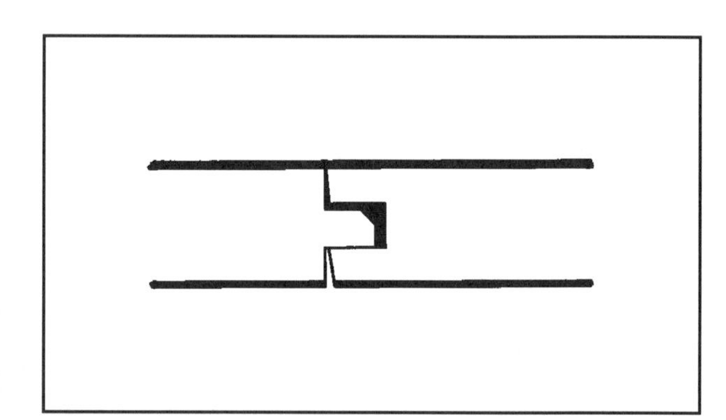

32-23 A patented edge joint used to secure laminate floor covering. *(Courtesy Wilsonart International)*

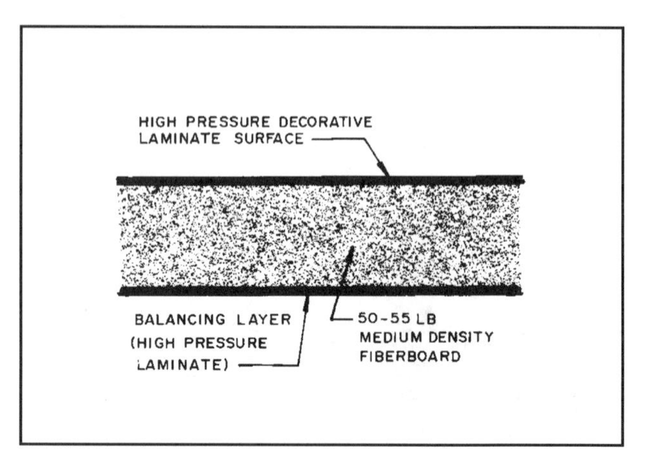

32-22 This laminate floor covering is made with high-pressure laminate outer surfaces and a medium-density fiberboard core. *(Courtesy Wilsonart International)*

32-24 Fill voids and low spots with leveling compound. When dry, clean the floor thoroughly. *(Courtesy Wilsonart International)*

32-25 Unroll the special foam padding in the same direction as the laminate floor planks will be laid. *(Courtesy Wilsonart International)*

The floor is installed using a floating concept; it is not nailed or glued to the subfloor. The pieces are glued together at the tongue-and-groove edges, forming a large, single floating floor covering. The manufacturer supplies a foam padding that is placed over the floor before installation begins.

Following is a brief discussion of the installation process for laminated plank flooring.

1. Clean the subfloor and fill and smooth any dents or recessed areas **(see 32-24)**.
2. Plan the layout. Measure the width of the room and see how wide the space will be when you lay the flooring from one wall across the room. If the space is less than two inches, cut the first strip a little narrow.
3. Unroll the foam padding in the same direction as the flooring planks will be laid. Usually they are laid with the planks parallel with the incoming light, except in long narrow halls where they parallel the wall **(see 32-25)**. Unroll only one width, lay over it, and then unroll another width.
4. Begin laying in a corner of the room along a wall. Plan so you can lay the floor without having to walk on it. If there is no wall, nail a straightedge on top of the padding and start against it **(32-26)**.
5. Lay the planks from right to left with the groove side facing the wall **(see 32-27)**.

32-26 If there is no wall on the starting side of the room, nail a straightedge on top of the padding and start the first row along it. *(Courtesy Wilsonart International)*

32-27 Place the first row next to a wall or a straightedge and begin in a corner. *(Courtesy Wilsonart International)*

6. Fill the end groove with a continuous bead of glue and slide the planks together **(see 32-28)**. Insert the plastic spacers between the ends of the planks and the wall. Theses spacers provide a space to allow for expansion. They can also be gently tapped down to move the assembled flooring if it needs straightening **(see 32-29)**.

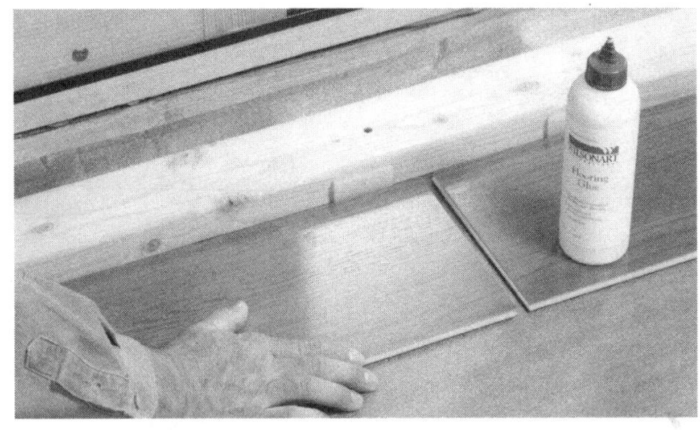

32-28 Fill the groove on the short end with a continuous bead of glue and slide the planks together. *(Courtesy Wilsonart International)*

32-29 These spacers provide a space for expansion and allow the floating floor to be adjusted in relation to the wall. *(Courtesy Wilsonart International)*

7. Wipe off excess glue after each piece is installed. Use a clean damp cloth **(see 32-30)**.
8. Measure and cut the last piece. Use a manufacturer-supplied pull bar to close the joint **(see 32-31)**.
9. Start each row with a plank 12 inches shorter than the previous one. This staggers the joints **(see 32-32)**.
10. A tapping block can be used to tap the rows together. After installing four rows of planks let the floor set for 45 minutes before you do the next four **(see 32-33)**.
11. After the first width of padding is covered unroll the next width. Butt the edges of the padding and seal with two-inch masking tape **(see 32-34)**.
12. If the door trim has already been installed, use a scrap piece of floor and pad and cut off the bottom before installing the planks **(see 32-35)**.
13. Carefully measure and cut pieces to fit around pipes and other protrusions **(see 32-36)**.
14. After the adhesive has set, install the shoe molding **(see 32-37)** and damp mop the surface with a cleaning solution provided by the manufacturer. Do not use the room for at least 12 hours after it has been finished.

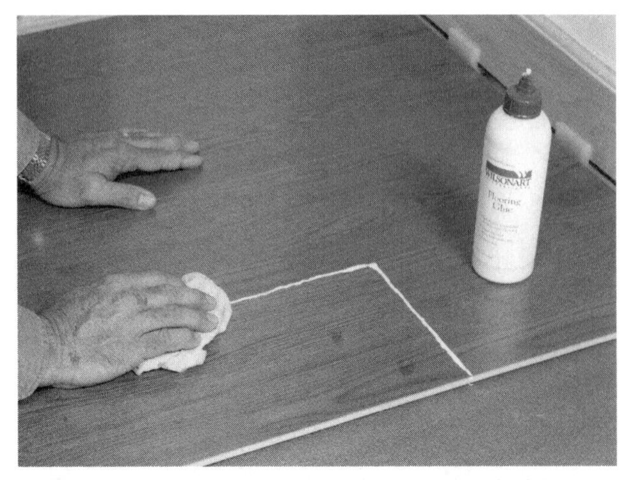

32-30 Wipe off the excess glue as soon as the plank is in place. Notice the plastic spacers at the wall. *(Courtesy Wilsonart International)*

32-31 Cut the last piece to length, allowing ¼ inch for expansion. Pull the joint closed with the pull bar. *(Courtesy Wilsonart International)*

32-32 Start each row with a plank 12 inches shorter than the previous one. This will stagger the joints, giving a pleasing appearrance to the installation. (*Courtesy Wilsonart International)*

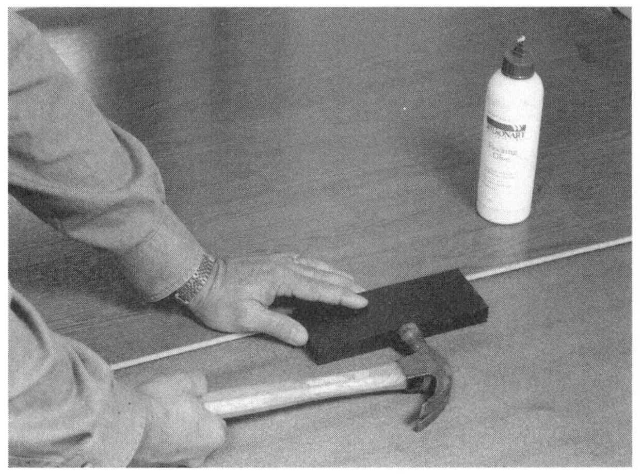

32-33 Use a tapping block and hammer and occasionally tap the rows together. *(Courtesy Wilsonart International)*

32-34 When the second row of padding is installed, butt the edges and seal with two-inch masking tape. *(Courtesy Wilsonart International)*

32-35 Undercut door trim with a jamb saw. Use a piece of scrap flooring as a spacer. *(Courtesy Wilsonart International)*

32-36 Locate pipes and cut the planks to fit around them. Cut the holes about ¼ inch larger than the pipe diameters. Apply glue to the butting edges. *(Courtesy Wilsonart International)*

32-37 Finish the installation by removing the spacers and installing the shoe molding. Do not allow anyone to walk on the floor for 12 hours. *(Courtesy Wilsonart International)*

CHAPTER 33

STAIRS

Stairs can be laid out and cut by the carpenter; or they can be bought completely cut to size, and then assembled and installed by the finish carpenter. If the carpenter builds stairs that are to be carpeted, they often have structurally sound softwood treads cut and fitted. If exposed, finished hardwood treads are required. These are usually bought from a building materials supplier and cut and fitted on the job. They are unfinished but completely machined, ready to install.

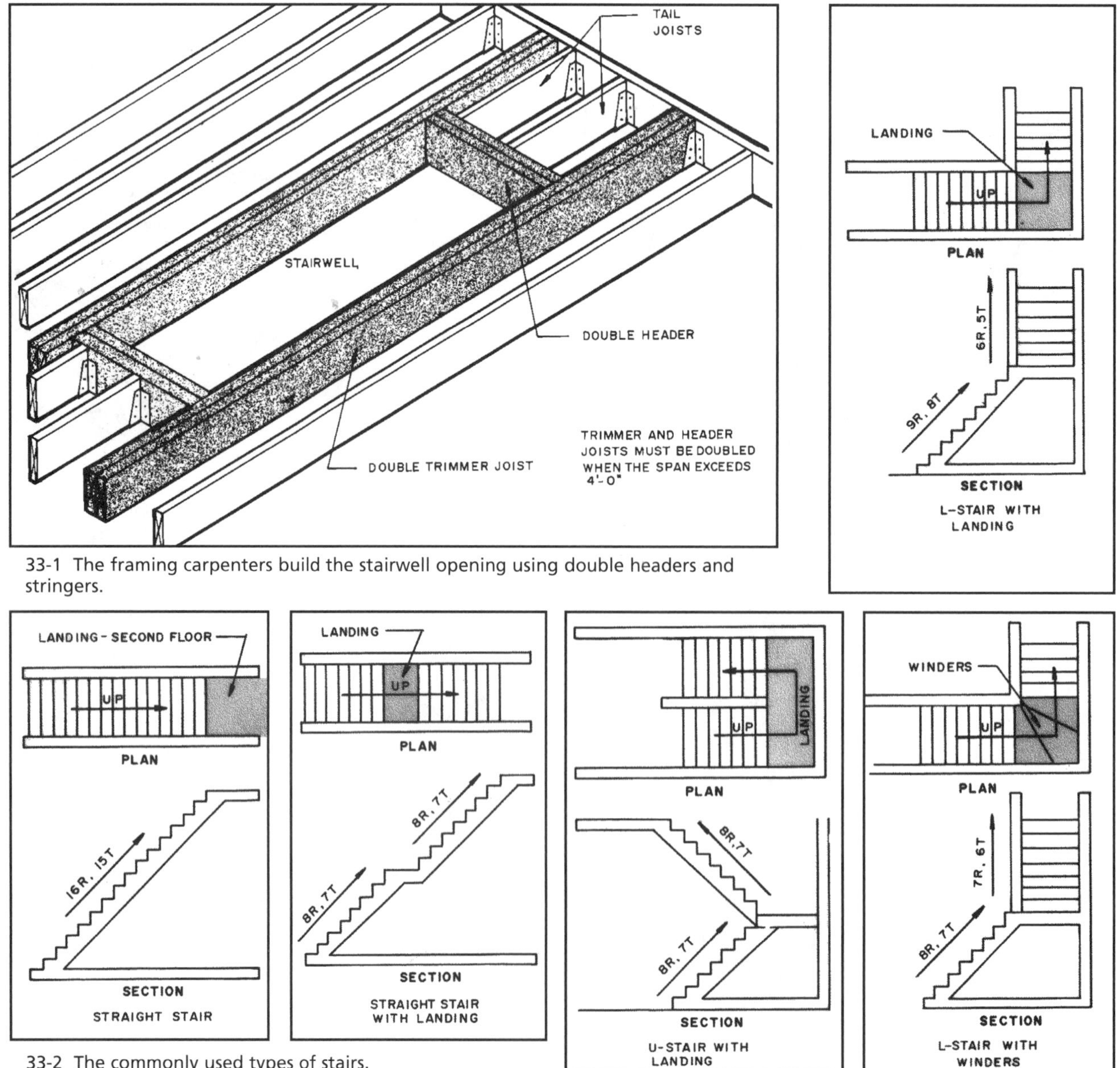

33-1 The framing carpenters build the stairwell opening using double headers and stringers.

33-2 The commonly used types of stairs.

THE STAIRWELL

When the framing carpenters finish their work, they should leave stairwell openings framed, as shown in **33-1**. Notice that the header and trimmer joists are doubled.

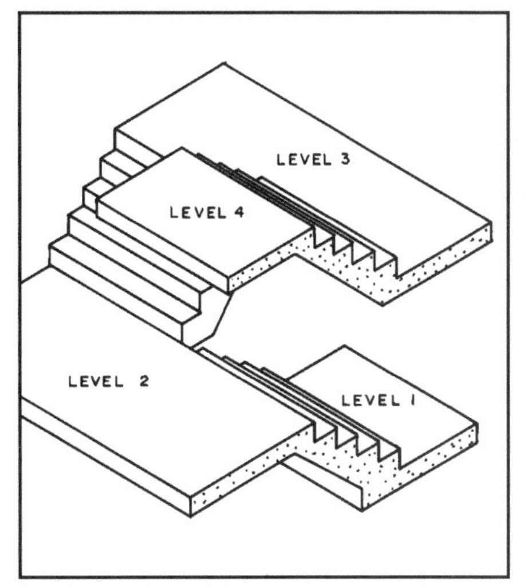

33-3 Stairs in split-level houses typically have short runs and are placed above each other.

33-4 This stair is open on one side.

TYPES OF STAIRS

The commonly used types of stairs are **straight**, stairs with **landings** (straight, **L-shaped,** and **U-shaped**), and stairs with winders **(see 33-2)**.

The straight stair is perhaps the most frequently used type. It consists of a straight run from one level to another. Codes typically require stairs with a rise above a specified vertical distance, such as 12 feet (3.6m), to place a landing about halfway up. Its purpose is to provide a break in the climb, giving a place to rest. Landings are also used to change the direction of a stair, as shown for the L-shaped stairs. Since landings take considerable floor space, winders are sometimes used instead when space is tight. Winders, however, are difficult to use and are avoided whenever possible. The design of winders must, as always, meet local codes. U-shaped stairs require the most floor space but permit the stairs to reverse direction. Split-level houses use a series of short straight stairs, each about half the height of a single story. The stair has a short run and has excellent ceiling height because of the stair that is generally placed directly above it **(see 33-3)**.

Stairs may be open on both sides, open on one side, closed on both sides, or partially open **(see 33-4)**.

BUILDING CODES

Often the stair design is detailed on the working drawings by the designer. It must be designed to meet local building codes. Sometimes the specifications for the stairs are not available on the drawing, and the owner relies on the carpenter to design and build them. Therefore it is important that the carpenter know the requirements specified for the stairs in the code and be able to consult the code for any changes or updates.

Following are examples of **typical** code requirements. As always, consult your local code.

1. **Stair Width**—If the occupant load is 50 occupants or less, 36 inches (914mm) is required. In private, single-family dwellings the 36-inch width is also the minimum. The width is defined as the clear distance inside the finished wall. To get a clear 36-inch width, the stairwell must be framed wide enough to allow for the finish wall **(see 33-5)**.

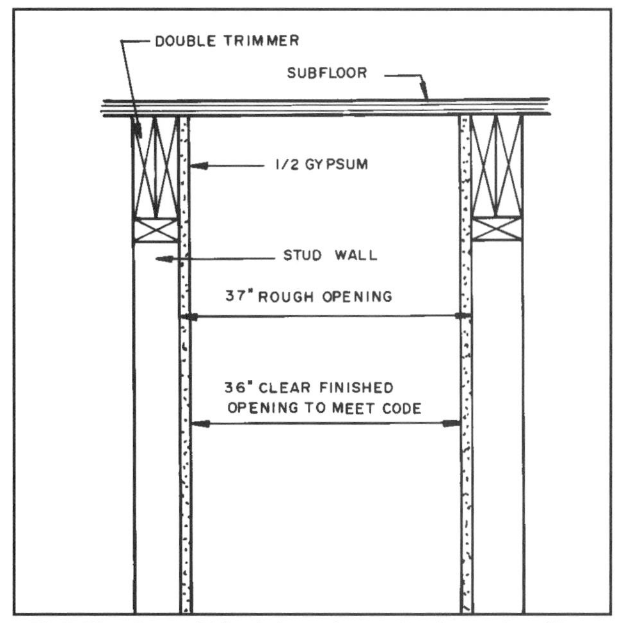

33-5 The clear width of the stair opening is regulated by the local building code.

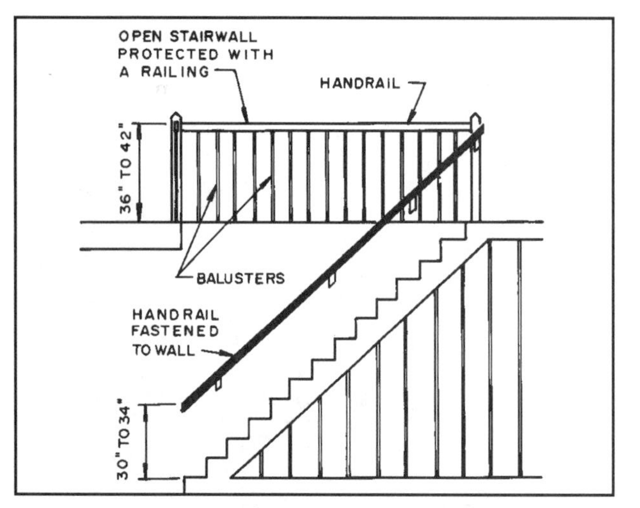

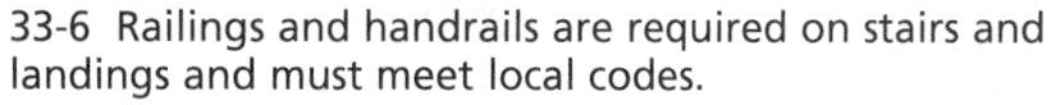

33-6 Railings and handrails are required on stairs and landings and must meet local codes.

2. **Handrail**—A handrail is required on one side of a stair in single-family dwellings. If a stair has less than three risers, no handrail is required. The handrail can project a maximum of 3½ inches (89mm) from each side of the stair.

   The handrail must be continuous for the full length of the stair and extend 6 inches (15.2cm) beyond the top and bottom risers.

   The handrail must be at least 30 inches (762mm) but not more than 34 inches (863mm) above the tread nosing. The handrail should be 1½ to 2 inches (38 to 50mm) in diameter and have at least a 1½-inch space between the handrail and the wall **(see 33-6)**.

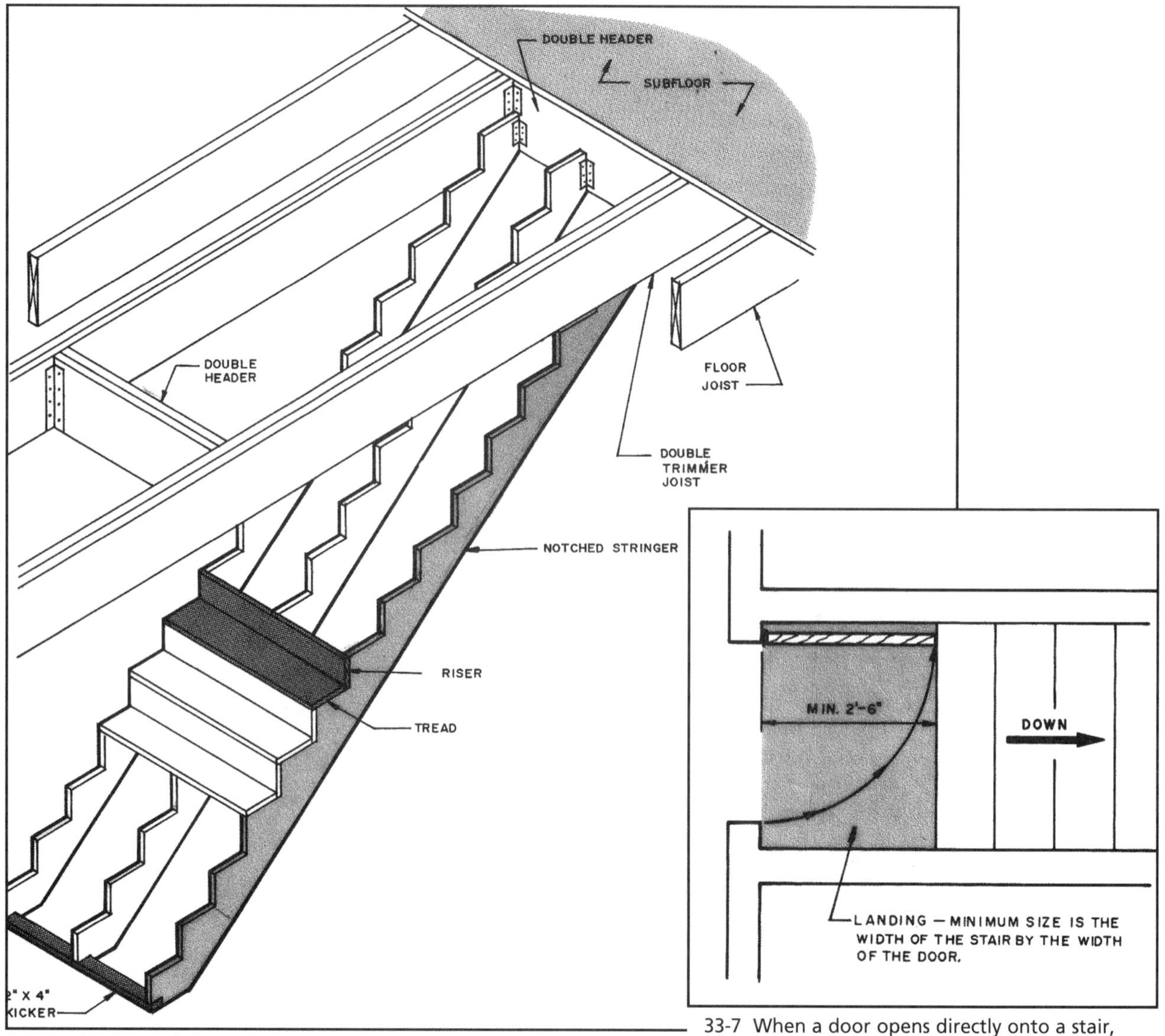

33-8 The parts of a straight stair.

33-7 When a door opens directly onto a stair, codes require a landing be provided.

3. **Railings**—For single-family dwelling units, railings should be 36 inches high **(see 33-6)**.
4. **Headroom**—For single-family dwellings, there must be a 6′-8″ (2030mm) clearance on major stairs. Some codes permit 6′-6″ (1980mm) on minor stairs, such as basement stairs.
5. **Landings**—The dimension in the direction of travel must be equal to the width of the stair but not exceed 4′-0″ (1220mm) on straight-run stairs. The vertical distance between landings should not be more than 12 feet (3.6m) unless a landing is placed halfway up the stairs.

   When a door opens directly onto a stair a landing is required immediately inside the door. It should be equal to the width of the door **(see 33-7)**.
6. **Treads and Risers**—Single-family dwellings with less than 10 occupants can have a riser of not less than 4 inches (101mm) or more than 7 to 7¾ inches (178 to 196mm). The minimum tread depth is 9 to 11 inches (229 to 279mm) depending on the code used. All the risers should be the same size, and all treads should be the same size. Any variation can cause accidents.

These requirements vary from one local code to another. Again, check your local building code.

The width of winders 12 inches (305mm) from the narrow side of the tread must equal the specified tread width. No part of the tread should be less than 6 inches (152mm) wide.

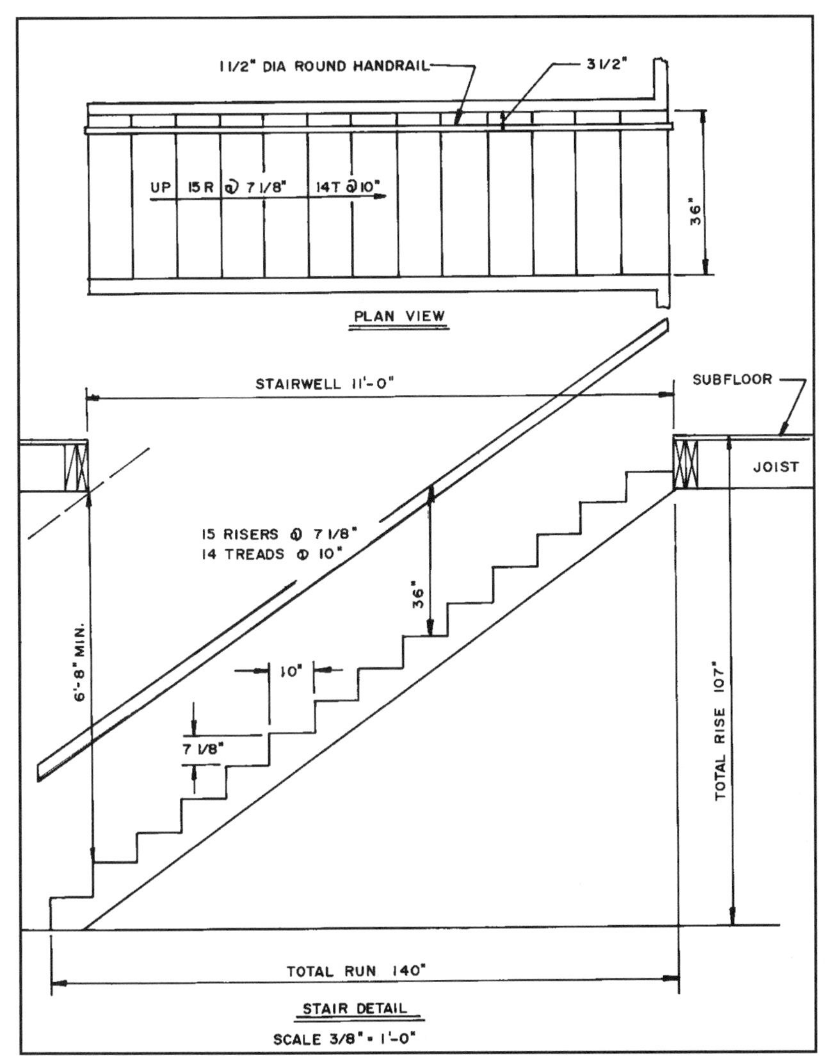

33-9 A typical detail drawing found on an architectural working drawing giving basic information on a stair to be built between two walls.

PARTS OF A STAIR

The parts of a typical stair are shown in **33-8**. The names of the parts vary sometimes in different sections of the country. The **stringer** (also called a **carriage**) is the main structural member and should be of quality stock and free of large knots and other defects. The **kicker** is nailed to the subfloor and resists movement at the bottom of the stair. The **tread** is the horizontal board on which you step. The **riser** is a vertical board that encloses the stair.

A TYPICAL STAIR WORKING DESIGN DRAWING

The drawing in **33-9** is typical of what is found on working design drawings. It gives all the information needed by the carpenter to build the stairs. Factory-manufactured stairs sometimes have instructions sent with them to assist the carpenter. The drawing shows the plan view and a side view giving total rise, total run, and tread and riser data. Notice that there is one less tread than riser. This is because the floor at the top of the stair replaces one tread.

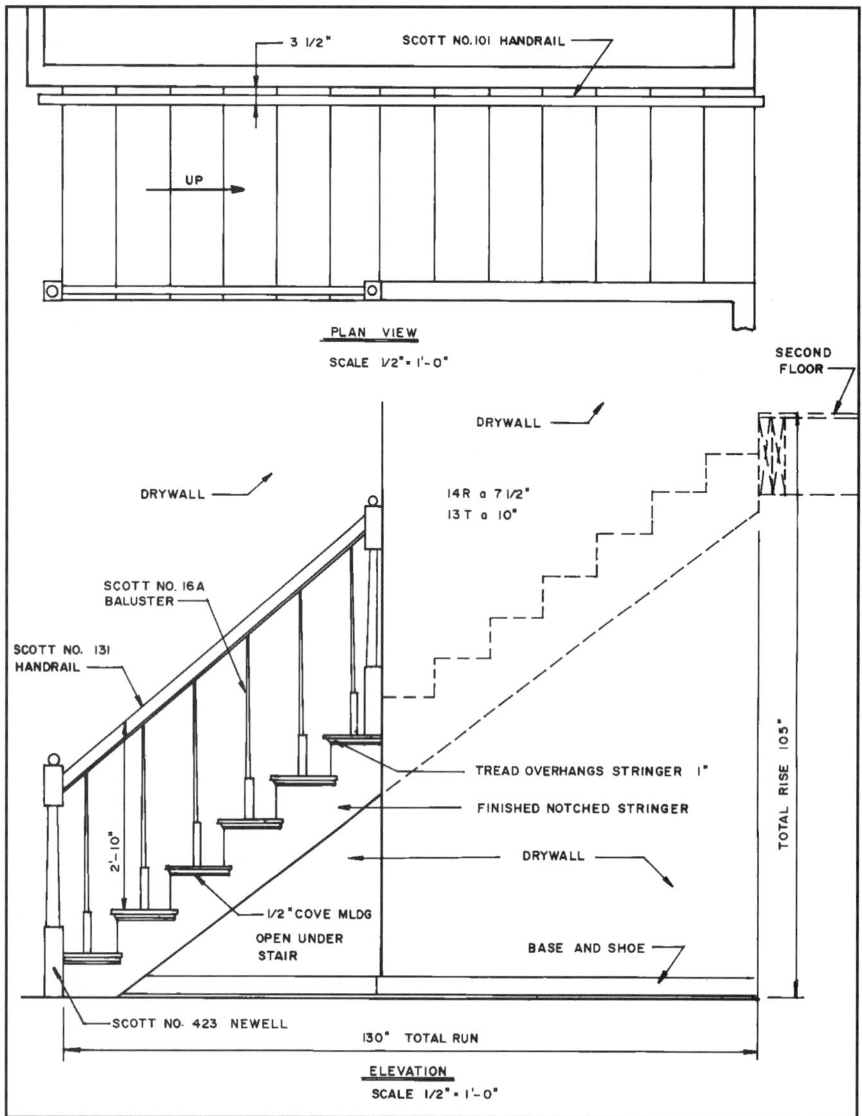

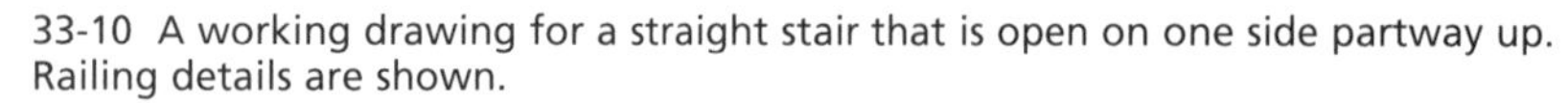
33-10 A working drawing for a straight stair that is open on one side partway up. Railing details are shown.

A typical working drawing for a stair with one side open is shown in **33-10**.

Occasionally the stair is designed with the top tread flush with the double header at the second floor as shown in **33-11**. When this is done, the stair stringer will have the same number of treads as risers and the total run will be longer because of the extra tread.

FIGURING TREADS & RISERS

If the working drawings do not give the tread and riser data, then they must be calculated. The restrictions set by the building code must, as always, be observed. Begin by measuring the actual total rise as shown in **33-9**. Be certain to measure the **actual distance** on the job rather than relying on the distance given on the drawing. Select the rise, such as 7 inches (178mm). For this example, divide the total rise by 7 inches. This will give the number of risers. Usually it does not work out to a whole number. In 33-9 the total rise was 107 inches and, when divided by the risers, 107 divided by 7, gave 15.28 risers; to get a whole number of risers, try dividing the whole numbers above and below the fractional number to see what rise they would require. Dividing 107 by 15 risers gives a rise of 7.13 inches; dividing 107 by 16 risers gives a rise of 6.68 inches. Most likely, 15 risers at 7.13 inches (7⅛ inches or 181mm) would be used. Since you have 15 risers you will have 14 treads. If the specified tread size is 10 inches (254mm) the total run will be 14 treads × 10 inches or 140 inches (3556mm).

Now check to be certain you have the required headroom. Measure from the double header to a distance equal to the height of the tread directly below it, as shown in **33-9**. If all is going well, the clearance will be greater than the minimum 6′-8″

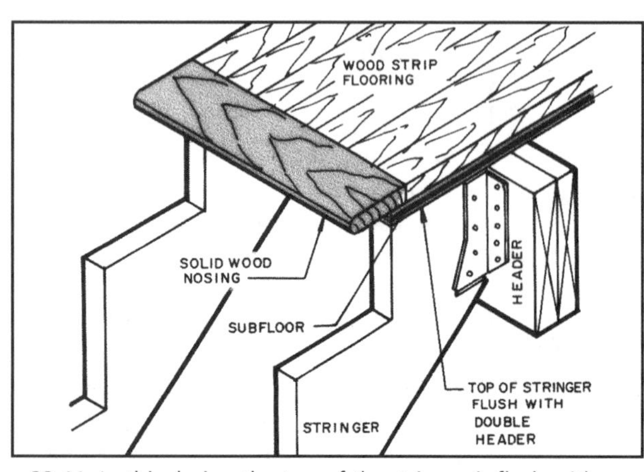

33-11 In this design the top of the stringer is flush with the header and the subfloor runs over the stringer. This stair will have the same number of treads as risers.

(2030mm). If it is not possible to open the stairwell enough to get the required headroom, the header can be set back and the edge sloped, as shown in **33-12**.

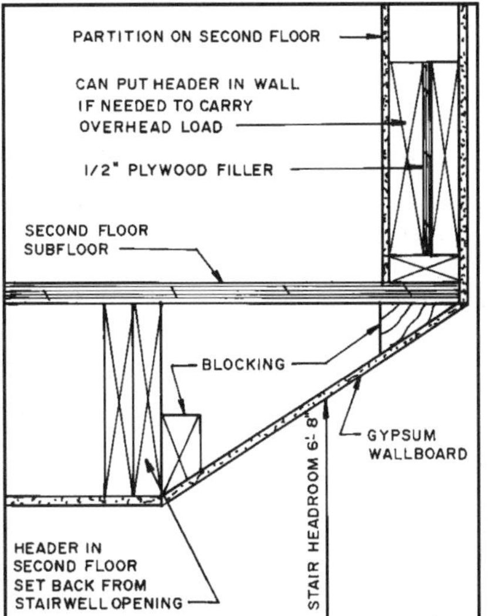

33-12 One way to provide additional headroom.

LAYING OUT THE STRINGER

First it is necessary to calculate the approximate length of the stringer so that stock of the proper length can be secured. This is done by using the Pythagorean theorem, which is used to find the hypotenuse of a triangle **(study 33-13)**. The Pythagorean theorem states that the square root of the sum of the squares of the sides of a right triangle equals the length of the hypotenuse. The hypotenuse in stair-building represents the length of the theoretical stringer.

To find the hypotenuse **(refer again to 33-13)** you take the square root of the sum of the squares of the other two sides of the triangle. Any pocket calculator will have a square root key to enable you to do this operation. In our example, the hypotenuse (carriage) was 14.68 feet, so you need 16-foot stock. It is always best to get stock a little long, because so often the ends have splits that must be cut away. Douglas fir and yellow pine are strong and suitable for this.

To lay out the stringer, use a framing square. Put stair gauges on the riser and tread sizes, as shown in **33-14**. Stair gauges are small clamp-like devices fastened to the framing square. Begin the layout at the bottom by marking the bottom cut, then move the square and mark the next riser and tread. Continue up the stock until all have been marked. Lay out the kicker notch at the bottom side. This is usually a 2 × 4 nominal (1½ × 3½-inch or 39 × 89mm) notch.

Most codes require that the stringer have 3½ inches of uncut stock between the back edge of the carriage and the finish cut.

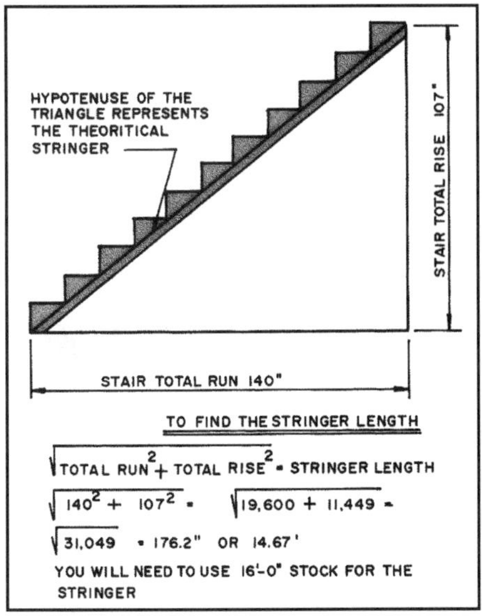

33-13 How to calculate the approximate length of the stringer.

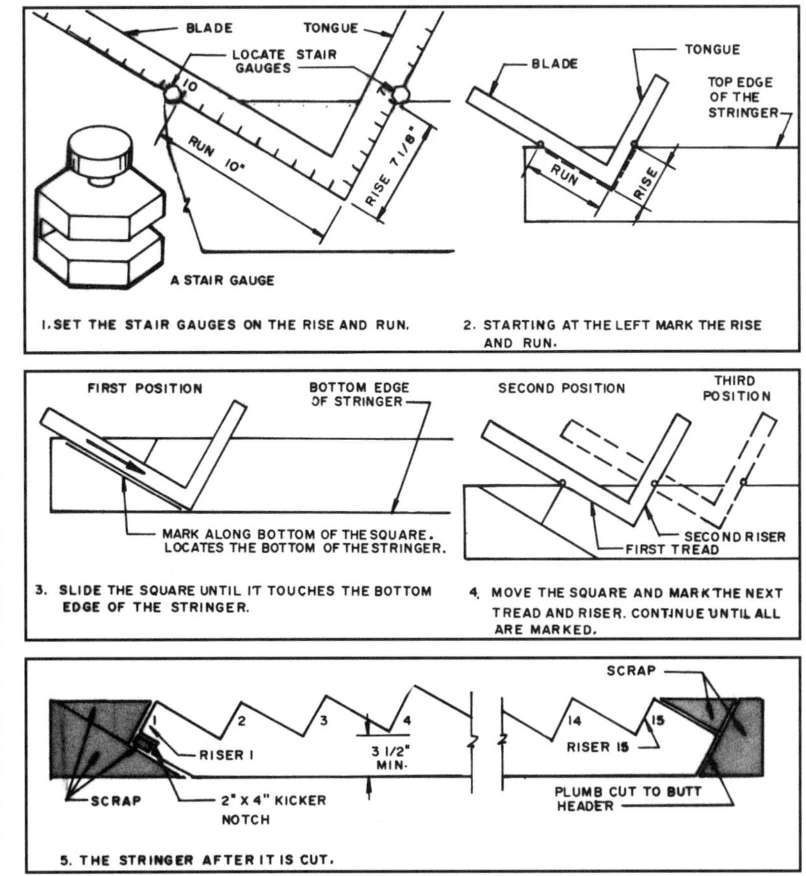

33-14 The steps to lay out a stair stringer.

DROPPING THE STRINGER

Before cutting the stringer, it is necessary to consider if and how much the carriage will need to be lowered to allow for conditions at the subfloor **(study 33-15)**.

Part 1 of 33-15 shows the stringer as laid out and cut. It is resting on the subfloor. The first rise must be reduced by an amount equal to the thickness of the tread board. If this were not done, the total rise for the first step would be 7 inches (178mm) plus the thickness of the tread. This makes the first rise greater than the rest and is **not acceptable**.

Part 2 of 33-15 shows the stringer shortened by a distance equal to the thickness of the tread board.

Part 3 of 33-15 shows the stringer on the subfloor with ¾-inch- (19mm) thick finished wood flooring butting it. The stringer must be shortened by a distance equal to the thickness of the tread board minus the thickness of the finished wood flooring.

CUTTING THE STRINGER

The cuts can be made with a portable circular saw. Some carpenters prefer to make all the tread cuts first and then do all the riser cuts. Be aware that this saw undercuts at the corner where the tread and riser meet. If this does not leave the required 3½ inches (89mm) of uncut stock, do not cut to the corner with the circular saw. Finish the cut with a saber saw or handsaw.

Considerations before Installing Stringers

Usually a stair will butt one or possibly two walls. These walls will eventually be covered with drywall or some other finish material. To protect the finished wall from damage, a notched or closed **skirt** is installed on each wall. The skirt not only protects the wall from damage but covers the joint between the wall and stringer.

A closed skirt is shown in **33-16**. The skirt fits between the drywall and the stringer and is nailed to the studs. The top edge of the closed skirt is placed about 3 to 4 inches (approx. 70 to 100mm) above the stair tread nosing. The end of the skirt at the floor is plumb-cut to butt the base.

The installation of such a construction often proceeds by installing the stringers long before any drywall is in place.

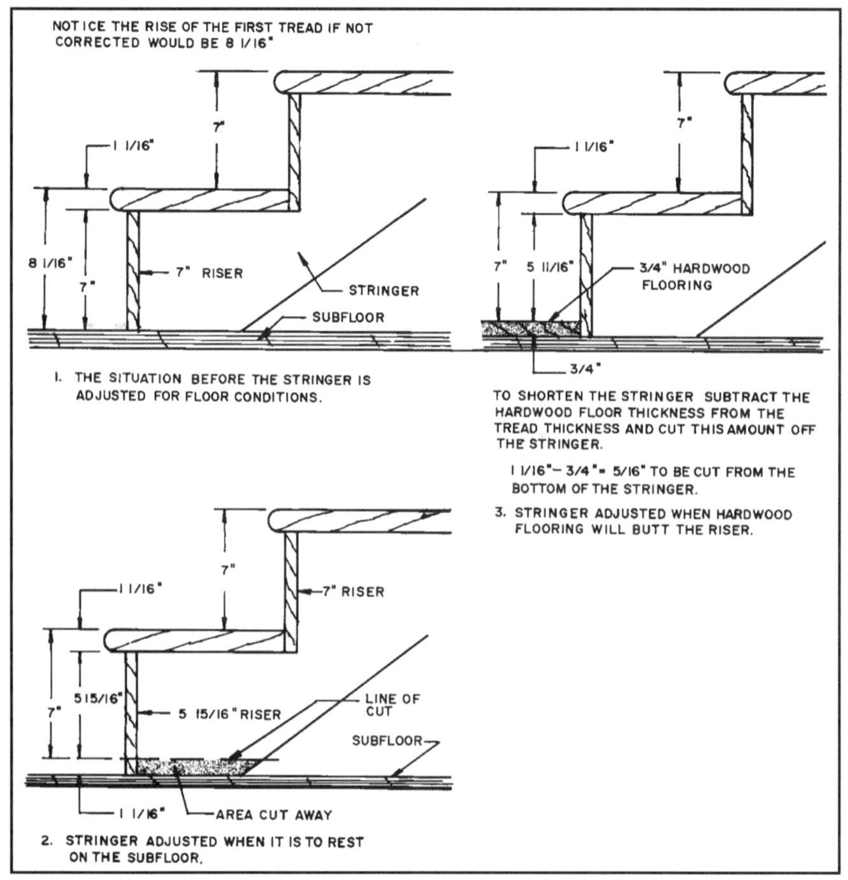

33-15 How to figure the amount that the stringer needs to be shortened.

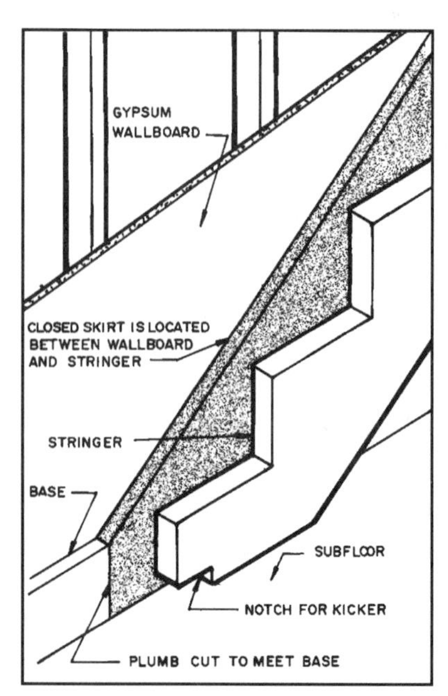

33-16 The closed skirt is installed between the drywall and the stringer.

The stringer is spaced away from the studs a distance needed to install drywall and the closed skirt at a later date. Usually a 1½-inch (38mm) space is adequate for ½-inch (13mm) drywall plus a ¾-inch (19mm) closed skirt **(see 33-17)**. The stringers are needed early during the construction to provide access to the second floor. Temporary treads of two-inch (51mm) stock are nailed in place. These are removed after the drywall is in place and the finish carpenter is ready to install the closed skirt, treads, and risers.

Another technique is to use a notched skirt that sits on top of the stringer as shown in **33-18**. A gap for the drywall is left between the studs and the stringer next to the wall. A gap of ¾ inch (19mm) is adequate for ½-inch (13mm) drywall. Adjust to suit other thicknesses of wall finish material.

INSTALLING THE STRINGERS

The assembly of a closed skirt and the stringers is shown in **33-19**. The skirt is nailed through the gypsum wallboard into the studs. The stringers are secured to the header. The following figures will show a number of ways this assembly can occur. While some omit the kicker, it is good insurance for a stable installation.

Notice in **33-19** there is a second stair directly below the one to the second floor. This is typically done to save floor space. Two stairs occupy the space needed for one. Notice also the use of three stringers. Codes usually require a center stringer if the stair is 30 inches wide or wider. Since codes require residential stairs to be 36 inches wide the center stringer is always installed.

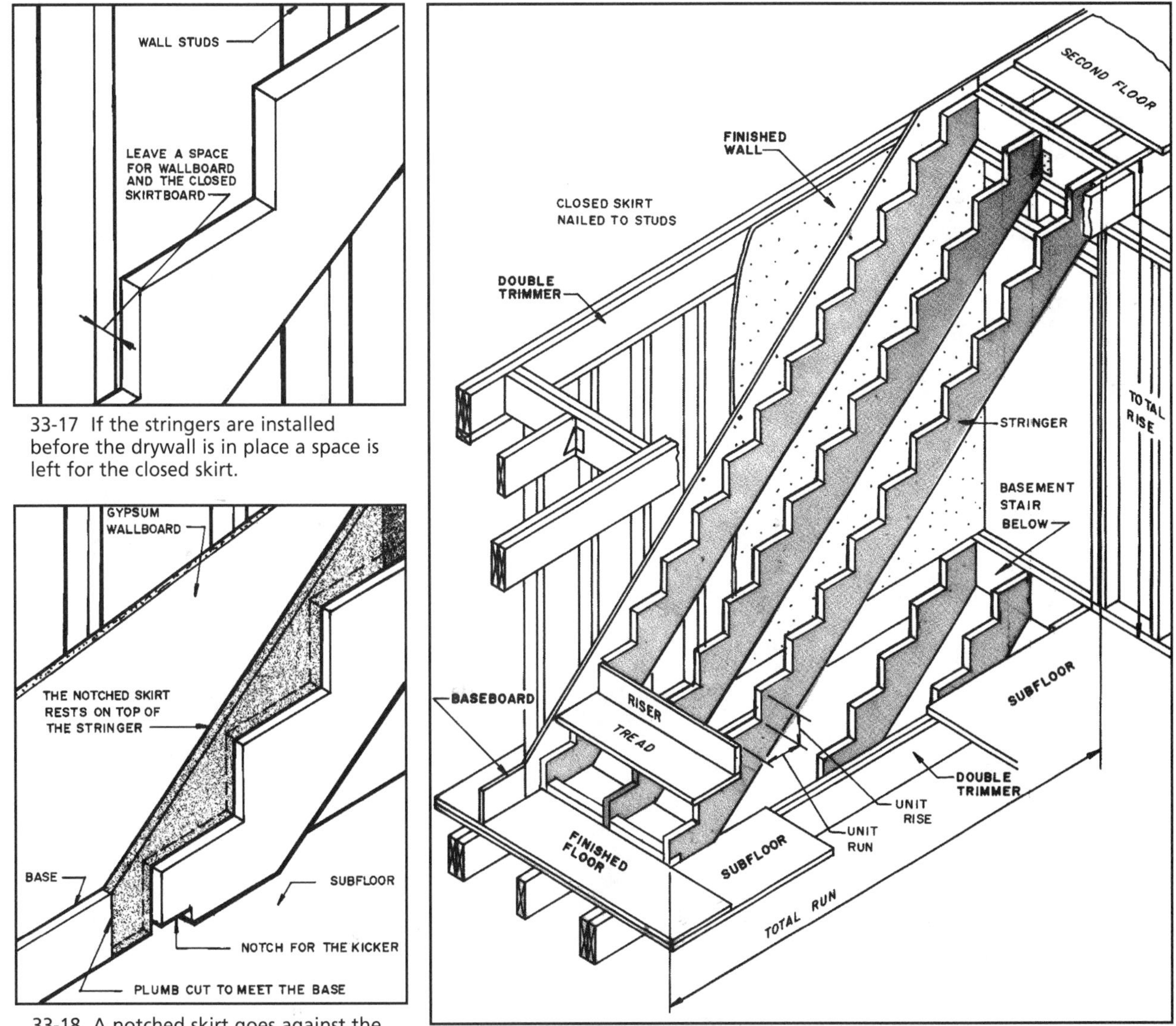

33-17 If the stringers are installed before the drywall is in place a space is left for the closed skirt.

33-18 A notched skirt goes against the drywall and on top of the stringer.

33-19 Skirt and stringer assembly for a straight stair.

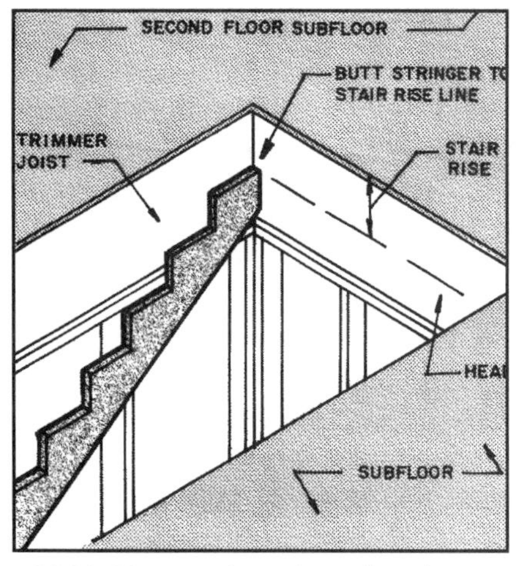

33-20 Measure down from the subfloor to locate a line representing the stair rise.

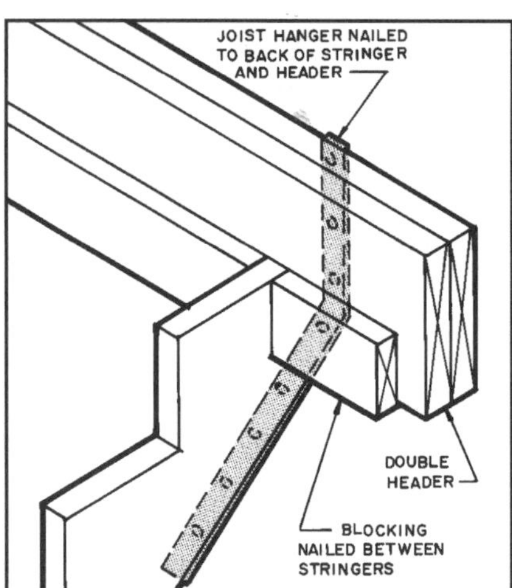

33-21 Stringers can be hung with metal straps.

Generally the stringer is designed to butt the double header on the stairwell opening and drop down a distance equal to one riser **(see 33-20)**. If the second floor is to have a wood flooring over the subfloor, this drop would have to allow for that even though the finished wood flooring has not yet been installed.

There are a number of ways to install the stringers. One method uses metal joist hanger straps that are nailed to the header and to the back edge of the stringer **(see 33-21)**. The strap is usually 18 to 24 inches (approx. 460 to 610mm) long. Since the

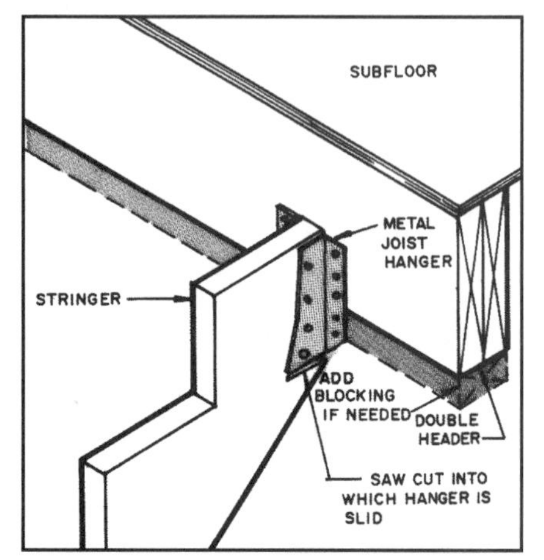

33-22 Stringers can be hung with metal joist hangers.

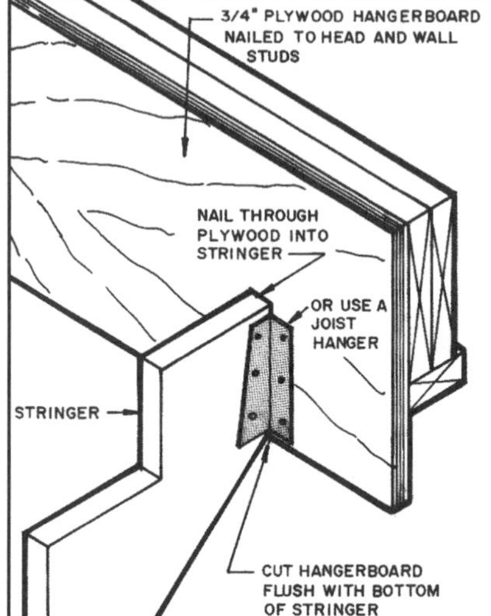

33-23 The header can be widened with ¾-inch plywood and the stringer hung on it.

33-26 This fully assembled temporary stair is used to provide access to the floor above until it is time to build the permanent stair.

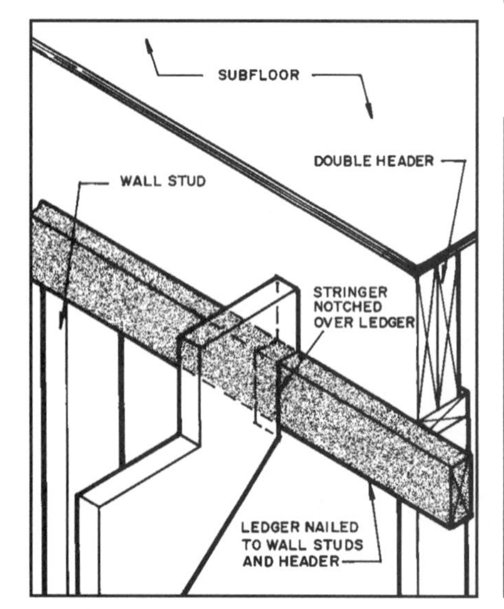

33-24 Stringers can be hung on a wood ledger.

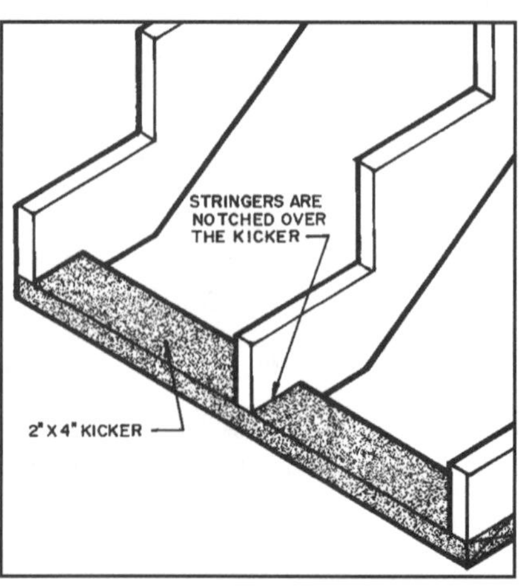

33-25 Kickers are used to hold the bottom of the stringer.

stringers have little side support, blocking is nailed between them and the header.

Another method uses metal joist hangers as shown in **33-22**. The stringer will need a relief cut to permit it to slip on the hanger. The hanger is nailed to the stringer and to the header.

A ¾-inch- (19mm) thick plywood hangerboard can be used instead of metal joist hangers as shown in **33-23**. It is nailed to the header and to the wall studs if there is a wall below the header. The stringer can be secured by nailing through the plywood into the butting edge of the stringer. Some prefer to use metal joist hangers. The hangerboard should run up to the top of the header and be cut flush with the bottom of the butting stringer.

Often there is a wall directly below the double header, as shown in **33-24**. In such a case the top plate and additional blocking serve to back up the stringers. In this instance a wood ledger can be nailed to the wall and the stringer notched around it.

There are times when the top tread is designed to be flush with the second floor as shown earlier in **33-11**. In this situation the top end of the stringer fully butts the header and can be secured with metal hangers or a ledger.

The bottom end of the stringer is notched to sit on a kicker that is nailed to the subfloor. Toenail the stringer to the kicker **(see 33-25)**.

STAIRS DURING CONSTRUCTION

Some builders prefer to install a temporary stair providing access to the second floor. This is an assembled stair unit which is set in place much like a ladder **(see 33-26)**. It is removed intact when the permanent stair is constructed and is used again on another job.

In **33-27** are photos of a finished stringer installation that has temporary two-inch-thick treads providing access to the second floor. What happens to the treads as the drywall is installed is shown in **33-28**. In this photo the mess is on the permanent plywood treads which will be covered by carpet. A thorough cleaning job is necessary.

33-27 Three views: These are permanent stringers with temporary treads that will be removed and replaced with permanent treads.

33-28 These winder treads are covered with gypsum dust and debris and must be thoroughly cleaned before the carpet is installed.

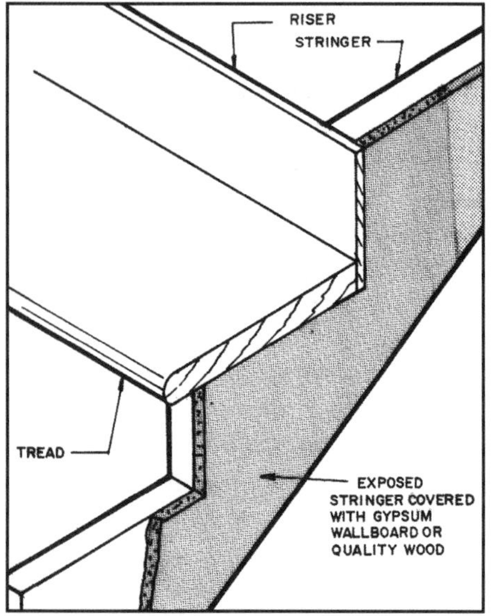

33-29 The exposed stringer is covered with a finishing material such as gypsum wallboard or a quality wood.

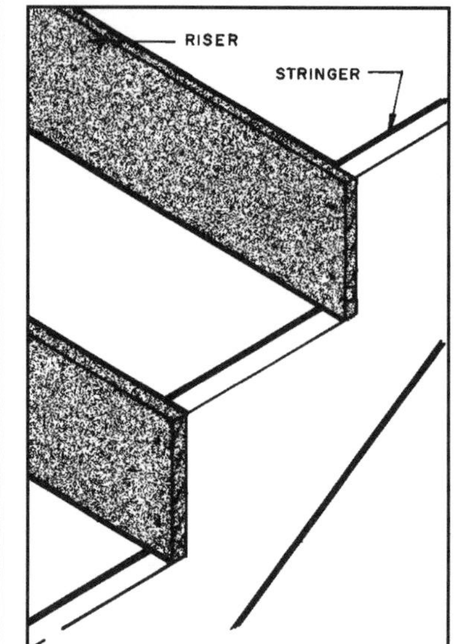

33-30 Install the risers with finishing nails.

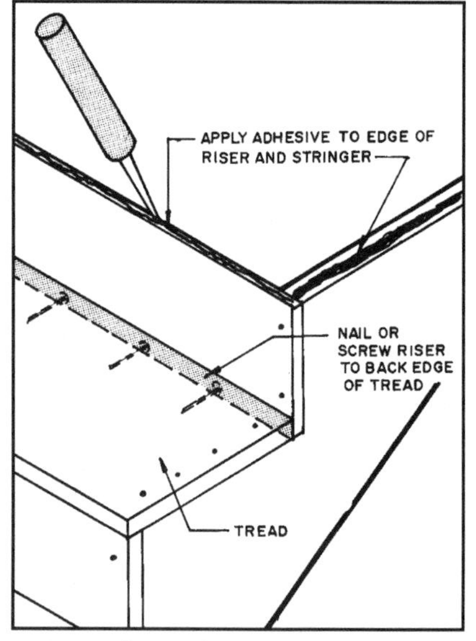

33-31 Glue and nail the treads to the stringer. Nail or screw the riser to the back edge of the tread.

TREADS & RISERS FOR CARPETED STAIRS

If stairs are to be carpeted, the treads and risers can be cut from ¾-inch (19mm) plywood. Some carpenters prefer to use 1⅛-inch (28mm) plywood for the treads. This decision must be made early on as the stair is planned, because it will influence the design of the stringer. Usually the drywall and skirt are in place before the treads and risers are installed. The treads and risers should be carefully cut square and fit close to, but not touch, the skirt. The small gap will be covered by the carpet and will eliminate a possible source of squeaks caused by the treads rubbing on the skirt board. If one side of the stair is to be exposed, the stringer will be covered by a finish skirt or drywall as shown in **33-29**.

Begin installation by gluing and nailing the risers first **(see 33-30)**. They can be nailed with 8d finishing nails that are set and covered. This is important as these are installed to take any bow out of the stringers. Some carpenters will install the bottom riser first, then the top riser, then move down a few to install several in the center to pull the stringers into line.

Finally install the tread boards with three 8d or 10d finishing nails into each carriage. Apply an adhesive to the top edge of the riser and the stringer before nailing. This will help reduce possible squeaks **(see 33-31)**.

Conditions at the top of the stair will vary. If the stair and second floor are to be carpeted, they should be framed as shown in **33-32**. The nosing will be the same as that used on the stair. However, sometimes the nosing is omitted on the stair.

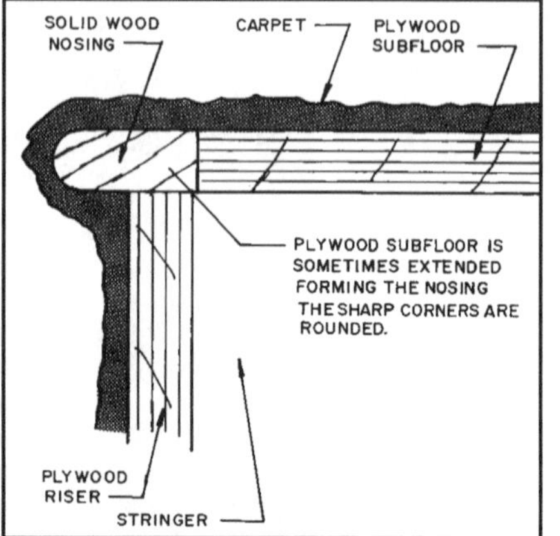

33-32 When the stair and second floor are to be carpeted, the carpet on the floor usually runs over the nosing and butts the carpet on the stair below the nosing.

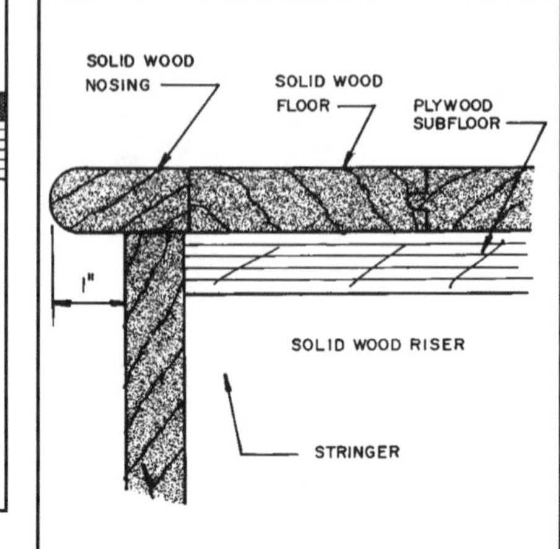

33-33 When the second floor is to be solid wood flooring, a nosing strip is used at the riser.

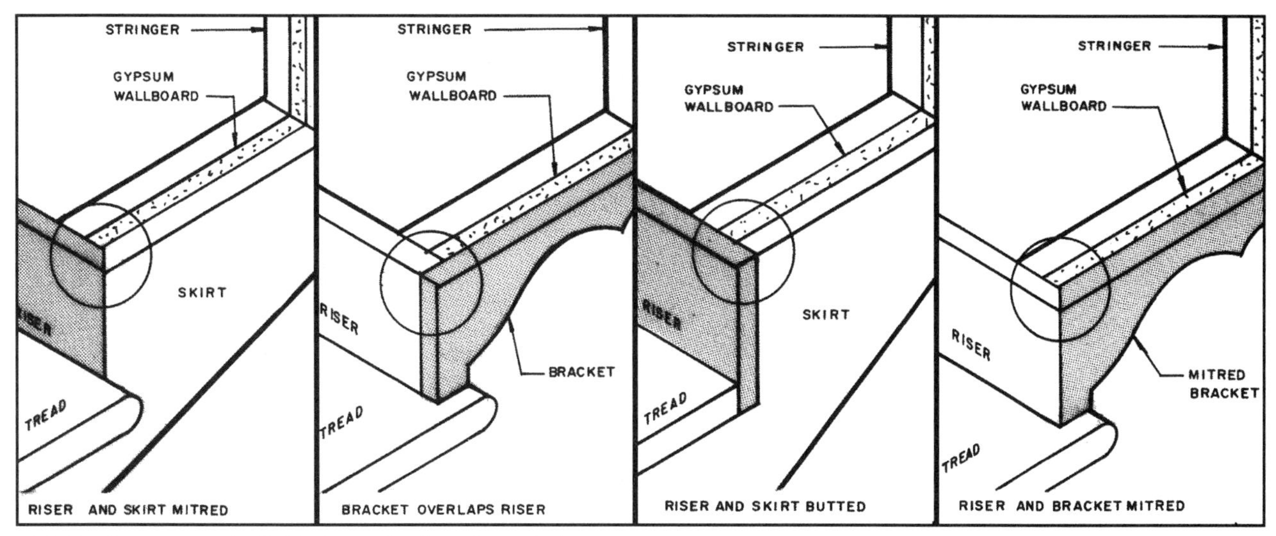

33-34 Various ways to finish the ends of solid wood risers.

INSTALLING TREADS & RISERS FOR SOLID WOOD STAIRS

Factory-manufactured treads and risers are available in a variety of hard- and softwoods. The risers are typically ¾ inch (19mm) thick and the treads 1¹⁄₁₆ inches (27mm) thick.

If the second floor is to have a solid wood floor, a nosing will be installed extending over the riser one inch (25mm) **(see 33-33)**.

There are several ways solid wood treads and risers can be installed. The riser may be nailed to the stringer with the end grain permitted to show, or the finished skirt could be mitered and the riser mitered, forming a corner where no end grain is visible. Another approach is to miter the riser and use a bracket to hide the end grain **(see 33-34)**.

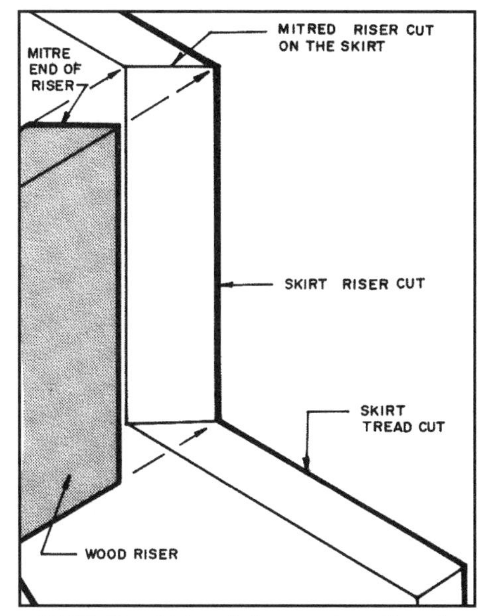

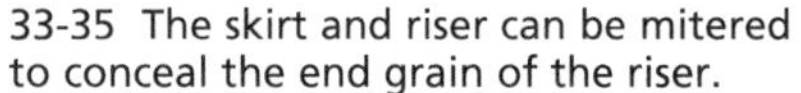
33-35 The skirt and riser can be mitered to conceal the end grain of the riser.

The mitered skirt has the riser cut on a 45-degree angle to meet the miter on the end of the riser. The skirt is laid out much the same as the stringer except that the riser is extended beyond the edge of the riser cut on the stringer at a 45-degree angle **(see 33-35)**. Therefore, the mitered skirt layout would be like that shown in **33-36**. Cutting the riser miter requires careful, accurate work. The exposed edge must be sharp and free from chips. Some cut the miter slightly less than 45 degrees. The gap on top is covered by the tread **(see 33-37)**.

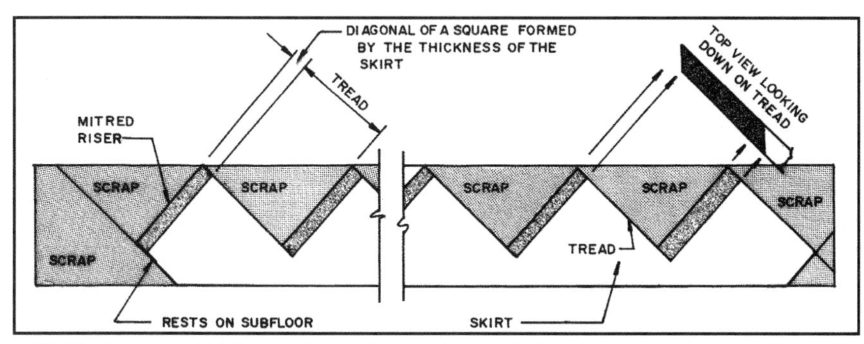

33-36 A typical skirt layout for a mitered riser and skirt corner.

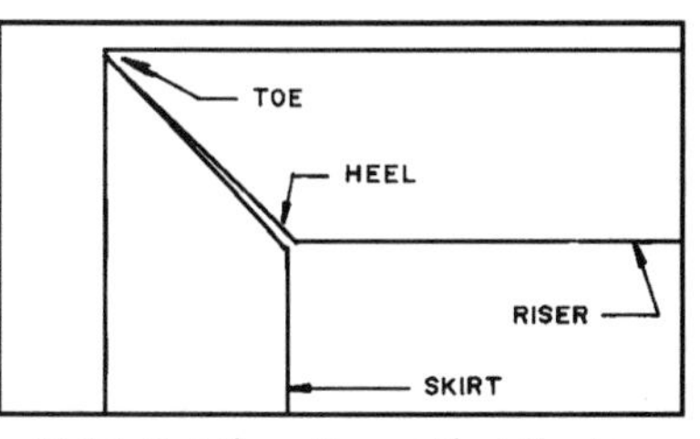

33-37 Cut the miter so that the toe touches and there is a small gap at the heel.

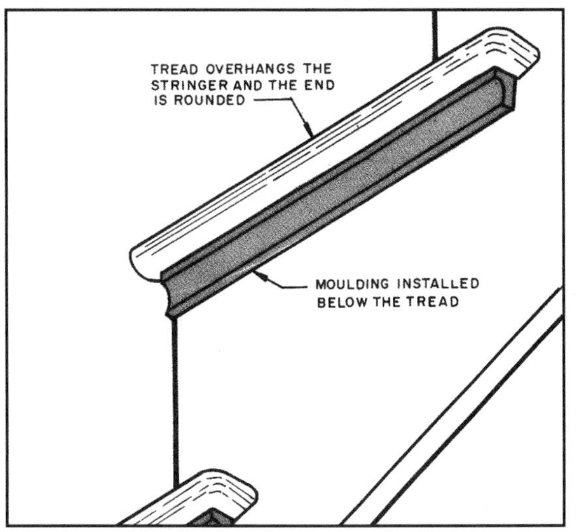

33-38 A tread on an exposed stringer can overhang the stringer and have a decorative molding or a bracket below the tread to cover any gap.

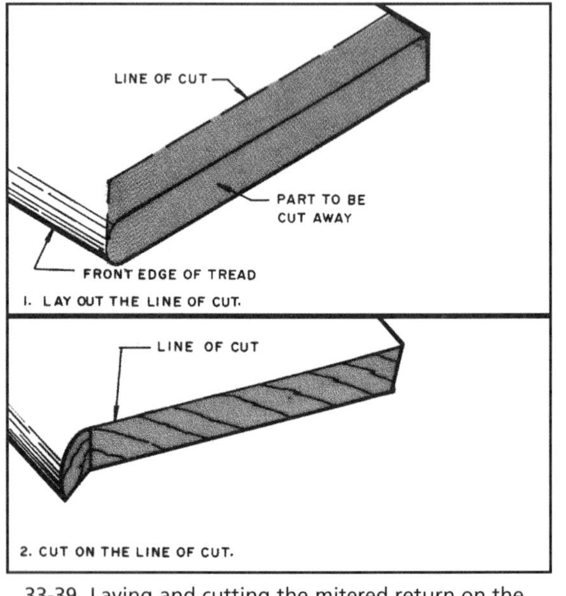

33-39 Laying and cutting the mitered return on the end of the tread.

33-41 This stair has been covered to protect it from damage during the final stages of construction.

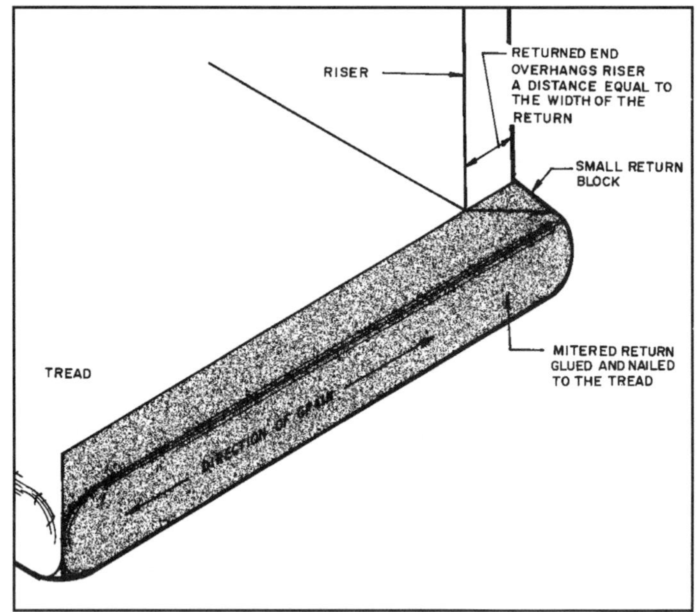

33-40 Cut the mitered return piece and glue and nail it to the end of the tread.

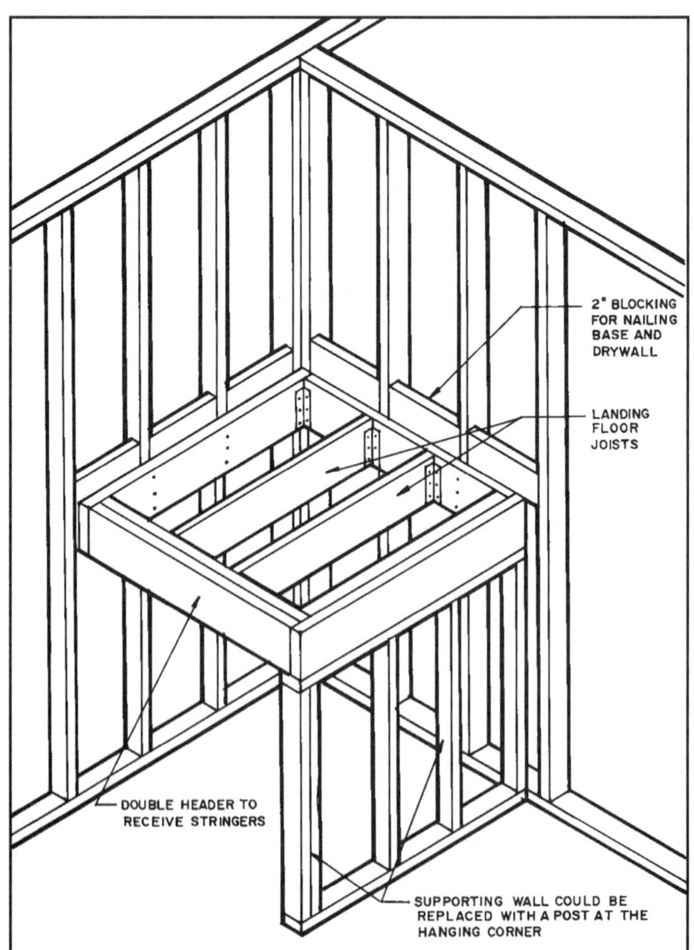

33-42 Typical framing for the landing of an L-shaped stair.

The tread should bear on the stringer, not the skirt, so that the skirt is set a little below the tread. The gap is covered with a small molding **(see 33-38)** or a bracket **(refer to 33-34)**.

The tread overhang can be return-mitered or shaped. The shaped return is cut with a router, and care must be taken to avoid splitting the end of the tread.

The mitered return is more difficult to make. It should be made before the tread is cut to length. Begin by laying out the cuts on the treads **(see 33-39)**. Carefully cut to the lines. A fine-toothed blade is required so that a smooth, clip-free edge is left. Usually the returns are not installed on the treads until they are secured to the stringer and the newel posts—the large posts at the bottom of the stair—are in place. The mitered returns are glued and nailed in place **(see 33-40)**. The back end is also mitered, and a small return block finishes off the installation.

When installing the hardwood treads and risers, butt them against the skirt. When notched stringers are used, the ends must be clearly cut and fit squarely against the skirt. Over time this may produce a small, visible gap. To avoid this, a housed stringer could be used. Glue and nail or screw the hardwood treads in place. The screw heads can be covered with wood plugs.

After hardwood treads and risers are installed they must be covered with some form of protective material. By now almost all interior work should have been completed, so traffic will be minimal. An instance where the builder covered a stair with scraps of rigid foam sheathing and overlaid the entire stair with plastic sheeting is shown in **33-41**.

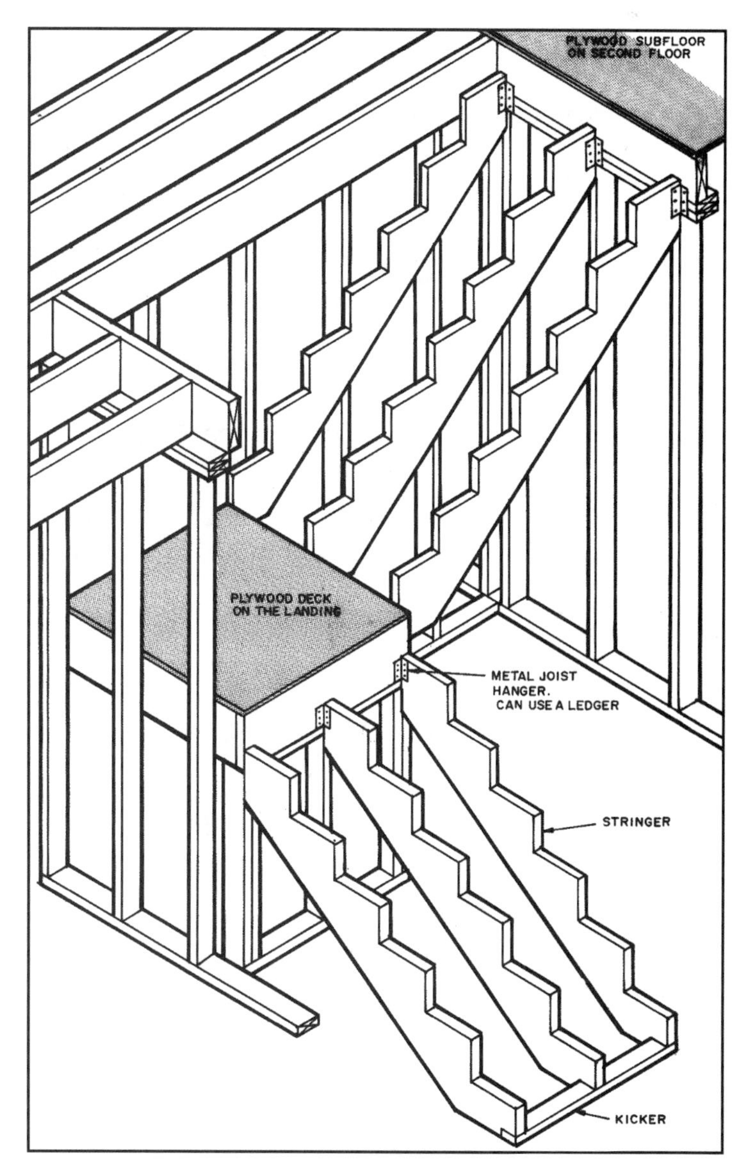

33-43 Typical framing for a stair with a landing.

BUILDING LANDINGS

The actual framing of a landing will vary some with the specific installation. In all cases it must be very sound and well constructed. The following examples are typical.

A landing for an L-shaped stair should be as wide and deep as the finished width of the stair. Generally it is in the middle of the rise, but this is not necessarily so. It can occur as needed, and this is determined by the floor space available.

The design of the landing and the two flights of stairs is the same as described for straight flights. Remember that the landing actually replaces a tread in the total distance. Therefore, figure the tread and riser sizes as described earlier for a straight stair. At one of the tread locations, position the landing. The treads and risers in both flights should be the same.

The landing is framed in much the same way as the floor **(see 33-42)**. Usually 2 × 8 joists 16 inches (406mm) O.C. are used on small landings. They are nailed to the studs and hung with metal joist hangers. The header against which the stringers rest is often doubled. Posts or a wall is used to support corners without a wall to hold them **(see 33-43)**.

The tops of the stringers are secured to the landing as described earlier. The bottom of the second set of stringers can be fastened to the landing as shown in **33-44**. Remember to allow space between the studs and stringer for the drywall and finish stringer, as described earlier.

Framing for a landing on a U-shaped stair is shown in **33-45**. It is framed as described for the smaller landing. A supporting wall or posts beneath the unsupported double joist add greatly to the stiffness of the landing.

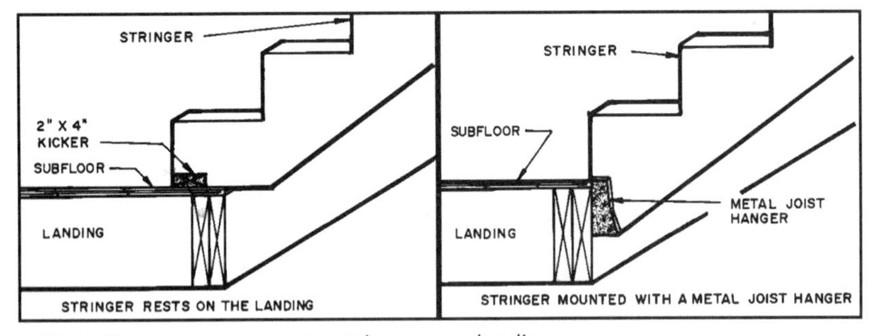

33-44 Two ways to mount a stringer on a landing.

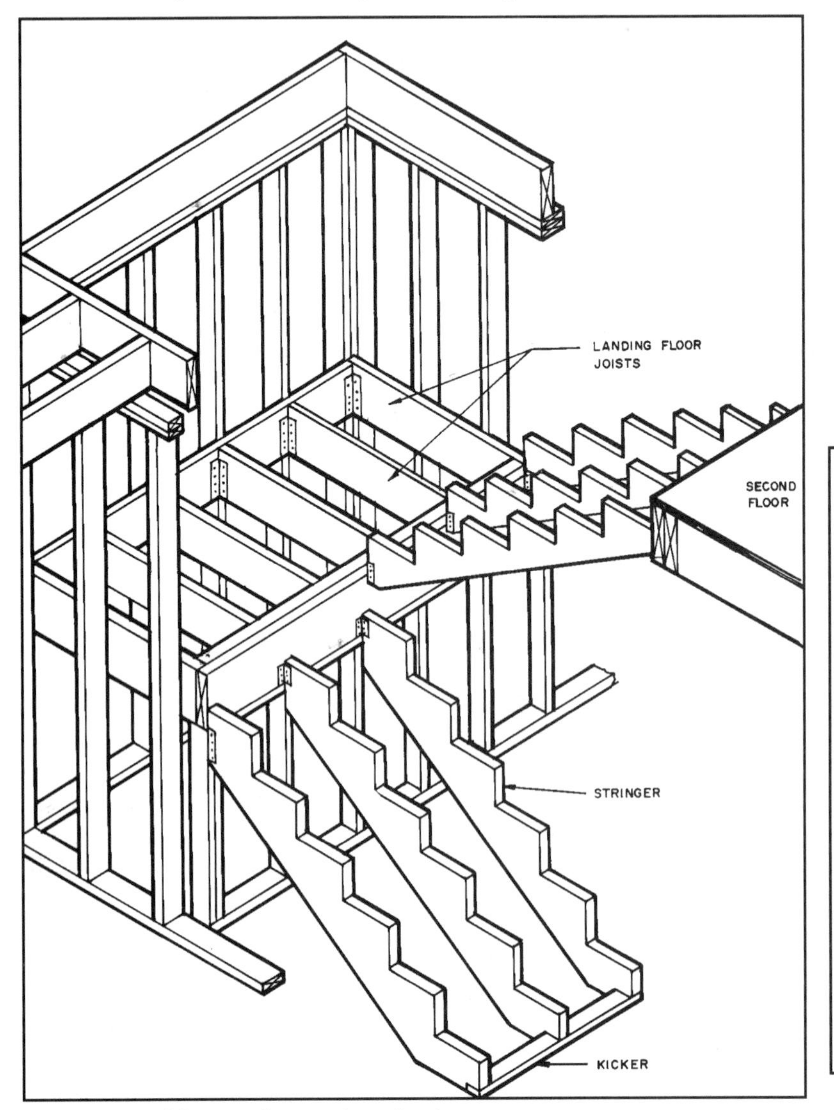

33-45 Typical framing for a U-shaped stair.

BUILDING WINDERS

Winders are used in the same way as landings, which is to enable a stair to make a 90-degree or 180-degree turn **(see 33-46)**. Before designing the winders be certain to check your local building code. Typical requirements are given earlier in this chapter.

Generally three triangular treads are used to turn a 90-degree corner, as shown in **33-47**. This permits the stair to rise the required distance, turn a corner, and take considerably less floor space than if a landing were used.

To begin, calculate the number of treads and risers as discussed earlier. Since you know it takes three treads to turn the corner, the remaining number are divided between the two flights. Usually space requirements are such that the first flight is

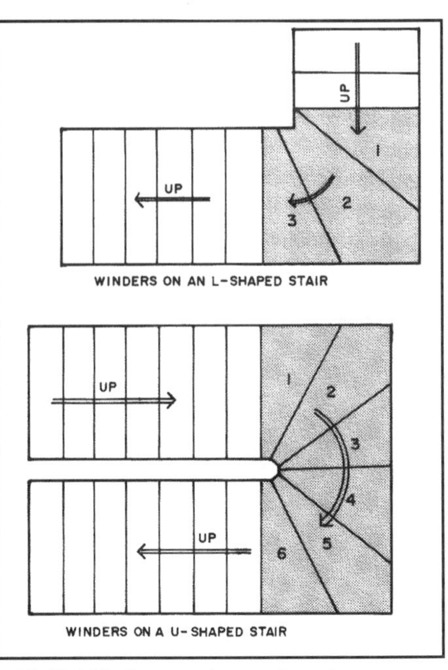

33-46 Winders can be used to change the direction of a stair.

shorter than the second, but this is not mandatory. Notice ahead in **33-49** that the installation begins by constructing a landing as described earlier. Then two triangular treads (the winders) are built on top of this, and the second flight joins the top or third winder. The basic construction is the same as described earlier in this chapter.

Laying Out the Winders

The layout of the winders is an important part of the job because, if not done properly, building codes may be violated.

After the platform is complete, it will have a ¾-inch (19mm) plywood subfloor. The winders can be laid out on this surface. The exact procedure will depend on your local building code. It should be noted that most codes require a winder to be a minimum of 6 inches (152mm) wide at the narrow end and equal to the width of the tread, 12 inches (305mm), in the middle. This will require that for a 36-inch- (915mm) wide stair the platform will have to be 46¼ inches (1174mm), as shown in **33-48**.

The width of the stair (36 inches) is laid out and the 6-inch (152mm) narrow end is drawn. To find the sides of the triangle forming the end of the second tread, use the Pythagorean theorem as follows:

$$\sqrt{(\text{side } \mathbf{A}^2 + \text{side } \mathbf{B}^2)} = 6 \text{ (required small end)}$$

$$\sqrt{36} = 6$$

$$\mathbf{A}^2 = 18, \mathbf{B}^2 = 18$$

$$\mathbf{A} \text{ and } \mathbf{B} = \sqrt{18}, \text{ or } 4\tfrac{1}{4} \text{ inches}$$

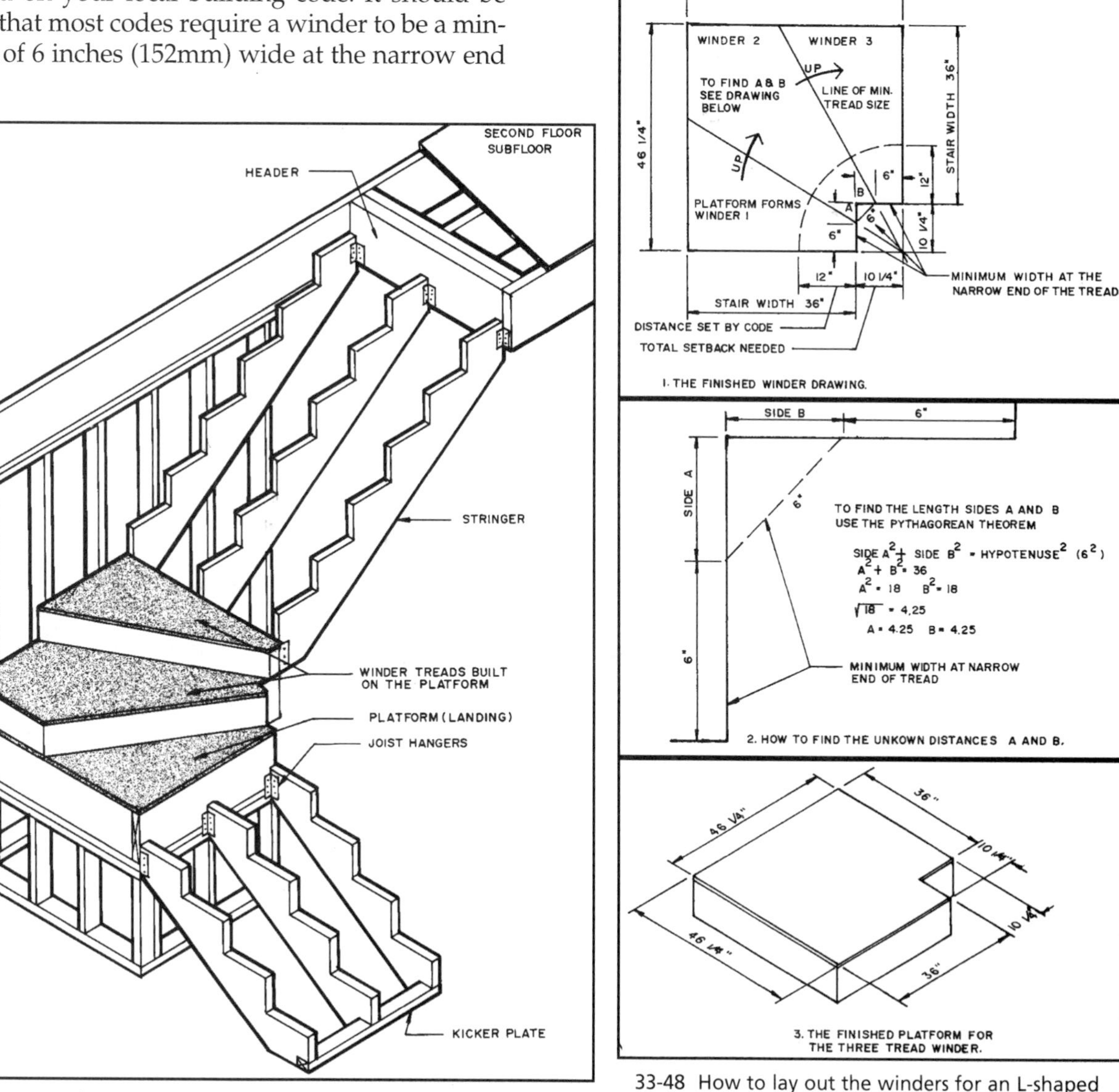

33-47 Typical framing for an L-shaped stair with winders.

33-48 How to lay out the winders for an L-shaped stair.

Building the Winders

While the winders can be built using carriages as explained for straight stairs, it is easier to build each as a separate unit and mount one on top of the other. Typical construction details are shown in **33-49**. The stock used to form each winder unit is two inches (50mm) thick and must be ripped to width to get the correct riser height. For example, if you want a 7-inch (178mm) rise and use a ¾-inch (19mm) plywood tread, the two-inch (50mm) stock must be cut 6¼ inches (158mm) wide. Assemble by gluing and nailing. Be certain the nails are flush or a little below the tread surface so that they do not penetrate the carpet.

Installing the Stringers

Install the stringers as explained earlier for straight stairs. Metal joist hangers or ledgers are commonly used **(refer back to 33-47)**.

HOUSED STRINGER STAIRS

Housed stringers have dadoes cut in their sides into which the treads and risers are inserted. Usually these are bought machined to fit a particular job. However, a carpenter could lay out and route the dadoes on the job. The dadoes are typically routed ⅜ inch (10mm) into the stringer.

A housed stringer is shown in **33-50**. It is used when the stair is enclosed by walls on both sides, or the housed stringer can be on the side butting the wall and open stringers used for the other two **(see 33-51)**.

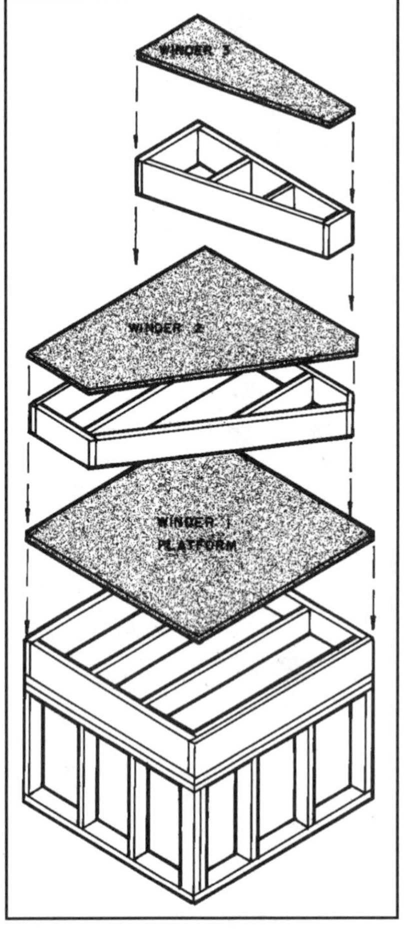

33-49 One way to frame the winders for an L-shaped stair.

Assembling Stairs with Housed Stringers

This discussion assumes the housed stringers are made by a stair manufacturer who also supplies the treads, risers, and wedges.

1. Determine the length of the treads and risers. The factors involved are shown in **33-52**.

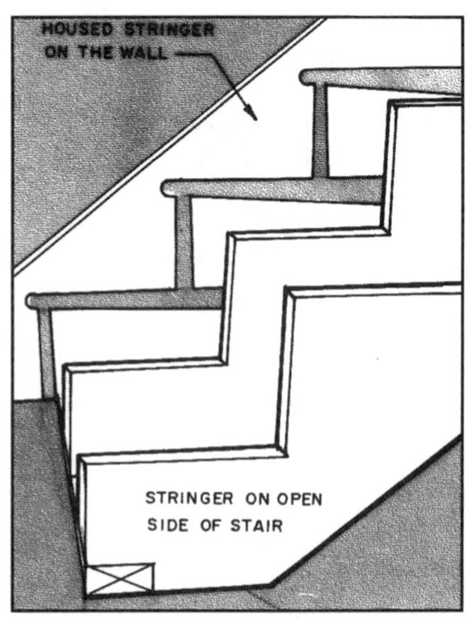

33-51 The housed stringer can be used on the wall when the other side of the stair is open.

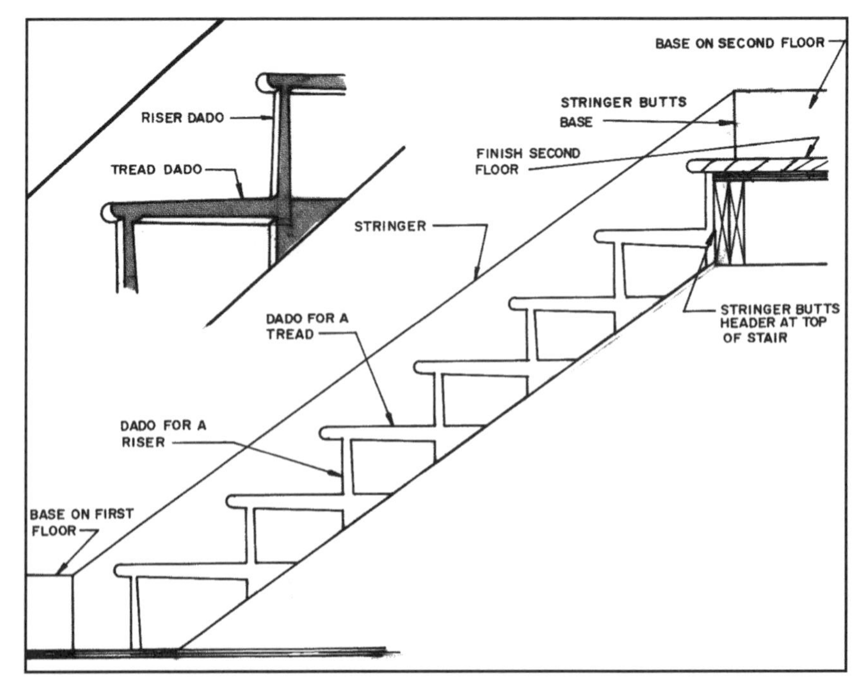

33-50 The housed stringer has the area for the tread and riser recessed.

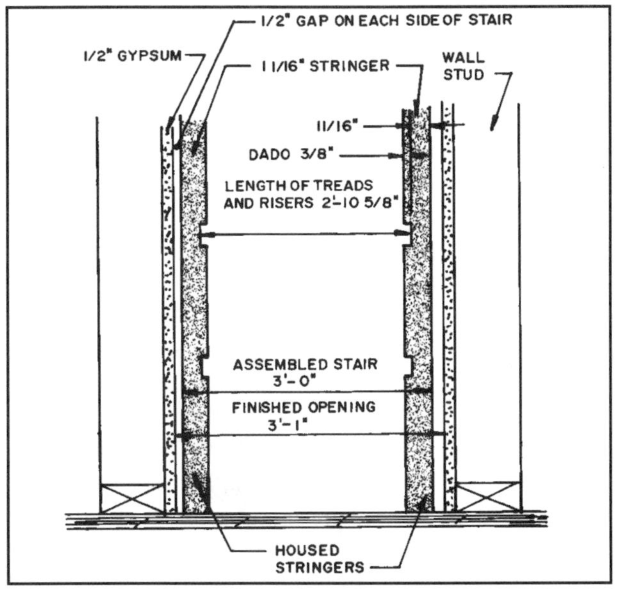

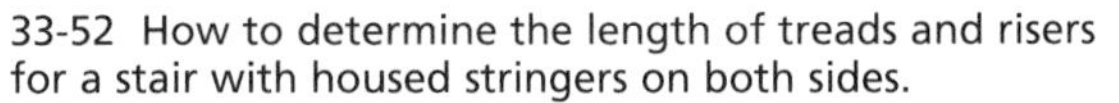

33-52 How to determine the length of treads and risers for a stair with housed stringers on both sides.

33-53 This tool lets you quickly find the length of treads and risers. (Courtesy Stairtool, Inc.)

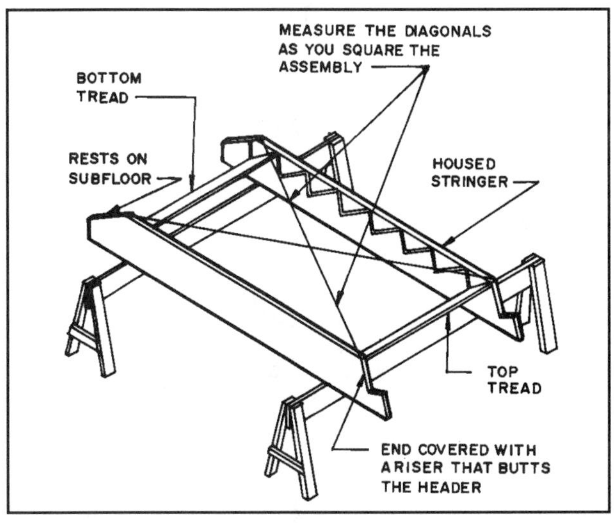

33-54 Begin assembly by installing top and bottom treads.

Measure the distance between the studs forming the stairwell. For our example we will use a three-foot- (915mm) wide stair. Then subtract twice the thickness of the finished wall covering material. Notice that the rough opening minus the gypsum drywall leaves a one-inch (25mm) space for positioning the 3′-0″ (914mm) assembled stair between the walls.

2. Examine the housed stringers. Typically they are $1\frac{1}{16}$ inch (27mm) thick and the dado is $\frac{3}{8}$ inch (10mm) deep. The length of the treads and risers is equal to 3′-0″ minus $1\frac{3}{8}$ inch (35mm) or 2′-10⅝″ (880mm). The amount subtracted is found by subtracting the depth of the dado from the thickness of the stringer and doubling it (two stringers). A tool used to measure tread and riser length is shown in **33-53**. It will fit between stringers, skirtboards, or the dadoes in a housed stringer.
3. Check the treads and risers for width and length. Cut to size as necessary. Notice that the top riser is $1\frac{3}{8}$ inch (35mm) longer than the rest because it is nailed to the face of the stringer at the top of the stair.
4. Place the stringers on sawhorses, as shown in **33-54**. Install the top tread first. Apply adhesive to the dado, and insert the tread so that the front nosing fits snugly into the rounded end of the dado **(see 33-55)**.

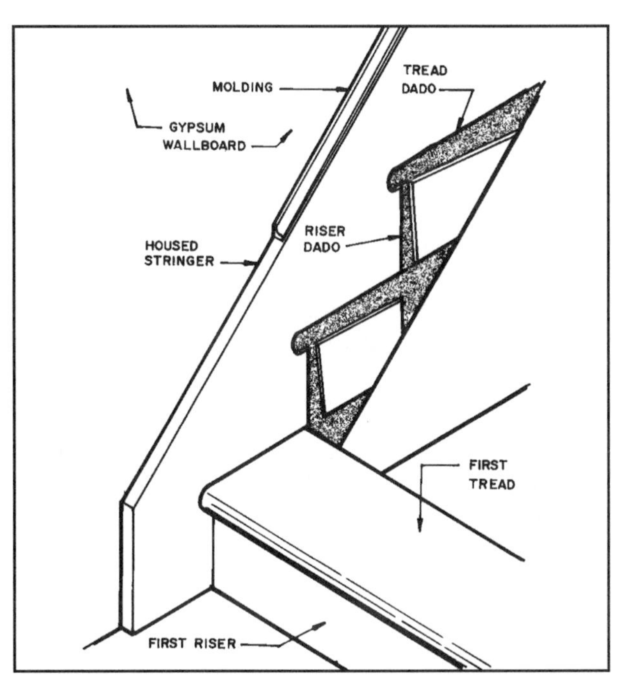

33-55 Insert the tread in the dado.

5. The treads are held in place with glued wedges **(see 33-56)**. Now apply glue to the wedge, and drive it in place between the bottom of the tread and the edge of the dado. Check the front side of the stair (which is facing the floor) to be certain the tread nosing is in place before you set the wedge tightly in place. This may involve tapping the tread some and then tapping the wedge until the joint on the front is closed. Be certain the back edge of the tread is in line with the front of the dado to receive the riser. Drive several 8d box nails through the stringer into the edge of the tread. Be certain one is in the front nosing. These serve to pull the tread fully into the dado.

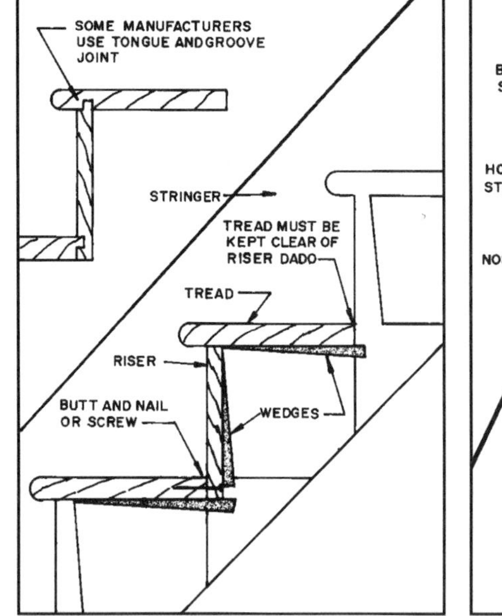

33-56 Glue and wedge the treads and risers in the dadoes in the stringer.

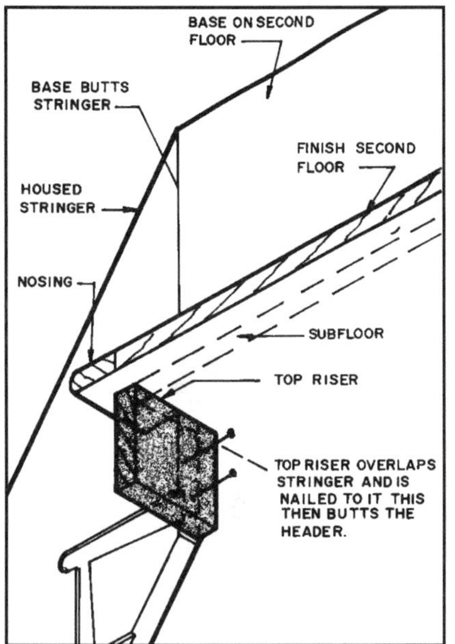

33-57 The top end of a housed stringer is cut square to receive the riser and butt the header.

6. Now install the bottom tread in the same manner.

7. Check to be certain that the stringers are straight and the rectangular unit formed is square. Squareness can be checked by measuring the diagonal from opposite corners **(refer to 33-54)**. If one diagonal is longer than the other, it is necessary to apply pressure on the long diagonal to move the unit square. It may be necessary to nail blocks to the sawhorses to hold the stringers in a square position.

8. Install the other treads as just described. Some carpenters prefer to install the next one in the center to help hold the unit square and keep the stringers straight.

9. Install the top riser. Remember that it is longer than the rest because it is face nailed to the back of the stringers **(see 33-57)**.

10. Install the bottom riser. It may be narrower than the rest because of conditions at the floor.

11. Install all the other risers. They are set in the dadoes, glued, and wedged the same as the treads. Nail through the bottom edge of the riser into the back edge of the tread and through the stringers into the end of the riser with 8d box nails. Space the nails about 8 inches (200mm) apart. Be certain that the joint between the back of the tread and riser is closed.

12. Nail or screw glue blocks about 6 inches (150mm) long joining the tread and riser in the center of the stair **(see 33-58)**. This helps stiffen the unit and reduces squeaking.

13. Turn the assembled stair over, and nail a small molding under the nosing. Nail with 4d wire brads **(see 33-59)**.

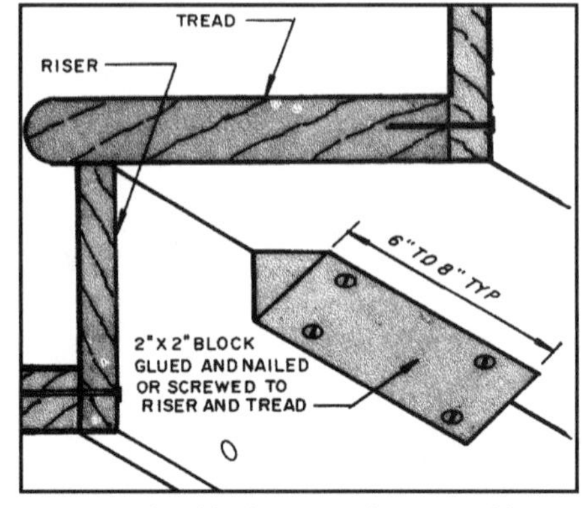

33-58 Glue blocks can reduce possible squeaks.

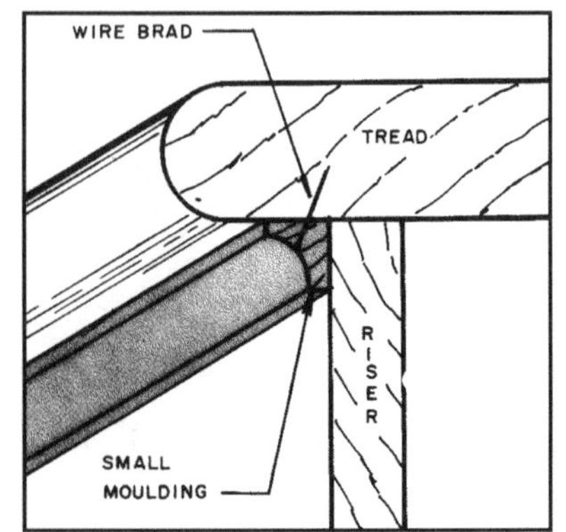

33-59 A small molding can be nailed below the tread.

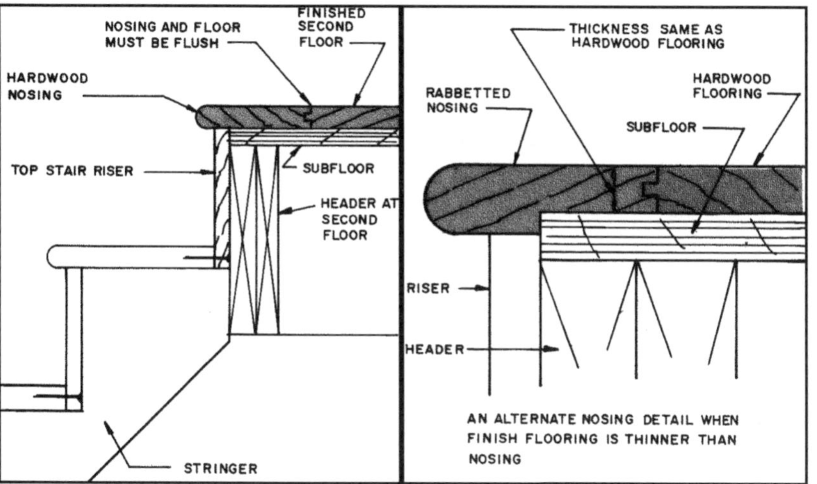

33-60 Adjust the length of the stringer so the nosing and finished floor are flush.

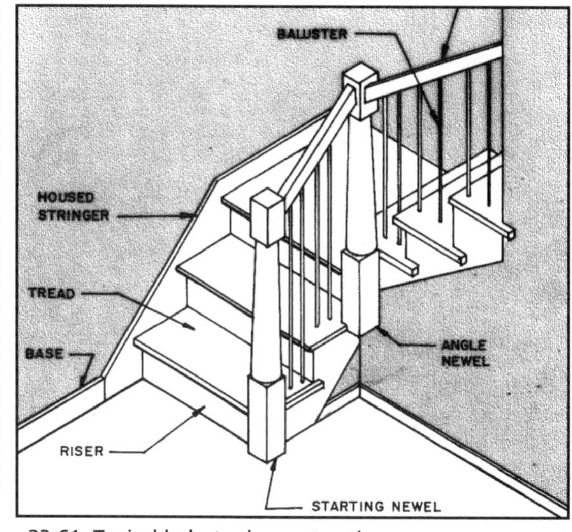

33-61 Typical balustrade construction.

Installing the Assembled Stair

The assembled stair is long and heavy. It will take several people to work it into position. If the gypsum drywall has not been applied to the walls yet, it enables workers to get underneath the stair to hold it or to provide temporary bracing. It would be very difficult to install the stair otherwise.

1. Slide the stair into the stair opening and raise it up, placing the top riser against the header on the second floor. Adjust it so that it is in the center of the rough opening, leaving room on each side for the gypsum drywall. The top edge of the top dado should be level with the top of the finished floor on the second level **(see 33-60)**. Shim the stringer or trim it as necessary to get this level. If the first floor is to have a hardwood floor that has not been installed yet, set the stringers on wood blocks that are the same thickness as the wood floor.
2. Join the stringers to the header with metal joist hangers or a ledger, as explained for notched carriage stairs.
3. Some carpenters also nail the stringers to the wall studs; this will stiffen the stair and is done after the drywall is in place. Use finishing nails and set the heads so that they can be concealed.
4. Secure the bottom end of the stringers to the finished hardwood floor or subfloor. This can be 2 · 4 blocking nailed to the carriages and the subfloor. This is located behind the riser.
5. Finally, install the top rabbeted nosing into the top dado. It should be joined securely to the top-floor subfloor or the top of the header (depending on how the second floor will finally be finished).
6. After the drywall or other finished wall covering is installed, cut and nail a molding on top of the stringer to cover the gap between the wall and the stringer.

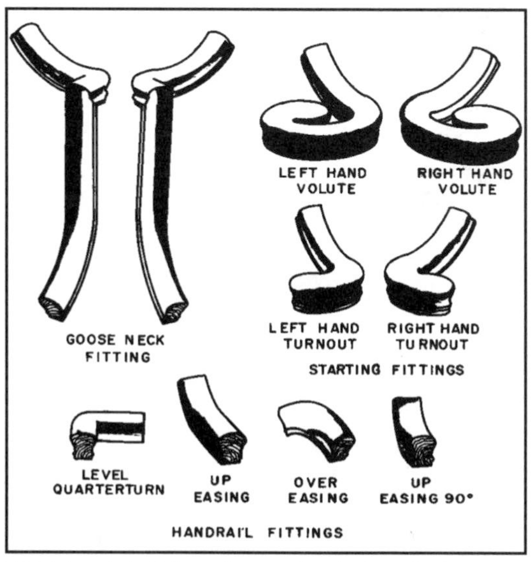

33-62 Typical fittings designed to be installed on top of newel posts to change the direction of the rail.

STOCK STAIR PARTS

Stock stair parts are available from lumber and millwork materials suppliers. They have available a wide range of sizes and designs from which to choose. This includes **dadoed stringers, treads, risers, rails, fittings, balusters,** and **newels**.

Stock Balustrades

A balustrade includes the rail, balusters, and newels **(see 33-61)**. The handrail is manufactured in straight lengths with a variety of fittings used to turn corners and change levels. Some of the fittings used to change direction are shown in **33-62**.

These balustrade fittings are designed to fit on top of the newel posts. Examples are shown in **33-63** and **33-64**.

The balusters are vertical rods running from the bottom of the rail to the treads. They are inserted into holes drilled in the treads and handrail. Typical designs are shown in **33-65**. The balustrade begins at the bottom of the stair with a large post called a newel. If the stair changes directions, a handrail fitting such as shown earlier in **33-61** is used. The use of a gooseneck is shown in **33-66**. These fittings are expensive and only the most experienced finish carpenters install them. Often this is an area of specialization, and a stair builder is called to install the finished stair parts. Typical handrail and wallrail profiles are shown in **33-67**.

Other Stock Parts

Various types of starting steps are available, as shown in **33-68**. Applications of these are illustrated in **33-69** and **33-70**.

33-63 This gooseneck fitting permits the handrail to drop several inches before it joins the handrail as it descends a flight of stairs.

33-64 This fitting turns the handrail 90 degrees.

33-66 The stair balustrade turns 90 degrees from a landing and uses a gooseneck fitting to make the turn and align the handrail.

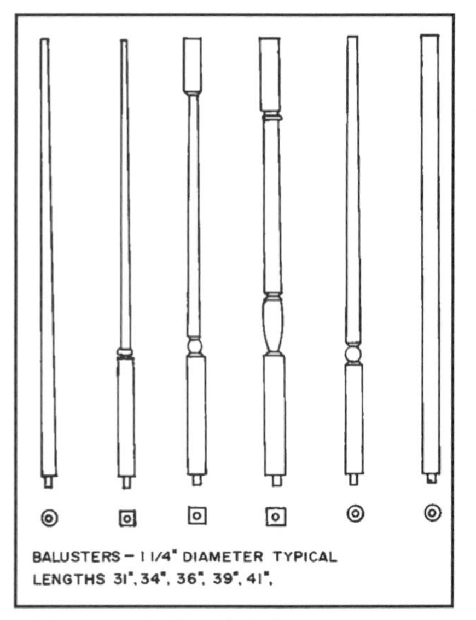

33-65 Typical stock balusters.

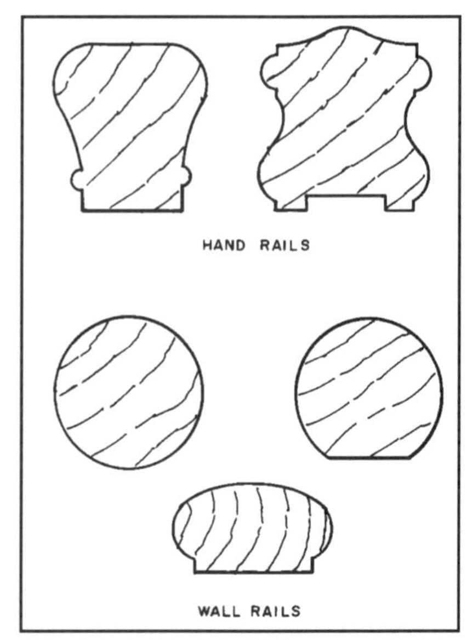

33-67 Typical stock handrail and wallrail profiles.

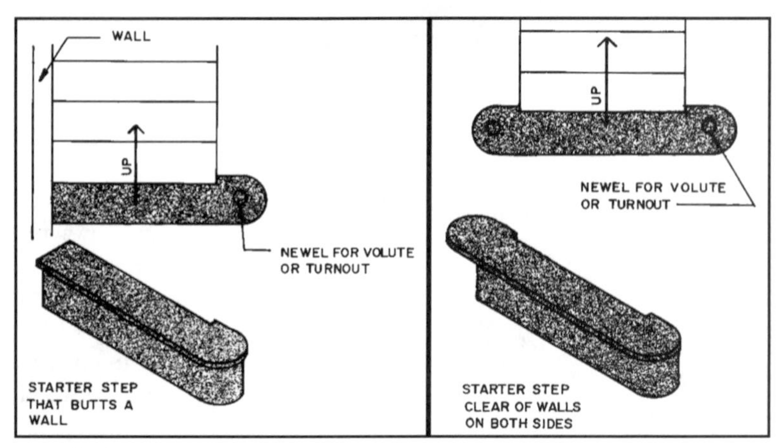

33-68 Typical starting steps.

33-69 A starting newel rests on a starting step.

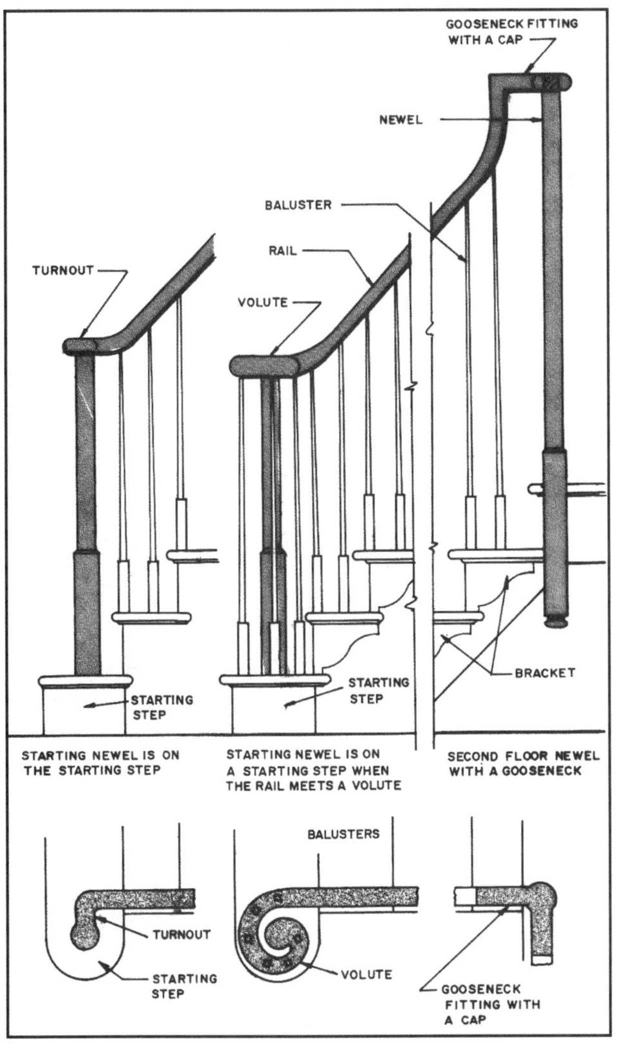

33-70 Typical over-the-post balustrade construction.

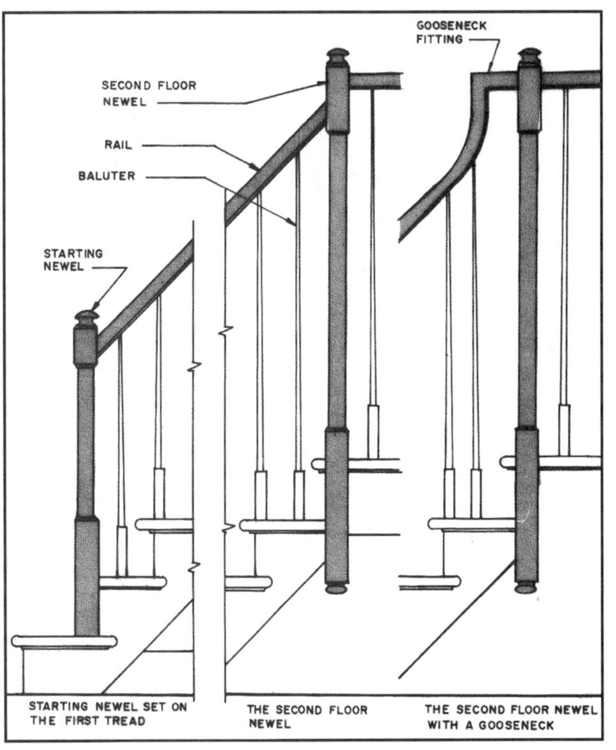

33-71 Typical post-to-post balustrade construction.

Two Balustrade Systems

Stock parts for balustrade construction are available for post-to-post and over-the-post construction. These are shown in **33-70** and **33-71**. Starting steps are required for over-the-post systems when the balustrade begins with a volute or turnout rail fitting. A finished over-the-post stair is shown in **33-72**.

33-72 A finished L-shaped stair with an over-the-post balustrade.

33-73 A folding stair opened up, providing access to the area above.

FOLDING STAIRS

Folding stairs are used to provide infrequent access to storage, such as in an attic above a garage **(see 33-73)**. They are not acceptable for access to living areas. The opening in the ceiling is framed with headers and trimmers as used with permanent stairs. When locating the opening be certain there is enough floor area below for the stair to fold down. If the house was built with roof trusses you should never cut the ceiling joists without the approval of an engineer. The size of the opening will vary depending upon the manufacturer of the stair.

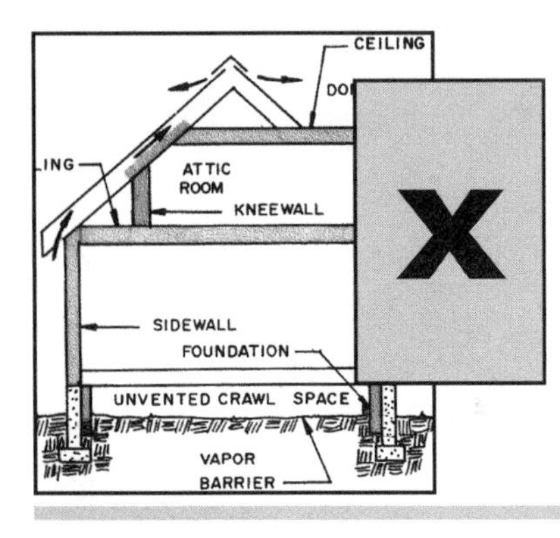

X INSULATION & INTERIOR WALL FINISH

CHAPTER 34

Thermal Insulation, Sound & Moisture Control

Thermal insulation is used to reduce the transmission of heat through ceilings, walls, and floors. In the summer it reduces the heat flow from the exterior, thus reducing air-conditioning costs. In the winter it helps hold the furnace heat inside the building, thus reducing heating costs. In both cases it enables the building to maintain a rather stable temperature, which contributes to the comfort of the occupants. It is an important factor in energy conservation.

TRANSFER OF HEAT

Heat moves through a material from the warm side toward the cold side. Therefore, in the summer the heated exterior walls and roof transfer heat to the cooler interior. In the winter the heat inside the building tends to move to the colder exterior **(see 34-1)**.

The goal of a well designed and constructed building is to control the transfer of heat into and out of the interior of the building. Thermal insulation is one material used to control heat transfer. It is placed in the walls, ceiling, and floor and on the foundation. Insulation is usually installed by contractors specializing in this area of construction.

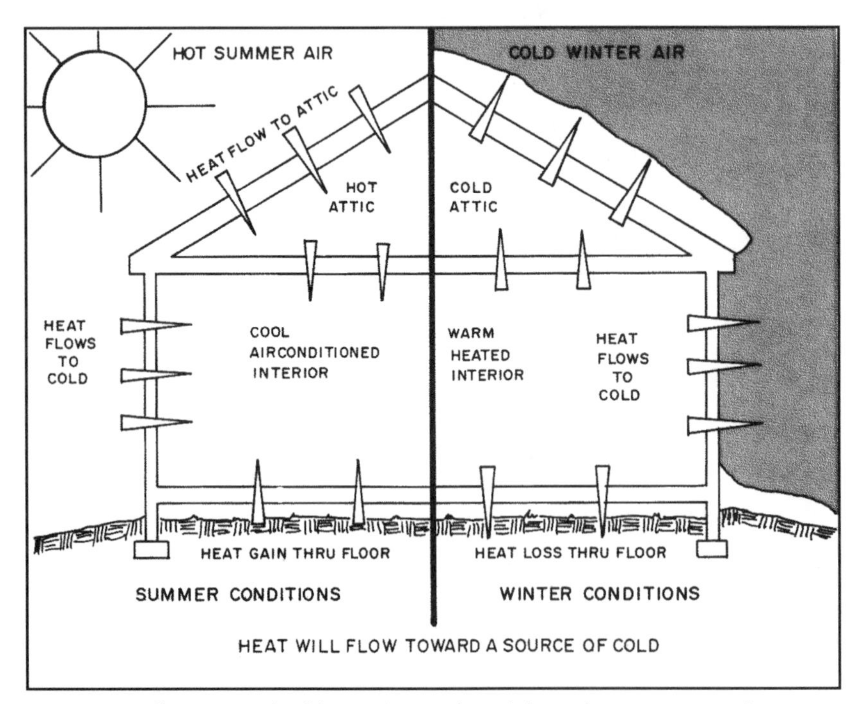

34-1 Heat flows toward cold. Heat is transferred through most construction materials and through leaks in the assembly.

How Heat Is Transferred

Heat is transferred by convection, conduction, and radiation.

Convection refers to the movement of heat through the motion of air or a liquid. A hot air furnace heats the air and blows it into the interior of the building. The heat is therefore transferred by convection. Hot water moving through the plumbing system also moves heat by convection **(see 34-2)**.

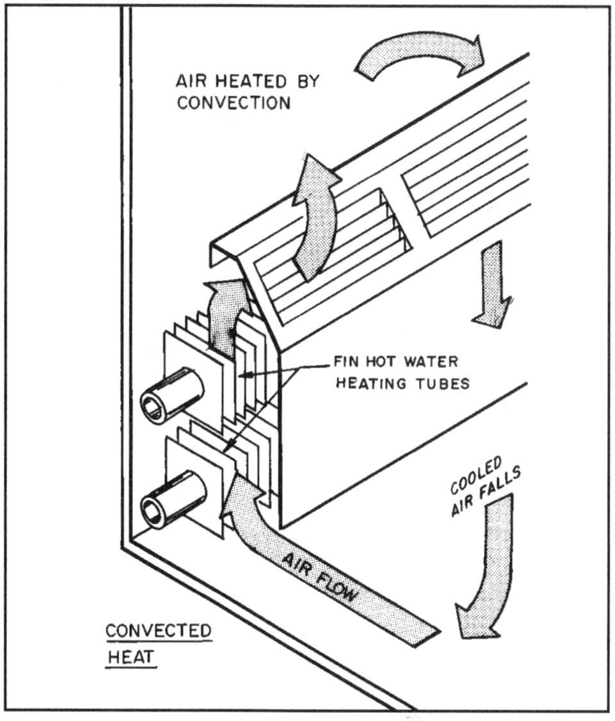

34-2 The hot air leaving this finned hot water unit rises and moves toward the cooler air in the room. As the air cools it falls to the floor and is reheated and recycled.

Conduction refers to the movement of heat through a solid material, such as metal or masonry. When heat strikes the material the heat is conducted through it by passing from one molecule to another. If you hold a flame to one end of a wire and hold the other end in your hand you will soon feel the heat moving by conduction through the wire **(see 34-3)**.

Radiation refers to the direct transmission of heat by invisible rays moving through space. For example, heat you feel when you stand near a fire is heat radiated to you and heats anything it may strike such as the furniture, walls, floor, or ceiling **(see 34-4)**. Radiant heat can be reflected off of anything it strikes and can warm the air by convection.

HEAT TRANSMISSION & RESISTANCE VALUES

The measurement of heat transmission and resistance is indicated by three factors, conductivity (*k*), conductance (*C*), and transmittance (*U*). The resistance (*R*) factor is determined by the *k* and *C* values.

Conductivity (*k*) Values Conductivity is a measure of the amount of heat that is transferred through a homogenous material such as wood or concrete. Homogenous materials are those that have a uniform composition throughout their thickness. The *k* value is a measurement of the number of Btu (British thermal units) per hour that will pass through one square foot of a material that is one-inch-thick and has a temperature difference of one degree F between its two surfaces. (A British thermal unit is the quantity of heat required to raise the temperature of one pound of water one degree F at a specified temperature—39 degrees F.)

Conductance (*C*) Values Conductance is similar to the *k* value except it applies to materials that are not homogenous or that have air cavities.

Transmittance (*U*) Values Transmittance is a measure of the number of Btu per hour that will pass through one square foot of an assembly of

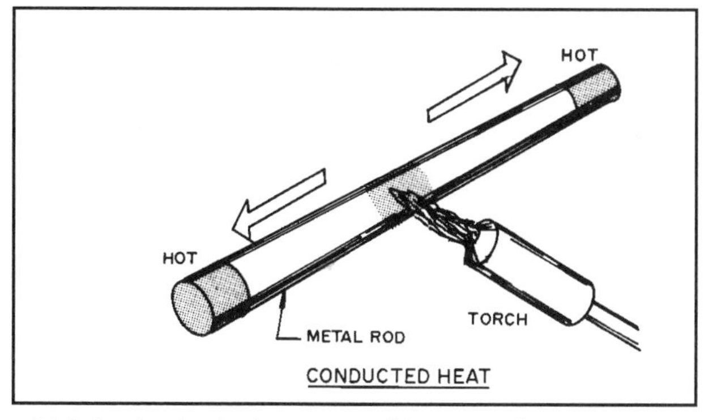

34-3 Conduction is the process of heat transfer that occurs through a solid material when one molecule transfers heat directly to another molecule.

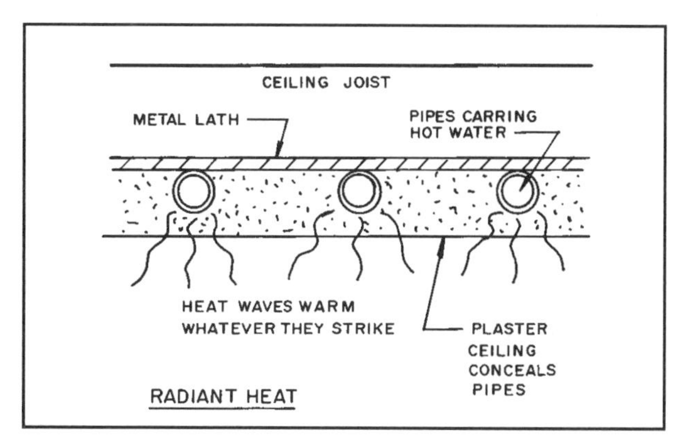

34-4 Radiant heat is directly transmitted from the source as heat waves which warm anything they strike.

34-5 How to calculate the *U*-value of a wall assembly.

	R-value
Outside film surface	0.17
Wood siding shingles	0.87
⅜-in. plywood sheathing	0.47
4-in. glass fiber insulation	11.00
½-in. gypsum wallboard	0.45
Inside air film	0.61
R-value	13.57

$$U = \frac{1}{R} = \frac{1}{13.57} = 0.07$$

materials when there is a one degree F temperature difference between its two sides. An example is a wood-framed wall. The *U* value includes the *R*-value of all of the materials plus air spaces and surface air films. The *U*-value is the reciprocal of the *R* value: $U=1/R$. In **34-5** the *R*-value of a wood-framed exterior wall has been determined to be 13.57; therefore the *U*-value is 1/13.57 or 0.07. The smaller the *U*-value, the higher the insulation value of the assembly.

Resistance (*R*) Values Resistance is a measure of a material's capacity to resist the flow of heat through it. Insulation materials are specified by their *R*-value. The higher the *R*-value the

34-6 *R*-values of selected construction materials.

Structural and Finish Materials	*R*-value
Wood bevel siding, ½ × 8 in., lapped	*R* 0.81
Wood siding shingles, 16 in., 7½-in. exposure	*R* 0.87
Asbestos-cement shingles	*R* 0.03
Stucco, per inch	*R* 0.20
Building paper	*R* 0.06
½-in. nail-base insul. board sheathing	*R* 1.14
½-in. insul. board sheathing, regular density	*R* 1.32
25⁄32-in. insul. board sheathing, regular density	*R* 2.04
¼-in. plywood	*R* 0.31
⅜-in. plywood	*R* 0.47
½-in. plywood	*R* 0.62
⅝-in. plywood	*R* 0.78
¼-in. hardboard	*R* 0.18
Softwood, per inch	*R* 1.25
Softwood board, ¾ in. thick	*R* 0.94
Concrete blocks, three oval cores	
Cinder aggregate, 4 in. thick	*R* 1.11
Cinder aggregate, 12 in. thick	*R* 1.89
Cinder aggregate, 8 in. thick	*R* 1.72
Sand and gravel aggregate, 8 in. thick	*R* 1.11
Lightweight aggregate (expanded clay, shale, slag, pumice, etc.), 8 in. thick	*R* 2.00
Concrete blocks, two rectangular cores	
Sand and gravel aggregate, 8 in. thick	*R* 1.04
Lightweight aggregate, 8 in. thick	*R* 2.18
Common brick, per inch	*R* 0.20
Face brick, per inch	*R* 0.11
Sand-and-gravel concrete, per inch	*R* 0.08
Sand-and-gravel concrete, 8 in. thick	*R* 0.64
½-in. gypsumboard	*R* 0.45
⅝-in. gypsumboard	*R* 0.56
½-in. lightweight-aggregate gypsum plaster	*R* 0.32
25⁄32-in. hardwood finish flooring	*R* 0.68
Asphalt, linoleum, vinyl, or rubber floor tile	*R* 0.05
Carpet and fibrous pad	*R* 2.08
Carpet and foam rubber pad	*R* 1.23
Asphalt roof shingles	*R* 0.44
Wood roof shingles	*R* 0.94
⅜-in. built-up roof	*R* 0.33
Glass	
Single glass (winter)	*U* 1.13
Single glass (summer)	*U* 1.06
Insulating glass (double)	
¼-in. air space (winter)	*U* 0.65
¼-in. air space (summer)	*U* 0.61
½-in. air space (winter)	*U* 0.58
½-in. air space (summer)	*U* 0.56
Storm windows	
1–4 in. air space (winter)	*U* 0.56
1–4 in. air space (summer)	*U* 0.54

greater the insulation value. Dead air space, as in a wall cavity or between two panes of glass, has some *R*-value. *R*-values for selected building materials and *U*-values for some types of glazing are in **34-6**.

TYPES OF THERMAL INSULATION

Thermal insulation is available made from a range of materials such as fiberglass, mineral fibers, mineral wool, various foams, foils, and a number of recycled materials such as paper. These vary in cost and *R*-value. The *R*-values of selected types of insulation are in **34-7**.

Insulation batts are available in 15- and 23-inch (380 and 584mm) widths and are 4 feet (1.2.m) in length. They are available in thicknesses from 3 to 10 inches (76 to 254mm).

Insulation blankets are the same as batts except they are in rolls often 40 feet or longer. Both are typically fiberglass or rock wool fibers. They are stapled to the studs as shown in **34-8**. Some types have an aluminum foil facing which faces the inside of the building. It reduces the transmission of vapor into the wall. They are also used to insulate floors, attics, and roofs as shown in **34-9** and **34-10**.

34-8 Insulation blankets are used to insulate the area between wall studs. *(Courtesy Owens Corning)*

34-9 Fiberglass batts are used between the floor joists to insulate the floor over a crawl space. *(Courtesy Owens Corning)*

34-7 *R*-values for various types of insulation.*

R-Value	Blankets or batts		Loose and Blown Fill
	Mineral Fiber	Fiberglass	Cellulosic Fiber
R-11	3¼ to 3¾	4 to 5¼	3¾
R-19	5¾ to 6¼	7 to 8¾	6½
R-30	9 to 9½	11¼	10½
R-38	11½ to 12	14 to 17¾	13
R-49	15 to 15½	18 to 23	17

Insulating Board Products

Type	*R*-value per inch of thickness
Polystyrene Expanded	4
Polystyrene Extruded	5
Polyisocyanurate	6 to 7
Polyurethane	6 to 7
Phenolic	8

* *R*-values may vary for the same thickness. Consult the manufacturer for specific data. *(U.S. Department of Energy)*

34-10 This attic insulation is being stapled to the ceiling joists before the gypsum drywall ceiling has been installed. *(Courtesy Owens Corning)*

Blown-in insulation is typically fiberglass or a cellulose fibrous product that is placed in wall cavities. The product is blown into a cavity created by stapling a strong plastic vapor barrier over the studs, as shown in **34-11**. The vapor barrier should be strong enough so that it will not bulge when the insulation fibers are compressed behind it. The cavity is filled by cutting a small opening in the vapor barrier and inserting the hose from the blowing machine into the wall cavity **(see 34-12)**. Consult the manufacturer of the system for specific details.

Loose-fill insulation is available as a loose fibrous or granular material. Fibrous insulation may be cellulosic fibers obtained from waste materials (wood chips and newsprint), glass fibers, mineral wool, and cotton fibers. It is available in bags and can be poured or sprayed over the area to be insulated **(see 34-13)**.

Granular insulation is typically perlite (expanded volcanic rock), vermiculite (expanded mica), cork, and expanded polystyrene. It is poured into the cavities to be insulated.

Reflective insulation is usually a copper or aluminum foil in sheets or rolls. The rolls are 24 and 48 inches wide and up to 500 feet long. Reflective insulation is available as a single thickness or a multilayer batt that has dead air spaces between the layers **(see 34-14)**. Reflective insulation is bonded to the face of some blankets and gypsum wallboard sheets.

Rigid insulation board is made from organic fibers (wood, cane), mineral wool fibers, glass fibers, corkboard, and various types of expanded plastics. It is used on all parts of the building such as the roof, ceiling, walls, floor, and foundation **(see 34-15)**.

Rigid insulation panels are available in thicknesses from one to three inches and up to 48 × 96-inch sheets. Actual sizes vary with the material used to form the panel.

Foamed-in-place insulation is a creamy mix that is pumped through a hose into cavities needing insulation. It is generally a polyurethane or phenol-based compound. After it is in place it hardens into a rigid cellular mass. The foam can also be sprayed onto surfaces such as roof decks, where it hardens. The finished roof is laid over the hardened foam.

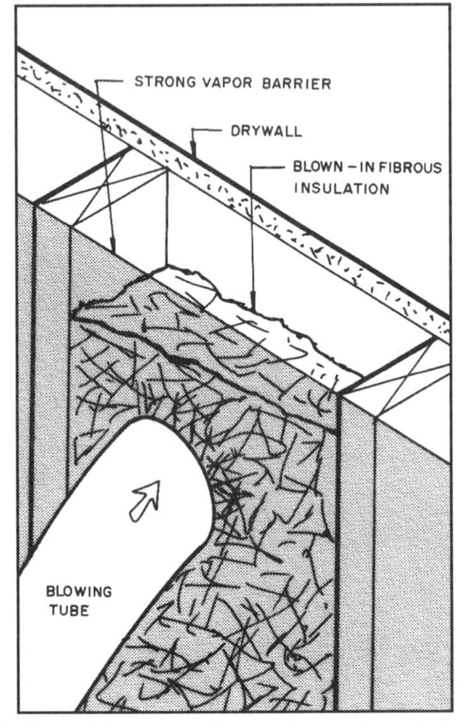

34-11 Blown-in fibrous wall insulation is placed between a vapor barrier and the finish wall material on the other side of the wall.

34-12 Blown-in fibrous wall insulation is set in place by blowing it through a slot cut in the vapor barrier. *(Courtesy Ark-Seal, Inc., International)*

34-13 Loose-fiber-type insulation can be poured or blown into place. *(Courtesy Owens Corning)*

34-14 Reflective insulation is stapled to the inside face of the wood studs. *(Courtesy Stanley Bostitch)*

34-15 Rigid insulation is available in large sheets and is used on walls, floors, foundations, and other locations. *(Courtesy Dow Chemical Company)*

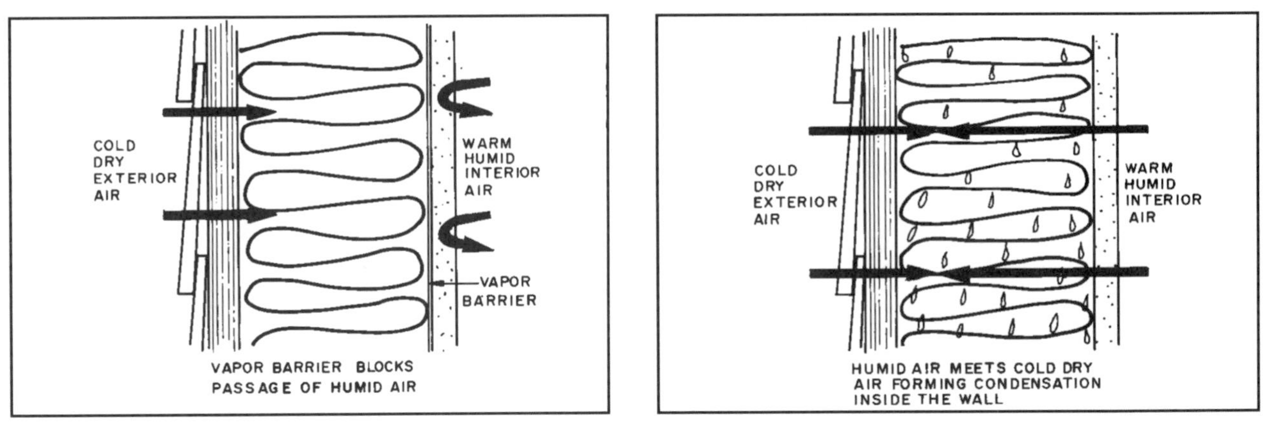

34-16 The vapor barrier blocks the passage of interior moisture through the wall where it would otherwise dampen the insulation and cause paint to peel on exterior siding.

VAPOR BARRIERS

A vapor barrier is a material used to keep moisture in the air from passing through the surface upon which it has been installed. It is applied to the walls, ceiling, and floor of rooms and faces the heated area **(see 34-16)**. Moisture generated inside the room (such as from cooking) is stopped from penetrating the wall, floor, or ceiling and getting into the insulation **(see 34-17)**. Materials frequently used are plastic films, aluminum foil, and asphalt-laminated paper.

Vapor barriers are compared based on their perm ratings. The perm rating is a measure of the material's resistance to vapor penetration. Codes require vapor barriers have a perm rating of 1.0. For example, 15-lb. asphalt felt paper has a perm rating of 1.0, 6-mil polyethylene sheeting has a perm rating of 0.06, and aluminum foil vapor barrier is rated at 0.00.

34-17 Applying a polyethylene vapor barrier over the unfaced insulation blankets on the interior side of a frame wall. *(Courtesy Owens Corning)*

MOISTURE CONTROL

Moisture in the air can cause considerable damage to a building. Various techniques are used to control and remove moisture, before damage occurs. One

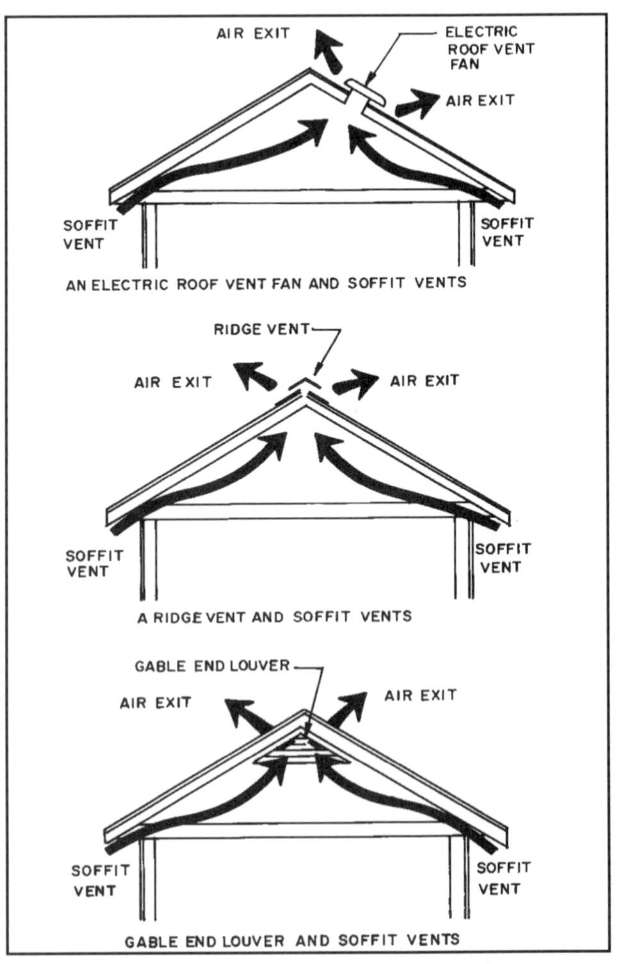

34-18 Commonly used methods for venting moisture out of the attic.

method is to use vapor barriers on the inside walls, ceilings, and floors. Buildings with crawl spaces should have the ground covered with a vapor barrier. This keeps moisture in the soil from raising the humidity in the crawl space. If this is not done, the joists and subfloor will mold and rot.

Another major consideration is moisture in the attic. The attic must be thoroughly ventilated all year long to remove moisture that might accumulate there. This is done by providing a means for the exterior air to flow into and out of the attic; typically with soffit vents and some type of exit vent near the ridge **(see 34-18)**. Additional information can be found in Chapter 19.

Building codes specify the amount of net free ventilating required for natural attic ventilation. For example, one square foot (0.092m^2) of free ventilation area is often required for each 150 square feet (13.8m^2) of the ceiling area. If approved vapor barriers are used on the ceiling, this can be reduced to one square foot (0.092m^2) for every 300 square feet (27.6m^2) of ceiling area.

Electric ventilation fans can also be used to remove attic moisture. Both natural and mechanical attic ventilation also remove excess heat that can cause damage to the building.

WHEN TO INSULATE

After the building is framed and the exterior has been made weather tight, the electricians, plumber, and heating and air-conditioning contractor move in to do their part. Once they have completed their work and the building inspector has approved it the insulation contractor can begin to work.

WHERE TO INSULATE

All heated areas should be surrounded by insulation **(see 34-19)**. This includes the walls, ceilings, floor, and foundation. Some prefer not to insulate the ceiling of a room having a heated area above it. Ventilated crawl spaces should have insulation between the floor joists **(refer to 34-1)**. Unvented spaces, such as basements should have insulation around the foundation and two feet below grade.

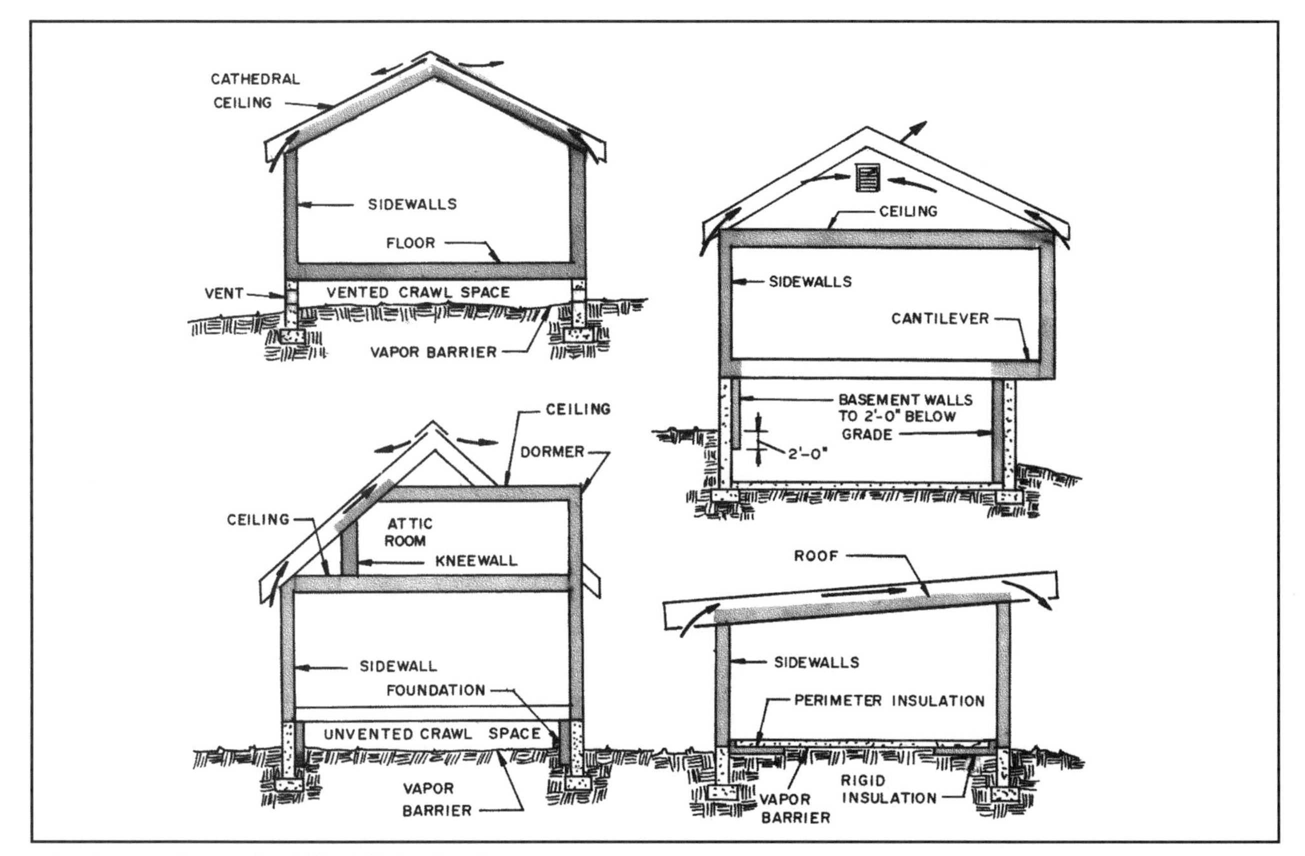

34-19 Heated areas should be fully insulated.

Insulation should be placed in all openings around doors and windows **(see 34-20)**. It should not be packed in tightly. The insulation value of fibrous insulation depends upon the air spaces existing between the fibers. A vapor barrier is installed over this insulation.

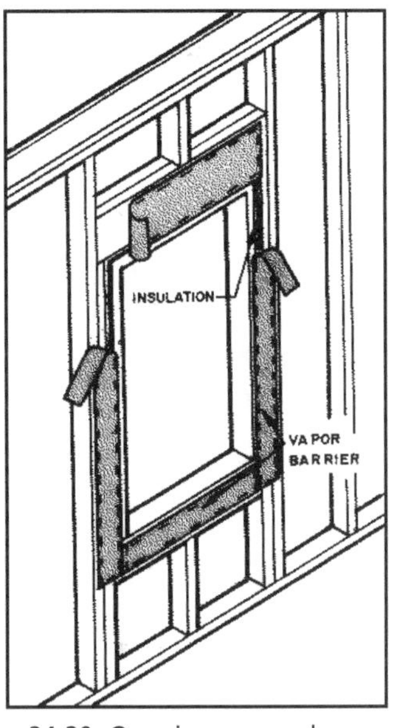

34-20 Openings around door and window frames should be filled with insulation and covered with a vapor barrier.

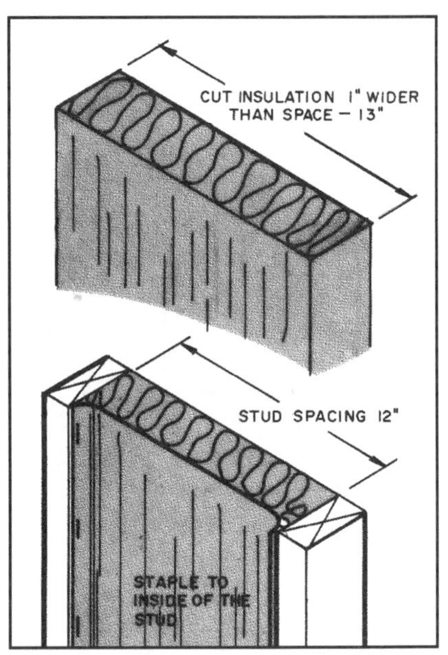

34-23 Cut insulation that must fit into narrow stud spacings one inch wider than the space between the studs.

HOW MUCH INSULATION

The amount of insulation depends upon such things as the cost of fuel, nature of the climate, desired comfort, and type of construction. Other factors include the extent of weatherstripping, the thermal properties of doors and windows, and to what degree the orientation of the building utilizes the sun for heating.

One recommendation for insulation is shown in **34-21**. Notice that even in the warm climates a ceiling should have an *R*-26 rating to handle air-conditioning requirements. In the most northern states the following *R*-values are recommended: ceiling, 38; walls, 19; and floor, 22. In the southern states the recommendations are ceiling, 26; walls, 13; and floor, 11.

HOW TO INSTALL INSULATION

Blanket and batt-type insulation have tabs that are stapled to the inside of the wood framing members **(see 34-22)**. Insulation with a vapor barrier already attached is installed with this barrier facing into the room.

When installing fibrous insulation in walls be certain it touches the top and bottom plates. If it is necessary to cut a blanket to fit a small space, cut it about 1 inch wider and use this as the tab for stapling **(see 34-23)**.

Insulation should fit tightly around electrical boxes, pipes, and other items protruding from the wall **(see 34-24)**.

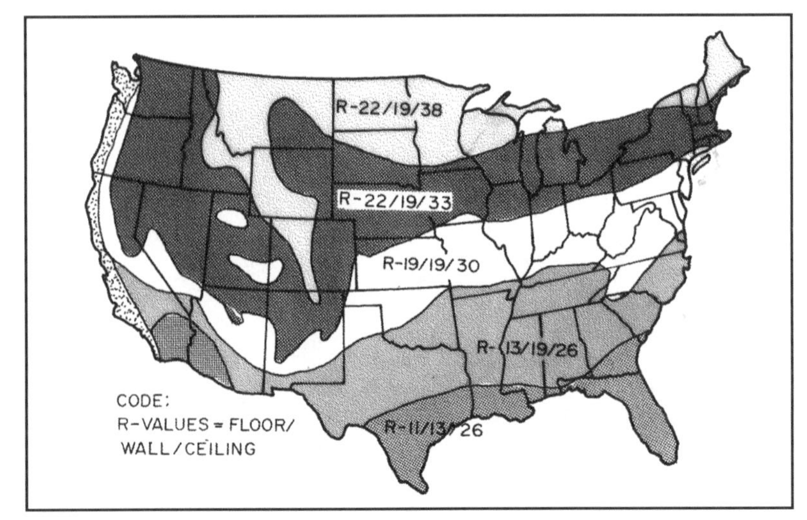

34-21 Recommended *R*-values for various parts of the United States.

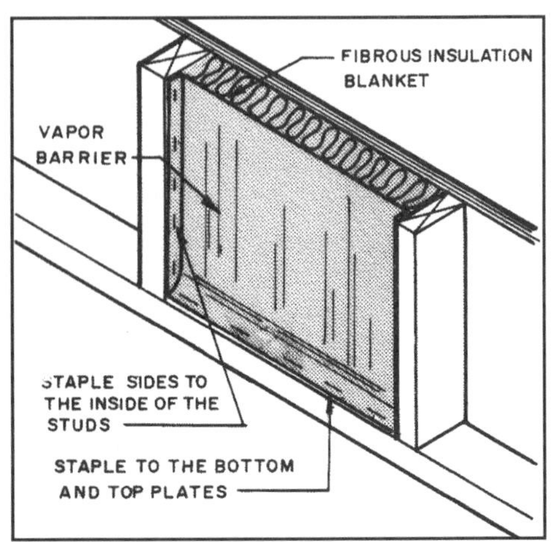

34-22 Staple fibrous insulation blankets to the inside face of the studs and to the top and bottom plates.

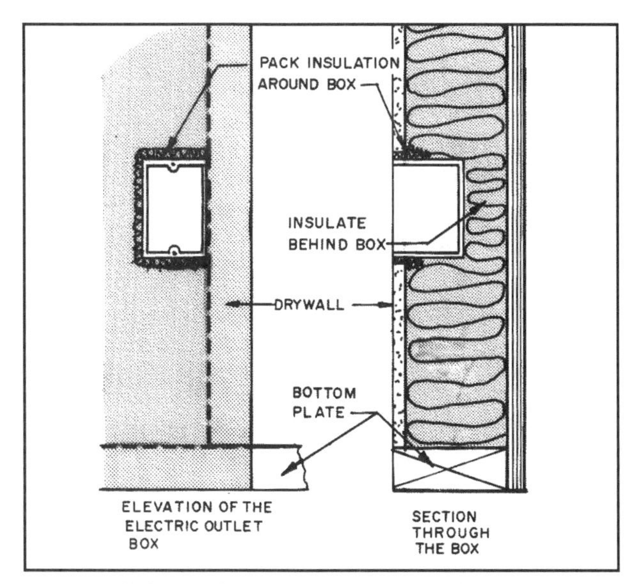

34-24 Fill the gap between the gypsum wallboard and electrical boxes and other protruding items with insulation.

While water and sewer pipes are not usually placed in exterior walls because they may freeze if they are there, it is important to fit insulation between them and the sheathing **(see 34-25)**.

Rigid insulation is cut to fit the space between members and pressed into place. It is held by friction. It must have a vapor barrier stapled over it **(see 34-26)**.

Masonry walls can be insulated by fastening 2 × 2-inch strips **(see 34-27)** or building a regular stud wall **(see 34-28)** against them and installing insulation in the normal manner.

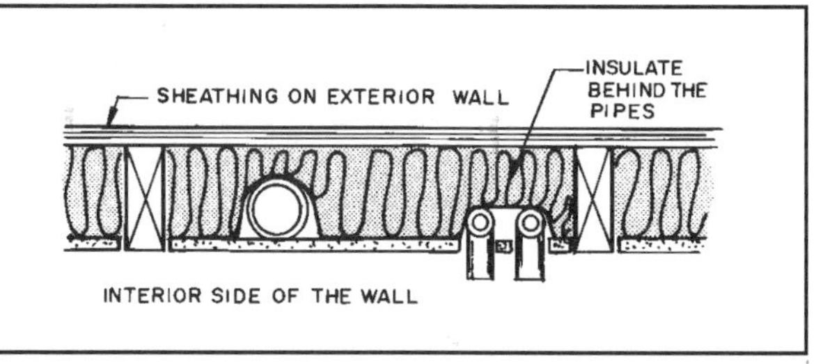

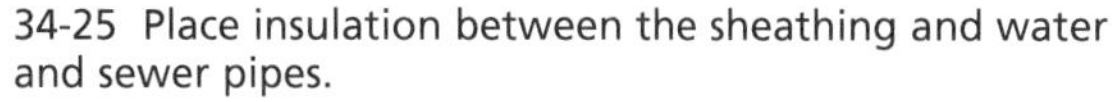
34-25 Place insulation between the sheathing and water and sewer pipes.

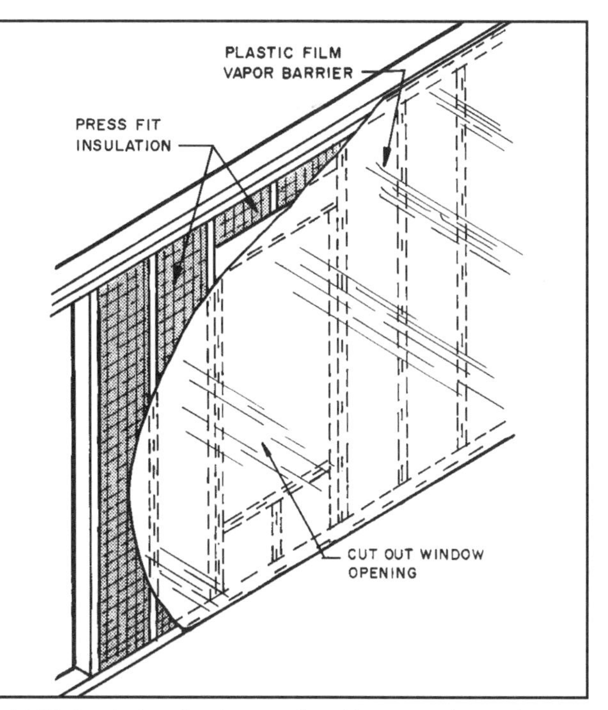

34-26 Insulation that is press-fitted between the studs must have a vapor barrier stapled over it.

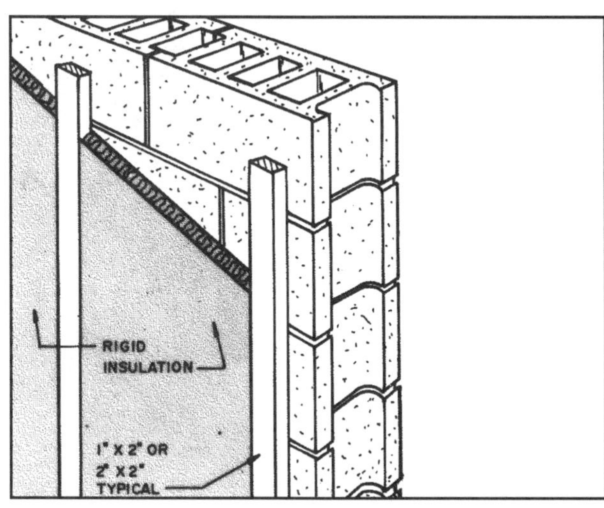

34-27 Masonry walls can be insulated with rigid insulation panels. Wood furring strips are mounted on the wall for nailing the finish wall covering.

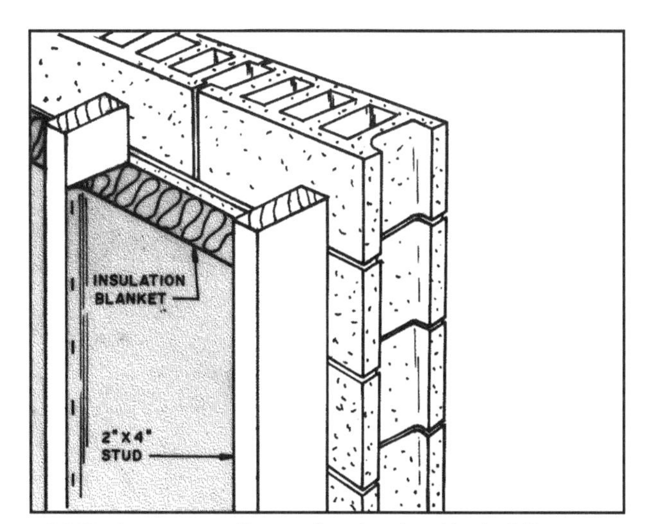

34-28 Basement walls are often insulated by building a 2 × 4 wall inside and insulating it in the same manner as stud exterior walls.

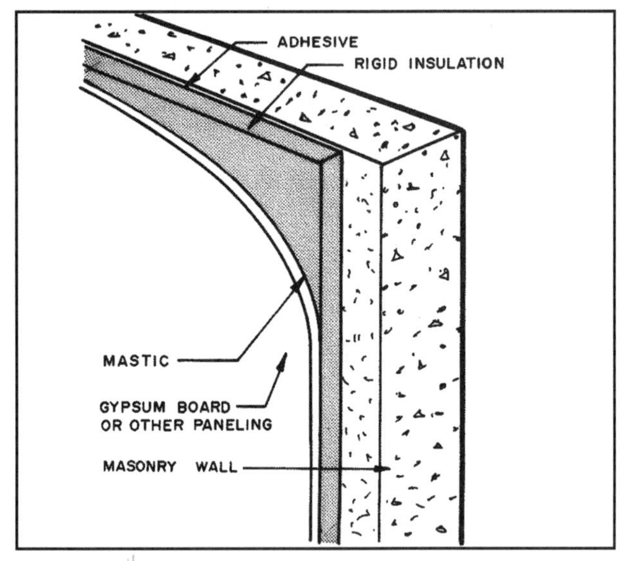

34-29 Rigid insulation panels can be glued to the masonry wall and the finish wall material is glued to the insulation.

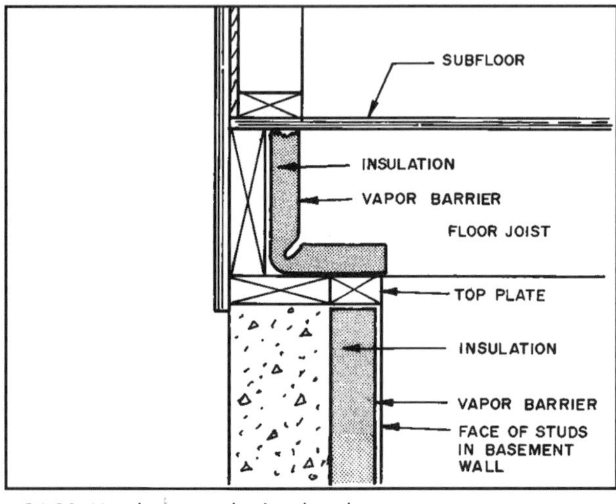

34-30 Headers must be insulated.

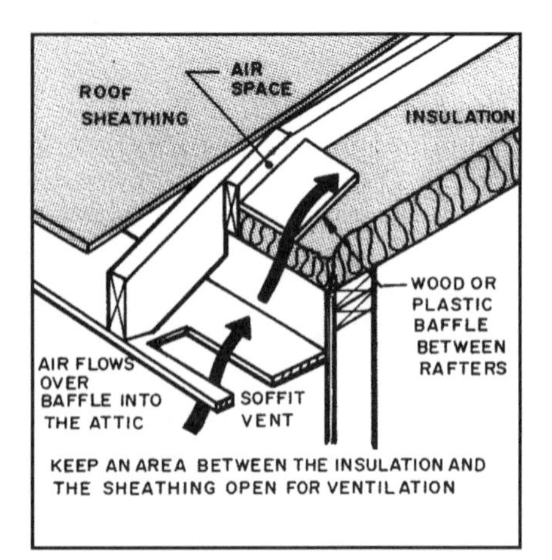

34-31 When soffit vents are used provision must be made so the insulation does not block the air flow into the attic.

Rigid insulation can be glued to the masonry wall and a finished wall panel glued to it **(see 34-29)**. In all cases be certain to insulate the floor header as shown in **34-30**.

When insulating the ceiling be certain the insulation does not block the flow of air from the soffit vent. If this appears to be a problem, a baffle can be installed between rafters to assure a passage for the air **(see 34-31)**.

Insulation should be kept at least three inches away from light fixtures recessed in the ceiling **(see 34-32)**. Place noncombustible insulation between the chimney and the ceiling framing.

Unvented crawl spaces can be insulated with fibrous or rigid insulation **(see 34-33)**.

When a floor cantilevers over the exterior wall, the header and cantilevered area must be insulated **(see 34-34)**.

Insulation can be held in place in floors that are not to have a finished ceiling below by using wires or stapling chicken wire over the bottom of the joists.

Concrete Slab Insulation

Concrete slab construction requires the use of rigid insulation around the edge of the slab. It is placed before the slab is poured. A few installation examples are in **34-35**. Before the slab is poured a polyethylene vapor barrier is placed over the insulation and gravel is spread to cover the bottom of the entire slab.

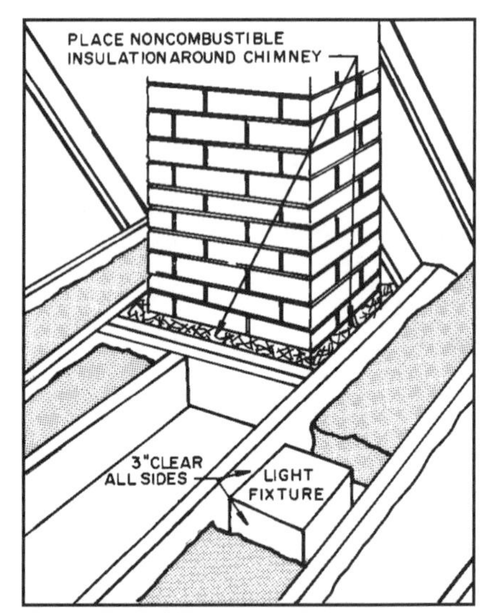

34-32 Keep insulation away from recessed light fixtures because they get very hot. Place noncombustible insulation between the headers, joists, and chimney.

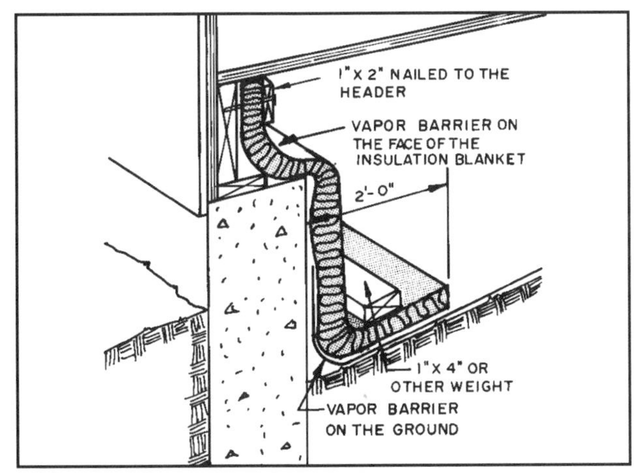

34-33 Headers and crawl space walls can be insulated with fibrous blanket insulation.

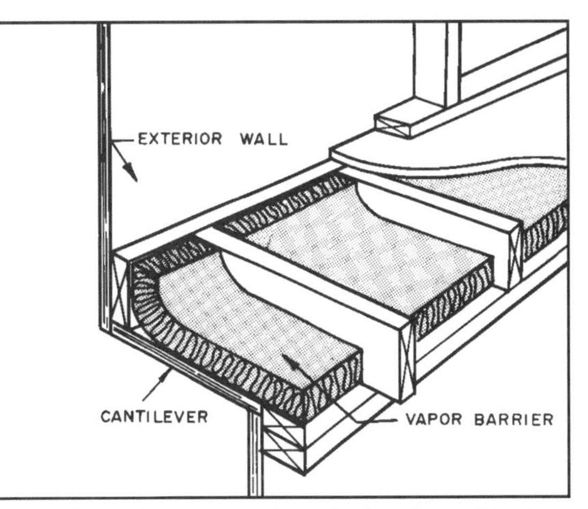

34-34 Cantilevered areas must have the header and horizontal area fully insulated.

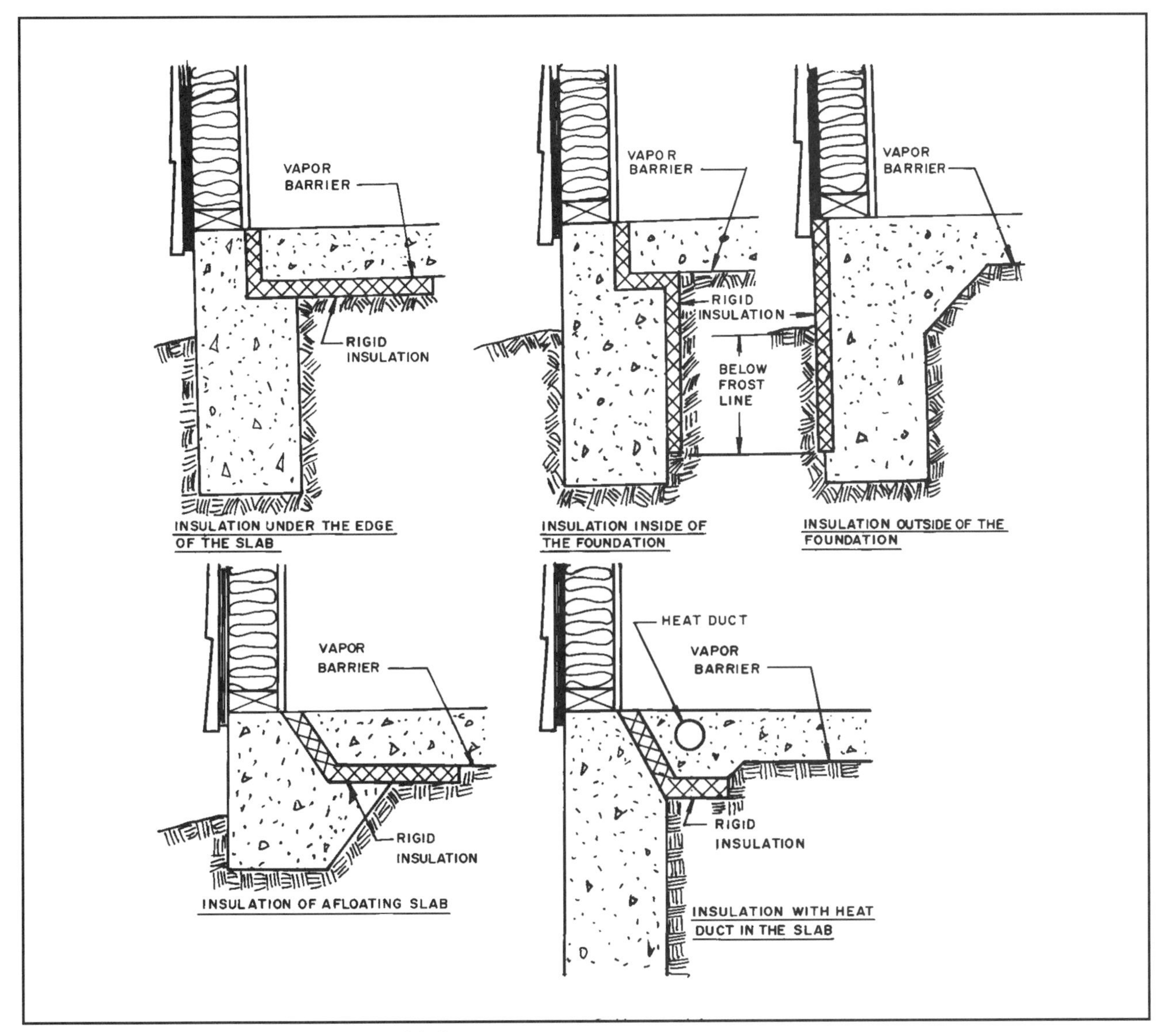

34-35 Various ways to insulate concrete slab floors.

AIR INFILTRATION

Air retarders are used to reduce energy loss by preventing interior heated or air-conditioned air from escaping through cracks in the building shell. They also block drafts of hot or cold outside air from passing into the building. Major sources of air infiltration occur around doors, windows, electrical outlets, plumbing, and heat ducts. The high quality of windows and doors today has greatly reduced this problem. If they are properly installed, little infiltration will occur. The space between outlets, pipes, and other protrusions through the finished interior wall should be sealed with high-quality caulking **(see 34-36)**.

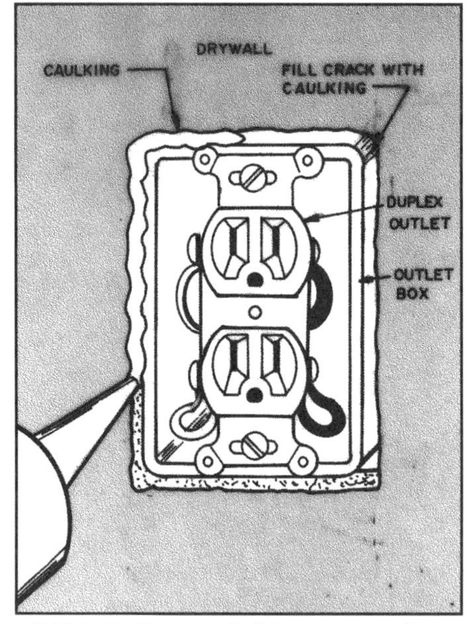

34-36 Caulk around all items protruding through the interior wall covering material. This reduces air infiltration and the transmission of sound through the wall.

Air infiltration can be greatly reduced by covering the sheathing with a plastic film commonly referred to as house wrap. This material seals cracks between panels of sheathing at the sill and is wrapped around the rough openings for doors and windows. Installation details are in Chapter 10.

SOUND INSULATION

A building is a more pleasant place and the occupants are happier if the transmission of sound between areas can be greatly reduced. Areas in a home that are especially critical are the bathrooms and bedrooms. Extensive sound control efforts should be made to isolate these areas.

Sound travels in waves that radiate from the source through the air until they strike a wall, ceiling, or floor. These surfaces are set in vibration by the fluctuating pressure of the sound waves. As the material vibrates, some of the sound is conducted to the other side **(see 34-37)**. The ability of a material or an assembly of materials to resist sound transmis-

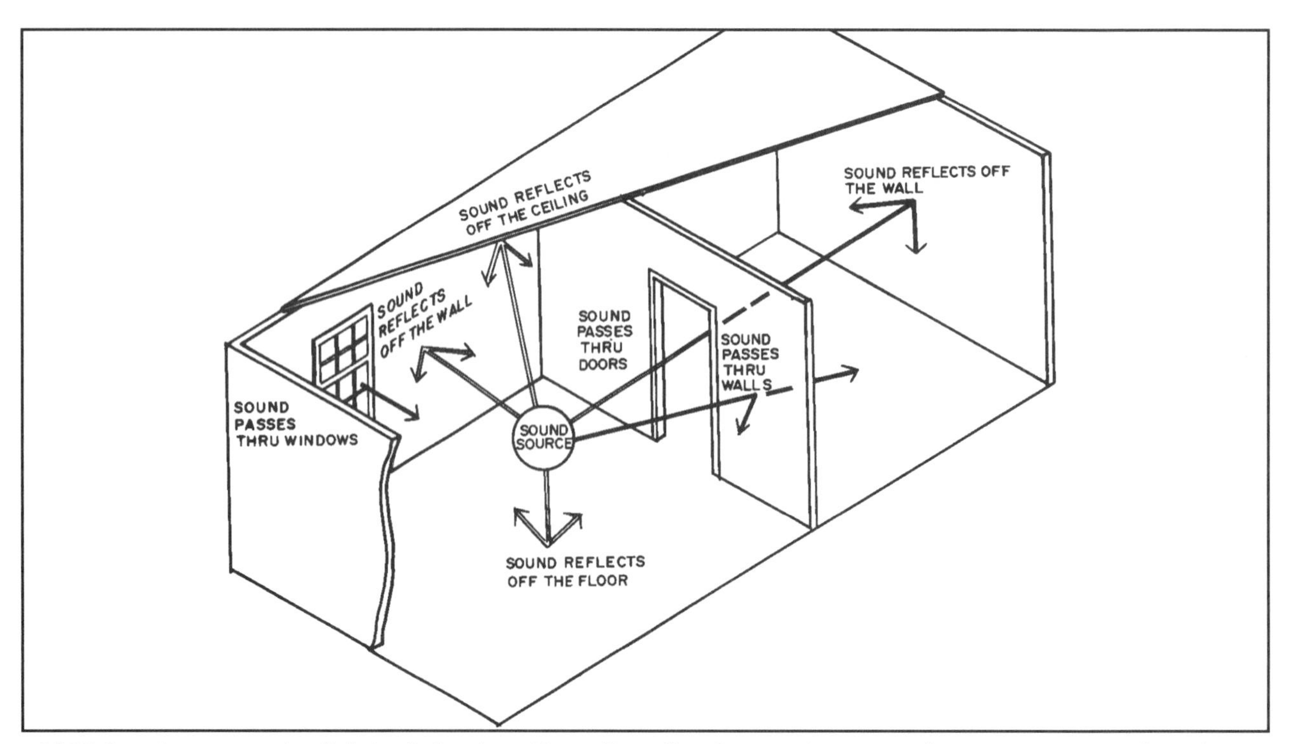

34-37 Sound waves travel radially in all directions. They reflect off surfaces, and some sound wave energy passes through some materials.

34-38 **Typical sound transmission characteristics for several STC ratings.**

STC Number	
25	Normal speech can be understood easily
35	Loud speech audible but not intelligible
45	You must strain to hear loud speech
50	Loud speech not audible
60 to 70	Practically no sound heard

sion is described as its Sound Transmission Class (STC). STC is expressed in decibels. The higher the STC rating, the greater the ability to limit the transmission of sound. The established values of sound transmission classes are shown in **34-38**. Generally STC ratings of around 50 are acceptable for residential construction.

Sound Control Materials

Acoustic batts are designed to be placed in wall, floor, and ceiling assemblies to reduce the amount of sound passing through them. While fiberglass insulation is effective, acoustic fiberglass batts are better. Acoustic batts are not intended to serve as thermal insulation.

Sound-deadening board is a panel made from glass fibers that is installed next to the studs or joists and covered with gypsum wallboard or some other wall finishing material.

Glass fiber wall panels are made from glass fibers and have a decorative plastic fabric covering. They are bonded to the walls and ceilings of the room.

There is a wide range of **acoustical ceiling tiles** and panels made from mineral fibers bonded by organic and inorganic materials. They have the surface perforated with holes or some irregular channels which increase the sound-deadening properties.

Constructions for Sound Control

The most commonly used material to control sound through partitions and ceiling/floor assemblies in residential construction is acoustical batts. They are installed in the wall or ceiling/floor assembly in the same manner as thermal insulation batts and blankets **(34-39)**.

34-39 Sound-attenuation batts are installed in the wall cavity in the same manner as thermal insulation. *(Courtesy Owens Corning)*

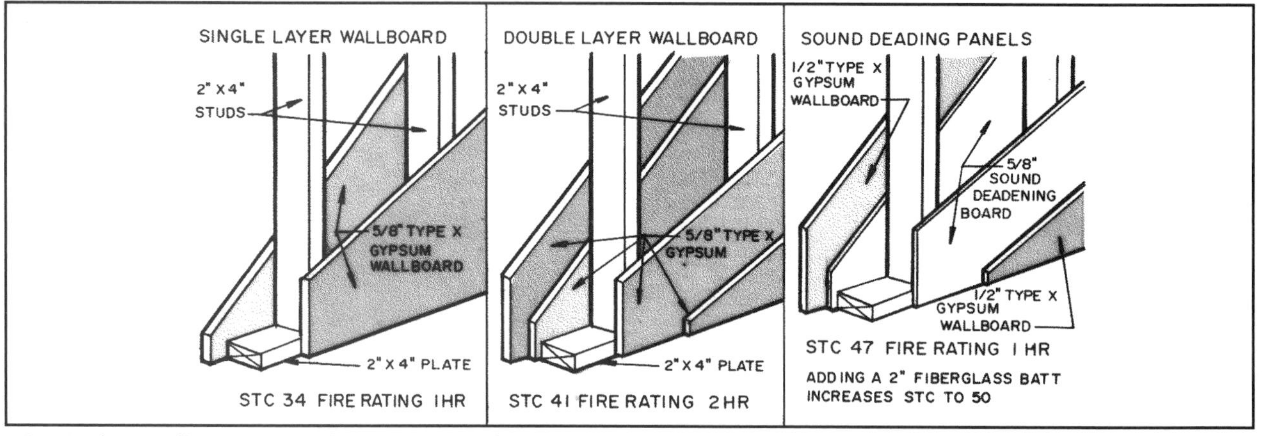

34-40 These wall constructions have STC ratings from 34 to 50 and one- and two-hour fire ratings.

Some typical wall constructions with their STC and fire ratings are shown in **33-40** through **34-43**. The fire rating is an indication of the time in hours the wall assembly can withstand exposure to fire.

Several designs using standard 2 × 4-inch studs are shown in **34-40** with various wall covering materials. The same type of construction but with the gypsum wallboard mounted on one side over a series of resilient metal furring channels is shown in **34-41**. Notice how this increases the STC and fire rating over the single layer wall seen earlier in **34-37**. Another technique is to stagger 2 × 4-inch studs on a 2 × 6-inch plate. One such assembly is in **34-42**. Still another assembly is shown in **34-43**. It completely separates the wall and utilizes fibrous insulation batts and sound-deadening board. The fastening recommendations of the gypsum wallboard manufacturer must be observed in order to get the full value of the design.

Sound also passes between floor and ceilings. They must resist **airborne noises** as well as **impact noises**, such as someone walking. Impact noise is expressed as the **impact isolation class** (IIC). Some typical floor-ceiling constructions with their IIC, STC, and fire ratings are in **34-44**.

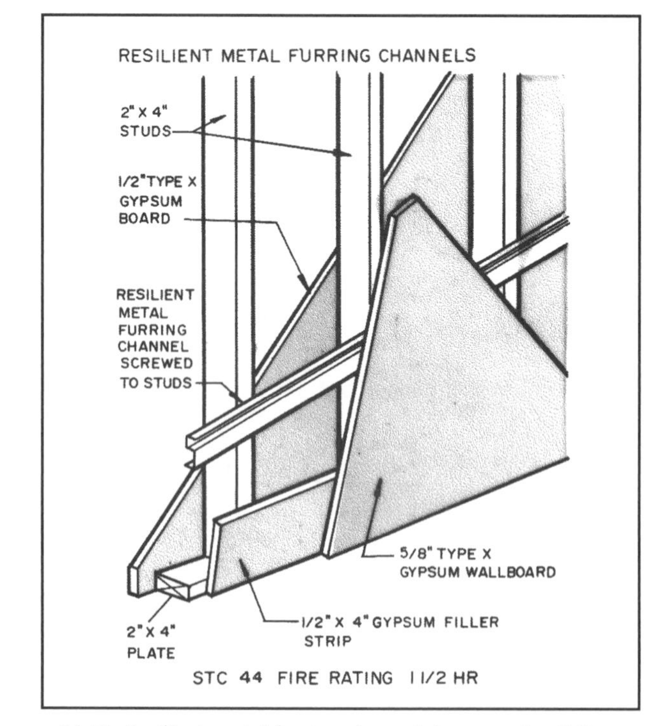

34-41 Resilient metal furring channels increase the STC and fire rating.

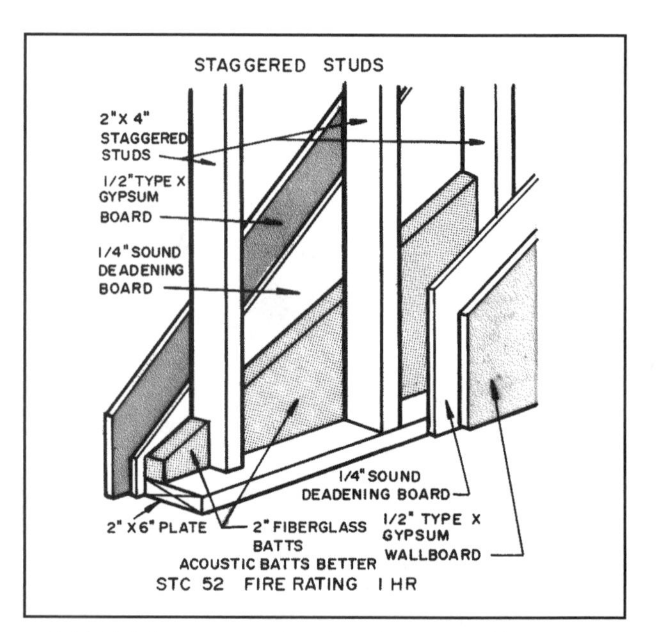

34-42 This is a staggered stud partition assembly using sound-deadening board below the gypsum wallboard.

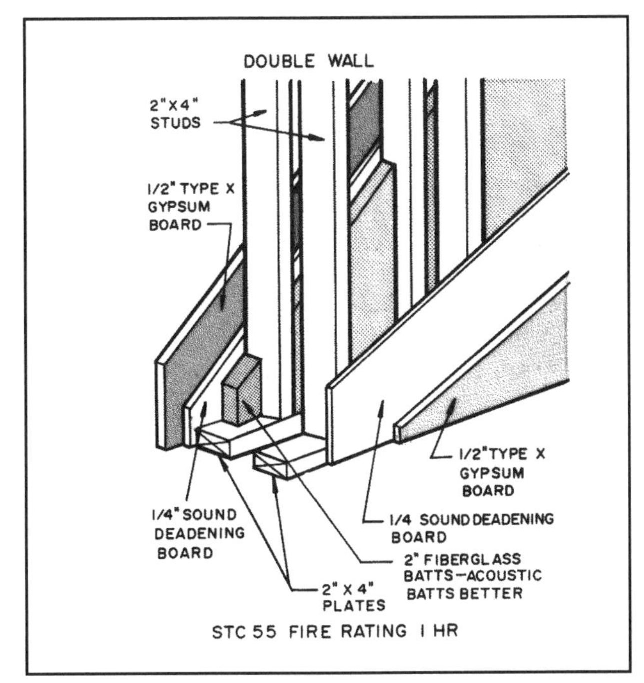

34-43 This is one design for a double-stud wall that completely separates the faces of the walls in the abutting rooms.

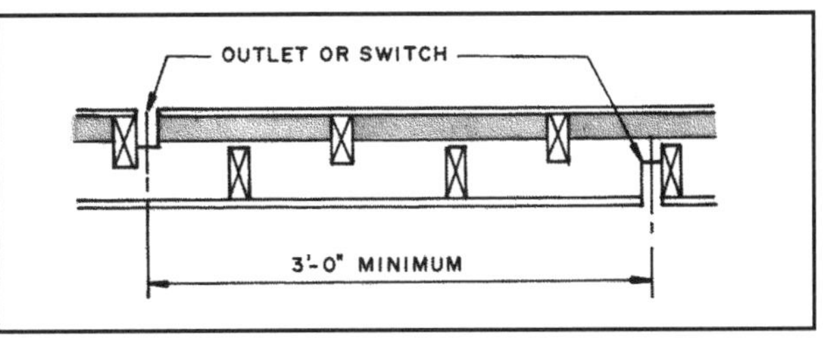

34-45 Staggered electrical switches and outlets interrupt any passage of sound around them and through the wall.

Sound passage between rooms can also be reduced by caulking around electrical outlets, plumbing, and other things penetrating the finished wall **(refer to 34-36)**. It is recommended that electrical switches and outlets not be installed back-to-back but be staggered **(see 34-45)**.

Sound can also be handled by absorption. **Sound-absorbing** materials minimize the amount of noise by stopping the reflection of sound back into the room. Various acoustical materials such as acoustical ceiling tile and wall panels, curtains, and carpets are good sound absorbers.

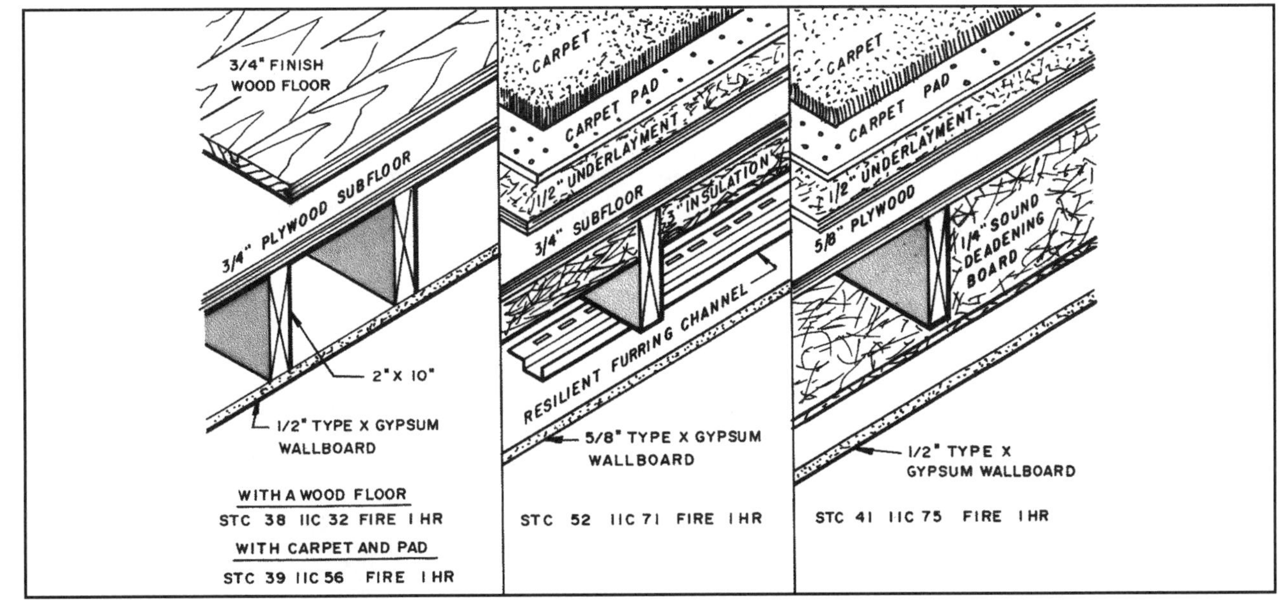

34-44 A couple of typical floor-ceiling assemblies that use various materials to reduce the transfer of sound and impact noise.

CHAPTER 35

INSTALLING GYPSUM DRYWALL

Gypsum sheet products are used to finish interior ceilings and walls, essentially replacing the labor-intensive plaster lath of an earlier era. They consist of a gypsum core sandwiched between two sheets of special paper. The product is strong and highly fire resistant.

The most frequently used drywall is in sheets 4 feet (1.2m) wide and 8 feet (2.4m) long. Sheets 54 inches wide are available, which helps when walls are 8′-6″ or 9′-0″ high. However 9-, 10-, 12-, and 14-foot (2.7, 3, 3.7, and 4.3m) lengths are available. The standard thicknesses are ⅜, ½, and ⅝ inch (10, 12, and 16mm). The thicker the sheet, the less likely it will sag after installation and the higher the fire-resistance rating. The long edges are tapered, forming a shallow valley when the sheets are butted **(see 35-1)**. This is filled with joint compound and tape to produce a smooth, concealed joint. The four-foot ends of the sheets are usually cut square, which, when used to form an exposed joint, are harder to conceal with joint compound and tape than the tapered sides.

TYPES OF GYPSUM DRYWALL

There are a variety of gypsum panels, each with special properties.

Regular panels which have tapered, long side edges are commonly used in finishing residential walls.

Type X panels have a specially formulated core that increases their ability to resist fire.

Vapor barrier panels have an aluminum foil backing serving as a vapor barrier when applied with the foil layer next to the framing.

Moisture-resistant panels are used as a base for the adhesive application of ceramic tile wall covering. They have a light green or blue face paper. The paper covering of this type is chemically treated to resist the penetration of water; it is typically used on walls in bathrooms and kitchens.

Predecorated standard panels are also available with a vinyl wall covering on the exposed face in a wide range of colors and designs. A variety of gypsum predecorated moldings are available to cover the joints in predecorated-panel installations.

Flexible panels are available for use on arches and curved walls. They are ¼ inch thick with heavy paper facing and will resist cracking when bent.

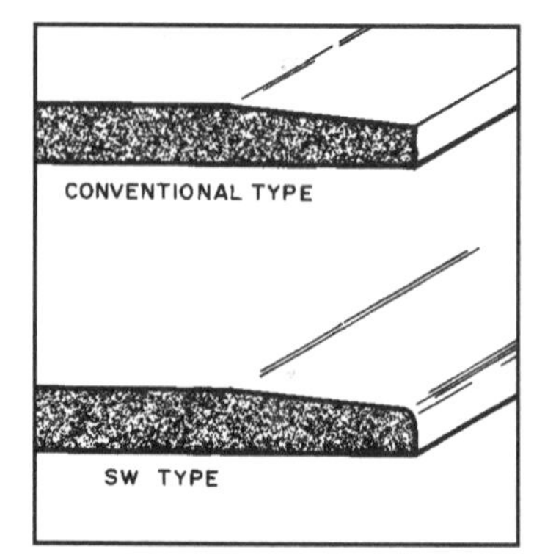

35-1 The long edges of some gypsum wallboard panels are tapered to permit the tape and compound covering the joint to be flush with the surface of the panel.

35-2 All-purpose joint compound is available ready-mixed.

35-3 Paper tape is commonly used with joint compound to cover joints between panels.

High-strength panels are used where sagging is a possibility, such as on a ceiling.

Abuse-resistant panels have a reinforced core, a strong liner paper on the back, and a heavy face paper. They are available in Type X fire-resistant panels. They resist dents from impact better than normal panels.

JOINT COMPOUNDS

Drywall joint compound is used to conceal the butt joints between panels and the heads of the nails or screws used to secure the panels to the studs or joists. It is available in powdered or premixed form. The powdered type is mixed with water to the proper consistency. Premixed compounds are ready to use. They have the advantage of immediate usability and correct consistency, and are free of lumps and air bubbles that are often found in site-mixed powdered compounds.

Powder-type compounds are available in a number of setting times ranging from 20 to 360 minutes. They can be stored for long periods in dry storage and cold conditions. However, they should be brought to normal room temperature before mixing.

Premixed compounds are available in several types. The **taping compound** is used for embedding the tape. The **topping compound** is used for the second and third finish coats over the taping compound. An **all-purpose compound** can be used for taping, finishing, and texturing the surface **(see 35-2)**.

JOINT-REINFORCING MATERIALS

The most commonly used joint reinforcement is **reinforced paper tape (see 35-3)**. It is two inches (50.8mm) wide and is available in rolls up to 500 feet (139.5m) long. It has a crease in the center making it easy to fold for reinforcing interior and exterior corners. It is bonded to the gypsum wallboard with joint compound.

Corners are reinforced with galvanized steel or plastic **corner beads**. A couple of metal corner beads are shown in **35-4**. The corner beads may be installed by nailing or stapling through the drywall into the stud **(see 35-5)**.

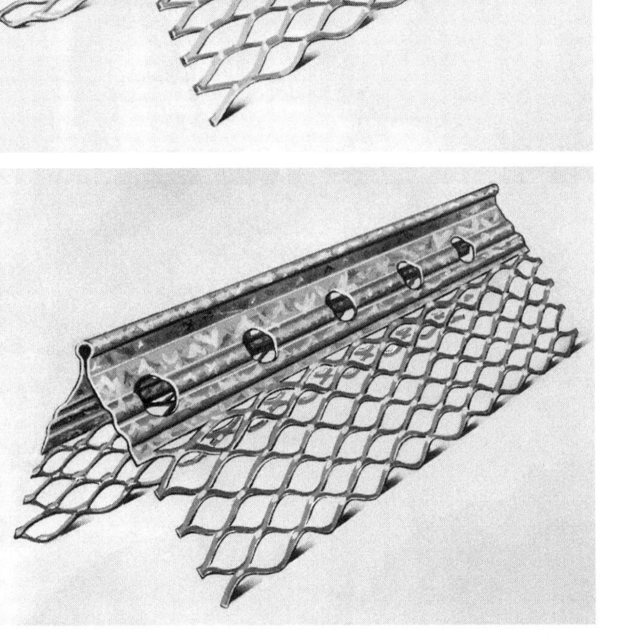

35-4 Two types of metal corner bead. *(Courtesy United States Gypsum Company)*

35-5 This metal corner bead has been nailed to the studs through pre-punched holes. It is being covered with joint compound. *(Courtesy United States Gypsum Company)*

A flexible metal corner tape is available that is folded, forming the bead. It can be used on curved corners. A few of the **plastic corner beads** that are available are shown in **35-6**. They are used for straight and curved corners and may be nailed or stapled. The reveal bead provides a ½-inch-(13mm) deep reveal. The width of the reveal can be ½, ¾, or one inch (13, 19, or 25mm). A hideaway expansion joint bead has a vinyl membrane over the V-shaped expansion area. Drywall compound is laid up to the edge but not over the vinyl membrane.

A bullnose corner bead gives the corner a round appearance. A bullnose archway corner bead has the nailing flange notched, permitting it to bend around the curved arch. The low-profile corner bead is most commonly used in residential construction. It provides a small but damage-resistant corner. The adjustable inside and outside corner beads will flex to cover corners on angles from 90 degrees to 135 degrees. Many other types of beads, moldings, and trim are also available.

After the corner beads are in place they are covered with joint compound, which is feathered out on the drywall, as shown earlier in **35-5**.

Exposed edges of gypsum drywall panels are reinforced and concealed with U-shaped metal trim, as shown in **35-7**. The trim is slid on the gypsum panel and nailed to the stud. It is concealed with joint compound. Several types of edges are available.

DRYWALL FINISHES

Possibly the greatest number of square feet of installed drywall are finished by painting with an interior paint. Wallpaper is also widely used. The ceilings and some walls are given a textured

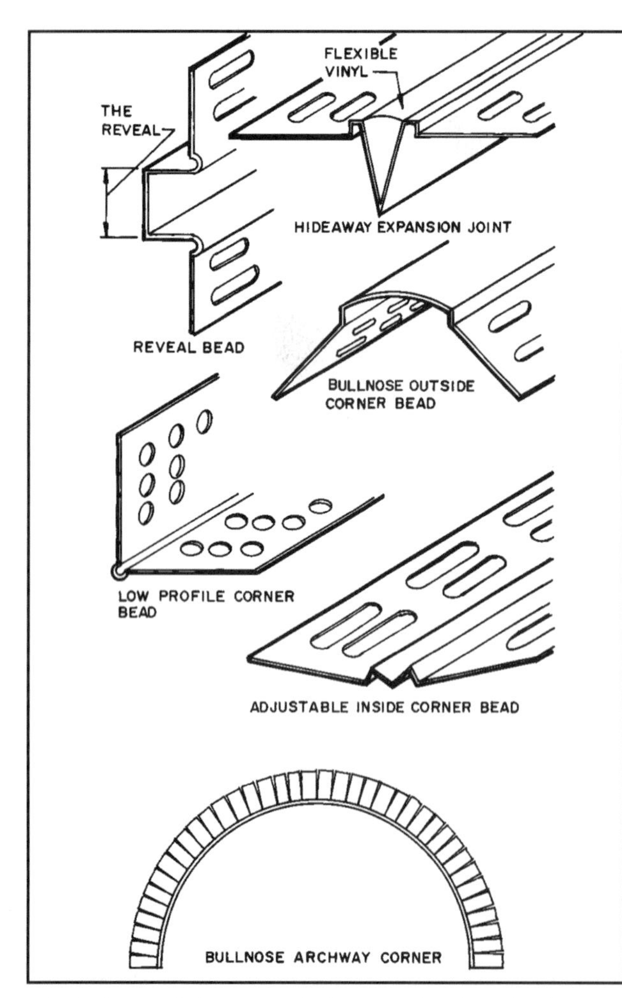

35-6 A few of the many plastic corner beads available. *(Courtesy Trim-Tex, Inc.)*

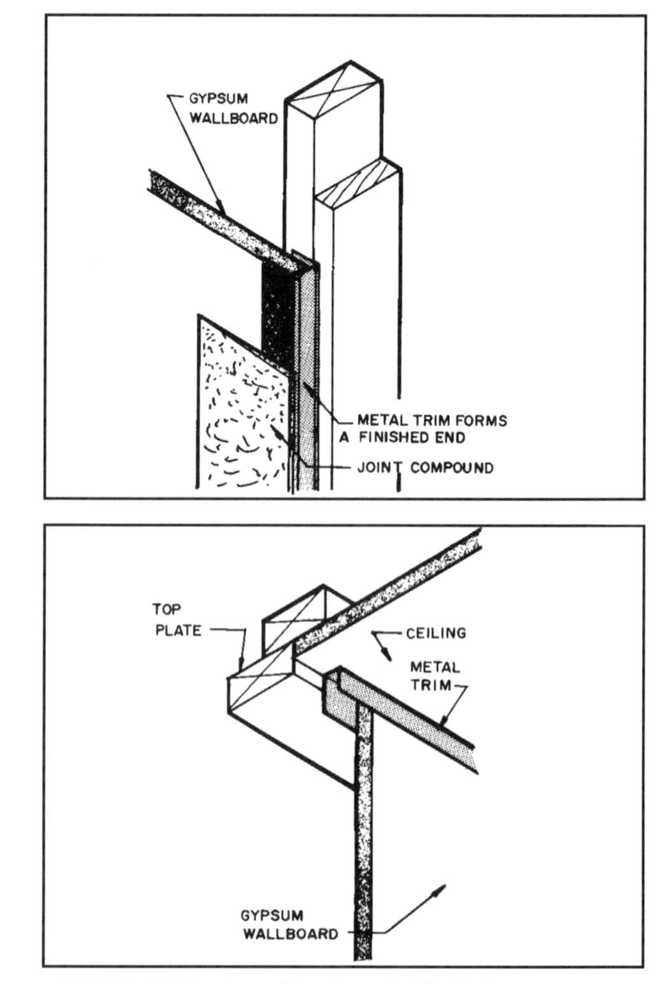

35-7 Metal trim is used to give a finished edge to gypsum wallboard that is exposed to view.

finish. Compound manufacturers have available a variety of materials that can be brushed, rolled, or sprayed over the panels to produce such a textured surface **(see 35-8)**.

TEMPERATURE CONTROL

The rooms in which gypsum panels are to be installed must be kept at 55 degrees F to 70 degrees F (13 to 21 degrees C), 24 hours a day, during periods when the outdoor temperature is below 55 degrees F (13 degrees C). Ventilation must be provided to remove excess moisture that builds up as the joint compound dries.

ON-SITE STORAGE

Gypsum drywall panels must be stored in a protected area. Typically this is inside the building where they are to be used. They should be stacked flat on a clean floor. Since the ceiling is installed first, the panels for this use must be on the top of the pile. Since gypsum panels are heavy, consideration must be given to the total weight that a pile will put on the floor joists.

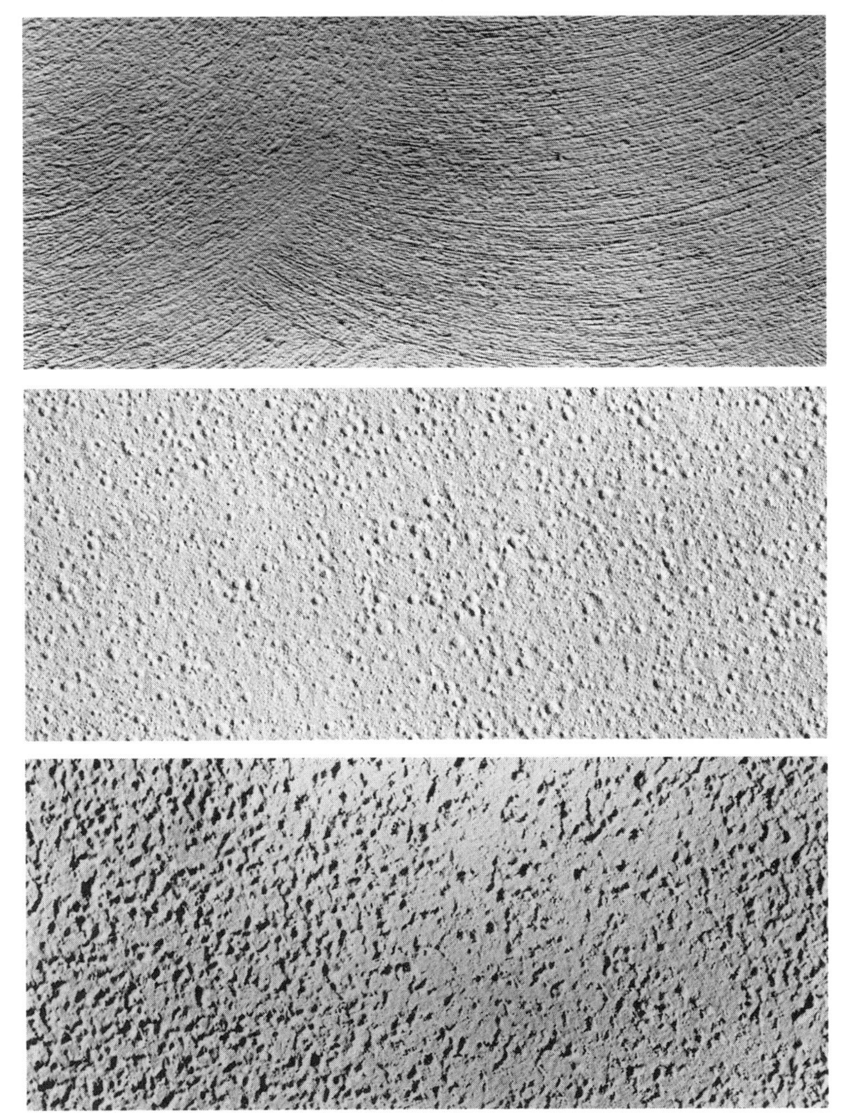

35-8 A few of the surface textures that may be used on gypsum wallboard. *(Courtesy United States Gypsum Company)*

PRELIMINARY PREPARATION

Before beginning to install gypsum drywall, the framing should be checked for squareness and spacing. The standard four-foot (1.2m) width panel will span four studs, and the eight-foot (2.4m) length will span seven studs spaced 16 inches (40.6cm) O.C. If they have not been carefully spaced, it will be necessary to nail scabs onto the misplaced studs so that the edges of the sheet rest firmly on a stud **(see 35-9)**.

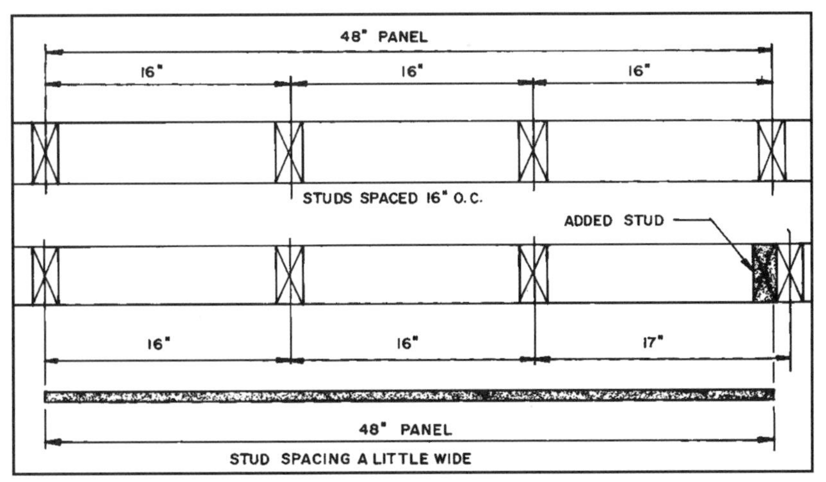

35-9 Improperly spaced studs require that extra studs be installed to support the edge of the wallboard.

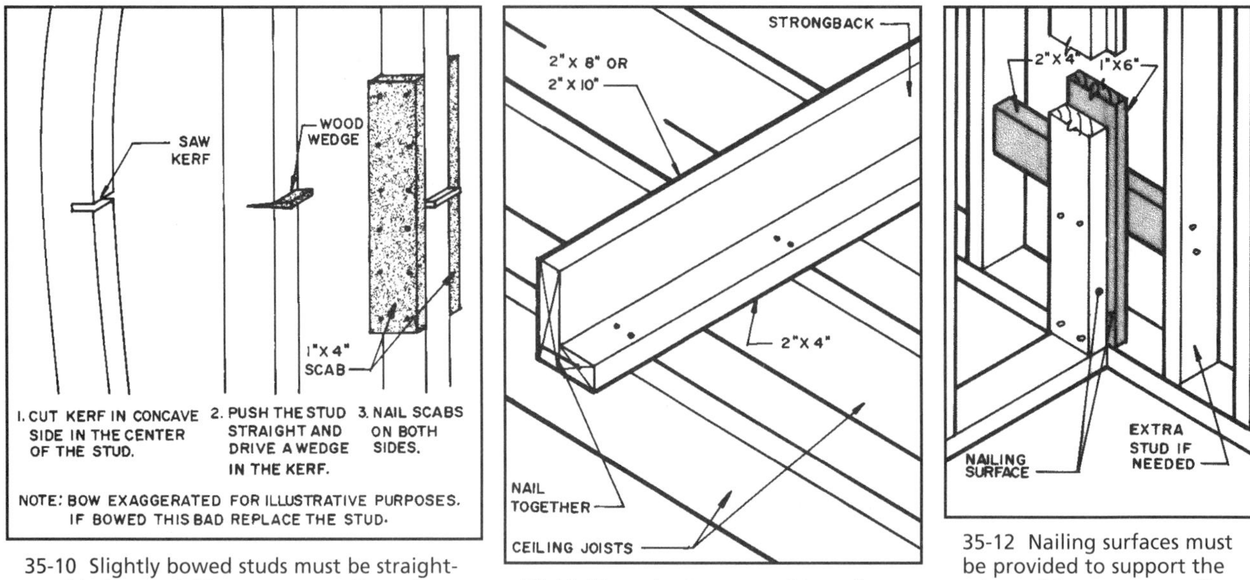

35-10 Slightly bowed studs must be straightened before installing gypsum wallboard. Many times the bowed stud is replaced.

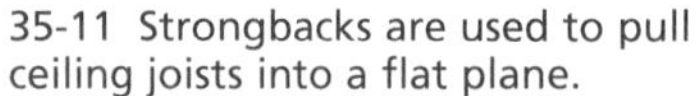

35-11 Strongbacks are used to pull ceiling joists into a flat plane.

35-12 Nailing surfaces must be provided to support the edges of the gypsum wallboard.

The studs and ceiling are checked for squareness by running a chalkline from one end of the wall to the other. Bowed studs can be straightened by sawing into the hollow sides at the middle of the stud, pulling the stud straight, and driving a thin wedge into the saw kerf until the stud remains straight. Then a two-foot- (610mm) long, 1 × 4-inch scab is nailed across the cut on both sides of the stud to hold it straight **(see 35-10)**. Trim off the part of the wedge extending beyond the stud after the scabs are nailed in place. If the bow is too great it is best to replace the stud.

Ceiling joists can often be straightened by installing some form of strongback across them in the attic. One example is shown in **35-11**. When the joists are nailed to the strongback, they are pulled into line. The material used to make the strongback should be straight. Check for straightness with a chalkline.

Check to see that the framing has provided the needed nailing surfaces. This is shown in Chapter 10. For example, a corner formed by the intersection of two partitions needs to be framed to provide nailing surfaces as shown in **35-12**.

A device that speeds up the installation of drywall panels is a plastic clip installed by the finish carpenters before the drywall crew arrives. The clips are nailed or stapled every two feet (61cm)

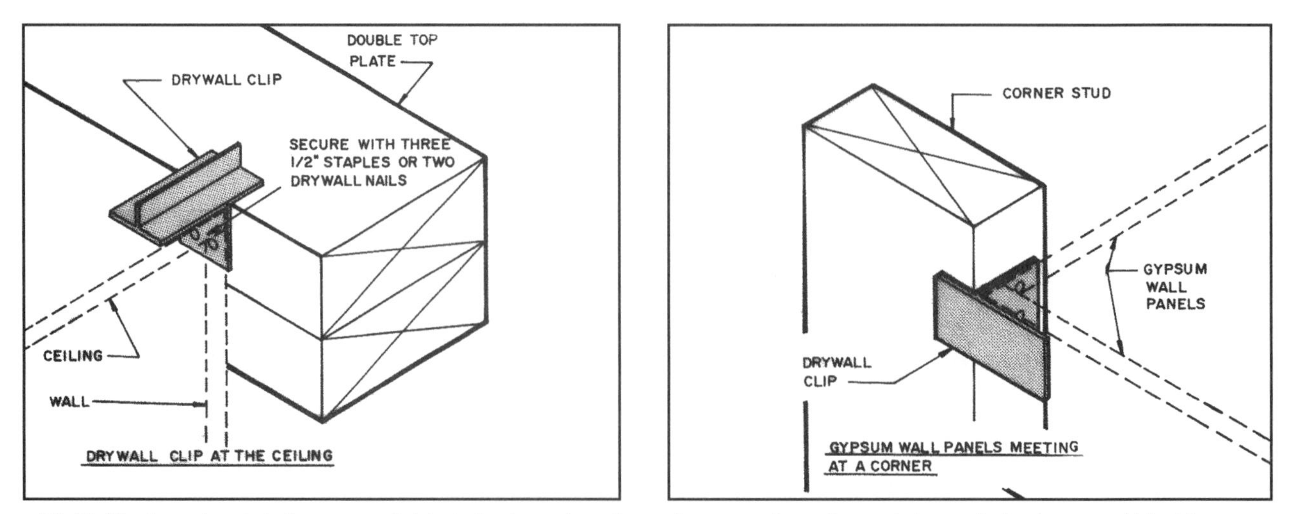

35-13 Plastic and metal clips are available to back up the edges of gypsum drywall panels instead of using wood blocking.

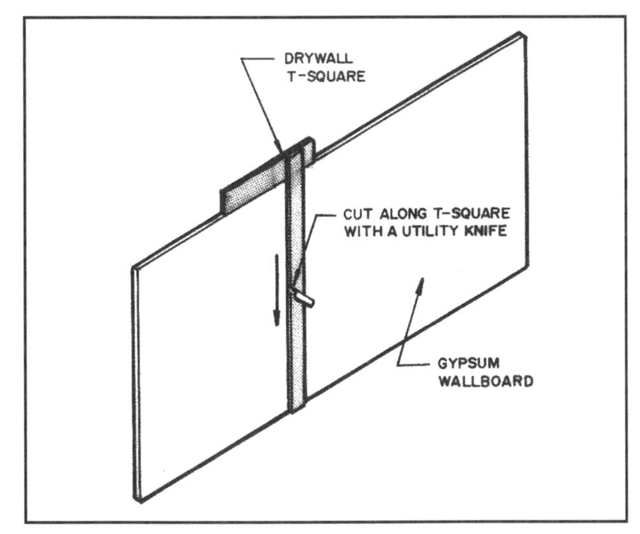

35-14 Straight cuts are made with a drywall T-square and a sharp utility knife.

along the vertical corners and every 16 inches (40.6cm) around the tops of the walls. These reduce the need to install the wood framing that is usually required for nailing the panels at the ceiling and in the corners. Once the panels are in place, they are nailed in the conventional manner **(see 35-13)**.

CUTTING GYPSUM PANELS

Straight cuts across a panel are made by marking the location of the cut, placing a drywall T-square on the mark, and scoring the panel with a utility knife **(see 35-14)**. The panel may be cut lying flat or in a vertical position. Then bend the board and it will snap along the scored line **(see 35-15)**. Cut through the paper on the back side of the panel **(see 35-16)**. The two pieces will separate **(see 35-17)**. Holes can be cut with a circular cutting tool or a keyhole saw **(see 35-18)**. Drywall saws are also used to make various cuts in gypsum panels. Ragged edges can be smoothed with a knife, file, multiblade forming tool, or coarse sandpaper.

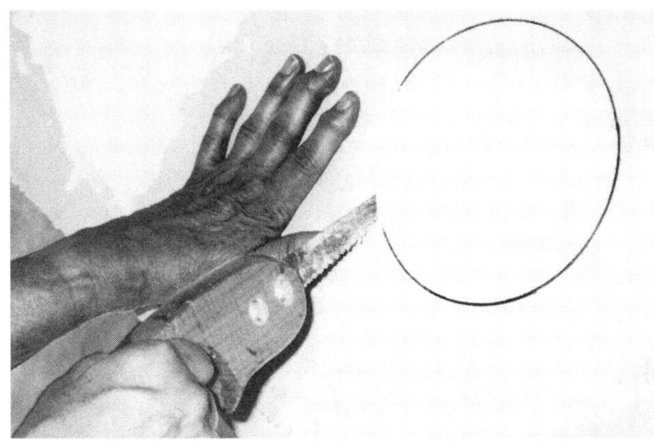

35-18 This circular opening is being cut with a keyhole saw.

35-15 After scoring the panel on one side with the knife, bend it to snap the gypsum core along the line that was cut. *(Courtesy United States Gypsum Company)*

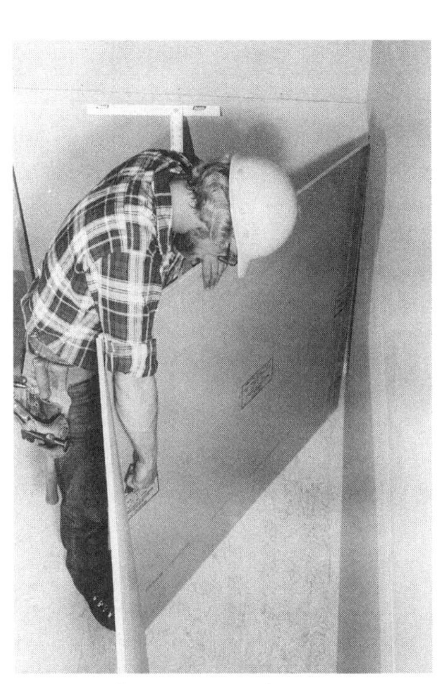

35-16 After snapping the gypsum core, cut the paper on the other side. *(Courtesy United States Gypsum Company)*

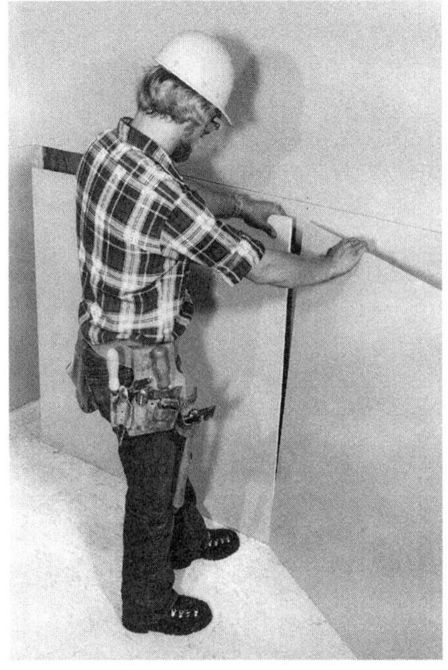

35-17 After the back paper has been cut, the wallboard parts will separate. *(Courtesy United States Gypsum Company)*

35-19 Adhesive is applied to the back of a gypsum panel that will be bonded to another panel that has been nailed or screwed to the studs. *(Courtesy United States Gypsum Company)*

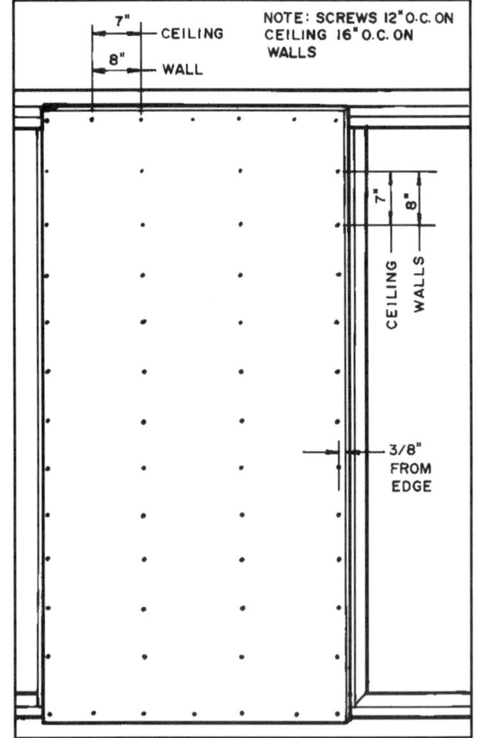

35-20 The recommended nailing pattern for single-layer wall and ceiling panels.

INSTALLING DRYWALL PANELS

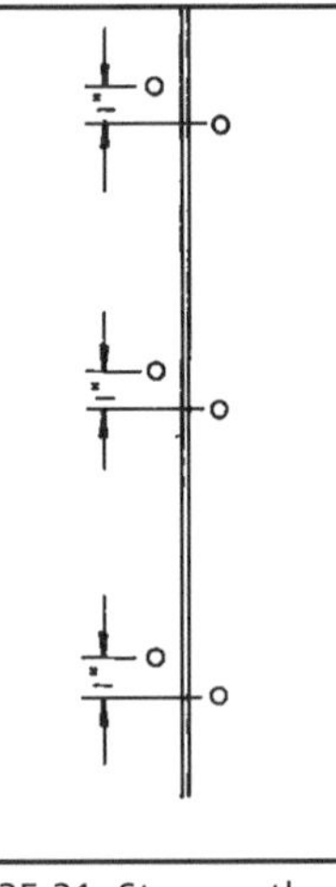

35-21 Stagger the nails along butting panels.

Panels may be installed using a **single-layer** or **double-layer application**. **Single-layer application** is used on interior ceilings and walls where economy is important. It provides for fast installation and fire protection. Most residential work uses single-layer installation. **Double-layer application** has one layer of gypsum drywall nailed or screwed directly to the studs, and this is overlaid with a second layer. It offers greater strength, greater fire protection, and reduced sound transmission. The second layer is bonded to the first layer with an adhesive **(see 35-19)**. Some nails are used to hold the panels as the adhesive sets. The base layer may be secured to the studs horizontally or vertically. The bonded face layer must be applied with the long joists perpendicular to the long joints in the base layer.

Panels may be nailed or screwed to wood studs and screwed to metal studs. They may be single-nailed, double-nailed, attached with screws, or bonded with adhesives. Single-nailed panels must have the nails located ⅜, inch (9.5mm) from the edge of the panel. On wood frame construction they should be spaced seven inches (178mm) apart on the ceiling and eight inches (203mm) apart on the walls **(see 35-20)**.

Screws are spaced 12 inches (305mm) apart on the ceiling and 16 inches (406mm) apart on the walls. The nails should be staggered by about one inch (25mm) on the butting joints **(see 35-21)**.

A double-nailed panel is shown in **35-22**. Double-nailing reduces the number of nail pops that occur as the wood framing shrinks, and helps keep the panels tight to the studs. The nails are spaced seven inches (178mm) around the perimeter of ceiling panels and 8 inches (203mm) on wall panels. Nails are doubled within the panel, spaced two inches (50mm) apart, and spaced 12 inches (304mm) O.C.

SINGLE-LAYER APPLICATION OVER A WOOD FRAME

The panels can be applied with the long side **perpendicular to** or **parallel with** the studs. Check local codes to see if fire regulations specify the method to be used **(see 35-23 and 35-24)**. Generally perpendicular application is used, because it greatly reduces the lineal feet of joints, provides a stronger wall, and tends to bridge over irregu-

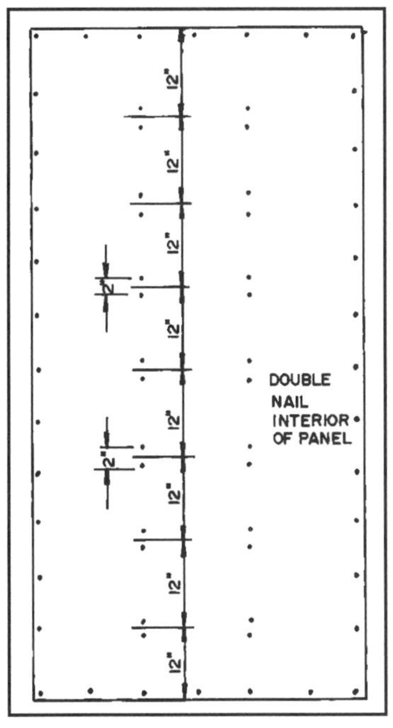

35-22 A double-nailed panel.

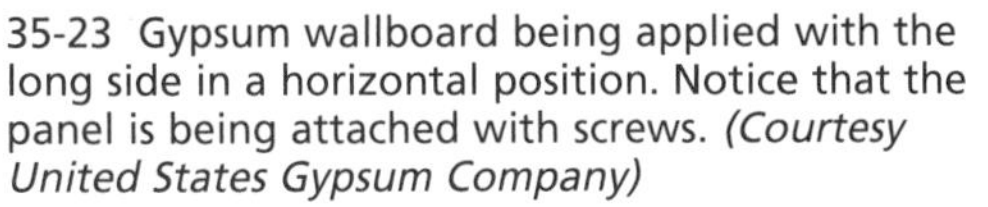

35-23 Gypsum wallboard being applied with the long side in a horizontal position. Notice that the panel is being attached with screws. *(Courtesy United States Gypsum Company)*

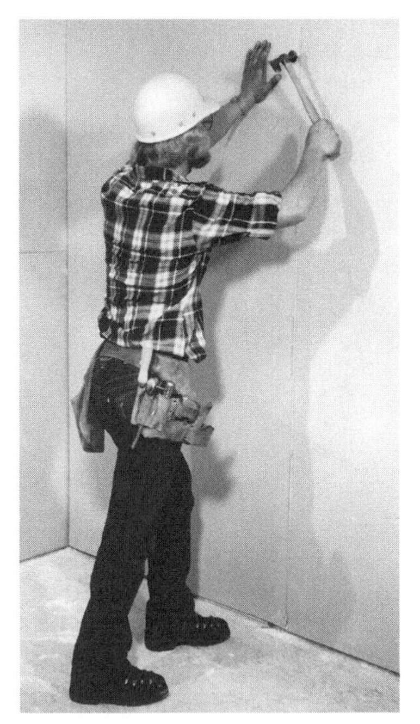

35-24 Gypsum wallboard being applied with the long side in a vertical position. *(Courtesy United States Gypsum Company)*

larities in the line of the studs. For ceiling heights greater than 8′-1″ (2.46m), parallel application is usually more practical.

All ends and edges of panels must occur over framing members, except when the joints are at right angles to the framing members.

Nails should be driven straight into the studs. If a nail misses the stud, it should be removed. The nail head should not break the paper cover. If it does break it, drive another nail abut two inches (50mm) away **(see 35-25)**. These same recommendations apply to screws. Use a drywall hammer to drive nails, because it has a crowned face designed for setting the nails and producing the required dimple around the head. This hammer is shown in **35-26**.

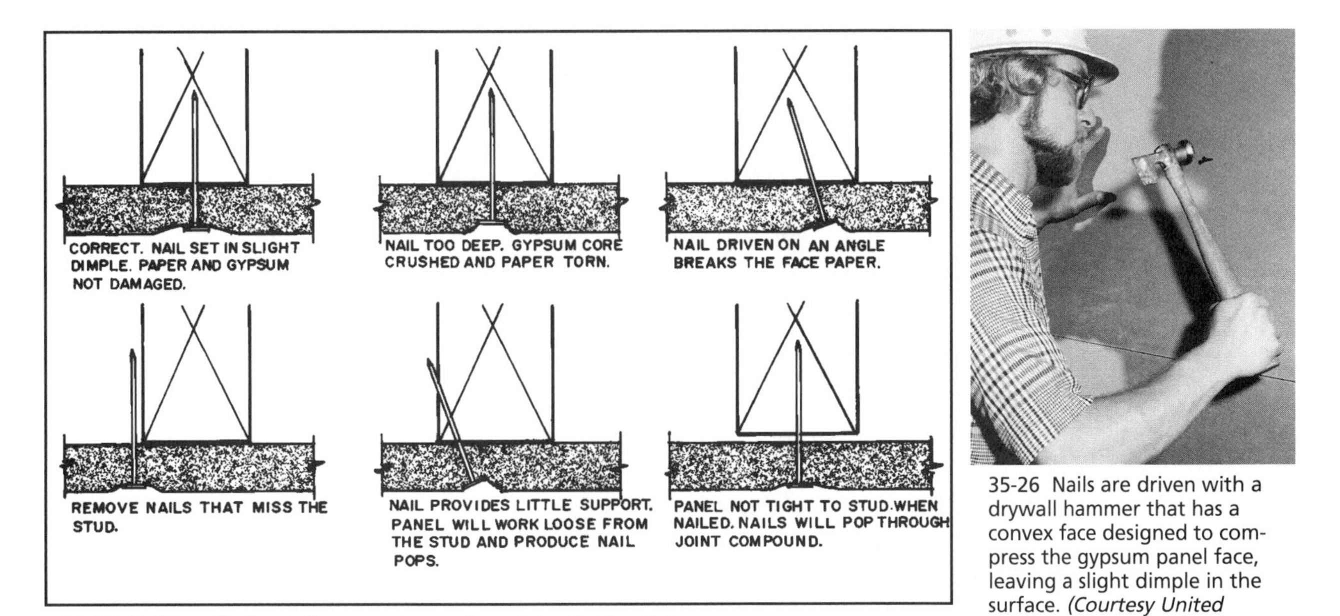

35-25 Nail installation influences the quality of the finished job.

35-26 Nails are driven with a drywall hammer that has a convex face designed to compress the gypsum panel face, leaving a slight dimple in the surface. *(Courtesy United States Gypsum Company)*

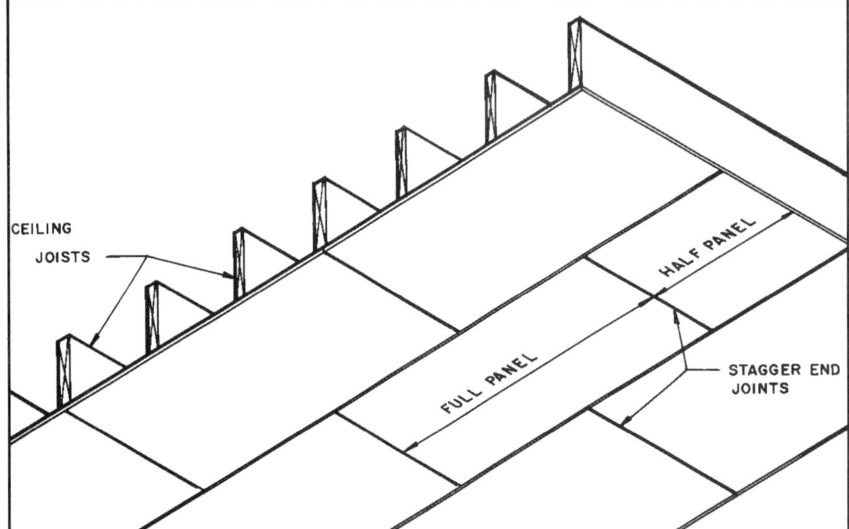

35-27 Ceiling panels are installed before the wall panels. The end joints are staggered by half a panel.

35-28 Gypsum panels on the ceiling can be held in place with an adjustable support. *(Courtesy Patterson Avenue Tool Company)*

Begin by installing the ceiling panels. Start nailing in the middle of the panel, and work towards the edges and ends. If there are end joints, stagger them about four feet (1220mm) apart **(see 35-27)**. The panel can be held in place by two or more workers. The use of T-shaped supports or mechanical jacks greatly assists the process **(see 35-28, 35-29, and 35-30)**. If possible, get panel lengths long enough to cross the ceiling without requiring end butt joints. End butt joints are difficult to conceal. It is especially desirable to double-nail the ceiling, since it is the area most likely to sag or pop nails.

Next, nail the wall panels. Keep the edge and end joints about 1⁄16 inch (1.5mm) apart to allow for possible expansion. They should meet the ceiling neatly so that a sharp corner can be taped. If a molding is to be used at the ceiling, the size of the

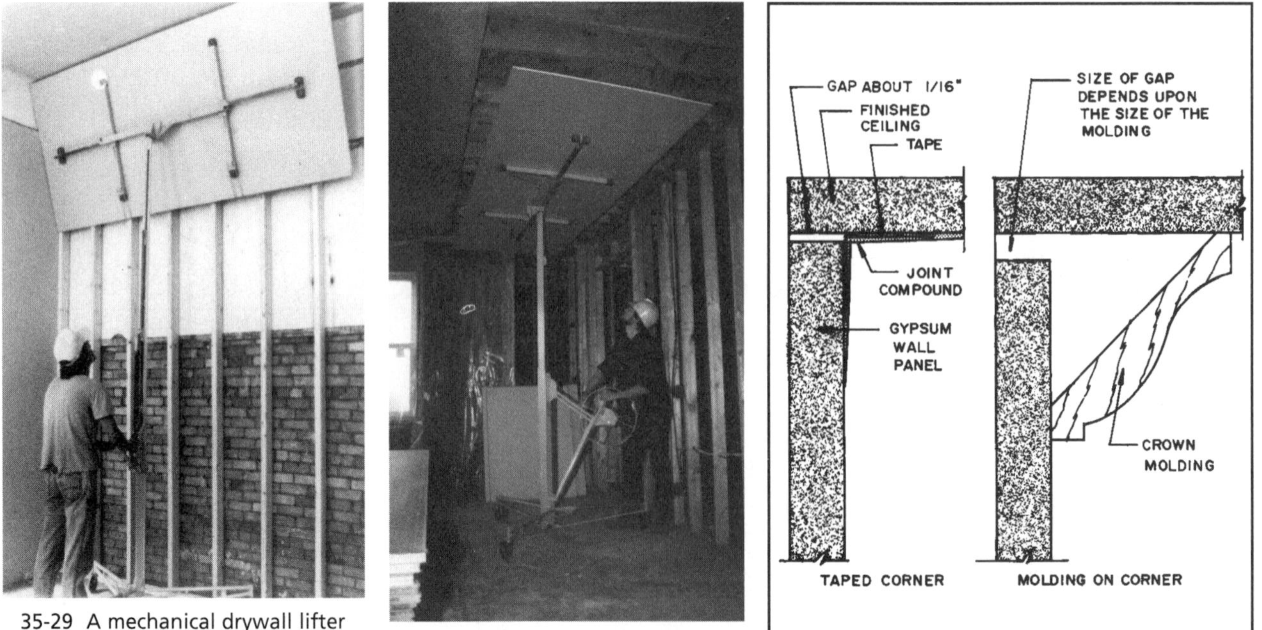

35-29 A mechanical drywall lifter such as this can be used to place a panel on a high wall. *(Courtesy Telpro, Inc.)*

35-30 This mechanical drywall lifter can place panels on the ceiling. *(Courtesy Telpro, Inc.)*

35-31 Leave a gap between the gypsum wall panel and the finished ceiling.

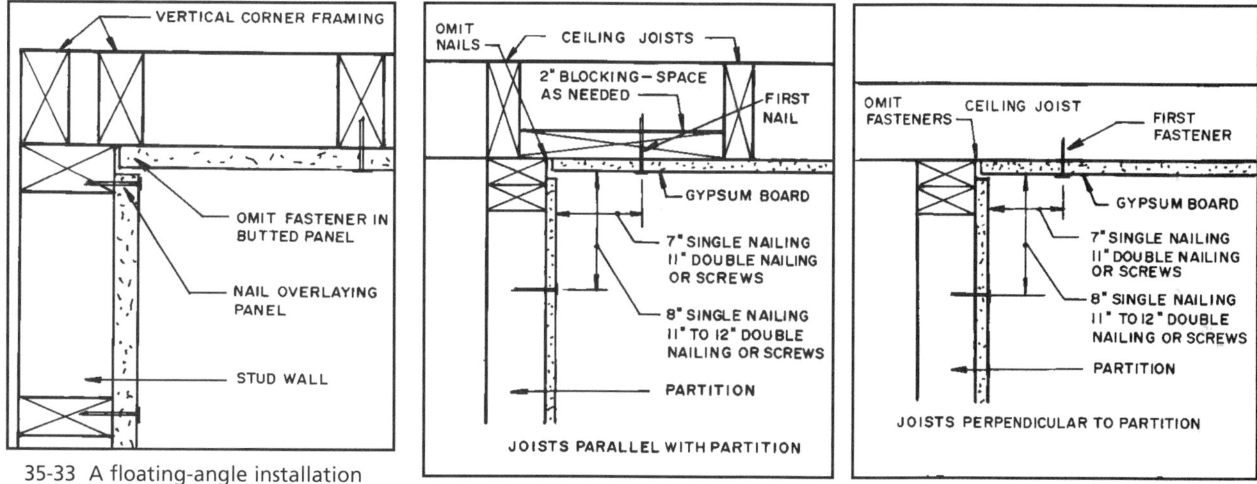

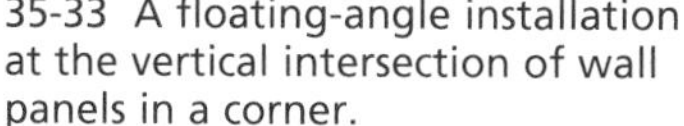

35-33 A floating-angle installation at the vertical intersection of wall panels in a corner.

35-32 A floating-angle installation between the ceiling and wall panels.

gap is not critical **(see 35-31)**. It is recommended that the corner at the ceiling be installed using floating-angle construction. This helps minimize fasteners popping at the wall and ceiling intersection. Recommended nailing procedures for an intersection parallel with and perpendicular to ceiling joists are in **35-32**. The same technique for vertical corners is shown in **35-33**.

When installing wall panels horizontally, hang the top panels first **(see 35-34)** and try to avoid end joints. Often a panel can butt a door or window frame, thus hiding the end with the wood casing to be installed around the door or window opening. Begin nailing the panel from the top edge in the middle of the panel. Work down and towards the ends. Next, install the bottom panel in the same way. If necessary, trim the bottom panel so that it clears the subfloor. This edge will be covered with a baseboard or carpeting. If a baseboard is used, it need not be a smooth cut. If the floor is to be carpeted it should be close to the subfloor and neatly trimmed.

Vertical panels are nailed by starting on the edge that butts another panel; nail towards the opposite ends and edges. Nail the interior of the panel and proceed towards the edges. The panels should be held tightly to the framing as they are nailed.

INSTALLING DRYWALL ON CURVED WALLS & ARCHES

It takes careful work and planning to hang curved walls and arches. First of all the carpenter should have properly prepared the framing. The space between wall studs and other framing material is very important. If the space is too great, the wall will have a series of flat surfaces.

Hanging Curved Walls

It is recommended that you use a special ¼-inch flexible gypsum panel designed for curved surface installation. This is a two-layer installation with a second ¼-inch panel overlaid on the first. Bending radii for ¼-inch flexible wallboard are specified by the manufacturer. When the panels are placed

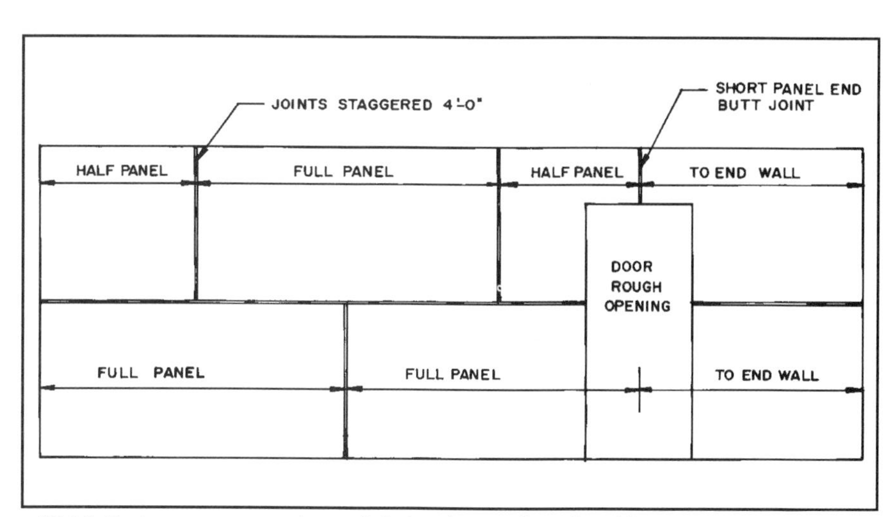

35-34 Try to arrange panels to minimize the number of end butt joints.

horizontally they typically have a bend radius of 30 to 32 inches, and if placed vertically with the bend across the width of the panel they have a bend radius of 18 to 20 inches. The manufacturer will also specify the maximum distance between studs, which is commonly 8 to 9 inches.

When covering **concave surfaces,** place a wood stop at one end of the curve and press the panel against it from the other edge. This helps bow the panel to the wall. Start installing the fasteners from the end against the stop **(see 35-35)**.

When installing on **convex surfaces,** first nail one end to the framing with nails or screws. Then push the panel against the framing nailing toward the loose end **(see 35-36)**.

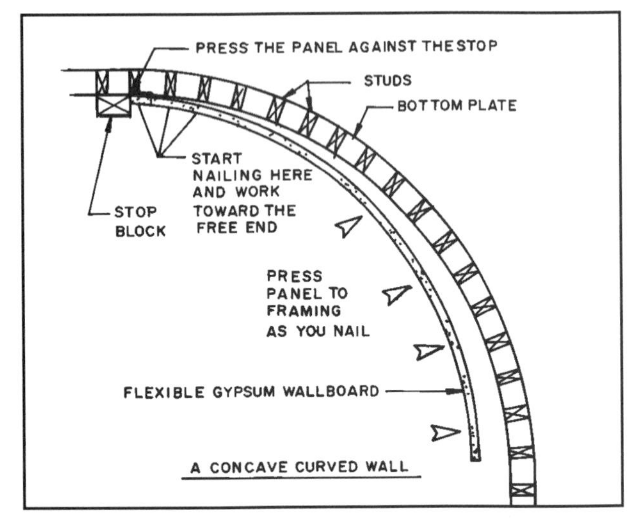

35-35 This is how to install flexible gypsum wallboard on the concave side of a curved wall.

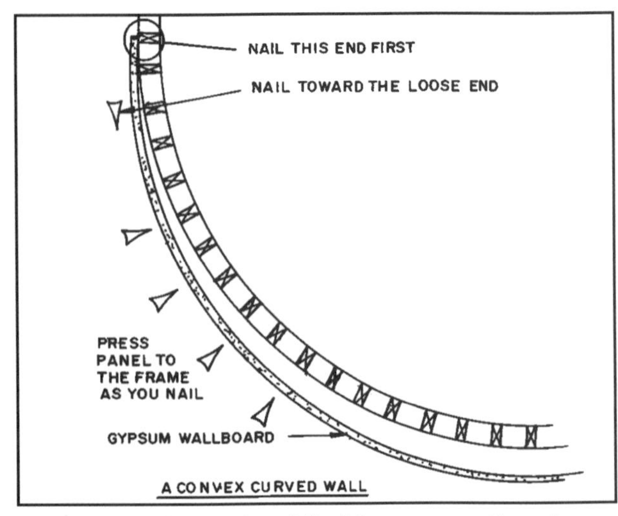

35-36 This is how to install flexible gypsum wallboard on the convex side of a curved wall.

This product may require wetting if smaller radii are required or if the temperature is lower than 67 degrees F and the humidity is below 45 percent. Apply 10 to 15 minutes before bending.

When regular gypsum wallboard is used on curved walls, the minimum radii are much larger than those with the flexible panels.

Either wood or metal studs can be used for curved walls. When wood is used the studs can be nailed to the floor, with curved wood blocking nailed between them serving as a bottom plate **(see 35-37)**. When metal studs are used the bottom runner is made to bend to the desired radius and the metal studs are set in the runner.

Install the panels in the same manner as other walls. Using screws can be a big help in stabilizing the installation. Be careful in nailing or screwing because the panel will not necessarily always be flat against the stud. Nail into the edge of the stud touching the panel.

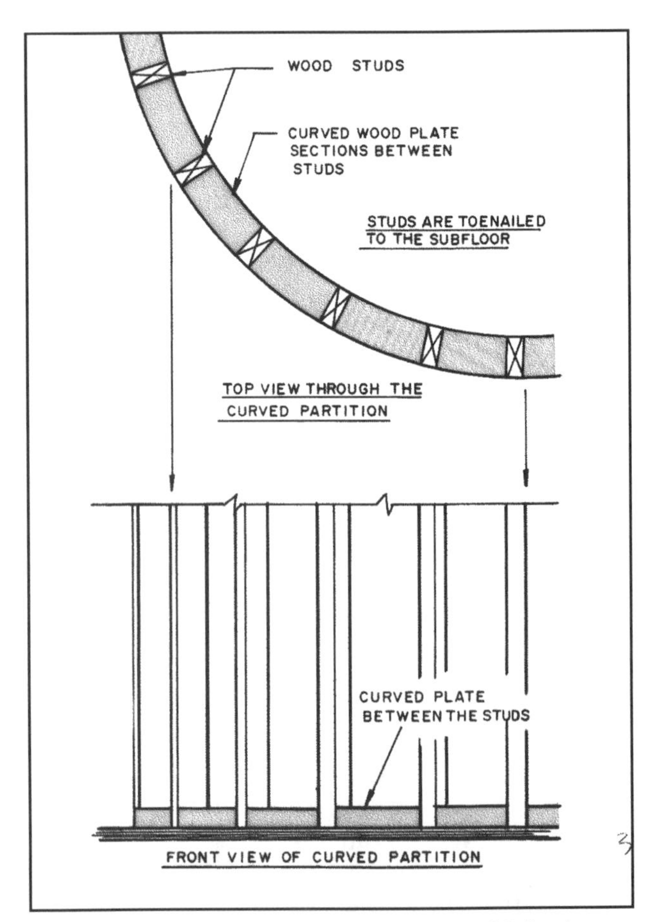

35-37 Wood studs in a curved wall are spaced following the recommendations of the gypsum wallboard manufacturer. Notice the curved sections forming the bottom plate.

Place the fasteners no farther than 12 inches O.C. Be certain to stagger the joints on the second layer away from those on the first layer. If possible use panels long enough to make the curve without end joints. End joints in a curve are very hard to finish.

Joints on curved walls are finished with paper tape and three coats of joint compound.

Installing Arches

Arches are widely used for interior partition openings and add a grace and beauty to the partition and the room **(see 35-38)**. It is important that the carpenter properly frame the arch so an adequate nailing surface is available **(see 35-39)**.

There are several materials used to cover arches. **Regular** gypsum board can be used. However it will not bend around the arch unless it is scored every ½, ¾, or one inch on the back. The smaller the arch, the closer together the score marks. As the panel is nailed in place the score marks open, allowing the panel to round the arch.

The use of **¼-inch high-flex** gypsum board permits most arches to be covered by simply bending the panel around the arch. Start fastening in the center and work toward each end. It can be lightly wetted, as explained earlier, if the arch is small **(see 35-40)**. Both the regular and high-flex panels require the edge to be covered with a corner bead.

35-38 Arches on interior walls present the drywall installer with a challenging installation.

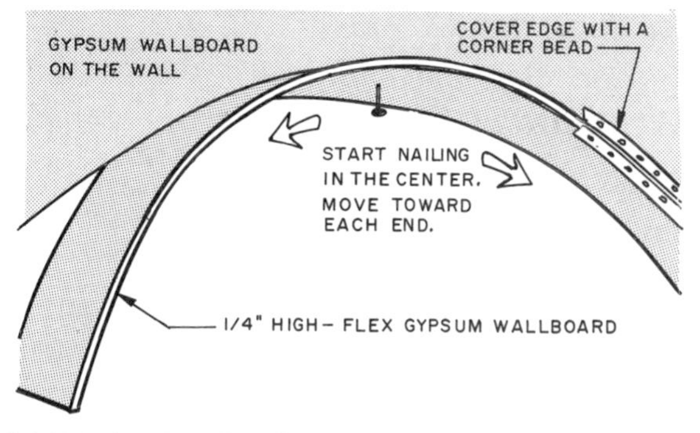

35-40 When installing flexible gypsum wallboard on the arch, begin nailing in the center and work toward each end.

35-39 This interior arch has been framed with 2 × 4-inch blocking and plywood sheathing. This provides a solid base for installing the gypsum drywall.

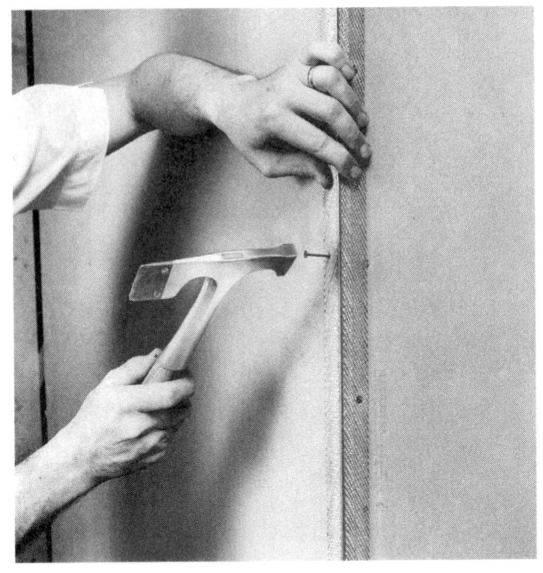

35-41 This corner bead is being nailed through the gypsum wallboard into the stud. *(Courtesy United States Gypsum Company)*

35-42 This gypsum drywall installer is cutting plastic corner beads with a tin snips. *(Courtesy Trim-Tex, Inc.)*

INSTALLING CORNER BEADS

Corner beads are installed by nailing or stapling every nine inches (228mm) **(see 35-41)**. Place the bead firmly against the panel and nail. The bead should stand out a little from the panel so that joint compound can penetrate through the holes in it to get behind the bead. Dimple the nail into the bead so that the head can be covered with joint compound. Try to avoid butting the ends of beads together to get a long piece; it is difficult to get a tight joint. The beads can be cut to length with tin snips **(see 35-42)**.

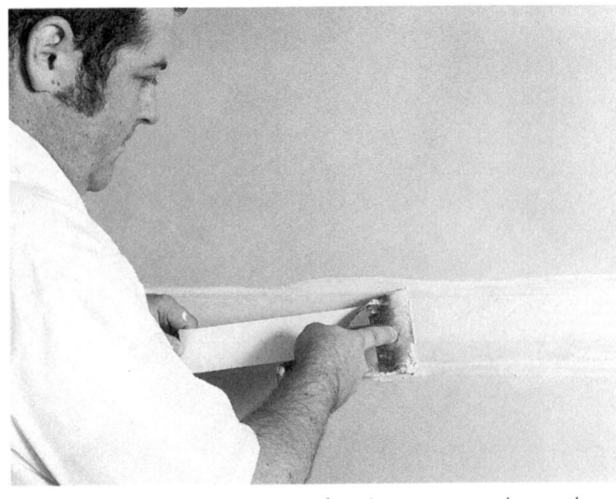

35-43 After applying a coat of taping compound over the joint, center the tape and lightly press it into the compound. *(Courtesy Georgia-Pacific Corporation.)*

TAPING THE JOINTS

Be certain all nails and screws are properly set. Apply a continuous coat of taping compound to fill the space formed by the edges of the tapered panels. Center the tape over the joint and lightly press it into the compound. Draw the finishing knife along the joint to remove excess compound, and cover the tape with a thin layer of compound (1⁄32 inch or 0.8mm), as shown in **35-43**. Allow to dry before applying topping coats. Do not use topping compound for this first step.

Apply a taping compound over the nail heads. This should be enough to fill the depressions and bring them level with the surface of the panel.

After the tape coating has dried, apply a coating of topping compound. Using a wide finishing knife, feather a thin coat out to about 7 to 10 inches

35-44 After the tape coating has dried, apply a topping coat over it that is wider than the taping coat. *(Courtesy Georgia-Pacific Corporation)*

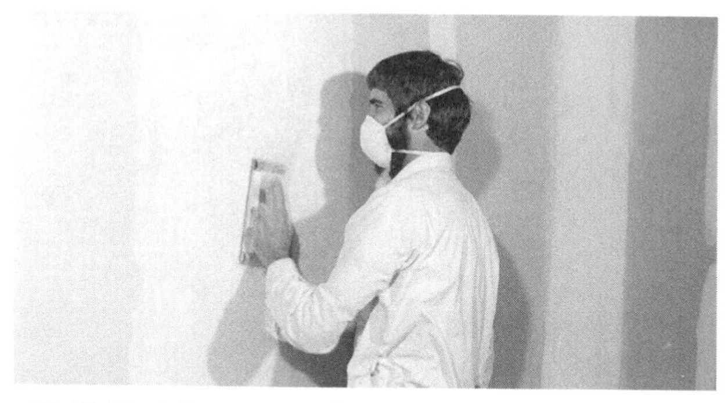

35-45 The joint compound can be hand-sanded using a wide, flat sanding pad and fine sandpaper or abrasive cloth. *(Courtesy Georgia-Pacific Corporation)*

(214 to 305mm) wide. It should be at least two inches (50mm) beyond the edge of the tape coat **(see 35-44)**.

After the second coat is dry, smooth out any tool marks or ridges with the finishing knife. Then apply the third coat using topping compound. Feather the edges until they are about two inches (50mm) or more beyond the edge of the second coat.

When the third coat is dry, the joint can be sanded. Use very-fine-grade sandpapers or abrasive cloths. Any depressions should be filled with topping compound, allowed to dry, and then be resanded. Dust is a severe problem, so respirators and safety glasses are required **(see 35-45)**. A sanding unit that uses a vacuum to remove much of the dust is shown in **35-46**. It not only provides protection for the worker but helps prevent the gypsum dust from spreading all over the house. Cleanup after a typical sanding operation is difficult and expensive.

35-46 While a drywall sander such as this uses a vacuum and filters to remove much of the dust created, you should always wear a dust mask. *(Courtesy PermaGlas Mesh, Inc.)*

Another technique for controlling dust during drywall sanding or other operations is to isolate the dust in the area of work with a curtain wall. One unit, shown in **35-47**, has an adjustable metal framework that carries the dust-resisting curtain. Another product is a dust door; this unit fully seals off doors from dust-producing areas, as shown in **35-48**.

35-47 A portable barrier such as this provides a flexible means for isolating dust to the area in which work is taking place. *(Courtesy Curtain Wall Co.)*

35-48 This dust control system seals off doors and other openings to areas in which dust will be produced. *(Courtesy Brophy Design, Inc., Manufacturers of the Dust Door and Wall System)*

1. Mechanically tape the ceiling joint.

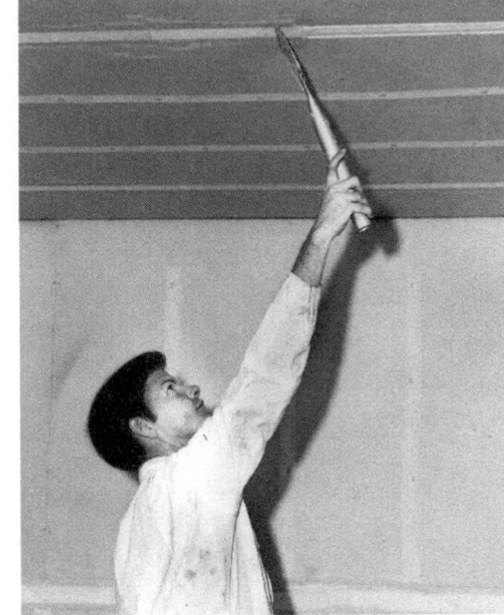

2. Then wipe down the joint with a broad knife.

3. Apply the finish coats after the compound has hardened.

35-49 Mechanical devices such as these make taping the ceiling easier. *(Courtesy United States Gypsum Company)*

A number of mechanical devices are used to apply joint compound and tape. A device that applies joint compound and tape is shown in **35-49**. After the joint is wiped down with a broad knife, finish coats can be mechanically applied. Interior corners can also have compound and tape applied as shown in **35-50**. Then the tape can be dressed with a mechanical finisher.

INSTALLING PREDECORATED GYPSUM PANELS

Predecorated gypsum panels have some form of finished surface applied during their manufacture. These are typically a vinyl or fabric covering, or a painted or other type of liquid coating applied to provide a smooth or textured surface.

Before installing the wall panels, finish the ceiling, and tape and sand it so all dust and tool work is complete. Some prefer to paint it so there is no danger of splattering the walls.

The easiest way for you to install predecorated panels is to place them in a vertical position and nail them to the studs with colored nails supplied by the manufacturer. Use a plastic-headed hammer. The color helps the nails blend into the panel. The panels are typically ½ inch thick and the nails supplied are 1⅜ inches long. Nail them ⅜ inch from the edge and eight inches O.C. on the edges and in the field of the panel.

1. Apply the compound and tape on the inside corner.

2. Dress the joint with a corner-finisher.

35-50 Devices such as these can be used to tape inside corners. *(Courtesy United States Gypsum Company)*

Since the panels are available in lengths up to 10 feet, when applied vertically no end joints occur. Cut the panels to length with a very sharp utility knife in the same manner as regular gypsum panels.

Since there can be some variation in the decorative surface, you should line up a series of panels against the wall and rearrange them to get the appearance you like.

Joints are covered with a variety of moldings supplied by the manufacturer. Some examples are in **35-51**. The snap-on panels have a base strip that is nailed or screwed every 12 inches O.C. to the framing. The colored trim is pressed against it and snaps over the sides. These vinyl trim pieces are best cut with a fine-toothed hacksaw. The cuts can be smoothed with a fine-toothed file or sandpaper.

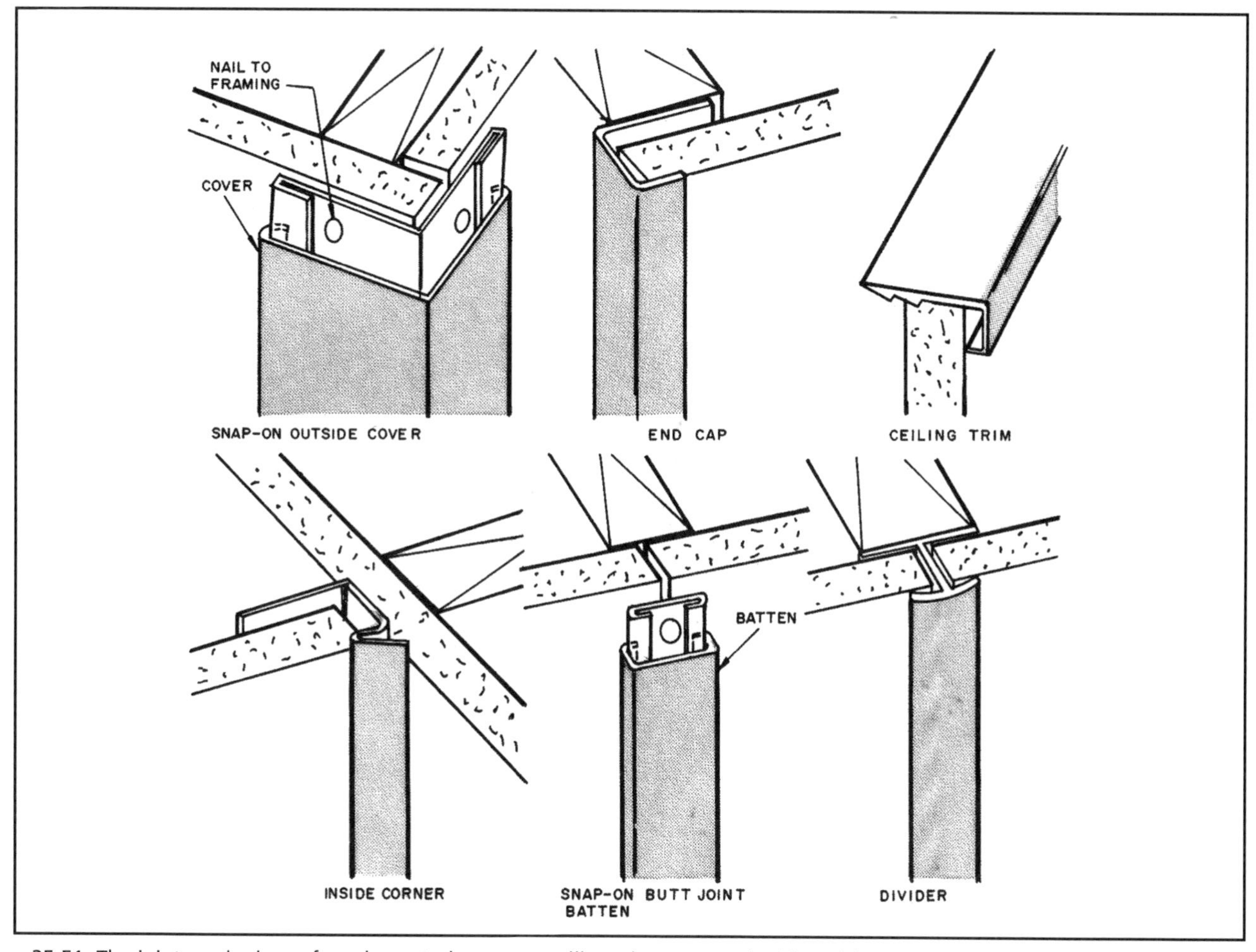

35-51 The joints and edges of predecorated gypsum wallboard are covered with moldings.

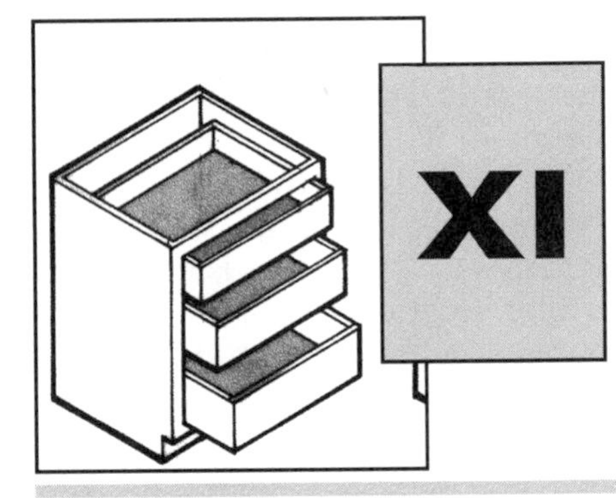

XI CABINET CONSTRUCTION

CHAPTER 36

CABINETS & COUNTERTOPS

Various built-in cabinets and countertops are installed by the staff of the company supplying the cabinets. They specialize in this work and do it rapidly and efficiently. There are times when the finish carpenter is expected to do this work.

Cabinet installation is one of the last things done as the house is nearing completion. The drywall should be taped, sanded, and painted before installing the cabinets. The finish floor is typically installed after the cabinets have been installed. The placement of the cabinets is shown on the architectural working drawings.

WORKING DESIGN DRAWINGS

The designer conceives and executes **working design drawings** for the kitchen, baths, and other areas in which cabinets are wanted. The **specifications** for the job detail technical information about the cabinets. The **floor plan** shows the plan view of the required cabinets, and **elevations** of each wall are drawn showing the locations of doors, drawers, shelves, and other features. An example of a typical floor plan of a kitchen is shown in **36-1** and the cabinet elevations are shown in **36-2**. Before the cabinets are ordered or custom built, the cabinetmaker or building materials supplier must visit the house and take measurements of the finished room to verify the exact sizes required. The dimensions given on the architectural drawings, while representative of the specifications and layout of the space, are not precise enough to use for ordering the cabinets. Also, things not on the plan may be found to exist, such as a heating duct, plumbing, or electrical outlets. As you measure the room make a layout similar to that shown in **36-3**. Notice the heat vent under the sink. Provision must be made to duct the heat out under the cabinet.

MASS-PRODUCED & CUSTOM-BUILT CABINETS

Cabinets are produced in three basic ways. These include **mass-produced** assembled and unassembled units, cabinets **built on the site** by the carpenter, and **custom-built** units made in a local cabinet shop.

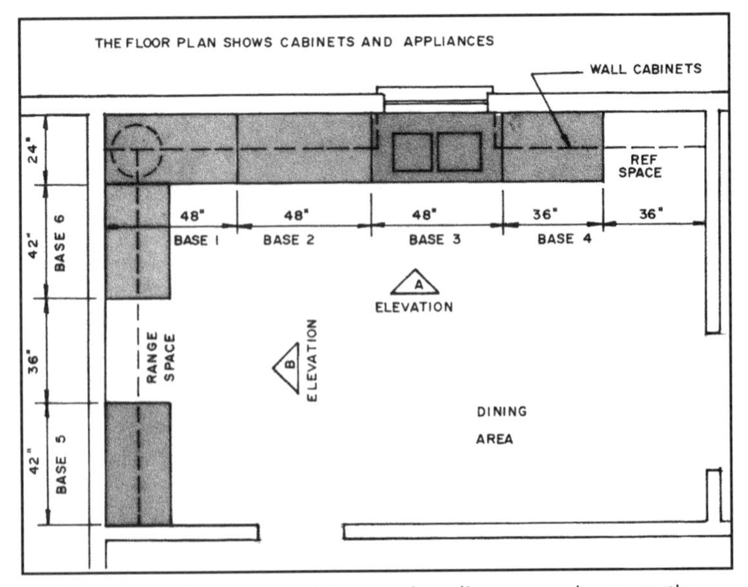

36-1 The locations of the cabinets and appliances are shown on the floor plan of the house. Notice it indicates that there are two elevations, A and B, of the cabinets. See 36-2.

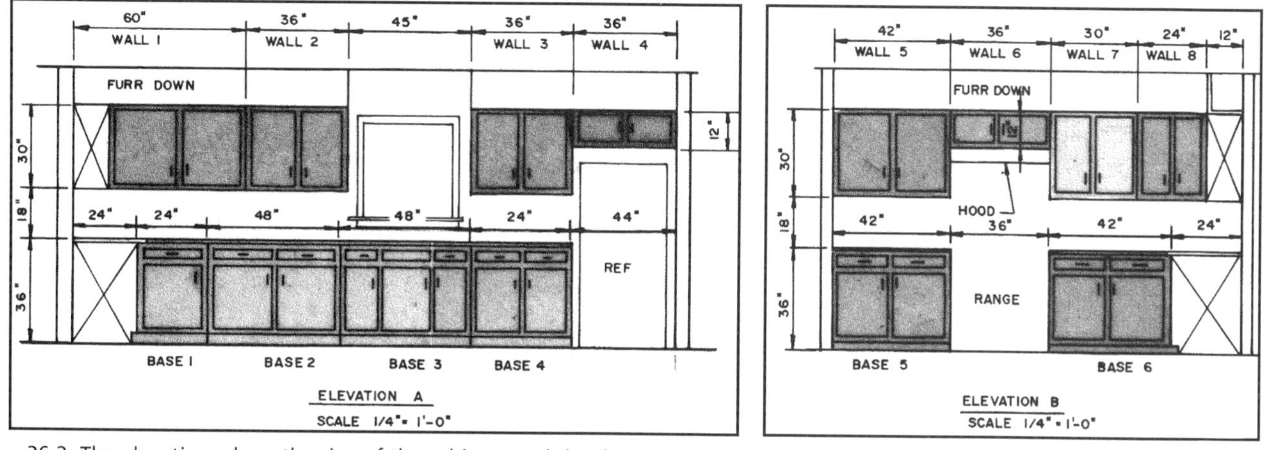

36-2 The elevations show the size of the cabinets and the door swing.

Mass-produced cabinets are manufactured in a factory by the same techniques used to make furniture. The use of production woodworking equipment permits the cabinet to be made using high-quality joinery and finishes applied in a controlled atmosphere **(see 36-4)**. While most mass-produced cabinets used are completely assembled and finished in the factory, they can be purchased precut but not assembled. The carpenter assembles these on the job. Most are designed so that they can be nailed or stapled together.

Manufactured cabinets, also referred to as **architectural mill**, are divided into three levels of quality. The **economy grade** is the lowest and least expensive. Joinery and construction is simple, and less costly wood products are used. **Custom-grade** cabinets have better-quality framing and joinery and are possibly the most commonly used grade. **Premium-grade** cabinets use the best joinery, and expensive woods. They will use extensive moldings and machined panels.

Site-built cabinets are constructed by the carpenter or a cabinetmaker on the job. Power tools, wood, and plywood are brought to the job. All the pieces are cut and assembled, and the units are then installed and finished. The joinery on site-built cabinets must be relatively simple. Since the finish is applied after they are installed, the conditions are less than ideal.

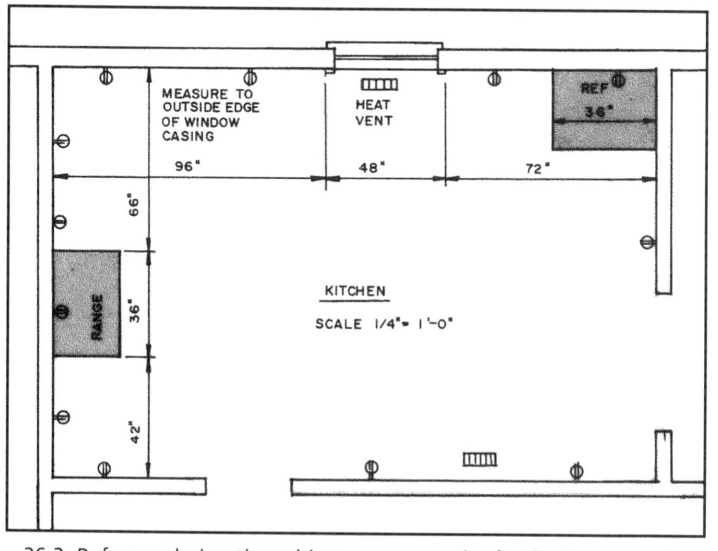

36-3 Before ordering the cabinets, measure the finished room and carefully locate all features that may be involved with the installation, such as windows and heat registers.

36-4 Factory-produced cabinets are available in a wide range of styles and sizes and have a quality finish applied in the factory under controlled conditions. *(Courtesy Wellborn Cabinet, Inc.)*

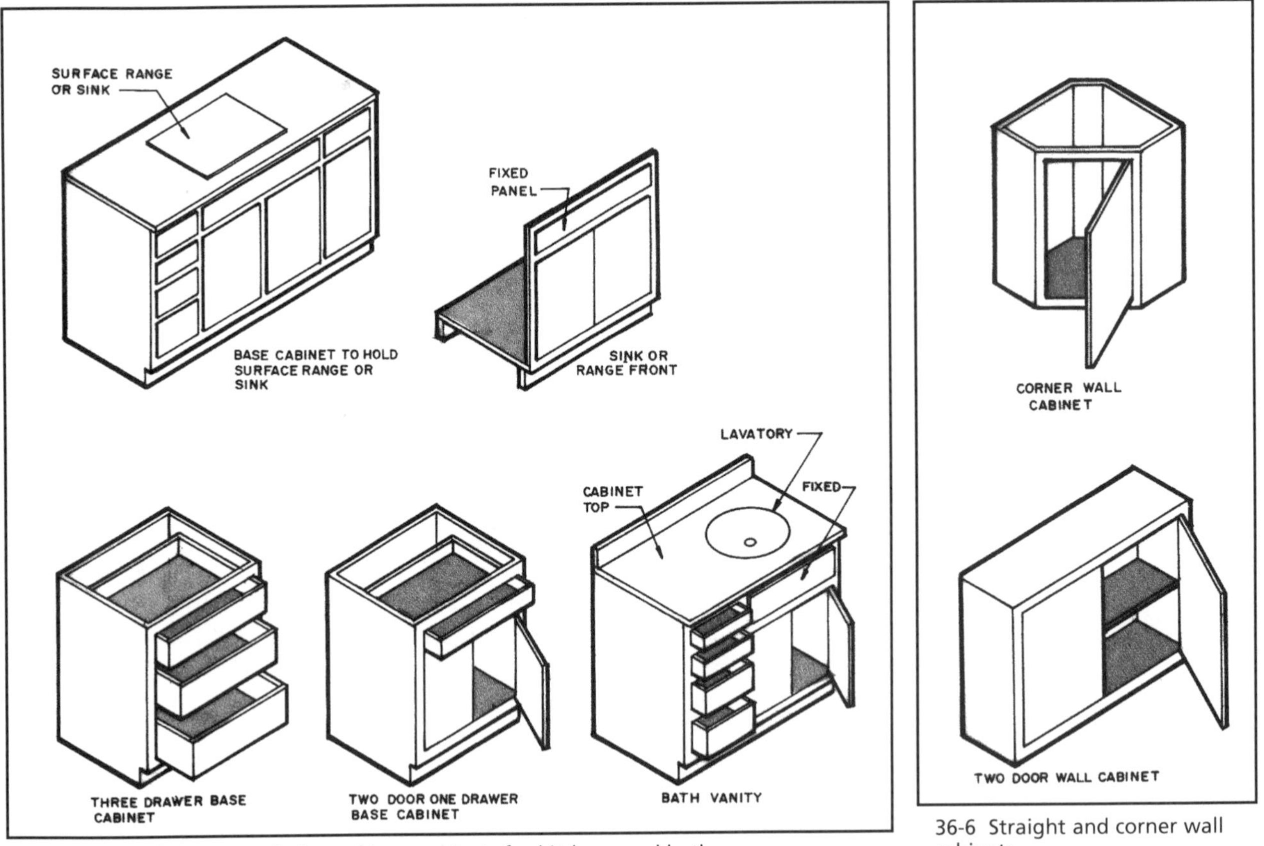

36-5 A few of the many designs of base cabinets for kitchens and baths.

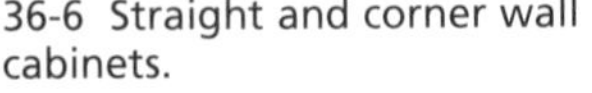

36-6 Straight and corner wall cabinets.

Custom-built cabinets are built by a cabinetmaker in a local woodworking shop. Their construction and finish are usually of higher quality than those of site-built units. Units not available from a mass-production supplier can be satisfactorily made this way. The quality depends on the skills of the cabinetmaker.

KINDS OF CABINET

Cabinets most commonly required in residential construction are in the kitchen, bath, utility room, and various built-in storage units such as bookcases and areas where articles are to be displayed.

In the kitchen the base cabinet rests on the floor and is designed for storage, and to hold a sink, cooking top, or range, with a top that serves as a work surface **(see 36-5)**. It has both drawer and door sections.

Wall cabinets are mounted on the wall about 18 inches (approx. 46cm) above the top of the base cabinet. They usually have adjustable shelves and are used for general storage **(see 36-6)**. Most wall cabinets are made 30 inches (approx. 76cm) high, which is about as high as most people can reach. The space between the top of the wall cabinet and the ceiling can be left open or closed with a drop ceiling **(see 36-7)**. If left open, the area can be used to display interesting articles.

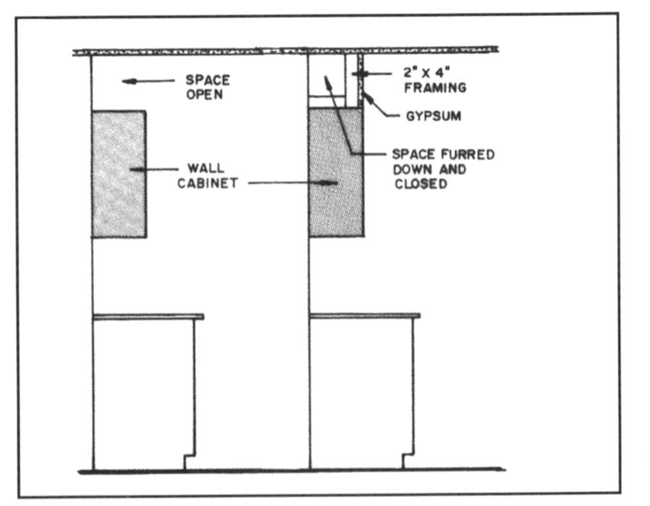

36-7 The space above wall cabinets may be left open or furred down and enclosed.

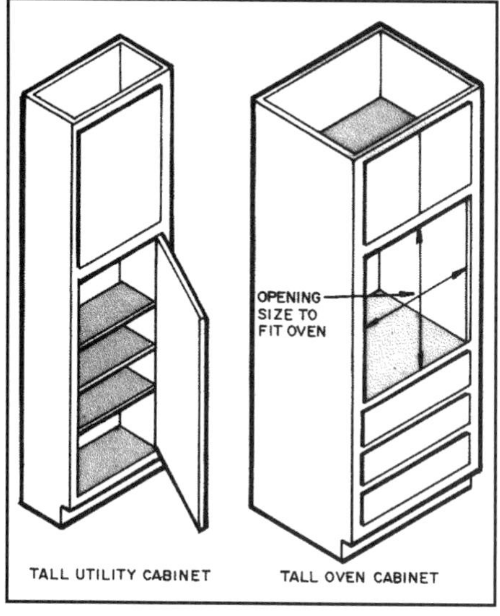

36-8 Typical tall cabinets.

36-9 This bath installation provides considerable storage for towels and linens plus a sink base cabinet. *(Courtesy Wellborn Cabinet, Inc.)*

Tall cabinets are generally used to store nonperishable foods. They are built so that the top lines up with the top of the wall cabinets **(see 36-8)**.

Vanity cabinets of all types are used in the bathroom to hold the wash basin. They have drawers and doors below providing storage, as shown in **36-9**.

CABINET CONSTRUCTION

While those who install the cabinets typically do not build them it is helpful if they have a basic knowledge of the various types of construction.

The various parts of a cabinet are identified in **36-10**. These terms are commonly used in the industry. Cabinets can be divided into two major types, based on their construction. These are units with a **face frame** and units **without a face frame** (often called **European style**).

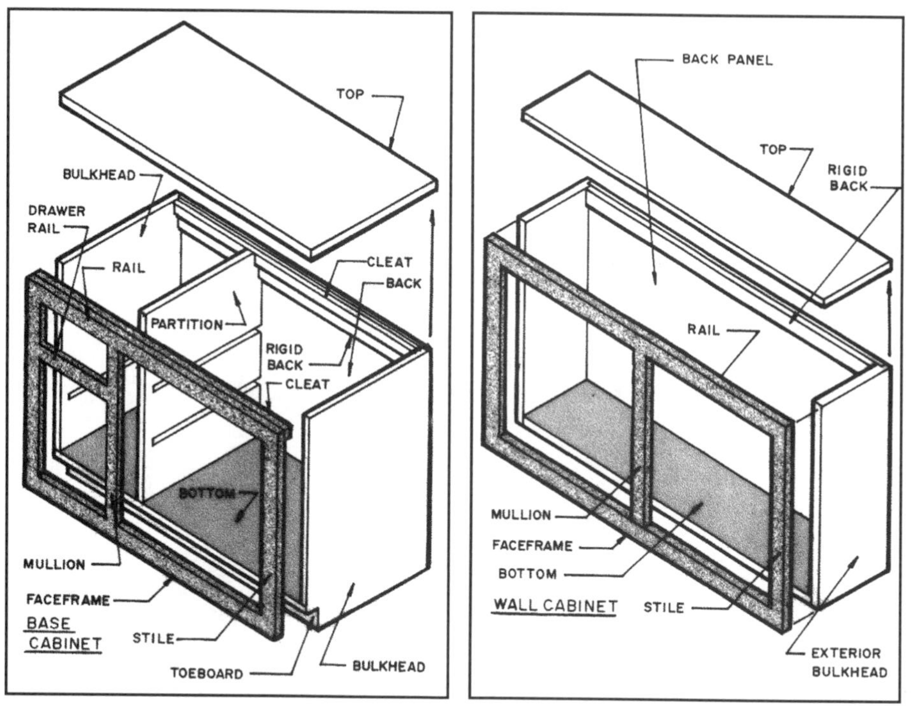

36-10 The parts of conventionally built cabinets.

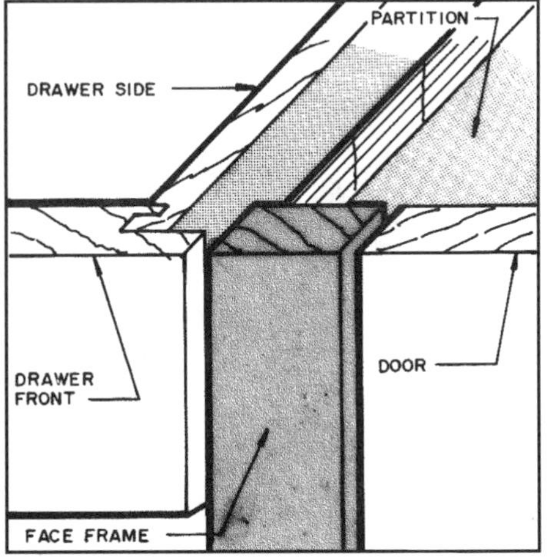

36-11 This type of conventional construction sets the doors and drawer fronts flush with the face frame.

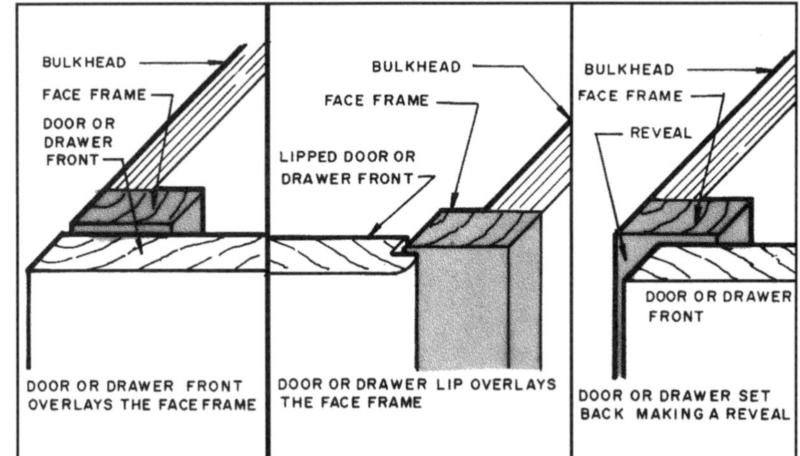

36-12 Conventional construction with doors and drawer fronts extending over the face frame.

Construction with a Face Frame

This type of construction can have doors and drawers flush with the face frame as shown in **36-11** or can overlap or overlay the face frame as shown in **36-12**. The flush construction is more expensive because of the necessity to fit the doors and drawers within the framed opening.

Construction without a Face Frame

This type of construction can have the doors set flush with the face of the bulkheads, top and bottom, or can have the doors overlay the bulkheads, top and bottom, or the overlay can be adjusted to provide a reveal which accents the separation between the doors and drawer fronts **(see 36-13)**.

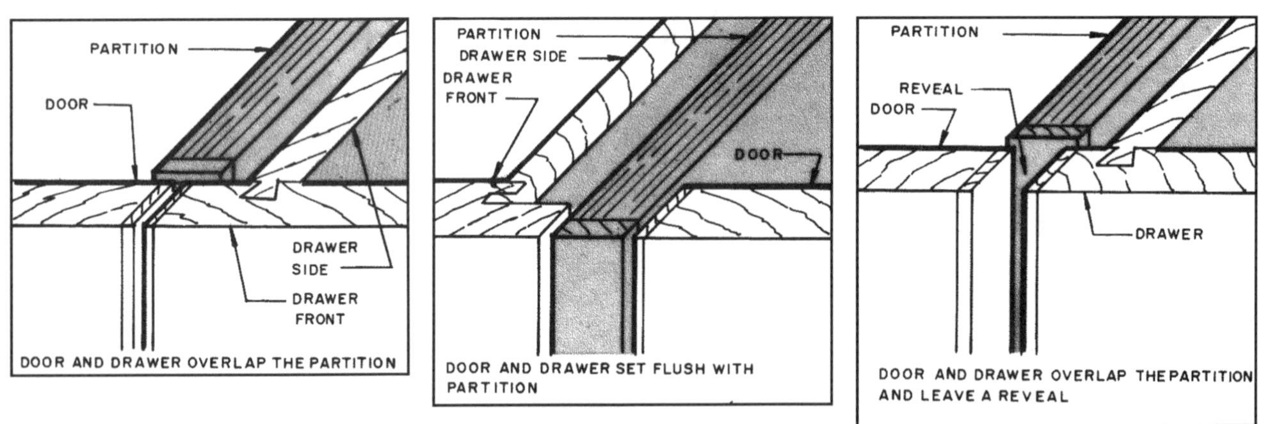

36-13 Typical cabinet construction when a face frame is not used.

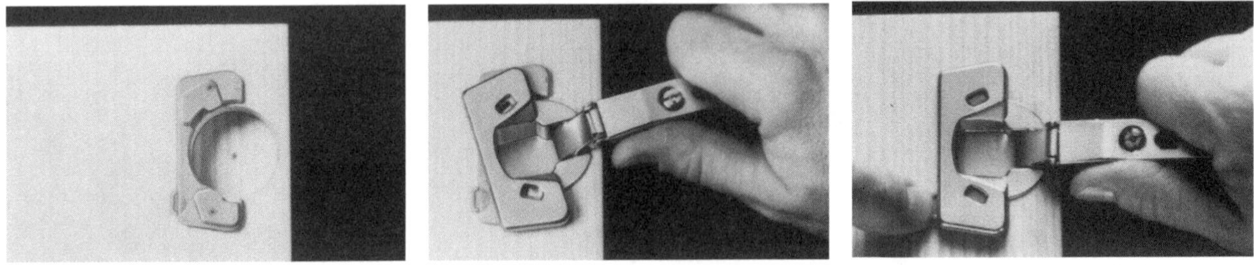

1. Cut recess into door and attach anchor plate.

2. Insert hinge in hole and rotate to lock in position.

3. Leaf on right of installed hinge is ready to be screwed to cabinet frame.

36-14 The steps to install one type of hinge for use on overlaid doors. *(Courtesy Melpa Furniture Fittings)*

36-15 This hinge permits the door to be removed and reinstalled by pressing the lock on the end of the leaf. *(Courtesy Melpa Furniture Fittings)*

1. Once hinge is mounted on door and anchor plate on cabinet side, insert hinges on anchors.

2. Press lock on end of hinge; door is locked in position. It can be removed by releasing the lock.

36-16 This hinge is mounted on the door and inside the cabinet. The door can be removed by pressing the lock on the end. *(Courtesy Melpa Furniture Fittings)*

There are a variety of hinges designed to hang cabinet doors on European-style cabinets. The steps to install one type of hinge are shown in **36-14**. The hinge permits the door to be installed and removed when desired by pressing the lock pad on the end of the hinge **(see 36-15)**. The installation of doors with this hinge is shown in **36-16**.

Another hinge has adjustments that permit the door to be repositioned slightly after the hinges have been installed. This enables the cabinetmaker to get the best position possible **(see 36-17)**.

Three-Way Independent Adjustment
(For use with the Blum 195H7100 mounting plate)

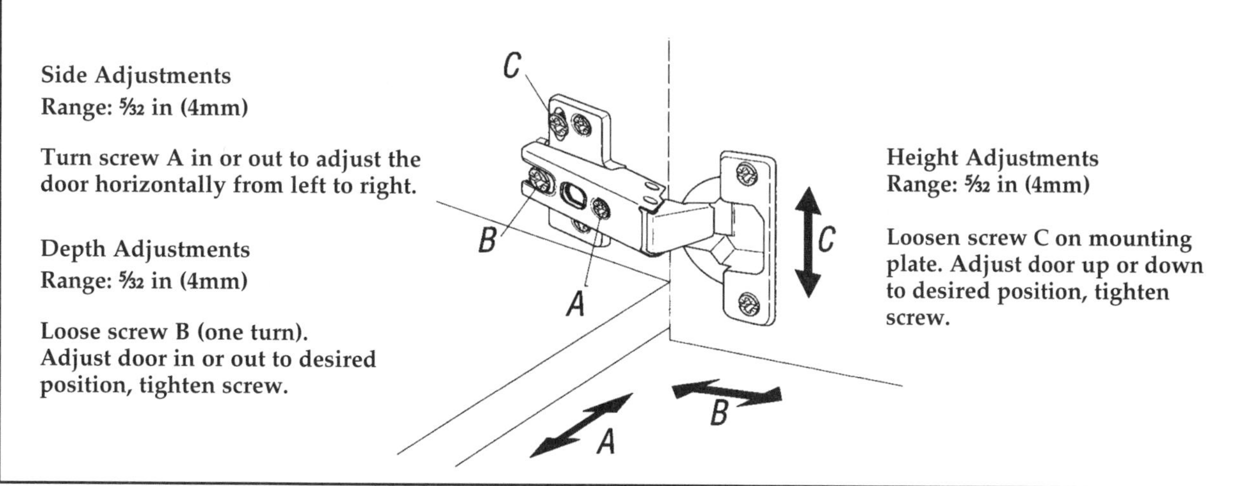

36-17 This hinge permits the door position to be adjusted in three directions. *(Courtesy Julius Blum, Inc.)*

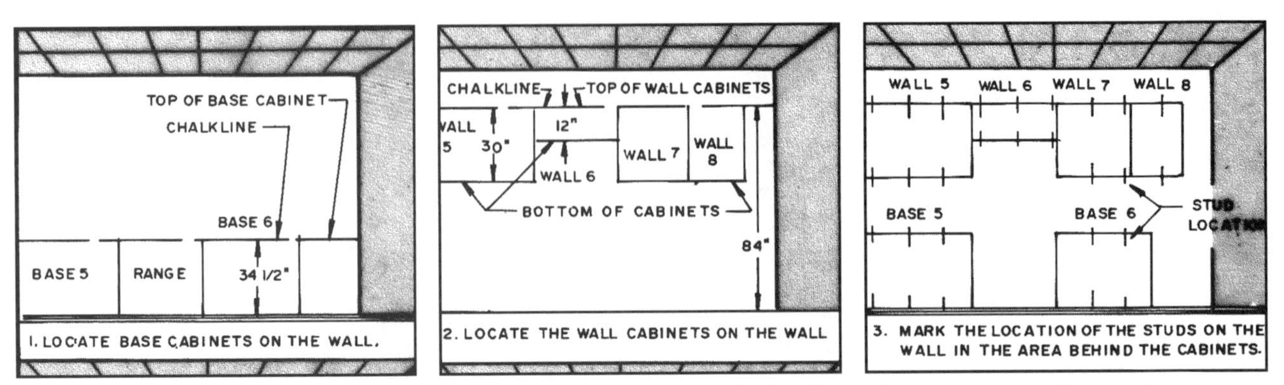

36-18 Locate the base and wall cabinets on the wall. Then mark the studs. This illustration shows the locations of the cabinets in elevation B in 36-2.

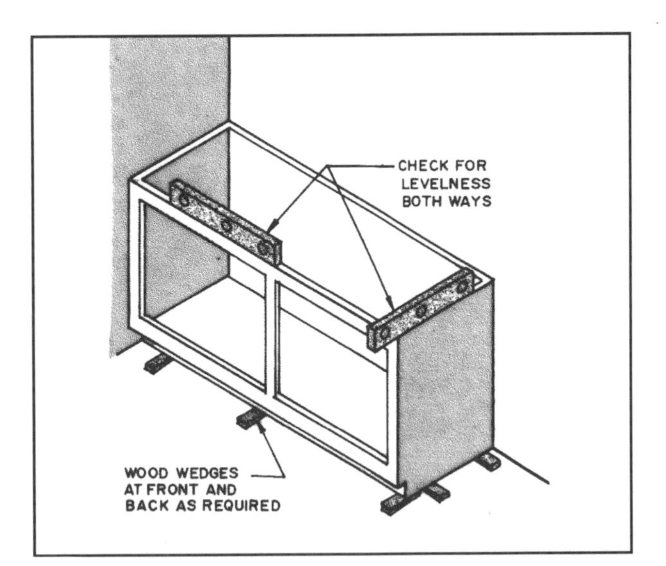

36-19 Set the first base cabinet in a corner and check for levelness.

PREPARING THE ROOM FOR INSTALLATION

Check the area in which the cabinets are to be installed to be certain that the walls are finished and the floor is clean. Also make certain the pipes, ductwork, and electrical wiring that will be required have been installed. Check the floor with a level to see if it is free of high spots and long irregularities. The drywall or plaster should be thoroughly dry so that the cabinets are not exposed to high levels of moisture.

Check the architectural drawings with the cabinets on site to verify the correct style and sizes are available. Be certain to measure the finished room. Sometimes changes are made that do not appear on the drawings.

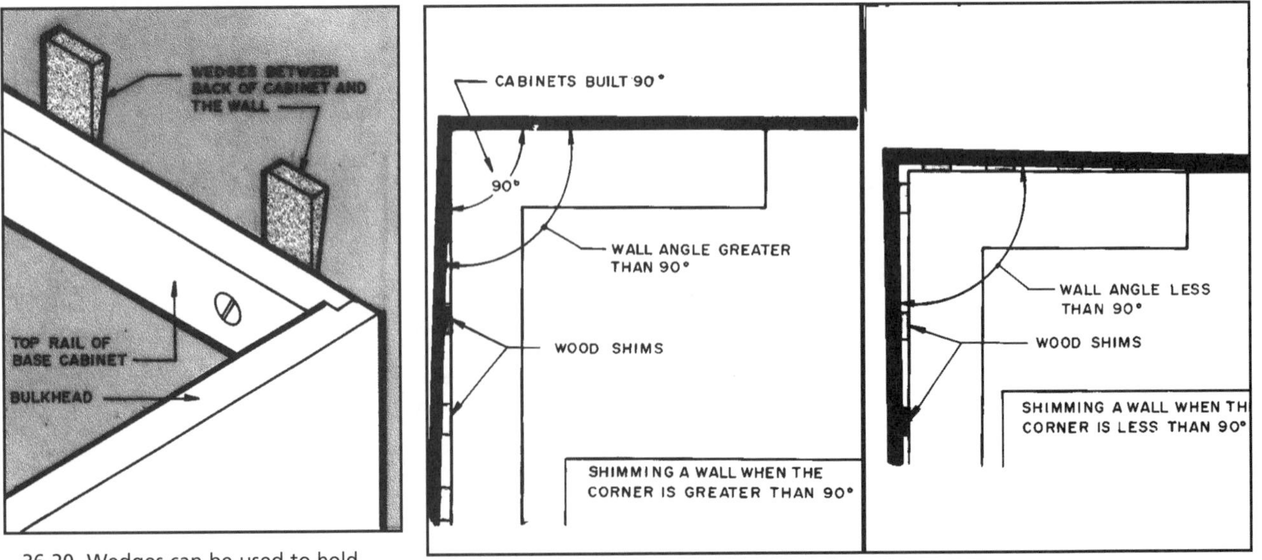

36-20 Wedges can be used to hold the base cabinets straight if the wall bows or is out of plumb.

36-21 When a wall is out of square, use wood shims to hold the cabinets in square.

ESTABLISH LEVEL LOCATION LINES

Begin installation by drawing level lines locating the top of the base cabinet without the countertop and the bottom of the wall cabinets as shown in **36-18**. The fastest way is to use a laser level as explained in Chapter 5. A chalkline can also be used.

1. Measure the height of the base cabinet on the wall. This is typically 34½ inches (876mm). This point should be above the highest point on the floor in the area where the cabinet is to be located. Locate the sides of each unit.
2. Measure up 84 inches (213cm) to locate the top of the wall cabinet and down as required to locate the bottom edge of each wall cabinet. Typical lengths are 12, 15, 18, and 30 inches (30.5, 38, 45.7, and 76cm). Locate the sides of each unit.
3. With the laser level project these points along the line and set the cabinet to them. If a laser level is not used you can run a chalkline along the wall. Check it carefully for levelness with a line level and mark the line on the wall. It is very important that these lines be level.

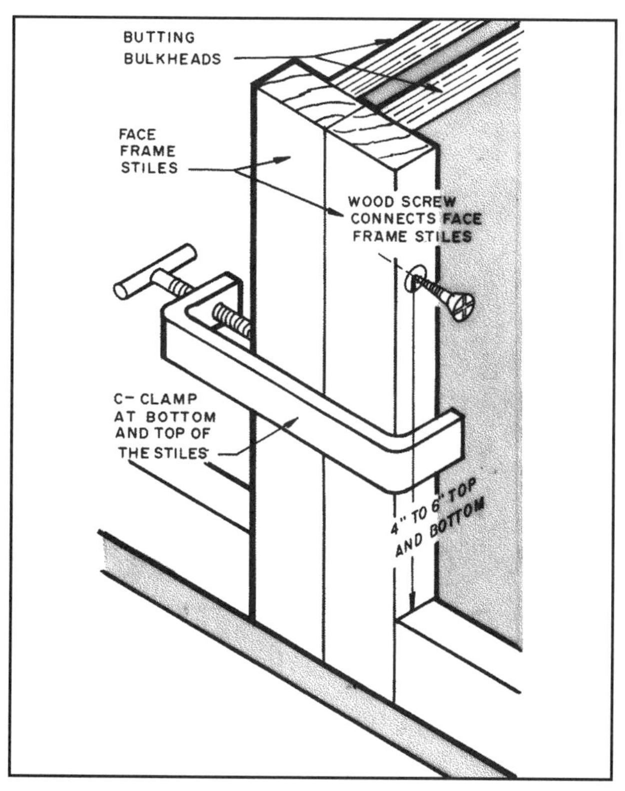

36-22 Butting cabinets are joined with wood screws through the stiles.

Finally mark the location of the studs so the cabinets can be screwed to them.

You may install either the wall or the base cabinets first.

INSTALLING THE BASE CABINETS

The following is a commonly used procedure for installing base cabinets.

1. Begin with a corner unit. If the floor is not level, shim it with wood wedges, keeping the top on the line on the wall, and checking to be certain it is level as well as parallel and perpendicular to the wall **(see 36-19)**.
2. Next, check the back of the cabinet as it butts the wall. If the wall has a bow, shim the back of the cabinets so that they line up straight **(see 36-20)**.
3. If a cabinet turns a corner, check to see whether the walls are perpendicular. If they are not, shim as shown in **36-21**.
4. The base is joined to each stud by wood screws through the rigid back. Drill holes through the rigid back and secure the base to the wall with No. 8 or 10 flathead wood screws that go into the stud at least ¾ inch (19mm).
5. Butt the second base unit to this first one. Clamp the butting ends together with C-clamps. Join the units with flathead wood screws through the frame of one unit into the next unit **(see 36-22)**. Then level as needed, and fasten to the wall stud. At the rear of the cabinet insert a wood strip and drive screws through the bulkhead from each side, as shown in **36-23**.

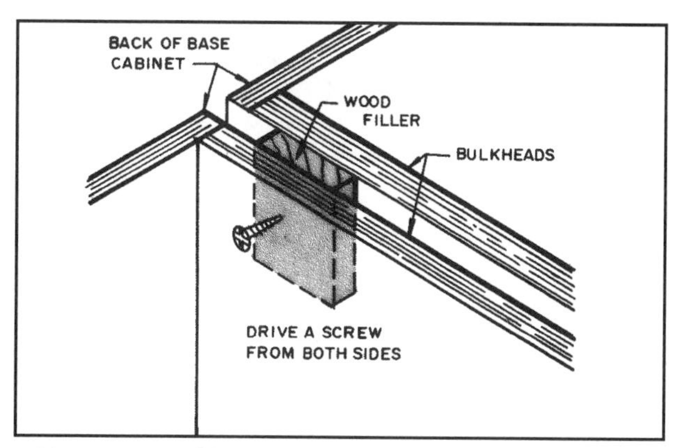

36-23 A wood filler block is screwed in place between the bulkheads at the rear of the cabinets.

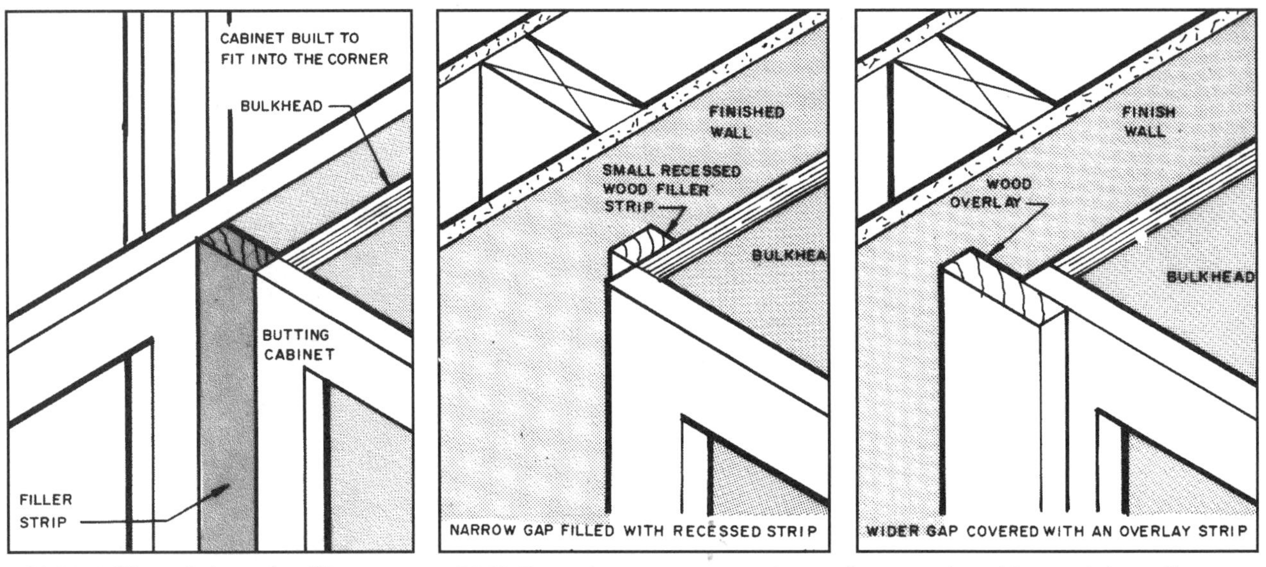

36-24 A filler strip is used to fill any gap between cabinets forming a corner or between a cabinet and a wall.

36-25 Two other ways to conceal a gap between the cabinet and the wall.

6. When a space is left in the row of base cabinets for an appliance, check the front of the cabinets so they are flush across the opening.
7. In many cases the stock base and wall cabinets will not provide the exact length required. When this happens, filler strips are added to the face frame to fill the gap. These strips are available prefinished to match the grain and color of manufactured cabinets. They must be cut to width and joined to the base or wall cabinet, as shown in **36-24**. Two other techniques are shown in **36-25**.

Some installers drive a couple of wood screws through the base at the floor and run the base used around the room along the end of the cabinet (see **36-26**). In this case, the toe board will have to extend out and be mitered where it meets this piece of base.

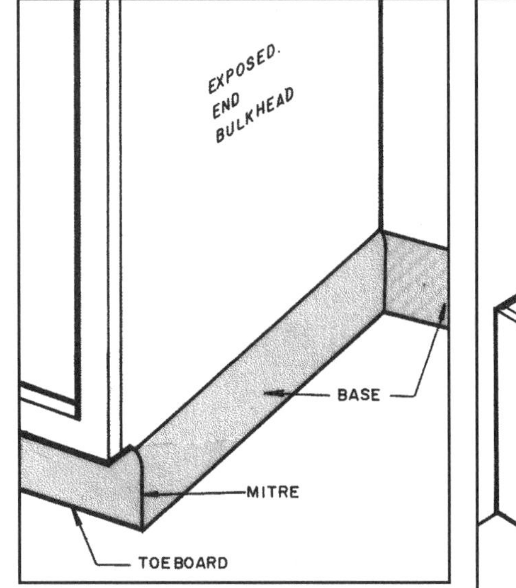

36-26 The exposed end bulkhead can have the base used in the room run along it at the floor.

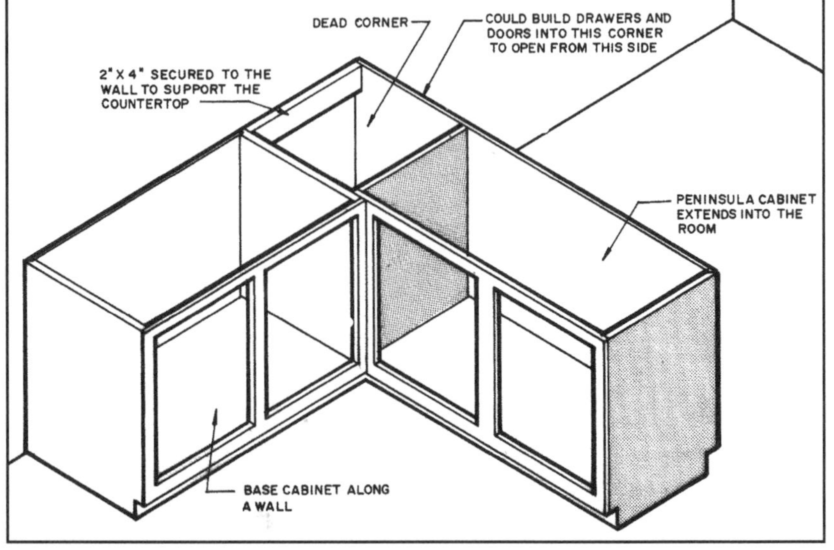

36-27 Peninsula base cabinets project into the room from the base cabinets along the wall.

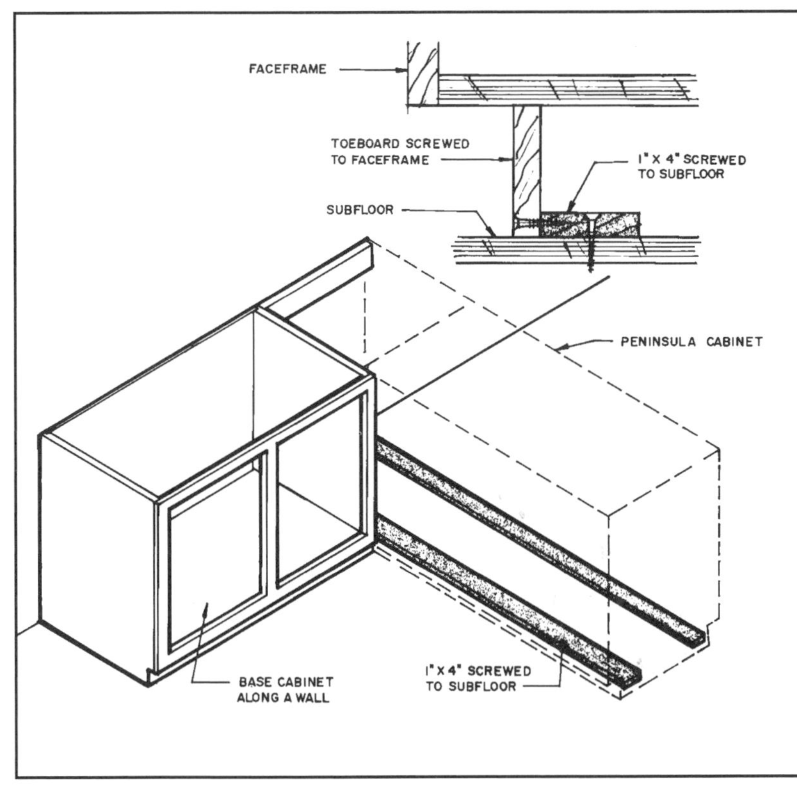

36-28 The peninsula cabinet is screwed to the wall cabinet and the subfloor.

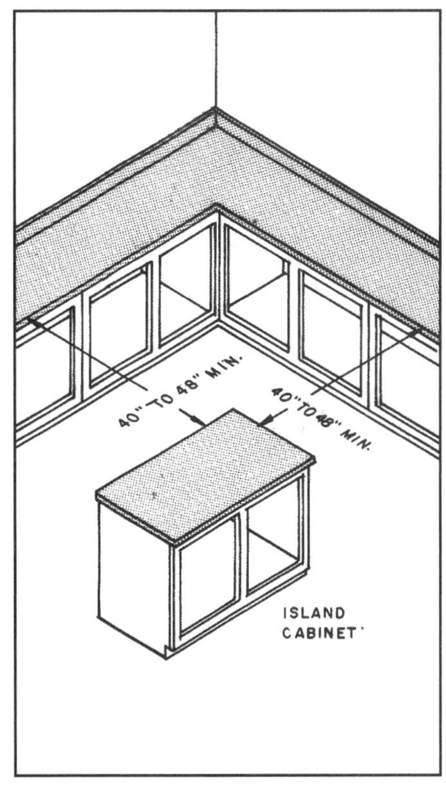

36-29 Island cabinets are freestanding and must have aisles specified by the building code.

INSTALLING PENINSULA & ISLAND BASE UNITS

Peninsular units connect to a wall base cabinet and project into the room, as shown in **36-27**. They are available in a variety of designs, but installation is much the same as with wall base cabinets. They are joined to the wall base unit by screws through the face frame. The toe board is secured to the floor by screwing or nailing it to a wood blocking screwed to the subfloor **(see 36-28)**. The island base unit is freestanding. It is secured to the subfloor in the same manner as the peninsula base unit **(see 36-29 and 36-30)**.

36-30 This island counter holds the cooking unit and provides storage space. *(Courtesy Wellborn Cabinet, Inc.)*

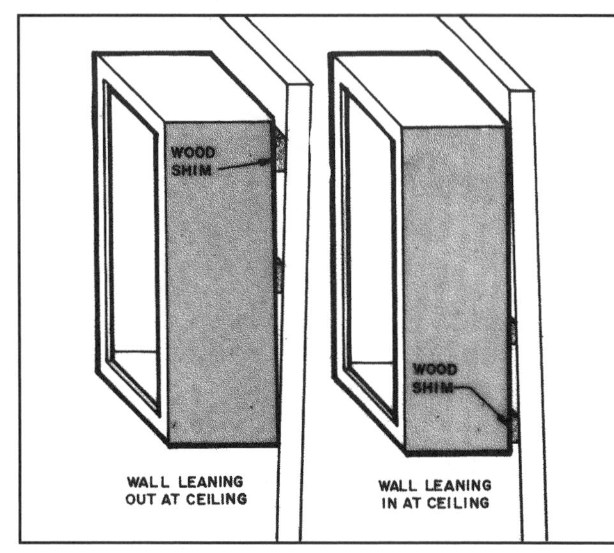

36-31 If the wall is not plumb, shim the cabinets so they are plumb.

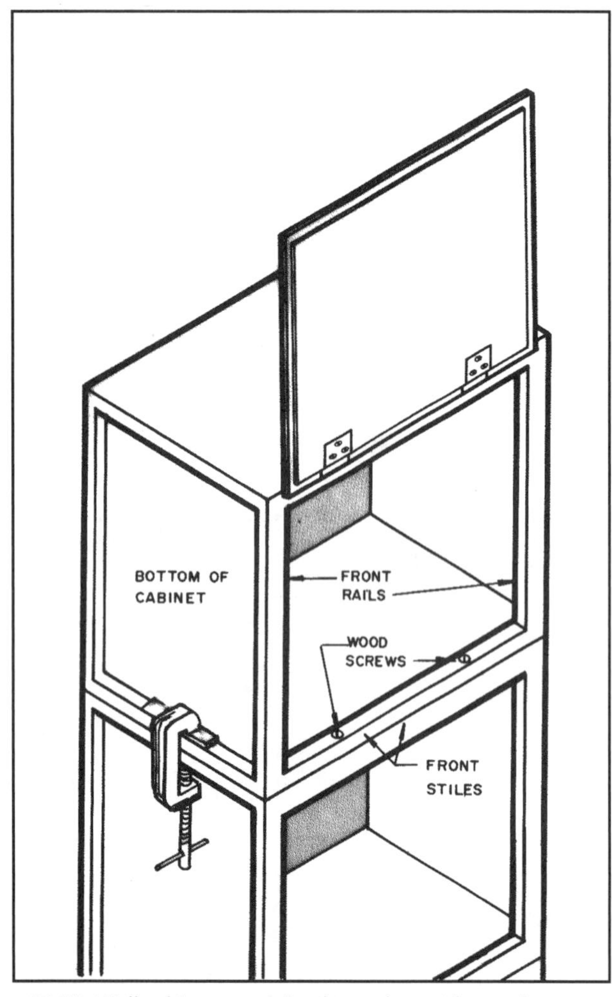

36-32 Wall cabinets are joined together with wood screws through the front stiles before they are raised up on the wall.

INSTALLING THE WALL CABINETS

1. As with base cabinets, check the wall to see if it has any bows. Make note of these, because they will require that the wall cabinets be shimmed to keep them straight. The wall may also be out of plumb and require shimming as shown in **36-31**.
2. Check the corners of the room to see whether the walls are square. If not, this may require shimming as shown in **36-21**.
3. Join the units together on the floor in the order in which they are to be on the wall. Clamp them with C-clamps. Put pieces of wood between the cabinet and the clamp to prevent marring the cabinet. Be certain that the bottom rails are in perfect alignment. Then join them together with wood screws through the front stiles. Round-head screws are preferred. At least two screws should be used to join each cabinet **(see 36-32)**.

Measure the location of any ducts or electrical services that must be put through the cabinet. Mark and cut the required openings in the cabinet.

4. Mark the stud locations on the wall. Measure and mark the stud locations on the back rails of the cabinet. Drill the holes for the screws through the upper and lower rear rails. Use at least four screws per cabinet. Large double-door cabinets should have six screws **(36-33)**.

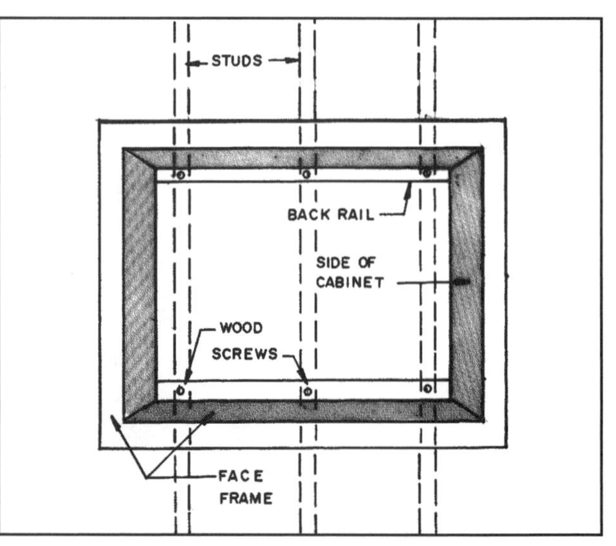

36-33 Wall cabinets are hung with wood screws that go through the back rails and into the studs.

5. Start the installation of wall cabinets in a corner. Raise the cabinets into position on the wall, lining them up with the location line. Some carpenters use wood supports that rest on a temporary plywood or particleboard top on the base cabinets. The wall cabinets can rest on these as they are leveled and plumbed **(see 36-34)**.

If the wall cabinets are installed before the base units, carpenter-built I-braces or commercial jacks can be used to support them **(see 36-35)**.

Another useful technique is to nail a temporary wood strip along the line of the bottom of the cabinets. This holds them steadier than if just the T-brace is used **(see 36-36)**.

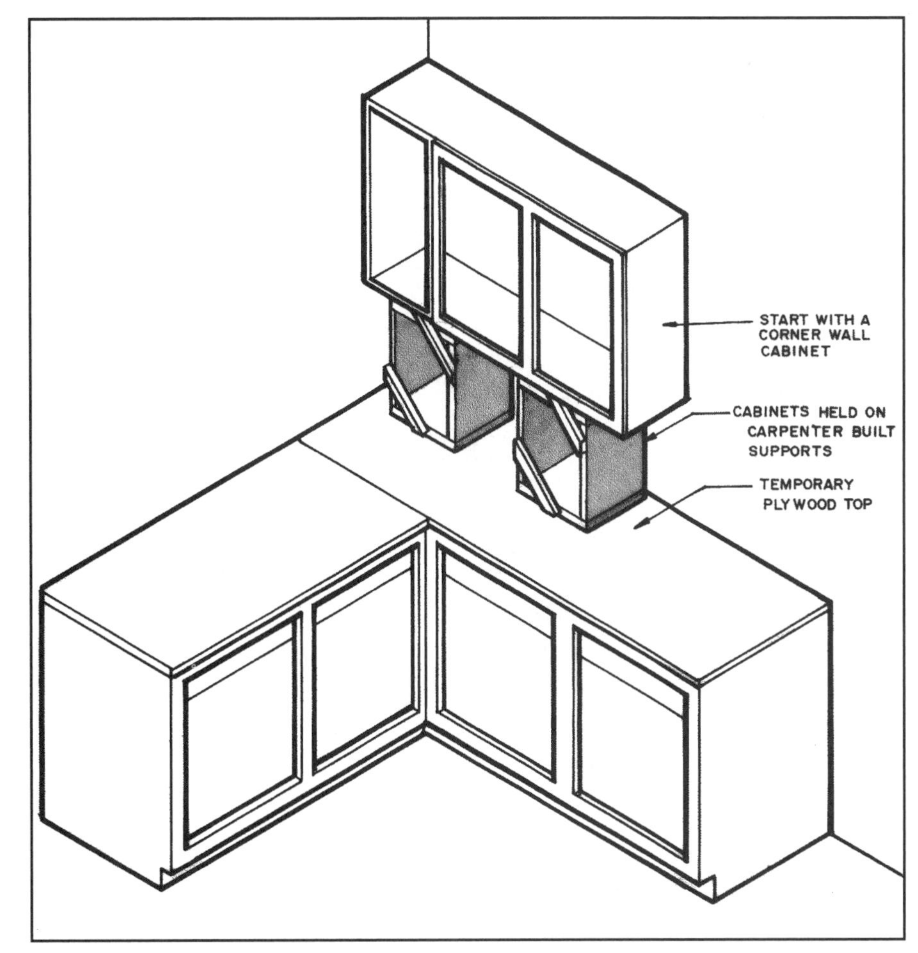

36-34 Wall cabinets can be supported while being installed by using carpenter-built supports that rest on a temporary top on the base cabinets.

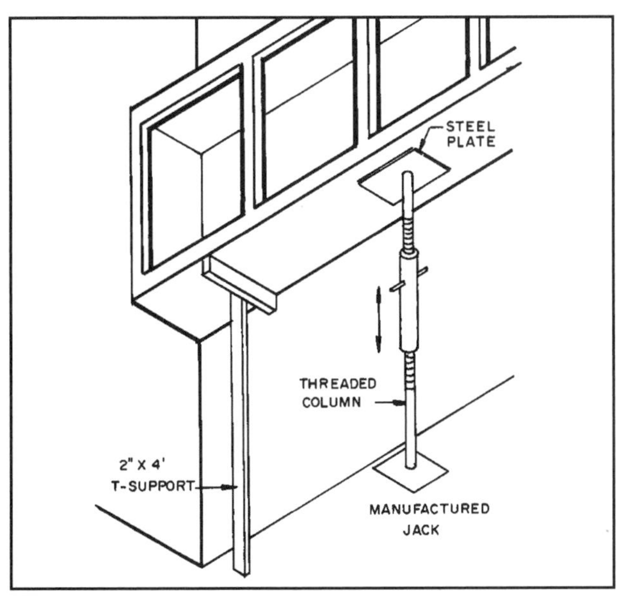

36-35 When the wall cabinets are hung first they can be supported with carpenter-built T-supports or commercially available jacks.

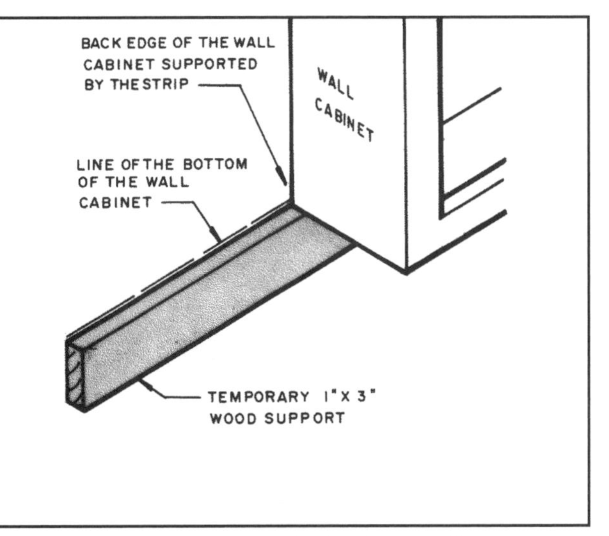

36-36 Nail a temporary wood strip along the line on the wall locating the bottom of the wall cabinets. Be certain it is level.

Another lifting tool **(see 36-37)** picks up the cabinet on the floor, rolls it into place along the wall, easily raises and lowers it so it lines up with the location line on the wall, and holds the cabinet in place as it is screwed to the wall framing **(see 36-38)**.

6. Drive the screws through the holes drilled in the top rear rail of the cabinet into the studs. They should penetrate the stud at least one inch (25mm) **(see 36-39)**. Wood screws are typically No. 9 or 10 round head and are generally 2¼ inches (57mm) long so they penetrate the ¾-inch (18mm) back rail, the ½-inch (12mm) gypsum wallboard, and go one inch (25mm) into the stud.
7. If the cabinet is to carry a heavy load, it is best to nail a 2 × 4-inch hanging strip between the studs before the finish wall is applied. Then screw the cabinets to the studs and the hanging strip.
8. If the wall is irregular, it may be necessary to place wood shims behind the cabinet before giving the screws their final tightening. If the wall is crooked and the cabinets are fastened tightly to it, they may become bowed. The rails may open up. The doors may not close. The cabinet must not be under any strain that throws it out of plumb.

36-37 The Gel-Lift holds the cabinet above the floor and rolls it against the wall. *(Courtesy Gel-Lift)*

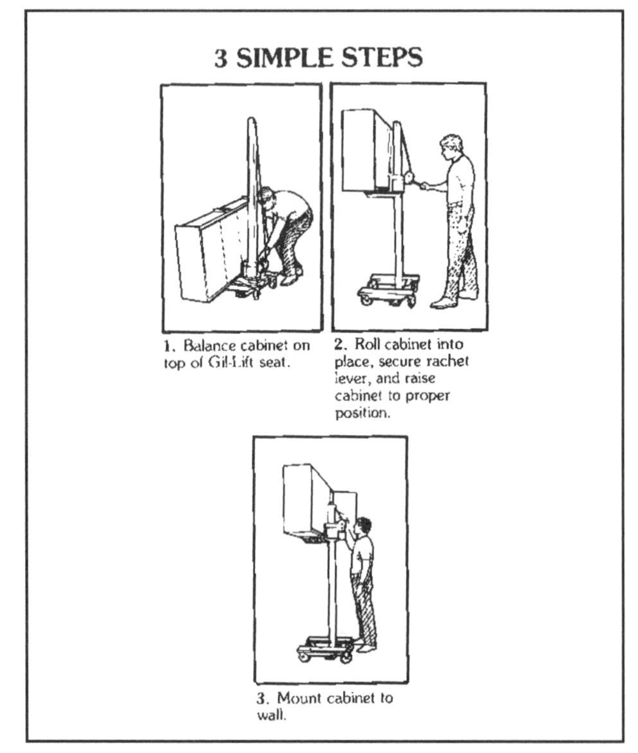

36-38 The steps to lift, locate, and hold a cabinet for installation. *(Courtesy Gel-Lift)*

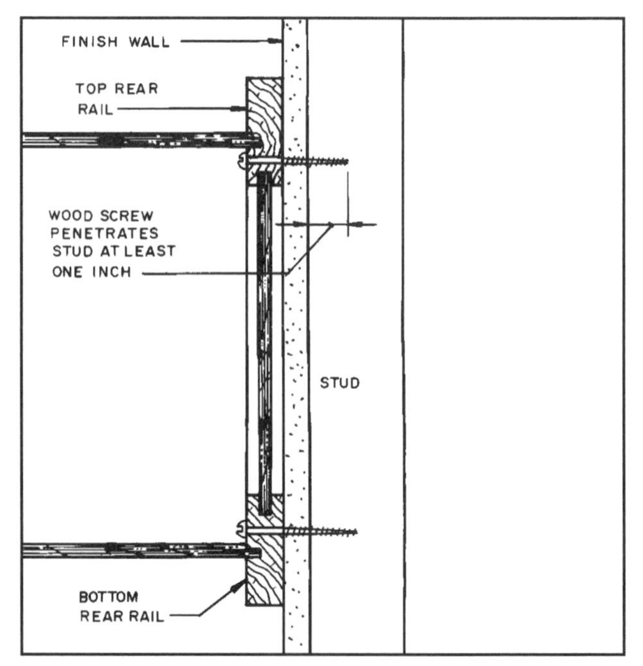

36-39 Wood screws holding wall cabinets to the wall should penetrate the stud at least one inch.

INSTALLING EUROPEAN-STYLE CABINETS

European-style cabinets have no face frame, as shown earlier in **36-13**. Some American cabinet manufacturers make European-style cabinets, but most prepare them to be installed as explained earlier in this chapter. The European-style cabinet installation system is fast and efficient. The upper cabinets are hung from a steel rail cleat or suspension fitting that is screwed to the wall studs. The base cabinets stand on metal legs secured to the bottom of the cabinet. The legs may be adjusted to level the base.

To install the cabinets, snap a chalkline on the wall locating the top of the base cabinet, the bottom of the wall cabinet, and the top of the wall cabinet hanging rail. This rail is generally a little above the top of the cabinet when it is hung **(see 36-40)**.

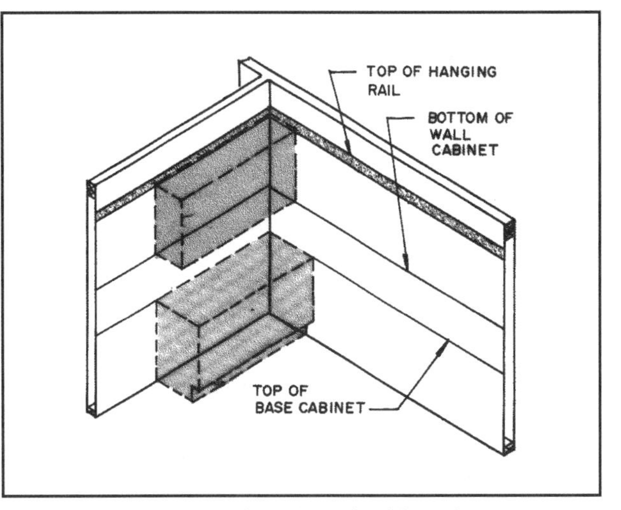

36-40 How to lay out the cabinet level lines for European-style cabinets.

The hanging rail that is screwed to the studs is usually about 1¼ inches (32mm) wide. It has an offset bend in the top into which hanging hooks screwed to the back of the cabinet will fit. A stronger installation is made by using two-inch (50mm) blocking between the studs where the hanging rail will be located. Then run screws through the rail and drywall into the blocking every two or three inches (50 to 76mm).

The upper cabinet has hanging hooks secured to the back rail. The hook has two adjusting screws. One is used to move the cabinet in and out from the wall. The other adjusts the cabinet up and down so that it can be leveled. If the hanging rail is exposed to view, a crown molding may be used to conceal it.

After the wall cabinets are hung and leveled, they are bolted together with bolts supplied with the cabinets. Clamp the cabinets together and finish drilling the holes located for the bolts inside the cabinets. If bolts are not available, the cabinets can be joined with wood screws.

The base cabinet has metal leveling legs fitting into plastic sockets secured to the bottom of the cabinet. A hole in the cabinet floor above each leg permits the length to be adjusted with a screwdriver, or the legs can be turned by hand **(see 36-41)**. A cap is supplied to cover the hole.

European-style cabinets are usually narrower than the standard 24-inch (61cm) American cabinet. This means that the countertop must be made narrower or the base cabinet held out from the wall with a wood strip nailed or screwed to the studs. Europeans do not secure their base cabinets to the wall. However, it is a good practice to screw them to the studs or the spacer strip.

The toe kick is covered with a plywood or particleboard strip covered with a plastic laminate. It can be cut to the required width. This depends on the finish floor and whether it has been installed. On the back of the toe kick board, mark the location of each leg. The manufacturer supplies clips that fit into a groove in the back of the toe kick board. The clips fit around the leg, holding the board in place.

It should be noted that this is only a general description; there may be differences in cabinets made by various manufacturers.

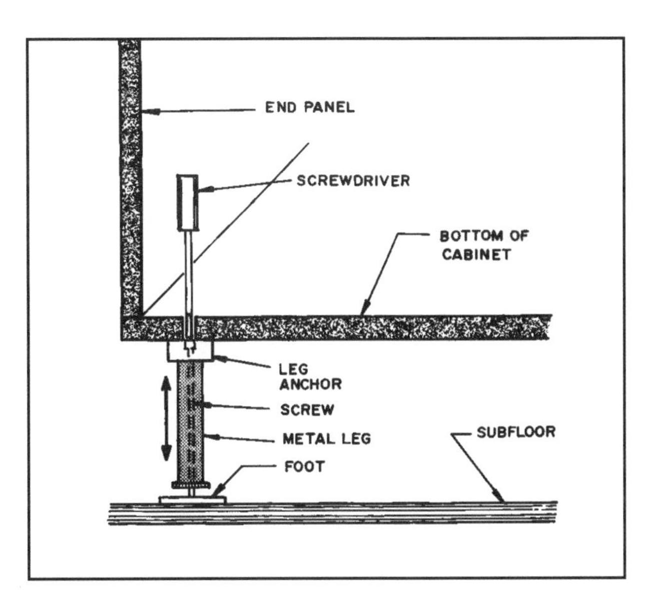

36-41 The base cabinet is set on metal legs that permit it to be leveled by adjusting their lengths.

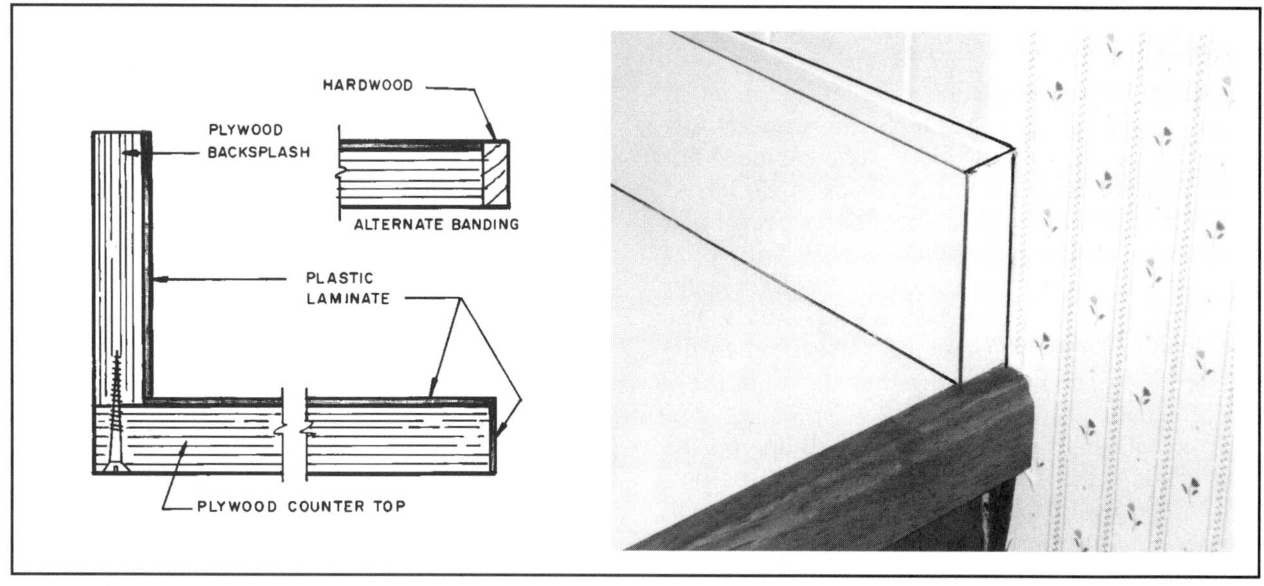

36-42 This countertop has a plastic laminate bonded to the backsplash, top, and exposed edges.

INSTALLING COUNTERTOPS

It is common practice for the company supplying the cabinets to also supply the finished countertops ready to install. Commonly used countertops include a plywood base and backsplash upon which a plastic laminate has been installed. The backsplash is covered with the plastic and then screwed to the top **(see 36-42)**. The edge may be covered with plastic laminate or a wood edging that matches the wood used for the cabinets.

Another type uses the same materials but has a cove where the backsplash meets the top and the countertop edge is rounded. The plastic laminate runs across all of this with no joints **(see 36-43)**.

Countertops are also available with the bowl and countertop cast as a single unit including the backsplash. They are cast with some type of plastic, such as an acrylic polymer, and selected minerals **(see 36-44)**. They are placed on the base cabinet and held

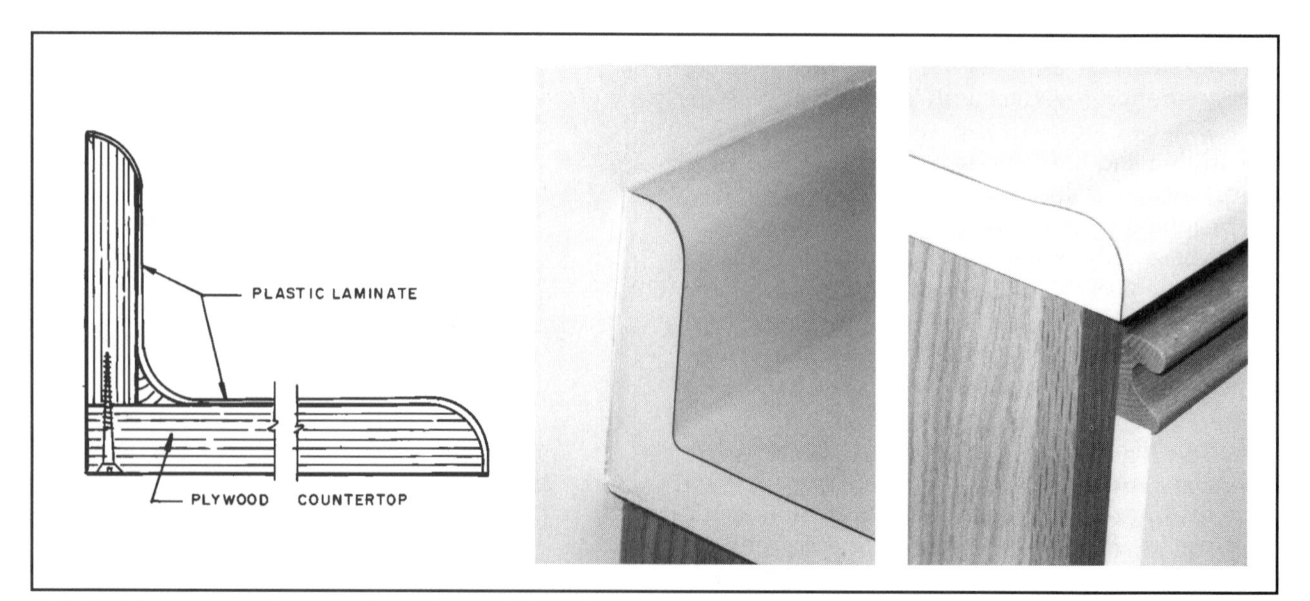

36-43 This countertop has a continuous layer of plastic laminate covering the backsplash, top, and exposed edges.

36-44 Molded plastic countertops have the sink bowl and backsplash cast with the top, forming a single unified assembly.

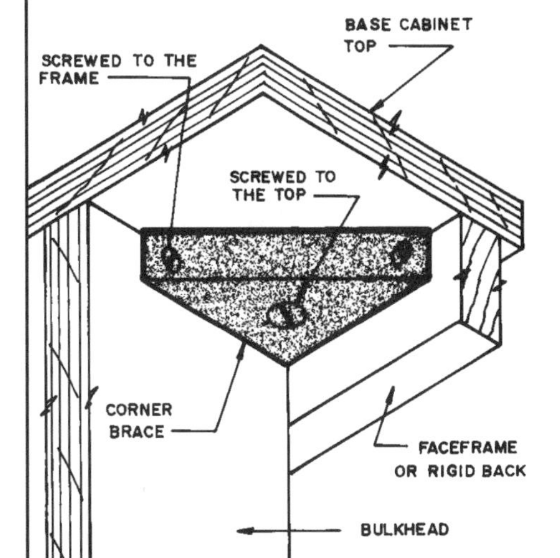

36-45 Countertops can be joined to the base with corner braces.

in place with several pads of adhesive along the top edge of the base cabinet.

The plywood-based tops are secured to the base cabinet by screws through braces located at each corner of the base cabinet **(see 36-45)**. Some cabinets have a one-inch (25mm) wood strip along the top edge used to secure the top.

If the finished wall has an irregular or bumpy surface, you may have to cut away small portions to enable the top to fit square and flush.

If the top is to have a sink and it is plywood with a plastic laminate, position the metal rim on the top in the exact location. Trace around the outside of the edge of the rim that touches the top. Be certain that there are at least two inches of clear space beyond the edge of the hole **(see 36-46)**. This area is needed to hold the metal clips that pull the rim, sink, and top together **(see 36-47)**.

Cut to the line with an electric saber saw. Insert the metal rim to see if it fits. A file can be used to remove high spots from the sawed opening.

Apply a thin bead of caulking to the bottom side of the metal rim. Set the sink and rim into the opening. Place the clips on the rim below the countertop and tighten. Wipe away any caulking that may be squeezed out from under the rim.

After the installation is complete, check the doors and drawers for proper operation. Adjust the door catches, if necessary.

Carefully wipe down the cabinets with a soft, clean cloth. Be certain that the drawers and shelves are free of sawdust and plaster.

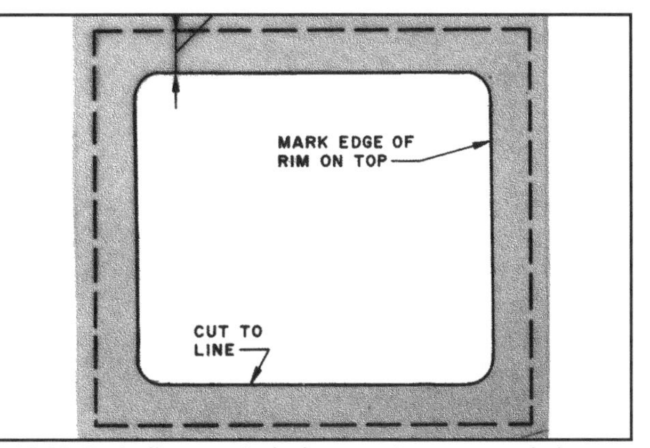

36-46 Typically a two-inch clearance is recommended around the hole to be cut for a sink. This is needed for the brackets that hold the sink in the cabinet.

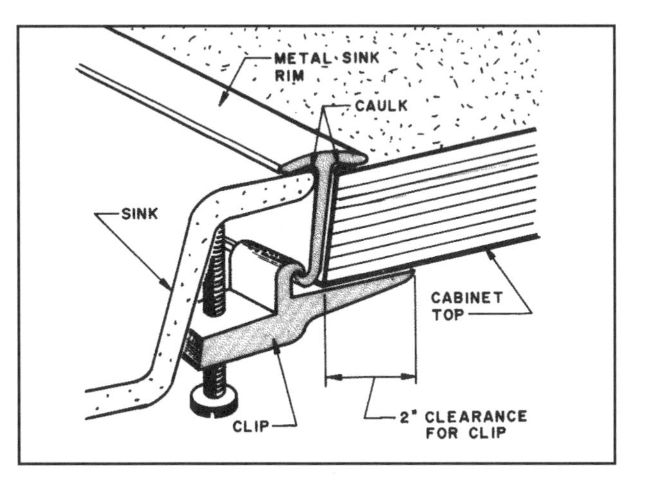

36-47 Metal clips clamp the sink and sink rim to the plywood countertop.

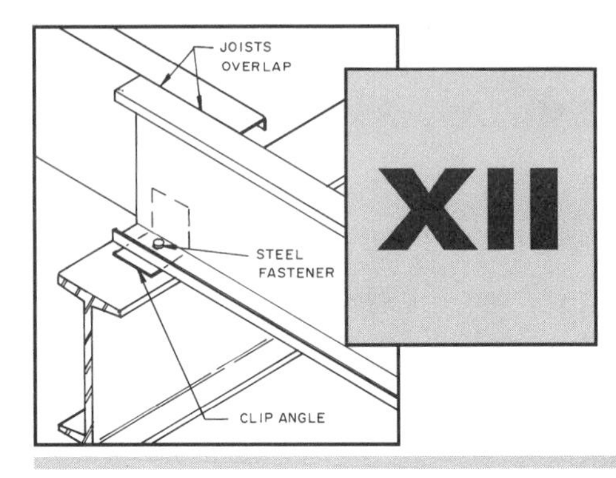

XII LIGHT-FRAME STEEL CONSTRUCTION

CHAPTER 37

Cold-Formed Structural Steel Products

Cold-formed structural steel products are designed and manufactured based on the American Iron and Steel Institute (AISI) publication, *AISI Specifications for the Design of Cold-Formed Steel Structural Members*. Manufacturers of these products may produce items that have different specifications so you must consult with the manufacturer to be certain of the properties of each product.

The load-bearing products, such as studs and joists, must have the properties to carry the imposed load as well as to take into account other factors such as height restrictions or allowable deflection.

The products shown in this chapter illustrate those typically available from various manufacturers. The use of these products is for illustrative purposes and does not indicate anything about their capacity for a specific application. Selection of structural members is made by the design engineer.

Table 37-1 Minimum Uncoated Material Thickness

Designation (mils)	Minimum Uncoated Thickness inches (mm)	Reference Gauge Number
18	0.018 (0.455)	25
27	0.027 (0.683)	22
33	0.033 (0.836)	20
43	0.043 (1.087)	18
54	0.054 (1.367)	16
68	0.068 (1.720)	14
97	0.097 (2.454)	12

For SI: 1" = 25.4mm

MATERIALS

Load-bearing steel framing members are cold formed from structural-quality sheet steel complying with ASTM A653, ASTM 4792, and ASTM A875. ASTM is the American Society for Testing and Materials. Non-structural members comply with ASTM C645.

Members may be uncoated or coated. The thickness of the uncoated products is shown in **Table 37-1**. Coated members are typically hot-dipped galvanized though other coating materials can be used. The thickness is specified by various ASTM specifications. It should be noted that products that exceed ASTM specifications are available.

FIELD IDENTIFICATION

All load-bearing steel framing members have a label on them every 48 inches (1219mm) that contains the following:

1. Name of the manufacturer.
2. Minimum uncoated steel thickness.
3. The abbreviation "ST" to indicate they are a structural members.
4. The weight of any metallic coating.
5. Minimum yield strength in kips per square inch.

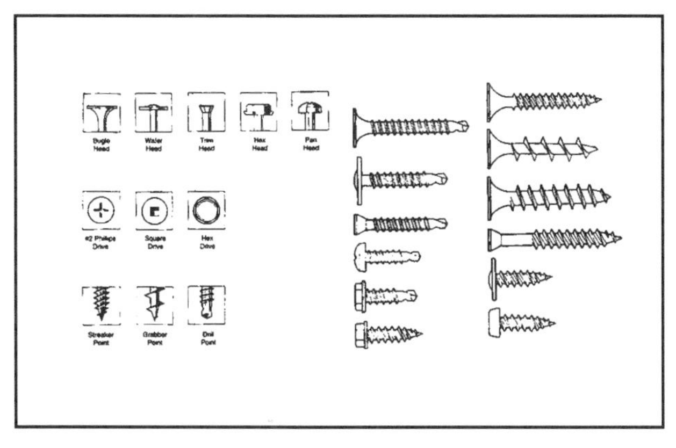

37-1 Some of the screw-type fasteners used in assembling a cold-formed lightweight steel frame. *(Courtesy Grabber Construction Products)*

A typical note would read: APEX MFG., 54 mills, ST. GF60, 33 ksi.

FASTENERS

While screws are the most commonly used fasteners, pneumatically driven fasteners, power-actuated fasteners, crimping, and welding may be used. Again the type and size of fasteners must be decided by a registered engineer.

Screws

Screws should be corrosion resistant and spaced center-to-center as per engineering design. Some of the self-drilling fasteners are shown in **37-1**.

Bolts

Bolts must meet the requirements of ASTM A307. Washers must be used below the head and nut. Bolt spacing and diameters are determined by the design engineer.

COLD-FORMED STEEL PRODUCTS

The C-section is a structural member used for studs, joists, headers, beams, girders, and rafters. The components of a C-section are in **37-2**. The web depth measurements are taken from the outside of the flanges. The width dimensions also use outside dimensions.

Track is used for the top and bottom plates for walls, and band joists for the flooring system. It is much like the C-section but has no lips, as shown in **37-3**.

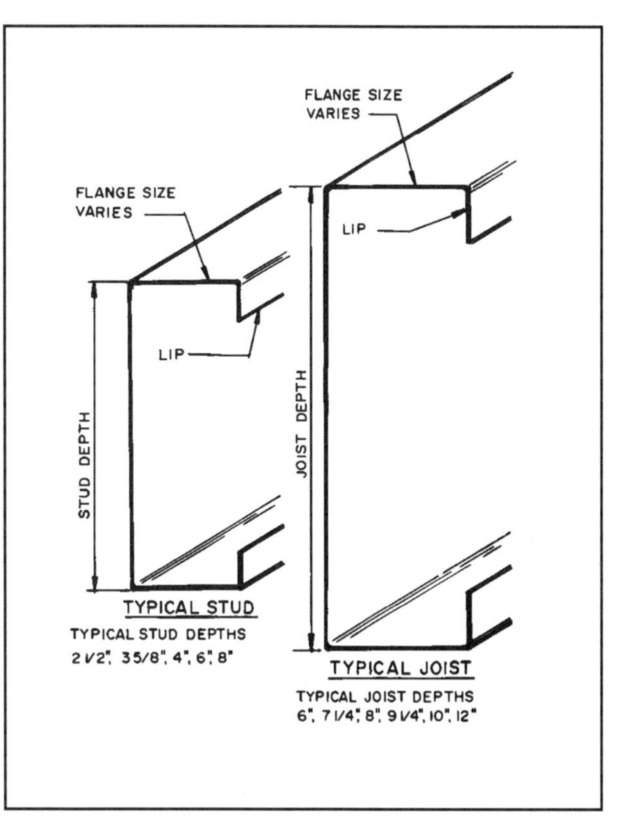

37-2 C-shaped cold-formed steel structural members are used for many purposes.

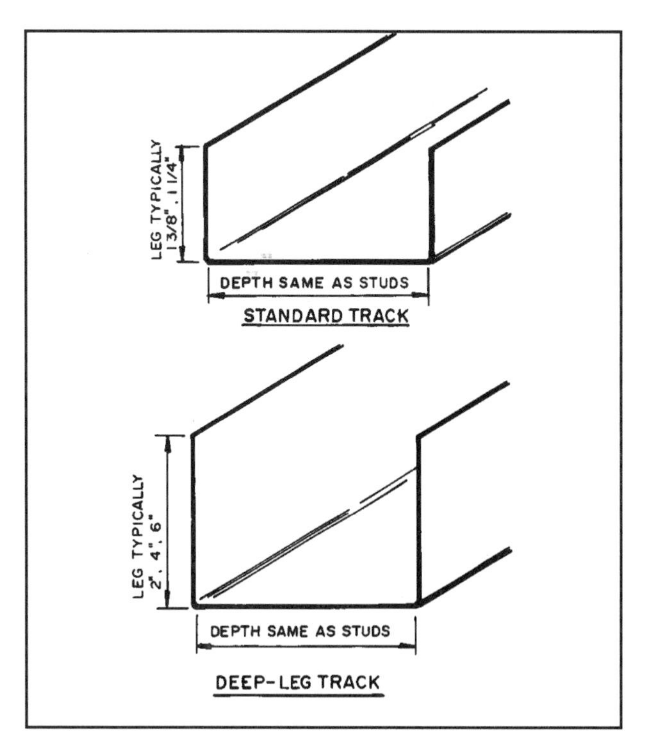

37-3 Commonly available track.

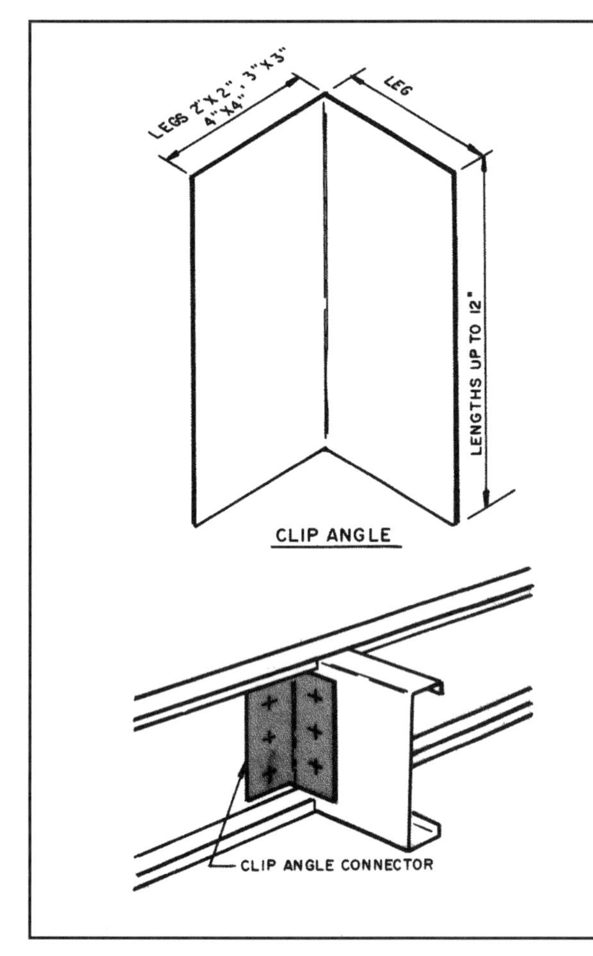

37-4 Clip angle connectors are used to join the structural members.

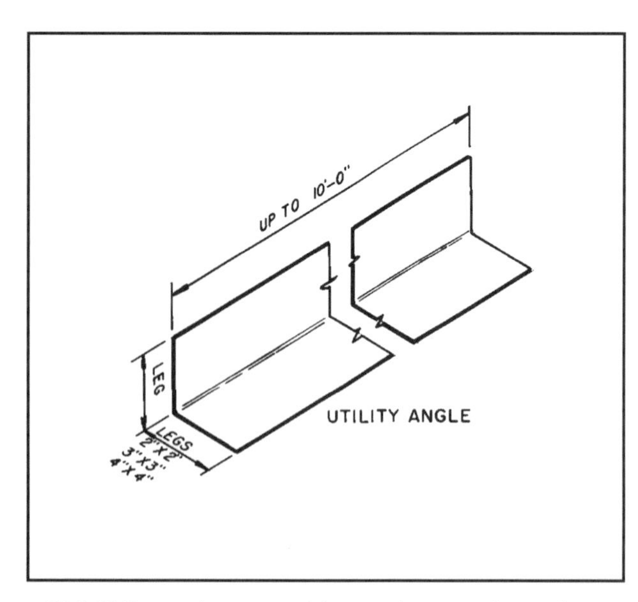

37-5 Utility angles are used for attachments where a long angle is needed.

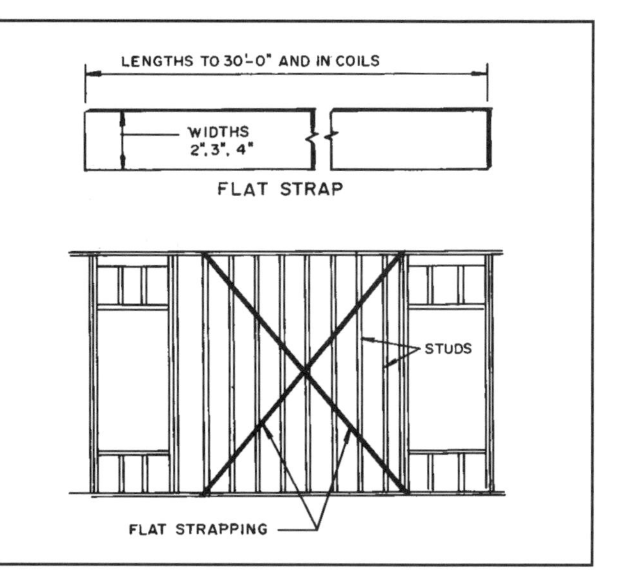

37-6 Flat strapping is used to resist racking in stud wall assemblies.

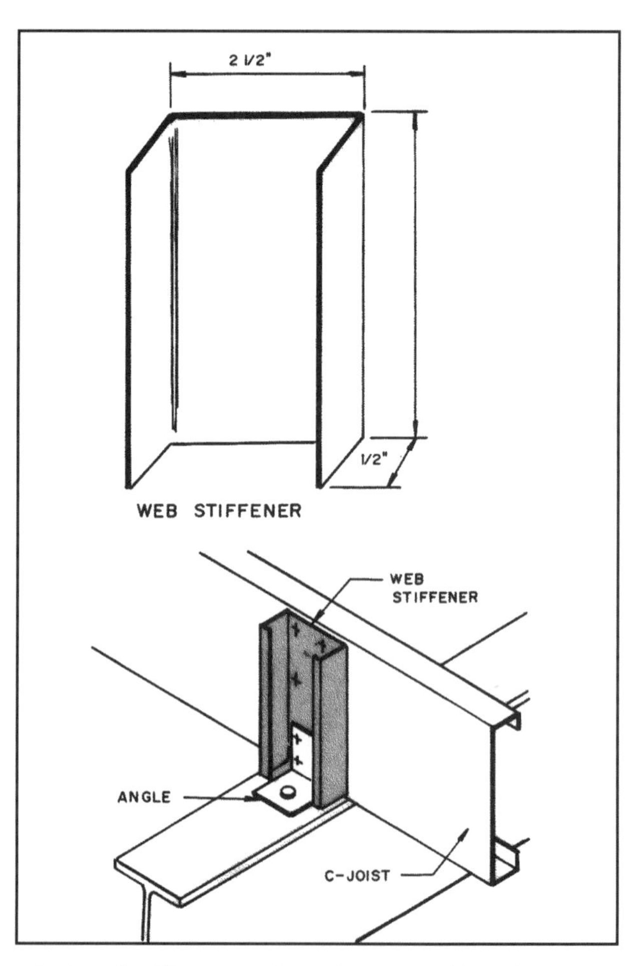

37-7 Web stiffeners reinforce the C-shaped joists by transferring axial loads to bearing members below.

Clip angles and utility angles are L-shaped members generally used to connect other steel members. They are generally formed on a 90-degree angle **(see 37-4 and 37-5)**. They are made in a number of sizes.

Flat strapping is a long, flat steel strip typically used to brace other members, such as a row of studs **(see 37-6)**.

Web stiffeners are C-shaped framing members added to the web of a structural member to strengthen it **(see 37-7)**.

V-bars are used for bridging between studs or joists. They are screwed to each member to impede rotation **(see 37-8)**.

Various types of joist hangers are available, as shown in **37-9**. They are connected to beams to carry floor and ceiling joists.

Joist track is used as a header across the ends of joists as shown in **37-10**. (Refer to **37-13** for an installation detail.)

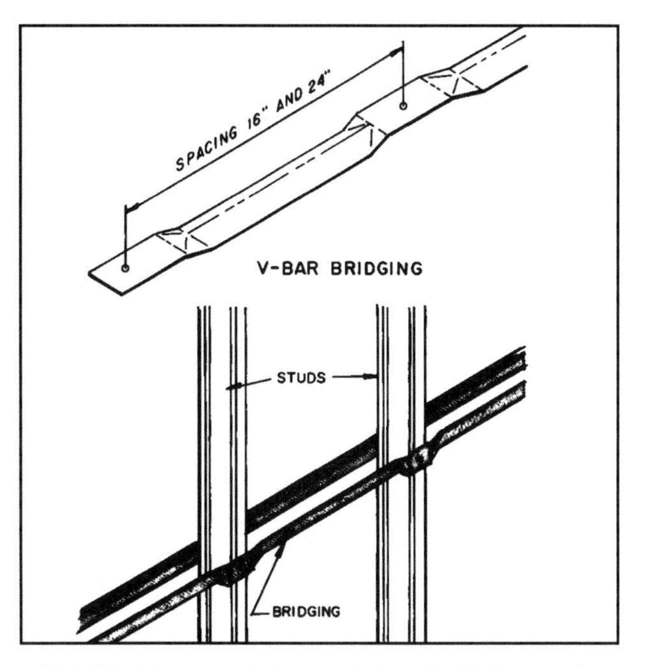

37-8 Bridging is used between studs and joists to provide lateral support.

PRE-PUNCHED HOLES

Studs and in some cases joists are provided with pre-punched holes through which electrical and plumbing systems are run. The shape of the punchout will vary depending upon the company making the member. The American Iron and Steel Institute specifications require these punchouts be at least 12 inches (30.5cm) from the end of the member and a minimum of 24 inches (61cm) apart center-to-center as shown in **37-11**.

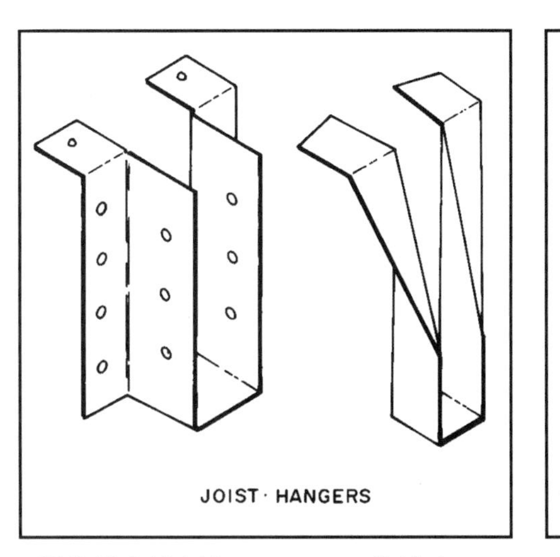

37-9 Metal joist hangers are available in a wide range of sizes.

37-10 Joist track is run around the ends of the joists on the perimeter of the building.

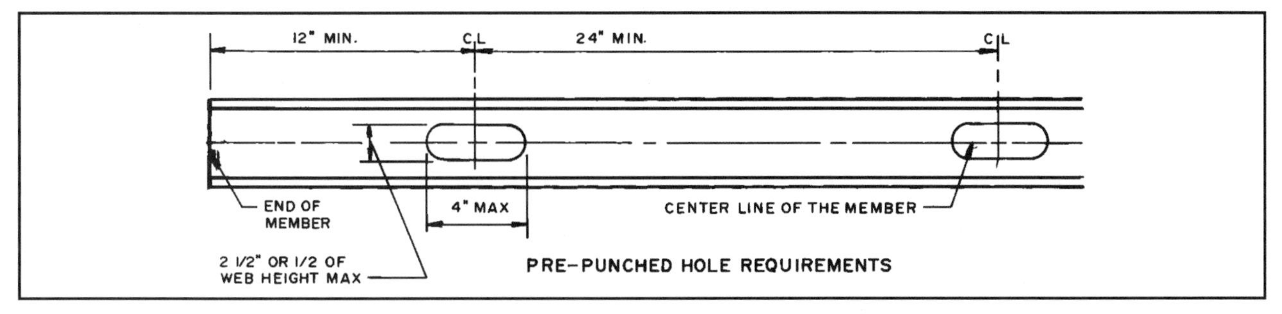

37-11 Requirements for pre-punched holes in cold-formed steel structural members. *(Courtesy American Iron and Steel Institute)*

TRUSSES

Steel roof trusses are much the same as wood trusses, described in Chapter 18. The major differences include steel structural members that have different properties than wood, steel connectors, and different fasteners and fastening techniques. These must be designed by a competent steel framing designer. They are manufactured in the same types as shown in Chapter 18. A finished truss is shown in **37-12**.

TIES

Cold-formed steel framing uses basically the same types of ties used for wood frame construction. These include such items as seismic and hurricane ties, straps, bridging, and hold-downs **(see 37-13 and 37-14)**.

TOOLS

When it is necessary to cut cold-formed lightweight steel members a chopsaw is generally

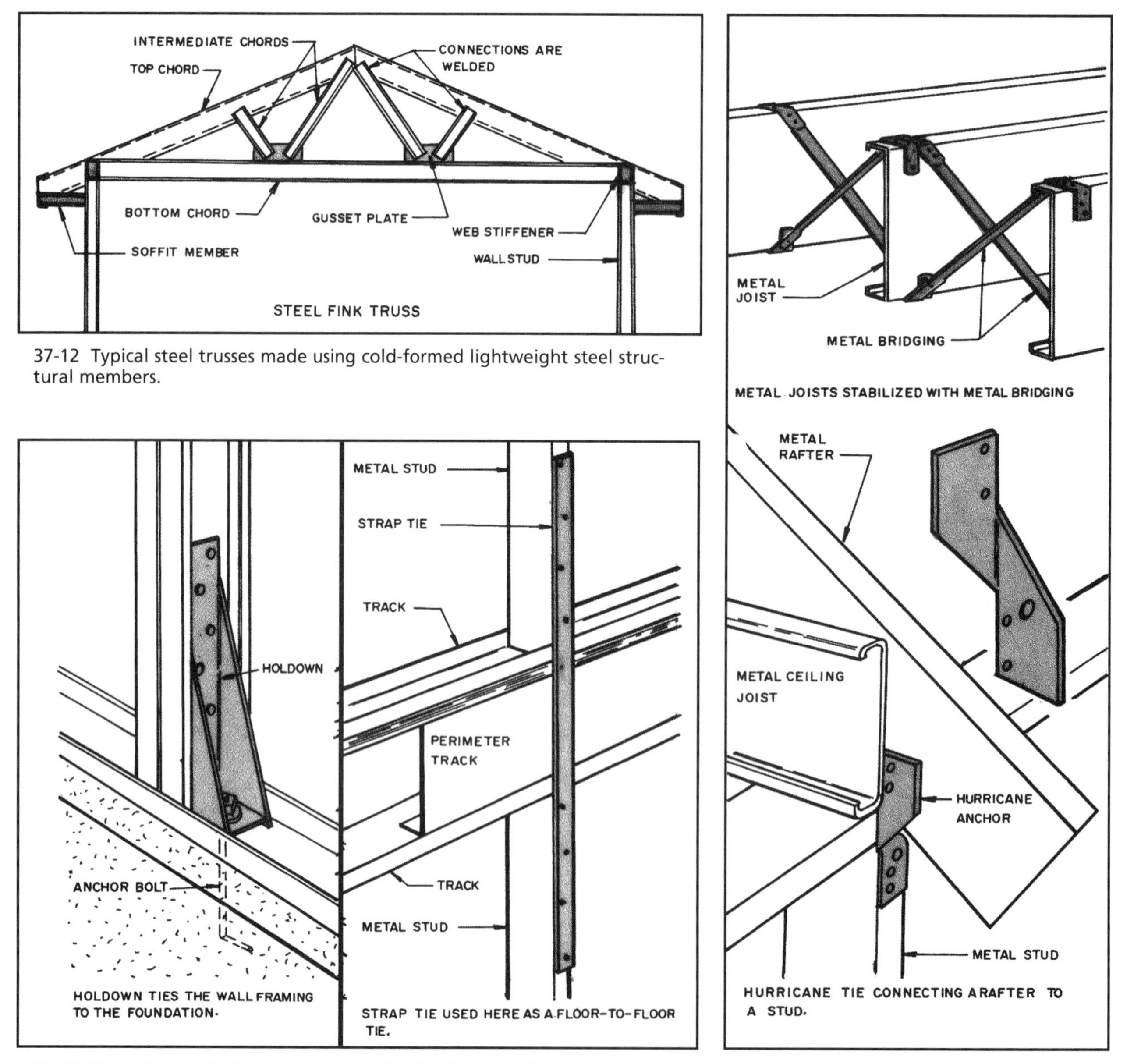

37-12 Typical steel trusses made using cold-formed lightweight steel structural members.

37-13 Strap ties and hold-downs are used in the framing of buildings using cold-formed steel structural members. *(Courtesy Simpson Strong-Tie Company, Inc.)*

37-14 Hurricane ties are used on rafters and bridging brace joists. *(Courtesy Simpson Strong-Tie Company, Inc.)*

used. Either a fine-toothed metal cutting blade or a reinforced abrasive cut-off wheel is used **(see 37-15)**. These can also be used on portable circular saws. Since sparks and flying metal particles are produced, the worker must have full-face shield protection. Portable electric circular saws are also used. Metal cutting blades and abrasive cut-off wheels are also used. These are used for light duty work. The chopsaw has more power (usually four or five horsepower).

There are portable electric shears that will cut up to 20-gauge steel **(see 37-16)**. A nibbler makes quick cuts in thinner stock **(see 37-17)**. Various types of hand operated metal snips are available.

Clamps are important tools because the various parts need to be held while they are being fastened.

A variety of screw fasteners are used to secure metal framing members. An electric screwdriver is probably the most frequently used tool **(see 37-18)**. The clutch engages, connecting the chuck with the screw to the rotating motor when the screw is pressed against the steel. The clutch releases when the pressure is released. Some drills have torque release clutches that automatically stop driving the screw when a preset torque is reached.

Tools used to level and plumb the structural frame are discussed in Chapter 5.

37-15 This chopsaw is cutting cold-formed lightweight steel studs with an abrasive wheel. *(Courtesy Delta International Machinery Corp.)*

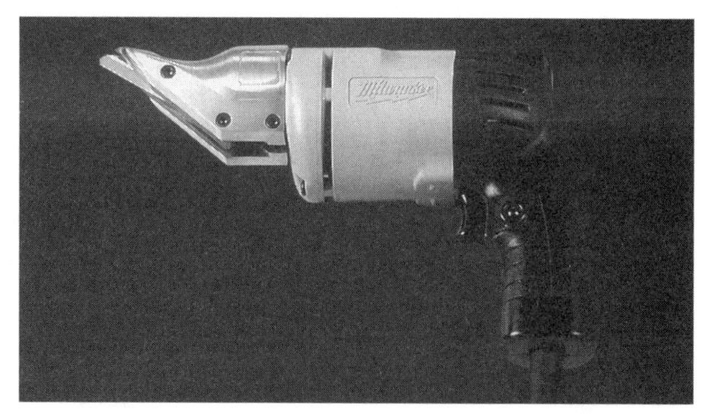

37-16 This portable electric shear can cut round, square, and irregular-shaped holes as well as all shapes of edges. *(Courtesy Milwaukee Electric Tool Corporation)*

37-17 This nibbler will cut flat, irregular, and corrugated metal. *(Courtesy Milwaukee Electric Tool Corporation)*

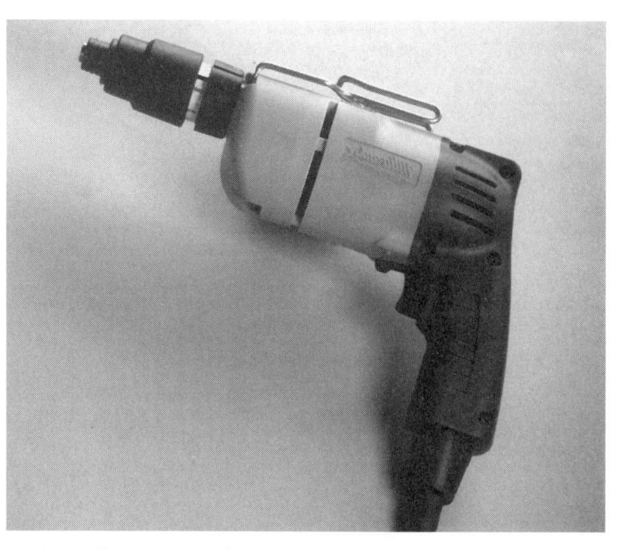

37-18 Electric screwdrivers are used to drive the threaded fasteners used to assemble the metal structural members. *(Courtesy Milwaukee Electric Tool Corporation)*

CHAPTER 38

COLD-FORMED STRUCTURAL STEEL FRAMING

Originally steel framing members were designed in the same sizes as wood so they could substitute for wood. Now they are designed with the recognition of their strength and uniformity of load-bearing properties so that an entire range of steel sizes and shapes are available, offering sizes different from wood members. The designer now utilizes the unique strengths and properties of the material as the structural frame is designed.

The details shown in this chapter are for illustrative purposes only and are not intended to be used without the advice of an engineer experienced in light steel framing construction design. Neither are the illustrations all-inclusive. There are many other ways the structural members could be used, connected, or applied to a structural frame.

A competent designer is needed to provide technical information and designs for other features, such as the foundation, thermal insulation, moisture control, sound attenuation, and special requirements, as needed, for seismic or high wind conditions.

BUILDING CODES

Most building codes have not yet developed recommended methods for cold-formed lightweight steel framing. Therefore it is imperative that the services of a competent engineer be used to design the steel structure. This is especially important in areas where particular situations occur such as frequent high winds, heavy snows, and seismic conditions.

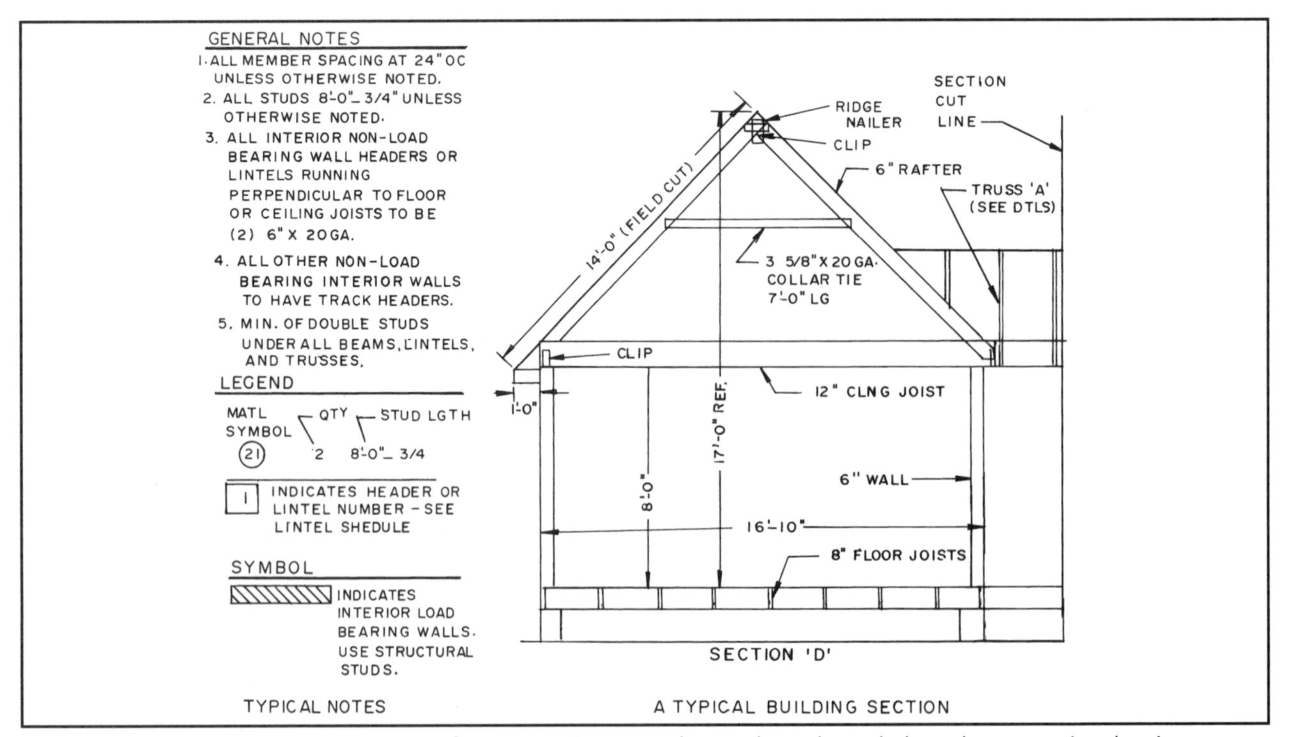

38-1 The architectural drawings contain information using general notes, legends, symbols, and many section drawings. *(Courtesy HSF Home Steel Framing, Inc.)*

Codes currently refer the builder to the *AISI Specifications for the Design of Cold-Formed Steel Structural Members* and the *AISI Load and Resistance Factor Design Specifications for Cold-Formed Steel Structural Members.* AISI refers to the American Iron and Steel Institute.

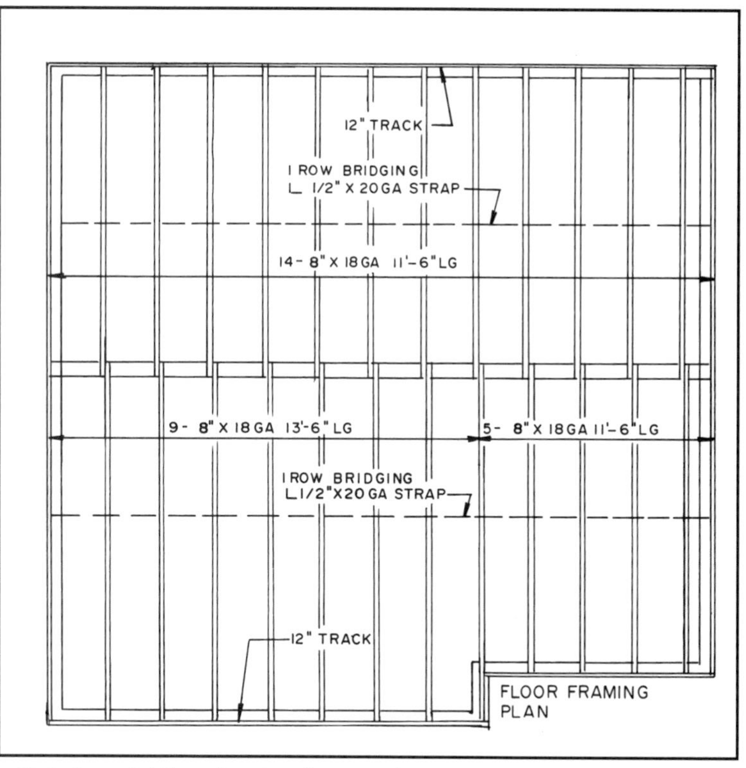

38-2 A partial floor framing drawing. *(Courtesy HSF Home Steel Framing, Inc.)*

FRAMING PLANS

The framing plans must be prepared by a competent designer or engineer. Included is information in the form of notes, legends, schedules, and a wide range of section drawings **(see 38-1)**. A partial floor framing plan is shown in **38-2**, a partial wall framing plan is shown in **38-3**, a partial ceiling framing plan is shown in **38-4** on the following page, and a partial roof framing plan is shown in **38-5** on the following page.

ASSEMBLY METHODS

There are three commonly used methods for assembling cold-formed light steel framed buildings: stick-built, panelized, and preengineered systems.

Stick-Built

Stick-built construction involves cutting the members to length on the job and joining them with screws or other recommended fasteners. The steel is received on the job in stock lengths in the same manner as wood for stick-built wood-framed buildings. After the structural framework has been erected, the sheathing, windows, doors, and finish exterior materials are installed.

Panelized Method

Panelized techniques involve assembling floor, wall, and roof sections into large assemblies. These are built in a shop where the

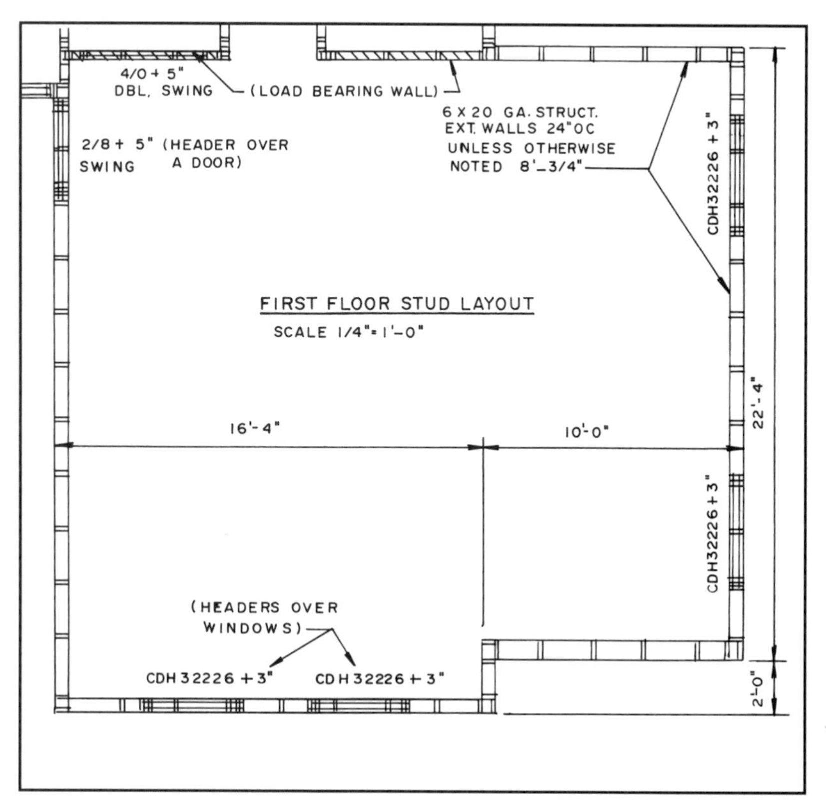

38-3 A partial wall framing plan. *(Courtesy HSF Home Steel Framing, Inc.)*

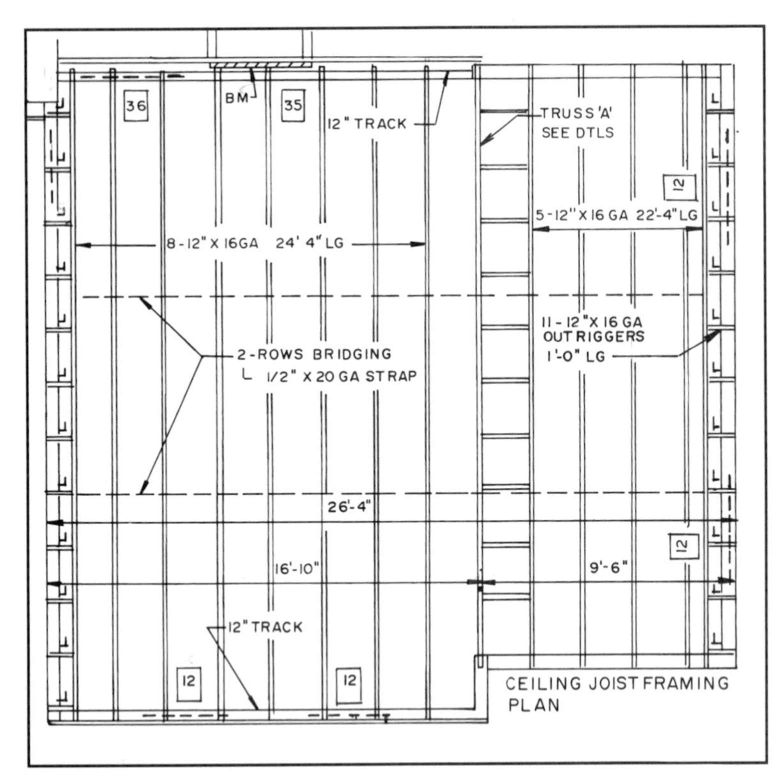

38-4 A partial ceiling framing plan. *(Courtesy HSF Home Steel Framing, Inc.)*

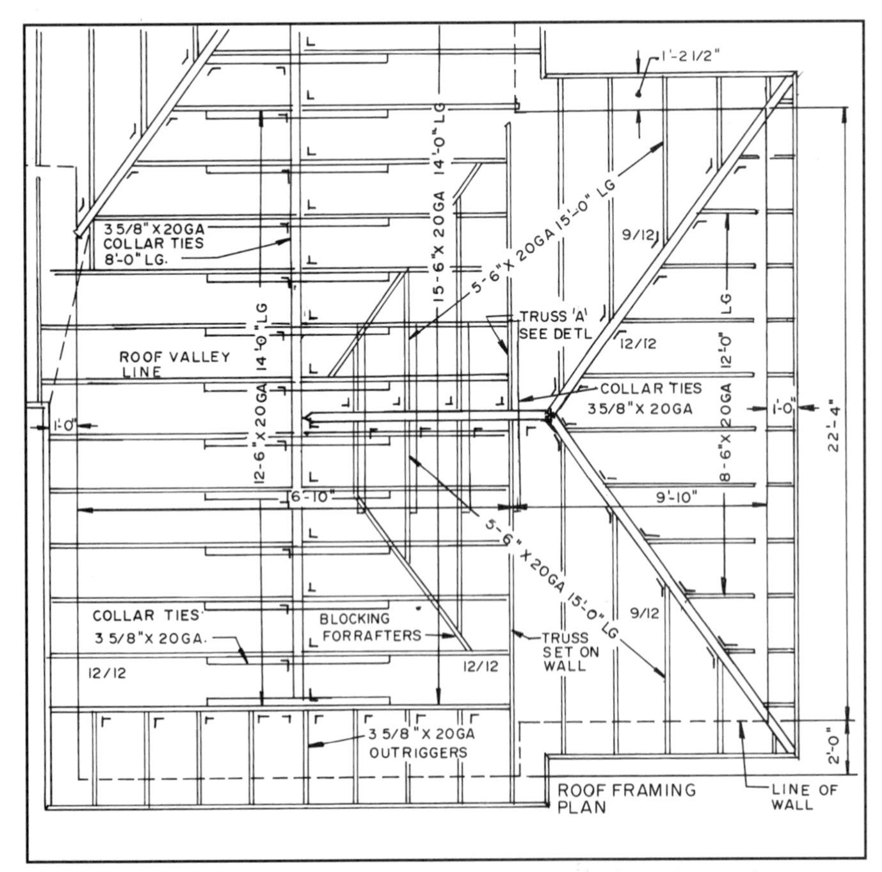

38-5 A partial roof framing plan. *(Courtesy HSF Home Steel Framing, Inc.)*

work can proceed regardless of the weather **(see 38-6)**. The assembled panels are moved to the site and set in place. A crane is used to lift heavy assemblies.

Preengineered Systems

A preengineered system uses a frame of load-carrying members to support secondary lighter-weight steel sections to fill in between the heavy structural frame. These buildings are designed and manufactured by companies specializing in preengineered building components. They are shipped to the site and erected on the foundation prepared by the general contractor.

FRAMING THE FLOOR

Floor framing can begin when the foundation is in place. The framing crew will need a set of architectural working drawings that show which members to use and where they belong. Connection details are also necessary.

The basic process is about the same as described for wood-framed floors. The joists rest on the foundation and on beams used to reduce the span. Floor joists bearing on the foundation and perpendicular to it can be installed as shown in **38-7**. This illustration shows the use of metal clip angles and anchor bolts. It also has used a web stiffener and the joist track on the perimeter of the floor. The joist track is connected to the joists with a clip angle.

When the joists run parallel with the foundation, the installation is usually made as shown in **38-8**. The outermost joist is set in from the edge of the foundation to allow the installation of the joist track.

When floor joists are supported by a beam, they are overlapped (side by side) and secured to the beam with a clip angle as shown in **38-9**. Continuous bridging is required between the joists and is placed directly over the beam.

If it is desired to keep the joists in line (end to end) when they rest on a beam, they can be installed as shown in **38-10**. This is important when a load-bearing wall is to be directly above the beam.

38-6 This assembly platform is used to assemble a wall panel. *(Courtesy American Iron and Steel Institute,* Jordan Commons Cold-Formed Steel Training Manual, RG95-09, 1995*)*

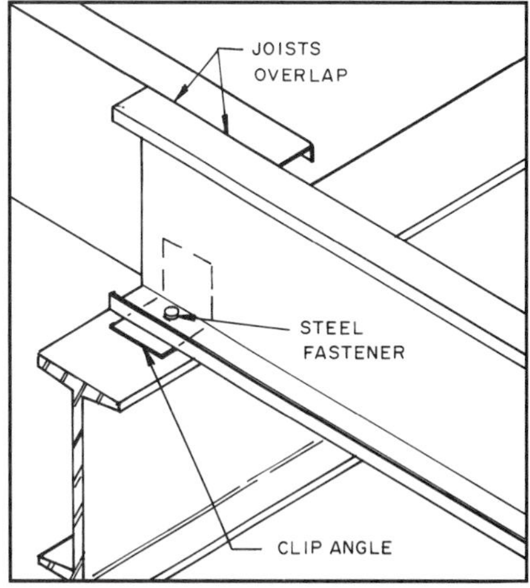

38-9 These floor joists are overlapped on a beam. *(Courtesy American Iron and Steel Institute,* Low-Rise Residential Construction Details, RG-934, 1993*)*

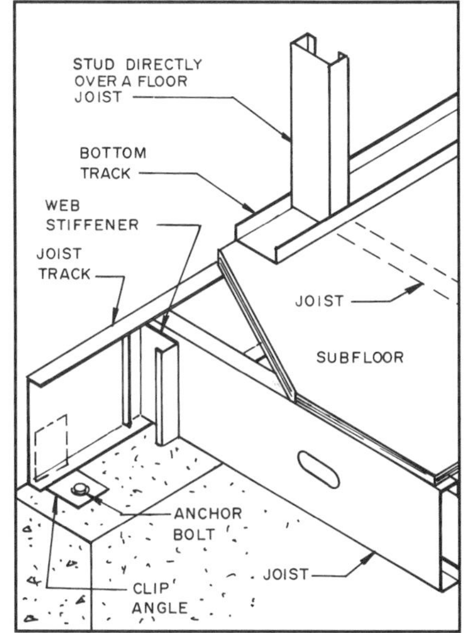

38-7 This assembly shows how floor joists that run perpendicular to the foundation are installed. *(Courtesy American Iron and Steel Institute,* Low-Rise Residential Construction Details, RG-934, 1993*)*

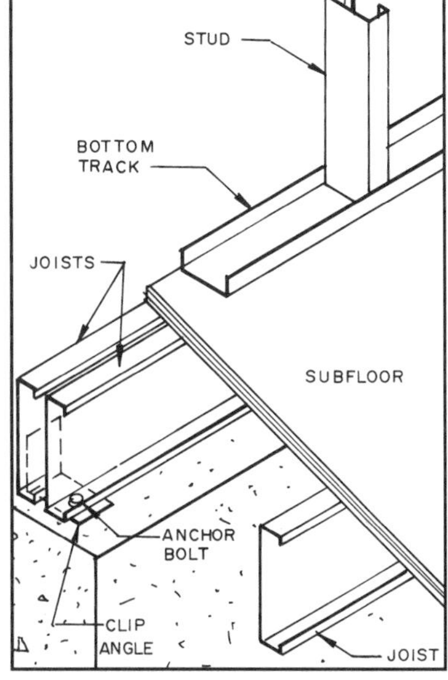

38-8 This is one way to install the end joists when they run parallel with the foundation. *(Courtesy American Iron and Steel Institute,* Low-Rise Residential Construction Details, RG-934, 1993*)*

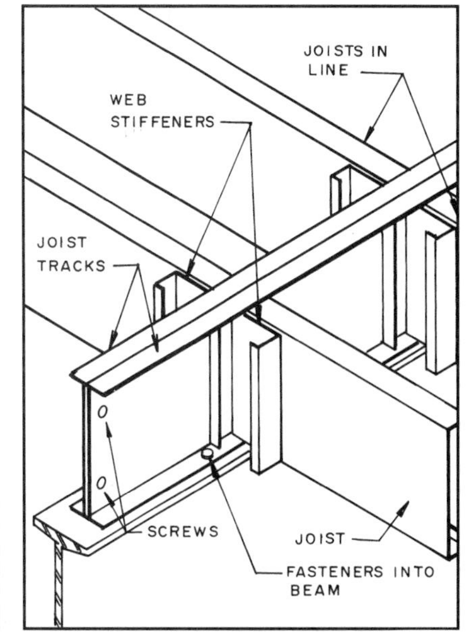

38-10 One way to frame joists on a beam so they are in line. *(Courtesy American Iron and Steel Institute,* Low-Rise Residential Construction Details, RG-934, 1993*)*

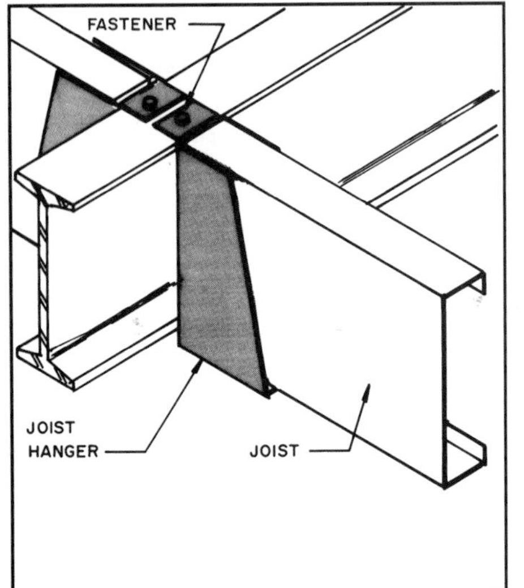

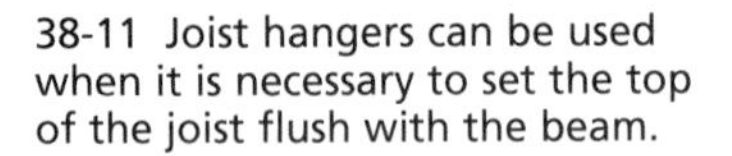

38-11 Joist hangers can be used when it is necessary to set the top of the joist flush with the beam.

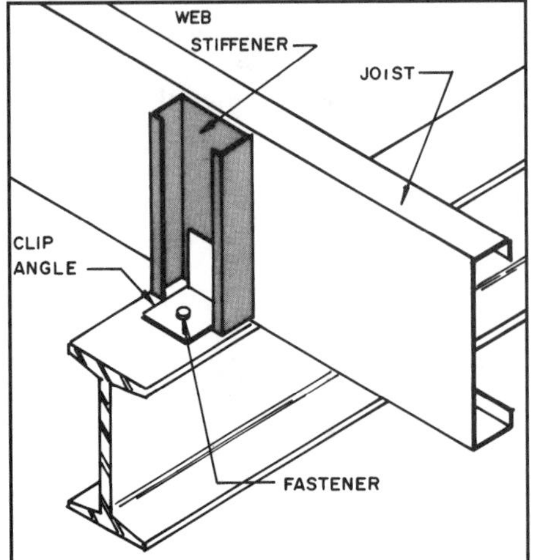

38-12 Joists that run across beams have a web stiffener added to provide needed support. *(Courtesy American Iron and Steel Institute,* Low-Rise Residential Construction Details, RG-934, 1993*)*

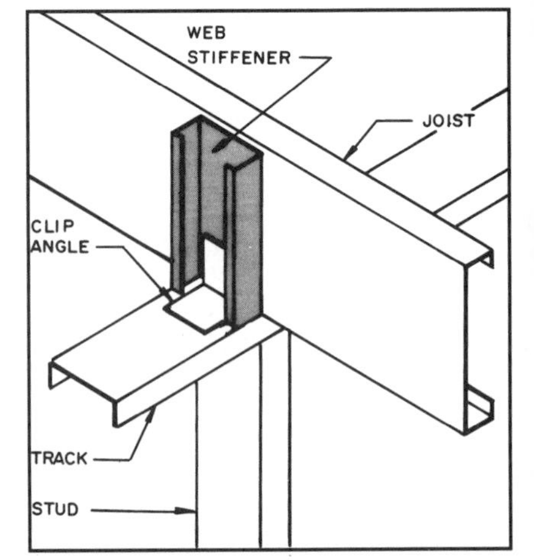

38-13 Web stiffeners are used to strengthen the joist if it runs across an interior load-bearing partition. *(Courtesy American Iron and Steel Institute,* Low-Rise Residential Construction Details, RG-934, 1993*)*

If it is desired to have the tops of the joists flush with the top of the beam, joist hangers can be used **(see 38-11)**. When the joists run over the beam they are secured to it with clip angles and a web stiffener is added **(see 38-12)**.

The joists could be ceiling joists supporting storage, or second floor joists. Two-story buildings often carry the second floor joists on interior partitions, as shown in **38-13**, and on exterior walls, as shown in **38-14**.

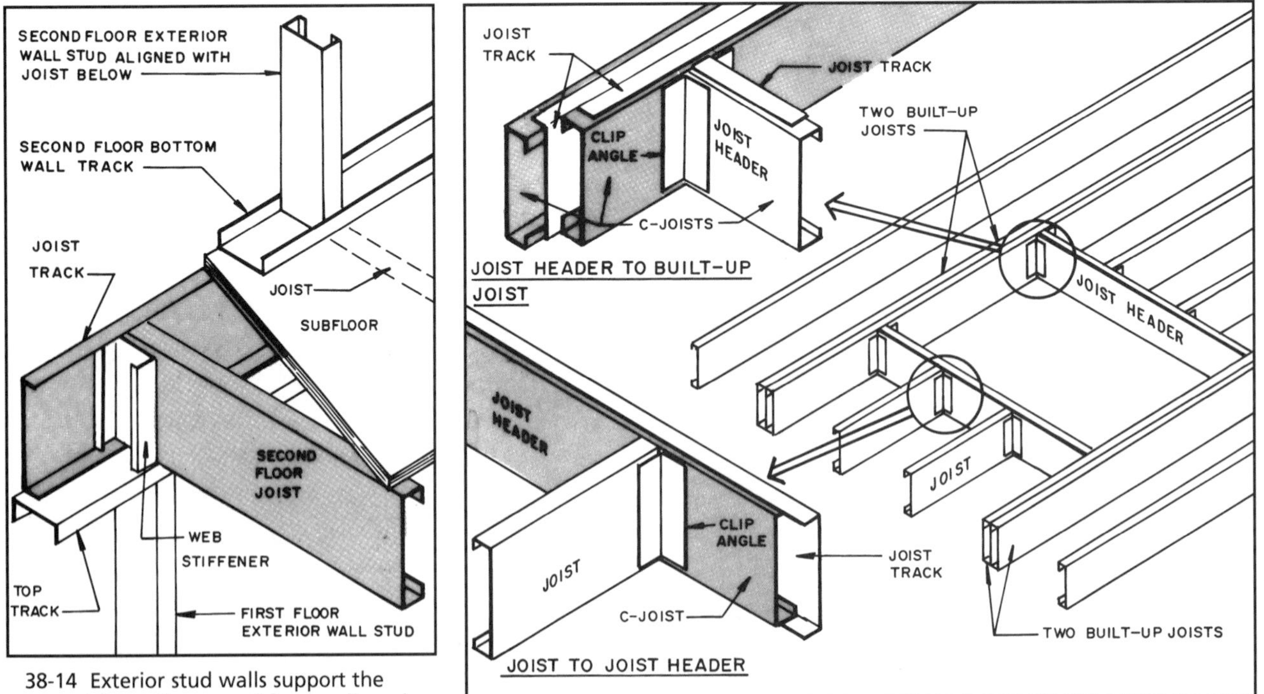

38-14 Exterior stud walls support the second floor, the second floor wall, and the roof. *(Courtesy American Iron and Steel Institute,* Low-Rise Residential Construction Details, RG-934, 1993*)*

38-15 Openings in floors are framed with built-up headers and side joists. *(Courtesy American Iron and Steel Institute,* Low-Rise Residential Construction Details, RG-934, 1993*)*

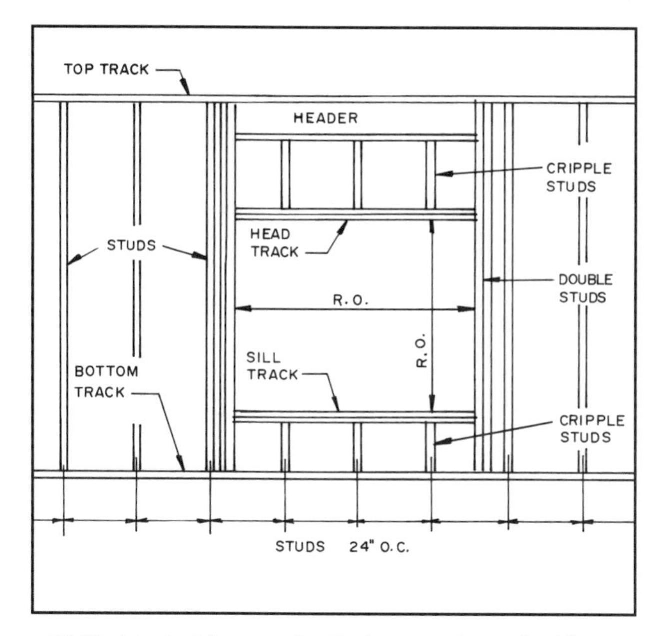

38-16 A typical framing detail of an exterior wall with a window opening. Notice there is a stud located every 24 inches (610mm).

Openings in floors require built-up side joists and headers made from a joist and a joist track. Additional joists can be used if span and loads require it **(see 38-15)**. A single joist can be secured to a header with a clip angle.

Plywood and oriented strand board are typically used as subflooring. They are secured to the steel joists with self-drilling screws.

Bridging is shown later in this chapter in **38-32**, **38-33**, and **38-34**.

Floor joists are doubled under load-bearing partitions. The web stiffener helps transmit the load to the structural members below the joist.

FRAMING WALLS

Steel frame walls may be load-bearing or nonload-bearing the same as for wood-framed walls. Studs should be aligned with the floor joists below **(see 38-14)**. This transmits any load through the structure. When assembling the wall, lay out the location of the studs on each side but do not break the center-to-center spacing **(see 38-16)**. The studs are doubled on each side of the opening and covered with an enclosure track needed for the sill to connect to the stud **(see 38-17)**. The studs are secured (usually with screws) to the bottom track, which has been screwed to the subfloor **(see 38-18)**. A complete assembly is shown in **38-17**.

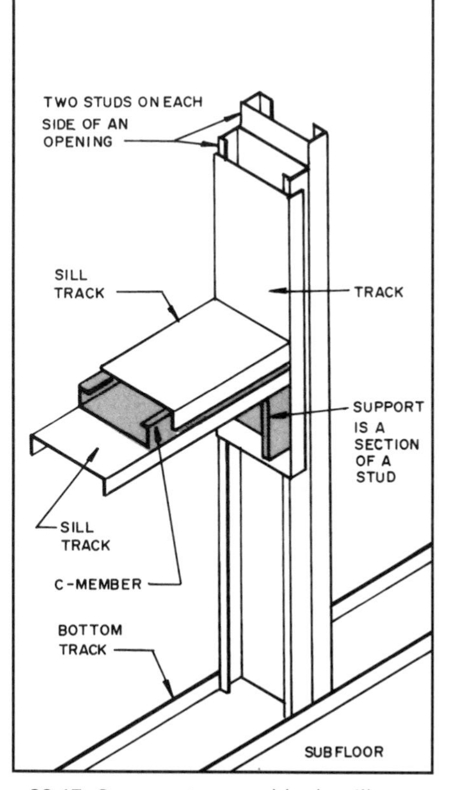

38-17 One way to assemble the sill below a window rough opening. *(Courtesy American Iron and Steel Institute, Low-Rise Residential Construction Details, RG-934, 1993)*

38-18 The bottom track is screwed to the subfloor.

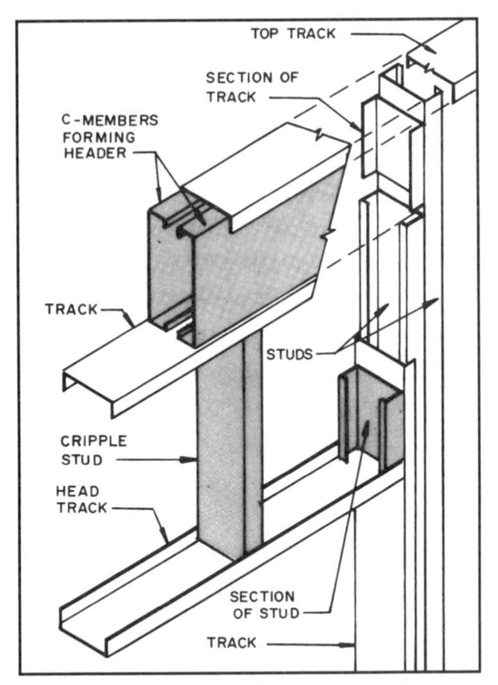

38-19 One header design used in load-bearing walls. *(Courtesy American Iron and Steel Institute,* Low-Rise Residential Construction Details, RG-934, 1993*)*

38-21 An assembled interior partition raised in place and braced to hold it plumb. Notice the horizontal bridging.

38-20 A finished framed window rough opening. *(Courtesy American Iron and Steel Institute,* Jordan Commons Cold-Formed Steel Training Manual, RG95-09, 1995*)*

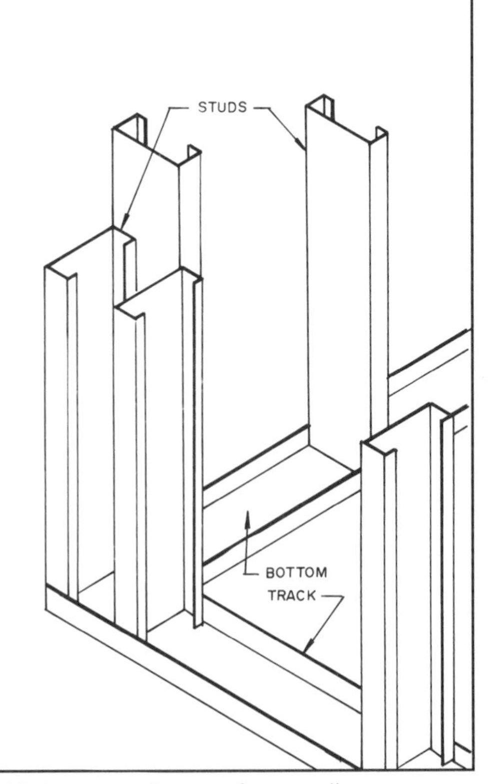

38-22 A suggested way to frame wall corners. *(Courtesy American Iron and Steel Institute,* Low-Rise Residential Construction Details, RG-934, 1993*)*

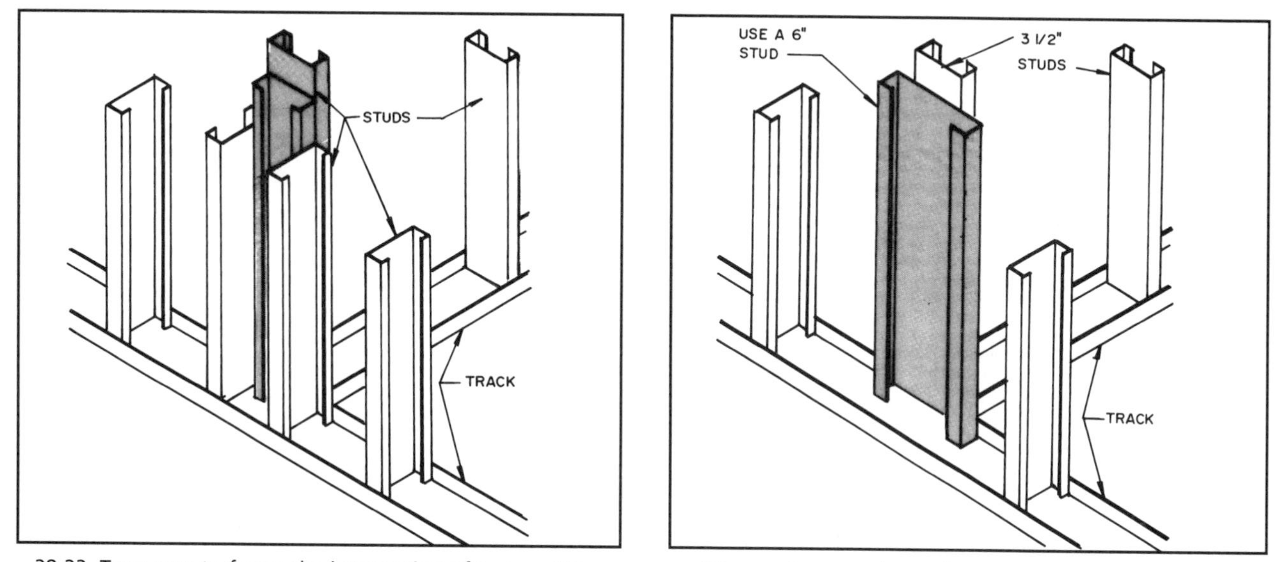

38-23 Two ways to frame the intersection of two partitions or a partition and an exterior wall.

A wall opening for a door or window requires a header (**see 38-19**). If it is a window opening, a sill is required as shown earlier in **38-16**. A finished window opening is shown in **38-20**.

The assembled wall section is raised and braced in the same manner as wood-framed walls (**see 38-21**). When walls meet forming a corner, an assembly such as shown in **38-22** is made. When a partition meets another partition or an exterior wall it can be framed as shown in **38-23**.

If the building has a concrete floor, the bottom track is secured to the concrete (**see 38-24**) with power-driven fasteners through the bottom plate into the concrete (**see 38-25**).

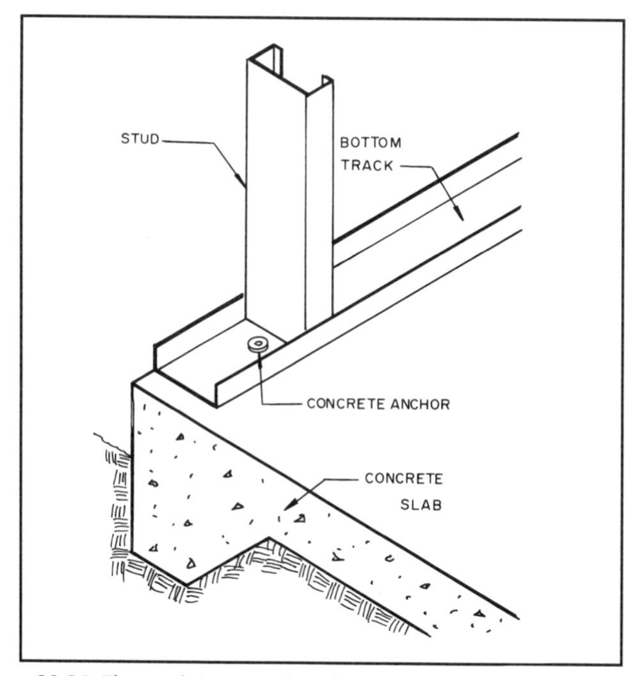

38-24 The track is secured to the concrete slab with expansion bolts or some other type of concrete fastener. *(Courtesy American Iron and Steel Institute,* Low-Rise Residential Construction Details, RG-934, 1993*)*

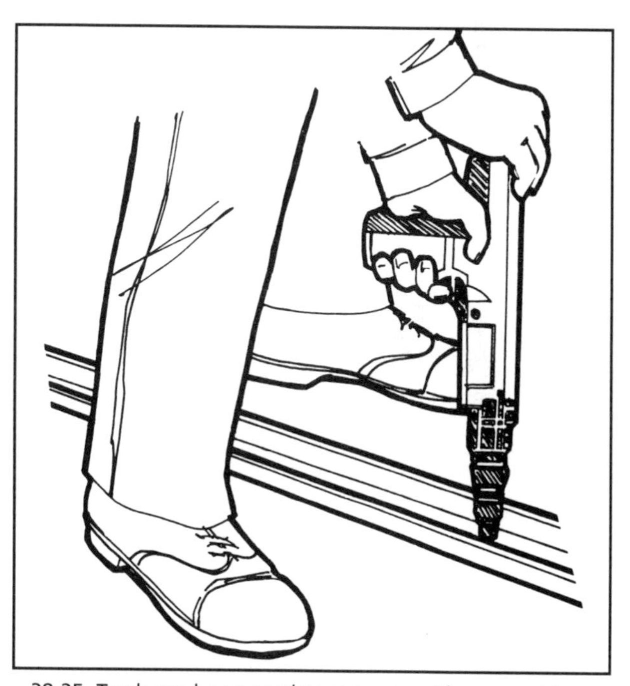

38-25 Track can be secured to concrete with power-driven concrete fasteners. *(Courtesy American Iron and Steel Institute,* Low-Rise Residential Construction Details, RG-934, 1993*)*

Plumbing, electrical wiring, and heating and air-conditioning are installed much the same as in wood-framed walls. Many studs are available with prepunched holes (refer to Chapter 37). If it is necessary to make a hole it can be cut with a hole punch and the sharp edges covered by inserting a plastic grommet. Electrical boxes are secured to the studs with screws as shown in **38-26**.

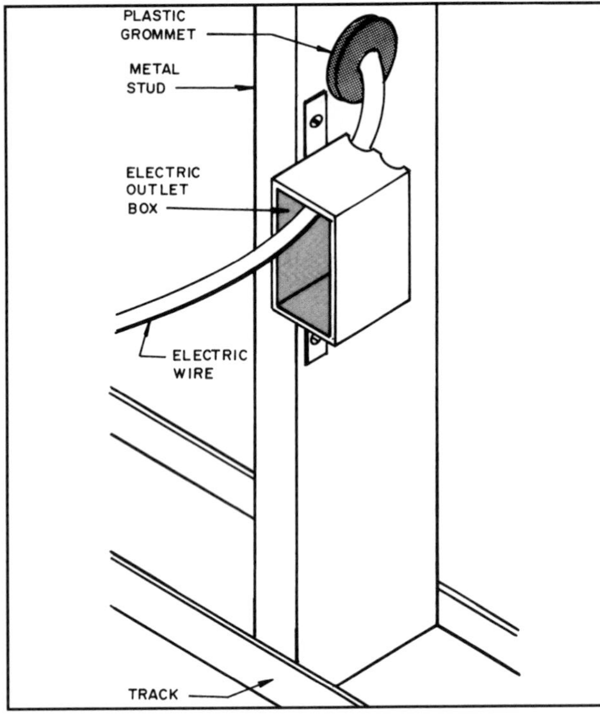

38-26 Electric outlet boxes have tabs that are screwed to the stud. Notice the plastic grommet used to protect the wire from the sharp edges of the hole in the stud.

FRAMING THE ROOF

The framing of steel roofs is essentially the same as those discussed for wood roofs in Chapter 12. They may be stick built or use steel trusses. The stick-built roof has the rafters (C-members) butt a ridge board made by assembling a C-joist and a joist track **(see 38-27)**. They are assembled with clip angles. The rafter rests on the top track and is joined to it with a clip angle. The ceiling joist is secured to the rafter as shown in **38-28**. The joists are spaced as specified on the working drawings and located as described for wood-framed roofs in Chapters 13 through 16.

Steel trusses are manufactured and trucked to the site. They are designed and assembled as directed by a competent structural designer **(see 38-29)**. They may bear on the top track as shown in **38-30** or be raised above it with a vertical web as shown in **38-31**.

Rafters and trusses should be placed directly above a stud. When this is not possible the top track needs to be reinforced with additional track

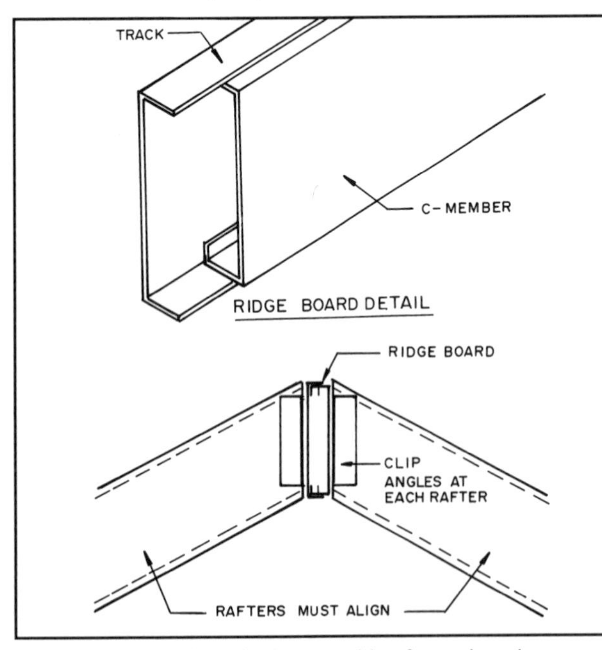

38-27 The ridge board is an assembly of a track and a C-shaped member. *(Courtesy American Iron and Steel Institute,* Low-Rise Residential Construction Details, RG-934*)*

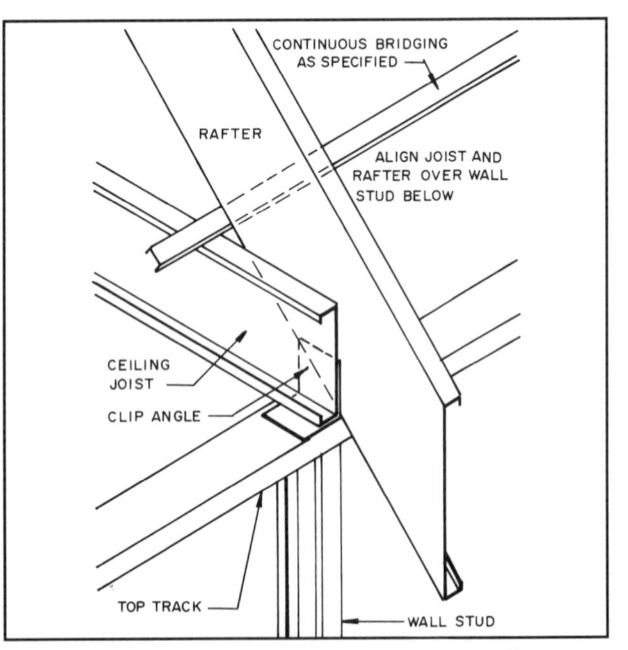

38-28 Typical roof construction at the exterior wall. *(Courtesy American Iron and Steel Institute,* Low-Rise Residential Construction Details, RG-934*)*

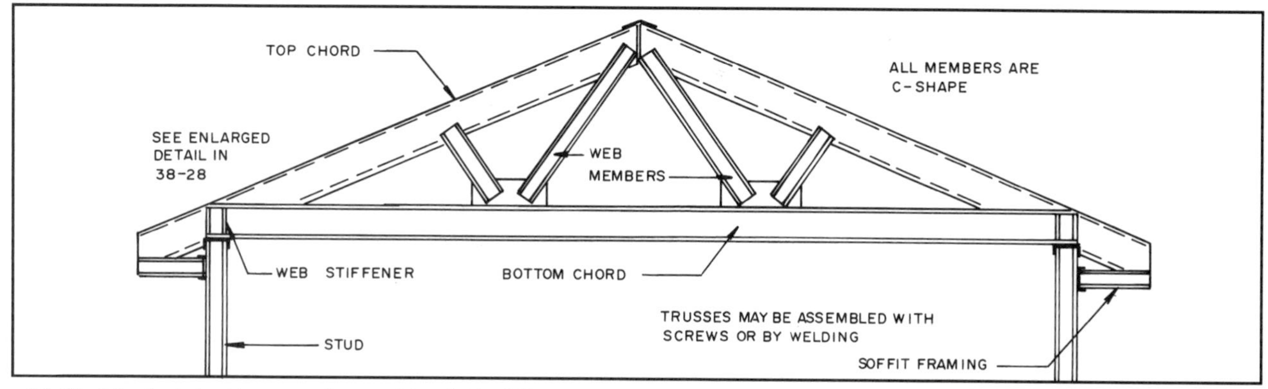

38-29 A typical steel truss used on residential buildings.

or a piece of a joist. Roof sheathing is usually plywood or oriented strand board. It is screwed to the rafters.

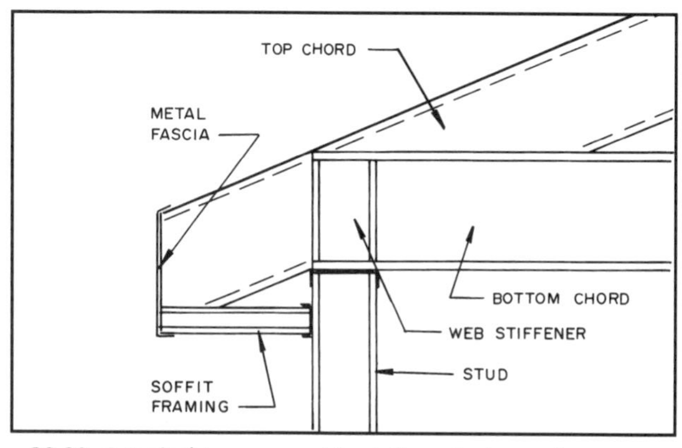

38-30 A typical truss assembly at the exterior wall.

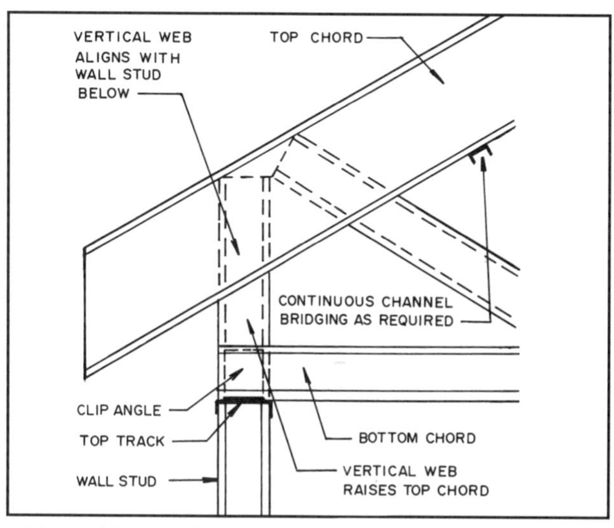

38-31 This truss design uses a vertical web member to raise the top chord above the top of the wall. *(Courtesy American Iron and Steel Institute, Low-Rise Residential Construction Details, RG-934, 1993)*

TIES, BLOCKING & BRIDGING

Floor joists are bridged to prevent them from twisting, moving sideways, or having bounce. The bridging reduces movement in the joists and stabilizes the floor. The subfloor also adds considerably to a stable floor. Bridging typically is X-bridging, solid blocking, or a flat strap. X-bridging is used to stabilize joists, rafters, and studs **(see 38-32)**. Some manufacturers make a V-shaped bridging that adds additional stiffness (see Chapter 37).

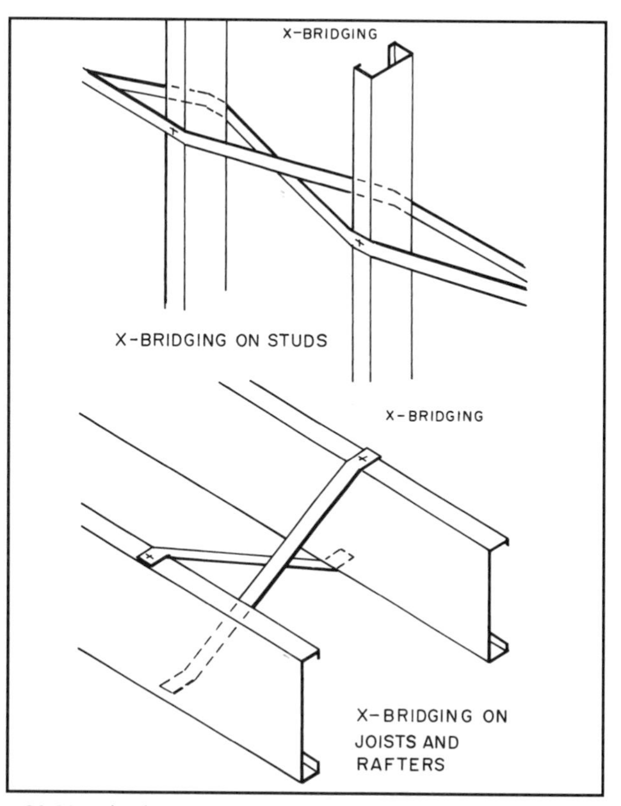

38-32 X-bridging is used to stabilize joists, rafters, and studs. *(Courtesy American Iron and Steel Institute, Low-Rise Residential Construction Details, RG-934, 1993)*

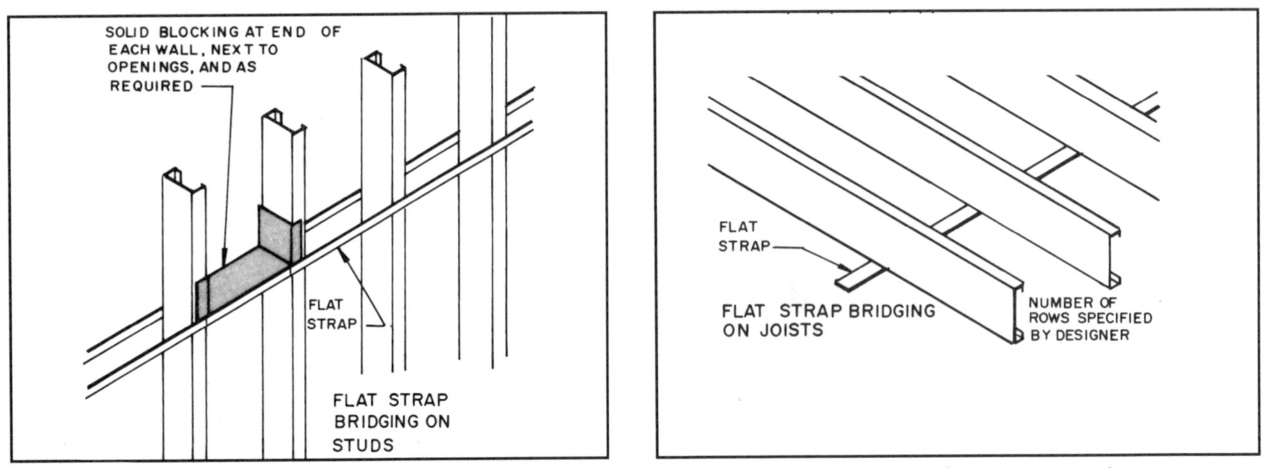

38-33 Flat-strap bridging is used to stabilize joists and studs. *(Courtesy American Iron and Steel Institute,* Low-Rise Residential Construction Details, RG-934, 1993*)*

Flat strapping is also used by securing it to the bottom of joists or horizontally on studs **(see 38-33)**. It is secured to studs on a diagonal, which reduces the chance of the wall racking **(see 38-34)**.

Flat strapping is also temporarily joined to panelized wall sections to hold them square and prevent racking as they are lifted and placed **(see 38-35)**.

Solid blocking is made using short pieces of studs, joists, or track. The blocking should be the same depth as the member being bridged. The pieces must fit firmly against the members and are connected with clip angles **(see 38-36)**. Floor joists and studs can be stabilized with solid blocking.

When a wall requires bracing against shear forces, wide steel straps are installed, forming an X-shaped reinforcement **(see 38-37)**. Shear reinforcement is designed to resist shear stresses or diagonal tension stresses as caused by wind. Plywood or OSB sheathing can also be used for this purpose.

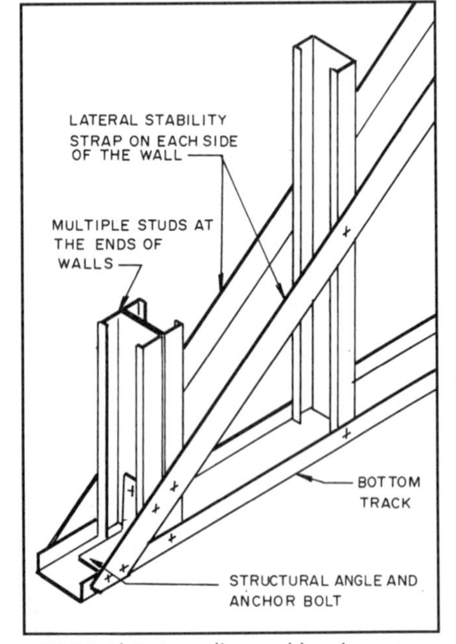

38-34 Flat-strap diagonal bracing provides lateral bracing of the wall structure and reduces racking. *(Courtesy American Iron and Steel Institute,* Low-Rise Residential Construction Details, RG-934, 1993*)*

38-35 Flat strapping is used to temporarily brace prefabricated panelized walls while they are moved to the site and raised into place *(Courtesy American Iron and Steel Institute,* Jordon Commons Cold-Formed Steel Training Manual, RG95-09, 1995*)*

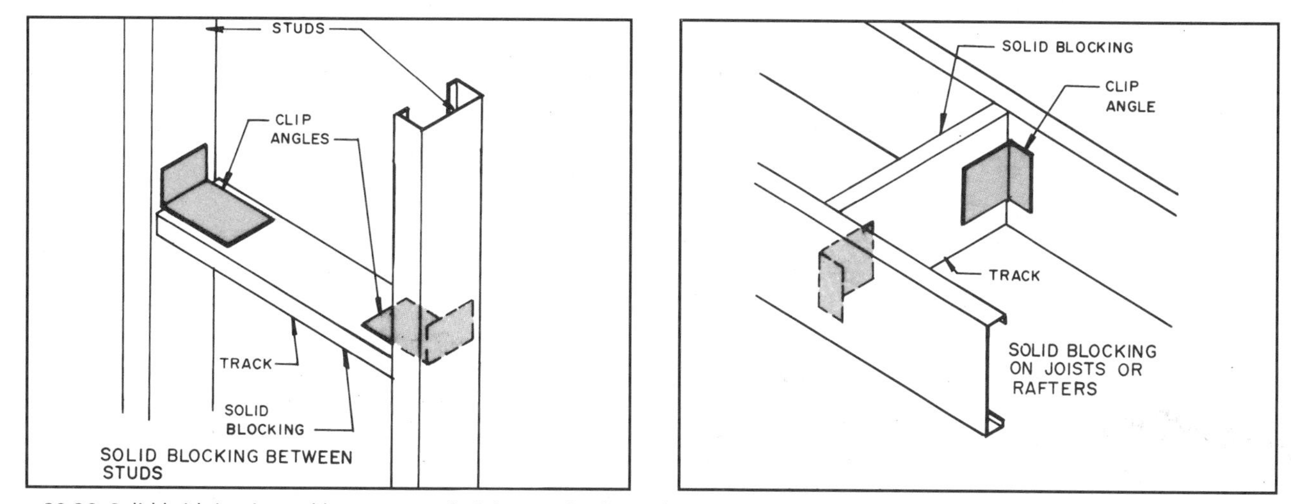

38-36 Solid bridging is used between studs, joists, and rafters. *(Courtesy American Iron and Steel Institute,* Low-Rise Residential Construction Details, RG-934, 1993*)*

38-37 Wide steel straps are on a diagonal across the studs, providing excellent shear bracing. *(Courtesy American Iron and Steel Institute,* Jordan Commons Cold-Formed Steel Training Manual, RG95-09, 1995*)*

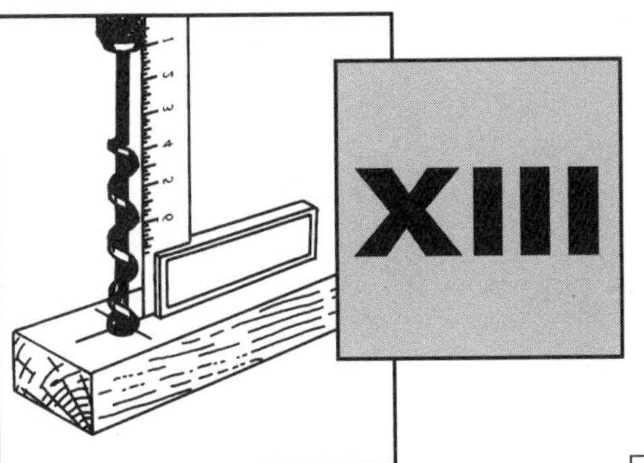

XIII TOOLS & MATERIALS

CHAPTER 39

Wood & Reconstituted Wood Products

Wood is an environmentally friendly material because it is a renewable resource. Large tracts of land are occupied by trees replanted after the mature timber was harvested. Other materials, such as aluminum, steel, and plastics, have a limited source of supply and require far more energy to produce than wood. Wood is also recyclable and biodegradable; therefore it does not contaminate the earth.

Table 39-1 Common species of softwoods and hardwoods.

Softwoods	
Alaska cedar	Western larch
Incense cedar	Jack pine
Port Orford cedar	Lodgepole pine
Eastern red cedar	Norway pine
Western red cedar	Ponderosa pine
Northern white cedar	Sugar pine
Southern white cedar	Idaho white pine
Cypress	Northern white pine
Balsam fir	Longleaf yellow pine
Douglas fir	Shortleaf pine
Noble fir	Eastern spruce
White fir	Engelmann spruce
Eastern hemlock	Sitka spruce
Mountain hemlock	Tamarack
West coast hemlock	Pacific yew
Western juniper	Redwood

Hardwoods	
Black walnut	Willow
Mahogany	American elm
Philippine mahogany	Sweet gum
Sugar maple	White ash
Yellow birch	American beech
Black cherry	Cottonwood
White oak	Rock elm
Red oak	Hickory
Yellow poplar	Bassword

WOOD IN CONSTRUCTION

The major use for wood in light frame construction is for structural framing and exposed, finished surfaces such as siding, trim, and flooring. Many factory-manufactured units, such as cabinets, doors, and windows, are of wood construction.

Wood is an excellent construction material because it is lightweight, strong, attractive, and a good insulator. Some woods are stronger than others and some can withstand exposure to the weather or soil or insects. The selection of the best species of wood for a particular application is important to producing a satisfactory end result.

SPECIES OF WOOD

While there are hundreds of species of trees, only a few are used extensively in construction. The species are divided into two major groups, softwoods and hardwoods, as shown in **Table 39-1**.

Softwoods have needle-like leaves and produce some form of cone. They retain the needles year round and so are green all the time. Typical examples are pine, fir, cedar, cypress, spruce, and redwood. **Hardwood** trees have broad fanlike leaves that drop off in the winter. Typical examples are walnut, oak, maple, poplar, and mahogany. The descriptors "hardwood" and "softwood" do not refer to the actual hardness, since some hardwoods are softer than some softwoods **(see 39-1)**.

Construction framing and finishing lumber is primarily from softwood trees **(see 39-2)**.

Hardwoods are used for cabinets, built-in furniture, floors, and sometimes interior trim that is left natural or stained to bring out the color and grain **(see 39-3)**.

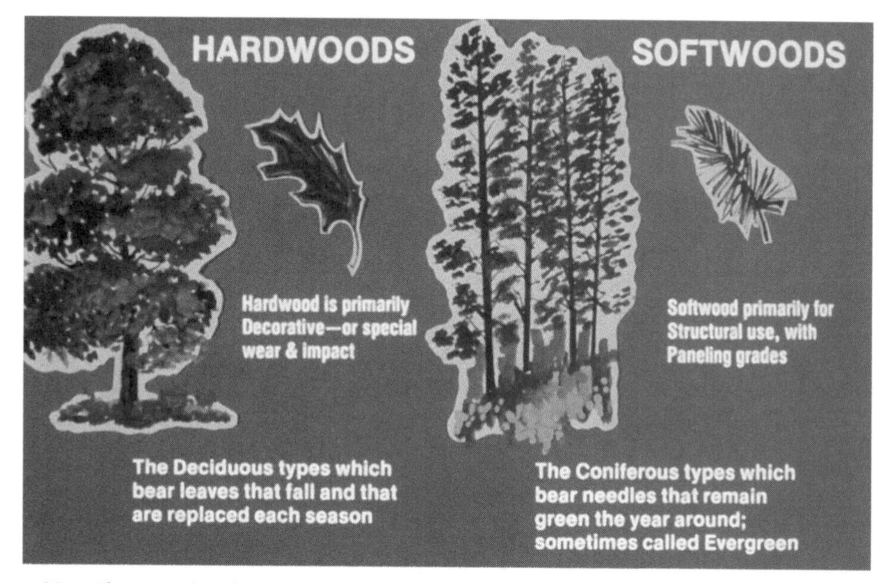

39-1 The two classifications of trees: deciduous and coniferous. *(Courtesy Fine Hardwood Veneer Association)*

39-2 This wall and roof framing used softwood lumber.

39-3 This cabinet was made from birch.

Some commonly used softwoods that are harvested in the United States include Douglas fir, white fir, southern yellow pine, white pine, Ponderosa pine, sugar pine, western larch, western red cedar, spruce, cypress, and redwood **(see 39-4)**. Some commonly used softwoods that are harvested in Canada include spruce, larch, hemlock, and red cedar.

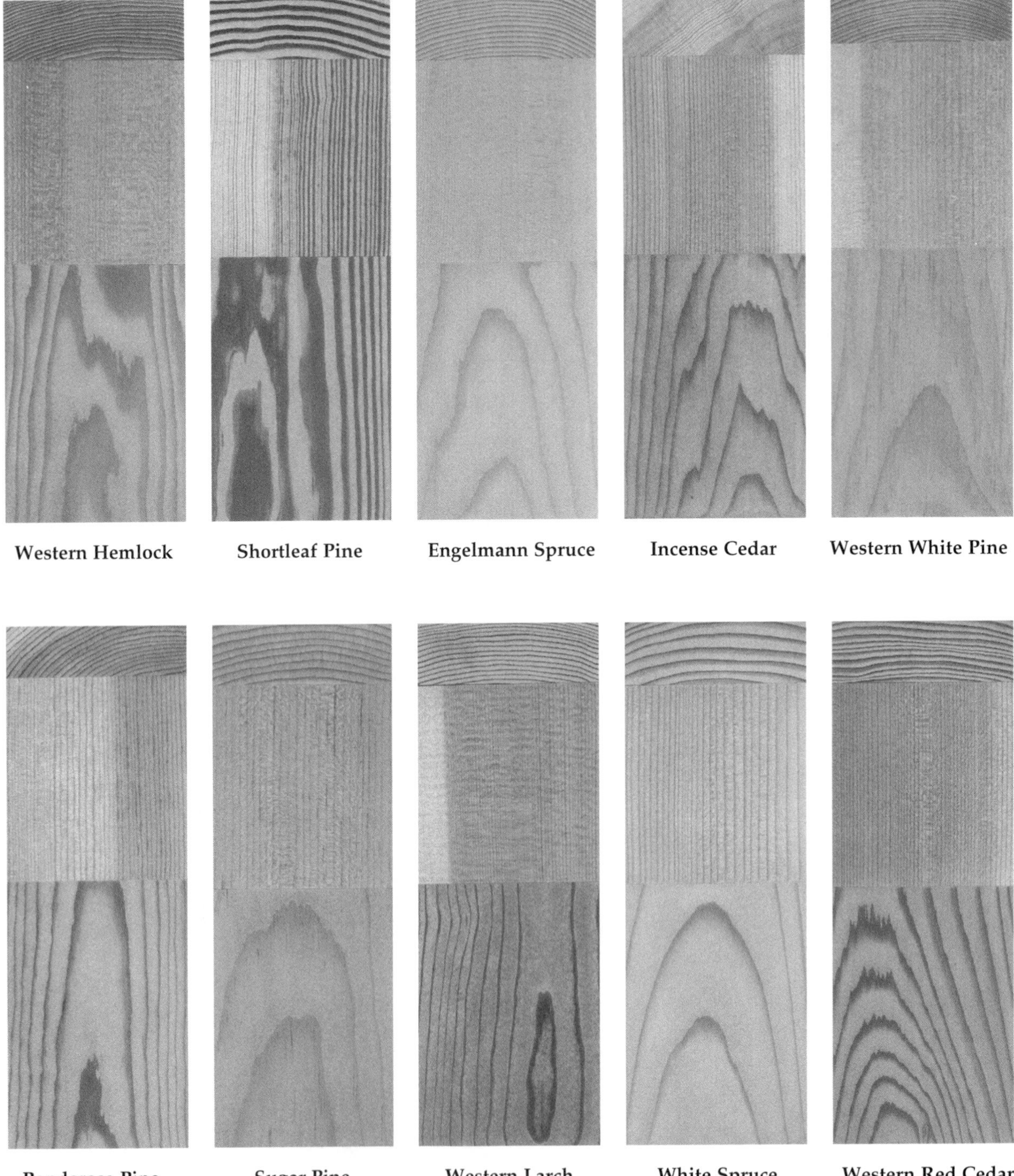

39-4 Commonly used softwoods: above and opposite. *(Courtesy U.S. Forest Products Laboratory, Forest Service, USDA)*

PROPERTIES OF WOOD

Wood has a number of desirable properties that make it excellent for light frame construction.

1. It is easily cut and shaped.
2. It is a good insulator. An inch of wood is 15 times as efficient an insulator as concrete, 400 times as efficient as steel, and 1770 times as efficient as aluminum.
3. It is a good electrical insulator.
4. It provides a barrier to sound transmission.
5. It can be glued to form large structural members and sheet products.
6. It will not rust or corrode.
7. It can be painted.
8. It can be joined with mechanical fasteners, such as nails, screws, staples, bolts, and metal truss plates.
9. It absorbs chemicals, enabling it to resist rot and fire.
10. It is lightweight and has a good strength-to-weight ratio.
11. It has some resistance to alkalis and acids.

Likewise wood has limitations that must be considered as material choices are being made.

1. It is readily combustible and special precautions must be taken if a potential fire danger exists. It can be impregnated with fire-retardant chemicals.
2. It is hygroscopic: It absorbs moisture and swells, and it dries, releasing moisture and shrinking. It must be protected from moisture.
3. When subjected to high moisture levels, wood is easily attacked by spores of wood-destroying fungi, causing it to decay. Most species of wood in this situation must be protected with preservative chemicals.
4. Wood is attacked by a variety of insects, such as termites, beetles, and borers. Proper construction practices and insect-controlling chemicals are necessary to prevent damage to most species of wood.
5. Wood is a naturally occurring material. Therefore, it is not of uniform quality or strength as are other materials such as steel. It will have defects such as knots, cracks, and checks that influence appearance and strength.
6. Many species of wood are used in light frame construction. Each has different physical and mechanical properties. The framer must consider the properties of the species being used so that the building is structurally sound.
7. If improperly installed, dried, or protected, wood will twist, cup, bow, and crook. Once this happens it cannot be restored to the original flat condition.

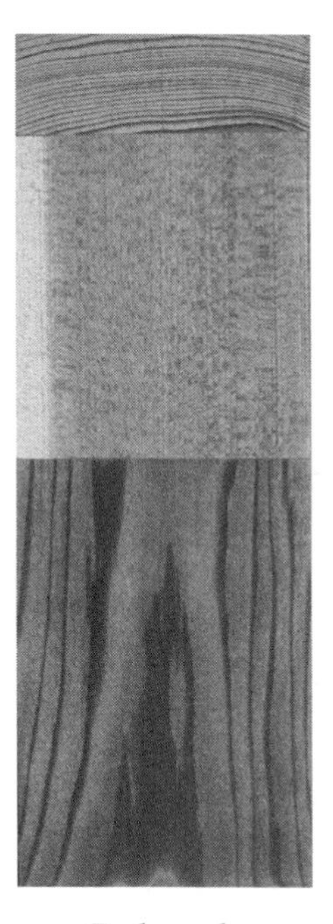

Redwood

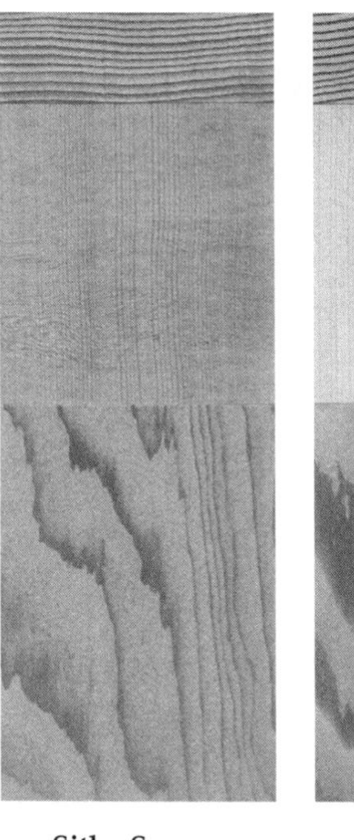

Sitka Spruce

Douglas Fir

39-5 The logs are moved into the mill and are debarked and washed. *(Courtesy Western Wood Products Association)*

MANUFACTURE OF LUMBER

After the logs have been harvested they are moved to a central area where they are stored and kept wet until they are to be cut into lumber. Then the logs are moved into the mill and onto the sawing deck, where they are debarked and washed **(see 39-5)**. Each log is then placed on the carriage of the headrig and moved across a large band saw that cuts them into boards, dimension lumber, and timbers **(see 39-6)**. They are then stacked and allowed to air-dry for a time before going into a kiln.

39-7 After they are dried, the rough-sawed boards are moved to the planing mill and dressed to finished sizes. *(Courtesy Western Wood Products Association)*

After the stock has been dried to the desired moisture content it goes to the planing mill, where it is surfaced on all four sides to standard dimensions **(see 39-7)**, and then moves to saws which cut it to the desired lengths **(see 39-8)**. Finally it is passed down a conveyor belt and past lumber graders, who note any defects and mark each piece according to standard lumber grades **(see 39-9)**.

39-6 The debarked logs are placed on the carriage and moved across a large band saw cutting them into boards and timbers. *(Courtesy Western Wood Products Association)*

39-8 The dressed boards are cut to length with a large circular saw. *(Courtesy Western Wood Products Association)*

CUTTING SOFTWOOD LUMBER

Lumber used for framing purposes is plain-sawed. If used for finished floors, it might be quarter-sawed **(see 39-10)**. Plain-sawed lumber is manufactured by first squaring a log and then cutting the boards tangent to the growth rings. Plain-sawed lumber is cheaper and produces wider boards than quarter-sawing. It does tend to warp and shrink more.

Quarter-sawed lumber is produced by cutting a log into quarters and cutting each of these into boards. The growth rings are cut so they form an angle of 45 to 90 degrees with the surface. This exposes the edges of the growth rings on the surface, and since they are very hard, they resist wear. This also reduces warping, shrinking, and twisting **(see 39-11)**.

BARK
PLAIN SAWED
QUARTER SAWED

39-10 This is how the log is cut to produce plain-sawed and quarter-sawed lumber.

SEASONING SAWED LUMBER

Before wood can be used for construction purposes the amount of moisture in it must be reduced. Often half the weight of a freshly sawed board is water. For general framing use, the lumber should have a moisture content of 12 to 15 percent. For furniture and cabinets it must be only 6 to 10 percent.

39-9 The finished boards move past a lumber grader who marks the grade on each piece. *(Courtesy Western Wood Products Association)*

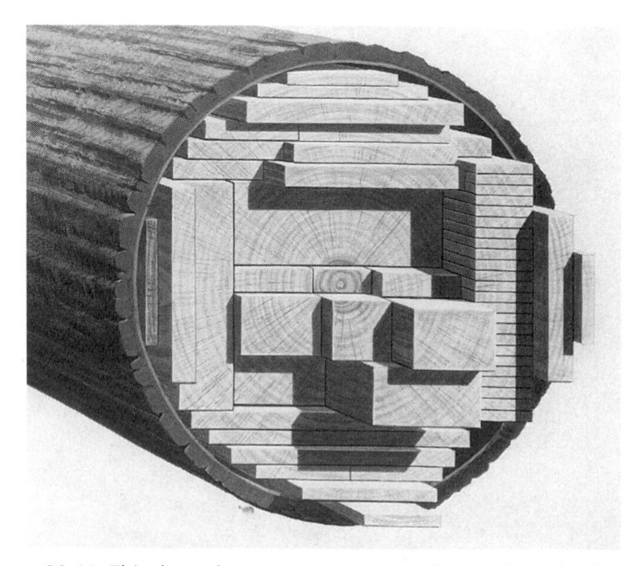

39-11 This shows how a sawyer can produce various-sized boards and timbers as the log is being sawed. *(Courtesy Western Wood Products Association)*

Lumber can be air dried or kiln dried. **Air-drying** is done by stacking the lumber in piles with wood sticks between the layers so that air can circulate between the layers. The lumber is then let to sit outdoors. It takes several months to lower the moisture content to 15 to 20 percent by air-drying. It is difficult to control moisture content by air-drying. Therefore, quality lumber is usually kiln-dried.

Kiln-dried lumber is stacked on carts with sticks between the layers. It is then run into long ovenlike buildings called kilns **(see 39-12)**. The moisture, temperature, and airflow are controlled. Steam is pumped in, saturating the wood. Then the steam is reduced, and heated air is passed over the wood, removing the moisture. Softwoods for construction purposes can be kiln dried to the required moisture content in three to five days. A wood preservative is applied to the ends to prevent splitting. Kiln-dried lumber should be covered with plastic sheets to protect it from rain.

As wood is seasoned, some surface checks, cracks, and splits occur. Kiln-drying reduces these defects to a minimum. It is important to remember that the moisture content of wood will vary with the environment in which it is used. Wood with 15 percent moisture content used in an area with a natural eight percent moisture content, as in a dry western U.S. state, will continue to lose moisture after it has been nailed in place. This causes the wood to shrink, which will produce, for instance, cracks in gypsum and plaster walls. Likewise, wood in a moist climate will swell, causing doors and windows to stick. It is important that the moisture content of the wood be in balance with the environment.

CHECKING MOISTURE CONTENT

The moisture content of wood can be checked on the job with a moisture meter. One type has two needles that are pressed into the wood. It measures the electrical resistance of the current flow through the wood between the two needles **(see 39-13)**. Another types uses metal plates that are pressed against the surface of the wood. It measures the relationship between the moisture content sensed and a fixed setting.

Moisture Content Designations

Most construction lumber is kiln dried to 15 to 19 percent. Generally the moisture content is included in the information found on the grade stamp placed

39-12 Lumber is dried in kilns using steam and heat. *(Courtesy Western Wood Products Association)*

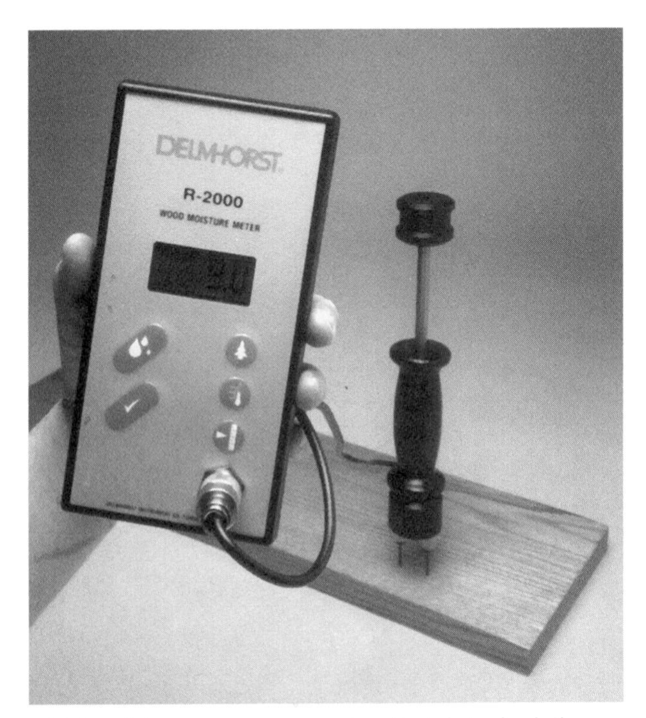

39-13 Moisture meters are used on the job to check the moisture content of lumber. *(Courtesy Delmhorst Instrument Company)*

on each piece. Typically designations include MC-15 (moisture content 15 percent), KD-15 or KD-19 (kiln dried to 15 percent or 19 percent), S-DRY (surfaced at 15 percent), and S-DRY-19 (surfaced at 19 percent). If the lumber is surfaced green (not dried) it will be stamped S-GRN. The moisture content will be somewhere above 19 percent. S-GRN lumber is surfaced larger than S-DRY so that when it dries and shrinks it will be the same size as the S-DRY lumber. It is not widely used.

The moisture content of wood is important because it influences the use of structural lumber and engineering designs, such as truss design. Wood becomes stronger and stiffer as the moisture content is **lowered**. Therefore, a member marked KD-15 is stronger than the same number at KD-19. Fasteners placed in wood at a high moisture content tend to loosen as the moisture content is lowered. This influences the strength of the structure.

DEFECTS IN LUMBER

Defects in lumber reduce its strength and affect its appearance. Defects are considered as lumber is graded. The defects are described in rules used to grade lumber. Some frequently listed defects and descriptions follow:

Bow is curvature flatwise from a straight line from one end of the board to the other.

Checks are separations of the grain fibers running lengthwise. The cracks go into the board but not through it **(see 39-14)**.

Crook is curvature edgewise from a straight line from one end of the board to the other.

Cup is curvature flatwise from a straight line across the width of the piece.

Dry rot is a condition in which the wood decays and crumbles into small granular particles. It is caused by a fungus that will live in wood that is wet a great deal of the time, such rafters wetted by a leaking roof.

Knots are growth defects caused by branches. The branches embedded in the trunk are cut when the boards are sawed. When grading lumber, the size, location, and firmness of the knots are considered.

Pitch is a heavy concentration of resin in wood cells in a single location.

Pitch pockets occur between growth rings; they are open cavities that are filled with pitch.

Shakes are lengthwise separations of the grain that occur between or through the annual rings.

Splits are separations of the grain fibers running lengthwise; these separations go completely through the board **(see 39-14)**.

Stains are discolorations of the board which damage the natural color but do not affect the strength.

Wane is the presence of bark or a lack of wood on the edge of a manufactured board. It causes the edge to be rounded **(see 39-15)**.

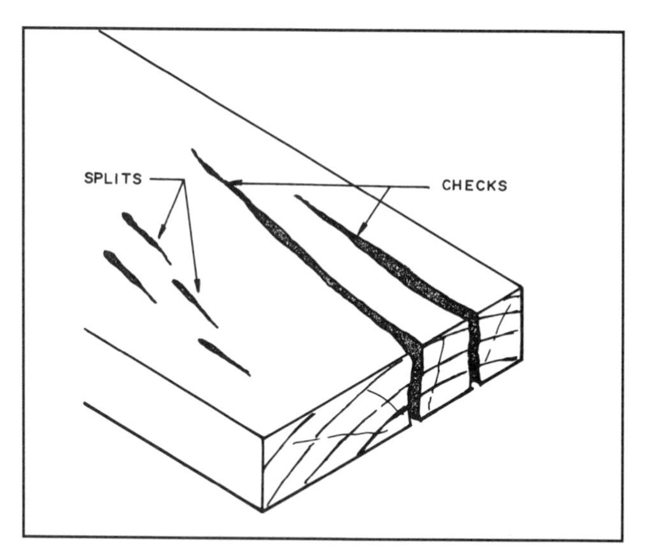

39-14 Splits and checks are frequently found defects.

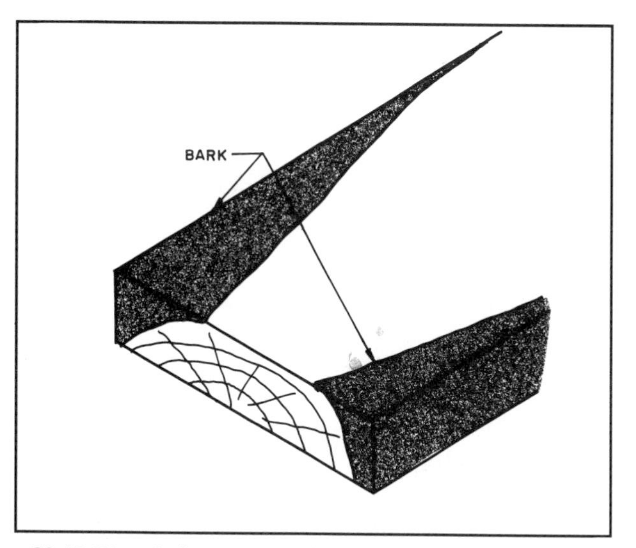

39-15 Wane is the presence of bark on the edges of lumber.

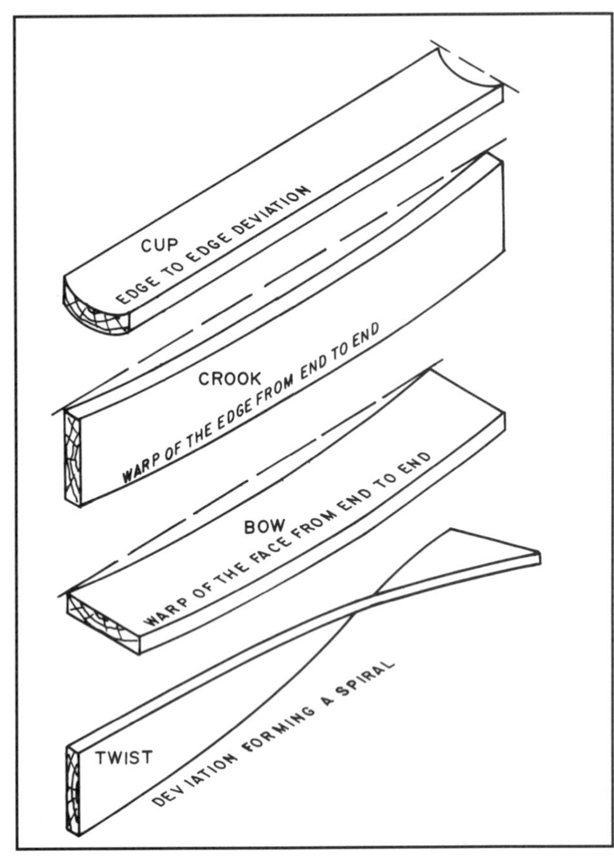

39-16 Wood can warp in various ways.

Warp describes variations from a straight, true plane surface. It includes bow, crook, cup, twist, and combinations of these **(see 39-16)**.

TREATED WOOD

Wood can be treated to make it resistant to decay, insect attack, and fire. Treated wood is protected from decay when exposed to the weather, placed in the ground, or used in places with high humidity and in areas where termites are common.

The wood is treated under pressure and controlled conditions in a pressure tank with chemicals which penetrate into the structure of the wood. Since wood is hydroscopic it readily accepts the liquid chemical solution. Wood is treated following the recommendations of the American Wood Preservers Association. They recommend specific concentrations of the chemical for wood to be used above the ground and in the ground. Do not let stock treated for above-ground use come in contact with the ground.

While treated wood tends to resist attack by termites, the most effective action you can take is to have the soil around the building treated by a licensed exterminator.

SOFTWOOD LUMBER GRADING

Grading specifications for lumber manufactured in the United States are recorded in the National Grading Rule for Softwood Dimension Lumber: Product Standard 20-70 (PS 20-70). This standard provides a common classification for sizes, grades, terminology, and moisture content for softwood dimension lumber. It provides standard rules for graded lumber regardless of the physical properties. National and regional agencies involved with the grading of lumber manufactured in their area are listed in Appendix T.

Canadian lumber grading rules are formulated and maintained by the National Lumber Grades Authority (NLGA). The NLGA rule is approved and enforced by the Canadian Lumber Standards Accreditation Board and by the American Lumber Standards Board of Review. This approval provides acceptance under all Canadian and U.S. building codes. Membership in the NLGA includes all the lumber manufacturers' associations in Canada that have approved grading agencies, as well as independent grading agencies. Lumber manufactured according to NLGA rules meets the provision of the Canadian standard, CSA 0141, and the American standard, PS 20-70 (see listing in Appendix).

Lumber is graded by two methods: visual inspection and machine evaluation. Most wood used for general construction is visually graded. The grading agency, grade classification, and moisture content are indicated on the grade stamp placed on each piece of lumber **(see 39-17)**.

MACHINE-EVALUATED LUMBER

Machine-evaluated lumber is used where physical properties must be more accurately determined, such as in the manufacture of trusses, laminated beams and columns, and I-joists.

Machine-evaluated lumber is tested to ascertain the bending stress (F_b) and the modulus of elasticity (E), which is a measure of the stiffness. Engineers designing structural components such as laminated beams or trusses need this information so that they can select a satisfactory species

SPIB—Southern Pine Inspection Bureau, the grading agency	**KD 19, KD 15—Kiln-dried moisture percentage**
No. 1, No. 2, No. 3—The Lumber Grade	**S-DRY—Surfaced after kiln-drying**
	7—Number of the mill manufacturing the lumber

39-17 Grade marks used by the manufacturers of southern pine lumber. *(Courtesy Southern Pine Inspection Bureau)*

and proper sizes to carry the design loads over the established spans.

Machine stress rating gives a more precise grade designation than visual stress rating. The process involves passing the piece of lumber through a machine that applies a load and measures the deflection. A computer records the average and minimum elasticity values, which are stamped on the lumber as it leaves the machine. The piece is then visually graded for any defects that may reduce its strength, and a grade stamp is printed on the piece **(see 39-18)**. Detailed information on lumber grades is available from lumber manufacturers' associations.

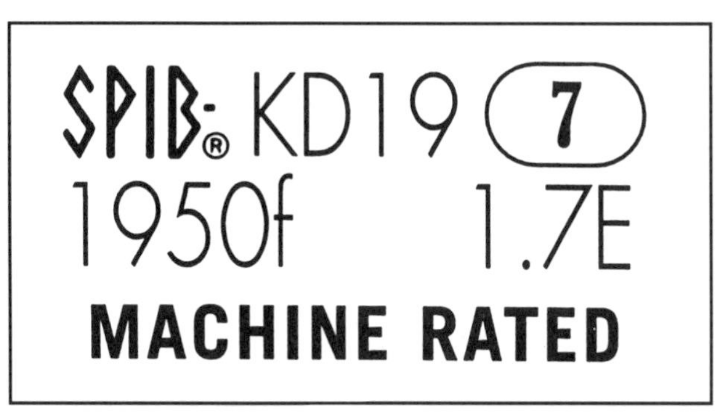

SPIB—Southern Pine Inspection Bureau, the grading agency
KD19—Kiln dried to 19% moisture
1950f—Extreme fiber stress in bending
1.7E—Modulus of elasticity in millions of pounds per square foot
MACHINE RATED—Machine tested for strength properties
7—Number of the mill manufacturing the lumber

39-18 Machine stress-rated lumber grade marks. *(Courtesy Southern Pine Inspection Bureau)*

USE CLASSIFICATIONS

Lumber is classified into five major areas related to its intended use. These are structural, industrial-and-shop, yard, select, and common lumber.

Structural lumber is two inches or more in thickness and width. It is machine stress rated for strength.

Industrial-and-shop lumber is used for the manufacture of various wood products such as doors and windows.

Yard lumber includes grades and sizes intended for construction and building purposes. It is classified as select and common.

Select lumber has good appearance and finishing qualities, and it is available in grades B & B, C, C & B, and D.

Common lumber is used for general construction and utility purposes. It is available in grades No. 1, No. 2, No. 3, and No. 4.

SIZE CLASSIFICATIONS

Lumber is classified into three groups by the size of the members. These are boards, dimension lumber, and timbers.

Boards are under two inches in nominal thickness and one-inch or more in width.

Dimension lumber is from two inches to, but not including, five inches thick and two inches or more in width, and it is identified as framing, joists, planks, rafters, studs, small timbers, etc.

Timbers are five inches or more in their least dimension, and they are identified as beams, stringers, posts, caps, girders, purling, sills, etc.

Dimension Lumber

Dimension lumber is used for framing members such as joists, studs, rafters, planks, and small timbers. It is classified into two categories by width: 2 × 4 inches wide, and five inches and wider. These are classified into four use categories: structural light framing, light framing, studs, and structural joists and planks **(Table 39-2)**.

Structural light framing grades are used for engineered applications where higher design values are needed. Light framing grades have good appearance but lower design values. Stud-grade lumber has design values making it suitable for stud uses including the construction of load-bearing walls. Structural joists and planks have qualities needed for engineered applications required of lumber five inches and wider.

Timbers

Timbers are wood members 5 × 5 inches and larger. There are two categories: stress rated and non-stress rated. Stress-rated timbers are used where high strength and stiffness values and good appearance are needed. Non-stress rated timbers have not been tested and permit a number of defects to be present. Timber grades are shown in **Table 39-3**. They are available in 10-, 16-, 18-, and 20-foot lengths.

Table 39-2 Dimension lumber classification.

Lumber	Grades
Structural light framing	Select Structural (Sel Str)
2" to 4" thick	No. 1
2" to 4" wide	No. 2
	No. 3
Light framing	Construction (Const)
2" to 4" thick	Standard (Stand)
2" to 4" wide	Utility (Util)
Studs	Stud
2" to 4" thick	
2" to 4" wide	
10'0" and shorter	
Structural joists and planks	Select Structural (Sel Str)
2" to 4" thick	No. 1
5" and wider	No. 2
	No. 3

Nominal Size

Boards and other lumber are typically designated by nominal size. Nominal means that the stated size approximately matches the size of the rough lumber before it is surfaced. The stated—nominal—size of the thickness and width of lumber differs from its actual size. Thus rough lumber 2 inches by 4 inches that is surfaced is actually closer to 1½ inches by 3½ inches, but it is still referred to as two-by-four lumber. In this book the nominal size "two-by-four" is written without units as 2 × 4. When a measurement is actual, the units are given. The same is true for one-by-six, four-by-four, etc. written as 1 × 6, 4 × 4, etc.

AMERICAN STANDARD LUMBER SIZES

The standard sizes of lumber used for light frame construction are shown in **Table 39-4** and **Table 39-5**. Other sizes, such as for siding, are available from lumber manufacturers' associations. The sizes shown are for lumber surfaced on four sides (S4S). However, for some applications it may be surfaced on one side (SIS), two edges (S2E), or some combination of sides and edges (SIS2E, S2S1E, or SIS1E). The nominal size, as stated above, matches the rough size to which the material is first sawed. Generally when speaking about lumber sizes the nominal size is

Table 39-3 Classification of timbers.

Timbers	Grades
Stress rated	Select Structural (Sel Str)
5" × 5" and larger	No. 1 SR (Stress Rated)
	No. 2 SR (Stress Rated)
Non-stress rated	Square Edges and Sound (SE & S) timbers
5" × 5" and larger	No. 1
	No. 2
	No. 3

referred to rather than the actual size. Surfaced lumber is smaller than the nominal size because material is removed in the planing operation. Notice that the pieces lose ½ inch in thickness and width as they are surfaced. This applies to lumber up to and including six-inch-wide stock. After this the width loses ¾ inch.

Green lumber is dressed slightly larger than dry lumber so that when it air dries it is the same size as dressed dry lumber. Green wood shrinks about 1⁄32 inch per inch of width across the grain and in thickness. Length is basically not changed by shrinkage.

Softwood lumber is manufactured in lengths from 6 to 24 feet in two-foot increments.

Table 39-4 Softwood lumber sizes for dressed dimensional and structural lumber.

Thickness[a] (inches)		
	Standard ALS Minimum Dressed	
Nominal	Dry	Green
2	1½	–
2½	2[b]	2⅟16
3	2½[b]	2⁹⁄16[c]
3½	3[b]	3⅟16
4	3½[b]	3⁹⁄16
5 and thicker	½ off nominal	½ off nominal

Width (inches)		
	Standard ALS Minimum Dressed	
Nominal	Dry	Green
2	1½	–
3	2½	2⁹⁄16[c]
4	3½	3⁹⁄16[c]
5	4½	4⅝[c]
6	5½	5⅝[c]
8	7¼	7½
10	9¼	9½
12	11¼	11½
14	13	13
16	15¼	15½
18	17¼	17½
20	19¼	19½
5 and wider	½ off nominal	½ off nominal

a A 2-inch dressed green thickness of 1⁹⁄16 applies to widths of 14 inches and over.
b Not required to be dry unless specified.
c These green widths apply to thicknesses of 3 and 4 inches only, except as provided in footnote **a**.

CANADIAN LUMBER

The species of Canadian lumber are about the same as those harvested in the United States. Canadian softwood species are combined into four main species groups: Spruce-Pine-Fir (S-P-F),

Table 39-5 Softwood lumber sizes for finish lumber and boards.

	Thickness (inches)		Width (inches)	
	Nominal	Dressed	Nominal	Dressed
Finish	⅜	5⁄16	2	1½
	½	7⁄16	3	2½
	⅝	9⁄16	4	3½
	¾	⅝	5	4½
	1	¾	6	5½
	1¼	1	7	6½
	1½	1¼	8	7¼
	1¾	1⅜	9	8¼
	2	1½	10	9¼
	2½	2	11	10¼
	3	2½	12	11¼
	3½	3	14	13¼
	4	3½	16	15¼
Boards	1	¾	2	1½
	1¼	1	3	2½
	1½	1¼	4	3½
			5	4½
			6	5½
			7	6½
			8	7¼
			9	8¼
			10	9¼
			11	10¼
			12	11¼
			over 12	off ¾

Douglas Fir-Larch (N), Hem-Fir (N), and Northern Species. The various species in each group are similar in strength and appearance. About three-fourths of the production of lumber is from the S-P-F group.

Grade stamps for the first three groups use the group name. Woods in the Northern Species group are identified by the actual name of the species, such as red pine. Grades of Canadian dimension lumber are identical to those in use in the United States and meet the requirements of the American Softwood Lumber Standards (ALS) PS 20-70 and are designated by Canadian lumber standards (CSA 0141). There are twelve Canadian lumber grading agencies. Each has its own grade stamp. All grade stamps contain the same information. The only difference is the identification of the grading agency.

SPAN DESIGN VALUES IN THE UNITED STATES & CANADA

New design values for lumber manufactured in the United States and Canada were published in 1991. These were developed over a period of years by a project entitled the In-Grade Testing Program. (New span values for southern pine were published in 1993, and are included in Appendices J, K, and L.)

The In-Grade Testing Program developed new design values for visually graded North American (U.S. and Canadian) softwood species. These design values are based on actual experience from tests with full-size lumber specimens. This research effort was conducted by the Southern Pine Inspection Bureau, the West Coast Lumber Inspection Bureau, the Western Wood Products Association, the Canadian Wood Council, and other rule-writing lumber associations. The U.S. Forest Products Laboratory conducted the research in cooperation with the various lumber associations. The new design values for dimension lumber are available in the *Supplement to the 1991 National Design Specifications® for Wood Construction (NDS®)*, which is available from the National Forest Products Association, 1250 Connecticut Ave., Washington, DC 20036.

The various lumber associations publish span tables for species available in their region. The NDS provides the basis for the design of wood structures adhering to United States building codes. The new data are reported in Southern Pine Inspection Bureau publications using a size-adjusted/repetitive-member-adjusted format and named by **empirical values**. The Western Wood Products Association, the Northeast Lumber Manufacturers Association, and the Canadian Wood Council present the new data as constants—for particular species combinations—that can be adjusted for the application to which the lumber is to be used. These new numbers are identified as **base values**. The base value is more accurate mathematically and fits into the design and engineering procedures as used for steel and concrete design.

The base values are the core of structural lumber safety, whereas design values are used for visually graded lumber safety. Design values for visually graded lumber are assigned to six basic properties of wood. The person ascertaining the performance of lumber can assign **conditions of use** to the base value to ascertain the strength of the member for the specified conditions. The six base values include (1) extreme fiber stress in bending (F_b), (2) tension parallel to grain (F_t), (3) horizontal shear (F_v), (4) compression parallel to grain ($F_{c//}$), (5) compression perpendicular to grain ($F_{c\perp}$), and (6) modulus of elasticity (E or MOE). An example of base values is shown in **Table 39-6**.

To find the final design value of the member, the base value must first be adjusted for size, and then adjusted for conditions of use.

There are seven conditions of use which are applied to the various base values to get the adjusted value. These adjustment factors include:

Size Factor (C_f)—applied to dimension lumber base values

Repetitive Member Factor (C_r)—applied to size-adjusted F_b (bending stress)

Duration of Load Adjustment (C_d)—applied to size-adjusted values

Horizontal Shear Adjustment (C_h)—applied to F_v (horizontal shear) values

Flat Use Factors (C_{fu})—applied to size-adjusted F_b (bending stress)

Adjustments for Compression Perpendicular to Grain (C_c)—applied to $F_{c\perp}$ (perpendicular compression) values

Wet Use Factors (C_m)—applied to size-adjusted values

The value derived after applying the conditions of use is the **adjusted value**. This is the actual design value for the piece of lumber under the prescribed conditions. Adjustment factors and detailed instructions may be obtained from lumber associations such as the Western Wood Products Association or the National Forest Products Association.

Following is an example of how to use base values. Assume the lumber is DF-S SS 2 × 4 inches (Douglas Fir-South, Select Structural grade, nominal 2 × 4-inch member). The designer selects a base value—for instance, the extreme fiber stress in bending, F_b—to be considered. The base value for the extreme fiber stress in bending for this species and grade is found in the base value table to be 1300 psi. Multiply the base value by the **size factor** for 2 × 4-inch stock, which is found in the size factor table to be 1.5. Therefore, 1300 × 1.5 equals 1950 psi. If the member is to be used in repetitive loading applications, such as for a floor joist, multiply 1950 psi by the repetitive factor, which is found in the **repetitive factor** table to be 1.15. Therefore, 1950 × 1.15 equals 2242.50 psi, which is the **adjusted bending stress** required for the member.

Lumber associations publish manuals giving span data for members such as joists and rafters. The person using these data must read the section in the front that explains what these span data represent. This section tells what factors have been applied to the base values to get the span. For example, typically joists and rafter data will be adjusted for bending values, repetitive loads, duration of load, and deflection. Procedures are given in the manuals for calculating other factors such as shear, tension, and compression.

When the In-Grade Testing Program was designed, the various softwood species used for structural dimension lumber were grouped to assist in the specification of their properties. Canadian softwood species were combined into four species groups: Spruce-Pine-Fir (SPF), Douglas Fir-Larch (North), Hem-Fir (North), and Northern Species (cedars, aspen, poplar, and several species of pine). The Western Wood Products Association grouped their species into six groupings. These groupings include Douglas Fir-Larch, Douglas Fir-South, Hem-Fir, Spruce-Pine-Fir (South), Western Cedars, and Western Woods (all of the above plus other species of pine, fir, and hemlock). The Southern Pine Inspection Bureau

Table 39-6 Base values for western dimensional lumber, sizes two inches to four inches thick by two inches and wider.*

Species or Group	Grade	Extreme Fiber Stress in Bending "F_b" Single	Tension Parallel to Grain "F_t"	Horizontal Shear "F_v"	Compression Perpendicular "$F_{c\perp}$"	Compression Parallel to Grain "$F_{c//}$"	Modulus of Elasticity "E"
Douglas Fir-Larch	Select Structural	1450	1000	95	625	1700	1,900,000
	No. 1 & Btr	1150	775	95	625	1500	1,800,000
	No. 1	1000	675	95	625	1450	1,700,000
	No. 2	875	575	95	625	1300	1,600,000
	No. 3	500	325	95	625	750	1,400,000
	Construction	1000	650	95	625	1600	1,500,000
	Standard	550	375	95	625	1350	1,400,000
	Utility	275	175	95	625	875	1,300,000
	Stud	675	450	95	625	825	1,400,000

* Use with appropriate adjustment factors. *(Courtesy Western Wood Products Association)*

has one group, Southern Pine. These groupings were used when the base values were developed. The base values are for all wood species in the specified group.

METRIC LUMBER SIZES

Lumber in the United States is still manufactured using feet and inches. In Canada some is manufactured in feet and inches because it is sold in the United States. Metric sizes used in Canada are shown in **Table 39-7** and **Table 39-8**. Metric lumber dressed green is two millimeters larger than the dressed-dry size to allow for shrinkage.

SPECIFYING LUMBER QUANTITIES

When lumber is ordered it is specified by giving the thickness, width, and length, and the number of pieces of this size. Lumber is priced by the board foot. A **board foot** is equal to 144 cubic inches of wood. For example, a piece 12 × 12 inches that is one inch thick contains 144 cubic inches **(see 39-19)**. Board feet are calculated by multiplying the nominal thickness in inches, width in inches, and length in feet and dividing this resultant by 12. Note the following example for finding the board feet for 10 pieces of 2 × 8 lumber 14 feet long:

No. of boards × T (inches) × W (inches) × L (feet)
divided by 12 = board feet

(10 × 2 × 8 × 14 = 2240)
divided by 12 = 186.7 board feet

Boards less than one inch thick are figured as one inch. If board feet are figured frequently, a table such as in Appendix R is used, or the data are part of a computer estimating program. The proportions shown in **Table 39-9** are useful to remember to get quick board-feet answers. These figures are for one-inch-thick stock. If thicker stock is to be figured, get the answer for one-inch stock and multiply by the thickness. Some framing squares have board-feet tables stamped on the blade. The figures are based on one-inch-thick stock.

The inch markings on the outer edge of the blade represent the width of the board in inches. The length is given in the vertical column under the figure "12" on the outer edge of the blade. The figures appearing where the two rows meet are board feet for a one-inch board. The number represents whole and fractional board feet **(see 39-20)**. For example, assume a board 10 inches wide and 14 feet long. Where these two rows meet is the figure 11.8. This means the board contains 11.8 or 11⅘ board feet.

VENEERED & RECONSTITUTED WOOD PRODUCTS

Wood product manufacturers produce a variety of products made using wood in various forms (as strands, veneers, chips, and flakes) to produce panels and related products. These are constantly being improved and new products appear on the market. The architect and the carpenter must be

Table 39-7 Metric dimension lumber.

Surfaced dry* (actual size in millimeters)	Surfaced dry* (equivalent size in inches)
38 × 89	1½ × 3½
38 × 140	1½ × 5½
38 × 184	1½ × 7¼
38 × 235	1½ × 9¼
38 × 286	1½ × 11¼
89 × 140	3½ × 5½
89 × 184	3½ × 7¼
89 × 235	3½ × 9¼
89 × 286	3½ × 11¼
89 × 337	3½ × 13¼
89 × 387	3½ × 15¼

* Lumber dressed green—add 2mm to dry thickness and width surfaced dimensions.

Table 39-8 Metric boards—thickness.*

Surfaced dry (actual size in millimeters)	Surfaced dry (equivalent size in inches)
17	43/64
19	¾

*Widths same as shown on Table 39-7 for dimension lumber.

Table 39-9 Quick estimate for the number of board feet in a piece of lumber.

Width (inches)	Thickness* (inches)	Board feet equals
3	1 or less	¼ of the length
4	1 or less	⅓ of the length
6	1 or less	½ of the length
9	1 or less	¾ of the length
12	1 or less	Same as the length
15	1 or less	1¼ of the length

*For thicker stock, find board feet or 1 inch of thickness and multiply by the thickness.

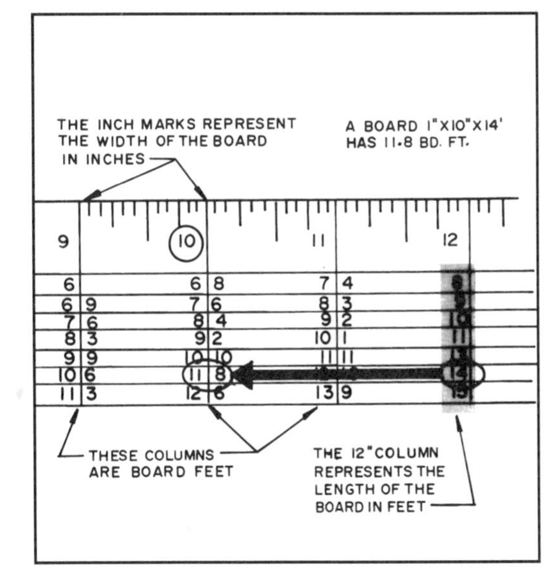

39-20 Some framing squares have board foot tables on the blade.

aware of the properties of these products so they are used for applications for which they will serve successfully.

These are divided into two major groups: veneered panels, and reconstituted panels, which are made from wood strands, chips, and flakes.

VENEERED STRUCTURAL PANELS

Veneered structural panels, better known as plywood, are made by bonding thin sheets of veneer, called plies, in layers forming a panel. The plies are always in odd numbers of layers, giving balanced construction. By alternating the direction of the grain in the layers, the grain runs perpendicular to the adjacent layer, producing a strong, stiff material that will not split like a solid wood board **(see 39-21)**. The following discussion refers to the standards of the American Plywood Association (APA)—The Engineered Wood Association.

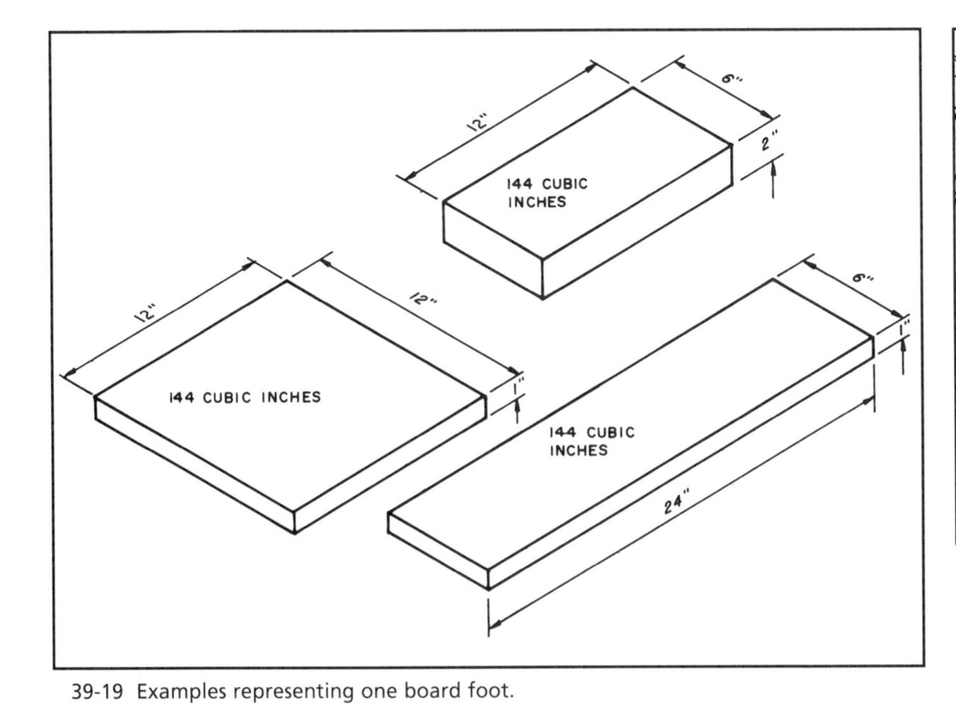

39-19 Examples representing one board foot.

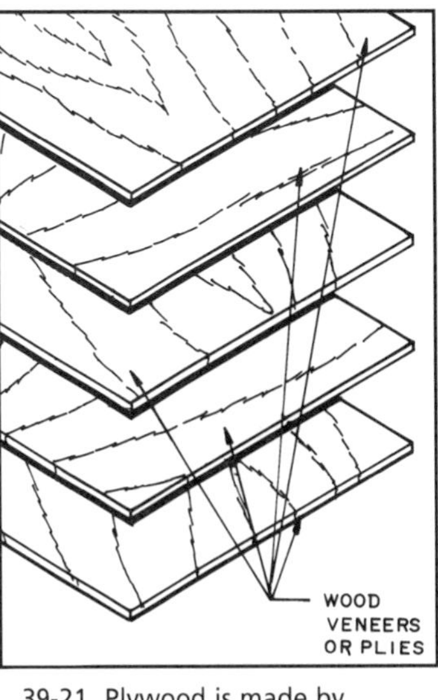

39-21 Plywood is made by bonding thin wood plies with the grain in each ply perpendicular to the plies on each side. There is always an odd number of plies in a panel.

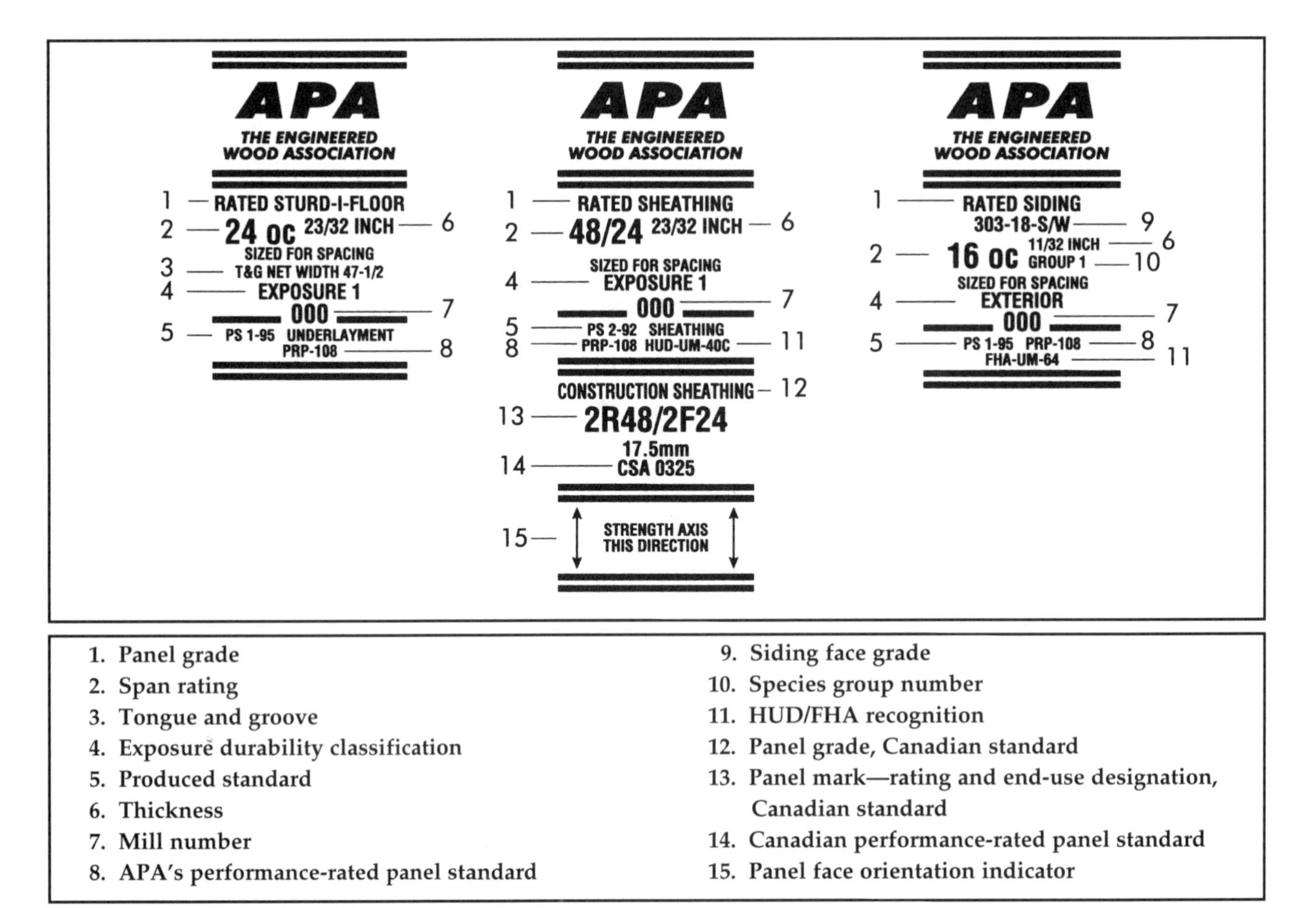

39-22 Examples of APA-grade stamps for sheathing, siding, and Sturd-I-Floor. *(Courtesy APA—The Engineered Wood Association)*

Grade	Description
A	Smooth, paintable. Not more than 18 neatly made repairs, boat, sled, or router type, and parallel to grain, permitted. Wood or synthetic repairs permitted. May be used for natural finish in less demanding applications.
B	Solid surface. Shims, sled or router repairs, and tight knots to one inch across grain permitted. Wood or synthetic repairs permitted. Some minor splits permitted.
C Plugged	Improved C veneer with splits limited to ⅛-inch width and knotholes or other open defects limited to ¼ × ½ inch. Admits some broken grain. Wood or synthetic repairs permitted.
C	Tight knots to 1½ inches. Knotholes to one inch across grain and some to 1½ inches if total width of knots and knotholes is within specified limits. Synthetic or wood repairs. Discoloration and sanding defects that do not impair strength permitted. Limited splits allowed. Stitching permitted.
D	Knots and knotholes to 2½-inches width across grain and ½ inch larger within specified limits. Limited splits allowed. Stitching permitted. Limited to Interior, Exposure 1, and Exposure 2 panels.

39-23 The quality of the face and back veneers is indicated by these veneer grades.

APA Trademarks

APA trademarks give considerable information about the properties of the panel. The trademarks for APA-Rated Structural panels are in **36-22**. Refer back to these as you read the following discussion.

Veneer Grades

Structural panel grades are identified in terms of the veneer grades used on the face and back of the panel, such as A-B grades. The grade designations are shown in **39-23**.

Exposure Durability

APA trademarked panels are produced in four exposure durability classifications: Exterior, Exposure 1, Exposure 2, and Interior. The exposure durability classification relates to the integrity of the glue bond.

Exterior panels have a fully waterproof bond and are used where subject to permanent exposure to the weather.

Exposure 1 panels have a fully waterproof bond and are designed to be used where they will be exposed to the weather during long construction delays.

Exposure 2 panels have an intermediate glue and are to be used for protected construction where there may be occasional high humidity or possible leaks.

Interior panels are intended for interior use only.

Group Number

Plywood is manufactured from over 70 species of wood. These have been divided on the basis of their strength and stiffness into five groups. Group 1 panels are made with the strongest species, Group 2 with the next strongest, and on to Group 5, which is made with the weakest species.

Span Ratings

APA-Rated Sheathing, Sturd-I-Floor, and Rated Siding have their span ratings indicated by numbers which indicate the maximum center-to-center spacing in inches of the supports over which the panels may be placed. The rating numbers for Rated Sheathing and Sturd-I-Floor apply when their long dimension is placed **across** the supports. The Rated Siding span rating applies when the panels are applied **vertically**.

Rated Sheathing spans are a two-number rating, such as 32/16. The first number (32), is when it is used as roof sheathing and the second (16) is for floor sheathing. Span ratings for Rated Sturd-I-Floor and Rated Siding are a single number, such as 16 or 24.

APA-RATED PRODUCTS

APA Performance-Rated products include APA-Rated Sheathing, APA Structural 1 Rated Sheathing, APA-Rated Sturd-I-Floor, and APA-Rated Siding.

APA-Rated Sheathing is used for wall and roof sheathing, subfloors, and other construction applications. It is available in the following thicknesses: 5/16, 3/8, 7/16, 15/32, 1/2, 19/32, 5/8, 23/32, and 3/4 inch. It has the following span ratings: 16/0, 20/0, 24/0, 24/16, 32/16, 40/20, and 48/24. The span rating indicates the maximum safe unsupported span when used as sheathing. The first number is for use as roof sheathing and the second number is the span when used as floor sheathing. It is available in square-edged panels. APA-Rated Sheathing is manufactured as plywood, oriented strand board, and Com-Ply®, and in Exposure 1, Exposure 2, and Exterior durability classifications.

APA Structural 1 Rated Sheathing is used where cross-panel strength and stiffness or shear properties are of maximum importance, such as in the construction of shear walls and panelized roofs. All plies are special improved grades, and panels marked PSI are limited to Group 1 species (the strongest species). They are made with span ratings of 20/0, 24/0, 24/16, 32/16, 40/20, and 48/24, and in Exterior and Exposure 1 durability classifications.

Structure 1 panels are manufactured as plywood, oriented strand board, and Com-Ply®. Typical allowable spans for selected wood panel products are given in **Table 39-10**. Always check the manufacturer's recommendations or the trademark for specific information.

APA-Rated Sturd-I-Floor is a structural panel used as a combination subfloor and underlayment over which carpet can be directly laid. It has high resistance to concentrated and impact loads. Usually an underlayment is placed over it if vinyl floor covering is to be laid. It is available as square-edged and tongue-and-groove-edged panels. Square-edged panels require that wood blocking be installed below unsupported joints between panels. It has span ratings of 16, 20, 24, 32, and 48 inches and durability classifications of Exterior, Exposure 1, and Exposure 2. Standard thicknesses are 19⁄32, 5⁄8, 23⁄32, 3⁄4, and 1 1⁄8 inches.

APA-Rated Siding is available in panel and lap siding. It includes a variety of products such as plywood, overlaid oriented strand board, and composite materials.

Panel siding can be applied directly to the studs or over sheathing. It is available in a variety of surface textures and patterns. Panel thicknesses vary with the particular pattern but range from 11⁄32 to 5⁄8 inch. Panel sizes are 48 × 96 inches, 48 × 108 inches, and 48 × 120 inches **(see 39-24)**.

Lap siding is available in rough- sawed or smooth overlaid surfaces with square or beveled edges. It is available in thicknesses of 11⁄32, 3⁄8, 15⁄32, 1⁄2, 19⁄32, and 5⁄8 inch, widths up to 12 inches, and lengths to 16 feet **(see 39-25).**

TREATED PLYWOOD

Plywood is an approved structural panel that can be pressure treated to resist rot and decay or treated with fire-retardant chemicals. Panels that are treated must have Exterior or Exposure 1 adhesive. Panels treated with waterborne preservative must be dried to a maximum moisture content of 18 percent after treatment. Those treated with fire-retardant chemicals must be dried to a moisture content of 15 percent after treatment.

FIRE-RETARDANT-TREATED WOOD & PLYWOOD

There are situations where the local building code requires the framer to use fire-retardant-treated wood (FRTW). Fire-retardant treatments differ for interior and exterior use.

There are two types of interior formulations in use, types A and B.

Interior Type B (AWPA-C20) is the older. (AWPA is the American Wood Preservers Association.) It is a formulation of mineral or organic water-soluble salts, such as zinc chloride, boric acid, ammonium sulfate, monobasic

Table 39-10 Typical allowable spans for structural panel products.*

Material	Wall sheathing (studs 16" O.C.)	Roof sheathing (framing 24" O.C.) edge support	Roof sheathing (framing 24" O.C.) no edge support	Subfloor (joists 16" O.C.)	Subfloor-underlayment (joists 24" O.C.)
plywood	5⁄16"	3⁄8"	1⁄2"	1⁄2"	3⁄4"
Com-Ply®	3⁄8"	3⁄8"	1⁄2"	1⁄2"	3⁄4"
waferboard	3⁄8"	7⁄16"	9⁄16"	5⁄8"	3⁄4"
oriented strand board	3⁄8"	3⁄8"	7⁄16"	7⁄16"	3⁄4"
particleboard	3⁄8"	3⁄8"	9⁄16"	5⁄8"	3⁄4"

*These are general spans. Consult with manufacturer's data for panels available in your locality.

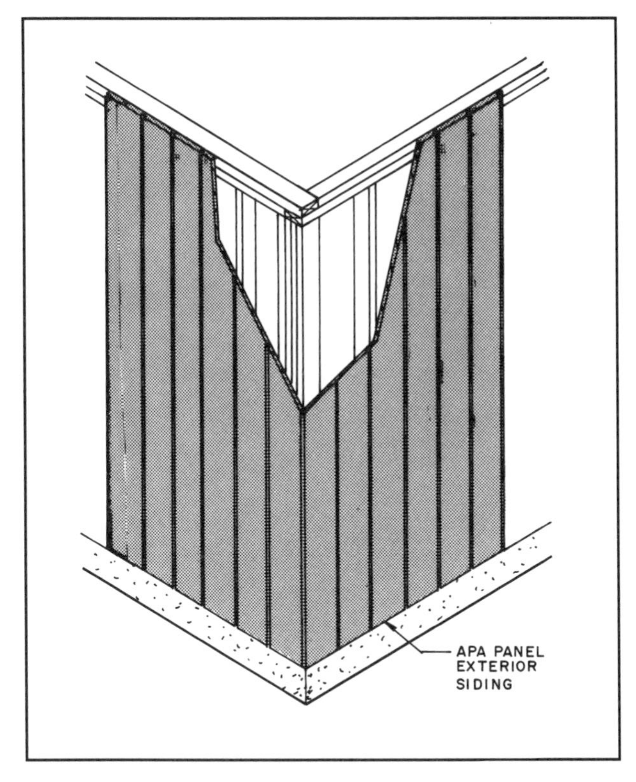

39-24 APA panel-type exterior siding can be applied over sheathing or directly to the studs.

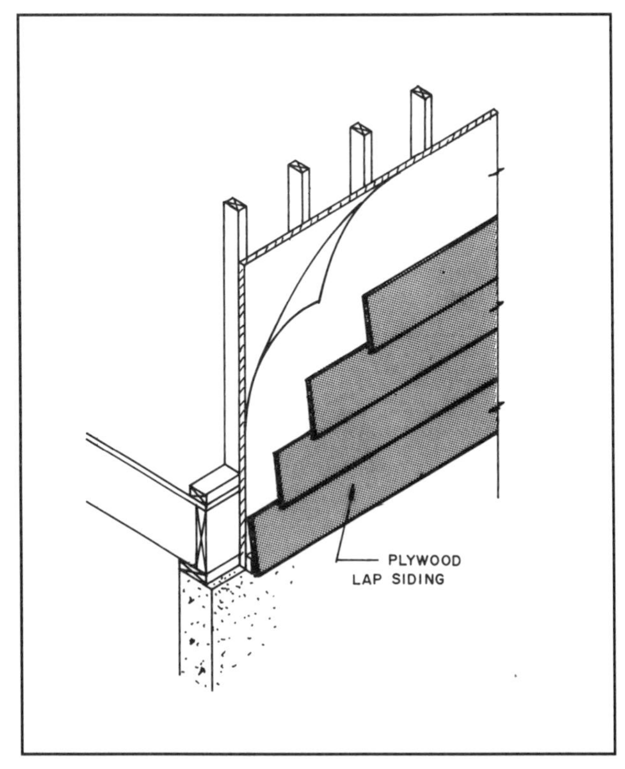

39-25 APA plywood lap siding.

ammonium phosphate, dibasic ammonium phosphate, and sodium tetraborate (borax).

Borax resists flaming but is not good at retarding glow. **Ammonium phosphates** will check flaming and glowing. **Boric acid** is not the best flame-resistant material but does check glowing. It is recommended that the manufacturer of the fire-retardant material be consulted before making a choice. Wood treated with one or more of these salts is more hygroscopic than untreated wood or than wood treated with an exterior-type retardant. When used in areas of high humidity, the moisture draws the salts in the wood to the surface, causing nails and other metal parts to corrode.

Interior Type A (AWPA-CA20), a newer interior formulation, is no more corrosive than untreated wood, but is not intended for exterior use. Treated wood must be redried to the required moisture content.

Exterior Types (EXT-FRTW) vary with the manufacturer but in general consist of water-soluble monomers or phenolic resins which, when polymerized, become insoluble in water.

The best way to apply fire-retardant chemicals is to use some form of pressure treatment. This provides a deeper penetration of the soluble salts into the wood. Some situations require a partial penetration while others require penetration completely through the wood. After the wood has been pressure treated it must be redried to the required moisture content and to set the chemicals.

Other Plywood Panels

There are many other plywood panels used for hundreds of applications. Contact APA—The Engineered Wood Association for details.

RECONSTITUTED WOOD PANELS

Reconstituted wood panels are made by bonding wood strands, chips, or flakes, forming panels in sizes the same as plywood. The properties vary and the manufacturer's recommendations must be observed. The bonding agent greatly influences where the panels can be used and their strength. Widely used panels include particleboard, waferboard, oriented strand board, and hardboard.

Particleboard

Particleboard is made by bonding wood chips using synthetic resin as a binder. It is widely used in the furniture and cabinetmaking industry and as a floor underlayment. Structural grades have been developed and accepted by some building codes as roof sheathing and subflooring. The grades and uses are shown in **Table 39-11**.

Table 39-11 General uses and grades of particleboard.

Use	Grade
Commercial	M-1, S
Industrial	M-2, 3
High-density industrial	H-1, 2, 3
Door core	LD-1, 2
Exterior construction	M-1, 2, 3 Exterior glue
Exterior industrial	M-1, 2,3 Exterior glue
High-density industrial	H-1, 2, 3 Exterior glue
Underlayment	PBU
Manufactured home decking	D-2, 3

(Courtesy Composite Panel Association)

Waferboard

Waferboard is made by bonding large wood flakes (called wafers) 1½ inches (39mm) and longer into panels having the same thicknesses and sizes as particleboard. The wafers are bonded with an exterior-type phenolic resin. The panels are used for subflooring and wall and roof sheathing. Waferboard is no longer widely used. It has been largely replaced with oriented strand board.

Oriented Strand Board (OSB)

Oriented strand board is a structural panel formed by bonding wood strands sliced from logs. The strands are mixed with wax and a waterproof exterior-type resin binder. As the panels are formed the strands are aligned in the direction of the long side of the panel.

OSB panels are widely used for roof and wall sheathing, subfloors, siding, soffits, and underlayment. The panels are marked with a trademark stamp by a certification board of the American Plywood Association (APA), PFS/TECO Quality Assurance Agency, Professional Services Industries, Inc. (PSI), or the Structural Board Association (SBA). The stamps give the same information shown earlier for plywood in **39-22**.

Hardboard

Hardboard is made from wood chips converted into fibers that are then bonded into panels under heat and pressure. It is made following the standards of the American Hardboard Association. It is made in panels from 1/12 to 1⅛ inch (2 to 28mm) and 4 × 8 foot (1220 × 2440mm) sheets. It is available in five classes as shown in **Table 39-12**.

WOOD PRESERVATIVES

Wood exposed to the soil, or exposed to or immersed in water, requires treatment by some type of wood preservative.

Creosote is an oil-borne preservative used to treat wood to be in contact with the soil, such as railroad ties, or in water, such as pilings for piers. The most effective type is coal-tar creosote, but oil-tar, water-tar, and water-gas-tar creosotes are available. Creosote gives the wood a black to dark brown appearance.

Pentachlorophenol is an oil-borne preservative that penetrates deeply into the wood and is long lasting. It is used on many products, including pilings for use in land or water, glued laminated beams, and utility poles.

Waterborne preservatives are odorless and paintable. They are infused by a pressure treatment process. Wood must be dried to the desired moisture content after treatment.

Table 39-12 Classifications of hardboard panels.

Class 1	Tempered
Class 2	Standard
Class 3	Service-tempered
Class 4	Service
Class 5	Industrialite

The types of inorganic arsenic waterborne preservative in common use include ammoniacal copper arsenate (ACA), chromated copper arsenate (CCA), and ammoniacal copper zinc arsenate (ACZA). These protect against insect attack and decay. They should never be used where they will come in contact with food or be burned. Those handling treated wood should wear gloves and wash their clothes separately from other clothes. Always wear a mask when cutting treated wood.

OTHER WOOD PRODUCTS

The number of products manufactured from wood is extensive and these products are described in the various chapters in this book. These range from structural members such as glulam beams, laminated-veneer lumber, and I-joists (introduced in Chapter 9) to finished products such as doors (Chapter 23), windows (Chapter 24), cabinets (Chapter 36), and moldings (Chapters 28, 29, and 30).

CHAPTER 40

FASTENERS

Considerable attention is given to the sizes and strengths of the various members making up the structural frame of a house. It is just as important to give careful consideration to the kinds and sizes of fasteners used to assemble these materials. These typically include nails, screws, bolts, metal connectors, specialty units, and bonding agents.

BUILDING CODES

Building codes have specific nailing and stapling requirements indicating the number and size of fasteners to be used. They also address the use of metal connections and bonding agents. The codes contain extensive tables giving very clear and detailed instructions. The carpenter should be thoroughly familiar with these requirements.

NAILS

There are many different types of nails produced for use in specific applications. The nail recommended for a particular use should be used to provide satisfactory results.

Nails are made from a variety of metals including steel, stainless steel, aluminum, and copper. Aluminum, copper, and stainless steel resist rust. Steel nails can be zinc coated to resist rust and improve holding properties. Cement-coated nails are used when increased holding power is needed. High-carbon steel nails are very hard and are used to nail into concrete and masonry. Nails used in structural framing have large-diameter heads, while those for finishing purposes have very small

40-1 Sizes and number of nails per pound for common nails.

Length	Penny	Gauge	Approx. Nails/lb
1	2	15	875
1¼	3	14	585
1½	4	12½	315
1¾	5	12½	270
2	6	11½	180
2¼	7	11½	160
2½	8	10¼	105
2¾	9	10¼	95
3	10	9	70
3¼	12	9	65
3½	16	8	50
4	20	6	30

COMMON NAIL (GENERAL CONSTRUCTION)

40-2 Sizes and number of nails per pound for box and casing nails.

Length	Penny	Gauge	Approx. Nails/lb
1	2	15½	1010
1¼	3	14½	635
1½	4	14	470
1¾	5	14	405
2	6	12½	235
2¼	7	12½	210
2½	8	11½	145
2¾	9	11½	130
3	10	10½	95
3¼	12	10½	86
3½	16	10	70
4	20	9	50

BOX NAIL (LIGHT CONSTRUCTION)

CASING NAIL (INTERIOR TRIM)

heads, which permits them to be set below the surface of the wood and allows the holes made to be hidden with a filler. Some nails have smooth shanks, while others have threaded, barbed, grooved, twisted, or ring shanks, which increase holding power.

SPECIFYING NAIL SIZES & QUANTITY

Nail sizes are indicated by the term "penny," which is represented by the symbol "d." For example, a 10d nail is a 10-penny nail. When the size is specified this also establishes the wire diameter used to make the nail. For example, a 10d common nail is made from No. 9 wire. The smaller the wire gauge number, the larger the wire diameter.

Nails are sold by the pound. A pound of 20d nails will contain fewer nails than a pound of 10d nails because they are larger and heavier. Nails that are coated, as with zinc, are heavier than uncoated (bright) nails so there will be a few fewer per pound.

The most frequently used nails are common, box, finish, and casing. These are shown in **40-1**, **40-2**, and **40-3**.

40-3 Sizes and number of nails per pound for finishing nails.

Length	Penny	Gauge	Approx. Nails/ lb
1	2	16½	135
1¼	3	15½	305
1½	4	15	545
1¾	5	15	500
2	6	13	310
2¼	7	12½	235
2½	8	12½	190
2¾	9	12½	170
3	10	11½	120
3¼	12	11½	110
3½	16	11	90
4	20	10	60

FINISHING NAIL (TRIM AND CABINETS)

NAILS FOR FRAMING

The common nail is most frequently used for structural framing. It has a large head and wire diameter. The box nail also has a large head, but it is thinner and the wire diameter is smaller than the common nail. It is used where the common nail might split the wood but where the strength of the smaller box nail is adequate for the application, such as when installing exterior wood siding.

Duplex nails (often called scaffold nails) are used in rough construction where they will eventually have to be removed. Scaffold construction is a typical example. They have a double head and are driven until the lower head hits the wood. The second head is used to pull the nail out **(see 40-4)**.

NAILS FOR FINISH APPLICATIONS

Finish and casing nails are used for securing interior trim, baseboards, and some types of cabinet work where the head is to be concealed. Casing nails have a larger head than finish nails, and they are used where greater holding power is needed. The wire brad is much like the finish nail but is available in very small diameters **(see 40-5)**. Its length is typically specified in inches.

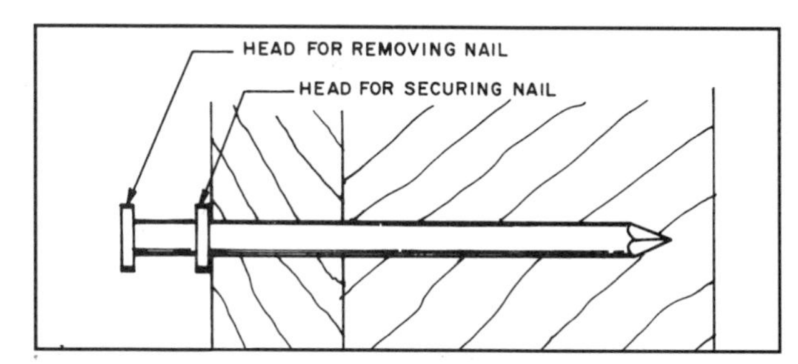

40-4 A duplex nail is used where it is installed only temporarily and will eventually have to be removed.

40-5 Sizes and gauges of wire brads.

Length (inches)	Gauge Number
½	20
¾	20
1	18
1¼	16
1½	14

WIRE BRADS (LIGHT MATERIAL)

There are many other types of nails used for specific applications **(see 40-6)**.

HOLDING POWER OF COMMON NAILS

The holding power of common nails is the withdrawal resistance (the force required to pull a nail straight out) when the nail is driven perpendicular to the grain of the wood. Holding power tests also measure the lateral resistance (a sideways pull to remove the nail). This varies for each size nail depending upon its diameter as well as the species and specific gravity of the wood. As the density and specific gravity increase, the holding power increases. For example, oak has a higher specific gravity than western cedar. Therefore it has a

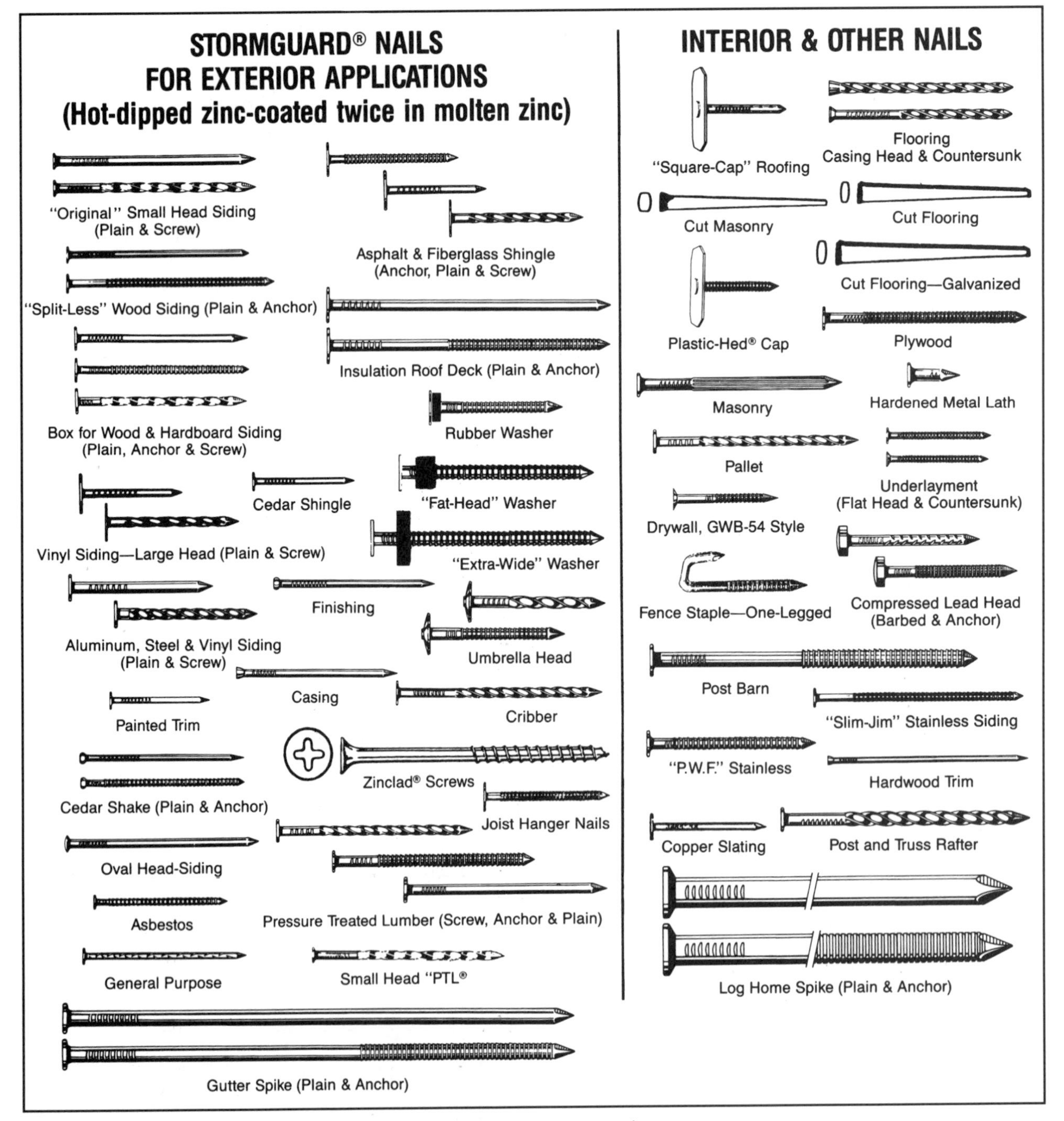

40-6 Nails used in light frame construction. *(Courtesy W.H. Maze Company)*

roughly four times greater holding power. The shank of the nail also influences holding power. Screw and annular ring shanks have greater holding power than smooth-shank nails **(see 40-7)**.

POWER STAPLERS & NAILERS

Power staplers and nailing devices are used considerably by framing carpenters **(see 40-8)**. These power nailing tools operate on compressed air and are available in a wide range of nailers and staplers.

Typical nailers use either coils or nested-head strips of nails **(see 40-9)**. A wide variety of nails can be power driven. These include small brads, finishing nails, trim and siding nails, sheathing and decking nails, roofing nails, concrete nails, and long heavy-duty spikes for timber construction. Nails are available with smooth, ring, and screw shanks.

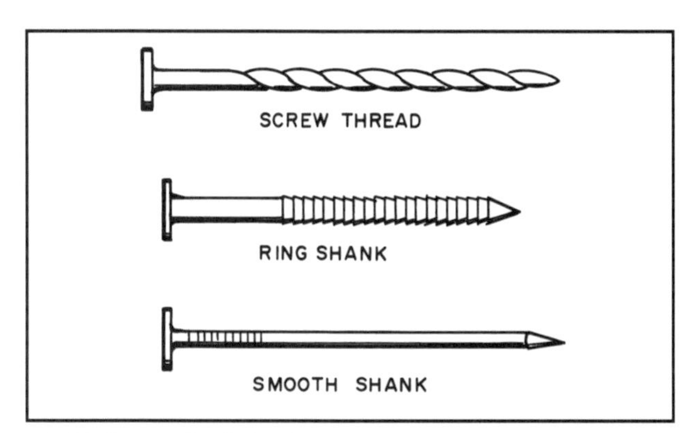

40-7 Types of nail shanks.

40-8 Power nailers are widely used by framing and finish carpenters. *(Courtesy Senco Fastening Systems)*

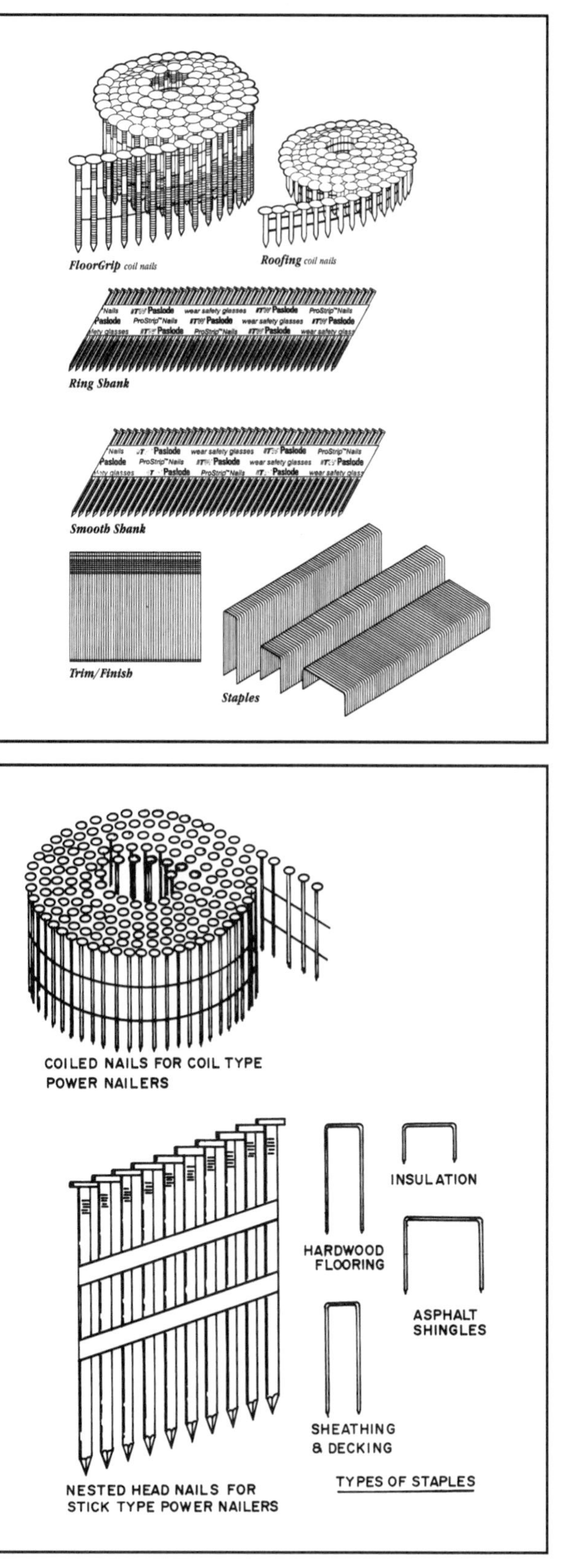

40-9 Power nailers use coil-and-stick-type nail cartridges. Power staplers can drive a wide variety of specialized staples. *(Courtesy Paslode, an Illinois Tool Company)*

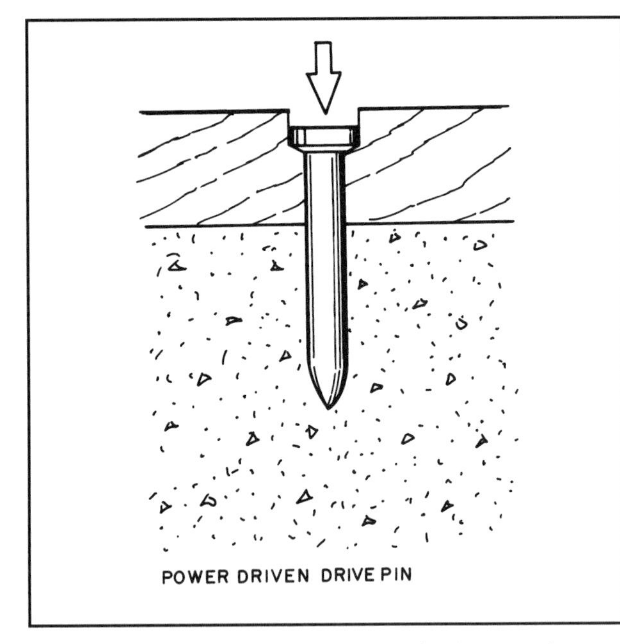

40-10 Power-driven drive pins are used to join wood, metal, and other materials to concrete.

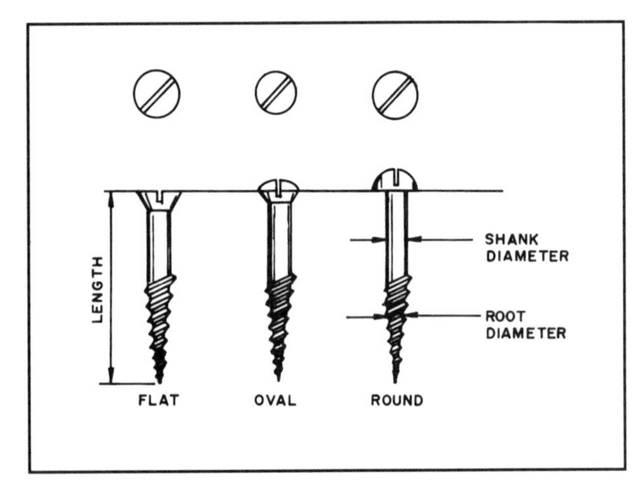

40-11 Common types of wood screws.

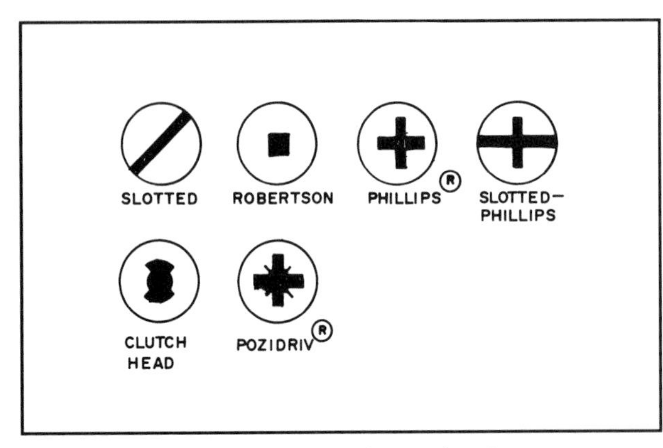

40-12 Types of recesses in wood screw heads.

Power staplers are sometimes used instead of nailers. They are especially useful when securing soft materials, such as foam insulation. Types of staples typically available include those for use on sheathing, decking, asphalt shingles, insulation, installing finish materials, and hardwood flooring.

There are power-driven tools using .22 caliber crimped loads that drive specially designed fasteners into concrete, steel, and masonry. Other pneumatic fasteners can join metal to metal, wood to metal, and gypsum board to metal **(see 40-10)**.

WOOD SCREWS

Wood screws are used for many applications in construction. Most commonly they are used in cabinet construction and for installing hardware. Gypsum wall panels are also frequently installed with screws. Screws have greater holding power than nails and are used where this extra strength is needed. They take longer to install than nails but can be easily removed if necessary.

The most common head types are flat, round, and oval. **Flathead screws** are set flush with the surface and are used to install hinges and hardware. The heads of **roundhead screws** remain above the surface. **Oval heads** are partially in the material and are used where an attractive exposed screw is required **(see 40-11)**. Screw heads are made in a variety of head recesses **(see 40-12)**. The slotted, Phillips, and slotted/Phillips recesses are most commonly found. The Phillips-type recess makes driving the screw easier and faster.

Table 40-1 Typical sizes of wood screws.

Length (inches)	Shank diameter (wire gauge number)
¼	2, 4
⅜	2, 3, 4, 5, 6
½ and ⅜	2, 3, 4, 5, 6, 7, 8
¾, ⅞, and 1	4, 5, 6, 7, 8, 9, 10, 11, 12
1¼	4, 5, 6, 7, 8, 9, 10, 11, 12, 14, 16
1½	6, 7, 8, 9, 10, 11, 12, 14, 16
1¾	6, 8, 9, 10, 12, 14, 16
2	8, 9, 10, 12, 14, 16
2¼	10, 12, 14
2½	8, 9, 10, 12, 14, 16
3	10, 12, 14, 16

The sizes of commonly used wood screws are given in **Table 40-1**. The length is the distance the screw will penetrate the wood, as shown in **40-11**. The diameter of the screw shank is indicated by the gauge number of the wire used to make it. Each length of screw is available in inches for each gauge number of the wire used to make it. Each length of screw is available in several gauge sizes. The approximate diameter in inches for each group is listed in **Table 40-2**.

A particleboard screw is designed for fastening into particleboard. It has widely spaced threads that are needed to set it into the particleboard **(see 40-13)**.

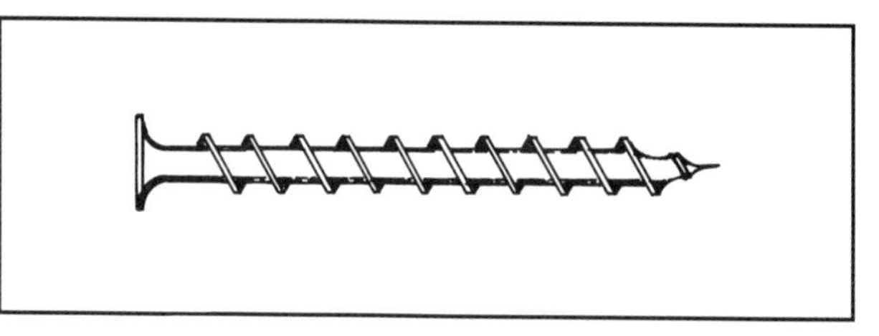

40-13 A special screw designed for use in particleboard.

BOLTS & HEAVY DUTY SCREWS

Bolts and screws are used to join heavy wood members such as those in heavy timber and pole building construction. The frequently used types are the lag screw, carriage bolt, and machine bolt **(see 40-14 and 40-15)**. Lag screws and machine bolts must have a large washer below the head to keep them from sinking into the wood. The nuts for the carriage and machine bolts must also have a washer under them. The square section of the carriage bolt penetrates the wood and the head rests on top of it.

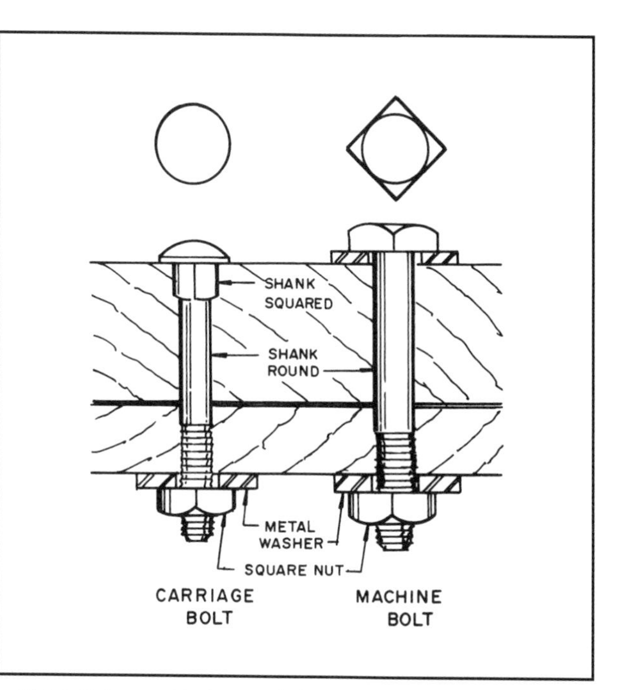

40-14 Bolts used in heavy timber and pole construction.

Table 40-2 Wood screw diameters in inches and by gauge number.

Wire gauge number	Decimal size	Fractional size
0	0.060	1/16
1	0.073	5/64
2	0.086	3/32
3	0.099	7/64
4	0.112	1/8
5	0.125	1/8
6	0.138	9/64
7	0.151	5/32
8	0.164	11/64
9	0.177	3/16
10	0.190	13/64
11	0.203	13/64
12	0.216	7/32
14	0.242	1/4
16	0.268	9/32
18	0.294	19/64
20	0.320	21/64

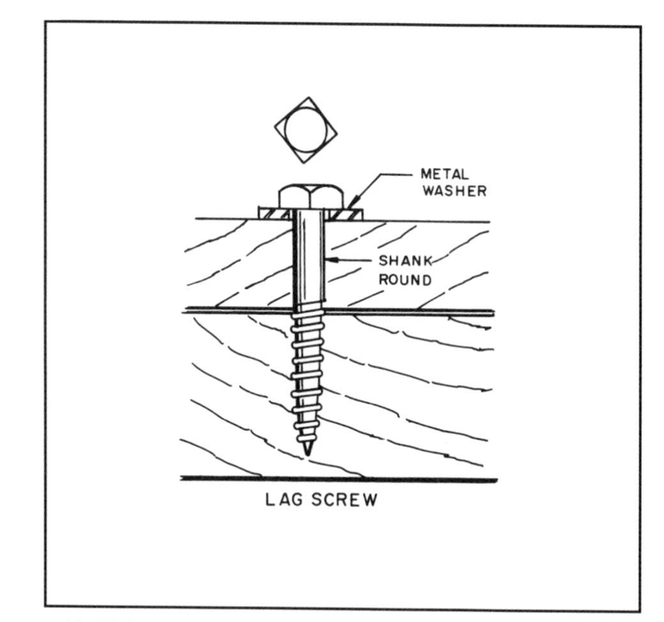

40-15 Lag screws are used in heavy timber and pole construction.

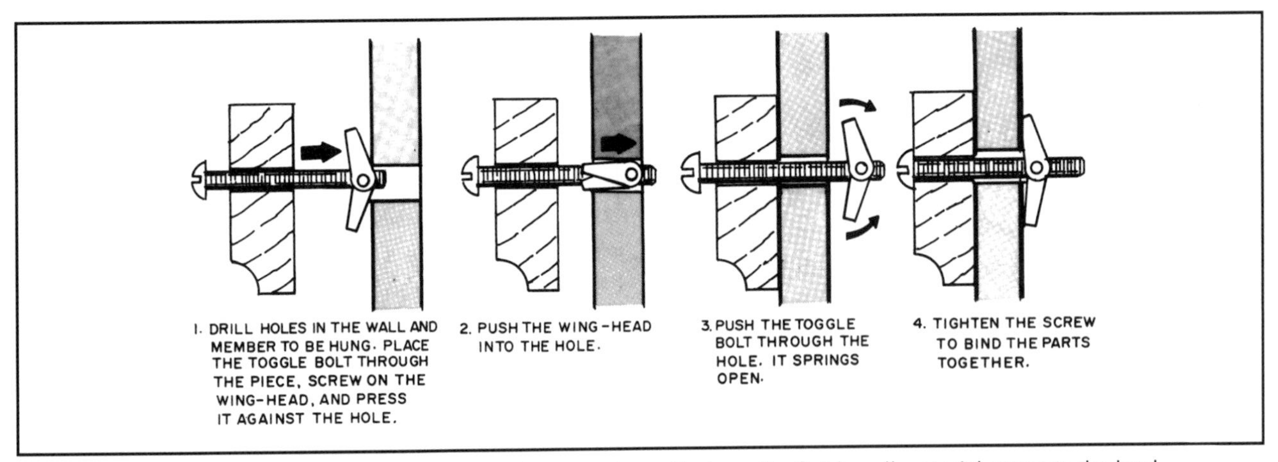

40-16 Toggle bolts are one device used to mount items on the wall where the finish wall material supports the load.

WALL FASTENERS

Typically there are things that need to be anchored to a wall. In some cases the wall is hollow and the anchoring device can be a toggle bolt expansion anchor or a small plastic anchor plug.

The toggle bolt has a spring-loaded head. A hole is drilled through the finished wall covering large enough for the head to pass through. When inside it springs open and the bolt through it is tightened as shown in **40-16.**

The hollow wall anchor is installed by drilling a hole through the finished wall and sliding the device into the opening. As the screw is tightened the metal shield flattens out against the inside of the wall **(see 40-17)**.

A plastic anchor is a small, tapered plug. A hole is drilled in the wall and the plug is inserted. When a screw is driven into the plug it expands and binds on the sides of the hole **(see 40-18)**. It is a light duty anchor and is used on drywall and masonry.

Each of these devices can be used to hang small cabinets, mirrors, towel bars, drapery hangers, and other lightweight items. The key to success is the strength of the wallboard into which they are inserted.

ANCHORING TO MASONRY & CONCRETE

There are a number of products available for mounting an item to a concrete or masonry wall. One type has a sleeve that expands as the bolt is tightened **(see 40-19)**. A hole is drilled in the concrete or masonry. As the screw is run into the sleeve it expands and binds to the sides of the hole.

Another type is installed in the hole and the drive screw is hit with a hammer, driving it into the expansion sleeve. This makes the sleeve expand and bind to the sides of the hole **(see 40-20)**.

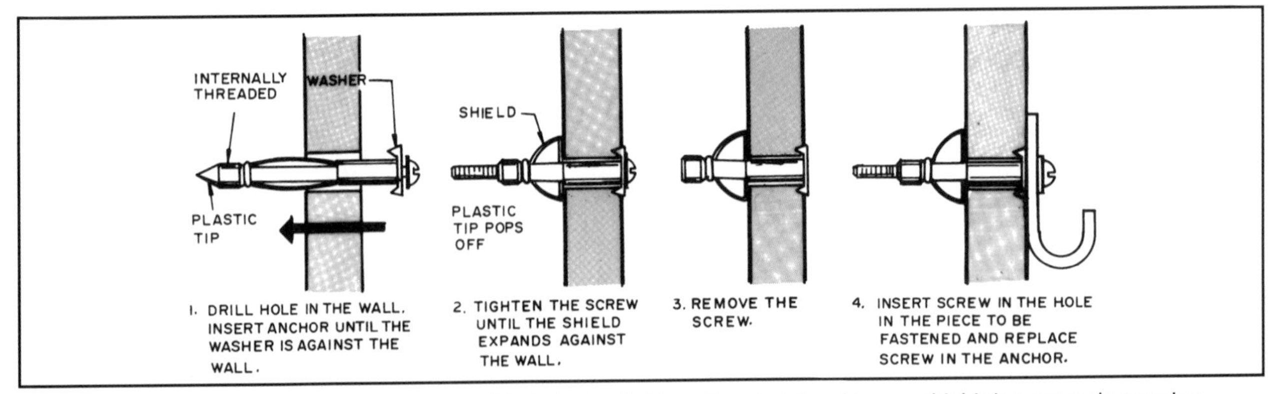

40-17 Hollow wall anchors slide through a small hole in the finish wall material and have a shield that expands, pressing against the wall as the screw is tightened.

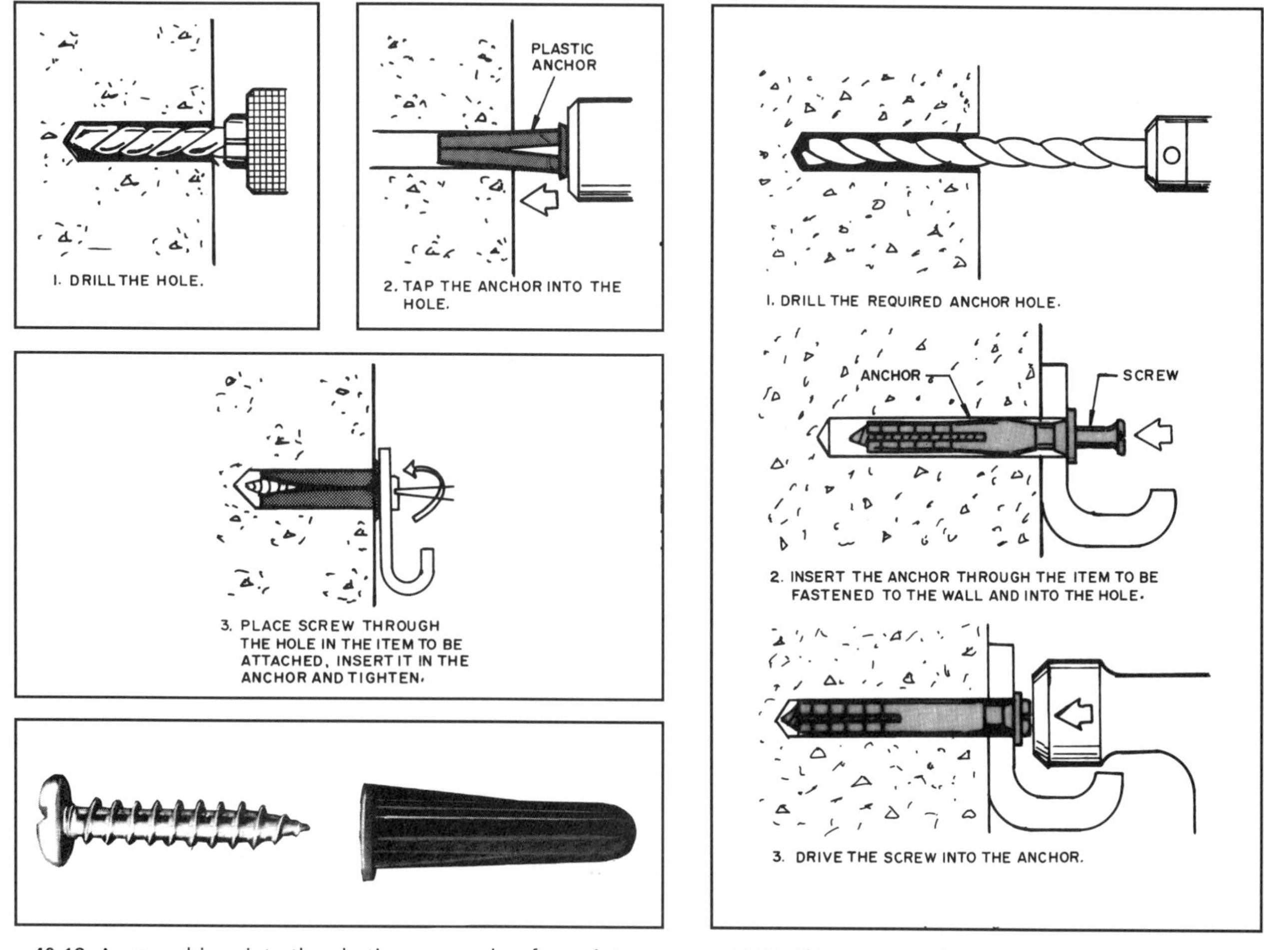

40-18 A screw driven into the plastic screw anchor forces it to bind against the sides of the hole in the masonry. *(Courtesy Hilti, Inc.)*

40-20 This impact anchor expands as the drive screw is driven into it. *(Courtesy Hilti, Inc.)*

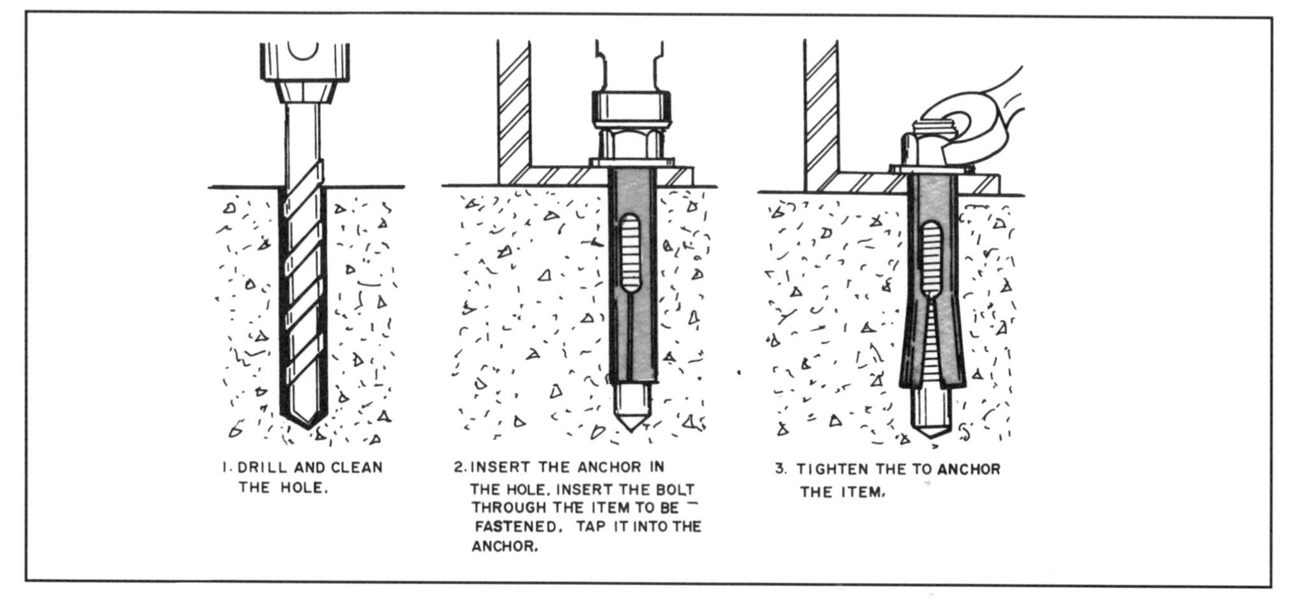

40-19 This masonry anchor has an expansion sleeve that is expanded as the bolt is screwed into it. *(Courtesy Hilti, Inc.)*

METAL CONNECTORS

A large number of metal connectors are used in the assembly of wood structural frames. Some of these are illustrated in other chapters of this book. Basically they range from simple flat plate and angle connectors **(see 40-21)** to a large number of connectors joining posts, beams, girders, joists, rafters, and other structural members. A few are shown in **40-22**.

BONDING AGENTS

Adhesives are finding increasing use in light frame construction. Uses include bonding the subfloor to the joists, laminating wood members together to form a larger member, installing drywall and paneling, and bonding rigid foam insulation.

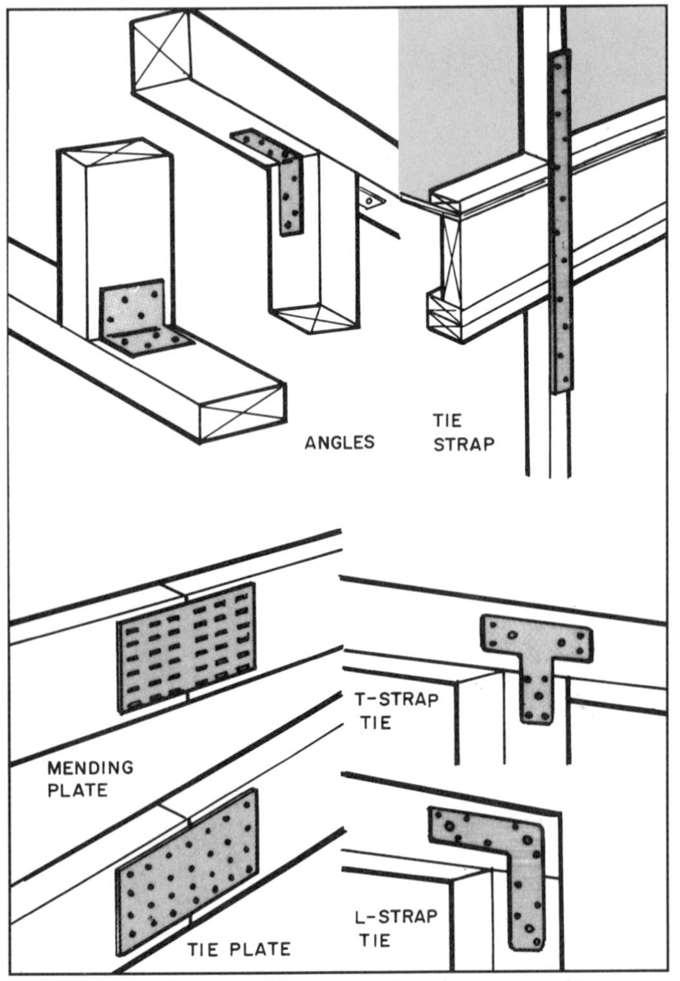

40-21 Some of the metal flat and angle connectors used to assemble wood structures.

There are hundreds of bonding agents on the market. You must study their specifications to be certain they are right for the job. They fall into three major classifications: adhesives, glues, and cements. Adhesives are made from synthetic materials. Glues are made from natural materials. Cements are rubber-based bonding agents.

ADHESIVES

Adhesives are made from synthetic materials and may be either thermosetting or thermoplastic. Thermosets produce a chemical reaction when mixed and are water and heat resistant.

THERMOSETS

Epoxy is used to bond nonporous materials such as glass, steel, and plastics.

Resorcinol-formaldehyde is used to produce a waterproof connection. It is effective when used on porous surfaces, such as wood.

Urethane-formaldehyde is sold as a powder and mixed on the job with water. It is water resistant but not waterproof. It has a long clamp time.

THERMOPLASTICS

Thermoplastic bonding agents are the most widely used for bonding wood. They are all moisture resistant but not waterproof.

Polyvinyl (also called white glue) is a good general-purpose bonding agent. Its strength is only moderate, but it serves many light-duty purposes.

Aliphatic agents (yellow carpenter's glue) is fast setting. It is stronger than and superior to white glue.

GLUES

Glues are made from vegetable and animal products, such as bones and hides. They are not widely used since the development of the many adhesives.

Casein glue is a powder made from dried milk curds. It is mixed with water to form a paste. It produces a waterproof joint.

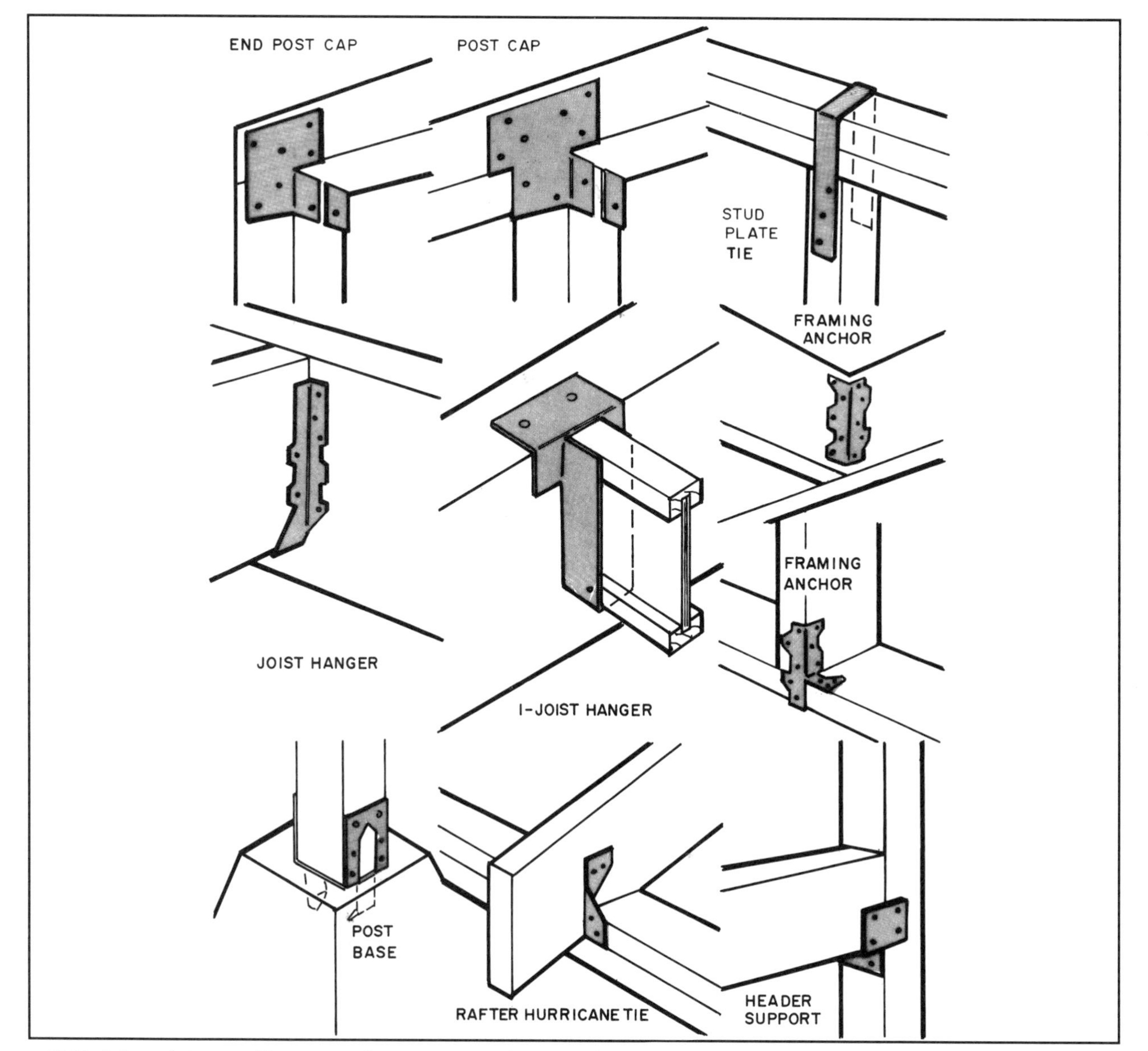

40-22 A few of the metal hangers and connectors used in wood frame construction.

CEMENTS

Cements are made from synthetic rubber. Some are highly flammable.

Contact cement is used to bond plastic laminate to counter tops. It bonds instantly and cannot be moved or adjusted once it is in place.

Mastic is a thick form of contact cement sold in tubes and applied with a caulking gun. It is used to join sheathing to studs and floor joists and paneling to walls. Mastic is also sold in several-gallon containers for applying ceramic tiles.

CHAPTER 41

THE CARPENTER'S TOOL BOX

A qualified carpenter will maintain an extensive set of hand tools. While power tools are used for many operations the fully qualified, skilled carpenter will be able to use all of the hand tools available for the trade. High-quality tools are essential for satisfactory work. While they cost more, they last longer, perform better, and reduce the likelihood of accidents and injury. These tools must be maintained (sharpened, cleaned, etc.) and stored in a manner to protect them from damage. A carefully constructed or quality commercial tool chest is important. There should be a place for each tool and each tool should be kept in its place.

The following discussion describes the most commonly used hand tools and appropriate working attire.

CLOTHING

Clothing should be comfortable and permit bending and other movement. Safety regulations require that items that could be caught in power tools be secured, such as long sleeves—roll them up—or scarves—remove them. For safety's sake **steel-toed** leather shoes are recommended. However, plastic athletic shoes, as used for jogging, are available with steel toes; these are more comfortable than leather work shoes. Some form of **knee pads** are useful when installing baseboards and flooring.

Safety equipment such as **eye** and **ear protection** is required. Some jobs require a **mask** to prevent dust from entering your lungs. This is especially important when sanding, cutting, or shaping stock.

The **tool belt** is used by many finish carpenters **(see 41-1)**. It may be made from leather or a synthetic material. The belt has a variety of large pouches and small pockets in which small tools are carried. Larger tools are hung on the outside. Some of the tools placed in it are those used most frequently such as pencils, rules, and tapes. Some finish carpenters carry only the tools they need for the job at hand rather than load themselves down with tools they will not likely need. There are work situations, such as in close quarters or where the tools might scratch a wall or cabinet, in which the belt is laid aside. Basically it is up to the craftsman to decide what to carry.

Various safety equipment should be part of the carpenter's personal belongings **(see 41-2)**. See Chapter 1 for more information.

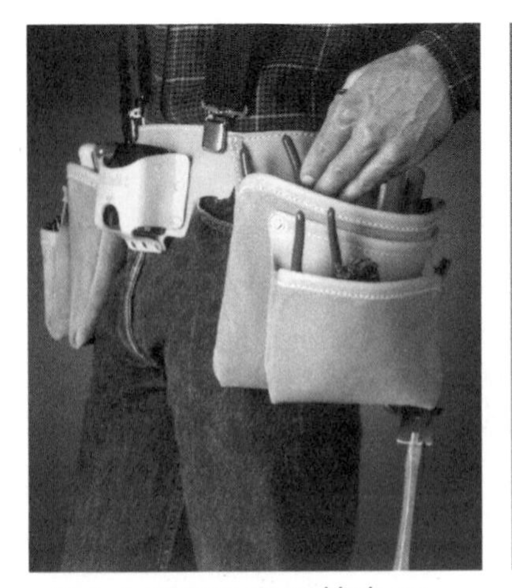

41-1 A carpenter's tool belt carries the tools needed for a specific job. *(Courtesy McGuire-Nicholas Manufacturing Co.)*

41-2 A hard hat and safety glasses similar to these should be in every carpenter's tool box. *(Courtesy Aearo Company)*

LAYOUT TOOLS

Finish carpenters do a lot of layout work. Commonly used layout tools are shown in **41-3**. The **folding rule** and **tapes** are used to measure linear distances. **Squares** are used for layout and checking for squareness. The **T-bevel** is used for laying out and checking angles.

Various marking tools are used. Most common is the **pencil**; however, a line scribed with a **knife** is more accurate **(see 41-4)**. Circles can be drawn with a **compass** or **trammel points**, and the exact locations of holes to be bored or drilled are located with an awl **(see 41-5)**.

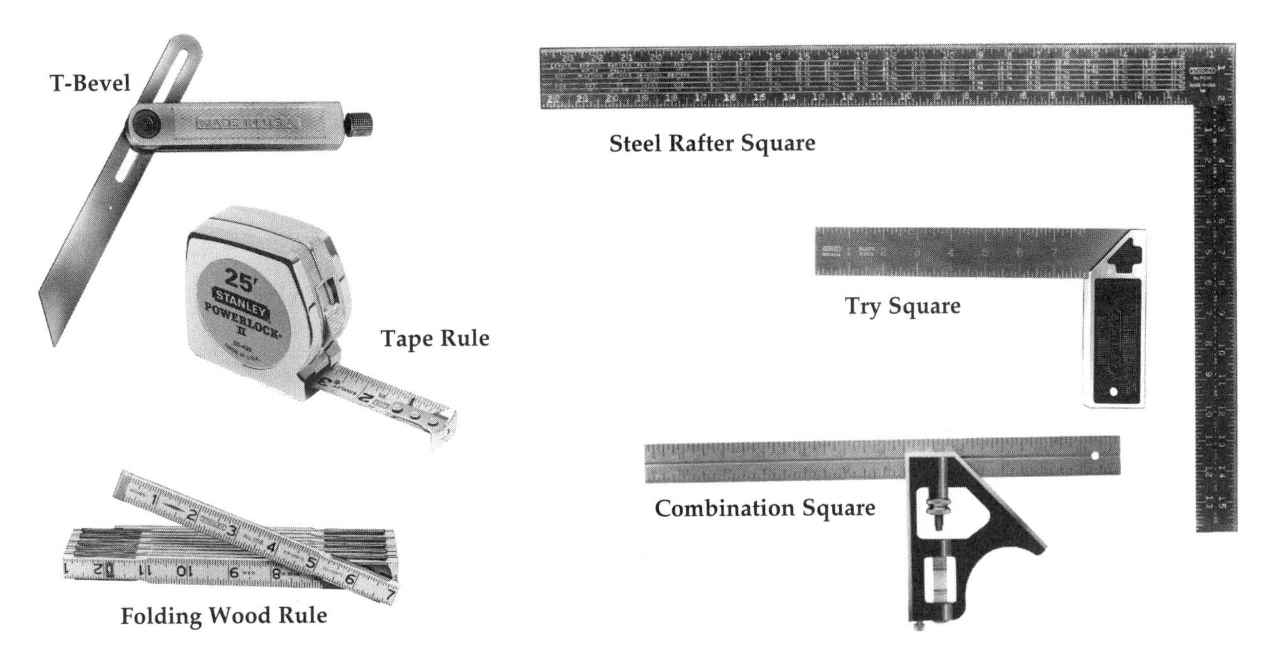

41-3 Commonly used layout tools. (© 1998 Stanley Tools)

LINEAR MEASUREMENT

Currently most buildings are still designed using feet and inches as the system of linear measurement. The rules and tapes are divided into units as shown in **41-6**. The United States government has indicated that federal buildings should be designed and built using metric units. This means that most of the measurements made by the carpenter will be either millimeters (mm) or centimeters (cm). A metric rule is shown in **41-6**. Notice that one centimeter is made up of 10 millimeters. Additional metric information is in Appendix S.

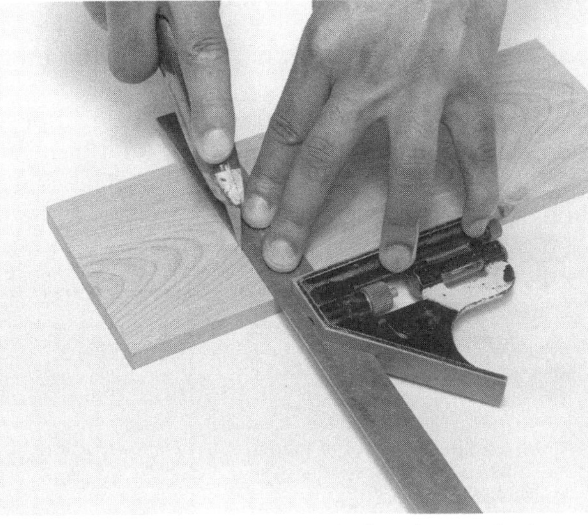

41-4 Accurate layouts can be marked on the wood using a utility knife.

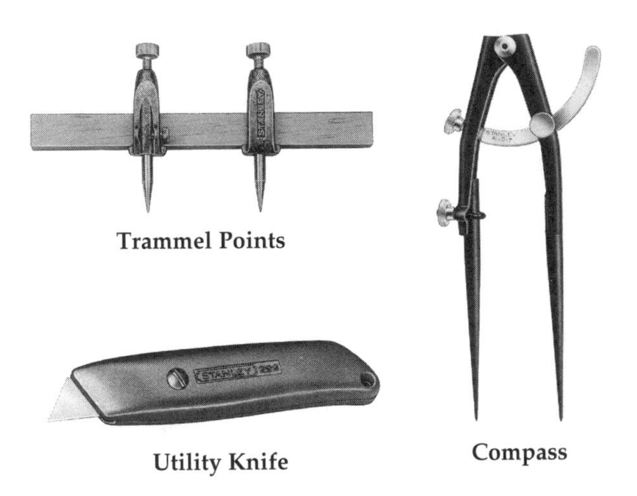

41-5 Commonly used marking tools. *(© 1998 Stanley Tools)*

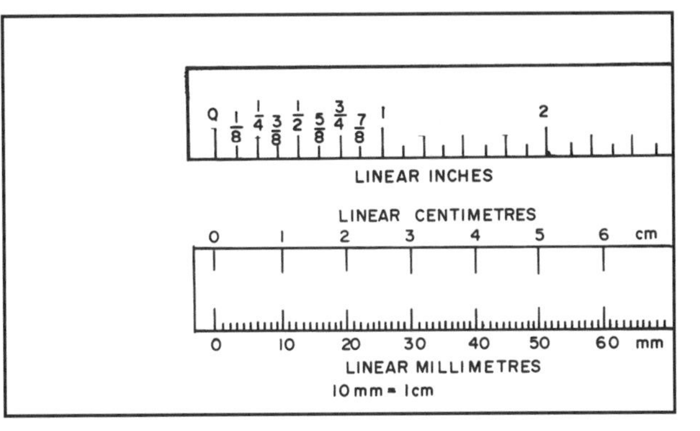

41-6 Inch and metric scales are used in carpentry work.

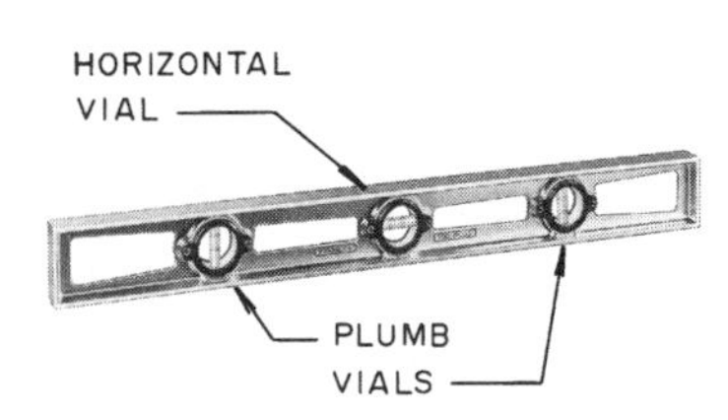

41-7 This spirit level made of aluminum or magnesium is used to check for levelness and plumb. *(© 1998 Stanley Tools)*

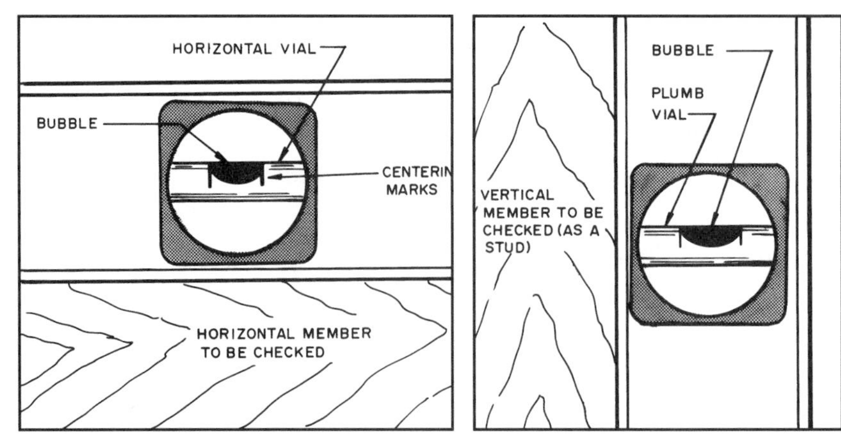

41-8 The horizontal vial is used to check for levelness (horizontal).

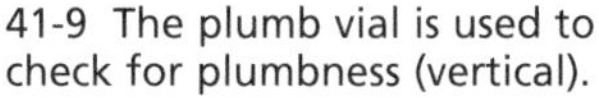
41-9 The plumb vial is used to check for plumbness (vertical).

LEVELS

Levels are tools used to set and check members that must be in a perfectly horizontal plane. They are also used to check plumb, which is the vertical position.

The **spirit level** is the most commonly used type. It has an aluminum or magnesium body and vials holding a bubble **(see 41-7)**. When the bubble is located between the lines on the horizontal vial the item is level (horizontal) **(see 41-8)**. Plumb (vertical) is checked with the plumb vials **(see 41-9)**. Levels are available in lengths from 24 inches (610mm) to 78 inches (1982mm).

When a chalkline is used to locate a level position, a **line level** is hung on it **(see 41-10)**. It has hooks which are placed over the line. The chalkline must be very tight.

41-10) A line level is used with a chalkline to check for levelness. *(© 1998 Stanley Tools)*

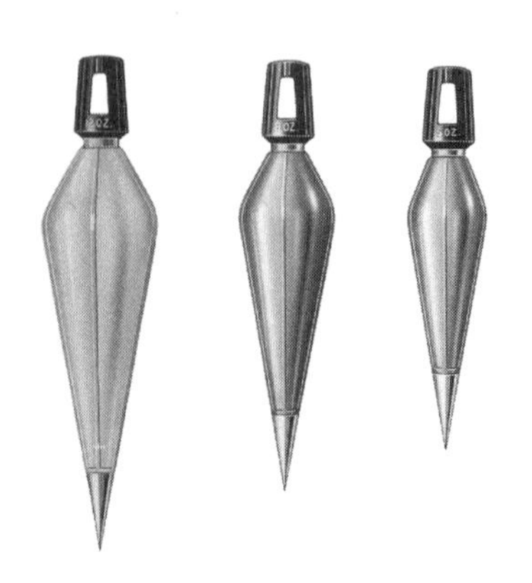
41-11 Plumb bobs are used to check the plumbness of structural members. *(© 1998 Stanley Tools)*

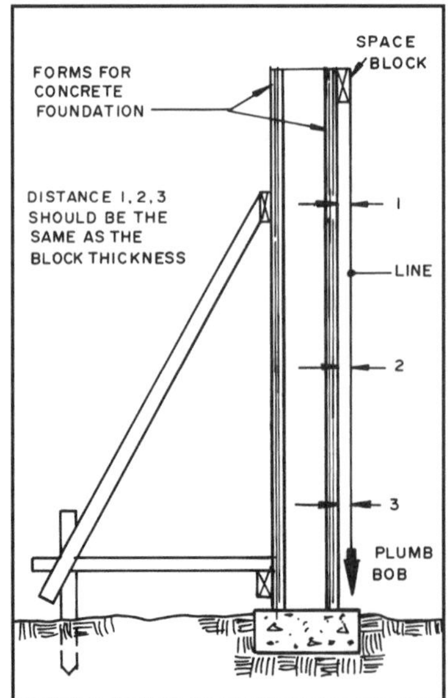

41-12 When using a plumb line measure the distance between the line and the feature being checked. It must be the same the entire length.

A **plumb bob** is used with a chalkline to check for plumb **(see 41-11)**.

The chalkline with the plumb bob is dropped along the feature to be checked for plumb. Measure the distance between the line and the feature. When the distance is the same at the top and bottom it is plumb **(see 41-12)**.

Other devices for checking for levelness and plumb can be seen in Chapter 5.

HANDSAWS

While much sawing is performed with various power saws, there are situations where a good, sharp handsaw is needed. **Crosscut** and **rip saws** are used for rough cutting. They leave rough edges that, if exposed or used in a joint, must be smoothed. **Backsaws** and **dovetail saws** have fine teeth and produce a relatively smooth surface. They are used when an accurate cut is needed. The **compass** and **keyhole saws** are used to cut internal openings, such as an opening in a sheet of paneling for an electrical outlet. The **coping saw** is used to make curved cuts in thin stock and to cope the ends of molding **(see 41-13)**. **Hand miter boxes** are lightweight and convenient to use **(see 41-14)**.

The rip and crosscut saws appear to be identical. The major difference is the layout of the teeth. **Crosscut saws** are designed to cut across the grain of a board. They are usually 26

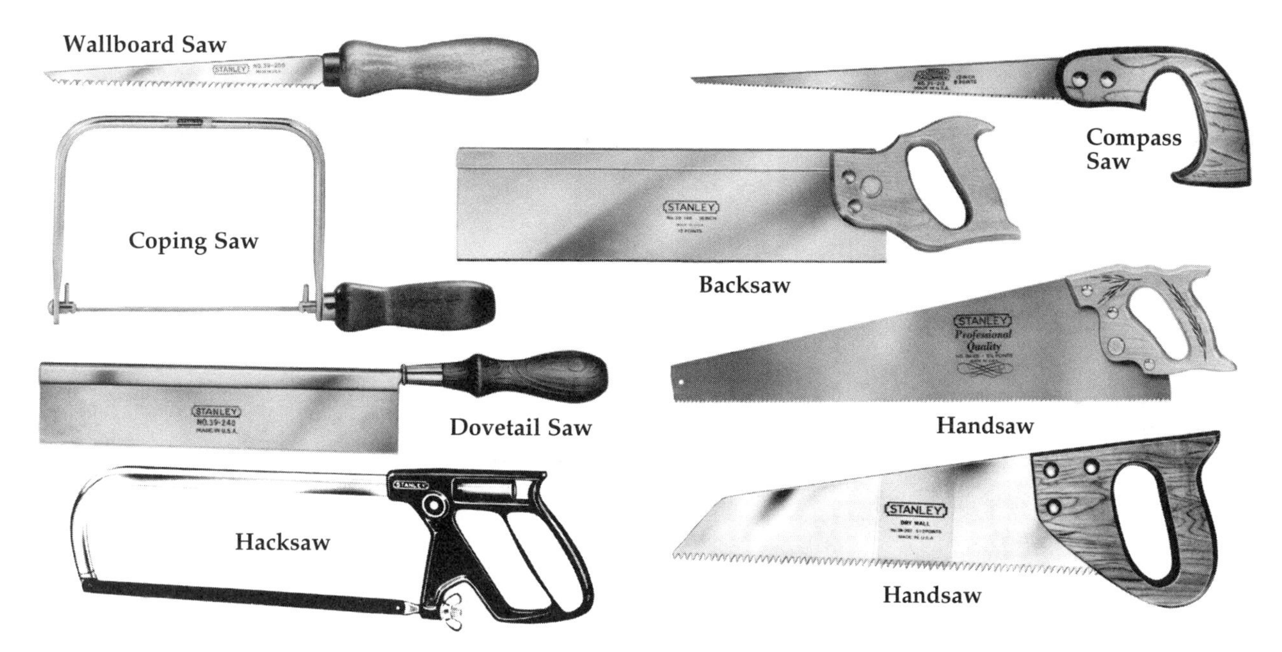

41-13 Commonly used woodworking saws. *(© 1998 Stanley Tools)*

inches long and have either 8 or 11 cutting points per inch on the cutting edge. A saw has one more point per inch than it has teeth **(see 41-15)**. The teeth on a saw are **set** (bent) alternately to the right and left. A saw is set so that the slot it cuts is wider than the thickness of the blade. This prevents the saw from binding on cut surfaces **(see 41-16)**. Some saws have a taper ground from the teeth to the top of the blade. The taper reduces the amount of set required.

The sharp points of the teeth on a crosscut saw sever the wood fibers as the saw moves across them. This produces a kerf wider than the thickness of the saw blade. A **kerf** is the slot formed by the saw as it cuts the board. Notice in **41-16** that the teeth are slanted on the bottom. The sharp outside corner cuts the wood fiber, and the slanted portion removes the wood behind this.

A **ripsaw** is used to cut with the grain of a board. The teeth are set differently from those on a crosscut saw. While the teeth have a set, they are also flat on the bottom. Actually, they are more like a series of small chisels which remove small chips of wood **(see 41-16)**. A common ripsaw size is 26 inches long with 5½ points per inch **(see 41-15)**.

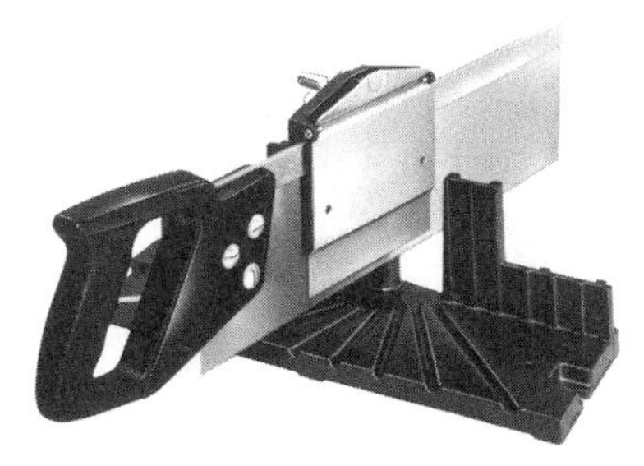
41-14 This hand-operated miter box is lightweight and easy to move about the job. *(© 1998 Stanley Tools)*

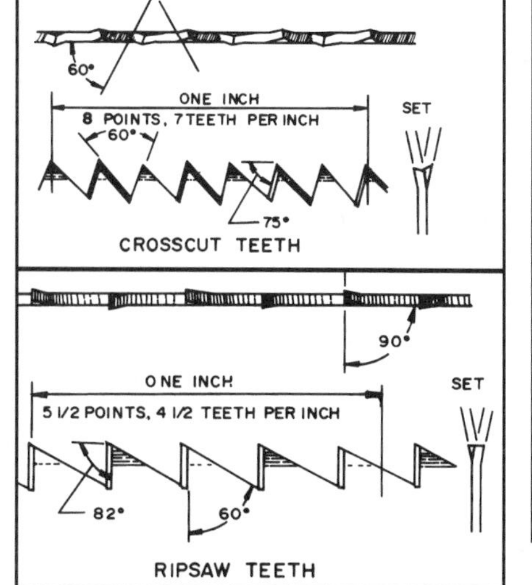

41-15 Crosscut and ripsaw teeth used on handsaws.

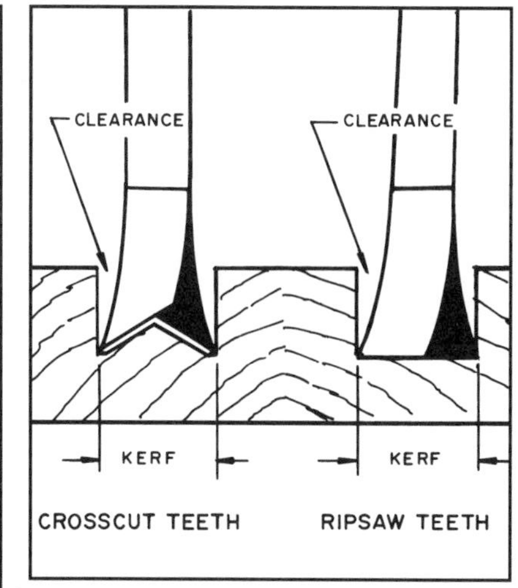

41-16 Saw teeth are set to provide a kerf wider than the thickness of the saw blade to prevent it from binding on the wood.

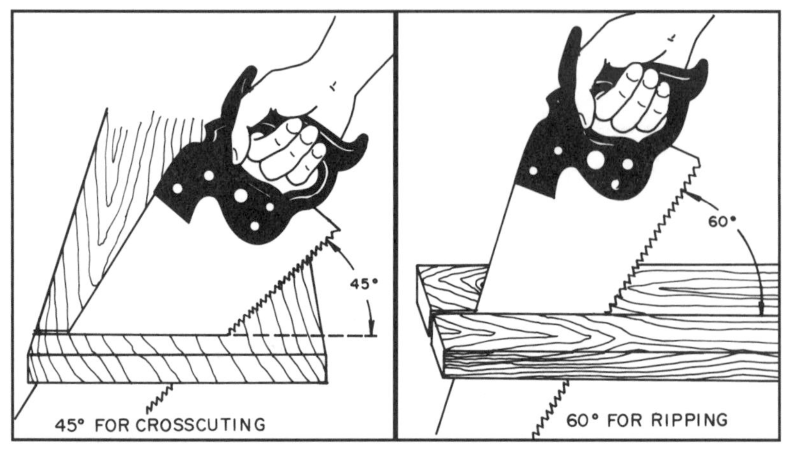

41-17 To crosscut, hold the crosscut saw at an angle of 45 degrees to the board. To rip, hold the ripsaw at an angle of 60 degrees

41-18 The two most frequently used hand planes. *(© 1998 Stanley Tools)*

When crosscutting stock with a handsaw keep the teeth on an angle of 45 degrees with the face of the board. When ripping hold the saw on an angle of about 60 degrees **(see 41-17)**.

HAND PLANES

The most common type of hand plane used is the **bench plane**. It is available in several lengths. The longest, the **jointer plane** (24 inches or 61cm) is used to smooth long edges. Other types include the **fore plane** (18 inches or 45.7cm) and **jack plane** (8 inches or 20.3cm) **(see 41-18)**. A **block plane** is a small plane (4 to 7 inches or 102 to 17.8cm) used to smooth the end grain of stock **(see 41-19)**. The blade is set on a low angle enabling it to shear off the fibers in the end grain of a board **(see 41-20)**.

Always plane the board in the direction of the grain **(see 41-21)**. The grain pattern shows this direction. If you go against the grain the surface will chip.

41-19 A block plane. *(© 1998 Stanley Tools)*

41-20 A block plane is small and is used to plane end grain.

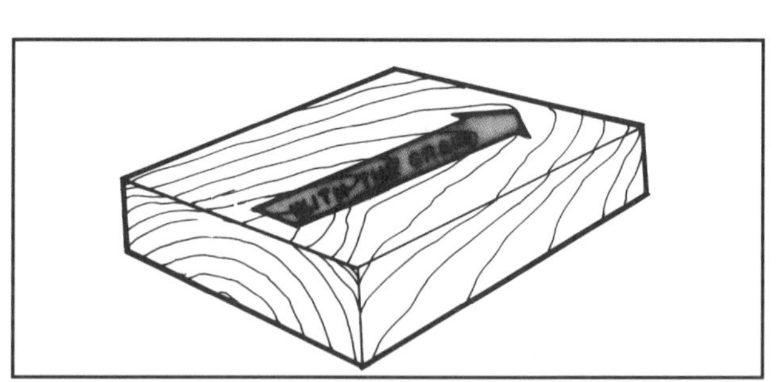

41-21 Always plane boards with the direction of the grain.

41-22 Two ways to plane end grain without splitting off the end of the board.

41-23 The procedure for planing the faces and edges of boards.

When planing end grain you must never plane all the way across the board. If you do, the end of the board will split off. Several ways to plane end grain are shown in **41-22**.

The steps for planing the surface of stock with the grain involve first securing it so it does not move. Set the plane to take a very thin cut. Begin by placing the toe of the plane on the stock and pushing down on the knob **(41-23A)**. Move the plane forward. As the bottom rests on the stock keep even pressure on the knob and the handle **(see 41-23B)**. As the end of the board is reached ease up on the knob and keep a firm pressure on the handle **(see 41-23C)**.

CHISELS & GOUGES

Chisels are used for a variety of jobs such as cutting rectangular openings or trimming joints. The best type of chisel has the blade and tang forged as a single piece. This permits it to be struck with a mallet.

A typical set will have chisels from ¼ inch (6mm) to 1½ inches (38mm) **(see 41-24)**.

A **gouge** is made like a wood chisel but the blade is formed into a V or U shape. It is used to make decorative cuts **(see 41-25)**.

When using a chisel or gouge keep both hands on the tool behind the cutting edge as shown in **41-25**. In this way you will never cut yourself if it slips. The stock should be firmly clamped to a bench or other support so it will not slide as you work on it.

SHARPENING EDGE TOOLS

The most commonly used edge tools requiring sharpening are plane irons, chisels, gouges, and knives. If the cutting edge is a little dull but still has the proper shape and is not nicked, it can be sharpened (honed) on an **oilstone** or **waterstone**. Oilstones are lightly coated with a sharpening oil as the tool is moved over them, whereas waterstones use water. The oil or water prevents the surface of the stone from being clogged with metal particles. Stones are available in various degrees of coarseness.

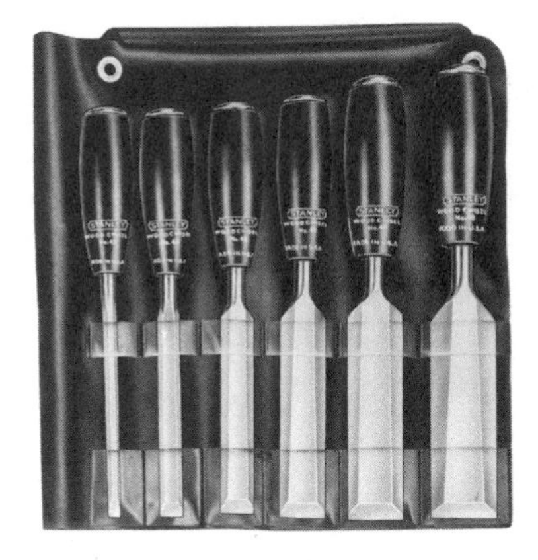

41-24 A quality set of wood chisels that have the tang and blade forged as a single piece. *(© 1998 Stanley Tools)*

41-25 A gouge is formed into a V or U shape so it cuts these shapes in the wood.

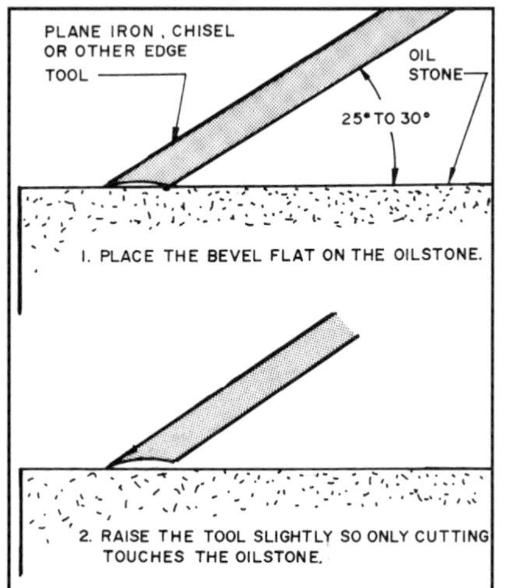

41-26 The steps for honing the cutting edge of the plane iron or chisel.

41-27 The cutting tool may be moved over the oil stone in a figure-eight pattern.

41-28 After honing the cutting edge, lay the back side of the honed cutting tool flat on the oilstone and move it back and forth easily to remove any burrs on the cutting edge.

The key to sharpening the edge tool is to hold it against the stone at the proper angle. Most plane irons, chisels, and gouges have a cutting edge of 25 degrees to 30 degrees.

To hone the cutting edge, place the bevel flat on the stone. Raise the tool about five degrees so the cutting edge touches the stone **(see 41-26)**. Move the tool back and forth across the stone. Some prefer to move it in a figure-eight pattern **(see 41-27)**. Start with the coarsest stone, which is usually an 800 grit. After the beveled side has been honed, move to a finer-grit stone (1200 grit) and finally to the finest grit (6000), which puts a shiny finish to the cutting edge. Then turn the tool over and lay it flat on the finest stone. Move it back and forth several times to remove any burrs left on the edge **(see 41-28)**.

Gouges are honed using a slip stone. It has one surface concave and the other convex. Hone the outside bevel of a gouge on the concave surface **(see 41-29)** and the inside bevel on the convex surface **(see 41-30)**. Knives are sharpened by placing the cutting bevel flat on the oilstone, raising the back edge slightly, and stroking it across the stone **(see 41-31)**.

If the cutting edge is nicked or the bevel is no longer concave, the edge must be ground before honing **(see 41-32)**. The tool is placed on the tool rest, which is adjusted so that the bevel on the tool is ground at the correct angle. Some grinders have tool-grinding attachments that hold the tool for grinding. A fine-abrasive cutting wheel must be used. After the grinder is started, the tool is moved up so that it just touches the wheel. It is immediately moved across the wheel in a parallel

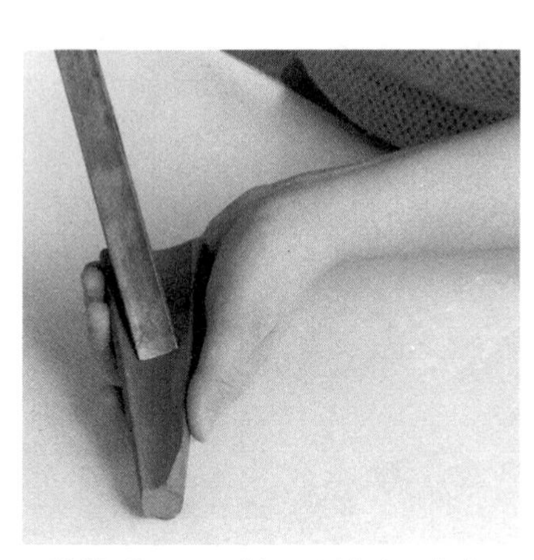

41-29 Gouges with outside beveled cutting edges are honed on the concave surface of the slipstone.

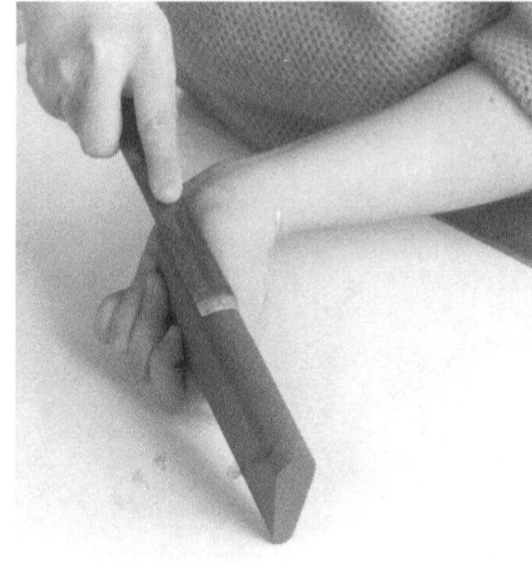

41-30 Gouges with inside beveled cutting edges are honed on the convex surface of the slipstone.

41-31 Knives are sharpened in the same manner as other edge tools.

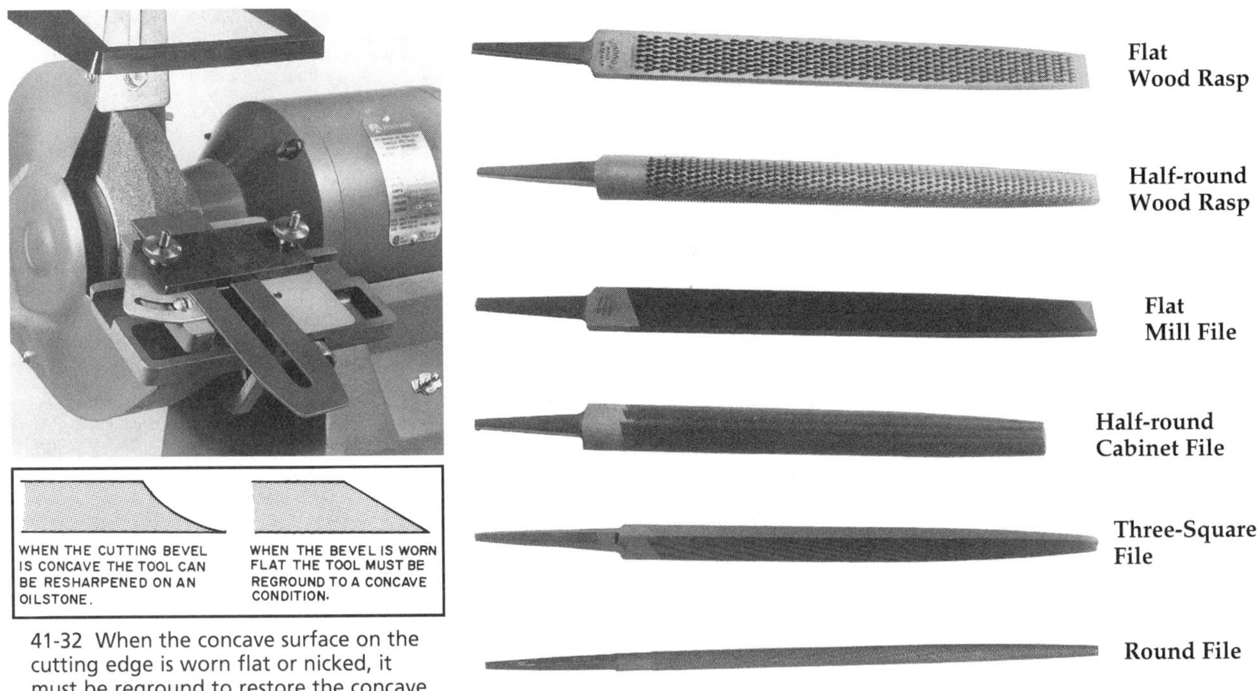

41-32 When the concave surface on the cutting edge is worn flat or nicked, it must be reground to restore the concave condition. *(Courtesy Delta International Machinery)*

41-33 Files commonly used in woodworking activities.

movement. Take very light cuts and keep the tool moving. Do not let the edge of the tool get too hot or it will turn blue, indicating it has lost its temper (hardness) and will not stay sharp very long. When the concave bevel is complete, hone to a fine cutting edge. Always wear safety glasses when using the power grinder **(see 41-32)**.

FINISHING TOOLS

The most frequently used finishing tools are **files, cabinet scrapers, rasps,** and **surface-forming tools**. There are a wide variety of files available. Those designed for use on wood include a **rasp, cabinet file,** and **wood file**. The rasp has very large teeth and removes wood rapidly, leaving a very rough surface. The cabinet file has finer teeth and can clean up a surface following the rasp. The wood file has the finest teeth but still leaves a fairly rough surface **(see 41-33)**. Metal-cutting files have very fine teeth and can be used on wood, but they clog easily and need frequent cleaning. Files are available in flat, half-round, round, and triangular shapes **(see 41-34)**.

Surface-forming tools have a thin, perforated metal blade and serve the same purpose as files. The teeth on the blade are open so that the wood chips pass through the blade, reducing clogging **(see 41-35)**. When they get dull a new blade is installed.

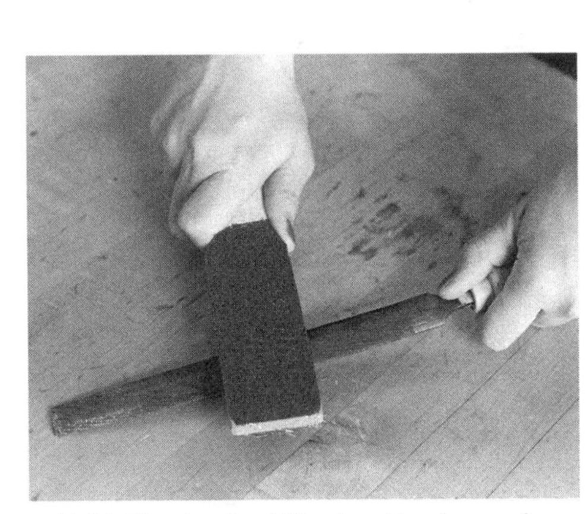

41-34 Fine-toothed files tend to clog and are cleaned with a wire brushlike tool called a file card.

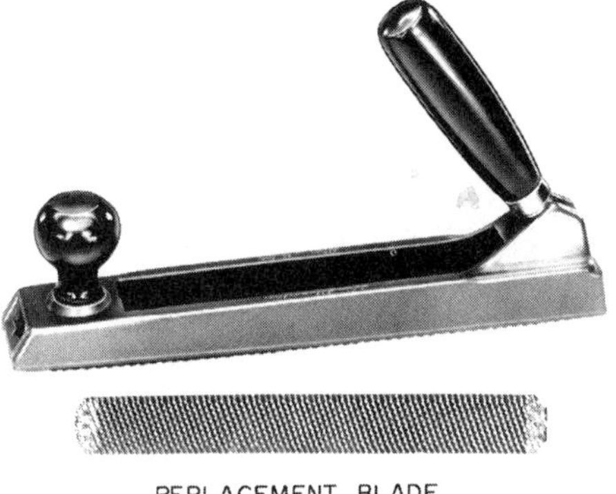

41-35 One of the many types of Surform forming tools. *(© 1998 Stanley Tools)*

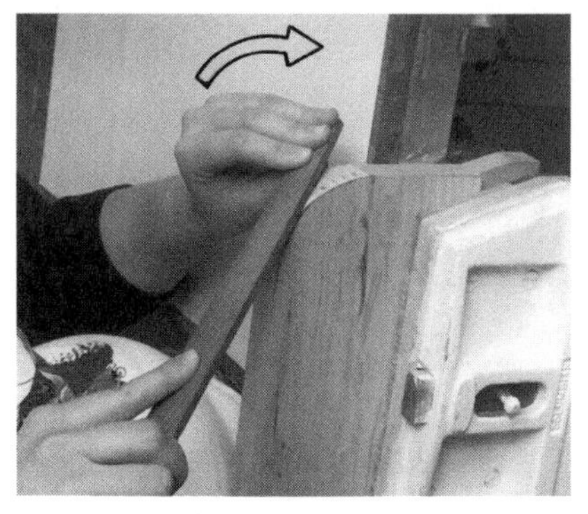

41-36 Flat files are used to smooth convex surfaces.

41-37 Half-round files are used to smooth concave surfaces. File down the curve to the bottom.

41-38 This rectangular handheld scraper provides a fine finish before sanding.

Flat files are used to smooth convex surfaces and half-round files are used for concave surfaces **(see 41-36 and 41-37)**.

Cabinet scrapers are useful to give a final smoothness to a wood surface before sanding. They are available in handheld rectangular **(see 41-38)** and curved types. Some have handles **(see 41-39)**. Hand scrapers are sharpened by filing the edge square and then burnishing each corner to form a cutting burr **(see 41-40)**.

41-39 This scraper blade is held in a metal frame and is easy to use.

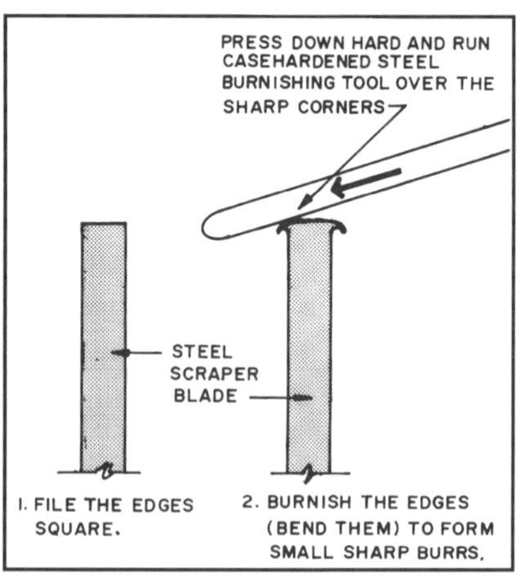

41-40 Scraper blades are sharpened by filing the corners square and then bending them over into a burr with a burnisher.

DRILLING & BORING TOOLS

Larger-diameter holes are often bored with a **brace** and **auger bit** or one of the special large-diameter bits. Small holes are drilled with a push drill.

The brace, shown in **41-41**, hold the bits in its chuck. The bits have a tapered, square tang that fits into the jaws in the chuck. An auger bit is shown in **41-42**. These are sold in sets beginning with a ¼-inch (6mm) diameter and going larger up to 1½ inches (38mm). The hole is bored more accurately if the center is punched with an awl, giving

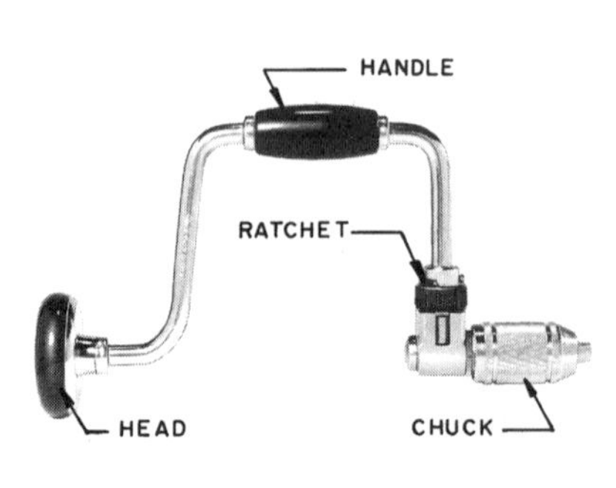

41-41 The brace is used to hold bits that bore large-diameter holes. *(© 1998 Stanley Tools)*

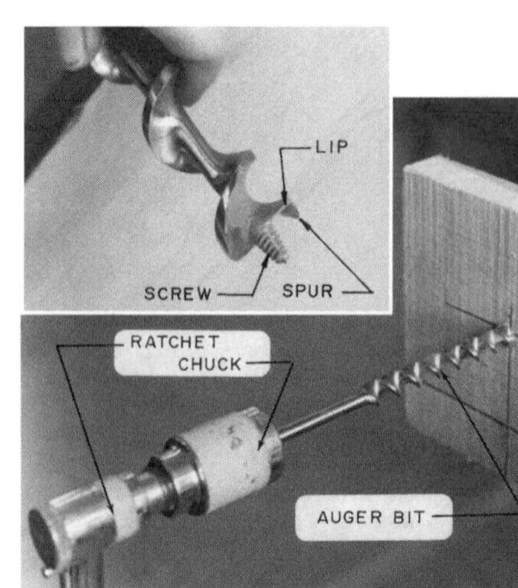

41-42 Auger bits are held in the chuck of the brace and are used to bore large-diameter holes.

41-43 An awl is used to punch the center of a hole to permit the auger bit to be located accurately.

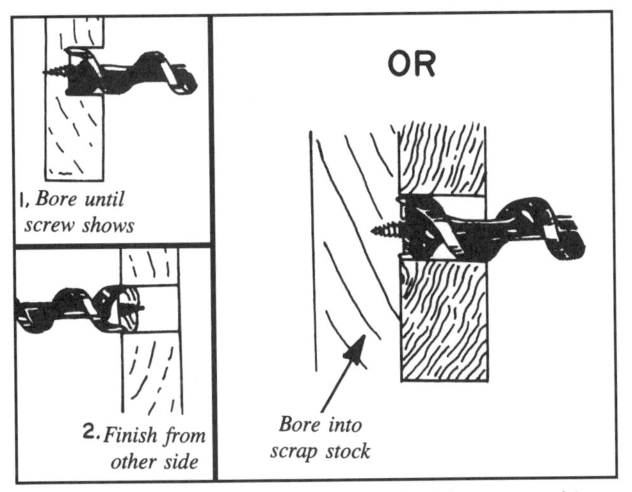

41-44 Two ways to bore through stock with an auger bit without splitting out the wood on the back side.

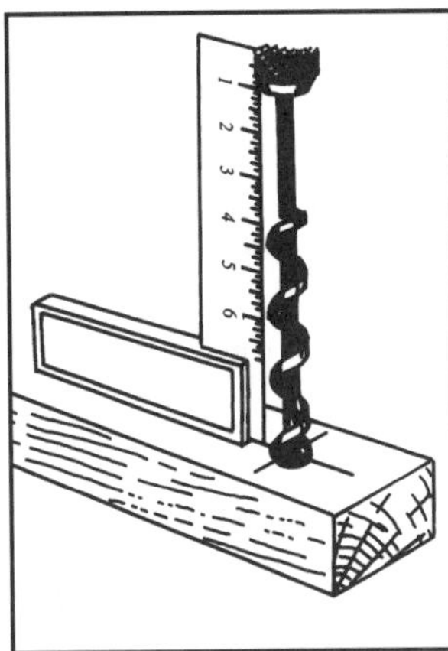

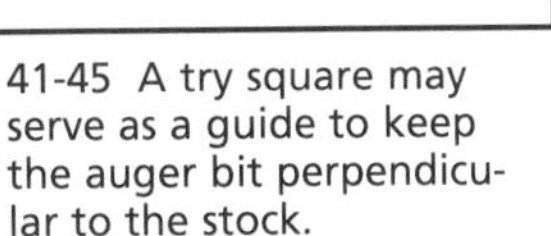

41-45 A try square may serve as a guide to keep the auger bit perpendicular to the stock.

41-46 A T-bevel may be used to line up an auger bit to bore a hole on an angle.

a recess for the screw to get started **(see 41-43)**. When boring with an auger bit or other special bits bore through from one side until the screw breaks through. Then remove the bit from the stock and finish the hole by boring through from the other side **(see 41-44)**.

When it is necessary to bore the hole perpendicular through the stock the auger bit can be aligned with a square **(see 41-45)**. Holes to be on a specific angle can be lined up with a T-bevel **(see 41-46)**. Holes requiring a flat bottom are bored with a **Forstner bit**. These are sold in sets and are available in a range of diameters. Since a **Forstner bit** does not have a screw it will bore close to the back face without breaking through **(see 41-47)**. It has a tang on the end and is used in a brace.

The **expansive bit** has an adjustable cutter and can bore holes from ⅝- to 1¾-inch (16 to 44mm) diameter and a larger size will go from ⅞ to 3 inches (22 to 76mm) diameter **(see 41-48)**.

The **push drill** uses a fluted drill point. The drill sets have diameters from 1/16 to 11/64 inch (1.5 to 4.4mm) **(see 41-49)**.

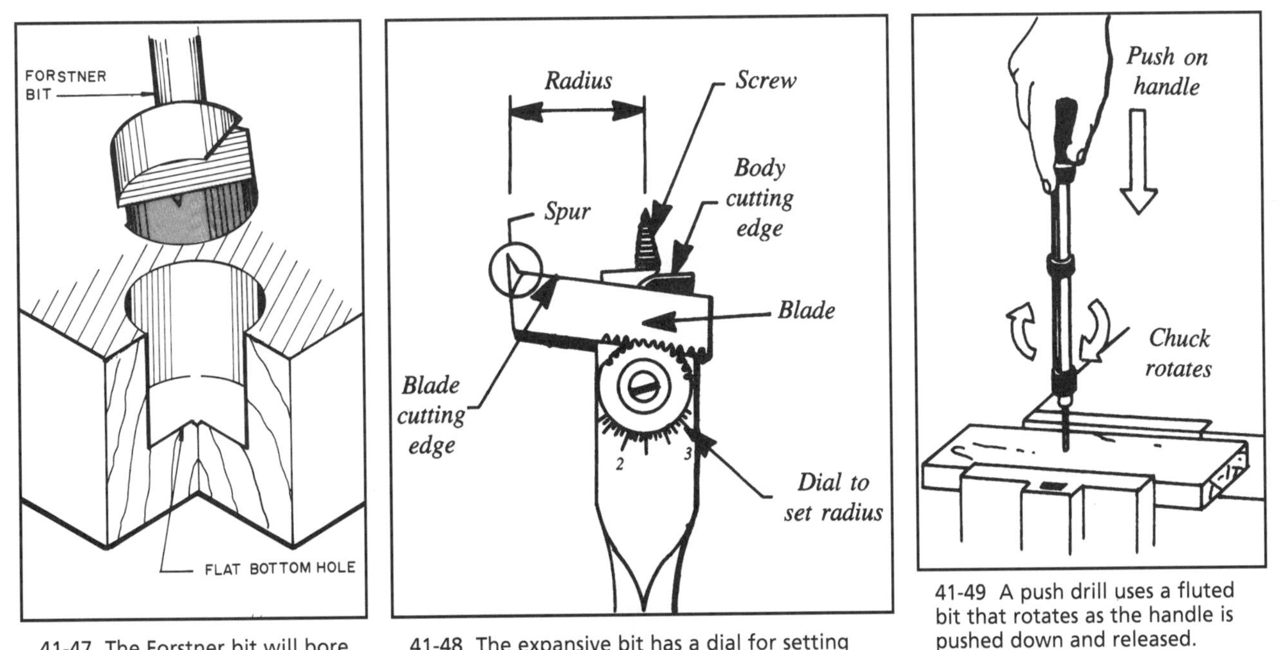

41-47 The Forstner bit will bore flat-bottomed holes.

41-48 The expansive bit has a dial for setting the radius of the adjustable cutter.

41-49 A push drill uses a fluted bit that rotates as the handle is pushed down and released. *(© 1998 Stanley Tools)*

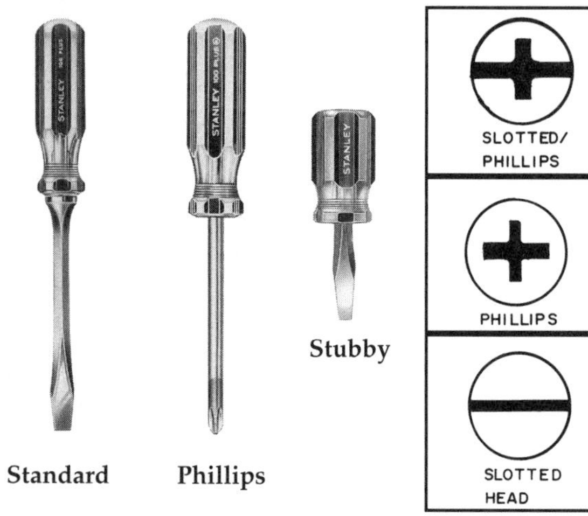

41-50 The commonly used screwdrivers. *(© 1998 Stanley Tools)*

41-51 The types of screw head slots.

41-52 A spiral-ratchet screwdriver turns the bit when you push down on the handle. *(© 1998 Stanley Tools)*

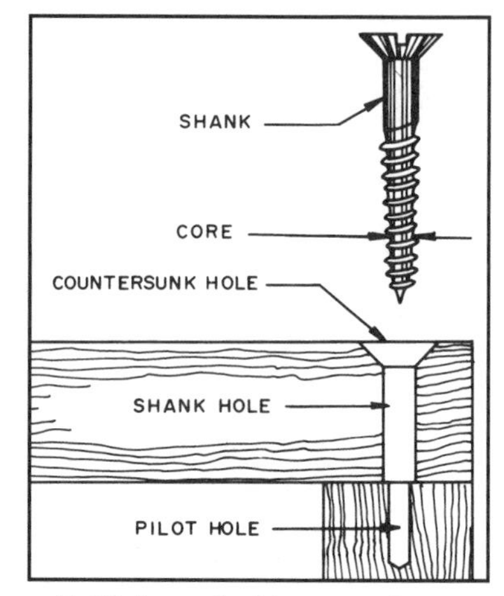

41-53 Properly driven wood screws require a shank and a pilot hole be drilled to relieve stress on the wood.

SCREWDRIVERS

The two most frequently used screwdrivers are the **standard** and the **Phillips (see 41-50)**. They are available with a range of tip sizes to fit various-sized screws; standard tips range from ⅛ to ⅜ inch (3 to 10mm) wide. Phillips tip sizes are indicated by a number and range from 0 to 24. Some screws have a slotted Phillips head enabling you to use either type of screwdriver **(see 41-51)**.

Be certain to select the tip that fully engages the opening in the head of the screw.

The **spiral-ratchet screwdriver** has a spiral-grooved spindle that turns when you push down on the handle. The direction of rotation can be reversed to remove screws. It has a chuck that holds the bits. The tool comes with a number of tip types and sizes. These are inserted in the chuck to fit the screw to be driven **(see 41-52)**.

Preparation for properly installing a wood screw is shown in **41-53**. A hole called a shank is drilled in the top board. It is about the same diameter as the screw. A smaller hole about the size of the core is drilled in the bottom board. If a flathead screw is used the shank hole is countersunk. Roundhead screws sit flat on top of the board. Properly installed flathead and roundhead screws are in **41-54**. A countersink is shown in **41-55**.

HAMMERS

A variety of hammers are available. They should be high quality with a tempered face. Since most finish work requires careful nailing, lighter-weight hammers are often used. A 13-ounce (370g) hammer is popular. Other weights available are 16 ounces (450g), 20 ounces (570g), and 22 ounces (625g). Framing carpenters use the heavier sizes because they drive larger nails, and hammer marks do not spoil the job because they are not seen.

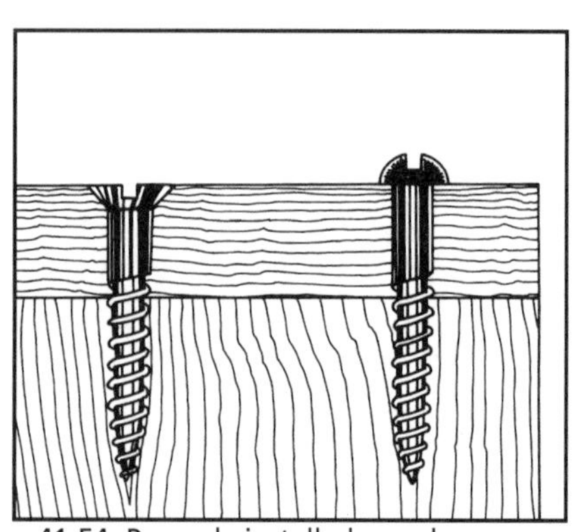

41-54 Properly installed wood screws.

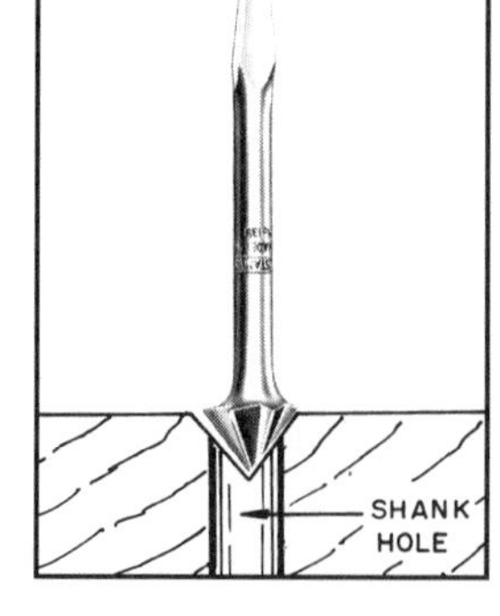

41-55 A countersink. *(© 1998 Stanley Tools)*

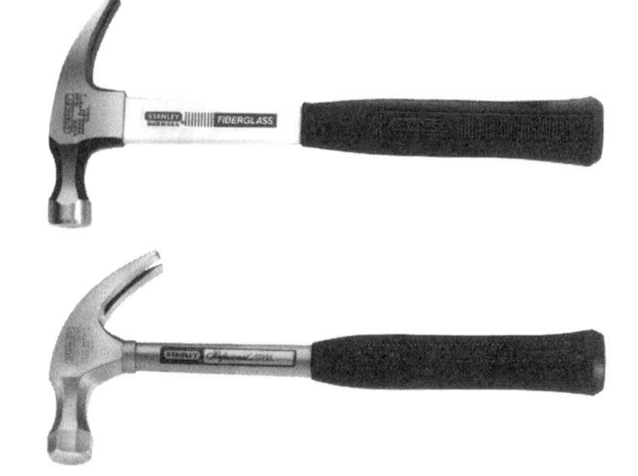

41-56 Carpenters' hammers. *(© 1998 Stanley Tools)*

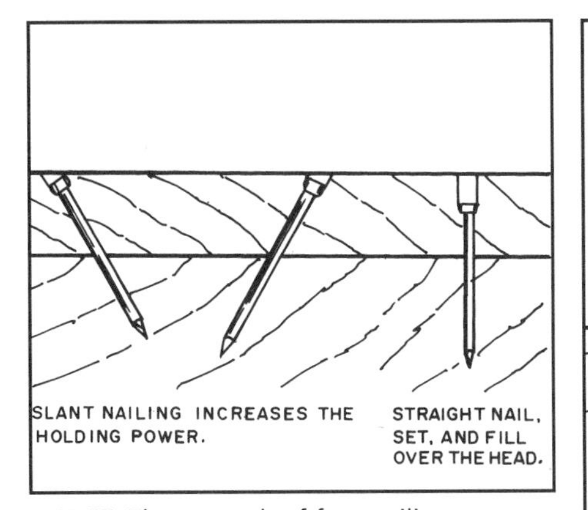

41-57 The strength of face-nailing can be increased by slanting the nails toward each other.

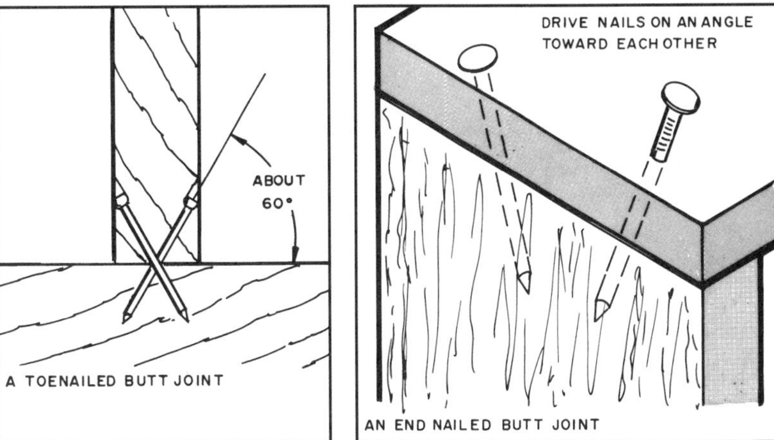

41-58 The butt joint can be toenailed or end-nailed.

The commonly used types are the curved-claw hammer and the straight-claw ripping hammer **(see 41-56)**. The curved-claw hammer is used for both finish carpentry and rough framing. Weights range from 13 ounces (370g) to 22 ounces (625g). Framing carpenters use the heavier hammers. Ripping hammers are used for rough work where it is often necessary to pry boards apart. A hammer holster is hung on the carpenter's belt and used to carry the hammer when it is not in use.

PROPER NAILING TECHNIQUES

The strength of a nailed wood joint depends upon the nails used and how they are installed. Nails are discussed in Chapter 40. Wood is joined by face-nailing, slant nailing, toenailing, and tacking. **face-nailing** involves driving the nail perpendicular to the wood. It is used when strength is not important. **Slant nailing** increases the holding power and involves driving the nails in opposite directions **(see 41-57)**. When a butt joint is to be nailed it may be **toenailed** or **end-nailed (see 41-58)**. When end-nailing, slant the nails toward the center of the stock. **Tacking** involves driving a nail partway into the joint to temporarily hold it. The nail is removed after its purpose has been served **(see 41-59)**.

Nails with small heads, such as finishing nails, can be set below the surface of the wood with a **nail set**. The holes can be filled with an appropriate filler **(see 41-60)**.

Clinching the nails greatly increases their holding power. Drive the nails through both pieces of wood, allowing them to stick out the other side. Bend them on a 90-degree angle as shown in **41-61**. Double bending is even stronger than single nailing. You can use pliers to make the first bend.

Extra holding power is available by using screw- or threaded-type nails. Examples of these nails are shown in Chapter 40.

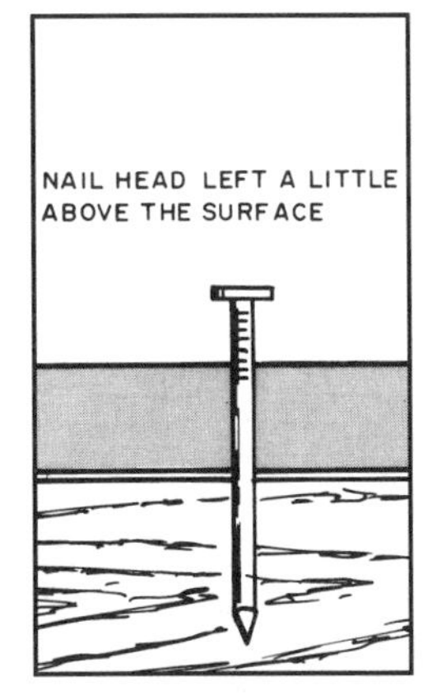

41-59 Tacking is used to temporarily hold two boards together.

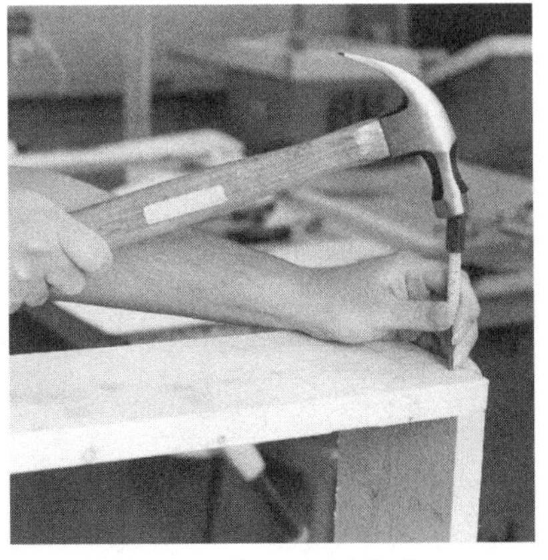
41-60 Use a nail set to sink the heads of finishing and casing nails below the surface of the wood. *(© 1998 Stanley Tools)*

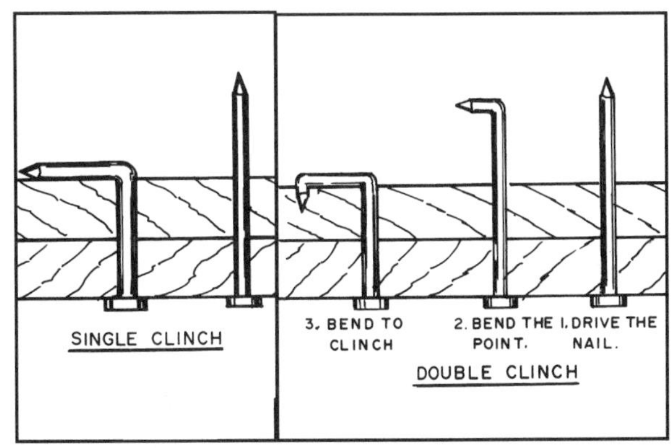

41-61 Clinching the nail increases the holding power.

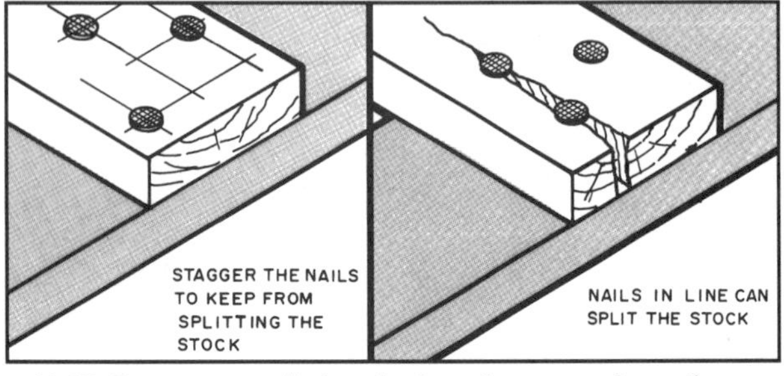

41-62 To prevent splitting the board, stagger the nails.

41-63 Whenever possible nail the thinner piece of wood to the thicker piece.

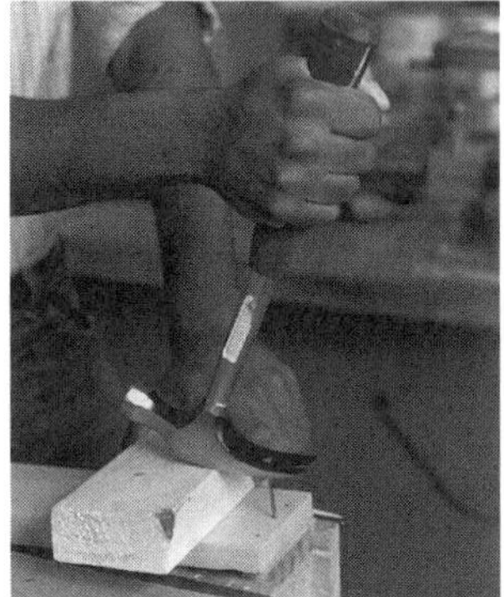

41-64 A scrap block under the hammer head makes it easier to remove long nails.

Often there is a danger that the nail will split the wood. In this case you can drill a small hole in the top piece, and even in the bottom piece when necessary. Consider staggering the nails so they do not line up with the grain of the wood **(see 41-62)**. When nailing a thin piece to a thicker piece, nail through the thin piece into the thick piece if possible **(see 41-63)**.

When it is necessary to remove a nail, it helps to put a block under the hammer as shown in **41-64**.

HATCHETS

Hatchets have a sharp cutting edge and a nailing face. Commonly used types are the half-hatchet, used for concrete form construction, cutting stakes, and other general uses; the wallboard hatchet, used to install gypsum drywall; and the shingle hatchet, used to split and nail wood shingles and shakes **(see 41-65)**.

CLAMPING TOOLS

The most frequently used clamping tools include hand screws **(41-66)**, C-clamps **(41-67)**, and bar clamps **(41-68)**. These are available in a number of designs and sizes. A miter vise is used to hold mitered stock while the miter is being nailed together **(41-69)**.

BONDING AGENTS

Glues, adhesives, and **cements** are bonding agents used in wood construction (also refer to Chapter 40). Glues are made from natural materials, adhesives from synthetic materials, and cements from rubber-based materials. Often bonding agents are used with mechanical fasteners such as nails or screws **(see 41-70)**.

The types of glues available are extensive and many are not of use to the finish carpenter. Following are some that are useful.

GLUES

Two types of glues often used for cabinetwork are **liquid hide glue** and **casein glue**. Liquid hide glue is an animal product in liquid form. It will fill small gaps and chips in the joint and is used for interior work. It will set up in about two hours.

Casein glue is made from dry milk curds and is in powder form. It is mixed with water to form a thick cream. It is water resistant and can be used on out-of-doors applications that are not directly exposed to the weather.

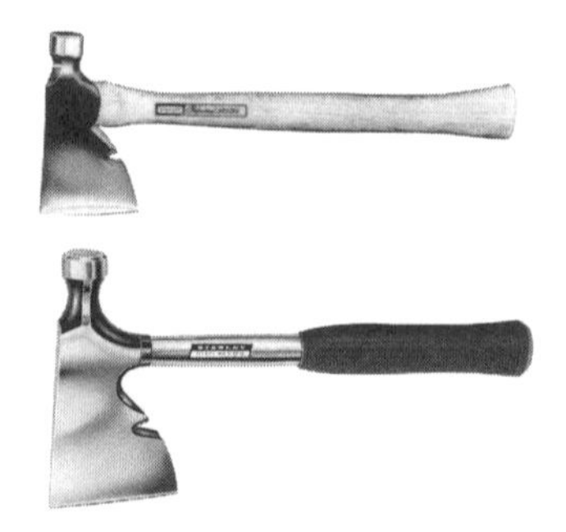

41-65 Two of the hatchets used by carpenters. *(© 1998 Stanley Tools)*

41-66 Hand screws have long parallel jaws.

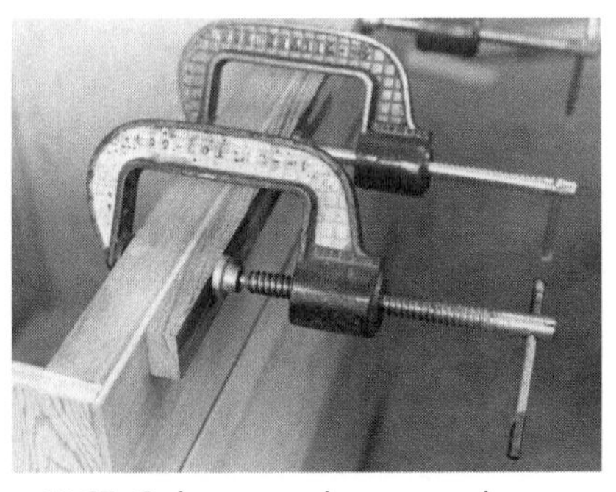

41-67 C-clamps apply pressure by tightening the screw.

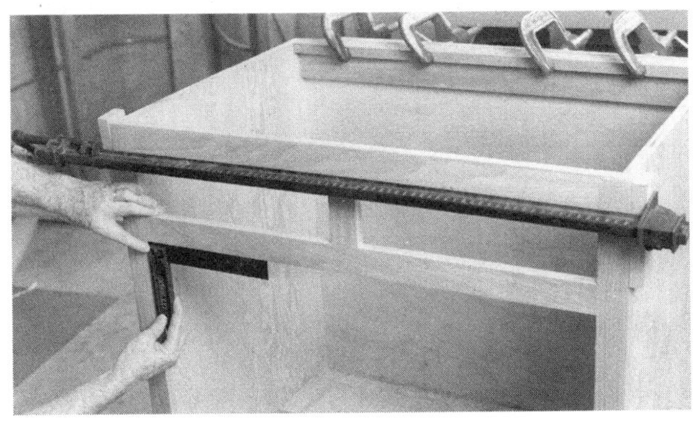

41-68 Bar clamps are used to clamp long items.

ADHESIVES

The most frequently used bonding agents are adhesives. They are classified under two types, **thermoplastics** and **thermosets.**

The major types of thermoplastics include **polyvinyl, aliphatic resin, alpha cyanoacrylate,** and **hot melts**. Polyvinyl is commonly called white glue. It is sold in squeeze bottles and is ready to use. It sets up in about 30 minutes but the joint requires 24 hours for a complete cure. Wood bonded with polyvinyl must have a moisture content of 6 to 10 percent.

Aliphatic resin is yellow and is stronger than the white polyvinyl adhesives.

Alpha cyanoacrylate is marketed as Super Glue. It is not used on porous materials such as wood but bonds metal, plastics, and ceramic materials.

Hot melts glue wood and almost any other type of material. They are heated and applied with a special gun-like tool. They are not very strong and are used where strength is not important, such as when bonding overlays to cabinet doors.

Thermoset adhesives are more resistant to heat and moisture than many bonding agents. The major types the finish carpenter may use are **urea-formaldehyde** and **resorcinol-formaldehydes**.

Urea-formaldehyde adhesives are sold as a powder. When mixed with water, a chemical reaction begins; therefore, do not mix more than will be used in a few hours. They are water resistant and requires 16 hours clamping time. In the store it is often called plastic resin.

Resorcinal-formaldehydes produce a waterproof joint and are used on roof trusses and exposed surfaces where failure cannot be tolerated. They are sold in two containers; one contains the liquid resin and the other the catalyst. These are mixed and stirred. The joint must be kept clamped for at least 16 hours. One type is identified as Elmer's waterproof glue.

CEMENTS

The two cements used in building construction are **contact cement** and **mastic**. Contact cement is used to bond wood veneers, plastic laminates, and other decorative materials to a wood substrate. It is applied to both surfaces, allowed to dry, and then the surfaces are brought together **(see 41-71)**.

Mastic is a thick contact cement usually applied with a caulking gun to bond plywood subfloors to the joists and wall sheathing to the studs.

TOOL STORAGE

Generally the finish carpenter will have a **tool-storage unit** built in the back of his/her truck or van. Many build their own storage boxes and shelves, while others buy metal units that fit in the back of the vehicle. Others use a metal tool chest of some kind that has large rollers. Many finish carpenters will not leave tool chests on the site overnight because of the danger of theft. Some form of small tool box with a handle can be used to move the tools needed that day from the truck to the building.

41-69 A miter vice holds the mitered stock as it is being glued and nailed.

41-70 A wide range of glues, adhesives, and cements are available.

41-71 Contact cement is used to bond plastic laminate to the substrate.

CHAPTER 42

Using Power Tools

Finish carpenters use a variety of power tools to saw, drill, shape, nail, staple, and set screws. Some are **stationary** and others are **portable**. Stationary power tools are mounted on a base resting on the floor. Portable power tools are lightweight and are carried as the work progresses. Portable power tools are operated by electricity, batteries, and compressed air.

It is important to purchase only high-quality tools; anything less will not produce the work expected and could be dangerous. Select only tools that have their electrical system double insulated to ensure that electrical shocks are not possible. All tools should have guards on the cutting edges providing good protection. Be certain the manufacturer has a reliable service and repair system.

The construction industry is one of the more hazardous. Follow the general safety rules governing the use of power tools and the recommendations for using specific tools.

STATIONARY CIRCULAR SAWS

The stationary circular saw is used for a variety of cuts and produces more accurate results than the portable circular saw. The most common operations include ripping and crosscutting. When it has table extensions, it will accurately cut large panels such as plywood and particleboard. This tool can cause many accidents, so observing safety rules is very important. The saw in **42-1** has a 10-inch (25.4cm) blade and will cut stock 3¼ inches (8.2cm) thick at 90 degrees and 2⅛ inches (5.4cm) thick at a 45-degree miter. It is light enough so that it can be easily moved around the construction job.

General Power-Tool Safety Rules

1. Wear some form of approved eye protection.
2. Use only tools with guards in place.
3. Be certain the electrical cord and extension cord are not worn or damaged.
4. If the tool has a three-wire plug, be certain the source of electricity has a third-wire ground and use a ground fault interrupter (GFI).
5. Remove all rings, bracelets, necklaces, etc., that may become caught. Long hair can also be a problem and should be confined with a hair net.
6. Do not use a tool that is not in good condition.
7. Be certain the switch operates properly.
8. Use only sharp saws and cutters. Dull tools can cause accidents.
9. After changing a saw blade or other cutter, recheck before starting the tool to make certain it is properly seated and securely locked in place.
10. Work being cut must rest on a secure surface so it will not slip while being cut.
11. Avoid working when you are excessively fatigued.
12. Let the tool cut at its normal pace. Do not force or overcrowd to speed up the cut. This can cause kickbacks.
13. After finishing a cut let the saw or cutter stop rotating before laying the tool down.

The Occupational Safety and Health Administration (OSHA) requirements for power tools are detailed in the publication *Hand and Power Tools, OSHA 3080.* Order from OSHA Publication Office, 200 Constitution Ave. NW, Room N-3647, Washington, DC 20210.

42-1 This lightweight stationary circular saw can be moved around the building as required by the carpenter. *(Courtesy Delta International Machinery Corp.)*

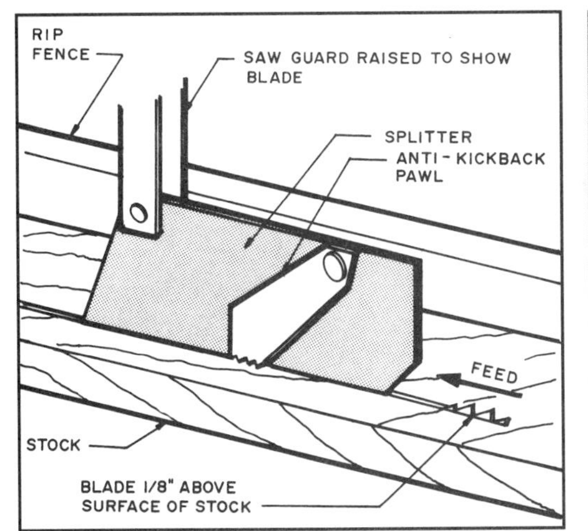

42-2 The circular saw blade should not extend more than ⅛ inch (3mm) above the stock to be cut.

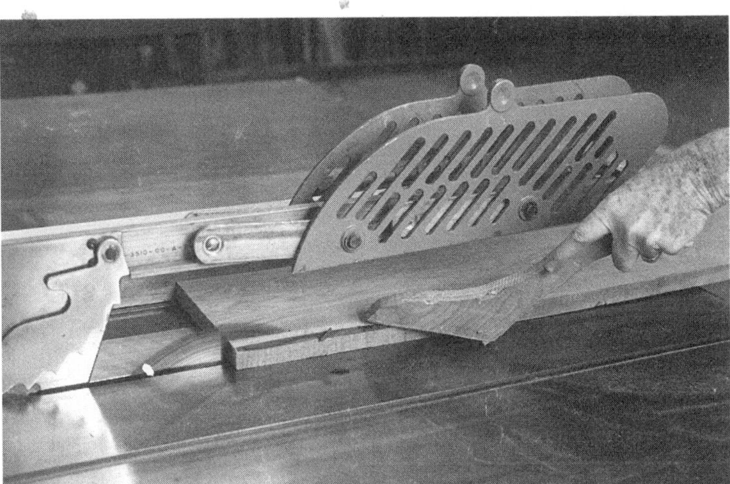

42-3 Use a wood push stick to control wood when ripping narrow stock. Note that the guard has been removed only to show the cutting process.

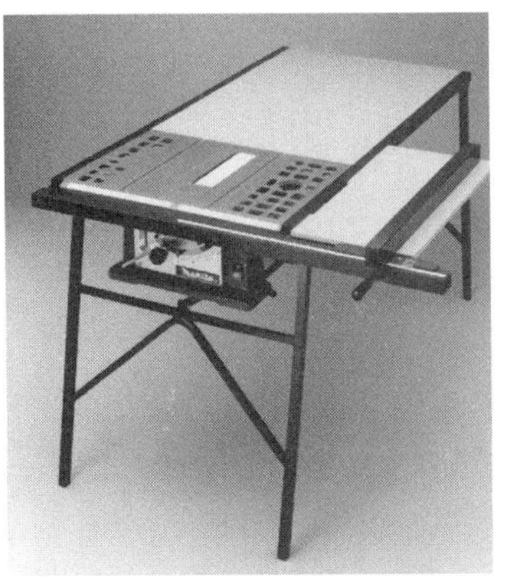

42-4 This stand holds a small, portable circular saw, expands the size of the table, and raises the saw to a convenient working height. *(Courtesy Trojan Manufacturing, Inc.)*

42-5 A material support stand used to carry the end of long stock as it is being ripped on a table saw. *(Courtesy Trojan Manufacturing, Inc.)*

Stationary Circular Saw Safety Recommendations

1. Do not adjust saw while it is running.
2. Always wear safety goggles or safety glasses with side shields.
3. In some operations a dust mask should be used.
4. The saw table and the floor around the saw should be kept free from sawdust and debris.
5. When cutting wide stock, such as a plywood panel, use an auxiliary support to keep the panel level.
6. Keep all guards in place. Repair them if they do not function properly.
7. The blade should project no more than ⅛ to ¼ inch (3 to 6mm) above the work **(see 42-2)**.
8. Never reach over the blade or place your hands within several inches of it.
9. Never cut a piece of stock freehand. Always use a miter gauge or the rip fence.
10. Do not stand directly behind the blade. Kickbacks do occur and can cause serious injury.
11. When ripping narrow stock use a push stick **(see 42-3)**.
12. Be certain the rip fence is parallel with the blade. This reduces binding and burning and possible kickbacks.
13. Do not use cracked or burned blades.
14. Blades that wobble or vibrate must be destroyed and discarded.
15. Be certain the blade is installed so that it rotates in the proper direction.
16. Place small, portable saws on a firm base at a convenient working height above the floor **(see 42-4)**.
17. Have someone help tail off the end when cutting long boards. A stand as shown in **42-5** can be used to support the end of long boards as they are ripped.
18. Do not cut wet, cupped, or warped boards.
19. Do not use the rip fence as a stop when cutting short pieces to length. Use a stop block instead **(see 42-6 on the following page)**.
20. Do not leave a running saw unattended. If you have to leave it, shut it off.
21. Use the proper blade for the job at hand.

Blade Selection

There are many varieties of circular saw blades available. The hole in the blade must match the diameter of the shaft on the saw. Following are the common types of blades.

A **cross blade** is used to cut stock across the grain. A **hollow-ground blade** is used for fine, smooth cuts. A **ripsaw blade** has large-set, chisel-like teeth and removes wood rapidly. It is used to cut with the grain. The **combination saw blade** has a combination of crosscut and rip teeth and is used for both ripping and crosscutting. The **plywood saw blade** is designed to cut plywood with a minimum of chipping and leaves a smooth surface.

Circular Saw Parts

A typical saw is shown in **42-1**. Before operating a saw, be certain you are familiar with the operating parts and adjustments. Usually the switch is located

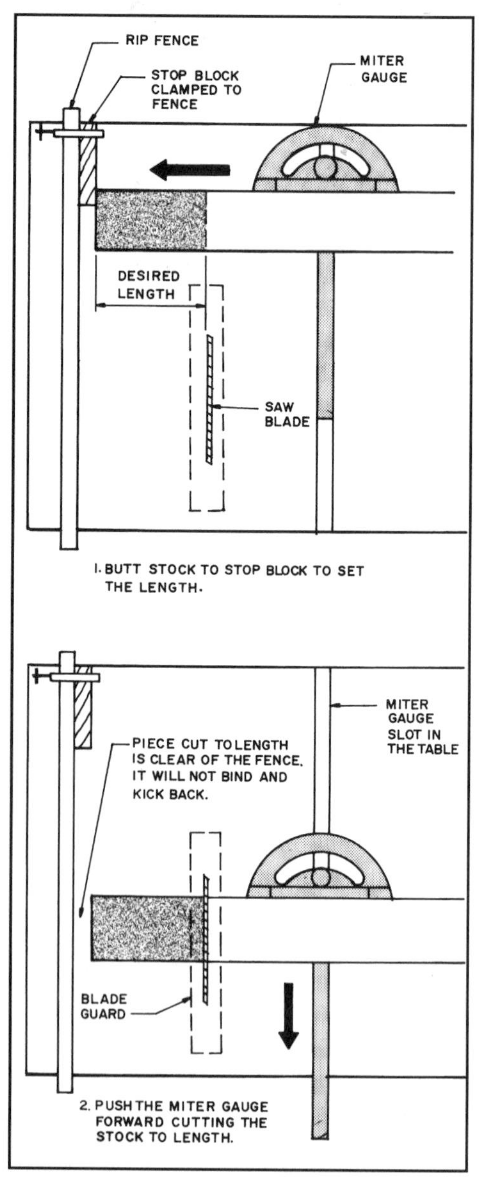

42-6 Use a stop block when cutting many small pieces to length. Always keep a guard over the blade when crosscutting.

Steps for Crosscutting Stock (see 42-7)

1. Set the miter gauge on the desired angle. Slide it in the groove in the table.
2. Crank the blade up so it protrudes ⅛ inch (3mm) above the stock to be cut.
3. Place the saw guard over the blade.
4. Be certain the rip fence is moved over, out of the way. Remove it from the saw if necessary.
5. Place the stock to be cut on the table and firmly against the miter gauge.
6. Start the saw. Do not begin to cut until it reaches its full speed.
7. Holding the board firmly against the miter gauge, slide it into the saw. Sight through the guard to see that the mark on the board lines up with the saw blade and that the blade will cut on the waste side of the line.
8. Push the board and miter gauge on past the saw. Then remove the board from the miter gauge.
9. If the scrap end cut off is small, use a stick to push it off the table or wait until the blade stops turning to remove it.

Steps for Ripping Stock

1. Measure the required distance between the saw blade and rip fence. Measure from the edge of the teeth bent toward the fence.
2. Raise the blade so that it is ⅛ inch (3mm) above the thickness of the stock to be cut.
3. Place the guard over the blade.
4. Start the saw. Do not begin to cut until the spinning blade has reached full speed.
5. Place the straight, smooth edge of the board against the fence. Do not try to rip stock with curved or warped edges.
6. Hold the stock against the fence and feed it into the saw. Keep an even pressure **(see 42-8)**.
7. Have someone on the other side of the saw to receive long boards. This helper must only support the stock and should not pull on it.
8. When the end of the stock nears the saw, use a push stick to move the stock past the saw. If the board is 12 inches (30.5cm) or wider, you can safely move the stock with your hand if you keep your hand against the rip fence.

on the front within quick and easy reach. Two handles below the table are used to raise and lower the blade and tilt it for cutting on angles.

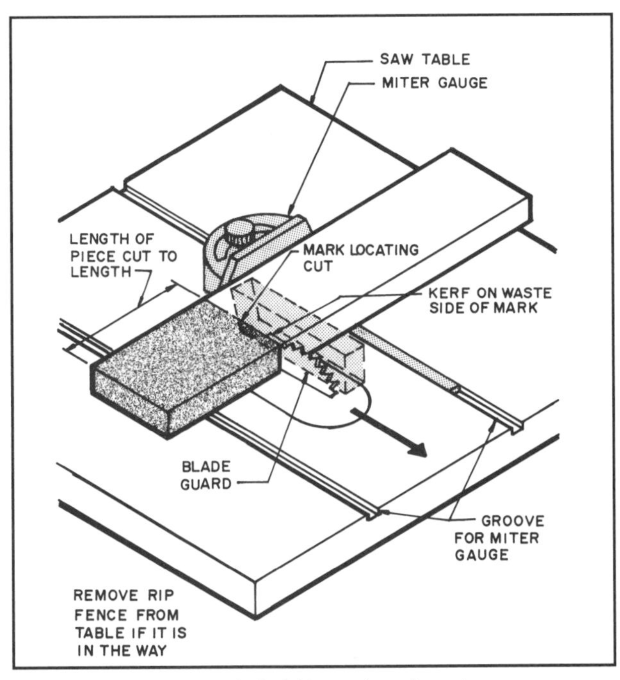

42-7 To crosscut stock, hold it against the miter gauge, and move them together toward the blade. Be certain the rip fence is well clear of the stock. Always have the guard in place.

RADIAL-ARM SAWS

A radial-arm saw is a type of circular saw that has the blade above the table, and it moves on an arm that extends over the table. A typical radial-arm saw is shown in **42-9**. This will rip and crosscut as well as cut miters.

The blade is mounted directly onto the motor. The motor and blade are raised and lowered by a crank on the top of the column. They also move horizontally by gliding along the arm. The blade can be tilted to cut on an angle. The blade must be installed with the teeth pointing down toward the table and rotating toward the fence. This action helps hold the stock being cut against the fence **(see 42-10)**.

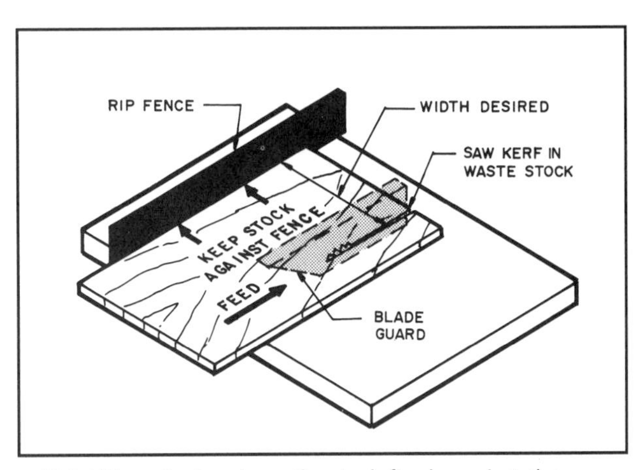

42-8 When ripping, keep the stock firmly against the fence. Always have the guard in place.

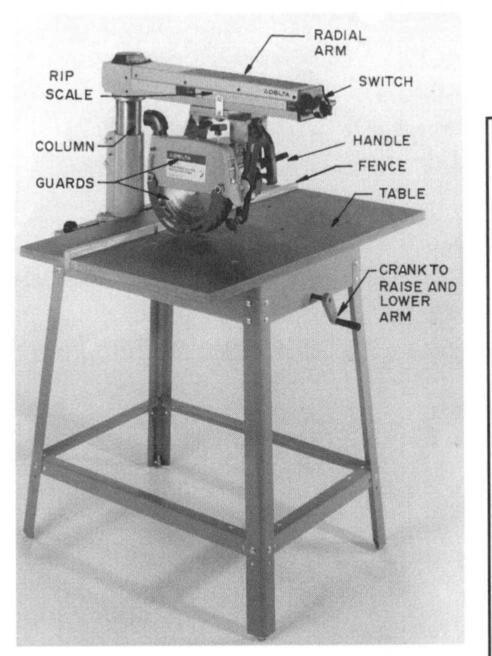

42-9 A lightweight radial-arm saw that can be easily moved about as the carpenter works. *(Courtesy Delta International Machinery Corp.)*

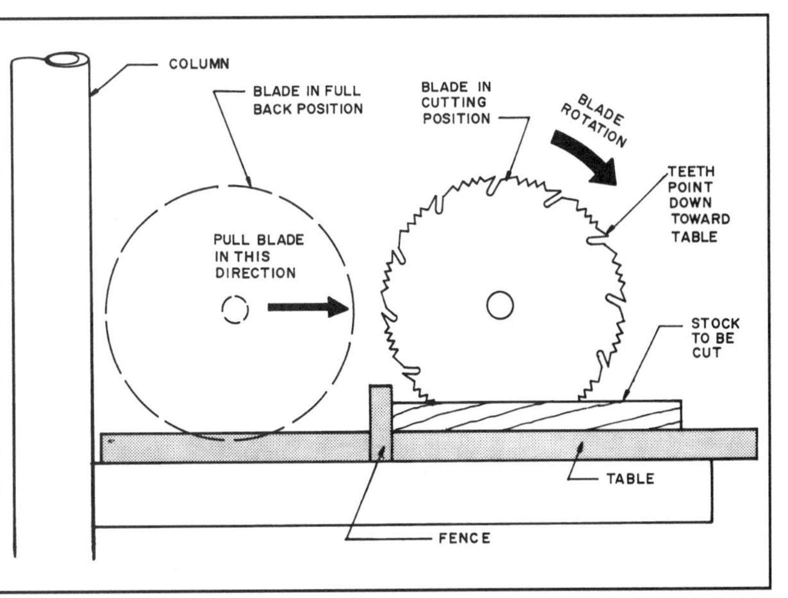

42-10 The teeth on the blade on a radial-arm saw must point down toward the table at the outside, opposite the fence.

Radial-Arm Saw Safety Recommendations

1. Never operate unless all guards are in place.
2. Be certain the blade is installed so that it rotates in the proper direction. If it is not installed properly, the saw will tend to run out toward the operator.
3. When ripping be certain to feed the stock into the blade from the proper direction **(see 42-11)**.
4. When crosscutting, hold the stock firmly until the blade is clear of the stock.
5. Keep both hands clear of the blade at all times. Never reach across the front of the saw.
6. Always return the blade/motor assembly to the full rear position after crosscutting a board.
7. Set the anti-kickback device so that it is slightly below the surface of the stock.
8. When ripping, be certain the blade is parallel with the fence. If not, binding will occur which will produce a damaging kickback.
9. Be certain to use the spreader device when you are ripping stock to keep the stock from binding behind the blade after it has been cut.
10. When ripping stock, use a push stick to feed stock past the saw.
11. Never cut anything freehand. Always keep the stock tight to the fence.
12. Do not cut wet, cupped, or warped wood.
13. When ripping make certain the blade is rotating toward you.

Ripping

1. Move the yoke out on the arm. Release the clamp and rotate it 90 degrees. Lock it to the arm so that the distance between the blade and the fence is the size desired. Set it so that the blade is rotating toward the board as it is fed into the saw **(refer to 42-11)**.
2. Be certain all adjustments are locked tight.
3. Lower the blade so that it just touches the wood table. It should cut a shallow kerf in the table when turned on.
4. Adjust the anti-kickback device so that it is slightly below the surface of the stock.
5. Adjust the blade guards so that the blade is covered.
6. Turn on the power. Recheck to be certain that the blade is rotating upwards toward you.
7. Place the smooth edge of the stock against the fence and feed into the blade. Do not rip curved, warped, or wet stock.
8. When the end of the stock approaches the blade, use a push stick to feed it past the blade.

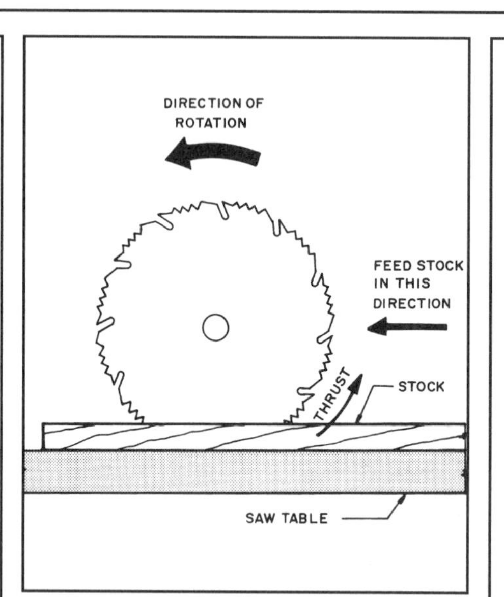

42-11 When ripping with a radial-arm saw, feed the stock into the blade from the side where the teeth are rotating upwards.

42-12 When crosscutting, hold the stock firmly against the fence, keep the guard in place, and keep your hands clear of the path of the saw.

Crosscutting

1. Adjust the arm to the desired angle of the cut. Lock it tightly in place.
2. Lower the blade so that it cuts a slight kerf in the wood table.
3. Adjust the anti-kickback device so that the fingers are just above the surface of the stock.
4. Place the blade-motor yoke back against the column so that the blade is clear of the front of the fence.
5. Place the stock firmly against the fence. Do not cut warped stock.
6. Turn on the saw.
7. Hold the stock to the fence with one hand. Be certain your hand is clear of the path of the saw. With the other hand move the saw forward into the stock. Hold it firmly so that you control the rate of cut **(see 42-12)**.
8. When the cut is complete, return the yoke to the rear position with the blade behind the fence. Then you can move the cut pieces of stock from the table.

Mitering

To cut a flat miter, rotate the arm right or left to the required number of degrees, which is usually 45 degrees. There is an automatic stop, setting the arm on this angle. Place the stock against the fence and cut in the same manner as crosscutting **(see 42-13)**.

A compound meter is cut with the radial-arm saw by adjusting the arm to the correct angle and tilting the motor unit to the correct number of degrees **(see 42-14)**. The stock is placed against the fence and cut the same as for a flat miter. To cut a compound miter on a 45-degree angle forming a 90-degree corner, tilt the blade to 30 degrees and place the arm at 35.25 degrees.

42-13 Miters are cut by pivoting the radial arm to the desired angle, and cutting in the same manner as crosscutting stock.

42-14 A compound miter requires that the saw be set on an angle (as for a regular miter) and tilted to the required angle to produce the desired cut.

MITER SAWS

The miter saw is used to cut miters and crosscut narrow stock. It resembles the radial-arm saw except that the blade-motor unit swings up and down on a pivot and may be adjusted to swing right and left on various angles **(see 42-15)**.

Miter saws are specified by the diameter of the blade. A 12-inch (30.5mm) saw will crosscut 2 × 8-inch stock at 90 degrees and 2 × 6- and 4 × 4-inch stock at a 45-degree miter. It will also miter 5¼-inch (133mm) crown molding and 5½-inch (140mm) base molding. A 10-inch (254mm) saw will crosscut 2 × 6-inch stock at 90 degrees and 2 × 4-inch stock at 45 degrees.

Miter Saw Safety Recommendations

1. Always leave all guards in place.
2. Be certain the blade revolves down toward the table.
3. Since the saw swings down, it is necessary to be especially alert and keep your hands well clear of the cutting area.
4. Hold the stock firmly against the fence.
5. As soon as a cut is complete, release the switch.
6. Always hold stock against the fence. Never try to cut freehand.
7. Be certain the spring that lifts the saw to an upright position is properly adjusted. Do not use the saw if it will not automatically rise to the upright position when released.
8. Secure the miter saw to a firm base that holds it at a comfortable working height **(see 42-16)**. Another miter saw stand is shown in **42-17**.

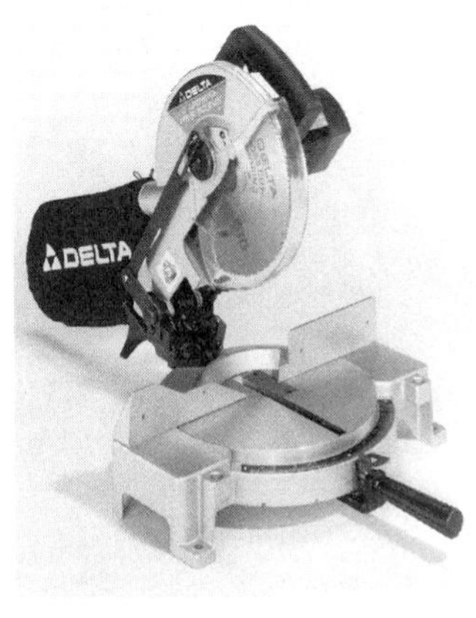

42-15 A power miter saw is used for making square and angled cuts. *(Courtesy Delta International Machinery Corp.)*

42-16 This portable lightweight stand holds the miter saw at a convenient height and provides portability. *(Courtesy Trojan Manufacturing, Inc.)*

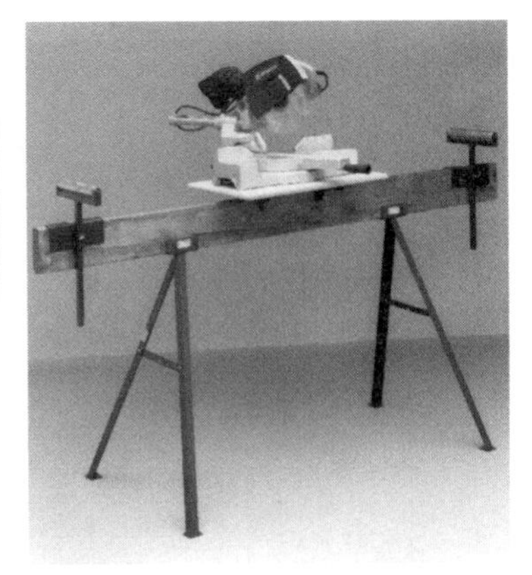

42-17 Steel sawhorse legs and a wood base can provide a simple work stand for a miter saw.

42-18 This molding is being cut on a miter with a power miter saw. *(Courtesy DeWalt Industrial Tool Company)*

Cutting Miters

1. Set the saw on the desired angle. Lock it tightly in place.
2. Place the stock firmly against the fence.
3. Turn on the saw.
4. Once you are certain your hands are clear, and the stock is firm against the fence, lower the blade and complete the cut **(see 42-18)**.
5. Release the switch and raise the saw. Be careful of the blade because on most saws it is still rotating. Some models have a brake to slow down the spin in a very short period of time.

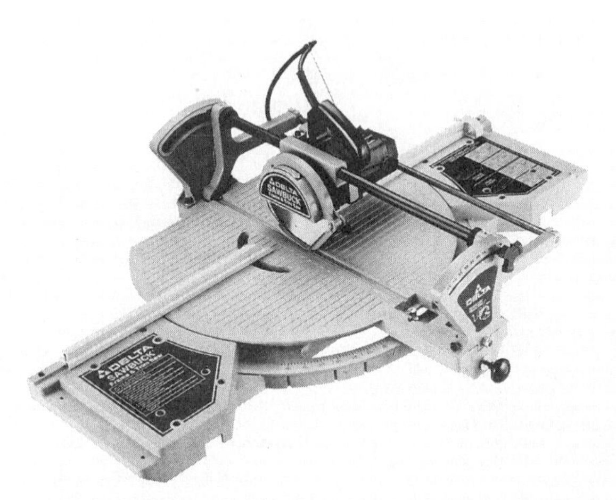

42-19 A framing and trimming saw. *(Courtesy Delta International Machinery Corp.)*

FRAMING & TRIMNMINGSAWS

A versatile saw that can crosscut and miter wide stock is the framing and trimming saw shown in **42-19**. It will crosscut stock up to 2¾ inches (7cm) thick and 16 inches (40.6cm) wide and miter 1¾-inch (4.4cm) stock 12 inches (30.5cm) wide at 45 degrees **(see 42-20 and 42-21)**.

It can also be used to cut dadoes commonly used for stair and cabinet construction **(42-22)**. The saw head can be tilted to cut compound miters.

When using the framing and trimming saw, observe the safety rules for the use of miter and radial-arm saws.

42-20 The framing and trimming saw can miter wide stock. *(Courtesy Delta International Machinery Corp.)*

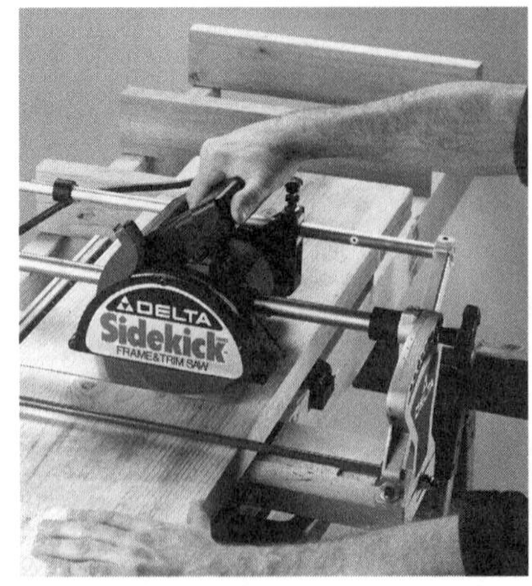

42-21 The framing and trimming saw produces accurate crosscuts. *(Courtesy Delta International Machinery Corp.)*

42-22 The framing and trimming saw can use a dado head to cut dadoes. *(Courtesy Delta International Machinery Corp.)*

SABER OR BAYONET SAWS

The saber or bayonet saw is primarily used to cut curves and internal cuts. It can also be used for crosscutting, ripping, and mitering but is not as accurate as other types of saws. A typical saber saw is shown in **42-23**. Light-duty saber saws are used to cut thin materials such as plywood sheets **(see 42-24)**. Heavy-duty saber saws can be used to cut two-inch (5.1cm) stock. Several brands will cut stock up to five inches (12.7cm) thick **(see 42-25)**.

Manufacturers provide a variety of blades for various purposes **(see 42-26)**. Thin wood is best cut with a fine-toothed blade such as one with 12 teeth per inch. A blade with 10 teeth per inch is used on thicker panels. If cutting two-inch stock, a blade with six teeth per inch will produce the fastest cut. Some blades will cut metal and plastics and can be used to cut through nails. Thin, hard metals and very thin woods are cut with a blade having 24 teeth per inch. A blade should always have two teeth in contact with the material. Thicker, softer metals require a blade with 14 teeth per inch.

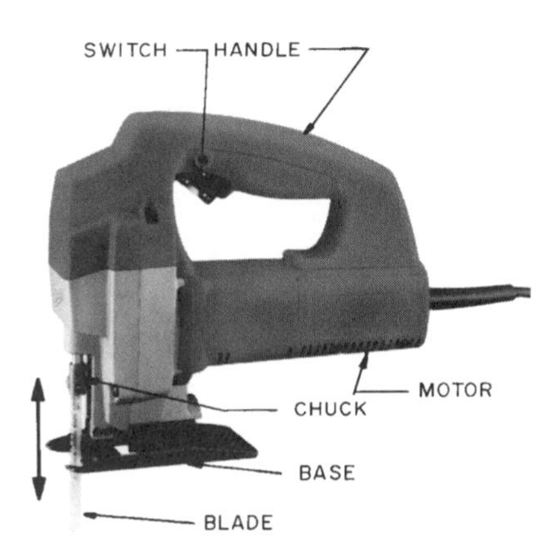

42-23 A saber or bayonet saw. *(Courtesy Milwaukee Electric Tool Corporation)*

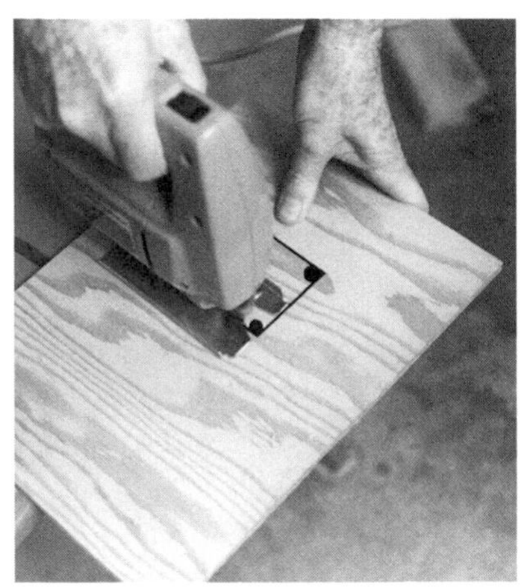

42-24 This saber saw is making an internal cut in a thin piece of plywood. *(Courtesy DeWalt Industrial Tool Company)*

Saber Saw Safety

1. Unplug the saw before changing blades.
2. Check to be certain the blade is securely held in the chuck.
3. As you cut, keep the cord out of the way.
4. When beginning a cut, place the base firmly on the wood surface before touching the blade to the wood.
5. If the blade gets stuck, turn off the power and slide the blade clear.
6. When plunge-cutting, use blades designed for that purpose.
7. After cutting, the blade is very hot, so do not touch it.
8. Do not force the saw to cut faster than normal.
9. Do not use dull blades.
10. Be sure the wood to be cut is held so that it will not slip.
11. Let the saw get to full speed before starting to cut.
12. Keep all fingers on top of the board.

42-25 Saber saws can cut stock up to 1½ inches (38mm) thick. *(Courtesy DeWalt Industrial Tool Company)*

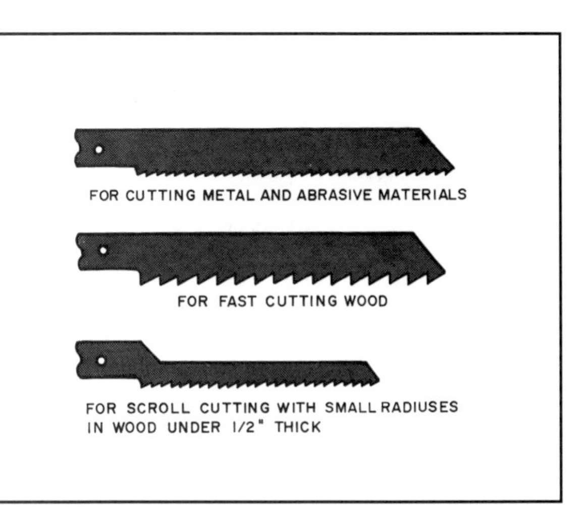

42-26 Several of the many types of blade available for use with a saber saw.

Internal cuts can be made by drilling holes at the corners of the area to be removed. The blade is lowered into the hole and the cut proceeds in a normal manner. Internal cuts in thin stock can also be made by plunge-cutting. The back edge of the base is placed on the panel surface. The saw is started and the blade slowly lowered to the wood surface. It cuts its way through the panel until the saw base is flat on the surface. The cut proceeds in the normal manner **(see 42-27)**.

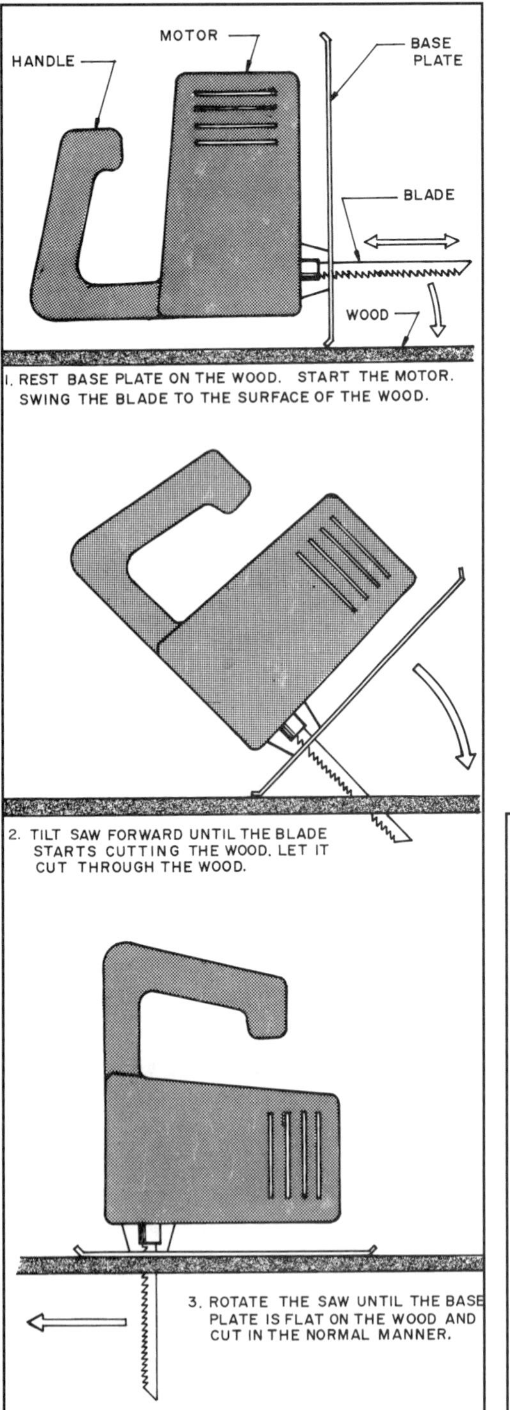

42-27 Saber saws are used to make plunge cuts through thin panels.

RECIPROCATING SAWS

A reciprocating saw operates much like a saber saw. A blade is held in a chuck and moves back and forth to produce a cutting action **(see 42-28)**. It will saw wood, metal, and plastic. A variety of blades are available for cutting various materials. The reciprocating saw does not produce accurate cuts but is used for rough-cutting operations such as cutting notches for pipes and holes for electric outlet boxes, cutting pipe and metal duct work, and sawing wood. It is a versatile saw and useful for many operations **(see 42-29)**.

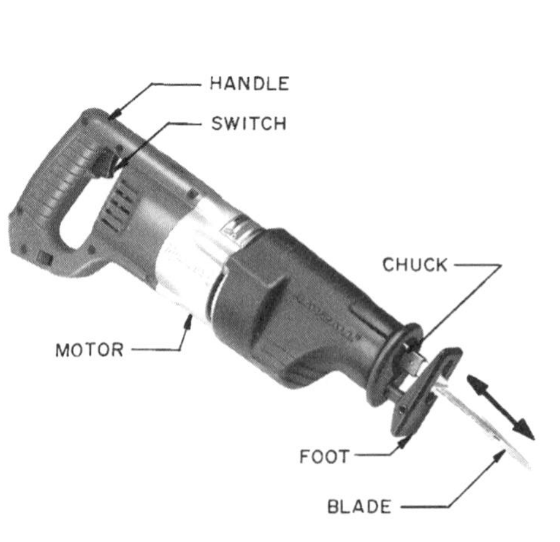

42-28 A reciprocating saw. *(Courtesy Milwaukee Electric Tool Corporation)*

Reciprocating Saw Safety

1. Observe the safety rules listed for the saber saw.
2. Use the shortest blade that will do the job.
3. When plunge-cutting use the foot piece and blade specified by the manufacturer.
4. Before making a blind cut, as into a wall, be certain no electric wires or plumbing is behind the surface.
5. Hold the tool by the insulated gripping surfaces.

42-29 A reciprocating saw can be used to cut openings in materials such as this subfloor for a heat register. *(Courtesy Milwaukee Electric Tool Corporation)*

PORTABLE CIRCULAR SAWS

The portable circular saw is probably the most frequently used power tool on a framing job. It is also the most dangerous. Carelessness and improper use cause serious accidents.

A typical saw is shown in **42-30**. Popular saw sizes are 7¼-inch (18.4cm) and 8¼-inch (20.9cm) diameters. However, they are available from five inches (12.7cm) to 10 inches (25.4cm) in diameter. They are used for crosscutting and ripping solid lumber, plywood, and other panel products. There are many types available, and the purchaser needs to consider possible uses in his/her selection. Factors to consider are the blade size, power available to the blade, guards, provisions for grounding, whether it has a brake on the blade, its weight, and its type of drive.

The saw cuts from the bottom of the material, so it produces a smoother cut at the bottom. This means the best surface of the stock should be placed down, and the cut made from the other side **(see 42-31)**.

The angle the blade makes with the surface can be adjusted from 90 degrees to 45 degrees. This permits the production of bevel cuts on ends and edges. To do this the base is unlocked and pivoted to the angle desired **(see 42-32)**.

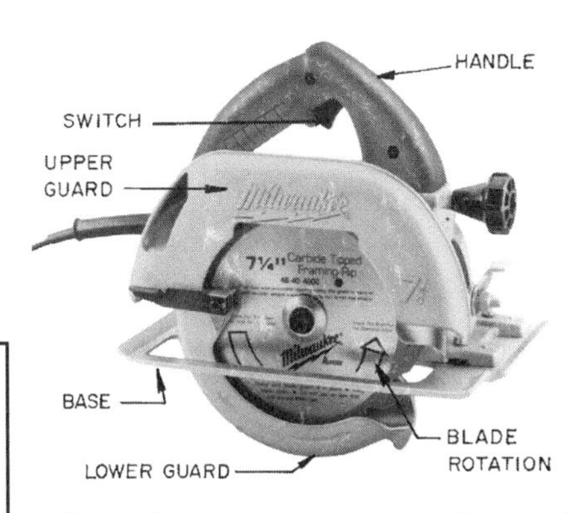

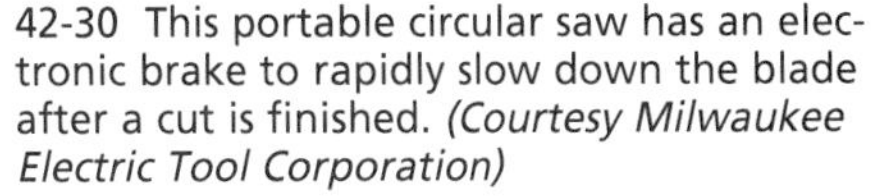
42-30 This portable circular saw has an electronic brake to rapidly slow down the blade after a cut is finished. *(Courtesy Milwaukee Electric Tool Corporation)*

Portable Circular Saw Safety

1. Do not use blades that are dull, that are cracked, or that wobble.
2. Keep the extension cord clear of the work. Be certain the cord is long enough to let you finish the cut. A short cord may jerk the saw back, causing a kickback.
3. The piece being cut must be clamped or otherwise held firmly in place.
4. Never hold a piece of wood in your hand while trying to cut it. It must be firmly held on some stationary object.
5. Do not try to cut small pieces of wood.
6. Set the depth of cut so that the blade protrudes no more than ⅛ inch (3mm) below the wood being cut.
7. Never, ever place your hands below the wood being cut.
8. Allow the blade to reach full speed before starting to cut.
9. A bind between the saw and wood could cause a kickback. If the saw binds, release the switch immediately. Put a wood wedge in the kerf to hold it open before trying to continue the cut.
10. Do not use the saw if the guards are not working.
11. After finishing a cut, let the guard close and the blade stop before moving it to another position.
12. Never use a dull blade. A dull blade will cause kickbacks and burning of the blade.
13. Do not force the saw to cut faster than its normal pace.
14. Remove pitch and resin buildup from the blade.
15. Do not cut wet wood. This increases friction and loads the blade with wet sawdust.
16. When cutting large pieces, have provision to keep the pieces from falling when the cut is completed. A helper is often a good solution.

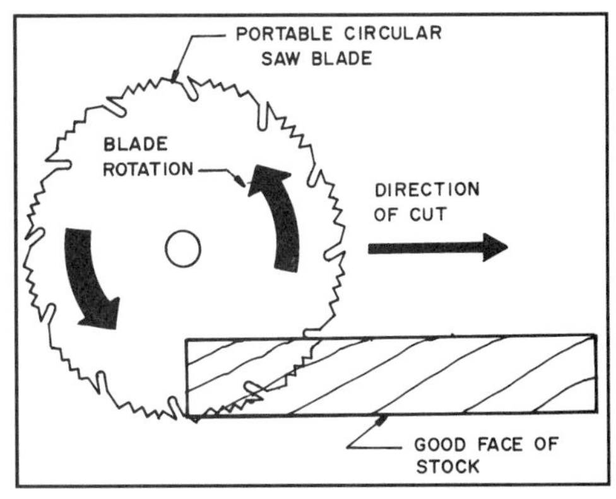

42-31 The blade of a portable circular saw cuts up from the bottom of the stock.

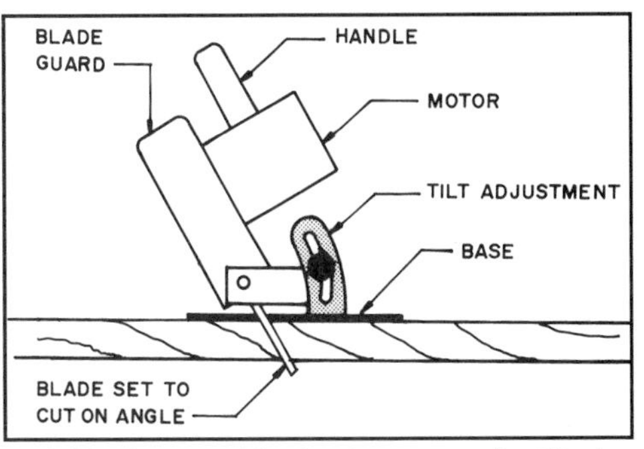

42-32 The portable circular saw can be tilted on its base to cut bevels.

The depth of cut can be adjusted by loosening a lock on the base and pivoting it up or down to expose more or less of the blade below the base **(see 42-33)**.

When installing a blade, be certain it is on the arbor, so that the teeth cut up from the bottom, as shown earlier in **42-31**. Usually the blade is marked with an arrow on the outside face, indicating the direction of rotation.

Crosscutting can be done freehand, if accuracy is not important **(see 42-34)**. A straightedge clamped to the stock will provide a fence against which the saw can slide. This provides a more accurate cut.

Ripping can also be done freehand, keeping the blade cutting along a line drawn on the surface. Long straightedges or straight stock can be clamped to the surface to produce a more accurate cut. One such device is shown in **42-35**. The aluminum track is clamped to the stock, and the saw is attached to a carriage that rolls along the track. A precision circular saw table is shown in **42-36**. The saw mounts on a carriage that slides on two round rails. It can be set up to make precision straight cuts, miter cuts, or cuts on a variety of angles. An adaptation of this uses round tracks and a mounted portable circular saw to function as a panel saw **(see 42-37)**.

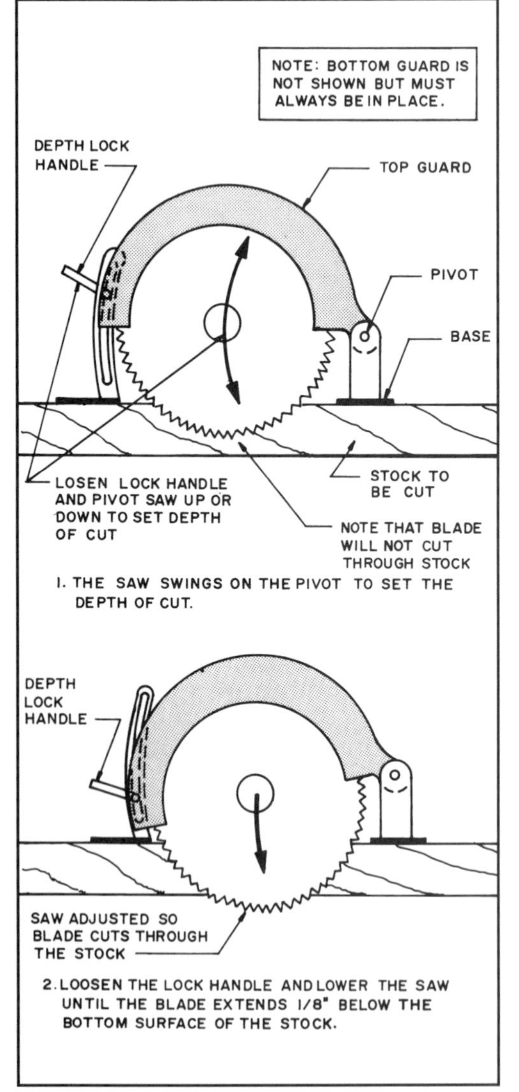

42-33 The depth of cut on a portable circular saw is set by raising or lowering the saw in relation to the base plate.

42-34 The portable circular saw can be used to crosscut stock. *(Courtesy DeWalt Industrial Tool Company)*

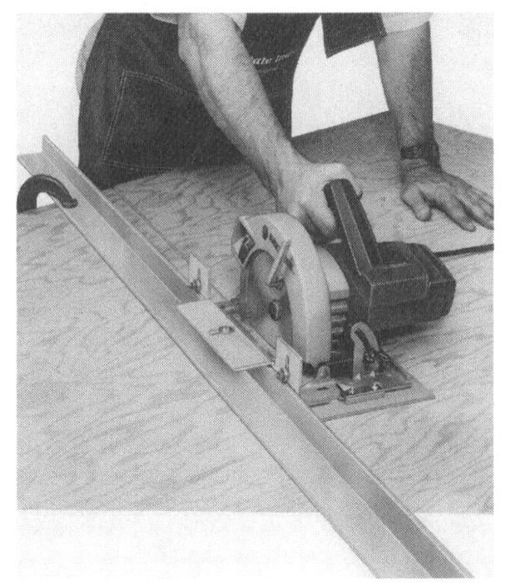

42-35 This portable panel saw system provides a guide so that large panels may be cut accurately. *(Courtesy Penn State Industries)*

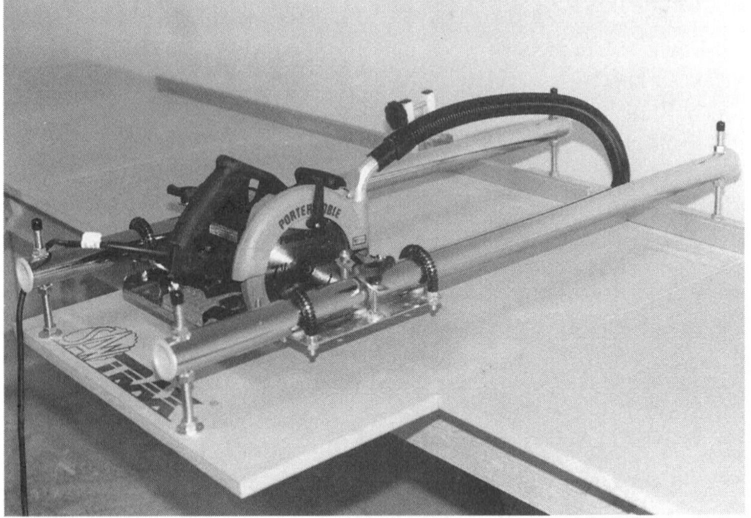

42-36 This unit makes precision straight cuts and may be set to cut a range of angles. *(Courtesy Saw Trax Manufacturing Co., Inc.)*

The same types of blade described for stationary circular saws are available for use on portable saws.

PORTABLE ELECTRIC PLANES

Portable electric planes are used to smooth edges of stock and faces of narrow boards. They will cut chamfers, rabbets, butt joints, and tenons. The width of the cutter is generally about three inches (76mm) and a typical depth of cut in a single pass is 11⁄64 inch (4.4mm). The depth of cut is regulated by raising or lowering the front shoe **(see 42-38)**.

To use the electric plane, set the desired depth of cut as specified by the manufacturer. Do not set it too deep. It is better to take two light cuts than one overly deep cut. The fence slides along the face of the board and keeps the plane moving in a straight line. Start the motor, and let it reach full speed. Place the front shoe on the board, and slide the plane along the surface. Hold the plane as specified by the manufacturer. At the start of the cut keep most of the downward pressure on the front shoe. As the end of the surface is reached, keep most of the pressure on the rear shoe **(see 42-39)**.

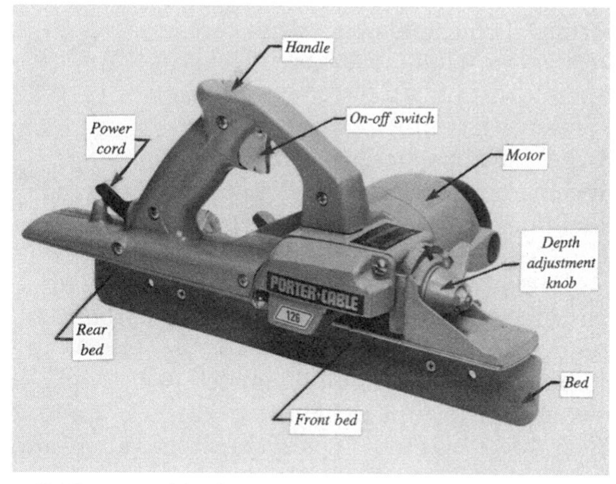

42-38 A portable electric plane. *(Courtesy Porter-Cable Corporation)*

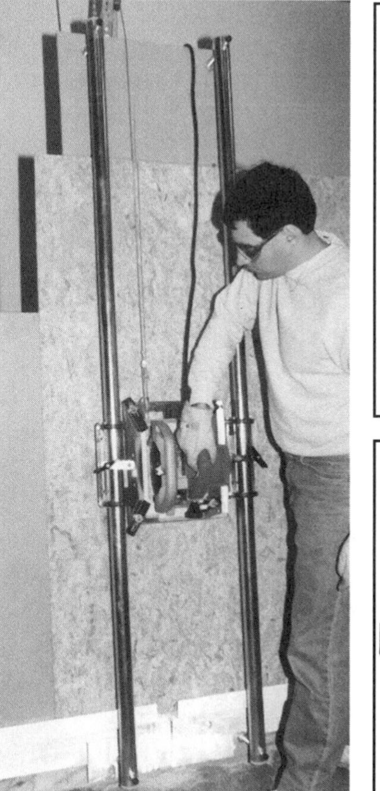

42-37 This unit turns a portable circular saw into a panel saw. *(Courtesy Saw Trax Manufacturing Co., Inc.)*

Portable Electric Plane Safety

1. Be certain the cutters are sharp and installed as recommended by the manufacturer.
2. Allow the cutter head to reach full speed before starting a cut.
3. Use thinner cuts on harder woods.
4. Hold the tool as recommended by the manufacturer. Large planes require two hands. Small block planes can be held with one hand.
5. Be certain the stock is held securely. Otherwise the plane will throw it back.
6. Keep both hands above the plane.
7. Do not feed material faster than the saw seems able to cut easily.
8. When possible, avoid cutting knots, especially with deep cuts.

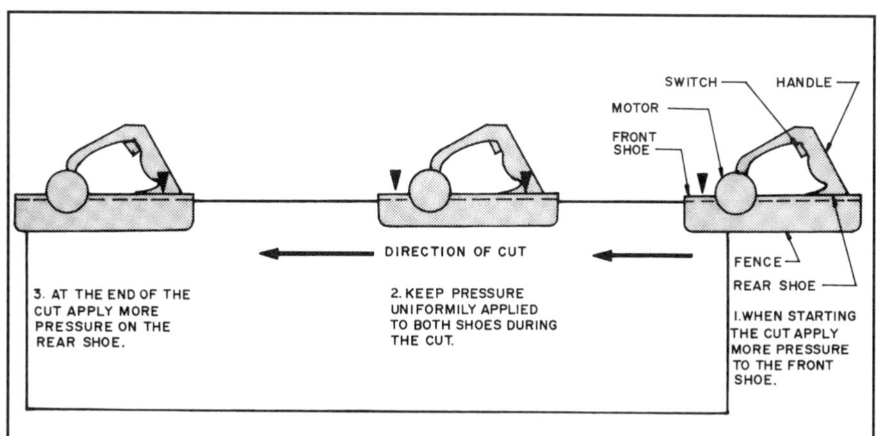

42-39 How to make a cut with a portable electric plane.

PORTABLE ELECTRIC DRILLS

Portable electric drills are available in a wide range of sizes and features. Some have electric motors connected to 120-V outlets by a cord, and others are battery operated and are called cordless drills. Typical electric drills are shown in **42-40** and **42-41**. Most small drills have a pistol-grip handle and can be operated with one hand. Larger, more powerful drills have a pistol grip and a second handle to control the torque produced by large-diameter drills.

42-40 A reversing, portable electric drill with a keyless chuck. *(Courtesy Milwaukee Electric Tool Corporation)*

42-41 This is a 12-volt, keyless, cordless, portable electric drill. *(Courtesy Milwaukee Electric Tool Corporation)*

Electric drills operating on 120-V current are specified by the chuck size and the amperage of the electric motor. The chuck size indicates the maximum-diameter drill shank it will hold. Most frequently used sizes are ⅜ and ½ inch (10 and 13mm). Typical amperage ratings run from 3 to 10 amps. Battery-operated cordless drills are specified by chuck size and the available voltage. Common chuck sizes are ⅜ and ½ inch, and voltages of 7 and 13 are typical. Both of the ones shown in **42-40** and **42-41** have a keyless chuck.

Bits are installed in key-type chucks which are then tightened with a key **(see 42-42)**. When using straight-shank twist drills, insert the round shank into the chuck but do not let any of the twisted flutes enter. Drills with no flutes, such as spade bits, are inserted into the chuck as far as they will go **(see 42-43)**. It is recommended that the chuck be tightened by inserting the key and tightening at all three holes in the chuck. The bit can be released by inserting the key in any one of the holes.

When drilling holes, be certain to keep the drill perpendicular to the work. If it drifts off to an angle, the bit may break or may bind in the work. When bits in large, powerful drills bind, the tool could twist out of your hands, causing injury.

Some drills can have the speed of rotation varied and the direction of rotation can be reversed.

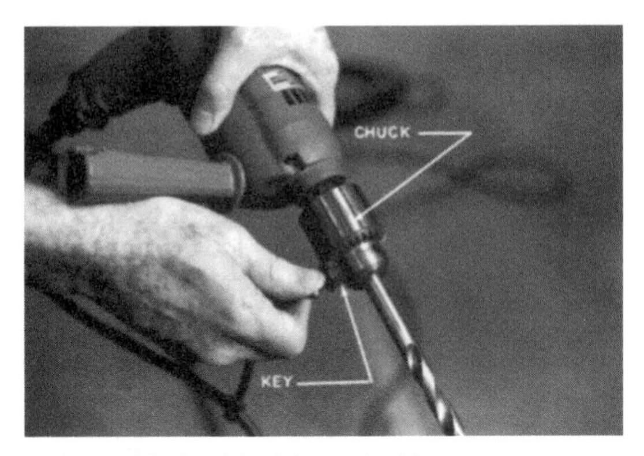

42-42 This chuck is tightened with a chuck key. *(Courtesy DeWalt Industrial Tool Company)*

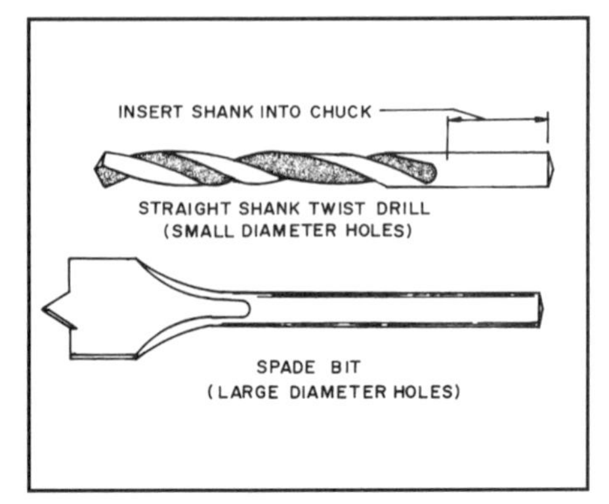

42-43 A commonly used drill and bit.

Portable Electric Drill Safety

1. Do not use a drill if the switch does not operate properly.
2. Remove the chuck key before starting the drill.
3. Always hold the tool securely.
4. Do not force the drill bit into the work.
5. Use sharp drill bits.
6. If the drill bit binds in the work, immediately release the switch. Free the bit from the work before proceeding.
7. When the drill is about to break through the back of the stock, reduce the pressure.
8. Drill bits get very hot, so do not touch them immediately after finishing a hole.
9. Keep drill bits clean and free of wood chips or resin deposits.
10. Be certain the work being drilled is firmly supported.

These can be equipped with a screwdriver bit and used to drive screws. A very low speed of rotation is used.

ELECTRIC DRYWALL SCREWDRIVERS

Gypsum drywall is commonly secured to wood and metal studs with special screws. These are driven with an electric screwdriver. When the screw is tight, the clutch releases the drive **(see 42-44)**. Cordless electric screwdrivers are also available **(see 42-45)**.

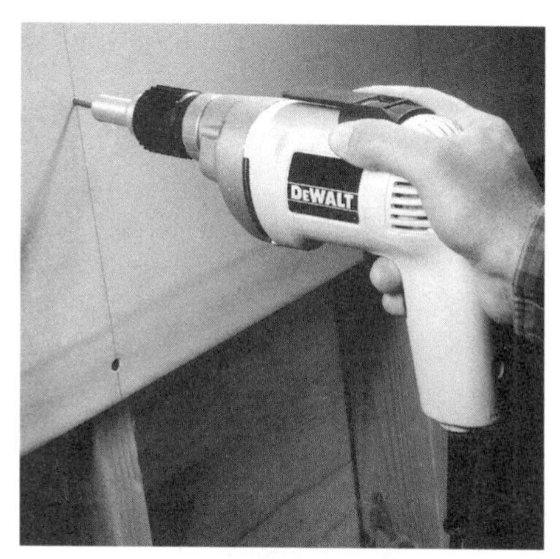

42-44 Gypsum wallboard can be installed with special screws driven with an electric screwdriver. *(Courtesy DeWalt International Tool Company)*

42-45 Cordless electric screwdrivers are convenient to use. *(Courtesy Milwaukee Electric Tool Corporation)*

PORTABLE ROUTERS

The portable router is used to make decorative cuts and joinery. It consists of two major parts, the motor and the base **(see 42-46)**. On the end of the motor shaft is a collet that holds the router bits. This adjustment sets the depth of cut **(see 42-47)**. A lock binds the motor and base together when the proper depth has been set. For general uses a one-horsepower router will be adequate. A finish carpenter will most likely want a two- or three-horsepower router so that deeper cuts may be made. In **42-48** is a router table. The router is mounted below the table with the cutter extending above it. This is operated like a shaper.

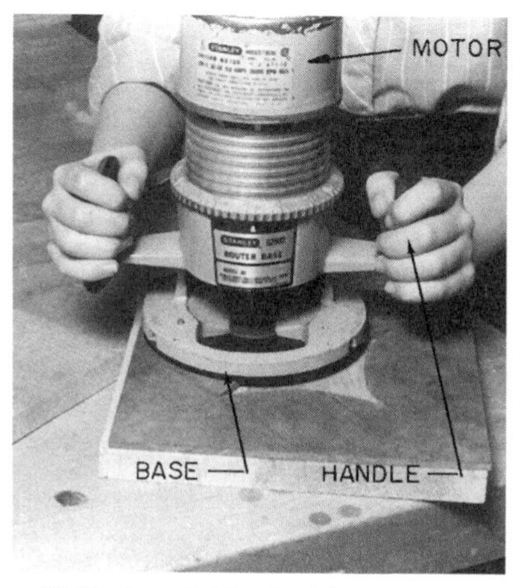

42-46 A portable electric router.

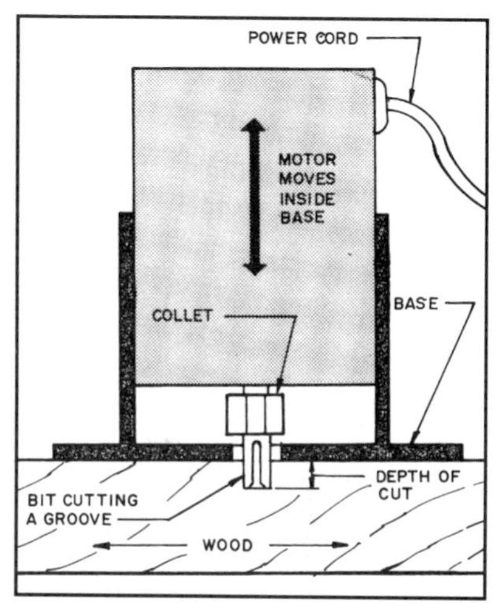

42-47 The router motor is raised and lowered to set the depth of cut.

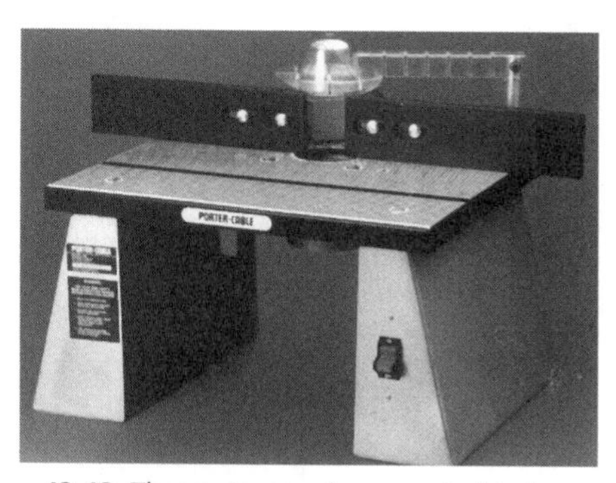

42-48 The router can be mounted below a router table and serve as a light-duty shaper. *(Courtesy Porter-Cable Corporation)*

Router Safety

1. Do not make deep cuts and overload the router. Limit cuts to ⅛ to 3⁄16 inch (3 to 5mm).
2. Before changing the cutter, unplug the power cord.
3. Stock to be routed should be clamped to sawhorses or a workbench.
4. Hold portable routers with two hands, using the handles provided on the machine.
5. Laminate trimmers are designed to be held with one hand.
6. Never place your hands near the rotating cutter.
7. When routing an outside edge of the stock, move the router in a counterclockwise direction.
8. Be certain the bit is correctly installed.
9. Do not let the rotating bit become tangled in your clothing. Remove rings and other jewelry, eliminate long hanging clothing, and put long hair in a hair net.
10. Wear eye protection and a face shield.
11. Do not use dull router bits.

A **laminate trimmer** is used to trim high-pressure plastic laminates, as used on countertops, so it is flush with the edge of the substrate. It is a small, lightweight unit that operates similarly to a router (**see 42-49**).

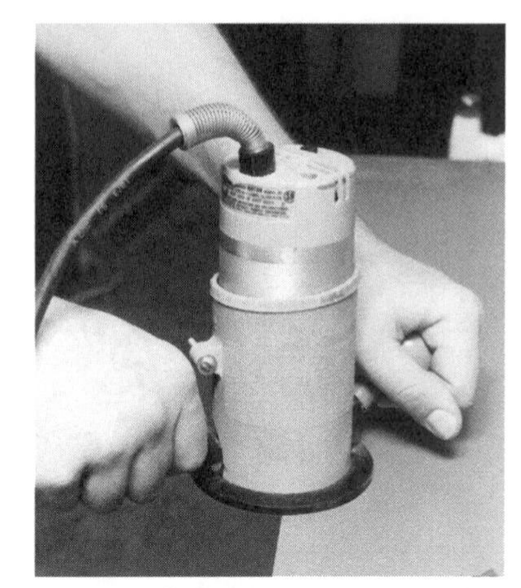

42-49 A laminate trimmer is used to finish cutting the edge of countertop plastic laminate after it has been bonded to the substrate.

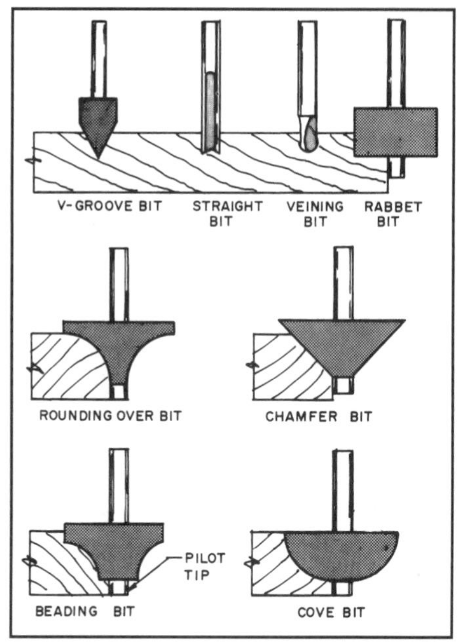

42-50 A few of the router bits available for routing edges and decorating surfaces.

Router Bits

There are dozens of types of router bits available. Some produce various shapes on the edges of stock; others produce decorative cuts on the surface; and others still are used to cut grooves (**see 42-50**). The cutting edges are made from carbon steel and are available carbide tipped. The carbon-steel bits are less expensive but the carbide-tipped will last much longer. Those who do a lot of routing will most likely use carbide-tipped cutters. Carbide cutters can also be used to trim plastic laminates, particleboard, and plywood. Some have a pilot tip, which is a small roller on the end of the bit. It rolls along the edge of the board and controls the width of the cut.

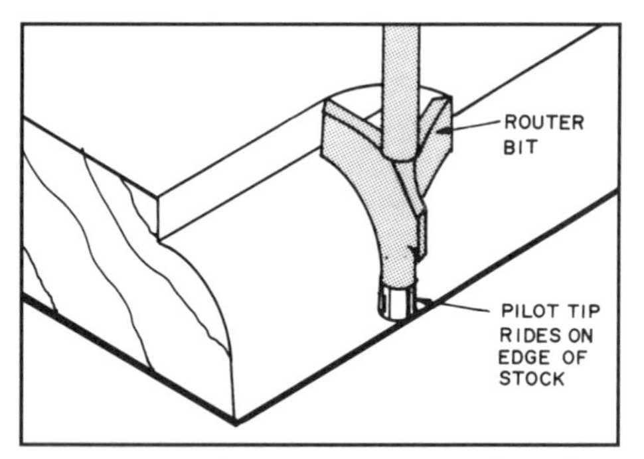

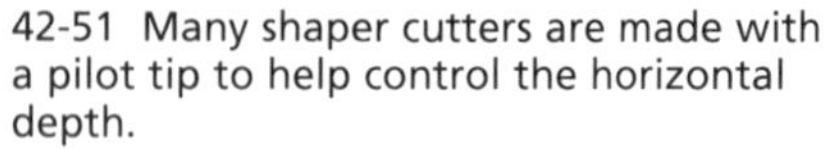

42-51 Many shaper cutters are made with a pilot tip to help control the horizontal depth.

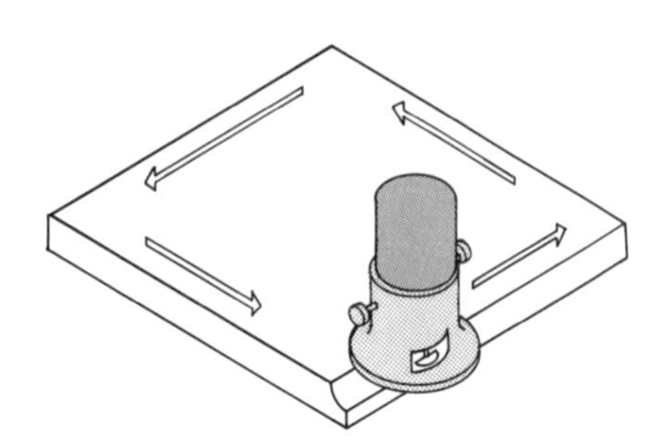

42-52 Move the router from left to right.

Routing an Edge

Begin by clamping the stock firmly to a workbench or sawhorses. Then set the depth of cut. If it appears to have a heavy cut, set it to take a lighter partial cut first, and then set the bit to take a full cut for the second pass. Light-duty routers may require more than two passes for a deep cut. Overloading may also damage the cutter.

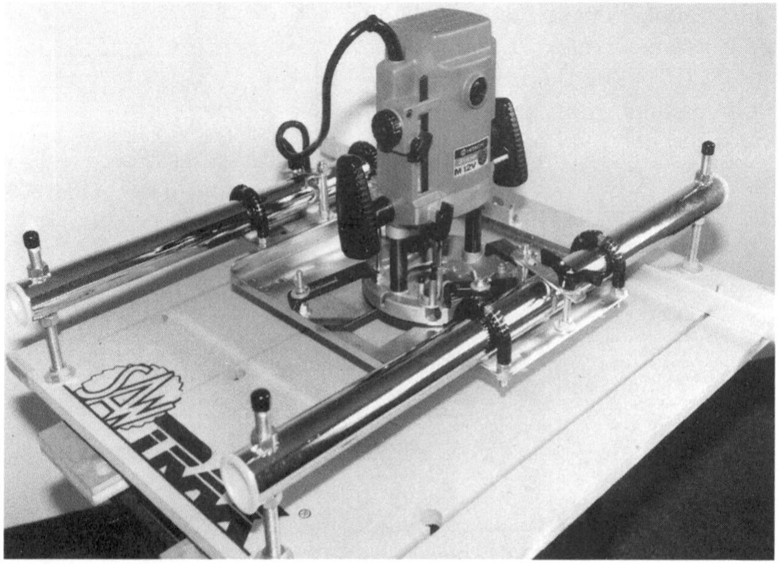

42-53 This router control system permits many edge moldings and decorative surface cuts to be accurately completed. *(Courtesy Saw Trax Manufacturing Co., Inc.)*

Once the stock is secure and the depth set, plug in the router. Most bits used to shape edges have a pilot tip riding along the edge of the stock. This controls the horizontal depth to which the bit will enter the wood **(see 42-51)**.

To begin the cut, start the motor and let it get to full speed. Place the base flat on the workpiece and slowly move the cutter into the wood until the pilot tip touches the edge. Move the router from left to right in a steady motion. Experience will show you how fast to move. If you go too slow the cutter will burn the wood. If you go too fast the cut may be bumpy rather than smooth. At the end of the cut lift the router **(see 42-52)**.

In **42-53** is a router control system that mounts the router on a moving carriage that slides on side bars. The stock is placed on the table below the router and clamped in place. The router can be moved to cut rabbets, dadoes, edge moldings, or decorative surface cuts.

Decorative Routing

The decorative router cutters do not have a pilot, so the router must be guided by a template or the router guide that is usually supplied with the router. The router guide is used to make straight cuts as shown in **42-54**. A circular decorative cut may be made by placing a circular guide on the rods from the straight guide **(see 42-55)**. Rotate the router counterclockwise.

Cutting Grooves

Grooves can be cut on the edges of the stock by following the setup shown in **42-56**. The commercial straight guide is used to keep the bit in the desired location on the stock. It helps if the stock is supported by wood support strips on the side. This makes it easier to adjust the guide.

BISCUIT JOINERS

A biscuit, or plate, joiner **(see 42-57)** uses a small cutter to cut short, circular kerfs in the edges of wood to be joined. Pressed-wood splines, often called biscuits or plates, are glued into these kerfs holding the wood parts together. They are commonly used on edge-to-edge, butt, and miter joints **(see 42-58)**.

42-54 This router guide is used to make accurate, straight cuts.

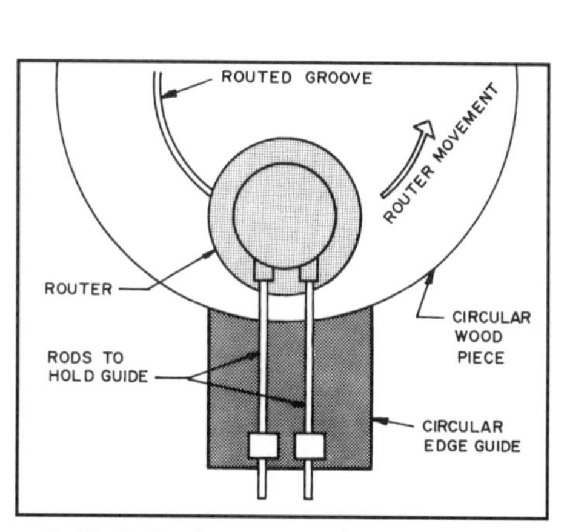

42-55 A circular router edge guide is used when routing circular edges.

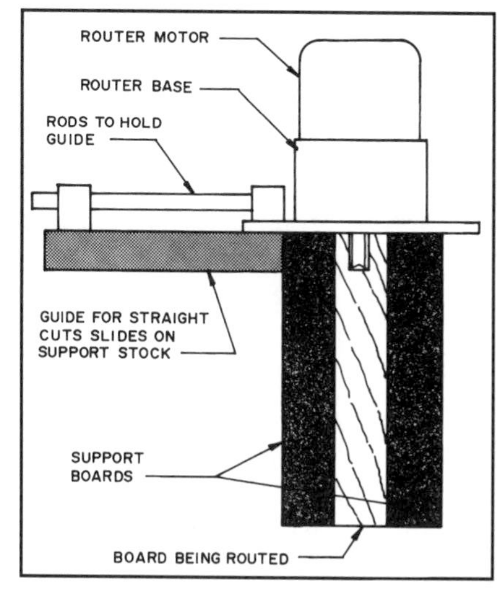

42-56 When routing grooves in the edges of stock, clamp support boards on each side to provide a broader surface.

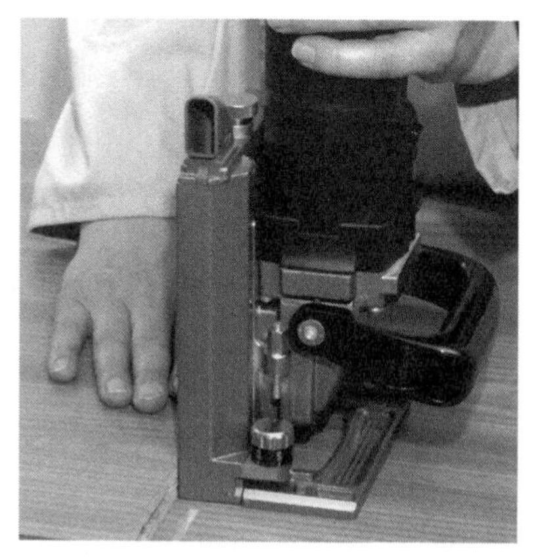

42-57 The biscuit, or plate, joiner cuts small circular kerfs in the edges of stock to be joined. *(Courtesy Colonial Saw Company)*

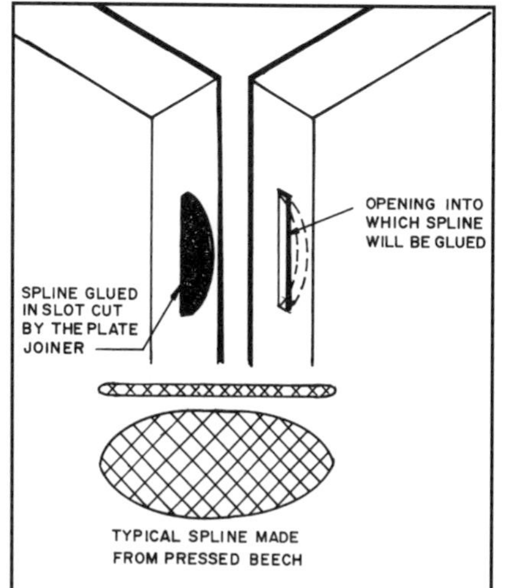

42-58 Pressed-wood biscuits (splines or plates) are glued in kerfs cut by the biscuit, or plate, joiner.

Belt Sander Safety

1. Unplug the power cord before changing the abrasive belt.
2. Be certain the switch is OFF before plugging in the power cord.
3. Clamp pieces to be sanded to a workbench or sawhorses.
4. Keep both hands on the handles provided on the sander.
5. Keep the power cord out of the way. Some run it over their shoulder.
6. Do not set the sander down while the belt is still turning.

42-59 This heavy-duty belt sander will remove wood rapidly. *(Courtesy Porter-Cable Corporation)*

PORTABLE SANDERS

The two commonly used portable sanders are for belt- and pad-type finishing. The belt sander is primarily used for removing large amounts of wood from flat surfaces **(see 42-59)**. It is available in a variety of sizes. The size is indicated by the length and width of the belt, with 3 × 21 inches and 4 × 24 inches (76 × 533mm and 102 × 610mm) being widely used. Some have a dust bag attached to reduce the dust in the air.

Using the Belt Sander

Before starting the belt sander be certain the belt is properly installed. The sander has tracking knobs on each side which need frequent adjustment to keep the belt running straight on the pulleys.

Clamp the stock to be sanded or it will be thrown across the room when the moving belt is placed on it. Place the sander flat on the piece, grip the handles firmly, and turn on the power. Immediately move the sander, following the pattern in **42-60**. The strokes should overlap and be short. Always sand with the grain.

Finishing Sanders

Finishing sanders are used to produce a smooth finished surface, after the heavier belt sander has already been applied. One type has a rectangular pad **(see 42-61)** that may move in an orbital, straight-line, or vibrating direction **(see 42-62)**. The pad is covered with abrasive paper held to the pad with clamps on each end. Since it is a finish sander, generally fine-grit abrasive papers are used.

When using the pad sander, work with the grain. Do not bear down on the sander to try to speed up removing a defect; use a coarser paper or go back to the belt sander instead. A vibrating sander and an orbital sander may be moved in any direction across the board. Straight-line sanders must be moved with the grain.

A very fine finish can be obtained using a random-orbit finishing sander **(see 42-63)**. Sanding pads are bonded to the pad with pressure-sensitive adhesive-backed paper. The sander can be moved in any direction without leaving any scratch marks. It is used for the final sanding operation.

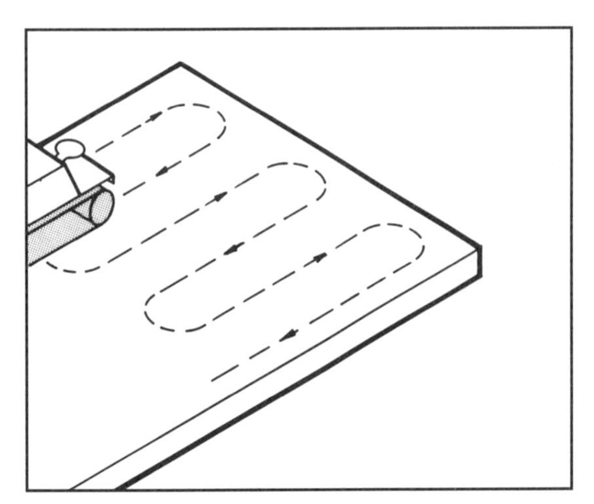

42-60 Follow this pattern when using a belt sander.

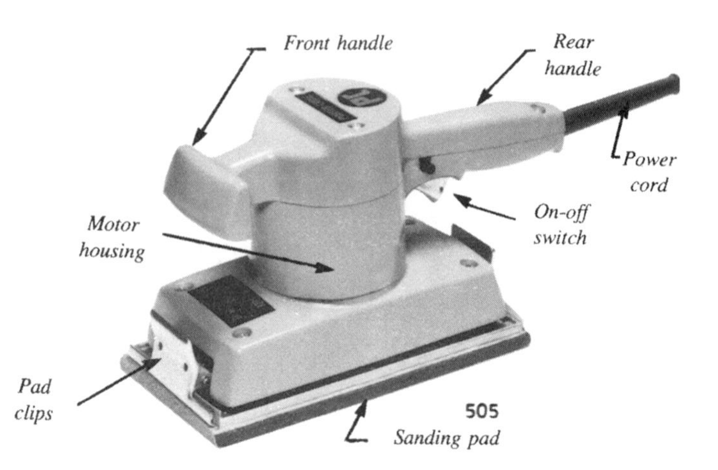

42-61 A finishing sander. *(Courtesy Porter-Cable Corporation)*

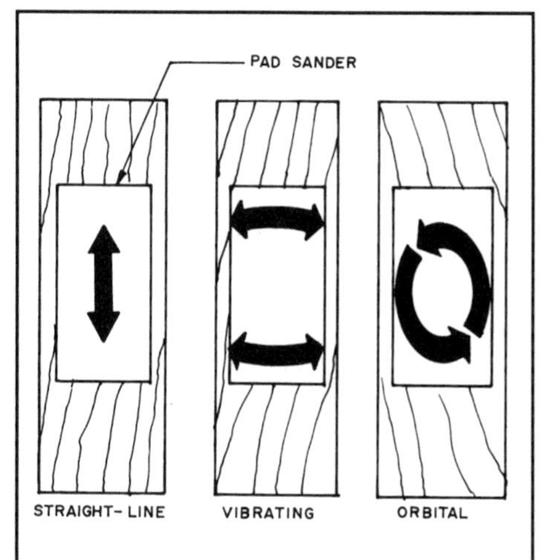

42-62 Finishing sanders are available that produce different sanding patterns.

42-63 A random-orbit finishing sander. *(Courtesy Porter-Cable Corporation)*

42-64 Power nailers can be used for many of the jobs performed by the finish carpenter. *(Courtesy Senco Products, Inc.)*

42-65 Portable air compressors are required for operating many of the tools used by finish carpenters. *(Courtesy Thomas Industries)*

Power Nailer & Stapler Safety

1. Before operating any power nailer or stapler, study the owner's manual and observe all operating recommendations.
2. Treat any power nailer or stapler as you would a gun. Never point it at anyone, even as a joke. The projected fastener can cause serious physical damage.
3. Keep both hands behind the nail-ejecting tube.
4. Keep your feet and legs behind and clear of the tool.
5. Wear safety glasses.
6. Use only the fasteners designed for the tool and recommended by the manufacturer.
7. Keep the tool tight against the surface being fastened. Do not let it bounce.
8. Maintain the recommended air pressure.
9. When making repairs or adjustments, disconnect the tool from the air line.
10. Do not use a nailer that will discharge a nail when the end of the tool is not against a surface.

POWER NAILERS & STAPLERS

Power nailers speed up the work of the finish carpenter. They can be used for almost any work that needs to be accomplished. Jobs such as the installation of doors, casing, baseboard, and cabinets are examples where power nailers and staplers are helpful **(see 42-64)**. Power nailers operate off of an air compressor. However, a cordless type is available. Some nailers drive small brads such as are required when installing small trim and moldings. Brad sizes range from ⅝ inch to 1⅝ inches (15 to 42mm). A finish nailer is used for installing light and heavy trim, paneling, and stair parts. Finish nails range from 1 to 2 inches (25.4 to 50.8mm) in length. Heavy-duty nailers are used by framing carpenters and drive nails up to 16d (3½ inches or 89mm).

Power staplers also find use on interior work but are used in places where the staples will not be seen.

Air Compressors

It is important to have an air compressor supplying a constant air stream at the required **pressure**. Air pressures from 80 to 100 psi (551 to 689 kPa) are common. The air supply must also provide the required **lubrication** to the power tool as specified by the manufacturer. Water condenses in the air tank and must be regularly drained. Filters are used to keep the air clean and remove abrasive sludge created by rust and dust in the air supply system. The air intake filter must be frequently cleaned **(see 42-65)**.

ELECTRICAL POWER

The carpenter will need a variety of **electric extension cords**. Frequently the building will be without electricity, and power has to be brought in from an electric drop pole. It will have a meter and box into which a long, heavy-gauge electric wire can be run into the house. Some use 12- or 14-gauge thermoplastic-covered exterior wire for this long extension. In the building the wire connects to an electrical box having one or more outlets. Smaller flexible wire cords run from the box to the area where lights or power tools are to be used. Extension cords must use wire with a large enough diameter to carry the load over the length of the wire **(see 42-66)**. Whenever there is any doubt about bringing power into the building, a qualified electrician should be consulted.

While portable power tools should always be doubled insulated to protect from electrical shock and the circuit being used should be grounded, further protection against a fault, or short circuit, should be obtained by installing a **ground-fault interrupter (GFI)**.

These ground-fault interrupter (GFI) units are plugged into the grounded circuit **(see 42-67)**. They sense an electrical fault current that would be too small to trip the normal circuit breaker but which could still flow through a person in contact with faulty equipment and a grounded surface. The GFI opens, or interrupts, the circuit, protecting the operator. It is especially needed when working on the ground or wet surfaces.

42-66 Recommended extension cord sizes for use with portable electric tools.

(For Rubber Types S, SO, SR, SJ, SJO, SV, SP & Thermoplastic Types ST, SRT, SJT, SVP, SPT)

Nameplate Amperes	Cord Length in Feet																			
	25	50	75	100	125	150	175	200	225	250	275	300	325	350	375	400	425	450	475	500
1	16	16	16	16	16	16	16	16	16	16	16	16	16	16	16	16	16	16	16	14
2	16	16	16	16	16	16	16	16	16	16	14	14	14	14	14	12	12	12	12	12
3	16	16	16	16	16	16	14	14	14	14	12	12	12	12	12	12	10	10	10	10
4	16	16	16	16	16	14	14	12	12	12	12	12	12	10	10	10	10	10	10	10
5	16	16	16	16	14	14	12	12	12	12	10	10	10	10	10	8	8	8	8	8
6	16	16	16	14	14	12	12	12	10	10	10	10	10	8	8	8	8	8	8	8
7	16	16	14	14	12	12	12	10	10	10	10	8	8	8	8	8	8	8	8	8
8	14	14	14	14	12	12	10	10	10	10	8	8	8	8	8	8	8	8	8	
9	14	14	14	12	12	10	10	10	8	8	8	8	8	8	8	8	8			
10	14	14	14	12	12	10	10	10	8	8	8	8	8	8	8					
11	12	12	12	12	10	10	10	8	8	8	8	8	8	8						
12	12	12	12	12	10	10	8	8	8	8	8	8	8							
13	12	12	12	12	10	10	8	8	8	8	8	8								
14	10	10	10	10	10	10	8	8	8	8	8									
15	10	10	10	10	10	8	8	8	8	8										
16	10	10	10	10	10	8	8	8	8	8										
17	10	10	10	10	10	8	8	8	8											
18	8	8	8	8	8	8	8	8	8											
19	8	8	8	8	8	8	8	8												
20	8	8	8	8	8	8	8	8												

Notes: Wire sizes are for 3-CDR cords, one CDR of which is used to provide a continuous grounding circuit from tool housing to receptacle.
Wire sizes shown are A.W.G. (American Wire Gauge).
Based on 115-V power supply; ambient temp. of 86 degrees F (30 degrees C).

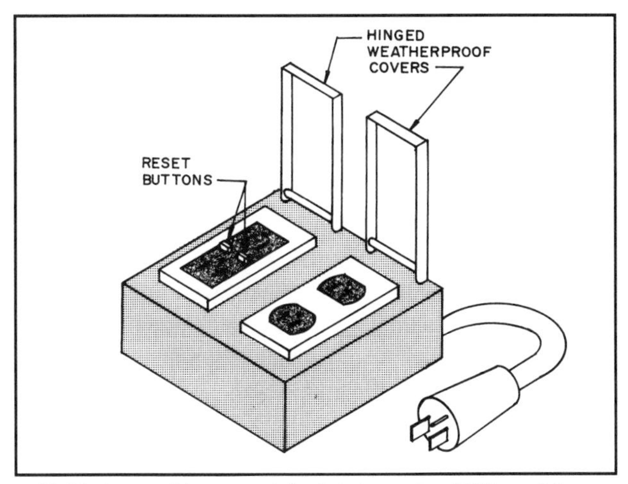

42-67 A portable ground-fault interrupter (GFI) used for protection against a fault or short circuit, where permanent GFI circuit protection has not been installed. Power tools are plugged into these outlets.

GROUND-FAULT PROTECTION ON CONSTRUCTION SITES

It is important that electrical equipment on the construction site be protected against 120-volt electrical hazards. This is accomplished by the use of ground-fault circuit interrupters (GFCI) or through an assured equipment grounding conductor program.

INSULATION & GROUNDING

Insulation and grounding are two means of preventing injury from electrical shock while operating electrically powered equipment. Conductor insulation must be an approved nonconductive material such as plastic. Grounding is achieved by using approved grounding electrodes. Any metal enclosures around a motor, or box around electrical switches, circuit breakers, and controls, must be connected to the ground. Conductors with damaged insulation, such as on an extension cord, should be replaced or professionally repaired. Conductors with old, cracked insulation should be cut up into short pieces and discarded. The use of a ground-fault circuit interrupter is another method of overcoming insulation and grounding deficiencies.

GROUND-FAULT CIRCUIT INTERRUPTERS

A GFCI is a fast-acting circuit breaker that senses small imbalances in the circuit caused by current leakage to ground and, in a fraction of a second, shuts off the electricity. The GFCI matches the amount of current going to an electrical device against the amount of current returning from the device along the normal electrical path. When the amount moving to the device differs from the amount returning in the circuit by 5 millamps, the GFCI interrupts the electrical power in about 1/40 of a second.

ADDITIONAL INFORMATION

Ground-Fault Protection on Construction Sites, U.S. Department of Labor, Occupational Safety and Health Administration, Washington, DC 20001

APPENDICES

APPENDIX A

STRUCTURAL FRAMING & WOOD FLOOR NAILING REQUIREMENTS

Structural Floor Framing Nailing Requirements

Type of Connection	Nailing Method	Number of Nails	Nail Size	Comments
Joist to sill or girder	toenail	3	8d	
Header to joist	end-nail	3	12d/16d	
Header or stringer joist to sill	toenail	———	8d	24 inches O.C.
Built-up beams and girders	face-nail	———	12d	12 inches O.C. on top and bottom, two 12d at each end
Wood bridging	toenail	2	8d	

Structural Wall Framing Nailing Requirements

Type of Connection	Nailing Method	Number of Nails	Nail Size	Comments
Bottom plate to stud	end-nail	2	12d/16d	or four 8d toenail
Top plate to stud	end-nail	2	12d/16d	
Bottom plate to joist	face-nail	———	12d/16d	16 inches O.C.
Double studs	face-nail	———	12d/16d	24 inches O.C., stagger
Double top plates	face-nail	———	12d/16d	16 inches O.C.
Top plate overlaps and intersections	face-nail	2	12d/16d	
Header, two pieces	face-nail	———	12d/16d	16 inches O.C., stagger on each edge
One-inch diagonal bracing	face-nail	2	8d	to each stud and plate

Structural Ceiling Framing Nailing Requirements

Type of Connection	Nailing Method	Number of Nails	Nail Size	Comments
Ceiling joists to top plate	toenail	3	8d	
Ceiling joist laps at partition	face-nail	3	12d/16d	

Structural Roof Framing Nailing Requirements

Type of Connection	Nailing Method	Number of Nails	Nail Size	Comments
Rafter to top plate	toenail	3	8d	
Rafter to parallel ceiling joist	face-nail	3	12d/16d	
Rafter to hip or valley rafter	toenail	3	8d	
Rafter to ridgeboard	toenail	3	12d	
One-inch collar beams to rafter	face-nail	3	8d	
Two-inch collar beams to rafter	face-nail	2	12d	

Wood Flooring Nailing Requirements

Tongue-and-Groove over Subfloor	Nailing Method	Number of Nails	Nail Size	Comments
⅜″		——	6d	
½″	all blind-nailed	——	7d	all spaced 10 to 12 inches apart
⅝″		——	8d	
1″		——	8d	
1¼″		——	8d	
1½″		——	9d	
2″ planks		2	16d	at each bearing or as recommended by manufacturer

Appendix B

Subfloor, Roof & Wall Sheathing Nailing Requirements

(Common or box nails may be used unless otherwise specified.)

Solid Wood

Size	Nailing
1″ x 8″ or less	Face-nail with two 8d
Wider than 1″ x 8″	Face-nail with three 8d
2″ planks	Face-nail with two 16d

Plywood Sheathing

Fasteners spaced 6 inches on center at edges and 12 inches on center on interior bearings except for floors, where 10 inches on center is required.

Thickness	Nailing
½″ or less	6d common or deformed shank
⅝″ to ¾″	8d common or 6d deformed shank
⅞″ to 1″	8d common or deformed shank
1⅛″ to 1¼″	10d common or 8d deformed shank

Plywood Combined Subfloor Underlayment

Thickness	Nailing
¾″ or less	6d deformed shank
⅞″ to 1″	8d deformed shank
1⅛″ to 1¼″	10d common or 8d deformed shank

Particleboard Wall Sheathing

Nails spaced 6 inches on center at edges, 12 inches on center on interior bearings except floors, where 10 inches on center is required.

Thickness	Nailing
⅜″ to ½″	6d common
⅝″ to ¾″	8d common

Fiberboard Sheathing

Fasteners spaced 7 inches on center at exterior edges and 6 inches on center at intermediate supports.

Thickness	Nailing
½″	6d common or 11-gauge corrosion-resistant roofing nails with 7/16″-diameter head and 1½″ long
25/32″	8d common or 11-gauge corrosion-resistant roofing nails with 7/16″-diameter head and 1¾″ long

Oriented Strand Board

Nails spaced 6 inches on center on panel edges and 12 inches on center on panel interior (floor, wall, roof).

Thickness	Nailing
½″	6d common
19/32″ to ¾″	8d common or 6d annular or spiral

Appendix C

Canadian Lumber Floor Joist Spans, Feet & Inches

Span data for Canadian lumber in Appendices C through I have been determined on the same basis as those published by the National Forest Products Association. Spans are based on the use of lumber in dry service conditions and the use of base values for strength and stiffness that have been modified for repetitive member use, size factor, and load duration.

Appendix D includes a table that gives a slope factor that must be used to determine the sloping distance.

The horizontal span in the table is multiplied by the slope factor to get acceptable sloping distances. For example, if a slope is 4 in 12 and the horizontal rafter span is 12'6", the sloping distance is 12.5 x 1.054 = 13.175 feet or 13'2".

(Spruce-Pine-Fir)

(Sleeping Rooms and Attics)

Maximum Allowable Spans (feet-inches)

	2 x 6			2 x 8			2 x 10			2 x 12		
	Grade											
Spacing (inches)	Sel. Str.	No.1/ No.2	No.3	Sel. Str.	No.1/ No.2	No.3	Sel. Str.	No.1/ No.2	No.3	Sel. Str.	No.1/ No.2	No.3
12	11-7	11-3	9-8	15-3	14-11	12-4	19-5	19-0	15-0	23-7	23-0	17-5
16	10-6	10-3	8-5	13-10	13-6	10-8	17-8	17-2	13-0	21-6	19-11	15-1
19.2	9-10	9-8	7-8	13-0	12-9	9-9	16-7	15-8	11-10	20-2	18-3	13-9
24	9-2	8-11	6-10	12-1	11-6	8-8	15-5	14-1	10-7	18-9	16-3	12-4

(Courtesy Canadian Wood Council)

(Spruce-Pine-Fir)

(All Rooms Except Sleeping Rooms and Attics)

Maximum Allowable Spans (feet-inches)

	2 x 6			2 x 8			2 x 10			2 x 12		
	Grade											
Spacing (inches)	Sel. Str.	No.1/ No.2	No.3	Sel. Str.	No.1/ No.2	No.3	Sel. Str.	No.1/ No.2	No.3	Sel. Str.	No.1/ No.2	No.3
12	10-6	10-3	8-8	13-10	13-6	11-0	17-8	17-3	13-5	21-6	20-7	15-7
16	9-6	9-4	7-6	12-7	12-3	9-6	16-0	15-5	11-8	19-6	17-10	13-6
19.2	9-0	8-9	6-10	11-10	11-6	8-8	15-1	14-1	10-7	18-4	16-3	12-4
24	8-4	8-1	6-2	11-0	10-3	7-9	14-0	12-7	9-6	17-0	14-7	11-0

(Courtesy Canadian Wood Council)

Appendix D

Canadian Lumber Rafter Spans, Feet & Inches

(Spruce-Pine-Fir)

(Flat or Low Slope,* No Ceiling Load, Live Load—20 psf)

	Maximum Allowable Spans (feet-inches)											
	2 x 6			2 x 8			2 x 10			2 x 12		
Spacing (inches)	Grade Sel. Str.	No.1/ No.2	No.3	Sel. Str.	No.1/ No.2	No.3	Sel. Str.	No.1/ No.2	No.3	Sel. Str.	No.1/ No.2	No.3
12	15-2	14-9	12-0	19-11	19-6	15-3	25-5	24-7	18-7	30-11	28-6	21-7
16	13-9	13-5	10-5	18-2	17-5	13-2	23-2	21-4	16-1	28-2	24-8	18-8
19.2	12-11	12-7	9-6	17-1	15-11	12-0	21-9	19-5	14-8	26-6	22-7	17-1
24	12-0	11-3	8-6	15-10	14-3	10-9	20-2	17-5	13-2	24-1	20-2	15-3

* Slope not over 3 in 12

(Courtesy Canadian Wood Council)

(Spruce-Pine-Fir)

(Flat or Low Slope,* No Ceiling Load, Live Load—30 psf)

	Maximum Allowable Spans (feet-inches)											
	2 x 6			2 x 8			2 x 10			2 x 12		
Spacing (inches)	Grade Sel. Str.	No.1/ No.2	No.3	Sel. Str.	No.1/ No.2	No.3	Sel. Str.	No.1/ No.2	No.3	Sel. Str.	No.1/ No.2	No.3
12	13-3	12-11	10-5	17-5	17-0	13-2	22-3	21-4	16-1	27-1	24-8	18-8
16	12-0	11-9	9-0	15-10	15-1	11-5	20-2	18-5	13-11	24-7	21-5	16-2
19.2	11-4	10-11	8-3	14-11	13-9	10-5	19-0	16-10	12-9	23-1	19-6	14-9
24	10-6	9-9	7-4	13-10	12-4	9-4	17-8	15-1	11-5	20-11	17-6	13-2

* Slope not over 3 in 12

(Courtesy Canadian Wood Council)

Rafter Slope Factor

Slope (in 12)	3	4	5	6	7	8	9	10	11	12	13	14	15	16	17	18	19	20
Slope Factor	1.031	1.054	1.083	1.118	1.158	1.202	1.25	1.302	1.357	1.414	1.474	1.537	1.601	1.667	1.734	1.803	1.873	1.944

Note: Sloping Distance = Rafter Span multiplied by Slope Factor

(Spruce-Pine-Fir)

(Medium or High Slope,* No Ceiling Load, Heavy Roofing, Live Load—20 psf)

Maximum Allowable Spans (feet-inches)

	2 x 4			2 x 6			2 x 8			2 x 10		
	Grade											
Spacing (inches)	Sel. Str.	No.1/ No.2	No.3	Sel. Str.	No.1/ No.2	No.3	Sel. Str.	No.1/ No.2	No.3	Sel. Str.	No.1/ No.2	No.3
12	10-7	10-1	7-7	16-8	14-9	11-2	21-11	18-8	14-1	27-3	22-9	17-3
16	9-8	8-9	6-7	15-2	12-9	9-8	19-4	16-2	12-2	23-7	19-9	14-11
19.2	9-1	7-11	6-0	13-11	11-8	8-10	17-7	14-9	11-2	21-6	18-0	13-7
24	8-5	7-1	5-5	12-5	10-5	7-10	15-9	13-2	10-0	19-3	16-1	12-2

* Slope over 3 in 12

(Courtesy Canadian Wood Council)

(Spruce-Pine-Fir)

(Medium or High Slope,* No Ceiling Load, Heavy Roofing, Live Load—30 psf)

Maximum Allowable Spans (feet-inches)

	2 x 4			2 x 6			2 x 8			2 x 10		
	Grade											
Spacing (inches)	Sel. Str.	No.1/ No.2	No.3	Sel. Str.	No.1/ No.2	No.3	Sel. Str.	No.1/ No.2	No.3	Sel. Str.	No.1/ No.2	No.3
12	9-3	8-10	6-8	14-7	13-0	9-10	19-2	16-5	12-5	24-0	20-1	15-2
16	8-5	7-8	5-10	13-3	11-3	8-6	17-0	14-3	10-9	20-9	17-5	13-2
19.2	7-11	7-0	5-4	12-3	10-3	7-9	15-6	13-0	9-10	19-0	15-11	12-0
24	7-4	6-3	4-9	11-0	9-2	6-11	13-11	11-8	8-9	17-0	14-2	10-9

* Slope over 3 in 12

(Courtesy Canadian Wood Council)

(Spruce-Pine-Fir)

(Medium or High Slope,* No Ceiling Load, Light Roofing, Live Load—20 psf)

Maximum Allowable Spans (feet-inches)

	2 x 4			2 x 6			2 x 8			2 x 10		
	Grade											
Spacing (inches)	Sel. Str.	No.1/ No.2	No.3	Sel. Str.	No.1/ No.2	No.3	Sel. Str.	No.1/ No.2	No.3	Sel. Str.	No.1/ No.2	No.3
12	10-7	10-4	8-8	16-8	16-3	12-8	21-11	21-3	16-1	28-0	25-11	19-7
16	9-8	9-5	7-6	15-2	14-6	11-0	19-11	18-5	13-11	25-5	22-5	17-0
19.2	9-1	8-10	6-10	14-3	13-3	10-0	18-9	16-9	12-8	23-11	20-6	15-6
24	8-5	8-1	6-1	13-3	11-10	9-0	17-5	15-0	11-4	21-11	18-4	13-10

* Slope over 3 in 12

(Courtesy Canadian Wood Council)

Appendix E

Canadian Lumber Ceiling Joist Spans, Feet & Inches

(Spruce-Pine-Fir)

(Drywall Finish, No Future Sleeping Rooms and No Attic Storage)

Maximum Allowable Spans (feet-inches)

	2 x 4			**2 x 6**			**2 x 8**			**2 x 10**		
	Grade											
Spacing (inches)	**Sel. Str.**	**No.1/ No.2**	**No.3**	**Sel. Str.**	**No.1/ No.2**	**No.3**	**Sel. Str.**	**No.1/ No.2**	**No.3**	**Sel. Str.**	**No.1/ No.2**	**No.3**
12	12-2	11-10	10-10	19-1	18-8	15-10	25-2	24-7	20-1	32-1	31-4	24-6
16	11-0	10-9	9-5	17-4	16-11	13-9	22-10	22-4	17-5	29-2	28-1	21-3
19.2	10-4	10-2	8-7	16-4	15-11	12-6	21-6	21-0	15-10	27-5	25-8	19-5
24	9-8	9-5	7-8	15-2	14-9	11-2	19-11	18-9	14-2	25-5	22-11	17-4

(Courtesy Canadian Wood Council)

(Spruce-Pine-Fir)

(Drywall Finish, No Future Sleeping Rooms and Limited Attic Storage Available)

Maximum Allowable Spans (feet-inches)

	2 x 4			**2 x 6**			**2 x 8**			**2 x 10**		
	Grade											
Spacing (inches)	**Sel. Str.**	**No.1/ No.2**	**No.3**	**Sel. Str.**	**No.1/ No.2**	**No.3**	**Sel. Str.**	**No.1/ No.2**	**No.3**	**Sel. Str.**	**No.1/ No.2**	**No.3**
12	9-8	9-5	7-8	15-2	14-9	11-2	19-11	18-9	14-2	25-5	22-11	17-4
16	8-9	8-7	6-8	13-9	12-10	9-8	18-2	16-3	12-4	23-2	19-10	15-0
19.2	8-3	8-0	6-1	12-11	11-9	8-10	17-1	14-10	11-3	21-8	18-2	13-8
24	7-8	7-2	5-5	12-0	10-6	7-11	15-10	13-3	10-0	19-5	16-3	12-3

(Courtesy Canadian Wood Council)

Appendix F

Canadian Lumber Floor & Ceiling Joist Spans, Metric

Floor Joist Spans: Living Quarters (Spruce-Pine-Fir) — **Maximum spans (m)**

Grade		38 x 140			38 x 184			38 x 235			38 x 286		
Nailed & glued subfloor thickness (mm)		Briding & strapping joist spacing (mm)											
		300	400	600	300	400	600	300	400	600	300	400	600
15.5	SS	3.24	2.95	2.57	4.26	3.87	3.38	5.26	4.91	4.32	5.90	5.51	5.13
	No. 1&2	3.14	2.85	2.49	4.12	3.75	3.27	5.09	4.74	4.18	5.71	5.32	4.26
	No. 3	3.08	2.80	2.43	4.05	3.61	2.95	5.00	4.42	3.61	5.61	5.13	4.19
18.5	SS	3.24	2.95	2.57	4.26	3.87	3.38	5.45	4.95	4.32	6.42	5.90	5.26
	No. 1&2	3.14	2.85	2.49	4.12	3.75	3.27	5.27	4.79	4.18	6.21	5.71	5.09
	No. 3	3.08	2.80	2.43	4.05	3.61	2.95	5.10	4.42	3.61	5.92	5.13	4.19

(Courtesy Canadian Wood Council)

Floor Joist Spans: Bedrooms & Attics (Spruce-Pine-Fir) — **Maximum spans (m)**

Grade		38 x 140			38 x 184			38 x 235			38 x 286		
Nailed & glued subfloor thickness (mm)		Briding & strapping joist spacing (mm)											
		300	400	600	300	400	600	300	400	600	300	400	600
15.5	SS	3.59	3.26	2.85	4.56	4.25	3.75	5.26	4.91	4.57	5.90	5.51	5.13
	No. 1&2	3.47	3.16	2.76	4.41	4.11	3.62	5.09	4.74	4.42	5.71	5.32	4.96
	No. 3	3.41	3.10	2.71	4.33	4.04	3.37	5.00	4.66	4.12	5.61	5.23	4.78
18.5	SS	3.59	3.26	2.85	4.72	4.29	3.75	5.72	5.26	4.79	6.42	5.90	5.43
	No. 1&2	3.47	3.16	2.76	4.57	4.15	3.62	5.53	5.09	4.63	6.21	5.71	5.25
	No. 3	3.41	3.10	2.71	4.48	4.07	3.37	5.44	5.00	4.12	6.10	5.61	4.78

(Courtesy Canadian Wood Council)

Ceiling Joist Spans (Spruce-Pine-Fir) — **Maximum spans (m)**

Grade	38 x 89			38 x 140			38 x 184			38 x 235		
	Joist Spacing (mm)											
	300	400	600	300	400	600	300	400	600	300	400	600
SS	3.22	2.92	2.55	5.06	4.60	4.02	6.65	6.05	5.28	8.50	7.72	6.74
No. 1&2	3.11	2.83	2.47	4.90	4.45	3.89	6.44	5.85	5.11	8.22	7.47	6.52
No. 3	3.06	2.78	2.43	4.81	4.37	3.82	6.32	5.74	5.02	8.07	7.33	6.34

(Courtesy Canadian Wood Council)

APPENDIX G

CANADIAN LUMBER ROOF RAFTER, SOLID 89-MM FLOOR BEAM & SUBFLOORING SPANS, METRIC

Roof Rafter Spans (Spruce-Pine-Fir) **Roof Snow Load—1.5kPa Maximum spans (m)**

Grade	38 x 89			38 x 140			38 x 184			38 x 235		
	Rafter Spacing (mm)											
	300	400	600	300	400	600	300	400	600	300	400	600
SS	2.81	2.55	2.23	4.42	4.02	3.51	5.81	5.28	4.61	7.42	6.74	5.89
No. 1&2	2.72	2.47	2.16	4.28	3.89	3.40	5.62	5.11	4.41	7.18	6.52	5.39
No. 3	2.67	2.39	1.95	3.95	3.42	2.79	4.80	4.16	3.40	5.87	5.08	4.15

(Courtesy Canadian Wood Council)

Solid 89-mm Floor Beam Spans (Spruce-Pine-Fir) **Maximum spans (m)**

Grade	Supported length (m)	Supporting One Floor in Houses				Supporting Two Floors in Houses			
		89 x 235	89 x 286	89 x 337	89 x 387	89 x 235	89 x 286	89 x 337	89 x 387
SS	**2.4**	4.13	4.81	5.41	5.89	3.01	3.40	3.70	3.91
	3.0	3.70	4.31	4.84	5.12	2.50	2.83	3.10	3.28
	3.6	3.37	3.85	4.18	4.40	2.16	2.46	2.69	2.86
	4.2	2.99	3.38	3.68	3.88	1.92	2.19	2.41	2.56
	4.8	2.67	3.03	3.30	3.49	1.74	1.99	2.19	2.34
No. 1&2	**2.4**	3.49	4.07	4.57	4.98	2.65	3.09	3.47	3.78
	3.0	3.13	3.64	4.09	4.46	2.37	2.76	3.10	3.28
	3.6	2.85	3.32	3.73	4.07	2.16	2.46	2.69	2.86
	4.2	2.64	3.08	3.46	3.77	1.92	2.19	2.41	2.56
	4.8	2.47	2.88	3.23	3.49	1.74	1.99	2.19	2.34

(Courtesy Canadian Wood Council)

Subflooring, Thickness **(mm)**

Maximum spacing of supports (mm)	Plywood	Waferboard and strand board		Lumber
		R-1, 0-1 grades	0-2 grade	
400	15.5	15.9	15.5	17.0
500	15.5	15.9	15.5	19.0
600	18.5	19.0	18.5	19.0

(Courtesy Canadian Wood Council)

APPENDIX H

CANADIAN LUMBER BUILT-UP FLOOR BEAM SPANS, METRIC

Built-Up Floor Beam Spans (Spruce-Pine-Fir) — **Maximum spans (m)**

Grade	Supported length (m)	Supporting One Floor in Houses, Size of beam (mm): 338 x 184	438 x 184	338 x 235	438 x 235	338 x 286	438 x 286
SS	**2.4**	3.84	4.43	4.70	5.42	5.45	6.29
	3.0	3.43	3.97	4.20	4.85	4.87	5.63
	3.6	3.14	3.62	3.79	4.43	4.19	5.14
	4.2	2.78	3.35	3.25	4.10	3.60	4.76
	4.8	2.43	3.14	2.84	3.79	3.15	4.19
No. 1&2	**2.4**	3.25	3.75	3.97	4.59	4.61	5.32
	3.0	2.90	3.35	3.55	4.10	4.12	4.76
	3.6	2.65	3.06	3.24	3.74	3.76	4.34
	4.2	2.45	2.83	3.00	3.47	3.48	4.02
	4.8	2.30	2.65	2.81	3.24	3.15	3.76

(Courtesy Canadian Wood Council)

Built-Up Floor Beam Spans (Spruce-Pine-Fir) — **Maximum spans (m)**

Grade	Supported length (m)	Supporting Two Floors in Houses, Size of beam (mm): 338 x 184	438 x 184	338 x 235	438 x 235	338 x 286	438 x 286
SS	**2.4**	2.80	3.36	3.27	4.11	3.62	4.77
	3.0	2.24	2.98	2.62	3.49	2.90	3.86
	3.6	1.86	2.49	2.18	2.91	2.42	3.22
	4.2	1.60	2.13	1.87	2.49	2.07	2.76
	4.8	1.40	1.86	1.64	2.18	1.81	2.42
No. 1&2	**2.4**	2.46	2.85	3.01	3.48	3.50	4.04
	3.0	2.20	2.55	2.62	3.11	2.90	3.61
	3.6	1.86	2.32	2.18	2.84	2.42	3.22
	4.2	1.60	2.13	1.87	2.49	2.07	2.76
	4.8	1.40	1.86	1.64	2.18	1.81	2.42

(Courtesy Canadian Wood Council)

APPENDIX I

CANADIAN LUMBER GLUED-LAMINATED FLOOR BEAMS, METRIC

Glued-Laminated Floor Beam Spans — **Maximum spans (m)**

Stress grade designation	Beam width (mm)	Supported length (m)	Supporting One Floor in Houses Beam depth mm) 228	266	304	342	380	412	456
20f-E	80	2.4	4.32	5.04	5.76	6.48	7.20	7.92	8.64
		3.0	3.87	4.51	5.15	5.80	6.44	7.09	7.73
		3.6	3.53	4.12	4.70	5.29	5.88	6.47	7.06
		4.2	3.27	3.81	4.36	4.90	5.44	5.99	6.53
		4.8	3.06	3.57	4.07	4.58	5.09	5.60	6.11
	130	2.4	5.51	6.43	7.35	8.26	9.18	10.10	11.02
		3.0	4.93	5.75	6.57	7.39	8.21	9.03	9.86
		3.6	4.50	5.25	6.00	6.75	7.50	8.25	9.00
		4.2	4.16	4.86	5.55	6.25	6.94	7.64	8.33
		4.8	3.90	4.54	5.19	5.84	6.49	7.14	7.79

(Courtesy Canadian Wood Council)

Glued-Laminated Floor Beam Spans — **Maximum spans (m)**

Stress grade designation	Beam width (mm)	Supported length (m)	Supporting Two Floors in Houses Beam depth mm) 228	266	304	342	380	412	456
20f-E	80	2.4	3.28	3.83	4.37	4.92	5.47	6.01	6.56
		3.0	2.93	3.42	3.91	4.40	4.89	5.38	5.87
		3.6	2.68	3.12	3.57	4.02	4.46	4.91	5.36
		4.2	2.48	2.89	3.31	3.72	4.13	4.54	4.96
		4.8	2.32	2.71	3.09	3.48	3.86	4.25	4.64
	130	2.4	4.18	4.88	5.57	6.27	6.97	7.66	8.36
		3.0	3.74	4.36	4.99	5.61	6.23	6.85	7.48
		3.6	3.41	3.98	4.55	5.12	5.69	6.26	6.83
		4.2	3.16	3.69	4.21	4.74	5.27	5.79	6.32
		4.8	2.96	3.45	3.94	4.43	4.93	5.42	5.91

Note: Supported length means ½ the sum of the joist spans on both sides of the beam. Straight interpolation may be used for other supported lengths. Spans are clear spans between supports. For total span, add two bearing lengths.

(Courtesy Canadian Wood Council)

Appendix J

Southern Pine Floor Joist Spans, Feet & Inches

Floor Joists—Sleeping Rooms and Attic Floors (Empirical Design Values), 30 psf Live Load, 10 psf Dead Load, *l*/360

Size (inches)	Spacing (inches on center)	Grade			
		Select Structural	No. 1	No. 2	No. 3
2 x 6	12	12-3	12-0	11-10	10-5
	16	11-2	10-11	10-9	9-1
	24	9-9	9-7	9-4	7-5
2 x 8	12	16-2	15-10	15-7	13-3
	16	14-8	14-5	14-2	11-6
	24	12-10	12-7	12-4	9-5
2 x 10	12	20-8	20-3	19-10	15-8
	16	18-9	18-5	18-0	13-7
	24	16-5	16-1	14-8	11-1
2 x 12	12	25-1	24-8	24-2	18-8
	16	22-10	22-5	21-1	16-2
	24	19-11	19-6	17-2	13-2

(Courtesy Southern Pine Marketing Council)

Floor Joists—All Rooms Except Sleeping Rooms and Attic Floors (Emperical Design Values), 40 psf Live Load, 10 psf Dead Load, *l*/360

These spans are based on the 1993 AFPA (formerly NFPA) Span Tables for Joists and Rafters and the 1991 SPIB Grading Rules. They are intended for use in covered structures or where the moisture content in use does not exceed 19 percent for an extended period of time. Loading conditions are expressed in psf (pounds per square foot). Deflection is limited to span in inches divided by 360 and is based on live load only. Check sources of supply for availability of lumber in lengths greater than 20′0″.

Size (inches)	Spacing (inches on center)	Grade			
		Select Structural	No. 1	No. 2	No. 3
2 x 6	12	11-2	10-11	10-9	9-4
	16	10-2	9-11	9-9	8-1
	24	8-10	8-8	8-6	6-7
2 x 8	12	14-8	14-5	14-2	11-11
	16	13-4	13-1	12-10	10-3
	24	11-8	11-5	11-0	8-5
2 x 10	12	18-9	18-5	18-0	14-0
	16	17-0	16-9	16-1	12-2
	24	14-11	14-7	13-2	9-11
2 x 12	12	22-10	22-5	21-9	16-8
	16	20-9	20-4	18-10	14-5
	24	18-1	17-5	15-4	11-10

(Courtesy Southern Pine Marketing Council)

Appendix K

Southern Pine Ceiling Joist Spans, Feet & Inches

Ceiling Joists—Drywall Ceiling No Attic Storage (Empirical Design Values), 10 psf Live Load 5 psf Dead Load, *l*/240

Size (inches)	Spacing (inches on center)	Grade			
		Select Structural	No. 1	No. 2	No. 3
2 x 4	12	12-11	12-8	12-5	11-7
	16	11-9	11-6	11-3	10-9
	24	10-3	10-0	9-10	8-2
2 x 6	12	20-3	19-11	19-6	17-1
	16	18-5	18-1	17-8	14-9
	24	16-1	15-9	15-6	12-1
2 x 8	12	26-0	26-0	25-8	21-8
	16	24-3	23-10	23-4	18-9
	24	21-2	20-10	20-1	15-4
2 x 10	12	26-0	26-0	26-0	25-7
	16	26-0	26-0	26-0	22-2
	24	26-0	26-0	24-0	18-1

(Courtesy Southern Pine Marketing Council)

Ceiling Joists—Drywall Ceiling, No Future Sleeping Rooms, but Limited Storage Available (Empirical Design Values), 20 psf Live Load, 10 psf Dead Load, *l*/240

These spans are based on the 1993 AFPA (formerly NFPA) Span Tables for Joists and Rafters and the 1991 SPIB Grading Rules. They are intended for use in covered structures or where the moisture content in use does not exceed 19 percent for an extended period of time. Loading conditions are expressed in psf (pounds per square foot). Deflection is limited to span in inches divided by 240 and is based on live load only. Check sources of supply for availability of lumber in lengths greater than 20′0″.

Size (inches)	Spacing (inches on center)	Grade			
		Select Structural	No. 1	No. 2	No. 3
2 x 4	12	10-3	10-0	9-10	8-2
	16	9-4	9-1	8-11	7-1
	24	8-1	8-0	7-8	5-9
2 x 6	12	16-1	15-9	15-6	12-1
	16	14-7	14-4	13-6	10-5
	24	12-9	12-6	11-0	8-6
2 x 8	12	21-2	20-10	20-1	15-4
	16	19-3	18-11	17-5	13-3
	24	16-10	15-11	14-2	10-10
2 x 10	12	26-0	26-0	24-0	18-1
	16	24-7	23-2	20-9	15-8
	24	21-6	18-11	17-0	12-10

(Courtesy Southern Pine Marketing Council)

APPENDIX L

SOUTHERN PINE RAFTER SPANS, FEET & INCHES

Rafters—Light Roofing; No Finished Ceiling; Snow Load (Empirical Design Values), 20 psf Live Load, 10 psf Dead Load, *l*/180, C_D = 1.15

Size (inches)	Spacing (inches on center)	Grade			
		Select Structural	No. 1	No. 2	No. 3
2 x 4	12	11-3	11-1	10-10	8-9
	16	10-3	10-0	9-10	7-7
	24	8-11	8-9	8-3	6-2
2 x 6	12	17-8	17-4	16-8	12-11
	16	16-1	15-9	14-5	11-2
	24	14-1	13-6	11-9	9-1
2 x 8	12	23-4	22-11	21-7	16-5
	16	21-2	20-10	18-8	14-3
	24	18-6	17-0	15-3	11-7
2 x 10	12	26-0	26-0	25-8	19-5
	16	26-0	24-9	22-3	16-10
	24	23-8	20-3	18-2	13-9

(Courtesy Southern Pine Marketing Council)

Rafters—Light Roofing; No Finished Ceiling; Snow Load (Empirical Design Values), 30 psf Live Load, 10 psf Dead Load, *l*/180, C_D = 1.15

These spans are based on the 1993 AFPA (formerly NFPA) Span Tables for Joists and Rafters and the 1991 SPIB Grading Rules. They are intended for use in covered structures or where the moisture content in use does not exceed 19 percent for an extended period of time. Loading conditions are expressed in psf (pounds per square foot). Deflection is limited to span in inches divided by 180 and is based on live load only. The load duration factor, C_D, is 1.15 for snow loads. Check sources of supply for availability of lumber in lengths greater than 20′0″.

Size (inches)	Spacing (inches on center)	Grade			
		Select Structural	No. 1	No. 2	No. 3
2 x 4	12	9-10	9-8	9-6	7-7
	16	8-11	8-9	8-7	6-7
	24	7-10	7-8	7-1	5-4
2 x 6	12	15-6	15-2	14-5	11-2
	16	14-1	13-9	12-6	9-8
	24	12-3	11-9	10-2	7-11
2 x 8	12	20-5	20-0	18-8	14-3
	16	18-6	18-0	16-2	12-4
	24	16-2	14-9	13-2	10-1
2 x 10	12	26-0	24-9	22-3	16-10
	16	23-8	21-5	19-3	14-7
	24	20-8	17-6	15-9	11-11

(Courtesy Southern Pine Marketing Council)

Rafters—Medium Roofing; No Finished Ceiling; Snow Load (Empirical Design Values), 20 psf Live Load, 15 psf Dead Load, l/180, C_D = 1.15

Size (inches)	Spacing (inches on center)	Grade			
		Select Structural	No. 1	No. 2	No. 3
2 x 4	12	11-3	11-1	10-9	8-1
	16	10-3	10-0	9-4	7-0
	24	8-11	8 -5	7-7	5-9
2 x 6	12	17-8	17-4	15-5	11-11
	16	16-1	15-4	13-4	10-4
	24	14-1	12-6	10-11	8-5
2 x 8	12	23-4	22-3	19-11	15-3
	16	21-2	19-3	17-3	13-2
	24	18-6	15-9	14-1	10-9
2 x 10	12	26-0	26-0	23-10	18-0
	16	26-0	22-11	20-7	15-7
	24	23-6	18-9	16-10	12-9

(Courtesy Southern Pine Marketing Council)

Rafters—Medium Roofing; No Finished Ceiling; Snow Load (Empirical Design Values), 30 psf Live Load, 15 psf Dead Load, l/180, C_D = 1.15

These spans are based on the 1993 AFPA (formerly NFPA) Span Tables for Joists and Rafters and the 1991 SPIB Grading Rules. They are intended for use in covered structures or where the moisture content in use does not exceed 19 percent for an extended period of time. Loading conditions are expressed in psf (pounds per square foot). Deflection is limited to span in inches divided by 180 and is based on live load only. The load duration factor, C_D, is 1.15 for snow loads. Check sources of supply for availability of lumber in lengths greater than 20′0″.

Size (inches)	Spacing (inches on center)	Grade			
		Select Structural	No. 1	No. 2	No. 3
2 x 4	12	9-10	9-8	9-6	7-2
	16	8-11	8-9	8-3	6-2
	24	7-10	7-5	6-8	5-0
2 x 6	12	15-6	15-2	13-7	10-6
	16	14-1	13-6	11-9	9-1
	24	12-3	11-1	9-7	7-5
2 x 8	12	20-5	19-8	17-7	13-5
	16	18-6	17-0	15-3	11-7
	24	16-2	13-11	12-5	9-6
2 x 10	12	26-0	23-4	21-0	15-10
	16	23-8	20-3	18-2	13-9
	24	20-8	16-6	14-10	11-3

(Courtesy Southern Pine Marketing Council)

Rafters—Heavy Roofing; No Finished Ceiling; Snow Load (Empirical Design Values), 20 psf Live Load, 20 psf Dead Load, $l/180$, $C_D = 1.15$

Size (inches)	Spacing (inches on center)	Grade			
		Select Structural	No. 1	No. 2	No. 3
2 x 4	12	11-3	11-1	10-1	7-7
	16	10-3	9-8	8-8	6-7
	24	8-11	7-11	7-1	5-4
2 x 6	12	17-8	16-7	14-5	11-2
	16	16-1	14-4	12-6	9-8
	24	14-1	11-9	10-2	7-11
2 x 8	12	23-4	20-10	18-8	14-3
	16	21-2	18-0	16-2	12-4
	24	18-3	14-9	13-2	10-1
2 x 10	12	26-0	24-9	22-3	16-10
	16	26-0	21-5	19-3	14-7
	24	22-0	17-6	15-9	11-11

(Courtesy Southern Pine Marketing Council)

Rafters—Heavy Roofing; No Finished Ceiling; Snow Load (Empirical Design Values), 30 psf Live Load, 20 psf Dead Load, $l/180$, $C_D = 1.15$

These spans are based on the 1993 AFPA (formerly NFPA) Span Tables for Joists and Rafters and the 1991 SPIB Grading Rules. They are intended for use in covered structures or where the moisture content in use does not exceed 19 percent for an extended period of time. Loading conditions are expressed in psf (pounds per square foot). Deflection is limited to span in inches divided by 180 and is based on live load only. The load duration factor, C_D, is 1.15 for snow loads. Check sources of supply for availability of lumber in lengths greater than 20′0″.

Size (inches)	Spacing (inches on center)	Grade			
		Select Structural	No. 1	No. 2	No. 3
2 x 4	12	9-10	9-8	9-0	6-9
	16	8-11	8-8	7-9	5-10
	24	7-10	7-1	6-4	4-9
2 x 6	12	15-6	14-10	12-11	10-0
	16	14-1	12-10	11-2	8-8
	24	12-3	10-6	9-1	7-1
2 x 8	12	20-5	18-8	16-8	12-9
	16	18-6	16-2	14-5	11-0
	24	16-2	13-2	11-10	9-0
2 x 10	12	26-0	22-2	19-11	15-1
	16	23-8	19-2	17-3	13-0
	24	19-8	15-8	14-1	10-8

(Courtesy Southern Pine Marketing Council)

Appendix M

Western Lumber Floor Joist Spans, Feet & Inches

Floor Joists—(Empirical Design Values) 30 psf Live Load, 10 psf Dead Load, *l*/360

Species or Group	Grade \ Span O.C.	2 x 6 12″	2 x 6 16″	2 x 6 24″	2 x 8 12″	2 x 8 16″	2 x 8 24″	2 x 10 12″	2 x 10 16″	2 x 10 24″	2 x 12 12″	2 x 12 16″	2 x 12 24″
Douglas Fir-Larch	Sel. Struc	12-6	11-4	9-11	16-6	15-0	13-1	21-0	19-1	16-8	25-7	23-3	20-3
	No. 1 & Btr.	12-3	11-2	9-9	16-2	14-8	12-10	20-8	18-9	16-1	25-1	22-10	18-8
	No. 1	12-0	10-11	9-7	15-10	14-5	12-4	20-3	18-5	15-0	21-8	21-4	17-5
	No. 2	11-10	10-9	9-1	15-7	14-1	11-6	19-10	17-2	14-1	23-0	19-11	16-3
	No. 3	9-8	8-5	6-10	12-4	10-8	8-8	15-0	13-0	10-7	17-5	15-1	12-4
Douglas Fir-South	Sel. Struc	11-3	10-3	8-11	14-11	13-6	11-10	19-0	17-3	15-1	23-1	21-0	18-4
	No. 1	11-0	10-0	8-9	14-6	13-2	11-6	18-6	16-10	14-3	22-6	20-3	16-6
	No. 2	10-9	9-9	8-6	14-2	12-10	11-2	18-0	16-5	13-8	21-11	19-4	15-10
	No. 3	9-6	8-2	6-8	12-0	10-5	8-6	14-8	12-8	10-4	17-0	14-8	12-0
Hem-Fir	Sel. Struc	11-10	10-9	9-4	15-7	14-2	12-4	19-10	18-0	15-9	24-2	21-11	19-2
	No. 1 & Btr.	11-7	10-6	9-2	15-3	13-10	12-1	19-5	17-8	15-5	23-7	21-6	17-10
	No. 1	11-7	10-6	9-2	15-3	13-10	12-0	19-5	17-8	14-8	23-7	20-9	17-0
	No. 2	11-0	10-0	8-9	14-6	13-2	11-4	18-6	16-10	13-10	22-6	19-8	16-1
	No. 3	9-8	8-5	6-10	12-4	10-8	8-8	15-0	13-0	10-7	17-5	15-1	12-4

Residential occupancy sleeping rooms (BOCA only). Attics with storage under the Standard Code. Does not apply in UBC areas.

(Courtesy Western Wood Products Association)

Floor Joists—(Empirical Design Values) 40 psf Live Load, 10 psf Dead Load, *l*/360

Species or Group	Grade \ Span O.C.	2 x 6 12″	2 x 6 16″	2 x 6 24″	2 x 8 12″	2 x 8 16″	2 x 8 24″	2 x 10 12″	2 x 10 16″	2 x 10 24″	2 x 12 12″	2 x 12 16″	2 x 12 24″
Douglas Fir-Larch	Sel. Struc	11-4	10-4	9-0	15-0	13-7	11-11	19-1	17-4	15-2	23-3	21-1	18-5
	No. 1 & Btr.	11-2	10-2	8-10	14-8	13-4	11-8	18-9	17-0	14-5	22-10	20-5	16-8
	No. 1	10-11	9-11	8-8	14-5	11-1	11-0	18-5	16-5	13-5	22-0	19-1	15-7
	No. 2	10-9	9-9	8-1	14-2	12-7	10-3	17-9	15-5	12-7	20-7	17-10	14-7
	No. 3	8-8	7-6	6-2	11-0	9-6	7-9	13-5	11-8	9-6	15-7	13-6	11-0
Douglas Fir-South	Sel. Struc	10-3	9-4	8-2	13-6	12-3	10-9	17-3	15-8	13-8	21-0	19-1	16-8
	No. 1	10-0	9-1	7-11	13-2	12-0	10-5	16-10	15-3	12-9	20-6	18-1	14-9
	No. 2	9-9	8-10	7-9	12-10	11-8	10-0	16-5	14-11	12-2	19-11	17-4	14-2
	No. 3	8-6	7-4	6-0	10-9	9-3	7-7	13-1	11-4	9-3	15-2	13-2	10-9
Hem-Fir	Sel. Struc	10-9	9-9	8-6	14-2	12-10	11-3	18-0	16-5	14-4	21-11	19-11	17-5
	No. 1 & Btr.	10-6	9-6	8-4	13-10	12-7	11-0	17-8	16-0	13-9	21-6	19-6	16-0
	No. 1	10-6	9-6	8-4	13-10	12-7	10-9	17-8	16-0	13-1	21-6	18-7	15-2
	No. 2	10-0	9-1	7-11	13-2	12-0	10-2	16-10	15-2	12-5	20-4	17-7	14-4
	No. 3	8-8	7-6	6-2	11-0	9-6	7-9	13-5	11-8	9-6	15-7	13-6	11-0

Residential occupancies include private dwelling, private apartment, and hotel guest rooms. Deck under CABO and Standard Codes.

(Courtesy Western Wood Products Association)

Appendix N

Western Lumber Ceiling Joist Spans, Feet & Inches

Ceiling Joists—(Empirical Design Values) 20 psf Live Load, 10 psf Dead Load, *l*/240

Species or Group	Grade \ Span O.C.	2 x 6			2 x 8			2 x 10			2 x 12		
		12″	16″	24″	12″	16″	24″	12″	16″	24″	12″	16″	24″
Douglas Fir-Larch	Sel. Struc	10-5	9-6	8-3	16-4	14-11	13-0	21-7	19-7	17-1	27-6	25-0	20-11
	No. 1 & Btr.	10-3	9-4	8-1	16-1	14-7	12-0	21-2	18-8	15-3	26-4	22-9	18-7
	No. 1	10-0	9-1	7-8	15-9	13-9	11-2	20-1	17-5	14-2	24-6	21-3	17-4
	No. 2	9-10	8-9	7-2	14-10	12-10	10-6	18-9	16-3	13-3	22-11	19-10	16-3
	No. 3	7-8	6-8	5-5	11-2	9-8	7 -11	14-2	12-4	10-0	17-4	15-0	12-3
Douglas Fir-South	Sel. Struc	9-5	8-7	7-6	14-9	13-5	11-9	19-6	17-9	15-6	24-10	22-7	19-9
	No. 1	9-2	8-4	7-3	14-5	13-0	10-8	19-0	16-6	13-6	23-3	20-2	16-5
	No. 2	8-11	8-1	7-0	14-1	12-6	10-2	18-3	15-9	12-11	22-3	19-3	15-9
	No. 3	7-6	6-6	5-3	10-11	9-6	7-9	13-10	12-0	9-9	16-11	14-8	11-11
Hem-Fir	Sel. Struc	9-10	8-11	7-10	15-6	14-1	12-3	20-5	18-6	16-2	26-0	23-8	20-6
	No. 1 & Btr.	9-8	8-9	7-8	15-2	13-9	11-6	19-11	17-10	14-7	25-2	21-9	17-9
	No. 1	9-8	8-9	7-6	15-2	13-5	10-11	19-7	16-11	13-10	23-11	20-8	16-11
	No. 2	9-2	8-4	7-1	14-5	12-8	10-4	18-6	16-0	13-1	22-7	19-7	16-0
	No. 3	7-8	6-8	5-5	11-2	9-8	7-11	14-2	12-4	10-0	17-4	15-0	12-3

Use these loading conditions for the following: Limited attic storage where development of future rooms is not possible. Ceilings where the roof pitch is steeper than 3 in 12. Where the clear height in the attic is greater than 30 inches. Drywall ceilings.

(Courtesy Western Wood Products Association)

Appendix O

Western Lumber Rafter Spans, Feet & Inches

Roof Rafters—(Empirical Design Values) 20 psf Live Load, 15 psf Dead Load, *l*/180

Species or Group	Grade	Span / O.C.	2 x 6 12″	2 x 6 16″	2 x 6 24″	2 x 8 12″	2 x 8 16″	2 x 8 24″	2 x 10 12″	2 x 10 16″	2 x 10 24″	2 x 12 12″	2 x 12 16″	2 x 12 24″
Douglas Fir-Larch	Sel. Struc		18-0	16-4	14-0	23-9	21-7	17-8	30-4	26-6	21-7	35-5	30-8	25-1
	No. 1 & Btr.		17-7	15-3	12-5	22-3	19-4	15-9	27-3	23-7	19-3	31-7	27-4	22-4
	No. 1		16-5	14-3	11-7	20-9	18-0	14-8	25-5	22-0	17-11	29-5	25-6	20-10
	No. 2		15-4	13-3	10-10	19-5	16-10	13-9	23-9	20-7	16-9	27-6	23-10	19-6
	No. 3		11-7	10-1	8-2	14-8	12-9	10-5	17-11	15-7	12-8	20-10	18-0	14-9
Douglas Fir-South	Sel. Struc		16-3	14-9	12-11	21-5	19-6	16-9	27-5	24-10	20-6	33-4	29-1	23-9
	No. 1		15-7	13-6	11-0	19-9	17-1	13-11	24-1	20-10	17-0	27-11	24-2	19-9
	No. 2		14-11	12-11	10-6	18-10	16-4	13-4	23-1	20-0	16-4	26-9	23-2	18-11
	No. 3		11-4	9-10	8-0	14-4	12-5	10-2	17-6	15-2	12-4	20-3	17-7	14-4
Hem-Fir	Sel. Struc		17-0	15-6	13-6	22-5	20-5	17-5	28-7	26-0	21-3	34-10	30-2	24-8
	No. 1 & Btr.		16-8	14-7	11-11	21-4	18-5	15-1	26-0	22-6	18-5	30-2	26-1	21-4
	No. 1		16-0	13-10	11-4	20-3	17-6	14-4	24-9	21-5	17-6	28-8	24-10	20-3
	No. 2		15-2	13-1	10-8	19-2	16-7	13-7	23-5	20-3	16-7	27-2	23-6	19-2
	No. 3		11-7	10-1	8-2	14-8	12-9	10-5	17-11	15-7	12-8	20-10	18-0	14-9

No snow load. Roof slope greater than 3 in 12. Heavy roof covering. No ceiling finish.

(Courtesy Western Wood Products Association)

Roof Rafters—(Empirical Design Values) 30 psf Snow Load, 15 psf Dead Load, *l*/180

Species or Group	Grade	Span / O.C.	2 x 6 12″	2 x 6 16″	2 x 6 24″	2 x 8 12″	2 x 8 16″	2 x 8 24″	2 x 10 12″	2 x 10 16″	2 x 10 24″	2 x 12 12″	2 x 12 16″	2 x 12 24″
Douglas Fir-Larch	Sel. Struc		15-9	14-4	11-10	20-9	18-4	15-0	25-10	22-5	18-3	30-0	26-0	21-2
	No. 1 & Btr.		14-11	12-11	10-6	18-10	16-4	13-4	23-0	19-11	16-3	26-8	23-1	18-11
	No. 1		13-11	12-0	9-10	17-7	15-3	12-5	21-6	18-7	15-2	24-11	21-7	17-7
	No. 2		13-0	11-3	9-2	16-5	14-3	11-8	20-1	17-5	14-2	23-3	20-2	16-6
	No. 3		9-10	8-6	6-11	12-5	10-9	8-9	15-2	13-2	10-9	17-7	15-3	12-5
Douglas Fir-South	Sel. Struc		14-3	12-11	11-2	18-9	17-0	14-2	23-11	21-2	17-4	28-5	24-7	20-1
	No. 1		13-2	11-5	9-4	16-8	14-5	11-9	20-4	17-8	14-5	23-7	20-5	16-8
	No. 2		12-7	10-11	8-11	16-0	13-10	11-3	19-6	16-11	13-9	22-7	19-7	16-0
	No. 3		9-7	8-3	6-9	12-1	10-6	8-7	14-10	12-10	10-6	17-2	14-10	12-2
Hem-Fir	Sel. Struc		14-10	13-6	11-7	19-7	17-10	14-8	25-0	22-0	18-0	29-6	25-6	20-10
	No. 1 & Btr.		14-3	12-4	10-1	18-0	15-7	12-9	22-0	19-1	15-7	25-6	22-1	18-0
	No. 1		13-6	11-9	9-7	17-2	14-10	12-1	20-11	18-1	14-10	24-3	21-0	17-2
	No. 2		12-10	11-1	9-1	16-2	14-0	11-6	19-10	17-2	14-0	22-11	19-11	16-3
	No. 3		9-10	8-6	6-11	12-5	10-9	8-9	15-2	13-2	10-9	17-7	15-3	12-5

Roof slope greater than 3 in 12. Heavy roof covering. No ceiling finish. *(Courtesy Western Wood Products Association)*

APPENDIX P

TYPICAL LOADS FOR GLULAM BEAMS

Typical Loads* for Glulam Roof Beams**

Span in feet	Roof Beam 3″ x 6	9	12	15
10	272	745	1104	1380
14	—	335	676	986
16	—	224	518	789
20	—	115	272	505

* Loads shown are pounds per lineal foot including weight of beam.
** Applicable for straight, simply supported beams. (Consult the manufacturer for specific load data.)

Typical Loads* for Glulam Floor Beams**

Span in feet	Floor Beam 3″ x 6	9	12	15
8	266	896	1200	1500
14	—	167	397	774
16	—	112	266	519
20	—	—	136	266

* Loads shown are pounds per lineal foot including weight of beam.
** Applicable for straight, simply supported beams. (Consult the manufacturer for specific load data.)

APPENDIX Q

SOLID & BUILT-UP WOOD BEAMS*

Width of Structure	Girder Size (in.) (Built-up or Solid)	Maximum Span When Supporting a Load-Bearing Interior Partition: 1 Story	1½ or 2 Story
Up to 26 ft.	6 x 8	7′–0″	6′–0″
	6 x 10	9′–0″	7′–6″
	6 x 12	10′–6″	9′–0″
26 ft. to 32 ft.	6 x 8	6′–6″	5′–6″
	6 x 10	8′–0″	7′–0″
	6 x 12	10′–0″	8′–0″

* Spans based on allowable fiber stress of 1500 psi. FHA Minimum Property Standards for One and Two Family Dwellings

Appendix R

Board Feet for Selected Lumber Sizes

Nominal size (inches)	Length (feet)								
	8	10	12	14	16	18	20	22	24
1 x 2	1⅓	1⅔	2	2⅓	2⅔	3	3⅓	3⅔	4
1 x 3	2	2½	3	3½	4	4½	5	5½	6
1 x 4	2⅔	3⅓	4	4⅔	5⅓	6	6⅔	7⅓	8
1 x 5	3⅓	4⅙	5	5⅚	6⅔	7½	8⅓	9⅙	10
1 x 6	4	5	6	7	8	9	10	11	12
1 x 7	4⅔	5⅚	7	8⅙	9⅓	10½	11⅔	12⅚	14
1 x 8	5⅓	6⅔	8	9⅓	10⅔	12	13⅓	14⅔	16
1 x 10	6⅔	8⅓	10	11⅔	13⅓	15	16⅔	18⅓	20
1 x 12	8	10	12	14	16	18	20	22	24
1¼ x 4	3⅓	4⅙	5	5⅚	6⅔	7½	8⅓	9⅙	10
1¼ x 6	5	6¼	7½	8¾	10	11¼	12½	13¾	15
1¼ x 8	6⅔	8⅓	10	11⅔	13⅓	15	16⅔	18⅓	20
1¼ x 10	8⅓	10 5/12	12½	14 7/12	16⅔	18¾	20⅚	22 11/12	25
1¼ x 12	10	12½	15	17½	20	22½	25	27½	30
1½ x 4	4	5	6	7	8	9	10	11	12
1½ x 6	6	7½	9	10½	12	13½	15	16½	18
1½ x 8	8	10	12	14	16	18	20	22	24
1½ x 10	10	12½	15	17½	20	22½	25	27½	30
1½ x 12	12	15	18	21	24	27	30	33	36
2 x 2	2⅔	3⅓	4	4⅔	5⅓	6	6⅔	7⅓	8
2 x 4	5⅓	6⅔	8	9⅓	10⅔	12	13⅓	14⅔	16
2 x 6	8	10	12	14	16	18	20	22	24
2 x 8	10⅔	13⅓	16	18⅔	21⅓	24	26⅔	29⅓	32
2 x 10	13⅓	16⅔	20	23⅓	26⅔	30	33⅓	36⅔	40
2 x 12	16	20	24	28	32	36	40	44	48
3 x 3	6	7½	9	10½	12	13½	15	16½	18
3 x 6	12	15	18	21	24	27	30	33	36
3 x 8	16	20	24	28	32	36	40	44	48
3 x 10	20	25	30	35	40	45	50	55	60
3 x 12	24	30	36	42	48	54	60	66	72
4 x 4	10⅔	13⅓	16	18⅔	21⅓	24	26⅔	29⅓	32
4 x 6	16	20	24	28	32	36	40	44	48
4 x 8	21⅓	26⅔	32	37⅓	42⅔	48	53⅓	58⅔	64
4 x 10	26⅔	33⅓	40	46⅔	53⅓	60	66⅔	73⅓	80
4 x 12	32	40	48	56	64	72	80	88	96

APPENDIX S

METRIC CONVERSION

Feet and Inch Conversions

1 inch	=	25.4 mm
1 foot	=	304.8 mm
1 psi	=	6.89 kPa
1 psf	=	0.048 kPa

Metric Conversions

1 mm	=	0.039 inches
1 m	=	3.28 feet
1 kPa	=	20.88 psf

mm	=	millimeter
m	=	meter
kPa	=	kilopascal
psi	=	pounds per square inch
psf	=	pounds per square foot

Inches to Millimeters and Centimeters

mm—millemeter cm—centimeter

inches	mm	cm	inches	mm	cm	inches	mm	cm
⅛	3	0.3	9	229	22.9	30	762	76.2
¼	6	0.6	10	254	25.4	31	787	78.7
⅜	10	1.0	11	279	27.9	32	813	81.3
½	13	1.3	12	305	30.5	33	838	83.8
⅝	16	1.6	13	330	33.0	34	864	86.4
¾	19	1.9	14	356	35.6	35	889	88.9
⅞	22	2.2	15	381	38.1	36	914	91.4
1	25	2.5	16	406	40.6	37	940	94.0
1¼	32	3.2	17	432	43.2	38	965	96.5
1½	38	3.8	18	457	45.7	39	991	99.1
1¾	44	4.4	19	483	48.3	40	1016	101.6
2	51	5.1	20	508	50.8	41	1041	104.1
2½	64	6.4	21	533	53.3	42	1067	106.7
3	76	7.6	22	559	55.9	43	1092	109.2
3½	89	8.9	23	584	58.4	44	1118	111.8
4	102	10.2	24	610	61.0	45	1143	114.3
4½	114	11.4	25	635	63.5	46	1168	116.8
5	127	12.7	26	660	66.0	47	1194	119.4
6	152	15.2	27	686	68.6	48	1219	121.9
7	178	17.8	28	711	71.1	49	1245	124.5
8	203	20.3	29	737	73.7	50	1270	127.0

Appendix T

Associations Involved in the Grading, Manufacturing & Promotion of Wood & Related Products

United States

American Lumber Standards Committee
P O Box 210
Germantown, MD 20875-0210

American Plywood Association
P O Box 11700
Tacoma, WA 98411-1700

California Lumber Inspection Service
1885 The Alameda
P O Box 6989
San Jose, CA 95150

California Redwood Association
405 Enfrente Drive, Suite 200
Novato, CA 94949

National Forest Products Association
1250 Connecticut Ave, NW
Washington, DC 20036

Northeastern Lumber Manufacturers Association
272 Tuttle Road
P O Box 87A
Cumberland Center, ME 04021

Northern Softwood Lumber Bureau
P O Box 87A
Cumberland Center, ME 04021

Pacific Lumber Inspection Bureau, Inc.
P O Box 7235
Bellevue, WA 98000-1235

Redwood Inspection Service
405 Enfrente Drive, Suite 200
Novato, CA 94949

Southern Forest Products Association
P O Box 64170
Kenner, LA 70064

Southern Pine Inspection Bureau
4709 Scenic Highway
Pensacola, FL 32504-9094

Timber Products Inspection
P O Box 919
Conyers, GA 30207

West Coast Lumber Inspection Bureau
P O Box 23145
Portland, OR 97223

Western Wood Products Association
Yeon Building
522 S.W. Fifth Ave.
Portland, OR 97204-2122

Canada

Alberta Forest Products Association
11710 Kingsway Ave., Suite 104
Edmonton, Alberta, Canada T5G 0X5

Canadian Lumberman's Association
27 Goulbourn Ave.
Ottawa, Ontario, Canada K1N 8C7

Canadian Wood Council
1730 St. Laurent Blvd., Suite 350
Ottawa, Ontario, Canada K1G 5L1

Cariboo Lumber Manufacturers Association
197 Second Ave. North, Suite 301
Williams Lake, British Columbia,
Canada V2G 1Z5

Central Forest Products Association, Inc.
P O Box 1169
Hudson Bay, Saskatchewan,
Canada S0E 0Y0

Council of Forest Industries of British Columbia
555 Burrard St., Suite 1200
Vancouver, British Columbia,
Canada V7X 1S7

Interior Lumber Manufacturers Association
1855 Kischner Road, Suite 350
Kelowna, British Columbia,
Canada V1Y 4N7

MacDonald Inspection
211 School House Street
Coquitlam, British Columbia,
Canada V3K 4X9

Maritime Lumber Bureau
P O Box 459
Amherst, Nova Scotia,
Canada B4H 4A1

National Lumber Grades Authority
260-1055 West Hastings St.
Vancouver, British Columbia,
Canada V6E 2E9

Ontario Lumber Manufacturers Association
55 University Ave., Suite 325
P O Box 8
Toronto, Ontario, Canada M5J 2H7

Pacific Lumber Inspection Bureau
1110-355 Burrard St.
Vancouver, British Columbia,
Canada V6C 2G8

Quebec Lumber Manufacturers Association
5055 Boulevard Hamel West, Suite 200
Quebec, Quebec,
Canada G2E 2G6

Western Red Cedar Lumber Association
1200-555 Burrard St.
Vancouver, British Columbia,
Canada V7X 1S7

INDEX

Adhesives, 640, 655
Air compressors, 673
Air infiltration, 556
Anchor clip, steel, 92
Anchors
  for shoring, 14
  sill, 92–93
  for walls, 638
Angle of repose, 13–14
Apron, window casing with, 364–366
Architect's rod, 60
Architectural drawings, 34–42
Attic
  access, on floor plan, 51
  truss, 255
  ventilation, 184–185, 274–277

Balloon framing, 140–141
Balustrades, 541–543
Barricades, 15
Basement, 48–49
Base shoes, 446
Bath, framing, 127
Batterboards, 74
Beam pocket, 90
Beams, 278, 293, 685-686
  box, 111–113
  collar, 186–187, 204
  floor, 108, 279
  glued laminated, 110, 695
  grade, 96–97, 679-694
  installing, 114–115
  notching, 128
  plywood-lumber, 111–113
  roof, 279
  steel, 109
  wood, solid and built-up, 109–110, 695
Bench mark, 47
Bird's mouth, 184–185, 196–197
Biscuit joiners, 671
Blocks, cold-formed steel, 607–609
Board foot, 624–625, 696
Bolts, 593, 637
  anchor, 92
  expansion, 430
  toggle, 638
Bonding agents, 640–641, 654–655
Bracing
  exterior walls, 157–159
  forms, 89
  let-in wood, 158
  permanent, for trusses, 260
  for pole construction, 302–303
Bridging
  cold-formed steel, 607–609
  wood, 133–134
Building codes, 30, 33, 338, 471
  cold-formed structural steel framing, 598–599
  fasteners, 632
  interior moldings, 456
  model, 30–31
  paneling and wainscoting, 472
  roof, 402
  stairs, 523–525
Building permits, 32–33
Building sections, 53

Cabinets
  base, 582–584
  construction of, 579–581
  European-style, 579, 589
  on floor plan, 50, 54
  level lines, establishing, 582–583
  mass-produced and custom-built, 576–578
  peninsula and island base units, 584–585
  preparing room for installation, 582
  types of, 578–579
  wall, 586–588
Cabinet scrapers, 650
Casings
  door, 332–335
  window, 360–364
Ceiling
  flat-roof, framing for, 181
  joists, 175–180
  openings, framing, 181
Cement, Portland, 100–101
Cements, adhesive, 641, 655
Ceramic tile installation, 517–518
Cheek cuts, 218
Chimney
  flashing, 409
  openings, sheathing for, 405
Chisels, 647
Clamping tools, 654–655
Clothing, protective, 642
Cold-formed steel products, 593–595
  assembly method, 599–601
  building codes, 598–599
  of floor, 600–603
  framing, 598
  of roof, 606–607
  structural, 592–593
  ties, blocks and bridging, 596, 607–609
  tools, 596–597
  trusses, 596
  of wall, 602–606
Column footings, 80–81
Columns, 293
Composite stressed-skin panels, 282
Concrete, 100
  admixtures, 102
  aggregates in, 101–102
  curing, 105–106
  formwork, 76
  insulation, 554–555
  mixing, 101–103
  placing, 104–105
  Portland cement and, 100–101
  reinforcing, 106–107
  slab floors, 95–96
  slab preparation, 504–505
  sources, 103–104
  stairs, 92f, 93
  tiles, 412
Conductance values (C), 545
Conduction, 545
Conductivity values (k), 545
Connectors, metal, 429
Contour lines, 66
Convection, 544
Corner boards, 375
Corners, inside and outside, 382–383
Cornices
  box, 266–268
  close, 266, 269–270
  framing, on gable roof, 208–209
  open, 266, 269
  returns, 273–274
Countertops
  installing, 590–591
  working design drawings, 576–577
Crane safety, 17
Crawl space, 48–49
Creosote, 630
Cricket, framing, 210
Cripple jack rafters, 184, 232
Cripples, 144–145

Crosscutting, 658–660
Dado, 468
Dead loads, 186
Deck/decking
connectors for, 429–430
design, 431–432
installing, 435–436
joists for, 428
posts for, 429
radius edge, 428
second floor, 438
span, single and double, 280
steps, 440–442
synthetic, 436–437
wood-framed, 433–434
Door frames
exterior, 310
garage, 332
interior, 314–318, 320
in solid masonry wall, 313–314
Doors
bifold, installation of, 326–330
bypass sliding, installation of, 326
casings, 332–335
exterior, 306–307, 310–313
on floor plan, 50
flush, 305
garage, 330–332
hands or swing of, 308
hanging, 320–321
hinges for, 321–325
interior, 304–305
lockset installation, 335–337
openings for, 144–145, 309
pocket, 325
sliding, bypass for, 326
symbols for, 41
weatherstripping for, 309
Door schedules, 51
Door stops, 337, 446
Dormers, 248, 401
finishing, 251
gable, 248, 250
hip, 251
shed, 248–249
Drilling and boring tools, 650–651, 668–669
Drip cap, 371, 377
Drip edge, 406
Drywall
cutting, 565
finishes, 562–563
installing, 566—575
joint compounds, 560–561
preliminary preparation, 563–565
storage, on-site, 563
taping joints, 572–574
temperature control, 563
types of, 560–561
Eave flashing, 406
Edge sander, 510
Electrical power, 42, 51, 674
Electronic distance-measuring devices (EDM), 56–57
Elevation
above-grade construction, finding, 66
finish floor, 46
grade, finding difference in, 65
reading, 52
site, 47
Engineer's rod, 60
Epoxy, 640
Excavations
for footings, 75
preparing, 13–15
shoring, 14–15
sloping requirements, 13–14
Exterior insulation and finish system (EIFS), 399–400
Eye safety, 8

Face frame, 579
Face protection equipment, 8
Fall protection, 10–11
Fascia, 371
Fascia surface, 184–185
Fasteners, 632–641. *See also specific fasteners*
Finished grade, 47
Finishes, 431, 511
Finishing tools, 649–650
Fink-truss, 253
Fireblocks, 142, 148
Fire extinguishers, types of, 16–17
Fireplace mantels and surrounds, 55, 471
Fire precautions, 16–17
Fire-retardant-treated wood (FRTW), 628–629
First aid, 11–12
Flashing, 406–410
Floor
beams, 108
cold-formed steel products framing, 600–603
construction, using post, plank and beam system, 283–284
decking, 293
framing, 108
joists, 118–125
openings, 15–16, 126
plans, reading, 50–51
for pole construction, 301–302
projections, framing, 127
surfacing, 508–510
trusses, wood, 131–133
Flooring. *See also* Wood flooring
cushion and carpet, 512–513
laminate, 496, 518–521
resilient, 513–518
sheet, 516–517
Floor plans, 50–51
Footings
excavating for, 75
grade stakes, setting, 65
forms, 77–81
Forms
bracing, 89
for foundation walls, 83–84
textured surfaces, 84
Forstner bit, 651
Foundation
for decks and porches, 431
for house, 44
locating, 72–73
permanent wood, 97–99
plans, reading, 48–49
for post, plank and beam framing, 282–283
sections, 54
Framing
balloon, 140–141
ceiling, 174–175, 180–181
gable-end, 206–208
gable roof, 192–193
plans, 54–55, 186–187
platform, 140–141
siding and, 383
square, 193–194, 216
structural, 676
wood stud wall, 478
Frieze, 469
Frost line, 76

Gable-end (rake), 377
finishing, 270–272
overhang, framing, 261
sheathing, 405
for siding, 377–379
truss, 255
Gable roof, 182, 183, 192
cornice, 208–210
cricket or saddle, 210
end projection, 204–208
frame, 192–193
framing
with I-joists, 210–212
on L-shaped building, 263–265
openings, 208
rafter layout, 193–196
sheathing, 212–213
valleys, 213
Gain (hinge mortise), 323
Gambrel roof, 183, 242–245

Garage doors, overhead, 330–332
Girders, 293
Glasses/goggles, safety, 8
Glazing, 339
Glues, 640, 654–655
Gouges, 647
Grade marks, on plywood sheathing, 165
Grooves, cutting, 671
Ground-fault circuit interrupters (GFCIs), 17, 674–675
Grounding, electrical, 675
Gussets, metal, for truss fabrication, 255
Gypsum sheet products. *See* Drywall

Hammers, 652–653
Hand planes, 646–647
Handrails, for stairs, 524
Handsaws, 644–646
Hardboard, 473, 630
Hard hats, 7–8
Hardwoods, 610–611
Hatchets, 654
Headers, 126, 146–149
Hearing protection devices, 8–9
Heat transfer/resistance, 544–547
Heavy timber construction, 278, 292–297
  flame resistance, 293
  framing members, 293–294
  information resources, 297
  structural members, 296–297
Hinges, for door, 321–325
Hip gable, 223
Hip roof, 182–183, 214
  Dutch, 182–183
  framing, 261–263
  one-half of 45-degree thickness of member, 217
  rafters, 214–215
  raising, 223–225
  ridgeboard, finding length of, 222–223
Hip-valley cripple jack, 232–234
Hip-valley jack rafters, 226–227
Horizontal angles, laying out, 66–67
Horizontal line of sight, 57
Housekeeping rules, general, 7
Houses, 44–45
Housewrap, 169

I-joist, 119
  floor system, 128–131
  for gable roof framing, 210–212
  standards, 131
Inspection, 32–33
Insulation
  electrical, 675
  installation of, 552–555
  for pole construction, 303
  sound, 556–559
  thermal, 544, 551–552
  types, 547–549
  for window casing, 360

Jamb extensions, 344–345, 360
Joint compound, drywall, 560–561
Joints
  coped, cutting, 458
  horizontal, 382
  isolation, 94
  reinforcing materials, for drywall, 560–562
  scarf, 460
  vertical, 382
Joist hangers, metal, 121
Joists. *See also* I-joists
  ceiling, 175–180, 682-683, 688, 693
  for decks and porches, 428
  floor, 118–125, 683, 687, 692
  notching, 128
  tail, 126
  trimmer, 126
Joist-to-beam connections, 119–121, 434

Kerf, 645
Keyway, 78
Kilogram, 36

Ladder jacks, 27
Ladders, 18–22
Laminated veneer lumber (LVL), 111
Laminate trimmer, 670
Laser levels, 68–71
Layout, checking, 73
Layout tools, 642–643
Leveling operations, on-site, 64
Leveling rod, 60
Levels, 57–58, 644
  automatic, 58
  laser, 69–70
  leveling, 62–63
  setting up, 62–63
Level transit, 57–59, 62–63
Live loads, 186
Loads, 687-695
Lockset installation, 335–337
Lumber
  air-drying, 616
  APA-rated products, 627–629
  Canadian, 621–622
  in construction, 610
  defects in, 617–618
  dimension, 620
  kiln-dried, 616
  machine-evaluated, 618–619
  manufacture of, 614–615
  metric sizes, 624
  parallel strand, 110–111
  quantities, specifying, 624–625
  quarter-sawed, 615
  seasoned sawed, 615–616
  sheathing, 402–403
  size, 619–621
  softwood, cutting, 615
  softwood grading, 618
  span design values, 622–624, 681-694
  use classifications, 619

Mansard roof, 183, 246–247
Masonry wall, 479
Mastic, 641
Measurement
  area, 36
  conversion factors, 36–37
  linear, 34–35, 56–57, 643, 696
  scales, for architectural drawings, 37–38
  volume, 36
Metric system, 35–37, 60, 683-686, 697
Mitering, 458–462, 661–662
Moisture control, 550–551
Molding, 444
  base, 460–462
  cap base, 460
  carpenter-made, 449–453
  chair rail, 469
  crown, 465–469
  custom-designed, 445
  door stop, 337
  interior, 456–471
  joinery, 466
  mitering outside corners, 467–468
  panel, 470
  picture, 470
  polymer, 447–448, 470–471
  quantities, estimating, 454–455
  securing and storing wood materials, 455
  shoe, 460–461
  stain-grade, 448–449
  stock, 445–447
  trim quantities, estimating, 469
  wood, grades, and species, 448–449
Monolithically cast slab, 95–96
Monolithically cast wall forms, 85
M-truss, 253, 255
Mudsill (sill plate), 115–116
Muntins, 344–345, 490

Nailers
  for ceiling materials, 265
  power, 635–636, 673
Nailing techniques, 653–654

Nails
  for cedar shingles and shakes, 418
  common, holding power of, 634–635
  for decks and porches, 430
  for finish applications, 633–634
  for framing, 633, 676-679
  sizes, 118–119, 632–633
  for wood shingles/siding, 369–370, 395

Occupational Safety and Health Administration (OSHA), 6–7, 10
  ladder requirements, 18–19
  planks requirements, 22–23
  scaffolding regulations, 22
Oil finishes, 511
Open valley, 407
Oriented strand board (OSB), 139, 164–167, 405, 630
OSHA. *See* Occupational Safety and Health Administration
Overhang, marking, 218–219

Paneling, 472
  building codes, 472
  hardboard, installing, 485–486
  interior wall, 473
  locating and fitting panels, 480–482
  nailing, 482–483
  on-site storage, 473–474
  plywood, 478, 480, 484–485
  preparing walls for, 478–480
  solid-wood, 473–477
  square-edged, 474
Panelized construction, 173–174, 599–601
Panels, 82–83
  reconstituted wood, 629–630
Parquet flooring, 495, 505–507
Particleboard, 630
Partitions
  assembling, 156–157
  constructing using post, plank and beam, 289
  interior, 162
  intersecting, 152–153
  load-bearing, 141
  nonload-bearing, 141, 180–181
  platform framing techniques, 141, 150–151
  supporting, 125
Patios, 94–95
Personal fall arrest system, 10–11
Phenolphthalein test, 504
Pier footings, 79–81
Piers, 82, 114
Pilasters, 89, 471
Pilings, 14–15
Pipe columns, 114–115
Plancher, 184, 185
Plancher line, locating, 198–199
Planes, portable electric, 667
Plank and beam system, 278–279
Planks
  for decking, 280–282, 405
  for scaffolding, 22–23
Plate layout, for walls/partitions, 153–155
Platform framing, 140–141
  door and window openings, 144–145
  exterior corners, 149
  exterior wall, 142–143
  fireblocks, 142, 148
  headers, 146–149
  interior partitions, 150–151
  solid and engineered studs, 143–144
  wall and partitions, 141
Plinth blocks, rosettes, 367, 463
Plumb bob, 58
Plumbing, on floor plan, 51
Plywood, 625
  APA trademarks, 626–627
  exposure durability, 627
  fire-retardant-treated, 628–629
  group number, 627
  paneling, 473–474
  sheathing, 164–167, 403–405
  span ratings, 627
  subfloor, installing, 134–136, 138
  treated, 628
  veneer grades, 626–627
Plywood-on-slab method, 504
Pole construction, 298
  bracing, 302–303
  design, 298
  embedment, 300–301
  floors, 301, 302
  frame construction, 302
  framing systems for, 298–299
  insulation, 303
  platform framing for, 301
Polyethylene film test, 504
Polyurethane, 511
Porches, 428
  connectors for, 429–430
  information resources, 443
  joists for, 428
  posts for, 429
  roof construction, 442–443
  second floor, 438
  steps, 440–442
Portland cement plaster (stucco), 396–398
Portland cements, 100–101
Posts, 278, 280
  for decks and porches, 429
  exterior wall construction, 284–286
  floor construction, 283–284
  for interior partition construction, 289
  roof construction, 286–289
  structure, assembly procedure for, 290–291
  slenderness ratio, 280
  steel, 114–115
  wood, 114
Post-to-beam connections, 434
Preengineered systems, for cold-formed structural steel framing, 600
Preservatives, waterborne, 630–631
Property lines, 46
Pump jacks, 28–29
Purlins, 188, 204–205

Radiation, 545
Rafters
  backing or dropping, 219
  common, 184, 192
  determining length of, 215–217
  duplicating and cutting, 198–201
  erecting, 201–204
  hip, 184, 214–215
  hip jack, 184, 220–223
  jack, 232, 234– 235
  laying out, 193—198, 218
  length, approximate, 193–195
  long valley, 232
  seat of, 184–185
  short valley, 232
  sizes, determining, 189–190
  step-off, 194–196
  tail of, 184–185
  types of, 184
  vs. valley jack rafters, 233
Rafter span tables, 193, 196, 216, 680-681, 684, 689-691
Railings, 438–441, 524–525
Rake, finishing, 270–272
Reinforcing bars (rebars), 78–79, 106–107
Resistance values (R), 546–547
Ridgeboard lengths, intersecting, 234–236
Ridge line, 184–185
Ridge vents, 416–417
Ripping, 658–660
Rise, 188–191
Riser, 525
Roof
  anchorage, 188
  blind valley construction, 229–230
  building codes, 402

cold-formed steel product
framing, 606–607
decking, 294
design considerations, 186
erection procedure, 236–237
flashing, 406
flat, 182, 238–239
framing, for heavy timber
construction, 293
having equal spans, 226–228
having unequal spans, 228–229
intersections, 226–227
laying out valley rafters for roofs
with unequal spans, 232
layout valley rafters for roofs with
equal spans, 230–231
openings, in gable roof, framing,
208
porch, 442–443
post, plank and beam system for,
286–289
ridge and hips, finishing, 416
sheathing, 402–405
shed, 182, 240–241
slope, 52, 266, 417
technical terms, 188–189
total rise, finding, 190–191
trusses, 252–255
types of, 182–183
windows, 354–357
Roofing
brackets, 27
built-up, 413
installation, safety, 413
materials for, 410–413
metal, 413, 425–427
tile, clay and concrete, 422–425
Ropes, securing of, 12–13
Routers, portable, 669–671
Run, 189

Saddle, framing, 210
Safety
around major equipment, 17
fall protection, 10–11
first aid and, 11–12
general housekeeping rules and, 7
information resources, 17
laser level, 71
for materials handling, 16
OSHA and, 6–7
power tool, 656. *See also specific power tools*
protective clothing for, 642
roofing installation, 413
scaffolding, 26–27
securing ropes for, 12–13
tool, 12
Sanders, portable, 672–673
Saws
circular, 656–659, 665–667
frame and trim, 662
handsaws, 644–646
miter, 661
radial-arm, 659–660
reciprocating, 664
saber or bayonet, 663–664
safety, 657, 660, 661, 663
Scaffolding, 22–27
Schedules, residential construction, 51
Screed, 94–95, 504–505
Screens, window, 345
Screwdrivers, 652, 669
Screws, 430, 593, 636–637
Sealers, penetrating, 511
Setbacks, 31, 46
Shakes, wood
for roof, 403, 417–423
for siding, 390–394
types of, 412
Sharpening edge tools, 647–649
Sheathing
at chimney openings, 405
exterior walls, 158–159, 164–165
fiberboard, 168
at gable-end, 405
gable roof, 212–213
gypsum, 167
plywood and OSB, 164–167
rigid foam plastic, 168
taping, 170
underlayment, 405
Shingles, 395, 395f
additional course, laying, 414–415
asphalt and fiberglass asphalt,
410–411, 413
installing, 392–395
lock-type, installing, 416
nails used with, 395
patterns, 391
starter strip for first course, 414
at valleys, 406–407, 416
wood, 403, 411, 417–420
Sidewalks, 94–95
Siding
corners, installation considerations
for, 375–376
fiber-cement, 386–387
gable-end treatment, 377–379
hardboard, 383–386
plywood, 379–383
preparation for installing, 368–369
steel, 396
types of, 368
vinyl and aluminum, 388–390
wood, 369–377
Sill anchors, 92–93
Sill plate (mudsill), 115–117
Site plan
building location on, 72
reading, 46–47
Skylights, 341, 357
Slope, 189
Soffit (box cornices), 266–268
Soffit and fascia system, 274
Softwoods, 610–613
Sound insulation, 556–559
Span, 175, 189
Specifications, 34
Squares, 642–643
Stains, 617
Stair gauge, 194–195
Stairs, 522
building codes, 523–525
carpeted, treads and risers for, 532
during construction, 530–531
figuring treads and risers, 525–527
floor plan details, 54
folding, 543
housed stringer, 538–541
landings, 524–525, 534–536
solid wood, installing treads and
risers for, 533–535
stock parts, 541–543
stringers, 527–531, 537–538
temporary, 29
types of, 522–523
width of, 523
windings, building, 536–538
working design drawing, 525–526
Stairwell, 522–523
Staplers, power, 635–636, 673
Steel square method, for determining
hip rafter length, 216
Steps, 440–442
Stick-built construction, 192, 599
Stiles, 490
Stool, window casing with, 364–365
Storm windows, 339
Straps, metal, 158
Stringer, 525
Stucco, 368, 396–398
Studs, 143–144
Subfloor, 134–139, 493, 678
Surveying, 38, 67
Surveyor-to-rodman signals, 64
Swanson Speed® square, 197–198
Sway braces, installing, 204
Symbols, on architectural drawings,
38, 39–42

Tacking, 653
Taping, sheathing, 170
Template, 335

Theodolites (level transits), 88
Thermoplastics and thermosets, 640, 655
Ties, cold-formed steel, 596, 607–609
Tile floor installation, 514–516
Tongue-and-groove paneling, 474
Tool belt, 642
Tools. *See also specific tools*
  cold-formed steel products, 596–597
  layout, 642–643
  power, 655–656
  safety, 12
  storage of, 655
Top plate, 163
Transmittance values (U), 545–546
Tresle jacks, 28
Trimmer, 144
Trim quantities, estimating, 469
Tripods, 59–60
Trusses, 252
  attic, 255
  cold-formed steel products, 596
  common jack, 261
  cornices, 274
  design and fabrication, 255–257
  erecting, 258–259
  gable roof, eave framing details, 264–265
  handling, 257
  hip, 255
  hip jack, 261
  Howe, 253, 255
  king-post, 253
  permanent bracing for, 260
  piggyback gable, 260
  scissors, 253
  storing, 257
Truss roof, 265

Underlayment, 405

Valley
  closed, 407
  closed-cut, 407
  flashing, 406–408
  gable roof, 213
Valley cripple jacks, 232
Valley rafters, 184
  cripple jack, 232, 234–235
  jack, 184, 226, 232–233
  vs. hip rafters, 233
Vapor barriers, 550
Varnishes, 511
V-bars, 595
Veneered structural panels. *See* Plywood
Veneers, masonry, 400–401
Ventilation, of attic, 184–185
Vernier scale, 61–62
Vertical angles, laying out, 67
Vertical arc, 62

Waferboard, 404, 630
Wainscoting, 472, 486–491
Wall/walls
  assembled, erecting, 159–160
  bracing, 157–159
  building/assembling, 154–157
  cold-formed steel framing, 602–606
  exterior, 368
  fasteners, 638–639
  forms, 84–88
  framing on concrete slabs, 163
  jacks, 162
  layout, on floor, 153
  openings for, 15–16, 90–92
  panels, manufactured, 171–172
  platform framing technique, 141–143
  preparing for paneling, 478–480
  post, plank and beam system, 284–286
  sheathing, 158–159, 164–165
Wax finishes, 511
Web stiffeners, 594–595
Windows
  awning and hopper, 340–341
  bay and bow, 342–343, 352–354
  building codes, 338
  casement, 340
  casings, 360–367
  double-hung, 339
  elliptical and circle top, installing, 351–352
  energy-efficient, 338–339
  fixed, 340–341
  on floor plan, 50
  frame and sash materials, 344
  installing, 346–351
  multiple-pane, 339
  openings for, platform framing technique, 144–145
  on plan, 343–344
  preparation for installation, 345
  replacement, 343, 358
  roof, 354–357
  schedules for, 51, 344, 346
  screens, 345
  single-hung, 340
  sliding, 340
  symbols, architectural drawings, 40
  trim, interior, 359–360
  types of, 339–343
Wood. *See also* Lumber
  beams, rafters, and joists, 679-695
  moisture content, checking, 616–617
  products, 624–625, 631
  properties of, 613
  species, 610–612
  timbers, 620
  treated, 618
Wood flooring, 493, 676
  installing hardwood floors over radiant heating, 507–508
  laminated, 503
  parquet block, 505–507
  plank, 503–504
  solid, 492–494
  strip, 494
  end grain, 496
  inlaid panels, 496
  installing over concrete slab, 504–505
  laminated, 494–495
  laying, 496–503
  parquet blocks, 495
  planks, 495
  solid, 678, 695
  subfloor, 493, 678
Wood preservatives, 630–631
W-truss, 253

Zoning regulations, 31–32